T0332689

Statistical Methods in the Atmospheric Sciences

Statistical Methods in the Atmospheric Sciences

Fourth Edition

Daniel S. Wilks
Professor Emeritus
Department of Earth & Atmospheric Sciences
Cornell University, Ithaca, NY, USA

ELSEVIER

Elsevier
Radarweg 29, PO Box 211, 1000 AE Amsterdam, Netherlands
The Boulevard, Langford Lane, Kidlington, Oxford OX5 1GB, United Kingdom
50 Hampshire Street, 5th Floor, Cambridge, MA 02139, United States

Notices

Knowledge and best practice in this field are constantly changing. As new research and experience broaden our understanding, changes in research methods, professional practices, or medical treatment may become necessary.

Practitioners and researchers must always rely on their own experience and knowledge in evaluating and using any information, methods, compounds, or experiments described herein. In using such information or methods they should be mindful of their own safety and the safety of others, including parties for whom they have a professional responsibility.

To the fullest extent of the law, neither the Publisher nor the authors, contributors, or editors, assume any liability for any injury and/or damage to persons or property as a matter of products liability, negligence or otherwise, or from any use or operation of any methods, products, instructions, or ideas contained in the material herein.

Library of Congress Cataloging-in-Publication Data
A catalog record for this book is available from the Library of Congress

British Library Cataloguing-in-Publication Data
A catalogue record for this book is available from the British Library

ISBN: 978-0-12-815823-4

For information on all Elsevier publications
visit our website at https://www.elsevier.com/books-and-journals

Publisher: Candice Janco
Acquisition Editor: Laura Kelleher
Editorial Project Manager: Katerina Zaliva
Production Project Manager: Prem Kumar Kaliamoorthi
Cover Designer: Matthew Limbert

Typeset by SPi Global, India

Contents

Contents

Part III
Multivariate Statistics   551

11. Matrix Algebra and Random Matrices   553

12. The Multivariate Normal Distribution   587

13. Principal Component (EOF) Analysis   617

Preface to the Fourth Edition

In preparing this fourth edition of *Statistical Methods in the Atmospheric* Sciences I have again tried to serve the needs both of instructors and students for a textbook, while also supporting researchers and operational practitioners needing a reasonably comprehensive but readable and not-too-cumbersome reference. The primary student audience will likely be upper-division undergraduates and beginning graduate students. These readers may wish to ignore the many literature references except in cases where additional information may be desired as a result of the material having been presented too briefly or otherwise inadequately. Researchers and other practitioners may see little use for the exercises at the ends of the chapters, but will find useful entry points into the broader literature in the references, which are more concentrated in the more research-oriented sections.

The most prominent change from the third edition is inclusion of a separate chapter on the burgeoning area of ensemble forecasting, including statistical (MOS) postprocessing of ensemble forecasts. However, all of the chapters have been updated, including but not limited to new and expanded treatments of extreme-value statistics; ANOVA, experimental design and comparisons among multiple means; regularization/shrinkage methods in regression and other techniques; nonparametric and "machine learning" regression methods; verification for probability density forecasts, ensemble forecasts, spatial structure field forecasts, and sampling and inference for verification statistics; regularization and missing data issues in PCA; expanded treatment of CCA and allied methods; and "machine learning" methods in discrimination, classification, and clustering.

The extensive reference list in this new edition includes more than 400 entries that were not listed in the previous edition. Of these, half are dated 2011 and later, having appeared since publication of the third edition. Also included in this edition are new examples and exercises, and a new appendix listing symbols and acronyms.

Of course it is unrealistic to expect that a work of this length could be produced without errors, and there are undoubtedly many in this book. Please take a moment to let me know about any of these that you might find, by contacting me at dsw5@cornell.edu. A list of errata for this and previous editions will be collected and maintained at https://tinyurl.com/WilksBookErrata, and at https://bit.ly/2DPeyPc.

Preface to the Third Edition

In preparing the third edition of *Statistical Methods in the Atmospheric* Sciences I have again tried to serve the needs of both instructors and students for a textbook, while also supporting researchers and operational practitioners who need a reasonably comprehensive but not-too-cumbersome reference.

All of the chapters have been updated from the second edition. This new edition includes nearly 200 new references, of which almost two-thirds are dated 2005 and later. The most prominent addition to the text is the new chapter on Bayesian inference. However, there are also new sections on trend tests and multiple testing, as well as expanded treatment of the Bootstrap; new sections on generalized linear modeling and developments in ensemble MOS forecasting; and six new sections in the forecast verification chapter, reflecting the large amount of attention this important topic has received during the past five years.

I continue to be grateful to the many colleagues and readers who have offered suggestions and criticisms that have led to improvements in this new edition, and who have pointed out errors in the second edition. Please continue to let me know about the errors that will be found in this revision, by contacting me at dsw5@cornell.edu. A list of these errata will be collected and maintained at http://atmos.eas.cornell.edu/~dsw5/3rdEdErrata.pdf.

I have been very gratified by the positive responses to the first edition of this book since it appeared about 10 years ago. Although its original conception was primarily as a textbook, it has come to be used more widely as a reference than I had initially anticipated. The entire book has been updated for this second edition, but much of the new material is oriented toward its use as a reference work. Most prominently, the single chapter on multivariate statistics in the first edition has been expanded to the final six chapters of the current edition. It is still very suitable as a textbook, but course instructors may wish to be more selective about which sections to assign. In my own teaching, I use most of Chapters 1 through 7 as the basis for an undergraduate course on the statistics of weather and climate data; Chapters 9 through 14 are taught in a graduate-level multivariate statistics course.

I have not included large digital data sets for use with particular statistical or other mathematical software, and for the most part I have avoided references to specific URLs (Web addresses). Even though larger data sets would allow examination of more realistic examples, especially for the multivariate statistical methods, inevitable software changes would eventually render these obsolete to a degree. Similarly, Web sites can be ephemeral, although a wealth of additional information complementing the material in this book can be found on the Web through simple searches. In addition, working small examples by hand, even if they are artificial, carries the advantage of requiring that the mechanics of a procedure must be learned firsthand, so that subsequent analysis of a real data set using software is not a black-box exercise.

Many, many people have contributed to the revisions in this edition by generously pointing out errors and suggesting additional topics for inclusion. I would like to thank particularly Matt Briggs, Tom Hamill, Ian Jolliffe, Rick Katz, Bob Livezey, and Jery Stedinger for providing detailed comments on the first edition and for reviewing earlier drafts of new material for the second edition. This book has been materially improved by all these contributions.

This text is intended as an introduction to the application of statistical methods to atmospheric data. The structure of the book is based on a course that I teach at Cornell University. The course primarily serves upper-division undergraduates and beginning graduate students, and the level of the presentation here is targeted to that audience. It is an introduction in the sense that many topics relevant to the use of statistical methods with atmospheric data are presented, but nearly all of them could have been treated at greater length and in more detail. The text will provide a working knowledge of some basic statistical tools sufficient to make accessible the more complete and advanced treatments available elsewhere.

This book assumes that you have completed a first course in statistics, but basic statistical concepts are reviewed before being used. The book might be regarded as a second course in statistics for those interested in atmospheric or other geophysical data. For the most part, a mathematical background beyond first-year calculus is not required. A background in atmospheric science is also not necessary, but it will help the reader appreciate the flavor of the presentation. Many of the approaches and methods are applicable to other geophysical disciplines as well.

In addition to serving as a textbook, I hope this will be a useful reference both for researchers and for more operationally oriented practitioners. Much has changed in this field since the 1958 publication of the classic *Some Applications of Statistics to Meteorology*, by Hans A. Panofsky and Glenn W. Brier, and no really suitable replacement has since appeared. For this audience, my explanations of statistical tools that are commonly used in atmospheric research will increase the accessibility of the literature and will improve your understanding of what your data sets mean.

Finally, I acknowledge the help I received from Rick Katz, Allan Murphy, Art DeGaetano, Richard Cember, Martin Ehrendorfer, Tom Hamill, Matt Briggs, and Pao-Shin Chu. Their thoughtful comments on earlier drafts have added substantially to the clarity and completeness of the presentation.

Part I

Preliminaries

Chapter 1

Introduction

1.1. WHAT IS STATISTICS?

"Statistics is the discipline concerned with the study of variability, with the study of uncertainty, and with the study of decision-making in the face of uncertainty" (Lindsay et al. 2004, p. 388). This book is concerned with the use of statistical methods in the atmospheric sciences, specifically in the various specialties within meteorology and climatology, although much of what is presented is applicable to other fields as well.

Students (and others) often resist statistics, and many perceive the subject to be boring beyond description. Before the advent of cheap and widely available computers, this negative view had some basis, at least with respect to applications of statistics involving the analysis of data. Performing hand calculations, even with the aid of a scientific pocket calculator, was indeed tedious, mind numbing, and time consuming. The capacity of an ordinary personal computer is now well beyond the fastest mainframe computers of just a few decades ago, but some people seem not to have noticed that the age of computational drudgery in statistics has long passed. In fact, some important and powerful statistical techniques were not even practical before the abundant availability of fast computing, and our repertoire of these "big data" methods continues to expand in parallel with ongoing increases in computing capacity. Even when liberated from hand calculations, statistics is sometimes still seen as uninteresting by people who do not appreciate its relevance to scientific problems. Hopefully, this book will help provide that appreciation, at least within the atmospheric sciences.

Fundamentally, statistics is concerned with uncertainty. Evaluating and quantifying uncertainty, as well as making inferences and forecasts in the face of uncertainty, are all parts of statistics. It should not be surprising, then, that statistics has many roles to play in the atmospheric sciences, since it is the uncertainty about atmospheric behavior that makes the atmosphere interesting. For example, many people are fascinated by weather forecasting, which remains interesting precisely because of the uncertainty that is intrinsic to the problem. If it were possible to make perfect forecasts or nearly perfect forecasts even one day into the future (i.e., if there were little or no uncertainty involved), the practice of meteorology would present few challenges, and would be similar in many ways to the calculation of tide tables.

1.2. DESCRIPTIVE AND INFERENTIAL STATISTICS

It is convenient, although somewhat arbitrary, to divide statistics into two broad areas: descriptive statistics and inferential statistics. Both are relevant to the atmospheric sciences.

The descriptive side of statistics pertains to the organization and summarization of data. The atmospheric sciences are awash with data. Worldwide, operational surface and upper-air observations are

Statistical Methods in the Atmospheric Sciences. https://doi.org/10.1016/B978-0-12-815823-4.00001-8

routinely taken at thousands of locations in support of weather forecasting activities. These are supplemented with aircraft, radar, profiler, and satellite data. Observations of the atmosphere specifically for research purposes are less widespread, but often involve very dense sampling in time and space. In addition, dynamical models of the atmosphere,[1] which undertake numerical integration of the equations describing the physics of atmospheric flow, produce yet more numerical output for both operational and research purposes.

As a consequence of these activities, we are often confronted with extremely large batches of numbers that, we hope, contain information about natural phenomena of interest. It can be a nontrivial task just to make some preliminary sense of such data sets. It is typically necessary to organize the raw data, and to choose and implement appropriate summary representations. When the individual data values are too numerous to be grasped individually, a summary that nevertheless portrays important aspects of their variations—a statistical model—can be invaluable in understanding them. It is worth emphasizing that it is not the purpose of descriptive data analyses to "play with numbers." Rather, these analyses are undertaken because it is known, suspected, or hoped that the data contain information about a natural phenomenon of interest, which can be exposed or better understood through the statistical analysis.

Inferential statistics is traditionally understood as consisting of methods and procedures used to draw conclusions regarding underlying processes that generate the data. For example, one can conceive of climate as the process that generates weather (Stephenson et al., 2012), so that one goal of climate science is to understand or infer characteristics of this generating process on the basis of the single sample realization of the atmospheric record that we have been able to observe. Thiébaux and Pedder (1987) express this point somewhat poetically when they state that statistics is "the art of persuading the world to yield information about itself." There is a kernel of truth here: Our physical understanding of atmospheric phenomena comes in part through statistical manipulation and analysis of data. In the context of the atmospheric sciences, it is sensible to interpret inferential statistics a bit more broadly as well and to include statistical forecasting of both weather and climate. By now this important field has a long tradition and is an integral part of operational forecasting at meteorological centers throughout the world.

1.3. UNCERTAINTY ABOUT THE ATMOSPHERE

The notion of uncertainty underlies both descriptive and inferential statistics. If atmospheric processes were constant, or strictly periodic, describing them mathematically would be easy. Weather forecasting would also be easy, and meteorology would be boring. Of course, the atmosphere exhibits variations and fluctuations that are irregular. This uncertainty is the driving force behind the collection and analysis of the large data sets referred to in the previous section. It also implies that weather forecasts are inescapably uncertain. The weather forecaster predicting a particular temperature on the following day is not at all surprised (and perhaps is even pleased) if the subsequent observation is different by a degree or two, and users of everyday forecasts also understand that forecasts involve uncertainty (e.g., Joslyn and Savelli, 2010). "Uncertainty is a fundamental characteristic of weather, seasonal climate, and

---

1. These are often referred to as NWP (numerical weather prediction) models, which term was coined in the middle of the last century (Charney and Eliassen, 1949) in order to distinguish dynamical from traditional subjective (e.g., Dunn, 1951) weather forecasting. However, as exemplified by the contents of this book, statistical methods and models are also numerical, so that the more specifically descriptive term "dynamical models" seems preferable.

hydrological prediction, and no forecast is complete without a description of its uncertainty" (National Research Council, 2006). Communicating this uncertainty promotes forecast user confidence, helps manage user expectations, and honestly reflects the state of the underlying science (Gill et al., 2008).

In order to deal quantitatively with uncertainty it is necessary to employ the tools of probability, which is the mathematical language of uncertainty. Before reviewing the basics of probability, it is worthwhile to examine why there is uncertainty about the atmosphere. After all, we have large, sophisticated dynamical computer models that represent the physics of the atmosphere, and such models are used routinely for forecasting its future evolution. Individually, these models have traditionally been formulated in a way that is deterministic, that is, without the ability to represent uncertainty. Once supplied with a particular initial atmospheric state (pressures, winds, temperatures, moisture content, etc., comprehensively through the depth of the atmosphere and around the planet) and boundary forcings (notably solar radiation, and sea- and land-surface conditions), each will produce a single particular result. Rerunning the model with the same inputs will not change that result.

In principle, dynamical atmospheric models could provide forecasts with no uncertainty, but they do not, for two reasons. First, even though the models can be very impressive and give quite good approximations to atmospheric behavior, they do not contain complete and true representations of the governing physics. An important and essentially unavoidable cause of this problem is that some relevant physical processes operate on scales too small and/or too fast to be represented explicitly by these models, and their effects on the larger scales must be approximated in some way using only the large-scale information. Although steadily improving computing capacity continues to improve the dynamical forecast models through increased resolution, Palmer (2014a) has noted that hypothetically achieving cloud-scale (<1 km) resolution would require exascale computing, which in turn would require hundreds of megawatts of electrical power to run the computing machinery!

Even if all the relevant physics could somehow be included in atmospheric models, however, we still could not escape the uncertainty caused by what has come to be known as *dynamical chaos*. The modern study of this phenomenon was sparked by an atmospheric scientist (Lorenz, 1963), who also has provided a very readable introduction to the subject (Lorenz, 1993). Smith (2007) provides another very accessible introduction to dynamical chaos. Simply and roughly put, the time evolution of a nonlinear, deterministic dynamical system (e.g., the equations of atmospheric motion, and presumably also the atmosphere itself) depends very sensitively on the initial conditions of the system. If two realizations of such a system are started from only very slightly different initial conditions, their two time evolutions will eventually diverge markedly. Imagine that one of these realizations is the real atmosphere and that the other is a perfect mathematical model of the physics governing the atmosphere. Since the atmosphere is always incompletely observed, it will never be possible to start the mathematical model in exactly the same state as the real system. So even if a computational model of the atmosphere could be perfect, it would still be impossible to calculate what the real atmosphere will do indefinitely far into the future.

Since forecasts of future atmospheric behavior will always be uncertain, probabilistic methods will always be needed to describe adequately that behavior. Some in the field have appreciated this fact since at least the beginning of practically realizable dynamical weather forecasting. For example, Eady (1951) observed that "forecasting is necessarily a branch of statistical physics in its widest sense: both our questions and answers must be expressed in terms of probabilities." Lewis (2005) nicely traces the history of probabilistic thinking in dynamical atmospheric prediction. The realization that the atmosphere exhibits chaotic dynamics has ended the dream of perfect (uncertainty-free) weather forecasts that formed the philosophical basis for much of 20th-century meteorology (an account of this history and scientific culture is provided by Friedman, 1989). Jointly, chaotic dynamics and the unavoidable errors in mathematical representations of the atmosphere imply that "all meteorological prediction problems, from weather forecasting to climate-change projection, are essentially probabilistic" (Palmer, 2001). Whether or not the atmosphere is fundamentally a random system, for most practical purposes it might as well be (e.g., Smith, 2007).

Finally, it is worth noting that randomness is not a state of complete unpredictability, or "no information," as is sometimes thought. Atmospheric predictability is typically defined with respect to the degree of statistical relatedness between forecasts and subsequent outcomes, characterized in terms of their probability distributions (DelSole and Tippett, 2018). A random process is not fully and precisely predictable or determinable, but may well be partially so.

To illustrate, the amount of precipitation that will occur tomorrow where you live is a random quantity, not known to you today. However, a simple statistical analysis of climatological precipitation records at your location would yield relative frequencies of past precipitation amounts providing substantially more information about tomorrow's precipitation at your location than I have as I sit writing this sentence. A still less uncertain idea of tomorrow's rain might be available to you in the form of a weather prediction that quantifies the uncertainty for the possible rainfall amounts in terms of probabilities. Uncertainty relates to how well something is known. Reducing uncertainty about random meteorological events is the purpose of weather forecasts, and reducing uncertainty about the nature of underlying natural phenomena is the purpose of much of scientific research.

Chapter 2

Review of Probability

2.1. BACKGROUND

This chapter presents a brief review of the basic elements of probability. More complete treatments of the basics of probability can be found in any good introductory statistics text.

Our uncertainty about the atmosphere, or about almost any other system for that matter, is of different degrees in different instances. For example, you cannot be completely certain whether or not rain will occur at your home tomorrow, or whether the average temperature next month will be greater or less than the average temperature this month. But you may be more sure about one or the other of these questions.

It is not sufficient, or even particularly informative, to say that an event is uncertain. Rather, we are faced with the problem of expressing or characterizing degrees of uncertainty. One approach is to use qualitative descriptors such as "likely," "unlikely," "possible," or "chance of." Conveying uncertainty through such phrases, however, is ambiguous and open to varying interpretations (Beyth-Marom, 1982; Murphy and Brown, 1983; National Research Council, 2006; Wallsten et al., 1986). For example, Figure 2.1 shows median endpoints for probability ranges corresponding to 10 qualitative uncertainty descriptors, elicited from twenty social science graduate students.

Because of the ambiguity associated with qualitative uncertainty descriptors, it is generally preferable to express uncertainty quantitatively, and this is done using numbers called *probabilities*. In a limited sense, probability is no more than an abstract mathematical system that can be developed logically from three premises called the *Axioms of Probability*. This system would be of no interest to many people, including perhaps yourself, except that the resulting abstract concepts are relevant to real-world problems involving uncertainty. Before presenting the axioms of probability and a few of their more important implications, it is necessary first to define some terminology.

2.2. THE ELEMENTS OF PROBABILITY

2.2.1. Events

An *event* is a set, or class, or group of possible uncertain outcomes. Events can be of two kinds: A *compound event* can be decomposed into two or more (sub)events, whereas an *elementary event* cannot. As a simple example, think about rolling an ordinary six-sided die. The event "an even number of spots comes up" is a compound event, since it will occur if either two, four, or six spots appear. The event "six spots come up" is an elementary event.

In simple situations like rolling a die, it is usually obvious which events are simple and which are compound. But more generally, just what is defined to be elementary or compound often depends on the situation at hand and the purposes for which an analysis is being conducted. For example, the event

Statistical Methods in the Atmospheric Sciences. https://doi.org/10.1016/B978-0-12-815823-4.00002-X

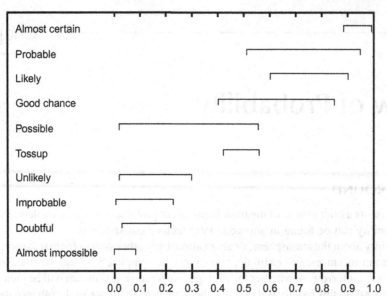

FIGURE 2.1 Median probability ranges corresponding to 10 qualitative uncertainty descriptors, as elicited from twenty social science graduate students. *Modified from Wallsten et al. (1986).*

"precipitation occurs tomorrow" could be an elementary event to be distinguished from the elementary event "precipitation does not occur tomorrow." But if it is important to distinguish further between forms of precipitation, "precipitation occurs" would be regarded as a compound event, possibly composed of the three elementary events: "liquid precipitation," "frozen precipitation," and "both liquid and frozen precipitation." If we were interested further in how much precipitation will occur, these three events would themselves be regarded as compound, each composed of at least two elementary events. In that case, for example, the compound event "frozen precipitation" could occur if either of the elementary events "frozen precipitation containing at least 0.01 in. water equivalent" or "frozen precipitation containing less than 0.01 in. water equivalent" were to occur.

2.2.2. The Sample Space

The *sample space* or *event space* is the set of all possible elementary events. Thus the sample space represents the universe of all possible outcomes or events. Equivalently, it is the largest possible compound event.

The relationships among events in a sample space can be represented geometrically, using what is called a *Venn Diagram*. Often the sample space is drawn as a rectangle and the events within it are drawn as circles, as in Figure 2.2a. Here the sample space is the rectangle labeled **S**, which might contain the set of possible precipitation outcomes for tomorrow. Four elementary events are depicted within the boundaries of the three circles. The "No precipitation" circle is drawn not overlapping the others because neither liquid nor frozen precipitation can occur if no precipitation occurs (i.e., in the absence of precipitation). The hatched area common to both "Liquid precipitation" and "Frozen precipitation" represents the event "both liquid and frozen precipitation." That part of **S** in Figure 2.2a not surrounded by circles is interpreted as representing the "null" event, which cannot occur.

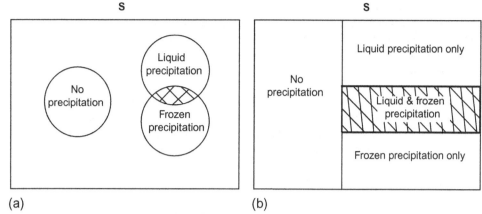

FIGURE 2.2   Venn diagrams representing the relationships of selected precipitation events. The hatched regions represent the event "both liquid and frozen precipitation." (a) Events portrayed as circles in the sample space. (b) The same events portrayed as space-filling rectangles.

It is not necessary to draw or think of circles in Venn diagrams to represent events. Figure 2.2b is an equivalent Venn diagram drawn using rectangles filling the entire sample space **S**. Drawn in this way, it is clear that **S** is composed of exactly four elementary events representing the full range of outcomes that may occur. Such a collection of all possible elementary (according to whatever working definition is current) events is called *mutually exclusive and collectively exhaustive* (MECE). Mutually exclusive means that no more than one of the events can occur. Collectively exhaustive means that at least one of the events will occur. A set of MECE events completely fills a sample space.

Note that Figure 2.2b could be modified to distinguish among precipitation amounts by adding a vertical line somewhere in the right-hand side of the rectangle. If the new rectangles on one side of this line were to represent precipitation of 0.01 in. or more, the rectangles on the other side would represent precipitation less than 0.01 in. The modified Venn diagram would then depict seven MECE events.

2.2.3. The Axioms of Probability

Once the sample space and its constituent events have been carefully defined, the next step is to associate probabilities with each of the events. The rules for doing so all flow logically from the three Axioms of Probability. Formal mathematical definitions of the axioms exist, but they can be stated qualitatively as follows:

1. The probability of any event is nonnegative.
2. The probability of the compound event **S** is 1.
3. The probability that one or the other of two mutually exclusive events occurs is the sum of their two individual probabilities.

2.3. THE MEANING OF PROBABILITY

The axioms are the essential logical basis for the mathematics of probability. That is, the mathematical properties of probability can all be deduced from the axioms. Some of these properties are listed later in this chapter.

However, the axioms are not very informative about what probability actually means. There are two dominant views of the meaning of probability—the Frequency view and the Bayesian view—and other interpretations exist as well (De Elia and Laprise, 2005; Gillies, 2000). Perhaps surprisingly, there has been no small controversy in the world of statistics as to which is correct. Passions have actually run so high on this issue that adherents of one interpretation or the other have been known to launch personal (verbal) attacks on those supporting a different view! Little (2006) presents a thoughtful and balanced assessment of the strengths and weaknesses of the two main perspectives.

It is worth emphasizing that the mathematics are the same for both Frequentist and Bayesian probability, because both follow logically from the same axioms. The differences are entirely in interpretation. Both of these dominant interpretations of probability have been accepted and found to be useful in the atmospheric sciences, in much the same way that the particle/wave duality of the nature of electromagnetic radiation is accepted and useful in the field of physics.

2.3.1. Frequency Interpretation

The Frequency interpretation is the mainstream view of probability. Its development in the 18th century was motivated by the desire to understand games of chance and to optimize the associated betting. In this view, the probability of an event is exactly its long-run relative frequency. This definition is formalized in the *Law of Large Numbers*, which states that the ratio of the number of occurrences of event $\{E\}$ to the number of opportunities for $\{E\}$ to have occurred converges to the probability of $\{E\}$, denoted $\Pr\{E\}$, as the number of opportunities increases. This idea can be written formally as

$$\Pr\left\{ \left| \frac{a}{n} - \Pr\{E\} \right| \geq \varepsilon \right\} \rightarrow 0 \text{ as } n \rightarrow \infty, \tag{2.1}$$

where a is the number of occurrences, n is the number of opportunities (thus a/n is the relative frequency), and ε is an arbitrarily small number. Equation 2.1 says that when there have been many opportunities, n, for the event $\{E\}$ to occur, the relative frequency a/n is likely to be close to $\Pr\{E\}$. In addition, the relative frequency and the probability are more likely to be close as n becomes progressively larger.

The Frequency interpretation is intuitively reasonable and empirically sound. It is useful in such applications as estimating climatological probabilities by computing historical relative frequencies. For example, in the last 50 years there have been $31 \times 50 = 1550$ August days. If rain had occurred at a location of interest on 487 of those days, a natural estimate for the climatological probability of precipitation at that location on an August day would be $487/1550 = 0.314$.

2.3.2. Bayesian (Subjective) Interpretation

Strictly speaking, employing the Frequency view of probability requires a long series of identical trials. For estimating climatological probabilities from a sufficiently long series of historical weather data this requirement presents essentially no problem. However, thinking about probabilities for events like {the football team at your college or alma mater will win at least half of their games next season} does present some difficulty in the relative frequency framework. Although abstractly we can imagine a hypothetical series of football seasons identical to the upcoming one, this series of fictitious football seasons is of no help in actually estimating a probability for the event.

The subjective interpretation is that probability represents the degree of belief, or quantified judgment, of a particular individual about the occurrence of an uncertain event. For example, there is now a long history of weather forecasters routinely (and very skillfully) assessing probabilities for events like precipitation occurrence on days in the near future. If your college or alma mater is a large enough school that professional gamblers take an interest in the outcomes of its football games, probabilities regarding those outcomes are also regularly assessed—subjectively.

Two individuals can assess different subjective probabilities for an event without either necessarily being wrong, and often such differences in judgment are attributable to differences in information and/or experience. However, the fact that different individuals may have different subjective probabilities for the same event does not mean that an individual is free to choose any numbers and call them probabilities. The quantified judgment must be a consistent judgment in order to be a legitimate subjective probability. This means, among other things, that subjective probabilities must be consistent with the axioms of probability, and thus with the mathematical properties of probability implied by the axioms.

2.4. SOME PROPERTIES OF PROBABILITY

One reason Venn diagrams can be so useful is that they allow probabilities to be visualized geometrically as areas. Familiarity with geometric relationships in the physical world can then be used to better grasp the more abstract world of probability. Imagine that the area of the rectangle in Figure 2.2b is 1, according to the second axiom. The first axiom says that no areas can be negative. The third axiom says that the total area of nonoverlapping parts is the sum of the areas of those parts.

Some of the mathematical properties of probability that follow logically from the axioms are listed in this section. The geometric analog for probability provided by a Venn diagram can be used to help visualize them.

2.4.1. Domain, Subsets, Complements, and Unions

Together, the first and second axioms imply that the probability of any event will be between zero and one, inclusive:

$$0 \leq \Pr\{E\} \leq 1. \tag{2.2}$$

If $\Pr\{E\}=0$ the event cannot occur. If $\Pr\{E\}=1$ the event is absolutely sure to occur.

If event $\{E_2\}$ necessarily occurs whenever event $\{E_1\}$ occurs, $\{E_1\}$ is said to be a *subset* of $\{E_2\}$. For example, $\{E_1\}$ and $\{E_2\}$ might denote occurrence of frozen precipitation, and occurrence of precipitation of any form, respectively. In this case the third axiom implies

$$\Pr\{E_1\} \leq \Pr\{E_2\}. \tag{2.3}$$

The *complement* of event $\{E\}$ is the (generally compound) event that $\{E\}$ does not occur. In Figure 2.2b, for example, the complement of the event "liquid and frozen precipitation" is the compound event "either no precipitation, or liquid precipitation only, or frozen precipitation only." Together the second and third axioms imply

$$\Pr\{E^C\} = 1 - \Pr\{E\}, \tag{2.4}$$

where $\{E^C\}$ denotes the complement of $\{E\}$. (Some authors use an overbar as an alternative notation to represent complements. This use of the overbar is very different from its most common statistical meaning, which is to denote an arithmetic average.)

The *union* of two events is the compound event that one or the other, or both, of the events occur. In set notation, unions are denoted by the symbol ∪. As a consequence of the third axiom, probabilities for unions can be computed using

$$\Pr\{E_1 \cup E_2\} = \Pr\{E_1 \text{ or } E_2 \text{ or both}\}$$
$$= \Pr\{E_1\} + \Pr\{E_2\} - \Pr\{E_1 \cap E_2\}. \tag{2.5}$$

The symbol ∩ is called the *intersection* operator, and

$$\Pr\{E_1 \cap E_2\} = \Pr\{E_1, E_2\} = \Pr\{E_1 \text{ and } E_2\} \tag{2.6}$$

is the event that both $\{E_1\}$ and $\{E_2\}$ occur. The notation $\{E_1, E_2\}$ is equivalent to $\{E_1 \cap E_2\}$. Another name for $\Pr\{E_1, E_2\}$ is the *joint probability* of $\{E_1\}$ and $\{E_2\}$. Equation 2.5 is sometimes called the *Additive Law of Probability*. It holds whether or not $\{E_1\}$ and $\{E_2\}$ are mutually exclusive. However, if the two events are mutually exclusive, the probability of their intersection (i.e., their joint probability) is zero, since mutually exclusive events cannot both occur.

The probability for the joint event, $\Pr\{E_1, E_2\}$ is subtracted in Equation 2.5 to compensate for its having been counted twice when the probabilities for events $\{E_1\}$ and $\{E_2\}$ are added. This can be seen most easily by thinking about how to find the total geometric area enclosed by the two overlapping circles in Figure 2.2a. The hatched region in Figure 2.2a represents the intersection event {liquid precipitation and frozen precipitation}, and it is contained within both of the two circles labeled "Liquid precipitation" and "Frozen precipitation."

The additive law, Equation 2.5, can be extended to the union of three or more events by thinking of $\{E_1\}$ or $\{E_2\}$ as a compound event (i.e., a union of other events), and recursively applying Equation 2.5. For example, if $\{E_2\}=\{E_3 \cap E_4\}$, substituting into Equation 2.5 yields, after some rearrangement,

$$\Pr\{E_1 \cup E_3 \cup E_4\} = \Pr\{E_1\} + \Pr\{E_3\} + \Pr\{E_4\}$$
$$- \Pr\{E_1 \cap E_3\} - \Pr\{E_1 \cap E_4\} - \Pr\{E_3 \cap E_4\}. \tag{2.7}$$
$$+ \Pr\{E_1 \cap E_3 \cap E_4\}.$$

This result may be difficult to grasp algebraically but is fairly easy to visualize geometrically. Figure 2.3 illustrates the situation. Adding together the areas of the three circles individually (the first line in

FIGURE 2.3 Venn diagram illustrating computation of the probability of the union of three intersecting events in Equation 2.7. The regions with two overlapping hatch patterns have been double-counted, and their areas must be subtracted to compensate. The central region with three overlapping hatch patterns has been triple-counted, but then subtracted three times when the double-counting is corrected. Its area must be added back again.

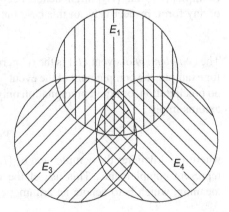

Equation 2.7) results in double-counting the areas with two overlapping hatch patterns, and triple-counting the central area contained in all three circles. The second line of Equation 2.7 corrects the double-counting, but subtracts the area of the central region three times. This area is added back a final time in the third line of Equation 2.7.

2.4.2. DeMorgan's Laws

Manipulating probability statements involving complements of unions or intersections, or statements involving intersections of unions or complements, is facilitated by the two relationships known as *DeMorgan's Laws*,

$$\Pr\left\{ (A \cup B)^C \right\} = \Pr\{A^C \cap B^C\} \tag{2.8a}$$

and

$$\Pr\left\{ (A \cap B)^C \right\} = \Pr\{A^C \cup B^C\}. \tag{2.8b}$$

The first of these laws, Equation 2.8a, expresses the fact that the complement of a union of two events is the intersection of the complements of the two events. In the geometric terms of the Venn diagram, the events outside the union of $\{A\}$ and $\{B\}$ (left-hand side) are simultaneously outside of both $\{A\}$ and $\{B\}$ (right-hand side). The second of DeMorgan's Laws, Equation 2.8b, says that the complement of an intersection of two events is the union of the complements of the two individual events. Here, in geometric terms, the events not in the overlap between $\{A\}$ and $\{B\}$ (left-hand side) are those either outside of $\{A\}$ or outside of $\{B\}$, or both (right-hand side).

2.4.3. Conditional Probability

It is often the case that we are interested in the probability of an event, given that some other event has occurred or will occur. For example, the probability of freezing rain, given that precipitation occurs, may be of interest; or perhaps we need to know the probability of coastal wind speeds above some threshold, given that a hurricane makes landfall nearby. These are examples of *conditional probabilities*. The event that must be "given" is called the *conditioning event*. The conventional notation for conditional probability is a vertical line, so denoting $\{E_1\}$ as the event of interest and $\{E_2\}$ as the conditioning event, conditional probability is denoted as

$$\Pr\{E_1 | E_2\} = \Pr\{E_1 \text{ given that } E_2 \text{ has occurred or will occur}\}. \tag{2.9}$$

If the event $\{E_2\}$ has occurred or will occur, the probability of $\{E_1\}$ is the conditional probability $\Pr\{E_1|E_2\}$. If the conditioning event has not occurred or will not occur, the conditional probability by itself gives no information on the probability of $\{E_1\}$.

More formally, conditional probability is defined in terms of the intersection of the event of interest and the conditioning event, according to

$$\Pr\{E_1 | E_2\} = \frac{\Pr\{E_1 \cap E_2\}}{\Pr\{E_2\}}, \tag{2.10}$$

provided that the probability of the conditioning event is not zero. Intuitively, it makes sense that conditional probabilities are related to the joint probability of the two events in question, $\Pr\{E_1 \cap E_2\}$. Again, this

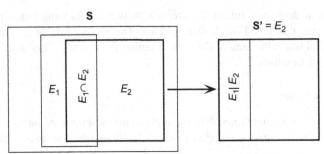

FIGURE 2.4 Illustration of the definition of conditional probability. The unconditional probability of $\{E_1\}$ is that fraction of the area of **S** occupied by $\{E_1\}$ on the left side of the figure. Conditioning on $\{E_2\}$ amounts to considering a new sample space, **S'** composed only of $\{E_2\}$, since this means we are concerned only with occasions when $\{E_2\}$ occurs. Therefore the conditional probability $\Pr\{E_1|E_2\}$ is given by the proportion of the area of the new sample space **S'** $= \{E_2\}$ that is occupied by $\{E_1\}$. This proportion is computed in Equation 2.10.

is easiest to understand through the analogy to areas in a Venn diagram, as shown in Figure 2.4. We understand the unconditional probability of $\{E_1\}$ to be represented by that proportion of the sample space S occupied by the rectangle labeled E_1. Conditioning on $\{E_2\}$ means that we are interested only in those outcomes including $\{E_2\}$. We are, in effect, throwing away any part of S not contained in $\{E_2\}$. This amounts to considering a new sample space, S', that is coincident with $\{E_2\}$. The conditional probability $\Pr\{E_1|E_2\}$ therefore is represented geometrically as that proportion of area of the new sample space (corresponding to $\{E_2\}$) that is occupied by $\{E_1\}$. If the conditioning event and the event of interest are mutually exclusive, the conditional probability clearly must be zero, since their joint probability will be zero.

2.4.4. Independence

Rearranging the definition of conditional probability, Equation 2.10, yields the form of this expression called the *Multiplicative Law of Probability*:

$$\Pr\{E_1 \cap E_2\} = \Pr\{E_1|E_2\}\,\Pr\{E_2\} = \Pr\{E_2|E_1\}\,\Pr\{E_1\}. \tag{2.11}$$

Two events are said to be *independent* if the occurrence or nonoccurrence of one does not affect the probability of the other. For example, if we roll a red die and a white die, the probability of an outcome on the red die does not depend on the outcome of the white die, and vice versa. The outcomes for the two dice are independent. Independence between $\{E_1\}$ and $\{E_2\}$ implies $\Pr\{E_1|E_2\}=\Pr\{E_1\}$ and $\Pr\{E_2|E_1\}=\Pr\{E_2\}$. Independence of events makes the calculation of joint probabilities particularly easy, since the multiplicative law then reduces to

$$\Pr\{E_1 \cap E_2\} = \Pr\{E_1\}\,\Pr\{E_2\}, \text{ for } \{E_1\} \text{ and } \{E_2\} \text{ independent}. \tag{2.12}$$

Equation 2.12 is extended easily to the computation of joint probabilities for more than two independent events, by simply multiplying all the probabilities of the independent unconditional events.

Example 2.1. Conditional Relative Frequency

Consider estimating climatological (i.e., long-run, or "population") probabilities using the data given in Table A.1 of Appendix A. Climatological probabilities conditional on other events can be computed. Such probabilities are sometimes referred to as conditional climatological probabilities, or *conditional climatologies*.

Suppose it is of interest to estimate the probability of at least 0.01 in. of liquid equivalent precipitation at Ithaca in January, given that the minimum temperature is at least 0°F. Physically, these two events would be expected to be related since very cold minimum temperatures typically occur on clear nights, and precipitation occurrence requires clouds. This physical relationship would lead us to expect that these two events would be statistically related (i.e., not independent) and that the conditional probabilities of precipitation given different minimum temperature conditions will be different from each other and from the unconditional probability. For example, on the basis of our understanding of the underlying physical processes, we expect the probability of precipitation given minimum temperature of 0°F or higher will be larger than the conditional probability given the complementary event of minimum temperature colder than 0°F.

To estimate the first of these probabilities using conditional relative frequency, we are interested only in those data records for which the Ithaca minimum temperature was at least 0°F. There are 24 such days in Table A.1. Of these 24 days, 14 show measurable precipitation (ppt), yielding the estimate $\Pr\{\text{ppt} \geq 0.01 \text{ in.} | T_{min} \geq 0°F\} = 14/24 \approx 0.58$. The precipitation data for the seven days on which the minimum temperature was colder than 0°F have been ignored. Since measurable precipitation was recorded on only one of these seven days, we could estimate the conditional probability of precipitation given the complementary conditioning event of minimum temperature colder than 0°F as $\Pr\{\text{ppt} \geq 0.01 \text{ in.} | T_{min} < 0°F\} = 1/7 \approx 0.14$. The corresponding estimate of the unconditional probability of precipitation would be $\Pr\{\text{ppt} \geq 0.01 \text{ in.}\} = 15/31 \approx 0.48$.   ◇

The difference between the conditional probability estimates calculated in Example 2.1 reflects statistical dependence. Since the underlying physical processes are well understood, we would not be tempted to speculate that relatively warmer minimum temperatures somehow cause precipitation. Rather, the temperature and precipitation events show a statistical relationship because of their (different) physical relationships to clouds. When dealing with statistically dependent variables whose physical relationships may not be known, it is well to remember that statistical dependence does not necessarily imply a physical cause-and-effect relationship, but may instead reflect more complex interactions within the physical data-generating process.

Example 2.2. Persistence as Conditional Probability

Atmospheric variables often exhibit statistical dependence with their own past and future values. In the terminology of the atmospheric sciences, this dependence through time is usually known as *persistence*. Persistence can be defined as the existence of (positive) statistical dependence among successive values of the same variable or among successive occurrences of a given event. Positive dependence means that large values of the variable tend to be followed by relatively large values, and small values of the variable tend to be followed by relatively small values.

Typically the source of persistence is that the measurement interval is shorter than (at least one of) the timescale(s) of the underlying physical process(es). Accordingly, it is usually the case that statistical dependence of meteorological variables in time is positive. For example, the probability of an above-average temperature tomorrow is higher if today's temperature was above average. Thus another name for persistence is *positive serial dependence*. When present, this frequently occurring characteristic has important implications for statistical inferences drawn from atmospheric data, as will be seen in Chapter 5.

Consider characterizing the persistence of the event {precipitation occurrence} at Ithaca, again using the small data set of daily values in Table A.1 of Appendix A. Physically, serial dependence would be expected in these data because the typical timescale for the midlatitude synoptic waves with which most winter precipitation is associated at this location is several days, and this is longer than the daily observation interval. The statistical consequence should be that days for which measurable precipitation is reported should tend to occur in runs, as should days without measurable precipitation.

To evaluate serial dependence for precipitation events, it is necessary to estimate conditional probabilities of the type Pr{ppt today|ppt yesterday}. Since data set A.1 contains no records for either December 31, 1986 or February 1, 1987, there are 30 yesterday/today data pairs to work with. To estimate Pr{ppt today|ppt yesterday} we need to only count the number of days reporting precipitation (as the conditioning, or "yesterday" event) that were followed by a day reporting precipitation (as the event of interest, or "today"). When estimating this conditional probability, we are not interested in what happened following days on which no precipitation was reported. Excluding January 31, there are 14 days on which precipitation was reported. Of these, 10 were followed by another day with nonzero precipitation, and four were followed by dry days. The conditional relative frequency estimate therefore would be Pr{ppt today|ppt yesterday}=10/14≈0.71. Similarly, conditioning on the complementary event (no precipitation "yesterday") yields the estimate Pr{ppt today|no ppt yesterday}=5/16≈0.31. The difference between these conditional probability estimates confirms the serial dependence in these data and quantifies the tendency of the wet and dry days to occur in runs. These two conditional probabilities also constitute a "conditional climatology." ◇

2.4.5. Law of Total Probability

Sometimes probabilities must be computed indirectly because of limited information. One relationship that can be useful in such situations is the *Law of Total Probability*. Consider a set of MECE events, {E_i}, $i=1, ..., I$ on a sample space of interest. Figure 2.5 illustrates this situation for $I=5$ events. If there is an event {A}, also defined on this sample space, its probability can be computed by summing the joint probabilities

$$Pr\{A\} = \sum_{i=1}^{I} Pr\{A \cap E_i\}. \tag{2.13}$$

The notation on the right-hand side of this equation indicates summation of terms defined by the mathematical template to the right of the uppercase sigma, for all integer values of the index i between 1 and I, inclusive. Substituting the multiplicative law of probability yields

FIGURE 2.5 Illustration of the Law of Total Probability. The sample space S contains the event {A}, represented by the oval, and five MECE events, {E_i}.

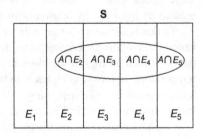

$$Pr\{A\} = \sum_{i=1}^{I} Pr\{A | E_i\} Pr\{E_i\}. \tag{2.14}$$

If the unconditional probabilities $Pr\{E_i\}$ and the conditional probabilities of $\{A\}$ given each of the MECE events $\{E_i\}$ are known, the unconditional probability of $\{A\}$ can be computed. It is important to note that Equation 2.14 is correct only if the events $\{E_i\}$ constitute a MECE partition of the sample space.

Example 2.3. Combining Conditional Probabilities Using the Law of Total Probability

Example 2.2 can also be viewed in terms of the Law of Total Probability. Consider that there are only $I = 2$ MECE events partitioning the sample space: $\{E_1\}$ denotes precipitation yesterday, and $\{E_2\} = \{E_1^C\}$ denotes no precipitation yesterday. Let the event $\{A\}$ be the occurrence of precipitation today. If the data were not available, we could compute $Pr\{A\}$ using the conditional probabilities through the Law of Total Probability. That is, $Pr\{A\} = Pr\{A|E_1\} Pr\{E_1\} + Pr\{A|E_2\} Pr\{E_2\} = (10/14)(14/30) + (5/16)(16/30) = 0.50$. Since the data are available in Appendix A, the correctness of this result can be confirmed simply by counting. $\diamond$

2.4.6. Bayes' Theorem

Bayes' Theorem is an interesting combination of the Multiplicative Law and the Law of Total Probability. In a relative frequency setting, Bayes' Theorem is used to "invert" conditional probabilities. That is, if $Pr\{E_1|E_2\}$ is known, Bayes' Theorem may be used to compute $Pr\{E_2|E_1\}$. In the Bayesian framework, developed in Chapter 6, it is used to optimally revise or update subjective probabilities consistent with new information.

Consider again a situation such as that shown in Figure 2.5, in which there is a defined set of MECE events $\{E_i\}$ and another event $\{A\}$. The Multiplicative Law (Equation 2.11) can be used to find two expressions for the joint probability of $\{A\}$ and any of the events $\{E_i\}$,

$$\begin{aligned} Pr\{A, E_i\} &= Pr\{A | E_i\} Pr\{E_i\} \\ &= Pr\{E_i|A\} Pr\{A\} \end{aligned} \tag{2.15}$$

Combining the two right-hand sides and rearranging yields

$$Pr\{E_i | A\} = \frac{Pr\{A | E_i\} Pr\{E_i\}}{Pr\{A\}} = \frac{Pr\{A | E_i\} Pr\{E_i\}}{\sum_{j=1}^{J} Pr\{A | E_j\} Pr\{E_j\}}. \tag{2.16}$$

The Law of Total Probability has been used to rewrite the denominator. Equation 2.16 is the expression for Bayes' Theorem. It is applicable separately for each of the MECE events $\{E_i\}$. Note, however, that the denominator is the same for each E_i, since $Pr\{A\}$ is obtained each time by summing over all the events, indexed in the denominator by the subscript j.

Example 2.4. Bayes' Theorem from a Relative Frequency Standpoint

Conditional probabilities for precipitation occurrence given minimum temperatures above or below 0°F were estimated in Example 2.1. Bayes' Theorem can be used to compute the converse conditional probabilities, concerning temperature events given that precipitation did or did not occur. Let $\{E_1\}$ represent minimum temperature of 0°F or above, and $\{E_2\}=\{E_1^C\}$ be the complementary event that minimum temperature is colder than 0°F. Clearly the two events are a MECE partition of the sample space. Recall that minimum temperatures of at least 0°F were reported on 24 of the 31 days, so that the unconditional climatological estimates of the probabilities for the temperature events would be Pr $\{E_1\}=24/31$ and $\Pr\{E_2\}=7/31$. Recall also that $\Pr\{A|E_1\}=14/24$ and $\Pr\{A|E_2\}=1/7$.

Equation 2.16 can be applied separately for each of the two events $\{E_i\}$. In each case the denominator is $\Pr\{A\}=(14/24)(24/31)+(1/7)(7/31)=15/31$. (This differs slightly from the estimate for the probability of precipitation obtained in Example 2.3, since there the data for December 31 could not be included.) Using Bayes' Theorem, the conditional probability for minimum temperature at least 0°F given precipitation occurrence is $(14/24)(24/31)/(15/31)=14/15$. Similarly, the conditional probability for minimum temperature below 0°F given nonzero precipitation is $(1/7)(7/31)/(15/31)=1/15$. Since all the data are available in Appendix A, these calculations can be verified directly by counting.   ◇

Example 2.5. Bayes' Theorem from a Subjective Probability Standpoint

A subjective (Bayesian) probability interpretation corresponding to the calculations in Example 2.4 can also be made. Suppose a weather forecast specifying the probability of the minimum temperature being at least 0°F is desired. If no more sophisticated information were available, it would be natural to use the unconditional climatological probability for the event, $\Pr\{E_1\}=24/31$, to represent the forecaster's uncertainty or degree of belief in the outcome. In the Bayesian framework this baseline state of information is known as the *prior probability*. Assume, however, that the forecaster could know whether or not precipitation will occur on that day. That information would affect the forecaster's degree of certainty in the temperature outcome. Just how much more certain the forecaster can become depends on the strength of the relationship between temperature and precipitation, expressed in the conditional probabilities for precipitation occurrence given the two minimum temperature outcomes. These conditional probabilities, $\Pr\{A|E_i\}$ in the notation of this example, are known as the *likelihoods*. If precipitation occurs, the forecaster is more certain that the minimum temperature will be at least 0°F, with the revised probability given by Equation 2.16 as $(14/24)(24/31)/(15/31)=14/15$. This modified or updated (in light of the additional information regarding precipitation occurrence) judgment regarding the probability of a very cold minimum temperature not occurring is called the *posterior probability*. Here the posterior probability is larger than the prior probability of 24/31. Similarly, if precipitation will not occur, the forecaster is more confident that the minimum temperature will not be 0°F or warmer. Note that the differences between this example and Example 2.4 are entirely in the interpretations, and that the computations and numerical results are identical.   ◇

2.5. EXERCISES

2.1 In the climate record for 60 winters at a given location, single-storm snowfalls greater than 35 cm occurred in nine of those winters (define such snowfalls as event "A"), and the coldest temperature was below −25°C in 36 of the winters (define this as event "B"). Both events "A" and "B" occurred in three of the winters.

a. Sketch a Venn diagram for a sample space appropriate to this data.

b. Write an expression using set notation for the occurrence of 35 cm snowfalls, −25°C temperatures, or both. Estimate the climatological probability for this compound event.

   c. Write an expression using set notation for the occurrence of winters with both 35 cm snowfalls and temperatures not falling below −25°C. Estimate the climatological probability for this compound event.

   d. Write an expression using set notation for the occurrence of winters having neither −25°C temperatures nor 35 cm snowfalls. Again, estimate the climatological probability.

2.2 Using the January 1987 data set in Table A.1, define event "A" as Ithaca $T_{max} > 32°F$, and event "B" as Canandaigua $T_{max} > 32°F$.

   a. Explain the meanings of $Pr(A)$, $Pr(B)$, $Pr(A, B)$, $Pr(A \cup B)$, $Pr(A|B)$, and $Pr(B|A)$.

   b. Estimate, using relative frequencies in the data, $Pr(A)$, $Pr(B)$, and $Pr(A, B)$.

   c. Using the results from part (b), calculate $Pr(A|B)$.

   d. Are events "A" and "B" independent? How do you know?

2.3 Again using the data in Table A.1, estimate probabilities of the Ithaca maximum temperature being at or below freezing (32°F), given that the previous day's maximum temperature was at or below freezing,

   a. Accounting for the persistence in the temperature data.

   b. Assuming (incorrectly) that sequences of daily temperatures are independent.

2.4 Three radar sets, operating independently, are searching for "hook" echoes (a radar signature associated with tornados). Suppose that each radar has a probability of 0.05 of failing to detect this signature when a tornado present.

   a. Sketch a Venn diagram for a sample space appropriate to this problem.

   b. What is the probability that a tornado will escape detection by all three radars?

   c. What is the probability that a tornado will be detected by all three radars?

2.5 Suppose the Probability-of-Precipitation (PoP) forecast for 12UTC (tomorrow morning) to 00UTC (tomorrow night) is 0.7, and the PoP forecast for 00UTC (tomorrow night) to 12UTC (the morning of the following day) is 0.4.

   a. Assuming that the precipitation occurrences in these two time periods are independent, what is the probability of precipitation for the 24 h period running from 12UTC tomorrow through 12UTC the following day?

   b. In the real world, there is positive statistical dependence between precipitation occurrences in consecutive 12 h time periods. For example, the conditional probability of precipitation in the second (00UTC to 12UTC) 12 h period, given that precipitation occurred in the first period, is higher than the corresponding unconditional probability. If this dependence were accounted for, would the probability of precipitation for the combined 24 h period be larger or smaller than the answer to part (a) of this question? Justify your answer mathematically.

2.6 The effect of cloud seeding on suppression of damaging hail is being studied in your area, by randomly seeding or not seeding equal numbers of candidate storms. Suppose the probability of damaging hail from a seeded storm is 0.10, and the probability of damaging hail from an unseeded storm is 0.40. If one of the candidate storms has just produced damaging hail, what is the probability that it was seeded?

Part II

Univariate Statistics

Chapter 3

Empirical Distributions and Exploratory Data Analysis

3.1. BACKGROUND

One very important application of statistical ideas in meteorology and climatology is in making sense of a new set of data. Ultimately, the goal is to extract insight about the processes underlying the generation of the numbers. As mentioned in Chapter 1, meteorological observing systems and computer models, supporting both operational and research efforts, produce torrents of numerical data. Many of these large data sets are easily available through the internet, on such websites as www.ncdc.noaa.gov, www. ecmwf.int/en/forecasts/datasets, www.data.gov, and www.data.gov.uk. Even when working with a rela tively small data set, it can be a significant task just to get a feel for a new batch of numbers, and to begin to make some sense of them.

Broadly speaking, this activity is known as *Exploratory Data Analysis*, or EDA. Its systematic use increased substantially following Tukey's (1977) pathbreaking and very readable book of the same name. EDA methods draw heavily on a variety of graphical tools to aid in the comprehension of the large batches of numbers that may confront an analyst. Graphics are a very effective means of com-pressing and summarizing data, portraying much in little space, and exposing unusual features of a data set. Sometimes unusual data points result from errors in recording or transcription, and it is well to know about these as early as possible in an analysis. Sometimes the unusual data are valid and may turn out to be the most interesting and informative parts of a data set.

Many EDA methods were designed originally to be applied by hand, with pencil and paper, to small (up to perhaps 200-point) data sets. Modern computing capabilities have greatly broadened the scope of statistical graphics, a large variety of which are easily available (e.g., R Development Core Team, 2017; Theus and Urbanek, 2009).

3.1.1. Robustness and Resistance

Many of the classical techniques of statistics work best when fairly stringent assumptions about the nature of the data are met. For example, it is often assumed that data will follow the familiar bell-shaped curve of the Gaussian distribution (Section 4.4.2). Classical procedures can behave very badly (i.e., produce quite misleading results) if their assumptions are not satisfied by the data to which they are applied.

The assumptions of classical statistics were not made out of ignorance, but rather out of necessity. Invocation of simplifying assumptions in statistics, as in other fields, has allowed progress to be made

Statistical Methods in the Atmospheric Sciences. https://doi.org/10.1016/B978-0-12-815823-4.00003-1

through the derivation of elegant analytic results, which are relatively simple but powerful mathematical formulas. As has also been the case in many quantitative fields, the advent of cheap computing power has freed the data analyst from sole dependence on such results, by allowing alternatives requiring less stringent assumptions to become practical. This does not mean that the classical methods are no longer useful. However, it is much easier to check that a given set of data satisfies particular assumptions before a classical procedure is used, and good alternatives are computationally feasible in cases where the classical methods may not be appropriate.

Two important properties of EDA methods are that they are *robust* and *resistant*. Robustness and resistance are two aspects of reduced sensitivity to assumptions about the nature of a set of data. A robust method is not necessarily optimal in any particular circumstance, but performs reasonably well in most circumstances. For example, the sample average is the best characterization of the center of a set of data if it is known that those data follow a Gaussian distribution. However, if those data are decidedly non-Gaussian (e.g., if they are a record of extreme rainfall events), the sample average may yield a misleading characterization of their center. In contrast, robust methods generally are not sensitive to particular assumptions about the overall nature of the data.

A resistant method is not unduly influenced by a small number of outliers, or "wild data." As indicated previously, such points often show up in a batch of data through errors of one kind or another. The results of a resistant method change very little if a small fraction of the data values are changed, even if they are changed drastically. In addition to not being robust, the sample average is not a resistant characterization of the center of a data set, either. Consider the small set {11, 12, 13, 14, 15, 16, 17, 18, 19}. Its average is 15. However, if instead the set {11, 12, 13, 14, 15, 16, 17, 18, 91} had resulted from a transcription error, the "center" of the data (erroneously) characterized using the sample average would be 23. Resistant measures of the center of a batch of data, such as those to be presented later, would be changed little or not at all by the substitution of "91" for "19" in this simple example.

3.1.2. Quantiles

Many common summary measures rely on the use of selected sample *quantiles* (also known as *fractiles*). Quantiles and fractiles are essentially equivalent to the more familiar term, *percentile*. A sample quantile, q_p, is a number having the same units as the data, which exceeds that proportion of the data given by the subscript p, with $0 < p < 1$. The sample quantile q_p can be interpreted approximately as that value expected to exceed a randomly chosen member of the data set, with probability p. Equivalently, the sample quantile q_p would be regarded as the $p \times 100$th percentile of the data set.

The determination of sample quantiles requires that a batch of data first be arranged in order. Sorting small sets of data by hand presents little problem. Sorting larger sets of data is best accomplished by computer. Historically, the sorting step presented a major bottleneck in the application of robust and resistant procedures to large data sets. Today the sorting can be done easily using either a spreadsheet or data analysis program on a desktop computer, or one of many sorting algorithms available in collections of general-purpose computing routines (e.g., Press et al., 1986).

The sorted, or ranked, data values from a particular sample are called the *order statistics* of that sample. Given a set of data {$x_1, x_2, x_3, x_4, x_5, ..., x_n$}, the order statistics for this sample would be the same numbers, sorted in ascending order. These sorted values are conventionally denoted using parenthetical subscripts, that is, by the set {$x_{(1)}, x_{(2)}, x_{(3)}, x_{(4)}, x_{(5)}, ..., x_{(n)}$}. Here the ith smallest of the n data values is denoted $x_{(i)}$.

Certain sample quantiles are used especially often in the exploratory summarization of data. The *median*, or $q_{0.5}$, or 50th percentile, is most commonly used. This is the value at the center of the data set, in the sense that equal proportions of the data fall above and below it. If a data set contains an odd number of values, the median is simply the middle order statistic. If there are an even number, however, the data set has two middle values. In this case the median is conventionally taken to be the average of these two middle values. Formally,

$$q_{0.5} = \begin{cases} x_{([n+1]/2)} & , \ n \ \text{odd} \\ \dfrac{x_{(n/2)} + x_{([n/2]+1)}}{2} & , \ n \ \text{even} \end{cases}. \tag{3.1}$$

The *quartiles*, $q_{0.25}$ and $q_{0.75}$, are almost as commonly used as the median. Usually these are called the lower (LQ) and upper quartiles (UQ), respectively. They are located half-way between the median, $q_{0.5}$, and the extremes, $x_{(1)}$ and $x_{(n)}$. In typically colorful terminology, Tukey (1977) calls $q_{0.25}$ and $q_{0.75}$ the "*hinges*," apparently imagining that the data set can be folded first at the median, and then at the quartiles. The quartiles are thus the two medians of the half-data sets between $q_{0.5}$ and the extremes. If n is odd, these half-data sets each consist of $(n+1)/2$ points, and both include the median. If n is even these half-data sets each contain $n/2$ points, and do not overlap. The upper and lower *terciles*, $q_{0.333}$ and $q_{0.667}$, separate a data set into thirds, although sometimes the term tercile is used also to refer to any of the three equally sized portions of the data set so defined. Other quantiles that also are used frequently enough to be named are the four *quintiles*, $q_{0.2}$, $q_{0.4}$, $q_{0.6}$, and $q_{0.8}$; the *eighths*, $q_{0.125}$, $q_{0.375}$, $q_{0.625}$, and $q_{0.875}$ (in addition to the quartiles and median); and the nine *deciles*, $q_{0.1}$, $q_{0.2}$, ..., $q_{0.9}$.

Example 3.1. Computation of Common Quantiles

If there are $n=9$ data values in a batch of data, the median is $q_{0.5}=x_{(5)}$, or the fifth largest of the nine. The lower quartile is $q_{0.25}=x_{(3)}$, and the upper quartile is $q_{0.75}=x_{(7)}$.

If $n=10$, the median is the average of the two middle values, and the quartiles are the single middle values of the upper and lower halves of the data. That is, $q_{0.25}$, $q_{0.5}$, and $q_{0.75}$ are $x_{(3)}$, $[x_{(5)}+x_{(6)}]/2$, and $x_{(8)}$, respectively.

If $n=11$ then there is a unique middle value, but the quartiles are determined by averaging the two middle values of the upper and lower halves of the data. That is, $q_{0.25}$, $q_{0.5}$, and $q_{0.75}$ are $[x_{(3)}+x_{(4)}]/2$, $x_{(6)}$, and $[x_{(8)}+x_{(9)}]/2$, respectively.

For $n = 12$ both quartiles and the median are determined by averaging pairs of middle values; $q_{0.25}$, $q_{0.5}$, and $q_{0.75}$ are $[x_{(3)}+x_{(4)}]/2$, $[x_{(6)}+x_{(7)}]/2$, and $[x_{(9)}+x_{(10)}]/2$, respectively. $\diamond$

Estimating quantiles other than the median and quartiles involves use of more elaborate formulas, several choices for which are available (Hyndman and Fan, 1996).

3.2. NUMERICAL SUMMARY MEASURES

Some simple robust and resistant numerical summary measures are available that can be used without hand plotting or computer graphic capabilities. Often these will be the first quantities to be computed from a new and unfamiliar set of data. The next three subsections describe numerical summary measures of *location*, *spread*, and *symmetry*. Location refers to the central tendency or general magnitude of the data values. Spread denotes the degree of variation or dispersion around the center. Symmetry describes the balance with which the data values are distributed about their center. Asymmetric data tend to spread more either on the high side (have a long right tail) or the low side (have a long left tail). These three types of numerical summary measures correspond to the first three statistical moments of a data sample, but the classical measures of these moments (i.e., the sample mean, sample variance, and sample coefficient of skewness, respectively) are neither robust nor resistant.

3.2.1. Location

The median, $q_{0.5}$, is the most common robust and resistant measure of central tendency. Consider again the data set $\{11, 12, 13, 14, 15, 16, 17, 18, 19\}$. The median and mean are both 15. If, as noted before, the "19" is replaced erroneously by "91," the *mean*

$$\bar{x} = \frac{1}{n} \sum_{i=1}^{n} x_i \tag{3.2}$$

(=23) is very strongly affected, illustrating its lack of resistance to outliers. The median is unchanged by this common type of data error.

The *trimean* is a slightly more complicated measure of location, which takes into account more information about the magnitudes of the data. It is a weighted average of the median and the quartiles, with the median receiving twice the weight of each of the quartiles:

$$\text{Trimean} = \frac{q_{0.25} + 2q_{0.5} + q_{0.75}}{4}. \tag{3.3}$$

The *trimmed mean* is another resistant measure of location, whose sensitivity to outliers is reduced by removing a specified proportion of the largest and smallest observations. If the proportion of observations omitted at each end of the data distribution is α, then the α-trimmed mean is

$$\bar{x}_\alpha = \frac{1}{n - 2k} \sum_{i=k+1}^{n-k} x_{(i)}, \tag{3.4}$$

where k is an integer rounding of the product αn, the number of data values "trimmed" from each tail. The trimmed mean reduces to the ordinary mean (Equation 3.2) for $\alpha = 0$.

Other methods of characterizing location can be found in Andrews et al. (1972), Goodall (1983), Rosenberger and Gasko (1983), and Tukey (1977).

3.2.2. Spread

The *Interquartile Range* (IQR) is the most common, and simplest, robust and resistant measure of spread (also known as dispersion or scale). The IQR is simply the difference between the upper and lower quartiles:

$$\text{IQR} = q_{0.75} - q_{0.25}. \tag{3.5}$$

The IQR is a good index of the spread in the central part of a data set, since it simply specifies the range of the central 50% of the data. The fact that it ignores the upper and lower 25% of the data makes it quite resistant to outliers. This quantity is sometimes also called the *fourth-spread*.

It is worthwhile to compare the IQR with the conventional measure of scale of a data set, the *sample standard deviation*

$$s = \sqrt{\frac{1}{n-1}\sum_{i=1}^{n}(x_i - \bar{x})^2}. \tag{3.6}$$

The square of the sample standard deviation, s^2, is known as the *sample variance*. Because of the square root in Equation 3.6, the standard deviation has the same physical dimensions as the underlying data. The standard deviation is neither robust nor resistant. It is very nearly just the square root of the average squared difference between the data points and their sample mean. (The division by $n-1$ rather than n often is done in order to compensate for the fact that the x_i are closer, on average, to their sample mean than to the true population mean: dividing by $n-1$ exactly counters the resulting tendency for the sample standard deviation to be too small, on average.) Even one very large data value will be felt very strongly because it will be especially far away from the mean and that difference will be magnified by the squaring process. Consider again the set $\{11, 12, 13, 14, 15, 16, 17, 18, 19\}$. The sample standard deviation is 2.74, but it is greatly inflated to 25.6 if "91" erroneously replaces "19." It is easy to see that in either case IQR$=4$.

The IQR is very easy to compute, but it does have the disadvantage of not making much use of a substantial fraction of the data. The *median absolute deviation* (MAD) is a more complete, yet reasonably simple alternative. The MAD is easiest to understand by imagining the transformation $y_i = |x_i - q_{0.5}|$. Each transformed value y_i is the absolute value of the difference between the corresponding original data value and the median. The MAD is then just the median of the transformed (y_i) values:

$$\text{MAD} = median\,|x_i - q_{0.5}|. \tag{3.7}$$

Although this process may seem a bit elaborate at first, a little thought illustrates that it is analogous to computation of the standard deviation, but using operations that do not emphasize outlying data. The median (rather than the mean) is subtracted from each data value, any negative signs are removed by the absolute value (rather than squaring) operation, and the center of these absolute differences is located by their median (rather than their mean).

The *trimmed variance* is a still more elaborate measure of spread. The idea, as for the trimmed mean (Equation 3.4), is to omit a proportion of the largest and smallest values and compute the analog of the sample variance (the square of Equation 3.6)

$$s_\alpha^2 = \frac{1}{n-2k}\sum_{i=k+1}^{n-k}\left(x_{(i)} - \bar{x}_\alpha\right)^2. \tag{3.8}$$

Again, k is the nearest integer to αn, and squared deviations from the consistent trimmed mean (Equation 3.4) are averaged. The trimmed variance is sometimes multiplied by an adjustment factor to make it more consistent with the ordinary sample variance, s^2 (Graedel and Kleiner, 1985).

Other measures of spread can be found in Hosking (1990) and Iglewicz (1983).

3.2.3. Symmetry

The *sample skewness coefficient* is the conventional moments-based measure of symmetry in a batch of data,

$$\gamma = \frac{\frac{1}{n-1}\sum_{i=1}^{n}(x_i - \bar{x})^3}{s^3}. \tag{3.9}$$

This statistic is neither robust nor resistant. The numerator is similar to the sample variance, except that the average is over cubed deviations from the mean. Thus the sample skewness coefficient is even more sensitive to outliers than is the standard deviation. The average cubed deviation in the numerator is divided by the cube of the sample standard deviation in order to standardize and nondimensionalize the skewness coefficient, so that comparisons of skewness among different data sets are more meaningful.

Notice that cubing differences between the data values and their mean preserves the signs of these differences. Since the differences are cubed, the data values farthest from the mean will dominate the sum in the numerator of Equation 3.9. If there are a few very large data values, the sample skewness will tend to be positive. Therefore batches of data with long right tails are referred to both as right skewed and positively skewed. Data that are physically constrained to lie above a minimum value (such as precipitation or wind speed, both of which must be nonnegative) are often positively skewed. Conversely, if there are a few very small (or large negative) data values, these will fall far below the mean. The sum in the numerator of Equation 3.9 will then be dominated by a few large negative terms, so that the sample skewness coefficient will tend to be negative. Data with long left tails are referred to as left skewed or negatively skewed. For essentially symmetric data, the skewness coefficient will be near zero.

The *Yule-Kendall index*,

$$\gamma_{YK} = \frac{(q_{0.75} - q_{0.5}) - (q_{0.5} - q_{0.25})}{IQR} = \frac{q_{0.25} - 2q_{0.5} + q_{0.75}}{IQR}, \tag{3.10}$$

is a robust and resistant alternative to the sample skewness. It is computed by comparing the distance between the median and each of the two quartiles. If the data are right skewed, at least in the central 50% of the data, the distance to the median will be greater from the upper quartile than from the lower quartile. In this case the Yule-Kendall index will be greater than zero, consistent with the usual convention of right skewness being positive. If the positive skewness is strong enough that $q_{0.25} = q_{0.50}$, then $\gamma_{YK} = 1$. Conversely, left-skewed data will be characterized by a negative Yule-Kendall index, and if the negative skewness is sufficiently strong that $q_{0.75} = q_{0.50}$, then $\gamma_{YK} = -1$. Analogously to Equation 3.9, division by the interquartile range nondimensionalizes γ_{YK} (i.e., scales it in a way that the physical dimensions, such as meters or millibars, cancel) and thus improves its comparability between data sets.

Alternative measures of skewness can be found in Brooks and Carruthers (1953) and Hosking (1990).

3.2.4. Kurtosis

Extending Equation 3.9 by increasing the exponent leads to the coefficient of *kurtosis*,

$$\kappa = \frac{\frac{1}{n-1} \sum_{i=1}^{n} (x_i - \bar{x})^4}{s^4} - 3. \tag{3.11}$$

Although this summary statistic is often characterized as reflecting "flatness" or "peakedness" of a data distribution, it is really a measure of the weights of the upper and lower tails of a distribution relative to the distribution center (Westfall, 2014). Distributions for which $\kappa > 0$ are termed *leptokurtic* and have relatively heavy tails. Distributions for which $\kappa < 0$ are termed *platykurtic* and have relatively light tails. The subtraction of 3 in Equation 3.11 is a convention that allows comparison with the kurtosis of the

Gaussian, or bell-curve distribution, for which $\kappa = 0$. Accordingly, Equation 3.11 is often also known as *excess kurtosis*.

3.3. GRAPHICAL SUMMARY DEVICES

Numerical summary measures are quick and easy to compute and display, but they can express only a small amount of detail. In addition, their visual impact is limited. Graphical displays for exploratory data analysis have been devised that require only slightly more effort to produce.

3.3.1. Stem-and-Leaf Display

The *stem-and-leaf display* is a very simple but effective tool for producing an overall view of a new set of data. At the same time it provides the analyst with an initial exposure to the individual data values. In its simplest form, the stem-and-leaf display groups the data values according to their all-but-least significant digits. These values are written in either ascending or descending order to the left of a vertical bar, constituting the "stems." The least significant digit for each data value is then written to the right of the vertical bar, on the same line as the more significant digits with which it belongs. These least significant values constitute the "leaves."

Figure 3.1a shows a stem-and-leaf display for the January 1987 Ithaca maximum temperatures in Table A.1. The data values are reported to whole degrees and range from 9°F to 53°F. The all-but-least significant digits are thus the tens of degrees, which are written to the left of the bar. The display is built up by proceeding through the data values one by one, and writing its least significant digit on the appropriate line. For example, the temperature for 1 January is 33°F, so the first "leaf" to be plotted is the first "3" on the stem of temperatures in the 30s. The temperature for 2 January is 32°F, so a "2" is written to the right of the "3" just plotted for 1 January.

The initial stem-and-leaf display for this particular data set is a bit crowded, since most of the values are in the 20s and 30s. In cases like this, better resolution can be obtained by constructing a second plot, like that in Figure 3.1b, in which each stem has been split to contain only the values 0–4 or 5–9. Sometimes the opposite problem will occur, and the initial plot is too sparse. In that case (if there are at least three significant digits), replotting can be done with stem labels omitting the two least significant digits. Less stringent groupings can also be used. Regardless of whether or not it may be desirable to split or

```
                                          5•  3
                                          4*  5
                                          4•
   5 │ 3                                  3*  6  7  7
                                          3•  0  0  0  2  2  2  3  3  4  4
   4 │ 5                                  2*  5  5  6  6  6  7  7  8  8  9  9
   3 │ 3 2 0 0 7 7 0 6 2 3 4 2 0 4        2•  2  4
                                          1*  7
   2 │ 9 5 9 5 8 7 6 8 4 6 2 6 7          1•
                                          0*  9
   1 │ 7
   0 │ 9

        (a)                                   (b)
```

FIGURE 3.1 Stem-and-leaf displays for the January 1987 Ithaca maximum temperatures in Table A.1. The plot in panel (a) results after a first pass through the data, using the 10s as "stem" values. Optionally, a bit more resolution is obtained in panel (b) by creating separate stems for least-significant digits from 0 to 4 (•), and from 5 to 9 (*). At this stage it is also easy to sort the data values before rewriting them.

consolidate stems, it can be useful to rewrite the display with the leaf values sorted, as has also been done in Figure 3.1b.

The stem-and-leaf display is much like a quickly plotted histogram of the data, placed on its side. In Figure 3.1, for example, it is evident that these temperature data are reasonably symmetrical, with most of the values falling in the upper 20s and lower 30s. Optionally, sorting the leaf values also facilitates extraction of quantiles of interest. In this case it is easy to count inward from the extremes in Figure 3.1b to find that the median is 30, and that the two quartiles are 26 and 33.

It can happen that there are one or more outlying data points that are far removed from the main body of the data set. Rather than plot many empty stems, it is usually more convenient to just list these extreme values separately at the upper and/or lower ends of the display, as in Figure 3.2. This display is of data of wind speeds in kilometers per hour (km/h) to the nearest tenth. Merely listing two extremely large values and two values of calm winds at the top and bottom of the plot has reduced the length of the display by more than half. It is quickly evident that the data are strongly skewed to the right, as often occurs for wind data.

The stem-and-leaf display in Figure 3.2 also reveals something that might have been missed in a tabular list of the daily data. All the leaf values on each stem are the same. Evidently a rounding process has been applied to the data, knowledge of which could be important to some subsequent analyses. In this case the rounding process consists of transforming the data from the original units (knots) to km/h. For example, the four observations of 16.6 km/h result from original observations of 9 knots. No values on the 17 km/h stem would be possible, since observations of 10 knots transform to 18.5 km/h.

3.3.2. Boxplots

The *boxplot*, or *box-and-whisker plot*, is a very widely used graphical tool introduced by Tukey (1977). It is a simple plot of five sample quantiles: the minimum, $x_{(1)}$; the lower quartile, $q_{0.25}$; the median, $q_{0.5}$; the upper quartile, $q_{0.75}$; and the maximum, $x_{(n)}$. Using these five numbers, the boxplot essentially presents a quick sketch of the distribution of the underlying data.

FIGURE 3.2 Stem-and-leaf display of 0100 local-time wind speeds (km/h) at the Newark, New Jersey, Airport during December 1974. Very high and very low values are written outside the plot itself to avoid having many blank stems. The striking grouping of repeated leaf values suggests that a rounding process has been applied to the original observations. *From Graedel and Kleiner (1985).*

```
High: 38.8, 51.9

25 | 9
24 | 0
23 |
22 |
21 |
20 |
19 |
18 | 55
17 |
16 | 6666
15 |
14 |
13 |
12 | 9999
11 | 1111111
10 |
 9 | 22222
 8 |
 7 | 4444

Low: 0.0, 0.0
```

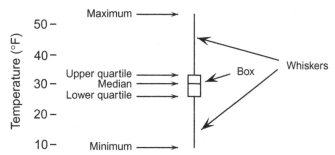

FIGURE 3.3   A simple boxplot, or box-and-whiskers plot, for the January 1987 Ithaca maximum temperature data. The upper and lower ends of the box are drawn at the quartiles, and the bar through the box is drawn at the median. The whiskers extend from the quartiles to the maximum and minimum data values. The arrows and associated labels are shown here only to define the plot attributes and are not included in practice.

Figure 3.3 shows a boxplot for the January 1987 Ithaca maximum temperature data in Table A.1. The box in the middle of the diagram is bounded by the upper and lower quartiles, and thus locates the central 50% of the data. The bar inside the box locates the median. The whiskers extend away from the box to the two extreme values.

Boxplots can convey a surprisingly large amount of information at a glance. It is clear from the small range occupied by the box in Figure 3.3, for example, that the data are concentrated quite near 30°F. Being based only on the median and the quartiles, this portion of the boxplot is highly resistant to any outliers that might be present. The full range of the data is also apparent at a glance. Finally, we can see easily that these data are nearly symmetrical, since the median is near the center of the box, and the whiskers are of comparable lengths.

3.3.3. Schematic Plots

A shortcoming of the boxplot is that information about the tails of the data distribution is highly generalized. The whiskers extend to the highest and lowest values, but there is no information about the distribution of data points within the upper and lower quarters of the data. For example, although Figure 3.3 shows that the highest maximum temperature is 53°F, it gives no information as to whether this is an isolated point (with the remaining warm temperatures cooler than, say, 40°F) or whether the warmer temperatures are more or less evenly distributed between the upper quartile and the maximum.

It is often useful to have some idea of the degree of unusualness of the extreme values. The *schematic plot*, which was also originated by Tukey (1977), is a refinement of the boxplot that presents more detail in the tails. The schematic plot is identical to the boxplot, except that extreme points deemed to be sufficiently unusual are plotted individually. Just how extreme is sufficiently unusual depends on the variability of the data in the central part of the sample, as reflected by the IQR. A given extreme value is regarded as being less unusual if the two quartiles are far apart (i.e., if the IQR is large), and more unusual if the two quartiles are near each other (the IQR is small).

The dividing lines between less- and more-unusual points are known in Tukey's idiosyncratic terminology as the *"fences."* Four fences are defined: inner and outer fences, above and below the data, according to

$$\text{Upper outer fence} = q_{0.75} + 3\,IQR$$
$$\text{Upper inner fence} = q_{0.75} + \frac{3\,IQR}{2}$$
$$\text{Lower inner fence} = q_{0.25} - \frac{3\,IQR}{2}$$
$$\text{Lower outer fence} = q_{0.25} - 3\,IQR.$$

(3.12)

Thus the two outer fences are located three times the distance of the interquartile range above and below the two quartiles. The inner fences are midway between the outer fences and the quartiles, being 1.5 times the distance of the interquartile range away from the quartiles.

In the schematic plot, points within the inner fences are called "inside." The range of the inside points is shown by the extent of the whiskers. Data points between the inner and outer fences are referred to as being "outside" and are plotted individually in the schematic plot. Points above the upper outer fence or below the lower outer fence are called "far out" and are plotted individually with a different symbol. These automatically generated boundaries, while somewhat arbitrary, have been informed by Tukey's experience and intuition. The resulting differences from the simple boxplot are illustrated in Figure 3.4. In common with the boxplot, the box in a schematic plot shows the locations of the quartiles and the median.

Example 3.2. Construction of a Schematic Plot

Figure 3.4 is a schematic plot for the January 1987 Ithaca maximum temperature data. As can be determined from Figure 3.1, the quartiles for these data are 33°F and 26°F, and the IQR $= 33 - 26 = 7$°F. From this information it is easy to compute the locations of the inner fences at $33+(3/2)(7)=43.5$°F and $26-(3/2)(7)=15.5$°F. Similarly, the outer fences are $33+(3)(7)=54$°F and $26-(3)(7)=5$°F. The dashed lines locating the fences are normally not included in schematic plots, but have been shown in Figure 3.4 for clarity.

The two warmest temperatures, 53°F and 45°F, are larger than the upper inner fence, and are shown individually by circles. The coldest temperature, 9°F, is less than the lower inner fence, and is also plotted individually. The whiskers are drawn to the most extreme temperatures inside the fences, 37°F

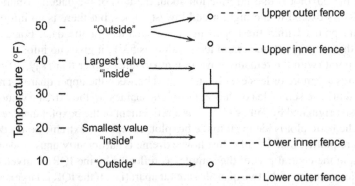

FIGURE 3.4 A schematic plot for the January 1987 Ithaca maximum temperature data. The central box portion of the figure is identical to the boxplot of the same data in Figure 3.3. The three values outside the inner fences are plotted separately. None of the values are beyond the outer fences, or "far out." Notice that the whiskers extend to the most extreme "inside" data values, and not to the fences. Dashed lines locating the fences are shown here for clarity, but are not normally included in a schematic plot. The arrows, dashed lines, and associated labels are shown only to define the plot attributes and are not included in practice.

and 17°F. If the warmest temperature had been 55°F rather than 53°F, it would have fallen outside the outer fence (far out) and would have been plotted individually with a different symbol. This separate symbol for the far out points is often an asterisk. ◇

One important use of schematic plots or boxplots is simultaneous graphical comparison of several batches of data. This use of schematic plots is illustrated in Figure 3.5, which shows side-by-side schematic plots for all four of the batches of temperature data in Table A.1. Of course it is known in advance that the maximum temperatures are warmer than the minimum temperatures, and comparing their schematic plots brings out this difference quite strongly. Apparently, Canandaigua was slightly warmer than Ithaca during this month, and more strongly so for the minimum temperatures. The Ithaca minimum temperatures were evidently more variable than the Canandaigua minimum temperatures. For both locations, the minimum temperatures are more variable than the maximum temperatures, especially in the central parts of the distributions represented by the boxes. The location of the median in the upper ends of the boxes in the minimum temperature schematic plots suggests a tendency toward negative skewness, as does the inequality of the whisker lengths for the Ithaca minimum temperature data. The maximum temperatures appear to be reasonably symmetrical for both locations. Note that none of the minimum temperature data are outside the inner fences, so that boxplots of the same data would be identical.

3.3.4. Other Boxplot Variants

Two variations on boxplots or schematic plots suggested by McGill et al. (1978) are sometimes used, particularly when comparing side-by-side plots. The first is to plot each box width proportional to $\sqrt{n}$. This simple variation allows plots for data having larger sample sizes to stand out and give a stronger visual impact.

The notched boxplot or schematic plot is a second variant. The boxes in these plots resemble hourglasses, with the constriction, or waist, located at the median. The lengths of the notched portions of the box differ from plot to plot, reflecting estimates of preselected confidence limits (Chapter 5) for the

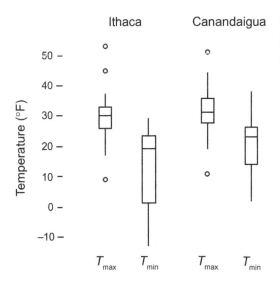

FIGURE 3.5 Side-by-side schematic plots for the January 1987 temperatures in Table A.1. The minimum temperature data for both locations are all "inside," so the schematic plots are identical to ordinary boxplots.

median. The details of constructing these intervals are given in Velleman and Hoaglin (1981). Combining both of these techniques, that is, constructing notched, variable-width plots, is straightforward. If the notched portion needs to extend beyond the quartiles, however, the overall appearance of the plot can begin to look a bit strange (an example can be seen in Graedel and Kleiner, 1985). An alternative to notching is to add shading or stippling in the box to span the computed interval, rather than deforming its outline with notches.

3.3.5. Histograms

The *histogram* is a very familiar graphical display device for representing the distribution of a single batch of data. The range of the data is divided into class intervals or *bins*, and the number of values falling into each interval is counted. The histogram then consists of a series of rectangles whose widths are defined by the class limits implied by the binwidths, and whose heights depend on the number of values in each bin. Example histograms are shown in Figure 3.6. Histograms quickly reveal such attributes of the data distribution as location, spread, and symmetry. If the data are multimodal (i.e., more than one "hump" in the distribution of the data), this is quickly evident as well.

Usually the widths of the bins are chosen to be equal. In that case the heights of the histogram bars are simply proportional to the numbers of counts. The vertical axis can be labeled to give either the number of counts represented by each bar (the absolute frequency) or the proportion of the entire sample represented by each bar (the relative frequency). More properly, however, it is the areas of the histogram bars (rather than their heights) that are proportional to probabilities. This point becomes important if the histogram bins are chosen to have unequal widths, or when a parametric probability function (Chapter 4) is to be superimposed on the histogram.

The main issue to be confronted when constructing a histogram is choice of the binwidth. Intervals that are too wide will result in important details of the data being masked (the histogram is too smooth). Intervals that are too narrow will result in a plot that is irregular and difficult to interpret (the histogram is

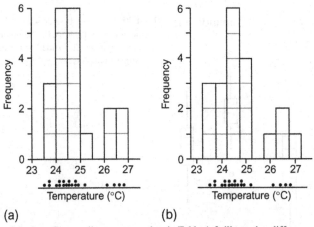

(a)   (b)

FIGURE 3.6   Histograms of the June Guayaquil temperature data in Table A.3, illustrating differences that can arise due to arbitrary shifts in the horizontal placement of the bins. Neither of these plots in panels (a) or (b) is more "correct" than the other. This figure also illustrates that each histogram bar can be viewed as being composed of stacked "building blocks" (gray) equal in number to the number of data values in the bin. Dotplots below each histogram locate the original data.

too rough). In general, narrower histogram bins are justified by larger data samples, but the nature of the data also influences the choice. A good approach to selecting the binwidth, w, is to begin by computing

$$w \approx \frac{c\,IQR}{n^{1/3}}, \tag{3.13}$$

where c is a constant in the range of perhaps 2.0–2.6. Results given in Scott (1992) indicate that $c = 2.6$ is optimal for Gaussian (bell-shaped) data, and that smaller values are more appropriate for skewed and/or multimodal data.

The initial binwidth computed using Equation 3.13, or arrived at according to any other rule, should be regarded as just a guideline or rule of thumb. Other considerations also will enter into the choice of the binwidth, such as the practical desirability of having the class boundaries fall on values that are natural with respect to the data at hand. (Computer programs that plot histograms must use rules such as that in Equation 3.13, and one indication of the care with which the software has been written is whether the resulting histograms have natural or arbitrary bin boundaries.) For example, the January 1987 Ithaca maximum temperature data has $IQR = 7°F$, and $n = 31$. A binwidth of 5.7°F would be suggested initially by Equation 3.13, using $c = 2.6$ since the schematic plot for these data (Figure 3.5) look at least approximately Gaussian. A natural choice in this case might be to choose 10 bins of width 5°F, yielding a histogram looking much like the stem-and-leaf display in Figure 3.1b.

3.3.6. Kernel Density Smoothing

One interpretation of the histogram is as a nonparametric estimator for the underlying probability distribution from which the data have been drawn. "Nonparametric" means that fixed mathematical forms of the kind presented in Chapter 4 are not assumed. However, the alignment of the histogram bins on the real line is an arbitrary choice, and construction of a histogram requires essentially that each data value is rounded to the center of the bin into which it falls. For example, in Figure 3.6a the bins have been aligned so that they are centered at integer temperature values ±0.25°C, whereas the equally valid histogram in Figure 3.6b has shifted these by 0.25°C. The two histograms in Figure 3.6 present somewhat different impressions of the data, although both indicate bimodality that can be traced (through the asterisks in Table A.3) to the occurrence of El Niño. Another, possibly less severe, difficulty with the histogram is that the rectangular nature of the histogram bars presents a rough appearance and appears to imply that any value within a given bin is equally likely.

Kernel density smoothing is an alternative to the histogram that does not require arbitrary rounding to bin centers, and which presents a smooth result. The application of kernel smoothing to the empirical frequency distribution of a data set produces the *kernel density estimate*, which is a nonparametric alternative to the fitting of a parametric probability density function (Chapter 4). It is easiest to understand kernel density smoothing as an extension of the histogram. As illustrated in Figure 3.6, after rounding each data value to its bin center the histogram can be viewed as having been constructed by stacking rectangular building blocks above each bin center, with the number of blocks equal to the number of data points in each bin. In Figure 3.6 the distribution of the data is indicated below each histogram in the form of *dotplots*, which locate each data value with a dot, and indicate instances of repeated data with stacks of dots.

The rectangular building blocks in Figure 3.6 each have area equal to the binwidth (0.5°F), because the vertical axis is just the raw number of counts in each bin. If instead the vertical axis had been chosen so the area of each building block was $1/n$ ($=1/20$ for these data), the resulting histograms would be

TABLE 3.1   Some Commonly Used Smoothing Kernels

Name	$h(t)$	Support [t for Which $h(t)>0$]	$1/\sigma_k$
Quartic (Biweight)	$(15/16)(1-t^2)^2$	$-1<t<1$	$\sqrt{7}$
Triangular	$1-|t|$	$-1<t<1$	$\sqrt{6}$
Quadratic (Epanechnikov)	$(3/4)(1-t^2)$	$-1<t<1$	$\sqrt{5}$
Gaussian	$(2\pi)^{-1/2}\exp[-t^2/2]$	$-\infty<t<\infty$	1

quantitative estimators of the underlying probability distribution, since the total histogram area would be 1 in each case, and total probability must sum to 1.

Kernel density smoothing proceeds in an analogous way, using characteristic shapes called *kernels*, that are generally smoother than rectangles. Table 3.1 lists four commonly used smoothing kernels, and Figure 3.7 shows their shapes graphically. These are all nonnegative functions with unit area, that is, $\int h(t)\,dt=1$ in each case, so each is a proper probability density function (discussed in more detail in Chapter 4). In addition, all are centered at zero. The *support* (value of the argument t for which $h(t)>0$) is $-1<t<1$ for the triangular, quadratic, and quartic kernels and covers the entire real line for the Gaussian kernel. The kernels listed in Table 3.1 are appropriate for use with continuous data (taking on values over all or some portion of the real line). Some kernels appropriate to discrete data (able to take on only a finite number of values) are presented in Rajagopalan et al. (1997).

Instead of stacking rectangular kernels centered on bin midpoints (which is one way of looking at histogram construction), kernel density smoothing is achieved by stacking kernel shapes, equal in number to the number of data values, with each stacked element being centered at the data value it represents. Of course in general kernel shapes do not fit together like building blocks, but kernel density smoothing is achieved through the mathematical equivalent of stacking, by adding the heights of all the kernel functions contributing to the smoothed estimate at a given value, x_0,

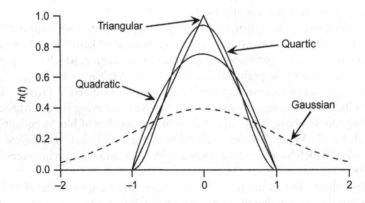

FIGURE 3.7   The four commonly used smoothing kernels defined in Table 3.1.

$$\hat{f}(x_0) = \frac{1}{nw} \sum_{i=1}^{n} h\left(\frac{x_0 - x_i}{w}\right). \tag{3.14}$$

The argument within the kernel function indicates that each of the kernels employed in the smoothing (corresponding to the data values x_i close enough to the point x_0 that the kernel height is not zero) is centered at its respective data value x_i and is scaled in width relative to the shapes as plotted in Figure 3.7 by the smoothing parameter w. Consider, for example, the triangular kernel in Table 3.1, with $t=(x_0-x_i)/w$. The function $h[(x_0-x_i)/w]=1-|(x_0-x_i)/w|$ is an isosceles triangle with support (i.e., nonzero height) for $x_i-w<x_0<x_i+w$, and the area within this triangle is w, because the area within $1-|t|$ over the support interval $-1<t<1$ is 1 and its base has been expanded (or contracted) by a factor of w. Therefore in Equation 3.14 the kernel heights stacked at the value x_0 will be those corresponding to any of the x_i at distances closer to x_0 than w. In order for the area under the entire function in Equation 3.14 to integrate to 1, which is desirable if the result is meant to estimate a probability density function, each of the n kernels to be superimposed should have area $1/n$. This is achieved by dividing each $h[(x_0-x_i)/w]$, or equivalently dividing their sum, by the product nw.

The choice of kernel type is usually less important than choice of the smoothing parameter. The Gaussian kernel is intuitively appealing, but it is computationally slower both because of the exponential function calls, and because its infinite support leads to all data values contributing to the smoothed estimate at any x_0 (none of the n terms in Equation 3.14 are ever zero). On the other hand, all the derivatives of the resulting function will exist, and nonzero probability is estimated everywhere on the real line, whereas these are not characteristics of probability density functions estimated using the other kernels listed in Table 3.1.

Example 3.3. Kernel Density Estimates for the Guayaquil Temperature Data
Figure 3.8 shows kernel density estimates for the June Guayaquil temperature data in Table A.3, corresponding to the histograms in Figure 3.6. The four probability density estimates have been constructed using the quartic kernel and four choices for the smoothing parameter w, which increase from panels (a) through (d). The role of the smoothing parameter is analogous to that of the histogram binwidth, also called w, in that larger values result in smoother shapes that progressively suppress details. Smaller values result in more irregular shapes that reveal more details, including more of the sampling variability. Figure 3.8b, plotted using $w=0.6$, also shows the individual kernels that have been summed to produce the smoothed density estimate. Since $w=0.6$ and the support of the quartic kernel is $-1<t<1$ (see Table 3.1) the width of each of the individual kernels in Figure 3.8b is 1.2°C. The five repeated data values 23.7, 24.1, 24.3, 24.5, and 24.8 (compare the dotplots at the bottom of Figure 3.6) are represented by the five taller kernels, the areas of which are each $2/n$. The remaining 10 data values are unique, and their kernels each have area $1/n$. ◇

Comparing the panels in Figure 3.8 emphasizes that a good choice for the smoothing parameter w is critical. Silverman (1986) suggests that a reasonable initial choice for use with the Gaussian kernel could be

$$w = \frac{\min\left\{0.9\,s, \frac{2}{3}IQR\right\}}{n^{1/5}}, \tag{3.15}$$

where s is the standard deviation of the data. Equation 3.15 indicates that less smoothing (smaller w) is justified for larger sample sizes n, although w should not decrease with sample size as quickly as does the

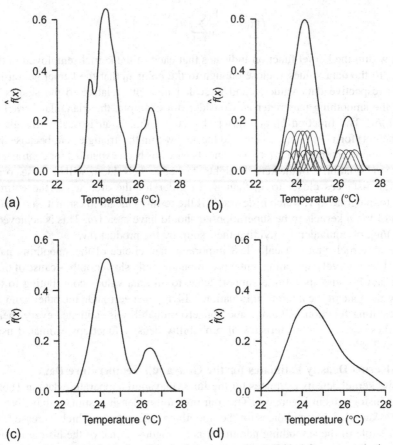

FIGURE 3.8 Kernel density estimates for the June Guayaquil temperature data in Table A.3, constructed using the quartic kernel and (a) $w=0.3$, (b) $w=0.6$, (c) $w=0.92$, and (d) $w=2.0$. Also shown in panel (b) are the individual kernels that have been added together to construct the estimate. These same data are shown as histograms in Figure 3.6.

histogram binwidth (Equation 3.13). Since the Gaussian kernel is intrinsically broader than the others listed in Table 3.1 (compare Figure 3.7), smaller smoothing parameters are appropriate for these, in proportion to the reciprocals of the kernel standard deviations (Scott, 1992), which are listed in the last column of Table 3.1. For the Guayaquil temperature data, $s=0.98$ and IQR$=0.95$, so 2/3 IQR is smaller than $0.9s$, and Equation 3.15 yields $w=(2/3)(0.95)/20^{1/5}=0.35$ for smoothing these data with a Gaussian kernel. But Figure 3.8 was prepared using the more compact quartic kernel, whose standard deviation is $1/\sqrt{7}$, yielding an initial choice for the smoothing parameter $w=(\sqrt{7})(0.35)=0.92$ (Figure 3.8c).

When kernel smoothing is used for an exploratory analysis or construction of an esthetically pleasing data display, a recommended smoothing parameter computed according to Equation 3.15 will often be the starting point for a subjective choice following some exploration through trial and error, and this process may even enhance the exploratory data analysis. In instances where the kernel density estimate will be used in subsequent quantitative analyses it may be preferable to estimate the smoothing parameter objectively using cross-validation methods similar to those presented in Chapter 7 (Scott, 1992; Sharma et al., 1998; Silverman, 1986). Adopting the exploratory approach, both $w=0.92$

(Figure 3.8c) and $w=0.6$ (Figure 3.8b) appear to produce reasonable balances between display of the main data features (here, the bimodality related to El Niño) and suppression of irregular sampling variability. Figure 3.8a, with $w=0.3$, is too rough for most purposes, as it retains irregularities that can probably be ascribed to sampling variations, and (almost certainly spuriously) indicates zero probability for temperatures near 25.5°C. On the other hand, Figure 3.8d is clearly too smooth, as it suppresses entirely the bimodality in the data.

Kernel smoothing can be extended to bivariate, and higher dimensional, data using the product-kernel estimator

$$\hat{f}(\boldsymbol{x}_0) = \frac{1}{n\, w_1\, w_2 \cdots w_K} \sum_{i=1}^{n} \left[\prod_{k=1}^{K} h\left(\frac{x_{0,k} - x_{i,k}}{w_k}\right) \right] .. \tag{3.16}$$

Here there are K data dimensions, $x_{0,k}$ denotes the point at which the smoothed estimate is produced in the kth of these dimensions, and the uppercase *pi* indicates multiplication of factors analogously to the summation of terms indicated by an uppercase sigma. The same (univariate) kernel $h(\bullet)$ is used in each dimension, although not necessarily with the same smoothing parameter w_k. In general the multivariate smoothing parameters w_k will need to be larger than for the same data smoothed alone (i.e., for a univariate smoothing of the corresponding kth variable in the vector $\boldsymbol{x}$) and should decrease with sample size in proportion to $n^{-1/(K+4)}$. Equation 3.16 can be extended to include also nonindependence of the kernels among the K dimensions by using a multivariate probability density (e.g., the multivariate normal distribution described in Chapter 12) for the kernel (Scott, 1992, Sharma et al., 1998, Silverman, 1986).

Kernel density estimates can be combined with boxplots to produce an informative graphic known as the *violin plot* (Hintze and Nelson, 1998). A violin plot usually consists of a central boxplot, with kernel density estimates based on the same underlying data plotted symmetrically on both sides of the boxplot. Figure 3.9 shows violin plots for the same maximum and minimum temperature data portrayed as

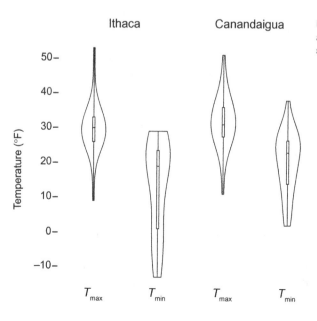

FIGURE 3.9 Violin plots for the January 1987 temperature data in Table A.1, which can be compared to the schematic plots for the same data shown in Figure 3.5.

schematic plots in Figure 3.5. Violin plots show more detail about the data distribution than the simpler boxplots alone, while the boxplots on the midlines allow identification of the median and quartiles of the smoothed distribution. Distributional features which might not be evident from a boxplot alone can be discerned in a violin plot. For example, the violin plot for a bimodal distribution would tend to take on the shape of the musical instrument for which it is named. The violin plot for the Ithaca minimum temperatures in Figure 3.9 begins to suggest this attribute.

Finally, note that kernel smoothing can be applied in settings other than estimation of probability distribution functions. When estimating a general smoothing function, which is not constrained to integrate to 1, a smoothed value of a function $y = f(x)$ at any point x_0 can be computed using the *Nadaraya-Watson kernel-weighted average*,

$$f(x_0) = \frac{\sum_{i=1}^{n} h\left(\frac{x_0 - x_i}{w}\right) y_i}{\sum_{i=1}^{n} h\left(\frac{x_0 - x_i}{w}\right)}, \tag{3.17}$$

where y_i is the raw value at x_i of the response variable to be smoothed. For example, Figure 3.10 shows mean numbers of tornado days per year, based on daily tornado occurrence counts in 80×80 km grid squares, for the period 1980–1999. The figure was produced after a three-dimensional smoothing using a Gaussian kernel, smoothing in time with $w = 15$ days, and smoothing in latitude and longitude with $w = 120$ km. The figure allows a straightforward interpretation of the underlying data, which in raw form are very erratic in both space and time.

More on kernel smoothing methods can be found in Chapter 6 of Hastie et al. (2009).

3.3.7. Cumulative Frequency Distributions

The *cumulative frequency distribution* is a display related to the histogram. It is also known as the *empirical cumulative distribution function*. The cumulative frequency distribution is a two-dimensional plot in which the vertical axis shows cumulative probability estimates associated with data values on the horizontal axis. That is, the plot represents relative frequency estimates for the probability that an arbitrary or random future datum will not exceed the corresponding value on the horizontal axis. Thus the cumulative frequency distribution is like the integral of a histogram with arbitrarily narrow binwidth. Figure 3.11 shows two empirical cumulative distribution functions, illustrating that they are step functions with probability jumps occurring at the data values. Just as histograms can be smoothed using kernel density estimators, smoothed versions of empirical cumulative distribution functions can be obtained by integrating the result of a kernel smoothing.

The vertical axes in Figure 3.11 show the empirical cumulative distribution function, $p(x)$, which can be expressed as

$$p(x) \approx \Pr\{X \leq x\}. \tag{3.18}$$

The notation on the right side of this equation can be somewhat confusing at first, but is standard in statistical work. The uppercase letter X represents the generic random variable or the "arbitrary or random future" value referred to in the previous paragraph. The lowercase x, on both sides of Equation 3.18, represents a specific value of the random quantity. In the cumulative frequency distribution, these specific values are plotted on the horizontal axis.

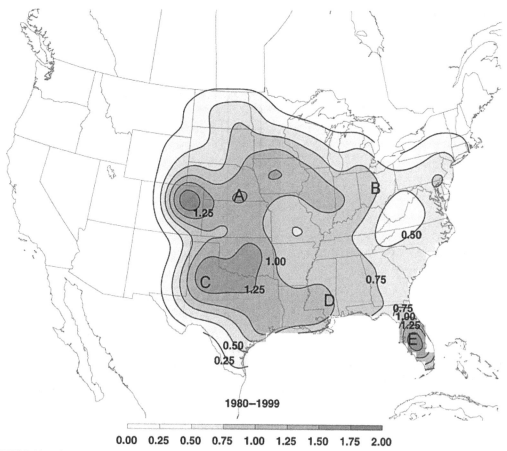

FIGURE 3.10 Mean numbers of tornado days per year in the United States, as estimated using a three-dimensional (time, latitude, longitude) kernel smoothing of daily, 80×80 km gridded tornado occurrence counts. *From Brooks et al. (2003). © American Meteorological Society. Used with permission.*

In order to construct a cumulative frequency distribution, it is necessary to estimate $p(x)$ using the ranks, i, of the order statistics, $x_{(i)}$. In the literature of hydrology these estimates are known as *plotting positions* (e.g., Harter, 1984), reflecting their historical use in graphically comparing the empirical distributions with candidate parametric functions (Chapter 4) that might be used to represent them. There is a substantial literature devoted to equations that can be used to calculate plotting positions, and thus to estimate cumulative probabilities from data sets. Most are particular cases of the formula

$$p\left(x_{(i)}\right) = \frac{i - a}{n + 1 - 2a} \quad , 0 \leq a \leq 1, \tag{3.19}$$

in which different values for the constant a result in different plotting position estimators, some of which are presented in Table 3.2. The names in this table relate to authors who proposed the various estimators and not to particular probability distributions that may be named for the same authors.

Several of the plotting positions in Table 3.2 are motivated by characteristics of the *sampling distributions* of the cumulative probabilities associated with the order statistics. The notion of a

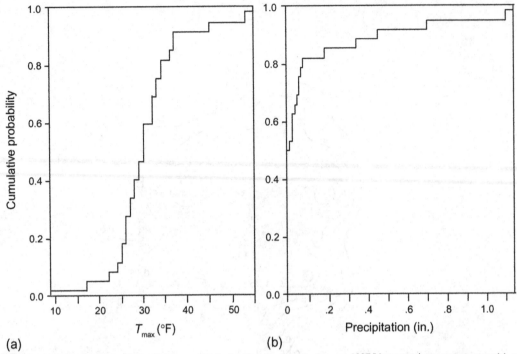

FIGURE 3.11   Empirical cumulative frequency distribution functions for the January 1987 Ithaca maximum temperature (a), and precipitation data (b). The S-shape exhibited by the temperature data is characteristic of reasonably symmetrical data, and the concave downward character exhibited by the precipitation data is characteristic of data that are skewed to the right.

TABLE 3.2   Some Common Plotting Position Estimators for Cumulative Probabilities Corresponding to the *i*th Order Statistic, $x_{(i)}$, and the Corresponding Values of the Parameter *a* in Equation 3.18

Name	Formula	a	Interpretation
Weibull	$i/(n+1)$	0	mean of sampling distribution
Benard & Bos-Levenbach	$(i-0.3)/(n+0.4)$	0.3	approximate median of sampling distribution
Tukey	$(i-1/3)/(n+1/3)$	1/3	approximate median of sampling distribution
Cunnane	$(i-2/5)/(n+1/5)$	2/5	subjective choice, commonly used in hydrology
Gringorten	$(i-0.44)/(n+0.12)$	0.44	consonance with the Gumbel distribution (Equation 4.67)
Hazen	$(i-1/2)/n$	1/2	midpoints of *n* equal intervals on [0, 1]
Gumbel	$(i-1)/(n-1)$	1	mode of sampling distribution

sampling distribution is considered in more detail in Chapter 5, but briefly think about hypothetically obtaining a large number of data samples of size n from some unknown distribution. The ith order statistics from these samples will differ somewhat from each other, but each will correspond to some cumulative probability in the distribution from which the data were drawn. In aggregate over the large number of hypothetical samples there will be a distribution—the sampling distribution—of cumulative probabilities corresponding to the ith order statistic. One way to imagine this sampling distribution is as a histogram of cumulative probabilities for, say, the smallest (or any of the other order statistics) of the n values in each of the batches. This notion of the sampling distribution for cumulative probabilities is expanded upon more fully in a climatological context by Folland and Anderson (2002).

The mathematical form of the sampling distribution of cumulative probabilities corresponding to the ith order statistic is known to be a Beta distribution (see Section 4.4.6), with parameters $\alpha = i$ and $\beta = n - i + 1$, regardless of the distribution from which the x's have been independently drawn (Gumbel, 1958). Thus the Weibull ($a = 0$) plotting position estimator is the mean of the cumulative probabilities corresponding to a particular $x_{(i)}$, averaged over many hypothetical samples of size n. Similarly, the Benard & Bos-Levenbach ($a = 0.3$) and Tukey ($a = 1/3$) estimators approximate the medians of these distributions. The Gumbel ($a = 1$) plotting position locates the modal (single most frequent) cumulative probability, although it ascribes zero and unit cumulative probability to $x_{(1)}$ and $x_{(n)}$, respectively, leading to the generally unwarranted implication that the probabilities of observing data more extreme than these are zero. It is possible also to derive plotting position formulas using the reverse perspective, thinking about the sampling distributions of data quantiles x_i corresponding to particular, fixed cumulative probabilities (e.g., Cunnane, 1978; Stedinger et al., 1993). Plotting positions resulting from this approach depend on the distribution from which the data have been drawn, although the Cunnane ($a = 2/5$) plotting position is a compromise approximation to many of them.

In practice most of the various plotting position formulas produce quite similar results, especially when judged in relation to the intrinsic variability (Equation 4.59b) of the sampling distribution of the cumulative probabilities, which is much larger than the differences among the various plotting positions in Table 3.2. Generally very reasonable results are obtained using moderate (in terms of the parameter a) plotting positions such as Tukey or Cunnane. Reviewing and comparing properties of various plotting positions, Hyndman and Fan (1996) conclude with an overall preference for the Tukey estimator.

Figure 3.11a shows the cumulative frequency distribution for the January 1987 Ithaca maximum temperature data, using the Tukey ($a = 1/3$) plotting position to estimate the cumulative probabilities. Figure 3.11b shows the Ithaca precipitation data displayed in the same way. For example, the coldest of the 31 temperatures in Figure 3.11a is $x_{(1)} = 9°F$, and $p(x_{(1)})$ is plotted at $(1 - 0.333)/(31 + 0.333) = 0.0213$. The steepness in the center of the plot reflects the concentration of data values in the center of the distribution, and the flatness at high and low temperatures results from data being more rare there. The S-shaped character of this plot is indicative of a reasonably symmetrical data distribution, with comparable numbers of observations on either side of the median at a given distance from the median. The cumulative distribution function for the precipitation data (Figure 3.11b) rises quickly on the left because of the high concentration of data values there, and then rises more slowly in the center and right of the figure because of the relatively fewer large observations. The concave downward character of this cumulative distribution function is thus indicative of positively skewed data. A plot of cumulative probability for a batch of negatively skewed data would show just the reverse characteristics: a very shallow slope in the left and center of the diagram, rising steeply toward the right, yielding a function that would be concave upward.

3.4. REEXPRESSION

It is possible that the original scale of measurement may obscure important features in a set of data. If so, an analysis can be facilitated, or may yield more revealing results, if the data are first subjected to a mathematical transformation. Such transformations can also be very useful for helping data conform to the assumptions of regression analysis (see Section 7.2), or allowing application of multivariate statistical methods that may assume Gaussian distributions (see Chapter 12). In the terminology of exploratory data analysis, such data transformations are known as *reexpression* of the data.

3.4.1. Power Transformations

Often data transformations are undertaken in order to make the distribution of values more nearly symmetrical, and the resulting symmetry may allow use of more familiar and traditional statistical techniques. Sometimes a symmetry-producing transformation can make exploratory analyses, such as those described in this chapter, more revealing. These transformations can also aid in comparing different batches of data, for example, by rendering the relationship between two variables more nearly linear. Another important use of transformations is to make the variations or dispersion (i.e., the spread) of one variable less dependent on the value of another variable, in which case the transformation is called *variance stabilizing*.

Undoubtedly the most commonly used (although not the only possible—see, e.g., Equation 12.12) symmetry-producing transformations are the *power transformations*, defined by one or the other of the two closely related functions

$$T_1(x) = \begin{cases} x^\lambda & , \lambda > 0 \\ \ln(x) & , \lambda = 0, \\ -(x^\lambda) & , \lambda < 0 \end{cases} \tag{3.20a}$$

and

$$T_2(x) = \begin{cases} \dfrac{x^\lambda - 1}{\lambda} & , \lambda \neq 0 \\ \ln(x) & , \lambda = 0 \end{cases} . \tag{3.20b}$$

These transformations are useful when dealing with unimodal (single-humped) distributions of strictly positive data variables. Each of these functions defines a family of transformations indexed by the single parameter λ. The name power transformation derives from the fact that the important work of these transformations—changing the shape of the data distribution—is accomplished by the exponentiation, or raising the data values to the power λ. Thus the sets of transformations in Equations 3.20a and 3.20b are actually quite comparable, and a particular value of λ produces the same effect on the overall shape of the data in either case. The transformations in Equation 3.20a are of a slightly simpler form and are often employed because of the greater ease. The transformations in Equation 3.20a are sometimes known as "*Tukey's ladder*". The transformations in Equation 3.20b, also known as the *Box-Cox transformations*, are simply shifted and scaled versions of Equation 3.20a, and are sometimes

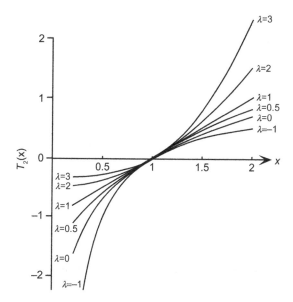

FIGURE 3.12 Graphs of the power transformations in Equation 3.20b for selected values of the transformation parameter λ. For $\lambda = 1$ the transformation is linear and produces no change in the shape of the data distribution. For $\lambda < 1$ the transformation reduces all data values, with larger values more strongly affected. The reverse effect is produced by transformations with $\lambda > 1$.

more useful when comparing among different transformations. Also, Equation 3.20b is mathematically "nicer" since the limit of the transformation in the upper equality as $\lambda \to 0$ is actually the function $\ln(x)$.

In both Equations 3.20a and 3.20b, adjusting the value of the parameter λ yields specific members of an essentially continuously varying set of smooth transformations. These transformations are sometimes referred to as the *ladder of powers*. A few of these transformation functions are plotted in Figure 3.12. The curves in this figure are functions specified by Equation 3.20b, although the corresponding curves from Equation 3.20a have the same shapes. Figure 3.12 makes it clear that use of the logarithmic transformation for $\lambda = 0$ fits neatly into the spectrum of the power transformations. This figure also illustrates another property of the power transformations, which is that they are all increasing functions of the original variable, x. This property is achieved in Equation 3.20a by the negative sign in the transformations with $\lambda < 1$. For the transformations in Equation 3.20b this sign reversal is achieved by dividing by λ. This strictly increasing property of the power transformations implies that they are order preserving, so that the smallest value in the original data set will correspond to the smallest value in the transformed data set, and likewise for the largest values. In fact, there will be a one-to-one correspondence between all the order statistics of the original and transformed distributions. Thus the median, quartiles, and so on, of the original data will be transformed to the corresponding quantiles of the transformed data.

Clearly for $\lambda = 1$ the shape of the data distribution remains unchanged. For $\lambda > 1$ the data values are increased (except for the subtraction of $1/\lambda$ and division by λ, if Equation 3.20b is used), with the larger values being increased more than the smaller ones. Therefore power transformations with $\lambda > 1$ will help produce symmetry when applied to negatively skewed data. The reverse is true for $\lambda < 1$, where larger data values are decreased more than smaller values. Power transformations with $\lambda < 1$ are therefore applied to data that are originally positively skewed, in order to produce more nearly symmetric distributions. Figure 3.13 illustrates the mechanics of this process for an originally positively

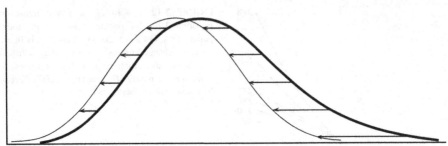

FIGURE 3.13   Effect of a power transformation with $\lambda < 1$ on a data distribution with positive skew (heavy curve). Arrows indicate that the transformation moves all the points to the left, with the larger values being moved much more. The resulting distribution (light curve) is reasonably symmetric.

skewed distribution (heavy curve). Applying a power transformation with $\lambda < 1$ reduces all the data values, but affects the larger values more strongly. An appropriate choice of λ can often produce at least approximate symmetry through this process (light curve). Choosing an excessively small or negative value for λ would yield an overcorrection, resulting in the transformed distribution being negatively skewed.

Initial inspection of an exploratory graphic such as a schematic plot can indicate quickly the direction and approximate magnitude of the skew in a batch of data. It is thus usually clear whether a power transformation with $\lambda > 1$ or $\lambda < 1$ is appropriate, but a specific value for the exponent will not be so obvious. A number of approaches to choosing an appropriate transformation parameter have been suggested. The simplest of these is the d_λ statistic (Hinkley, 1977),

$$d_\lambda = \frac{|\,\mathrm{mean}(\lambda) - \mathrm{median}(\lambda)\,|}{\mathrm{spread}(\lambda)}. \tag{3.21}$$

Here, spread is some resistant measure of dispersion, such as the IQR or MAD. Each value of λ will produce a different mean, median, and spread in a particular set of data, and these functional dependencies on λ are indicated in the equation. The Hinkley d_λ is used to decide among power transformations essentially by trial and error, by computing its value for each of a number of different choices for λ. Usually these trial values of λ are spaced at intervals of 1/2 or 1/4. That choice of λ producing the smallest d_λ is then adopted to transform the data. One very easy way to do the computations is with a spreadsheet program on a desk computer.

The basis of the d_λ statistic is that the mean and median will be very close for symmetrically distributed data. Therefore as successively stronger power transformations (values of λ increasingly far from 1) move the data toward symmetry, the numerator in Equation 3.21 will move toward zero. As the transformations become too strong, the numerator will begin to increase relative to the spread measure, resulting in the d_λ increasing again.

Equation 3.21 is a simple and direct approach to finding a power transformation that produces symmetry or near symmetry in the transformed data. A more sophisticated approach was suggested in the original Box and Cox (1964) paper, which is particularly appropriate when the transformed data should have a distribution as close as possible to the bell-shaped Gaussian, for example, when the results of multiple transformations will be summarized simultaneously through the multivariate Gaussian or multivariate normal distribution (see Chapter 12). In particular, Box and Cox suggested choosing the power transformation exponent to maximize the log-likelihood function (see Section 4.6) for the Gaussian distribution

$$L(\lambda) = -\frac{n}{2} \ln\left[s^2(\lambda)\right] + (\lambda - 1) \sum_{i=1}^{n} \ln[x_i].$$ (3.22)

Here n is the sample size, and $s^2(\lambda)$ is the sample variance (computed with a divisor of n rather than $n-1$, see Equation 4.84b) of the data after transformation with the exponent λ. It is important to realize that the sum of the logarithms in the second term of Equation 3.22 pertains to the *un*transformed data. As was the case for using the Hinkley statistic (Equation 3.21), different values of λ may be tried, and the one yielding the largest value of $L(\lambda)$ is chosen as most appropriate. It is possible that the two criteria will yield different choices for λ since Equation 3.21 addresses only symmetry of the transformed data, whereas Equation 3.22 tries to accommodate all aspects of the Gaussian distribution, including but not limited to its symmetry. Note, however, that choosing λ by maximizing Equation 3.22 does not necessarily produce transformed data that are close to Gaussian if the original data are not well suited to the transformations in Equation 3.20.

Equations 3.20 and 3.22 are valid only if zero or negative values of the variable x cannot be realized. For transformation of data that include some zero or negative values, the original recommendation by Box and Cox (1964) was to modify the transformation by adding a positive constant to each data value, with the magnitude of the constant being large enough for all the data to be shifted onto the positive half of the real line. This easy approach is often adequate, but it is somewhat arbitrary and fails entirely if a future negative value of x is larger in absolute value than this constant. Yeo and Johnson (2000) have proposed a unified extension of the Box-Cox transformations that accommodate data anywhere on the real line:

$$T_3(x) = \begin{cases} \left[(x+1)^\lambda - 1\right]/\lambda & , x \geq 0 \text{ and } \lambda \neq 0 \\ \ln(x+1) & , x \geq 0 \text{ and } \lambda = 0 \\ -\left[(-x+1)^{2-\lambda} - 1\right]/(2-\lambda) & , x < 0 \text{ and } \lambda \neq 2 \\ -\ln(-x+1) & , x < 0 \text{ and } \lambda = 2 \end{cases}.$$ (3.23)

For $x > 0$, Equation 3.23 achieves the same effect as Equation 3.20b, although with the curves shifted to the left by one unit. The graphs of $T_3(x)$ resemble those in Figure 3.12, except that they pass through the origin. The simplest approach to choosing the transformation parameter λ for Equation 3.23 is again the Hinkley statistic (Equation 3.21), although Yeo and Johnson (2000) also provide a maximum likelihood estimation procedure.

Example 3.4. Choosing an Appropriate Power Transformation
Table 3.3 presents the 1933–1982 January Ithaca precipitation data from Table A.2 in Appendix A, sorted in ascending order and subjected to the power transformations $T_2(x)$ in Equation 3.20b, with $\lambda = 1$, $\lambda = 0.5$, $\lambda = 0$, and $\lambda = -0.5$. For $\lambda = 1$ this transformation amounts only to subtracting 1 from each data value. Note that even for the negative exponent $\lambda = -0.5$ the ordering of the original data is preserved in all the transformations, so that it is easy to determine the medians and the quartiles of the original and transformed data.

Figure 3.14 shows schematic plots for the data in Table 3.3. The untransformed data (leftmost plot) are clearly positively skewed, which is usual for distributions of precipitation amounts. All three of the

TABLE 3.3 Ithaca January Precipitation 1933–1982, From Table A.2 ($\lambda=1$)

Year	$\lambda=1$	$\lambda=0.5$	$\lambda=0$	$\lambda=-0.5$	Year	$\lambda=1$	$\lambda=0.5$	$\lambda=0$	$\lambda=-0.5$
1933	−0.56	−0.67	−0.82	−1.02	1948	0.72	0.62	0.54	0.48
1980	−0.48	−0.56	−0.65	−0.77	1960	0.75	0.65	0.56	0.49
1944	−0.46	−0.53	−0.62	−0.72	1964	0.76	0.65	0.57	0.49
1940	−0.28	−0.30	−0.33	−0.36	1974	0.84	0.71	0.61	0.53
1981	−0.13	−0.13	−0.14	−0.14	1962	0.88	0.74	0.63	0.54
1970	0.03	0.03	0.03	0.03	1951	0.98	0.81	0.68	0.58
1971	0.11	0.11	0.10	0.10	1954	1.00	0.83	0.69	0.59
1955	0.12	0.12	0.11	0.11	1936	1.08	0.88	0.73	0.61
1946	0.13	0.13	0.12	0.12	1956	1.13	0.92	0.76	0.63
1967	0.16	0.15	0.15	0.14	1965	1.17	0.95	0.77	0.64
1934	0.18	0.17	0.17	0.16	1949	1.27	1.01	0.82	0.67
1942	0.30	0.28	0.26	0.25	1966	1.38	1.09	0.87	0.70
1963	0.31	0.29	0.27	0.25	1952	1.44	1.12	0.89	0.72
1943	0.35	0.32	0.30	0.28	1947	1.50	1.16	0.92	0.74
1972	0.35	0.32	0.30	0.28	1953	1.53	1.18	0.93	0.74
1957	0.36	0.33	0.31	0.29	1935	1.69	1.28	0.99	0.78
1969	0.36	0.33	0.31	0.29	1945	1.74	1.31	1.01	0.79
1977	0.36	0.33	0.31	0.29	1939	1.82	1.36	1.04	0.81
1968	0.39	0.36	0.33	0.30	1950	1.82	1.36	1.04	0.81
1973	0.44	0.40	0.36	0.33	1959	1.94	1.43	1.08	0.83
1941	0.46	0.42	0.38	0.34	1976	2.00	1.46	1.10	0.85
1982	0.51	0.46	0.41	0.37	1937	2.66	1.83	1.30	0.95
1961	0.69	0.60	0.52	0.46	1979	3.55	2.27	1.52	1.06
1975	0.69	0.60	0.52	0.46	1958	3.90	2.43	1.59	1.10
1938	0.72	0.62	0.54	0.48	1978	5.37	3.05	1.85	1.21

The data have been sorted, with the power transformations $T_2(x)$ in Equation 3.20b applied for $\lambda=1$, $\lambda=0.5$, $\lambda=0$, and $\lambda=-0.5$. For $\lambda=1$ the transformation subtracts 1 from each data value. Schematic plots of these data are shown in Figure 3.14.

values outside the fences are large amounts, with the largest being far out. The three other schematic plots show the results of progressively stronger power transformations with $\lambda<1$. The logarithmic transformation ($\lambda=0$) both minimizes the Hinkley d_λ statistic (Equation 3.21) with IQR as the measure of spread and maximizes the Gaussian log-likelihood (Equation 3.22). The near symmetry exhibited by the schematic plot for the logarithmically transformed data supports the conclusion that it is best among the possibilities considered according to both criteria. The more extreme inverse square-root transformation ($\lambda=-0.5$) has evidently overcorrected for the positive skewness, as the three smallest amounts are now outside the lower fence. ◇

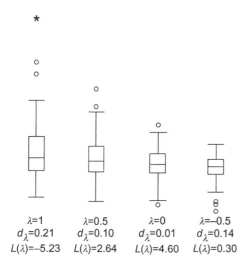

FIGURE 3.14 The effect of the power transformations $T_2(x)$ in Equation 3.20b on the January total precipitation data for Ithaca, 1933–1982 (Table A.2). The original data ($\lambda=1$) are skewed to the right, with the largest value being far out. The square-root transformation ($\lambda=0.5$) improves the symmetry somewhat. The logarithmic transformation ($\lambda=0$) produces a reasonably symmetric distribution. When subjected to the more extreme inverse square-root transformation ($\lambda=-0.5$) the data begin to exhibit negative skewness. The logarithmic transformation would be chosen as best by both the Hinkley d_λ statistic (Equation 3.21), and the Gaussian log-likelihood $L(\lambda)$ (Equation 3.22).

3.4.2. Some Other Nonlinear Transformations

In order for the shape of a transformed data distribution to be different from its untransformed counterpart, the transformation must be nonlinear. Equations 3.20 and 3.23 are by far the most frequently used such transformations, but they are not the only possibilities. This section presents a few others that can be useful in particular circumstances.

When the original data consist of proportions, probabilities, or other quantities p on the unit interval, $0 < p < 1$, it can be useful to transform them using

$$x = \ln\left(\frac{p}{1-p}\right), \tag{3.24}$$

which is known as the *log-odds, logit,* or *logistic* transformation. The bracketed quantity is called the *odds ratio.* The log-odds transformation yields transformed data on the full real line, $-\infty < x < \infty$.

Data that consist of correlations, so that $-1 < r < 1$, can be transformed to the full real line using the *Fisher Z,* or inverse hyperbolic tangent, transform,

$$Z = \frac{1}{2} \ln\left[\frac{1+r}{1-r}\right]. \tag{3.25}$$

Other nonlinear transformations can be designed to achieve particular goals. For example, Wang et al. (2012) propose the log-hyperbolic sine transformation for positively skewed data y,

$$z = \ln\left(\frac{e^y - e^{-y}}{2}\right). \tag{3.26}$$

This transformation was designed specifically for use with hydrological data which simultaneously exhibit positive skewness, and the tendency for variance to initially increase with y but stabilize for larger y. In contrast, Box-Cox transformations (Equations 3.20 or 3.23) with exponent $\lambda < 1$ will be effective at improving symmetry for positively skewed data, but will increasingly suppress variance as the data values become larger. The transformation in Equation 3.26 can be further elaborated by adjusting the tapering of the variance suppression by working with the linearly transformed variable $y = a + bx$, where now x is the original untransformed quantity, and a and b represent tuning constants.

3.4.3. Standardized Anomalies

Linear transformations, which do not change the shape of the data distribution, can nevertheless be useful when we are interested in working simultaneously with batches of data that are related but not strictly comparable because of differences in location and/or scale. One instance of this situation occurs when the data are subject to seasonal variations. Direct comparison of raw monthly temperatures, for example, will usually show little more than the dominating influence of the seasonal cycle: at most northern mid-latitude locations a record warm January will still be much colder than a record cool July. In situations of this sort, reexpression of the data in terms of *standardized anomalies* can be very helpful.

The standardized anomaly, z, is computed simply by subtracting the sample mean of the raw data x and dividing by the corresponding sample standard deviation:

$$z = \frac{x - \bar{x}}{s_x} = \frac{x'}{s_x}. \tag{3.27}$$

In the jargon of the atmospheric sciences, an *anomaly* x' is understood to mean the subtraction from a data value of a relevant average, as in the numerator of Equation 3.27. The term anomaly does not connote a data value or event that is abnormal or necessarily even unusual. The standardized anomaly in Equation 3.27 is produced by dividing the anomaly in the numerator by the corresponding standard deviation. This transformation is sometimes also referred to as a *normalization*. It would also be possible to construct standardized anomalies using resistant measures of location and spread, for example, subtracting the median and dividing by IQR, but this is rarely done. Use of standardized anomalies is motivated by ideas related to the bell-shaped Gaussian distribution, which are explained in Section 4.4.2. However, it is not necessary to assume that a batch of data follows any particular distribution in order to reexpress them in terms of standardized anomalies, and transforming non-Gaussian data according to Equation 3.27 will not make their distribution shape be any more Gaussian, because linear transformations do not change the shape of a data distribution.

The idea behind the standardized anomaly is to try to remove the influences of location and spread from a data sample. The physical units of the original data cancel, so standardized anomalies are always dimensionless quantities. Subtracting the mean produces a series of anomalies, x', located somewhere near zero. Division by the standard deviation puts excursions from the mean in different batches of data on equal footings. Collectively, a data sample that has been transformed to a set of standardized anomalies will exhibit a mean of zero and a standard deviation of 1.

To illustrate, it is often the case that summer temperatures are less variable than winter temperatures. We might find that the standard deviation for average January temperature at some location is around 3°C, but that the standard deviation for average July temperature at the same location is close to 1°C. An average July temperature 3°C colder than the long-term mean for July would then be quite unusual, corresponding to a standardized anomaly of −3. An average January temperature 3°C warmer than the long-term mean January temperature at the same location would be a fairly ordinary occurrence, corresponding to a standardized anomaly of only +1. Another way to look at the standardized anomaly is as a measure of distance, in standard deviation units, between a data value and its mean.

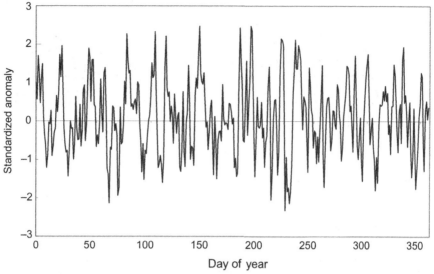

FIGURE 3.15 Standardized anomalies for the annual cycle of average daily temperatures at Boston, Massachusetts (USA), 1920–1999. *Adapted from Godfrey et al. (2002). © American Meteorological Society. Used with permission.*

Example 3.5. Standardizing a Nonstationary Time Series

Figure 3.15 demonstrates the idea of data standardization, comparing average daily temperatures throughout the year, at Boston, USA. Of course the raw temperatures exhibit warmer means in summer and colder means in winter, but also exhibit greater variability in winter than summer. The 365 daily standardized anomalies plotted in Figure 3.15 have been constructed by subtracting means for each day over the 80 years 1920–1999, and then dividing by the corresponding day-by-day standard deviations computed over the same period. That is, $z(t) = [x(t) - \bar{x}(t)]/s(t)$, where $\bar{x}(t)$ and $s(t)$ have been computed as smooth functions of the date t, as described in Section 11.4. The procedure achieves comparability of the values across seasons even though the means and standard deviations vary strongly throughout the year. Overall, the resulting standardized anomalies are centered near zero, and nearly all are smaller than 2 in absolute value. It is sometimes suggested that the peak around day 22 (the "January thaw") is somehow unusual, but Figure 3.15 indicates that this interpretation is simply an artifact of temperatures in winter being more variable than in other seasons (Godfrey et al., 2002).

Figure 3.15 illustrates the use of standardized anomalies to allow comparability across time, but the same idea can be applied for spatial comparisons by computing the standardized anomalies using location-specific means and standard deviations (e.g., Dabernig et al., 2017).   ◇

Example 3.6. Expressing Climatic Data in Terms of Standardized Anomalies

Figure 3.16 illustrates the use of standardized anomalies in an operational context, where Equation 3.27 is applied twice. The plotted points are values of the *Southern Oscillation Index*, which is an index of the atmospheric component of El-Niño-Southern Oscillation (ENSO) phenomenon, that is used by the Climate Prediction Center of the U.S. National Centers for Environmental Prediction (Ropelewski and Jones, 1987). The values of this index in the figure are derived from month-by-month differences in the standardized anomalies of sea-level pressure at two tropical locations: Tahiti, in the central Pacific Ocean; and Darwin, in northern Australia. In terms of Equation 3.27 the first step toward generating

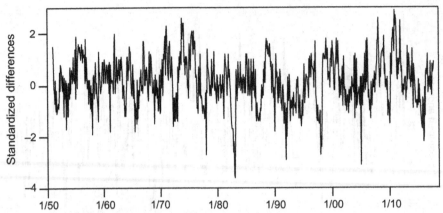

FIGURE 3.16 Standardized differences between the standardized monthly sea level pressure anomalies at Tahiti and Darwin (Southern Oscillation Index), January 1951–October 2017.

Figure 3.16 is to calculate the difference $\Delta z = z_{Tahiti} - z_{Darwin}$ for each month during the years plotted. The standardized anomaly z_{Tahiti} for January 1951, for example, is computed by subtracting the average pressure for all Januaries at Tahiti from the observed monthly pressure for January 1951. This difference is then divided by the standard deviation characterizing the year-to-year variations of January atmospheric pressure at Tahiti.

The curve in Figure 3.16 is based on monthly values that are themselves standardized anomalies of this difference of standardized anomalies Δz, so that Equation 3.27 has been applied three times to the original data. The first two of the standardizations are undertaken to minimize the influences of seasonal changes in the average monthly pressures and the year-to-year variability of the monthly pressures, separately at the two locations. The third standardization, calculating the standardized anomalies of the differences Δz, ensures that the resulting index will have unit standard deviation. For reasons that will be made clear in the discussion of the Gaussian distribution in Section 4.4.2, this attribute aids qualitative judgments about the unusualness of a particular index value.

Physically, during El Niño events the center of tropical Pacific precipitation activity shifts eastward from the western Pacific (near Darwin) to the central Pacific (near Tahiti). This shift is associated with higher than average surface pressures at Darwin and lower than average surface pressures at Tahiti, which together produce a negative value for the index plotted in Figure 3.16. The exceptionally strong El-Niño event of 1982–1983 is especially prominent in this figure. ◇

3.5. EXPLORATORY TECHNIQUES FOR PAIRED DATA

The ideas presented so far in this chapter have pertained mainly to the manipulation and investigation of single batches of data. Some comparisons have been made, such as the side-by-side schematic plots in Figure 3.5. There, several distributions of data from Appendix A were plotted, but potentially important aspects of the structure of those data were not shown. In particular, the relationships between variables observed on a given day were masked when the data from each batch were separately ranked prior to construction of the schematic plots. However, for each observation in one batch there is a corresponding observation from the same date in any one of the others. In this sense, these data are *paired*. Elucidating relationships among sets of data pairs often yields important insights.

3.5.1. Scatterplots

The nearly universal format for graphically displaying paired data is the familiar *scatterplot* or *x–y plot*. Geometrically, a scatterplot is simply a collection of points in the plane whose two Cartesian coordinates are the values of each member of the data pair. Scatterplots allow easy examination of such features in the data as trends, curvature in the relationship, clustering of one or both variables, changes of spread of one variable as a function of the other, and extraordinary points or outliers.

Figure 3.17 is a scatterplot of the maximum and minimum temperatures for Ithaca during January 1987. It is immediately apparent that very cold maxima are associated with very cold minima, and there is a tendency for the warmer maxima to be associated with the warmer minima. This scatterplot also shows that the central range of maximum temperatures is not strongly associated with minimum temperature, since maxima near 30°F occur with minima anywhere in the range of −5° to 20°F, or warmer.

Two optional but often useful embellishments on the scatterplot are also illustrated in Figure 3.17. The first of these is to sketch the individual data distributions, called the *marginal distributions*, using the lines on the upper and right-hand boundaries of the frame. These lines give the visual impression of fringe on a floor rug, and so Figure 3.17 is an example of a *rug plot*. The second embellishment is

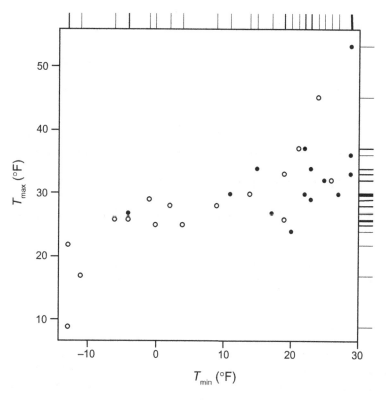

FIGURE 3.17 Scatterplot for daily maximum and minimum temperatures during January 1987 at Ithaca, New York. "Fringes" along the margins separately indicate the individual data distributions, with repeated data represented by heavier lines. Closed circles represent days with at least 0.01 in. of precipitation (liquid equivalent).

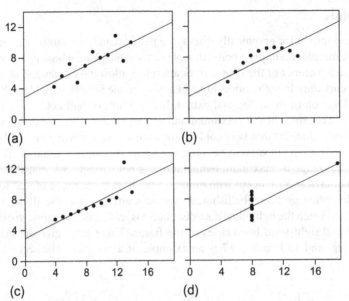

FIGURE 3.18 "Anscombe's quartet," illustrating the ability of graphical EDA to discern data features more powerfully than can a few numerical summaries. Each horizontal ("x") variable in panels (a) – (d) has the same mean (9.0) and standard deviation (11.0), as does each of the vertical ("y") variables (mean 7.5, standard deviation 4.12). Both the ordinary (Pearson) correlation coefficient ($r_{x,y}=0.816$) and the conventional least-squares regression relationship ($y=3+x/2$) are the same for all four of the panels.

the use of more than one type of plotting symbol. Here points representing days on which at least 0.01 in. (liquid equivalent) of precipitation were recorded are plotted using the filled circles. As was evident in Example 2.1 concerning conditional probability, precipitation days tend to be associated with warmer minimum temperatures. The scatterplot indicates that the maximum temperatures tend to be warmer as well on wet days, but that the effect is not as pronounced.

The scatterplots in Figure 3.18, known as *Anscombe's quartet* (Anscombe, 1973), illustrate the power of graphical EDA relative to computation of a few simple numerical summaries. The four sets of x–y pairs have been designed to have the same means and standard deviations in each panel, and the same ordinary (Pearson) correlation coefficient (Section 3.5.2) and the same least-squares linear regression relationship (Section 7.2.1). However, it is clear only from the graphical expositions that the relationships between the pairs of variables are very different in each case.

3.5.2. Pearson (Ordinary) Correlation

Often an abbreviated, single-valued measure of association between two variables, say x and y, is needed. In such situations, data analysts almost automatically (and sometimes somewhat uncritically) calculate a correlation coefficient. Usually, the term correlation coefficient is used to mean the "Pearson product-moment coefficient of linear correlation" between two variables x and y.

One way to view the *Pearson correlation* is as the ratio of the sample covariance between the two variables to the product of the two standard deviations,

$$r_{x,y} = \frac{Cov(x,y)}{s_x s_y} = \frac{\dfrac{1}{n-1}\sum\limits_{i=1}^{n}[(x_i - \bar{x})(y_i - \bar{y})]}{\left[\dfrac{1}{n-1}\sum\limits_{i=1}^{n}(x_i - \bar{x})^2\right]^{1/2}\left[\dfrac{1}{n-1}\sum\limits_{i=1}^{n}(y_i - \bar{y})^2\right]^{1/2}}$$

$$= \frac{\sum\limits_{i=1}^{n}(x_i' y_i')}{\left[\sum\limits_{i=1}^{n}(x_i')^2\right]^{1/2}\left[\sum\limits_{i=1}^{n}(y_i')^2\right]^{1/2}}, \tag{3.28}$$

where the primes denote anomalies, or subtraction of mean values, as before. Note that the sample variance is a special case of the covariance (numerator in Equation 3.28), with $x = y$. One application of the covariance is in the mathematics used to describe turbulence, where the average product of, for example, the horizontal velocity anomalies u' and v' is called the *eddy covariance*, and is used in the framework of Reynolds averaging (e.g., Stull, 1988).

The Pearson product-moment correlation coefficient is neither robust nor resistant. It is not robust because strong but nonlinear relationships between the two variables x and y may not be recognized. It is not resistant since it can be extremely sensitive to one or a few outlying point pairs. Nevertheless it is often used, both because its form is well suited to mathematical manipulation, and because it is closely associated with regression analysis (see Section 7.2), and the bivariate (Equation 4.31) and multivariate (see Chapter 12) Gaussian distributions.

The Pearson correlation has two important properties. First, it is bounded by -1 and 1; that is, $-1 \le r_{x,y} \le 1$. If $r_{x,y} = -1$ there is a perfect, negative linear association between x and y. That is, the scatterplot of y versus x consists of points all falling along one line, and that line has negative slope. Similarly if $r_{x,y} = 1$ there is a perfect positive linear association. Note, however, that $|r_{x,y}| = 1$ says nothing about the slope of the perfect linear relationship between x and y except that it is not zero, and it says nothing about any vertical offset in their relationship that may be evident in a scatterplot. The second important property is that the square of the Pearson correlation, $r_{x,y}^2$, specifies the proportion of the variability of one of either x or y that is linearly accounted for, or described, by the other. It is sometimes said that $r_{x,y}^2$ is the proportion of the variance of one variable "explained" by the other, but this interpretation is imprecise at best and is sometimes misleading. The correlation coefficient provides no explanation at all about the relationship between the variables x and y, at least not in any physical or causative sense. It may be that x physically causes y or vice versa, but often both result physically from some other or many other quantities or processes.

The heart of the Pearson correlation coefficient is the covariance between x and y in the numerator of Equation 3.28. The denominator is in effect just a scaling constant and is always positive. Thus the Pearson correlation is essentially a nondimensionalized covariance. Consider the hypothetical cloud of (x, y) data points in Figure 3.19, recognizable immediately as exhibiting positive correlation. The two perpendicular lines passing through the two sample means define four quadrants, labeled conventionally using Roman numerals. For points in quadrant I, both the x and y values are larger than their respective means ($x' > 0$ and $y' > 0$), so that both factors being multiplied will be positive. Therefore points in quadrant I contribute positive terms to the sum in the numerator of Equation 3.28. Similarly, for points in quadrant III, both x and y are smaller than their respective means ($x' < 0$ and $y' < 0$), and again the product of their anomalies will be positive. Thus points in quadrant III will also contribute positive terms

FIGURE 3.19 A hypothetical cloud of points in two dimensions, illustrating the mechanics of the Pearson correlation coefficient (Equation 3.28). The two sample means divide the plane into four quadrants, numbered I–IV.

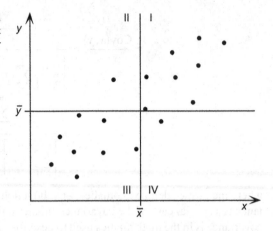

to the sum in the numerator. For points in quadrants II and IV one of the two variables x and y is above its mean and the other is below. Therefore the product in the numerator of Equation 3.28 will be negative for points in quadrants II and IV, and these points will contribute negative terms to the sum.

Most of the points in Figure 3.19 are in either quadrants I or III, and therefore most of the terms in the numerator of Equation 3.28 are positive. Only the two points in quadrants II and IV contribute negative terms, and these are small in absolute value since the x and y values are relatively close to their respective means. The result is a positive sum in the numerator and therefore a positive covariance. The two standard deviations in the denominator of Equation 3.28 must always be positive, which yields a positive correlation coefficient overall for the points in Figure 3.19. If most of the points had been in quadrants II and IV, the point cloud would slope downward rather than upward, and the correlation coefficient would be negative. If the point cloud were more or less evenly distributed among the four quadrants the correlation coefficient would be near zero, since the positive and negative terms in the sum in the numerator of Equation 3.28 would tend to cancel.

Another way of looking at the Pearson correlation coefficient is produced by moving the scaling constants in the denominator (the standard deviations), inside the summation of the numerator. This operation yields

$$r_{x,y} = \frac{1}{n-1} \sum_{i=1}^{n} \left[\frac{(x_i - \bar{x})}{s_x} \frac{(y_i - \bar{y})}{s_y} \right] = \frac{1}{n-1} \sum_{i=1}^{n} z_{x_i} z_{y_i}, \quad (3.29)$$

showing that the Pearson correlation is (nearly) the average product of the variables after conversion to standardized anomalies.

From the standpoint of computational economy, the formulas presented so far for the Pearson correlation are awkward. This is true whether or not the computation is to be done by hand or by a computer program. In particular, they all require two passes through a data set before the result is achieved: the first to compute the sample means, and the second to accumulate the terms involving deviations of the data values from their sample means (the anomalies). Passing twice through a data set requires twice the effort and provides double the opportunity for keying errors when using a hand calculator, and can amount to substantial increases in computer time, especially when using small computing systems such as in data loggers (Farrugia and Micallef, 2006). Therefore it is often useful to know the *computational form* of the Pearson correlation, which allows it to be calculated with only one pass through a data set.

The computational form arises through an easy algebraic manipulation of the summations in the correlation coefficient. Consider the numerator in Equation 3.28. Carrying out the indicated multiplication yields

$$
\begin{aligned}
\sum_{i=1}^{n}[(x_i - \bar{x})(y_i - \bar{y})] &= \sum_{i=1}^{n}[x_i y_i - x_i \bar{y} - y_i \bar{x} - \bar{x}\bar{y})] \\
&= \sum_{i=1}^{n}(x_i y_i) - \bar{y}\sum_{i=1}^{n}x_i - \bar{x}\sum_{i=1}^{n}y_i + \bar{x}\bar{y}\sum_{i=1}^{n}(1) \\
&= \sum_{i=1}^{n}(x_i y_i) - n\bar{x}\bar{y} - n\bar{x}\bar{y} + n\bar{x}\bar{y} \\
&= \sum_{i=1}^{n}(x_i y_i) - \frac{1}{n}\left[\sum_{i=1}^{n}x_i\right]\left[\sum_{i=1}^{n}y_i\right]
\end{aligned}
\tag{3.30}
$$

The second line in Equation 3.30 is arrived at through the realization that the sample means are constant, once the individual data values are determined, and therefore can be moved (factored) outside the summations. In the last term on this line there is nothing left inside the summation but the number 1, and the sum of n of these is simply n. The third step recognizes that the sample size multiplied by the sample mean yields the sum of the data values, which follows directly from the definition of the sample mean (Equation 3.2). The fourth step simply substitutes again the definition of the sample mean, to emphasize that all the quantities necessary for computing the numerator of the Pearson correlation can be known after one pass through the data. These are the sum of the x's, the sum of the y's, and the sum of their products.

It should be apparent from the similarity in form between Equation 3.30 and the summations in the denominator of the Pearson correlation that analogous formulas can be derived for them or, equivalently, for the sample standard deviation. The mechanics of the derivation are exactly as followed in Equation 3.30, with the result being

$$
s_x = \left[\frac{\sum x_i^2 - n\bar{x}^2}{n-1}\right]^{1/2} = \left[\frac{\sum x_i^2 - \frac{1}{n}(\sum x_i)^2}{n-1}\right]^{1/2}.
\tag{3.31}
$$

A similar result, of course, is obtained for y. Mathematically, Equation 3.31 is exactly equivalent to the formula for the sample standard deviation in Equation 3.6. Thus Equations 3.30 and 3.31 can be substituted into the form of the Pearson correlation given in Equations 3.28 or 3.29, to yield the computational form for the correlation coefficient

$$
r_{x,y} = \frac{\sum_{i=1}^{n}x_i y_i - \frac{1}{n}\left(\sum_{i=1}^{n}x_i\right)\left(\sum_{i=1}^{n}y_i\right)}{\left[\sum_{i=1}^{n}x_i^2 - \frac{1}{n}\left(\sum_{i=1}^{n}x_i\right)^2\right]^{1/2}\left[\sum_{i=1}^{n}y_i^2 - \frac{1}{n}\left(\sum_{i=1}^{n}y_i\right)^2\right]^{1/2}}.
\tag{3.32}
$$

Analogously, a computational form for the sample skewness coefficient (Equation 3.9) is

$$\gamma = \frac{\frac{1}{n-1}\left[\Sigma x_i^3 - \frac{3}{n}(\Sigma x_i)(\Sigma x_i^2) + \frac{2}{n^2}(\Sigma x_i)^3\right]}{s^3}. \tag{3.33}$$

It is important to mention a cautionary note regarding the computational forms just derived. There is a potential problem inherent in their use, which stems from the fact that they are very sensitive to roundoff errors. The problem arises because these formulas involve differences of two numbers that may be of comparable magnitude. To illustrate, suppose that the two terms on the last line of Equation 3.30 have each been saved to five significant digits. If the first three of these digits are the same, their difference will then be known only to two significant digits rather than five. The remedy to this potential problem is to retain as many as possible (preferably all) of the significant digits in each calculation, for example, by using the double-precision representation when programming floating-point calculations on a computer.

Example 3.7. Some Limitations of Linear Correlation

Consider the two artificial data sets in Table 3.4. The data values are few and small enough that the computational form of the Pearson correlation can be used without discarding any significant digits. For Set I, the Pearson correlation is $r_{x,y} = +0.88$, and for Set II the Pearson correlation is $r_{x,y} = +0.61$. Thus moderately strong linear relationships appear to be indicated for both sets of paired data.

The Pearson correlation is neither robust nor resistant, and these two small data sets have been constructed to illustrate these deficiencies. Figure 3.20 shows scatterplots for the two data sets, with Set I in panel (a) and Set II in panel (b). For Set I the relationship between x and y is actually stronger than indicated by the linear correlation of 0.88. The data points all fall very nearly on a smooth curve, but since that curve is not a straight line the Pearson coefficient underestimates the strength of the relationship. It is not robust to deviations from linearity in a relationship.

TABLE 3.4   Artificial Paired Data Sets for Correlation Examples

Set I		Set II	
x	y	x	y
0	0	2	8
1	3	3	4
2	6	4	9
3	8	5	2
5	11	6	5
7	13	7	6
9	14	8	3
12	15	9	1
16	16	10	7
20	16	20	17

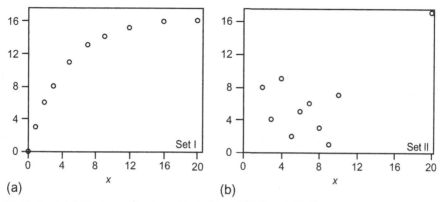

FIGURE 3.20 Scatterplots of the two artificial sets of paired data in Table 3.4. The Pearson correlation for the data in panel (a) (Set I in Table 3.4) of only 0.88 underrepresents the strength of the relationship, illustrating that this measure of correlation is not robust to nonlinearities. The Pearson correlation for the data in panel (b) (Set II) is 0.61, reflecting the overwhelming influence of the single outlying point, and illustrating lack of resistance.

Figure 3.20b illustrates that the Pearson correlation coefficient is not resistant to outlying data. Except for the single outlying point, the data in Set II exhibit very little structure. If anything these remaining nine points are weakly negatively correlated. However, the data pair $x=20$ and $y=17$ are so far from their respective sample means that the product of the resulting two large positive differences in the numerator of Equation 3.28 or Equation 3.29 dominate the entire sum, which erroneously indicates a moderately strong positive relationship among the ten data pairs overall.   ◇

3.5.3. Spearman Rank Correlation and Kendall's τ

Robust and resistant alternatives to the Pearson product-moment correlation coefficient are available. The first of these is known as the *Spearman rank correlation* coefficient. The Spearman correlation is simply the Pearson correlation coefficient computed using the ranks of the data. Conceptually, either Equation 3.28 or Equation 3.29 is applied, but to the ranks of the data rather than to the data values themselves. For example, consider the first data pair, (2, 8), in Set II of Table 3.4. Here $x=2$ is the smallest of the 10 values of x and therefore has rank 1. Being the eighth smallest of the 10, $y=8$ has rank 8. Thus this first data pair would be transformed to (1, 8) before computation of the correlation. Similarly, both x and y values in the outlying pair (20,17) are the largest of their respective batches of 10 and would be transformed to (10, 10).

In practice it is not necessary to use Equation 3.28, 3.29, or 3.32 to compute the Spearman rank correlation. Rather, the computations are simplified because we know in advance what the transformed values will be. Because the data are ranks, they consist simply of all the integers from 1 through the sample size n. For example, the average of the ranks of any of the four data batches in Table 3.4 is $(1+2+3+4+5+6+7+8+9+10)/10=5.5$. Similarly, the standard deviation (Equation 3.31) of these first ten positive integers is about 3.028. More generally, the average of the integers from 1 to n is $(n+1)/2$, and their variance is $n(n+1)/12$. Taking advantage of this information, computation of the Spearman rank correlation can be simplified to

$$r_{rank} = 1 - \frac{6 \sum_{i=1}^{n} D_i^2}{n(n^2-1)}, \qquad (3.34)$$

where D_i is the difference in ranks between the ith pair of data values. In cases of ties, where a particular data value appears more than once, all of these equal values are assigned their average rank before computing the D_i's.

Kendall's τ is a second robust and resistant alternative to the conventional Pearson correlation. Kendall's τ is calculated by considering the relationships among all possible matchings of the data pairs (x_i, y_i), of which there are $n(n-1)/2$ in a sample of size n. Any such matching in which both members of one pair are larger than their counterparts in the other pair are called *concordant*. For example, the pairs $(3, 8)$ and $(7, 83)$ are concordant because both numbers in the latter pair are larger than their counterparts in the former. Matchups in which each pair has one of the larger values, for example $(3, 83)$ and $(7, 8)$, are called *discordant*. The slope of the line segment connecting a concordant pair will be positive, and the slope of the segment connecting a discordant pair will be negative. Kendall's τ is calculated by subtracting the number of discordant pairs, N_D, from the number of concordant pairs, N_C, and dividing by the number of possible matchups among the n observations,

$$\tau = \frac{N_C - N_D}{n(n-1)/2}. \qquad (3.35)$$

Pairs for which one or both elements are identical contribute 1/2 to both N_C and N_D.

Example 3.8. Comparison of Spearman and Kendall Correlations for the Table 3.4 Data
In Set I of Table 3.4, there is a monotonic relationship between x and y, so that each of the two batches of data is already arranged in ascending order. Therefore both members of each of the n pairs have the same rank within its own batch, and the differences D_i are all zero. Actually, because of rounding the two largest y values are equal, so that each would be assigned the rank 9.5. Other than this tie, the sum in the numerator of the second term in Equation 3.34 is zero, and the Spearman rank correlation is essentially 1. This result better characterizes the strength of the relationship between x and y than does the Pearson correlation of 0.88. The Pearson correlation coefficient reflects the strength of linear relationships, but the Spearman rank correlation reflects the strength of monotone relationships.

Because the data in Set I exhibit an essentially perfect positive monotone relationship, all of the 10 $(10-1)/2=45$ possible matchups between data pairs yield concordant relationships. For data sets with perfect negative monotone relationships (one of the variables is strictly decreasing as a function of the other), all comparisons among data pairs yield discordant relationships. Except for the one tie, all comparisons for Set I are concordant relationships. $N_C=44.5$, so that Equation 3.31 would produce $\tau=(44.5-0.5)/45=0.978$.

For the data in Set II, the x values are presented in ascending order, but the y values with which they are paired are jumbled. The difference of ranks for the first record is $D_1=1-8=-7$. There are only three data pairs in Set II for which the ranks match (the fifth, sixth, and the outliers of the tenth pair). The remaining seven pairs will contribute nonzero terms to the sum in Equation 3.34, yielding $r_{rank}=0.018$ for Set II. This result reflects much better the very weak overall relationship between x and y in Set II than does the Pearson correlation of 0.61.

Calculation of Kendall's τ for Set II is facilitated by their being sorted according to increasing values of the x variable. Given this arrangement, the number of concordant combinations can be determined by counting the number of subsequent y variables that are larger than each of the first through $(n-1)$st listings in the table. Specifically, there are two y variables larger than 8 in $(2, 8)$ among the nine values below it, five y variables larger than 4 in $(3, 4)$ among the eight values below it, one y variable larger than 9 in $(4, 9)$ among the seven values below it, ..., and one y variable larger than 7 in $(10, 7)$ in the single value below it. Together there are $2+5+1+5+3+2+2+2+1=23$ concordant combinations, and $45-23=22$ discordant combinations, yielding $\tau=(23-22)/45=0.022$.   ◇

3.5.4. Serial Correlation

In Chapter 2 meteorological persistence, or the tendency for weather in successive time periods to be similar, was illustrated in terms of conditional probabilities for the two discrete events "precipitation" and "no precipitation." For continuous variables (e.g., temperature), persistence typically is characterized in terms of *serial correlation* or *temporal autocorrelation*. The prefix "auto" in autocorrelation denotes the correlation of a variable with itself, so that temporal autocorrelation indicates the correlation of a variable with its own future and past values. Sometimes such correlations are referred to as *lagged correlations*. Almost always, autocorrelations are computed as Pearson product-moment correlation coefficients, although there is no reason why other forms of lagged correlation cannot be computed as well.

The process of computing autocorrelations can be visualized by imagining two copies of a sequence of data values being written, with one of the series shifted by one unit of time. This shifting is illustrated in Figure 3.21, using the January 1987 Ithaca maximum temperature data from Table A.1. This data series has been rewritten, with the middle part of the month represented by ellipses, on the first line. The same record has been recopied on the second line, but shifted to the right by one day. This process results in 30 pairs of temperatures within the box, which are available for the computation of a correlation coefficient.

Autocorrelations are computed by substituting the lagged data pairs into the formula for the Pearson correlation (Equation 3.28). For the lag-1 autocorrelation there are $n-1$ such pairs. The only real confusion arises because the mean values for the two series will in general be slightly different. In Figure 3.21, for example, the mean of the 30 boxed values in the upper series is 29.77°F, and the mean for the boxed values in the lower series is 29.73°F. This difference arises because the upper series does not include the temperature for 1 January, and the lower series does not include the temperature for 31 January. Denoting the sample mean of the first $n-1$ values with the subscript "−" and that of the last $n-1$ values with the subscript "+," the lag-1 autocorrelation is

$$r_1 = \frac{\sum\limits_{i=1}^{n-1}[(x_i - \bar{x}_-)(x_{i+1} - \bar{x}_+)]}{\left[\sum\limits_{i=1}^{n-1}(x_i - \bar{x}_-)^2\right]^{1/2}\left[\sum\limits_{i=2}^{n}(x_i - \bar{x}_+)^2\right]^{1/2}} \tag{3.36}$$

For the January 1987 Ithaca maximum temperature data, for example, $r_1 = 0.52$.

```
33 | 32 30 29 25 30 53  · · ·  17 26 27 30 34 |
    | 33 32 30 29 25 30 53  · · ·  17 26 27 30 | 34
```

FIGURE 3.21 Construction of a shifted time series of January 1987 Ithaca maximum temperature data. Shifting the data by one day leaves 30 data pairs (enclosed in the box) with which to calculate the lag-1 autocorrelation coefficient.

The lag-1 autocorrelation is the most commonly computed measure of persistence, but it is also sometimes of interest to compute autocorrelations at longer lags. Conceptually, this is no more difficult than the procedure for the lag-1 autocorrelation, and computationally the only difference is that the two series are shifted by more than one time unit. Of course, as a time series is shifted increasingly relative to itself there is progressively less overlapping data to work with. Equation 3.36 can be generalized to the lag-k autocorrelation coefficient using

$$r_k = \frac{\sum_{i=1}^{n-k}[(x_i - \bar{x}_-)(x_{i+k} - \bar{x}_+)]}{\left[\sum_{i=1}^{n-k}(x_i - \bar{x}_-)^2\right]^{1/2}\left[\sum_{i=k+1}^{n}(x_i - \bar{x}_+)^2\right]^{1/2}}. \tag{3.37}$$

Here the subscripts "$-$" and "$+$" indicate sample means over the first and last $n-k$ data values, respectively. Equation 3.37 is valid for $0 \le k < n-2$, although it is usually only the lowest few values of k that will be of interest. So much data is lost at large lags that lagged correlations for roughly $k > n/2$ or $k > n/3$ rarely are computed.

In situations where a long data record is available it is sometimes acceptable to use an approximation to Equation 3.37, which simplifies the calculations and allows use of a computational form. In particular, if the data series is sufficiently long, the overall sample mean will be very close to the subset averages of the first and last $n-k$ values. The overall sample standard deviation will be close to the two subset standard deviations for the first and last $n-k$ values as well. Invoking these assumptions leads to the commonly used approximation

$$r_k \approx \frac{\sum_{i=1}^{n-k}[(x_i - \bar{x})(x_{i+k} - \bar{x})]}{\sum_{i=1}^{n}(x_i - \bar{x})^2} = \frac{\sum_{i=1}^{n-k}(x_i x_{i+k}) - \frac{n-k}{n^2}\left(\sum_{i=1}^{n}x_i\right)^2}{\sum_{i=1}^{n}x_i^2 - \frac{1}{n}\left(\sum_{i=1}^{n}x_i\right)^2}. \tag{3.38}$$

3.5.5. Autocorrelation Function

Together, the collection of autocorrelations computed for various lags is called the *autocorrelation function*. Often autocorrelation functions are displayed graphically, with the autocorrelations plotted as a function of lag. Figure 3.22 shows the first seven values of the sample autocorrelation function for the January 1987 Ithaca maximum temperature data. An autocorrelation function always begins with $r_0 = 1$, since any unshifted series of data will exhibit perfect correlation with itself. It is typical for an autocorrelation function to exhibit a more or less gradual decay toward zero as the lag k increases, reflecting the generally weaker statistical relationships between pairs of data points further removed from each other in time. It is instructive to relate this observation to the context of weather forecasting. If the autocorrelation function did not decay toward zero after a few days, making reasonably accurate forecasts at that range would be very easy: simply forecasting today's observation (the persistence forecast) or some modification of today's observation would give good results.

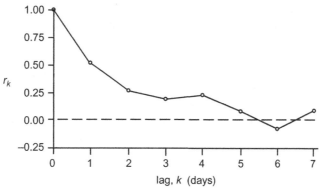

FIGURE 3.22   Sample autocorrelation function for the January 1987 Ithaca maximum temperature data. The correlation is 1 for $k=0$, since the unlagged data are perfectly correlated with themselves. The autocorrelation function decays to essentially zero for $k \geq 5$.

Sometimes it is useful to rescale the autocorrelation function, by multiplying all the autocorrelations by the variance of the data. The result, which is proportional to the numerators of Equations 3.37 and 3.38, is called the *autocovariance function*,

$$\gamma_k = \sigma^2 r_k, \ k = 0, 1, 2, \cdots. \tag{3.39}$$

The existence of autocorrelation in meteorological and climatological time series has important implications regarding the applicability of some standard statistical methods to atmospheric data. In particular, uncritical application of classical methods requiring independence of data within a sample will often give badly misleading results when applied to strongly persistent series. In some cases it is possible to successfully modify these techniques, by accounting for the temporal dependence using sample autocorrelations. This topic will be discussed in Chapter 5.

3.6. VISUALIZATION FOR HIGHER-DIMENSIONAL DATA

Graphical methods are essential when exploration, analysis, or comparison of matched data consisting of more than two variables is required. The methods presented so far can be applied only to pairwise subsets of the variables. Simultaneous display of three or more variables is more difficult due to a combination of geometric and cognitive problems. The geometric problem is that most available display media (e.g., paper and computer screens) are two-dimensional, so that directly plotting higher-dimensional data requires a geometric projection onto the plane, during which process information is inevitably lost. The cognitive problem derives from the fact that our brains have evolved to deal with life in a three-dimensional world, and visualizing four or more dimensions simultaneously is difficult or impossible.

Systematic study and articulation of principles for data visualization are relatively recent. In addition to Tukey's (1977) pathbreaking book, notable contributions include Tufte (1983, 1990), Cleveland (1994), and Wilkinson (2005). Some broad principles emerge from the work of these authors. Notably, Tufte (1983) calls for maximization of the ratio of "data-ink" to "total ink," and minimization of *chartjunk*, or gratuitous decoration. Cleveland (1994) expresses these two ideas as making the data stand out and avoiding superfluity. These books are also full of detailed suggestions for better graphic design, such as (from Cleveland, 1994) drawing tick-marks on frame exteriors in order not to obscure the data, adjusting plot aspect ratios to best allow assessment of rates of change, and using logarithmic scales when understanding proportional changes or multiplicative factors is important, to name a few.

A variety of clever graphical tools have been devised for multivariate (three or more variables simultaneously) EDA, and there is plenty of room for the design of new graphical tools suited to particular purposes. In addition to the ideas presented in this section, some multivariate graphical EDA devices designed particularly for ensemble forecasts are described in Section 8.5, and a high-dimensional EDA approach based on principal component analysis is described in Section 13.7.3.

3.6.1. The Star Plot

If the number of variables, K, is not too large, each of a set of n K-dimensional observations can be displayed graphically as a *star plot*. The star plot is based on K coordinate axes sharing the same origin, spaced $360°/K$ apart on the plane. For each of the n observations, the value of the kth of the K variables is proportional (with perhaps some minimum value subtracted) to the radial plotting distance on the corresponding axis. The "star" consists of line segments connecting these points to their counterparts on adjacent radial axes.

For example, Figure 3.23 shows star plots for the last 5 (of $n=31$) days of the January 1987 data in Table A.1. Since there are $K=6$ variables, the six axes are separated by $360°/6=60°$, and each is identified with one of the variables as indicated in the plot for 27 January. In general the scales of proportionality on star plots are different for different variables and are designed so the smallest value (or some value near but below it) corresponds to the origin and the largest value (or some value near and above it) corresponds to the full length of the axis. Because the variables in Figure 3.23 are matched in type, the scales for the three types of variables have been chosen identically in order to better compare them. For example, the origin for both the Ithaca and Canandaigua maximum temperature axes corresponds to 10°F, and the ends of these axes correspond to 40°F. The precipitation axes have zero at the origin and 0.15 in. at the ends, so that the double-triangle shapes for 27 and 28 January indicate zero precipitation at both locations for those days. The near symmetry of the stars suggests strong correlations for the pairs of like variables (since their axes have been plotted 180 degree apart), and the tendency for the stars to get larger through time indicates warmer and wetter days at the end of the month.

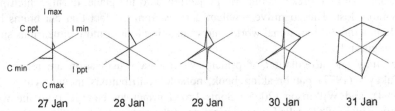

FIGURE 3.23   Star plots for the last five days in the January 1987 data in Table A.1, with axes labeled for the 27 January star only. Approximate radial symmetry in these plots reflects correlation between like variables at the two locations, and expansion of the stars through the time period indicates warmer and wetter days at the end of the month.

3.6.2. The Glyph Scatterplot

The *glyph scatterplot* is an extension of the ordinary scatterplot, in which the simple dots locating points on the two-dimensional plane defined by two variables are replaced by "glyphs," or more elaborate symbols that encode the values of additional variables in their sizes, shapes, and/or colors. Figure 3.15 is a primitive glyph scatterplot, with the filled/open circular glyphs indicating the binary precipitation/no-precipitation variable.

Figure 3.24 is a simple glyph scatterplot displaying three variables relating to evaluation of a small set of winter maximum temperature forecasts. The two scatterplot axes are the forecast and observed temperatures, rounded to 5°F bins, and the circular glyphs are drawn so that their areas are proportional to the numbers of forecast-observation pairs in a given 5 × 5°F square bin. Choosing area to be proportional to the third variable (here, counts in each bin) is preferable to radius or diameter because the glyph areas correspond better to the visual impression of size.

Figure 3.24 is essentially a two-dimensional histogram for this bivariate set of temperature data, but is more effective than a direct generalization to three dimensions of a conventional two-dimensional histogram for a single variable. Figure 3.25 shows such a perspective-view bivariate histogram for the same data, which is usually ineffective because projection of the three dimensions onto the two-dimensional page has introduced ambiguities about the locations of individual points. This is so, even though each point in Figure 3.25 is tied to its location on the forecast-observed plane at the apparent base of the plot through the vertical tails, and the points falling exactly on the diagonal are indicated by open plotting symbols. Figure 3.24 speaks more clearly than Figure 3.25 about the data, for example, showing immediately that there is an overforecasting bias (forecast temperatures systematically warmer than the corresponding observed temperatures, on average), particularly for the colder forecasts.

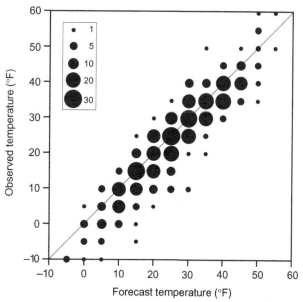

FIGURE 3.24 Glyph scatterplot of the bivariate frequency distribution of forecast and observed winter daily maximum temperatures for Minneapolis, 1980–1981 through 1985–1986. Temperatures have been rounded to 5°F intervals, and the circular glyphs have been scaled to have areas proportional to the counts (inset).

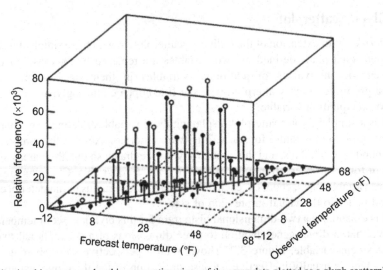

FIGURE 3.25   Bivariate histogram rendered in perspective view, of the same data plotted as a glyph scatterplot in Figure 3.24. Even though data points are located on the forecast-observation plane by the vertical tails and points on the 1:1 diagonal are further distinguished by open circles, the projection from three dimensions to two makes the figure difficult to interpret. *From Murphy et al. (1989). © American Meteorological Society. Used with permission.*

Effective alternatives to the glyph scatterplot in Figure 3.24 for displaying the bivariate frequency distribution might be a contour plot of the bivariate kernel density estimate (see Section 3.3.6) for these data; or a *hexbin plot*, which divides the Cartesian plane into a tessellation of hexagons and indicates the third dimension (in the present case, numbers of forecast-observation pairs) using either shading or color.

More elaborate glyphs than the circles in Figure 3.24 can be used to simultaneously display multivariate data with more than three variables. For example, star glyphs as described in Section 3.6.1 could be used as the plotting symbols in a glyph scatter plot. Virtually any shape that might be suggested by the data or the scientific context can be used in this way as a glyph. For example, Figure 3.26 shows a glyph that simultaneously displays seven meteorological quantities: wind direction, wind speed, sky cover, temperature, dew point temperature, pressure, and current weather condition. When these glyphs are plotted as a scatterplot defined by longitude (horizontal axis) and latitude (vertical axis), the result is a raw weather map, which is, in effect, a graphical EDA depiction of a nine-dimensional data set describing the spatial distribution of weather at a particular time. Similarly, Figure 3.27 shows a glyph map of observed linear temperature trends for the period 1950–2010 over much of North America.

FIGURE 3.26   An elaborate glyph, known as a meteorological station model, simultaneously depicting seven quantities. When plotted on a map, two location variables (latitude and longitude) are added as well, increasing the dimensionality of the depiction to nine, in what amounts to a glyph scatterplot of the weather data.

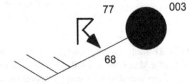

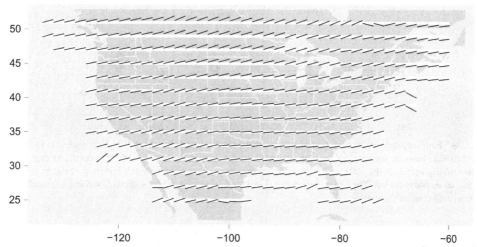

FIGURE 3.27 Glyph map of observed linear temperature trends for the period 1950–2010 over much of North America. *From Wickham et al. (2012).*

3.6.3. The Rotating Scatterplot

Figure 3.25 illustrates that it is generally unsatisfactory to attempt to extend the two-dimensional scatterplot to three dimensions by rendering it as a perspective view. The problem occurs because the three-dimensional counterpart of the scatterplot consists of a point cloud located in a volume rather than on the plane, and geometrically projecting this volume onto any one plane results in ambiguities about distances perpendicular to that plane. One solution to this problem is to draw larger and smaller symbols that are respectively closer to and further from the front in the direction of the projection, in a way that mimics the change in apparent size of physical objects with distance.

More effective, however, is to view the three-dimensional data in a computer animation known as a *rotating scatterplot*. At any instant the rotating scatterplot is a projection of the three-dimensional point cloud, together with its three coordinate axes for reference, onto the two-dimensional surface of the computer screen. But the plane onto which the data are projected can be changed smoothly in time, typically using the computer mouse, in a way that produces the illusion that we are viewing the points and their axes rotating around the three-dimensional coordinate origin, "inside" the computer monitor. The apparent motion can be rendered quite smoothly, and it is this continuity in time that allows a subjective sense of the shape of the data in three dimensions to be developed as we watch the changing display. In effect, the animation substitutes time for the missing third dimension.

It is not really possible to convey the power of this approach in the static form of a book page. However, an idea of how this works can be had from Figure 3.28, which shows four snapshots from a rotating scatterplot sequence, using the June Guayaquil data for temperature, pressure, and precipitation in Table A.3, with the 5 El Niño years indicated by open circles. Initially (Figure 3.28a) the temperature axis is oriented out of the plane of the page, so what appears is a simple two-dimensional scatterplot of precipitation versus pressure. In Figure 3.28b–d, the temperature axis is rotated into the plane of the page, which allows a gradually changing perspective on the arrangement of the points relative to each other and relative to the projections of the coordinate axes. Figure 3.28 shows only about 90

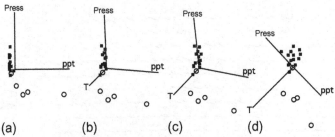

FIGURE 3.28 Four snapshots of the evolution of a three-dimensional rotating plot of the June Guayaquil data in Table A.3, in which the 5 El Niño years are shown as circles. The temperature axis is perpendicular to, and extends out of, the page in panel (a), and the three subsequent panels (b)–(d) show the changing perspectives as the temperature axis is rotated into the plane of the page, in a direction down and to the left. The visual illusion of a point cloud suspended in a three-dimensional space is much greater in a live rendition with continuous motion.

degree of rotation. A "live" examination of these data with a rotating plot usually would consist of choosing an initial direction of rotation (here, down, and to the left), allowing several full rotations in that direction, and then possibly repeating the process for other directions of rotation until an appreciation of the three-dimensional shape of the point cloud had developed.

3.6.4. The Correlation Matrix

The *correlation matrix* is a very useful device for simultaneously displaying correlations among more than two batches of matched data. For example, the data set in Table A.1 contains matched data for six variables. Correlation coefficients can be computed for each of the 15 distinct pairings of these six variables. In general, for K variables, there are $(K)(K-1)/2$ distinct pairings, and the correlations between them can be arranged systematically in a square array, with as many rows and columns as there are matched data variables whose relationships are to be summarized. Each entry in the array, $r_{i,j}$, is indexed by the two subscripts, i and j, that point to the identity of the two variables whose correlation is represented. For example, $r_{2,3}$ would denote the correlation between the second and third variables in a list. The rows and columns in the correlation matrix are numbered correspondingly, so that the individual correlations are arranged as shown in Figure 3.29.

FIGURE 3.29 The layout of a correlation matrix, $[R]$. Correlations $r_{i,j}$ between all possible pairs of variables are arranged so that the first subscript, i, indexes the row number, and the second subscript, j, indexes the column number.

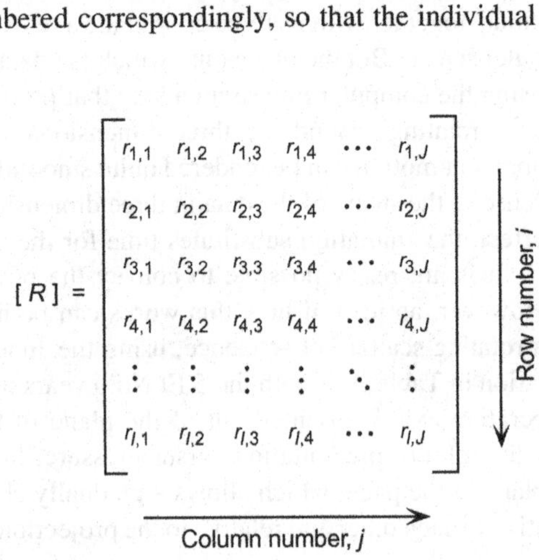

$$[R] = \begin{bmatrix} r_{1,1} & r_{1,2} & r_{1,3} & r_{1,4} & \cdots & r_{1,J} \\ r_{2,1} & r_{2,2} & r_{2,3} & r_{2,4} & \cdots & r_{2,J} \\ r_{3,1} & r_{3,2} & r_{3,3} & r_{3,4} & \cdots & r_{3,J} \\ r_{4,1} & r_{4,2} & r_{4,3} & r_{4,4} & \cdots & r_{4,J} \\ \vdots & \vdots & \vdots & \vdots & \ddots & \vdots \\ r_{I,1} & r_{I,2} & r_{I,3} & r_{I,4} & \cdots & r_{I,J} \end{bmatrix}$$

Row number, i

Column number, j

The correlation matrix was not designed for exploratory data analysis, but rather as a notational shorthand that allows mathematical manipulation of the correlations in the framework of linear algebra (see Chapter 11). As a format for an organized exploratory arrangement of correlations, parts of the correlation matrix are redundant, and some are simply uninformative. Consider first the diagonal elements of the matrix, arranged from the upper left to the lower right corners, that is, $r_{1,1}, r_{2,2}, r_{3,3}, \ldots, r_{K,K}$. These are the correlations of each of the variables with themselves and are always equal to 1. Realize also that the correlation matrix is symmetric. That is, the correlation $r_{i,j}$ between variables i and j is exactly the same number as the correlation $r_{j,i}$, between the same pair of variables, so that the correlation values above and below the diagonal of 1's are mirror images of each other. Therefore as noted earlier, only $(K)(K-1)/2$ of the K^2 entries in the correlation matrix provide distinct information.

Table 3.5 presents correlation matrices for the data in Table A.1. The matrix on the left contains Pearson product-moment correlation coefficients, and the matrix on the right contains Spearman rank correlation coefficients. As is consistent with usual practice when using correlation matrices for display rather than computational purposes, only one of the upper and lower triangles of each matrix actually is printed. Omitted are the uninformative diagonal elements and the redundant upper triangular elements. Only the $(6)(5)/2 = 15$ distinct correlation values are presented.

Important features in the underlying data can be discerned by studying and comparing these two correlation matrices. First, notice that the six correlations involving only temperature variables have comparable values in both matrices. The strongest Spearman correlations are between like temperature variables at the two locations. Correlations between maximum and minimum temperatures at the same location are moderately large, but weaker. The correlations involving one or both of the precipitation variables differ substantially between the two correlation matrices. There are only a few very large precipitation amounts for each of the two locations, and these tend to dominate the Pearson correlations, as explained previously. On the basis of this comparison between the correlation matrices, we therefore would suspect that the precipitation data contained some outliers, even without the benefit of knowing the type of data, or of having seen the individual numbers. The rank correlations would be expected to better reflect the degree of association for data pairs involving one or both of the precipitation variables. Subjecting the precipitation variables to a monotonic transformation appropriate to reducing the skewness would produce no changes in the matrix of Spearman correlations, but would be expected to improve the agreement between the Pearson and Spearman correlations.

TABLE 3.5   Correlation Matrices for the Data in Table A.1

	Ith. Ppt	Ith. Max	Ith. Min	Can. Ppt	Can. Max	Ith. Ppt	Ith. Max	Ith. Min	Can. Ppt	Can. Max
Ith. Max	−0.024					0.319				
Ith. Min	0.287	0.718				0.597	0.761			
Can. Ppt	0.965	0.018	0.267			0.750	0.281	0.546		
Can. Max	−0.039	0.957	0.762	−0.015		0.267	0.944	0.749	0.187	
Can. Min	0.218	0.761	0.924	0.188	0.810	0.514	0.790	0.916	0.352	0.776

Only the lower triangle of the matrices is shown, to omit redundancies and the uninformative diagonal values. The left matrix contains Pearson product-moment correlations, and the right matrix contains Spearman rank correlations.

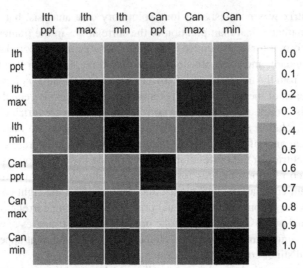

FIGURE 3.30   Heatmap of the Spearman correlations in Table 3.5, using a grayscale to indicate the magnitudes.

3.6.5. Heatmaps

When there are a large number of variables being related through their correlations, the very large number of pairwise comparisons can be overwhelming, in which case this arrangement of the numerical values in a correlation matrix is not particularly effective as an EDA device. However, different colors or shading levels can be assigned to particular ranges of correlation, and then plotted in the same two-dimensional arrangement as the numerical correlations on which they are based, in order to more directly gain a visual appreciation of the patterns of relationship. Figure 3.30 shows such a plot for the Spearman correlations presented in Table 3.5. This is an example of what is known as a *heatmap*, which can also be plotted using color instead of a grayscale. Here the full correlation matrix has been plotted, including the unit correlations along the main diagonal. The shading shows clearly that the strongest correlations are between like temperature variables and that the weakest correlations are between precipitation variables and temperature variables.

3.6.6. The Scatterplot Matrix

The *scatterplot matrix* is a graphical extension of the correlation matrix. The physical arrangement of the correlation coefficients in a correlation matrix is convenient for quick comparisons of relationships between pairs of variables, but distilling these relationships to a single number such as a correlation coefficient inevitably hides important details. A scatterplot matrix is an arrangement of individual scatterplots according to the same logic governing the placement of individual correlation coefficients in a correlation matrix.

Figure 3.31 is a scatterplot matrix for the January 1987 data in Table A.1, with the scatterplots arranged in the same pattern as the correlation matrices in Table 3.5. The complexity of a scatterplot matrix can be bewildering at first, but a large amount of information about the joint behavior of the data is displayed very compactly. For example, quickly evident from a scan of the precipitation rows and columns in Figure 3.31 is the fact that there are just a few large precipitation amounts at each of the

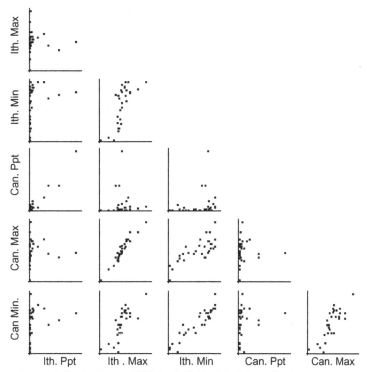

FIGURE 3.31   Scatterplot matrix for the January 1987 data in Table A.1 of Appendix A.

two locations. Looking vertically along the column for Ithaca precipitation, or horizontally along the row for Canandaigua precipitation, the eye is drawn to the largest few data values, which appear to line up. Most of the precipitation points correspond to small amounts and therefore cluster along the opposite axes. Focusing on the plot of Canandaigua versus Ithaca precipitation, it is apparent that the two locations received most of their precipitation for the month on the same few days. Also evident is the association of precipitation with milder minimum temperatures that was seen in previous looks at these same data. The closer relationships between maximum and maximum, or minimum and minimum temperature variables at the two locations—as compared to the maximum versus minimum temperature relationships at one location—can also be seen clearly.

The scatterplot matrix in Figure 3.31 has been drawn without the diagonal elements in the positions that correspond to the unit correlation of a variable with itself in a correlation matrix. A scatterplot of any variable with itself would be equally uninteresting, consisting only of a straight-line collection of points at a 45-degree angle. However, it is possible to use the diagonal positions in a scatterplot matrix to portray useful univariate information about the variable corresponding to that matrix position. One simple choice would be schematic plots of each of the variables in the diagonal positions. Another potentially useful choice is the Q-Q plot (Section 4.5.2) for each variable, which graphically compares the data with a reference distribution; for example, the bell-shaped Gaussian distribution. Sometimes the diagonal positions are used merely to contain labels for the respective variables.

The scatterplot matrix can be even more revealing if constructed using software allowing *brushing* of data points in related plots. When brushing, the analyst can select a point or set of points in one plot, and the corresponding points in the same data record then also light up or are otherwise differentiated in all the other plots then visible. For example, when preparing Figure 3.17, the differentiation of Ithaca temperatures occurring on days with measurable precipitation was achieved by brushing another plot (that plot was not reproduced in Figure 3.17) involving the Ithaca precipitation values. The solid circles in Figure 3.17 thus constitute a temperature scatterplot conditional on nonzero precipitation. Brushing can also sometimes reveal surprising relationships in the data by keeping the brushing action of the computer mouse in motion. The resulting "movie" of brushed points in the other simultaneously visible plots essentially allows the additional dimension of time to be used for differentiating relationships in the data.

3.6.7. Correlation Maps

Correlation matrices such as those in Table 3.5 are understandable and informative, so long as the number of quantities represented (six, in the case of Table 3.5) remains reasonably small. When the number of variables becomes large it may not be possible to easily make sense of the individual values, or even to fit their correlation matrix on a single page. A frequent cause of atmospheric data being excessively numerous for effective display in a correlation or scatterplot matrix is the necessity of working simultaneously with data from a large number of locations. In this case the geographical arrangement of the locations can be used to organize the correlation information in map form.

Consider, for example, summarization of the correlations among surface pressure at perhaps 200 locations around the world. This would be only a modestly large set of data by modern standards. However, this many batches of pressure data would lead to $(200)(199)/2 = 19,100$ distinct station pairs, and as many correlation coefficients. A technique that has been used successfully in such situations is construction of a series of *one-point correlation maps*.

Figure 3.32, taken from the classic paper by Bjerknes (1969), is a one-point correlation map for annual surface pressure data. Displayed on this map are contours of Pearson correlations between the pressure data at roughly 200 locations with those at Djakarta, Indonesia. Djakarta is thus the "one point" in this one-point correlation map. Essentially, the quantities being contoured are the values in the row or

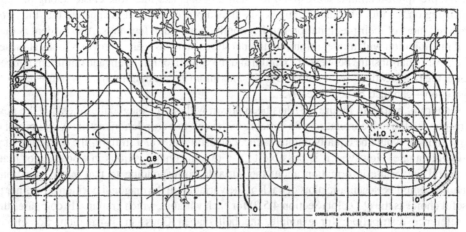

FIGURE 3.32   One-point correlation map of annual surface pressures at locations around the globe with those at Djakarta, Indonesia. The strong negative correlation of −0.8 at Easter Island reflects the atmospheric component of the El Niño-Southern Oscillation phenomenon. *From Bjerknes (1969). © American Meteorological Society. Used with permission.*

column corresponding to Djakarta in the very large correlation matrix containing all the 19,100 or so correlation values. A complete representation of that large correlation matrix in terms of one-point correlation maps would require as many maps as stations, or in this case about 200. However, not all the maps would be as interesting as Figure 3.32, although the maps for nearby stations (e.g., Darwin, Australia) would look very similar.

Clearly Djakarta is located under the +1.0 on the map, since the pressure data there are perfectly correlated with themselves. Not surprisingly, pressure correlations for locations near Djakarta are quite high, with gradual declines toward zero at locations somewhat further away. This pattern is the spatial analog of the tailing off of the (temporal) autocorrelation function indicated in Figure 3.22. The surprising feature in Figure 3.32 is the region in the eastern tropical Pacific, centered on Easter Island, for which the correlations with Djakarta pressure are strongly negative. These negative correlations imply that in years when average pressures at Djakarta (and nearby locations, such as Darwin) are high, pressures in the eastern Pacific are low, and vice versa. This correlation pattern is an expression in the surface pressure data of the ENSO phenomenon, sketched earlier in this chapter, and is an example of what has come to be known as a *teleconnection* pattern. In the ENSO warm phase, the center of tropical Pacific convection moves eastward, producing lower than average pressures near Easter Island and higher than average pressures at Djakarta. When the precipitation shifts westward during the cold phase, pressures are low at Djakarta and high at Easter Island.

Not all geographically distributed correlation data exhibit teleconnection patterns such as the one shown in Figure 3.32. However, many large-scale fields, especially pressure (or geopotential height) fields, show one or more teleconnection patterns. A device used to simultaneously display these aspects of the large underlying correlation matrix is the *teleconnectivity* map. To construct a teleconnectivity map, the row (or column) for each station or gridpoint in the correlation matrix is searched for the largest negative value. The teleconnectivity value for location i, T_i, is the absolute value of that most negative correlation,

$$T_i = \left| \min_j \left(r_{i,j} \right) \right|. \tag{3.40}$$

Here the minimization over j (the column index for $[R]$) implies that all correlations $r_{i,j}$ in the ith row of $[R]$ are searched for the smallest (most negative) value. For example, in Figure 3.32 the largest negative correlation with Djakarta pressures is with Easter Island, is -0.80. The teleconnectivity for Djakarta surface pressure would therefore be 0.80, and this value would be plotted on a teleconnectivity map at the location of Djakarta. To construct the full teleconnectivity map for surface pressure, the other 199 or so rows of the correlation matrix, each corresponding to another station, would be examined for the largest negative correlation (or, if none were negative, then the smallest positive one), and its absolute value would be plotted at the map position of that station.

Figure 3.33 shows the teleconnectivity map for northern hemisphere winter 500 mb heights. The density of the shading indicates the magnitude of the individual gridpoint teleconnectivity values. The locations of local maxima of teleconnectivity are indicated by the positions of the numbers, ×100. The arrows in Figure 3.33 point from the teleconnection centers (i.e., the local maxima in T_i) to the location with which each maximum negative correlation is exhibited. The unshaded regions indicate gridpoints for which the teleconnectivity is relatively low. The one-point correlation maps for locations in these unshaded regions would tend to show gradual declines toward zero at increasing distances, analogously to the time correlations in Figure 3.22, but without declining much further to large negative values.

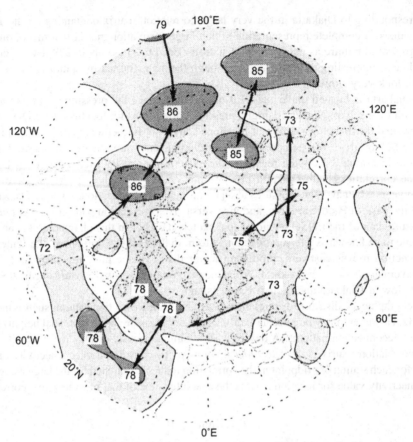

FIGURE 3.33   Teleconnectivity, or absolute value of the strongest negative correlation from each of many one-point correlation maps plotted at the base grid point, for winter 500mb heights. *From Wallace and Blackmon (1983).*

It has become apparent that a fairly large number of these teleconnection patterns exist in the atmosphere, and the many double-headed arrows in Figure 3.33 indicate that these group naturally. Especially impressive is the four-center pattern arcing from the central Pacific to the southeastern United States, known as the *Pacific-North America*, or *PNA pattern*. Notice, however, that these patterns emerged here from a statistical, exploratory analysis of a large quantity of atmospheric data. This type of work actually had its roots in the early part of the 20th century (see Brown and Katz, 1991) and is a good example of exploratory data analysis in the atmospheric sciences turning up interesting features in very large data sets.

3.6.8. On the Use of Color

Modern computing interfaces allow easy manipulation of aspects of color and pattern that can be used to enhance quantitative communication. Color can be very effective at graphically conveying information if used intelligently. Unfortunately color can be distracting and ineffective if used poorly.

Both principles and tools for choosing effective color schemes in graphical displays are available. Cleveland (1994) distinguishes two distinct uses of color for statistical graphics. The first is to aid in the discrimination among groups or categories of graphed elements. An example would be use of colored dots as glyphs in a glyph scatterplot. For this purpose, choosing a small number of distinct hues (basic colors) is most effective. However, be aware that many readers will be unable to distinguish red and green hues (e.g., Light and Bartlein, 2004).

In contrast, visual estimation of continuously varying quantitative information is better achieved using a range of saturation (lightness) together with one or a small number of hues, whereas a more traditional "rainbow" palette is often less effective for this purpose. Further exposition and elaboration on these ideas are available in Stauffer et al. (2015) and Retchless and Brewer (2016). Online tools to aid in the selection of appropriate and effective color schemes are available at www.colorbrewer2.org, and www.hclwizard.org.

3.7. EXERCISES

3.1. Compare the median, trimean, and the mean of the precipitation data in Table A.3.

3.2. Compute the MAD, the IQR, and the standard deviation of the pressure data in Table A.3.

3.3. Draw a stem-and-leaf display for the temperature data in Table A.3.

3.4. Compute the Yule-Kendall Index and the skewness coefficient using the temperature data in Table A.3.

3.5. Draw the empirical cumulative frequency distribution for the pressure data in Table A.3. Compare it with a histogram of the same data.

3.6. Compare the boxplot and the schematic plot representing the precipitation data in Table A.3.

3.7. Use Hinkley's d_λ to find an appropriate power transformation for the precipitation data in Table A.2 using Equation 3.19a, rather than Equation 3.19b as was done in Example 3.4. Use IQR in the denominator of Equation 3.20.

3.8. Construct side-by-side schematic plots for the candidate, and final, transformed distributions derived in Exercise 3.7. Compare the result to Figure 3.13.

3.9. Express the June 1951 temperature in Table A.3 as a standardized anomaly.

3.10. Plot the autocorrelation function up to lag 3, for the Ithaca minimum temperature data in Table A.1.

3.11. Construct a scatterplot of the temperature and pressure data in Table A.3.

3.12. Construct correlation matrices for the data in Table A.3 using
   a. The Pearson correlation.
   b. The Spearman rank correlation.

3.13. Draw and compare star plots of the data in Table A.3 for each of the years 1965 through 1969.

Chapter 4

Parametric Probability Distributions

4.1. BACKGROUND

4.1.1. Parametric vs. Empirical Distributions

In Chapter 3, methods for exploring and displaying variations in data sets were presented. These methods had at their core the expression of how, empirically, a particular set of data are distributed through their range. This chapter presents an approach to the summarization of data distributions that involves imposition of particular mathematical forms, called *parametric distributions*, to represent variations in the underlying data. These mathematical forms are theoretical constructs, and so yield idealizations of real data.

It is worth taking a moment to understand why we would want to force real data to fit an abstract mold. The question is worth considering because parametric distributions *are* abstractions. They will represent real data only approximately, although in many cases the approximation can be very good indeed. There are three ways in which employing parametric probability distributions may be useful.

- *Compactness.* Particularly when dealing with large data sets, repeatedly manipulating the raw data can be cumbersome, or even severely limiting. A well-fitting parametric distribution reduces the number of quantities required for characterizing properties of the data from the full n order statistics $(x_{(1)}, x_{(2)}, x_{(3)}, \dots, x_{(n)})$ to a small number of distribution parameters.
- *Smoothing and interpolation.* Real data are subject to sampling variations that lead to gaps or rough spots in their empirical distributions. For example, in Figures 3.1 and 3.11a there are no maximum temperature values between 10°F and 16°F, although certainly maximum temperatures in this range can and do occur during January at Ithaca. A parametric distribution imposed on these data would represent the possibility of these temperatures occurring, as well as allowing estimation of their probabilities of occurrence.
- *Extrapolation.* Estimating probabilities for events outside the range of a particular data set requires assumptions about as-yet-unobserved behavior. Again referring to Figure 3.11a, the empirical cumulative probability associated with the coldest temperature, 9°F, was estimated as 0.0213 using the Tukey plotting position. The probability of a maximum temperature this cold or colder could be estimated as 0.0213, but nothing further can be said quantitatively about the probability of January maximum temperatures colder than 5°F or 0°F without the imposition of a probability model such as that produced by a parametric distribution.

The distinction is being drawn between empirical and parametric data representations, but it should be emphasized that use of parametric probability distributions is not independent of empirical considerations. In particular, before embarking on the representation of data using parametric functions, we must

Statistical Methods in the Atmospheric Sciences. https://doi.org/10.1016/B978-0-12-815823-4.00004-3

decide among candidate distribution forms, fit parameters of the chosen distribution, and check that the resulting function does, indeed, provide a reasonable fit. All three of these steps require use of real data.

4.1.2. What Is a Parametric Distribution?

A parametric distribution is an abstract mathematical form, or characteristic shape, or a family of characteristic shapes. Some of these mathematical forms arise naturally as a consequence of certain kinds of data-generating processes, and when applicable these are especially plausible candidates for concisely representing variations in a set of data. Even when there is not a strong natural justification behind the choice of a particular parametric distribution, it may be found empirically that the distribution represents a set of data very well.

The specific nature of a parametric distribution is determined by particular values for entities called *parameters* of that distribution. For example, the Gaussian (or "normal") distribution has as its characteristic shape the familiar symmetric bell. However, merely asserting that a particular batch of data, say average September temperatures at a location of interest, is well represented by the Gaussian distribution is not very informative about the nature of the data, without specifying *which* Gaussian distribution represents the data. There are infinitely many particular examples of the Gaussian distribution, corresponding to all possible values of the two distribution parameters μ and σ. But knowing, for example, that the monthly temperature for September is well represented by the Gaussian distribution with $\mu = 60°F$ and $\sigma = 2.5°F$ conveys a large amount of information about the nature and magnitudes of the variations of September temperatures at that location.

4.1.3. Parameters vs. Statistics

There is potential for confusion between distribution parameters and *sample statistics*. Distribution parameters are abstract characteristics of a particular parametric distribution. They succinctly represent underlying population, or data-generating process, properties. By contrast, a statistic is any quantity computed from a sample of data. Usually, the notation for sample statistics uses Roman (i.e., ordinary) letters, and parameters are typically written using Greek letters.

The confusion between parameters and statistics arises because, for some common parametric distributions, certain sample statistics are good estimators for the distribution parameters. For example, the sample standard deviation, s (Equation 3.6), a statistic, can be confused with the parameter σ of the Gaussian distribution because the two often are equated when finding a particular Gaussian distribution to best match a data sample. Distribution parameters are estimated (fitted) using sample statistics. However, it is not always the case that the fitting process is as simple as that for the Gaussian distribution, where usually the sample mean is equated to the parameter μ and the sample standard deviation is equated to the parameter σ.

4.1.4. Discrete vs. Continuous Distributions

There are two distinct types of parametric distributions, corresponding to different types of data, or random variables. *Discrete distributions* describe random quantities (i.e., the data of interest) that can take on only particular values. That is, the allowable values are finite, or at least countably infinite. For example, a *discrete random variable* might take on only the values 0 or 1; or any of the nonnegative integers; or one of the colors red, yellow, or blue. A *continuous random variable* typically can take on any value within a specified range of the real numbers. For example, a continuous random variable might be defined on the real numbers between 0 and 1, or the nonnegative real numbers, or, for some distributions, the entire real line.

Strictly speaking, using a *continuous distribution* to represent observable data implies that the underlying observations are known to an arbitrarily large number of significant figures. Of course this is never true, but it is convenient and not too inaccurate to represent as continuous those variables that are continuous conceptually but reported discretely. Temperature and precipitation are two obvious examples that really range over some portion of the real number line, but which are usually reported to discrete multiples of 1°F and 0.01 in. in the United States. Little is lost when treating these discrete observations as samples from continuous distributions.

4.2. DISCRETE DISTRIBUTIONS

A large number of parametric distributions exist that are applicable to discrete random variables. Many of these are listed in the encyclopedic volume by Johnson et al. (1992), together with results concerning their properties. Only five of these, the binomial distribution, the geometric distribution, the negative binomial distribution, the multinomial distribution, and the Poisson distribution, are presented here.

4.2.1. Binomial Distribution

The *binomial distribution* is one of the simplest parametric distributions, and therefore is employed often in textbooks to illustrate the use and properties of parametric distributions more generally. This distribution pertains to outcomes of situations where, on some number of occasions (sometimes called "trials"), one or the other of two MECE (mutually exclusive and collectively exhaustive) events will occur. Classically the two events have been called "success" and "failure," but these are arbitrary labels. More generally, one of the events (say, the success) is assigned the number 1, and the other (the failure) is assigned the number zero.

The random variable of interest, X, is the number of event occurrences (given by the sum of 1s and 0s) in some number of trials. The number of trials, N, can be any positive integer, and the variable X can take on any of the nonnegative integer values from 0 (if the event of interest does not occur at all in the N trials) to N (if the event occurs on each occasion). The binomial distribution can be used to calculate probabilities for each of these $N + 1$ possible values of X if two conditions are met: (1) the probability of the event occurring does not change from trial to trial (i.e., the occurrence probability is *stationary*), and (2) the outcomes on each of the N trials are mutually independent. These conditions are rarely strictly met, but real situations can be close enough to this ideal that the binomial distribution provides sufficiently accurate representations.

One implication of the first restriction, relating to constant occurrence probability, is that events whose probabilities exhibit regular cycles must be treated carefully. For example, the event of interest might be thunderstorm or dangerous lightning occurrence, at a location where there is a diurnal or annual variation in the probability of the event. In cases like these, subperiods (e.g., hours or months, respectively) with approximately constant occurrence probabilities usually would be analyzed separately.

The second necessary condition for applicability of the binomial distribution, relating to event independence, is often more troublesome for atmospheric data. For example, the binomial distribution usually would not be directly applicable to daily precipitation occurrence or nonoccurrence. As illustrated by Example 2.2, such events often exhibit substantial day-to-day dependence. For situations like this the binomial distribution can be generalized to a theoretical stochastic process called a Markov chain, discussed in Section 10.2. On the other hand, the year-to-year statistical dependence in atmospheric behavior is usually weak enough that occurrences or nonoccurrences of an event in consecutive annual periods can be considered to be effectively independent (12-month climate forecasts would be much easier if they were not!). The first four examples in this chapter take advantage of this fact.

The usual first illustration of the binomial distribution is in relation to coin flipping. If the coin is fair, the probability of either heads or tails is 0.5, and does not change from one coin-flipping occasion (or, equivalently, from one coin) to the next. If $N > 1$ coins are flipped simultaneously or in sequence, the outcome on one of the coins does not affect the other outcomes. The coin-flipping situation thus satisfies all the requirements for description by the binomial distribution: dichotomous, independent events with constant probability.

Consider a game where $N = 3$ fair coins are flipped simultaneously, and we are interested in the number, X, of heads that result. The possible values of X are 0, 1, 2, and 3. These four values are a MECE partition of the sample space for X, and their probabilities must therefore sum to 1. In this simple example, you may not need to think explicitly in terms of the binomial distribution to realize that the probabilities for these four events are 1/8, 3/8, 3/8, and 1/8, respectively.

In the general case, probabilities for each of the $N+1$ values of X are given by the *probability distribution function*, or *probability mass function*, for the binomial distribution,

$$\Pr\{X = x\} = \binom{N}{x} p^x (1-p)^{N-x}, \quad x = 0, 1, ..., N. \tag{4.1}$$

Here, consistent with the usage in Equation 3.18, the uppercase X indicates the random variable whose precise value is unknown, or has yet to be observed. The lowercase x denotes a specific, particular value that the random variable can take on. The binomial distribution has two parameters, N and p. The parameter p is the probability of occurrence of the event of interest (the success) on any one of the N independent trials. For a given pair of the parameters N and p, Equation 4.1 is a function associating a probability with each of the discrete values $x = 0, 1, 2, ..., N$, such that $\sum_x \Pr\{X = x\} = 1$. That is, the probability distribution function distributes probability over all events in the sample space. Note that the binomial distribution is unusual in that both of its parameters are conventionally represented by Roman letters.

The right-hand side of Equation 4.1 consists of two parts: a combinatorial part and a probability part. The combinatorial part specifies the number of distinct ways of realizing x success outcomes from a collection of N trials. It is pronounced "N choose x" and is computed according to

$$\binom{N}{x} = \frac{N!}{x!(N-x)!}. \tag{4.2}$$

By convention, $0! = 1$. For example, when tossing $N = 3$ coins, there is only one way that $x = 3$ heads can be achieved: all three coins must come up heads. Using Equation 4.2, "three choose three" is given by $3!/(3!0!) = (1 \cdot 2 \cdot 3)/(1 \cdot 2 \cdot 3 \cdot 1) = 1$. There are three ways in which $x = 1$ can be achieved: either the first, the second, or the third coin can come up heads, with the remaining two coins coming up tails; using Equation 4.2 we obtain $3!/(1!2!) = (1 \cdot 2 \cdot 3)/(1 \cdot 1 \cdot 2) = 3$.

The probability part of Equation 4.1 follows from the multiplicative law of probability for independent events (Equation 2.12). The probability of a particular sequence of exactly x independent event occurrences and $N-x$ nonoccurrences is simply p multiplied by itself x times, and then multiplied by $1-p$ (the probability of nonoccurrence) $N-x$ times. The number of these particular sequences of exactly x event occurrences and $N-x$ nonoccurrences is given by the combinatorial part, for each x, so that the product of the combinatorial and probability parts in Equation 4.1 yields the probability for x event occurrences, regardless of their locations in the sequence of N trials.

Example 4.1. Binomial Distribution and the Freezing of Cayuga Lake, I

Consider the data in Table 4.1, which lists years during which the surface of Cayuga Lake, in central New York State, was observed to have frozen. Cayuga Lake is rather deep and will freeze only after a long period of exceptionally cold and cloudy weather. In any given winter, the lake surface either freezes or it does not. Whether or not the lake freezes in a given winter is essentially independent of whether or not it froze in recent years. Unless there has been appreciable climate change in the region over the past 200 years, the probability that the lake will freeze in a given year has been effectively constant through the period of the data in Table 4.1. This assumption is increasingly questionable as the planet progressively warms, but if we can assume near-stationarity of the annual freezing probability, p, we expect the binomial distribution to provide a good statistical description of the freezing of this lake.

In order to use the binomial distribution as a representation of the statistical properties of the lake-freezing data, we need to *fit the distribution* to the data. Fitting the distribution simply means finding particular values for the distribution parameters, p and N in this case, for which Equation 4.1 will behave as much as possible like the data in Table 4.1. The binomial distribution is somewhat unique in that the parameter N depends on the question we want to ask, rather than on the data per se. If we want to compute the probability of the lake freezing next winter, or in any single winter in the future, $N = 1$. (The special case of Equation 4.1 with $N = 1$ is called the *Bernoulli distribution*, and one realization of a success or failure is called a *Bernoulli trial*.) If we want to compute probabilities for the lake freezing at least once during some decade in the future, $N = 10$.

The binomial parameter p in this application is the probability that the lake freezes in any given year. It is natural to estimate this probability using the relative frequency of the freezing events in the data. This is a straightforward task here, except for the small complication of not knowing exactly when the climate record starts. The written record clearly starts no later than 1796, but probably began some years before that. Suppose that the data in Table 4.1 represent a 230-year record. The 10 observed freezing events then lead to the relative frequency estimate for the binomial p of $10/230 = 0.0435$.[1]

We are now in a position to use Equation 4.1 to compute probabilities for a variety of events relating to the freezing of this lake. The simplest kinds of events to work with have to do with the lake freezing exactly a specified number of times, x, in a specified number of years, N. For example, the probability of the lake freezing exactly once in 10 years is

TABLE 4.1 Years in which Cayuga Lake Has Frozen, as of 2018

1796	1904
1816	1912
1856	1934
1875	1961
1884	1979

---

1. Cayuga lake nearly froze again in 2015. The 8 March visible satellite image from that year shows approximately 1/3 of the lake surface as open water. Incorporating this event into the data set of Table 4.1 would yield the estimate $p \approx 11/230 = 0.0478$ for the annual freezing probability.

$$\Pr\{X=1\} = \binom{10}{1}(.0435)^{1}(1-.0435)^{10-1} = \frac{10!}{1!9!}(.0435)(.9565)^{9} = 0.292. \tag{4.3}$$

In reality the result in Equation 4.3 is likely an overestimate, because of the ongoing climate warming.   ◇

Example 4.2. Binomial Distribution and the Freezing of Cayuga Lake, II

A somewhat harder class of events to deal with is exemplified by the problem of calculating the probability that the lake freezes at least once in 10 years. It is clear from Equation 4.3 that this probability will be no smaller than 0.292, since the probability for the compound event will be given by the sum of the probabilities $\Pr\{X=1\}+\Pr\{X=2\}+\cdots+\Pr\{X=10\}$. This result follows from Equation 2.5, and the fact that these events are mutually exclusive: the lake cannot freeze both exactly once and exactly twice in the same decade.

The brute-force approach to this problem is to calculate all 10 probabilities in the sum and then add them up. However, this approach is rather tedious, and quite a bit of effort can be saved by giving the problem a bit more thought. Consider that the sample space here is composed of 11 MECE events: that the lake freezes exactly 0, 1, 2, ..., or 10 times in a decade. Since the probabilities for these 11 events must sum to 1, it is much easier to proceed using

$$\Pr\{X \geq 1\} = 1 - \Pr\{X=0\} = 1 - \frac{10!}{0!10!}(.0435)^{0}(.9565)^{10} = 0.359. \tag{4.4}$$

               ◇

It is worth noting that the binomial distribution can be applied to situations that are not intrinsically binary, through a suitable redefinition of events. For example, temperature is not intrinsically binary and is not even intrinsically discrete. However, for some applications it is of interest to consider the probability of frost, that is, $\Pr\{T \leq 32°F\}$. Together with the probability of the complementary event, $\Pr\{T > 32°F\}$, the situation is one concerning dichotomous events, and therefore could be a candidate for representation using the binomial distribution.

4.2.2. Geometric Distribution

The *geometric distribution* is related to the binomial distribution, describing a different aspect of the same data-generating situation. Both distributions pertain to a collection of independent trials in which one or the other of a pair of dichotomous events occurs. The trials are independent in the sense that the probability of the "success" occurring, p, does not depend on the outcomes of previous trials, and the sequence is stationary in the sense that p does not change over the course of the sequence (as a consequence of, e.g., an annual cycle or a changing climate). For the geometric distribution to be applicable, the collection of trials must occur in a sequence.

The binomial distribution pertains to probabilities that particular numbers of successes will be realized in a fixed number of trials. The geometric distribution specifies probabilities for the number of trials that will be required to observe the next success. For the geometric distribution, this number of trials is the random variable X, and the probabilities corresponding to its possible values are given by the probability distribution function of the geometric distribution,

$$\Pr\{X=x\} = p(1-p)^{x-1}, \quad x = 1, 2, \ldots \tag{4.5}$$

Here X can take on any positive integer value, since at least one trial will be required in order to observe a success, and it is possible (although vanishingly probable) that we would have to wait indefinitely for this

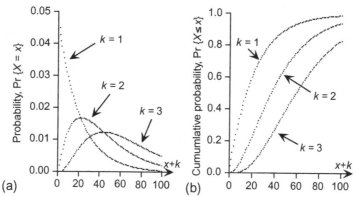

FIGURE 4.1   Probability distribution functions (a), and cumulative probability distribution functions (b), for the waiting time $x + k$ years for Cayuga Lake to freeze k times, using the negative binomial distribution, Equation 4.6.

outcome. Equation 4.5 can be viewed as an application of the multiplicative law of probability for independent events, as it multiplies the probability for a success by the probability of observing a sequence of $x - 1$ consecutive failures. The function $k = 1$ in Figure 4.1a shows an example geometric probability distribution, for the Cayuga Lake freezing probability $p = 0.0435$.

Usually the geometric distribution is applied to trials that occur consecutively through time, so it is sometimes called the *waiting distribution*. The distribution has been used to describe lengths of weather regimes or spells. One application of the geometric distribution is description of sequences of dry time periods (where we are waiting for a wet event) and wet periods (during which we are waiting for a dry event), when the time dependence of events follows the first-order Markov process (Waymire and Gupta 1981; Wilks 1999a), described in Section 10.2.

4.2.3. Negative Binomial Distribution

The *negative binomial distribution* is closely related to the geometric distribution, although this relationship is not indicated by its name, which comes from a technical derivation with parallels to a similar derivation for the binomial distribution. The probability distribution function for the negative binomial distribution is defined for nonnegative integer values of the random variable x,

$$\Pr\{X = x\} = \frac{\Gamma(k+x)}{x!\,\Gamma(k)} p^k \,(1-p)^x, \quad x = 0, 1, 2, \ldots \tag{4.6}$$

The distribution has two real-valued parameters, p, $0 < p < 1$ and k, $k > 0$. For integer values of k the negative binomial distribution is called the *Pascal distribution* and has an interesting interpretation as an extension of the geometric distribution of waiting times for the first success in a sequence of independent Bernoulli trials with probability p. In this case, the negative binomial X pertains to the number of failures until the kth success, so that $x + k$ is the total waiting time required to observe the kth success.

The notation $\Gamma(k)$ on the right-hand side of Equation 4.6 indicates a standard mathematical function known as the *gamma function*, defined by the definite integral

TABLE 4.2 Values of the Gamma Function, $\underline{\Gamma}(k)$ (Equation 4.7), for $1.00 \le k \le 1.99$

k	0.00	0.01	0.02	0.03	0.04	0.05	0.06	0.07	0.08	0.09
1.0	1.0000	0.9943	0.9888	0.9835	0.9784	0.9735	0.9687	0.9642	0.9597	0.9555
1.1	0.9514	0.9474	0.9436	0.9399	0.9364	0.9330	0.9298	0.9267	0.9237	0.9209
1.2	0.9182	0.9156	0.9131	0.9108	0.9085	0.9064	0.9044	0.9025	0.9007	0.8990
1.3	0.8975	0.8960	0.8946	0.8934	0.8922	0.8912	0.8902	0.8893	0.8885	0.8879
1.4	0.8873	0.8868	0.8864	0.8860	0.8858	0.8857	0.8856	0.8856	0.8857	0.8859
1.5	0.8862	0.8866	0.8870	0.8876	0.8882	0.8889	0.8896	0.8905	0.8914	0.8924
1.6	0.8935	0.8947	0.8959	0.8972	0.8986	0.9001	0.9017	0.9033	0.9050	0.9068
1.7	0.9086	0.9106	0.9126	0.9147	0.9168	0.9191	0.9214	0.9238	0.9262	0.9288
1.8	0.9314	0.9341	0.9368	0.9397	0.9426	0.9456	0.9487	0.9518	0.9551	0.9584
1.9	0.9618	0.9652	0.9688	0.9724	0.9761	0.9799	0.9837	0.9877	0.9917	0.9958

$$\Gamma(k) = \int_0^\infty t^{k-1} e^{-t} dt. \tag{4.7}$$

In general, the gamma function must be evaluated numerically (e.g., Abramowitz and Stegun, 1984; Press et al., 1986) or approximated using tabulated values, such as those given in Table 4.2. It satisfies the factorial recurrence relationship,

$$\Gamma(k+1) = k\,\Gamma(k), \tag{4.8}$$

allowing Table 4.2 to be extended indefinitely. For example, $\Gamma(3.50) = (2.50)\,\Gamma(2.50) = (2.50)(1.50)$ $\Gamma(1.50) = (2.50)(1.50)(0.8862) = 3.323$. Similarly, $\Gamma(4.50) = (3.50)\,\Gamma(3.50) = (3.50)\,(3.323) = 11.631$. The gamma function is also known as the *factorial function*, the reason for which is especially clear when its argument is an integer (e.g., in Equation 4.6 when k is an integer), that is, $\Gamma(k+1) = k!$.

With this understanding of the gamma function, it is straightforward to see the connection between the negative binomial distribution with integer k as a waiting distribution for k successes, and the geometric distribution (Equation 4.5) as a waiting distribution for the first success, in a sequence of independent Bernoulli trials with success probability p. Since X in Equation 4.6 is the number of failures before observing the kth success, and the total number of trials to achieve k successes will be $x + k$, then for $k = 1$, Equations 4.5 and 4.6 pertain to the same situation. The numerator in the first factor on the right-hand side of Equation 4.6 is $\Gamma(x+1) = x!$, canceling the $x!$ in the denominator. Realizing that $\Gamma(1) = 1$ (see Table 4.2), Equation 4.6 reduces to Equation 4.5 except that Equation 4.6 pertains to $k = 1$ additional trial since it also includes that $k = 1$st success.

Example 4.3. Negative Binomial Distribution and the Freezing of Cayuga Lake, III

Assuming again that the freezing of Cayuga Lake is well represented statistically by a series of annual Bernoulli trials with $p = 0.0435$, what can be said about the probability distributions for the number of years required to observe k winters in which the lake freezes? As noted earlier, these probabilities will be those pertaining to X in Equation 4.6.

Figure 4.1a shows three of these negative binomial distributions, for $k = 1$, 2, and 3, shifted to the right by k years in order to show the distributions of waiting times, $x + k$. That is, the leftmost points in the three functions in Figure 4.1a all correspond to $X = 0$ in Equation 4.6. For $k = 1$ the probability distribution function is the same as for the geometric distribution (Equation 4.5), and the figure shows that the probability of freezing in the next year is simply the Bernoulli $p = 0.0435$. The probabilities that year $x + 1$ will be the next freezing event decrease smoothly at a fast enough rate that probabilities for the first freeze being more than a century away are quite small. It is impossible for the lake to freeze $k = 2$ times before next year, so the first probability plotted in Figure 4.1a for $k = 2$ is at $x + k = 2$ years, and this probability is $p^2 = 0.0435^2 = 0.0019$. These probabilities rise through the most likely waiting time for two freezes at $x + 2 = 23$ years before falling again, although there is a nonnegligible probability that the lake still will not have frozen twice within a century. When waiting for $k = 3$ freezes, the probability distribution of waiting times is flattened more and shifted even further into the future.

An alternative way of viewing these distributions of waiting times is through their cumulative probability distribution functions,

$$\Pr\{X \leq x\} = \sum_{t=0}^{x} \Pr\{X = t\}, \qquad (4.9)$$

which are plotted in Figure 4.1b. Here all the probabilities for waiting times t less than or equal to a waiting time x of interest have been summed, analogously to Equation 3.18 for the empirical *cumulative distribution function* (CDF). For $k = 1$, the CDF rises rapidly at first, indicating that the probability of the first freeze occurring within the next few decades is quite high, and that it is nearly certain that the lake will freeze next within a century (assuming that the annual freezing probability p is stationary so that, e.g., it is not decreasing through time as a consequence of a changing climate). These functions rise more slowly for the waiting times for $k = 2$ and $k = 3$ freezes; and indicate a probability around 0.93 that the lake will freeze at least twice, and a probability near 0.82 that the lake will freeze at least three times, during the next century, again assuming that the climate is stationary.   ◇

Use of the negative binomial distribution is not limited to integer values of the parameter k, and when k is allowed to take on any positive value the distribution may be appropriate for flexibly describing variations in data on counts. For example, the negative binomial distribution has been used (in slightly modified form) to represent the distributions of spells of consecutive wet and dry days (Wilks 1999a), and annual numbers of landfalling Atlantic hurricanes (Hall and Jewson 2008) in a way that is more flexible than Equation 4.5 because values of k different from 1 produce different shapes for the distribution, as in Figure 4.1a. In general, appropriate parameter values must be determined by the data to which the distribution will be fit. That is, specific values for the parameters p and k must be determined that will allow Equation 4.6 to look as much as possible like the empirical distribution of the data that it will be used to represent.

The simplest way to find appropriate values for the parameters in the more general situation of non-integer k, that is, to fit the distribution, is to use the *method of moments*. To use the method of moments we mathematically equate the sample moments and the distribution (or population) moments. Since there are two parameters, it is necessary to use two distribution moments to define them. The first moment is the mean and the second moment is the variance. In terms of the distribution parameters, the mean of the negative binomial distribution is $\mu = k(1-p)/p$, and the variance is $\sigma^2 = k(1-p)/p^2$. Estimating p and k using the method of moments involves simply setting these expressions equal to the corresponding sample moments and solving the two equations simultaneously for the parameters.

That is, each data value x is an integer, and the mean and variance of these x's are calculated, and substituted into the equations

$$p = \frac{\bar{x}}{s^2},$$

(4.10a)

and

$$k = \frac{\bar{x}^2}{s^2 - \bar{x}}.$$

(4.10b)

4.2.4. Multinomial Distribution

The *multinomial distribution* extends the binomial distribution to situations where the MECE partition of the sample space consists of more than two discrete events. In common with the binomial distribution, the multinomial distribution relates to a sequence or collection of N independent trials, but on each of these trials one and only one of $K > 2$ outcomes $X_k, k = 1, \ldots, K$, occurs. The probabilities $p_k, k = 1, \ldots, K$ for each of these outcomes are unchanging (i.e., the process is stationary), and of course $\sum_k p_k = 1$.

The probability distribution function for the multinomial distribution is

$$\Pr\{X_1 = x_1, X_2 = x_2, \ldots, X_K = x_K\} = \frac{N!}{x_1! x_2! \ldots x_K!} p_1^{x_1} p_2^{x_2} \ldots p_K^{x_K},$$

(4.11)

which distributes probability over the K outcome variables. Note that for $K = 2$, Equation 4.11 reduces to the probability distribution function for the binomial distribution (Equation 4.1), and that each of the X_k individually follow the binomial distribution with parameters p_k and N.

Because the K outcomes x_k must sum to N, they can jointly be represented geometrically on a $K - 1$-dimensional hyperplane. Binomial outcomes, for example, can be represented on the real line (i.e., in one dimension), since either one of the numbers of the complementary events "success" and "failure" is sufficient to define the other. For $K = 3$ the outcomes can be jointly represented on a two-dimensional surface (see Section 9.4.9 for an example). Distribution fitting (i.e., estimation of the p_ks) is most easily achieved by substituting the respective outcome relative frequencies.

4.2.5. Poisson Distribution

The *Poisson distribution* describes the numbers of discrete events occurring in a series, or a sequence, and so pertains to data on counts that can take on only nonnegative integer values. Usually the sequence is understood to be in time, for example, the occurrence of storms in a particular geographic region over the course of a year. However, it is also possible to apply the Poisson distribution to counts of events occurring in one or more spatial dimensions, such as the number of gasoline stations along a particular stretch of highway, or the distribution of hailstones over a small area.

Poisson events occur randomly, but at a constant average rate. That is, the average rate at which Poisson events are generated is stationary. The individual events being counted must be independent, in the sense that their occurrences do not depend on whether or how many other events may have occurred elsewhere in nonoverlapping portions of the sequence. Given the average rate of event occurrence, the probabilities of particular numbers of events in a given interval depend only on the size of the interval over which events will be counted. A sequence of such events is sometimes said to have been generated by a *Poisson process*. As was the case for the binomial distribution, strict adherence to this independence condition is often difficult to demonstrate in atmospheric data, but the Poisson distribution can still yield a useful representation if the degree of dependence is not too strong. Ideally, Poisson events should be rare enough that the probability of more than one occurring simultaneously is very small. One way of motivating the Poisson distribution mathematically is as the limiting case of the binomial distribution, as p approaches zero and N approaches infinity.

The Poisson distribution has a single parameter, μ, that specifies the average occurrence rate. The Poisson parameter is sometimes called the *intensity* and has physical dimensions of occurrences per unit time. The probability distribution function for the Poisson distribution is

$$\Pr\{X=x\} = \frac{\mu^x e^{-\mu}}{x!}, \quad x=0,1,2,\ldots, \tag{4.12}$$

which associates probabilities with all possible numbers of occurrences, X, from zero to infinitely many. Here $e \approx 2.718$ is the base of the natural logarithms. The sample space for Poisson events therefore contains (countably) infinitely many elements. Clearly the summation of Equation 4.12 for x running from zero to infinity must be convergent and equal to 1. The probabilities associated with very large numbers of counts are vanishingly small, since the denominator in Equation 4.12 is $x!$

To use the Poisson distribution it must be fit to a sample of data. Again, fitting the distribution means finding the specific value for the single parameter μ that makes Equation 4.12 behave as similarly as possible to a data set at hand. For the Poisson distribution, a good way to estimate the parameter μ is by using the method of moments. Fitting the Poisson distribution is thus especially easy, since its one parameter is the mean number of occurrences per unit time, which can be estimated directly as the sample average of the number occurrences per unit time.

Example 4.4. Poisson Distribution and Annual U.S. Hurricane Landfalls
The Poisson distribution is a natural and commonly used statistical model for representing hurricane occurrence statistics (e.g., Parisi and Lund 2008). Consider the Poisson distribution in relation to the annual number of hurricanes making landfall on the U.S. coastline, from Texas through Maine, for 1899–1998, shown as the dashed histogram in Figure 4.2. During the 100 years covered by these data, 170 hurricanes made landfall on the U.S. coastline (Neumann et al. 1999). The counts range from zero U.S. hurricane landfalls in 16 of the 100 years, through six U.S. hurricane landfalls in two of the years (1916 and 1985). The average, or mean, rate of U.S. hurricane landfall occurrence is simply 170/100 = 1.7 landfalls/year, so this average is the method-of-moments estimate of the Poisson intensity for these data. Having fit the distribution by estimating a value for its parameter, the Poisson distribution can be used to compute probabilities that particular numbers of hurricanes will make landfall on the U.S. coastline annually. The first 8 of these probabilities (pertaining to zero through seven hurricane landfalls per year) are plotted in the form of the solid histogram in Figure 4.2.

The Poisson distribution allocates probability smoothly (within the limitation that the data are discrete) among the possible outcomes, with the most probable numbers of landfalls being near the mean rate of 1.7 per year. The distribution of the data shown by the dashed histogram resembles that of the fitted Poisson distribution but is more irregular, especially for the more active years, due at least in part to sampling variations. For example, there does not seem to be a physically based reason why five hurricanes per year should be less likely than six. Fitting a distribution to these data provides a sensible way to smooth out such variations, which is desirable if the irregular variations in the data histogram are not

FIGURE 4.2 Histogram of numbers of U.S. landfalling hurricanes for 1899–1998 *(dashed)* and fitted Poisson distribution with $\mu = 1.7$ hurricanes/year *(solid)*.

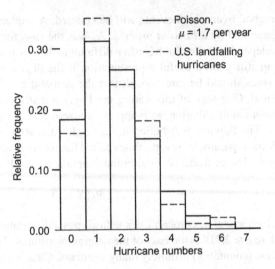

physically meaningful. Similarly, using the Poisson distribution to summarize the data allows quantitative estimation of probabilities for large numbers of landfalls in a year. Even though none of the years in this 100-year record had more than six U.S. hurricane landfalls, even more active years are not physically impossible and the fitted Poisson distribution allows probabilities for such events to be estimated. For example, according to this Poisson model, the probability of seven U.S. hurricane landfalls occurring in a given year would be estimated as $\Pr\{X = 7\} = 1.7^7 e^{-1.7}/7! = 0.00149$.   ◇

4.3. STATISTICAL EXPECTATIONS

4.3.1. Expected Value of a Random Variable

The *expected value* of a random variable or function of a random variable is simply the probability-weighted average of that variable or function. This weighted average is called the expected value, although we do not necessarily expect this outcome to occur in the informal sense of an "expected" event being likely. It can even happen that the statistical expected value is an impossible outcome. Statistical expectations are closely tied to probability distributions, since the distributions will provide the weights or weighting function for the weighted average. The ability to work easily with statistical expectations can be a strong motivation for choosing to represent data using parametric distributions rather than empirical distribution functions.

It is easiest to see expectations as probability-weighted averages in the context of a discrete probability distribution, such as the binomial. Conventionally, the expectation operator is denoted $E[]$, so that the expected value for a discrete random variable is

$$E[X] = \sum_x x \Pr\{X = x\}. \tag{4.13}$$

The equivalent notation $\langle X \rangle = E[X]$ is sometimes used for the expectation operator. The summation in Equation 4.13 is taken over all allowable values of X. For example, the expected value of X when X follows the binomial distribution is

TABLE 4.3 Expected Values (Means) and Variances for Probability Distribution Functions Described in Section 4.2, in Terms of Their Distribution Parameters

Distribution	Probability Distribution Function	$\mu = E[X]$	$\sigma^2 = Var[X]$
Binomial	Equation 4.1	Np	$Np(1-p)$
Geometric	Equation 4.5	$1/p$	$(1-p)/p^2$
Negative Binomial	Equation 4.6	$k(1-p)/p$	$k(1-p)/p^2$
Poisson	Equation 4.12	μ	μ

$$E[X] = \sum_{x=0}^{N} x \binom{N}{x} p^x (1-p)^{N-x}. \tag{4.14}$$

Here the allowable values of X are the nonnegative integers up to and including N, and each term in the summation consists of the specific value of the variable, x, multiplied by the probability of its occurrence from Equation 4.1.

The expectation $E[X]$ has a special significance, since it is the mean of the distribution of X. Distribution (or population, or data-generating process) means are conventionally denoted using the symbol μ. It is possible to analytically simplify Equation 4.14 to obtain, for the binomial distribution, the result $E[X] = Np$. Thus the mean of any binomial distribution is given by the product $\mu = Np$. Expected values for discrete probability distributions described in Section 4.2 are listed in Table 4.3, in terms of their distribution parameters. The U.S. hurricane landfall data in Figure 4.2 provide an example of the expected value $E[X] = 1.7$ landfalls being impossible to realize in any year.

4.3.2. Expected Value of a Function of a Random Variable

It can be very useful to compute expectations, or probability-weighted averages, of functions of random variables, $E[g(x)]$. Since the expectation is a linear operator, expectations of functions of random variables have the following properties·

$$E[c] = c, \tag{4.15a}$$

$$E[c\,g_1(x)] = c\,E[g_1(x)], \tag{4.15b}$$

$$E\left[\sum_{j=1}^{J} g_j(x)\right] = \sum_{j=1}^{J} E\left[g_j(x)\right], \tag{4.15c}$$

where c is any constant and $g_j(x)$ is any function of x. Because the constant c does not depend on x, $E[c] = \sum_x c \Pr\{X = x\} = c \sum_x \Pr\{X = x\} = c \cdot 1 = c$. Equations 4.15a and 4.15b reflect the fact that constants can be factored out of the summations when computing expectations. Equation 4.15c expresses the important property that the expectation of a sum is equal to the sum of the separate expected values.

Use of the properties expressed in Equation 4.15 can be illustrated with the expectation of the function $g(x) = (x - \mu)^2$. The expected value of this function is called the *variance* and is conventionally denoted by σ^2. Applying the properties in Equations 4.15 to this expectation yields

$$
\begin{aligned}
Var[X] = E\left[(X-\mu)^2\right] &= \sum_x (x-\mu)^2 \Pr\{X=x\} \\
&= \sum_x (x^2 - 2\mu x + \mu^2)\Pr\{X=x\} \\
&= \sum_x x^2 \Pr\{X=x\} - 2\mu \sum_x x\Pr\{X=x\} + \mu^2 \sum_x \Pr\{X=x\} \\
&= E\left[X^2\right] - 2\mu E[X] + \mu^2 \cdot 1 \\
&= E\left[X^2\right] - \mu^2.
\end{aligned}
\tag{4.16}
$$

Notice the similarity of the last equality on the first line of Equation 4.16 to the sample variance, given by the square of Equation 3.6. Similarly, the final equality in Equation 4.16 is analogous to the computational form for the sample variance, given by the square of Equation 3.31. Notice also that combining the first line of Equation 4.16 with the properties in Equation 4.15 yields

$$
Var[c\, g(x)] = c^2\, Var[g(x)]. \tag{4.17}
$$

Variances for four of the univariate discrete distributions described in Section 4.2 are listed in Table 4.3.

Example 4.5. Expected Value of a Function of a Binomial Random Variable

Table 4.4 presents the computation of statistical expectations for the binomial distribution with $N=3$ and $p=0.5$. These parameters correspond to the situation of simultaneously flipping three coins and counting $X=$ the number of heads. The first column shows the possible outcomes of X, and the second column shows the probabilities for each of the outcomes, computed according to Equation 4.1.

The third column in Table 4.4 shows the individual terms in the probability-weighted average $E[X]=\sum_x [x\,\Pr(X=x)]$. Adding these four values yields $E[X]=1.5$, as would be obtained by multiplying the two distribution parameters $\mu = Np$, in Table 4.3

The fourth column in Table 4.4 similarly shows the construction of the expectation $E[X^2]=3.0$. We might imagine this expectation in the context of a hypothetical game, in which the player receives $\$X^2$, that is, nothing if zero heads come up, $1 if one head comes up, $4 if two heads come up, and $9 if three heads come up. Over the course of many rounds of this game, the long-term average payout would be $E[X^2]=\$3.00$. An individual willing to pay more than $3 to play this game would be either foolish or inclined toward taking risks.

TABLE 4.4 Binomial Probabilities for $N=3$ and $p=0.5$, and the Construction of the Expectations $E[X]$ and $E[X^2]$ as Probability-Weighted Averages

X	$\Pr(X=x)$	$x \cdot \Pr(X=x)$	$x^2 \cdot \Pr(X=x)$
0	0.125	0.000	0.000
1	0.375	0.375	0.375
2	0.375	0.750	1.500
3	0.125	0.375	1.125
		$E[X]=1.500$	$E[X^2]=3.000$

Notice that the final equality in Equation 4.16 can be verified for this particular binomial distribution using Table 4.4. Here $E[X^2] - \mu^2 = 3.0 - (1.5)^2 = 0.75$, agreeing with $Var[X] = Np(1-p) = 3(0.5)(1-0.5) = 0.75$.   ◇

4.4. CONTINUOUS DISTRIBUTIONS

Most atmospheric variables can take on any of a continuum of values. Temperature, precipitation amount, geopotential height, wind speed, and other quantities are at least conceptually not restricted to integer values of the physical units in which they are measured. Even though the nature of measurement and reporting systems is such that measurements are rounded to discrete values, the set of reportable values is large enough that most such variables can still be treated as continuous quantities.

Many continuous parametric distributions exist. Those used most frequently in the atmospheric sciences are discussed in subsequent sections. Encyclopedic information on these and many other continuous distributions can be found in Johnson et al. (1994, 1995).

4.4.1. Distribution Functions and Expected Values

The mathematics of probability for continuous variables are somewhat different, although analogous, to those for discrete random variables. In contrast to probability calculations for discrete distributions, which involve summation over a discontinuous probability distribution function (e.g., Equation 4.1), probability calculations for continuous random variables involve integration over continuous functions called *probability density functions* (PDFs). A PDF is sometimes referred to more simply as a *density*.

Conventionally, the PDF for a random variable X is denoted $f(x)$. Just as summation of a discrete probability distribution function over all possible values of the random quantity must equal 1, the integral of any PDF over all allowable values of x must equal 1:

$$\int_x f(x)dx = 1. \tag{4.18}$$

A function cannot be a PDF unless it satisfies this condition. Furthermore, a PDF $f(x)$ must be nonnegative for all values of x. No specific limits of integration have been included in Equation 4.18, because different probability densities are defined over different ranges of the random variable (i.e., have different *support*).

Probability density functions are the continuous parametric analogs of the familiar histogram (see Section 3.3.5) and of the nonparametric kernel density estimate (see Section 3.3.6). However, the meaning of the PDF is often initially confusing precisely because of the analogy with the histogram. In particular, the height of the density function $f(x)$, obtained when it is evaluated at a particular value of the random variable, is not in itself meaningful in the sense of defining a probability. The confusion arises because often it is not realized that probability is proportional to area, and not to height, in both the PDF and the histogram.

Figure 4.3 shows a hypothetical PDF, defined for a nonnegative random variable X. A PDF can be evaluated for a specific value of the random variable, say $x = 1$, but by itself $f(1)$ is not meaningful in terms of probabilities for X. In fact, since X varies continuously over some portion of

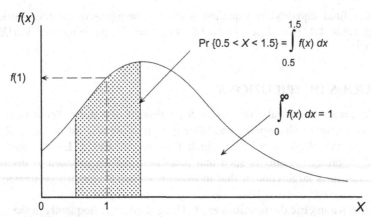

FIGURE 4.3   Hypothetical PDF $f(x)$ for a nonnegative random variable, X. Evaluation of $f(x)$ is not, by itself, meaningful in terms of probabilities for specific values of X. Probabilities are obtained by integrating portions of $f(x)$.

the real numbers, the probability of *exactly* $X = 1$ is infinitesimally small. It is meaningful, however, to think about and compute probabilities for values of a random variable in finite neighborhoods around $X = 1$. Figure 4.3 shows the probability of X being between 0.5 and 1.5 as the integral of the PDF between these limits.

An idea related to the PDF is that of the CDF. The CDF is a function of the random variable X, given by the integral of the PDF up to a particular value of x. Thus the CDF specifies probabilities that the random quantity X will not exceed particular values. It is therefore the continuous counterpart to the empirical CDF, Equation 3.18, and the discrete CDF, for example, Equation 4.9. Conventionally, CDFs are denoted $F(x)$:

$$F(x) = \Pr\{X \le x\} = \int_{X \le x} f(x)dx. \tag{4.19}$$

Again, specific integration limits have been omitted from Equation 4.19 to indicate that the integration is performed from the minimum allowable value of X to the particular value, x, that is the argument of the function. Since the values of $F(x)$ are probabilities, $0 \le F(x) \le 1$.

Equation 4.19 transforms a particular value of the random variable to a cumulative probability. The value of the random variable corresponding to a particular cumulative probability is given by the inverse of the CDF,

$$F^{-1}(p) = x(F), \tag{4.20}$$

where p is the cumulative probability. That is, Equation 4.20 specifies the upper limit of the integration in Equation 4.19 that will yield a particular cumulative probability $p = F(x)$. Since this inverse of the CDF specifies the data quantile corresponding to a particular probability, Equation 4.20 is also called the *quantile function.*

Statistical expectations are defined for continuous as well as for discrete random variables. As is the case for discrete variables, the expected value of a variable or a function is the probability-weighted average of that variable or function. Since probabilities for continuous random variables are computed by integrating their density functions, the expected value of a function of a random variable is given by the integral

TABLE 4.5 Expected Values (Means) and Variances for Continuous Probability Density Functions Described in This Section, in Terms of Their Parameters

Distribution	PDF	$E[X]$	$Var[X]$
Gaussian	Equation 4.24	μ	σ^2
Lognormal[1]	Equation 4.36	$\exp[\mu + \sigma^2/2]$	$(\exp[\sigma^2] - 1) \exp[2\mu + \sigma^2]$
Zero-truncated Gaussian	Equation 4.39	Equation 4.41a	Equation 4.41b
Logistic	Equation 4.42	μ	$\sigma^2 \pi^2/3$
Gamma	Equation 4.45	$\alpha\beta$	$\alpha\beta^2$
Exponential	Equation 4.52	β	β^2
Chi-square	Equation 4.54	ν	2ν
Pearson III	Equation 4.55	$\zeta + \alpha\beta$	$\alpha\beta^2$
Beta	Equation 4.58	$\alpha/(\alpha + \beta)$	$(\alpha\beta)/[(\alpha+\beta)^2(\alpha+\beta+1)]$
Gumbel[2]	Equation 4.57	$\zeta + \gamma\beta$	$\beta\pi/\sqrt{6}$
GEV[3]	Equation 4.63	$\zeta - \beta[1 - \Gamma(1 - \kappa)]/\kappa$	$\beta^2(\Gamma[1 - 2\kappa] - \Gamma^2[1 - \kappa])/\kappa^2$
Weibull	Equation 4.66	$\beta\Gamma[1 + 1/\alpha]$	$\beta^2(\Gamma[1 + 2/\alpha] - \Gamma^2[1 + 1/\alpha])$
Mixed Exponential	Equation 4.78	$w\beta_1 + (1 - w)\beta_2$	$w\beta_1^2 + (1 - w)\beta_2^2 + w(1 - w)(\beta_1 - \beta_2)^2$

[1]For the lognormal distribution, μ and σ^2 refer to the mean and variance of the log-transformed variable $y = \ln(x)$.
[2]$\gamma = 0.57721 \ldots$ is Euler's constant.
[3]For the GEV the mean exists (is finite) only for $\kappa < 1$, and the variance exists only for $\kappa < 1/2$.

$$E[g(x)] = \int_x g(x) f(x) \, dx. \tag{4.21}$$

Expectations of continuous random variables also exhibit the properties in Equations 4.15 and 4.17. For $g(x) = x$, $E[X] = \mu$ is the mean of the distribution whose PDF is $f(x)$. Similarly, the variance of a continuous variable is given by the expectation of the function $g(x) = (x - E[X])^2$,

$$
\begin{aligned}
Var[X] = E\left[(x - E[X])^2\right] &= \int_x (x - E[X])^2 f(x) \, dx \\
&= \int_x x^2 f(x) \, dx - (E[X])^2 = E[X^2] - \mu^2.
\end{aligned}
\tag{4.22}
$$

Note that, depending on the particular functional form of $f(x)$, some or all of the integrals in Equations 4.19, 4.21, and 4.22 may not be analytically computable, and for some distributions the integrals may not even exist.

Table 4.5 lists means and variances for the distributions to be described in this section, in terms of the distribution parameters.

4.4.2. Gaussian Distributions

The *Gaussian distribution* plays a central role in classical statistics and has many applications in the atmospheric sciences as well. It is sometimes also called the *normal* distribution, although this name carries the unwarranted connotation that it is in some way universal, or that deviations from it are in some way unnatural. Its PDF is the bell-shaped curve, familiar even to people who have not studied statistics.

The breadth of applicability of the Gaussian distribution follows in large part from a very powerful theoretical result, known as the *Central Limit Theorem*. Informally, the Central Limit Theorem states that in the limit, as the sample size becomes large, the sum (or, equivalently because it is proportional, the arithmetic mean) of a set of independent observations will have a Gaussian *sampling distribution*. That is, a histogram of the sums or sample means of a large number of different batches of the same kind of data, each of size n, will look like a bell curve if n is large enough. This is true regardless of the distribution from which the original data have been drawn. The data need not even be from the same distribution! Actually, the independence of the observations is not really necessary for the shape of the resulting distribution to be Gaussian either (see Section 5.2.4), which considerably broadens the applicability of the Central Limit Theorem for atmospheric data.

What is not clear for particular data sets is just how large the sample size must be for the Central Limit Theorem to apply. In practice this sample size depends on the distribution from which the summands are drawn. If the summed observations are themselves taken from a Gaussian distribution, the sum of any number of them (including, of course, $n = 1$) will also be Gaussian. For underlying distributions not too unlike the Gaussian (unimodal and not too asymmetrical), the sum of a modest number of observations will be nearly Gaussian. Summing daily temperatures to obtain a monthly averaged temperature is a good example of this situation. Daily temperature values can exhibit noticeable asymmetry (e.g., Figure 3.5), but are usually much more symmetrical than daily precipitation values. Conventionally, average daily temperature is approximated as the average of the daily maximum and minimum temperatures, so that the average monthly temperature is computed as

$$\bar{T} = \frac{1}{30} \sum_{i=1}^{30} \frac{T_{max}(i) + T_{min}(i)}{2}, \tag{4.23}$$

for a month with 30 days. Here the average monthly temperature is computed from the sum of 60 numbers drawn from two more or less symmetrical distributions. It is not surprising, in light of the Central Limit Theorem, that monthly temperature values are often very successfully represented by Gaussian distributions.

A contrasting situation is that of the monthly total precipitation, constructed as the sum of, say, 30 daily precipitation values. There are fewer numbers going into this sum than is the case for the average monthly temperature in Equation 4.23, but the more important difference has to do with the distribution of the underlying daily precipitation amounts. Typically most daily precipitation values are zero, and most of the nonzero amounts are small. That is, the distributions of daily precipitation amounts are usually very strongly skewed to the right (e.g., Figure 3.11b). Generally, the distribution of sums of 30 such values is also skewed to the right, although not so extremely. The schematic plot for $\lambda = 1$ in Figure 3.14 illustrates this asymmetry for total January precipitation at Ithaca. Note, however, that the distribution of Ithaca January precipitation totals in Figure 3.14 is much more symmetrical than the corresponding distribution for the underlying daily precipitation amounts in Figure 3.11b. Even though the summation of 30 daily values has not produced a Gaussian distribution for the monthly totals, the shape of the distribution of monthly precipitation is much closer to the Gaussian than the very strongly skewed distribution of the daily precipitation amounts. In humid climates, the distributions of seasonal (i.e., 90-day) precipitation totals begin to approach the Gaussian, but even annual precipitation totals at arid locations can exhibit substantial positive skewness. Whether the sample size is adequate for invocation of CLT for a particular data set can

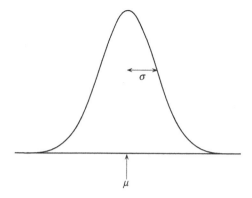

FIGURE 4.4 Probability density function for the Gaussian distribution, Equation 4.24. The mean, μ, locates the center of this symmetrical distribution, and the standard deviation, σ, controls the degree to which the distribution spreads out. Nearly all of the probability is within $\pm 3\sigma$ of the mean.

be assessed through bootstrapping (Section 5.3.5), examining the shape of the resulting bootstrap distribution for the sample mean.

The PDF for the Gaussian distribution is

$$f(x) = \frac{1}{\sigma\sqrt{2\pi}} \exp\left[-\frac{(x-\mu)^2}{2\sigma^2}\right], \quad -\infty < x < \infty. \tag{4.24}$$

The two distribution parameters are the mean, μ, and the standard deviation, σ; and π is the mathematical constant 3.14159.... Gaussian random variables are defined on the entire real line, so Equation 4.24 is valid for $-\infty < x < \infty$. Graphing Equation 4.24 results in the familiar bell-shaped curve shown in Figure 4.4. This figure shows that the mean locates the center of this symmetrical distribution, and the standard deviation controls the degree to which the distribution spreads out. Nearly all the probability is within $\pm 3\sigma$ of the mean.

In order to use the Gaussian distribution to represent a set of data, it is necessary to fit the two distribution parameters. Good parameter estimates for this distribution are easily obtained using the method of moments. Again, the method of moments amounts to nothing more than equating as many sample moments and distribution moments as there are parameters to be estimated. In the case of the Gaussian distribution the first two moments correspond exactly to the distribution parameters: the first moment is the mean, μ, and the second moment is the variance, σ^2. Therefore we simply estimate μ as the sample mean (Equation 3.2), and σ as the sample standard deviation (Equation 3.6).

If a data sample follows at least approximately a Gaussian distribution, these parameter estimates will make Equation 4.24 behave similarly to the data. Then, in principle, probabilities for events of interest can be obtained by integrating Equation 4.24. Practically, however, analytic integration of Equation 4.24 is impossible, so that a formula for the CDF, $F(x)$, for the Gaussian distribution does not exist. Rather, Gaussian probabilities are obtained in one of two ways. If the probabilities are needed as part of a computer program, the integral of Equation 4.24 can be economically approximated (e.g., Abramowitz and Stegun 1984) or computed by numerical integration (e.g., Press et al. 1986) to precision that is more than adequate. If only a few probabilities are needed, it is practical to compute them by hand using tabulated values such as those in Table B.1 in Appendix B.

In either of these two situations, a data transformation will nearly always be required. This is because Gaussian probability tables and algorithms pertain to the *standard Gaussian distribution*, that is, the Gaussian distribution having $\mu = 0$ and $\sigma = 1$. Conventionally, the random variable described by the standard Gaussian distribution is denoted as z. Its PDF simplifies to

$$\phi(z) = \frac{1}{\sqrt{2\pi}} \exp\left[-\frac{z^2}{2}\right]. \tag{4.25}$$

The notation $\phi(z)$ is often used for the PDF of the standard Gaussian distribution, rather than $f(z)$. Similarly, $\Phi(z)$ is the conventional notation for the CDF of the standard Gaussian distribution, which is related to the "error function," erf(z) according to erf(z)=2$\Phi[z\sqrt(2)]$-1. Any Gaussian random variable, x, can be transformed to standard form, z, simply by subtracting its mean and dividing by its standard deviation,

$$z = \frac{x - \mu}{\sigma}. \tag{4.26}$$

In practical settings, the mean and standard deviation usually need to be estimated using the corresponding sample statistics, so that we use

$$z = \frac{x - \bar{x}}{s}. \tag{4.27}$$

Note that whatever physical units characterize x will cancel in this transformation, so that the standardized variable, z, is always dimensionless.

Equation 4.27 is exactly the same as the standardized anomaly of Equation 3.27. Any batch of data can be transformed by subtracting the mean and dividing by the standard deviation, and this transformation will produce transformed values having a sample mean of zero and a sample standard deviation of one. However, the transformed data will not follow a Gaussian distribution unless the untransformed data do. Use of the standardized variable in Equation 4.27 to obtain Gaussian probabilities is illustrated in the following example.

Example 4.6. Evaluating Gaussian Probabilities

Consider a Gaussian distribution characterized by $\mu = 22.2°F$ and $\sigma = 4.4°F$. These parameters were fit to a set of average January temperatures at Ithaca. Suppose we are interested in evaluating the probability that an arbitrarily selected, or future, January will have average temperature as cold as or colder than 21.4°F, the value observed in 1987 (see Table A.1). Transforming this temperature using the standardization in Equation 4.26 yields $z = (21.4°F - 22.2°F)/4.4°F = -0.18$. Thus the probability of a temperature as cold as or colder than 21.4°F is the same as the probability of a value of z as small as or smaller than -0.18: $\Pr\{X \leq 21.4°F\} = \Pr\{Z \leq -0.18\}$.

Evaluating $\Pr\{Z \leq -0.18\}$ is easy, using Table B.1 in Appendix B, which contains cumulative probabilities for the standard Gaussian distribution, $\Phi(z)$. Looking across the row in Table B.1 labeled -0.1 to the column labeled 0.08 yields the desired probability, 0.4286. Evidently, there is a substantial probability that an average temperature this cold or colder will occur in January at Ithaca.

Notice that Table B.1 contains no rows for positive values of z. These are not necessary because the Gaussian distribution is symmetric. This means, for example, that $\Pr\{Z \geq +0.18\} = \Pr\{Z \leq -0.18\}$, since there will be equal areas under the curve in Figure 4.4 to the left of $z = -0.18$, and to the right of $z = +0.18$. Therefore Table B.1 can be used more generally to evaluate probabilities for $z > 0$ by applying the relationship

$$\Pr\{Z \leq z\} = 1 - \Pr\{Z \leq -z\}, \tag{4.28a}$$

or, equivalently,

$$\Phi(z) = 1 - \Phi(-z), \tag{4.28b}$$

which follows from the fact that the total area under the curve of any PDF is 1 (Equation 4.18).

Using Equation 4.28 it is straightforward to evaluate $\Pr\{Z \leq +0.18\} = 1 - 0.4286 = 0.5714$. The average January temperature at Ithaca to which $z = +0.18$ corresponds is obtained by inverting Equation 4.26,

$$x = \sigma z + \mu. \tag{4.29}$$

The probability is 0.5714 that an average January temperature at Ithaca will be no greater than $(4.4°F)$ $(0.18) + 22.2°F = 23.0°F$.

It is only slightly more complicated to compute probabilities for outcomes between two specific values, say Ithaca January temperatures between 20°F and 25°F. Since the event $\{X \leq 20°F\}$ is a subset of the event $\{X \leq 25°F\}$, the desired probability, $\Pr\{20°F < T \leq 25°F\}$ can be obtained by the subtraction $\Phi(z_{25}) - \Phi(z_{20})$. Here $z_{25} = (25.0°F - 22.2°F)/4.4°F = 0.64$ and $z_{20} = (20.0°F - 22.2°F)/4.4°F = -0.50$. Therefore (from Table B.1), $\Pr\{20°F < T \leq 25°F\} = \Phi(z_{25}) - \Phi(z_{20}) = 0.739 - 0.309 = 0.430$.

It is also sometimes required to evaluate the inverse of the standard Gaussian CDF, that is, the standard Gaussian quantile function, $\Phi^{-1}(p)$. This function specifies values of the standard Gaussian variate, z, corresponding to particular cumulative probabilities, p. Again, an explicit formula for this function cannot be written, but Φ^{-1} can be evaluated using Table B.1 in reverse. For example, to find the average January Ithaca temperature defining the lowest decile (i.e., the coldest 10% of Januaries), the body of Table B.1 would be searched for $\Phi(z) = 0.10$. This cumulative probability corresponds almost exactly to $z = -1.28$. Using Equation 4.29, $z = -1.28$ corresponds to a January temperature of $(4.4°F)(-1.28) + 22.2°F = 16.6°F$.   ◇

When high precision is not required for Gaussian probabilities, a "pretty good" approximation to the standard Gaussian CDF can be used,

$$\Phi(z) \approx \frac{1}{2}\left[1 \pm \sqrt{1 - \exp\left(\frac{-2z^2}{\pi}\right)}\right]. \tag{4.30}$$

The positive root is taken for $z > 0$ and the negative root is used for $z < 0$. The maximum errors produced by Equation 4.30 are about 0.003 (probability units) in magnitude, which occur at $z = \pm 1.65$. Equation 4.30 can be inverted to yield an approximation to the Gaussian quantile function, but the approximation is poor for the tail (i.e., for extreme) probabilities that are often of greatest interest.

Bivariate Normal Distribution

In addition to the power of the Central Limit Theorem, another reason that the Gaussian distribution is used so frequently is that it easily generalizes to higher dimensions. That is, it is usually straightforward to represent joint variations of multiple Gaussian variables through what is called the *multivariate Gaussian* or *multivariate normal distribution*. This distribution is discussed more extensively in Chapter 12, since in general the mathematical development for the multivariate Gaussian distribution requires use of matrix algebra.

However, the simplest case of the multivariate Gaussian distribution, describing the joint variations of two Gaussian variables, can be presented without vector notation. This two-variable distribution is known as the *bivariate Gaussian* or *bivariate normal distribution*. It is sometimes possible to use this distribution to describe the behavior of two non-Gaussian distributions if the variables are first subjected to transformations such as those in Equation 3.21 or 3.24. In fact the opportunity to use the bivariate normal can be a major motivation for using such transformations.

Let the two variables considered be x and y. The bivariate normal distribution is defined by the joint PDF

$$f(x, y) = \frac{1}{2\pi\sigma_x\sigma_y\sqrt{1-\rho^2}} \exp\left\{-\frac{1}{2(1-\rho^2)}\left[\left(\frac{x-\mu_x}{\sigma_x}\right)^2 + \left(\frac{y-\mu_y}{\sigma_y}\right)^2 - 2\rho\left(\frac{x-\mu_x}{\sigma_x}\right)\left(\frac{y-\mu_y}{\sigma_y}\right)\right]\right\}.$$

(4.31)

As a generalization of Equation 4.24 from one to two dimensions, this function defines a surface above the x–y plane rather than a curve above the x-axis. For continuous bivariate distributions, including the bivariate normal, probability corresponds geometrically to the volume under the surface defined by the PDF so that, analogously to Equation 4.18, necessary conditions to be fulfilled by any bivariate PDF are

$$\iint\limits_{x,y} f(x, y)\, dy\, dx = 1, \quad \text{and } f(x, y) \geq 0.$$

(4.32)

The bivariate normal distribution has five parameters: the two means and standard deviations for the variables x and y, and the correlation between them, ρ. The two marginal distributions for the variables x and y (i.e., the univariate PDFs $f(x)$ and $f(y)$) must both be Gaussian distributions and have parameters μ_x, σ_x, and μ_y, σ_y, respectively. It is usual, although not guaranteed, for the joint distribution of any two Gaussian variables to be bivariate normal. Fitting the bivariate normal distribution is very easy. The means and standard deviations are estimated using their sample counterparts for the x and y variables separately, and the parameter ρ is estimated as the Pearson product–moment correlation between x and y, Equation 3.28.

Figure 4.5 illustrates the general shape of the bivariate normal distribution. It is mound shaped in three dimensions, with properties that depend on the five parameters. The function achieves its maximum height above the point (μ_x, μ_y). Increasing σ_x stretches the density in the x direction and increasing σ_y stretches it in the y direction. For $\rho = 0$ the density is symmetric around the point (μ_x, μ_y) with respect to both the x- and y-axes. Curves of constant height (i.e., intersections of $f(x,y)$ with planes parallel to the x–y plane) are concentric circles if $\rho = 0$ and $\sigma_x = \sigma_y$, and are concentric ellipses otherwise. As ρ increases in absolute value the density function is stretched diagonally, with the curves of constant height becoming increasingly elongated ellipses. For negative ρ the orientation of these

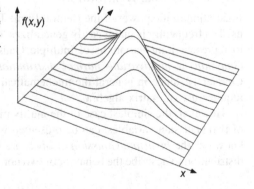

FIGURE 4.5 Perspective view of a bivariate normal distribution with $\sigma_x = \sigma_y$, and $\rho = -0.75$. The individual lines depicting the hump of the bivariate distribution have the shapes of (univariate) Gaussian distributions, illustrating that conditional distributions of x given a particular value of y are themselves Gaussian.

ellipses is as depicted in Figure 4.5: larger values of x are more likely to occur simultaneously with smaller values of y, and smaller values of x are more likely with larger values of y. The ellipses have the opposite orientation (positive slope) for positive values of ρ.

Probabilities for joint outcomes of x and y are given by the double integral of Equation 4.31 over the relevant region in the plane, for example

$$\Pr\{(y_1 < Y \le y_2) \cap (x_1 < X \le x_2)\} = \int_{x_1}^{x_2} \int_{y_1}^{y_2} f(x, y)\, dy\, dx. \tag{4.33}$$

This integration cannot be done analytically, and in practice numerical methods usually are used. Probability tables for the bivariate normal distribution do exist (National Bureau of Standards 1959), but they are lengthy and cumbersome. It is possible to compute probabilities for elliptically shaped regions, called probability ellipses, centered on (μ_x, μ_y) using the method illustrated in Example 12.1. When computing probabilities for other regions, it can be more convenient to work with the bivariate normal distribution in standardized form. This is the extension of the standardized univariate Gaussian distribution (Equation 4.25) and is achieved by subjecting both the x and y variables to the transformation in Equation 4.26 or 4.27. Thus $\mu_{z_x} = \mu_{z_y} = 0$ and $\sigma_{z_x} = \sigma_{z_y} = 1$, leading to the bivariate density

$$\phi\left(z_x, z_y\right) = \frac{1}{2\pi\sqrt{1 - \rho^2}} \exp\left[-\frac{z_x^2 - 2\rho\, z_x z_y + z_y^2}{2\,(1 - \rho^2)}\right]. \tag{4.34}$$

A very useful property of the bivariate normal distribution is that the conditional distribution of one of the variables, given any particular value of the other, is univariate Gaussian. This property is illustrated graphically in Figure 4.5, where the individual lines defining the shape of the distribution in three dimensions themselves have Gaussian shapes. Each indicates a function proportional to a conditional distribution of x given a particular value of y. The parameters for these conditional Gaussian distributions can be calculated from the five parameters of the bivariate normal distribution. For the conditional distribution of x given a particular value of y, the conditional Gaussian density function $f(x|Y = y)$ has parameters

$$\mu_{x|y} = \mu_x + \rho\frac{\sigma_x}{\sigma_y}\left(y - \mu_y\right) \tag{4.35a}$$

and

$$\sigma_{x|y} = \sigma_x\sqrt{1 - \rho^2}. \tag{4.35b}$$

Equation 4.35a relates the mean of x to the distance of y from its mean, scaled according to the product of the correlation and the ratio of the standard deviations. It indicates that the conditional mean $\mu_{x|y}$ is larger than the unconditional mean μ_x if y is greater than its mean and ρ is positive, or if y is less than its mean and ρ is negative. If x and y are uncorrelated, knowing a value of y gives no additional information about y, and $\mu_{x|y} = \mu_x$ since $\rho = 0$. Equation 4.35b indicates that, unless the two variables are uncorrelated, $\sigma_{x|y} < \sigma_x$, regardless of the sign of ρ. Knowing y provides some information about x if the two have nonzero correlation, and the diminished uncertainty about x is reflected by the smaller standard deviation. In this sense, ρ^2 is often interpreted as the proportion of the variance in x that is accounted for by y.

FIGURE 4.6 Gaussian distributions representing the unconditional distribution for daily January maximum temperature at Canandaigua, and the conditional distribution given that the Ithaca maximum temperature was 25°F. The large correlation between maximum temperatures at the two locations results in the conditional distribution being much sharper, reflecting substantially diminished uncertainty.

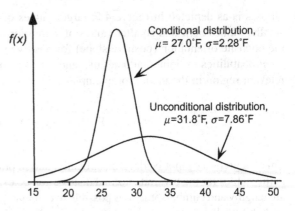

Example 4.7. Bivariate Normal Distribution and Conditional Probability

Consider the maximum temperature data for January 1987 at Ithaca and Canandaigua, in Table A.1. Figure 3.5 indicates that these data are fairly symmetrical, so that it may be reasonable to model their joint behavior as bivariate normal. A scatterplot of these two variables is shown in one of the panels of Figure 3.31. The average maximum temperatures are 29.87°F and 31.77°F at Ithaca and Canandaigua, respectively. The corresponding sample standard deviations are 7.71°F and 7.86°F. Table 3.5 indicates their Pearson correlation to be 0.957.

With such a large correlation, knowing the temperature at one location should give very strong information about the temperature at the other. Suppose it is known that the Ithaca maximum temperature is 25°F, and probability information about the Canandaigua maximum temperature is needed, perhaps for the purpose of estimating a missing value there. Using Equation 4.35a, the conditional mean for the distribution of maximum temperature at Canandaigua, given that the Ithaca maximum temperature is 25°F, is 27.0°F—substantially lower than the unconditional mean of 31.77°F. Using Equation 4.35b, the conditional standard deviation is 2.28°F. This would be the conditional standard deviation regardless of the particular value of the Ithaca temperature chosen, since Equation 4.35b does not depend on the value of the conditioning variable. The conditional standard deviation is so much lower than the unconditional standard deviation because of the high correlation of maximum temperatures between the two locations. As illustrated in Figure 4.6, this reduced uncertainty means that any of the conditional distributions for Canandaigua temperature given the Ithaca temperature will be much sharper than the unmodified, unconditional distribution for Canandaigua maximum temperature.

Using these parameters for the conditional distribution of maximum temperature at Canandaigua, we can compute such quantities as the probability that the Canandaigua maximum temperature is at or below freezing, given that the Ithaca maximum is 25°F. The required standardized variable is $z = (32 - 27.0)/2.28 = 2.19$, which corresponds to a probability of 0.986. By contrast, the corresponding climatological probability (without benefit of knowing the Ithaca maximum temperature) would be computed from $z = (32 - 31.8)/7.86 = 0.025$, corresponding to the much lower probability 0.510. ◇

4.4.3. Some Gaussian Distribution Variants

Although it is mathematically possible to fit Gaussian distributions to data that are nonnegative and positively skewed, the results are generally not useful. For example, the January 1933–1982 Ithaca precipitation data in Table A.2 can be characterized by a sample mean of 1.96 in. and a sample standard deviation of 1.12 in. These two statistics are sufficient to fit a Gaussian distribution to these data, and this distribution is shown as the dashed PDF in Figure 4.16, but applying this fitted distribution leads to non-sense. In particular, using Table B.1, we can compute the probability of negative precipitation as $\Pr\{Z < (0.00 - 1.96)/1.12\} = \Pr\{Z < -1.75\} = 0.040$. This computed probability is not especially large, but neither is it vanishingly small. The true probability is exactly zero: observing negative precipitation is impossible.

Several variants of the Gaussian distribution exist that can be useful for representing these kinds of data. Three of these are presented in this section: the lognormal distribution, the truncated Gaussian distribution, and the censored Gaussian distribution.

Lognormal Distribution

As noted in Section 3.4.1, one approach to dealing with skewed data is to subject them to a power transformation that produces an approximately Gaussian distribution. When that power transformation is logarithmic (i.e., $\lambda = 0$ in Equation 3.21), the (original, untransformed) data are said to follow the *lognormal distribution*, with PDF

$$f(x) = \frac{1}{x \sigma_y \sqrt{2\pi}} \exp\left[-\frac{\left(\ln x - \mu_y\right)^2}{2\sigma_y^2}\right], \quad x > 0. \tag{4.36}$$

Here μ_y and σ_y are the mean and standard deviation, respectively, of the transformed variable, $y = \ln(x)$. Actually, the lognormal distribution is somewhat confusingly named, since the random variable x is the *anti*log of a variable y that follows a Gaussian distribution.

Parameter fitting for the lognormal distribution is simple and straightforward: the mean and standard deviation of the log-transformed data values y—that is, μ_y and σ_y, respectively—are estimated by their sample counterparts. The relationships between these parameters, in Equation 4.36, and the mean and variance of the original variable X are

$$\mu_x = \exp\left[\mu_y + \frac{\sigma_y^2}{2}\right] \tag{4.37a}$$

and

$$\sigma_x^2 = \left(\exp\left[\sigma_y^2\right] - 1\right) \exp\left[2\mu_y + \sigma_y^2\right]. \tag{4.37b}$$

Lognormal probabilities are evaluated simply by working with the transformed variable $y = \ln(x)$, and using computational routines or probability tables for the Gaussian distribution. In this case the standard Gaussian variable

$$z = \frac{\ln(x) - \mu_y}{\sigma_y}, \tag{4.38}$$

follows a Gaussian distribution with $\mu_z = 0$ and $\sigma_z = 1$.

The lognormal distribution is sometimes somewhat arbitrarily assumed for positively skewed data. In particular, the lognormal too frequently is used without checking whether a different power transformation might produce more nearly Gaussian behavior. In general, it is recommended that other candidate power transformations also be investigated as explained in Section 3.4.1 before the lognormal distribution is adopted to represent a particular data set.

Truncated Gaussian Distribution

The *truncated Gaussian distribution* provides another approach to modeling positive and positively skewed data using the Gaussian bell-curve shape. Although it is mathematically possible to fit a conventional Gaussian distribution to such data in the usual way, unless both the mean is positive and the standard deviation is much smaller than the mean, the resulting Gaussian distribution will specify substantial probability for impossible negative outcomes. A zero-truncated Gaussian distribution is a Gaussian distribution having nonzero probability only for positive values of the random variable, with the portion of the PDF corresponding to negative values cut off, or truncated. Zero-truncated Gaussian distributions have been used successfully to represent distributions of forecast wind speeds (e.g., Thorarinsdottir and Gneiting 2010, Baran 2014).

Figure 4.3 illustrates a typical zero-truncated Gaussian shape, for which $\mu > 0$ and σ is comparable in magnitude to μ. The truncated Gaussian PDF is written in terms of the PDF and CDF of the standard Gaussian distribution, and has the form

$$f(x) = \phi\left(\frac{x-\mu}{\sigma}\right)\left[\sigma\,\Phi\left(\frac{\mu}{\sigma}\right)\right]^{-1}, \quad x > 0. \tag{4.39}$$

The factor $\Phi(\mu/\sigma)$ in the denominator is necessary in order that Equation 4.39 will integrate to unity. In effect, the probability for $x \leq 0$ has been spread proportionally across the rest of the distribution. The corresponding CDF is then

$$F(x) = \left[\Phi\left(\frac{x-\mu}{\sigma}\right) - \Phi\left(\frac{-\mu}{\sigma}\right)\right] / \Phi\left(\frac{\mu}{\sigma}\right), \tag{4.40}$$

which converges to $\Phi[(x-\mu)/\sigma]$ (i.e., an ordinary Gaussian CDF) for $\mu \gg \sigma$.

Estimating the distribution location parameter μ and scale parameter σ is best done using iterative fitting methods (e.g., Thorarinsdottir and Gneiting 2010) such as maximum likelihood (Section 4.6). Because Equation 4.39 specifies zero probability for $x \leq 0$, the mean of a zero-truncated Gaussian distribution is necessarily larger than the location parameter μ, and the variance is necessarily smaller than the scale parameter σ^2. Specifically,

$$E[x] = \mu + \sigma\frac{\phi(\mu/\sigma)}{\Phi(\mu/\sigma)} \tag{4.41a}$$

and

$$Var[x] = \sigma^2\left\{1 - \frac{\mu}{\sigma}\frac{\phi(\mu/\sigma)}{\Phi(\mu/\sigma)} - \left[\frac{\phi(\mu/\sigma)}{\Phi(\mu/\sigma)}\right]^2\right\}. \tag{4.41b}$$

Analogously to Equation 4.40, Equations 4.41a and 4.41b converge to μ and σ^2, respectively, for $\mu \gg \sigma$.

Censored Gaussian Distribution

The *censored Gaussian distribution* is an alternative to truncation that may be more appropriate for representing precipitation data, where there is a discrete probability mass at exactly zero. Zero censoring, in contrast to zero truncation, allows a probability distribution to represent values falling hypothetically

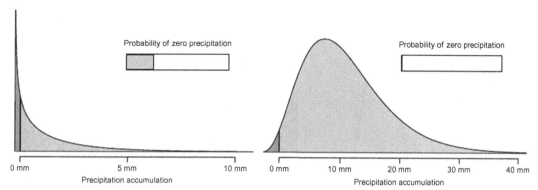

FIGURE 4.7 Illustration of the use of censoring to represent the discontinuous probability spike for zero precipitation, based on shifted-gamma (Pearson III) distributions. Probability for negative amounts is attributed to exactly zero precipitation, as indicated by the inset bars. *From Scheuerer and Hamill (2015a). © American Meteorological Society. Used with permission.*

below a censoring threshold, even though those values have not been observed. The censoring threshold is generally zero for precipitation data, and any probability corresponding to negative values is assigned to exactly zero, yielding a probability spike there. The idea is distinct from truncation, where probability for any negative values is spread proportionally across the positive values.

Bárdossy and Plate (1992) used zero-censored Gaussian distributions to represent (power-transformed) precipitation data, enabling a spatial precipitation model using the multivariate Gaussian distribution (Chapter 12). The censored distribution idea may also be applied to distributions other than the Gaussian in order to represent the discontinuous probability spike at zero precipitation, including the generalized extreme-value distribution (Section 4.4.7) (Scheuerer 2014), the gamma distribution (Section 4.4.5) (Wilks 1990), the shifted-gamma (Pearson III, Section 4.4.5) (Scheuerer and Hamill 2015a, Baran and Nemoda 2016), and the logistic distribution (Section 4.4.4) (Stauffer et al. 2017). Figure 4.7 illustrates the idea for two censored distributions. Probability for nonzero precipitation amounts is distributed according to the fitted distributions. Any probability that those distributions attribute to negative amounts is assigned to exactly zero precipitation, as indicated by the shaded inset bars.

4.4.4. Logistic Distributions

The *logistic distribution* shape is quite similar to that of the Gaussian distribution, but with somewhat heavier tails. Its two parameters are the location, μ, and scale, σ. The PDF and CDF for the logistic distribution are, respectively,

$$f(x) = \exp\left(\frac{x-\mu}{\sigma}\right)\left(1 + \exp\left(\frac{x-\mu}{\sigma}\right)\right)^{-2} \tag{4.42}$$

and

$$F(x) = \frac{\exp\left(\frac{x-\mu}{\sigma}\right)}{1 + \exp\left(\frac{x-\mu}{\sigma}\right)}. \tag{4.43}$$

The mean of the logistic distribution is equal to the location parameter μ, but the variance is $\sigma^2\pi^2/3$, so that logistic distributions with unit variance have $\sigma = \sqrt{3}/\pi$. Figure 4.8 compares the standard (zero mean and unit variance) logistic distribution (solid curves) and the corresponding standard Gaussian distribution

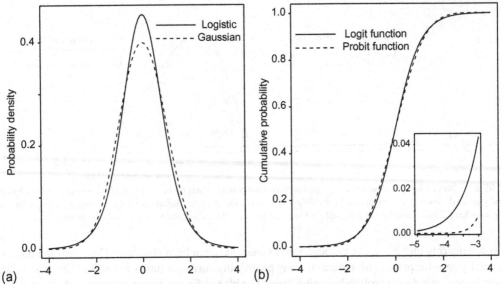

FIGURE 4.8 Comparison of logistic *(solid)* and Gaussian *(dashed)* probability densities (a), and CDFs (b). Inset in panel (b) shows detail of left-tail behavior. Both distributions shown have zero mean and unit variance.

(dashed curves), in terms of (a) their PDFs and (b) CDFs. As indicated in the inset in Figure 4.8b, the heavier tails of the logistic distribution are most evident for more extreme values. For example, $\Pr\{x \le -4\}$ is 0.00003 for the standard Gaussian distribution (cf. Table B.1), but 0.0007 for the logistic distribution according to Equation 4.43. The CDF for the logistic distribution is called the *logit function*, and the very similar CDF for standard Gaussian distribution is called the *probit function*.

Since the logistic CDF can be written in closed form, so also can its quantile function, which is

$$F^{-1}(p) = \mu + \sigma \ln\left(\frac{p}{1-p}\right). \tag{4.44}$$

For $\mu = 0$ and $\sigma = 1$, Equation 4.44 is equivalent to the log-odds transformation (Equation 3.24).

4.4.5. Gamma Distributions

The statistical distributions of many atmospheric variables are distinctly asymmetric and skewed to the right. Often the skewness occurs when there is a physical limit on the left that is relatively near the range of the data. Common examples are precipitation amounts or wind speeds, which are physically constrained to be nonnegative. The *gamma distribution* is a common choice for representing such data.

The gamma distribution is defined by the PDF

$$f(x) = \frac{(x/\beta)^{\alpha-1} \exp(-x/\beta)}{\beta \, \Gamma(\alpha)} \quad , \quad x, \alpha, \beta > 0. \tag{4.45}$$

The two parameters of the distribution are α, the shape parameter; and β, the scale parameter. The quantity $\Gamma(\alpha)$ is the gamma function, defined in Equation 4.7, evaluated at α. The PDF of the gamma

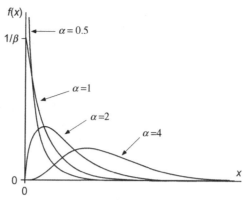

FIGURE 4.9   Gamma distribution density functions for four values of the shape parameter, α.

distribution takes on a wide range of shapes depending on the value of the shape parameter, α. As illustrated in Figure 4.9, for $\alpha < 1$ the distribution is very strongly skewed to the right, with $f(x) \rightarrow \infty$ as $x \rightarrow 0$. For $\alpha = 1$ the function intersects the vertical axis at $1/\beta$ for $x = 0$ (this special case of the gamma distribution is called the *exponential distribution*, which is described more fully later in this section). For $\alpha > 1$ the gamma distribution density function begins at the origin, $f(0) = 0$. Progressively larger values of α result in less skewness, and a shifting of probability density to the right. For very large values of α (larger than perhaps 50–100) the gamma distribution approaches the Gaussian distribution in form.

The parameter α is always dimensionless. The role of the scale parameter, β, effectively is to stretch or squeeze (i.e., to scale) the gamma density function to the right or left, depending on the overall magnitudes of the data values represented. Notice that the random quantity x in Equation 4.45 is divided by β in both places where it appears. The scale parameter β has the same physical dimensions as x. As the distribution is stretched to the right by larger values of β, its height must drop in order to satisfy Equation 4.18, and conversely as the density is squeezed to the left its height must rise. These adjustments in height are accomplished by the β in the denominator of Equation 4.45. The versatility in shape of the gamma distribution makes it an attractive candidate for representing precipitation data. It is often used for this purpose, although observed precipitation data often exhibit heavier tails than corresponding fitted gamma distributions, especially at smaller spatial scales and for shorter accumulation periods (e.g., Cavanaugh and Gershunov 2015, Katz et al., 2002, Wilks 1999a).

The gamma distribution is more difficult to work with than the Gaussian distribution because obtaining good parameter estimates from particular batches of data is not as straightforward. The simplest (although certainly not best) approach to fitting a gamma distribution is to use the method of moments. Even here, however, there is a complication, because the two parameters for the gamma distribution do not correspond exactly to moments of the distribution, as is the case for the Gaussian distribution. The mean of the gamma distribution is given by the product $\alpha\beta$, and the variance is $\alpha\beta^2$. Equating these expressions with the corresponding sample quantities yields a set of two equations in two unknowns, which can be solved to yield the moments estimators

$$\hat{\alpha} = \bar{x}^2 / s^2 \tag{4.46a}$$

and

$$\hat{\beta} = s^2 / \bar{x}. \tag{4.46b}$$

The moments estimators for the gamma distribution are usually reasonably accurate for large values of the shape parameter, perhaps $\alpha > 10$, but can yield poor results for small values of α (Thom 1958, Wilks 1990). The moments estimators in this case are said to be inefficient, in the technical sense of not making maximum use of the information in a data set. The practical consequence of this inefficiency is that particular values of the parameters calculated using Equation 4.46 are erratic, or unnecessarily variable, from data sample to data sample.

A much better approach to parameter fitting for the gamma distribution is to use the method of *maximum likelihood*. For many distributions, including the gamma distribution, maximum likelihood fitting requires an iterative procedure that is really only practical using a computer. Section 4.6 presents the method of maximum likelihood for fitting parametric distributions, including the gamma distribution in Example 4.13.

Approximations to the *maximum likelihood estimators* (MLEs) for the gamma distribution are available that do not require iterative computation. Two of these employ the sample statistic

$$D = \ln(\bar{x}) - \frac{1}{n}\sum_{i=1}^{n} \ln(x_i), \tag{4.47}$$

which is the difference between the natural log of the sample mean and the mean of the logs of the data. Equivalently, the sample statistic D is the difference between the logs of the arithmetic and geometric means. Notice that the sample mean and standard deviation are not sufficient to compute the statistic D, since each data value must be used to compute the second term in Equation 4.47.

A simple approximation to maximum likelihood estimation for the gamma distribution is due to Thom (1958). The Thom estimator for the shape parameter is

$$\hat{\alpha} = \frac{1 + \sqrt{1 + 4D/3}}{4D}, \tag{4.48}$$

after which the scale parameter is obtained from

$$\hat{\beta} = \frac{\bar{x}}{\hat{\alpha}}. \tag{4.49}$$

An alternative approach is a polynomial approximation to the shape parameter (Greenwood and Durand 1960). One of two equations is used,

$$\hat{\alpha} = \frac{0.5000876 + 0.1648852D - 0.0544274D^2}{D}, \quad 0 \le D \le 0.5772, \tag{4.50a}$$

or

$$\hat{\alpha} = \frac{8.898919 + 9.059950D + 0.9775373D^2}{17.79728D + 11.968477D^2 + D^3}, \quad 0.5772 \le D \le 17.0, \tag{4.50b}$$

depending on the value of D. The scale parameter is again subsequently estimated using Equation 4.49.

Another approximation to maximum likelihood for the gamma distribution is presented in Ye and Chen (2017). Because all three of these approximations require computing logarithms of the data, none are suitable if some values are exactly zero. In that case a workaround using the censoring concept can be used (Wilks 1990).

As was the case for the Gaussian distribution, the gamma density function is not analytically integrable. Gamma distribution probabilities must therefore be obtained either by computing approximations to the CDF (by numerically integrating Equation 4.45) or from tabulated probabilities. Formulas and computer routines for this purpose can be found in Abramowitz and Stegun (1984) and Press et al. (1986), respectively. A table of gamma distribution probabilities is included as Table B.2 in Appendix B.

In any of these cases, gamma distribution probabilities will be available for the *standard gamma distribution*, with $\beta = 1$. Therefore it is nearly always necessary to transform by rescaling the variable X of interest (characterized by a gamma distribution with arbitrary scale parameter β) to the standardized variable

$$\xi = x/\beta, \tag{4.51}$$

which follows a gamma distribution with $\beta = 1$. The standard gamma variate ξ is dimensionless. The shape parameter, α, will be the same for both x and ξ. The procedure is analogous to the transformation to the standardized Gaussian variable, z, in Equations 4.26 and 4.27.

Cumulative probabilities for the standard gamma distribution are given by a mathematical function known as the *incomplete gamma function*, $P(\alpha, \xi) = \Pr\{\Xi \leq \xi\} = F(\xi)$. It is this function that was used to compute the probabilities in Table B.2. The cumulative probabilities for the standard gamma distribution in Table B.2 are arranged in an inverse sense to the Gaussian probabilities in Table B.1. That is, quantiles (transformed data values, ξ) of the distributions are presented in the body of the table, and cumulative probabilities are listed as the column headings. Different probabilities are obtained for different shape parameters, α, which appear in the first column.

Example 4.8. Evaluating Gamma Distribution Probabilities

Consider the data for January precipitation at Ithaca during the 50 years 1933–1982 in Table A.2. The average January precipitation for this period is 1.96 in. and the mean of the logarithms of the monthly precipitation totals is 0.5346, so Equation 4.47 yields $D = 0.139$. Both Thom's method (Equation 4.48) and the Greenwood and Durand formula (Equation 4.50a) yield $\alpha = 3.76$ and $\beta = 0.52$ in. This result agrees well with the maximum likelihood values obtained using the method outlined in Example 4.13. In contrast, the moments estimators (Equation 4.46) yield $\alpha = 3.09$ and $\beta = 0.64$ in.

Adopting the approximate MLEs, the unusualness of the January 1987 precipitation total at Ithaca can be evaluated with the aid of Table B.2. That is, by representing the climatological variations in Ithaca January precipitation by the fitted gamma distribution with $\alpha = 3.76$ and $\beta = 0.52$ in., the cumulative probability corresponding to 3.15 in. (the sum of the daily values for Ithaca in Table A.1) can be computed.

First, applying Equation 4.51, the standard gamma variate $\xi = 3.15$ in./0.52 in. = 6.06. Adopting $\alpha = 3.75$ as the closest tabulated value to the fitted $\alpha = 3.76$, it can be seen that $\xi = 6.06$ lies between the tabulated values for $F(5.214) = 0.80$ and $F(6.354) = 0.90$. Interpolation yields $F(6.06) = 0.874$, indicating that there is approximately one chance in eight for a January this wet or wetter to occur at Ithaca. The probability estimate could be refined slightly by interpolating between the rows for $\alpha = 3.75$ and $\alpha = 3.80$ to yield $F(6.06) = 0.873$, although this additional calculation would probably not be worth the effort.

Table B.2 can also be used to invert the gamma CDF to find precipitation values corresponding to particular cumulative probabilities, $\xi = F^{-1}(p)$, that is, to evaluate the quantile function. Dimensional precipitation values are then recovered by reversing the transformation in Equation 4.51. Consider estimation of the median January precipitation at Ithaca. This will correspond to the value of ξ satisfying $F(\xi) = 0.50$ which, in the row for $\alpha = 3.75$ in Table B.2, is 3.425. The corresponding dimensional precipitation amount is given by the product $\xi\beta = (3.425)(0.52 \text{ in.}) = 1.78$ in. By comparison, the sample median of the precipitation data in Table A.2 is 1.72 in. It is not surprising that the median is less than the mean of 1.96 in., since the distribution is positively skewed. A (perhaps surprising, but often unappreciated) fact exemplified by this comparison is that below "normal" (i.e., below average) precipitation is typically more likely than above normal precipitation, as a consequence of the positive skewness of the distribution of precipitation. ◇

Example 4.9. Gamma Distribution in Operational Climatology, I. Reporting Seasonal Outcomes
The gamma distribution can be used to report monthly and seasonal precipitation amounts in a way that allows comparison with locally applicable climatological distributions. Figure 4.10 shows an example of this format for global precipitation for March–May 2014. The precipitation amounts for this 3-month period are not shown as accumulated depths, but rather as quantiles corresponding to local climatological gamma distributions. Five categories are mapped: less than the 10th percentile $q_{0.1}$, between the 10th and 30th percentile $q_{0.3}$, between the 30th and 70th percentile $q_{0.7}$, between the 70th and 90th percentile $q_{0.9}$, and wetter than the 90th percentile.

It is immediately clear which regions received substantially less, slightly less, about the same, slightly more, or substantially more precipitation during this period as compared to the underlying climatological distributions, even though the shapes of these distributions vary widely. One of the advantages of expressing precipitation amounts in terms of climatological gamma distributions is that these very strong differences in the shapes of the precipitation climatologies do not confuse comparisons between locations. Figure 4.11 illustrates the definition of the percentiles using a gamma PDF with $\alpha = 2$. The distribution is divided into five categories corresponding to the five shading levels in Figure 4.10, with the precipitation amounts $q_{0.1}$, $q_{0.3}$, $q_{0.7}$, and $q_{0.9}$ separating regions of the distribution containing 10%, 20%, 40%, 20%, and 10% of the probability, respectively. ◇

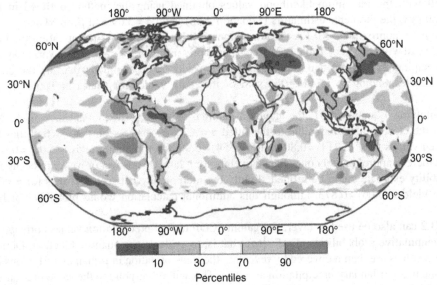

FIGURE 4.10 Global precipitation totals for the period March–May 2014, expressed as percentile values of local gamma distributions. *From Blunden and Arndt (2015). © American Meteorological Society. Used with permission.*

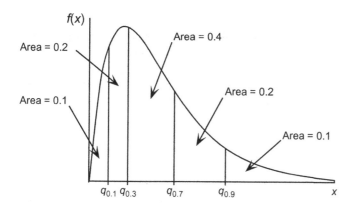

FIGURE 4.11 Illustration of the precipitation categories in Figure 4.10 in terms of a gamma distribution density function with $\alpha = 2$. Outcomes drier than the 10th percentile lie to the left of $q_{0.1}$. Areas with precipitation between the 30th and 70th percentiles (between $q_{0.3}$ and $q_{0.7}$) would be unshaded on the map. Precipitation in the wettest 10% of the climatological distribution lies to the right of $q_{0.9}$.

Exponential Distribution

There are several important special cases of the gamma distribution, which result from particular restrictions on the parameters α and β. For $\alpha = 1$, the gamma distribution reduces to the *exponential distribution*, with PDF

$$f(x) = \frac{1}{\beta} \exp\left(-\frac{x}{\beta}\right), \quad x \geq 0. \tag{4.52}$$

The shape of this density is simply an exponential decay, as indicated in Figure 4.7, for $\alpha = 1$. Equation 4.52 is analytically integrable, so the CDF for the exponential distribution exists in closed form,

$$F(x) = 1 - \exp\left(-\frac{x}{\beta}\right). \tag{4.53}$$

The quantile function is easily derived by solving Equation 4.53 for x (Equation 4.94). Since the shape of the exponential distribution is fixed by the restriction $\alpha = 1$, it is usually not suitable for representing variations in quantities like precipitation, although mixtures of two exponential distributions (see Section 4.4.9) can represent daily nonzero precipitation values quite well.

An important use of the exponential distribution in atmospheric science is in the characterization of the size distribution of raindrops, called drop-size distributions (e.g., Sauvageot 1994). When the exponential distribution is used for this purpose, it is called the *Marshall–Palmer distribution*, and generally denoted $N(D)$, which indicates a distribution over the numbers of droplets as a function of their diameters. Drop-size distributions are particularly important in radar applications where, for example, reflectivities are computed as expected values of a quantity called the backscattering cross-section, with respect to a drop-size distribution such as the exponential.

Erlang Distribution

The second special case of the gamma distribution is the *Erlang distribution*, in which the shape parameter α is restricted to integer values. One application of the Erlang distribution is as the distribution of waiting times until the αth Poisson event, for the Poisson rate $\mu = 1/\beta$.

Chi-Square Distribution

The *chi-square* (χ^2) distribution is a yet another special case of the gamma distribution. Chi-square distributions are gamma distributions with scale parameter $\beta = 2$. Chi-square distributions are conventionally written in terms of an integer-valued parameter called the *degrees of freedom*, denoted v. The relationship to the gamma distribution more generally is that the degrees of freedom are twice the gamma distribution shape parameter, or $\alpha = v/2$, yielding the chi-square PDF

$$f(x) = \frac{x^{(v/2-1)} \exp(-x/2)}{2^{v/2}\, \Gamma(v/2)}, \quad x > 0. \tag{4.54}$$

Since it is the gamma scale parameter that is fixed at $\beta = 2$ to define the chi-square distribution, Equation 4.54 is capable of the same variety of shapes as the full gamma distribution, as the shape parameter α varies. Because there is no explicit horizontal scale in Figure 4.9, it could be interpreted as showing chi-square densities with $v = 1, 2, 4,$ and 8 degrees of freedom. The chi-square distribution arises as the distribution of the sum of v squared independent standard Gaussian variates and is used in several ways in the context of statistical inference (see Chapters 5 and 12). Table B.3 lists right-tail quantiles for chi-square distributions.

Pearson III Distribution

The gamma distribution is also sometimes generalized to a three-parameter distribution by moving the PDF to the left or right according to a shift parameter ζ. This three-parameter gamma distribution is also known as the Pearson Type III, or simply *Pearson III distribution*, and has PDF

$$f(x) = \frac{\left(\dfrac{x-\zeta}{\beta}\right)^{\alpha-1} \exp\left(-\dfrac{x-\zeta}{\beta}\right)}{|\beta\,\Gamma(\alpha)|}, \quad x > \zeta \text{ for } \beta > 0, \text{ or } x < \zeta \text{ for } \beta < 0. \tag{4.55}$$

This distribution is also sometimes called the *shifted gamma distribution*. Usually the scale parameter β is positive, which results in the Pearson III being a gamma distribution shifted to the right if $\zeta > 0$, with support $x > \zeta$. However, Equation 4.55 also allows $\beta < 0$, in which case the PDF is reflected (and so has a long left tail and negative skewness), and the support is $x < \zeta$.

Generalized Gamma Distribution

Sometimes, analogously to the lognormal distribution, the random variable x in Equation 4.55 has been log-transformed, in which case the distribution of the original variable $[=\exp(x)]$ is said to follow the *log-Pearson III distribution*, which is commonly used in hydrologic applications (e.g., Griffis and Stedinger 2007). Other power transformations besides the logarithm can also be used, and in that case the resulting distributions typically are formulated without the shift parameter ζ in Equation 4.55, yielding the generalized gamma distribution (Stacy 1962), with PDF

$$f(x) = \frac{\lambda x^{\alpha\lambda-1}}{\beta^{\alpha\lambda}\,\Gamma(\alpha)} \exp\left[-(x/\beta)^{\lambda}\right], \quad x > 0. \tag{4.56}$$

Here the power transformation parameter λ functions in the same way as in Equation 3.20.

Example 4.10. Gamma Distribution in Operational Climatology, II. The Standardized Precipitation Index

The *Standardized Precipitation Index* (SPI) is a popular approach to characterizing drought or wet-spell conditions, by expressing precipitation for monthly or longer periods in terms of the corresponding climatological distribution. McKee et al. (1993) originally proposed using gamma distributions for this purpose, and Guttman (1999) has suggested using the Pearson III distribution.

Computation of the SPI is accomplished through the *normal quantile transform*,

$$z = \Phi^{-1}[F(x)], \tag{4.57}$$

which is an instance of the operation sometimes also referred to as *quantile mapping*. The SPI is equal to z in Equation 4.57, so that a precipitation value x is characterized in terms of the standard Gaussian variate z that yields the same cumulative probability as the original data that follows the distribution $F(x)$. The SPI is thus a probability index that expresses precipitation deficits (SPI <0) or excesses (SPI >0) in a standardized way, accounting for differences in precipitation climatologies due to geographic and/or timescale differences. Precipitation accumulations characterized by $|SPI| > 1.0$, >1.5, and >2.0 are qualitatively and somewhat arbitrarily characterized as being dry or wet, moderately dry or wet, and extremely dry or wet, respectively (Guttman 1999).

Consider computing the SPI for the January 1987 Ithaca precipitation accumulation of 3.15 in. Example 4.8 showed that approximate maximum likelihood estimates for the gamma distribution characterizing January Ithaca precipitation, 1933–1982, are $\alpha = 3.76$ and $\beta = 0.52$ in., and that the cumulative probability corresponding to $x = 3.15$ in. in the context of this distribution is $F(3.15 \text{ in.}) = 0.873$. The SPI for the January 1987 precipitation at Ithaca is then the normal quantile transform (Equation 4.57) of the precipitation amount, SPI $= \Phi^{-1}[F(3.15 \text{ in.})] = \Phi^{-1}[0.873] = +1.14$. That is, the standard Gaussian variate having the same cumulative probability as does the 1987 January precipitation within its own climatological distribution is $z = +1.14$.

The SPI is routinely computed for timescales ranging from 1 month, as was just done for January 1987, through 2 years. For any timescale, the accumulated precipitation is characterized in terms of the corresponding cumulative probability within the distribution fitted to historical data over the same timescale and same portion of the annual cycle. So, for example, a 2-month SPI for January–February at a location of interest would involve fitting a gamma (or other suitable) distribution to the historical record of January plus February precipitation at that location. Similarly, annual SPI values are computed with respect to probability distributions for total annual precipitation. ◇

4.4.6. Beta Distributions

Some variables are restricted to segments of the real line that are bounded on two sides. Often such variables are restricted to the interval $0 \leq x \leq 1$. Examples of physically important variables subject to this restriction are cloud amount (observed as a fraction of the sky) and relative humidity. An important, more abstract, variable of this type is probability, where a parametric distribution can be useful in summarizing the frequency of use of forecasts, for example, of daily rainfall probability. The parametric distribution usually chosen to represent variations in these types of data is the *beta distribution*.

The PDF of the beta distribution is

$$f(x) = \left[\frac{\Gamma(\alpha + \beta)}{\Gamma(\alpha)\,\Gamma(\beta)} \right] x^{\alpha-1} (1-x)^{\beta-1}, \quad 0 \leq x \leq 1, \ \alpha, \beta > 0. \tag{4.58}$$

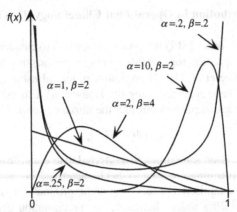

FIGURE 4.12 Five example PDFs for beta distributions. Mirror images of these distributions are obtained by reversing the parameters α and β.

This is a very flexible function, taking on many different shapes depending on the values of its two parameters, α and β. Figure 4.12 illustrates five of these. In general, for $\alpha \leq 1$ probability is concentrated near zero (e.g., $\alpha = 0.25$ and $\beta = 2$, or $\alpha = 1$ and $\beta = 2$, in Figure 4.12), and for $\beta \leq 1$ probability is concentrated near 1. If both parameters are less than one the distribution is U-shaped. For $\alpha > 1$ and $\beta > 1$ the distribution has a single mode (hump) between 0 and 1 (e.g., $\alpha = 2$ and $\beta = 4$, or $\alpha = 10$ and $\beta = 2$, in Figure 4.12), with more probability shifted to the right for $\alpha > \beta$, and more probability shifted to the left for $\alpha < \beta$. Beta distributions with $\alpha = \beta$ are symmetric. Reversing the values of α and β in Equation 4.58 results in a density function that is the mirror image (horizontally flipped) of the original.

Beta distribution parameters usually are fit using the method of moments. Using the expressions for the first two moments of the distribution,

$$\mu = \alpha / (\alpha + \beta) \tag{4.59a}$$

and

$$\sigma^2 = \frac{\alpha \beta}{(\alpha + \beta)^2 (\alpha + \beta + 1)}, \tag{4.59b}$$

the moments estimators

$$\hat{\alpha} = \frac{\bar{x}^2 (1 - \bar{x})}{s^2} - \bar{x} \tag{4.60a}$$

and

$$\hat{\beta} = \frac{\hat{\alpha} (1 - \bar{x})}{\bar{x}} \tag{4.60b}$$

are easily obtained. Alternatively, Mielke (1975) presents algorithms for maximum likelihood estimation.

The *uniform*, or *rectangular distribution*, is an important special case of the beta distribution, with $\alpha = \beta = 1$, and PDF $f(x) = 1$. The uniform distribution plays a central role in the computer generation of random numbers (see Section 4.7.1).

Use of the beta distribution is not limited only to variables having support on the unit interval [0, 1]. A variable, say y, constrained to any interval $[a, b]$ can be represented by a beta distribution after subjecting it to the transformation

$$x = \frac{y - a}{b - a}.$$ (4.61)

In this case parameter fitting is accomplished using

$$\bar{x} = \frac{\bar{y} - a}{b - a}$$ (4.62a)

and

$$s_x^2 = \frac{s_y^2}{(b - a)^2},$$ (4.62b)

which are then substituted into Equation 4.60.

The integral of the beta probability density does not exist in closed form except for a few special cases, for example, the uniform distribution. Probabilities can be obtained through numerical methods (Abramowitz and Stegun 1984, Press et al. 1986), where the CDF for the beta distribution is known as the *incomplete beta function*, $I_x(\alpha, \beta) = \Pr\{0 \leq X \leq x\} = F(x)$. Tables of beta distribution probabilities are given in Epstein (1985) and Winkler (1972b).

4.4.7. Extreme-Value Distributions, I. Block-Maximum Statistics

The statistics of *extreme values* can be approached through description of the behavior of the largest of m values. Such data are extreme in the sense of being unusually large, and by definition are also rare. Often extreme-value statistics are of interest because the physical processes generating extreme events, and the societal impacts that occur because of them, are also large and unusual. A typical example of block-maximum extreme-value data is the collection of *annual maximum* daily precipitation values. In each of n years there is a wettest day of the $m = 365$ days, and the collection of these n wettest days is an extreme-value data set. Table 4.6 presents a small example annual maximum data set, for daily precipitation at Charleston, South Carolina. For each of the $n = 20$ years, the precipitation amount for the wettest of its $m = 365$ days is shown in the table.

A basic result from the theory of extreme-value statistics states (e.g., Coles 2001, Leadbetter et al. 1983) that the largest of m independent observations from a fixed distribution will follow a known

TABLE 4.6 Annual Maxima of Daily Precipitation Amounts (Inches) at Charleston, South Carolina, 1951–1970

1951	2.01	1956	3.86	1961	3.48	1966	4.58
1952	3.52	1957	3.31	1962	4.60	1967	6.23
1953	2.61	1958	4.20	1963	5.20	1968	2.67
1954	3.89	1959	4.48	1964	4.93	1969	5.24
1955	1.82	1960	4.51	1965	3.50	1970	3.00

distribution increasingly closely as m increases, regardless of the (single, fixed) distribution from which the observations have come. This result is called the *Extremal Types Theorem* and is the analog within the statistics of extremes of the Central Limit Theorem for the distribution of sums converging to the Gaussian distribution. The Extremal Types Theorem is valid even if the data within blocks are not independent, although in that case effectively the blocks are shorter (Coles 2001). The theory and approach are equally applicable to distributions of extreme minima (smallest of m observations) by analyzing the variable $-X$.

The distribution toward which the sampling distributions of largest-of-m values converges is called the *generalized extreme value*, or GEV, distribution, with PDF

$$f(x) = \frac{1}{\beta}\left[1 + \frac{\kappa(x-\zeta)}{\beta}\right]^{-1-\frac{1}{\kappa}} \exp\left\{-\left[1 + \frac{\kappa(x-\zeta)}{\beta}\right]^{-\frac{1}{\kappa}}\right\}, \quad 1 + \kappa(x-\zeta)/\beta > 0. \quad (4.63)$$

Here there are three parameters: a location (or shift) parameter ζ, a scale parameter β, and a shape parameter κ. Equation 4.63 can be integrated analytically, yielding the CDF

$$F(x) = \exp\left\{-\left[1 + \frac{\kappa(x-\zeta)}{\beta}\right]^{-\frac{1}{\kappa}}\right\}, \quad (4.64)$$

and this CDF can be inverted to yield an explicit formula for the quantile function,

$$F^{-1}(p) = \zeta + \frac{\beta}{\kappa}\{[-\ln(p)]^{-\kappa} - 1\}. \quad (4.65)$$

Particularly in the hydrological literature, Equations 4.63 through 4.65 are often written with the sign of the shape parameter κ reversed.

Because the moments of the GEV (see Table 4.5) involve the gamma function, estimating GEV parameters using the method of moments is no more convenient than alternative methods that yield more accurate results. The distribution usually is fit using either the method of maximum likelihood (see Section 4.6), or a method known as *L-moments* (Hosking 1990, Stedinger et al. 1993) that is used frequently in hydrological applications. L-moments fitting tends to be preferred for small data samples (Hosking 1990). Maximum likelihood methods can be adapted easily to include effects of *covariates* or additional influences. For example, the possibility that one or more of the distribution parameters may have a trend due to climate changes can be represented easily (Cooley 2009, Katz et al. 2002, Kharin and Zwiers 2005, Lee et al. 2014, Zhang et al. 2004). For moderate and large sample sizes the results of the two parameter estimation methods are usually similar. Using the data in Table 4.6, the maximum likelihood estimates for the GEV parameters are $\zeta = 3.50$, $\beta = 1.11$, and $\kappa = -0.29$, and the corresponding L-moment estimates are $\zeta = 3.49$, $\beta = 1.18$, and $\kappa = -0.32$.

Gumbel Distribution

Three special cases of the GEV are recognized, depending on the value of the shape parameter κ. The limit of Equation 4.63 as κ approaches zero yields the PDF

$$f(x) = \frac{1}{\beta} \exp\left\{ -\exp\left[-\frac{(x-\zeta)}{\beta} \right] - \frac{(x-\zeta)}{\beta} \right\}, \tag{4.66}$$

known as the *Gumbel*, or *Fisher–Tippett Type I*, *distribution*. The Gumbel distribution is the limiting form of the GEV for extreme data drawn independently from distributions with well-behaved (i.e., exponential) tails, such as the Gaussian and the gamma. However, it is not unusual to find that the right tail of the Gumbel distribution may be too thin for this distribution to appropriately represent probabilities for daily rainfall extremes (e.g., Brooks and Carruthers 1953). The Gumbel distribution is so frequently used to represent the statistics of extremes that it is sometimes incorrectly called "the" extreme-value distribution. The Gumbel PDF is skewed to the right and exhibits its maximum at $x = \zeta$. Gumbel distribution probabilities can be obtained from the CDF

$$F(x) = \exp\left\{ -\exp\left[-\frac{(x-\zeta)}{\beta} \right] \right\}. \tag{4.67}$$

Gumbel distribution parameters can be estimated through maximum likelihood or L-moments, as described earlier for the more general case of the GEV, but the simplest way to fit this distribution is to use the method of moments. The moments estimators for the two Gumbel distribution parameters are computed using the sample mean and standard deviation. The estimation equations are

$$\hat{\beta} = \frac{s\sqrt{6}}{\pi} \tag{4.68a}$$

and

$$\hat{\zeta} = \bar{x} - \gamma\hat{\beta}, \tag{4.68b}$$

where $\gamma = 0.57721\ldots$ is *Euler's constant*.

Fréchet Distribution

For $\kappa > 0$ the Equation 4.63 is called the *Fréchet*, or *Fisher–Tippett Type II distribution*.

These distributions exhibit what are called "heavy" tails, meaning that the PDF decreases rather slowly for large values of x. One consequence of heavy tails is that some of the moments of Frechet distributions are not finite. For example, the integral defining the variance (Equation 4.22) is infinite for $\kappa > 1/2$, and even the mean (Equation 4.21 with $g(x) = x$) is not finite for $\kappa > 1$. The notable practical consequence of heavy tails is that quantiles associated with large cumulative probabilities (i.e., Equation 4.65 with $p \approx 1$) will be quite large. It is often found that annual maxima for precipitation and streamflow data exhibit heavy tails (e.g., Morrison and Smith 2002, Papalexiou and Koutsoyiannis 2013, Serinaldi and Kilsby 2014).

Weibull Distribution

The third special case of the GEV distribution occurs for $\kappa < 0$ and is known as the *Weibull*, or *Fisher–Tippett Type III distribution*. In addition to use in extreme-value statistics, Weibull distributions are often chosen to represent wind data (e.g., Conradsen et al. 1984, He et al. 2010, Justus et al., 1978). Typically Weibull distributions are written with the shift parameter $\zeta = 0$, and a parameter transformation, yielding the PDF

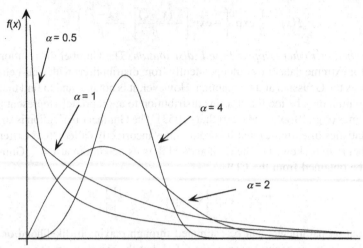

FIGURE 4.13 Weibull distribution PDFs for four values of the shape parameter, α.

$$f(x) = \left(\frac{\alpha}{\beta}\right)\left(\frac{x}{\beta}\right)^{\alpha-1} \exp\left[-\left(\frac{x}{\beta}\right)^{\alpha}\right], \quad x, \alpha, \beta > 0. \tag{4.69}$$

As is the case for the gamma distribution, the two parameters α and β are called the shape and scale parameters, respectively. The form of the Weibull distribution also is controlled similarly by the two parameters. The response of the shape of the distribution to different values of α is shown in Figure 4.13. In common with the gamma distribution, $\alpha \leq 1$ produces reverse "J" shapes and strong positive skewness, and for $\alpha = 1$ the Weibull distribution also reduces to the exponential distribution (Equation 4.52) as a special case. Also in common with the gamma distribution, the scale parameter acts similarly to either stretch or compress the basic shape along the x-axis, for a given value of α. For $\alpha \approx 3.6$ the Weibull is very similar to the Gaussian distribution. However, for shape parameters larger than this the Weibull density exhibits negative skewness, which is visible in Figure 4.13 for $\alpha = 4$.

The PDF for the Weibull distribution is analytically integrable, resulting in the CDF

$$F(x) = 1 - \exp\left[-\left(\frac{x}{\beta}\right)^{\alpha}\right]. \tag{4.70}$$

This equation can easily be solved for x to yield the quantile function. As is the case for the GEV more generally, the moments of the Weibull distribution involve the gamma function (see Table 4.5), so there is no computational advantage to parameter fitting by the method of moments. Usually Weibull distributions are fit using either maximum likelihood (see Section 4.6) or L-moments (Stedinger et al. 1993).

Average Return Periods

The result of an extreme-value analysis is often simply a summary of quantiles corresponding to large cumulative probabilities, for example, the event with an annual probability of 0.01 of being exceeded. Unless n is rather large, direct empirical estimation of these extreme quantiles will not be possible (cf. Equation 3.19), and a well-fitting extreme-value distribution provides a reasonable and objective way to extrapolate to probabilities that may be substantially larger than $1 - 1/n$. Often these extreme probabilities are expressed as average *return periods*,

$$R(x) = \frac{1}{\omega\,[1 - F(x)]}. \tag{4.71}$$

The return period $R(x)$ associated with a quantile x typically is interpreted to be the average time between occurrences of events of that magnitude or greater. The return period is a function of the CDF evaluated at x and the average sampling frequency ω. For annual maximum data $\omega = 1$ per year, in which case the event x corresponding to a cumulative probability $F(x) = 0.99$ will have probability $1 - F(x)$ of being exceeded in any given year. This value of x would be associated with a return period of 100 years and would be called the 100-year event. Extreme-event risks expressed in terms of return periods appear to be less well understood by the general public than are annual event probabilities (Grounds et al., 2018). In addition, the computation of return periods in Equation 4.71 implicitly assumes climate stationarity, but as the climate changes the (e.g.) 100-year event may become larger or smaller over the next century. It may thus be better to express extreme-value risk as the annual occurrence probability $F(x)$, changes in which might be modeled using time trends for the distribution parameters (e.g., Chen and Chu, 2014; Katz et al. 2002; Kharin and Zwiers, 2005).

Example 4.11. Return Periods and Cumulative Probability
As noted earlier, a maximum likelihood fit of the GEV distribution to the annual maximum precipitation data in Table 4.6 yielded the parameter estimates $\zeta = 3.50$, $\beta = 1.11$, and $\kappa = -0.29$. Using Equation 4.65 with cumulative probability $p = 0.5$ yields a median of 3.89 in. This is the daily precipitation amount that has a 50% chance of being exceeded in a given year. This amount will therefore be exceeded on average in half of the years in a hypothetical long climatological record, and so the average time separating daily precipitation events of this magnitude or greater is 2 years (Equation 4.71).

Because $n = 20$ years for these data, the median can be well estimated directly as the sample median. But consider estimating the 100-year 1-day precipitation event from these data. According to Equation 4.71 this corresponds to the cumulative probability $F(x) = 0.99$, whereas the empirical cumulative probability corresponding to the most extreme precipitation amount in Table 4.6 might be estimated as $p \approx 0.967$, using the Tukey plotting position (see Table 3.2). However, using the GEV quantile function (Equation 4.65) together with Equation 4.71, a reasonable estimate for the 100-year maximum daily amount is calculated to be 6.32 in. The corresponding 2- and 100-year precipitation amounts derived from the L-moment parameter estimates, $\zeta = 3.49$, $\beta = 1.18$, and $\kappa = -0.32$, are 3.90 and 6.33 in., respectively.

It is worth emphasizing that the T-year event is in no way guaranteed to occur within a particular period of T years, and indeed the probability distribution for the waiting time until the next occurrence of an extreme event will be quite broad (e.g., Wigley 2009). The probability that the T-year event occurs in any given year is $1/T$, for example, $1/T = 0.01$ for the $T = 100$-year event. In any particular year, the occurrence of the T-year event is a Bernoulli trial, with $p = 1/T$. Therefore the geometric distribution (Equation 4.5) can be used to calculate probabilities of waiting particular numbers of years for the event. Another interpretation of the return period is as the mean of the geometric distribution for the waiting time. The probability of the 100-year event occurring in an arbitrarily chosen century can be calculated as $\Pr\{X \le 100\} = 0.634$ using Equation 4.5. That is, there is more than a 1/3 chance that the 100-year event will not occur in any particular 100 years. Similarly, the probability of the 100-year event not occurring in 200 years is approximately 0.134. ◇

One important motivation for studying and modeling the statistics of extremes is to estimate annual probabilities of rare and potentially damaging events, such as extremely large daily precipitation amounts that might cause flooding, or extremely large wind speeds that might cause damage to structures. In applications like these, the assumptions of classical extreme-value theory, namely, that the underlying events are independent and come from the same distribution, and that the number of individual (usually daily) values m is sufficient for convergence to the GEV, may not be met. Most problematic for the application of extreme-value theory is that the underlying data often will not be drawn from the same distribution, for example, because of an annual cycle in the statistics of the m ($=365$, usually) values, and/or because the largest of the m values are generated by different processes in different blocks (years). For example, some of the largest daily precipitation values may occur because of hurricane landfalls, some may occur because of large and slowly moving thunderstorm complexes, and some may occur as a consequence of near-stationary frontal boundaries. The statistics of (i.e., the underlying PDFs corresponding to) the different physical processes may be different (e.g., Walshaw 2000).

These considerations do not invalidate the GEV (Equation 4.63) as a candidate distribution to describe the statistics of extremes, and empirically this distribution often is found to be an excellent choice even when the assumptions of extreme-value theory are not met. However, in the many practical settings where the classical assumptions are not valid the GEV is not guaranteed to be the most appropriate distribution to represent a set of extreme-value data. The appropriateness of the GEV should be evaluated along with other candidate distributions for particular data sets (Madsen et al. 1997, Wilks 1993), possibly using approaches presented in Section 4.6 or 5.2.6.

4.4.8. Extreme-Value Distributions, II. Peaks-Over-Threshold Statistics

Approaching the statistics of extremes using the block-maximum approach of Section 4.4.7 can be wasteful of data, when there are values that are not largest in their block (e.g., year of occurrence) but may be larger than the maxima in other blocks. An alternative approach to assembling a set of extreme-value data is to choose the largest n values regardless of their year of occurrence. The result is called *partial-duration* data in hydrology. This approach is known more generally as *peaks-over-threshold*, or POT, since any values larger than a (large) minimum level are chosen, and we are not restricted to choosing the same number of extreme values as there may be years in the climatological record. Because the underlying data may exhibit substantial serial correlation, some care is required to ensure that selected partial-duration data represent distinct events.

When the data underlying an extreme-value analysis have been abstracted using POT sampling, there is some theoretical support for characterizing them using the *generalized Pareto distribution*, with PDF

$$f(x) = \frac{1}{\sigma^*}\left[1 + \frac{\kappa(x-u)}{\sigma^*}\right]^{-\frac{1}{\kappa}-1} \tag{4.72}$$

and CDF

$$F(x) = 1 - \left[1 + \frac{\kappa(x-u)}{\sigma^*}\right]^{-1/\kappa}. \tag{4.73}$$

This distribution arises as an approximation to the distribution of POT data taken from a GEV distribution, with the numbers of these peaks in a given time period following a Poisson distribution (Coles 2001, Hosking and Wallis 1987, Katz et al. 2002). Here u is the threshold for the POT sampling, which should be relatively high. The shape parameter κ in Equations 4.72 and 4.73 has the same value as that for the related GEV distribution. Also in common with the GEV, generalized Pareto distributions with $\kappa > 0$ exhibit heavy tails. The scale parameter σ^* in Equations 4.72 and 4.73 is related to the parameters of the corresponding GEV distribution and to the sampling threshold u according to

$$\sigma^* = \beta + \kappa(u - \zeta). \tag{4.74}$$

Coles (2001) suggests that u can be optimized by fitting the distribution using a range of thresholds, and choosing a value slightly above the point at which the fitted shape parameters κ are fairly constant, and the scale parameters σ^* vary linearly according to Equation 4.74. An alternative empirical choice that has been suggested in the context of daily data is to choose u such that $1.65n$ extreme values are chosen, given n years of data (Madsen et al. 1997, Stedinger et al. 1993), yielding $\omega = 1.65\,\text{year}^{-1}$ in Equation 4.71.

The foregoing theory assumes statistical independence of the POT data, whereas often serial correlation in the underlying data series will give rise to clusters of consecutive values larger than u. Each of these clusters might reasonably be interpreted as a single event. Coles (2001) describes a process of *declustering*, whereby a subjective empirical rule defines clusters of exceedances and only the largest value within each cluster is incorporated into the POT data set. Nonstationarity (e.g., trends due to climate change) can be dealt with in POT settings with the same approach taken for block-maximum data, namely, by expressing one or more of the distribution parameters as a function of time (e.g., Katz 2013).

4.4.9. Mixture Distributions

The parametric distributions presented so far in this chapter may be inadequate for data that arise from more than one generating process or physical mechanism. An example is the Guayaquil temperature data in Table A.3, for which histograms are shown in Figure 3.6. These data are clearly bimodal; with the smaller, warmer hump in the distribution associated with El Niño years, and the larger, cooler hump consisting mainly of the non-El Niño years. Although the Central Limit Theorem suggests that the Gaussian distribution should be a good model for monthly averaged temperatures, the clear differences in the Guayaquil June temperature climate associated with El Niño make the Gaussian a poor choice to represent these data overall. However, separate Gaussian distributions for El Niño years and non-El Niño years might provide a good probability model for these data.

Cases like this are natural candidates for representation with *mixture distributions*, which are weighted averages of two or more PDFs. Any number of PDFs can be combined to form a mixture distribution (Everitt and Hand 1981, McLachlan and Peel 2000, Titterington et al. 1985), but by far the most commonly used mixture distributions are weighted averages of two component PDFs,

$$f(x) = wf_1(x) + (1 - w)f_2(x). \tag{4.75}$$

The component PDFs $f_1(x)$ and $f_2(x)$ can be any distributions, and usually although not necessarily they are of the same parametric form. The weighting parameter w, $0 < w < 1$, determines the contribution of each component density to the mixture PDF and can be interpreted as the probability that a realization of the random variable X will have come from $f_1(x)$.

Of course the properties of a mixture distribution depend on the properties of the component distributions and on the weighting parameter. The mean is simply the weighted average of the two component means,

FIGURE 4.14 Probability density function for the mixture (Equation 4.75) of two Gaussian distributions fit to the June Guayaquil temperature data (Table A.3). The result is very similar to the kernel density estimate derived from the same data, Figure 3.8b.

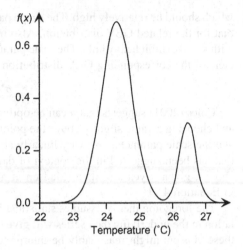

$$\mu = w\mu_1 + (1-w)\mu_2. \tag{4.76}$$

On the other hand, the variance

$$\sigma^2 = \left[w\sigma_1^2 + (1-w)\sigma_2^2\right] + \left[w(\mu_1 - \mu)^2 + (1-w)(\mu_2 - \mu)^2\right]$$
$$= w\sigma_1^2 + (1-w)\sigma_2^2 + w(1-w)(\mu_1 - \mu_2)^2, \tag{4.77}$$

has contributions from the weighted variances of the two distributions (first square-bracketed terms on the first line), plus additional dispersion deriving from the difference of the two means (second square-bracketed terms). Mixture distributions are clearly capable of representing bimodality (or, when the mixture is composed of three or more component distributions, multimodality), but mixture distributions can also be unimodal if the differences between component means are small enough relative to the component standard deviations or variances.

Usually mixture distributions are fit using maximum likelihood, using the EM algorithm (see Section 4.6.3). Figure 4.14 shows the PDF for a maximum likelihood fit of a mixture of two Gaussian distributions to the June Guayaquil temperature data in Table A.3, with parameters $\mu_1 = 24.34°C$, $\sigma_1 = 0.46°C$, $\mu_2 = 26.48°C$, $\sigma_2 = 0.26°C$, and $w = 0.80$ (see Example 4.14). Here μ_1 and σ_1 are the parameters of the first (cooler and more probable) Gaussian distribution, $f_1(x)$, and μ_2 and σ_2 are the parameters of the second (warmer and less probable) Gaussian distribution, $f_2(x)$. The mixture PDF in Figure 4.14 results as a simple (weighted) addition of the two component Gaussian distributions, in a way that is similar to the construction of the kernel density estimates for the same data in Figure 3.8, as a sum of scaled kernels that are themselves PDFs. Indeed, the Gaussian mixture in Figure 4.14 resembles the kernel density estimate derived from the same data in Figure 3.8b. The means of the two component Gaussian distributions are well separated relative to the dispersion characterized by the two standard deviations, resulting in the mixture distribution being strongly bimodal.

Gaussian distributions are the most common choice for components of mixture distributions, but mixtures of exponential distributions (Equation 4.52) are also important and frequently used. In particular, the mixture distribution composed of two exponential distributions is called the *mixed exponential distribution* (Smith and Schreiber 1974), with PDF

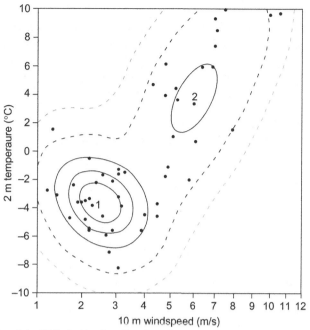

FIGURE 4.15 Contour plot of the PDF of a bivariate Gaussian mixture distribution, fit to an ensemble of 51 forecasts for 2 m temperature and 10 m wind speed, made at 180 h lead time. The wind speeds have first been square root transformed to make their univariate distribution more Gaussian. Dots indicate individual forecasts made by the 51 ensemble members. The two constituent bivariate Gaussian densities $f_1(x)$ and $f_2(x)$ are centered at "1" and "2," respectively, and the smooth lines indicate level curves of their mixture $f(x)$, formed with $w = 0.57$. Solid contour interval is 0.05, and the heavy and light dashed lines are 0.01 and 0.001, respectively. Adapted from Wilks (2002b).

$$f(x) = \frac{w}{\beta_1} \exp\left(-\frac{x}{\beta_1}\right) + \frac{1-w}{\beta_2} \exp\left(-\frac{x}{\beta_2}\right). \tag{4.78}$$

The mixed exponential distribution has been found to be well suited for representing nonzero daily precipitation data (Woolhiser and Roldan 1982, Foufoula-Georgiou and Lettenmaier 1987, Wilks 1999a) and is especially useful for simulating (see Section 4.7) spatially correlated daily precipitation amounts (Wilks 1998).

   Mixture distributions are not limited to combinations of univariate continuous PDFs. Equation 4.75 can as easily be used to form mixtures of discrete probability distribution functions or mixtures of multivariate joint distributions. For example, Figure 4.15 shows the mixture of two bivariate Gaussian distributions (Equation 4.31) fit to a 51-member ensemble forecast (see Section 8.2) for temperature and wind speed. The distribution was fit using the maximum likelihood algorithm for multivariate Gaussian mixtures given in Smyth et al. (1999) and Hannachi and O'Neill (2001). Although multivariate mixture distributions are quite flexible in accommodating unusual-looking data, this flexibility comes at the cost of needing to estimate a large number of parameters, so use of relatively elaborate probability models of this kind may be limited by the available sample size. The mixture distribution in Figure 4.15 requires 11 parameters to characterize it: two means, two variances, and one correlation for each of the two component bivariate distributions, plus the weight parameter w.

Other Combination Distributions

Combinations of component distributions of different parametric types have been used especially to represent hydrologic variables, in order to better model their extreme tail behavior. Although not a mixture distribution in the sense of Equation 4.75, Furrer and Katz (2008) combined gamma and generalized Pareto distributions to represent daily precipitation amounts, by stitching together a gamma distribution for most of the data with a generalized Pareto distribution for the extreme amounts. The generalized Pareto scale parameter was adjusted in order to achieve a continuous (although not smooth) transition between the two. Carreau and Bengio (2009) take a similar approach when augmenting Gaussian distributions with generalized Pareto tails, and Li et al. (2012) analogously combine exponential distributions with generalized Pareto tails.

Vrac and Naveau (2007) employed a "dynamic" mixture of gamma and generalized Pareto distributions, in which the weight w in Equation 4.75 depends on the amount x, so that the mixture density is defined as

$$f(x) = \frac{[1 - w(x)]f_\Gamma(x) + w(x)f_{gP}(x)}{c}, \qquad (4.79)$$

where $f_\Gamma(x)$ denotes a gamma density (Equation 4.45), $f_{gP}(x)$ denotes a generalized Pareto density (Equation 4.72) with threshold $u = 0$, and c is a normalizing constant which has been defined explicitly by Li et al. (2012). The weight function takes the form

$$w(x) = \frac{1}{2} + \frac{1}{\pi}\arctan\left(\frac{x - m}{\tau}\right), \qquad (4.80)$$

where m and τ are fitted parameters. Equation 4.80 obviates the need to choose a threshold u, progressively downweights the light-tailed gamma distribution in favor of the generalized Pareto tail distribution as x increases, and ensures a smooth and continuous transition between the two.

4.5. QUALITATIVE ASSESSMENTS OF THE GOODNESS OF FIT

Having fit a parametric distribution to a batch of data, it is worthwhile to verify that the theoretical probability model provides an adequate description. Fitting an inappropriate distribution can lead to erroneous conclusions being drawn. Quantitative methods for evaluating the closeness of fitted distributions to underlying data rely on ideas from formal hypothesis testing, and a few such methods will be presented in Section 5.2.5. This section describes some qualitative, graphical methods useful for subjectively discerning the goodness of fit. These methods are instructive even if a formal goodness of fit test of fit is to be computed also. A formal test may indicate an inadequate fit, but it may not inform the analyst as to the specific nature of the discrepancies. Graphical comparison of the data and the fitted distribution allow diagnosis of where and how the parametric representation may be inadequate.

4.5.1. Superposition of a Fitted Parametric Distribution and Data Histogram

Superposition of the fitted distribution and the histogram is probably the simplest and most intuitive means of comparing a fitted parametric distribution to the underlying data. Gross departures of the parametric model from the data can readily be seen in this way. If the data are sufficiently numerous, irregularities in the histogram due to sampling variations will not be too distracting.

For discrete data, the probability distribution function is already very much like the histogram. Both the histogram and the probability distribution function assign probability to a discrete set of outcomes. Comparing the two requires only that the same discrete data values, or ranges of the data values, are plotted, and that the histogram and distribution function are scaled comparably. This second condition is met by plotting the histogram in terms of relative, rather than absolute, frequency on the vertical axis. Figure 4.2 is an example, showing the superposition of a Poisson probability distribution function on the histogram of observed annual numbers U.S. hurricane landfalls.

The procedure for superimposing a continuous PDF on a histogram is entirely analogous. The fundamental constraint is that the integral of any PDF, over the full range of the random variable, must be one. That is, Equation 4.18 is satisfied by all PDFs. One approach to matching the histogram and the density function is to rescale the density function. The correct scaling factor is obtained by computing the area occupied collectively by all the bars in the histogram plot. Denoting this area as A, it is easy to see that multiplying the fitted density function $f(x)$ by A produces a curve whose area is also A because, as a constant, A can be taken out of the integral: $\int_x A \cdot f(x)\, dx = A \cdot \int_x f(x)\, dx = A \cdot 1 = A$. Note that it is also possible to rescale the histogram heights so that the total area contained in the bars is 1. This latter approach is more traditional in statistics, since the histogram is regarded as an estimate of the density function.

Example 4.12. Superposition of PDFs onto a Histogram

Figure 4.16 illustrates the procedure of superimposing fitted distributions and a histogram, for the 1933–1982 January precipitation totals at Ithaca from Table A.2. Here $n = 50$ years of data, and the bin width for the histogram (consistent with Equation 3.13) is 0.5 in., so the area occupied by the histogram rectangles is $A = (50)(0.5) = 25$. Superimposed on this histogram are PDFs for the gamma distribution fit

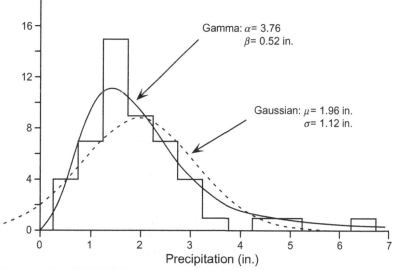

FIGURE 4.16 Histogram of the 1933–1982 Ithaca January precipitation data from Table A.2, with the fitted gamma *(solid)* and Gaussian *(dashed)* PDFs. Each of the two density functions has been multiplied by $A = 25$, since the bin width is 0.5 in. and there are 50 observations. Apparently the gamma distribution provides a reasonable representation of the data. The Gaussian distribution underrepresents the right tail and implies nonzero probability for negative precipitation.

using Equation 4.48 or 4.50a *(solid curve)*, and the Gaussian distribution fit by matching the sample and distribution moments *(dashed curve)*. In both cases the PDFs (Equations 4.45 and 4.24, respectively) have been multiplied by 25 so that their areas are equal to that of the histogram. It is clear that the symmetrical Gaussian distribution is a poor choice for representing these positively skewed precipitation data, since too little probability is assigned to the largest precipitation amounts and nonnegligible probability is assigned to impossible negative precipitation amounts. The gamma distribution represents these data much more closely and provides a quite plausible summary of the year-to-year variations in the data. The fit appears to be worst for the 0.75–1.25 in. and 1.25–1.75 in. bins, although this easily could have resulted from sampling variations. This same data set will also be used in Section 5.2.5 to test formally the fit of these two distributions.　　　　　　　　　　　　　　　　　　　　　　　◇

4.5.2. Quantile–Quantile (Q–Q) Plots

Quantile–quantile (Q–Q) plots compare empirical (data) and fitted CDFs in terms of the dimensional values of the variable (the empirical quantiles). The link between observations of the random variable X and the fitted distribution is made through the quantile function, or inverse of the CDF (Equation 4.20), evaluated at estimated levels of cumulative probability.

The Q–Q plot is a scatterplot. Each coordinate pair defining the location of a point consists of a data value and the corresponding estimate for that data value derived from the quantile function of the fitted distribution. Adopting the Tukey plotting position formula (see Table 3.2) as the estimator for empirical cumulative probability (although others could reasonably be used), each point in a Q–Q plot would have the Cartesian coordinates $(F^{-1}[(i-1/3)/(n+1/3)], x_{(i)})$. Thus the ith point on the Q–Q plot is defined by the ith smallest data value, $x_{(i)}$, and the value of the random variable corresponding to the sample cumulative probability $p = (i-1/3)/(n+1/3)$ in the fitted distribution. A Q–Q plot for a fitted distribution representing the data perfectly would have all points falling on the 1:1 diagonal line.

Figure 4.17 shows Q–Q plots comparing the fits of gamma and Gaussian distributions to the 1933–1982 Ithaca January precipitation data in Table A.2 (the parameter estimates are shown in Figure 4.16). Figure 4.17 indicates that the fitted gamma distribution corresponds well to the data through most of its range, since the quantile function evaluated at the estimated empirical cumulative probabilities is quite close to the observed data values, yielding points very close to the diagonal 1:1 line. The fitted distribution seems to underestimate the largest few points, suggesting that the tail of the fitted gamma distribution may be too thin.

On the other hand, Figure 4.17 shows the Gaussian fit to these data is clearly inferior. Most prominently, the left tail of the fitted Gaussian distribution is too heavy, so that the smallest theoretical quantiles are too small, and in fact the smallest two are actually negative. Through the bulk of the distribution the Gaussian quantiles are further from the 1:1 line than the gamma quantiles, indicating a less accurate fit, and on the right tail the Gaussian distribution underestimates the largest quantiles even more than does the gamma distribution.

It is possible also to compare fitted and empirical distributions by reversing the logic of the Q–Q plot, and producing a scatterplot of the empirical cumulative probability (estimated using a plotting position, Table 3.2) as a function of the fitted CDF, $F(x)$, evaluated at the corresponding data value. Plots of this kind are called *probability–probability*, or *P–P plots*. P–P plots seem to be used less frequently than Q–Q plots, perhaps because comparisons of dimensional data values can be more intuitive than comparisons of cumulative probabilities. Also, because P–P plots converge to the (0,0) and (1,1) points at the extremes, regardless of the closeness of correspondence, they are less sensitive to differences in the tails of a

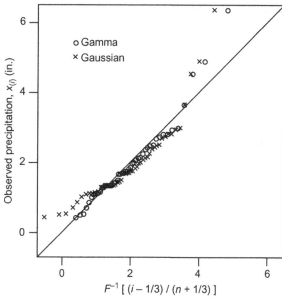

FIGURE 4.17 Quantile–quantile plots for gamma (∘) and Gaussian (×) fits to the 1933–1982 Ithaca January precipitation data in Table A.2. Observed precipitation amounts are on the vertical, and amounts inferred from the fitted distributions using the Tukey plotting position are on the horizontal. Diagonal line indicates 1:1 correspondence.

distribution, which are often of most interest. Both Q–Q and P–P plots belong to a broader class of plots known as *probability plots*.

4.6. PARAMETER FITTING USING MAXIMUM LIKELIHOOD

4.6.1. The Likelihood Function

For many distributions, fitting parameters using the simple method of moments produces inferior results that can lead to misleading inferences and extrapolations. The *method of maximum likelihood* is a versatile and important alternative. As the name suggests, the method seeks to find values of the distribution parameters that maximize the *likelihood function*. The procedure follows from the notion that the likelihood is a measure of the degree to which the data support particular values of the parameter(s) (e.g., Lindgren 1976). As explained more fully in Chapter 6, a Bayesian interpretation of the procedure (except for small sample sizes) would be that the maximum likelihood estimators (MLEs) are the most probable values for the parameters, given the observed data.

Notationally, the likelihood function for a single observation, x, looks identical to the probability density (or, for discrete variables, the probability distribution) function, and the difference between the two can be confusing initially. The distinction is that the PDF is a function of the data for fixed values of the parameters, whereas the likelihood function is a function of the unknown parameters for fixed values of the (already observed) data. Just as the joint PDF of n independent variables is the product of the n individual PDFs, the likelihood function for the parameters of a distribution given a sample of n independent data values is the product of the n individual likelihood functions. For example, the likelihood function for the Gaussian parameters μ and σ, given a sample of n observations, x_i, $i = 1$, ..., n, is

$$\Lambda(\mu, \sigma) = \sigma^{-n} \left(\sqrt{2\pi} \right)^{-n} \prod_{i=1}^{n} \exp \left[-\frac{(x_i - \mu)^2}{2\sigma^2} \right]. \tag{4.81}$$

Here the $\prod$ indicates multiplication of terms of the form indicated to its right. Actually, the Gaussian likelihood can be any function proportional to Equation 4.81, so the constant factor involving the square root of 2π could have been omitted because it does not depend on either of the two parameters. It has been included here to emphasize the relationship between Equations 4.81 and 4.24. The right-hand side of Equation 4.81 looks exactly the same as the joint PDF for n independent Gaussian variates from the same distribution, except that the parameters μ and σ are the variables, and the x_i denote fixed constants. Geometrically, Equation 4.81 describes a surface above the μ–σ plane that takes on a maximum value above a specific pair of parameter values, which depend on the particular data set given by the x_i values.

Usually it is more convenient to work with the logarithm of the likelihood function, known as the *log-likelihood*. Since the logarithm is a strictly increasing function, the same parameter values will maximize both the likelihood and log-likelihood functions. The log-likelihood function for the Gaussian parameters, corresponding to Equation 4.81 is

$$L(\mu, \sigma) = \ln[\Lambda(\mu, \sigma)] = -n \ln(\sigma) - n \ln\left(\sqrt{2\pi}\right) - \frac{1}{2\sigma^2} \sum_{i=1}^{n} (x_i - \bar{x})^2, \tag{4.82}$$

where, again, the term involving 2π is not strictly necessary for locating the maximum of the function because it does not depend on the parameters μ or σ.

Conceptually, at least, maximizing the log-likelihood is a straightforward exercise in calculus. For the Gaussian distribution the exercise really is straightforward, since the maximization can be done analytically. Taking derivatives of Equation 4.82 with respect to the parameters yields

$$\frac{\partial L(\mu, \sigma)}{\partial \mu} = \frac{1}{\sigma^2} \left[\sum_{i=1}^{n} x_i - n\mu \right] \tag{4.83a}$$

and

$$\frac{\partial L(\mu, \sigma)}{\partial \sigma} = -\frac{n}{\sigma} + \frac{1}{\sigma^3} \sum_{i=1}^{n} (x_i - \mu)^2. \tag{4.83b}$$

Setting each of these derivatives equal to zero and solving yields, respectively,

$$\hat{\mu} = \frac{1}{n} \sum_{i=1}^{n} x_i \tag{4.84a}$$

and

$$\hat{\sigma} = \sqrt{\frac{1}{n} \sum_{i=1}^{n} (x_i - \hat{\mu})^2}. \tag{4.84b}$$

These are the MLEs for the Gaussian distribution, which are readily recognized as being very similar to the moments estimators. The only difference is the divisor in Equation 4.84b, which is n rather than $n - 1$. The divisor $n - 1$ is often adopted when computing the sample standard deviation, because that choice yields an unbiased estimate of the population value. This difference points out the fact that the MLEs for a particular distribution may not be unbiased. In this case the estimated standard deviation (Equation 4.84b) will tend to be too small, on average, because the x_i are on average closer to the sample mean computed from them in Equation 4.84a than to the true mean, although these differences are small for large n.

4.6.2. The Newton–Raphson Method

The MLEs for the Gaussian distribution are somewhat unusual, in that they can be computed analytically. It is more usual for approximations to the MLEs to be calculated iteratively. One common approach is to think of the maximization of the log-likelihood as a nonlinear rootfinding problem to be solved using the multidimensional generalization of the *Newton–Raphson method* (e.g., Press et al. 1986). This approach follows from the truncated Taylor expansion of the derivative of the log-likelihood function

$$L'(\boldsymbol{\theta}^*) \approx L'(\boldsymbol{\theta}) + (\boldsymbol{\theta}^* - \boldsymbol{\theta}) L''(\boldsymbol{\theta}), \tag{4.85}$$

where $\boldsymbol{\theta}$ denotes a generic vector of distribution parameters and $\boldsymbol{\theta}^*$ are the true values to be approximated. Since it is the *derivative* of the log-likelihood function, $L'(\boldsymbol{\theta}^*)$, whose roots are to be found, Equation 4.85 requires computation of the second derivatives of the log-likelihood, $L''(\boldsymbol{\theta})$. Setting Equation 4.85 equal to zero (to find a maximum in the log-likelihood, L) and rearranging yields the expression describing the algorithm for the iterative procedure,

$$\boldsymbol{\theta}^* = \boldsymbol{\theta} - \frac{L'(\boldsymbol{\theta})}{L''(\boldsymbol{\theta})}. \tag{4.86}$$

Beginning with an initial guess, $\boldsymbol{\theta}$, an updated set of estimates $\boldsymbol{\theta}^*$ are computed by subtracting the ratio of the first to second derivatives, which are in turn used as the guesses for the next iteration.

Example 4.13. Algorithm for Maximum Likelihood Estimation of Gamma Distribution Parameters
In practice, use of Equation 4.86 is somewhat complicated by the fact that usually more than one parameter must be estimated simultaneously, so that $L'(\boldsymbol{\theta})$ is a vector of first derivatives, and $L''(\boldsymbol{\theta})$ is a matrix of second derivatives. To illustrate, consider the gamma distribution (Equation 4.45). For this distribution, Equation 4.86 becomes

$$\begin{bmatrix} \alpha^* \\ \beta^* \end{bmatrix} = \begin{bmatrix} \alpha \\ \beta \end{bmatrix} - \begin{bmatrix} \partial^2 L/\partial \alpha^2 & \partial^2 L/\partial \alpha \partial \beta \\ \partial^2 L/\partial \beta \partial \alpha & \partial^2 L/\partial \beta^2 \end{bmatrix}^{-1} \begin{bmatrix} \partial L/\partial \alpha \\ \partial L/\partial \beta \end{bmatrix}$$

$$= \begin{bmatrix} \alpha \\ \beta \end{bmatrix} - \begin{bmatrix} -n\Gamma''(\alpha) & -n/\beta \\ -n/\beta & \dfrac{n\alpha}{\beta^2} - \dfrac{2\Sigma x}{\beta^3} \end{bmatrix}^{-1} \begin{bmatrix} \sum \ln(x) - n \ln(\beta) - n\Gamma'(\alpha) \\ \sum x/\beta^2 - n\alpha/\beta \end{bmatrix}, \tag{4.87}$$

where $\Gamma'(\alpha)$ and $\Gamma''(\alpha)$ are the first and second derivatives of the gamma function (Equation 4.7), which must be evaluated or approximated numerically (e.g., Abramowitz and Stegun 1984). The matrix-algebra notation in this equation is explained in Chapter 11. Equation 4.87 would be implemented by starting with initial guesses for the parameters α and β, perhaps using the moments estimators (Equations 4.46). Updated values, α^* and β^*, would then result from a first application of Equation 4.87. The updated values would then be substituted into the right-hand side of Equation 4.87, and the process repeated until convergence of the algorithm. Convergence might be diagnosed by the parameter estimates changing sufficiently little, perhaps by a small fraction of a percent, between iterations. Note that in practice the Newton–Raphson algorithm may overshoot the likelihood maximum on a given iteration, which could result in a decline from one iteration to the next in the current approximation to the log-likelihood. Often the Newton–Raphson algorithm is programmed in a way that checks for such likelihood decreases and tries smaller changes in the estimated parameters (although in the same direction specified by, in this case, Equation 4.87).   ◇

4.6.3. The EM Algorithm

Maximum likelihood calculations using the Newton–Raphson method are generally fast and effective in applications where estimation of relatively few parameters is required. However, for problems involving more than perhaps three parameters, the computations required can expand dramatically. Even worse, the iterations can be quite unstable (sometimes producing "wild" updated parameters θ^* well away from the maximum likelihood values being sought) unless the initial guesses are so close to the correct values that the estimation procedure itself is almost unnecessary.

The *EM*, or *Expectation–Maximization algorithm* (McLachlan and Krishnan 1997), is an alternative to Newton–Raphson that does not suffer these problems. It is actually somewhat imprecise to call the EM algorithm an "algorithm," in the sense that there is not an explicit specification (like Equation 4.86 for the Newton–Raphson method) of the steps required to implement it in a general way. Rather, it is more of a conceptual approach that needs to be tailored to particular problems.

The EM algorithm is formulated in the context of parameter estimation given "incomplete" data. Accordingly, on one level, it is especially well suited to situations where some data may be missing (censored), or unobserved (truncated) above or below known thresholds, or recorded imprecisely because of coarse binning. Such situations are handled easily by the EM algorithm when the estimation problem would be easy (e.g., reducing to an analytic solution such as Equation 4.84) if the data were "complete." More generally, an ordinary (i.e., not intrinsically "incomplete") estimation problem can be approached with the EM algorithm if the existence of some additional unknown (and possibly hypothetical or unknowable) data would allow formulation of a straightforward (e.g., analytical) maximum likelihood estimation procedure. Like the Newton–Raphson method, the EM algorithm requires iterated calculations, and therefore an initial guess for the parameters to be estimated. When the EM algorithm can be formulated for a maximum likelihood estimation problem, some of the difficulties experienced by the Newton–Raphson approach do not occur, and in particular the updated log-likelihood will not decrease from iteration to iteration, regardless of how many parameters are being estimated simultaneously. For example, the bivariate distribution shown in Figure 4.15, which required simultaneous estimation of 11 parameters, was fit using the EM algorithm. This problem would have been numerically impractical with the Newton–Raphson approach unless the correct answer had been known to good approximation initially.

Just what will constitute the sort of "complete" data allowing the machinery of the EM algorithm to be used smoothly will differ from problem to problem and may require some creativity to define. Accordingly, it is not practical to outline the method here in enough generality to serve as stand-alone instruction in its use, although the following example illustrates the nature of the process. Further examples of its use in the atmospheric science literature include Hannachi and O'Neill (2001), Katz and Zheng (1999), Sansom and Thomson (1992), and Smyth et al. (1999). The original source paper is Dempster et al. (1977), and the authoritative book-length treatment is McLachlan and Krishnan (1997).

Example 4.14. Fitting a Mixture of Two Gaussian Distributions with the EM Algorithm

Figure 4.14 shows a PDF fit to the Guayaquil temperature data in Table A.3, assuming a mixture distribution in the form of Equation 4.75, where both component PDFs $f_1(x)$ and $f_2(x)$ have been assumed to be Gaussian (Equation 4.24). As noted in connection with Figure 4.14, the fitting method was maximum likelihood, using the EM algorithm.

One interpretation of Equation 4.75 is that each datum x has been drawn from either $f_1(x)$ or $f_2(x)$, with overall probabilities w and $(1 - w)$, respectively. It is not known which x's might have been drawn from which PDF, but if this more complete information were somehow to be available, then fitting the mixture of two Gaussian distributions indicated in Equation 4.75 would be straightforward: the parameters μ_1 and σ_1 defining the PDF $f_1(x)$ could be estimated using Equation 4.84 on the basis of the $f_1(x)$ data only, the parameters μ_2 and σ_2 defining the PDF $f_2(x)$ could be estimated using Equation 4.85 on the basis of the $f_2(x)$ data only, and the mixing parameter w could be estimated as the sample proportion of $f_1(x)$ data.

Even though the labels identifying particular x's as having been drawn from either $f_1(x)$ or $f_2(x)$ are not available (so that the data set is "incomplete"), the parameter estimation can proceed using the expected values of these hypothetical identifiers at each iteration step. If the hypothetical identifier variable would have been binary (equal to 1 for $f_1(x)$, and equal to 0 for $f_2(x)$), its expected value given each data value x_i would correspond to the probability that x_i was drawn from $f_1(x)$. The mixing parameter w would be equal to the average of these n hypothetical binary variables.

Equation 15.35 specifies the expected values of the hypothetical indicator variables (i.e., the n conditional probabilities) in terms of the two PDFs $f_1(x)$ and $f_2(x)$, and the mixing parameter w:

$$P(f_1 \mid x_i) = \frac{w f_1(x_i)}{w f_1(x_i) + (1 - w) f_2(x_i)}, \quad i = 1, \dots, n. \tag{4.88}$$

Equation 4.88 defines the E- (or expectation-) part of this implementation of the EM algorithm, where statistical expectations have been calculated for the unknown (and hypothetical) binary group membership data. Having calculated these n posterior probabilities, the updated maximum-likelihood estimate for the mixing parameter is

$$w = \frac{1}{n} \sum_{i=1}^{n} P(f_1 \mid x_i). \tag{4.89}$$

The remainder of the -M (or -maximization) part of the EM algorithm consists of ordinary maximum-likelihood estimation (Equations 4.84, for Gaussian-distribution fitting), using the expected quantities from Equation 4.88 in place of their unknown "complete-data" counterparts:

$$\hat{\mu}_1 = \frac{1}{nw} \sum_{i=1}^{n} P(f_1 \mid x_i) x_i, \tag{4.90a}$$

$$\hat{\mu}_2 = \frac{1}{n(1-w)} \sum_{i=1}^{n} [1 - P(f_1 \mid x_i)] x_i, \tag{4.90b}$$

$$\hat{\sigma}_1 = \left[\frac{1}{nw} \sum_{i=1}^{n} P(f_1 \mid x_i)(x_i - \hat{\mu}_1)^2 \right]^{1/2}, \tag{4.90c}$$

TABLE 4.7 Progress of the EM Algorithm Over the Seven Iterations Required to Fit the Mixture of Gaussian PDFs Shown in Figure 4.13.

Iteration	w	μ_1	μ_2	σ_1	σ_2	Log-Likelihood
0	0.50	22.00	28.00	1.00	1.00	−79.73
1	0.71	24.26	25.99	0.42	0.76	−22.95
2	0.73	24.28	26.09	0.43	0.72	−22.72
3	0.75	24.30	26.19	0.44	0.65	−22.42
4	0.77	24.31	26.30	0.44	0.54	−21.92
5	0.79	24.33	26.40	0.45	0.39	−21.09
6	0.80	24.34	26.47	0.46	0.27	−20.49
7	0.80	24.34	26.48	0.46	0.26	−20.48

and

$$\hat{\sigma}_2 = \left[\frac{1}{n(1-w)} \sum_{i=1}^{n} [1 - P(f_1 \mid x_i)](x_i - \hat{\mu}_2)^2 \right]^{1/2}. \tag{4.90d}$$

That is, Equation 4.90 implements Equation 4.84 for each of the two Gaussian distributions $f_1(x)$ and $f_2(x)$, using expected values for the hypothetical indicator variables, rather than sorting the x's into two disjoint groups. If these hypothetical labels could be known, such a sorting would correspond to the $P(f_1|x_i)$ values being equal to the corresponding binary indicators, so that Equation 4.89 would be the relative frequency of $f_1(x)$ observations; and each x_i would contribute to either Equations 4.90a and 4.90c, or to Equations 4.90b and 4.90d, only.

This implementation of the EM algorithm, for estimating parameters of the mixture PDF for two Gaussian distributions in Equation 4.75, begins with initial guesses for the five distribution parameters μ_1, σ_1, μ_2 and σ_2, and w. These initial guesses are used in Equations 4.88 and 4.89 to obtain the initial estimates for the posterior probabilities $P(f_1|x_i)$. Updated values for the mixing parameter w and the two means and two standard deviations are then obtained using Equations 4.89 and 4.80, and the process is repeated until convergence. For many problems, including this one, it is not necessary for the initial guesses to be particularly good ones. For example, Table 4.7 outlines the progress of the EM algorithm in fitting the mixture distribution that is plotted in Figure 4.14, beginning with the rather poor initial guesses $\mu_1 = 22°C$, $\mu_2 = 28°C$, $\sigma_1 = \sigma_2 = 1°C$, and $w = 0.5$. Note that the initial guesses for the two means are not even within the range of the data (although good initial guesses would be required for complicated higher-dimensional likelihoods having multiple local maxima). Nevertheless, Table 4.7 shows that the updated means are quite near their final values after only a single iteration, and that the algorithm has converged after seven iterations. The final column in this table illustrates that the log-likelihood increases monotonically with each iteration. ◇

4.6.4. Sampling Distribution of Maximum Likelihood Estimates

Even though maximum likelihood estimates may require elaborate computations, they are still sample statistics that are functions of the underlying data. As such, they are subject to sampling variations for the same reasons and in the same ways as more ordinary statistics, and so have sampling distributions that characterize the precision of the estimates. For sufficiently large sample sizes, these sampling distributions are approximately Gaussian, and the joint sampling distribution of simultaneously estimated parameters is approximately multivariate Gaussian (e.g., the joint sampling distribution of the estimates for α and β in Equation 4.87 is approximately bivariate normal, see Example 12.2).

Let $\boldsymbol{\theta} = [\theta_1, \theta_2, ..., \theta_K]$ represent a K-dimensional vector of parameters to be estimated. For example in Equation 4.87, $K = 2$, $\theta_1 = \alpha$, and $\theta_2 = \beta$. The estimated variance–covariance matrix for the approximate multivariate Gaussian ([Σ], in Equation 12.1) sampling distribution is given by the inverse of the *information matrix*,

$$Var\left(\hat{\boldsymbol{\theta}}\right) = [I(\boldsymbol{\theta})]^{-1} \tag{4.91}$$

(the matrix algebra notation is defined in Chapter 11).

The information matrix is the negative expectation of the matrix of second derivatives of the log-likelihood function with respect to the vector of parameters. In the setting of maximum likelihood the information matrix is generally estimated by the negative of the matrix $L''(\boldsymbol{\theta})$ in Equation 4.86, evaluated at the estimated parameter values $\hat{\boldsymbol{\theta}}$, which is called the *observed Fisher information*,

$$[I(\boldsymbol{\theta})] \approx - \begin{bmatrix} \partial^2 L/\partial\hat{\theta}_1^2 & \partial^2 L/\partial\hat{\theta}_1\partial\hat{\theta}_2 & \cdots & \partial^2 L/\partial\hat{\theta}_1\partial\hat{\theta}_K \\ \partial^2 L/\partial\hat{\theta}_2\partial\hat{\theta}_1 & \partial^2 L/\partial\hat{\theta}_2^2 & \cdots & \partial^2 L/\partial\hat{\theta}_2\partial\hat{\theta}_K \\ \vdots & \vdots & \ddots & \vdots \\ \partial^2 L/\partial\hat{\theta}_K\partial\hat{\theta}_1 & \partial^2 L/\partial\hat{\theta}_K\partial\hat{\theta}_2 & \cdots & \partial^2 L/\partial\hat{\theta}_K^2 \end{bmatrix}. \tag{4.92}$$

The approximation is generally close for large sample sizes. Note that the inverse of the observed Fisher information matrix appears as part of the Newton–Raphson iteration for the estimation itself, for example, for parameter estimation for the gamma distribution in Equation 4.87. One advantage of using this algorithm is therefore that the estimated variances and covariances for the joint sampling distribution of the estimated parameters will already have been calculated at the final iteration. The EM algorithm does not automatically provide these quantities, but they can, of course, be computed from the estimated parameters: either by substitution of the parameter estimates into analytical expressions for the second derivatives of the log-likelihood function, or through a finite-difference approximation to the derivatives.

4.7. STATISTICAL SIMULATION

An underlying theme of this chapter is that uncertainty in physical processes can be described by suitable probability distributions. When a component of a physical phenomenon or process of interest is uncertain, that phenomenon or process can still be studied through computer simulations, using algorithms that generate numbers that can be regarded as random samples from the relevant probability distribution(s). The generation of these apparently random numbers is called *statistical simulation*.

This section describes algorithms that are used in statistical simulation. These algorithms consist of deterministic recursive functions, so their output is not really random at all. In fact, their output can be duplicated exactly if desired, which can help in the debugging of code and is an advantage when executing controlled replication of numerical experiments. Although these algorithms are sometimes called *random-number generators*, the more correct name is *pseudo-random-number generator*, since their deterministic output only appears to be random. However, quite useful results can be obtained by regarding them as being effectively random.

Essentially all random-number generation begins with simulation from the uniform distribution, with PDF $f(u) = 1, 0 \leq u \leq 1$, which was described in Section 4.7.1. Simulating values from other distributions involves transformation of one or more uniform variates. Much more on this subject than can be presented here, including computer code and pseudocode for many particular algorithms, can be found in such references as Boswell et al. (1993), Bratley et al. (1987), Dagpunar (1988), Press et al. (1986), Tezuka (1995), and the encyclopedic Devroye (1986).

The material in this section pertains to generation of scalar, independent random variates. The discussion emphasizes generation of continuous variates, but the two general methods described in Sections 4.7.2 and 4.7.3 can be used for discrete distributions as well. Extension of statistical simulation to correlated sequences is included in Sections 10.2.4 and 10.3.7 on time-domain time series models. Extensions to multivariate simulation are presented in Section 12.4.

4.7.1. Uniform Random Number Generators

As noted earlier, statistical simulation depends on the availability of a good algorithm for generating apparently random and uncorrelated samples from the uniform [0, 1] distribution, which can be transformed to simulate random sampling from other distributions. Arithmetically, uniform random number generators take an initial value of an integer, called the *seed*, operate on it to produce an updated seed value, and then rescale the updated seed to the interval [0, 1]. The initial seed value is chosen by the programmer, but usually subsequent calls to the uniform generating algorithm operate on the most recently updated seed. The arithmetic operations performed by the algorithm are fully deterministic, so restarting the generator with a previously saved seed will allow exact reproduction of the resulting "random" number sequence.

The *linear congruential generator* is the most commonly encountered algorithm for uniform random number generation, defined by

$$S_n = a S_{n-1} + c, \quad \text{Mod } M \tag{4.93a}$$

and

$$u_n = S_n / M. \tag{4.93b}$$

Here S_{n-1} is the seed brought forward from the previous iteration; S_n is the updated seed; and a, c, and M are integer parameters called the multiplier, increment, and modulus, respectively. The quantity u_n in Equation 4.93b is the uniform variate produced by the iteration defined by Equation 4.93. Since the updated seed S_n is the remainder when $aS_{n-1} + c$ is divided by M, S_n is necessarily smaller than M, and the quotient in Equation 4.93b will be <1. For $a > 0$ and $c \geq 0$ Equation 4.93b will be >0. The parameters in Equation 4.93a must be chosen carefully if a linear congruential generator is to work at all well. The sequence S_n repeats with a period of at most M, and it is common to choose the modulus as a prime number that is nearly as large as the largest integer that can be represented by the computer on which the algorithm will be run. Many computers use 32-bit (i.e., 4-byte) integers, and $M = 2^{31} - 1$ is a usual choice in that case, often in combination with $a = 16,807$ and $c = 0$.

Linear congruential generators can be adequate for some purposes, particularly in low-dimensional applications. In higher dimensions, however, their output is patterned in a way that is not space filling. In particular, pairs of successive u's from Equation 4.93b fall on a set of parallel lines in the $u_n - u_{n+1}$ plane,

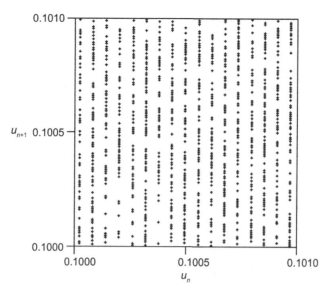

FIGURE 4.18   1000 nonoverlapping pairs of uniform random variates in a small portion of the square defined by $0 < u_n < 1$ and $0 < u_{n+1} < 1$; generated using Equation 4.93, with $a = 16,807$, $c = 0$, and $M = 2^{31} - 1$. This small domain contains 17 of the parallel lines onto which the successive pairs fall over the whole unit square.

triples of successive u's from Equation 4.93b fall on a set of parallel planes in the volume defined by the $u_n - u_{n+1} - u_{n+2}$ axes, and so on, with the number of these parallel features diminishing rapidly as the dimension K increases, approximately according to $(K! \, M)^{1/K}$. Here is another reason for choosing the modulus M to be as large as reasonably possible, since for $M = 2^{31} - 1$ and $K = 2$, $(K! \, M)^{1/K}$ is approximately 65,000.

Figure 4.18 shows a magnified view of a portion of the unit square, onto which 1000 nonoverlapping pairs of uniform variates generated using Equation 4.93 have been plotted. This small domain contains 17 of the parallel lines onto which successive pairs from this generator fall, which are spaced at an interval of 0.000059. Note that the minimum separation of the points in the vertical is much closer, indicating that the spacing of the nearly vertical lines of points does not define the resolution of the generator. The relatively close horizontal spacing in Figure 4.18 suggests that simple linear congruential generators may not be too crude for some low-dimensional purposes, although see Section 4.7.4 for a pathological interaction with a common algorithm for generating Gaussian variates in two dimensions. However, in higher dimensions the number of hyperplanes onto which successive groups of values from a linear congruential generator are constrained decreases rapidly, so that it is impossible for algorithms of this kind to generate many of the combinations that should be possible: for $K = 3, 5, 10$, and 20 dimensions, the number of hyperplanes containing all the supposedly randomly generated points is smaller than 2350, 200, 40, and 25, respectively, even for the relatively large modulus $M = 2^{31} - 1$. Note that the situation can be very much worse than this if the generator parameters are chosen poorly: a notorious but formerly widely used generator known as RANDU (Equation 4.93 with $a = 65,539$, $c = 0$, and $M = 2^{31}$) is limited to only 15 planes in three dimensions.

Direct use of linear congruential uniform generators cannot be recommended because of their patterned results in two or more dimensions. Better algorithms can be constructed by combining two or more independently running linear congruential generators, or by using one such generator to shuffle the output of another. Examples are given in Bratley et al. (1987) and Press et al. (1986). The *Mersenne twister* (Matsumoto and Nishimura 1998), which is freely available and easily found through a Web search on that name, is an attractive alternative with apparently very good properties.

4.7.2. Nonuniform Random Number Generation by Inversion

Inversion is the easiest method of nonuniform variate generation to understand and program, when the quantile function $F^{-1}(p)$ (Equation 4.20) exists in closed form. It follows from the fact that, regardless of the functional form of the CDF $F(x)$, the distribution of the variable defined by that transformation, $u = F(x)$ follows the distribution that is uniform on [0, 1]. This relationship is called the *probability integral transform* (PIT). The converse is also true (i.e., the inverse PIT), so that the CDF of the transformed variable $x(F) = F^{-1}(u)$ is $F(x)$, if the distribution of u is uniform on [0, 1]. Therefore to generate a variate with CDF $F(x)$, for which the quantile function $F^{-1}(p)$ exists in closed form, we need only to generate a uniform variate as described in Section 4.7.1, and invert the CDF by substituting that value into the quantile function.

Inversion also can be used for distributions without closed-form quantile functions, by using numerical approximations, iterative evaluations, or interpolated table lookups. Depending on the distribution, however, these workarounds might be insufficiently fast or accurate, in which case other methods would be more appropriate.

Example 4.15. Generation of Exponential Variates Using Inversion
The exponential distribution (Equations 4.52 and 4.53) is a simple continuous distribution, for which the quantile function exists in closed form. In particular, solving Equation 4.53 for the cumulative probability p yields

$$F^{-1}(p) = -\beta \ln(1-p). \tag{4.94}$$

Generating exponentially distributed variates requires only that a uniform variate be substituted for the cumulative probability p in Equation 4.94, so $x(F) = F^{-1}(u) = -\beta \ln(1-u)$. Figure 4.19 illustrates the

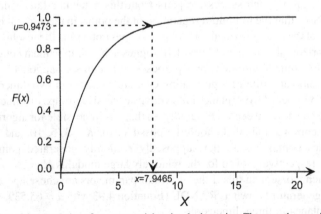

FIGURE 4.19   Illustration of the generation of an exponential variate by inversion. The smooth curve is the CDF (Equation 4.53) with mean $\beta = 2.7$. The uniform variate $u = 0.9473$ is transformed, through the inverse of the CDF, to the generated exponential variate $x = 7.9465$. This figure also illustrates that inversion produces a monotonic transformation of the underlying uniform variates.

process for an arbitrarily chosen u, and the exponential distribution with mean $\beta = 2.7$. Note that the numerical values in Figure 4.19 have been rounded to a few significant figures for convenience, but in practice all the significant digits would be retained in a computation.

Since the uniform distribution is symmetric around its middle value 0.5, the distribution of $1 - u$ is also uniform on [0, 1], so that exponential variates can be generated just as easily using $x(F) = F^{-1}(1 - u)$ $= -\beta \ln(u)$. Even though this is somewhat simpler computationally, it may be worthwhile to use $-\beta \ln(1 - u)$ anyway in order to maintain the monotonicity of the inversion method. In that case the quantiles of the underlying uniform distribution correspond exactly to the quantiles of the distribution of the generated variates, so the smallest u's correspond to the smallest x's, and the largest u's correspond to the largest x's. One instance where this property can be useful is in the comparison of simulations that might depend on different parameters or different distributions. Maintaining monotonicity across such a collection of simulations (and beginning each with the same random number seed) can allow more precise comparisons among the different simulations, because a greater fraction of the variance of differences between simulations is then attributable to differences in the simulated processes, and less is due to sampling variations in the random number streams. This technique is known as *variance reduction* in the simulation literature.   ◇

4.7.3. Nonuniform Random Number Generation by Rejection

The inversion method is mathematically and computationally convenient when the quantile function can be evaluated simply, but it can be awkward otherwise. The *rejection method*, or *acceptance–rejection method*, is a more general approach which requires only that the PDF, $f(x)$, of the distribution to be simulated can be evaluated explicitly. However, in addition, an envelope PDF, $g(x)$, must also be found. The envelope density $g(x)$ must have the same support as $f(x)$ and should be easy to simulate from (e.g., by inversion). In addition, a constant $c > 1$ must be found such that $f(x) \leq c\,g(x)$, for all x having nonzero probability. That is, $f(x)$ must be dominated by the function $c\,g(x)$ for all relevant x. The difficult part of designing a rejection algorithm is finding an appropriate envelope PDF with a shape similar to that of the distribution to be simulated, so that the constant c can be as close to 1 as possible.

Once the envelope PDF and a constant c sufficient to ensure domination have been found, simulation by rejection proceeds in two steps, each of which requires an independent call to the uniform generator. First, a candidate variate is generated from $g(x)$ using the first uniform variate u_1, perhaps by inversion as $x = G^{-1}(u_1)$. Second, the candidate x is subjected to a random test using the second uniform variate: the candidate x is accepted if $u_2 \leq f(x)/[c\,g(x)]$, otherwise the candidate x is rejected and the procedure is tried again with a new pair of uniform variates.

Figure 4.20 illustrates use of the rejection method, to simulate from the quartic density (see Table 3.1). The PDF for this distribution is a fourth-degree polynomial, so its CDF could be found easily by integration to be a fifth-degree polynomial. However, explicitly inverting the CDF (solving the fifth-degree polynomial) could be problematic, so rejection is a plausible method to simulate from this distribution. The triangular distribution (also given in Table 3.1) has been chosen as the envelope distribution $g(x)$, and the constant $c = 1.12$ is sufficient for $c\,g(x)$ to dominate $f(x)$ over $-1 \leq x \leq 1$. The triangular function is a reasonable choice for the envelope density because it dominates $f(x)$ with a relatively small value for the stretching constant c, so that the probability for a candidate x to be rejected is relatively small. In addition, it is simple enough that we easily can derive its quantile function, allowing simulation through inversion. In particular, integrating the triangular PDF yields the CDF

FIGURE 4.20 Illustration of simulation from the quartic (biweight) density, $f(x) = (15/16)(1-x^2)^2$ (Table 3.1), using a triangular density (Table 3.1) as the envelope, with $c = 1.12$. Twenty-five candidate xs have been simulated from the triangular density, of which 21 have been accepted (+) because they also fall under the distribution $f(x)$ to be simulated, and four have been rejected ($\circ$) because they fall outside it. Light gray lines point to the values simulated, on the horizontal axis.

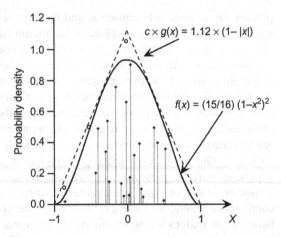

$$G(x) = \begin{cases} \dfrac{x^2}{2} + x + \dfrac{1}{2}, & -1 \leq x \leq 0, \\ -\dfrac{x^2}{2} + x + \dfrac{1}{2}, & 0 \leq x \leq 1, \end{cases} \tag{4.95}$$

which can be inverted to obtain the quantile function

$$x(G) = G^{-1}(p) = \begin{cases} \sqrt{2p} - 1, & 0 \leq p \leq 1/2, \\ 1 - \sqrt{2(1-p)}, & 1/2 \leq p \leq 1. \end{cases} \tag{4.96}$$

Figure 4.20 indicates 25 candidate points, of which 21 have been accepted (+), with light gray lines pointing to the corresponding generated values on the horizontal axis. The horizontal coordinates of these points are $G^{-1}(u_1)$, that is, random draws from the triangular density $g(x)$ using the uniform variate u_1. Their vertical coordinates are $u_2 c\, g[G^{-1}(u_1)]$, which is a uniformly distributed distance between the horizontal axis and $c\, g(x)$, evaluated at the candidate x using the second uniform variate u_2. Essentially, the rejection algorithm works because the two uniform variates define points distributed uniformly (in two dimensions) under the function $c\, g(x)$, and a candidate x is accepted according to the conditional probability that it is also under the PDF $f(x)$. The rejection method is thus very similar to Monte Carlo integration of $f(x)$. An illustration of simulation from this distribution by rejection is included in Example 4.16.

One drawback of the rejection method is that some pairs of uniform variates are wasted when a candidate x is rejected, and this is the reason that it is desirable for the constant c to be as small as possible: the probability that a candidate x will be rejected is $1 - 1/c$ ($= 0.107$ for the situation in Figure 4.20). Another property of the method is that an indeterminate, random number of uniform variates is required for one call to the algorithm, so that the synchronization of random number streams that is possible when using the inversion method is more difficult to achieve when using rejection.

4.7.4. Box–Muller Method for Gaussian Random Number Generation

One of the most frequently needed distributions in simulation is the Gaussian (Equation 4.24). Since the CDF for this distribution does not exist in closed form, neither does its quantile function, so generation of

Gaussian variates by inversion can be done only approximately. Alternatively, standard Gaussian (Equation 4.25) variates can be generated in pairs using a clever transformation of a pair of independent uniform variates, through an algorithm called the *Box–Muller method*. Corresponding dimensional (nonstandard) Gaussian variables can then be reconstituted using the distribution mean and variance, according to Equation 4.29.

The Box–Muller method generates pairs of independent standard bivariate normal variates z_1 and z_2, that is, a random sample from the bivariate PDF in Equation 4.34, with correlation $\rho = 0$ so that the level contours of the PDF are circles. Because the level contours are circles, any direction away from the origin is equally likely, implying that in polar coordinates the PDF for the angle of a random point is uniform on $[0, 2\pi]$. A uniform angle on this interval can be easily simulated from the first of the pair of independent uniform variates as $\theta = 2\pi u_1$. The CDF for the radial distance of a standard bivariate Gaussian variate is

$$F(r) = 1 - \exp\left[-\frac{r^2}{2}\right], \quad 0 \le r \le \infty, \tag{4.97}$$

which is known as the *Rayleigh distribution*. Equation 4.97 is easily invertible to yield the quantile function $r(F) = F^{-1}(u_2) = -2\ln(1 - u_2)$. Transforming back to Cartesian coordinates, the generated pair of independent standard Gaussian variates is

$$\begin{aligned} z_1 &= \cos(2\pi u_1)\sqrt{-2\ln(u_2)}, \\ z_2 &= \sin(2\pi u_1)\sqrt{-2\ln(u_2)}. \end{aligned} \tag{4.98}$$

The Box–Muller method is very common and popular, but caution must be exercised in the choice of a uniform generator with which to drive it. In particular, the lines in the u_1–u_2 plane produced by simple linear congruential generators, illustrated in Figure 4.18, are operated upon by the polar transformation to yield spirals in the z_1–z_2 plane, as discussed in more detail by Bratley et al. (1987). This patterning is clearly undesirable, and more sophisticated uniform generators are essential when generating Box–Muller Gaussian variates.

4.7.5. Simulating from Mixture Distributions and Kernel Density Estimates

Simulation from mixture distributions (Equation 4.75) is only slightly more complicated than simulation from one of the component PDFs. It is a two-step procedure, in which a component distribution is selected according to the weights, w_i, which can be regarded as probabilities with which the component distributions will be chosen. Having randomly selected a component distribution, a variate from that distribution is generated and returned as the simulated sample from the mixture.

Consider, for example, simulation from the mixed exponential distribution, Equation 4.78, which is a probability mixture of two exponential PDFs, so that $w_1 = w$ in Equation 4.78 and $w_2 = 1 - w$. Two independent uniform variates are required in order to produce one realization from the mixture distribution: one uniform variate to choose one of the two exponential distributions, and the other to simulate from that distribution. Using inversion for the second step (Equation 4.94) the procedure is simply

$$x = \begin{cases} -\beta_1 \ln(1 - u_2), & u_1 \le w, \\ -\beta_2 \ln(1 - u_2), & u_1 > w. \end{cases} \tag{4.99}$$

Here the exponential distribution with mean β_1 is chosen with probability w, using u_1, and the inversion of whichever of the two distributions is chosen is implemented using the second uniform variate u_2.

The kernel density estimate, described in Section 3.3.6 is an interesting instance of a mixture distribution. Here the mixture consists of n equiprobable PDFs, each of which corresponds to one of n observations of a variable x, so that $w_i = 1/n$, $i = 1, \ldots, n$. These PDFs are often of one of the forms listed in Table 3.1. Again, the first step is to choose which of the n data values on which the kernel to be simulated from in the second step will be centered, which can be done according to

$$\text{choose } x_i \text{ if } \frac{i-1}{n} \leq u < \frac{i}{n}, \tag{4.100a}$$

yielding

$$i = \text{int}[nu + 1]. \tag{4.100b}$$

Here int[·] indicates retention of the integer part only, or truncation of fractions.

Example 4.16. Simulation from the Kernel Density Estimate in Figure 3.8b

Figure 3.8b shows a kernel density estimate representing the Guayaquil temperature data in Table A.3, constructed using Equation 3.14, the quartic kernel (see Table 3.1), and smoothing parameter $w = 0.6$. Using rejection to simulate from the quartic kernel density, at least three independent uniform variates will be required to simulate one random sample from this distribution. Suppose these three uniform variates are generated as $u_1 = 0.257990$, $u_2 = 0.898875$, and $u_3 = 0.465617$.

The first step is to choose which of the $n = 20$ temperature values in Table A.3 will be used to center the kernel to be simulated from. Using Equation 4.100b, this will be x_i, where $i = \text{int}[20 \cdot 0.257990 + 1]$ $= \text{int}[6.1598] = 6$, yielding $T_6 = 24.3°\text{C}$, because $i = 6$ corresponds to the year 1956 in Table A.3.

The second step is to simulate from a quartic kernel, which can be done by rejection, as illustrated in Figure 4.20. First, a candidate x is generated from the dominating triangular distribution by inversion (Equation 4.96) using the second uniform variate, $u_2 = 0.898875$. This calculation yields $x(G) = 1 - [2 (1 - 0.898875)]^{1/2} = 0.550278$. Will this value be accepted or rejected? This question is answered by comparing u_3 to the ratio $f(x)/[c\,g(x)]$, where $f(x)$ is the quartic PDF, $g(x)$ is the triangular PDF, and $c = 1.12$ in order for $c\,g(x)$ to dominate $f(x)$. We find, then, that $u_3 = 0.465617 < 0.455700/[1.12 \cdot 0.449722]$ $= 0.904726$, so the candidate $x = 0.550278$ is accepted.

The value x just generated is a random draw from a standard quartic kernel, centered on zero and having unit smoothing parameter. Equating it with the argument of the kernel function h in Equation 3.14 yields $x = 0.550278 = (T - T_6)/w = (T - 24.3°\text{C})/0.6$, which centers the kernel on T_6 and scales it appropriately, so that the final simulated value is $T = (0.550278)\,(0.6) + 24.3 = 24.63°\text{C}$. ◇

4.8. EXERCISES

4.1. Using the binomial distribution as a model for the freezing of Cayuga Lake as presented in Examples 4.1 and 4.2, calculate the probability that the lake will freeze at least once during the four-year stay of a typical Cornell undergraduate in Ithaca.

4.2. Compute probabilities that Cayuga Lake will freeze next.
   a. In exactly 5 years.
   b. In 25 or more years.

4.3. In an article published in the journal *Science*, Gray (1990) contrasts various aspects of Atlantic hurricanes occurring in drought vs. wet years in sub-Saharan Africa. During the 18-year drought period 1970–1987, only one strong hurricane (intensity 3 or higher) made landfall on the east coast of the United States, but 13 such storms hit the eastern United States during the 23-year wet period 1947–1969.

a. Assume that the number of hurricanes making landfall in the eastern United States follows a Poisson distribution whose characteristics depend on African rainfall. Fit two Poisson distributions to Gray's data (one conditional on drought, and one conditional on a wet year, in West Africa).

b. Compute the probability that at least one strong hurricane will hit the eastern United States, given a dry year in West Africa.

c. Compute the probability that at least one strong hurricane will hit the eastern United States, given a wet year in West Africa.

4.4. Assume that a strong hurricane making landfall in the eastern United States causes, on average, $5 billion in damage. What are the expected values of annual hurricane damage from such storms, according to each of the two conditional distributions in Exercise 4.3?

4.5. Derive a general expression for the quantity $E[X^2]$ when X is distributed according to the binomial distribution, in terms of the distribution parameters p and N.

4.6. For any Gaussian distribution,
a. What is the probability that a randomly selected data point will fall between the two quartiles?
b. What is the probability that a randomly selected data point will fall outside the inner fences (either above or below the main body of the data)?

4.7. Using the June temperature data for Guayaquil, Ecuador, in Table A.3,
a. Fit a Gaussian distribution.
b. Without converting the individual data values, determine the two Gaussian parameters that would have resulted if this data had been expressed in°F.
c. Construct a histogram of this temperature data, and superimpose the density function of the fitted distribution on the histogram plot.

4.8. Using the Gaussian distribution with $\mu = 19°C$ and $\sigma = 1.7°C$:
a. Estimate the probability that January temperature (for Miami, Florida) will be colder than 18°C.
b. What temperature will be higher than all but the warmest 1% of Januaries at Miami?

4.9. The distribution of total summer (June, July, and August) precipitation at Montpelier, Vermont, can be represented by a gamma distribution with shape parameter $\alpha = 40$ and scale parameter $\beta = 0.24$ in. Gamma distributions with shape parameters this large are well approximated by Gaussian distributions having the same mean and variance. Compare the probabilities computed using gamma and Gaussian distributions that this location will receive no more than 7 in. of precipitation in a given summer.

4.10. For the Ithaca July rainfall data given in Table 4.8,
a. Fit a gamma distribution using Thom's approximation to the MLEs.
b. Without converting the individual data values, determine the values of the two parameters that would have resulted if the data had been expressed in mm.

c. Construct a histogram of this precipitation data and superimpose the fitted gamma density function.

TABLE 4.8 July Precipitation at Ithaca, New York, 1951–1980 (inches)

1951	4.17	1961	4.24	1971	4.25
1952	5.61	1962	1.18	1972	3.66
1953	3.88	1963	3.17	1973	2.12
1954	1.55	1964	4.72	1974	1.24
1955	2.30	1965	2.17	1975	3.64
1956	5.58	1966	2.17	1976	8.44
1957	5.58	1967	3.94	1977	5.20
1958	5.14	1968	0.95	1978	2.33
1959	4.52	1969	1.48	1979	2.18
1960	1.53	1970	5.68	1980	3.43

4.11. Use the result from Exercise 4.10 to compute:
   a. The 30th and 70th percentiles of July precipitation at Ithaca.
   b. The difference between the sample mean and the median of the fitted distribution.
   c. The probability that Ithaca precipitation during any future July will be at least 7 in.

4.12. Using the lognormal distribution to represent the data in Table 4.8, recalculate Exercise 4.11.

4.13. What is the lower quartile of an Exponential distribution with $\beta = 15$ cm?

4.14. The average of the greatest snow depths for each winter at a location of interest is 80 cm, and the standard deviation (reflecting year-to-year differences in maximum snow depth) is 45 cm.
   a. Fit a Gumbel distribution to represent these data, using the method of moments.
   b. Derive the quantile function for the Gumbel distribution, and use it to estimate the snow depth that will be exceeded in only one year out of 100, on average.

4.15. Using the GEV distribution to represent the annual maximum precipitation data in Table 4.6,
   a. What is the daily precipitation amount corresponding to the 10-year return period?
   b. What is the probability that at least 2 years in the next decade will have daily precipitation amounts exceeding this value?

4.16. Consider the bivariate normal distribution as a model for the Canandaigua maximum and Canandaigua minimum temperature data in Table A.1.
   a. Fit the distribution parameters.
   b. Using the fitted distribution, compute the probability that the maximum temperature will be as cold or colder than 20°F, given that the minimum temperature is 0°F.

4.17. Construct a Q–Q plot for the temperature data in Table A.3, in comparison to the corresponding fitted Gaussian distribution.

4.18. a. Derive a formula for the MLE for the exponential distribution (Equation 4.52) parameter, β.
b. Derive a formula for the standard deviation of the sampling distribution for β, assuming n is large.

4.19. The frequency distribution for hourly averaged wind speeds at a particular location is well described by a Weibull distribution with shape parameter $\alpha = 1.2$ and scale parameter $\beta = 7.4\,\text{m/s}$. What is the probability that, during an arbitrarily selected hour, the average wind speed will be between $10\,\text{m/s}$ and $20\,\text{m/s}$?

4.20. Design an algorithm to simulate from the Weibull distribution by inversion.

Chapter 5

Frequentist Statistical Inference

5.1. BACKGROUND

Statistical inference refers broadly to the process of drawing conclusions from a limited data sample about the characteristics of a (possibly hypothetical) "population," or generating process, from which the data were drawn. Put another way, inferential methods are meant to extract information from data samples about the process or processes that generated them.

The most familiar instance of statistical inference is in the formal testing of statistical hypotheses, also known as *significance testing*. In their simplest form, these tests yield a binary decision that a particular hypothesis about the phenomenon generating the data may be true or not, so that this process is also known as *hypothesis testing*. However, limiting statistical inferences to such binary conclusions is unnecessarily restrictive and potentially misleading (e.g., Nicholls, 2001). It is usually better to consider and communicate elements of the inferential procedures beyond just a binary result in order to address degrees of confidence in the inferences. The most familiar of these procedures are based in the frequentist, or relative frequency, view of probability, and are usually covered extensively in introductory courses in statistics. Accordingly, this chapter will review only the basic concepts behind these most familiar formal hypothesis tests, and subsequently emphasize aspects of inference that are particularly relevant to applications in the atmospheric sciences. A different approach to characterizing confidence about statistical inferences is provided by Bayesian statistics, based on the subjective view of probability, an introduction to which is provided in Chapter 6.

5.1.1. Parametric Versus Nonparametric Inference

There are two contexts in which frequentist statistical inferences are addressed; broadly, there are two types of tests and inferences. *Parametric tests* and inferences are those conducted in situations where we know or assume that a particular parametric distribution is an appropriate representation for the data and/or the test statistic. *Nonparametric tests* are conducted without assumptions that particular parametric forms are appropriate in a given situation.

Very often, parametric tests consist essentially of making inferences about particular distribution parameters. Chapter 4 presented a selection of parametric distributions that have been found to be useful for describing atmospheric data. Fitting such a distribution amounts to distilling the information contained in a sample of data, so that the distribution parameters can be regarded as representing (at least some aspects of) the nature of the underlying data-generating process of interest. Thus a parametric statistical test concerning a physical process of interest can reduce to a test pertaining to a distribution parameter, such as a Gaussian mean μ.

Statistical Methods in the Atmospheric Sciences. https://doi.org/10.1016/B978-0-12-815823-4.00005-5

Nonparametric, or *distribution-free tests* and inferences proceed without the necessity of assumptions about what, if any, parametric distribution can well describe the data at hand. Nonparametric inferential methods proceed along one of two basic lines. One approach is to construct the procedure similarly to parametric procedures, but in such a way that the distribution of the data is unimportant, so that data from any distribution can be treated in the same way. In the following, procedures following this approach are referred to as *classical* nonparametric methods, since they were devised before the advent of cheap and abundant computing power. In the second approach, crucial aspects of the relevant distribution are inferred directly from the data, by repeated computer manipulations of the data themselves. These nonparametric methods are known broadly as *resampling* procedures.

5.1.2. The Sampling Distribution

The concept of the *sampling distribution* is fundamental to both parametric and nonparametric inferential methods. Recall that a statistic is some numerical quantity computed from a batch of data. The sampling distribution for a statistic is the probability distribution describing batch-to-batch variations of that statistic. Since the batch of data from which any sample statistic (including the test statistic for a hypothesis test) has been computed is subject to sampling variations, sample statistics are subject to sampling variations as well. The value of a statistic computed from a particular batch of data will in general be different from that for the same statistic computed using a different batch of the same kind of data. For example, average January temperature is obtained by averaging daily temperatures during that month at a particular location for a given year. This statistic is different from year to year.

The random batch-to-batch variations of sample statistics can be described using probability distributions just as the random variations of the underlying data can be described using probability distributions. Thus sample statistics can be viewed as having been drawn from probability distributions, and these distributions are called sampling distributions. The sampling distribution provides a probability model describing the relative frequencies of possible values of the statistic.

5.1.3. The Elements of Any Hypothesis Test

Any hypothesis test proceeds according to the following five steps:

1. Identify a *test statistic* that is appropriate to the data and question at hand. The test statistic is the quantity computed from the data values that will be the subject of the test. In parametric settings the test statistic will often be the sample estimate of a parameter of a relevant distribution. In nonparametric resampling tests there is nearly unlimited freedom in the definition of the test statistic.
2. Define a *null hypothesis*, usually denoted H_0. The null hypothesis defines a specific logical frame of reference against which to judge the observed test statistic. Often the null hypothesis will be a "straw man" that we hope to reject.
3. Define an *alternative hypothesis*, H_A. Many times the alternative hypothesis will be as simple as "H_0 is not true," although more complex and specific alternative hypotheses are also possible.
4. Obtain the *null distribution*, which is simply the sampling distribution for the test statistic, if the null hypothesis is true. Depending on the situation, the null distribution may be an exactly known parametric distribution, a distribution that is well approximated by a known parametric distribution, or an empirical distribution obtained by resampling the data. Identifying the null distribution is the crucial step in the construction of a hypothesis test.

5. Compare the observed test statistic to the null distribution. If the test statistic falls in a sufficiently improbable region of the null distribution, H_0 is rejected as too implausible to have been true given the observed evidence. If the test statistic falls within the range of ordinary values described by the null distribution, the test statistic is seen as consistent with H_0, which is then not rejected. Note that not rejecting H_0 does not necessarily mean that the null hypothesis is true, only that there is insufficient evidence to reject this hypothesis. Not rejecting H_0 does not mean that it is "accepted." Rather, when H_0 is not rejected, it is more precise to say that it is "not inconsistent" with the observed data.

5.1.4. Test Levels and *p* Values

The sufficiently improbable region of the null distribution just referred to is defined by the *rejection level*, or simply the *level*, of the test. The null hypothesis is rejected if the probability (according to the null distribution) of the observed test statistic, *and all other results at least as unfavorable to the null hypothesis*, is less than or equal to the test level. The test level is chosen in advance of the computations, but it depends on the particular investigator's judgment and taste, so that there is usually a degree of arbitrariness about its specific value. Table 5.1 lists qualitative characterizations of the strength of evidence provided by various test levels (i.e., ceilings on *p* values), as formulated by Fisher (Efron et al., 2001). Commonly the 5% level is chosen, although tests conducted at the 10% level or the 1% level are not unusual. In situations where penalties can be associated quantitatively with particular test errors (e.g., erroneously rejecting H_0), however, the test level can be optimized (see Winkler, 1972b).

The *p* value is the specific probability that the observed value of the test statistic, together with all other possible values of the test statistic that are at least as unfavorable to the null hypothesis, will occur (according to the null distribution). Thus the null hypothesis is rejected if the *p* value is less than or equal to the test level, and is not rejected otherwise.

Unfortunately, *p* values are commonly misused and misinterpreted, which has led the American Statistical Association (ASA) to formulate an official statement on their use and purposes (Wasserstein and Lazar, 2016). Among the six principles listed in this statement is the important caution that a *p* value is *not* the probability that H_0 is true (see also Ambaum, 2010). The ASA statement also notes that it is more informative to report the *p* value for a hypothesis test rather than simply a reject/not-reject decision at a particular test level, because the *p* value also communicates the confidence with which a null hypothesis has or has not been rejected; and that a *p* value does not measure the size of an effect or the importance of a result.

Another important principle from the ASA statement is that conducting multiple analyses and reporting only nominally significant results (colloquially known as "*p*-hacking") renders the values so derived to be essentially uninterpretable. Relatedly, it is necessary for the hypotheses being examined

TABLE 5.1 Fisher's Scale of Evidence Against a Null Hypothesis, Mapped to Test Levels (i.e., Ceilings on *p* Values)

Test level	0.20	0.10	0.05	0.025	0.01	0.005	0.001
Strength of evidence	Null	Borderline	Moderate	Substantial	Strong	Very strong	Overwhelming

From Efron et al. (2001).

to have been formulated without having seen the specific data that will be used to evaluate them. This separation may be problematic especially in settings like climate research, in which new data accumulate slowly. A somewhat fanciful counterexample illustrating this point has been provided by Von Storch (1995) and Von Storch and Zwiers (1999), Chapter 6).

5.1.5. Error Types and the Power of a Test

Another way of looking at the level of a test is as the probability of falsely rejecting the null hypothesis, given that it is true. This false rejection is called a *Type I error*, and its probability (the level of the test) is often denoted α. Type I errors are defined in contrast to *Type II errors*, which occur if H_0 is not rejected when it is in fact valid. The probability of a Type II error usually is denoted β.

Figure 5.1 illustrates the relationship of Type I and Type II errors for a test conducted at the 5% level. A test statistic falling to the right of a critical value, corresponding to the quantile in the null distribution yielding the test level as a tail probability, results in rejection of the null hypothesis. Since the area under the probability density function of the null distribution to the right of the critical value in Figure 5.1 (horizontal hatching) is 0.05, this is the probability of a Type I error. The portion of the horizontal axis corresponding to H_0 rejection is sometimes called the *rejection region* or the *critical region*. Outcomes in this range are not impossible under H_0, but rather have some small probability α of occurring. It is clear from this illustration that, although we would like to minimize the probabilities of both Type I and Type II errors, this is not possible. Their probabilities, α and β, can be adjusted by adjusting the level of the test, which corresponds to moving the critical value to the left or right; but decreasing α in this way necessarily increases β, and vice versa.

The level of the test, α, can be prescribed, but the probability of a Type II error, β, usually cannot. This is because alternative hypotheses are defined more generally than the null hypothesis, and typically consists of the union of many specific alternative hypotheses. The probability α depends on the null distribution, which must be known in order to conduct a test, but β depends on which specific alternative

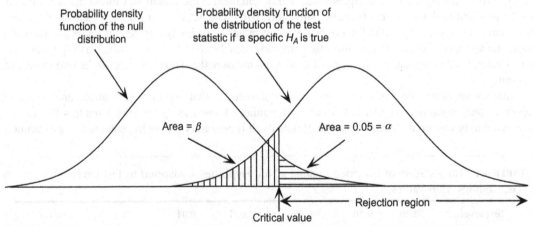

FIGURE 5.1　Illustration of the relationship of the rejection level, α, corresponding to the probability of a Type I error (horizontal hatching); and the probability of a Type II error, β (vertical hatching); for a test conducted at the 5% level. The horizontal axis represents possible values of the test statistic. Decreasing the probability of a Type I error necessarily increases the probability of a Type II error, and vice versa.

hypothesis would be applicable, and this is generally not known. Figure 5.1 illustrates the relationship between α and β for only one of a potentially infinite number of possible alternative hypotheses.

It is sometimes useful, however, to examine the behavior of β over a range of the possibilities for H_A. This investigation usually is done in terms of the quantity $1 - \beta$, which is known as the *power* of a test against a specific alternative. Geometrically, the power of the test illustrated in Figure 5.1 is the area under the sampling distribution on the right (i.e., for a particular H_A) that does not have vertical hatching. The relationship between the power of a test and a continuum of specific alternative hypotheses is called the *power function*. The power function expresses the probability of rejecting the null hypothesis, as a function of how far wrong it is. One reason why we might like to choose a less stringent test level (say, $\alpha = 0.10$) would be to better balance error probabilities for a test known to have low power.

5.1.6. One-Sided Versus Two-Sided Tests

A statistical test can be either *one-sided* or *two-sided*. This dichotomy is sometimes expressed in terms of tests being either *one-tailed* or *two-tailed*, since it is the probability in the extremes (tails) of the null distribution that governs whether a test result is interpreted as being significant. Whether a test is one-sided or two-sided depends on the nature of the hypothesis being tested.

A one-sided test is appropriate if there is a prior (e.g., a physically based) reason to expect that violations of the null hypothesis will lead to values of the test statistic on a particular side of the null distribution. This situation is illustrated in Figure 5.1, which has been drawn to imply that alternative hypotheses producing smaller values of the test statistic have been ruled out on the basis of prior information. In such cases the alternative hypothesis would be stated in terms of the true value being larger than the null hypothesis value (e.g., H_A: $\mu > \mu_0$), rather than the more vague alternative hypothesis that the true value is not equal to the null value (H_A: $\mu \neq \mu_0$). In the one-sided test situation illustrated in Figure 5.1, any test statistic larger than the $100 \cdot (1 - \alpha)$ percentile of the null distribution results in the rejection of H_0 at the α level, whereas very small values of the test statistic would not lead to a rejection of H_0.

A one-sided test is also appropriate when only values on one tail or the other of the null distribution are unfavorable to H_0 because of the way the test statistic has been constructed. For example, a test statistic involving a squared difference will be near zero if the difference is small, but will take on large positive values if the difference is large. In this case, results on the left tail of the null distribution could be quite supportive of H_0, so that only right-tail probabilities would result in H_0 rejection.

Two-sided tests are appropriate when either very large or very small values of the test statistic are unfavorable to the null hypothesis. Usually such tests pertain to the very general alternative hypothesis "H_0 is not true." The rejection region for two-sided tests consists of both the extreme left and extreme right tails of the null distribution. These two portions of the rejection region are delineated in such a way that the sum of their two probabilities under the null distribution yields the level of the test, α. That is, the null hypothesis is rejected at the α level if the test statistic is larger than $100 \cdot (1 - \alpha/2)$ percentile of the null distribution on the right tail, or is smaller than the $100 \cdot (\alpha/2)$ percentile of this distribution on the left tail. Thus a test statistic must be further out on the tail (i.e., more unusual with respect to H_0) to be declared significant in a two-tailed test as compared to a one-tailed test, at a specified test level. That the test statistic must be more extreme to reject the null hypothesis in a two-tailed test is appropriate, because generally one-tailed tests are used when additional (i.e., external to the test data) information exists, which then allows stronger inferences to be made.

5.1.7. Confidence Intervals: Inverting Hypothesis Tests

Hypothesis testing ideas can be used to construct *confidence intervals* around sample statistics. A typical use of confidence intervals is to construct *error bars* around plotted sample statistics in a graphical display, but more generally they allow a fuller appreciation of possible magnitudes of effects being investigated.

In essence, a confidence interval is derived from a hypothesis test in which the value of an observed sample statistic plays the role of the population parameter value under a hypothetical null hypothesis. The confidence interval around this sample statistic then consists of other possible values of the statistic for which that hypothetical H_0 would not be rejected. Whereas hypothesis tests evaluate probabilities associated with an observed test statistic in the context of a null distribution, conversely confidence intervals are constructed by finding the values of the test statistic that would not fall into the rejection region. In this sense, confidence interval construction is the inverse operation to hypothesis testing. That is, there is a duality between a one-sample hypothesis test and the computed confidence interval around the observed statistic, such that the $100 \cdot (1 - \alpha)\%$ confidence interval around an observed statistic will not contain the null-hypothesis value of the test if the test is significant at the α level, and will contain the null value if the test is not significant at the α level. Expressing the results of a hypothesis test in terms of the corresponding confidence interval will typically be more informative than simply reporting a reject/not-reject decision, because the width of the confidence interval and the distance of its endpoint from the null-hypothesis value will also communicate information about the degree of uncertainty in the sample estimate and about the strength of the inference.

It is tempting to think of a $100 \cdot (1 - \alpha)\%$ confidence interval as being wide enough to contain the true value with probability $1 - \alpha$, but this interpretation is not correct. The reason is that, in the frequentist view, a population parameter is a fixed if unknown constant. Therefore once a confidence interval has been constructed, the true value is either inside the interval or not. The correct interpretation is that $100 \cdot (1 - \alpha)\%$ of a large number of hypothetical similar confidence intervals, each computed on the basis of a different batch of data of the same kind (and therefore each being somewhat different from each other), will contain the true value.

Example 5.1. A Hypothesis Test Involving the Binomial Distribution

The hypothesis testing framework can be illustrated with a simple, although artificial, example. Suppose that advertisements for a tourist resort in the sunny desert southwest claim that, on average, 6 days out of 7 are cloudless during winter. To examine this claim, we would need to observe the sky conditions in the area on a number of winter days, and then compare the fraction observed to be cloudless with the claimed proportion of $6/7 = 0.857$. Assume that we could arrange to take observations on 25 independent occasions. (These would not be consecutive days, because of the serial correlation of daily weather values.) If cloudless skies are observed on 15 of those 25 days, is this observation consistent with, or does it justify questioning, the claim?

This problem fits neatly into the parametric setting of the binomial distribution. A given day is either cloudless or it is not, and observations have been taken sufficiently far apart in time that they can be considered to be independent. By confining observations to only a relatively small portion of the year, we can expect that the probability, p, of a cloudless day is approximately constant from observation to observation.

The first of the five hypothesis testing steps has already been completed, since the test statistic of $X = 15$ out of $N = 25$ days has been dictated by the form of the problem. The null hypothesis is that the resort advertisement was correct in claiming $p = 0.857$. Understanding the nature of advertising, it is reasonable to anticipate that, should the claim be false, the true probability will be lower. Thus the alternative hypothesis is that $p < 0.857$. That is, the test will be one-tailed, since results indicating $p > 0.857$ are not of interest with respect to possibly rejecting the truth of the claim. Our prior information regarding the nature of advertising claims will allow stronger inference than would have been the case if we were to have regarded alternatives with $p > 0.857$ as plausible.

Now the crux of the problem is to find the null distribution. That is, what is the sampling distribution of the test statistic X if the true probability of cloudless conditions is 0.857? This X can be thought of as the sum of 25 independent 0's and 1's, with the 1's having some constant probability of occurring on each of the 25 occasions. These are the conditions for the binomial distribution. Thus for this test the null distribution is binomial, with parameters $p = 0.857$ and $N = 25$.

It remains to compute the probability that 15 or fewer cloudless days would have been observed on 25 independent occasions if the true probability p is in fact 0.857. This probability is the p value for the test, which is a different usage for this symbol than the binomial distribution parameter, p. The direct, but tedious, approach to this computation is summation of the terms given by

$$\Pr\{X \le 15\} = \sum_{x=0}^{15} \binom{25}{x} 0.857^x (1 - 0.857)^{25-x}. \tag{5.1}$$

Here the terms for the outcomes for $X < 15$ must be included in addition to $\Pr\{X = 15\}$, since observing, say, only 10 cloudless days out of 25 would be even more unfavorable to H_0 than is $X - 15$. The p value for this test as computed from Equation 5.1 is only 0.0015. Thus $X \le 15$ is a highly improbable result if the true probability of a cloudless day is 6/7, and this null hypothesis would be resoundingly rejected. According to this test, the observed data provide very convincing evidence that the true probability is smaller than 6/7.

A much easier approach to the p-value computation in this example is to use the *Gaussian approximation to the binomial distribution*. This approximation follows from the Central Limit Theorem since, as the sum of some number of 0's and 1's, the random variable X will follow approximately the Gaussian distribution if N is sufficiently large. Here sufficiently large means roughly that $0 < p \pm 3[p(1 - p)/N]^{1/2} < 1$, in which case the binomial X can be characterized to good approximation using a Gaussian distribution with

$$\mu \approx Np \tag{5.2a}$$

and

$$\sigma \approx \sqrt{Np(1 - p)}. \tag{5.2b}$$

In the current example these parameters are $\mu \approx (25)(0.857) = 21.4$ and $\sigma \approx [(25)(0.857)(1 - 0.857)]^{1/2} = 1.75$. However, $p + 3[p(1 - p)/N]^{1/2} = 1.07$, which suggests that use of the Gaussian approximation is questionable in this example. Figure 5.2 compares the exact binomial null distribution with its Gaussian approximation. The correspondence is close, although the Gaussian approximation ascribes nonnegligible probability to the impossible outcomes $\{X > 25\}$, and correspondingly too little probability is assigned to the left tail. Nevertheless, the Gaussian approximation will be carried forward here to illustrate its use.

One small technical issue that must be faced here relates to the representation of discrete probabilities using a continuous probability density function. The p value for the exact binomial test is given by the discrete sum in Equation 5.1 yielding $\Pr\{X \le 15\}$, but its Gaussian approximation is given by the integral

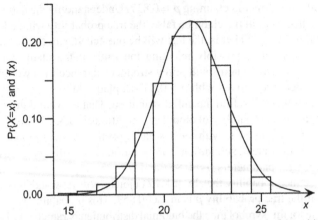

FIGURE 5.2 Relationship of the binomial null distribution (histogram bars) for Example 5.1, and its Gaussian approximation (*smooth curve*). The observed $X = 15$ falls on the far left tail of the null distribution. The exact p value from Equation 5.1 is Pr $\{X \leq 15\} = 0.0015$. Its approximation using the Gaussian distribution, including the continuity correction, is $\Pr\{X \leq 15.5\} = \Pr\{Z \leq -3.37\} = 0.00038$.

of the Gaussian PDF over the corresponding portion of the real line. This integral should include values >15 but closer to 15 than 16, since these also approximate the discrete $X = 15$. Thus the relevant Gaussian probability will be $\Pr\{X \leq 15.5\} = \Pr\{Z \leq (15.5 - 21.4)/1.75\} = \Pr\{Z \leq -3.37\} = 0.00038$, again leading to rejection but with too much confidence (too small a p value) because the Gaussian approximation puts insufficient probability on the left tail. The additional increment of 0.5 between the discrete $X = 15$ and the continuous $X = 15.5$ is called a *continuity correction*.

The Gaussian approximation to the binomial, Equations 5.2a and 5.2b, can also be used to construct a confidence interval (error bars) around the observed estimate of the binomial $\hat{p} = 15/25 = 0.6$. To do this, imagine a test whose null hypothesis is that the true binomial probability for this situation is 0.6. This test is then solved in an inverse sense to find the values of the test statistic defining the boundaries of the rejection region. That is, how large or small a value of x/N would be tolerated before this new null hypothesis would be rejected?

If a 95% confidence region is desired, the test to be inverted will be at the 5% level. Since the true binomial p could be either larger or smaller than the observed x/N, a two-tailed test (rejection regions for both very large and very small x/N) is appropriate. Referring to Table B.1, since this null distribution is approximately Gaussian, the standardized Gaussian variable cutting off probability equal to $0.05/2 = 0.025$ at the upper and lower tails is $z = \pm 1.96$. (This is the basis of the useful rule of thumb that a 95% confidence interval consists approximately of the mean value ± 2 standard deviations.) Using Equation 5.2a, the mean number of cloudless days should be $(25)(0.6) = 15$, and from Equation 5.2b the corresponding standard deviation is $[(25)(0.6)(1 - 0.6)]^{1/2} = 2.45$. Using Equation 4.28 with $z = \pm 1.96$ yields $x = 10.2$ and $x = 19.8$, leading to the 95% confidence interval bounded by $p = x/N = 0.408$ and 0.792. Notice that the claimed binomial p of $6/7 = 0.857$ falls outside this interval. The confidence interval computed exactly from the binomial probabilities is $[0.40, 0.76]$, with which the Gaussian approximation agrees very nicely. For the Gaussian approximation used to construct this confidence interval, $p \pm 3[p(1 - p)/N]^{1/2}$ ranges from 0.306 to 0.894, which is comfortably within the range $[0, 1]$.

Finally, what is the power of this test? That is, we might like to calculate the probability of rejecting the null hypothesis as a function of the true but unknown binomial p. As illustrated in Figure 5.1 the answer to this

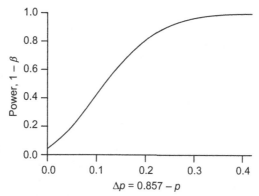

FIGURE 5.3 Power function for the test in Example 5.1. The vertical axis shows the probability of rejecting the null hypothesis, as a function of the difference between the true (and unknown) binomial p and the binomial p for the null distribution (=0.857).

question will depend on the level of the test, since it is more likely (with probability $1 - \beta$) to correctly reject a false null hypothesis if α is relatively large. Assuming a test at the 5% level, and again assuming the Gaussian approximation to the binomial distribution for simplicity, the (one-sided) critical value will correspond to $z = -1.645$ relative to the null distribution; or $-1.645 = (Np - 21.4)/1.75$, yielding $Np = 18.5$. The power of the test for a given alternative hypothesis is the probability observing the test statistic $X = \{$number of cloudless days out of $N\}$ less than or equal to 18.5, given the true binomial p corresponding to that alternative hypothesis, and will equal the area to the left of 18.5 in the approximate Gaussian sampling distribution for X defined by that binomial p and $N = 25$. Collectively, these probabilities for a range of alternative hypotheses constitute the power function for the test.

Figure 5.3 shows the resulting power function. Here the horizontal axis indicates the difference between the true binomial p and that assumed by the null hypothesis (=0.857). For $\Delta p = 0$ the null hypothesis is true, and Figure 5.3 indicates a 5% chance of rejecting it, which is consistent with the test being conducted at the 5% level. We do not know the true value of p, but Figure 5.3 shows that the probability of rejecting the null hypothesis increases as the true p is increasingly different from 0.857, until we are virtually assured of rejecting H_0 with a sample size of $N = 25$ if the true probability is smaller than about 0.5. If $N > 25$ days had been observed, the resulting power curve would be above that shown in Figure 5.3, so that probabilities of rejecting false null hypotheses would be greater (i.e., their power functions would climb more quickly toward 1), indicating more sensitive tests. Conversely, corresponding tests involving fewer samples would be less sensitive, and their power curves would lie below the one shown in Figure 5.3. ◇

5.2. SOME COMMONLY ENCOUNTERED PARAMETRIC TESTS

5.2.1. One-Sample *t*-Test

By far, the most commonly encountered parametric tests in classical statistics relate to the Gaussian distribution. Tests based on the Gaussian are so pervasive because of the strength of the Central Limit Theorem. As a consequence of this theorem, many non-Gaussian problems can be treated at least approximately in the Gaussian framework. The example test for the binomial parameter p in Example 5.1 is one such case.

Probably the most familiar statistical test is the *one-sample t-test*, which examines the null hypothesis that an observed sample mean has been drawn from a population or generating process centered at some previously specified mean, μ_0. If the number of data values making up the sample mean is large enough for its sampling distribution to be essentially Gaussian (by the Central Limit Theorem), then the test statistic

$$t = \frac{\bar{x} - \mu_0}{[\text{Vâr}(\bar{x})]^{1/2}} \tag{5.3}$$

follows a distribution known as *Student's t*, or simply the *t distribution*. Equation 5.3 resembles the standard Gaussian variable z (Equation 4.26), except that a sample estimate of the variance of the sample mean (denoted by the "hat" accent) has been substituted in the denominator.

The t distribution is a symmetrical distribution that is very similar to the standard Gaussian distribution, although with more probability assigned to the tails. That is, the t distribution has heavier tails than the Gaussian distribution. The t distribution is controlled by a single parameter, v, called the *degrees of freedom*. The parameter v can take on any positive integer value, with the largest differences from the Gaussian being produced for small values of v. For the test statistic in Equation 5.3, $v = n - 1$, where n is the number of independent observations being averaged in the sample mean in the numerator.

Tables of t distribution probabilities are available in almost any introductory statistics textbook. However, for even moderately large values of n (and therefore of v) the variance estimate in the denominator becomes sufficiently precise that the t distribution is closely approximated by the standard Gaussian distribution. The differences in tail quantiles are about 4% and 1% for $v = 30$ and 100, respectively, so for sample sizes of this magnitude and larger it is usually quite acceptable to evaluate probabilities associated with the test statistic in Equation 5.3 using standard Gaussian probabilities.

Use of the standard Gaussian PDF (Equation 4.25) as the null distribution for the test statistic in Equation 5.3 can be understood in terms of the Central Limit Theorem, which implies that the sampling distribution of the sample mean in the numerator will be approximately Gaussian if n is sufficiently large. Subtracting the mean μ_0 in the numerator will center that Gaussian distribution on zero if the null hypothesis, to which μ_0 pertains, is true. If n is also large enough that the standard deviation of the sampling distribution of the sample mean (the denominator) can be estimated sufficiently precisely, then the sampling distribution of the quantity in Equation 5.3 will also have unit standard deviation to good approximation. A Gaussian distribution with zero mean and unit standard deviation is the standard Gaussian distribution.

The variance of the sampling distribution of a mean of n independent observations, in the denominator of Equation 5.3, is estimated according to

$$\text{Vâr}[\bar{x}] = s^2/n, \tag{5.4}$$

where s^2 is the sample variance (the square of Equation 3.6) of the individual x's being averaged. Equation 5.4 is clearly true for the simple case of $n = 1$, but also makes intuitive sense for larger values of n. We expect that averaging together, say, pairs ($n = 2$) of x's will give quite irregular results from pair to pair. That is, the sampling distribution of the average of two numbers will have a high variance. On the other hand, averaging together batches of $n = 1000$ x's will give very consistent results from batch to batch, because the occasional very large x will tend to be balanced by the occasional very small x: a sample of $n = 1000$ will tend to have nearly equally many very large and very small values. The variance of the sampling distribution (i.e., the batch-to-batch variability) of the average of 1000 independent numbers will thus be small.

For small (absolute) values of t in Equation 5.3, the difference in the numerator is small in comparison to the standard deviation of the sampling distribution of the difference, suggesting a quite ordinary sampling fluctuation for the sample mean, which should not trigger rejection of H_0. If the difference in the numerator is more than about twice as large as the denominator in absolute value, the null hypothesis would usually be rejected, corresponding to a two-sided test at the 5% level (cf. Table B.1).

5.2.2. Tests for Differences of Mean Under Independence

Another common statistical test involves the difference between two independent sample means. Plausible atmospheric examples of this situation might be differences of average winter 500 mb heights when one or the other of two synoptic regimes had prevailed, or perhaps differences in average July temperature at a location as represented in a climate model under a doubling vs. no doubling of atmospheric carbon dioxide concentration.

In general, two sample means calculated from different batches of data, even if they are drawn from the same population or generating process, will be different. The usual test statistic in this situation is a function of the difference of the two sample means being compared, and the actual observed difference will almost always be some number other than zero. The null hypothesis is usually that the true difference is zero. The alternative hypothesis is either that the true difference is not zero (the case where no a priori information is available as to which underlying mean should be larger, leading to a two-tailed test), or that one of the two underlying means is larger than the other (leading to a one-tailed test). The problem is to find the sampling distribution of the difference of the two sample means, given the null hypothesis assumption about the difference between their population counterparts. It is in this context that the observed difference of means can be evaluated for unusualness.

Nearly always—and sometimes quite uncritically—the assumption is tacitly made that the sampling distributions of the two sample means being differenced are Gaussian. This assumption will be valid either if the data composing each of the sample means are Gaussian, or if the sample sizes are sufficiently large that the Central Limit Theorem can be invoked. If both of the two sample means have Gaussian sampling distributions their difference will be Gaussian as well, since any linear combination of Gaussian variables will itself follow a Gaussian distribution. Under these conditions the test statistic

$$z = \frac{(\bar{x}_1 - \bar{x}_2) \quad E[\bar{x}_1 - \bar{x}_2]}{(\text{Vâr}[\bar{x}_1 - \bar{x}_2])^{1/2}} \tag{5.5}$$

will be distributed as standard Gaussian (Equation 4.25) for large samples. Note that this equation has a form similar to both Equations 5.3 and 4.27.

If the null hypothesis is equality of means of the two populations from which values of x_1 and x_2 have been drawn, then

$$E[\bar{x}_1 - \bar{x}_2] = E[\bar{x}_1] - E[\bar{x}_2] = \mu_1 - \mu_2 = 0. \tag{5.6}$$

Thus a specific hypothesis about the magnitude of the two equal means is not required. If some other null hypothesis is appropriate to a problem at hand, that difference of underlying means would be substituted in the numerator of Equation 5.5.

The variance of a difference (or sum) of two independent random quantities is the sum of the variances of those quantities. Intuitively this makes sense since contributions to the variability of the difference are made by the variability of each the two quantities being differenced. With reference to the denominator of Equation 5.5,

$$\text{Vâr}[\bar{x}_1 - \bar{x}_2] = \text{Vâr}[\bar{x}_1] + \text{Vâr}[\bar{x}_2] = \frac{s_1^2}{n_1} + \frac{s_2^2}{n_2}, \tag{5.7}$$

where the last equality is achieved using Equation 5.4. Thus if the batches making up the two averages are independent, and the sample sizes are sufficiently large, Equation 5.5 can be transformed to good approximation to the standard Gaussian z by rewriting the test statistic as

$$z = \frac{\bar{x}_1 - \bar{x}_2}{\left[s_1^2/n_1 + s_2^2/n_2\right]^{1/2}}, \tag{5.8}$$

when the null hypothesis is that the two underlying means μ_1 and μ_2 are equal. This expression for the test statistic is appropriate when the variances of the two distributions from which the x_1's and x_2's are drawn are not equal. For relatively small sample sizes its sampling distribution is (approximately, although not exactly) the t distribution, with $v = \min(n_1, n_2) - 1$. For moderately large samples the sampling distribution is close to the standard Gaussian, for the same reasons presented in relation to its one-sample counterpart, Equation 5.3.

When it can be assumed that the variances of the distributions from which the x_1's and x_2's have been drawn are equal, that information can be used to calculate a single, *"pooled,"* estimate for that variance. Under this assumption of equal population variances, Equation 5.5 becomes instead

$$z = \frac{\bar{x}_1 - \bar{x}_2}{\left[\left(\frac{1}{n_1} + \frac{1}{n_2}\right)\left\{\frac{(n_1 - 1)s_1^2 + (n_2 - 1)s_2^2}{n_1 + n_2 - 2}\right\}\right]^{1/2}}. \tag{5.9}$$

The quantity in curly brackets in the denominator is the pooled estimate of the population variance for the data values, which is just a weighted average of the two sample variances, and has in concept been substituted for both s_1^2 and s_2^2 in Equations 5.7 and 5.8. The sampling distribution for Equation 5.9 is the t distribution with $v = n_1 + n_2 - 2$. However, when both n_1 and n_2 are moderately large it is again usually quite acceptable to evaluate probabilities associated with the test statistic in Equation 5.9 using the standard Gaussian distribution.

For small (absolute) values of z in either Equations 5.8 or 5.9, the difference of sample means in the numerator is small in comparison to the standard deviation of the sampling distribution of their difference in the denominator, indicating a quite ordinary value in terms of the null distribution. As before, if the difference in the numerator is more than about twice as large as the denominator in absolute value, and the sample size is moderate or large, the null hypothesis would be rejected at the 5% level for a two-sided test.

As is also the case for one-sample tests, Equation 5.8 or 5.9 for two-sample t-tests can be worked backwards to yield a confidence interval around an observed difference of the sample means, $\bar{x}_1 - \bar{x}_2$. A rejection of $H_0:\{\mu_1 = \mu_2\}$ at the α level would correspond to the $100 \cdot (1 - \alpha)\%$ confidence interval for this difference to not include zero. However, counterintuitively, the individual $100 \cdot (1 - \alpha)\%$ confidence intervals for $\bar{x}_1$ and $\bar{x}_2$ could very well overlap in that case (Lanzante, 2005; Schenker and Gentleman, 2001). That is, overlapping $100 \cdot (1 - \alpha)\%$ confidence intervals for two individual sample statistics can very easily be consistent with the two statistics being significantly different according to an appropriate two-sample α-level test. The discrepancy between the results of this so-called *overlap method* and a correct two-sample test is greatest when the two sample variances s_1^2 and s_2^2 are equal or nearly so, and progressively diminishes as the magnitudes of the two sample variances diverge. Conversely, nonoverlapping $(1 - \alpha) \cdot 100\%$ confidence intervals does imply a significant difference at the α-level, at least.

5.2.3. Tests for Differences of Mean for Paired Samples

Equation 5.7 is appropriate when the x_1's and x_2's are observed independently. An important form of nonindependence occurs when the data values making up the two averages are *paired*, or observed simultaneously. In this case, necessarily, $n_1 = n_2$. For example, the daily temperature data in Table A.1 of Appendix A are of this type, since there is an observation of each variable at both locations on each day. When paired data of this kind are used in a two-sample *t*-test, the two averages being differenced are generally correlated. When this correlation is positive, as will often be the case, Equation 5.7 or the denominators of Equations 5.8 or 5.9 will overestimate the variance of the sampling distribution of the difference in the numerators. The result is that the test statistic will be too small (in absolute value), on average, so that the calculated *p* values will be too large, and null hypotheses that should be rejected will not be.

We should expect the sampling distribution of the difference in the numerator of the test statistic to be affected if pairs of *x*'s going into the averages are strongly correlated. For example, the appropriate panel in Figure 3.31 indicates that the daily maximum temperatures at Ithaca and Canandaigua are strongly correlated, so that a relatively warm average monthly maximum temperature at one location would likely be associated with a relatively warm average at the other. A portion of the variability of the monthly averages is thus common to both, and that portion cancels in the difference in the numerator of the test statistic. That cancellation must also be accounted for in the denominator if the sampling distribution of the test statistic is to be approximately standard Gaussian.

The easiest and most straightforward approach to dealing with the *t*-test for paired data is to analyze differences between corresponding members of the $n_1 = n_2 = n$ pairs, which transforms the problem to the one-sample setting. That is, consider the sample statistic

$$\Delta = x_1 - x_2, \tag{5.10a}$$

with sample mean

$$\overline{\Delta} = \frac{1}{n}\sum_{i=1}^{n} \Delta_i = \overline{x}_1 - \overline{x}_2. \tag{5.10b}$$

The corresponding population mean will be $\mu_\Delta = \mu_1 - \mu_2$, which is often zero under H_0. The resulting test statistic is then of the same form as Equation 5.3,

$$z = \frac{\overline{\Delta} - \mu_\Delta}{\left(s_\Delta^2/n\right)^{1/2}}, \tag{5.11}$$

where s_Δ^2 is the sample variance of the n differences in Equation 5.10a. Joint variation in the pairs making up the difference $\Delta = x_1 - x_2$ is also automatically accounted for in the sample variance s_Δ^2 of those differences.

Equation 5.11 is an instance where positive correlation in the data is beneficial, in the sense that a more sensitive test can be achieved. Here a positive correlation results in a smaller standard deviation for the sampling distribution of the difference of means being tested, implying less underlying uncertainty. This sharper null distribution produces a more powerful test and allows smaller differences in the numerator to be detected as significantly different from zero.

Intuitively this effect on the sampling distribution of the difference of sample means makes sense as well. Consider again the example of Ithaca and Canandaigua temperatures for January 1987, which will be revisited in Example 5.3. The positive correlation between daily temperatures at the two locations will result in the batch-to-batch (i.e., January-to-January, or interannual) variations in the two monthly averages moving together for the two locations: months when Ithaca is warmer than usual tend also to be months when Canandaigua is warmer than usual. The more strongly correlated are x_1 and x_2, the less likely are the pair of corresponding averages from a particular batch of data to differ because of sampling variations. To the extent that the two sample averages are different, then, the evidence against their underlying means not being the same is stronger, as compared to the situation when their correlation is near zero.

5.2.4. Tests for Differences of Mean Under Serial Dependence

The material in the previous sections is essentially a recapitulation of well-known tests for comparing sample means, presented in almost every elementary statistics textbook. A key assumption underlying these tests is the independence among the individual observations composing each of the sample means in the test statistic. That is, it is assumed that all the x_1 values are mutually independent and that the x_2 values are mutually independent, whether or not the data values are paired. This assumption of independence leads to the expression in Equation 5.4 that allows estimation of the variance of the null distribution.

Atmospheric data often do not satisfy the independence assumption. Frequently the averages to be tested are time averages, and the persistence, or time dependence, often exhibited is the cause of the violation of the assumption of independence. Lack of independence invalidates Equation 5.4. In particular, meteorological persistence implies that the variance of a time average is larger than specified by Equation 5.4. Ignoring the time dependence thus leads to underestimation of the variance of sampling distributions of the test statistics in Sections 5.2.2 and 5.2.3. This underestimation leads in turn to an inflated value of the test statistic, and consequently to p values that are too small, and to overconfidence regarding the significance of the difference in the numerator. Equivalently, properly representing the effect of persistence in the data will require a larger sample size to reject a null hypothesis for a given magnitude of the difference in the numerator.

Figure 5.4 illustrates why serial correlation leads to a larger variance for the sampling distribution of a time average. The upper panel of this figure is an artificial time series of 100 independent Gaussian variates drawn from a generating process with $\mu = 0$, as described in Section 4.7.4. The series in the lower panel also consists of Gaussian variables having $\mu = 0$, but in addition this series has a lag-1 autocorrelation (Equation 3.36) of $\rho_1 = 0.6$. This value of the autocorrelation was chosen here because it is typical of the autocorrelation exhibited by daily temperatures (e.g., Madden, 1979; Wilks and Wilby, 1999). Both panels have been scaled to produce unit (population) variance. The two plots look similar because the autocorrelated series was generated from the independent series according to what is called a first-order autoregressive process (Equation 10.16).

The outstanding difference between the independent and autocorrelated pseudo-data in Figure 5.4 is that the correlated series is smoother, so that adjacent and nearby values tend to be more alike than in the independent series. The autocorrelated series exhibits longer runs of points away from the

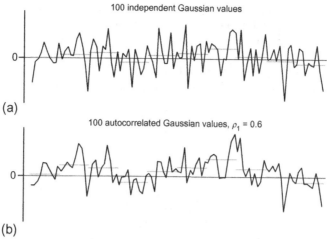

FIGURE 5.4 Comparison of artificial time series of (a) independent Gaussian variates, and (b) autocorrelated Gaussian variates having $\rho_1 = 0.6$. Both series were drawn from a generating process with $\mu = 0$, and the two panels have been scaled to have unit variances for the data points. Nearby values in the autocorrelated series tend to be more alike, with the result that averages over segments with $n = 10$ (*horizontal gray bars*) of the autocorrelated time series are more likely to be further from zero than are averages from the independent series. The sampling distribution of averages computed from the autocorrelated series accordingly has larger variance: the sample variances of the 10 subsample averages in panels (a) and (b) are 0.0825 and 0.2183, respectively.

(generating-process) mean value of zero. As a consequence, averages computed over subsets of the auto-correlated record are less likely to contain compensating points with large absolute value but of opposite sign, and those averages are therefore more likely to be far from zero than their counterparts computed using the independent values. That is, the averages over segments of the autocorrelated series will be less consistent from batch to batch. This is just another way of saying that the sampling distribution of an average of autocorrelated data has a higher variance than that of independent data. The gray horizontal lines in Figure 5.4 are subsample averages over consecutive sequences of $n = 10$ points, and these are visually more variable in Figure 5.4b. The sample variances of the 10 subsample means are 0.0825 and 0.2183 in panels (a) and (b), respectively.

Not surprisingly, the problem of estimating the variance of the sampling distribution of a time average has received considerable attention in the meteorological literature (e.g., Jones, 1975; Katz, 1982; Madden, 1979; Zwiers and Thiébaux, 1987; Zwiers and von Storch, 1995). One convenient and practical approach to dealing with the problem is to think in terms of the *effective sample size*, or *equivalent number of independent samples*, n'. That is, imagine that there is a fictitious sample size, $n' < n$ of independent values, for which the sampling distribution of the average has the same variance as the sampling distribution of the average over the n autocorrelated values at hand. Then, n' could be substituted for n in Equation 5.4, and the classical tests described in the previous section could be carried through as before.

Estimation of the effective sample size is most easily approached if it can be assumed that the underlying data follow a first-order autoregressive process (Equation 10.16). It turns out that first-order auto-regressions are often reasonable approximations for representing the persistence of daily meteorological values. This assertion can be appreciated informally by looking at Figure 5.4b. This plot consists of random numbers exhibiting first-order autoregressive dependence, but resembles statistically the day-to-day fluctuations in a meteorological variable like surface temperature.

In general the effective sample size depends on the full autocorrelation function for the data-generating process, according to Equation 10.38. The persistence in a first-order autoregression is completely characterized by the single parameter ρ_1, the lag-1 autocorrelation coefficient, because in that case the autocorrelation function is given by $\rho_k = \rho_1^k$. The lag-1 autocorrelation can be estimated from a data series using the sample estimate, r_1 (Equation 3.36). Using this value, the effective sample size for the sampling variance of a mean can be estimated using the approximation

$$n' \approx n \frac{1 - \rho_1}{1 + \rho_1}, \tag{5.12}$$

which is valid for moderate and large n. When there is no time correlation, $\rho_1 = 0$ and $n' = n$. As ρ_1 increases the effective sample size becomes progressively smaller. When a more complicated time-series model is necessary to describe the persistence, appropriate but more complicated expressions for the effective sample size can be derived (see Katz, 1982, 1985; and Section 10.3.5).

Note that Equation 5.12 is applicable only to sampling distributions of the mean, and different expressions will be appropriate for use with different statistics (e.g., Ebisuzaki, 1997; Faes et al., 2009; Thiébaux and Zwiers, 1984). For example, the approximate effective sample size for the variance, again assuming an underlying first-order autoregressive generating process, is $n(1 - \rho_1^2)/(1 + \rho_1^2)$ (e.g., Preisendorfer et al., 1981), and so is relatively less affected by autocorrelation than the corresponding value for the mean in Equation 5.12.

Using Equation 5.12, the counterpart to Equation 5.4 for the variance of a time average over a sufficiently large sample becomes

$$\mathrm{V\hat{a}r}[\bar{x}] = \frac{s^2}{n'} \approx \frac{s^2}{n}\left(\frac{1 + \rho_1}{1 - \rho_1}\right). \tag{5.13}$$

The ratio $(1 + \rho_1)/(1 - \rho_1)$ acts as a *variance inflation factor*, adjusting the variance of the sampling distribution of the time average upward to reflect the influence of the serial correlation. Sometimes this variance inflation factor is called the *time between effectively independent samples*, T_0 (e.g., Leith, 1973). Equation 5.4 can be seen as a special case of Equation 5.13, with $\rho_1 = 0$. Use of Equation 5.13 in the denominator of the test statistic in Equations 5.11 is in effect a special case of the *Diebold–Mariano test* (Diebold and Mariano, 1995, Hering and Genton 2011), which uses Equation 10.41, and so does not make a restrictive assumption about the correlation structure in the paired data, as the variance inflation factor.

Example 5.2. Unpaired Two-Sample *t*-Test for Autocorrelated Data

Magnitudes of autocorrelation typically exhibited by daily temperature data can have quite strong effects on statistical inferences based on such data. As an example, consider comparing the average 1988 temperatures for June and July at Ithaca, the values for which are listed in the second line of Table 5.2. The June temperature was slightly below average for the month, whereas the July temperature was several degrees warmer.

Considering also that July is climatologically warmer than June, a null hypothesis of equality for the means of the two data-generating processes that produced the monthly means in Table 5.2 seemingly should be easy to reject. However, because of the autocorrelation in the daily data that is near 0.5, Equation 5.12 yields effective sample sizes n' of 9.5 and 10.9 days for June and July respectively, which are reductions of about 2/3. Accordingly Equation 5.8 becomes $z = (83.5 - 75.2)/(7.75 + 11.8)^{1/2} = 1.88$.

TABLE 5.2 Summary Statistics Comparing Temperatures for June and July 1988 at Ithaca

	June	July
n	30	31
$\bar{x}$	75.2°F	83.5°F
s_x	10.6°F	9.2°F
r_1	0.52	0.48
n'	9.5	10.9
$\hat{\mathrm{Var}}(\bar{x})$	11.8°F^2	7.75°F^2

Considering that we know enough about the climatology of the area to expect that if there are differences July will be warmer, then a one-tailed test is appropriate and the p value is 0.030. The null hypothesis of equality for the data-generating process means might be rejected, but not especially strongly. If we did not know in advance that July should be warmer, then the (two-tailed) p value is 0.060 and the inference is weaker still. In contrast, ignoring the autocorrelation and assuming serial independence of the underlying daily data leads to $z = 3.18$ in Equation 5.8, corresponding to the 1-tailed p value of 0.00074, which is wildly overconfident.   ◇

Example 5.3. Paired Two-Sample t-Test for Autocorrelated Data

Consider testing whether the average maximum temperatures at Ithaca and Canandaigua for January 1987 (Table A.1 in Appendix A) are significantly different. This is equivalent to testing whether the difference of the two sample means is significantly different from zero, so that Equation 5.6 will hold for the null hypothesis. It has been shown previously (see Figure 3.5) that these two batches of daily data are reasonably symmetric and well behaved, so the sampling distributions of the monthly averages should be nearly Gaussian under the Central Limit Theorem.

The data for each location were observed on the same 31 days in January 1987, so the two batches are paired samples. Equation 5.11 is therefore the appropriate choice for the test statistic. Furthermore, we know that the daily data underlying the two time averages exhibit serial correlation (Figure 3.22 for the Ithaca data) so it may be expected that the effective sample size corrections in Equations 5.12 and 5.13 will be necessary as well.

Table A1 also shows the mean January 1987 temperatures, so the difference (Ithaca–Canandaigua) in mean maximum temperature is $29.87 - 31.77 = -1.9$°F. Computing the standard deviation of the differences between the 31 pairs of maximum temperatures yields $s_\Delta = 2.285$°F. The lag-1 autocorrelation for these differences is 0.076, yielding $n' = 31(1 - 0.076)/(1 + 0.076) = 26.6$. Since the null hypothesis is that the two population means are equal, $\mu_\Delta = 0$, and Equation 5.11 (using the effective sample size n' rather than the actual sample size n) yields $z = -1.9/(2.285^2/26.6)^{1/2} = -4.29$. This is a sufficiently extreme value not to be included in Table B.1, although Equation 4.30 estimates $\Phi(-4.29) \approx 0.000002$. The two-tailed p-value would be 0.000004, which is clearly significant. This extremely strong result is possible in part because much of the variability of the two temperature series is shared (the correlation between them is

0.957), and removing shared variance results in a rather small denominator for the test statistic. The magnitude of this shared variability is quantified in Example 5.18.

Notice that the lag-1 autocorrelation for the paired temperature differences is only 0.076, which is much smaller than the autocorrelations in the two individual series: 0.52 for Ithaca and 0.61 for Canandaigua. Much of the temporal dependence is also exhibited jointly by the two series, and so is removed when calculating the differences Δ_i. Here is another advantage of using the series of differences to conduct this test, and another major contribution to the strong result. The relatively low autocorrelation of the difference series translates into an effective sample size of 26.6 rather than only 9.8 (Ithaca) and 7.5 (Canandaigua), which produces an even more sensitive test.

Finally, consider the confidence interval for the mean difference μ_Δ in relation to the confidence intervals that would be calculated for the individual means μ_{Ith} and μ_{Can}. The 95% confidence interval around the observed mean difference of $\bar{x}_\Delta = -1.9°F$ is $-1.9°F \pm (1.96)(2.285)/(\sqrt{26.6})$, yielding the interval $[-2.77, -1.03]$. Consistent with the extremely low p value for the paired comparison test, this interval does not include zero, and indeed its maximum is well away from zero in standard error (standard deviation of the sampling distribution of $\bar{\Delta}$) units. In contrast, consider the 95% confidence intervals around the individual sample means for Ithaca and Canandaigua. For Ithaca, this interval is $29.9°F \pm (1.96)(7.71)/(\sqrt{9.8})$, or $[25.0, 34.7]$, whereas for Canandaigua it is $31.8°F \pm (1.96)(7.86)/(\sqrt{7.5})$, or $[26.2, 37.4]$. Not only do these two intervals overlap substantially, their length of overlap is greater than the sum of the lengths over which they do not overlap. Thus evaluating the significance of this difference using the so-called overlap method leads to a highly erroneous conclusion. This example provides a nearly worst case for the overlap method, since in addition to the two variances being nearly equal, members of the two data samples are strongly correlated, which also exacerbates the discrepancy with a correctly computed test (Jolliffe, 2007; Schenker and Gentleman, 2001).   ◇

Autocorrelation of values in a time series is not the only kind of statistical dependence that may affect statistical inferences. Director and Bornn (2015) address effective sample size adjustments in the context of spatial data.

5.2.5. Goodness-of-Fit Tests

When discussing the fitting of parametric distributions to data samples in Chapter 4, methods for visually and subjectively assessing the *goodness of fit* were presented. Formal, quantitative evaluations of the goodness of fit also exist, and these are carried out within the framework of hypothesis testing. The graphical methods can still be useful when formal tests are conducted, for example, in pointing out where and how a lack of fit is manifested. Many goodness-of-fit tests have been devised, but only a few common ones are presented here.

Assessing goodness of fit presents an atypical hypothesis test setting, in that these tests usually are computed to obtain evidence in favor of H_0, that the data at hand were drawn from a hypothesized distribution. The interpretation of confirmatory evidence is then that the data are "not inconsistent" with the hypothesized distribution, so the power of these tests is an important consideration. Unfortunately, because there are any number of ways in which the null hypothesis can be wrong in this setting, it is usually not possible to formulate a single best (most powerful) test. This problem accounts in part for the large number of goodness-of-fit tests that have been proposed (D'Agostino and Stephens, 1986), and for the ambiguity about which might be most appropriate for a particular problem. Note also that the tests described in this section assume mutually independent data, and as is the case for other hypothesis tests their uncorrected implementation using autocorrelated data will yield erroneously small p values.

Chi-Square Test

The *chi-square* (χ^2) *test* is a simple and common goodness-of-fit test. It essentially compares a data histogram with the probability distribution (for discrete variables) or probability density (for continuous variables) function. The χ^2 test actually operates more naturally for discrete random variables, since to implement it the range of the data must be divided into discrete classes, or bins. When alternative tests are available for continuous data they are usually more powerful, presumably at least in part because the rounding of data into bins, which may be severe, discards information. However, the χ^2 test is easy to implement and quite flexible, being for example, very straightforward to implement for multivariate data.

For continuous random variables, the probability density function is integrated over each of some number of MECE classes to obtain probabilities for data values in each class. Regardless of whether the data are discrete or continuous, the test statistic involves the counts of data values falling into each class in relation to the computed theoretical probabilities,

$$
\chi^2 = \sum_{classes} \frac{(\# \text{ Observed} - \# \text{ Expected})^2}{\# \text{ Expected}}
$$
$$
= \sum_{classes} \frac{(\# \text{ Observed} - n\Pr\{\text{data in class}\})^2}{n\Pr\{\text{data in class}\}}.
$$

(5.14)

In each class, the number (#) of data values expected to occur, according to the fitted distribution, is simply the probability of occurrence in that class multiplied by the sample size, n. This number of expected occurrences need not be an integer value. If the fitted distribution is very close to the data distribution, the expected and observed counts will be very close for each class, and the squared differences in the numerator of Equation 5.14 will all be very small, yielding a small χ^2. If the fit is not good, at least a few of the classes will exhibit large discrepancies. These will be squared in the numerator of Equation 5.14 and lead to large values of χ^2. It is not necessary for the classes to be of equal width or equal probability, but classes with small numbers of expected counts should be avoided. Sometimes a minimum of five expected events per class is imposed.

Under the null hypothesis that the data were drawn from the fitted distribution, the sampling distribution for the test statistic is the χ^2 distribution with parameter $v = (\#$ of classes $- \#$ of parameters fit $- 1)$ degrees of freedom. The test will be one-sided, because the test statistic is confined to positive values by the squaring process in the numerator of Equation 5.14, and small values of the test statistic support H_0. Right-tail quantiles for the χ^2 distribution are given in Table B.3.

Example 5.4. Comparing Gaussian and Gamma Distribution Fits Using the χ^2 Test

Consider the gamma and Gaussian distributions as candidates for representing the 1933–82 Ithaca January precipitation data in Table A.2. The approximate maximum likelihood estimators for the gamma distribution parameters (Equations 4.48 or 4.50a, and Equation 4.49) are $\alpha = 3.76$ and $\beta = 0.52$ in. The sample mean and standard deviation (i.e., the Gaussian parameter estimates) for these data are 1.96 in. and 1.12 in., respectively. The two fitted distributions are illustrated in relation to the data in Figure 4.16. Table 5.3 contains the information necessary to conduct the χ^2 tests for these two distributions. The precipitation amounts have been divided into six classes, or bins, the limits of which are indicated in the first

TABLE 5.3 The χ^2 Goodness-of-Fit Test Applied to Gamma and Gaussian Distributions for the 1933–82 Ithaca January Precipitation Data

Class	<1"	1–1.5"	1.5–2"	2–2.5"	2.5–3"	>3"
Observed #	5	16	10	7	7	5
Gamma:						
Probability	0.161	0.215	0.210	0.161	0.108	0.145
Expected #	8.05	10.75	10.50	8.05	5.40	7.25
Gaussian:						
Probability	0.195	0.146	0.173	0.178	0.132	0.176
Expected #	9.75	7.30	8.65	8.90	6.60	8.80

Expected numbers of occurrences in each bin are obtained by multiplying the respective probabilities by $n = 50$.

row of the table. The second row indicates the number of years in which the January precipitation total was within each class. Both distributions have been integrated over these classes to obtain probabilities for precipitation in each class. These probabilities were then multiplied by $n = 50$ to obtain the expected number of counts.

Applying Equation 5.14 yields $\chi^2 = 5.05$ for the gamma distribution and $\chi^2 = 14.96$ for the Gaussian distribution. As was also evident from the graphical comparison in Figure 4.16, these test statistics indicate that the Gaussian distribution fits these precipitation data substantially less well. Under the respective null hypotheses, these two test statistics are drawn from a χ^2 distribution with degrees of freedom $v = 6 - 2 - 1 = 3$; because Table 5.3 contains six classes, and two parameters (α and β, or μ and σ, for the gamma or Gaussian, respectively) were fit for each distribution.

Referring to the $v = 3$ row of Table B.3, $\chi^2 = 5.05$ is smaller than the 90th percentile value of 6.251, so the null hypothesis that the data have been drawn from the fitted gamma distribution would not be rejected even at the 10% level. For the Gaussian fit, $\chi^2 = 14.96$ is between the tabulated values of 11.345 for the 99th percentile and 16.266 for the 99.9th percentile, so this null hypothesis would be rejected at the 1% level, but not at the 0.1% level. ◇

Kolmogorov–Smirnov and Lilliefors Tests

The one-sample *Kolmogorov–Smirnov (K–S) test* is another very frequently used test of the goodness of fit. The χ^2 test essentially compares the histogram and the PDF or discrete distribution function, whereas the K–S test compares the empirical and fitted CDFs. Again, the null hypothesis is that the observed data were drawn from the distribution being tested, and a sufficiently large discrepancy will result in the null hypothesis being rejected. For continuous distributions the K–S test usually will be more powerful than the χ^2 test, and so usually will be preferred.

In its original form, the K–S test is applicable to any distributional form (including but not limited to any of the distributions presented in Chapter 4), provided that the parameters have *not* been estimated from the data sample being compared. In practice this provision can be a serious limitation to the use of the original K–S test, since it is often the correspondence between a fitted distribution and the particular batch of data used to estimate its parameters that is of interest. This may seem like a trivial problem, but it can have serious consequences, as has been pointed out by Crutcher (1975) and Steinskog et al. (2007). Estimating the parameters from the same batch of data used to test the goodness of fit results in the fitted distribution parameters being "tuned" to the data sample. When erroneously using K–S critical values that assume independence between the test data and the estimated parameters, it will often be the case that the null hypothesis (that the distribution fits well) will not be rejected when in fact it should be.

With modification, the K–S framework can be used in situations where the distribution parameters have been fit to the same data used in the test. In this situation, the K–S test is often called the *Lilliefors test*, after the statistician who did much of the early work on the subject (Lilliefors, 1967). Both the original K–S test and the Lilliefors test use the test statistic

$$D_n = \max_x |F_n(x) - F(x)|, \tag{5.15}$$

where $F_n(x)$ is the empirical cumulative probability, estimated as $F_n(x_{(i)}) = i/n$ for the ith smallest data value; and $F(x)$ is the theoretical cumulative distribution function evaluated at x (Equation 4.19). Thus the K–S test statistic D_n looks for the largest difference, in absolute value, between the empirical and fitted cumulative distribution functions. Any real and finite batch of data will exhibit sampling fluctuations resulting in a nonzero value for D_n, even if the null hypothesis is true and the theoretical distribution fits very well. If D_n is sufficiently large, the null hypothesis can be rejected. How large is large enough depends on the level of the test, of course; but also on the sample size, whether or not the distribution parameters have been fit using the test data, and if so also on the particular distribution form being fit.

When the parametric distribution to be tested has been specified completely externally to the data—the data have not been used in any way to fit the parameters—the original K–S test is appropriate. This test is distribution free, in the sense that its critical values are applicable to any distribution. These critical values can be obtained to good approximation (Stephens, 1974) using

$$C_\alpha = \frac{K_\alpha}{\sqrt{n} + 0.12 + 0.11/\sqrt{n}}, \tag{5.16}$$

where $K_\alpha = 1.224$, 1.358, and 1.628, for $\alpha = 0.10$, 0.05, and 0.01, respectively. The null hypothesis is rejected for $D_n \geq C_\alpha$. Alternatively, for n larger than perhaps 50, these critical levels can be obtained as continuous functions of α, using

$$C_\alpha = \sqrt{-\frac{\ln(\alpha/2)}{2n}}, \tag{5.17}$$

which in turn can be inverted to calculate p values for particular D_n.

Usually the original K–S test (and therefore Equations 5.16 and 5.17) is not appropriate because the parameters of the distribution being tested have been fit using the test data. But even in this case bounds

TABLE 5.4 Critical Values for the K–S Statistic D_n Used in the Lilliefors Test to Assess Goodness of Fit of Gamma Distributions, as a Function of the Estimated Shape Parameter, α, When the Distribution Parameters have been Fit Using the Data to be Tested

	20% level			10% level			5% level			1% level		
α	$n=25$	$n=30$	Large n	$n=25$	$n=30$	Large n	$n=25$	$n=30$	Large n	$n=25$	$n=30$	Large n
1	0.165	0.152	$0.84/\sqrt{n}$	0.185	0.169	$0.95/\sqrt{n}$	0.204	0.184	$1.05/\sqrt{n}$	0.241	0.214	$1.20/\sqrt{n}$
2	0.159	0.146	$0.81/\sqrt{n}$	0.176	0.161	$0.91/\sqrt{n}$	0.190	0.175	$0.97/\sqrt{n}$	0.222	0.203	$1.16/\sqrt{n}$
3	0.148	0.136	$0.77/\sqrt{n}$	0.166	0.151	$0.86/\sqrt{n}$	0.180	0.165	$0.94/\sqrt{n}$	0.214	0.191	$1.08/\sqrt{n}$
4	0.146	0.134	$0.75/\sqrt{n}$	0.164	0.148	$0.83/\sqrt{n}$	0.178	0.163	$0.91/\sqrt{n}$	0.209	0.191	$1.06/\sqrt{n}$
8	0.143	0.131	$0.74/\sqrt{n}$	0.159	0.146	$0.81/\sqrt{n}$	0.173	0.161	$0.89/\sqrt{n}$	0.203	0.187	$1.04/\sqrt{n}$
∞	0.142	0.131	$0.736/\sqrt{n}$	0.158	0.144	$0.805/\sqrt{n}$	0.173	0.161	$0.886/\sqrt{n}$	0.200	0.187	$1.031/\sqrt{n}$

The row labeled $\alpha = \infty$ pertains to the Gaussian distribution with parameters estimated from the data.
From Crutcher (1975). © American Meteorological Society. Used with permission.

on the true CDF, whatever its form, can be computed and displayed graphically using $F_n(x) \pm C_\alpha$ as limits covering the actual cumulative probabilities, with probability $1 - \alpha$. Values of C_α can also be used in an analogous way to calculate probability bounds on empirical quantiles consistent with a particular theoretical distribution (Loucks et al., 1981). Because the D_n statistic is a maximum over the entire data set, these bounds are valid jointly, for the entire distribution.

When the distribution parameters have been fit using the data at hand, Equation 5.16 is not sufficiently stringent, because the fitted distribution "knows" too much about the data to which it is being compared, and the Lilliefors test is appropriate. Here, however, the critical values of D_n depend on the distribution that has been fit. Table 5.4, from Crutcher (1975), lists critical values of D_n (above which the null hypothesis would be rejected) for four test levels, for the gamma distribution. These critical values depend on both the sample size and the estimated shape parameter, α. Larger samples will be less subject to irregular sampling variations, so the tabulated critical values decline for larger n. That is, smaller maximum deviations from the fitted distribution (Equation 5.15) are tolerated for larger sample sizes. Critical values in the last row of the table, for $\alpha = \infty$, pertain to the Gaussian distribution, since as the gamma shape parameter becomes very large the gamma distribution converges toward the Gaussian.

It is interesting to note that critical values for Lilliefors tests are usually derived through statistical simulation (see Section 4.7). The procedure is that a large number of samples from a known distribution are generated, and estimates of the distribution parameters are calculated from each of these samples. The agreement, for each synthetic data batch, between data generated from the known distribution and the distribution fit to it is then assessed using Equation 5.15. Since the null hypothesis is true in this protocol by construction, the α-level critical value is approximated as the $(1 - \alpha)$ quantile of that collection of synthetic D_n's. Thus Lilliefors-test critical values for any distribution that may be of interest can be computed using the methods described in Section 4.7. These methods can also be adapted to represent data having known forms and magnitude of autocorrelation.

Example 5.5. Comparing Gaussian and Gamma Fits Using the K–S Test

Again consider the fits of the gamma and Gaussian distributions to the 1933–82 Ithaca January precip-itation data, from Table A.2, shown in Figure 4.16. Figure 5.5 illustrates the Lilliefors tests for these two fitted distributions. In each panel of Figure 5.5, the black dots are the empirical cumulative probability estimates, $F_n(x)$, and the smooth curves are the fitted theoretical CDFs, $F(x)$, both plotted as functions of the observed monthly precipitation. Coincidentally, the maximum differences between the empirical and fitted theoretical cumulative distribution functions occur at the same (highlighted) point, yielding $D_n = 0.068$ for the gamma distribution (a) and $D_n = 0.131$ for the Gaussian distribution (b).

In each of the two tests to be conducted the null hypothesis is that the precipitation data were drawn from the fitted distribution, and the alternative hypothesis is that they were not. These will necessarily be one-sided tests, because the test statistic D_n is the absolute value of the largest difference between the parametric and empirical cumulative probabilities. Therefore values of the test statistic on the far right tail of the null distribution will indicate large discrepancies that are unfavorable to H_0, whereas values of the test statistic on the left tail of the null distribution will indicate $D_n \approx 0$, or near-perfect fits that are very supportive of the null hypothesis.

The critical values in Table 5.4 are the minimum D_n necessary to reject H_0. That is, they are leftmost bounds of the relevant rejection or critical regions. The sample size of $n = 50$ is sufficient to evaluate the tests using critical values from the large-n columns. In the case of the Gaussian distribution, the relevant row of the table is for $\alpha = \infty$. Since $0.886/\sqrt{50} = 0.125$ and $1.031/\sqrt{50} = 0.146$ bound the observed $D_n = 0.131$, the null hypothesis that the precipitation data were drawn from this Gaussian distribution would be rejected at the 5% level, but not the 1% level. For the fitted gamma distribution the nearest row in Table 5.4 is for $\alpha = 4$, where even at the 20% level the critical value of $0.75/\sqrt{50} = 0.106$ is substantially larger than the observed $D_n = 0.068$. Thus these data are quite consistent with the proposition of their having been drawn from this gamma distribution.

Regardless of the distribution from which these data were drawn, it is possible to use Equation 5.16 to calculate confidence intervals on its CDF. Using $K_\alpha = 1.358$, the *gray dots* in Figure 5.5 show the 95% confidence intervals for $n = 50$ as $F_n(x) \pm 0.188$. The intervals defined by these points cover the true CDF with 95% confidence, throughout the range of the data, because the K–S statistic pertains to the largest difference between $F_n(x)$ and $F(x)$, regardless of where in the distribution that maximum discrepancy may occur for a particular sample.   ◇

The related two-sample K–S test, or *Smirnov test*, compares two batches of data to one another under the null hypothesis that they were drawn from the same (but unspecified) distribution or generating process. The Smirnov test statistic,

$$D_S = \max_x |F_n(x_1) - F_n(x_2)|, \tag{5.18}$$

looks for the largest (in absolute value) difference between the empirical cumulative distribution func-tions of samples of n_1 observations of x_1 and n_2 observations of x_2. Unequal sample sizes can be accom-modated by the Smirnov test because the empirical CDFs are step functions (e.g., Figure 3.11), so that this maximum can occur at any of the values of x_1 or x_2. Again, the test is one-sided because of the absolute values in Equation 5.18, and the null hypothesis that the two data samples were drawn from the same distribution is rejected at the $\alpha \cdot 100\%$ level if

$$D_S > \left[-\frac{1}{2} \left(\frac{1}{n_1} + \frac{1}{n_2} \right) \ln \left(\frac{\alpha}{2} \right) \right]^{1/2}. \tag{5.19}$$

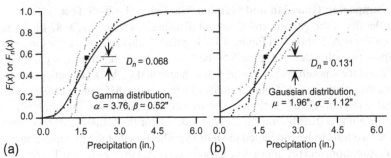

FIGURE 5.5 Illustration of the Kolmogorov–Smirnov D_n statistic in Lilliefors tests applied to the 1933–82 Ithaca January precipitation data fitted to fitted (a) gamma and (b) Gaussian distributions. *Solid curves* indicate cumulative distribution functions, and *black dots* show the corresponding empirical estimates. The maximum difference between the empirical and theoretical CDFs occurs for the *highlighted square point*, and is substantially greater for the Gaussian distribution. *Gray dots* show limits of the 95% confidence interval for the true CDF from which the data were drawn (Equation 5.16).

Filliben Q–Q Test

A good test for Gaussian distribution is often needed, for example, when the multivariate Gaussian distribution (see Chapter 12) will be used to represent the joint variations of (possibly power-transformed, Section 3.4.1) multiple variables. The Lilliefors test (Table 5.4, with $\alpha = \infty$) is an improvement in terms of power over the chi-square test for this purpose, but tests that are generally better (D'Agostino, 1986; Razali and Wah, 2011) can be constructed on the basis of the correlation between the empirical quantiles (i.e., the data), and the Gaussian quantile function based on their ranks. This approach was introduced by Shapiro and Wilk (1965), and both their original test formulation and its subsequent variants are known as *Shapiro–Wilk tests*. A computationally simple variant that is nearly as powerful as the original Shapiro–Wilk formulation was proposed by Filliben (1975). The test statistic is simply the correlation (Equation 3.32) between the empirical quantiles $x_{(i)}$ and the Gaussian quantile function $\Phi^{-1}(p_i)$, with p_i estimated using a plotting position (see Table 3.2) approximating the median cumulative probability for the ith order statistic (e.g., the Tukey plotting position, although Filliben (1975) used Equation 3.19 with $a = 0.3175$). That is, the test statistic is simply the correlation computed from the points on a Gaussian Q–Q plot. If the data are drawn from a Gaussian distribution these points should fall on a straight line, apart from sampling variations.

Table 5.5 presents critical values for the *Filliben test* for Gaussian distribution. The test is one-tailed, because high correlations are favorable to the null hypothesis that the data are Gaussian, so the null hypothesis is rejected if the correlation is smaller than the appropriate critical value. Because the points on a Q–Q plot are necessarily nondecreasing, the critical values in Table 5.5 are much larger than would be appropriate for testing the significance of the linear association between two independent (according to a null hypothesis) variables. Notice that, since the correlation will not change if the data are first standardized (Equation 3.27), this test does not depend in any way on the accuracy with which the distribution parameters may have been estimated. That is, the test addresses the question of whether the data were drawn from a Gaussian distribution but does not address, and is not confounded by, the question of what the parameters of that distribution might be.

TABLE 5.5 Critical Values for the Filliben (1975) Test for Gaussian Distribution, Based on the Q–Q Plot Correlation

n	0.5% Level	1% Level	5% Level	10% Level
10	0.860	0.876	0.917	0.934
20	0.912	0.925	0.950	0.960
30	0.938	0.947	0.964	0.970
40	0.949	0.958	0.972	0.977
50	0.959	0.965	0.977	0.981
60	0.965	0.970	0.980	0.983
70	0.969	0.974	0.982	0.985
80	0.973	0.976	0.984	0.987
90	0.976	0.978	0.985	0.988
100	0.9787	0.9812	0.9870	0.9893
200	0.9888	0.9902	0.9930	0.9942
300	0.9924	0.9935	0.9952	0.9960
500	0.9954	0.9958	0.9970	0.9975
1000	0.9973	0.9976	0.9982	0.9985

H_0 is rejected if the correlation is smaller than the appropriate critical value.

Example 5.6. Filliben Q–Q Correlation Test for Gaussian Distribution

The Q–Q plots in Figure 4.17 showed that the Gaussian distribution fits the 1933–82 Ithaca January precipitation data in Table A.2 less well than the gamma distribution. That Gaussian Q–Q plot is reproduced in Figure 5.6 (x's), with the horizontal axis scaled to correspond to standard Gaussian quantiles, z, rather than to dimensional precipitation amounts. Using the Tukey plotting position (see Table 3.2), estimated cumulative probabilities corresponding to (for example) the smallest and largest of these $n = 50$ precipitation amounts are $0.67/50.33 = 0.013$ and $49.67/50.33 = 0.987$. Standard Gaussian quantiles, z, corresponding to these cumulative probabilities (see Table B.1) are ± 2.22. The correlation for these $n = 50$ untransformed points is $r = 0.917$, which is smaller than all of the critical values in that row of Table 5.5. Accordingly, the Filliben test would reject the null hypothesis that these data were drawn from a Gaussian distribution, at the 0.5% level. The fact that the horizontal scale is the nondimensional z rather than dimensional precipitation (as in Figure 4.17) is immaterial, because the correlation is unaffected by linear transformations of either or both of the two variables being correlated.

Figure 3.14, in Example 3.4, indicated that a logarithmic transformation of these data was effective in producing approximate symmetry. Whether this transformation is also effective at producing a plausibly Gaussian shape for these data can be addressed with the Filliben test. Figure 5.6 also shows the standard Gaussian Q–Q plot for the log-transformed Ithaca January precipitation totals (o's). This relationship

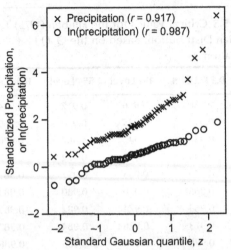

FIGURE 5.6 Standard Gaussian Q–Q plots for the 1933–82 Ithaca January precipitation in Table A.2 (x's), and for the log-transformed data (o's). Using Table 5.5, null hypotheses that these data were drawn from Gaussian distributions would be rejected for the original data ($p < 0.005$), but not rejected for the log-transformed data ($p > 0.10$).

is substantially more linear than for the untransformed data and is characterized by a correlation of $r = 0.987$. Again looking on the $n = 50$ row of Table 5.5, this correlation is larger than the 10% critical value, so the null hypothesis of Gaussian distribution would not be rejected.

Notice that Example 5.4 found that these data were also not inconsistent with a fitted gamma distribution. The goodness-of-fit tests cannot tell us whether these data were drawn from a gamma distribution, a lognormal distribution, or some other distribution that is similar to both the gamma and lognormal distributions that have been fit to these data. This ambiguity illustrates the situation that logically weaker statistical inferences result from nonrejection of null hypotheses.   ◇

Using statistical simulation (see Section 4.7), tables of critical Q–Q correlations can be obtained for other distributions, by generating large numbers of batches of size n from the distribution of interest, computing Q–Q plot correlations for each of these batches, and defining the critical value as that delineating the extreme $\alpha \cdot 100\%$ smallest of them. Results of this approach have been tabulated for the Gumbel distribution (Vogel, 1986), the uniform distribution (Vogel and Kroll, 1989), the GEV distribution (Chowdhury et al., 1991), and the Pearson III distribution (Vogel and McMartin, 1991). Heo et al. (2008) present regression-based critical values for Q–Q correlation tests pertaining to Gaussian, Gumbel, gamma, GEV, and Weibull distributions. The underlying simulations can also be adapted to represent data having known forms and magnitudes of autocorrelation.

5.2.6. Likelihood Ratio Tests

Sometimes we need to construct a test in a parametric setting, but the hypothesis is sufficiently complex that the simple, familiar parametric tests cannot be brought to bear. A flexible alternative, known as the *likelihood ratio test*, can be used if two conditions are satisfied. First, it must be possible to cast the problem in such a way that the null hypothesis pertains to some number, k_0 of free (i.e., fitted) parameters, and the alternative hypothesis pertains to some larger number, $k_A > k_0$, of parameters. Second, it must be possible to regard the k_0 parameters of the null hypothesis as a special case of the full parameter set of k_A parameters. That is, the null hypothesis is "nested" within the alternatives. Examples of this second condition on H_0 could include forcing some of the k_A parameters to have fixed values (often zero), or imposing equality between two or more of them. As the name implies, the likelihood ratio test compares the likelihoods associated with H_0 vs. H_A, when the k_0 and k_A parameters, respectively, have been fit using the method of maximum likelihood (see Section 4.6).

Even if the null hypothesis is true, the likelihood associated with H_A will always be at least as large as that for H_0. This is because the greater number of parameters $k_A > k_0$ allows the maximized likelihood function for the former greater freedom in accommodating the observed data. The null hypothesis is therefore rejected only if the likelihood associated with the alternative is sufficiently large that the difference is unlikely to have resulted from sampling variations.

The test statistic for the likelihood ratio test is

$$\Lambda^* = 2 \ln \left[\frac{\Lambda(H_A)}{\Lambda(H_0)} \right] = 2 \left[L(H_A) - L(H_0) \right]. \tag{5.20}$$

This quantity is also known as the *deviance*. Here $\Lambda(H_0)$ and $\Lambda(H_A)$ are the likelihood functions (see Section 4.6) associated with the null and alternative hypothesis, respectively. The second equality, involving the difference of the log-likelihoods $L(H_0) = \ln[\Lambda(H_0)]$ and $L(H_A) = \ln[\Lambda(H_A)]$, is used in practice since it is generally the log-likelihoods that are maximized (and thus computed) when fitting the parameters.

Under H_0, and given a large sample size, the sampling distribution of the statistic in Equation 5.20 is χ^2, with degrees of freedom $v = k_A - k_0$. That is, the degrees-of-freedom parameter is given by the difference between H_A and H_0 in the number of empirically estimated parameters. Since small values of Λ^* are not unfavorable to H_0, the test is one-sided and H_0 is rejected only if the observed Λ^* is in a sufficiently improbable region on the right tail.

Example 5.7. Testing for Climate Nonstationarity Using the Likelihood Ratio Test

Suppose there is a reason to suspect that the first 25 years (1933–57) of the Ithaca January precipitation data in Table A.2 have been drawn from a different gamma distribution than the second half (1958–82). This question can be tested against the null hypothesis that all 50 precipitation totals were drawn from the same gamma distribution using a likelihood ratio test. To perform the test it is necessary to fit gamma distributions separately to the two halves of the data, and compare these two distributions with the single gamma distribution fit using the full data set.

The relevant information is presented in Table 5.6, which indicates some differences between the two 25-year periods. For example, the average January precipitation ($=\alpha\beta$) for 1933–57 was 1.87 in., and the

TABLE 5.6 Gamma Distribution Parameters (MLEs) and Log-Likelihoods for Fits to the First and Second Halves of the 1933–82 Ithaca January Precipitation Data in Table A.2, and to the Full Data Set

	Dates	α	β	$\Sigma_i L(\alpha, \beta; x_i)$
H_A	1933–57	4.525	0.4128	−30.2796
	1958–82	3.271	0.6277	−35.8965
H_0	1933–82	3.764	0.5209	−66.7426

corresponding average for 1958–82 was 2.05 in. The year-to-year variability ($=\alpha\beta^2$) of January precipitation was greater in the second half of the period as well. Whether the extra two parameters required to represent the January precipitation using two gamma distributions rather than one are justified by the data can be evaluated using the test statistic in Equation 5.20. For this specific problem the test statistic is

$$\Lambda^* = 2\left\{\left[\sum_{i=1933}^{1957} L(\alpha_1, \beta_1; x_i)\right] + \left[\sum_{i=1958}^{1982} L(\alpha_2, \beta_2; x_i)\right] - \left[\sum_{i=1933}^{1982} L(\alpha_0, \beta_0; x_i)\right]\right\}, \qquad (5.21)$$

where the subscripts 1, 2, and 0 on the parameters refer to the first half, the second half, and the full period (null hypothesis), respectively, and the log-likelihood for the gamma distribution given a single observation, x_i, is (compare Equation 4.45)

$$L(\alpha, \beta; x_i) = (\alpha - 1)\ln(x_i/\beta) - x_i/\beta - \ln(\beta) - \ln[\Gamma(\alpha)]. \qquad (5.22)$$

The three terms in square brackets in Equation 5.21 are given in the last column of Table 5.6.

Using the information in Table 5.6, $\Lambda^* = 2(-30.2796 - 35.8965 + 66.7426) = 1.130$. Since there are $k_A = 4$ parameters under H_A ($\alpha_1, \beta_1, \alpha_2, \beta_2$) and $k_0 = 2$ parameters under H_0 (α_0, β_0), the null distribution is the χ^2 distribution with $v = 2$. Looking on the $v = 2$ row of Table B.3, we find $\chi^2 = 1.130$ is smaller than the median value, leading to the conclusion that the observed Λ^* is quite ordinary in the context of the null hypothesis that the two data records were drawn from the same gamma distribution, which would not be rejected. More precisely, recall that the χ^2 distribution with $v = 2$ is itself a gamma distribution with $\alpha = 1$ and $\beta = 2$, which in turn is the exponential distribution with $\beta = 2$. The exponential distribution has the closed-form CDF in Equation 4.53, which yields the right-tail probability (p value) $1 - F(1.130) = 0.5684$.   ◇

Example 5.8. Likelihood Ratio Tests comparing Simpler Versus More Elaborate Fitted Distributions

The annual maximum daily precipitation amounts in Table 4.6 can be represented by a GEV distribution (Equation 4.63) with parameters $\zeta = 3.49$, $\beta = 1.18$, and $\kappa = -0.32$, for which the minimized log-likelihood is $L_{GEV} = -30.273$. Since the fitted shape parameter κ is relatively small in absolute value it may be of interest to investigate whether the three-parameter GEV is justified by these data, in preference to fitting a two-parameter Gumbel distribution (Equation 4.66), effectively forcing $\kappa = 0$. Maximum likelihood estimates for the two Gumbel distribution parameters for these data are $\zeta = 3.33$ and $\beta = 1.05$, which yield the log-likelihood $L_{Gumbel} = -31.519$. Of course $L_{GEV} > L_{Gumbel}$ since the GEV has an additional free parameter, but whether this additional complexity is supported by the data can be investigated using a likelihood ratio test because the Gumbel distribution is a special case of the GEV. Here the null hypothesis is that the Gumbel distribution is adequate, and the alternative hypothesis is that these data support use of the GEV. The resulting test statistic (Equation 5.20) is therefore $\Lambda^* = 2(-30.273 + 31.519) = 2.49$. If the null hypothesis is true then this statistic follows the χ^2 distribution with $v = 3 - 2 = 1$ degree of freedom. Consulting the $v = 1$ row of Table B.3 we find that 2.49 is a bit smaller than the 90th percentile of this distribution, so that the right-tail p value is larger than 0.10. It appears that additional complexity of the GEV is not justified by these data, since the null hypothesis that they are Gumbel distributed has not been rejected.

Figure 4.14 shows a two-component Gaussian mixture distribution (Equation 4.75) representing the Guayaquil temperature data in Table A.3. The final line of Table 4.8 shows the five parameters that have been estimated using maximum likelihood, together with the maximized log-likelihood $L_{mix} = -20.48$. Might these data be represented adequately by a single Gaussian distribution? The mean and standard deviation (Equation 4.84a and b) of the 20 temperature values in Table A.3 are 24.76°C and 0.93°C, respectively, corresponding to the maximized log-likelihood (sum of 20 terms of the form of Equation 4.82) $L_{Gauss} = -27.47$. Since this latter distribution is a special case of the two-component Gaussian mixture, a likelihood ratio test can be computed in order to gauge how well the data justify the more elaborate model. The null hypothesis is that the single Gaussian distribution is adequate, and the alternative is that the five-parameter Gaussian mixture is supported by the data. The test statistic is $\Lambda^* = 2$ $(-20.48 + 27.47) = 13.98$. The appropriate χ^2 null distribution has $v = 5 - 2 = 3$ degrees of freedom, and the $v = 3$ row of Table B.3 shows that the null hypothesis is rejected, with $0.001 < p < 0.01$, supporting the use of the mixture model. ◇

5.3. NONPARAMETRIC TESTS

Not all formal hypothesis tests rest on assumptions involving specific parametric distributions for the data or for the sampling distributions of the test statistics. Tests not requiring such assumptions are called *nonparametric* or *distribution free*. Nonparametric methods are appropriate if either or both of the following conditions apply:

1. We know or suspect that the parametric assumption(s) required for a particular test are not met, for example, grossly non-Gaussian data in conjunction with the *t*-test for the difference of means in Equation 5.5.
2. A test statistic that is suggested or dictated by the scientific problem at hand is a complicated function of the data, and its sampling distribution is unknown and/or cannot be derived analytically.

The same hypothesis testing ideas apply to both parametric and nonparametric tests. In particular, the five elements of the hypothesis test presented at the beginning of this chapter apply also to nonparametric tests. The difference between parametric and nonparametric tests is in the way the null distribution is obtained in Step 4.

There are two branches of nonparametric testing. The first, called *classical nonparametric testing* in the following, consists of tests based on mathematical analysis of selected hypothesis test settings. These are older methods, devised before the advent of cheap and widely available computing. They employ analytic mathematical results (formulas) that are applicable to data drawn from any distribution. Only a few classical nonparametric tests will be presented here, although the range of classical nonparametric methods is much more extensive (e.g., Conover, 1999; Daniel, 1990; Sprent and Smeeton, 2001).

The second branch of nonparametric testing includes procedures collectively called *resampling tests*. A resampling test builds up a discrete approximation to the null distribution using a computer, by repeatedly operating on (resampling) the data set at hand. Since the null distribution is arrived at empirically, the analyst is free to use virtually any computable test statistic that may be relevant, regardless of how mathematically complicated it may be.

5.3.1. Classical Nonparametric Tests for Location

Two classical nonparametric tests for the difference in location between two data samples are especially common and useful. These are the *Wilcoxon–Mann–Whitney*, or *rank-sum test* for two independent samples (analogous to the parametric test in Equation 5.8) and the *Wilcoxon signed-rank test* for paired samples (corresponding to the parametric test in Equation 5.11).

Rank-Sum Test (Unpaired Data)

The Wilcoxon–Mann–Whitney rank-sum test was devised independently in the 1940s by Wilcoxon, and by Mann and Whitney, although in different forms. The notations from both forms of the test are commonly used, and this can be the source of some confusion. However, the fundamental idea behind the test is not difficult to understand. The test is resistant, in the sense that a few wild data values that would completely invalidate the *t*-test of Equation 5.8 will have little or no influence. It is robust in the sense that, even if all the assumptions required for the *t*-test in Equation 5.8 are met, the rank-sum test is almost as good (i.e., nearly as powerful). However, unlike the *t*-test, it is not invertible in a way that can yield a confidence interval computation.

Given two samples of independent (i.e., both serially independent and unpaired) data, the aim is to test for a possible difference in location. It is often erroneously stated that the null hypothesis for this test pertains to the difference between the two sample medians. However, the actual null hypothesis is that the two data samples have been derived from the same distribution, so that a random draw from the population underlying the first sample is equally likely to be larger or smaller than a counterpart from the second (Devine et al., 2018). Both one-sided (the center of one sample is expected in advance to be larger or smaller than the other if the null hypothesis is not true) and two-sided (no prior information on which sample should be larger) alternative hypotheses are possible. Importantly, the effect of serial correlation on the Wilcoxon–Mann–Whitney test is qualitatively similar to its effect on the *t*-test: the variance of the sampling distribution of the test statistic is inflated by serial correlation in the data, possibly leading to unwarranted rejection of H_0 if the problem is ignored (Yue and Wang, 2002). The same effect occurs in other classical nonparametric tests as well (Von Storch, 1995).

Under the null hypothesis that the two data samples are from the same distribution, the labeling of each data value as belonging to one group or the other is entirely arbitrary. That is, if the two data samples have really been drawn from the same population, each observation is as likely as the next to have been placed in one sample or the other by the process that generated the data. Under the null hypothesis, then, there are not n_1 observations in Sample 1 and n_2 observations in Sample 2, but rather $n = n_1 + n_2$ observations making up a single empirical distribution. The notion that the data labels are arbitrary because all the data have all been drawn from the same distribution under H_0 is known as the principle of *exchangeability*, which also underlies permutation tests, as discussed in Section 5.3.4.

The rank-sum test statistic is a function not of the data values themselves, but of their ranks within the n observations that are pooled under the null hypothesis. It is this feature that makes the underlying distribution(s) of the data irrelevant. Define R_1 as the sum of the ranks held by the members of Sample 1 in this pooled distribution, and R_2 as the sum of the ranks held by the members of Sample 2. Since there are n members of the pooled empirical distribution implied by the null distribution, $R_1 + R_2 = 1 + 2 + 3 + 4 + \ldots + n = (n)(n+1)/2$. Therefore the mean of this pooled distribution of ranks is $(n+1)/2$, and its variance is the variance of n consecutive integers $= n(n+1)/12$. If the two samples really have been drawn from the same distribution (i.e., if H_0 is true), then R_1 and R_2 will be similar in magnitude if $n_1 = n_2$. Regardless of whether or not the sample sizes are equal, however, R_1/n_1 and R_2/n_2 should be similar in magnitude if the null hypothesis is true.

The null distribution for R_1 and R_2 is obtained in a way that exemplifies the approach of nonparametric tests more generally. If the null hypothesis is true, the observed partitioning of the data into two groups of size n_1 and n_2 is only one of very many equally likely ways in which the n values could have been split and labeled. Specifically, there are $(n!)/[(n_1!)(n_2!)]$ such equally likely partitions of the data under the null hypothesis. For example, if $n_1 = n_2 = 10$, this number of possible distinct pairs of samples is 184,756. Conceptually, imagine the statistics R_1 and R_2 being computed for each of these 184,756 possible arrangements of the data. It is simply this very large collection of (R_1, R_2) pairs, or, more specifically, the collection of 184,756 scalar test statistics computed from these pairs, that constitutes the null distribution. If the observed test statistic characterizing the closeness of R_1 and R_2 falls comfortably near the middle this large empirical distribution, then that particular partition of the n observations is quite consistent with H_0. If, however, the observed R_1 and R_2 are more different from each other than under most of the other possible partitions of the data, H_0 would be rejected.

It is not actually necessary to compute the test statistic for all $(n!)/[(n_1!)(n_2!)]$ possible arrangements of the data. Rather, the Mann–Whitney U-statistic,

$$U_1 = R_1 - \frac{n_1}{2}(n_1 + 1) \tag{5.23a}$$

or

$$U_2 = R_2 - \frac{n_2}{2}(n_2 + 1), \tag{5.23b}$$

is computed for one or the other of the two Wilcoxon rank-sum statistics, R_1 or R_2. Both U_1 and U_2 carry the same information, since $U_1 + U_2 = n_1 n_2$, although some tables of null distribution probabilities for the rank-sum test evaluate unusualness of only the smaller of U_1 and U_2.

A little thought shows that the rank-sum test is a test for location in a way that is analogous to the conventional t-test. The t-test sums the data and equalizes the effects of different sample sizes by dividing by the sample size. The rank-sum test operates on sums of the ranks of the data, and the effects of possible differences in the sample sizes n_1 and n_2 are equalized using the Mann–Whitney transformation in Equation 5.23. This comparison is developed more fully by Conover and Iman (1981).

For even moderately large values of n_1 and n_2 (both larger than about 10), a simple method for evaluating null distribution probabilities is available. In this case, the null distribution of the Mann–Whitney U-statistic is approximately Gaussian, with

$$\mu_U = \frac{n_1 n_2}{2} \tag{5.24a}$$

and

$$\sigma_U = \left[\frac{n_1 n_2 (n_1 + n_2 + 1)}{12} \right]^{1/2}. \tag{5.24b}$$

Equation 5.24b is valid if all n data values are distinct and is approximately correct when there are few repeated values. If there are many tied values, Equation 5.24b overestimates the sampling variance, and a more accurate estimate is provided by

$$\sigma_U = \left[\frac{n_1 n_2 (n_1 + n_2 + 1)}{12} - \frac{n_1 n_2}{12(n_1 + n_2)(n_1 + n_2 - 1)} \sum_{j=1}^{J} \left(t_j^3 - t_j \right) \right]^{1/2}, \tag{5.25}$$

where J indicates the number of groups of tied values, and t_j indicates the number of members in group j.

For samples too small for application of the Gaussian approximation to the sampling distribution of U, tables of critical values (e.g., Conover, 1999), or an exact recurrence relation (Mann and Whitney, 1947; Mason and Graham, 2002) can be used.

Example 5.9. Evaluation of a Cloud Seeding Experiment Using the Wilcoxon–Mann–Whitney Test

Table 5.7 contains data from a weather modification experiment investigating the effect of cloud seeding on lightning strikes (Baughman et al., 1976). It was suspected in advance that seeding the storms would reduce lightning. The experimental procedure involved randomly seeding or not seeding candidate thunderstorms, and recording a number of characteristics of the lightning, including the counts of strikes presented in Table 5.7. There were $n_1 = 12$ seeded storms, exhibiting an average of 19.25 cloud-to-ground lightning strikes; and $n_2 = 11$ unseeded storms, with an average of 69.45 strikes.

Inspecting the data in Table 5.7, it is apparent that the distribution of lightning counts for the unseeded storms is distinctly non-Gaussian. In particular, the set contains one very large outlier of 358 strikes. We suspect, therefore, that uncritical application of the t-test (Equation 5.8) to test the significance of the difference in the observed mean numbers of lightning strikes could produce misleading results. This is because the single very large value of 358 strikes leads to a sample standard deviation for the unseeded storms of 98.93 strikes, which is larger even than the mean number. This large sample standard deviation would lead us to attribute a very large spread to the assumed t-distributed sampling distribution of the difference of means, so that even rather large values of the test statistic would be judged to be fairly ordinary.

TABLE 5.7 Counts of Cloud-to-Ground Lightning for Experimentally Seeded and Unseeded Storms

Seeded		Unseeded	
Date	Lightning Strikes	Date	Lightning Strikes
7/20/65	49	7/2/65	61
7/21/65	4	7/4/65	33
7/29/65	18	7/4/65	62
8/27/65	26	7/8/65	45
7/6/66	29	8/19/65	0
7/14/66	9	8/19/65	30
7/14/66	16	7/12/66	82
7/14/66	12	8/4/66	10
7/15/66	2	9/7/66	20
7/15/66	22	9/12/66	358
8/29/66	10	7/3/67	63
8/29/66	34		

From Baughman et al. (1976). © American Meteorological Society. Used with permission.

The mechanics of applying the rank-sum test to the data in Table 5.7 are presented in Table 5.8. In the left-hand portion of the table, the 23 data points are pooled and ranked, consistent with the null hypothesis that all the data came from the same population, regardless of the labels S (for seeded) or N (for not seeded). There are two observations of 10 lightning strikes, and as is conventional each has been assigned the average rank $(5+6)/2=5.5$. In the right-hand portion of the table, the data are segregated according to their labels, and the sums of the ranks of the two groups are computed. It is clear from this portion of Table 5.8 that the smaller numbers of strikes tend to be associated with the seeded

TABLE 5.8 Illustration of the Procedure of the Rank-Sum Test Using the Cloud-to-Ground Lightning Data in Table 5.7

	Pooled Data			Segregated Data		
Strikes	Seeded	Rank	Seeded	Rank	Seeded	Rank
0	N	1			N	1
2	S	2	S	2		
4	S	3	S	3		
9	S	4	S	4		
10	N	5.5			N	5.5
10	S	5.5	S	5.5		
12	S	7	S	7		
16	S	8	S	8		
18	S	9	S	9		
20	N	10			N	10
22	S	11	S	11		
26	S	12	S	12		
29	S	13	S	13		
30	N	14			N	14
33	N	15			N	15
34	S	16	S	16		
45	N	17			N	17
49	S	18	S	18		
61	N	19			N	19
62	N	20			N	20
63	N	21			N	21
82	N	22			N	22
358	N	23			N	23
		Sums of ranks:	$R_1=108.5$		$R_2=167.5$	

In the left portion of this table, the $n_1+n_2=23$ counts of lightning strikes are pooled and ranked. In the right portion of the table, the observations are segregated according to their labels of seeded (S) or not seeded (N) and the sums of the ranks for the two categories (R_1 and R_2) are computed.

storms, and the larger numbers of strikes tend to be associated with the unseeded storms. These differences are reflected in the differences in the sums of the ranks: R_1 for the seeded storms is 108.5, and R_2 for the unseeded storms is 167.5. The null hypothesis that seeding does not affect the number of lightning strikes can be rejected if this difference between R_1 and R_2 is sufficiently unusual against the backdrop of all possible $(23!)/[(12!)(11!)] = 1,352,078$ distinct arrangements of these data under H_0.

The Mann–Whitney U-statistic, Equation 5.23, corresponding to the sum of the ranks of the seeded data, is $U_1 = 108.5 - (6)(12+1) = 30.5$. The null distribution of all 1,352,078 possible values of U_1 for this data is closely approximated by the Gaussian distribution having (Equation 5.24) $\mu_U = (12)(11)/2 = 66$ and $\sigma_U = [(12)(11)(12+11+1)/12]^{1/2} = 16.2$. Within this Gaussian distribution, the observed $U_1 = 30.5$ corresponds to a standard Gaussian $z = (30.5 - 66)/16.2 = -2.19$. Table B.1 shows the (one-tailed) p value associated with this z to be 0.014, indicating that approximately 1.4% of the 1,352,078 possible values of U_1 under H_0 are smaller than the observed U_1. Accordingly, H_0 usually would be rejected.   ◇

Signed-Rank Test (Paired Data)

There is also a classical nonparametric test, the *Wilcoxon signed-rank test*, analogous to the paired two-sample parametric test of Equation 5.11. As is the case for its parametric counterpart, the signed-rank test takes advantage of positive correlation between the members of data pairs in assessing possible differences in location. In common with the unpaired rank-sum test, the signed-rank test statistic is based on ranks rather than the numerical values of the data. Therefore this test also does not depend on the distribution of the underlying data and is resistant to outliers.

Denote the data pairs (x_i, y_i), for $i = 1, ..., n$. The signed-rank test is based on the set of n differences, Δ_i, between the n data pairs. If the null hypothesis is true, and the two data sets represent paired samples from the same population or generating process, roughly equally many of these differences will be positive and negative, and the overall magnitudes of the positive and negative differences should be comparable. The comparability of the positive and negative differences is assessed by ranking them in absolute value. That is the n differences Δ_i are transformed to the series of ranks,

$$T_i = \text{rank} |\Delta_i| = \text{rank} |x_i - y_i|. \tag{5.26}$$

Data pairs for which $|\Delta_i|$ are equal are assigned the average rank of the tied values of $|\Delta_i|$, and pairs for which $x_i = y_i$ (implying $\Delta_i = 0$) are not included in the subsequent calculations. Denote as n^* the number of pairs for which $x_i \neq y_i$.

If the null hypothesis is true, the labeling of a given data pair as (x_i, y_i) could just as well have been reversed, so that the ith data pair is been just as likely to have been labeled (y_i, x_i). Changing the ordering reverses the sign of Δ_i, but yields the same $|\Delta_i|$. The unique information in the pairings that actually were observed is captured by separately summing the ranks, T_i, corresponding to pairs having positive or negative values of Δ_i, denoting as T either the statistic

$$T^+ = \sum_{\Delta_i > 0} T_i \tag{5.27a}$$

or

$$T^- = \sum_{\Delta_i < 0} T_i, \tag{5.27b}$$

respectively. Tables of null distribution probabilities sometimes require choosing the smaller of Equations 5.27a and 5.27b. However, knowledge of one is sufficient for the other, since $T^+ + T^- = n^*(n^*+1)/2$.

The null distribution of T is arrived at conceptually by considering again that H_0 implies the labeling of one or the other of each datum in a pair as x_i or y_i is arbitrary. Therefore under the null hypothesis there are $2n^*$ equally likely arrangements of the $2n^*$ data values at hand, and the resulting $2n^*$ possible values of T constitute the relevant null distribution. As before, it is not necessary to compute all possible values of the test statistic, since for moderately large n^* (greater than about 20) the null distribution is approximately Gaussian, with parameters

$$\mu_T = \frac{n^*(n^*+1)}{4} \tag{5.28a}$$

and

$$\sigma_T = \left[\frac{n^*(n^*+1)(2n^*+1)}{24}\right]^{1/2}. \tag{5.28b}$$

For smaller samples, tables of critical values for T^+ (e.g., Conover, 1999) can be used. Under the null hypothesis, $T (=T^+ \text{ or } T^-)$ will be close to μ_T because the numbers and magnitudes of the ranks T_i will be comparable for the negative and positive differences Δ_i. If there is a substantial difference between the x and y values in location, most of the large ranks will correspond to either the negative or positive Δ_i's, implying that T will be either very large or very small.

Example 5.10. Comparing Thunderstorm Frequencies Using the Signed Rank Test

The procedure for the Wilcoxon signed-rank test is presented in Table 5.9. Here the paired data are counts of thunderstorms reported in the northeastern United States (x) and the Great Lakes states (y) for the $n = 21$ years 1885–1905. Since the two areas are relatively close geographically, we expect that large-scale flow conducive to thunderstorm formation in one of the regions would be generally conducive in the other region as well. It is thus not surprising that the reported thunderstorm counts in the two regions are substantially positively correlated.

For each year the difference in reported thunderstorm counts, Δ_i, is computed, and the absolute values of these differences are ranked. None of the $\Delta_i = 0$, so $n^* = n = 21$. Years having equal differences, in absolute value, are assigned the average rank (e.g., 1892, 1897, and 1901 have the eighth, ninth, and tenth smallest $|\Delta_i|$, and are all assigned the rank 9). The ranks for the years with positive and negative Δ_i, respectively, are added in the final two columns, yielding $T^+ = 78.5$ and $T^- = 152.5$.

If the null hypothesis that the reported thunderstorm frequencies in the two regions are equal is true, then labeling of counts in a particular year as being Northeastern or Great Lakes is arbitrary and thus so is the sign of each Δ_i. Consider, arbitrarily, the test statistic T as the sum of the ranks for the positive differences, $T^+ = 78.5$. Its unusualness in the context of H_0 is assessed in relation to the $2^{21} = 2,097,152$ values of T^+ that could result from all the possible permutations of the data under the null hypothesis. This null distribution is closely approximated by the Gaussian distribution having $\mu_T = (21)(22)/4 = 115.5$ and $\sigma_T = [(21)(22)(42+1)/24]^{1/2} = 28.77$. The p value for this test is then obtained by computing the standard Gaussian $z = (78.5 - 115.5)/28.77 = -1.29$. If there is no reason to expect one or the other of the two regions to have had more reported thunderstorms, the test is two-tailed (H_A is simply "not H_0"), so the p value is $\Pr\{z \leq -1.29\} + \Pr\{z > +1.29\} = 2\,\Pr\{z \leq -1.29\} = 0.197$. The null hypothesis would not be rejected in this case. Note that the same result would be obtained if the test statistic $T^- = 152.5$ had been chosen instead. $\diamond$

TABLE 5.9 Illustration of the Procedure of the Wilcoxon Signed-Rank Test Using Data for Counts of Thunderstorms Reported in the Northeastern United States (x) and the Great Lakes States (y) for the Period 1885–1905

Year	Paired Data		Differences		Segregated Ranks	
	x	y	Δ_i	Rank $|\Delta_i|$	$\Delta_i > 0$	$\Delta_i < 0$
1885	53	70	−17	20		20
1886	54	66	−12	17.5		17.5
1887	48	82	−34	21		21
1888	46	58	−12	17.5		17.5
1889	67	78	−11	16		16
1890	75	78	−3	4.5		4.5
1891	66	76	−10	14.5		14.5
1892	76	70	+6	9	9	
1893	63	73	−10	14.5		14.5
1894	67	59	+8	11.5	11.5	
1895	75	77	−2	2		2
1896	62	65	−3	4.5		4.5
1897	92	86	+6	9	9	
1898	78	81	−3	4.5		4.5
1899	92	96	−4	7		7
1900	74	73	+1	1	1	
1901	91	97	−6	9		9
1902	88	75	+13	19	19	
1903	100	92	+8	11.5	11.5	
1904	99	96	+3	4.5	4.5	
1905	107	98	+9	13	13	
				Sums of ranks:	$T^+ = 78.5$	$T^- = 152.5$

Analogously to the procedure of the rank-sum test (see Table 5.8), the absolute values of the annual differences, $|\Delta_i|$, are ranked and then segregated according to whether Δ_i is positive or negative. The sum of the ranks of the segregated data constitutes the test statistic.
From Brooks and Carruthers (1953).

5.3.2. Mann–Kendall Trend Test

Investigating a possible trend through time of the central tendency of a data series is of interest in the context of a changing underlying climate, among other settings. The usual parametric approach to this kind of question is through conventional least-squares regression analysis (Section 7.2) with a time index as the predictor, and the associated test for the null hypothesis is that a regression slope is zero. The regression slope itself is proportional to the correlation between the time-series variable and the time index.

The *Mann–Kendall trend test* is a popular nonparametric alternative for testing for the presence of a trend, or nonstationarity of the central tendency, of a time series. In a parallel to the alternative parametric regression approach, the Mann–Kendall test arises as a special case of Kendall's τ (Equation 3.35), reflecting a tendency for monotone association between two variables. Accordingly it can accommodate nonlinear as well as linear trends. In the context of examining the possibility of trend underlying a time series x_i, $i = 1, \ldots, n$, the time index i (e.g., the year of observation of each datum) is by definition monotonically increasing, which simplifies the calculations.

The test statistic for the Mann–Kendall trend test is

$$S = \sum_{i=1}^{n-1} \sum_{j=i+1}^{n} \mathrm{sgn}(x_j - x_i) = \sum_{i<j} \mathrm{sgn}(x_j - x_i), \tag{5.29a}$$

where

$$\mathrm{sgn}(\Delta x) = \begin{cases} +1, & \Delta x > 0 \\ 0, & \Delta x = 0 \,. \\ -1, & \Delta x < 0 \end{cases} \tag{5.29b}$$

That is, the statistic in Equation 5.29a counts the number of all possible data pairs in which the first value is smaller than the second, and subtracts the number of data pairs in which the first is larger than the second. If the data x_i are serially independent and drawn from the same distribution (in particular, if the generating process has the same mean throughout the time series), then the numbers of data pairs for which $\mathrm{sgn}(\Delta x)$ is positive and negative should be nearly equal.

For moderate (n about 10) or larger series lengths, the sampling distribution of the test statistic in Equation 5.29 is approximately Gaussian, and if the null hypothesis of no trend is true this Gaussian null distribution will have zero mean. The variance of this distribution depends on whether all the x's are distinct, or if some are repeated values. If there are no ties, the variance of the sampling distribution of S is

$$\mathrm{Var}(S) = \frac{n(n-1)(2n+5)}{18}, \tag{5.30a}$$

otherwise the variance is

$$\mathrm{Var}(S) = \frac{n(n-1)(2n+5) - \sum_{j=1}^{J} t_j (t_j - 1)(2t_j + 5)}{18}. \tag{5.30b}$$

Analogously to Equation 5.25, J indicates the number of groups of repeated values, and t_j is the number of repeated values in the jth group. The test p value is evaluated using the standard Gaussian value

$$z = \begin{cases} \dfrac{S-1}{[\mathrm{Var}(S)]^{1/2}}, & S > 0 \\[2ex] \dfrac{S+1}{[\mathrm{Var}(S)]^{1/2}}, & S < 0 \end{cases} \,. \tag{5.31}$$

The mean of S under the null hypothesis is zero; however, the ± 1 in the numerator of Equation 5.31 represents a continuity correction. Alternatively for small sample sizes, Hamed (2009) has suggested use of beta sampling distributions for S.

If all n data values in the series are distinct, the relationship of Equation 5.29a to the value of Kendall's τ characterizing the relationship between the x's and the time index i is

$$S = \binom{n}{2} \tau = \frac{n(n-1)}{2} \tau. \tag{5.32}$$

Example 5.11. Testing for Climate Change Using the Mann–Kendall Test

In Example 5.7, the possibility of a change in the distribution of Ithaca January precipitation between the 1933–57 and 1958–82 periods was examined using a likelihood ratio test. A similar question can be addressed by examining the data in Table A.2 against a null hypothesis of no trend, using the Mann–Kendall Test.

Of the $(50)(49)/2 = 1225$ distinct pairs of individual data values, the earlier of the two is smaller for 580 of the pairs, and it is larger in 638 pairs, yielding $S = -58$. There are $J = 5$ groups of repeated precipitation values in Table A.2, four of which consist of pairs ($t_j = 2$) and one of which is a triple ($t_j = 3$). The subtracted correction term in the numerator of Equation 5.30b is therefore 138, yielding Var$(S) = [(50)(49)(105) - 138]/18 = 14,284$. Using the lower option in Equation 5.31, $z = (-58 + 1)/(14,284)^{1/2} = -0.477$, which is quite ordinary in the context of the null distribution, and associated with a rather large p value. Neglecting the effect of the repeated values, the corresponding Kendall τ characterizing the association between the precipitation data and the year labels in Table A.2 would be, according to Equation 5.32, $\tau = -57/[(50)(49)/2] = -0.0465$, which also indicates a very weak degree of association.

This result is quite consistent with the likelihood ratio test in Example 5.7, which also provided only extremely weak evidence against a null hypothesis of no climate change. However, it is important to keep in mind that the tests do not and cannot prove that no changes are occurring. Because of the logical structure of the hypothesis testing paradigm one can only conclude that any changes in Ithaca January precipitation would be occurring too slowly over the 50 years of data to be discerned against the very considerable year-to-year background variability in precipitation. ◇

Although the Mann–Kendall test can reject a null hypothesis of no trend, it does not return an estimate of the magnitude of any trend so detected. An approach for doing so is provided by the nonparametric regression method of Sen (1968, see also Section 7.7.2), which estimates the linear trend slope as the median of slopes between all $n(n-1)/2$ distinct data pairs. This procedure is thus naturally aligned with Kendall's τ, and with the Mann–Kendall trend test.

When annual time series values are the subject of a trend analysis the Mann–Kendall assumption of serial independence will usually be well met, but other series may exhibit sufficiently strong serial dependence that the results are adversely affected. As is the case in hypothesis testing more generally, positive serial correlation in the data series leads to underestimation of the sampling variance, with the result that the statistic in Equation 5.31 will be too large in absolute value, yielding p values that are too small. Lettenmaier (1976) proposed use of an effective-sample-size correction for first-order autoregressive time series that is approximately equivalent to those in Equations 5.12 and 5.13 for the t-test. In particular, assuming first-order autoregressive structure (Section 10.3.1) for the time dependence, Var[S] would be replaced in Equation 5.31 with Var*[S] = Var[S]$(1 + r_1)/(1 - r_1)$. Yue and Wang (2004) obtained good results with this modification. Here r_1 is the lag-1 autocorrelation in the data apart from contributions to positive autocorrelation induced by any trend, and so is generally estimated after detrending the data series, often by subtracting a linear regression function (see Section 7.2.1). Alternatively, relaxing the assumption of first-order autoregressive time dependence, Equation 10.41 rather than Equation 10.40 could be used to estimate the variance inflation factor. Cabilio et al. (2013) use the block bootstrap (Section 5.3.5) to estimate the sampling variance of the Mann–Kendall statistic for autocorrelated series.

5.3.3. Introduction to Resampling Tests

Since the advent of inexpensive and fast computing, another approach to nonparametric testing has become practical. This approach is based on the construction of artificial data sets from a given collection of real data, by resampling the observations in a manner consistent with the null hypothesis. Sometimes such methods are also known as *resampling tests, randomization tests, rerandomization tests*, or *Monte Carlo tests*. Resampling methods are highly adaptable to different testing situations, and there is considerable scope for the analyst to creatively design new tests to meet particular needs.

The basic idea behind resampling tests is to build up a collection of artificial data batches of the same size as the actual data at hand using a procedure that is consistent with the null hypothesis, and then to compute the test statistic of interest for each artificial batch. The result is as many artificial values of the test statistic as there are artificially generated data batches. Taken together, these reference test statistics constitute an estimated null distribution against which to compare the test statistic computed from the original data.

As a practical matter, a computer is programmed to do the resampling. Fundamental to this process are the uniform [0,1] random number generators described in Section 4.7.1. These algorithms produce streams of numbers that resemble independent values drawn independently from the probability density function $f(u) = 1$, $0 \leq u \leq 1$. The synthetic uniform variates are used to draw random samples from the data to be tested.

In general, resampling tests have two very appealing advantages. The first is that no assumptions regarding underlying parametric distributions for the data or the sampling distribution for the test statistic are necessary, because the procedures consist entirely of operations on the data themselves. The second is that any statistic that may be suggested as important by the scientific context of the problem can form the basis of the test, so long as it can be computed from the data. For example, when investigating location (i.e., overall magnitudes) of a sample of data, we are not confined to the conventional tests involving the arithmetic mean or sums of ranks; because it is just as easy to use alternative measures such as the median, the geometric mean, or more exotic statistics if any of these are more meaningful to the problem at hand. The data being tested can be scalar (each data point is one number) or vector-valued (data points are composed of pairs, triples, etc.), as dictated by the structure of each particular problem. Resampling procedures involving vector-valued data can be especially useful when the effects of spatial correlation must be captured by a test, in which case each element in the data vector corresponds to a different location, so that each data vector can be thought of as a "map."

Any computable statistic (i.e., any function of the data) can be used as a test statistic in a resampling test, but not all will be equally good. In particular, some choices may yield tests that are more powerful than others. Good (2000) suggests the following desirable attributes for candidate test statistics.

1. *Sufficiency*: All the relevant information about the distribution attribute or physical phenomenon of interest contained in the data is also reflected in the chosen statistic. Given a sufficient statistic, the data have nothing additional to say about the question being addressed.
2. *Invariance*: A test statistic should be constructed in a way that the test result does not depend on arbitrary transformations of the data, for example, from °F to °C.
3. *Loss*: The mathematical penalty for discrepancies that is expressed by the test statistic should be consistent with the problem at hand, and the use to which the test result will be put. Often squared-error losses are assumed in parametric tests because of mathematical tractability and connections with the Gaussian distribution, although squared-error loss is disproportionately sensitive to large differences relative to an alternative like absolute error. In a resampling test there is no reason to avoid absolute-error loss or other loss functions if these make more sense in the context of a particular problem.

In addition, Hall and Wilson (1991) point out that better results are obtained when the resampled statistic does not depend on unknown quantities, for example, unknown parameters. Such statistics are called *pivotal*. Equation 5.8 is an example of a pivotal statistic, since the differencing eliminates its dependence on an unknown location parameter, and the division by the estimated standard deviation removes dependence on an unknown scale parameter.

5.3.4. Permutation Tests

Two- (or more-) sample problems can often be approached using *permutation tests*. Early examples of permutation tests in the atmospheric science literature were provided by Mielke Jr. et al. (1981) and Preisendorfer and Barnett (1983). The concept behind permutation tests is not new (Fisher, 1935; Pitman, 1937), but the approach did not become practical until the advent of fast and abundant computing. (Fisher (1935) noted "The arithmetical procedure of such an examination is tedious …").

Permutation tests are a natural generalization of the Wilcoxon–Mann–Whitney test described in Section 5.3.1, and also depend on the principle of exchangeability. Exchangeability implies that, according to the null hypothesis, all the data were drawn from the same distribution. Therefore the labels identifying particular data values as belonging to one sample or another are arbitrary. Under H_0 these data labels are exchangeable.

The key difference between permutation tests generally, and the Wilcoxon–Mann–Whitney test as a special case, is that any test statistic that may be meaningful can be employed, including but certainly not limited to the particular function of the ranks given in Equation 5.23. Among other advantages, lifting restrictions on the mathematical form of possible test statistics expands the range of applicability of permutation tests to vector-valued data. For example, Mielke Jr. et al. (1981) provide a simple illustrative example using two batches of bivariate data ($x = [x, y]$) and the Euclidian distance measure to examine the tendency of the two batches to cluster in the [x, y] plane. Zwiers (1987a) gives an example of a permutation test that uses higher-dimensional multivariate Gaussian variates.

The exchangeability principle leads logically to the construction of the null distribution using samples drawn by computer from a pool of the combined data. As was the case for the Wilcoxon–Mann–Whitney test, if two batches of size n_1 and n_2 are to be compared, the pooled set to be resampled contains $n = n_1 + n_2$ points. However, rather than computing the test statistic using all possible $n!/(n_1!n_2!)$ groupings (i.e., permutations) of the pooled data, the pool is merely sampled some large number (perhaps 10,000) of times.

(An exception can occur when n is small enough for a *full enumeration* of all possible permutations to be practical, and some authors reserve the term "permutation test" for this situation.) For permutation tests the samples are drawn *without replacement*, so that on a given iteration each of the individual n observations is represented once and once only in one or the other of the artificial samples of size n_1 and n_2. In effect, the data labels are randomly permuted for each resample. For each of these pairs of synthetic samples the test statistic is computed, and the resulting distribution of (perhaps 10,000) outcomes forms the null distribution against which the observed test statistic can be compared.

An efficient permutation algorithm can be implemented in the following way. Assume for convenience that $n_1 \geq n_2$. The data values (or vectors) are first arranged into a single array of size $n = n_1 + n_2$. Initialize a reference index $m = n$. The algorithm proceeds by implementing the following steps n_2 times:

- Randomly choose x_i, $i = 1, \ldots, m$; using Equation 4.100 (i.e., randomly draw from the first m array positions).
- Exchange the array positions of (or, equivalently, the indices pointing to) x_i and x_m (i.e., each of the chosen x's will be placed in the bottom section of the n-dimensional array).
- Decrement the reference index by 1 (i.e., $m = m - 1$).

At the end of this process there will be a random selection of the n pooled observations in the first n_1 positions, which can be treated as Sample 1. The remaining n_2 data values at the end of the array can be treated as Sample 2. The scrambled array can be operated upon directly for subsequent random permutations—it is not necessary first to restore the data to their original ordering.

Example 5.12. Two-Sample Permutation Test for a Complicated Statistic

Consider again the lightning data in Table 5.7. Assume that their dispersion is best (from the standpoint of some criterion external to the hypothesis test) characterized by the *L-scale* statistic (Hosking, 1990),

$$\lambda_2 = \frac{2}{n(n-1)} \sum_{i=1}^{n-1} \sum_{j=i+1}^{n} |x_i - x_j|. \tag{5.33}$$

Equation 5.33 amounts to half the average difference, in absolute value, between all possible pairs of points in the sample of size n. For a tightly clustered sample of data each term in the sum will be small, and therefore λ_2 will be small. For a data sample that is highly variable, some of the terms in Equation 5.33 will be very large, and λ_2 will be correspondingly large.

To compare sample values of λ_2 from the seeded and unseeded storms in Table 5.7, we probably would use either the ratio or the difference of λ_2 for the two samples. A resampling test procedure provides the freedom to choose the one (or some other) making more sense for the problem at hand. Suppose the most relevant test statistic is the ratio $[\lambda_2(\text{seeded})]/[\lambda_2(\text{unseeded})]$. Under the null hypothesis that the two samples have the same L-scale, this statistic should be near one. If the seeded storms are more variable with respect to the numbers of lightning strikes, the ratio statistic should be greater than one. If the seeded storms are less variable, the statistic should be less than one. The ratio of L-scales has been chosen for this example arbitrarily, to illustrate that any computable function of the data can be used as the basis of a permutation test, regardless of how unusual or complicated it may be.

The null distribution of the test statistic is built up by sampling some (say 10,000) of the $23!/(12! \ 11!) = 1,352,078$ distinct partitions, or permutations, of the $n = 23$ data points into two batches of size $n_1 = 12$ and $n_2 = 11$. For each partition, λ_2 is computed according to Equation 5.33 for each of the two

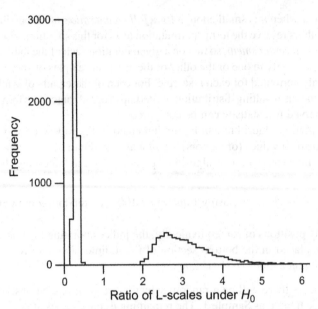

FIGURE 5.7   Histogram for the null distribution of the ratio of the L-scales for lightning counts of seeded versus unseeded storms in Table 5.7. The observed ratio of 0.188 is smaller than all but 49 of the 10,000 permutation realizations of the ratio, which provides very strong evidence that the lightning production by seeded storms was less variable than by unseeded storms. This null distribution is bimodal because the one outlier (353 strikes on 9/12/66) produces a very large L-scale in whichever of the two partitions it has been randomly assigned.

synthetic samples, and their ratio (with the value for the $n_1 = 12$ batch in the numerator) is computed and stored. The observed value of the ratio of the L-scales, 0.188, is then evaluated with respect to this empirically generated null distribution.

Figure 5.7 shows a histogram for the null distribution of the ratios of the L-scales constructed from 10,000 permutations of the original data. The observed value of 0.188 is smaller than all except 49 of these 10,000 values, which would lead to the null hypothesis being soundly rejected. Depending on whether a one-sided or two-sided test would be appropriate on the basis of prior external information, the p value would be either 0.0049 or 0.0098, respectively. Notice that this null distribution has the unusual feature of being bimodal, having two humps. This characteristic results from the large outlier in Table 5.7, 358 lightning strikes on 9/12/66, producing a very large L-scale in whichever partition it has been assigned. Partitions for which this observation has been assigned to the unseeded (n_2) group are in the left hump, and those for which the outlier has been assigned to the seeded (n_1) group are in the right hump.

The conventional test for differences in dispersion involves the ratio of sample variances, the null distribution for which would be the F distribution if the two underlying data samples are both Gaussian, but the F test is not robust to violations of the Gaussian assumption. Computing a permutation test on the basis the variance ratio $s^2(\text{seeded})/s^2(\text{unseeded})$ would be as easy if not easier than computing the L-scale ratio permutation test, and in that case the permutation null distribution would also be bimodal (and probably exhibit larger variance because the sample variance is not resistant to outliers). It is likely that the results of such a test would be similar to those for the permutation test based on the L-scale ratio. However, the corresponding parametric test, which would examine the observed $s^2(\text{seeded})/s^2(\text{unseeded}) = 0.0189$ in relation to the F distribution rather than the corresponding resampling distribution would likely be misleading, since the F distribution would not resemble the counterpart of Figure 5.7 for resampled variance ratios. $\diamond$

Permutation tests can also be applied in paired-data situations, analogously to the parametric paired t-test described in Section 5.2.3, even though the pairing induces reduction to a one-sample test. Indeed, Fisher's (1935) original description of the permutation method was applied in exactly this setting. Again consistent with the null hypothesis that the labels for each pair as belonging to one group or the other are arbitrary, the permutations are achieved by randomly (with probability 1/2) switching those labels for each pair. After each randomization of labels has been implemented, the test statistic of interest (which may be, but is not limited to, the difference of means) is computed. The procedure is repeated a large number of times, and the resulting collection of randomized test statistics provides the needed null distribution.

5.3.5. The Bootstrap

Permutation schemes are very useful in multiple-sample settings where the exchangeability principle applies. In one-sample settings permutation procedures are useless because there is nothing to permute: there is only one way to resample a single data batch with replacement, and that is to replicate the original sample by choosing each of the original n data values exactly once. When the exchangeability assumption cannot be supported, the justification for pooling multiple samples before permutation disappears, because the null hypothesis no longer implies that all data, regardless of their labels, were drawn from the same population.

In either of these situations an alternative computer-intensive resampling procedure called the *bootstrap* is available. The bootstrap is a newer idea than permutation, dating from Efron (1979). The idea behind the bootstrap is known as the *plug-in principle*, under which we estimate any function of the underlying (population) distribution by using (plugging into) the same function, but using the empirical distribution, which puts probability $1/n$ on each of the n observed data values. Put another way, the idea behind the bootstrap is to treat a finite sample at hand as similarly as possible to the unknown distribution from which it was drawn. This perspective leads to resampling *with replacement*, since an observation of a particular value from an underlying distribution does not preclude subsequent observation of an equal data value. In general the bootstrap is less accurate than the permutation approach when permutation is appropriate, but can be used in instances where permutation cannot. Fuller exposition of the bootstrap than is possible here can be found in Efron and Gong (1983), Efron and Tibshirani (1993), and Leger et al. (1992), among others. Some examples of its use in climatology are given in Downton and Katz (1993) and Mason and Mimmack (1992).

Resampling with replacement is the primary distinction in terms of mechanics between the bootstrap and the permutation approach, where the resampling is done without replacement. Conceptually, the resampling process is equivalent to writing each of the n data values on separate slips of paper and putting all n slips of paper in a hat. To construct one bootstrap sample, n slips of paper are drawn from the hat and their data values recorded, but each slip is put back in the hat and mixed (this is the meaning of "with replacement") before the next slip is drawn. Generally some of the original data values will be drawn into a given bootstrap sample multiple times, and others will not be drawn at all. On average, the fraction of the original n values that are omitted from a bootstrap sample is $(1 - 1/n)^n$, which for large

n is about $1/e \approx 0.368$. If n is small enough, all possible distinct bootstrap samples can be fully enumerated. However, the number of possible bootstrap samples increases rapidly with n, according to

$$\#\text{possible bootstrap samples} = \binom{2n-1}{n}. \tag{5.34}$$

For example, for $n = 5$, Equation 5.34 yields only 126 possible bootstrap samples, but the number increases to 92,378 for $n = 10$, and to nearly 69×10^9 for $n = 20$.

In practice, we usually program a computer to perform the resampling, using Equation 4.100 in conjunction with a uniform random number generator (Section 4.7.1). This process is repeated a large number, perhaps $n_B = 10,000$ times, yielding n_B bootstrap samples, each of size n. The statistic of interest is computed for each of these n_B bootstrap samples. The resulting frequency distribution is then used to approximate the true sampling distribution of that statistic.

Example 5.13. One-Sample Bootstrap: Confidence Interval for a Complicated Statistic

The bootstrap is often used in one-sample settings to estimate confidence intervals around observed values of a test statistic. Because we do not need to know the analytical form of its sampling distribution, the procedure can be applied to any test statistic, regardless of how mathematically complicated it may be. To take a hypothetical yet not especially complicated example, consider the standard deviation of the logarithms, $s_{\ln(x)}$, of the 1933–82 Ithaca January precipitation data in Table A.2 of Appendix A. This statistic has been chosen for this example arbitrarily, to illustrate that any computable sample statistic can be bootstrapped. Here scalar data are used, but Efron and Gong (1983) illustrate the bootstrap using vector-valued (paired) data, for which a confidence interval around the sample Pearson correlation coefficient was estimated.

The value of $s_{\ln(x)}$ computed from the $n = 50$ data values is 0.537, but in order to make inferences about the true value, we need to know or estimate its sampling distribution. Figure 5.8 shows a histogram of the sample standard deviations computed from $n_B = 10,000$ bootstrap samples of size $n = 50$ from the logarithms of this data set. This empirical distribution approximates the sampling distribution of $s_{\ln(x)}$ for these data.

Confidence regions for $s_{\ln(x)}$ are most easily approached using the straightforward and intuitive *percentile method* (Efron and Gong, 1983; Efron and Tibshirani, 1993). To form a $(1 - \alpha) \cdot 100\%$ confidence interval using this approach, we simply find the values of the estimates defining largest and smallest $n_B \cdot \alpha/2$ of the n_B bootstrap estimates. These values also define the central $n_B \cdot (1 - \alpha)$ of the estimates, which is the region of interest. In Figure 5.8, for example, the estimated 95% confidence interval for $s_{\ln(x)}$ using the percentile method is between 0.410 and 0.648.   ◇

The previous example illustrates use of the bootstrap in a one-sample setting where permutations are not possible. Bootstrapping is also applicable in multiple-sample situations where the data labels are not exchangeable, so that pooling and permutation of data is not consistent with the null hypothesis. These kinds of data can still be resampled with replacement using the bootstrap, while maintaining the separation of samples having meaningfully different labels.

To illustrate, consider investigating differences of means using the test statistic in Equation 5.5. Depending on the nature of the underlying data and the available sample sizes, we might not trust the Gaussian approximation to the sampling distribution of this statistic, in which case an attractive alternative would be to approximate it through resampling. If the data labels were exchangeable, it would be natural to compute a pooled estimate of the variance and use Equation 5.9 as the test statistic, estimating its sampling distribution through a permutation procedure because both the means and variances would

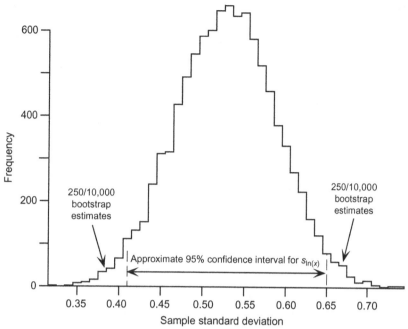

FIGURE 5.8 Histogram of $n_B = 10,000$ bootstrap estimates of the standard deviation of the logarithms of the 1933–82 Ithaca January precipitation data in Table A.2. The sample standard deviation computed directly from the data is 0.537. The 95% confidence interval for the statistic, as estimated using the percentile method, is also shown.

be equal under the null hypothesis. On the other hand, if the null hypothesis did not include equality of the variances, Equation 5.8 would be the correct test statistic, but it would not be appropriate to estimate its sampling distribution through permutation, because in this case the data labels would be meaningful, even under H_0. However, the two samples could be separately resampled with replacement (i.e., bootstrapped individually) to build up a bootstrap approximation to the sampling distribution of Equation 5.8. We would need to be careful in generating the bootstrap distribution for Equation 5.8 to construct the bootstrapped quantities consistent with the null hypothesis of equality of means. In particular, we could not bootstrap the raw data directly, because they have different means (whereas the two population means are equal according to the null hypothesis). One option would be to center each of the data batches at the overall mean (which would equal the estimate of the common, pooled mean, according to the plug-in principle). A more straightforward approach would be to estimate the sampling distribution of the test statistic directly, and then exploit the duality between hypothesis tests and confidence intervals to address the null hypothesis. This second approach is illustrated in the following example.

Example 5.14. Two-Sample Bootstrap Test for a Complicated Statistic

Consider again the situation in Example 5.12, in which we were interested in the ratio of L-scales (Equation 5.33) for the numbers of lightning strikes in seeded vs. unseeded storms in Table 5.7. The permutation test in Example 5.12 was based on the assumption that, under the null hypothesis, *all* aspects of the distribution of lightning strikes were the same for the seeded and unseeded storms. But pooling and permutation would not be appropriate if we wish to allow for the possibility that, even if the L-spread does not depend on seeding, other aspects of the distributions (e.g., the median numbers of lightning strikes) may be different.

Less restrictive null hypotheses can be accommodated by separately and repeatedly bootstrapping the $n_1 = 12$ seeded and $n_2 = 11$ unseeded lightning counts, and forming $n_B = 10,000$ samples of the ratio of one bootstrap realization of each, yielding bootstrap realizations of the test statistic λ_2(seeded)/ λ_2(unseeded). The result, shown in Figure 5.9 is a bootstrap estimate of the sampling distribution of this ratio for the data at hand. Its center is near the observed ratio of 0.188, which is the $q_{.4835}$ quantile of this bootstrap distribution. Even though this is not the bootstrap null distribution—which would be the sampling distribution if λ_2(seeded)/λ_2(unseeded) $= 1$—it can be used to evaluate the null hypothesis by examining the unusualness of λ_2(seeded)/λ_2(unseeded) $= 1$ with respect to this sampling distribution. The horizontal gray arrow indicates the 95% confidence interval for the L-scale ratio, estimated using the percentile method, which ranges from 0.08 to 0.75. Since this interval does not include 1, H_0 would be rejected at the 5% level (two-sided). The bootstrap L-scale ratios are >1 for only 33 of the $n_B = 10,000$ resamples, so the actual p value would be estimated as either 0.0033 (one-sided) or 0.0066 (two-sided), and so H_0 could be rejected at the 1% level as well. ◇

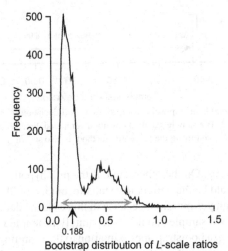

FIGURE 5.9 Bootstrap distribution for the ratio of L-scales for lightning strikes in seeded and unseeded storms, Table 5.7. The ratio is >1 for only 33 of 10,000 bootstrap samples, indicating that a null hypothesis of equal L-scales would be rejected. Also shown (gray arrows) is the 95% confidence interval for the ratio computed using the percentile method, which ranges from 0.08 to 0.75.

The percentile method is straightforward and easy to use, and gives generally good results in large-sample situations where the statistic being considered is unbiased and its sampling distribution is symmetrical or nearly so. For more moderate sample sizes and more general statistics, a better and more sophisticated method of bootstrap confidence interval construction is available, called bias-corrected and accelerated, or BC_a *intervals* (Efron, 1987; Efron and Tibshirani, 1993). BC_a intervals are more accurate than bootstrap confidence intervals based on the percentile method in the sense that the fraction of $(1 - \alpha) \cdot 100\%$ confidence intervals including the true value of the underlying statistic will be closer to $(1 - \alpha)$ for BC_a intervals.

In common with the percentile method, BC_a confidence intervals are based on quantiles of the bootstrap distribution. Denote the sample estimate of the statistic of interest, around which a confidence interval is to be constructed, as S. In Example 5.13, $S = s_{\ln(x)}$, the sample standard deviation of the log-transformed data. Denote the ith order statistic of the n_B bootstrap resamples of S as $S^*_{(i)}$. The percentile method estimates the lower and upper bounds of the $(1 - \alpha) \cdot 100\%$ confidence interval as $S^*_{(L)}$ and $S^*_{(U)}$, where $L = n_B \cdot \alpha_L = n_B \cdot \alpha/2$ and $U = n_B \cdot \alpha_U = n_B \cdot (1 - \alpha/2)$. BC_a confidence intervals are computed similarly, except that different quantiles of the bootstrap distribution are chosen, typically yielding $\alpha_L \neq \alpha/2$ and $\alpha_U \neq (1 - \alpha/2)$. Instead, the estimated confidence interval limits are based on

$$\alpha_L = \Phi\left[\hat{z}_0 + \frac{\hat{z}_0 + z(\alpha/2)}{1 - \hat{a}(\hat{z}_0 + z(\alpha/2))}\right] \tag{5.35a}$$

and

$$\alpha_U = \Phi\left[\hat{z}_0 + \frac{\hat{z}_0 + z(1 - \alpha/2)}{1 - \hat{a}(\hat{z}_0 + z(1 - \alpha/2))}\right]. \tag{5.35b}$$

Here $\Phi[\cdot]$ denotes the CDF of the standard Gaussian distribution, the parameter $\hat{z}_0$ is the bias correction, and the parameter $\hat{a}$ is the "acceleration." For $\hat{z}_0 = \hat{a} = 0$, Equations 5.35a and 5.35b reduce to the percentile method since, for example, $\Phi[z(\alpha/2)] = \alpha/2$.

The bias correction parameter $\hat{z}_0$ reflects median bias of the bootstrap distribution, or the difference between the estimated statistic S and the median of the bootstrap distribution, in units of Gaussian standard deviations. It is estimated using

$$\hat{z}_0 = \Phi^{-1}\left[\frac{\#\{S_i^* < S\}}{n_B}\right], \tag{5.36}$$

where the numerator inside the square brackets denotes the number of bootstrap estimates S_i^* that are smaller than the estimate computed using each of the n data values exactly once, S. Equation 5.36 is thus the normal quantile transform (Equation 4.57) of the relative frequency of bootstrap samples smaller than S. If exactly half of the S^* estimates are smaller than S, then the median bias is zero, because $\hat{z}_0 = \Phi^{-1}[1/2] = 0$.

The acceleration parameter $\hat{a}$ is conventionally computed using a statistic related to the *jackknife* estimate of the skewness of the sampling distribution of S. The jackknife (e.g., Efron and Hastie, 2016; Efron and Tibshirani, 1993) is a relatively early and therefore less computationally intensive resampling algorithm, in which the statistic S of interest is recomputed n times, each time omitting one of the n data values that were used to compute the original S. Denote the ith jackknife estimate of the statistic S, which has been computed after removing the ith data value, as S_{-i}; and denote the average of these n jackknife estimates as $\overline{S}_{jack} = (1/n)\Sigma_i S_{-i}$. The conventional estimate of the acceleration is then

$$\hat{a} = \frac{-\sum_{i=1}^{n}(S_{-i} - \overline{S}_{jack})^3}{6\left[\sum_{i=1}^{n}(S_{-i} - \overline{S}_{jack})^2\right]^{3/2}}. \tag{5.37}$$

Typical magnitudes for both $\hat{z}_0$ and $\hat{a}$ are on the order of $n^{-1/2}$ (Efron, 1987), so that as the sample size increases the BC_a and percentile methods yield increasingly similar results.

Example 5.15. A BC$_a$ Confidence Interval: Example 5.13 Revisited

In Example 5.13, a 95% confidence interval for the standard deviation of the log-transformed Ithaca January precipitation data from Table A.2 was computed using the straightforward percentile method. A 95% BC$_a$ confidence interval is expected to be more accurate, although it is more difficult to compute. The same $n_B = 10,000$ bootstrap samples of $S = s_{\ln(x)}$ are used in each case, but the difference will be that for the BC$_a$ confidence interval Equation 5.35 will be used to compute α_L and α_U, which will differ from $\alpha/2 = 0.025$ and $(1 - \alpha/2) = 0.975$, respectively.

The particular $n_B = 10,000$-member bootstrap distribution computed for these two examples contained 5552 bootstrap samples with $S_i^* < S = 0.537$. Using Equation 5.36, the bias correction is therefore estimated as $\hat{z}_0 = \Phi^{-1}[0.5552] = 0.14$. The acceleration in Equation 5.37 requires computation of the $n = 50$ jackknife values of the sample statistic, S_{-i}. The first three of these (i.e., standard deviations of the batches of 49 log-transformed precipitation omitting in turn the data for 1933, 1934, and 1935) are $S_{-1} = 0.505$, $S_{-2} = 0.514$, and $S_{-3} = 0.516$. The average of the 50 jackknifed values is 0.537, the sum of squared deviations of the jackknife values from this average is 0.004119, and the sum of cubed deviations is -8.285×10^{-5}. Substituting these values into Equation 5.37 yields $\hat{a} = (8.285 \times 10^{-5})/[6(0.004119)^{3/2}] = 0.052$. Using these values with $z(0.025) = -1.96$ in Equation 5.35a yields $\alpha_L = \Phi[-1.52] = 0.0643$, and similarly Equation 5.35b with $z(0.975) = +1.96$ yields $\alpha_U = \Phi[2.50] = 0.9938$.

The lower endpoint for the BC$_a$ estimate of the 95% confidence interval around $S = 0.537$ is thus the bootstrap quantile corresponding to $L = n_B \alpha_L = (10,000)(0.0643)$, or $S^*_{(643)} = 0.437$, and the upper endpoint is the bootstrap quantile corresponding to $U = n_B \alpha_U = (10,000)(0.9938)$, or $S^*_{(9938)} = 0.681$. This interval [0.437, 0.681] is slightly wider and shifted upward, relative to the interval [0.410, 0.648] computed in Example 5.13.   ◇

Parametric Bootstrap

Use of the bootstrap relies on having enough data for the underlying population or generating process to have been reasonably well sampled. A small sample may exhibit too little variability for bootstrap samples drawn from it to adequately represent the variability of the generating process that produced the data. For such relatively small data sets, an improvement over ordinary nonparametric bootstrapping may be provided by the *parametric bootstrap*, if a good parametric model can be identified for the data. The parametric bootstrap operates in the same way as the ordinary, nonparametric, bootstrap except that each of the n_B the bootstrap samples is a synthetic sample of size n that has been generated (Section 4.7) through random draws from a parametric distribution that has been fit to the size-n data sample. Kysely (2008) has compared the performance of parametric and nonparametric bootstrap confidence intervals in settings simulating extreme-value distributions for precipitation and temperature, and reports better results for the parametric bootstrap when $n \leq 40$.

Bootstrapping Autocorrelated Data

It is important to note that direct use of either bootstrap or permutation methods only makes sense when the underlying data to be resampled are independent. If the data are mutually correlated (exhibiting, e.g., time correlation or persistence) the results of these approaches will be misleading (Zwiers, 1987a, 1990), in the same way and for the same reason that autocorrelation adversely affects parametric tests. The random sampling used in either permutation or the bootstrap shuffles the original data, destroying the ordering that produces the autocorrelation.

One approach to respecting data correlation in bootstrapping, called *nearest-neighbor bootstrapping* (Lall and Sharma, 1996), accommodates serial correlation by resampling according to probabilities that depend on similarity to the previous few data points, rather the unvarying $1/n$ implied by the independence assumption. Essentially, the nearest-neighbor bootstrap resamples from relatively close analogs rather than from the full data set. The closeness of the analogs can be defined for both scalar and vector (multivariate) data.

The *sieve bootstrap* (Bickel and Bühlmann, 1999; Bühlmann, 1997; Choi and Hall, 2000) extends the idea of the parametric bootstrap to time-series data. In this method, the time dependence of a data series is modeled with a fitted autoregressive process (see Section 10.3), the complexity of which depends on the underlying data series. A large number n_B synthetic time series of length n are then generated (Section 10.3.7) from the fitted model, from which the desired inferences are then computed. Bühlmann (2002) concludes that the sieve bootstrap is the preferred method for resampling time-series data if it is known that the underlying data-generating process is a linear time-series process.

Time-series data are most commonly bootstrapped using the modification known as the *moving-blocks bootstrap* (Efron and Tibshirani, 1993; Künsch, 1989; Lahiri, 2003; Leger et al., 1992; Wilks, 1997b). Instead of resampling individual data values or data vectors, contiguous sequences of length L are resampled in order to build up a synthetic sample of size n. Figure 5.10 illustrates resampling a data series of length $n = 12$ by choosing $b = 3$ contiguous blocks of length $L = 4$, with replacement. The resampling works in the same way as the ordinary bootstrap, except that instead of resampling from a collection of n individual, independent values, the objects to be resampled with replacement are all the $n - L + 1$ contiguous subseries of length L.

The idea behind the moving-blocks bootstrap is to choose the blocklength L to be large enough for data values separated by this time period or more to be nearly independent. A good choice for the blocklength b is important to the success of the procedure, but a single best method for defining it has not emerged. Theoretical work (Bühlmann, 2002) suggests choosing $b \approx n^{1/3}$, although the optimality of this choice for the smaller samples encountered in practice is not clear. An alternative approach is to choose the blocklength from the middle of a range in which the results (e.g., a confidence interval width) change little, which is called the *minimum volatility method* (Politis et al., 1999). Intuitively, in addition to the blocklength increasing with increasing sample size n, it should also increase as the strength of dependence in the underlying time series increases, and should tend toward $b = 1$ (i.e., ordinary bootstrapping) as the time dependence progressively weakens (Braverman et al., 2011; Wilks, 1997b). If it can be assumed that the data follow a first-order autoregressive process (Equation 10.16), good results are achieved by choosing the blocklength according to the implicit equation (Wilks, 1997b)

$$L = (n - L + 1)^{(2/3)(1 - n'/n)}, \tag{5.38}$$

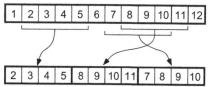

FIGURE 5.10 Schematic illustration of the moving-blocks bootstrap. Beginning with a time series of length $n = 12$ (above), $b = 3$ blocks of length $L = 4$ are drawn with replacement. The resulting time series (below) is one of $(n - L + 1)^b = 729$ equally likely bootstrap samples. *From Wilks (1997b).*

where n' is defined by Equation 5.12. Regardless of how L has been chosen, if n/L is not an integer, the number of blocks b is chosen as the next larger integer, but the final block in each bootstrap sample is truncated so each bootstrap series has length n.

One weakness of the block-bootstrap procedure just described is that data values contained in blocks originating near the beginning and end of the original data series will be undersampled. This undersampling can be avoided through use of the *circular block bootstrap* (Politis and Romano, 1992). Here the first $b-1$ values are appended to the end of the underlying data series, so that n blocks of length b can be resampled with equal probability.

5.4. MULTIPLICITY AND "FIELD SIGNIFICANCE"

Special problems occur when the results of multiple statistical tests must be evaluated simultaneously, which is known as the problem of test *multiplicity*. The multiplicity problem arises in many settings, but in meteorology and climatology it is most usually confronted when analyses involving atmospheric fields must be performed. Accordingly the multiplicity problem is sometimes conceptualized in terms of assessing *"field significance"* (Livezey and Chen, 1983). In this context the term atmospheric field often connotes a two-dimensional (horizontal) array of geographical locations at which data are available. It may be, for example, that two atmospheric models (one, perhaps, reflecting an increase of the atmospheric carbon dioxide concentration) both produce realizations of surface temperature at each of many gridpoints, and the question is whether the average temperatures portrayed by the two models are significantly different.

In principle, multivariate methods of the kind described in Section 12.5 would be preferred for this kind of problem, but often in practice the data are insufficient to implement them effectively if at all. Accordingly, statistical inference for these kinds of data is often approached by first conducting individual tests at each of the gridpoints, computing perhaps a collection of two-sample t-tests (Equation 5.8). If appropriate, a correction for serial correlation of the underlying data such as that in Equation 5.13 would be part of each of these local tests. Having conducted the local tests, however, it still remains to evaluate, collectively, the overall significance of the differences between the fields, or the field significance. This evaluation of overall significance is sometimes called determination of *global significance* or *pattern significance*. There are two major difficulties associated with this step. These derive from the problems of test multiplicity and from spatial correlation of the underlying data.

5.4.1. The Multiplicity Problem for Independent Tests

Consider first the problem of evaluating the collective significance of N independent hypothesis tests. If all their null hypotheses are true then the probability of falsely rejecting any one of them, picked at random, will be α. But we naturally tend to focus attention on the tests with the smallest p values, which would be a distinctly nonrandom sample of these N tests, and the bias deriving from this mental process needs to be accounted for in the analysis procedure.

The issue has been amusingly framed by Taleb (2001) in terms of the so-called *infinite monkeys theorem*. This is actually an allegory and not a theorem at all. If we could somehow put an infinite number of monkeys in front of keyboards and allow them to type random characters, it is virtually certain that one would eventually reproduce the *Iliad*. But it would not be reasonable to conclude that this particular monkey is special, in the sense, for example, that it would have a higher chance than any of the others of subsequently typing the *Odyssey*. Given the limitless number of monkeys typing, the fact that one has produced something recognizable does not provide strong evidence against a null hypothesis that this is just an ordinary monkey, whose future literary output will be as incoherent as that of any other monkey. In the realistic and less whimsical counterparts of this kind of setting, we must be careful to guard against *survivorship bias*, or focusing attention on the few instances or individuals surviving some test, and regarding them as typical or representative. That is, when we cherry-pick the (nominally) most significant results from a collection of tests, we must hold them to a higher standard (e.g., require smaller p values) than would be appropriate for any single test, or for a randomly chosen test from the same collection.

It has been conventional in the atmospheric sciences since publication of the Livezey and Chen (1983) paper to frame the multiple testing problem as a meta-test, where the data being tested are the results of N individual or "local" tests, and the "global" null hypothesis is that all the local null hypotheses are true. The Livezey–Chen approach was to compute the number of local tests exhibiting significant results, sometimes called the *counting norm* (Zwiers, 1987a), necessary to reject the global null hypothesis at a level α_{global}. Usually this global test level is chosen to be equal to the local test level, α. If there are $N = 20$ independent tests it might be naively supposed that, since 5% of 20 is 1, finding that any one of the 20 tests indicated a significant difference at the 5% level would be grounds for declaring the two fields to be significantly different, and that by extension, three significant tests out of 20 would be very strong evidence.

Although this reasoning sounds superficially plausible, because of survivorship bias it is really only even approximately true if there are very many, perhaps 1000, tests, and these tests are statistically independent (Livezey and Chen, 1983; Von Storch, 1982). Recall that declaring a significant difference at the 5% level means that, if the null hypothesis is true and there are really no significant differences, there is a probability no larger than 0.05 that evidence against H_0 as strong as or stronger than observed would have appeared by chance. For a single test conducted at the 5% level, the situation is analogous to rolling a 20-sided die (Figure 5.11), and observing that the side with the "20" on it has come up. However,

FIGURE 5.11   An icosahedron, or regular 20-sided geometrical solid, with distinct integer labels on each face. Participants in role-playing board games know this object as a "d20."

conducting $N = 20$ independent tests is probabilistically equivalent to rolling this die 20 times: there is a substantially higher chance than 5% that the side with "20" on it comes up at least once in 20 throws, and it is this latter situation that is analogous to the evaluation of the results from $N = 20$ independent hypothesis tests. If the die is rolled hundreds of times, it is virtually certain that the outcome that is rare for a single throw will occur many times.

Thinking about this analogy between multiple tests and multiple rolls of the 20-sided die suggests that we can quantitatively analyze the multiplicity problem for independent tests in the context of the binomial distribution and conduct a global hypothesis test based on the number of the N individual independent hypothesis tests that are nominally significant. Recall that the binomial distribution specifies probabilities for X successes out of N independent trials if the probability of success on any one trial is p. In the testing multiplicity context, X is the number of significant individual tests out of N tests conducted, and p is the level of the local tests.

Example 5.16. Illustration of the Livezey–Chen Approach for Independent Tests
In the hypothetical example just discussed, there are $N = 20$ independent tests, and $\alpha = 0.05$ is the level of each of these tests. Suppose the local tests pertain to inferences about means at N spatial locations, and $x = 3$ of the 20 tests have yielded significant differences. The question of whether the differences are (collectively) significant at the $N = 20$ gridpoints thus reduces to evaluating $\Pr\{X \geq 3\}$, given that the null distribution for the number of significant tests is binomial with $N = 20$ and $p = 0.05$. Using the binomial probability distribution function (Equation 4.1) with these two parameters, we find $\Pr\{X = 0\} = 0.358$, $\Pr\{X = 1\} = 0.377$, and $\Pr\{X = 2\} = 0.189$. Thus $\Pr\{X \geq 3\} = 1 - \Pr\{X < 3\} = 0.076$, and the null hypothesis that the two mean fields, as represented by the $N = 20$ gridpoints, are equal would not be rejected at the $\alpha_{global} \cdot 100\% = 5\%$ level. Since $\Pr\{X = 3\} = 0.060$, finding four or more significant local tests would result in a declaration of global (field) significance, at the (global) 5% level.

Even if there are no real differences, the chances of finding at least one significant test result out of 20 are almost 2 out of 3, since $\Pr\{X = 0\} = 0.358$. Until we are aware of and accustomed to the issue of test multiplicity, results such as these seem counterintuitive.   ◇

Livezey and Chen (1983) pointed out some instances in the literature of the atmospheric sciences where a lack of awareness of the multiplicity problem had led to conclusions that were not supported by the data. Wilks (2016b) found that this unfortunate situation has not improved in the intervening years, counting more than one-third of a representative sample of papers published in the *Journal of Climate* that ignored the problem of test multiplicity, while only 1% of these papers had addressed it (the remaining papers contained no multiple statistical inferences). Such results are usually displayed on maps that locate positions of nominally significant tests with black dots, leading to the visual impression of stippling over portions of the domain. By not addressing statistical test multiplicity and thereby producing unwarranted rejections of valid null hypotheses, this naive *stippling method* overstates research results even if the multiple tests are mutually independent (which is rarely the case). The result is that this naive procedure may lead researchers to overinterpret their data, and possibly to construct fanciful rationalizations for nonreproducible sampling fluctuations (Wilks, 2016b).

5.4.2. Field Significance and the False Discovery Rate

The Livezey–Chen approach to addressing test multiplicity by counting numbers of rejected null hypotheses is straightforward and attractive in its simplicity, but suffers from two important drawbacks.

The first is that the binary view of the local test results can reduce the global test power (i.e., can yield poor sensitivity for rejecting false null hypotheses). Local null hypotheses that are very strongly rejected (local p values that are very much smaller than α) carry no greater weight in the global test than do local tests for which the p values are only slightly smaller than α. That is, no credit is given for rejecting one or more local null hypotheses with near certainty when evaluating the plausibility of the global null hypothesis that all local null hypotheses are valid. The poor power of Livezey–Chen test is especially problematic if only a small fraction of the N local null hypotheses are not valid.

The second major problem is that the Livezey–Chen test is very sensitive to the effects of positive correlations among the data underlying the different tests, and therefore to positive correlation among the local test results. This situation occurs commonly when the local tests pertain to data at a collection of correlated spatial locations. The issue of spatially correlated tests will be taken up more fully in Section 5.4.3. The naive stippling approach shares this problematic attribute.

These shortcomings of the Livezey–Chen approach to addressing test multiplicity can in general be improved upon through the use of a global test statistic that depends on the magnitudes of the individual p values of the N local tests, rather than on simply counting the numbers of local tests having p values smaller than the chosen α. An attractive choice is to jointly analyze the results of the N multiple tests in a way that controls the *false discovery rate*, or FDR (Benjamini and Hochberg, 1995; Ventura et al., 2004), which is the expected fraction of nominally significant tests whose null hypotheses are actually valid. This terminology derives from the medical statistics literature, where rejection of an individual null hypothesis might correspond to a medical discovery, and survivorship bias in multiple testing is accounted for by controlling the maximum expected rate of erroneous null hypothesis rejection. Within the field significance paradigm, this ceiling on the FDR is numerically equal to the global test level, α_{global} (Wilks, 2006a).

The FDR approach to evaluating multiple hypothesis tests begins with the order statistics for the p values of the N tests, $p_{(1)}, p_{(2)}, p_{(3)}, \ldots, p_{(N)}$. The smallest (nominally most significant) of these is $p_{(1)}$, and the largest (least significant) is $p_{(N)}$. Results of individual tests are regarded as significant if the corresponding p value is no greater than

$$p_{\text{FDR}} = \max_{j=1,\cdots,N} \left\{ p_{(j)} : p_{(j)} \leq \frac{j}{N} \alpha_{\text{global}} \right\}. \tag{5.39}$$

That is, the sorted p values are evaluated with respect to a sliding scale, so that if the largest of them, $p_{(N)}$, is no greater than $\alpha_{\text{global}} = \text{FDR}$, then all N tests are regarded as statistically significant at that level. If $p_{(N)} > \alpha_{\text{global}}$ (i.e., if the largest p value does not lead to rejection of the corresponding null hypothesis) then survivorship bias is compensated by requiring $p_{(N-1)} \leq (N-1)\alpha_{\text{global}}/N$ in order for the second-least-significant test and all others with smaller p values to have their null hypotheses rejected. The general rule is that the null hypothesis for the test having the largest p value satisfying Equation 5.39 is rejected, as are the null hypotheses for all other tests with smaller p values. Survivorship bias is addressed by requiring a more stringent standard for declaring statistical significance as progressively smaller p values are considered in turn. If p values for none of the N tests satisfy Equation 5.39, then none are deemed to be statistically significant at the α_{global} level, and in effect the global null hypothesis that all N local tests have valid null hypotheses is not rejected at the α_{global} level. Equation 5.39 is valid regardless of the form of the hypothesis tests that produce the p values, provided that those tests operate at the correct level.

Example 5.17. Illustration of the FDR Approach to Multiple Testing
Consider again the hypothetical situation in Example 5.16, where $N = 20$ independent tests have been computed, of which 3 have p values smaller than $\alpha = 0.05$. The FDR approach accounts for how much smaller than $\alpha = 0.05$ each of these p values is, and thus provides greater test power than does the Livezey–Chen approach. Figure 5.12 plots the magnitudes of $N = 20$ hypothetical ranked p values of which, consistent with the calculations in Example 5.16, three are smaller than $\alpha = 0.05$ (*dotted horizontal line*). According to the calculations in Example 5.16, none of the corresponding hypothesis tests would be regarded as significant according to the Livezey–Chen approach, because $\Pr\{X \geq 3\} = 0.076 > 0.05 = \alpha_{global}$.

However, the Livezey–Chen approach does not consider how much smaller each of these p values is relative to $\alpha = 0.05$. As drawn in Figure 5.12, these smallest p values are $p_{(1)} = 0.001$, $p_{(2)} = 0.004$, and $p_{(3)} = 0.034$. Since all of the remaining p values are larger than $\alpha_{global} = 0.05$, none can satisfy Equation 5.39. Neither does $p_{(3)} = 0.034$ satisfy Equation 5.39, since $0.034 > (3/20)(0.05) = 0.0075$. However, $p_{(2)} = 0.004$ does satisfy Equation 5.39, so that the individual tests corresponding to both of the smallest two p values would be regarded as significant. Both would also be declared significant even if it were the case that $p_{(1)} > \alpha_{global}/N = 0.05/20 = 0.0025$. Both $p_{(1)}$ and $p_{(2)}$ are small enough that it is unlikely that they arose by chance from valid null hypotheses, even after accounting for the survivorship bias inherent in focusing on the most nominally significant of the N tests. Geometrically, the FDR approach rejects the null hypothesis for the largest ranked p value below the sloping dashed line in Figure 5.12, corresponding to $p_{FDR} = \alpha_{global}(\mathrm{rank}(p_{(j)})/N)$, and likewise for any other tests having smaller p values.

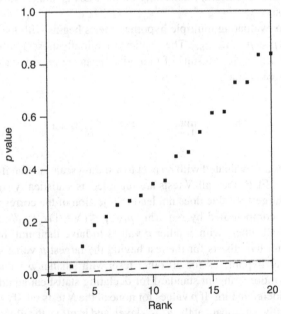

FIGURE 5.12   A hypothetical collection of 20 ranked p values, three of which are smaller than $\alpha = 0.05$ (*dotted horizontal line*). As detailed in Example 5.16, none would be considered statistically significant according to the Livezey–Chen procedure. Using the FDR approach, the largest p value below the sloping dashed line, and any other smaller p values, (in this case, the smallest two p values) would correspond to significant tests. All three would be declared significant according to the naive stippling procedure.

In contrast, the naive stippling approach ignores the effects of test multiplicity altogether, and would consider all three p values falling below the dotted horizontal line to be significant at the 5% level. ◇

5.4.3. Field Significance and Spatial Correlation

When a collection of multiple tests are performed using data from spatial fields, the positive spatial correlation of the underlying data induces statistical dependence among the local tests. Informally, we can imagine that positive correlation between data at two locations would result in the probability of a Type I error (falsely rejecting H_0) at one location being larger if a Type I error had occurred at the other location. This is because a hypothesis test statistic is a statistic like any other—a function of the data—and, to the extent that the underlying data are correlated, the statistics calculated from them will be also. Thus false rejections of the null hypothesis tend to cluster in space, possibly leading (if we are not careful) to the erroneous impression that a spatially coherent and physically meaningful feature may exist.

The binomial distribution underpinning the traditional Livezey–Chen procedure is very sensitive to positive correlation among the outcomes of the N tests, yielding too many spurious rejections of null hypotheses that are true. The same is true of the naive stippling approach, and indeed this is a usual response of conventional testing procedures to correlated data, as was illustrated in a different setting in Section 5.2.4. One approach suggested by Livezey and Chen (1983) was to hypothesize and estimate some number $N' < N$ of effectively independent gridpoint tests, as a spatial analog of Equation 5.12. A variety of approaches for estimating these "spatial degrees of freedom" have been proposed. Some of these have been reviewed by Van den Dool (2007), although even if this quantity can be estimated meaningfully the Livezey–Chen procedure still lacks power.

Another virtue of the FDR approach is that its sensitivity to correlation among the multiple tests is modest, and if anything results in tests that are more conservative (i.e., rejection levels that are smaller than the nominal α). Figure 5.13 illustrates this characteristic of the FDR approach for synthetic data generated on a $2° × 3°$ latitude–longitude grid (Figure 5.14) mimicking much of the northern hemisphere, when each of the local null hypotheses on the 3720-point grid is valid. The horizontal axis in Figure 5.13 corresponds to increasing spatial correlation, with the vertical arrow at $1.54 × 10^3$ km indicating a typical value for the 500 mb height field. The figure shows that the FDR approach is somewhat conservative when the underlying data are spatially correlated, but that approximately correct results can be achieved by choosing $\alpha_{FDR} = 2\alpha_{global}$.

By placing a control limit on the fraction of nominally significant gridpoint test results that are spurious, the FDR method enhances scientific interpretability of the patterns of significant results by focusing attention only on those results unlikely to have been inflated by survivorship bias. Figure 5.14 illustrates this point for the 3720-point grid of synthetic hypothesis tests to which Figure 5.13 also pertains, constructed with the level of spatial correlation indicated by the arrow in Figure 5.13. Here the 156 gridpoint null hypotheses in the central octagonal region are not true, and the remaining 3654 null hypotheses are valid. Figure 5.14a shows results for the FDR method with $\alpha_{FDR} = 0.10$, which is expected to yield FDR control at approximately the 5% level as indicated by Figure 5.13. The octagonal "signal" is clearly detected even though p values for 12 of the gridpoint tests in its interior (circles) fall above the threshold defined by Equation 5.39. Only six gridpoints near the leftmost edge of the figure yield rejections of valid null hypotheses, yielding the false discovery proportion $6/(144 + 6) = 0.04$. In contrast, Figure 5.14b shows corresponding results for the naive stippling method, where any gridpoint test with a p value smaller than 0.05 is flagged as nominally significant. In

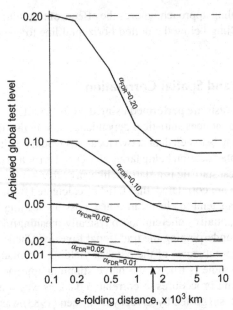

FIGURE 5.13 Achieved global test levels, which are probabilities of rejecting at least one gridpoint null hypothesis using the FDR sliding scale of Equation 5.39 when all null hypotheses are valid, for 3720 gridpoint t-tests as functions of the strength of the spatial correlation. For moderate and strong spatial correlation, approximately correct results can be achieved by choosing $\alpha_{FDR} = 2\alpha_{global}$. Arrow locates spatial correlation corresponding to northern hemisphere 500mb height fields. *From Wilks (2016b).*

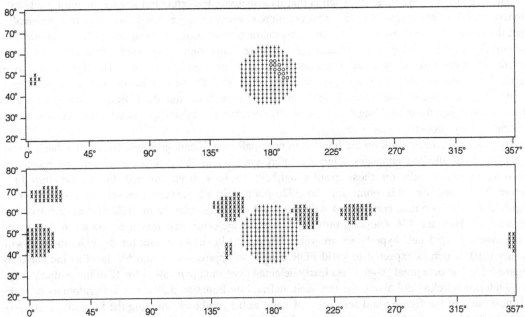

FIGURE 5.14 Maps of local test decisions made by (a) the FDR procedure with $\alpha_{FDR} = 0.10$ and (b) the naive stippling approach using $\alpha = 0.05$. Correct gridpoint null hypotheses are indicated by plus signs, failures to reject false null hypotheses are indicated by circles, and erroneous rejections of valid null hypotheses are indicated by x's. *From Wilks (2016b).*

addition to detecting the octagonal region of 156 incorrect null hypotheses, this method also indicates several regions of nominally but falsely significant tests, composed of 189 gridpoints in total. The false discovery proportion for the naive stippling method is therefore $189/(156+189)=0.55$, so that a majority of the gridpoint rejections are erroneous. The incorrectly rejected null hypotheses cluster spatially because of the spatial correlation of the underlying data, and accordingly this result might tempt an investigator to waste effort in constructing misleading rationales for their existence.

5.5. ANALYSIS OF VARIANCE AND COMPARISONS AMONG MULTIPLE MEANS

Section 5.2.1 reviewed the traditional approach to statistical inferences about a single mean, and Sections 5.2.2 and 5.2.3 considered the conventional methods for comparing two means. It is possible to extend these procedures to comparisons among multiple means, using a procedure known as Analysis of Variance (ANOVA). ANOVA was developed early in the last century (Fisher, 1925, 1935), mainly to support analysis of data from scientific (mainly agricultural and biological) experiments, and so different forms of ANOVA relate to different experimental designs. Such controlled-experiment applications in meteorology and climatology have been relatively few (e.g., Hingray et al., 2007; Räisänen, 2001; Sain et al., 2011; Sansom et al., 2013), because observational meteorological and climatological data cannot be manipulated in an experimental setting, and comprehensive computational experiments with dynamical weather and climate models are lengthy and expensive. However with increasing computing capacity and multi-institutional coordination such studies are becoming more common (e.g., Taylor et al., 2012).

5.5.1. One-Way ANOVA, and the Completely Randomized Experimental Design

One-way ANOVA is the simplest approach to multimean comparisons. It was developed to support the simple and straightforward experimental approach known as the completely randomized experimental design. In this design there are some number $J \geq 2$ experimental treatments, to which n_j experimental units have been assigned at random. The subscript on the sample sizes indicates that these need not be equal across the treatments. The ith experimental unit in the jth treatment exhibits the quantitative response $x_{i,j}$, and the scientific interest is comparison of the mean responses $\bar{x}_j, j = 1, ..., J$, under the null hypothesis that all J of these have been drawn from a population or generating process with the same mean, μ. For $J = 2$ the analysis and inference are equivalent to the (two-tailed) two-sample t-test in Equation 5.9.

The underlying statistical model for the one-way ANOVA is

$$X_{i,j} = \mu + \alpha_j + \varepsilon_{i,j}. \tag{5.40}$$

Here the coefficients $\alpha_j = \mu_j - \mu$ are the treatment effects, so that the null hypothesis implies $\alpha_1 = \alpha_2 = ... = \alpha_J = 0$. The treatment effect coefficients are constrained according to

$$\sum_{j=1}^{J} n_j \alpha_j = 0 \tag{5.41}$$

in order for the sum of the individual treatment means μ_j to yield the overall mean μ. The errors, or residuals, $\varepsilon_{i,j}$ are assumed to be independently Gaussian distributed, with mean zero and variance σ^2.

The ANOVA proceeds by deriving two independent variance estimates, which will both estimate the true common variance σ^2 if the null hypothesis is true. These are derived from computation of the total sum of squares around the overall mean

$$\text{SST} = \sum_{j=1}^{J} \sum_{i=1}^{n_j} \left(x_{i,j} - \bar{x}\right)^2, \tag{5.42a}$$

the (weighted) sum of the squares of the treatment means around the overall mean

$$\text{SSA} = \sum_{j=1}^{J} n_j \left(\bar{x}_j - \bar{x}\right)^2 \tag{5.42b}$$

and the sum of the squares of the data values around their respective treatment means

$$\text{SSE} = \sum_{j=1}^{J} \sum_{i=1}^{n_j} \left(x_{i,j} - \bar{x}_j\right)^2. \tag{5.42c}$$

These three quantities are related according to $\text{SST} = \text{SSA} + \text{SSE}$. Here the J treatment means are

$$\bar{x}_j = \frac{1}{n_j} \sum_{i=1}^{n_j} x_{i,j} \tag{5.43a}$$

and the overall mean is simply

$$\bar{x} = \frac{1}{n} \sum_{j=1}^{J} \sum_{i=1}^{n_j} x_{i,j}, \tag{5.43b}$$

where the total sample size $n = n_1 + n_2 + \ldots + n_J$. The two variance estimates are computed from equations 5.42b and 5.42c, by dividing by their respective degrees of freedom,

$$\text{MSA} = \frac{\text{SSA}}{J-1} \tag{5.44a}$$

and

$$\text{MSE} = \frac{\text{SSE}}{n-J}. \tag{5.44b}$$

The degrees of freedom are $J - 1$ in the denominator of Equation 5.44a because only $J - 1$ of the treatment means $\bar{x}_j$ need to be defined in order for their weighted sum to yield the overall mean $\bar{x}$. Similarly, only $n - J$ of the data values $x_{i,j}$ need be known if the J treatment means $\bar{x}_j$ in Equation 5.42c are given.

If the previously stated assumptions and the null hypothesis are true, both of the quantities in Equations 5.44 estimate the common variance σ^2. To the extent that one or more of the α_j in Equation 5.40 are not zero, then SSA in Equation 5.42b (and therefore also MSA in Equation 5.44a) will be inflated, and SSE in Equation 5.42c (and therefore also MAE in Equation 5.44b) will be deflated. Possible violations of the null hypothesis are accordingly tested by computing the ratio

$$F = \text{MSA}/\text{MSE}. \tag{5.45}$$

TABLE 5.10 Generic One-Way ANOVA Table

Source	df	SS	MS	F
Total	$n - 1$	SST (Equation 5.42a)		
Treatment	$J - 1$	SSA (Equation 5.42b)	MSA (Equation 5.44a)	F (Equation 5.45)
Error	$n - J$	SSE (Equation 5.42c)	MSE (Equation 5.44b)	

Typically these results for a one-way ANOVA are presented in tabular form, which is illustrated generically in Table 5.10. If the null hypothesis that all μ_j are equal is true, then the sampling distribution of the statistic in Equation 5.45 is the F distribution, with degrees-of-freedom parameters $v_1 = J - 1$ and $v_2 = n - J$. Violations of the null hypothesis will tend to lead to larger values of this ratio, so that H_0 is rejected if Equation 5.45 is sufficiently large. Right-tail critical values for the F distribution can be found in most introductory statistics textbooks.

Rejecting the null hypothesis in the one-way ANOVA indicates that at least one of the α_j in Equation 5.40 is unlikely to be zero, but this does not specify how many or which of these are responsible for the rejection result. A one-sample test addressing the significance of one of the treatment means can be computed using the conventional t test using the statistic

$$t = \frac{\bar{x}_j - \bar{x}}{\sqrt{\mathrm{MSE}/n_j}},$$

(5.46)

corresponding to Equation 5.3; and similarly a two-sample t-test addressing the equality of two of the treatment means can be computed using

$$t = \frac{\bar{x}_j - \bar{x}_k}{\sqrt{\mathrm{MSE}\left(\frac{1}{n_j} + \frac{1}{n_k}\right)}},$$

(5.47)

corresponding to Equation 5.9. In both of these cases the appropriate null distribution is the t distribution with $v = n - J$ degrees of freedom, and confidence intervals can be computed in the usual way.

The tests defined by Equations 5.46 and 5.47 are valid if the question being investigated has been defined in advance of seeing the data underlying the ANOVA calculations. If the means being tested are instead chosen on the basis of "interesting" post hoc results (e.g., the largest $\bar{x}_j$ or the largest difference between two treatment means) the inferential results will be optimistic due to survivorship bias (Section 5.4). Similarly, computing two or more of the J one-sample tests in Equation 5.46, or two or more of the $J(J - 1)/2$ possible two-sample tests in Equation 5.47, will require adjustments for test multiplicity. Possible approaches to dealing with these problems are to use the Bonferroni adjustment to the tail probabilities (i.e., require $p \leq \alpha/m$, where m is the number of multiple tests), employ false discovery rate control (Section 5.4.2), or compute a range test such as Tukey's "honest significant difference" (e.g., Steel and Torrie, 1960, pp. 109–110; available as the R routine TukeyHSD).

5.5.2. Two-Way ANOVA, and the Randomized Block Experimental Design

Just as better inferential precision can be achieved for paired data using the paired two-sample t-test outlined in Section 5.2.3, more powerful ANOVA testing can be achieved when the data can be grouped a priori into blocks within which the data are sufficiently strongly correlated. In the classical setting of agricultural trials, where the experimental treatments could be different plant varieties or different fertilization regimes, results from treatments imposed at multiple locations would form natural blocks because of the correlations in crop yield induced by the same weather sequences and similar soil conditions. In computer experiments that might look at responses of dynamical climate models to different boundary conditions and/or parameterization schemes, results obtained with models from different research centers would likely be good candidates for grouping into blocks. Such experiments would be examples of what are called randomized block designs.

In the simplest implementation of the randomized block design, there are J treatments as before, with experimental units grouped into K blocks that each contain one of the J experimental units. Thus the total sample size is $n = JK$. The underlying statistical model, corresponding to Equation 5.40, is

$$X_{j,k} = \mu + \alpha_j + \beta_k + \varepsilon_{j,k},\tag{5.48}$$

where as before the α_j are the treatment effects, and now in addition the β_k, $k = 1, 2, ..., K$, are the block effects. Both the block and treatment effects specify possible adjustments to the overall mean μ so that now $\Sigma_j \alpha_j = \Sigma_k \beta_k = 0$. As before, the null hypothesis of no treatment effects implies $\alpha_1 = \alpha_2 = ... = \alpha_J = 0$.

Again the ANOVA table is based on sums of squared deviations around the overall mean $\bar{x}$, which are the total sum of squares

$$\text{SST} = \sum_{k=1}^{K} \sum_{j=1}^{J} \left(x_{j,k} - \bar{x} \right)^2,\tag{5.49a}$$

the sum of squares of the block means around the overall mean is

$$\text{SSB} = J \sum_{k=1}^{K} \left(\bar{x}_k - \bar{x} \right)^2,\tag{5.49b}$$

the sum of squares of the treatment means around the overall mean is

$$\text{SSA} = K \sum_{j=1}^{J} \left(\bar{x}_j - \bar{x} \right)^2,\tag{5.49c}$$

and the error sum of squares is most easily computed as the residual

$$\text{SSE} = \text{SST} - \text{SSB} - \text{SSA}.\tag{5.49d}$$

Because there is a single experimental unit for each combination of treatment and block, the block and treatment means in Equations 5.49b and 5.49c are simply

$$\bar{x}_k = \frac{1}{J} \sum_{j=1}^{J} x_{j,k}\tag{5.50a}$$

TABLE 5.11 Generic Two-Way ANOVA Table

Source	df	SS	MS	F
Total	$n-1$	SST (Equation 5.49a)		
Blocks	$K-1$	SSB (Equation 5.49b)	MSB (Equation 5.51a)	MSB/MSE
Treatment	$J-1$	SSA (Equation 5.49c)	MSA (Equation 5.44a)	F (Equation 5.45)
Error	$n-K-J$ $+1$	SSE (Equation 5.49d)	MSE (Equation 5.51b)	

and

$$\bar{x}_j = \frac{1}{K}\sum_{K=1}^{K} x_{j,k},\tag{5.50b}$$

respectively. Equation 5.49d shows that, to the extent that there are strong block effects (i.e., the block means differ strongly from the overall mean in Equation 5.49b), the error sum of squares SSE will be reduced when this variation is subtracted, leading to more precise inferences.

The two-way ANOVA table is an extension of the one-way ANOVA table in Table 5.10, but with an additional line for the block effects, as shown in Table 5.11. The sums of squares in Equations 5.49 are arranged in the SS column. Under the null hypothesis of no treatment effects and no block effects on the mean (all $\alpha_j = 0$ and all $\beta_k = 0$ in Equation 5.48), SSB, SSA, and SSE are all independent estimates of the assumed common variance σ^2. Therefore the presence of such effects should be detectable in the ratios of the respective MS entries, given by the corresponding SS values divided by their respective degrees of freedom in the *df* column. These entries are given by Equation 5.44a for MSA,

$$MSB = \frac{SSB}{K-1}\tag{5.51a}$$

and

$$MSE = \frac{SSE}{n-K-J+1}.\tag{5.51b}$$

The primary interest is generally in possible treatment effects, examined using the F ratio MSA/MSE (- Equation 5.45), the null distribution for which is F, now with degrees-of-freedom parameters $v_1 = J-1$ and $v_2 = n - K - J + 1$. Similarly the null hypothesis of no block effects can be examined using the statistic $F = MSB/MSE$, the null distribution for which is F with parameters $v_1 = K-1$ and $v_2 = n - K - J + 1$.

Example 5.18. Comparing One-Way and Two-Way ANOVA, and the Respective *t*-Tests
Table 5.12 contains maximum temperature data extracted from Table A.1, but sampled every third day in order to suppress the serial correlation. Considering the two locations as $J = 2$ "treatments," Table 5.13a presents the one-way ANOVA table for examining the null hypothesis that the mean temperatures are equal. The treatment MSA (Equation 5.44a) is of comparable magnitude to the MSE (Equation 5.44b),

TABLE 5.12 Ithaca and Canandaigua Maximum Temperature Data (°F) from Table A.1, Sampled Every Third Day of January 1987 to Suppress Serial Correlation

Date	Ithaca T_{max}	Canandaigua T_{max}
1	33	34
4	29	29
7	37	44
10	30	33
13	33	34
16	45	44
19	32	36
22	28	29
25	9	11
28	26	26
31	34	38
Average	30.54	32.54
Std. Dev.	8.78	9.17

suggesting that the null hypothesis of equal means is plausible according to these calculations. Comparing the computed $F = 0.273$ to the F distribution with $v_1 = 1$ and $v_2 = 20$ degrees of freedom yields the p value 0.607, so that the null hypothesis is not rejected by this test.

Because the temperatures at these two nearby locations are strongly positively correlated (Table 3.5), much of the variance estimated by the MSE in Table 5.13a is shared by the two temperature data sets, and so does not contribute to uncertainty regarding their average difference. Accordingly the two-way ANOVA in Table 5.13b, with each of the $K = 11$ days regarded as a block, is expected to provide more precise and powerful inferences about any difference in the $J = 2$ means. Here SSB is quite large because the two temperature values in each block are strongly correlated. This shared variability is quantified by MSB, and is subtracted in Equation 5.49d to yield the much smaller MSE in Table 5.13b, even though the blocking has consumed an additional 10 degrees of freedom relative to the one-way analysis in Table 5.13a. These lost 10 degrees of freedom are therefore not available in the denominator of the computation of MSE in Table 5.13b. The primary interest is usually in the treatment effect, reflected by $F = 8.148$. Comparing this value with the F distribution having $v_1 = 1$ and $v_2 = 10$ degrees of freedom yields the $p = 0.017$, providing strong evidence against the null hypothesis. Because of the strong spatial correlation for the data at these two locations, the null hypothesis of no block effect is rejected even more strongly, since $F = 58.69$ with $v_1 = v_2 = 10$ degrees of freedom yields $p = 1.6 \times 10^{-7}$.

Since there are $J = 2$ treatment means being compared, the two ANOVAs in Table 5.13 are equivalent to the corresponding two-sample t-tests. In particular, quantiles of F distributions with $v_1 = 1$ degree of freedom are equal to the squares of (two-tailed) t distribution quantiles having v_2 degrees of freedom. The ANOVA in Table 5.13a, which ignores the spatial correlation that defines the paired

TABLE 5.13 One-Way (a) and Two-Way (b) ANOVA Tables for the Temperature data in Table 5.12, Considering the Locations as $J=2$ Treatments

(a) Source	df	SS	MS	F	(b) Source	df	SS	MS	F
Total	21	1633.45			Total	21	1633.45		
Treatment	1	22.00	22.00	0.273	Blocks	10	1588.45	158.45	58.69
Error	20	1611.45	80.57		Treatment	1	22.00	22.00	8.148
					Error	10	27.00	2.70	

The two-way ANOVA in (b) has been constructed using individual days as the $K=11$ blocks because of the spatial correlation between these two nearby locations.

nature of these data, is therefore equivalent to the two-sample t-test in Equation 5.9, because equal variances for the two samples has been assumed. The resulting test statistic is

$$t = \frac{30.54 - 32.54}{\left[\left(\frac{1}{11} + \frac{1}{11}\right)\left(\frac{8.78^2 + 9.17^2}{2}\right)\right]^{1/2}} = -0.5226, \tag{5.52}$$

and $t^2 = -0.5226^2 = 0.273$ matches the F ratio in Table 5.13a.

Similarly, the two-way ANOVA with individual days as blocks is equivalent to the paired two-sample t-test in Equation 5.11 and Example 5.3, which operate on the daily temperature differences and thereby subtract the shared variance. Using the smaller data set in Table 5.12, the resulting test statistic is

$$t = \frac{-2}{\left(2.324^2/11\right)^{1/2}} = -2.8545. \tag{5.53}$$

Again $t^2 = -2.8545^2 = 8.148$, matching the corresponding F ratio in Table 5.13b. The inference here is weaker than in the corresponding test in Example 5.3 because of the reduced sample size. ◇

One of the assumptions used for ANOVA is that the data are independent realizations from their population or generating process. For atmospheric data separated by sufficiently long sampling times, such as annual values, this assumption is generally valid. Zwiers (1987b) has proposed an approach to ANOVA when the data are autocorrelated, based on frequency-domain calculations (Section 10.5). Lund et al. (2016) describe an alternative approach, based on time-domain time series methods (Section 10.3).

The ANOVAs and corresponding experimental designs presented here are the simplest examples from a much larger class of possibilities. One such extension allows multiple experimental units for each treatment-block combination in the two-way ANOVA, in which case it is possible to include interaction coefficients $\gamma_{j,k}$ in Equation 5.48. Another variation is use of two-way ANOVA for a completely randomized design, examining different levels of two treatment types, which is suggested by the symmetry of SSB and SSA in Equations 5.49b and 5.49c, so that the "blocks" are just levels of the second treatment type. Much more on this subject can be found in such sources as Casella (2008), Montgomery (2013), and Wu and Hamada (2009).

5.5.3. Partitioning of Variance Contributions

One use to which ANOVA has been put in the atmospheric sciences literature does not involve hypothesis testing for means, but rather focuses on the variance partition within the ANOVA table. Zwiers (1987b) and Zwiers and Kharin (1998) consider potential predictability in dynamical climate models in the one-way ANOVA setting, with the error variance representing unpredictable internal variability that the analysis distinguishes from potentially predictable interannual variability.

Wang and Zwiers (1999) describe a similar analysis using two-way ANOVA, where responses in different years, and from different dynamical climate model formulations, represent different treatments. Similarly, Yip et al. (2011) and Northrop and Chandler (2014) use two-way ANOVA to partition variance attributable to different dynamical models, different climate-change trajectory scenarios, and internal climate variability, over a century of projected future climate evolution.

5.6. EXERCISES

5.1 For the June temperature data in Table A.3,
   a. Use a two-sample t-test to investigate whether the average June temperatures in El Niño and non-El Niño years are significantly different. Assume the variances are unequal and that the Gaussian distribution is an adequate approximation to the distribution of the test statistic.
   b. Construct a 95% confidence interval for the difference in average June temperature between El Niño and non-El Niño years.

5.2 Calculate n', the equivalent number of independent samples, for the two sets of minimum air temperatures in Table A.1.

5.3 Consider a data series for which $n = 100$, $\Sigma x = 25.00$, $\Sigma x^2 = 2481.25$, and for which the lag-1 autocorrelation is +0.6. Using an appropriately modified t-test, investigate the null hypothesis that the true mean is zero, versus the alternative that the true mean is greater than zero. Assume the Gaussian distribution is an adequate approximation to the true (t) distribution, and report a p value.

5.4 Use the data set in Table A.1 to test the null hypothesis that the average minimum temperatures for Ithaca and Canandaigua in January 1987 are equal. Compute p values, assuming the Gaussian distribution is an adequate approximation to the null distribution of the test statistic, and
   a. $H_A =$ the minimum temperatures are different for the two locations.
   b. $H_A =$ the Canandaigua minimum temperatures are warmer.

5.5 Suppose you are testing for significantly different means between a pair of time series. Assume that the standard deviations for both of these series are 2.7 units, the lag-1 autocorrelation for each is +0.37, and the sample sizes for the two series are equal but the data are not paired. How big do these samples need to be to reject a null hypothesis of no difference at the 10% level, if the observed difference in the means is 1.0 units? You have no prior reason to expect one mean to be larger than the other. Assume the standard Gaussian distribution is an adequate approximation for your test statistic.

5.6 You have 30 years' of daily January temperature data at a location of interest. The standard deviation of this set of $30 \times 31 = 930$ daily temperature values is 14.14°F. When you compute the 30 monthly mean temperatures, and then calculate the standard deviation of these 30 values, you find it is 4.40°F. Assuming the statistical properties of the climatology of temperature do not change appreciably from one part of January to another or from year to year,
   a. Estimate the lag-1 autocorrelation for the daily temperature data.
   b. Determine the standard deviation of 5-day means of the daily data.

5.7 The Fisher Z-transform (Equation 3.25) can be used for statistical inferences regarding correlation coefficients. For a null hypothesis of zero correlation, the sampling distribution of Z is Gaussian, with mean zero and variance $1/(n-3)$. How large a correlation coefficient computed from $n = 30$ independent data pairs is required to reject a null hypothesis of zero correlation at the 5% level, versus the alternative that $\rho \neq 0$?

5.8 Test the fit of the Gaussian distribution to the July precipitation data in Table 4.8, using.
    a. A K–S (i.e., Lilliefors) test.
    b. A Chi-square test.
    c. A Filliben Q–Q correlation test.

5.9 Test whether the 1951–80 July precipitation data in Table 4.8 might have been drawn from the same distribution as the 1951–80 January precipitation comprising part of Table A.2, using a likelihood ratio test, assuming gamma distributions.

5.10 Use the Wilcoxon–Mann–Whitney test to investigate whether the magnitudes of the pressure data in Table A.3 are lower in El Niño years,
    a. Using the exact one-tailed critical values 18, 14, 11, and 8 for tests at the 5%, 2.5%, 1%, and 0.5% levels, respectively, for the smaller of U_1 and U_2.
    b. Using the Gaussian approximation to the sampling distribution of U.

5.11 Discuss how the sampling distribution of the skewness coefficient (Equation 3.9) of June precipitation at Guayaquil could be estimated using the data in Table A.3, by bootstrapping. How could the resulting bootstrap distribution be used to estimate a 95% confidence interval for this statistic? If the appropriate computing resources are available, implement your algorithm.

5.12 Discuss how to construct a resampling test to investigate whether the variance of June precipitation at Guayaquil is different in El Niño versus non-El Niño years, using the data in Table A.3.
    a. Assuming that the precipitation distributions are the same under H_0.
    b. Allowing other aspects of the precipitation distributions to be different under H_0.
    If the appropriate computing resources are available, implement your algorithms.

5.13 Consider the following sorted p values from $N = 10$ independent hypothesis tests: 0.007, 0.009, 0.052, 0.057, 0.072, 0.089, 0.119, 0.227, 0.299, 0.533.
    a. Do these results support a conclusion of "field significance" (i.e., at least one of the 10 local null hypotheses can be rejected) at the $\alpha_{global} = 0.05$ level using either the Livezey–Chen binomial "counting" test with $\alpha = 0.05$, or the FDR approach?
    b. Which if any of the p values would lead to rejection of the respective local null hypotheses according to the calculations in part a, using each of the methods?

5.14 Augment the data in Table 5.12 with the minimum temperatures for Ithaca and Canandaigua from Table A.1, for the same 11 days. Then, considering the four temperature variables as "treatments,"
    a. Construct the one-way ANOVA table for testing equality of the four means.
    b. Construct the two-way ANOVA table for testing equality of the four means, considering the 11 days as blocks.

Chapter 6

Bayesian Inference

6.1. BACKGROUND

The Bayesian, or subjective, view of probability leads to a framework for statistical inference that is different from the more familiar "frequentist" methods that are the subject of Chapter 5. Bayesian inference is parametric, in that the subjects of the inferences are the parameters of probability distributions, of the kinds described in Chapter 4. A parametric distribution is assumed in order to characterize quantitatively the nature of the data-generating process, and its mathematical dependence on the parameter(s) about which inferences are being drawn. For example, if the data at hand have resulted from N independent and identical Bernoulli trials, then it would be natural to adopt the binomial distribution (Equation 4.1) as the data-generating model. The target of statistical inference would then be the binomial parameter, p, and inferences about p could then be used to more fully characterize the nature of the data-generating process.

Regarding probability as a quantitative expression of subjective degree of belief leads to two distinctive differences between the structures of Bayesian and frequentist inference. The first is that prior information (i.e., information available before the current data have been obtained or seen) about the parameter(s) of interest, often reflecting the analyst's subjective judgment, is quantified by a probability distribution. This distribution may or may not be of a familiar parametric form, such as one of the distributions discussed in Chapter 4. The calculations underlying Bayesian inference combine this prior information with the information provided by the data, in an optimal way.

The second difference between the two modes of inference has to do with the ways in which the parameters that are the targets of inference are viewed. In the frequentist view, parameters of a data-generating model are fixed, if unknown, constants. Accordingly in this view it makes no sense to think about or try to characterize uncertainty about them, since they are unvarying. Rather, frequentist inferences about parameters are made on the basis of the distribution of data statistics under (possibly hypothetical) repeated sampling. In contrast, the Bayesian approach allows the parameter(s) being studied to be regarded as being subject to uncertainty that can be quantified using a probability distribution, which is derived by combining the prior information with the data, in light of the chosen data-generating model.

The relative merits of frequentist and Bayesian inference continue to be debated within the statistics profession. A summary of the recent state of these discussions is provided by Little (2006).

6.2. THE STRUCTURE OF BAYESIAN INFERENCE

6.2.1. Bayes' Theorem for Continuous Variables

The computational algorithm for Bayesian inference is provided by Bayes' Theorem, which was presented for discrete variables in Equation 2.16. However, even if the data upon which inferences will be based are discrete, the parameters that are the subject of inference are generally continuous, in which

Statistical Methods in the Atmospheric Sciences. https://doi.org/10.1016/B978-0-12-815823-4.00006-7

case the probability distributions characterizing their uncertainty (the analyst's degrees of belief) may be represented as probability density functions. Analogously to Equation 2.16, Bayes' Theorem for continuous probability models can be expressed as

$$f(\theta|x) = \frac{f(x|\theta)f(\theta)}{f(x)} = \frac{f(x|\theta)f(\theta)}{\int_\theta f(x|\theta)f(\theta)\,d\theta}. \tag{6.1}$$

Here θ represents the parameter(s) about which inferences are to be drawn (e.g., a binomial probability p or a Poisson rate μ), and x represents the available data.

Equation 6.1 expresses the optimal combination of prior information and the available data for inference regarding the parameter(s) θ. Prior subjective beliefs and/or objective information regarding θ is quantified by the *prior distribution*, $f(\theta)$, which will be a continuous PDF when θ is a continuous parameter. It may be nontrivial to make a good assessment of the prior distribution for a given problem, and different analysts may reasonably reach different conclusions regarding it. The impact of the prior distribution, and consequences of different choices for it for inferences regarding θ, will be presented in more detail in Section 6.2.3.

The general nature of the data-generating process, and the quantitative influence of different values of θ on it, are represented by the *likelihood*, $f(x|\theta)$. Notationally, the likelihood appears to be identical to the probability distribution function representing the data-generating process for discrete data or to the PDF for continuous data. However, the distinction is that the likelihood is a function of the parameter(s) θ for fixed values of the data x, as was the case in Section 4.6.1, rather than a function of the data for fixed parameters. The function $f(x|\theta)$ expresses the relative plausibility ("likelihood") of the data at hand as a function of (given) different possible values for θ.

If the data are discrete, then the likelihood will look notationally like the probability distribution function chosen to represent the data-generating process, for example, Equation 4.1 for the binomial distribution or Equation 4.12 for the Poisson distribution. Both of these likelihoods would be functions of a continuous variable, that is, $\theta = p$ for the binomial, and $\theta = \mu$ for the Poisson. However, the likelihood $f(x|\theta)$ is generally not a PDF. Even though (for discrete x) $\sum_x \Pr\{X = x\} = 1$, in general $\int_\theta f(x|\theta)\,d\theta \neq 1$. If the data x are continuous, so that the likelihood looks notationally like the data PDF, the likelihood will in general also be a continuous function of the parameter(s) θ, but will typically also not itself be a PDF, again because $\int_\theta f(x|\theta)\,d\theta \neq 1$ even though $\int_x f(x|\theta)\,dx = 1$.

The optimal combination of the prior information, $f(\theta)$, and the information provided by the data in the context of the assumed character of the data-generating process, $f(x|\theta)$, is achieved through the product in the numerator on the right-hand side of Equation 6.1. The result is the *posterior distribution*, $f(\theta|x)$, which is the PDF for the parameter(s) θ characterizing the current best information regarding uncertainty about θ. The posterior distribution results from the process of updating the prior distribution in light of the information provided by the data, as seen through model provided by the likelihood for representing the data-generating process.

For settings in which all the data do not become available at the same time, this Bayesian updating can be computed sequentially. In such cases the analyst's assessment of the parameter uncertainty in the prior distribution $f(\theta)$ is first updated using whatever data is available initially, to yield a first iteration of the posterior distribution. That posterior distribution can then be further updated as new data become available, by applying Bayes' theorem with that initially calculated posterior distribution now playing the role of the prior distribution. The result of iterating Bayes' theorem in this way will be identical to what would be obtained if all the data had been used at the same time for a single updating of the initial prior distribution.

In order for the posterior distribution to be a proper PDF (i.e., integrating to 1), the product of the likelihood and the prior is scaled by the value $f(x) = \int_\Theta f(x|\theta) f(\theta) \, d\theta$ for the available data x, in the denominator of Equation 6.1. Because the important work of Equation 6.1 occurs in the numerator of the right-hand side, that equation is sometimes expressed simply as

$$f(\theta|x) \propto f(x|\theta) f(\theta),\tag{6.2}$$

or "the posterior is proportional to the likelihood times the prior."

Example 6.1. Iterative Use of Bayes' Theorem

Consider the simple but instructive situation in which data for the number of "successes" x in a sequence of N independent and identical Bernoulli trials are to be used to estimate the success probability for future trials. This parameter p controls the nature of the data-generating process in this setting, and clearly the relationship of the success probability to possible realizations of the data (i.e., the data-generating process) is provided by the binomial distribution (Equation 4.1). Accordingly the natural choice for the likelihood is

$$f(x|p) = \binom{N}{x} p^x (1-p)^{N-x} \propto p^x (1-p)^{N-x},\tag{6.3}$$

where the success probability p is the parameter θ about which inferences are to be made. The proportionality indicated in the second part of Equation 6.3 is appropriate because the combinatorial part of the binomial probability distribution function does not involve p, and so will factor out of the integral in the denominator of Equation 6.1 and cancel, for any choice of the prior distribution $f(p)$. Equation 6.3 is notationally identical to the discrete probability distribution function for the binomial distribution, Equation 4.1. However, unlike Equation 4.1, Equation 6.3 is not a discrete function of x, but rather is a continuous function of p, for a fixed number of successes x over the course of N independent trials.

An appropriate prior distribution $f(p)$ characterizing an analyst's initial uncertainty regarding possible values for p will depend on what, if any, information about p might be available before new data will be observed, as will be discussed more fully in Section 6.2.3. However, since $0 \le p \le 1$, any reasonable choice for $f(p)$ will have support on this interval. If the analyst has no initial idea regarding which values of p might be more or less likely, a reasonable prior might be the uniform distribution, $f(p) = 1$ (Section 4.4.6), which expresses the judgment that no value of p on the interval $0 \le p \le 1$ seems initially more plausible than any other.

Suppose now that the results of $N = 10$ Bernoulli trials from a process of interest become available, and of these $x = 2$ are successes. Bayes' Theorem provides the recipe for updating the initial indifference among possible values for p that is expressed by $f(p) = 1$, in the light of the results of these $N = 10$ observations. According to Equation 6.1, the posterior distribution is

$$f(p|x) = \frac{\binom{10}{2} p^2 (1-p)^8 \cdot 1}{\binom{10}{2} \displaystyle\int_0^1 p^2 (1-p)^8 \, dp} = \frac{\Gamma(12)}{\Gamma(3)\,\Gamma(9)} p^2 (1-p)^8.\tag{6.4}$$

Alternatively, using Equation 6.2,

$$f(p|x) \propto p^2 (1-p)^8 \cdot 1,\tag{6.5}$$

which achieves the same result, because the integral in the denominator of Equation 6.4 yields $\Gamma(3)\Gamma(9)/\Gamma(12)$, which is exactly the factor required for the posterior distribution $f(p|x)$ to integrate to 1 over $0 \le p \le 1$ and thus to be a PDF. In this case the posterior distribution is a beta distribution (Equation 4.58), with parameters $\alpha = x+1 = 3$ and $\beta = N - x + 1 = 9$. Posterior distributions will not always turn out to be recognizable and familiar parametric forms, but a beta distribution has resulted here because of the nature of the chosen prior distribution and its interaction with the specific mathematical form of the likelihood in Equation 6.3, as will be explained in Section 6.3.

The posterior distribution in Equation 6.4 is the result of updating the initial prior distribution in light of having observed $x = 2$ successes in $N = 10$ Bernoulli trials. It thus quantitatively expresses the degree of belief regarding the possible values for p after having observed these data, for an analyst whose prior beliefs had been well represented by $f(p) = 1$.

Consider now how these beliefs should change if data from additional realizations of the same Bernoulli process become available. Bayes' Theorem will be iterated again, updating the current state of knowledge or belief in light of the new data. The prior distribution for this next iteration of Bayes' Theorem is not the initial prior $f(p) = 1$, but rather the posterior distribution from the most recent probability updating, that is, the beta distribution from Equation 6.4. Suppose the next data observed are the results of $N = 5$ Bernoulli trials, of which $x = 3$ are successes. The second application of Equation 6.2 yields

$$f(p|x) \propto p^x(1-p)^{N-x} p^2 (1-p)^8 = p^{x+2}(1-p)^{N-x+8} = p^5(1-p)^{10}. \tag{6.6}$$

Neither the combinatorial part of the likelihood in Equation 6.3 or the ratio of gamma functions in the new prior distribution (Equation 6.4) depend on p, and so both cancel in the quotient of Equation 6.1. This updated posterior distribution is also a beta distribution, now with parameters $\alpha = 6$ and $\beta = 11$.

Figure 6.1 illustrates this probability updating process, by comparing the initial prior distribution $f(p) = 1$; the first posterior distribution (the beta distribution, Equation 6.4), with $\alpha = 3$, and $\beta = 9$, which becomes the next prior distribution; and the final posterior distribution (the beta distribution with $\alpha = 6$,

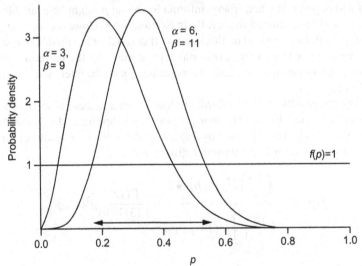

FIGURE 6.1 The prior distribution $f(p) = 1$, and two posterior beta distributions obtained after one ($\alpha = 3$, $\beta = 9$), and two ($\alpha = 6$, $\beta = 11$) applications of Equation 6.1, reflecting information contained in the two data installments. The double-headed arrow indicates the 95% CCI for p according to the second ($\alpha = 6$, $\beta = 11$) posterior distribution.

and $\beta = 11$, Equation 6.6). After the first application of Bayes' Theorem, it is evident that the most likely values for p are near the success relative frequency $x/N = 2/10$, and that values of $p > 0.7$ are associated with very small probability. After the second installment of data has been processed the most likely values for p are near the success relative frequency for all 15 realizations, or 5/15. If a single application of Bayes' Theorem had been made, using all of these data at once (i.e., $N = 15$ and $x = 5$), exactly the same posterior distribution (Equation 6.6) would have resulted from updating the original uniform prior distribution $f(p) = 1$. Similarly, if Equation 6.1 had been iterated 15 times, each using one of the Bernoulli realizations, the same posterior distribution would have resulted, regardless of the order in which the $x = 5$ successes and $N - x = 10$ nonsuccesses had been presented. $\diamond$

6.2.2. Inference and the Posterior Distribution

The posterior distribution, $f(\theta \mid x)$, provides the basis for statistical inference in the Bayesian framework. It is the result, through application of Bayes' Theorem, of the combination of prior beliefs about θ with information about θ contained in the data x. Thus communication of the posterior density fully expresses the analyst's beliefs regarding θ. When the posterior distribution is of a conventional parametric form (e.g., the beta distributions in Example 6.1), quoting the parameters of the posterior distribution is a compact and convenient way to communicate the analyst's degree of belief and uncertainty regarding θ. The parameters of the posterior distribution (and also of the prior distribution) are known as *hyper-parameters*, in order to more easily distinguish them from the parameter(s) that are the subjects of the statistical inference. In Example 6.1, inferences about the binomial parameter p were computed and expressed in terms of a beta posterior distribution, whose hyperparameters were $\alpha = 6$ and $\beta = 11$.

Especially if the posterior distribution is not of a familiar parametric form, for some purposes one might want to provide a point estimate for the parameter θ that is the subject of inference. There are several plausible choices for this characterization, provided by the various measures of central tendency of the posterior distribution. In particular, the mean, median, or mode of the posterior distribution might be chosen to communicate a point estimate for θ. In the case of the beta posterior distribution in Equation 6.6, the posterior mean is $6/17 = 0.353$ (Equation 4.59a), the median (which could be found through numerical integration, or tables such as those in Winkler, 1972b) is 0.347, and the posterior mode (value of p maximizing the posterior distribution) is 0.333.

The posterior mode can be an especially attractive point estimate because of its relationship to the maximum likelihood estimate for θ (Section 4.6). The influence of the prior distribution on the posterior distribution becomes quite small for problems where large amounts of data are available, so that the posterior distribution becomes nearly proportional to the likelihood alone. In that case the posterior mode is nearly the same as the value of θ maximizing the likelihood. In the case of a uniform prior distribution the posterior distribution is exactly proportional to the likelihood (Equation 6.2, with $f(\theta) = 1$), so that the posterior mode in Example 6.1 is exactly the maximum likelihood estimate for the binomial probability: $\hat{p} = 5/15 = 0.333$, having observed $x = 5$ successes in $N = 15$ trials.

Of course summarizing the posterior distribution using probabilities is more informative than is a single-number expression of central tendency. Most commonly this is done using a *central credible interval* (CCI), which will span a range for θ corresponding (in probability) to the middle portion of the posterior distribution. For example, the 95% CCI for the beta posterior distribution with $\alpha = 6$ and $\beta = 11$ in Figure 6.1 is [0.152, 0.587], as indicated by the double-headed arrow. These endpoints are calculated as the $q_{.025}$ and the $q_{.975}$ quantiles of the posterior distribution. The interpretation of this interval is that there is a 0.95 probability that θ lies within it. For many people this is a more natural inferential interpretation than the repeated-sampling concept associated with the $(1 - \alpha) \times 100\%$ frequentist confidence interval (Section 5.1.7), and indeed many people incorrectly ascribe this meaning of the Bayesian credible interval to frequentist confidence intervals (e.g., Ambaum, 2010).

An alternative, although generally more computationally difficult, credible interval is the *highest posterior density* (HPD) interval. The HPD interval also spans a specified amount of probability, but is defined with respect to the largest possible corresponding values of the posterior distribution. Imagine a horizontal line intersecting the posterior density, and thus defining an interval. The HPD interval corresponding to a given probability is defined by the two points of intersection of that horizontal line with the posterior density, for which the given probability is just encompassed by the interval. An HPD interval can thus be viewed as a probabilistic extension of the posterior mode. For a symmetric posterior distribution the HPD interval will coincide with the simpler CCI. For a skewed posterior distribution (such as in Figure 6.1), the HPD interval will be somewhat shifted and compressed relative to the CCI.

In some settings the probability that θ may be above or below some physically meaningful level could be of interest. In such cases the most informative summary of the posterior distribution might simply be a computation of the probability that θ is above or below the threshold.

6.2.3. The Role of the Prior Distribution

The prior distribution $f(\theta)$ quantitatively characterizes the analyst's uncertainty or degree of belief about possible values of the parameter θ, before new data become available. It is a potentially controversial element of Bayesian inference, because different people can reasonably differ in their judgments, and thus can reasonably hold prior beliefs that are different from each other. If the available data are relatively few, then different priors may lead to quite different posterior distributions, and thus to quite different inferences about θ. On the other hand, in data-rich settings the influence of the prior distribution is relatively much less important, so that inferences derived from most reasonable priors will be very similar to each other.

Accurately quantifying prior beliefs may be a difficult task, depending on the circumstances and the experience of the analyst. It is not necessary for a prior distribution to be of a known or familiar parametric form, for example, one of the distributions presented in Chapter 4. One approach to assessing subjective probability is through the use of hypothetical betting or "lottery" games in order to refine one's judgments about probabilities for discrete events (Section 7.10.4) or quantiles of continuous probability distributions (Section 7.10.5). In the continuous case, the subjectively elicited quantiles may provide a basis for constructing a continuous mathematical function representing the relative prior beliefs. Because of the equivalence between Equations 6.1 and 6.2 such functions need not necessarily be proper probability densities, although depending on the form of the function the normalizing constant in the denominator of Equation 6.1 may be difficult to compute.

Sometimes it is both conceptually and mathematically convenient to adopt a known parametric form for the prior distribution, and then to subjectively elicit its parameters (i.e., the prior hyperparameters) based on the properties of the chosen distributional form. For example, if one is able to form a judgment regarding the mean or median of one's prior distribution, this can provide a useful constraint on the prior hyperparameters. Certain parametric forms that are compatible with a particular data-generating model (i.e., the likelihood appropriate to a given problem) may greatly simplify the subsequent calculations, as discussed in Section 6.3, although a mathematically convenient prior that is a poor approximation to one's subjective judgments should not be chosen.

Another important aspect of the prior distribution relates to specification of zero probability for some of the mathematically allowable values of θ. This quite strong condition will usually not be justified, because any range of values for θ assigned zero probability by the prior cannot have nonzero probability in the posterior distribution, regardless of the strength of the evidence provided by the data. This point can be appreciated by examining Equations 6.1 or 6.2: any values of θ for which $f(\theta) = 0$ will necessarily yield $f(\theta|x) = 0$, for all possible data x. Any values of θ that cannot absolutely be ruled out by prior information (e.g., by constraints implied by the underlying physics, such as negative Kelvin temperatures) should be assigned nonzero (although possibly extremely small) probability in the prior distribution.

In situations where there is very little prior information with which to judge relative plausibility for different values of θ, it is natural to choose a prior distribution that does not favor particular values over others to an appreciable degree; that is, a prior distribution expressing as nearly as possible a state of ignorance. Such prior distributions are called *diffuse priors, vague priors, flat priors,* or *noninformative priors*. The prior distribution $f(p) = 1$ in Example 6.1 is an example of such a prior distribution.

Diffuse prior distributions are sometimes seen as being more objective, and therefore less controversial than priors expressing specific subjective judgments. In part this conclusion derives from the fact that a diffuse prior influences the posterior distribution to a minimum degree, by giving maximum weight in Bayes' Theorem to the (data-controlled) likelihood. In general the evidence provided in the data will overwhelm a diffuse prior unless the data sample is fairly small. As has already been noted, Bayesian inference with a diffuse prior will then usually be similar to inferences based on maximum likelihood.

When the parameter θ of interest is not bounded, either above or below or both, it may be difficult to construct a diffuse prior that is consistent with an analyst's subjective judgments. For example, if the parameter of interest is a Gaussian mean, its possible values include the entire real line. One possibility for a diffuse prior in this case could be a Gaussian distribution with zero mean and a very large but finite variance. This prior distribution is nearly flat, but still slightly favors values for the mean near zero. Alternatively, it might be useful to use an *improper prior*, having the property $\int_\theta f(\theta)\,d\theta \neq 1$, such as $f(\theta) = $ constant for $-\infty < \theta < \infty$. Surprisingly, improper priors do not necessarily lead to nonsense inferences, because of the equivalence of Equations 6.1 and 6.2. In particular, an improper prior is permissible if the integral in the denominator of Equation 6.1 yields a finite nonzero value, so that the resulting posterior distribution is a proper probability distribution, with $\int_\theta f(\theta|x)\,d\theta = 1$.

6.2.4. The Predictive Distribution

The ultimate goal of some inferential analyses will be to gain insight about future, yet-unobserved values of the data x^+, which in turn will be informed by the quantification of uncertainty regarding the parameter (s) θ. That is, we may wish to make probability forecasts for future data values that account both for the way their generating process varies for different values of θ, and for the relative plausibility of different values of θ provided by the posterior distribution.

The *predictive distribution* is a probability density function for future data that is derived from a combination of the parametric data-generating process and the posterior distribution for θ,

$$f(x^+) = \int_\theta f(x^+|\theta) f(\theta|x)\,d\theta. \tag{6.7}$$

Here x^+ denotes the future, yet-unobserved data, and x represents the data that has already been used in Bayes' Theorem to produce the current posterior distribution $f(\theta|x)$. Since Equation 6.7 expresses the unconditional PDF (if x is continuous) or probability distribution function (if x is discrete), $f(x|\theta)$ quantifies the data-generating process. It is the PDF (or probability distribution function) for the data given a particular value of θ, not the likelihood for θ given a fixed data sample x, although as before the two are notationally the same. The posterior PDF $f(\theta|x)$ quantifies uncertainty about θ according to the most recently available probability updating, and accordingly Equation 6.7 is sometimes called the *posterior predictive distribution*. If Equation 6.7 is to be applied before observing any data, $f(\theta|x)$ will be the prior distribution, in which case Equation 6.7 will be notationally equivalent to the denominator in Equation 6.1.

Equation 6.7 yields an unconditional PDF for future data x^+ that accounts both for the uncertainty about θ and uncertainty about x for each possible value of θ. It is in effect a weighted average of the PDFs $f(x^+|\theta)$ for all possible values of θ, where the weights are provided by posterior distribution. If θ could somehow be known with certainty, then $f(\theta|x)$ would put probability 1 on that value, and Equation 6.7 would simply be equal to the data-generating PDF $f(x^+|\theta)$ evaluated at that θ. However, Equation 6.7 explicitly accounts for the effects of uncertainty about θ, yielding increased uncertainty about future values of x consistent with the uncertainty about θ.

6.3. CONJUGATE DISTRIBUTIONS

6.3.1. Definition of Conjugate Distributions

An appropriate mathematical form for the likelihood in Equations 6.1 and 6.2, which characterizes the data-generating process, is often clearly dictated by the nature of the problem at hand. However, the form of the prior distribution is rarely so well defined, depending as it does on the judgment of the analyst. In this general case, where the form of the prior distribution is not constrained by the form of the likelihood, evaluation of Equations 6.1 and 6.7 may require numerical integration or other computationally intensive methods, and the difficulty is compounded if the probability updating must be computed iteratively rather than only once.

For certain mathematical forms of the likelihood, however, the computations of Bayes' Theorem can be greatly simplified if choice of a *conjugate distribution* for the prior to be used with that likelihood can be justified. A prior distribution that is conjugate to a particular likelihood is a parametric distribution that is similar mathematically to that likelihood, in a way that yields a posterior distribution that has the same parametric form as the prior distribution. Use of a conjugate distribution that is compatible with a given data-generating process greatly simplifies the computations associated with Bayesian inference by allowing closed-form expressions for the posterior PDF. In addition, simple relationships between the hyperparameters of the prior and posterior distributions can provide insights into the relative importance to the posterior distribution of the prior distribution and the available data. Use of conjugate distributions also facilitates iterative updating of Bayes' Theorem, since the previous posterior distribution, which becomes the new prior distribution when additional data become available, is of the same conjugate parametric form.

Choosing to work with a conjugate prior is convenient, but represents a strong constraint on how the analyst's prior beliefs can be expressed. When the parametric form of the conjugate distribution is very flexible there can be broad scope to approximate the analyst's actual prior beliefs, but an adequate representation is not guaranteed. On the other hand, representation of subjective beliefs using any mathematically explicit PDF will nearly always be an approximation, and the degree to which a nonconjugate prior might be a better approximation may be balanced against the advantages provided by conjugate distributions.

The following sections outline Bayesian inference using conjugate distributions, for three simple but important data-generating processes: the binomial, Poisson, and Gaussian distributions.

6.3.2. Binomial Data-Generating Process

When the data of interest consist of the numbers of "successes" x obtained from N independent and identically distributed Bernoulli trials, their probability distribution will be binomial (Equation 4.1). In this setting the inferential question typically pertains to the value of the success probability (p in Equation 4.1). The appropriate likelihood is then given by the first equality in Equation 6.3, which is notationally identical to Equation 4.1, but is a function of the success probability p given a fixed number of successes x in N independent realizations.

The conjugate prior distribution for the binomial data-generating process is the beta distribution (Equation 4.58). According to Equation 6.2, we can ignore the scaling constants $\binom{N}{x}$ and $\Gamma(\alpha + \beta)/$ $[\Gamma(\alpha)\Gamma(\beta)]$ in Equations 4.1 and 4.58, respectively, so that Bayes' Theorem for the binomial data-generating process and a beta prior distribution becomes

$$f(p\,|\,x) \propto p^x(1-p)^{N-x}\,p^{\alpha-1}p^{\beta-1} = p^{x+\alpha-1}(1-p)^{N-x+\beta-1}. \tag{6.8}$$

Here p is the Bernoulli success probability about which inferences are being computed, and α and β are the hyperparameters of the beta prior distribution. Because of the similarity in mathematical form (apart from the terms not involving p) between the binomial likelihood and the beta prior distribution, their product simplifies to the final equality in Equation 6.8. This simplification shows that the posterior distribution for the success probability, $f(p\,|\,x)$, is also a beta distribution, with hyperparameters

$$\alpha' = x + \alpha \tag{6.9a}$$

and

$$\beta' = N - x + \beta. \tag{6.9b}$$

Adopting the conjugate prior has allowed evaluation of Equation 6.1 using just these two simple relationships, rather than requiring a potentially difficult integration or some other computationally demanding procedure. Including the scaling constant to ensure that the posterior PDF integrates to 1,

$$f(p\,|\,x) = \frac{\Gamma(\alpha+\beta+N)}{\Gamma(x+\alpha)\,\Gamma(N-x+\beta)}p^{x+\alpha-1}(1-p)^{N-x+\beta-1}. \tag{6.10}$$

The relationship between the hyperparameters of the prior beta distribution, α and β, to the hyperparameters of the posterior beta distribution, Equations 6.9, illustrates a more general attribute of Bayesian inference. As more data accumulate, the posterior distribution depends progressively less on whatever choice has been made for the prior distribution (assuming that possible ranges of θ have not been assigned zero prior probability). In the present case of binomial inference with a conjugate prior, $x \gg \alpha$ and $N - x \gg \beta$ if a sufficiently large amount of data can be collected. Therefore the posterior density approaches the binomial likelihood (again apart from the scaling constants), since in that case $x \approx x + \alpha - 1$ and $N - x \approx N - x + \beta - 1$.

Although not mentioned at the time, Example 6.1 was computed using a conjugate prior distribution, because the uniform distribution $f(p) = 1$ is a special case of the beta distribution with hyperparameters

$\alpha = \beta = 1$, and this is exactly the reason that Equations 6.4 and 6.6 are also beta distributions. Equation 6.10 also illustrates clearly why the posterior distribution in Equation 6.6 was achieved regardless of whether Bayes' Theorem was applied individually for each of the two data batches as was done in Example 6.1, or only once after having observed $x = 5$ successes in the overall total of $N = 15$ realizations. In the latter case, the hyperparameters of the posterior beta distribution are also $x + \alpha = 5 + 1 = 6$ and $N - x + \beta = 15 - 5 + 1 = 11$. Since $\alpha = \beta = 1$ yields $f(p) = 1$ for the prior distribution, the posterior distributions in Example 6.1 are exactly proportional to the corresponding binomial likelihoods, which is why the posterior modes are equal to the corresponding maximum likelihood estimates for p. For beta distributions where $\alpha > 1$ and $\beta > 1$, the mode occurs at $(\alpha - 1)/(\alpha + \beta - 2)$.

The influence of uncertainty about the binomial success probability on the probability distribution for future numbers of successes, x^+, among N^+ future realizations, is quantified through the predictive distribution. In the setting of binomial likelihood and a conjugate beta prior distribution, Equation 6.7 is evaluated after substituting the binomial probability distribution function (Equation 4.1) with success probability p, for $f(x^+|\theta)$, and the posterior beta distribution from Equation 6.10 for $f(\theta|x)$. The result is the discrete probability distribution function

$$\Pr\{X^+ = x^+\} = \binom{N^+}{x^+}\left[\frac{\Gamma(N + \alpha + \beta)}{\Gamma(x + \alpha)\,\Gamma(N - x + \beta)}\right]\frac{\Gamma(x^+ + x + \alpha)\Gamma(N^+ + N - x^+ - x + \beta)}{\Gamma(N^+ + N + \alpha + \beta)} \tag{6.11a}$$

$$= \binom{N^+}{x^+}\left[\frac{\Gamma(\alpha' + \beta')}{\Gamma(\alpha')\,\Gamma(\beta')}\right]\frac{\Gamma(x^+ + \alpha')\Gamma(N^+ - x^+ + \beta')}{\Gamma(N^+ + \alpha' + \beta')}, \tag{6.11b}$$

known as the *beta-binomial*, or *Polya distribution*. This function distributes probability among the possible integer outcomes $0 \leq x^+ \leq N^+$. In Equation 6.11a, α and β are the hyperparameters for the prior beta distribution pertaining to p, and x indicates the number of successes in the N data realizations used to update that prior to the posterior distribution in Equation 6.10. The beta-binomial distribution in Equation 6.11b can be thought of as the probability distribution function for a binomial variable, when the success probability p is drawn randomly for each realization from the posterior beta distribution with hyperparameters α' and β'. The mean and variance for the beta-binomial distribution are

$$\mu = \frac{N^+(x + \alpha)}{N + \alpha + \beta} \tag{6.12a}$$

$$= \frac{N^+ \alpha'}{\alpha' + \beta'} \tag{6.12b}$$

and

$$\sigma^2 = \frac{N^+(x + \alpha)(N - x + \beta)(N^+ + N + \alpha + \beta)}{(N + \alpha + \beta)^2(N + \alpha + \beta + 1)} \tag{6.13a}$$

$$= \frac{N^+ \alpha' \beta'(N^+ + \alpha' + \beta')}{(\alpha' + \beta')^2(\alpha' + \beta' + 1)}. \tag{6.13b}$$

Example 6.2 Bayesian Reanalysis of Example 5.1

Example 5.1 considered a hypothetical situation in which the claim that the climatological probability of a cloudless day in winter is 6/7 was examined, after observing $x = 15$ cloudless days on $N = 25$

independent occasions. Analysis of this situation in a Bayesian framework is straightforward if the ana-lyst's prior uncertainty about the winter sunshine climatology at this location can be characterized with a beta distribution. Because beta distributions are able to represent a wide variety of shapes on the unit interval, they can often provide good approximations to an individual's subjective degree of belief about the true value of a probability, such as the binomial success probability, p.

Consider the effects of two possible prior distributions for this probability. First, someone with little or no knowledge of the context of this analysis might reasonably adopt the diffuse uniform prior distri-bution, equivalent to the beta distribution with $\alpha = \beta = 1$. Someone who is more sophisticated about the nature of advertising claims might use this prior knowledge to form the judgment that there might only be a 5% chance of this binomial p being above the claimed 6/7 value. If in addition this second individual thought that values of p above and below 0.5 were equally plausible (i.e., thinking the median of their prior distribution is 0.5), these two conditions together would fully determine a beta prior with $\alpha = \beta = 4$.

Because both of these two priors are beta distributions, it is straightforward to use Equation 6.10 to compute the posterior distributions after having observed $x = 15$ successes in $N = 25$ independent Ber-noulli trials. Because the beta distribution is conjugate to the binomial likelihood, both of these posterior distributions are also beta distributions. The uniform prior is updated by Equation 6.10 to the beta dis-tribution with $\alpha' = 16$ and $\beta' = 11$, and the $\alpha = \beta = 4$ prior distribution is updated by these same data to the posterior beta distribution with $\alpha' = 19$ and $\beta' = 14$.

These two posterior distributions and their corresponding priors are shown in Figure 6.2a and b. Although the two prior distributions are quite different from each other, even the modest amount of data used to update them has been sufficient for the two posterior distributions to be quite similar. For the posterior distribution in Figure 6.2a, the mode $[= (16 - 1)/(16 + 11 - 2) = 15/25 = 0.600]$ is exactly the maximum likelihood estimate for p because the prior $f(p) = 1$, so that the posterior is exactly propor-tional to the likelihood. In Figure 6.2b the posterior mode is 0.581, which is different from but still similar to the posterior mode in Figure 6.2a. Although the two posterior distributions in Figure 6.2 are similar, the sharper prior information in Figure 6.2b leads to a somewhat more concentrated (lower-variance) posterior distribution. This difference is reflected by the corresponding 95% CCIs, which are [0.406, 0.776] in Figure 6.2a and [0.406, 0.736] in Figure 6.2b. The claimed probability of $p = 6/7$ is quite implausible according to both of these posterior analyses, with $\Pr\{p \geq 6/7\} = 0.00048$

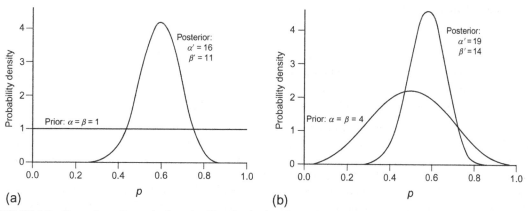

(a)   (b)

FIGURE 6.2   Comparison of posterior beta densities after having observed $x = 15$ successes in $N = 25$ Bernoulli trials, when (a) the prior beta density is uniform ($\alpha = \beta = 1$), and (b) the prior beta density has parameters $\alpha = \beta = 4$.

FIGURE 6.3 Beta-binomial predictive distribution with $\alpha' = 16$ and $\beta' = 11$ for the number of cloudless days X^+ in the next $N^+ = 5$ independent observations (solid histogram), compared to binomial probabilities obtained with $p = 0.6$ (dashed).

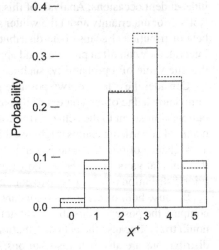

according to the posterior distribution in Figure 6.2a, and $\Pr\{p \geq 6/7\} = 0.000054$ in Figure 6.2b. Both results are generally consistent with the conclusion reached in Example 5.1.

In addition to inferences regarding the parameter p of the binomial data-generating process, in many situations it might also be of interest to make inferences about the probability distribution for future data values, which are quantified by the predictive distribution. For inferences about the binomial data-generating process that have been computed using conjugate beta distributions, the predictive distributions are beta-binomial distributions, Equation 6.11.

Suppose we are interested in the possible numbers of cloudless days, X^+, in the next $N^+ = 5$ independent observations of the sky condition at this desert resort, according to the posterior distribution in Figure 6.2a, with $\alpha' = 16$ and $\beta' = 11$. This will be a discrete distribution with $N^+ + 1 = 6$ possible outcomes, as indicated by the solid histogram bars in Figure 6.3. Not surprisingly the most likely outcome is $X^+ = 3$ cloudless days out of $N^+ = 5$. However, there are nonzero probabilities for the other 5 outcomes also, and the distribution of probability among the outcomes reflects both sampling variability deriving from the 5 Bernoulli trials, as well as uncertainty about the actual value of the Bernoulli success probability, p, that is quantified by the posterior distribution. The effect of this latter source of uncertainty can be appreciated by comparing the dotted histogram in Figure 6.3, which portrays the probabilities from the binomial distribution with $p = 0.6$ and $N = 5$. This binomial distribution would be the predictive distribution if it could be known with certainty that the success probability is 0.6, but uncertainty about p leads to additional uncertainty about X^+, so that the beta-binomial predictive distribution in Figure 6.3 allocates less probability to the middle values of X^+ and more probability to the extreme values.   ◇

Both the geometric distribution (Equation 4.5) and the negative binomial (Equation 4.6) distribution are closely related to the binomial data-generating process, since all three pertain to outcomes of independent Bernoulli trials. Looking more closely at these two probability distribution functions, it can be seen that the corresponding likelihood functions (again, apart from scaling constants not depending on the success probability p) are notationally analogous to the PDF for the beta distribution (again, apart from the scaling constants involving the gamma functions). As would be suggested by this similarity, beta distributions provide conjugate priors for these data-generating processes as well, allowing convenient Bayesian inference in these settings. Epstein (1985) provides the predictive distribution for the

Pascal (negative binomial distribution with integer parameter) data-generating process when a beta prior distribution is used, called the *beta-Pascal distribution*.

6.3.3. Poisson Data-Generating Process

The Poisson data-generating process (Section 4.2.4) is also amenable to simplification of Bayesian inference using conjugate prior distributions. In this case the parameter that is the subject of inference is the Poisson mean, μ, which specifies the average rate of event occurrences per unit interval (usually, a time interval). Rewriting the form of Equation 4.12 as a function of μ, and omitting the denominator that does not depend on it, the Poisson likelihood is proportional to

$$f(x\,|\,\mu) \propto \mu^x \exp[-\mu]. \tag{6.14}$$

This likelihood is mathematically similar to the PDF of the gamma distribution (Equation 4.45, with the mean μ as the random variable) which, again excluding factors not depending on μ, is proportional to

$$f(\mu) \propto \mu^{\alpha-1} \exp[-\mu/\beta]. \tag{6.15}$$

The two factors on the right-hand sides of Equations 6.14 and 6.15 combine when multiplied together in Equation 6.2, so that the gamma distribution is conjugate to the Poisson likelihood. Therefore when a gamma prior distribution for μ with hyperparameters α and β can be reasonably assumed (i.e., is consistent with a particular analyst's judgments, to good approximation), the resulting posterior distribution will also be gamma, and proportional to

$$f(\mu\,|\,x) \propto f(x\,|\,\mu)f(\mu) \propto \mu^x \exp[-\mu]\,\mu^{\alpha-1} \exp[-\mu/\beta] = \mu^{x+\alpha-1} \exp[-(1+1/\beta)\mu]. \tag{6.16}$$

The likelihood in Equation 6.14 pertains to the number of observed events, x, in a single unit time interval. Often the available data will consist of the total number of event counts over multiple (say, n) independent time intervals. In such cases the likelihood for the total number of events during the n time units will be the product of n likelihoods of the form of Equation 6.14. Denoting now the total number of events in these n time intervals as x, that Poisson likelihood is proportional to

$$f(x\,|\,\mu) \propto \mu^x \exp[-n\mu], \tag{6.17}$$

which when combined with a gamma prior distribution for μ (Equation 6.15) yields the posterior distribution

$$f(\mu\,|\,x) \propto f(x\,|\,\mu)f(\mu) \propto \mu^x \exp[-n\mu]\,\mu^{\alpha-1} \exp[-\mu/\beta] = \mu^{x+\alpha-1} \exp[-(n+1/\beta)\mu]. \tag{6.18}$$

Comparing the final expression in Equation 6.18 with Equation 4.45 it is clear that this posterior distribution is also a gamma distribution, with hyperparameters

$$\alpha' = \alpha + x \tag{6.19a}$$

and, since $1/\beta' = 1/\beta + n$,

$$\beta' = \frac{\beta}{1+n\beta}. \tag{6.19b}$$

The resulting posterior gamma PDF can therefore be expressed either in terms of the prior hyperparameters and the data,

$$f(\mu \mid x) = \frac{\left[\left(\frac{1}{\beta} + n\right)\mu\right]^{\alpha+x-1} \exp\left[-\left(\frac{1}{\beta} + n\right)\mu\right]}{\left(\frac{\beta}{1+n\beta}\right)\Gamma(\alpha+x)}, \tag{6.20a}$$

or in terms of the posterior hyperparameters in Equation 6.19,

$$f(\mu \mid x) = \frac{(\mu/\beta')^{\alpha'-1}\exp(-\mu/\beta')}{\beta'\,\Gamma(\alpha')}. \tag{6.20b}$$

As could also be seen in Equation 6.9 for the conjugate hyperparameters for the binomial data-generating process, Equation 6.19 shows that progressively larger amounts of data yield posterior gamma distributions that are less influenced by the prior hyperparameters α and β. In particular, as x and n both become large, $\alpha' \approx x$ and $\beta' \approx 1/n$. The dependence on the prior distribution is further lessened when the prior is diffuse. One possibility for a diffuse prior gamma distribution is $f(\mu) \propto 1/\mu$, which is uniform in $\ln(\mu)$. This is an improper prior distribution, but corresponds formally to the prior hyperparameters $\alpha = 1/\beta = 0$, so that Equation 6.19b yields $\alpha' = x$ and $\beta' = 1/n$, exactly, for the resulting posterior hyperparameters.

The predictive distribution, Equation 6.7, for (the discrete) numbers of future Poisson events $x^+ = 0$, 1, 2, ... in a given future unit interval, is the negative binomial distribution

$$\Pr\{X^+ = x^+\} = \frac{\Gamma(x^+ + \alpha')}{\Gamma(\alpha')\,x^+!}\left(\frac{1}{1+\beta'}\right)^{\alpha'}\left(\frac{\beta'}{1+\beta'}\right)^{x^+}. \tag{6.21}$$

This is of the same form as Equation 4.6, where the probability p has been parameterized in Equation 6.21 as $1/(1 + \beta')$. This result for the predictive distribution points out another interpretation for the negative binomial distribution, namely, that it describes a Poisson distribution with a random rate parameter μ, that is drawn anew for each time interval from the gamma distribution with parameters α' and β'. That is, the predictive distribution in Equation 6.21 accounts both for the interval-to-interval variability in the number of Poisson events for a particular value of the rate parameter μ, and for uncertainty about μ that is quantified by its gamma posterior distribution.

Example 6.3 Poisson Mean for U.S. Landfalling Hurricanes

It was noted in Example 4.4 that the Poisson is a natural distribution for characterizing the data-generating process for annual numbers of hurricanes making landfall in the United States. However, sample estimates of the Poisson rate μ must be based on the available data for annual U.S. hurricane counts, and are therefore subject to some uncertainty. These data are available from 1851 onward, but estimation of the Poisson rate for this phenomenon is complicated by the fact that the earlier data are generally believed to be less reliable.

One approach to dealing with the uneven reliability of the historical annual hurricane count data might be to focus only on the more recent years and ignore the older values. Elsner and Bossak (2001) suggested an alternative approach that makes use of the earlier data without assuming that it is of the same quality as the later data. Their approach was to use the earlier (1851–99) and less reliable data to estimate a prior distribution for the Poisson mean, and then to revise this prior distribution in light of the remaining (1900–2000) data, using Bayes' Theorem.

To specify their prior distribution, Elsner and Bossak (2001) bootstrapped (Section 5.3.5) the 1851–99 annual U.S. landfalling hurricane counts to estimate the sampling distribution for the mean annual number,

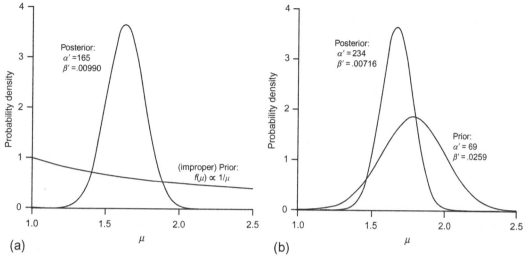

FIGURE 6.4   Posterior gamma PDFs for the Poisson mean characterizing annual numbers of U.S. landfalling hurricanes, resulting from updating (a) the diffuse, improper prior proportional to $1/\mu$, and (b), a gamma prior derived from bootstrapping hurricane landfall counts from the years 1851–99.

μ. The 5th and 95th percentiles of this estimated sampling distribution are 1.45 and 2.16 hurricanes per year, respectively, which quantiles are consistent with a gamma prior distribution with $\alpha = 69$ and $\beta = 0.0259$. The mean of this distribution (Table 4.5) is $\alpha\beta = (69)(0.0259) = 1.79$ hurricanes per year, which agrees well with the sample mean of 1.76 hurricanes per year for the years 1851–99.

For the $n = 101$ years 1900–2000, there were $x = 165$ U.S. landfalling hurricanes. Substituting these values into Equation 6.19, together with the prior hyperparameters $\alpha = 69$ and $\beta = 0.0259$, yields the gamma posterior hyperparameters $\alpha' = 234$ and $\beta' = 0.00716$. Alternatively, adopting the diffuse prior $\alpha = 1/\beta = 0$ leads to the gamma posterior distribution with $\alpha' = 165$ and $\beta' = 0.00990$. Figure 6.4 compares these prior-posterior pairs. Because both have large shape parameter α', each is closely approximated by a Gaussian distribution. The Elsner and Bossak (2001) posterior distribution in Figure 6.4b has a posterior mode of 1.668 (the mode of the gamma distribution, for $\alpha > 1$, is $\beta(\alpha - 1)$), and its 95% CCI is [1.46, 1.89]. The posterior distribution computed from the diffuse prior (Figure 6.4a) is similar but somewhat less sharp, having its mode at $(0.00990)(164-1) = 1.614$, with a 95% CCI of [1.38, 1.88]. The additional information in the nondiffuse prior distribution in Figure 6.4b has resulted in a lower-variance posterior distribution, exhibiting somewhat less uncertainty about the Poisson rate.

The probability distribution for numbers of U.S. landfalling hurricanes in some future year, accounting for both year-to-year differences in numbers of realized Poisson events, and uncertainty about their mean rate characterized by a gamma posterior distribution, is the negative binomial predictive distribution in Equation 6.21. Direct evaluation of Equation 6.21 for the present example is problematic, because the large arguments in the gamma functions will lead to numerical overflows. However, this problem can be circumvented by first computing the logarithms of the probabilities for each of the x^+ of interest, using series representations for the logarithm of the gamma function (e.g., Abramowitz and Stegun 1984, Press et al. 1986).

Figure 6.5 compares the negative binomial predictive distribution (solid histogram), computed using the posterior distribution in Figure 6.4b, to the Poisson distribution (dashed) with mean $\mu = 165/101 = 1.634$ (the annual average number of U.S. hurricane landfalls, 1900–2000). The two

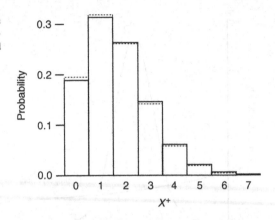

FIGURE 6.5 Negative binomial predictive distribution with $\alpha' = 234$ and $\beta' = 0.00716$ for the number of U.S. landfalling hurricanes (*solid histogram*), compared to Poisson probabilities obtained with $\mu = 165/101 = 1.634$ (*dashed*).

distributions are quite close, reflecting the rather compact character of the posterior distribution in Figure 6.4b, although the negative binomial predictive distribution has a slightly larger variance ($\sigma^2 = 1.687$, cf. Table 4.3) than the Poisson distribution ($\sigma^2 = \mu = 1.634$), which is reflected by the longer right tail. ◇

6.3.4. Gaussian Data-Generating Process

Bayesian inference for the mean μ of a Gaussian (Equation 4.24) data-generating process is also amenable to analytic treatment using conjugate prior and posterior distributions. The general case, where both the mean μ and variance σ^2 of the generating process are unknown, becomes quite complicated because the joint posterior distribution of the two parameters must be considered, even if their univariate prior distributions $f(\mu)$ and $f(\sigma^2)$ can reasonably be regarded as independent. Treatments of that case can be found in Epstein (1985) and Lee (1997), for example.

The more restricted case, where inferences about a Gaussian μ, assuming that the variance of the data-generating process is known, is much more straightforward. Instances where this assumption may be justified include analysis of data produced by an instrument whose measurement precision is well known, or in large-sample settings where the sample variance is known to estimate the variance of the generating process very closely.

An interesting aspect of Bayesian inference for the mean of a Gaussian generating process, assuming known variance, is that the conjugate prior and posterior distributions are also Gaussian. Furthermore, when the posterior distribution is Gaussian, then the predictive distribution is Gaussian as well. This situation is computationally convenient, but notationally confusing because four sets of means and variances must be distinguished. In the following, the symbol μ will be used for the mean of the data-generating process, about which inferences are to be made. The known variance of the data-generating process will be denoted as σ_*^2. The hyperparameters of the prior Gaussian distribution will be denoted as μ_h and σ_h^2, respectively, and will be distinguished from the posterior hyperparameters μ_h' and $\sigma_h^{2'}$. The parameters of the Gaussian predictive distribution will be represented by μ_+ and σ_+^2.

Using this notation the prior distribution is proportional to

$$f(\mu) \propto \frac{1}{\sigma_h} \exp\left[-\frac{(\mu - \mu_h)^2}{2\sigma_h^2}\right],$$ (6.22)

and the likelihood, given a data sample of n independent values x_i from the data-generating process, is proportional to

$$f(x|\mu) \propto \prod_{i=1}^{n} \exp\left[-\frac{(x_i - \mu)^2}{2\sigma_*^2}\right]. \tag{6.23a}$$

However, the sample mean carries all the relevant information in the data pertaining to μ (the sample mean is said to be sufficient for μ), so that the likelihood can be expressed more compactly as

$$f(\bar{x}|\mu) \propto \exp\left[-\frac{n(\bar{x} - \mu)^2}{2\sigma_*^2}\right], \tag{6.23b}$$

because the distribution for a sample mean of n data values from a Gaussian distribution with parameters μ and σ_*^2 is itself Gaussian, with mean μ and variance σ_*^2/n. Combining Equations 6.22 and 6.23b using Bayes' Theorem leads to the Gaussian posterior distribution for μ,

$$f(\mu|\bar{x}) = \frac{1}{\sqrt{2\pi}\sigma_{h'}} \exp\left[-\frac{(\mu - \mu_h')^2}{2\sigma_{h'}^2}\right], \tag{6.24}$$

where the posterior hyperparameters are

$$\mu_h' = \frac{\mu_h/\sigma_h^2 + n\bar{x}/\sigma_*^2}{1/\sigma_h^2 + n/\sigma_*^2} \tag{6.25a}$$

and

$$\sigma_{h'}^2 = \left(\frac{1}{\sigma_h^2} + \frac{n}{\sigma_*^2}\right)^{-1}. \tag{6.25b}$$

That is, the posterior mean is a weighted average of the prior mean and the sample mean, with progressively greater weight given to the sample mean as n increases. The reciprocal of the posterior variance is the sum of the reciprocals of the prior variance and the (known) data-generating variance, so that the posterior variance is necessarily smaller than both the prior variance and the data-generating variance, and decreases as n increases. Only the sample mean, and not the sample variance, appears in Equation 6.25 for the posterior parameters because of the assumption that σ_*^2 is known, so that no amount of additional data can improve our knowledge about it.

For analyses where diffuse prior distributions are appropriate, the most common approach when using a Gaussian prior distribution is to specify an extremely large prior variance, so that the prior distribution is nearly uniform over a large portion of the real line around the prior mean. In the limit of $\sigma_h^2 \to \infty$, the resulting diffuse prior distribution is uniform on the real line, and therefore improper. However, this choice yields $1/\sigma_h^2 = 0$ in Equation 6.25, so the posterior distribution is proportional to the likelihood, with $\mu_h' = \bar{x}$ and $\sigma_{h'}^2 = \sigma_*^2/n$.

Uncertainty about future data values x^+ from the Gaussian data-generating process results from the combination of sampling variability from the data-generating process itself in combination with uncertainty about μ that is expressed by the posterior distribution. These two contributions are quantified by the predictive distribution, which is also Gaussian, with mean

$$\mu_+ = \mu_h' \tag{6.26a}$$

and variance

$$\sigma_+^2 = \sigma_*^2 + \sigma_h^{2\prime}. \tag{6.26b}$$

Example 6.4 Bayesian Inference for Wind power Suitability

Before wind turbines for generation of electricity are purchased and installed at a location, an evaluation of the suitability of the local climate for wind power generation is prudent. A quantity of interest in this evaluation may be the average *wind power density* at 50 m height. Suppose a wind farm will be economically viable if the average annual wind power density is at least 400 W/m². Ideally a long climatological record of wind speeds would be very helpful in evaluating the suitability of a candidate site, but practically it may be possible to set up an anemometer to make wind measurements at a potential wind power site for only a year or two before the decision is made. How might such measurements be used to evaluate the wind power suitability?

The wind power density depends on the cube of wind speed, the distribution of which is usually positively skewed. However, when averaged over a long time period such as a year, the Central Limit Theorem suggests that the distribution of the annual average will be at least approximately Gaussian. Suppose previous experience with other wind farms is that the year-to-year variability in the annually averaged wind power density can be characterized by a standard deviation of 50 W/m². These conditions suggest a Gaussian data-generating process for the annual average wind power density at a location, with unknown mean μ and known standard deviation $\sigma_* = 50$ W/m².

Someone contemplating construction of a new wind power site will have some prior belief regarding possible values for μ. Suppose this person's prior distribution for μ is Gaussian, with mean $\mu_h = 550$ W/m². If in addition this person's judgment is that there is only a 5% chance that μ will be smaller than 200 W/m², the implied prior standard deviation is $\sigma_h = 212$ W/m².

Suppose now that it is possible to collect $n = 2$ years of wind data before deciding whether or not to begin construction, and that the average wind power densities for these 2 years are 420 and 480 W/m². These are certainly consistent with the degree of interannual variability implied by the standard deviation of the data-generating process, $\sigma_* = 50$ W/m² and yield $\bar{x} = 450$ W/m².

Modification of the prior distribution in light of the two annual data values using Bayes' Theorem yields the Gaussian posterior distribution in Equation 6.24, with posterior mean $\mu_h' = (550/212^2 + (2)$ $(450)/50^2)/(1/212^2 + 2/50^2) = 453.4$ W/m², and posterior standard deviation $\sigma_h' = (1/212^2 + 2/50^2)^{-1/2}$ $= 34.9$ W/m². The prior and posterior PDFs are compared in Figure 6.6. Having observed the wind power density for 2 years, uncertainty about its annual average value has decreased substantially. Even though the sample size of $n = 2$ is small, knowing that the generating-process standard deviation is 50 W/m², which is much smaller than the standard deviation of the prior distribution, has allowed these few data values to strongly constrain the location and spread of plausible values for μ in the posterior distribution. The probability, according to the posterior distribution, that the average annual wind power density is smaller than 400 W/m² is $\Pr\{z < (400 - 453.4)/34.9\} = \Pr\{z < -1.53\} = 0.063$.

The probability distribution for a future year's average wind power density, which would be if interest if the wind generation facility were to be built, is the Gaussian predictive distribution with parameters calculated using Equation 6.26, which are $\mu_+ = 453.4$ and $\sigma_+ = 61.0$ W/m². This distribution reflects uncertainty due both to the intrinsic interannual variability of the wind power density,

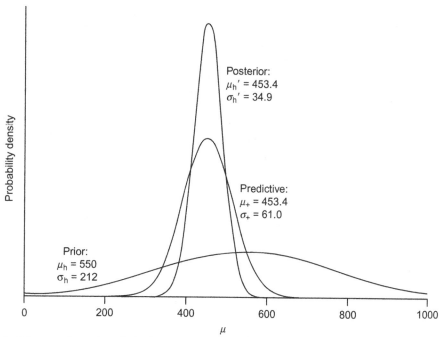

FIGURE 6.6 Prior, posterior, and predictive Gaussian distributions for the annually averaged wind power density, W/m^2.

characterized by σ_*^2, as well as uncertainty about the underlying climatological mean value μ that is expressed by the posterior distribution.   ◇

6.4. DEALING WITH DIFFICULT INTEGRALS

6.4.1. Markov Chain Monte Carlo (MCMC) Methods

Not all data-generating processes can be characterized by likelihood functions having conjugate prior and posterior distributions. Nor is it always the case that the form of a conjugate prior distribution is capable of adequately representing an analyst's beliefs about the parameter or parameters of the data-generating process, so that a nonconjugate prior distribution must be used. In either of these cases, the normalizing integral in the denominator of Equation 6.1 may not exist in closed form, and its explicit numerical integration may be difficult. The same problems often occur for the integral of the posterior distribution, on the left-hand sides of Equations 6.1 and 6.2, evaluation of which are necessary for computation of inferential quantities such as credible intervals.

The usual approach to Bayesian inference in such settings is the use of *Markov chain Monte Carlo*, or MCMC, methods. Rather than attempting to compute explicit expressions for, or numerical approximations to, the relevant integrals, MCMC methods operate through statistical simulation, or generation of (pseudo-) random samples from the distributions of interest, using "Monte Carlo" methods of the kinds described in Section 4.7. MCMC algorithms yield sequences of simulated values from a target distribution that constitute what is called a Markov chain, which means that these sequences of random numbers are not independent but rather exhibit a particular form of serial dependence. Markov chains for sequences of discrete variables are discussed in Section 10.2.

Given two conditions that are usually met when using MCMC methods, namely, that the Markov chain is aperiodic (never repeats exactly) and irreducible (cannot reach a point where some of the allowable values can never again be simulated), a very large sample of these simulated values approaches the target distribution. If the target distribution from which the random values have been drawn is the posterior distribution for a Bayesian analysis, then attributes of this distribution (e.g., posterior moments, credible intervals, etc.) can be well approximated using sample counterparts from a large collection of simulated values.

Convergence of the empirical distribution of random values from a MCMC algorithm to the actual underlying distribution as $n \to \infty$ occurs even though these samples from the target distribution are not mutually independent. Therefore the serial correlation in the simulated values does not present a problem if we are interested only in computing selected quantiles or moments of the target distribution. However, if (approximately) independent samples from the target distribution are needed, or if computer storage must be minimized, the chain may be "thinned." *Thinning* simply means that most of the simulated values are discarded, and only every mth simulated value is retained. An appropriate value of m depends on the nature and strength of the serial correlation in the simulated values, and might be estimated using the variance inflation factor, or "time between effectively independent samples" in Equation 5.13. Because simulated MCMC sequences may exhibit quite strong serial correlation, appropriate values of m can be 100 or larger.

Another practical issue to be considered is ensuring the convergence of the simulated values to the target distribution. Depending on the value used to initialize a Markov chain, the early portion of a simulated sequence may not be representative of the target distribution. It is usual practice to discard this first portion of a simulated sequence, called the *burn-in* period. Sometimes the length of the burn-in period is chosen arbitrarily (e.g., discard the first 1000 values), although a better practice is to create at a scatterplot of the simulated values as a function of their position number in the sequence, and look for a place after which the point scatter appears to "level off" and fluctuate with unchanging variance around a fixed value. Similarly, it is good practice to ensure that the simulations are being generated from an irreducible Markov chain, by initializing multiple simulated sequences from different starting points, and checking that the resulting distributions are the same, following the burn-in period.

Two approaches to constructing MCMC sequences are in general use. These are described in the next two sections.

6.4.2. The Metropolis–Hastings Algorithm

The *Metropolis–Hastings algorithm* is a procedure for random number generation that is similar to the rejection method (Section 4.7.3). In both cases it is necessary to know only the mathematical form of the PDF of the target distribution, and not its CDF (so the PDF need not be analytically integrable). Also in common with the rejection method, candidates for the next simulated value are drawn from a different distribution that is easy to sample from, and each candidate value may be accepted or not, depending on an additional random draw from the uniform [0,1] distribution. The Metropolis–Hastings algorithm is especially attractive for Bayesian inference because only a function proportional to the target PDF (Equation 6.2) needs to be known, rather than the complete PDF of the posterior distribution (Equation 6.1). In particular, the integral in the denominator of Equation 6.1 need never be computed.

To simulate from a posterior distribution $f(\theta|x)$, it is first necessary to choose a candidate-generating distribution $g(\theta)$ that is easy to simulate from, and which has the same support as $f(\theta|x)$. That is, $g(\theta)$ and $f(\theta|x)$ must be defined over the same range of the random argument θ.

The Metropolis–Hastings algorithm begins by drawing a random initial value, θ_0, from $g(\theta)$ for which $f(\theta_0|x) > 0$. Then, for each iteration, i, of the algorithm a new candidate value, θ_C, is drawn from the candidate-generating distribution, and used to compute the ratio

$$R = \frac{f(\theta_C|x)/f(\theta_{i-1}|x)}{g(\theta_C)/g(\theta_{i-1})}, \tag{6.27}$$

where θ_{i-1} denotes the simulated value from the previous iteration. Notice that the target density $f(\theta|x)$ appears as a ratio in Equation 6.27, so that whatever the normalizing constant in the denominator of Equation 6.1 might be, it cancels in the numerator of Equation 6.27.

Whether or not the candidate value θ_C is accepted as the next value, θ_i, in the Markov chain depends on the ratio in Equation 6.27. It will be accepted if $R \geq 1$,

$$\text{For } R \geq 1, \ \theta_i = \theta_C, \tag{6.28a}$$

otherwise

$$\text{For } R < 1, \ \theta_i = \begin{cases} \theta_C & \text{if } u_i \leq R \\ \theta_{i-1} & \text{if } u_i > R \end{cases}. \tag{6.28b}$$

That is, if $R \geq 1$ then θ_C is automatically accepted as the next value in the chain. If $R < 1$, then θ_C is accepted if u_i, which is an independent draw from the uniform [0,1] distribution, is no greater than R. Importantly, and differently from the rejection method described in Section 4.7.3, the previous value θ_{i-1} is repeated if the candidate value is not accepted.

The algorithm based on the ratio in Equation 6.27 is called "independence" Metropolis–Hastings sampling, but the resulting sequence of simulated values θ_1, θ_2, θ_3, ... is nevertheless a Markov chain exhibiting serial correlation, and that serial correlation may be quite strong. The procedure generally works best if the candidate-generating distribution $g(x)$ has heavier tails than the target distribution, which suggests that the prior distribution $f(\theta)$ may often be a good choice for the candidate-generating distribution, particularly if a straightforward algorithm is available for simulating from it.

Example 6.5 Gaussian Inference Without a Conjugate Prior Distribution
Example 6.4 considered evaluation of a hypothetical site for its wind power potential, using a Gaussian data-generating function to represent interannual variations in wind power density, and a conjugate prior distribution with mean $500\,\text{W/m}^2$ and standard deviation $212\,\text{W/m}^2$. This formulation was convenient, but the Gaussian prior distribution might not adequately represent an evaluator's prior beliefs about the wind power potential of the site, particularly as this prior distribution specifies a small but nonzero ($= 0.0048$) probability of impossible negative wind power densities.

Alternatively, the analyst might prefer to use a functional form for the prior distribution with support only on the positive part of the real line, such as the Weibull distribution (Equation 4.69). If, as before, the median and 5th percentile of the analyst's subjective distribution are 550 and $200\,\text{W/m}^2$, respectively, Equation 4.70 can be used to find that the consistent Weibull distribution parameters are $\alpha = 2.57$ and $\beta = 634\,\text{W/m}^2$.

The likelihood consistent with a Gaussian data-generating process is, as before, Equation 6.23b, and the Weibull prior distribution is proportional to

$$f(\mu) \propto \left(\frac{\mu}{\beta}\right)^{\alpha-1} \exp\left[-\left(\frac{\mu}{\beta}\right)^{\alpha}\right], \tag{6.29}$$

because the factor α/β in Equation 4.69 does not depend on μ. Accordingly, the posterior density is proportional to the product of Equations 6.23b and 6.29,

$$f(\mu|\bar{x}) \propto \exp\left[\frac{-n}{2\sigma_*^2}(\bar{x}-\mu)^2\right] \left(\frac{\mu}{\beta}\right)^{\alpha-1} \exp\left[-\left(\frac{\mu}{\beta}\right)^{\alpha}\right], \tag{6.30}$$

where as before $\sigma_* = 50\,\text{W/m}^2$ is the known standard deviation of the Gaussian data-generating process, and the sample mean of $\bar{x} = 450\,\text{W/m}^2$ was computed on the basis of $n = 2$ years of exploratory wind measurements.

The posterior PDF in Equation 6.30 is not a familiar form, and it is not clear that the normalizing constant (denominator in Equation 6.1) for it could be computed analytically. However, the Metropolis–Hastings algorithm allows simulation from this PDF, using a candidate-generating distribution $g(\mu)$ with the same support (positive real numbers) from which it is easy to simulate. A plausible choice for this candidate-generating distribution is the prior Weibull distribution $f(\mu)$, which clearly has the same support. Weibull variates can be generated easily using the inversion method (Section 4.7.4), as illustrated in Exercise 4.16.

Table 6.1 presents the results of the first 10 iterations of a realization of the Metropolis–Hastings algorithm. The algorithm has been initialized at $\mu_0 = 550$, which is the median of the prior distribution and which corresponds to nonzero density in the posterior distribution: $f(\mu_0|\bar{x}) = 0.00732$. The draw from the candidate-generating distribution on the first iteration is $\mu_C = 529.7$, yielding $R = 4.310$ in Equation 6.27, so that this candidate value is accepted as the simulated value for the first iteration, μ_1. This value becomes μ_{i-1} in the second iteration, in which the new candidate value $\mu_C = 533.6$ is generated. This value for the candidate in the second iteration yields $R = 0.773 < 1$, so it is necessary to generate the uniform [0,1] random number $u_2 = 0.3013$. Since $u_2 < R$ the candidate value is accepted

TABLE 6.1 Values for the Quantities in Equations 6.27 and 6.28, for the First 10 Iterations of a Realization of the Metropolis–Hastings Algorithm, Beginning With the Initial Value $\mu_0 = 550\,\text{W/m}^2$

It., i	μ_{i-1}	μ_C	$f(\mu_C|\bar{x})$	$f(\mu_{i-1}|\bar{x})$	$g(\mu_C)$	$g(\mu_{i-1})$	R	u_i	μ_i
1	550.0	529.7	0.03170	0.00732	0.00163	0.00162	4.310	–	529.7
2	529.7	533.6	0.02449	0.03170	0.00163	0.00163	0.773	0.3013	533.6
3	533.6	752.0	0.00000	0.02449	0.00112	0.00163	0.000	0.7009	533.6
4	533.6	395.7	0.10889	0.02449	0.00144	0.00163	5.039	–	395.7
5	395.7	64.2	0.00000	0.10889	0.00011	0.00144	0.000	0.9164	395.7
6	395.7	655.5	0.00000	0.10889	0.00144	0.00144	0.000	0.4561	395.7
7	395.7	471.2	0.32877	0.10889	0.00160	0.00144	2.717	–	471.2
8	471.2	636.6	0.00000	0.32877	0.00149	0.00160	0.000	0.0878	471.2
9	471.2	590.0	0.00015	0.32877	0.00158	0.00160	0.000	0.4986	471.2
10	471.2	462.3	0.36785	0.32877	0.00158	0.00160	1.128	–	462.3

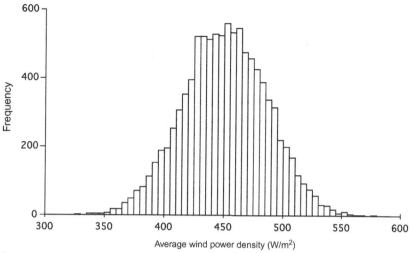

FIGURE 6.7   Histogram of 10,000 random draws from the posterior distribution in Equation 6.30, generated by the Metropolis–Hastings algorithm. The mean and standard deviation of this distribution are 451.0 and 35.4 W/m², respectively.

as $\mu_2 = 533.6$. In the third iteration, the candidate value of 752.0 is an extreme tail value in the posterior distribution, which yields $R = 0.000$ (to three decimal places). Since $u_3 -0.7009 > R$, the candidate value for the third iteration is rejected, and the generated value is the same as that from the second iteration, $\mu_3 = \mu_2 = 533.6$.

The process begun in Table 6.1 can be continued indefinitely, and for this simple example the necessary computations are very fast. Figure 6.7 shows a histogram of 10,000 of the resulting values generated from the posterior distribution, which are the results of every $m = 100$th of 1,000,000 iterations. Since the Weibull prior distribution used to arrive at this posterior distribution is very similar to the Gaussian prior distribution shown in Figure 6.6, it is not surprising that the histogram in Figure 6.7 is similar to the posterior distribution in Figure 6.6. The mean and standard deviation of the histogram in Figure 6.7 are 451.0 and 35.4 W/m², which are similar to the mean and standard deviation of 453.4 and 34.9 W/m², respectively, of the posterior distribution in Figure 6.6, $\Pr\{\mu < 400 \text{ W/m}^2\} = 0.076$ according to Figure 6.7, as compared to $\Pr\{\mu < 400 \text{ W/m}^2\} = 0.063$ for the posterior distribution in Figure 6.6.

The result in Figure 6.7 was produced with essentially no burn-in, other than having discarded results from the first $m - 1 = 99$ iterations. However, a scatterplot of the 10,000 values in Figure 6.7 as a function of their iteration number showed no apparent trends, either in location or dispersion.   ◇

6.4.3. The Gibbs Sampler

The Metropolis–Hastings algorithm is usually the method of choice for MCMC Bayesian inference in one-parameter problems, when a prior distribution conjugate to the form of the data-generating process is either not available or not suitable. It can also be implemented in higher-dimensional problems (i.e., those involving simultaneous inference about multiple parameters) when an appropriate higher-dimensional candidate-generating distribution is available. However, when simultaneous inferences regarding two or more parameters are to be computed, an alternative MCMC approach called the *Gibbs sampler* is more typically used. Casella and George (1992) present a gentle introduction to this algorithm.

The Gibbs sampler produces samples from a K-dimensional posterior distribution, where K is the number of parameters being considered, by simulating from the K univariate conditional distributions for each of the parameters, given fixed values for the remaining $K-1$ parameters. That is, a given K-dimensional joint posterior distribution $f(\theta_1, \theta_2, \theta_3, ..., \theta_K | x)$ can be characterized using the K univariate conditional distributions $f(\theta_1 | \theta_2, \theta_3, ..., \theta_K, x), f(\theta_2 | \theta_1, \theta_3, ..., \theta_K, x), ..., f(\theta_K | \theta_2, \theta_3, ..., \theta_{K-1}, x)$. Simulating from these individually will generally be easier and faster than simulating from the full joint posterior distribution. Denoting the simulated value for the kth parameter on the ith iteration as $\theta_{i,k}$, the ith iteration of the Gibbs sampler consists of the K steps:

1. Generate $\theta_{i,1}$ from $f(\theta_1 | \theta_{i-1,2}, \theta_{i-1,3}, ..., \theta_{i-1,K}, x)$
2. Generate $\theta_{i,2}$ from $f(\theta_2 | \theta_{i,1}, \theta_{i-1,3}, ..., \theta_{i-1,K}, x)$
$\vdots$
k. Generate $\theta_{i,k}$ from $f(\theta_k | \theta_{i,1}, \theta_{i,2}, \theta_{i,k-1}, ... \theta_{i-1,k+1}, ..., \theta_{i-1,K}, x)$
$\vdots$
K. Generate $\theta_{i,K}$ from $f(\theta_K | \theta_{i,1}, \theta_{i,2}, ..., \theta_{i,K-1}, x)$

The ith realization for θ_1 is simulated, conditional on values for the other $K-1$ parameters generated on the previous, $(i-1)^{st}$, iteration. The ith realization for θ_2 is simulated conditionally on the value $\theta_{i,1}$ just generated, and values for the remaining $K-2$ parameters from the previous iteration. In general, for each step within each iteration, values for the conditioning variables are the ones that have most recently become available. The procedure begins with initial ("0th iteration") values $\theta_{0,1}, \theta_{0,2}, \theta_{0,3}, ..., \theta_{0,K}$, drawn perhaps from the prior distribution.

Occasionally, analysis of the joint posterior distribution $f(\theta_1, \theta_2, \theta_3, ..., \theta_K | x)$ may yield explicit expressions for the K conditional distributions to be simulated from. More typically, Gibbs sampling is carried out numerically using freely available software such as BUGS (Bayesian inference Using Gibbs Sampling), or JAGS (Just Another Gibbs Sampler), which can be found through web searches on these acronyms. Regardless of whether the K conditional distributions are derived analytically or evaluated with software, the results are serially correlated Markov chains for simulated values of the parameters θ_k. The same burn-in and possible thinning considerations discussed in the previous sections are applicable to Gibbs samplers as well.

Gibbs sampling is especially well suited to Bayesian inference for *hierarchical models*, where the hyperparameters of a prior distribution are themselves endowed with their own prior distributions, called *hyperpriors*. Such models arise naturally when the parameter(s) of the data-generating process depend on yet other parameters that are not themselves explicit arguments of the likelihood.

Example 6.6 Hierarchical Bayesian Model for Hurricane Occurrences

Elsner and Jagger (2004) have investigated the relationship between annual numbers of U.S. land-falling hurricanes and two well-known features of the climate system, using a hierarchical Bayesian model. The first of these features is the El Niño-Southern Oscillation (ENSO) phenomenon, which they represented using the "cold tongue index", or average sea-surface temperature anomaly in the equatorial Pacific region bounded by 6°N–6°S and 180°–90°W. The second of these features is the *North Atlantic Oscillation* (NAO), which is represented by an index reflecting the strength and orientation of the pair of mutual teleconnectivity features over the Atlantic Ocean in Figure 3.33.

The data-generating process responsible for the number of hurricanes, x_i, in year i is assumed to be Poisson, with mean μ_i that may be different from year to year, depending on the state of the climate system as represented in terms of indices of ENSO and NAO,

$$\ln(\mu_i) = \beta_0 + \beta_1\,CTI_i + \beta_2\,NAO_i + \beta_3\,CTI_i\,NAO_i. \tag{6.31}$$

This hierarchical model is a Bayesian Poisson regression model, similar to the Poisson regression solved using maximum likelihood in Section 7.6.3. The logarithmic transformation on the left-hand side of Equation 6.31 ensures that the modeled μ_i will be strictly positive, as required. The resulting likelihood for the data-generating function, including the implicit expression for the μ_i in Equation 6.31, is (compare Equations 6.14 and 6.17)

$$f(x|\,\beta_0, \beta_1, \beta_2, \beta_3) \propto \prod_{i=1}^{n} \{\,[\exp(\beta_0 + \beta_1 CTI_i + \beta_2 NAO_i + \beta_3 CTI_i NAO_i)]^{x_i} \times$$

$$\exp[-\exp(\beta_0 + \beta_1 CTI_i + \beta_2 NAO_i + \beta_3 CTI_i NAO_i)]\,\}. \tag{6.32}$$

Inferences in this hierarchical model focus on the posterior distributions for the β's, and begin with specification of a (hyper-) prior distribution for them. The multivariate normal distribution (Equation 12.1) is a straightforward and usual choice in models like this, which characterizes initial uncertainty about each β individually as a distinct Gaussian distribution. Elsner and Jagger (2004) considered both a vague prior, and an informative prior based on 19th-century hurricane counts (as in Example 6.3).

Equation 6.31 is a complicated function, and when it is multiplied by the prior distribution (Equation 12.1) it yields an even more complicated posterior distribution for the four β's. However, simulations from it can be made using Gibbs sampling, and these were generated using BUGS. Using data for U.S. landfalling hurricane numbers, CTI, and NAO for the years 1900–2000, and vague priors for the four β's, Elsner and Jagger (2004) simulated the marginal posterior distributions for them in Figure 6.8. These are actually kernel density estimates (Section 3.3.6) computed with Gaussian kernels and smoothing parameter 0.17.

The posterior means and standard deviations in Figure 6.8 are (a) −0.380 and 0.125, (b) −0.191 and 0.078, and (c) 0.200 and 0.102. Panels (a) and (b) in Figure 6.8 suggest strongly that average annual U.S. landfalling hurricane numbers are meaningfully related to both CTI (more landfalling hurricanes on average for negative CTI, or La Niña conditions) and NAO (more U.S. landfalling hurricanes on average for negative NAO, or relatively lower pressures in the subtropical Atlantic), since in both cases values

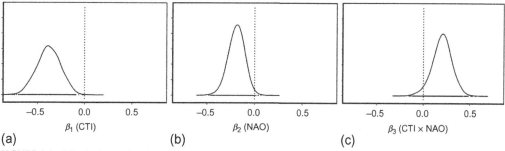

FIGURE 6.8 Marginal posterior distributions for the parameters (a) β_1, (b) β_2, and (c) β_3 in Equation 6.31. *From Elsner and Jagger (2004). © American Meteorological Society. Used with permission.*

near zero are unlikely and there is nearly zero probability that either coefficient is positive. The corresponding inference for β_3 in Figure 6.8c is not as strong, but assuming an approximately Gaussian shape for this posterior distribution implies the estimate $\Pr\{\beta_3 \leq 0\} \approx \Pr\{z \leq -2.00/0.102\} = \Pr\{z \leq -1.96\} = 0.025$. ◇

6.5. EXERCISES

6.1 Suppose a different analyst considering the data in Example 6.2 concludes that a reasonable prior distribution for the binomial p in this situation is Gaussian, with mean 2/3 and standard deviation 1/10.
   a. Find the parameters of a beta distribution that approximates this Gaussian prior distribution.
   b. Using the results of part (a), find the posterior distribution for p.
   c. Find the resulting predictive distribution for the number of "successes" in the next $N^+ = 5$ independent observations. (Use a computer to calculate the logs of the gamma function.)

6.2 Suppose you have concluded that your prior distribution for a parameter of interest is well represented by a Gumbel distribution. Evaluate the parameters of this distribution if
   a. The interquartile range of your prior distribution is (100, 400).
   b. The mean and standard deviation of your prior distribution are 270 and 200, respectively.
   c. What do these two distributions imply about your beliefs about the magnitude of the 100-year event?

6.3 Assume the annual numbers of tornados occurring in a particular county is well described by the Poisson distribution. After observing two tornados in this county during 10 years, a Bayesian analysis yields a posterior distribution for the Poisson rate that is a gamma distribution, with $\alpha' = 3.5$ and $\beta' = 0.05$.
   a. What was the prior distribution?
   b. What is the probability of the county experiencing at least one tornado next year?

6.4 Recalculate Example 6.4 if the analyst has less uncertainty about the eventual suitability of the site for wind power generation, so that an appropriate prior distribution is Gaussian with mean $\mu_h = 550\,\mathrm{W/m^2}$ and standard deviation $\sigma_h = 100\,\mathrm{W/m^2}$.
   a. Find the posterior distribution.
   b. Find the predictive distribution.

6.5 Consider how the analysis in Example 6.4 would change if a third year of wind measurements had been obtained, for which the average annual wind power density was $375\,\mathrm{W/m^2}$.
   a. Find the updated posterior distribution.
   b. Find the updated predictive distribution.

6.6 What value would be generated for μ_{11} in Table 6.1 after the 11th iteration if.
   a. $\mu_C = 350$ and $u_{11} = 0.135$?
   b. $\mu_C = 400$ and $u_{11} = 0.135$?
   c. $\mu_C = 450$ and $u_{11} = 0.135$?

Chapter 7

Statistical Forecasting

7.1. BACKGROUND

Much of operational weather and long-range (seasonal, or "climate") forecasting has a statistical basis. As a nonlinear dynamical system, the atmosphere is not perfectly predictable in a deterministic sense. Consequently, statistical methods are useful, and indeed necessary, parts of the forecasting enterprise. This chapter provides an introduction to statistical forecasting of scalar (single-number) quantities. Some methods suited to statistical prediction of vector (multiple values simultaneously) quantities, for example, spatial patterns, are presented in Sections 14.2.3, 14.3.2, and 15.4. The forecasting emphasis here is consistent with the orientation of this book, but the methods presented in this chapter are applicable to other settings as well.

Some statistical forecast methods operate without information from the fluid-dynamical forecast models that have become the mainstay of weather forecasting for lead times ranging from one day to a week or so in advance. Such pure statistical forecast methods are sometimes referred to as Classical, reflecting their prominence in the years before dynamical forecast information was available. These methods continue to be viable and useful at very short lead times (hours in advance), or very long lead times (weeks or more in advance), for which the dynamical forecast information is not available with sufficient promptness or accuracy, respectively.

Another important application of statistical methods to weather forecasting is in conjunction with dynamical forecast information. Statistical forecast equations routinely are used to postprocess and enhance the results of dynamical forecasts at operational weather forecasting centers throughout the world and are essential as guidance products to aid weather forecasters. The combined statistical and dynamical approaches are especially important for providing forecasts for quantities and locations (e.g., particular cities rather than gridpoints) not represented by the dynamical models.

The types of statistical forecasts mentioned so far are objective, in the sense that a given set of inputs always produces the same particular output. However, another important aspect of statistical weather forecasting is in the subjective formulation of forecasts, particularly when the forecast quantity is a probability or set of probabilities. Here the Bayesian interpretation of probability as a quantified degree of belief is fundamental. Subjective probability assessment forms the basis of many operationally important forecasts and is a technique that could be used more broadly to enhance the information content of operational forecasts.

7.2. LINEAR REGRESSION

Much of statistical weather forecasting is based on the procedure known as linear, least-squares regression. In this section, the fundamentals of linear regression are reviewed. Much more complete treatments can be found in standard texts such as Draper and Smith (1998) and Neter et al. (1996).

Statistical Methods in the Atmospheric Sciences. https://doi.org/10.1016/B978-0-12-815823-4.00007-9

7.2.1. Simple Linear Regression

Regression is most easily understood in the case of *simple linear regression*, which describes the linear relationship between two variables, say x and y. Conventionally the symbol x is used for the *independent*, or *predictor variable*, and the symbol y is used for the *dependent variable*, or *predictand*. The terms dependent and independent variable are in common in the literature of statistics and other disciplines, whereas the terms predictor and predictand are used primarily in the atmospheric and related sciences, having apparently been introduced by Gringorten (1949). More than one predictor ("x") variable is very often required in practical forecast problems, but the ideas for simple linear regression generalize easily to this more complex case of *multiple linear regression*. Therefore most of the important ideas about regression can be presented in the context of simple linear regression.

Essentially, simple linear regression seeks to summarize the relationship between x and y, shown graphically in their scatterplot, using a single straight line. The regression procedure chooses the line producing the least error for predictions of y given observations of x, within the (x, y) data set used to define that relationship. Exactly what is defined to be least error can depend on context, but the most usual error criterion is minimization of the sum (or, equivalently, the average) of the squared errors. It is the choice of the squared-error criterion that is the basis of the name *least-squares regression* or *ordinary least squares* (OLS) regression. Choosing the squared-error criterion is conventional not because it is necessarily best, but rather because it makes the mathematics analytically tractable. Adopting the squared-error criterion results in the line-fitting procedure being fairly tolerant of small discrepancies between the line and the points. However, the fitted line will adjust substantially to avoid very large discrepancies, and so the method is not resistant to outliers.

Figure 7.1 illustrates the situation. Given a data set of (x, y) pairs, the problem is to find the particular straight line,

$$\hat{y} = a + bx, \tag{7.1}$$

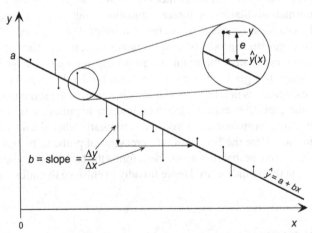

FIGURE 7.1 Schematic illustration of simple linear regression. The regression line, $\hat{y} = a + bx$, is chosen as the one minimizing some measure of the vertical differences (the residuals) between the points and the line. In least-squares regression that measure is the sum of the squared vertical distances. The inset shows a residual, e, as the difference between a data point and the regression line.

minimizing the squared vertical distances (thin lines) between it and the data points. The circumflex ("hat") accent signifies that the equation specifies a predicted value of y. The inset in Figure 7.1 indicates that the vertical distances between the data points and the line, also called errors or *residuals*, are defined as

$$e_i = y_i - \hat{y}(x_i). \tag{7.2}$$

There is a separate residual e_i for each data pair (x_i, y_i). Note that the sign convention implied by Equation 7.2 is for points above the line to be regarded as positive errors and points below the line to be negative errors. This is the usual convention in statistics, but is opposite to what often is seen in the atmospheric sciences, where forecasts smaller than the observations (the line being below the point) are regarded as having negative errors, and vice versa. However, the sign convention for the residuals is unimportant, since it is the minimization of the sum of squared residuals that defines the best-fitting line. Combining Equations 7.1 and 7.2 yields the regression equation,

$$y_i = \hat{y}_i + e_i = a + bx_i + e_i, \tag{7.3}$$

which says that the true value of the predictand is the sum of the predicted value (Equation 7.1) and the residual.

Finding analytic expressions for the least-squares intercept, a, and the slope, b, is a straightforward exercise in calculus. In order to minimize the sum of squared residuals,

$$\sum_{i=1}^{n} (e_i)^2 = \sum_{i=1}^{n} (y_i - \hat{y}_i)^2 = \sum_{i=1}^{n} (y_i - [a + bx_i])^2, \tag{7.4}$$

it is only necessary to set the derivatives of Equation 7.4 with respect to the parameters a and b to zero and solve. These derivatives are

$$\frac{\partial \sum_{i=1}^{n} (e_i)^2}{\partial a} = \frac{\partial \sum_{i=1}^{n} (y_i - a - bx_i)^2}{\partial a} = -2\sum_{i=1}^{n} (y_i - a - bx_i) = 0 \tag{7.5a}$$

and

$$\frac{\partial \sum_{i=1}^{n} (e_i)^2}{\partial b} = \frac{\partial \sum_{i=1}^{n} (y_i - a - bx_i)^2}{\partial b} = -2\sum_{i=1}^{n} x_i[(y_i - a - bx_i)] = 0. \tag{7.5b}$$

Rearranging Equations 7.5 leads to the so-called *normal equations*,

$$\sum_{i=1}^{n} y_i = na + b\sum_{i=1}^{n} x_i \tag{7.6a}$$

and

$$\sum_{i=1}^{n} x_i y_i = a\sum_{i=1}^{n} x_i + b\sum_{i=1}^{n} (x_i)^2. \tag{7.6b}$$

Dividing Equation 7.6a by n leads to the conclusion that the fitted regression line must pass through the point located by the two sample means of x and y. Finally, solving the normal equations for the regression parameters yields

$$b = \frac{\sum_{i=1}^{n}[(x_i - \bar{x})(y_i - \bar{y})]}{\sum_{i=1}^{n}(x_i - \bar{x})^2} = \frac{n\sum_{i=1}^{n}x_i y_i - \left(\sum_{i=1}^{n}x_i\right)\left(\sum_{i=1}^{n}y_i\right)}{n\sum_{i=1}^{n}(x_i)^2 - \left(\sum_{i=1}^{n}x_i\right)^2} \tag{7.7a}$$

and

$$a = \bar{y} - b\bar{x}. \tag{7.7b}$$

Equation 7.7a, for the slope, is proportional to the Pearson correlation coefficient between x and y, where the proportionality constant is given by the ratio of the two standard deviations

$$b = \frac{s_y}{s_x} r_{x,y}. \tag{7.8}$$

Accordingly the regression slope can be computed with a single pass through the data, using the computational form given as the second equality of Equation 7.7a. Note that, as was the case for the correlation coefficient, careless use of the computational form of Equation 7.7a can lead to roundoff errors since the numerator may be the difference between two large numbers.

7.2.2. Distribution of the Residuals

Thus far, fitting the straight line has involved no statistical ideas at all. All that has been required was to define least error to mean minimum squared error. The rest has followed from straightforward mathematical manipulation of the data, namely, the (x, y) pairs. To bring in statistical ideas, it is conventional to assume that the quantities e_i are independent random variables with zero mean and constant variance. Often, the additional assumption is made that these residuals follow a Gaussian distribution.

Assuming that the residuals have zero mean is not at all problematic. In fact, one convenient property of the least-squares fitting procedure is the guarantee that

$$\sum_{i=1}^{n} e_i = 0, \tag{7.9}$$

from which it is clear that the sample mean of the residuals (dividing this equation by n) is also zero.

Imagining that the residuals can be characterized in terms of a variance is really the point at which statistical ideas begin to come into the regression framework. Implicit in their possessing a variance is the idea that the residuals scatter randomly about some mean value (Equations 4.22 or 3.6). Equation 7.9 says that the mean value around which they will scatter is zero, so it is the regression line around which the data points will scatter. We then need to imagine a series of distributions of the residuals *conditional* on the x values, with each observed residual regarded as having been drawn from one of these conditional distributions. The constant variance assumption really means that the variance of the residuals is constant in x, or that all of these conditional distributions of the residuals have the same variance. Therefore a

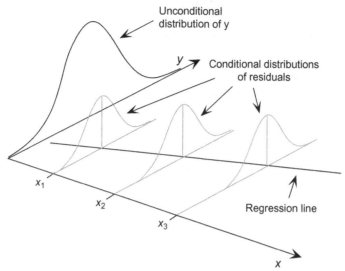

FIGURE 7.2 Schematic illustration of distributions (gray) of residuals around the regression line, conditional on these values of the predictor variable, x. The actual residuals are regarded as having been drawn from these distributions.

given residual (positive or negative, large or small) is by assumption equally likely to occur at any part of the regression line.

Figure 7.2 is a schematic illustration of the idea of a collection of conditional distributions centered on the regression line. The three small gray distributions are identical, except that their means are shifted higher or lower depending on the level of the regression line (predicted value of y) for each x. Extending this thinking slightly, it is not difficult to see that the regression equation can be regarded as specifying the conditional mean of the predictand, given a specific value of the predictor. Also shown by the large black distribution in Figure 7.2 is a schematic representation of the unconditional distribution of the predictand, y. The distributions of residuals are less spread out (have smaller variance) than the unconditional distribution of y, indicating that there is less uncertainty about y if a corresponding x value is known.

Central to the making of statistical inferences in the regression setting is estimation of this (constant) residual variance from the sample of residuals. Since the sample average of the residuals is guaranteed by Equation 7.9 to be zero, the square of Equation 3.6 becomes

$$s_e^2 = \frac{1}{n-2} \sum_{i=1}^{n} e_i^2,$$

(7.10)

where the sum of squared residuals is divided by $n-2$ because two parameters (a and b) have been estimated. Substituting Equation 7.2 then yields

$$s_e^2 = \frac{1}{n-2} \sum_{i=1}^{n} [y_i - \hat{y}(x_i)]^2.$$

(7.11)

Rather than compute the estimated residual variance using 7.11, however, it is more usual to use a computational form based on the relationship,

$$SST = SSR + SSE, \tag{7.12}$$

which proved in most regression texts. The notation in Equation 7.12 consists of acronyms describing the variation in the predictand, y (SST), and a partitioning of that variation between the portion represented by the regression (SSR), and the unrepresented portion ascribed to the variation of the residuals (SSE). The term SST is an acronym for sum of squares, total, which has the mathematical meaning of the sum of squared deviations of the y values around their mean,

$$SST = \sum_{i=1}^{n} (y_i - \bar{y})^2 = \sum_{i=1}^{n} y_i^2 - n\bar{y}^2. \tag{7.13}$$

This term is proportional (by the factor $n-1$) to the sample variance of y, and thus measures the overall variability of the predictand. The term SSR stands for the regression sum of squares or the sum of squared differences between the regression predictions and the sample mean of y,

$$SSR = \sum_{i=1}^{n} [\hat{y}(x_i) - \bar{y}]^2, \tag{7.14a}$$

which relates to the regression equation according to

$$SSR = b^2 \sum_{i=1}^{n} (x_i - \bar{x})^2 = b^2 \left[\sum_{i=1}^{n} x_i^2 - n\bar{x}^2 \right] = (n-1) \, b^2 \, s_x^2. \tag{7.14b}$$

Equation 7.14a indicates that a regression line differing little from the sample mean of the y values will have a small slope and produce a very small SSR, whereas one with a large slope will exhibit some large differences from the sample mean of the predictand and therefore produce a large SSR.

Finally, SSE refers to the sum of squared errors, or sum of squared differences between the residuals and their mean, which is zero,

$$SSE = \sum_{i=1}^{n} e_i^2. \tag{7.15}$$

Since this differs from Equation 7.10 only by the factor of $n-2$, rearranging Equation 7.12 yields the computational form

$$s_e^2 = \frac{1}{n-2} \{ SST - SSR \} = \frac{1}{n-2} \left\{ \sum_{i=1}^{n} y_i^2 - n\bar{y}^2 - b^2 \left[\sum_{i=1}^{n} x_i^2 - n\bar{x}^2 \right] \right\}. \tag{7.16}$$

An additional assumption that is implicit in the usual regression framework is that the predictor variable(s) is observed without error, so that the regression uncertainty relates only to the value of the predictand. Defining the residuals as vertical distances to the regression function, as in Figure 7.1, is consistent with this assumption. Although zero predictor uncertainty is rarely literally true, often uncertainty about the x values is much smaller than uncertainty about the predictions (or, equivalently, about the magnitudes of the residuals), in which case the assumption is reasonable from a practical perspective.

7.2.3. The Analysis of Variance Table

In practice, regression analysis is now almost universally done using computer software. A central part of the regression output from these software packages is a summary of the foregoing information in an

TABLE 7.1 Generic Analysis of Variance (ANOVA) Table for Simple Linear Regression

Source	df	SS	MS	F
Total	$n-1$	SST (7.13)		
Regression	1	SSR (7.14)	MSR $=$ SSR/1	($F=$ MSR/MSE)
Residual	$n-2$	SSE (7.15)	MSE $= s_e^2$	

The column headings df, SS, and MS stand for degrees of freedom, sum of squares, and mean square, respectively. Regression df $=1$ is particular to simple linear regression (i.e., a single predictor x). Parenthetical references are to equation numbers in the text.

analysis of variance, or ANOVA table. The ANOVA table was introduced in Section 5.5 as a vehicle for making inferences about mean responses in designed experiments. Such analyses can be viewed as special cases of regression, where the predictor variables are binary, $x \in \{0, 1\}$ (e.g., Draper and Smith, 1998), which are sometimes referred to as "dummy variables."

Usually, not all the information in an ANOVA table will be of interest, but it is such a universal form of regression output that you should understand its components. Table 7.1 outlines the arrangement of an ANOVA table for simple linear regression and indicates where the quantities described in the previous section are reported. The three rows correspond to the partition of the variation of the predictand as expressed in Equation 7.12. Accordingly, the Regression and Residual entries in the df (degrees of freedom) and SS (sum of squares) columns will sum to the corresponding entry in the Total row. Therefore the ANOVA table contains some redundant information, and as a consequence the output from some regression packages will omit the Total row entirely.

The entries in the MS (mean squared) column are given by the corresponding quotients of SS/df. For simple linear regression, the regression df $= 1$, and SSR $=$ MSR. Comparing with Equation 7.16, it can be seen that the MSE (mean squared error) is the estimated sample variance of the residuals. The total mean square, left blank in Table 7.1 and in the output of most regression packages, would be SST/$(n-1)$, or simply the sample variance of the predictand.

7.2.4. Goodness-of-Fit Measures

The ANOVA table also presents (or provides sufficient information to compute) three related measures of the fit of a regression, or the correspondence between the regression line and a scatterplot of the data. The first of these is the MSE. From the standpoint of forecasting, the MSE is perhaps the most fundamental of the three measures, since it indicates the variability of, or the uncertainty about, the observed y values (the quantities being forecast) around the forecast regression line. As such, it directly reflects the average accuracy of the resulting forecasts. Referring again to Figure 7.2, since MSE $= s_e^2$ this quantity indicates the degree to which the distributions of residuals cluster tightly (small MSE), or spread widely (large MSE) around a regression line. In the limit of a perfect linear relationship between x and y, the regression line coincides exactly with all the point pairs, the residuals are all zero, SST will equal SSR, SSE will be zero, and the variance of the residual distributions is also zero. In the opposite limit of absolutely no linear relationship between x and y, the regression slope will be zero, the SSR will be zero, SSE will equal SST, and the MSE will very nearly equal the sample variance of the predictand itself. In this unfortunate case, the three conditional distributions in Figure 7.2 would be indistinguishable from the unconditional distribution of y.

The relationship of the MSE to the strength of the regression fit is also illustrated in Figure 7.3. Panel (a) shows the case of a reasonably good regression, with the scatter of points around the regression line being fairly small. Here SSR and SST are nearly the same. Panel (b) shows an essentially useless regression, for values of the predictand spanning the same range as in panel (a). In this case the SSR is nearly zero since the regression has nearly zero slope, and the MSE is essentially the same as the sample variance of the y values themselves.

The second usual measure of the fit of a regression is the *coefficient of determination*, or R^2. This can be computed from

$$R^2 = \frac{SSR}{SST} = 1 - \frac{SSE}{SST},$$
(7.17)

which is often also displayed as part of standard regression output. The SSR is nearly equal to SST if each predicted value is close to its respective y, so that the corresponding residual is near zero. Therefore MSE and R^2 are different but related ways of expressing the closeness of or discrepancy between SST and SSR. The R^2 can be interpreted as the proportion of the variation of the predictand (proportional to SST) that is described or accounted for by the regression (SSR). Sometimes we see this concept expressed as the proportion of variation "explained," although this claim is misleading: a regression analysis can quantify the nature and strength of a relationship between two variables, but can say nothing about which variable (if either) causes the other. This is the same caveat that was offered in the discussion of the correlation coefficient in Chapter 3. For the case of simple linear regression, the square root of the coefficient of determination is exactly (the absolute value of) the Pearson correlation between x and y.

For a perfect regression, SSR=SST and SSE=0, so $R^2 = 1$. For a completely useless regression, SSR=0 and SSE=SST, so that $R^2 = 0$. Again, Figure 7.3b shows something close to this latter case. Comparing Equation 7.14a, the least-squares regression line is almost indistinguishable from the sample mean of the predictand, so SSR is very small. In other words, little of the variation in y can be ascribed to the regression so the proportion SSR/SST is nearly zero.

The third commonly used measure of the strength of the regression is the F ratio, generally given in the last column of the ANOVA table. The ratio MSR/MSE increases with the strength of the regression, since a strong relationship between x and y will produce a large MSR and a small MSE. Assuming that the residuals are independent and follow the same Gaussian distribution, and under the null hypothesis of no real linear

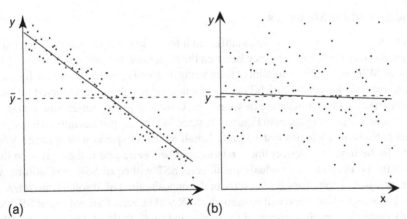

FIGURE 7.3 Illustration of the distinction between a fairly good regression relationship (a) and an essentially useless relationship (b). The points in panel (a) cluster closely around the regression line (solid), indicating small MSE, and the line deviates strongly from the average value of the predictand (dashed), producing a large SSR. In panel (b) the scatter around the regression line is large, and the regression line is almost indistinguishable from the mean of the predictand.

relationship, the sampling distribution of the F ratio has a known parametric form. Analogously to Equation 5.45, this distribution forms the basis of a test that is applicable in the case of simple linear regression if the single predictor has been chosen in advance of seeing the data subject to the analysis, but in the more general case of multiple regression (more than one x variable) problems of test multiplicity, to be discussed later, usually invalidate it. However, even if the F ratio cannot be used for quantitative statistical inference, it is still a valid qualitative index of the strength of a regression. See, for example, Draper and Smith (1998) or Neter et al. (1996) for discussions of the F test for overall significance of the regression.

7.2.5. Sampling Distributions of the Regression Coefficients

Another important use of the estimated residual variance is to obtain estimates of the sampling distributions of the regression coefficients. As statistics computed from a finite set of data subject to sampling variations, the computed regression intercept and slope, a and b, also exhibit sampling variability. That is, different batches of size n from the same data-generating process will yield different pairs of regression slopes and intercepts, and their sampling distributions characterize this batch-to-batch variability. Estimation of these sampling distributions allows construction of confidence intervals for the true population counterparts around the sample intercept and slope values, and provides a basis for hypothesis tests about the corresponding population values.

Under the assumptions listed previously, the sampling distributions for both intercept and slope are Gaussian. On the strength of the Central Limit Theorem, this result also holds at least approximately for any regression when n is large enough, because the estimated regression parameters (Equation 7.7) are obtained as the sums of large numbers of random variables. For the intercept the sampling distribution has parameters

$$\mu_a = a \tag{7.18a}$$

and

$$\sigma_a = s_e \left[\frac{\sum_{i=1}^{n} x_i^2}{n \sum_{i=1}^{n} (x_i - \bar{x})^2} \right]^{1/2}. \tag{7.18b}$$

For the slope the parameters of the sampling distribution are

$$\mu_b = b \tag{7.19a}$$

and

$$\sigma_b = \frac{s_e}{\left[\sum_{i=1}^{n} (x_i - \bar{x})^2 \right]^{1/2}}. \tag{7.19b}$$

Equations 7.18a and 7.19a indicate that the least-squares regression parameter estimates are unbiased. Equations 7.18b and 7.19b show that the precision with which the intercept and slope can be estimated from the data depends directly on the estimated standard deviation of the residuals, s_e, which is the square root of the MSE from the ANOVA table (Table 7.1). Additionally, the estimated slope and intercept are not independent, having correlation

$$r_{a,b} = \frac{-\bar{x}}{\frac{1}{n}\left(\sum_{i=1}^{n} x_i^2\right)^{1/2}}. \tag{7.20}$$

Taken together with the (at least approximately) Gaussian sampling distributions for a and b, Equations 7.18–7.20 define their joint bivariate normal (Equation 4.31) distribution. Equations 7.18b, 7.19b, and 7.20 are valid only for simple linear regression. With more than one predictor variable, analogous (vector) equations (Equation 11.40) must be used.

The output from regression packages will almost always include the standard errors (Equations 7.18b and 7.19b) in addition to the parameter estimates themselves. Some packages also include the ratios of the estimated parameters to their standard errors in a column labeled "t ratio." When this is done, a one-sample t-test (Equation 5.3) is implied, with the null hypothesis being that the underlying (population) mean for the parameter is zero. Sometimes a p value associated with this test is also automatically included in the regression output.

For the regression slope, this implicit t-test bears directly on the meaningfulness of the fitted regression. If the estimated slope is small enough that its true value could plausibly (with respect to its sampling distribution) be zero, then the regression is not informative, or useful for forecasting. If the slope is actually zero, then the value of the predictand specified by the regression equation is always the same and equal to its sample mean (cf. Equations 7.1 and 7.7b). If the assumptions regarding the regression residuals are satisfied, we would reject this null hypothesis at the 5% level if the estimated slope is, roughly, at least twice as large (in absolute value) as its standard error.

The same hypothesis test for the regression intercept often is offered by computerized statistical packages as well. Depending on the problem at hand, however, this test for the intercept may or may not be meaningful. Again, the t ratio is just the parameter estimate divided by its standard error, so the implicit null hypothesis is that the true intercept is zero. Occasionally, this null hypothesis is physically meaningful, and if so the test statistic for the intercept is worth looking at. On the other hand, it often happens that there is no physical reason to expect that the intercept might be zero. It may even be that a zero intercept is physically impossible. In such cases this portion of the automatically generated computer output is meaningless.

Example 7.1. A Simple Linear Regression

To concretely illustrate simple linear regression, consider the January 1987 minimum temperatures at Ithaca and Canandaigua from Table A.1 in Appendix A. Let the predictor variable, x, be the Ithaca minimum temperature, and the predictand, y, be the Canandaigua minimum temperature. The scatterplot of this data is shown the middle panel of the bottom row of the scatterplot matrix in Figure 3.31, and as part of Figure 7.10. A fairly strong, positive, and reasonably linear relationship is indicated.

Table 7.2 presents what the output from a typical statistical computer package would look like for this regression. The data set is small enough that the computational formulas can be worked through to verify the results. (A little work with a hand calculator will verify that $\Sigma x = 403$, $\Sigma y = 627$, $\Sigma x^2 = 10{,}803$, $\Sigma y^2 = 15{,}009$, and $\Sigma xy = 11{,}475$.) The upper portion of Table 7.2 corresponds to the template in Table 7.1, with the relevant numbers filled in. Of particular importance is MSE $= 11.780$, yielding as its square root the estimated sample standard deviation for the residuals, $s_e = 3.43°$F. This standard deviation addresses directly the precision of specifying the Canandaigua temperatures on the basis of the concurrent Ithaca temperatures, since we expect about 95% of the actual predictand values to be within $\pm 2s_e = \pm 6.9°$F of the temperatures given by the regression. The coefficient of determination is easily

TABLE 7.2 Example Output Typical of that Produced by Computer Statistical Packages, for Prediction of Canandaigua Minimum Temperature (y) using Ithaca Minimum Temperature (x) as the Predictor, from the January 1987 Data Set in Table A.1

Source	df	SS	MS	F
Total	30	2327.419		
Regression	1	1985.798	1985.798	168.57
Residual	29	341.622	11.780	

Variable	Coefficient	s.e.	t ratio
Constant	12.4595	0.8590	14.504
Ithaca Min	0.5974	0.0460	12.987

computed as $R^2 = 1985.798/2327.419 = 85.3\%$. The Pearson correlation is $\sqrt{0.853} = 0.924$, as was given in Table 3.5. The value of the F statistic is very high, considering that the 99th percentile of its distribution under the null hypothesis of no real relationship is about 7.5. We also could compute the sample variance of the predictand, which would be the total mean square cell of the table, as $2327.419/30 = 77.58°F^2$.

The lower portion of Table 7.2 gives the regression parameters, a and b, their standard errors, and the ratios of these parameter estimates to their standard errors. The specific regression equation for this data set, corresponding to Equation 7.1, would be

$$T_{Can.} = \underset{(0.859)}{12.46} + \underset{(0.046)}{0.597}\, T_{Ith.}. \tag{7.21}$$

Thus the Canandaigua temperature would be estimated by multiplying the Ithaca temperature by 0.597 and adding 12.46°F. The intercept $a = 12.46°F$ has no special physical significance except as the predicted Canandaigua temperature when the Ithaca temperature is 0°F. Notice that the standard errors of the two coefficients have been written parenthetically below the coefficients themselves. Although this is not a universal practice, it is very informative to someone reading Equation 7.21 without the benefit of the information in Table 7.2. In particular, it allows the reader to get a sense for the significance of the slope (i.e., the parameter b). Since the estimated slope is about 13 times larger than its standard error it is almost certainly not really zero. This conclusion speaks directly to the question of the meaningfulness of the fitted regression. On the other hand, the corresponding implied hypothesis test for the intercept is much less interesting, because the possibility of a zero intercept is of no physical interest. ◇

7.2.6. Examining Residuals

It is not sufficient to feed data to a computer regression package and uncritically accept the results. Some of the results can be misleading if the assumptions underlying the computations are not satisfied. Since these assumptions pertain to the residuals, it is important to examine the residuals for consistency with the assumptions made about their behavior.

One easy and fundamental check on the residuals can be made by examining a scatterplot of the residuals as a function of the predicted value $\hat{y}$. Many statistical computer packages provide this capability as a standard regression option. Figure 7.4a shows the scatterplot of a hypothetical data set, with the least-squares regression line, and Figure 7.4b shows a plot for the resulting residuals as a function of the predicted values. The residual plot presents the impression of "fanning," or exhibition of increasing spread as $\hat{y}$ increases. That is, the variance of the residuals appears to increase as the predicted value increases. This condition of nonconstant residual variance is called *heteroscedasticity*. Since the computer program that fit the regression has assumed constant residual variance, the MSE given in the ANOVA table is an overestimate for smaller values of x and y (where the points cluster closer to the regression line), and an underestimate of the residual variance for larger values of x and y (where the points tend to be further from the regression line). If the regression is used as a forecasting tool, we would be overconfident about forecasts for larger values of y, and underconfident about forecasts for smaller values of y. In addition, the sampling distributions of the regression parameters will be more variable than implied by Equations 7.18 and 7.19. That is, the parameters will not have been estimated as precisely as the standard regression output would lead us to believe.

Often nonconstancy of residual variance of the sort shown in Figure 7.4b can be remedied by transforming the predictand y, perhaps by using a power transformation (Equations 3.20 or 3.23). Figure 7.5 shows the regression and residual plots for the same data as in Figure 7.4 after logarithmically transforming the predictand. Recall that the logarithmic transformation reduces all the data values, but reduces the larger values more strongly than the smaller ones. Thus the long right tail of the predictand has been pulled in relative to the shorter left tail, as in Figure 3.13. As a result, the transformed data points

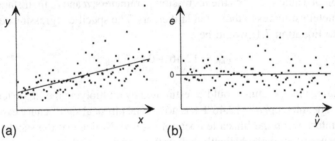

FIGURE 7.4 Hypothetical linear regression (a), and plot of the resulting residuals against the predicted values (b), for a case where the variance of the residuals is not constant. The scatter around the regression line in (a) increases for larger values of x and y, producing a visual impression of "fanning" in the residual plot (b). A transformation of the predictand is indicated.

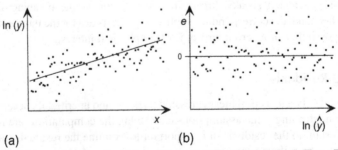

FIGURE 7.5 Scatterplot with regression (a), and resulting residual plot (b), for the same data in Figure 7.4, after logarithmically transforming the predictand. The visual impression of a horizontal band in the residual plot supports the assumption of constant variance of the residuals.

appear to cluster more evenly around the new regression line. Instead of fanning, the residual plot in Figure 7.5b gives the visual impression of a horizontal band, indicating appropriately constant variance of the residuals (*homoscedasticity*). Note that if the fanning in Figure 7.4b had been in the opposite sense, with greater residual variability for smaller values of $\hat{y}$ and lesser residual variability for larger values of $\hat{y}$, a transformation that stretches the right tail relative to the left tail (e.g., y^2) would have been appropriate.

It can also be informative to look at scatterplots of residuals as a function of the predictor variable. Figure 7.6 illustrates some of the forms such plots can take and their diagnostic interpretations. Figure 7.6a is similar to Figure 7.4b, in that the fanning of the residuals indicates nonconstancy of variance. Figure 7.6b illustrates a different form of heteroscedasticity that might be more challenging to remedy through a variable transformation. The type of residual plot in Figure 7.6c, with a linear dependence on the predictor of the linear regression, indicates that either the intercept a has been omitted or that the calculations have been done incorrectly. Deliberately omitting a regression intercept, called "forcing through the origin," is useful in some circumstances, but may not be appropriate even if it is known beforehand that the true relationship should pass through the origin. Particularly if data are available over only a restricted range, or if the actual relationship is nonlinear, a linear regression including an intercept term may yield better predictions. In this latter case a simple linear regression would be analogous to a first-order Taylor approximation about the mean of the training data.

Figure 7.6d shows a form for the residual plot that can occur when additional predictors would improve a regression relationship. Here the variance is reasonably constant in x, but the (conditional) average residual exhibits a dependence on x. Figure 7.6e illustrates the kind of behavior that can occur when a single outlier in the data has undue influence on the regression. Here the regression line has been pulled toward the outlying point in order to avoid the large squared error associated with it, leaving a trend in the other residuals. If the outlier were determined not to be a valid data point, it should either be corrected if possible or otherwise discarded. If it is a valid data point, a resistant approach such as LAD

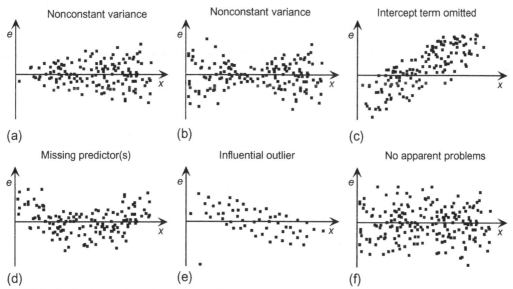

FIGURE 7.6   Idealized scatterplots of regression residuals vs. a predictor x, with corresponding diagnostic interpretations.

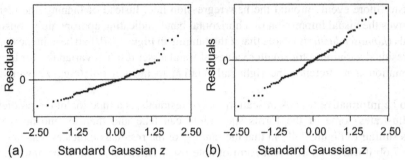

FIGURE 7.7 Gaussian quantile-quantile plots of the residuals for predictions of (a) the untransformed predictand in Figure 7.4a, and (b) the logarithmically transformed predictand in Figure 7.5. In addition to producing essentially constant residual variance, logarithmic transformation of the predictand has rendered the distribution of the residuals effectively Gaussian.

regression (Section 7.7.3) or median-slope regression (Section 7.7.2) might be more appropriate. Figure 7.6f again illustrates the desirable horizontally banded pattern of residuals, similar to Figure 7.5b.

A graphical impression of whether the residuals follow a Gaussian distribution can be obtained through a Q-Q plot. Such plots are often a standard option in statistical computer packages. Figure 7.7a and b show Q-Q plots for the residuals from Figures 7.4b and 7.5b, respectively. The residuals are plotted on the vertical, and the standard Gaussian variables corresponding to the empirical cumulative probability of each residual are plotted on the horizontal. The curvature apparent in Figure 7.7a indicates that the residuals from the regression involving the untransformed predictand are positively skewed relative to the (symmetric) Gaussian distribution. The Q-Q plot of residuals from the regression involving the logarithmically transformed predictand is very nearly linear. Evidently the logarithmic transformation has produced residuals that are close to Gaussian, in addition to stabilizing the residual variances. Similar conclusions could have been reached using a goodness-of-fit test (see Section 5.2.5).

It is also possible and desirable to investigate the degree to which the residuals are uncorrelated. This question is of particular interest when the underlying data are serially correlated, which is a common condition for atmospheric variables. A simple graphical evaluation can be obtained by plotting the regression residuals as a function of time. If groups of positive and negative residuals tend to cluster together (qualitatively resembling Figure 5.4b) rather than occurring more irregularly (as in Figure 5.4a), then time correlation can be suspected.

The *Durbin-Watson test* is a popular formal test for serial correlation of regression residuals that is included in many computer regression packages. This test examines the null hypothesis that the residuals are serially independent, against the alternative that they are consistent with a first-order autoregressive process (Equation 10.16). The Durbin-Watson test statistic,

$$d = \frac{\sum\limits_{i=2}^{n}(e_i - e_{i-1})^2}{\sum\limits_{i=1}^{n}e_i^2}, \tag{7.22}$$

computes the squared differences between pairs of consecutive residuals, divided by a scaling factor proportional to the residual variance. If the residuals are positively correlated, adjacent residuals will tend to be similar in magnitude, so the Durbin-Watson statistic will be relatively small. If the sequence

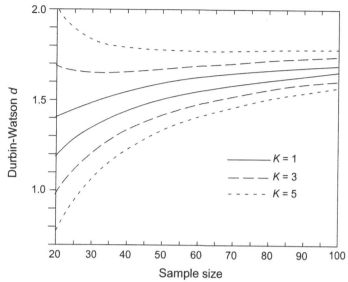

FIGURE 7.8 5%-level critical values for the Durbin-Watson statistic as a function of the sample size, for $K = 1, 3,$ and 5 predictor variables. A test statistic d below the relevant lower curve results in a rejection of the null hypothesis of zero serial correlation. If the test statistic is above the relevant upper curve the null hypothesis is not rejected. If the test statistic is between the two curves the test is indeterminate without additional calculations.

of residuals is randomly distributed, the sum in the numerator will tend to be larger. Therefore the null hypothesis that the residuals are independent is rejected if the Durbin-Watson statistic is sufficiently small.

Figure 7.8 shows critical values for Durbin-Watson tests at the 5% level. These vary depending on the sample size, and the number of predictor (x) variables, K. For simple linear regression, $K = 1$. For each value of K, Figure 7.8 shows two curves. If the observed value of the test statistic falls below the lower curve, the null hypothesis is rejected and we conclude that the residuals exhibit significant serial correlation. If the test statistic falls above the upper curve, we do not reject the null hypothesis that the residuals are serially uncorrelated. If the test statistic falls between the two relevant curves, the test is indeterminate. The reason behind the existence of this unusual indeterminate condition is that the null distribution of the Durban-Watson statistic depends on the data set being considered. In cases where the test result is indeterminate according to Figure 7.8, some additional calculations (Durban and Watson, 1971) can be performed to resolve the indeterminacy, that is, to find the specific location of the critical value between the appropriate pair of curves, for the particular data at hand.

Example 7.2. Examination of the Residuals from Example 7.1
A regression equation constructed using autocorrelated variables as predictand and predictor(s) does not necessarily exhibit strongly autocorrelated residuals. Consider again the regression between Ithaca and Canandaigua minimum temperatures for January 1987 in Example 7.1. The lag-1 autocorrelations (Equation 3.36) for the Ithaca and Canandaigua minimum temperature data are 0.651 and 0.672, respectively. The residuals for this regression are plotted as a function of time in Figure 7.9. A strong serial correlation for these residuals is not apparent, and their lag-1 autocorrelation as computed using Equation 3.36 is only 0.191.

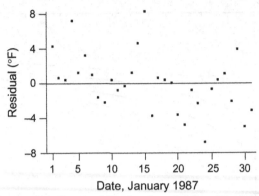

FIGURE 7.9 Residuals from the regression, Equation 7.21, plotted as a function of the date. A strong serial correlation is not apparent, but the tendency for a negative slope suggests that the relationship between Ithaca and Canandaigua temperatures may be changing through the month.

Having obtained the residuals for the Canandaigua vs. Ithaca minimum temperature regression, it is straightforward to compute the Durbin-Watson d (Equation 7.22). In fact, the denominator is simply the SSE from the ANOVA Table 7.2, which is 341.622. The numerator in Equation 7.22 must be computed from the residuals and is 531.36. These yield $d = 1.55$. Referring to Figure 7.8, the point at $n = 31$, $d = 1.55$ is well above the upper solid (for $K = 1$, since there is a single predictor variable) line, so the null hypothesis of uncorrelated residuals would not be rejected at the 5% level. Evidently the strong serial correlation exhibited by the predictand has been captured in the regression, through the strong serial correlation in the predictor.   ◇

When regression residuals are autocorrelated, statistical inferences based upon their variance are degraded in the same way, and for the same reasons, that were discussed in Section 5.2.4 (e.g., Matalas and Sankarasubramanian, 2003; Santer et al., 2000; Zheng et al., 1997). In particular, positive serial correlation of the residuals leads to inflation of the variance of the sampling distribution of their sum or average, because these quantities are less consistent from batch to batch of size n. When a first-order autoregression (Equation 10.16) is a reasonable representation for these correlations (characterized by r_1) it is appropriate to apply the same variance inflation factor, $(1 + r_1)/(1 - r_1)$ (the bracketed quantity in Equation 5.13), to the variance s_e^2 in, for example, Equations 7.18b and 7.19b (Matalas and Sankarasubramanian, 2003; Santer et al., 2000). The net effect is that the variance of the resulting sampling distribution is (appropriately) increased, relative to what would be calculated assuming independent regression residuals.

7.2.7. Prediction Intervals

Many times it is of interest to calculate *prediction intervals* around forecast values of the predictand (i.e., around the regression function), which are meant to bound a future value of the predictand with specified probability. When it can be assumed that the residuals follow a Gaussian distribution, it is natural to approach this problem using the unbiasedness property of the residuals (Equation 7.9), together with their estimated variance $MSE = s_e^2$. In terms of Figure 7.2, suppose the regression is being evaluated at a predictor value that is equal to x_3. The uncertainty associated with a corresponding future unknown value of the predictand would be represented approximately by the rightmost gray conditional residual distribution in Figure 7.2. Using Gaussian probabilities (Table B.1), we expect a 95% prediction interval for a future residual, or specific future forecast, to be approximately bounded by $\hat{y} \pm 2s_e$.

The $\pm 2s_e$ rule of thumb is often a quite good approximation to the width of a true 95% prediction interval, especially when the sample size is large. However, because both the sample mean of the predictand and the slope of the regression are subject to sampling variations, the prediction variance for future data (i.e., for data not used in the fitting of the regression) is somewhat larger than the regression MSE. For a forecast of y using the predictor value x_0, this prediction variance is given by

$$s_{\hat{y}}^2 = s_e^2 \left[1 + \frac{1}{n} + \frac{(x_0 - \bar{x})^2}{\sum\limits_{i=1}^{n}(x_i - \bar{x})^2} \right]. \tag{7.23}$$

That is, the prediction variance is proportional to the regression MSE, but is larger to the extent that the second and third terms inside the square brackets are appreciably larger than zero. The second term derives from the uncertainty in estimating the true mean of the predictand from a finite sample of size n (compare Equation 5.4), and becomes very small for large sample sizes. The third term derives from the uncertainty in estimation of the slope (it is similar in form to Equation 7.19b), and indicates that predictions far removed from the center of the data used to fit the regression will be more uncertain than predictions made near the sample mean. However, even if the numerator in this third term is fairly large, the term itself will tend to be small if a large data sample was used to construct the regression equation, since there are n nonnegative terms of generally comparable magnitude in the denominator.

It is sometimes also of interest to compute *confidence intervals* for the regression function itself. These will be narrower than the prediction intervals for future individual data values, reflecting a smaller variance in a way that is analogous to the variance of a sample mean being smaller than the variance of its underlying data values. The variance for the sampling distribution of the regression function, or equivalently the variance of the conditional mean of the predictand given a particular predictor value x_0, is

$$s_{\bar{y}|x_0}^2 = s_e^2 \left[\frac{1}{n} + \frac{(x_0 - \bar{x})^2}{\sum\limits_{i=1}^{n}(x_i - \bar{x})^2} \right]. \tag{7.24}$$

This expression is similar to Equation 7.23, but is smaller by the amount s_e^2. That is, there are contributions to this variance due to uncertainty in the mean of the predictand (or, equivalently the vertical position of the regression line, or the intercept), corresponding to the first of the two terms in the square brackets; and to uncertainty in the slope, corresponding to the second term. There is no contribution to Equation 7.24 reflecting scatter of data around the regression line, which is the difference between Equations 7.23 and 7.24.

Figure 7.10 compares prediction and confidence intervals computed using Equations 7.23 and 7.24, in the context of the regression from Example 7.1. Here the regression (Equation 7.21) fit to the 31 data points (dots) is shown by the heavy solid line. The 95% prediction interval around the regression computed as $\pm 1.96 \, s_{\hat{y}}$, using the square root of Equation 7.23, is indicated by the pair of slightly curved solid black lines. As noted earlier, these bounds are only slightly wider than those given by the simpler approximation $\hat{y} \pm 1.96 \, s_e$ (dashed lines), because the second and third terms in the square brackets of

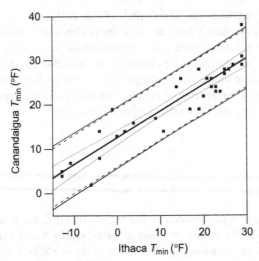

FIGURE 7.10 Prediction and confidence intervals around the regression derived in Example 7.1 (thick black line). Light solid lines indicate 95% prediction intervals for future data, computed using Equation 7.23, and the corresponding dashed lines simply locate the predictions $\pm 1.96\, s_e$. Light gray lines locate 95% confidence intervals for the regression function (Equation 7.24). Data to which the regression was fit are also shown.

Equation 7.23 are relatively small, even for moderate n. The pair of gray curved lines locates the 95% confidence interval for the conditional mean of the predictand. These are much narrower than the prediction interval because they account only for sampling variations in the regression parameters, without direct contributions from the prediction variance s_e^2.

Equations 7.18 through 7.20 define the parameters of a bivariate normal distribution for the two regression parameters. Imagine using the methods outlined in Section 4.7 to generate pairs of intercepts and slopes according to that distribution, and therefore to generate realizations of plausible regression lines. One interpretation of the gray curves in Figure 7.10 is that they would contain 95% of those regression lines (or, equivalently, 95% of the regression lines computed from different samples of data of this kind, each with size $n = 31$). The minimum separation between the gray curves (at the average Ithaca $T_{min} = 13°F$) reflects the uncertainty in the intercept. Their spreading at more extreme temperatures reflects the fact that uncertainty in the slope (i.e., uncertainty in the angle of the regression line) will produce more uncertainty in the conditional expected value of the predictand at the extremes than near the mean, because Equations 7.6a and 7.7b show that any regression line must pass through the point located by the two sample means.

The result in Example 7.2 indicates that the residuals for this regression can reasonably be regarded as independent. Also, some of the sample lag-1 autocorrelation of $r_1 = 0.191$ can be attributable to the time trend evident in Figure 7.9. However, if the residuals are significantly correlated, and the nature of that correlation is plausibly represented by a first-order autoregression (Equation 10.16), it would be appropriate to increase the residual variances s_e^2 in Equations 7.23 and 7.24 by multiplying them by the variance inflation factor $(1 + r_1)/(1 - r_1)$.

Special care is required when computing prediction and confidence intervals for regressions involving transformed predictands. For example, if the relationship shown in Figure 7.5a (involving a log-transformed predictand) were to be used in forecasting, dimensional values of the predictand would need to be recovered in order to make the forecasts interpretable. That is, the predictand $\ln(\hat{y})$ would need to be back-transformed, yielding the forecast $\hat{y} = \exp.[\ln(\hat{y})] = \exp.[a + bx]$. Similarly, the limits of the prediction intervals would also need to be back-transformed. For example, the 95% prediction interval would be approximately $\ln(\hat{y}) \pm 1.96\, s_e$, because the regression residuals and their assumed Gaussian distribution pertain to the transformed predictand values. The lower and upper limits of this interval, when expressed on the original untransformed scale of the predictand, would be approximately $\exp[a + bx - 1.96\, s_e]$ and $\exp[a + bx + 1.96 s_e]$. These limits would not be symmetrical around $\hat{y}$, and would extend further for the larger values, consistent with the longer right tail of the predictand distribution.

Equations 7.23 and 7.24 are valid for simple linear regression. The corresponding equations for multiple regression are similar, but are more conveniently expressed in matrix algebra notation (see Example 11.2). However, as is the case for simple linear regression, the prediction variance is quite close to the MSE for moderately large samples.

7.3. MULTIPLE LINEAR REGRESSION

7.3.1. Extending Simple Linear Regression

Multiple linear regression is the more general (and more common) implementation of linear regression. As in the case of simple linear regression, there is still a single predictand, y, but in distinction there is more than one predictor (x) variable. The preceding treatment of simple linear regression was relatively lengthy, in part because most of what was presented generalizes readily to the case of multiple linear regression.

Let K denote the number of predictor variables. Simple linear regression is then the special case of $K = 1$. The prediction equation (corresponding to Equation 7.1) becomes

$$\hat{y} = b_0 + b_1 x_1 + b_2 x_2 + \cdots + b_K x_K. \tag{7.25}$$

Each of the K predictor variables has its own coefficient, analogous to the slope, b, in Equation 7.1. For notational convenience, the intercept (or *regression constant*) is denoted as b_0 rather than a, as in Equation 7.1. These $K + 1$ regression coefficients often are called the *regression parameters*.

Equation 7.2 for the residuals is still valid, if it is understood that the predicted value $\hat{y}$ is a function of a vector of predictors, x_k, $k = 1, \ldots, K$. If there are $K = 2$ predictor variables, the residual can still be visualized as a vertical distance. In that case the regression function (Equation 7.25) is a surface rather than a line, and the residual corresponds geometrically to the distance above or below this surface along a line perpendicular to the (x_1, x_2) plane. The geometric situation is analogous for $K \geq 3$, but is not easily visualized. Also in common with simple linear regression, the average residual is guaranteed to be zero, so that the residual distributions are centered on the predicted values $\hat{y}_i$. Accordingly, these predicted values can be regarded as conditional means given particular values for a set of K predictors.

The $K + 1$ parameters in Equation 7.25 are found, as before, by minimizing the sum of squared residuals. This is achieved by simultaneously solving $K + 1$ equations analogous to Equation 7.5. This minimization is most conveniently done using matrix algebra, the details of which can be found in standard regression texts (e.g., Draper and Smith, 1998; Neter et al., 1996). The basics of the process are outlined in Example 11.2. In practice, the calculations usually are done using statistical software. They are again summarized in an ANOVA table, of the form shown in Table 7.3. As before, SST is

TABLE 7.3 Generic Analysis of Variance (ANOVA) Table for Multiple Linear Regression

Source	df	SS	MS	F
Total	$n-1$	SST		
Regression	K	SSR	$MSR = SSR/K$	$F = MSR/MSE$
Residual	$n-K-1$	SSE	$MSE = SSE/(n-K-1) = s_e^2$	

Table 7.1 for simple linear regression can be viewed as a special case, with $K=1$.

computed using Equation 7.13, SSR is computed using Equation 7.14a, and SSE is computed using the difference SST − SSR. The sample variance of the residuals is $MSE = SSE/(n-K-1)$.

The coefficient of determination is computed according to Equation 7.17, although in the more general case of multiple linear regression it is the square of the Pearson correlation coefficient between the observed and predicted values of the predictand y. This generalizes the meaning of R^2 as the square of the Pearson correlation between x and y for simple linear regression, because in that setting the predicted value is a linear function of the single predictor x, and correlation is insensitive to linear transformations.

Expressions for the variances characterizing prediction uncertainty and conditional-mean predictand uncertainty in multiple regression, generalizing Equations 7.23 and 7.24, are given by Equations 11.43 and 11.42, respectively. The procedures presented previously for examination of residuals are applicable to multiple regression as well.

7.3.2. Derived Predictor Variables in Multiple Regression

Multiple regression opens up the possibility of an essentially unlimited number of potential predictor variables. An initial list of potential predictor variables can be expanded manyfold by also considering nonlinear mathematical transformations of these variables as potential predictors. The derived predictors must be nonlinear functions of the primary predictors in order for the computations (in particular, for the matrix inversion indicated in Equation 11.39) to be possible. Such *derived predictors* can be very useful in producing a good regression equation.

In some instances the most appropriate forms for predictor transformations may be suggested by a physical understanding of the data-generating process. In the absence of a strong physical rationale for particular predictor transformations, the choice of a transformation or set of transformations may be made purely empirically, perhaps by subjectively evaluating the general shape of the point cloud in a scatterplot, or the nature of the deviation of a residual plot from its ideal form. For example, the curvature in the residual plot in Figure 7.6d suggests that addition of the derived predictor $x_2 = x_1^2$ might improve the regression relationship. It may happen that the empirical choice of a transformation for a predictor variable in regression leads to a greater physical understanding, which is a highly desirable outcome in a research setting. This outcome would be less important in a purely forecasting setting, where the emphasis is on producing good forecasts rather than knowing precisely why the forecasts are good.

Transformations such as $x_2 = x_1^2$, $x_2 = \sqrt{x_1}$, $x_2 = 1/x_1$, or any other power transformation of an available predictor, can be adopted as potential predictors. Similarly, trigonometric (sine, cosine, etc.), exponential or logarithmic functions, or combinations of these are useful in some situations. Another commonly used transformation is to a *binary variable*, also known as a *dummy variable* or *indicator variable.* Binary variables take on one of two values (usually 0 and 1, although the particular choices do not affect subsequent use of the regression equation), depending on whether the variable being transformed is above or below a threshold or cutoff, c. That is, a binary variable x_2 could be constructed from another predictor x_1 according to the transformation

$$x_2 = \begin{cases} 1, & \text{if } x_1 > c \\ 0, & \text{if } x_1 \leq c \end{cases}. \tag{7.26}$$

More than one binary predictor can be constructed from a single x_1 by choosing different values of the cutoff, c, for x_2, x_3, x_4, and so on.

Even though transformed variables may be nonlinear functions of other variables, the overall framework is still known as multiple linear regression. Once a derived variable has been defined it is just another variable, regardless of how the transformation was made. More formally, the "linear" in multiple linear regression refers to the regression equation being linear in the parameters, b_k.

Example 7.3. A Multiple Regression with Derived Predictor Variables

Figure 7.11 shows a scatterplot of the famous Keeling monthly averaged carbon dioxide (CO_2) concentration data from Mauna Loa in Hawaii, for the period March 1958 through December 2017. Representing the obvious time trend as a straight line yields the regression results presented in Table 7.4a, and the regression line is also plotted (dashed) in Figure 7.11. The results indicate a strong time trend, with the calculated standard error for the slope being much smaller than the estimated slope. The intercept merely estimates the CO_2 concentration at $t = 0$, or February 1958, so the implied test for

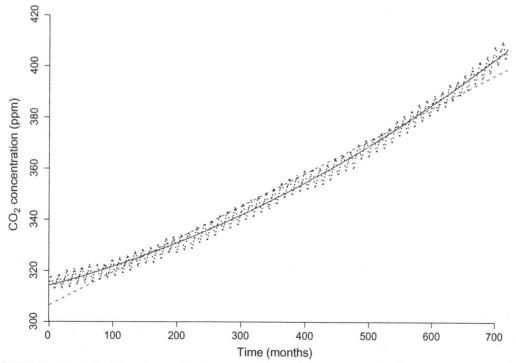

FIGURE 7.11   The Keeling Mauna Loa monthly CO_2 concentration data (March 1958–December 2017), with linear (dashed) and quadratic (solid) least-squares fits.

its difference from zero is of no interest. A literal interpretation of the MSE would suggest that a 95% prediction interval for measured CO_2 concentrations around the regression line would be about $\pm 2 \sqrt{MSE} = \pm 8$ ppm.

However, examination of a plot of the residuals versus time for this linear regression would reveal a bowing pattern similar to that in Figure 7.6d, with a tendency for positive residuals at the beginning and end of the record, and with negative residuals being more common in the central part of the record. This can be discerned from Figure 7.11 by noticing that most of the points fall above the dashed line early and late in the record, and fall below the line toward the middle.

This problem with the residuals can be alleviated (and the regression consequently improved) by fitting a quadratic curve to the time trend. To do this, a second predictor is added to the regression, and that predictor is simply the square of the time variable. That is, a multiple regression with $K = 2$ is fit using the predictors $x_1 = t$ and $x_2 = t^2$. Once defined, x_2 is just another predictor variable, taking on values between 1^2 and $718^2 = 515{,}524$. The resulting least-squares quadratic regression is shown by the solid curve in Figure 7.11, and the corresponding regression statistics are summarized in Table 7.4b.

Of course the SST in Tables 7.4a and 7.4b are the same since both pertain to the same predictand, the CO_2 concentrations. For the quadratic regression, both the coefficients $b_1 = 0.0657$ and $b_2 = 0.0000870$ are substantially larger than their respective standard errors. The value of $b_0 = 314.3$ is again just the estimate of the CO_2 concentration at $t = 0$, and judging from the scatterplot this intercept is a better estimate of its true value than was obtained from the simple linear regression. The data points are fairly evenly scattered around the quadratic trend line throughout the time period, so the residual plot would exhibit the desired horizontal banding. Using this analysis, an approximate 95% prediction interval of $\pm 2 \sqrt{MSE} = \pm 4.4$ ppm for CO_2 concentrations around the quadratic regression would be inferred throughout the range of these data.

The quadratic function of time provides a reasonable approximation of the annual-average CO_2 concentration for the 60 years represented by the regression, although we can find periods of time where the center of the point cloud wanders away from the curve. More importantly, however, a close inspection of the data points in Figure 7.11 reveals that they are not scattered randomly around the quadratic time trend. Rather, they execute a regular, nearly sinusoidal variation around the quadratic curve that is evidently an annual cycle. The resulting serial correlation in the residuals can easily be detected using the Durbin-Watson statistic, $d = 0.313$ (compare Figure 7.8). The CO_2 concentrations are lower in late summer and higher in late winter as a consequence of the annual cycle of photosynthetic carbon uptake by northern hemisphere land plants and carbon release from the decomposing dead plant parts. As will be shown in Section 10.4.2, this regular 12-month variation can be represented by introducing two more derived predictor variables into the equation, $x_3 = \cos(2\pi t/12)$ and $x_4 = \sin(2\pi t/12)$. Notice that both of these derived variables are functions only of the time variable t.

Table 7.4c indicates that, together with the linear and quadratic predictors included previously, these two harmonic predictors produce a very close fit to the data. The resulting prediction equation is

$$[CO_2] = \underset{(.1085)}{314.3} + \underset{(.0007)}{0.0657\,t} + \underset{(.0000)}{.0000872\,t^2} + \underset{(.0503)}{1.149}\,\cos\left(\frac{2\pi t}{12}\right) + \underset{(.0502)}{2.582}\,\sin\left(\frac{2\pi t}{12}\right), \qquad (7.27)$$

with all regression coefficients being much larger than their respective standard errors. The near equality of SST and SSR in Table 7.4c indicates that the predicted values are nearly coincident with the observed CO_2 concentrations (compare Equations 7.13 and 7.14a). The resulting coefficient of determination is $R^2 = 511{,}723/512356 = 99.88\%$, and the approximate 95% prediction interval implied by $\pm 2 \sqrt{MSE}$ is only ± 1.9 ppm. A graph of Equation 7.27 would wiggle up and down around the solid curve in Figure 7.11, passing rather close to each of the data points.   ◇

TABLE 7.4 ANOVA Tables and Regression Summaries for Three Regressions Fit to the 1958–2017 Keeling CO_2 Data in Figure 7.11

Source	df	SS	MS	F
(a) *Linear fit*				
Total	710	512,356		
Regression	1	501,014	501,014	31,317
Residual	709	11,342	15.998	
Variable	Coefficient	s.e.	t-ratio	
Constant	306.7	0.3027	1013	
t	0.1285	0.0007	177.0	
(b) *Quadratic fit*				
Total	710	512,356		
Regression	2	508,885	254,443	51,895
Residual	708	3471	4.903	
Variable	Coefficient	s.e.	t-ratio	
Constant	314.3	0.2536	1240	
t	0.0657	0.0016	40.6	
t^2	0.0000870	0.0000	40.1	
(c) *Including quadratic trend, and harmonic terms to represent the annual cycle*				
Total	710	512,356		
Regression	4	511,723	127,931	142,631
Residual	706	633.2	0.8969	
Variable	Coefficient	s.e.	t-ratio	
Constant	314.3	0.1085	2898	
t	0.0657	0.0007	94.9	
t^2	0.0000872	0.0000	93.8	
$\cos(2\pi t/12)$	1.149	0.0503	22.9	
$\sin(2\pi t/12)$	2.582	0.0502	51.4	

The variable t (time) is a consecutive numbering of the months, with March 1958 = 1 and December 2017 = 718. There are $n = 711$ data points and 7 missing months.

7.4. PREDICTOR SELECTION IN MULTIPLE REGRESSION

7.4.1. Why is Careful Predictor Selection Important?

There are almost always more potential predictors available than can be used in a statistical prediction procedure, and finding good subsets of these in particular cases is more difficult than might at first be imagined. The process is definitely not as simple as adding members of a list of potential predictors until an apparently good relationship is achieved. Perhaps surprisingly, there are dangers associated with including too many predictor variables in a forecast equation.

Example 7.4. An Overfit Regression

To illustrate the dangers of too many predictors, Table 7.5 presents total winter snowfall at Ithaca (inches) for the seven winters beginning in 1980 through 1986 and four arbitrary potential predictors: the U.S. federal deficit (in billions of dollars), the number of personnel in the U.S. Air Force, the sheep population of the United States (in thousands), and the average Scholastic Aptitude Test (SAT) scores of college-bound high-school students. Obviously these are nonsense predictors, which bear no real relationship to the snowfall at Ithaca.

Regardless of their lack of relevance, we can blindly offer these predictors to a computer regression package, and it will produce a regression equation. For reasons that will be made clear shortly, assume that the regression will be fit using only the six winters beginning in 1980 through 1985. That portion of available data used to produce the forecast equation is known as the *developmental sample, dependent sample,* or *training sample.* For the developmental sample of 1980–1985, the resulting equation is

$$Snow = 1161771 - 601.7\,yr - 1.733\,deficit + 0.0567\,AF\,pers. - 0.3799\,sheep + 2.882\,SAT$$

The ANOVA table accompanying this equation (not reproduced here) indicated MSE$=0.0000$, $R^2 = 100.00\%$, and $F = \infty$, that is, a perfect fit!

Figure 7.12 shows a plot of the regression-specified snowfall totals (line segments) and the observed data (circles). For the developmental portion of the record, the regression does indeed represent the data exactly, as indicated by the ANOVA statistics, even though it is obvious from the nature of the predictor variables that the specified relationship is not physically meaningful. In fact, essentially any five predictors would have produced exactly the same perfect fit (although with different regression coefficients, b_k) to the six developmental data points. More generally, any $K = n - 1$ predictors will produce a perfect regression fit to any predictand for which there are n observations. This concept is easiest to see for the case of $n = 2$, where a straight line can be fit using any $K = 1$ predictor (simple linear regression), since a line can be found that will pass through any two points in the plane, and only an intercept and a slope are necessary to define a line. The problem, however, generalizes to any sample size.

TABLE 7.5 A small Data Set Illustrating the Dangers of Overfitting

Winter Beginning	Ithaca Snowfall (in.)	U.S. Federal Deficit ($\times 10^9$)	U.S. Air Force Personnel	U.S. Sheep ($\times 10^3$)	Average SAT Scores
1980	52.3	59.6	557,969	12,699	992
1981	64.9	57.9	570,302	12,947	994
1982	50.2	110.6	582,845	12,997	989
1983	74.2	196.4	592,044	12,140	963
1984	49.5	175.3	597,125	11,487	965
1985	64.7	211.9	601,515	10,443	977
1986	65.6	220.7	606,500	9932	1001

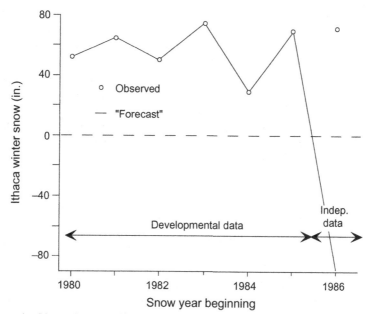

FIGURE 7.12 Forecasting Ithaca winter snowfall using the data in Table 7.5. The number of predictors is one fewer than the number of observations of the predictand in the developmental data, yielding perfect correspondence between the values specified by the regression and the predictand data for this portion of the record. The relationship falls apart completely when used with the 1986 data, which was not used in equation development. The regression equation has been grossly overfit.

Figure 7.12 indicates that the equation performs very poorly outside of the training sample, producing a meaningless forecast for negative snowfall during 1986–1987. Clearly, issuing forecasts equal to the climatological average total snowfall, or the snowfall for the previous winter, would yield better results than this overfit regression equation. ◇

Example 7.4 illustrates an extreme case of *overfitting* the data. Silver (2012) characterizes overfitting as "providing an overly specific solution to a general problem," and "the name given to the act of mistaking noise for a signal." That is, so many predictors have been used that an excellent fit has been achieved on the dependent data, but the fitted relationship falls apart when used with independent, or *verification data*—data not used in the development of the equation. In Example 7.4 the data for 1986 were reserved as a verification sample. Note that the problem of overfitting is *not* limited to cases where nonsense predictors are used in a forecast equation and will be a problem when too many meaningful predictors are included as well.

As ridiculous as it may seem, several important lessons can be drawn from Example 7.4:

- Begin development of a regression equation by choosing only physically reasonable or meaningful potential predictors. If the predictand of interest is surface temperature, for example, then temperature-related predictors such as the 1000–700 mb thickness (reflecting the mean virtual temperature in the layer), the 700 mb relative humidity (perhaps as a proxy for clouds), or the climatological average temperature for the forecast date (as a representation of the annual cycle of temperature) could be sensible candidate predictors. Understanding that clouds will form only in saturated air, a binary variable based on the 700 mb relative humidity also might be expected to contribute meaningfully to the regression. One consequence of this lesson is that a statistically literate person with insight into the physical problem ("domain expertise") may be more successful than a statistician at devising a forecast equation.

- A tentative regression equation needs to be tested on a sample of data not involved in its development. One way to approach this important step is simply to reserve a portion (perhaps a quarter, a third, or half) of the available data as the independent verification set and fit the regression using the remainder as the training set. The performance of the resulting equation will nearly always be better for the dependent than the independent data, since (in the case of least-squares regression) the coefficients have been chosen specifically to minimize the squared residuals in the developmental sample. A very large difference in performance between the dependent and independent samples would lead to the suspicion that the equation had been overfit. Often sufficient data are not available to reserve a separate verification set, in which case methods based on cross-validation (Section 7.4.4) are typically employed.

- A reasonably large developmental sample is needed if the resulting equation is to be "stable." Stability is usually understood to mean that the fitted coefficients are also applicable to independent (in particular, future) data, so that the resulting regression would be substantially unchanged if based on a different sample of the same kind of data. Stability thus relates to the precision of estimation of the regression coefficients, with smaller standard errors (Equations 7.18b and 7.19b, or square roots of the diagonal elements of Equation 11.40) being preferred. The number of coefficients that can be estimated with reasonable accuracy increases as the sample size increases, although in weather forecasting practice it has been found that there is little to be gained from including more than about a dozen predictor variables in a final regression equation (Glahn, 1985). In that kind of forecasting application there may be thousands of observations of the predictand in the developmental sample. Unfortunately, there is not a firm rule specifying a minimum ratio of sample size (number of observations of the predictand) to the number of predictor variables that can be supported in a final equation. Rather, testing on an independent data set is relied upon in practice to ensure stability of the regression.

7.4.2. Screening Predictors

Suppose the set of potential predictor variables for a particular problem could be assembled in a way that all physically relevant predictors were included, with exclusion of all irrelevant ones. This ideal can rarely, if ever, be achieved. Even if it could be, however, it generally would not be useful to include all the potential predictors in a final equation. This is because the predictor variables are almost always mutually correlated, so that the full set of potential predictors contains redundant information. Table 3.5, for example, shows substantial correlations among the six variables in Table A.1. Inclusion of predictors with strong mutual correlation is worse than superfluous, because this condition leads to poor estimates (high-variance sampling distributions) for the regression parameters. As a practical matter, then, we need a method to choose among potential predictors, and of deciding how many and which of them are sufficient to produce a good prediction equation.

In the jargon of statistical weather forecasting, the problem of selecting a good set of predictors from a pool of potential predictors is called *screening regression*, since the potential predictors must be subjected to some kind of screening, or filtering procedure. The most commonly used screening procedure is known as *forward selection* or *stepwise regression* in the broader statistical literature.

Given some number, M, candidate potential predictors for a least-squares linear regression, we begin the process of forward selection with the uninformative prediction equation $\hat{y} = b_0$. That is, only the intercept term is "in the equation," and this intercept is necessarily equal to the sample mean of the predictand. On the first forward selection step, all M potential predictors are examined for the strength of their linear relationship to the predictand. In effect, all the possible M simple linear regressions between the available predictors and the predictand are computed, and that predictor whose linear regression is best among all candidate predictors is chosen as x_1. At this stage of the screening procedure, then, the prediction equation is $\hat{y} = b_0 + b_1 x_1$. Note that in general the intercept b_0 no longer will be the average of the y values.

At the next stage of the forward selection, trial regressions are again constructed using all remaining $M - 1$ predictors. However, all these trial regressions also contain the variable selected on the previous step as x_1. That is, given the particular x_1 chosen on the previous step, that predictor variable yielding the best regression $\hat{y} = b_0 + b_1 x_1 + b_2 x_2$ is chosen as x_2. This new x_2 will be recognized as best because it produces that regression equation with $K = 2$ predictors that also includes the previously chosen x_1, having the highest R^2, the smallest MSE, and the largest F ratio.

Subsequent steps in the forward selection procedure follow this pattern exactly: at each step, that member of the potential predictor pool not yet in the regression is chosen that produces the best regression in conjunction with the $K-1$ predictors chosen on previous steps. In general, when these regression equations are recomputed the regression coefficients for the intercept and for the previously chosen predictors will change. These changes will occur because the predictors usually are correlated to a greater or lesser degree, so that information about the predictand is spread among the predictors differently as more predictors are added to the equation.

Example 7.5. Equation Development Using Forward Selection

The concept of predictor selection can be illustrated with the January 1987 temperature and precipitation data in Table A.1. As in Example 7.1 for simple linear regression, the predictand is Canandaigua minimum temperature. The potential predictor pool consists of maximum and minimum temperatures at Ithaca, maximum temperature at Canandaigua, the logarithms of the precipitation amounts plus 0.01 in. (in order for the logarithm to be defined for zero precipitation) for both locations, and the day of the month. The date predictor is included on the basis of the trend in the residuals apparent in Figure 7.9. Note that this example is somewhat artificial with respect to statistical weather forecasting, since in general the predictors (other than the date) will not be known in advance of the time that the predictand (minimum temperature at Canandaigua) will be observed. However, this small data set serves to illustrate the principles.

Figure 7.13 diagrams the process of choosing predictors using forward selection. The numbers in each table summarize the comparisons being made at each step. For the first ($K = 1$) step, no predictors are yet in the equation, and all six potential predictors are under consideration. At this stage the predictor producing the best simple linear regression is chosen, as indicated by the smallest MSE, and the largest R^2 and F ratio among the six. This best predictor is the Ithaca minimum temperature, so the tentative regression equation is exactly Equation 7.21.

Having chosen the Ithaca minimum temperature in the first stage there are five potential predictors remaining, and these are listed in the $K = 2$ table. Of these five, the one producing the best predictions in an equation that also includes the Ithaca minimum temperature is chosen. Summary statistics for these five possible two-predictor regressions are also shown in the $K = 2$ table. Of these, the equation including Ithaca minimum temperature and the date as the two predictors is clearly best, producing MSE $= 9.2°\text{F}^2$ for the dependent data.

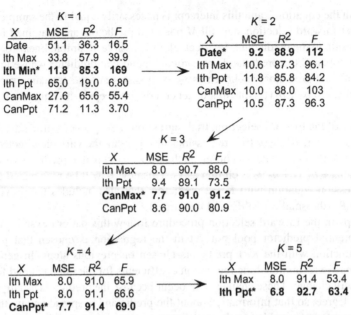

FIGURE 7.13 Diagram of the forward selection procedure for development of a regression equation for Canandaigua minimum temperature using as potential predictors the remaining variables in data set A.1, plus the date. At each step the variable is chosen (bold, starred) whose addition would produce the largest decrease in MSE or, equivalently, the largest increase in R^2 or F. At the final ($K = 6$) stage, only Ith. Max remains to be chosen, and its inclusion would produce MSE $= 6.8$, $R^2 = 93.0\%$, and $F = 52.8$.

With these two predictors now in the equation, there are only four potential predictors left at the $K = 3$ stage. Of these, the Canandaigua maximum temperature produces the best predictions in conjunction with the two predictors already in the equation, yielding MSE $= 7.7°F^2$ for the dependent data. Similarly, the best predictor at the $K = 4$ stage is Canandaigua precipitation, and the better predictor at the $K = 5$ stage is Ithaca precipitation. For $K = 6$ (all predictors in the equation) the MSE for the dependent data is $6.8°F^2$, with $R^2 = 93.0\%$.   ◇

An approach called *backward elimination* is an alternative to forward selection. The process of backward elimination is analogous but opposite to that of forward selection. Here the initial stage is a regression containing all M potential predictors, $\hat{y} = b_0 + b_1 x_1 + b_2 x_2 + \ldots + b_M x_M$, so backward elimination will not be computationally feasible if $M \geq n$. Usually this initial equation will be grossly overfit, containing many redundant and some possibly useless predictors. At each step of the backward elimination procedure, the least important predictor variable is removed from the regression equation. That variable will be the one whose coefficient is smallest in absolute value, relative to its estimated standard error, so that in terms of the sample regression output tables presented earlier, the removed variable will exhibit the smallest (absolute) t ratio. As in forward selection, the regression coefficients for the remaining variables require recomputation if (as is usually the case) the predictors are mutually correlated.

The processes of both forward selection and backward elimination must be stopped at some intermediate stage, as discussed in the subsequent two sections. However, there is no guarantee that forward selection and backward elimination will choose the same subset of the potential predictor pool for the final regression equation. Other predictor selection procedures for multiple regression also exist, and these might select still different subsets. The possibility that a chosen selection procedure might not select the "right" set of predictor variables might be unsettling at first, but as a practical matter this is not usually an important problem in the context of producing an equation for use as a forecast tool. Correlations among the predictor variables result in the situation that nearly the same information about the predictand can be extracted from different subsets of the potential predictors. Therefore if the aim of the regression analysis is only to produce reasonably accurate forecasts of the predictand, the black box approach of empirically choosing a workable set of predictors is quite adequate. However, we should not be so complacent in a research setting, where one aim of a regression analysis could be to find specific predictor variables most directly responsible for the physical phenomena associated with the predictand.

7.4.3. Stopping Rules

Both forward selection and backward elimination require a stopping criterion or stopping rule. Without such a rule, forward selection would continue until all M candidate predictor variables were included in the regression equation, and backward elimination would continue until all predictors had been eliminated. It might seem that finding the stopping point would be a simple matter of evaluating the test statistics for the regression parameters and their nominal p values as supplied by the computer regression package. Unfortunately, because of the way the predictors are selected, these implied hypothesis tests are not quantitatively applicable, for two reasons. First, a sequence of tests is being computed, and the nominal p values do not account for this multiple testing. In addition, at each step (either in selection or elimination) predictor variables are not chosen randomly for entry or removal. Rather, the best or worst, respectively, among the available choices is selected.

The problem is illustrated in Figure 7.14, taken from the study of Neumann et al. (1977). The specific problem represented in this figure is the selection of exactly $K = 12$ predictor variables from pools of

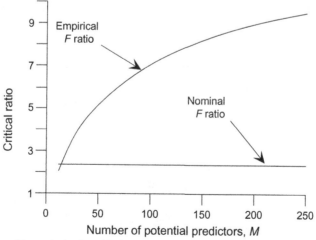

FIGURE 7.14   Comparison of the nominal and empirically (resampling-) estimated critical $(p = 0.01)$ F ratios for overall significance in a particular regression problem, as a function of the number of potential predictor variables, M. The sample size is $n = 127$, with the best $K = 12$ predictor variables to be included in each final regression equation. The nominal F ratio of 2.35 is applicable only for the case of $M = K$. When the forward selection procedure can choose from among more than K potential predictors the true critical F ratio is substantially higher. The difference between the nominal and actual values widens as M increases. *From Neumann et al. (1977).* © *American Meteorological Society. Used with permission.*

potential predictors of varying sizes, M, when there are $n = 127$ observations of the predictand. Ignoring the problems of multiple testing and nonrandom predictor selection would lead us to declare as significant any regression for which the F ratio in the ANOVA table is larger than the nominal critical value of 2.35. Naïvely, this value would correspond to the minimum F ratio necessary to reject the null hypothesis of no real relationship between the predictand and the twelve predictors, at the 1% level. The curve labeled empirical F ratio was arrived at using a resampling test, in which the same meteorological predictor variables were used in a forward selection procedure to predict 100 artificial data sets of $n = 127$ independent Gaussian random numbers each. This procedure simulates a situation consistent with the null hypothesis that the predictors bear no real relationship to the predictand, while automatically preserving the correlations among this particular set of predictors.

Figure 7.14 indicates that the nominal regression diagnostics give the correct answer only in the case of $K = M$, for which there is no ambiguity in the predictor selection since all the $M = 12$ potential predictors must be used to construct the $K = 12$ predictor equation, and only a single test is being computed. When the forward selection procedure has available some larger number $M > K$ potential predictor variables to choose from, the true critical F ratio is higher, and sometimes by a substantial amount. Even though none of the potential predictors in the resampling procedure bears any real relationship to the artificial (random) predictand, the forward selection procedure chooses those predictors exhibiting the highest chance correlations with the predictand, and these relationships result in apparently large F ratio statistics. Put another way, the p value associated with the nominal critical $F = 2.35$ is too large (less significant), by an amount that increases as more potential predictors are offered to the forward selection procedure. To emphasize the seriousness of the problem, the nominal F ratio in the situation of Figure 7.14 for the very stringent 0.01% level test is only about 3.7. The practical result of relying literally on the nominal critical F ratio is to allow more predictors into the final equation than are meaningful, with the danger that the regression will be overfit. The F ratio in Figure 7.14 is a single-number regression diagnostic convenient for illustrating the effects of overfitting, but these effects would be reflected in other aspects of the ANOVA table also. For example, most if not all of the nominal t ratios for the individual cherry-picked predictors when $M >> K$ would be larger than 2 in absolute value, incorrectly suggesting meaningful relationships with the (random) predictand.

Unfortunately, the results in Figure 7.14 apply only to the specific data set from which they were derived. In order to employ this approach to estimate the true critical F-ratio using resampling methods it would need to be repeated for each regression to be fit, since the statistical relationships among the potential predictor variables will be different in different data sets. In practice, other less rigorous stopping criteria usually are employed. For example, we might stop adding predictors in a forward selection when none of the remaining predictors would reduce the R^2 by a specified amount, perhaps 0.05%.

The stopping criterion can also be based on the MSE. This choice is intuitively appealing because, as the standard deviation of the residuals around the regression function, $\sqrt{\text{MSE}}$ directly reflects the anticipated precision of a regression. For example, if a regression equation were being developed to forecast surface temperature, little would be gained by adding more predictors if the MSE were already $0.01°F^2$, since this would indicate a $\pm 2s_e$ (i.e., approximately 95%) prediction interval around the forecast value of about $\pm 2\sqrt{0.01°F^2} = 0.2°F$. So long as the number of predictors K is substantially less than the sample size n, adding more predictor variables (even meaningless ones) will decrease the MSE for the developmental sample. This concept is illustrated schematically in Figure 7.15. Ideally the stopping criterion would be activated at the point where the MSE does not decline appreciably with the addition of more predictors, at perhaps $K = 12$ predictors in the hypothetical case shown in Figure 7.15.

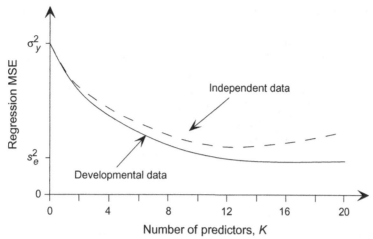

FIGURE 7.15   Schematic illustration of the regression MSE as a function of the number of predictor variables in the equation, K, for developmental data (solid), and for an independent verification set (dashed). *After Glahn (1985).*

Figure 7.15 indicates that the MSE for an independent data set will be larger than that achieved for the developmental data. This result should not be surprising, since the least-squares fitting procedure operates by optimizing the parameter values to minimize MSE for the developmental data. This under-estimation of the independent-data MSE provided by the MSE for a forecast equation on developmental data is an expression of what is sometimes called *artificial skill* (Davis, 1976; Michaelson, 1987). The precise magnitudes of the differences in MSE between developmental and independent data sets are not determinable solely from the regression output using the developmental data. That is, having seen only the regressions fit to the developmental data, we cannot know the value of the minimum MSE for independent data. Neither can we know if it will occur at a similar point (at around $K = 12$ in Figure 7.15), or whether the equation has been overfit and the minimum MSE for the independent data will be for a substantially smaller K. This situation is unfortunate, because the purpose of developing a forecast equation is to specify future, unknown values of the predictand using observations of the predictors that have yet to occur.

Figure 7.15 also indicates that, for forecasting purposes, the exact stopping point is usually not critical as long as it is approximately right. Again, this is because the MSE tends to change relatively little through a range of K near the optimum, and for purposes of forecasting it is the minimization of the MSE rather than the specific identities of the predictors that is important. By contrast, if the purpose of the regression analysis is scientific understanding, the specific identities of chosen predictor variables can be of primary interest, and the magnitudes of the resulting regression coefficients may lead to significant physical insight. In this case it is not reduction of prediction MSE, per se, that is desired, but rather that causal relationships between particular variables be suggested by the analysis.

7.4.4. Cross-Validation

Sometimes regression equations to be used for weather forecasting are tested on a sample of independent data that has been held back during the development of the forecast equation. In this way, once the number K and specific identities of the predictors have been fixed, an estimate of the distances between the solid and dashed MSE lines in Figure 7.15 can be estimated directly from the reserved data. If the deterioration in forecast precision (i.e., the unavoidable increase in MSE) is judged to be acceptable, the equation can be used operationally.

This procedure of reserving an independent verification data set is actually a restricted case of a technique known as *cross-validation* (Efron and Gong, 1983; Efron and Tibshirani, 1993; Elsner and Schmertmann, 1994; Michaelson, 1987). Cross-validation simulates prediction for future, unknown data by repeating the entire fitting procedure on data subsets, and then examining the predictions made for the data portions left out of each of these subsets. The most frequently used procedure is known as *leave-one-out cross-validation*, in which the fitting procedure is repeated n times, each time with a sample of size $n-1$, because one of the predictand observations and its corresponding predictor set are left out in each replication of the fitting process. The result is n (often only slightly) different prediction equations.

In leave-one-out cross-validation, the estimate of the prediction MSE is computed by forecasting each of the omitted observation using the equation developed from the remaining $n-1$ data values, computing the squared difference between the prediction and predictand for each of these equations, and averaging the n squared differences. Thus leave-one-out cross-validation uses all n observations of the predictand to estimate the prediction MSE in a way that allows each observation to be treated, one at a time, as independent data.

More generally, J-fold cross-validation leaves out groups of size J sequentially, so that the fitting process is repeated n/J (approximately, unless this ratio is an integer) times, each with reduced sample size (approximately) $n-J$. Leave-one-out cross-validation thus corresponds to $J=1$. Hastie et al. (2009) suggest that choosing $J=5$ or $J=10$ is often a good rule of thumb. These choices produce data subsets that differ from each other more than in leave-one-out cross-validation, leading to a less variable, more stable, estimate of the MSE, and which also require substantially less computation. When the sample size n is small and the predictions will be evaluated using a correlation measure, leaving out $J>1$ values at a time can be especially advantageous (Barnston and van den Dool, 1993).

It should be emphasized that each repetition of the cross-validation exercise is a repetition of the entire fitting algorithm, not a refitting of the specific statistical model derived from the full data set using $n-J$ data values. In particular, different prediction variables must be allowed to enter for different cross-validation subsets. DelSole and Shukla (2009) provide a cautionary analysis showing that failure to respect this precept can lead to random-number predictors exhibiting apparently real, cross-validated predictive ability. Any data transformations (e.g., standardizations with respect to climatological values) also need to be defined (and therefore possibly recomputed) without any reference to the withheld data in order for them to have no influence on the equation that will be used to predict them in the cross-validation exercise. However, the ultimate product equation, to be used for operational forecasts, would be fit using all the data after we are satisfied with the cross-validation results.

Cross-validation requires some special care when the data are serially correlated. In particular, data records adjacent to or near the omitted observation(s) will tend to be more similar to them than randomly selected ones, so the omitted observation(s) will be more easily predicted than the future observations they are meant to simulate. A solution to this problem is to leave out blocks of an odd number of consecutive observations, L, so the fitting procedure is repeated $n-L+1$ times on samples of size $n-L$ (Burman et al., 1994; Elsner and Schmertmann, 1994). The blocklength L is chosen to be large enough for the correlation between its middle value and the nearest data used in the cross-validation fitting to be small, and the cross-validation prediction is made only for that middle value. For $L=1$ this moving-blocks cross-validation reduces to leave-one-out cross-validation.

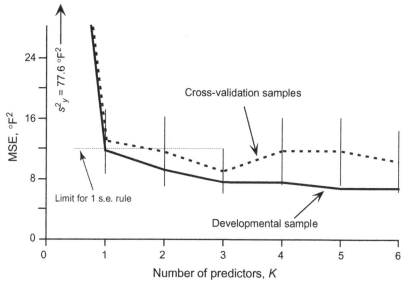

FIGURE 7.16   Plot of residual mean-squared error as a function of the number of regression predictors specifying Canandaigua minimum temperature, using the January 1987 data in Table A.1. Solid line shows MSE for developmental data (starred predictors in Figure 7.13). Dashed line shows estimated MSE achievable on independent data, with the same numbers of (possibly different) predictors, as computed through cross-validation, leaving out blocks of seven consecutive days. Whiskers show +1 standard errors around the cross-validated MSE estimates, and the horizontal dotted line locates the one-s.e. rule threshold. This plot is a real-data example corresponding to the idealization in Figure 7.15.

Example 7.6. Protecting against Overfitting Using Cross-Validation

Having used all the available developmental data to fit the regressions in Example 7.5, what can be done to ensure that these prediction equations have not been overfit? Fundamentally, what is desired is a measure of how the regressions will perform when used on data not involved in the fitting. Cross-validation is an especially appropriate tool for this purpose in the present example, because the small ($n = 31$) sample would be inadequate if a substantial portion of it had to be reserved for a validation sample.

Figure 7.16 evaluates MSEs for six regression equations obtained with forward selection. This figure shows real results in the same form as the idealization of Figure 7.15. The solid line indicates the MSE achieved on the developmental sample, obtained by adding the predictors in the order shown in Figure 7.13. Because a regression chooses precisely those coefficients minimizing MSE for the developmental data, this quantity is expected to be higher when the equations are applied to independent data. An estimate of how much higher is given by the average MSE from the cross-validation samples (dashed line). Because these data are autocorrelated, a simple leave-one-out cross-validation is expected to underestimate the prediction MSE. Here the cross-validation has been carried out omitting blocks of length $L = 7$ consecutive days, and repeating the entire forward selection procedure $n - L + 1 = 25$ times. Since the lag-1 autocorrelation for the predictand is approximately $r_1 = 0.6$ and the autocorrelation function exhibits approximately exponential decay (similar to that in Figure 3.22), the correlation between the predictand in the centers of the seven-day moving blocks and the nearest data used for equation fitting is $0.6^4 = 0.13$, corresponding to $R^2 = 1.7\%$, indicating near-independence.

Each cross-validation point in Figure 7.16 represents the average of 25 squared differences between an observed value of the predictand at the center of a block, and the forecast of that value produced by a regression equation fit to all the data except those in that block. Predictors are added to each of these equations according to the usual forward selection algorithm. The order in which the predictors are added in one of these 25 regressions is often the same as that indicated in Figure 7.13 for the full data set, but this order is not forced onto the cross-validation samples, and indeed is different for some of the data partitions.

The differences between the dashed and solid lines in Figure 7.16 are indicative of the expected prediction errors for future independent data (dashed), and those that would be inferred from the MSE on the dependent data as provided by the ANOVA table (solid). The minimum cross-validation MSE at $K = 3$ suggests that the best regression for these data may be the one with three predictors, and that it should produce prediction MSE on independent data of around $9.1°F^2$, yielding $\pm 2s_e$ confidence limits of $\pm 6.0°F$. $\diamond$

Recognizing that the cross-validated estimates of prediction MSE are subject to sampling variations, and that the cross-validated MSE estimates may be very similar near the minimum, Hastie et al. (2009) note that the possible resulting overfitting can be addressed using the so-called *one-s.e. rule*. The idea is to choose the simplest model (here, the regression with smallest K) whose accuracy is comparable to that of the best model, in the sense that its accuracy is no more than one standard error worse than that of the best model. The vertical whiskers in Figure 7.16 show ± 1 s.e. around each of the average cross-validated MSE estimates, where the standard errors are computed as the standard deviation of the 25 cross-validated MSE estimates divided by $\sqrt{25}$ (i.e., the square root of Equation 5.4). The light dotted horizontal line in Figure 7.16 locates this threshold level. The cross-validated MSE for $K = 1$ is above the one s.e. limit, but the value for $K = 2$ is below it, so the two-predictor regression would be chosen by the one-s.e. rule in this case.

A potentially interesting but as yet little-used twist on cross-validation has been proposed by Hothorn et al. (2005). The idea is to repeatedly bootstrap the available data sample, and then use the data values not included in the current bootstrap sample (which will number approximately $0.368\,n$, on average) as the independent test data. Results over all bootstrap data splits are then aggregated to yield the final cross-validation estimate. Hastie et al. (2009) suggest a similar idea, where predictions for each data value are averaged over the bootstrap iterations that did not draw that sample, and these average results are then averaged over all underlying n data values.

Before leaving the topic of cross-validation it is worthwhile to note that the procedure is sometimes mistakenly referred to as the *jackknife*, a relatively simple resampling procedure that was introduced in Section 5.3.5. The confusion is understandable because the jackknife is computationally analogous to leave-one-out cross-validation. Its purpose, however, is to estimate the bias and/or standard deviation of a sampling distribution nonparametrically, and using only the data in a single sample. Given a sample of n independent observations, the idea in jackknifing is to recompute a statistic of interest n times, omitting a different one of the data values each time. Attributes of the sampling distribution for the statistic can then be inferred from the resulting n-member jackknife distribution (Efron, 1982; Efron and Tibshirani, 1993). The jackknife and leave-one-out cross-validation share the mechanics of repeated recomputation on reduced samples of size $n - 1$, but cross-validation seeks to infer future forecasting performance, whereas the jackknife seeks to nonparametrically characterize the sampling distribution of a sample statistic.

7.5. REGULARIZATION/SHRINKAGE METHODS FOR MULTIPLE REGRESSION

One reason for the widespread use of the regression methods described in Sections 7.2 and 7.3 flows from the *Gauss-Markov theorem*, which shows that the least-squares parameter estimates b_k have the smallest variance (i.e., narrowest sampling distributions, and so are most consistent from batch to batch of training data) of any linear and unbiased estimates. (Unbiasedness implies that the mean of the sampling distribution locates the true value, on average.) However, in many instances the predictor variables in a multiple regression may exhibit some strong correlations among themselves, which leads to these sampling-distribution variances, although minimum conditional on the unbiasedness, being comparatively large and indeed large enough to negatively impact the accuracy of the predictions made with the resulting equations.

Although lack of bias sounds like a virtuous attribute because of the nonstatistical connotations of the word, better predictions may be possible when regression predictors are strongly correlated if the unbiasedness restriction is relaxed. As indicated by Equation 7.23, prediction accuracy is directly related to the regression MSE, which in turn depends on both the sampling variance and the bias:

$$MSE = variance + (bias)^2. \tag{7.28}$$

For an unbiased procedure, MSE is minimized by minimizing variance. But more generally MSE results from a trade-off between variance and bias, so that nonzero bias can be associated with smaller MSE if a reduction in variance can be achieved that is greater than the squared bias. Figure 7.17 illustrates the trade-off in terms of the sampling distributions for a hypothetical regression parameter. The dashed PDF represents the sampling distribution for the unbiased estimate, which is centered on the true value, E[b]. A possible sampling distribution for a biased estimator of the same parameter is shown as the solid density. Even though the biased estimates here will be too small on average, they will typically be closer to the true value because of the reduced sampling variance, and so usually would exhibit the smaller MSE. Bias and sampling error in parameter estimation propagates to bias and random errors in the resulting forecasts.

Regularization is the general term for allowing bias, with the aim of reducing MSE. It may be useful in regression when the correlations among the predictors are sufficiently strong by yielding regression equations that produce more accurate predictions, and parameter estimates that can be more meaningfully interpreted. When regression predictors are strongly correlated among themselves, a large positive estimated coefficient for one predictor can be balanced by a large negative coefficient for another. The magnitude of

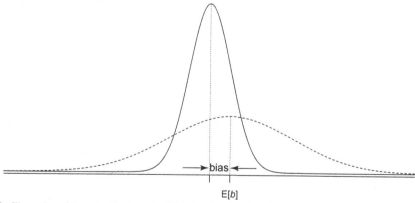

FIGURE 7.17 Illustration of the potential benefit of biased estimation, by reducing sampling variability sufficiently to yield a smaller MSE. Dashed density represents the sampling distribution of the unbiased estimator, and solid density represents sampling distribution of the biased estimator. Even though the biased estimator is too small on average, the estimation error is typically reduced.

such a trade-off between estimated coefficients for strongly correlated predictors can depend on the random sampling variations of the training data at hand, and so will not be reproducible in future independent data, leading to a lack of stability in the sense used in Section 7.4.1, and therefore to poor predictions. That is, small changes in the training data may lead to substantial changes in the estimated regression coefficients, which is symptomatic of large estimation variance. Another consequence can be that the estimated coefficients may not make sense physically, exhibiting the wrong sign relative to physical considerations, for example. In addition, the estimated regression coefficients tend to be too large in absolute value (Hoerl and Kennard, 1970), as is suggested by the breadth of the dashed probability density in Figure 7.17.

Regularization methods suppress these problems by allowing some bias in estimation of the regression parameters. In the process they also tend to reduce, or "shrink," the absolute magnitudes of the estimated coefficients. The interpretability of the resulting regression parameters may be improved in the sense that a relatively small subset of the predictors exhibiting the largest effects might be identified. Another advantage of regularization methods is that, unlike conventional least-squares regression, they can be applied when there are more potential predictors K than data values n.

The possible need for regularization methods in multiple regression can be diagnosed by looking at the squared multiple correlation coefficients for each of the predictor variables, in relation to all of the others. These will correspond to the R^2 value in a multiple regression in which the x variable of interest is predicted by the remaining $K–1$ predictors. Alternatively, these squared multiple correlations can be computed from the elements that appear on the diagonal of the inverse (Section 11.3.2) of the correlation matrix (Section 3.6.4) for the K predictor variables, according to $R_k^2 = 1–1/r_{kk}^{-1}$ for the kth element of the inverse correlation matrix. Marquardt (1970) suggests the rule of thumb that use of regularization in regression is appropriate if the largest of these squared multiple correlations is larger than 0.9 in absolute value, corresponding to a diagonal element in the inverse-correlation matrix larger than 10. Equivalently, the smallest eigenvalue (Section 11.3.3) of the correlation matrix for the x variables will be nearly zero.

7.5.1. Ridge Regression

Ridge regression, also known as *Tikhonov regularization*, is an early regularization procedure, dating from Hoerl and Kennard (1970). The basic idea is to control the potentially excessive magnitudes of estimated regression coefficients by imposing a limit, or budget, jointly on their magnitudes, according to

$$\sum_{k=1}^{K} b_k^{*2} \leq c, \tag{7.29}$$

where the b_k^* denote the regularized parameter estimates. In order to avoid undue emphasis on predictors that happen to exhibit large variance, perhaps arbitrarily because of the measurement scales of the corresponding predictors, the predictor variables in ridge regression are first standardized to have zero (sample) mean and unit variance according to Equation 3.27. That is, they are expressed as standardized anomalies. Often the predictand y is either centered (expressed as anomalies, with zero mean) or fully standardized, in which cases there is no intercept parameter b_0^*, but in any case the magnitude of the intercept is not penalized in Equation 7.29.

The ridge regression parameter estimates are obtained by minimizing a penalized residual sum of squares,

$$SSE_p = \sum_{i=1}^{n} e_i^2 + \lambda \sum_{k=1}^{K} b_k^{*2}, \tag{7.30}$$

where λ is the nonnegative regularization (or biasing) parameter. The regression parameters can be obtained analytically through the modification to Equation 11.39

$$\boldsymbol{b}^* = \left([X]^T[X] + \lambda[I]\right)^{-1} [X]^T \boldsymbol{y}. \tag{7.31}$$

The parameter c in Equation 7.29 is a decreasing function of λ, but their particular functional relationship depends on the training data being used. Estimation bias increases with increasing λ but estimation variance decreases, so that an optimum value yields minimum MSE. Unless all the predictor variables are uncorrelated there is a nonzero value of λ for which the biased MSE* is smaller than its conventional unconstrained counterpart (Hoerl and Kennard, 1970).

It is conventional to display the results of ridge regression by plotting the estimated regression parameters b_k^* as functions of the regularization parameter λ, which is known as the *ridge trace*. For sufficiently large λ the regression parameters will all have been forced to shrink to zero, which is generally a noninformative result. Although early practitioners would typically choose λ according to subjective rules of thumb (Hoerl and Kennard, 1970), the more modern approach is to optimize the regularization parameter using cross-validation (Section 7.4.4) to minimize the estimated prediction MSE, often in conjunction with the one-s.e. rule.

Example 7.7. Equation Development Using Ridge Regression

In Example 7.5, the use of forward selection was illustrated using the data in Appendix A.1, with the Canandaigua minimum temperature as the predictand. The remaining meteorological variables and the date were used as potential predictors, with the precipitation values transformed logarithmically. The largest of the six squared multiple correlation coefficients for these predictors is 0.94, exhibited by the Canandaigua maximum temperature, mainly because of its large correlation with the Ithaca maxima (Table 3.5). The squared multiple correlation for the Ithaca maxima is 0.93. Regularization may therefore be advantageous.

Figure 7.18 shows the ridge trace for the resulting regressions. For small values of the regularization parameter the estimated regression coefficients are nearly the same as for the unconstrained full model in Example 7.5. Although these coefficients have been computed using standardized variables, the scale on the vertical axis pertains to the variables on their original scales. As the regularization parameter is increased the regression coefficients are progressively shrunk toward zero. This effect is most pronounced for the two precipitation variables, which are strongly correlated with each other, and only modestly correlated with the predictand. In addition, the coefficients for the Canandaigua precipitation predictor changes sign to more physically plausible positive values as the regularization parameter increases.

A subjective interpretation of the ridge trace in Figure 7.18 would likely lead to removing the two precipitation predictors from the regression, both because they exhibit the greatest shrinkage, and because the coefficient for the Canandaigua precipitation changes sign, whereas the traces for the other predictors are nearly flat (e.g., Hoerl and Kennard, 1970; Marquardt and Snee, 1975). However, the procedure will not force estimated coefficients to zero unless the regularization parameter is very large. The dashed vertical line locates the regularized regression, with $\lambda = 3.96$, chosen by 10-fold cross-validation, using the one-s.e. rule. $\diamond$

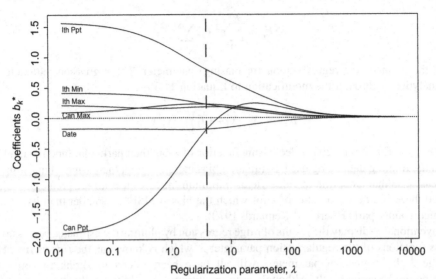

FIGURE 7.18 Ridge trace for regressions predicting Canandaigua minimum temperatures, using the other elements in Appendix A.1 as predictors. Regressions have been computed using standardized values, but the scale for the parameter estimates on the vertical axis pertains to the original dimensional values. Vertical dashed line locates the value of the regularization parameter chosen using 10-fold cross-validation with the one-s.e. rule.

7.5.2. The Lasso

The *Lasso* (least absolute shrinkage and selection operator, Tibshirani, 1996), or *L1 regularization*, is an alternative regularization procedure that constrains the magnitudes of the estimated regression coefficients according to

$$\sum_{k=1}^{K} |b_k^*| \leq c. \tag{7.32}$$

Comparing to Equation 7.29 for ridge regression, the Lasso parameter magnitude budget is formulated in terms of the absolute values rather than the squares of the estimated coefficients. In common with ridge regression, standardized values of the predictors, and often the predictand as well, are used. The penalty in Equation 7.32 does not involve an intercept, if any.

   Computation of Lasso regularization proceeds by minimizing

$$SSE_p = \sum_{i=1}^{n} e_i^2 + \lambda \sum_{k=1}^{K} |b_k^*|, \tag{7.33}$$

which is the counterpart of Equation 7.30 for ridge regression. As was also the case for ridge regression, the limit c of the parameter budget in Equation 7.32 is a decreasing function of λ, with the details of the relationship depending on the training data used. However, there is no closed-form expression for the parameter estimates as functions of the regularization parameter λ, corresponding to Equation 7.31 for ridge regression, and numerical methods must be used to evaluate them.

   Lasso regularization is computed for a range of the regularization parameter λ, with the results displayed in graphical form using a *coefficient profile graph*, which is the counterpart of the ridge trace in ridge regression. The estimated regression parameters shrink toward zero as λ increases so that the ceiling c progressively constrains their magnitudes. Also in common with ridge regression, a best value for λ is usually chosen to minimize estimated prediction MSE using cross-validation, often in conjunction with the one-s.e. rule. However, unlike ridge regression, Lasso regularization will progressively force the regularized coefficient to exactly zero as the parameter λ is increased, so that it may lead automatically to a subset of "best" predictors.

Example 7.8. Equation Development using the Lasso

The regression situation described in Examples 7.5, 7.6, and 7.7, where the Canandaigua minimum temperatures were predicted using the other quantities in Appendix A.1, can also be addressed with Lasso regularization. As before, the precipitation predictors are log-transformed, and all quantities are standardized.

Figure 7.19 shows the coefficient profile graph for this analysis, where the vertical axis pertains to regularized regression coefficients relating to the original untransformed variables. Qualitatively, the coefficient profile graph is similar to the ridge trace in Figure 7.18 for the ridge regressions based on the same data. For very small values of λ the parameter estimates approach their conventional least-squares counterparts. They are shrunk toward zero with increasing λ, and similarities in shape in the two figures for the various predictors can be discerned. Most notably, the coefficients for the two precipitation predictors begin with opposite signs and large absolute values, but decrease quickly to zero.

A notable difference in Figure 7.19 is that the predictors are progressively shrunk to exactly zero as λ is increased. The dashed vertical line locates the value of $\lambda = 1.37$ that is chosen using 10-fold cross-validation with the one-s.e. rule. At this point the nonzero predictors are Ithaca minima, Canandaigua maxima, and the date. Because the Lasso forces predictors to zero, moving from right to left in Figure 7.19 in effect traces out the results of a forward selection algorithm. Unlike in the forward

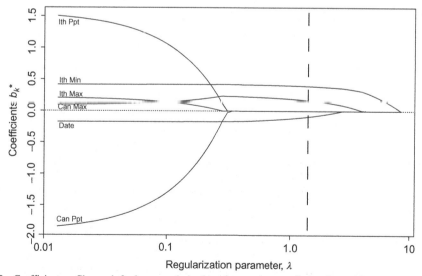

FIGURE 7.19 Coefficient profile graph for lasso regularization of regressions predicting Canandaigua minimum temperatures using the other elements of Appendix A.1 as predictors. Regressions have been computed using standardized values, but the scale for the parameter estimates on the vertical axis pertains to the original dimensional values. Vertical dashed line locates the value of the regularization parameter chosen using 10-fold cross-validation with the one-s.e. rule.

selection exercise in Example 7.5, here the Canandaigua maxima appear to be more important than the date predictor.   ◇

7.6. NONLINEAR REGRESSION

7.6.1. Generalized Linear Models

Although linear least-squares regression accounts for the overwhelming majority of regression applications, it is also possible to fit regression functions that are nonlinear in the regression parameters. Nonlinear regression can be appropriate when a nonlinear relationship is dictated by nature of the physical problem at hand, and/or the usual assumptions of Gaussian residuals with constant variance are untenable. In these cases the fitting procedure is usually iterative and based on maximum likelihood methods (see Section 4.6).

This section introduces two such regression structures, both of which are important examples of a class of nonlinear statistical models known as *generalized linear models* (GLM) (McCullagh and Nelder, 1989). Generalized linear models extend linear statistical models, such as multiple linear regression, by representing the predictand as a nonlinear transformation of a linear regression function. The nonlinearity is represented by a 1-to-1 (and therefore invertible) function known as the *link function*, $g(\hat{y})$. Accordingly, the GLM extension of the ordinary linear multiple regression (Equation 7.25) is

$$g(\hat{y}) = b_0 + b_1 x_1 + b_2 x_2 + \cdots + b_K x_K, \tag{7.34}$$

where the specific form of the link function is chosen according to the nature of the predictand data. Comparing Equation 7.34 and 7.25 shows that ordinary linear regression is a special case of a GLM model, with the identity link, i.e., $g(\hat{y}) = \hat{y}$. Because the link function will be invertible, GLM equations are often written equivalently as

$$\hat{y} = g^{-1}(b_0 + b_1 x_1 + b_2 x_2 + \cdots + b_K x_K). \tag{7.35}$$

7.6.2. Logistic Regression

One important advantage of statistical over (deterministic) dynamical forecasting methods is the capacity to produce probability forecasts. Inclusion of probability elements into the forecast format is advantageous because it provides an explicit expression of the inherent uncertainty or state of knowledge about the future weather, and because probabilistic forecasts allow users to extract more value from them when making decisions (e.g., Katz and Murphy, 1997a; Krzysztofowicz, 1983; Murphy, 1977; Thompson, 1962). In a sense, ordinary linear regression produces probability information about a predictand, for example, through the 95% prediction interval around the regression function approximated by the $\pm 2\sqrt{MSE}$ rule. More narrowly, however, probability forecasts are forecasts for which the predictand is expressed explicitly in terms of probability, rather than as the value of a physical variable.

The simplest setting for probability forecasting produces single probabilities for binary outcomes, for example, rain tomorrow or not. Systems for producing this kind of probability forecast are developed in a regression setting by first transforming the predictand to a binary (or dummy) variable, taking on the values zero (e.g., for no rain) and one (for any nonzero rain amount). That is, regression procedures are implemented after applying Equation 7.26 to the predictand, y, rather than to a predictor. In a sense, zero and one can be viewed as probabilities of the dichotomous event not occurring or occurring, respectively, after it has been observed.

The easiest approach to regression when the predictand is binary is to use the machinery of ordinary multiple regression as described previously. In the meteorological literature this is called Regression Estimation of Event Probabilities (REEP) (Glahn, 1985; Miller, 1964). The main justification for the use of REEP is that it is no more computationally demanding than the fitting of any other linear regression, and so historically has been used when computational resources have been limiting. The resulting predicted values are usually between zero and one, and it has been found through operational experience that these predicted values can usually be treated as specifications of probabilities for the event $\{Y = 1\}$. However, one obvious problem with REEP is that some of the resulting forecasts may not lie on the unit interval, particularly when the regression relationship is relatively strong (so that the slope parameter is relatively large in absolute value, Brelsford and Jones, 1967), there are relatively few predictors (so there is limited scope for compensating influences among them), or the predictands are near the limits or outside of their ranges in the training data. This logical inconsistency usually causes little difficulty in an operational setting because multiple-regression forecast equations with many predictors rarely produce such nonsense probability estimates. When the problem does occur the forecast probability is usually near zero or one, and the operational forecast can be issued as such.

Two other difficulties associated with forcing a linear regression onto a problem with a binary predictand are that the residuals are clearly not Gaussian, and their variances are not constant. Because the predictand can take on only one of two values, a given regression residual can also take on only one of two values, and so the residual distributions are Bernoulli (i.e., binomial, Equation 4.1, with $N = 1$). Furthermore, the variance of the residuals is not constant, but depends on the ith predicted probability p_i according to $(p_i)(1 - p_i)$.

It is possible to simultaneously bound the regression estimates for binary predictands on the interval $(0, 1)$, and to accommodate the Bernoulli distributions for the regression residuals, using a more theoretically satisfying technique known as log*istic regression*. Some examples of logistic regression in the atmospheric science literature are Applequist et al. (2002), Bröcker (2010), Hamill et al., 2004, Hilliker and Fritsch (1999), Lehmiller et al. (1997), and Lemcke and Kruizinga (1988).

Logistic regressions are fit to binary predictands using the log-odds, or *logit*, link function $g(p) = \ln [p/(1-p)]$ (Equation 3.24), yielding the generalized linear model

$$\ln \left(\frac{p}{1-p} \right) = b_0 + b_1 x_1 + \cdots + b_K x_K, \tag{7.36a}$$

which can be expressed also in the form of Equation 7.35 as

$$p = \frac{\exp (b_0 + b_1 x_1 + \cdots + b_K x_K)}{1 + \exp (b_0 + b_1 x_1 + \cdots + b_K x_K)} = \frac{1}{1 + \exp (-b_0 - b_1 x_1 - \cdots - b_K x_K)}. \tag{7.36b}$$

Here the predicted value p results from the ith set of predictors $(x_1, x_2, ..., x_K)$ of n such sets. Geometrically, logistic regression is most easily visualized for the single-predictor case $(K = 1)$, for which Equation 7.36b is an S-shaped curve that is a function of x_1. In the limit of $b_0 + b_1 x_1 \rightarrow + \infty$ the exponential function in the first equality of Equation 7.36b becomes arbitrarily large so that the predicted value p_i approaches one. As $b_0 + b_1 x_1 \rightarrow -\infty$, the exponential function approaches zero and thus so does the predicted value. Depending on the parameters b_0 and b_1, the function rises gradually or abruptly from zero to one (or falls, for $b_1 < 0$, from one to zero) at intermediate values of x_1. Thus it is guaranteed that logistic regression will produce properly bounded probability estimates. The logistic function is convenient mathematically, but it is not the only function that could be used in this context. Another alternative yielding a very similar shape involves using the inverse Gaussian CDF for the link function, yielding $p = \Phi(b_0 + b_1x_1 + \ldots + b_Kx_K)$, which is known as *probit regression*.

Equation 7.36a shows that logistic regression can be viewed as linear in terms of the logarithm of the odds ratio, $p/(1-p)$. Superficially it appears that Equation 7.36a could be fit using ordinary linear regression, except that the predictand is binary, so the left-hand side will be either $\ln(0)$ or $\ln(\infty)$. However, fitting the regression parameters can be accomplished using the method of maximum likelihood, recognizing that the residuals are Bernoulli variables. Assuming that Equation 7.36 is a reasonable model for the smooth changes in the probability of the binary outcome as a function of the predictors, the probability distribution function for the ith residual is Equation 4.1, with $N = 1$, and p_i as specified by Equation 7.36b. The corresponding likelihood is of the same functional form, except that the values of the predictand y and the predictors x are fixed, and the probability p_i is the variable. If the ith residual corresponds to a "success" (i.e., the event occurs, so $y_i = 1$), the likelihood is $\Lambda = p_i$ (as specified in Equation 7.36b), and otherwise $\Lambda = 1 - p_i = 1/(1 + \exp[b_0 + b_1x_1 + \ldots + b_Kx_K])$. If the n sets of observations (predictand and predictor(s)) are independent, the joint likelihood for the $K + 1$ regression parameters is simply the product of the n individual likelihoods, or

$$\Lambda(\boldsymbol{b}) = \prod_{i=1}^{n} \frac{y_i \exp(b_0 + b_1x_1 + \cdots + b_Kx_K) + (1 - y_i)}{1 + \exp(b_0 + b_1x_1 + \cdots + b_Kx_K)}. \tag{7.37}$$

Since the y's are binary $[0, 1]$ variables, each factor in Equation 7.37 for which $y_i = 1$ is equal to p_i (Equation 7.36b), and the factors for which $y_i = 0$ are equal to $1 - p_i$. As usual, it is more convenient to estimate the regression parameters by maximizing the log-likelihood

$$L(\boldsymbol{b}) = \ln[\Lambda(\boldsymbol{b})] = \sum_{i=1}^{n} \{y_i(b_0 + b_1x_1 + \cdots b_Kx_K) - \ln[1 + \exp(b_0 + b_1x_1 + \cdots b_Kx_K)]\}. \tag{7.38}$$

The combinatorial factor in Equation 4.1 has been omitted because it does not involve the unknown regression parameters, and so will not influence the process of locating the maximum of the function. Usually statistical software will be used to find the values of the b's maximizing this function, using iterative methods such as those outlined in Sections 4.6.2 or 4.6.3.

Some software will display information relevant to the strength of the maximum likelihood fit using what is called the *analysis of deviance* table, which is analogous to the ANOVA table for linear regression (Table 7.3). The idea underlying an analysis of deviance table is the likelihood ratio test (Equation 5.20). As more predictors and thus more regression parameters are added to Equation 7.36, the log-likelihood will progressively increase as more latitude is provided to accommodate the data. Whether that increase is sufficiently large to reject the null hypothesis that a particular, smaller, regression equation is adequate, is judged in terms of twice the difference of the log-likelihoods relative to the χ^2 distribution, with degrees-of-freedom ν equal to the difference in numbers of parameters between the null-hypothesis regression and the more elaborate regression being considered. More about analysis of deviance can be learned from sources such as Healy (1988) or McCullagh and Nelder (1989).

The likelihood ratio test is appropriate when a single candidate logistic regression is being compared to a null model. Often H_0 will specify that all the regression parameters except b_0 are zero, in which case the question being addressed is whether the predictors x being considered are justified in favor of the constant (no-predictor) model with $b_0 = \ln [\Sigma y_i /n/(1 - \Sigma y_i/n)]$. However, if multiple alternative logistic regressions are being entertained, computing the likelihood ratio test for each alternative raises the problem of test multiplicity (see Section 5.4.1). In such cases it is better to compute either the *Bayesian Information Criterion* (BIC) statistic (Schwarz, 1978)

$$BIC = -2L(\boldsymbol{b}) + (K+1) \ \ln(n) \tag{7.39}$$

or the *Akaike Information Criterion* (AIC) (Akaike, 1974)

$$AIC = -2L(\boldsymbol{b}) + 2(K+1), \tag{7.40}$$

for each candidate model. Both the AIC and BIC statistics consist of twice the negative of the log-likelihood plus a penalty for the number of parameters fit, and the preferred regression will be the one minimizing the chosen criterion. The BIC statistic will generally be better for large-n problems since its probability of selecting the proper member of the class of models considered approaches 1 as $n\longrightarrow\infty$. For smaller sample sizes BIC often chooses models that are simpler than justified by the data, in which cases AIC may be preferred. The AIC is biased for small sample size relative to the number of parameters estimated, and so will tend to yield overfit models. In such cases the corrected AIC (Hurvich and Tsai, 1989)

$$AIC_C = -2L(\boldsymbol{b}) + 2(K+1) + \frac{2(K+1)(K+2)}{n-K-2}, \tag{7.41}$$

will generally be preferred, where again $K+1$ is the number of parameters (including intercept) estimated for a given model. Equation 7.41 converges to Equation 7.40 for $n >> K$.

The Lasso (Section 7.5.2) is another alternative for selecting among potential predictors in a logistic regression (Bröcker, 2010). In that case predictors whose regression coefficients have not been shrunk to zero when the regularization parameter has been optimized through cross-validation, possibly using the one-s.e. rule, would be retained.

Example 7.9. Comparison of REEP and Logistic Regression

Figure 7.20 compares the results of REEP (dashed) and logistic regression (solid) for some of the January 1987 data from Table A.1. The predictand is daily Ithaca precipitation, transformed to a binary variable using Equation 7.26 with $c = 0$. That is, $y = 0$ if the precipitation is zero, and $y = 1$ otherwise. The predictor is the Ithaca minimum temperature for the same day. The REEP (linear regression) equation has been fit using ordinary least squares, yielding $b_0 = 0.208$ and $b_1 = 0.0212$. This equation specifies negative probability of precipitation if the temperature predictor is less than about $-9.8°F$, and specifies probability of precipitation greater than one if the minimum temperature is greater than about $37.4°F$. The parameters for the logistic regression, fit using maximum likelihood, are $b_0 = -1.76$ and $b_1 = 0.117$. The logistic regression curve produces probabilities that are similar to the REEP specifications through most of the temperature range, but are constrained by the functional form of Equation 7.36 to lie between zero and one, even for extreme values of the predictor.

Maximizing Equation 7.38 for logistic regression with a single ($K = 1$) predictor is simple enough that the Newton-Raphson method (see Section 4.6.2) can be implemented easily and is reasonably robust to poor initial guesses for the parameters. The counterpart to Equation 4.87 for this problem is

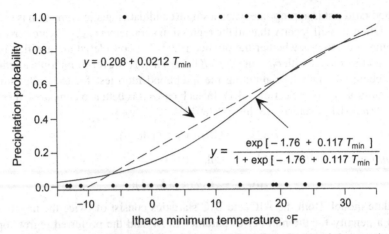

FIGURE 7.20 Comparison of regression probability forecasting using REEP (dashed) and logistic regression (solid) using the January 1987 data set in Table A.1. The linear function was fit using least squares, and the logistic curve was fit using maximum likelihood, to the data shown by the dots. The binary predictand $y = 1$ if Ithaca precipitation is greater than zero, and $y = 0$ otherwise.

$$\begin{bmatrix} b_0^* \\ b_1^* \end{bmatrix} = \begin{bmatrix} b_0 \\ b_1 \end{bmatrix} - \begin{bmatrix} \sum_{i=1}^{n}(p_i^2 - p_i) & \sum_{i=1}^{n}x_i(p_i^2 - p_i) \\ \sum_{i=1}^{n}x_i(p_i^2 - p_i) & \sum_{i=1}^{n}x_i^2(p_i^2 - p_i) \end{bmatrix}^{-1} \begin{bmatrix} \sum_{i=1}^{n}(y_i - p_i) \\ \sum_{i=1}^{n}x_i(y_i - p_i) \end{bmatrix}, \tag{7.42}$$

where p_i is a function of the regression parameters b_0 and b_1, and depends also on the predictor data x_i, as shown in Equation 7.36b. The first derivatives of the log-likelihood (Equation 7.38) with respect to b_0 and b_1 are in the vector enclosed by the rightmost square brackets, and the second derivatives are contained in the matrix to be inverted. Beginning with an initial guess for the parameters (b_0, b_1), updated parameters (b^*_0, b^*_1) are computed and then resubstituted into the right-hand side of Equation 7.42 for the next iteration. For example, assuming initially that the Ithaca minimum temperature is unrelated to the binary precipitation outcome, so $b_0 = -0.0645$ (the log of the observed odds ratio, for constant $p = 15/31$) and $b_1 = 0$; the updated parameters for the first iteration are $b^*_0 = -0.0645-(-0.251)(-0.000297)-(0.00936)$ $(118.0) = -1.17$, and $b^*_1 = 0-(0.00936)(-0.000297)-(-0.000720)(118.0) = 0.085$. These updated parameters increase the log-likelihood from -21.47 for the constant model (calculated using Equation 7.38, imposing $b_0 = -0.0645$ and $b_1 = 0$), to -16.00. After four iterations the algorithm has converged, with a final (maximized) log-likelihood of -15.67.

Is the logistic relationship between Ithaca minimum temperature and the probability of precipitation statistically significant? This question can be addressed using the likelihood ratio test (Equation 5.20). The appropriate null hypothesis is that $b_1 = 0$, so $L(H_0) = -21.47$, and $L(H_A) = -15.67$ for the fitted regression. If H_0 is true then the observed test statistic $\Lambda^* = 2\,[L(H_A) - L(H_0)] = 11.6$ is a realization from the χ^2 distribution with $\nu = 1$ (the difference in the number of parameters between the two regressions), and the test is 1-tailed because small values of the test statistic are favorable to H_0. Referring to the first row of Table B.3, it is clear that the regression is significant at the 0.1% level.

Another approach to assessing the significance of this regression is through use of the estimated sampling distributions of the fitted coefficients, as outlined in Section 4.6.4. For sufficiently large sample sizes, these sampling distributions are Gaussian, with variances estimated by the diagonal elements of the information matrix, which in the present example is the negative of the inverted matrix in Equation 7.42. The estimated sampling variance for the b_1 coefficient will be in the lower-right corner of that matrix, which is 0.00190. Accordingly the t statistic addressing the null hypothesis that $b_1 = 0$ would be $0.117/\sqrt{0.00190} = 2.68$, yielding a one-tailed p value of 0.0037. ◇

7.6.3. Poisson Regression

Another regression setting where the residual distribution may be poorly represented by the Gaussian is the case where the predictand consists of counts, that is, each of the y's is a nonnegative integer. Particularly if these counts tend to be small, the residual distribution is likely to be asymmetric, and we would like a regression predicting these data to be incapable of implying nonzero probability for negative counts.

A natural probability model for count data is the Poisson distribution (Equation 4.12). Recall that one interpretation of a regression function is as the conditional mean of the predictand, given specific value (s) of the predictor(s). If the outcomes to be predicted by a regression are Poisson-distributed counts, but the Poisson parameter μ may depend on one or more predictor variables, we can structure a regression to specify the Poisson mean as a nonlinear function of those predictors using the link function $g(\mu) = \ln(\mu)$. The resulting GLM can then be written as

$$\ln(\mu) = b_0 + b_1 x_1 + \cdots + b_K x_K, \tag{7.43a}$$

or

$$\mu = \exp\left[b_0 + b_1 x_1 + \cdots + b_K x_K\right]. \tag{7.43b}$$

Equation 7.43 is not the only function that could be used for this purpose, but framing the problem in this way makes the subsequent mathematics quite tractable, and the logarithmic link function ensures that any predicted Poisson mean is nonnegative. Some applications of Poisson regression are described in Elsner and Schmertmann (1993), Paciorek et al. (2002), Parisi and Lund (2008), Solow and Moore (2000), and Tippet et al. (2011).

Having framed the regression in terms of Poisson distributions for the y_i conditional on the corresponding set of predictor variables $x_i = \{x_1, x_2, \ldots, x_K\}$, the natural approach to parameter fitting is to maximize the Poisson log-likelihood, written in terms of the regression parameters. Again assuming independence among the n data values, the log-likelihood is

$$L(\boldsymbol{b}) = \sum_{i=1}^{n} \{y_i \left(b_0 + b_1 x_1 + \cdots + b_K x_K\right) - \exp\left(b_0 + b_1 x_1 + \cdots + b_K x_K\right)\}, \tag{7.44}$$

where the term involving $y!$ from the denominator of Equation 4.12 has been omitted because it does not involve the unknown regression parameters and so will not influence the process of locating the maximum of the function. An analytic maximization of Equation 7.44 in general is not possible, so that statistical software will approximate the maximum iteratively, typically using one of the methods outlined in Sections 4.6.2 or 4.6.3. For example, if there is a single ($K = 1$) predictor, the Newton-Raphson method (see Section 4.6.2) iterates the solution according to

$$\begin{bmatrix} b_0^* \\ b_1^* \end{bmatrix} = \begin{bmatrix} b_0 \\ b_1 \end{bmatrix} - \begin{bmatrix} -\sum_{i=1}^{n} \mu_i & -\sum_{i=1}^{n} x_i \mu_i \\ -\sum_{i=1}^{n} x_i \mu_i & -\sum_{i=1}^{n} x_i^2 \mu_i \end{bmatrix}^{-1} \begin{bmatrix} \sum_{i=1}^{n} (y_i - \mu_i) \\ \sum_{i=1}^{n} x_i (y_i - \mu_i) \end{bmatrix}, \tag{7.45}$$

where μ_i is the conditional mean as a function of the ith set of regression parameters as defined in Equation 7.43b. Equation 7.45 is the counterpart of Equation 4.87 for fitting the gamma distribution, and Equation 7.42 for logistic regression.

Example 7.10. A Poisson Regression

Consider the annual counts of tornados reported in New York state for 1959–1988, in Table 7.6. Figure 7.21 shows a scatterplot of these as a function of average July temperatures at Ithaca in the corresponding years. The solid curve is a Poisson regression function, and the dashed line shows the ordinary linear least-squares linear fit. The nonlinearity of the Poisson regression is quite modest over the range of the training data, although the regression function would remain strictly positive regardless of the magnitude of the predictor variable.

The relationship is weak, but slightly negative. The significance of the Poisson regression usually would be judged using the likelihood ratio test (Equation 5.20). The maximized log-likelihood (Equation 7.44) is 74.26 for $K = 1$, whereas the log-likelihood with only the intercept $b_0 = \ln(\Sigma y / n) = 1.526$ is 72.60. Comparing $\Lambda^* = 2 (74.26 - 72.60) = 3.32$ to the χ^2 distribution quantiles in Table B.3 with $\nu = 1$ (the difference in the number of fitted parameters) indicates that b_1 would be judged significantly different from zero at the 10% level, but not at the 5% level. Alternatively the estimated sampling variance, given by the lower-right element of the negative of the inverse matrix in Equation 7.45 (Section 4.6.4) is 0.00325, so that the t statistic for the null hypothesis that $b_1 = 0$ is $-0.104 / \sqrt{0.00325} = -1.82$, corresponding to the 2-tailed p value 0.069. For the linear regression, the t ratio

TABLE 7.6 Numbers of Tornados Reported Annually in New York State, 1959–1988

1959	3	1969	7	1979	3
1960	4	1970	4	1980	4
1961	5	1971	5	1981	3
1962	1	1972	6	1982	3
1963	3	1973	6	1983	8
1964	1	1974	6	1984	6
1965	5	1975	3	1985	7
1966	1	1976	7	1986	9
1967	2	1977	5	1987	6
1968	2	1978	8	1988	5

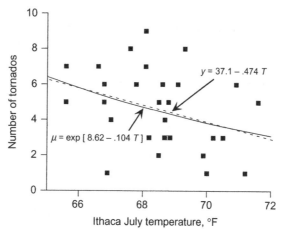

FIGURE 7.21 Annual New York tornado counts, 1959–1988 (Table 7.6), as a function of average Ithaca July temperature in the same year. Solid curve shows the Poisson regression fit using maximum likelihood (Equation 7.45), and dashed line shows ordinary least-squares linear regression.

for the slope parameter b_1 is -1.86, implying a two-tailed p value of 0.068, which is an essentially equivalent result.

The primary difference between the Poisson and linear regressions in Figure 7.21 is in the residual distributions, and therefore in the probability statements about the specified predicted values. Consider, for example, the number of tornados specified when $T = 70°F$. For the linear regression, $\hat{y} = 3.92$ tornados, with a Gaussian $s_e = 2.1$. Rounding to the nearest integer (i.e., using a continuity correction), the linear regression assuming Gaussian residuals implies that the probability for a negative number of tornados is $\Phi[(-0.5-3.92)/2.1] = \Phi[-2.10] = 0.018$, rather than the true value of zero. On the other hand, conditional on a temperature of 70°F, the Poisson regression specifies that the number of tornados will be distributed as a Poisson variable with mean $\mu = 3.82$. Using this mean, Equation 4.12 yields $\Pr\{Y < 0\} = 0$, $\Pr\{Y = 0\} = 0.022$, $\Pr\{Y = 1\} = 0.084$, $\Pr\{Y = 2\} = 0.160$, and so on. ◇

7.7. NONPARAMETRIC REGRESSION

7.7.1. Local Polynomial Regression and Smoothing

Regression settings where we may be unwilling to assume a specific mathematical form, such as Equation 7.25, can be approached nonparametrically using *local polynomial regression*. This method operates by fitting a separate low-order polynomial regression model in a limited neighborhood around each of many target points, building up the regression function as the collection of predictions at these target points. These local regressions are usually linear or quadratic in the predictor variable. Most often the approach is used as a smoothing technique, for example, as an aid to guiding the eye through a scatterplot, in which case it is often referred to as *loess smoothing*. It is also used to interpolate the predictand variable between observed data points, as well as providing the basis for predictions of future data.

Given n developmental data pairs x_i and y_i, the goal is to estimate or predict $\hat{y}(x_0)$ at a target point x_0. The target point need not be one of the values x_i in the training data. For each weighted local quadratic regression around a target point x_0, the regression parameters b_0, b_1, and b_2 are estimated by minimizing

$$\sum_{i=1}^{n} w_i(x_0)\left[y_i - b_0 - b_1(x_0 - x_i) - b_2(x_0 - x_i)^2\right]^2. \tag{7.46}$$

Weighted local linear regressions are obtained by minimizing Equation 7.46 while constraining $b_2 = 0$. Having fit these parameters for a given target point x_0, the estimate or prediction there is $\hat{y}(x_0) = b_0$.

Training-data values further away from the target point are deemphasized or neglected entirely using the weights

$$w_i(x_0) = \begin{cases} h\left(\dfrac{x_i - x_0}{\eta(x_0)}\right), & x_0 - \eta(x_0) < x_i < x_0 + \eta(x_0) \\ 0, & \text{otherwise} \end{cases} \tag{7.47}$$

where $\eta(x_0)$ is the bandwidth around the target point. Most frequently the tricube kernel,

$$h(u) = \left(1 - |u|^3\right)^3, \tag{7.48}$$

is used as the weighting function, although others such as those in Table 3.1 could be chosen instead.

The results of local regressions depend strongly on the bandwidth, which may be defined using the nearest-neighbor fraction λ. For each target-point regression the nearest λn training-data points receive nonzero weights in Equation 7.47. Specifically,

$$\eta(x_0) = |x_0 - d_{(\lambda n)}|, \tag{7.49}$$

where

$$d_i = |x_0 - x_i|, \quad i = 1, \ldots, n \tag{7.50}$$

are the distances of the training predictors to the target point, and the parenthetical subscript in Equation 7.49 (possibly rounded to a nearest integer) denotes the order statistic (Section 3.1.2), or the λnth smallest of the n distances. Chosen in this way, the bandwidth will tend to be narrower in regions of higher data density or away from the boundaries and tend to be wider in regions of lower data density or nearer the boundaries.

Larger values of the tuning parameter λ produce smoother results whereas smaller values allow the method to adapt more easily to local variations in the data. When the purpose of the local regression is data smoothing, the nearest-neighbor fraction could be chosen subjectively after viewing results of a range of possibilities. On the other hand, an objective method such as cross validation is more appropriate when the purpose is to obtain a predictive model.

Example 7.11. Smoothing the Mauna Loa CO$_2$ data

Figure 7.22 repeats the time series of monthly CO$_2$ measurements from Mauna Loa for 1958–2017, and the least-squares quadratic fit to those data from Figure 7.11 (light line). The heavy line shows the loess smooth of these data using weighted local linear regressions with nearest-neighbor fraction $\lambda = 0.10$, so that 10% of the training data are used in the local regressions for each of the many target points x_0. The two curves are very similar, although the local regression approach captures the dips below

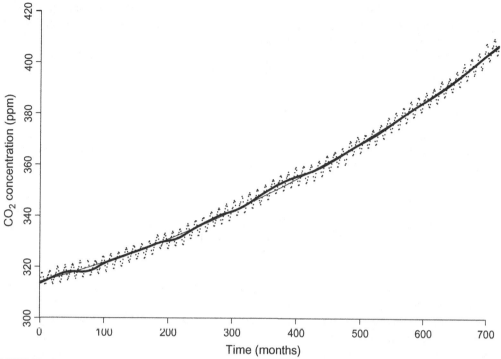

FIGURE 7.22 Loess smooth of the Mauna Loa CO_2 data using weighted local linear regressions with nearest-neighbor fraction $\lambda = 0.10$ (heavy curve). The light curve is the conventional quadratic least-squares fit reproduced from Figure 7.11.

the least-squares function around months 75 and 225, and the increase above the smoother function around month 375. Repeating the calculations with wider bandwidths defined by $\lambda = 0.30$ yields results nearly identical to the conventional regression. $\diamond$

Local linear regressions are effective in defining the long-term trend in Figure 7.22 because its curvature is modest. Local linear regressions may be less effective for data exhibiting stronger curvature, as they tend to undershoot peaks and overshoot valleys (Hastie et al., 2009; Loader, 1999), and in such situations local quadratic regressions usually provide better results.

The development in this section has been in terms of a single predictor variable, but the method extends naturally to settings with more than one predictor (Loader, 1999). For example, with 2 predictors the result is a surface above the plane defined by those two predictors. However, the method is generally not useful when there are more than three predictors, because there will be few training-data values in high-dimensional local neighborhoods (Hastie et al., 2009).

7.7.2. Theil-Sen (Median-Slope) Regression

Section 5.3.2 described the nonparametric Mann-Kendall trend test, which involves differencing all distinct pairs of the data values, and so relates to Kendall's τ (Equation 3.35) through Equation 5.32. Although the Mann-Kendall test can address the null hypothesis of no monotonic association, it cannot provide an estimate of the magnitude of any association or address the sampling distribution of such an estimate. Theil-Sen median-slope regression (Sen, 1968; Theil, 1950) provides these nonparametrically, assuming that the form of the association is linear.

As the name suggests, the method operates by computing the median of a collection of slope estimates. These estimates are all the pairwise combinations of

$$b_{i,j} = \frac{y_i - y_j}{x_i - x_j}, \tag{7.51}$$

for pairs in which $x_i \neq x_j$. Here y is the response variable and x is the independent variable (time index if the setting is a time series trend). If there are no repeated x values there are $n(n-1)/2$ such pairs. The median-slope regression is then defined by the slope

$$b = \text{median}\left(b_{i,j}\right) \tag{7.52a}$$

over the number of pairs for which $x_i \neq x_j$. The estimate for the intercept is

$$a = \text{median}\left(y_i - bx_i\right), \tag{7.52b}$$

over the n pairs in the training data, which result in a rank correlation between the x's and residuals from the regression line being approximately zero.

Median-slope regression is robust to non-Gaussian distribution and resistant to high fractions of outlying data, and yet performs reasonably well when the conventional assumptions in least-squares regression are met. Pairwise slopes for points involving outlying data will typically be extreme in absolute value, but because the slope estimate in Equation 7.51 is the median of all pairwise slopes it will be little affected by a modest number of erroneously large values. The slope estimate in Equation 7.52a is unbiased for an underlying true linear slope. Inferences regarding the estimated slope can be approached through bootstrapping.

Example 7.12. Comparison of Theil-Sen and Ordinary Linear Regression

Example 7.1 considered conventional least-squares regression between the Ithaca minimum temperature in Appendix A.1 as the predictor and the Canandaigua minimum temperature as predictand. There are six Ithaca temperatures appearing twice, and one Ithaca temperature appearing three times, yielding nine combinations of the 31 days with equal predictor values, so that there are $(31)(30)/2-9=456$ date pairs for which $x_i \neq x_j$ in Equation 7.51.

The median of these pairwise slopes is 0.600, which compares well with the least-squares slope estimate of 0.597 in Table 7.2 and Equation 7.21. The median of the 31 intercept estimates (Equation 7.52b with $b=0.600$) is 12.80, which also compares closely with the least-squares value of 12.46. It is unsurprising that these pairs of estimated regression parameters correspond so closely, because the relationship between the two maximum temperatures is rather well behaved, as can be seen in the appropriate panel of Figure 3.31.

Bootstrapping these data, and applying Equations 7.51, 7.52a, and 7.52b to each bootstrap sample, yields estimated (simple percentile-method) confidence intervals of (9.36, 14.12) for the intercept a, and (0.500, 0.724) for the slope b. Corresponding results using ± 2 standard error intervals for these parameters based on the conventional least-squares fitting in Table 7.2 and Equation 7.21 yields the intervals (10.74, 14.17) for the intercept and (0.505, 0.689) for the slope. It is unsurprising that the conventionally

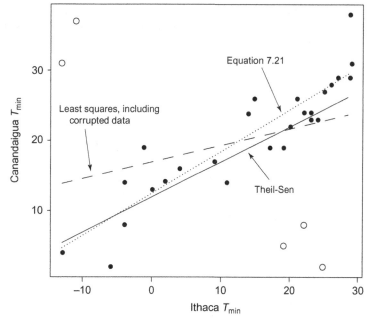

FIGURE 7.23 Scatterplot of January 1987 Canandaigua and Ithaca minimum temperatures, with five of the Canandaigua temperatures corrupted to outlying values (open circles). Solid line shows Theil-Sen median-slope regression, and dashed line shows least-squares fit to the data including the outliers. Dotted line shows least-squares fit to the uncorrupted data (Equation 7.21).

computed intervals are narrower in this instance because the data are plausibly Gaussian, contain no outliers, and appear to be homoscedastic. However, the degradation in estimation precision for the Theil-Sen estimators is quite modest.

The differences between the Theil-Sen and least-squares regressions are more substantial if the data set is contaminated with erroneous outlying values. Figure 7.23 shows a scatterplot in which five of the Canandaigua temperatures (open circles) have been artificially corrupted to be outliers. The least-squares fit to these data ($a = 16.8$, $b = 0.236$, dashed line) is adversely affected by the outlying data, whereas the Theil-Sen equation ($a = 11.8$, $b = 0.50$, solid line) differs only modestly from the least-squares fit (Equation 7.21) to the full correct data set ($a = 12.5$, $b = 0.60$, solid line). ◇

Examples of use of Theil-Sen regression in climatology and hydrology can be found in Hirsch et al. (1982), Huth and Pokorná (2004), Lettenmaier et al. (1994), and Romanic et al. (2015). Some alternative approaches to robust regression are considered by Muhlbauer et al. (2009).

7.7.3. Quantile Regression

Ordinary least-squares regression aims to find a function specifying the conditional mean of the predictand, given particular value(s) of the predictor(s). Alternatively, one can imagine seeking a function to specify the conditional median, or indeed functions specifying other conditional quantiles. This is the setting of *quantile regression* (Koenker and Bassett, 1978), which was introduced into the meteorological literature by Bremnes (2004).

Recall that the quantile q_p is the magnitude of the random variable exceeding p x 100% of the probability in its distribution. For example, $q_{.50}$ denotes the median or 50th percentile. In quantile regression a relatively small set of quantiles is selected; for example, Bremnes (2004) considered the five predictand quantiles $q_{.05}$, $q_{.25}$, $q_{.50}$, $q_{.75}$, and $q_{.95}$. Each selected quantile is represented by a distinct regression equation,

$$q_p(\boldsymbol{x}_i) = b_{q,0} + \sum_{k=1}^{K} b_{q,k} x_{i,k} \tag{7.53}$$

where $\boldsymbol{x}_i$ denotes the vector of K predictor variables $x_{i,1}, x_{i,2}, \ldots, x_{i,K}$. The predictors $x_{i,k}$ can be different for regression equations pertaining to different quantiles, and the considerations regarding predictor selection that were outlined in Section 7.4 apply also to quantile regression. The coefficients $b_{q,0}$ and $b_{q,k}$ are estimated separately for each quantile by numerically minimizing

$$\sum_{i=1}^{n} \rho_p \left[y_i - q_p(\boldsymbol{x}_i) \right], \tag{7.54}$$

where n is the training-data size, and

$$\rho_p(e_i) = \begin{cases} e_i p & , \ e_i \geq 0 \\ e_i(p-1) & , \ e_i < 0 \end{cases} \tag{7.55}$$

is called the check function. Because $0 < p < 1$ in Equation 7.55 the check function is nonnegative, so that Equation 7.54 is effectively a weighted sum of absolute values of the errors in the square brackets. Unlike least-squares regression, an analytic minimization of Equation 7.54 does not exist, so that parameter estimates must be computed numerically, usually using a linear programming algorithm.

When the target quantile is the median $q_{0.5}$, quantile regression is equivalent to *least absolute deviation* (LAD) *regression* (Bloomfield and Steiger, 1980; Mielke Jr. et al., 1996; Narula and Wellington, 1982). That is, computing a quantile regression for the median is equivalent to minimizing the sum over the n training samples of the absolute value of the square-bracketed quantity in Equation 7.54, because in that case $\rho_{0.5}[\varepsilon_i] = 0.5 |\varepsilon_i|$. With this perspective it can be easily seen that LAD regression, and quantile regression more broadly, is resistant to outliers because the residuals e_i are not squared. Interestingly, the idea behind LAD regression predates least-squares regression, and apparently Gauss developed least-squares regression as an analytically tractable alternative (Mielke, 1991).

When the assumptions of least-squares regression are met, namely, homoscedastic Gaussian residual distributions, conditional predictand quantile functions can instead be obtained easily and analytically using Equation 7.23. In addition to being resistant to outliers, quantile regression can also be useful when the residual distributions are asymmetric, or exhibit dispersion that depends on the predictor variable(s), or both. Figure 7.24, showing maximum winter Northern Hemisphere sea-ice extent for 1979–2010, illustrates the point by comparing linear quantile regression functions to the ordinary least-squares fit. The quantile-regression median function is very near the line for $q_{0.8}$ and much further from the $q_{0.2}$ line, and the least-squares mean function is substantially below the quantile-regression median, all of which indicate very substantial negative skewness for the residuals. In addition, the dispersion indicated by the distances between the $q_{0.8}$ and $q_{0.2}$ lines shows increasing variability of winter sea-ice extent later in the record (i.e., nonconstant residual variance), illustrating that quantile regression can easily represent heteroscedasticity.

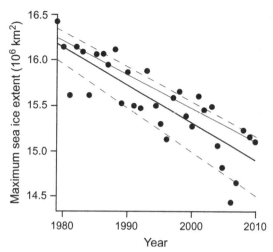

FIGURE 7.24 Linear quantile regressions representing the 20th and 80th percentiles (dashed), and median (light solid) of Northern Hemisphere maximum ice cover extent, 1979–2010. Heavy solid line shows the least-squares linear fit. *After Tareghian and Rasmussen (2013).*

One potential problem with quantile regressions arises as a consequence of the prediction equations for each target quantile being derived independently of one another. Especially in settings where training data are limited, or in instances where equations for a large number of densely spaced quantiles are estimated, the procedure may yield probabilistically incoherent results, meaning that the forecast cumulative probability for a higher quantile may be smaller than the probability for a lower quantile. Such cases would be reflected in plots like those in Figure 7.24 exhibiting functions that cross. For example, extrapolating much beyond 2010 in Figure 7.24 would imply the impossible $q_{0.5} > q_{0.8}$.

Some additional recent examples of the use of quantile regression include the studies of Ben Bouallègue (2016), Bentzien and Friederichs (2012), Jagger and Elsner (2009), Nielsen et al. (2006), and Wasko and Sharma (2014).

7.8. "MACHINE-LEARNING" METHODS

The names "machine-learning" (e.g., Efron and Hastie, 2016; Hsieh, 2009), or "statistical learning" (e.g., Hastie et al., 2009), or "artificial intelligence" methods (e.g., Haupt et al. 2009) refer to computationally intensive algorithms that have been developed relatively recently, enabled by the ongoing exponential increases in computing capacity. These nonparametric and nonlinear methods are extremely flexible and data adaptive. Since they typically involve large numbers of parameters, their estimation usually must involve large training data sets, and their effective use generally requires vigilant attention to potential overfitting. Unlike many traditional statistical methods, they can be deployed in settings where the number of fitted parameters is larger than the training-sample size. On the other hand, their structures are often less easily interpretable than are more conventional statistical methods, so that they are sometimes disparaged as "black boxes." These nonlinear methods may or may not outperform traditional linear statistical methods in atmospheric prediction, even as they are substantially more expensive (Mao and Monahan, 2018).

Machine learning methods are only beginning to be applied to meteorological and climatological forecasting problems, and this section outlines a few of the more common approaches.

7.8.1. Binary Regression Trees and Random Forests

Regression Trees

As the name suggests, *binary regression trees* (Breiman et al., 1984) are regression models, in the sense that each specifies a conditional mean for the predictand variable y conditional on a particular set of predictor variables $x_k, k = 1, ..., K$. An individual tree (i.e., a particular regression model) is built through a sequence of binary splittings of the training data made on the basis of the predictor variables, where each split progressively decreases the sum of squares of the training-sample predictand across the groups defined by the binary splits.

At the first step of the algorithm (the "base of the tree trunk"), all possible definitions for two groups g_1 and g_2 of the predictand, defined by binary splits of each of the predictors at each of its possible values, are examined to find the minimum of the combined sum of squares

$$SS = \sum_{y \in g_1} \left(y - \bar{y}_{g_1} \right)^2 + \sum_{y \in g_2} \left(y - \bar{y}_{g_2} \right)^2. \tag{7.56}$$

Thus $K(n-1)$ potential binary splits are evaluated at the initial step in order to find the one minimizing Equation 7.56. This number increases as the algorithm progresses, because the number of groups that can be potentially split increases. The minimization of Equation 7.56 defines the binary split that makes the resulting two (sub)groups of predictand values as different from each other as possible. The search for minimum sum-of-squares binary splits of one of the previously defined groups based on the predictors continues for the second and subsequent steps of the algorithm, yielding a sequence of branches for the tree that ends with a stopping criterion being satisfied. Definition of this endpoint (the "pruning") typically involves the cross-validated minimization of some function of the prediction error, often the mean-squared error (Hastie et al., 2009).

At the final stage, $s + 1$ groups have been defined on the basis of s algorithm steps, and each of the $s + 1$ terminal branches will be associated with some number of the training-data predictand values (the "leaves"). Using the tree for forecasting involves simply following the branching pattern according to the particular set of K predictor variables at hand, and then estimating the conditional predictand mean, given the predictor set, as the sample mean of the leaves at the end of the terminal branch. Unless the tree is rather large, the identities and specific values of the predictor variables that define each binary split can be easily interpreted.

When the predictand variable is not a continuous quantity, but rather consists of a finite number of discrete classes or groups, the same tree-building approach can be taken to build classification trees (see Section 15.6.2).

Example 7.13. A Simple Binary Regression Tree

Examples 7.5–7.8 illustrated specifications of the January 1987 Canandaigua minimum temperatures as the predictand, on the basis of the other variables listed in Table A.1 as predictors. Figure 7.25 shows the binary regression tree for the same problem.

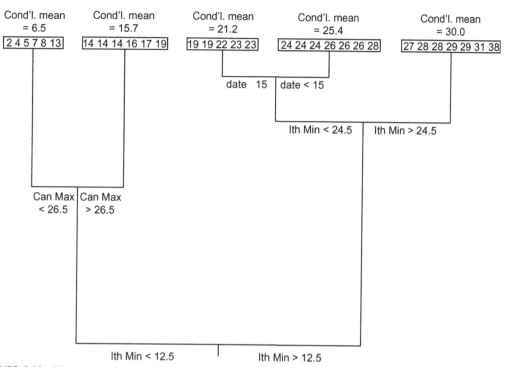

FIGURE 7.25 Binary regression tree for prediction of Canandaigua minimum temperatures, using the other variables in Table A.1 as potential predictors. Boxed values are training-data predictands at each of the five terminal branches, and the five corresponding conditional means are the possible predicted values.

The first binary split, at the base of the tree, indicates that the groups of predictand values most different from each other are defined by whether the Ithaca minimum temperature is colder or warmer than 12.5°F, with colder Ithaca temperatures predicting colder Canandaigua temperatures. The next most effective split, indicated by the next-highest horizontal bar, is based on whether the Canandaigua maximum temperature is below 26.5°F or not, again with colder predictor temperatures indicating colder predictand values. If the Ithaca minimum temperature is warmer than 12.5°F, the most effective binary split is defined by whether the Ithaca minimum temperature is colder or warmer than 24.5°F, and if colder the final split is defined by the date.

In common with the regression models developed in Examples 7.5 and 7.8, the Ithaca minima, Canandaigua maxima, and the date have been chosen as predictors here, but their use in prediction of the Canandaigua minima is nonlinear and discontinuous. The boxed values above the terminal branches show the training-data predictand values defined by the tree structure below (although these are not usually included in the portrayal of a regression tree). Evidently the binary tree structure has been very effective in segregating relatively homogeneous groups of training-data predictand temperatures, with only the 28° value at the second-warmest terminal branch being apparently out of place. The five corresponding conditional means are the possible values that the regression could return as predictions. For example, on any day for which the Ithaca minimum temperature is at least 25°, the corresponding conditional mean value for Canandaigua minimum temperature would be 30°. If the Ithaca minimum is at most 12° and the Canandaigua maximum temperature is at least 27°, the regression tree predicts 15.7° for the Canandaigua minimum temperature. ◇

Random Forests

Binary regression trees are simple to fit, easy to interpret, provide automatic selection of predictor variables, and can be used even when $n < K$. On the other hand, individual trees provide only discontinuous forecasts (e.g., five possible values in Figure 7.25). In addition, they can be unstable, in the sense of their structure varying strongly for different training data sets, and so may perform poorly on independent data. However, these problems can be ameliorated by averaging the predictions of many different regression trees. Such a collection of trees is known as a *random forest* (Breiman, 2001; Hastie et al., 2009).

The collection of regression trees in a random forest differ from each other in that each is computed based on a different bootstrap sample of the training data, which is called bootstrap aggregation, or *bagging* (Breiman, 1996; Hastie et al., 2009). Differences among trees in the random forest are further increased (and the required computations also reduced) by allowing only a random subset of the predictors to be considered as candidates for defining the binary splitting at each new branch. Typical choices for the number of predictors selected for consideration at each step are $K/3$ or $\sqrt{K}$. Forecasts are derived from the random forest by following the branches for each tree as directed by the predictor variables pertaining to the tree in question to their terminal branches, and then averaging the conditional means for the terminal branches of each tree in the forest.

A random forest delivers an estimate of the conditional mean of the predictand, given a particular set of predictor variables. The idea can be extended to estimation of the full predictive distribution using *quantile regression forests* (Meinshausen, 2006). Despite the similarity of the names, these are quite different from the quantile regressions described in Section 7.7.3. Forecast distributions from quantile random forests are based on the subsets of training-sample predictands at each of the terminal branches in the random forest (i.e., the leaves), which define an empirical predictive distribution for that terminal branch. The quantile-regression-forest predictive distribution is then formed as the average of the empirical distributions at the terminal leaves, over all trees in the random forest. Herman and Schumacher (2018), Taillardat et al. (2017), and Whan and Schmeits (2018) provide operationally oriented MOS (Section 7.9.2) examples. Because the resulting predictive distribution is the average of the conditional training-data empirical distributions for the terminal leaves, incoherent probability forecasts (such as negative event probabilities or nonnegative probabilities for impossible outcomes) cannot be produced. On the other hand, neither can nonzero probability be assigned to predictand values outside their range in the training data, so that extreme-value forecasts may be problematic.

7.8.2. Artificial Neural Networks

Artificial Neural Networks (ANNs) provide another nonlinear and nonparametric regression approach. The structure of ANNs is meant to emulate a particular conceptual model of brain functioning (McCulloch and Pitts, 1943). The idea is that an individual neuron sums the signals it receives from other neurons, and if that sum is sufficiently large it "fires" a signal of its own to other neurons. If the received signals are not strong enough the neuron is silent and does not signal to other neurons in the network.

Structures of particular ANNs, representing highly simplified abstractions of this underlying biological neural model, are typically represented by diagrams such as those in Figure 7.26, called feedforward *perceptrons*. The circles in this figure represent individual neurons, often called nodes, which are arranged in layers. Each node is connected to all other nodes in adjacent layers, as indicated by the lines, and the information flow is exclusively from left to right (i.e., feeding forward). The predictor data, consisting of $K = 4$ predictor variables in both panels of Figure 7.26, provide the initial signal to the first

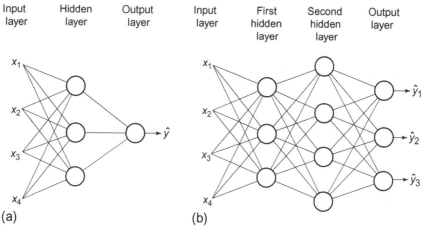

FIGURE 7.26   Two example ANN diagrams, both of which operate on $K = 4$ predictor variables. Panel (a) shows a network with a single hidden layer and one predictand variable. The network in panel (b) has two hidden layers and three related predictand outputs.

of possibly multiple "hidden" layers. The ANN represented in Figure 7.26a has a single hidden layer, and the ANN in Figure 7.26b has two hidden layers. The final, or output, layer transforms the outputs from the last hidden layer to the predicted value or values. For the ANN in Figure 7.26a this predictand is a single scalar value, and in Figure 7.26b the outputs are three related predictands.

Diagrams such as those in Figure 7.26 actually represent a sequence of parallel computations on the predictor variables. In particular, each circle indicates a transformation of a linear combination of the values it receives. For example, consider the simpler ANN represented in Figure 7.26a, and define z_j, $j = 1, 2, 3$, to be the output of the jth neuron in the hidden layer,

$$z_j = h\left(b_{j,0} + \sum_{k=1}^{K} b_{j,k} x_k\right), \tag{7.57}$$

where $h(\bullet)$ is called an *activation function*, which in general will be nonlinear (with the possible exception of the operations in the output layer). The b parameters are regression coefficients that need to be estimated using the training data.

Adhering to the original on/off concept of neuron interaction would suggest choosing a step function for the activation function. More commonly, smooth sigmoid (S-shaped) and bounded functions are chosen for this purpose. The logistic function

$$h(u) = \frac{1}{1 + e^{-u}} \tag{7.58}$$

(see also Equation 7.36b and Figure 7.20) is a very common choice, which is bounded by $0 < h(u) < 1$, and so approximates a step function in a way that is smooth and differentiable. For large positive values of its argument it is nearly fully "on," and for large negative argument values it is nearly fully "off." Another common choice for the activation function is the hyperbolic tangent function,

$$h(u) = \frac{e^u - e^{-u}}{e^u + e^{-u}}, \tag{7.59}$$

which is also smooth, differentiable, sigmoid in shape, and bounded as $-1 < h(u) < 1$. For u in the neighborhood of 0, Equations 7.58 and 7.59 approximate linear responses.

The activation function for the output layer is often different from that chosen for the hidden layers. In a conventional regression setting where there is a single scalar predictand, such as the ANN diagrammed in Figure 7.26a, the activation function in the output layer is usually just the identity $h(u) = u$. If the ANN is meant to yield probabilities for multiple discrete outcomes, such as diagrammed in Figure 7.26b, a reasonable choice is the *softmax* function

$$h(u_j) = \frac{e^{u_j}}{\sum_{m=1}^{J} e^{u_m}}, \tag{7.60}$$

which yields a collection of J positive values that sum to one. If the predictand is the probability for a single binary outcome, using the logistic function (Equation 7.58) at the single node of the output layer is a natural choice. These latter two cases are essentially probabilistic discrimination and classification (Chapter 15) applications.

The foregoing discussion indicates that ANNs are elaborations on the structure of generalized linear models (Section 7.6.1). Indeed, abstracting Figure 7.26a to a "network" with no hidden layer, and a single output node employing the logistic activation function, yields an ordinary logistic regression (Section 7.6.2). On the other hand, the complexity, and thus the flexibility, of ANN models expands very rapidly as layers and nodes are added. For example, possibly the simplest ANN network structure consists of a single, two-node hidden layer, and a single output node containing the identity activation function, which would be represented mathematically as

$$\hat{y} = b_0 + b_1\,h\left(b_{1,0} + \sum_{k=1}^{K} b_{1,k}x_k\right) + b_2\,h\left(b_{2,0} + \sum_{k=1}^{K} b_{2,k}x_k\right). \tag{7.61}$$

This is a highly flexible equation, but even this extremely simple ANN would require estimation of $2(K+1)+3$ parameters whose interpretation in terms of the underlying physical variables would be very difficult.

Parameter estimation typically begins using random, near-zero initial guesses for the parameters (and so a near-linear initial model), and proceeds iteratively by minimizing a penalty function such as squared prediction error or the negative of maximized likelihood. Because the number of parameters to be estimated may be quite large, regularization methods (Section 7.5), which penalize a model for large or numerous nonzero values for the parameters multiplying the inputs (the "intercept" parameters are usually not restricted), are generally needed to prevent overfitting. Accordingly, the predictor variables should be standardized (converted to standardized anomalies) to have comparable magnitudes, so that the regularization constraints are equitably distributed among them. Efron and Hastie (2016) and Hastie et al. (2009) provide details on fitting algorithms and their regularization.

There appear to be no clear-cut rules or guidelines for choosing the number of hidden layers or the numbers of nodes in each hidden layer. Hsieh (2009) notes that accurate representation of more complicated and irregular functional relationships requires more hidden nodes, whereas smoother relationships can be captured using fewer parameters. Hastie et al. (2009) state that typical ANN models usually contain somewhere between 5 and 100 hidden nodes in total. Efron and Hastie (2016) suggest that it is better to have too many rather than too few hidden nodes, since overall model complexity can be controlled through regularized parameter estimation.

7.8.3. Support Vector Regression

Consider again the conventional multiple regression model in Equation 7.25. Some of the possible approaches to estimating the parameters $b_0, b_1, ..., b_K$, can be generalized as finding the parameter set that minimizes the function

$$\sum_{i=1}^{n} g(y_i - \hat{y}) + \lambda \sum_{k=1}^{K} b_k^2 \qquad (7.62)$$

over the n training-data values. For example, if the error penalty $g(u) = u^2$ then Equation 7.62 represents ordinary least-squares regression (Section 7.3) if $\lambda = 0$, and ridge regression (Section 7.5.1) if $\lambda \neq 0$. Similarly, if $g(u) = |u|$ then Equation 7.62 represents LAD regression (quantile regression for the median, Section 7.7.3), with $(\lambda \neq 0)$ or without $(\lambda = 0)$ regularization.

In *support vector regression*, the parameters are estimated by minimizing Equation 7.62 using the "δ-insensitive" error penalty

$$g(u) = \begin{cases} 0 & , |u| \leq \delta \\ |u| - \delta, & |u| > \delta \end{cases}, \qquad (7.63)$$

that ignores regression residuals smaller than δ, and imposes an absolute-error penalty otherwise. Thus support vector regression seeks to orient the regression function in a way that maximizes the number of training-data points with residuals smaller than δ in absolute value and is fairly robust to outlying data because of the absolute error rather than squared-error loss specified by Equation 7.63. The parameter estimates depend only on the predictor values yielding residuals larger than δ in absolute value, which are the "support vectors." The two adjustable parameters δ and λ might be estimated through cross-validation.

The foregoing development described linear support vector regression. More typically nonlinear support vector regression is of interest, for which the regression equation can be written

$$\hat{y} = b_0 + \sum_{m=1}^{M} b_m h_m(x_1, x_2, ..., x_K). \qquad (7.64)$$

Here each of the basis functions h_m depends in general nonlinearly on all K of the predictor variables. Linear support vector regression is then a special case, with $M = K$, and $h_m(\mathbf{x}) = x_m$. The computational details are somewhat elaborate and can be found in Hastie et al. (2009) and Hsieh (2009). When the predictand values indicate memberships in discrete classes or groups, the same general approach can be used for classification (Section 15.6.1).

7.9. OBJECTIVE FORECASTS USING TRADITIONAL STATISTICAL METHODS

7.9.1. Classical Statistical Forecasting

Construction of weather forecasts through purely statistical means—that is, without the benefit of information from fluid-dynamical weather prediction models—has come to be known as classical statistical forecasting. This name reflects the long history of the use of purely statistical forecasting methods, dating from the time before the availability of dynamical forecast information. The accuracy of dynamical forecasts has advanced sufficiently that pure statistical forecasting is currently used in practical settings only for very short lead times or for fairly long lead times.

Very often classical forecast products are based on multiple regression equations of the kinds described in Sections 7.2 and 7.3. These statistical forecasts are objective in the sense that a particular set of inputs or predictors will always produce the same forecast for the predictand, once the forecast equation has been developed. However, many subjective decisions necessarily go into the development of the forecast equations.

The construction of a classical statistical forecasting procedure follows from a straightforward implementation of the ideas presented in previous sections of this chapter. Required developmental data consist of past values of the quantity to be forecast, and a matching collection of potential predictors whose values will be known prior to the forecast time. A forecasting procedure is developed using this set of historical data, which can then be used to forecast future values of the predictand on the basis of future observations of the predictor variables. It is thus a characteristic of classical statistical weather forecasting that the time lag is built directly into the forecast equation through the time-lagged relationships between the predictors and the predictand.

For lead times up to a few hours, purely statistical forecasts still find productive use. This short-lead forecasting niche is known as *nowcasting*. Dynamically based forecasts are not practical for nowcasting because of the delays introduced by the processes of gathering weather observations, data assimilation (calculation of initial conditions for the dynamical model), the actual running of the forecast model, and the postprocessing and dissemination of the results.

One very simple statistical approach that can produce competitive nowcasts is use of *conditional climatology*, that is, historical statistics subsequent to (conditional on) analogous weather situations in the past. The result could be a conditional frequency distribution for the predictand, or a single-valued forecast corresponding to the expected value (mean) of that conditional distribution. A more sophisticated approach is to construct a regression equation to forecast a few hours ahead. For example, Vislocky and Fritsch (1997) compare these two approaches for forecasting airport ceiling and visibility at lead times of one, three, and six hours.

At lead times beyond perhaps two weeks, purely statistical forecasts are again competitive with dynamical forecasts. At these longer lead times the sensitivity of dynamical models to the unavoidable small errors in their initial conditions, described in Chapter 8, makes explicit forecasting of specific weather events problematic. This estimated limit of approximately two weeks for explicit dynamical predictability of the atmosphere has remained basically unchanged since the question began being investigated in the 1960s (e.g., Buizza and Leutbecher, 2015; Lorenz, 1982; Simmons and Hollingsworth, 2002).

Although long-lead forecasts for seasonally averaged quantities currently are made using dynamical models (e.g., Barnston et al., 2003; Kirtman et al., 2014; Stockdale et al., 2011), comparable or even better predictive accuracy at substantially lower cost is still obtained through statistical methods (e.g., Hastenrath et al., 2009; Moura and Hastenrath, 2004; Quan et al., 2006; Van den Dool, 2007; Wilks, 2008; Zheng et al., 2008). Possibly the failure of dynamical methods to consistently outperform relatively simple statistical methods for seasonally averaged quantities is due to the linearization of these prediction problems that is induced by the inherent long time averaging (Yuval and Hsieh, 2002; Hsieh, 2009), through effects described by the Central Limit Theorem (Sections 4.4.2 and 12.5.1).

Often the predictands in these seasonal forecasts are spatial patterns, and so the forecasts involve multivariate statistical methods that are more elaborate than those described in Sections 7.2 and 7.3 (e.g., Barnston, 1994; Mason and Mimmack, 2002; Ward and Folland, 1991; Wilks, 2008, 2014a, 2014b; see Sections 14.2.3, 14.3.2, and 15.4). However, univariate regression methods are still appropriate and useful for single-valued predictands. For example, Knaff and Landsea (1997) used ordinary least-squares regression for seasonal forecasts of tropical sea-surface temperatures with observed sea-surface temperatures as predictors, and Elsner and Schmertmann (1993) used Poisson regression for seasonal prediction of hurricane numbers.

Example 7.14. A Set of Classical Statistical Forecast Equations
The flavor of classical statistical forecast methods can be appreciated by looking at the NHC-67 procedure for forecasting hurricane movement (Miller et al., 1968). This relatively simple set of regression equations was used as part of the operational suite of forecast models at the U.S. National Hurricane Center until 1988 (Sheets, 1990). Since hurricane movement is a vector quantity, each forecast consists of two equations: one for northward movement and one for westward movement. The two-dimensional forecast displacement is then computed as the vector sum of the northward and westward forecasts.

The predictands were stratified according to two geographical regions: north and south of 27.5°N latitude. That is, separate forecast equations were developed to predict storms located on either side of this latitude at the time of forecast initialization. This division was based on the subjective experience of the developers regarding the responses of hurricane movement to the larger-scale flow, and in particular that storms moving in the trade winds in the lower latitudes tend to behave less erratically. Separate forecast equations were also developed for "slow" vs. "fast" storms. The choice of these two stratifications was also made subjectively, on the basis of the experience of the developers. Separate equations are also needed for each forecast lead time (0–12 h, 12–24 h, 24–36 h, and 36–48 h, yielding a total of 2 (displacement directions) × 2 (regions) × 2 (speeds) × 4 (lead times) = 32 separate regression equations in the NHC-67 package.

The available developmental data set consisted of 236 northern cases (initial position for hurricanes) and 224 southern cases. Candidate predictor variables were derived primarily from 1000-, 700-, and 500-mb heights at each of 120 gridpoints in a 5° x 5° coordinate system that follows the storm. Predictors derived from these 3 × 120 = 360 geopotential height predictors, including 24-h height changes at each level, geostrophic winds, thermal winds, and Laplacians of the heights, were also included as candidate predictors. Additionally, two persistence predictors, observed northward and westward storm displacements in the previous 12 h, were included.

With vastly more potential predictors than observations, some screening procedure is clearly required. Here forward selection was used, with the (subjectively determined) stopping rule that no more than 15 predictors would be in any equation, and new predictors would be only included to the extent that they increased the regression R^2 by at least 1%. This second criterion was apparently sometimes relaxed for regressions with few predictors.

TABLE 7.7 Regression Results for the NHC-67 Hurricane Forecast Procedure, for the 0–12 h Westward Displacement of Slow Southern-Zone Storms, Indicating the Order in Which the Predictors were Selected and the Resulting R^2 at Each Step

Predictor	Coefficient	Cumulative R^2
Intercept	−2709.5	–
P_X	0.8155	79.8%
Z_{37}	0.5766	83.7%
P_Y	−0.2439	84.8%
Z_3	−0.1082	85.6%
P_{51}	−0.3359	86.7%

The meanings of the symbols for the predictors are P_X=westward displacement in the previous 12 h, Z_{37}=500 mb height at the point 10° north and 5° west of the storm, P_Y=northward displacement in the previous 12 h, Z_3=500 mb height at the point 20° north and 20° west of the storm, and P_{51}=1000 mb height at the point 5° north and 5° west of the storm. Distances are in nautical miles, and heights are in meters.
From Miller et al. (1968). © American Meteorological Society. Used with permission.

Table 7.7 presents the results for the 0–12 h westward displacement of slow southern storms in NHC-67. The five predictors are shown in the order they were chosen by the forward selection procedure, together with the R^2 value achieved on the developmental data at each step. The coefficients are those for the final ($K=5$) equation. The most important single predictor was the persistence variable (P_X), reflecting the tendency of hurricanes to change speed and direction fairly slowly. The 500 mb height at a point north and west of the storm (Z_{37}) corresponds physically to the steering effects of midtropospheric flow on hurricane movement. Its coefficient is positive, indicating a tendency for westward storm displacement given relatively high heights to the northwest, and slower or eastward (negative westward) displacement of storms located southeast of 500 mb troughs. The final two or three predictors appear to improve the regression only marginally—the predictor Z_3 increases the R^2 by <1%—and it is quite possible that the $K=2$ or $K=3$ predictor models might have been chosen, and might have been equally accurate for independent data, if cross-validation had been known to and computationally feasible for the developers. Remarks in Neumann et al. (1977) in relation to Figure 7.14, concerning the fitting of the similar NHC-72 regressions, are also consistent with the idea that the equation represented in Table 7.7 may have been overfit.   ◇

7.9.2. Model Output Statistics (MOS)

Pure classical statistical weather forecasts for lead times in the range of a few days to a week or two are generally no longer employed, since dynamical models now allow more accurate forecasts at these time scales. However, raw dynamical forecast outputs typically exhibit systematic errors which can be corrected through statistical postprocessing. Generally this process is carried out using large multiple regression equations in a way that is analogous to the classical approach, so that many of the same technical considerations pertaining to equation fitting apply. The difference has to do with the range of available predictor variables. In addition to conventional predictors such as current meteorological observations, the date, or climatological values of a particular meteorological element, predictor variables taken from the outputs of the dynamical models are also used.

There are three reasons why statistical reinterpretation of dynamical forecast output is useful for practical weather forecasting:

- There are important differences between the real world and its representation in the dynamical models, and these differences have important implications for the forecast enterprise (e.g., Gober et al., 2008). Dynamical models necessarily simplify and homogenize surface conditions, by representing the world as an array of gridpoints to which the forecast output pertains. Small-scale effects (e.g., of topography or small bodies of water) important to local weather may not be represented in a dynamical model. Also, locations and variables for which forecasts are needed may not be represented explicitly. However, statistical relationships can be developed between the information provided by the dynamical models and desired forecast quantities and locations to help alleviate these problems.
- Dynamical models are not complete and true representations of the workings of the atmosphere, particularly at the smaller time and space scales, and they are inevitably initialized at states that differ from the true initial state of the atmosphere. For both of these reasons, their forecasts are subject to errors. To the extent that these errors are systematic, statistical postprocessing can compensate and correct the resulting forecast biases.
- The dynamical models are deterministic. That is, even though the future state of the weather is inherently uncertain, a single integration is capable of producing only a single forecast for any meteorological element, given a particular set of initial model conditions. Using dynamical forecast information in conjunction with statistical methods allows quantification and expression of the uncertainty associated with different forecast situations. For example, it is possible to derive probability forecasts, using methods such as REEP or logistic regression, using predictors taken from even a single deterministic dynamical integration. Although ensemble forecasting (Chapter 8) is increasingly used to represent forecast uncertainty, these dynamical forecasts also benefit from statistical postprocessing (Sections 8.3 and 8.4).

The first statistical approach to be developed for taking advantage of deterministic dynamical forecasts was called *perfect prog* (Klein et al., 1959), which is short for perfect prognosis. As the name implies, the perfect prog technique made no attempt to correct for possible dynamical model errors or biases, but rather took their forecasts for future atmospheric variables at face value—assuming them to be perfect. The perfect prog method involved developing regression equations relating predictands of interest to simultaneously observed predictor variables. At first, it might seem that this would not be a productive approach to forecasting. For example, tomorrow's 1000–850 mb thickness may be an excellent predictor for tomorrow's maximum temperature, but tomorrow's thickness will not be known until tomorrow. However, in implementing the perfect-prog approach, it is the dynamical forecasts of the predictors (e.g., today's dynamical forecast for tomorrow's thickness) that are substituted into the regression equation as predictor values. Therefore the forecast time lag in the perfect-prog approach is contained entirely in the dynamical model. A key advantage of the perfect prog approach in the early days of dynamical weather forecasting was that it did not require an archive of past forecasts to fit and implement it.

The *Model Output Statistics* (MOS) approach (Carter et al., 1989; Glahn and Lowry, 1972) began to be used in preference to perfect prog after a sufficient historical archive of dynamical forecasts had been developed. The MOS approach extends classical statistical forecasting by including dynamical forecasts available at the time the forecast must be issued, but which pertain to the future time being forecast. Preference for the MOS approach over perfect prog derives from its capacity to include directly in the regression equations the influences of specific characteristics of particular dynamical models. Separate MOS forecast equations must be developed for different forecast lead times. This is because the error characteristics of the dynamical forecasts are different at different lead times, producing, for example, different statistical relationships between observed temperature and forecast thicknesses for 24 h versus 48 h in the future. In addition, since the MOS equations are tuned to the particular error characteristics of the model for which they were developed, different MOS equations will, in general, be required for use with different dynamical models.

The classical, perfect-prog, and MOS approaches are most commonly based on multiple linear regression, exploiting correlations between a predictand and available predictors (although nonlinear regressions can also be used: e.g., Lemcke and Kruizinga, 1988; Marzban et al., 2007; Vislocky and Fritsch, 1995b; Wilks, 2009, 2018b). In the classical approach it is the correlations between today's values of the predictors and tomorrow's predictand that forms the basis of the forecast. For the perfect-prog approach it is the simultaneous correlations between today's values of both predictand and predictors that are the statistical basis of the prediction equations. In the case of MOS forecasts, the prediction equations are constructed on the basis of correlations between dynamical forecasts as predictor variables, and the subsequently observed value of tomorrow's predictand.

Fitting MOS equations requires an archived record of forecasts from the same dynamical model that will ultimately be used to provide input to the MOS equations. Typically, several years of archived dynamical forecasts are required to develop a stable set of MOS forecast equations (e.g., Jacks et al., 1990). This requirement can be a substantial limitation, because the dynamical models are not static. Rather, these models regularly undergo changes aimed at improving their performance. Minor changes in a dynamical model leading to reductions in the magnitudes of its random errors will not substantially degrade the performance of a set of MOS equations (e.g., Erickson et al., 1991). However, modifications to the model that change—even substantially reducing—systematic errors will require redevelopment of accompanying MOS forecast equations. Since it is a change in the dynamical model that will have necessitated the redevelopment of a set of MOS forecast equations, it is often the case that a sufficiently long developmental sample of predictors from the improved dynamical model will not be immediately available. However, as the quality of dynamical forecast models continues to improve, fewer dynamical predictors are required to achieve good results using MOS regressions, so that shorter training periods can be used (Glahn, 2014).

The MOS approach to statistical forecasting has two advantages over the perfect prog approach that make MOS the method of choice when practical. The first of these is that model-calculated, but unobserved, quantities such as vertical velocity can be used as predictors. However, the dominating advantage of MOS over perfect prog is that systematic errors exhibited by the dynamical model are accounted for in the process of developing the MOS equations. Since the perfect-prog equations are developed without reference to the characteristics of any particular dynamical model, they cannot account for or correct their forecast errors. The MOS development procedure allows compensation for these systematic errors when the forecasts are computed. Systematic errors include such problems as progressive cooling or warming biases in the dynamical model with increasing forecast lead time, a tendency for synoptic features to move too slowly or too quickly in the dynamical model, and even the unavoidable decrease in forecast accuracy at increasing lead times.

The compensation for systematic errors in a dynamical model that is accomplished by MOS forecast equations is easiest to see in relation to a simple bias in an important predictor. Figure 7.27 illustrates

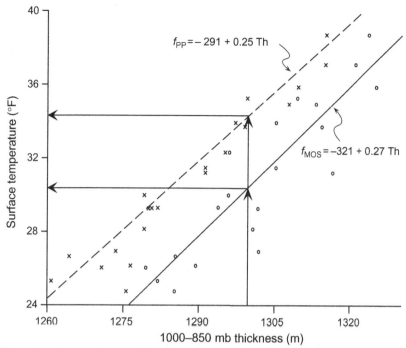

FIGURE 7.27 Illustration of the capacity of a MOS equation to correct for systematic bias in a hypothetical dynamical model. The x's represent observed, and the circles represent dynamically forecast 1000–850 mb thicknesses, in relation to hypothetical surface temperatures. The bias in the dynamical model is such that the forecast thicknesses are too large by about 15 m, on average. The MOS equation (solid line) is calibrated for this bias and produces a reasonable temperature forecast (lower horizontal arrow) when the forecast thickness is 1300 m. The perfect-prog equation (dashed line) incorporates no information regarding the attributes of the dynamical model and produces a surface temperature forecast (upper horizontal arrow) that is too warm as a consequence of the thickness bias.

a hypothetical case, where surface temperature is to be forecast using the 1000–850 mb thickness. The x's in the figure represent the (unlagged, or simultaneous) relationship of a set of observed thicknesses with observed temperatures, and the circles represent the relationship between thicknesses previously forecast by a dynamical model, with the same temperature data. As drawn, the hypothetical dynamical model tends to forecast thicknesses that are too large by about 15 m. The scatter around the perfect-prog regression line (dashed) derives from the fact that there are influences on surface temperature other than those captured by the concurrent 1000–850 mb thickness. The scatter around the MOS regression line (solid) is greater, because in addition it reflects errors in the dynamical model.

The observed thicknesses (x's) in Figure 7.27 appear to forecast the simultaneously observed surface temperatures reasonably well, yielding an apparently good perfect-prog regression equation (dashed line). The relationship between forecast thickness and observed temperature represented by the MOS equation (solid line) is substantially different, because it includes the tendency for this dynamical model to systematically overforecast thickness. If this model produces a thickness forecast of 1300 m (vertical arrows), the MOS equation corrects for the bias in the forecast thickness and produces a reasonable temperature forecast of about 30°F (lower horizontal arrow). Loosely speaking, the MOS knows that when

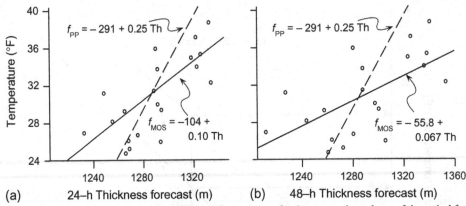

FIGURE 7.28 Illustration of the capacity of a MOS equation to account for the systematic tendency of dynamical forecasts to become less accurate at longer lead times. The points in these panels are simulated thickness forecasts, constructed from the x's in Figure 7.27 by adding random errors to the thickness values. As the forecast accuracy degrades at longer lead times, the perfect-prog equation (dashed line, reproduced from Figure 7.27) is increasingly overconfident and tends to forecast extreme temperatures too frequently. At longer lead times (b) the MOS equations increasingly provide forecasts near the climatological average temperature (30.8°F in this example).

this dynamical model forecasts 1300 m, a more reasonable expectation for the true future thickness is closer to 1285 m, which in the climatological data (x's) corresponds to a temperature of about 30°F. The perfect-prog equation, on the other hand, operates under the assumption that a dynamical model will forecast the future thickness perfectly. It therefore yields a temperature forecast that is too warm (upper horizontal arrow) when supplied with a thickness forecast that is too large.

A more subtle systematic error exhibited by all dynamical weather forecasting models is the degradation of forecast accuracy at increasing lead time. The MOS approach accounts for this type of systematic error as well. The situation is illustrated in Figure 7.28, which is based on the hypothetical observed data in Figure 7.27. The panels in Figure 7.28 simulate the relationships between forecast thicknesses from an unbiased dynamical model at 24- and 48-h lead time and the surface temperature, and have been constructed by adding random errors to the observed thickness values (x's) in Figure 7.27. These random errors exhibit $\sqrt{MSE} = 20\,m$ for the 24-h lead time and $\sqrt{MSE} = 30\,m$ at the 48-h lead time. The increased scatter of points for the simulated 48-h lead time illustrates that the regression relationship is weaker when the dynamical model is less accurate.

The MOS equations (solid lines) fit to the two sets of points in Figure 7.28 reflect the progressive loss of predictive accuracy of the dynamical model at longer lead times. As the scatter of points increases the slopes of the MOS forecast equations become more horizontal, leading to temperature forecasts that are more like the climatological mean temperature, on average. This characteristic is reasonable and desirable, since as the dynamical model provides less information about the future state of the atmosphere at longer lead times, temperature forecasts differing substantially from the climatological average temperature are progressively less well justified. In the limit of a few weeks lead time, a dynamical model will really provide no more information than will the climatological value of the predictand, so that the slope of the corresponding MOS equation would be zero, and the appropriate temperature forecast consistent with this (lack of) information would simply be the climatological average temperature. Thus it is sometimes said that MOS "converges to the climatology." By contrast, the perfect-prog equation (dashed lines, reproduced from Figure 7.27) take no account of the decreasing accuracy of the dynamical model at longer lead times and continue to produce temperature forecasts as if the thickness forecasts were perfect. Figure 7.28 emphasizes that the result is overconfident temperature forecasts, with both very warm and very cold temperatures forecast much too frequently.

Although MOS postprocessing of dynamical forecasts is strongly preferred to perfect prog and to the raw dynamical forecasts themselves, the pace of changes made to dynamical models continues to accelerate as computing capabilities progressively increase. Operationally it would not be practical to wait for two or three years of new dynamical forecasts to accumulate before deriving a new MOS system, even if the dynamical model were to remain static for that period of time. One option for maintaining MOS systems in the face of this reality is to retrospectively *reforecast* weather for previous years using the current updated dynamical model (Hagedorn, 2008; Hamill et al., 2004, 2013). Because daily weather data are typically strongly autocorrelated, the reforecasting process is more efficient if several days are omitted between the reforecast days (Hamill et al., 2004). Even if the computing capacity to reforecast is not available, a significant portion of the benefit of fully calibrated MOS equations can be achieved using a few months of training data (Mao et al., 1999; Neilley et al., 2002). Alternative common approaches include using longer developmental data records together with whichever version of the dynamical model was current at the time, and weighting the more recent forecasts more strongly. This can be done either by downweighting forecasts made with older model versions (Wilson and Vallée, 2002, 2003), or by gradually downweighting older data, usually using an algorithm called the *Kalman filter* (Cheng and Steenburgh, 2007, Crochet, 2004, Cui et al. 2012, Galanis and Anadranistakis, 2002, Homleid, 1995, Kalnay, 2003, Mylne et al., 2002b, Valée et al., 1996), although other approaches are also possible (Yuval and Hsieh, 2003).

7.9.3. Operational MOS Forecasts

Interpretation and extension of dynamical forecasts using MOS systems has a long-standing history at a number of national meteorological centers, including those in the Netherlands (Lemcke and Kruizinga, 1988), Britain (Francis et al., 1982), Italy (Conte et al., 1980), China (Lu, 1991), Spain (Azcarraga and Ballester, 1991), Canada (Brunet et al., 1988), and the United States (Carter et al., 1989; Glahn et al., 2009a), among others. Most of MOS applications have been oriented toward ordinary weather forecasting, but the method is equally well applicable in areas such as postprocessing of dynamical seasonal forecasts (e.g., Lepore et al., 2017; Shongwe et al., 2006; Vigaud et al., 2017).

MOS forecast products can be quite extensive, as exemplified by Table 7.8, which shows a collection of MOS forecasts for LaGuardia airport, New York City, for the 0600 UTC forecast cycle on 23 February 2018. This is one of hundreds of such panels for locations in the United States, for which these forecasts are issued four times daily and posted on the internet by the U.S. National Weather Service. Forecasts for a wide variety of weather elements are provided, at lead times up to 60 h and at intervals as close as 3 h. After the first few lines indicating the dates and times (UTC), are forecasts for daily maximum and minimum temperatures; temperatures, dew point temperatures, cloud coverage, wind speed, and wind direction at 3 h intervals; probabilities of measurable precipitation at 6- and 12-h intervals; forecasts for precipitation amount; thunderstorm probabilities; and forecast ceiling, visibility, and obstructions to visibility. Similar panels, based on several other dynamical models, are also produced and posted.

The MOS equations underlying forecasts such as those presented in Table 7.8 are seasonally stratified, usually with separate forecast equations for the "warm season" (April through September) and "cool season" (October through March). This two-season stratification allows the MOS forecasts to incorporate different relationships between predictors and predictands at different times of the year. A finer stratification (three-month seasons, or separate month-by-month equations) would probably be preferable if sufficient developmental data were available.

TABLE 7.8 Example MOS Forecasts Produced by the U.S. National Meteorological Center for LaGuardia Airport, New York City, Shortly after 0600 UTC on 23 February 2018

```
KLGA   GFS MOS GUIDANCE    2/23/2018   0600 UTC
DT /FEB 23                   /FEB 24                   /FEB 25      /
HR   12 15 18 21 00 03 06 09 12 15 18 21 00 03 06 09 12 15 18   00 06
X/N               44          59          25          53
TMP  37 38 39 40 41 42 44 45 46 51 56 56 52 46 43 43 44 48 52   52 50
DPT  28 28 32 37 40 40 41 41 40 37 36 36 36 35 35 36 37 38 39   43 37
CLD  OV OV OV OV OV OV OV OV OV BK OV OV OV OV OV OV OV OV OV   OV BK
WDR  07 09 09 10 13 22 24 27 30 33 36 36 05 07 07 07 08 08 06   01 30
WSP  08 09 08 06 04 06 07 06 07 07 07 07 07 07 10 14 16 15 08   06 06
P06        41    70    60     9     3    33    70    97    91    66  8
P12              70          60          38          97    91
Q06         0     1     1     0     0     0     1     2     3     3  2
Q12               1           1           0           3     4
T06   1/ 0/ 0/ 0/ 0/ 0/ 0/ 1/ 0/ 0/ 0/ 0/ 0/ 0/ 0/ 1/ 0/ 0/ 3/  1/ 3/
T12               0           5           0           1           3  8
POZ   6 11  7  6  3  2  2  0  1  0  0  1  2  3  0  0  1  0  0    0  2
POS  21 10  0  0  0  0  0  0  0  0  0  0  0  0  0  1  1  0  0    0  0
TYP   R  R  R  R  R  R  R  R  R  R  R  R  R  R  R  R  R  R  R    R  R
SNW                         0                       0
CIG   4  3  3  3  4  6  8  7  6  4  4  2  2  2  3  2  3  3  8    8
VIS   7  7  5  3  2  2  6  5  7  7  7  5  4  4  3  3  3  3  5    7
OBV   N  N BR BR BR BR BR FG BR BR BR  N  N  N BR BR BR BR BR   BR  N
```

A variety of weather elements are forecast, at lead times up to 60h and at intervals as close as 3h.

The forecast equations for all elements except temperatures, dew points, and winds are "regionalized." That is, developmental data from groups of nearby and climatically similar stations were composited in order to increase the sample size when deriving the forecast equations. For each regional group, then, forecasts are made with the same equations and the same regression coefficients. This does not mean that the forecasts for all the stations in the group are the same, however, since interpolation of the dynamical output to the different forecast locations yields different predictor values. Some of the MOS equations also contain predictors representing local climatological values, which introduces further differences in the forecasts for the different stations. Regionalization is especially valuable for producing good forecasts for relatively rare events.

In order to enhance consistency among the forecasts for different but related weather elements, some of the MOS equations are developed "simultaneously." This means that the same predictor variables, although with different regression coefficients, are forced into prediction equations for related predictands in order to enhance the consistency of the forecasts. This is an instance of *multivariate multiple regression* (e.g., Johnson and Wichern, 2007). For example, it would be physically unreasonable and clearly undesirable for the forecast dew point to be higher than the forecast temperature. To help ensure that such inconsistencies appear in the forecasts as rarely as possible, the MOS equations for maximum temperature, minimum temperature, and the 3-h temperatures and dew points all contain the same

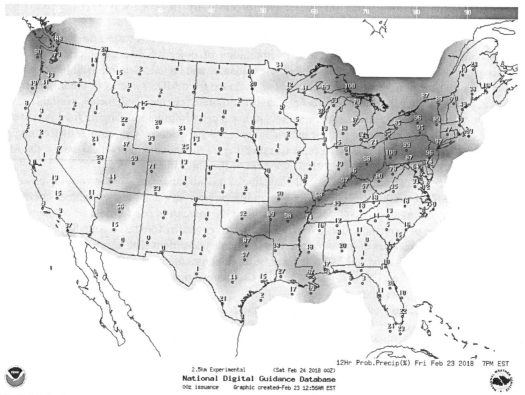

FIGURE 7.29 Example MOS forecasts in map form. The predictand is the probability (× 100) of at least 0.01 in. of precipitation during the 12-h period 7 AM–7 PM EST, 23 February 2018. *From sats.nws.noaa.gov/~mos/gmos/conus25/view_maps_js.php.*

predictor variables. Similarly, the four groups of forecast equations for wind speeds and directions, the 6- and 12-h precipitation probabilities, the 6- and 12-h thunderstorm probabilities, and the probabilities for precipitation types, were also developed simultaneously to enhance their consistency.

Because MOS forecasts are made for a large number of locations, it is possible to view them as maps, which are also posted on the internet. Some of these maps display selected quantities from the MOS panels such as the one presented in Table 7.8. Figure 7.29 shows a forecast map for the variable listed as "P12" in Table 7.8, meaning probabilities of at least 0.01 in. of (liquid-equivalent) precipitation, accumulated over a 12-h period.

7.10. SUBJECTIVE PROBABILITY FORECASTS

7.10.1. The Nature of Subjective Forecasts

Most of this chapter has dealt with objective forecasts, or forecasts produced by means that are automatic. Objective forecasts are determined unambiguously by the nature of the forecasting procedure and the values of the variables that are used to drive it. However, objective forecasting procedures necessarily rest on a number of subjective judgments made during their development. Nevertheless, some people feel more secure with the results of objective forecasting procedures, seemingly taking comfort from their lack of contamination by the vagaries of human judgment. Apparently, such individuals feel that objective forecasts are in some way less uncertain than human-mediated forecasts.

One very important—and perhaps irreplaceable—role of human forecasters in the forecasting process is in the subjective integration and interpretation of objective forecast information. These objective forecast products often are called forecast guidance, and include deterministic forecast information from dynamical integrations, and statistical guidance from MOS systems or other interpretive statistical products. Human forecasters also use, and incorporate into their judgments, available atmospheric observations (surface maps, radar images, etc.), and prior information ranging from persistence or simple climatological statistics, to their individual previous experiences with similar meteorological situations. The result is (or should be) a forecast reflecting, to the maximum practical extent, the forecaster's state of knowledge about the future evolution of the atmosphere.

Human forecasters can rarely, if ever, fully describe or quantify their personal forecasting processes (Stuart et al., 2007). Thus the distillation by a human forecaster of disparate and sometimes conflicting information is known as *subjective* forecasting. A subjective forecast is one formulated on the basis of the judgment of one or more individuals. Making a subjective weather forecast is a challenging process precisely because future states of the atmosphere are inherently uncertain. The uncertainty will be larger or smaller in different circumstances—some forecasting situations are more difficult than others—but it will never really be absent. Doswell (2004) provides some informed perspectives on the formation of subjective judgments in weather forecasting.

Since the future states of the atmosphere are inherently uncertain, a key element of a good and complete subjective weather forecast is the reporting of some measure of the forecaster's uncertainty. This point has been recognized since at least the 19th century (Murphy, 1998). It is the forecaster who is most familiar with the atmospheric situation, and it is therefore the forecaster who is in the best position to evaluate the uncertainty associated with a given forecasting situation. Although it is common for nonprobabilistic forecasts (i.e., forecasts containing no expression of uncertainty) to be issued, such as "tomorrow's maximum temperature will be 27°F," an individual issuing this forecast would not seriously expect the temperature to be exactly 27°F. Given a forecast of 27°F, temperatures of 26°F or 28°F would generally be regarded as nearly as likely, and in this situation the forecaster would usually not really be surprised to see tomorrow's maximum temperature anywhere between 25° and 30°F.

Although uncertainty about future weather can be reported verbally using phrases such as "chance" or "likely," such qualitative descriptions are open to different interpretations by different people (e.g., Murphy and Brown, 1983). Even worse, however, is the fact that such qualitative descriptions do not precisely reflect the forecaster's uncertainty about, or degree of belief in, the future weather. The forecaster's state of knowledge is most accurately reported, and the needs of the forecast user are best served, if the intrinsic uncertainty is quantified in probability terms. Thus the Bayesian view of probability as the degree of belief of an individual holds a central place in subjective forecasting. Note that since different forecasters have somewhat different information on which to base their judgments (e.g., different sets of experiences with similar past forecasting situations), it is perfectly reasonable to expect that their probability judgments may differ somewhat as well.

7.10.2. The Subjective Distribution

Before a forecaster reports a subjective degree of uncertainty as part of a forecast, he or she needs to have a mental image of that uncertainty. The information about an individual's uncertainty can be thought of as residing in their *subjective distribution* for the event in question. The subjective distribution is a probability distribution in the same sense as the parametric distributions described in Chapter 4. Sometimes, in fact, one of the distributions specifically described in Chapter 4 may provide a very good approximation to an individual's subjective distribution. Subjective distributions are interpreted from a Bayesian perspective as the quantification of an individual's degree of belief in each of the possible outcomes for the variable being forecast.

Each time a forecaster prepares to make a forecast, he or she internally develops a subjective distribution. The possible weather outcomes are subjectively weighed, and an internal judgment is formed as to their relative likelihoods. This process occurs whether or not the forecast is to be a probability forecast, or indeed whether or not the forecaster is even consciously aware of the process. However, unless we believe that uncertainty can somehow be expunged from the process of weather forecasting, it should be clear that better forecasts will result when forecasters think explicitly about their subjective distributions and the uncertainty that those distributions describe.

It is easiest to approach the concept of subjective probabilities with a familiar but simple example. Subjective probability-of-precipitation (PoP) forecasts have been routinely issued in the United States since 1965. These forecasts specify the probability that measurable precipitation (i.e., at least 0.01 in.) will occur at a particular location during a specified time period. The forecaster's subjective distribution for this event is so simple that we might not notice that it is a probability distribution. However, the events "precipitation" and "no precipitation" divide the sample space into two MECE events. The distribution of probability over these events is discrete and consists of two complementary elements: one probability for the event "precipitation" and another probability for the event "no precipitation," and either one of these two probabilities can be easily computed from the other. This distribution will be different for different forecasting situations, and perhaps for different forecasters assessing the same situation. However, the only thing about a forecaster's subjective distribution for the PoP that can change from one forecasting occasion to another is the probability, and this will be different to the extent that the forecaster's degree of belief regarding the future precipitation occurrence is different. The PoP ultimately issued by the forecaster should be the forecaster's subjective probability for the event "precipitation," or perhaps a suitably rounded version of that probability. That is, it is the forecaster's job to evaluate the uncertainty associated with the possibility of future precipitation occurrence and to report that uncertainty to the users of the forecasts.

7.10.3. Central Credible Interval Forecasts

It has been argued here that inclusion of some measure of the forecaster's uncertainty should be included in any weather forecast. Forecast users can use the added uncertainty information to make better, economically more favorable, decisions (e.g., Roulston et al., 2006). Historically, resistance to the idea of probability forecasting has been based in part on the practical consideration that the forecast format should be compact and easily understandable. In the case of PoP forecasts, the subjective distribution is sufficiently simple that it can be reported with a single number, and is no more cumbersome than issuing a nonprobabilistic forecast of "precipitation" or "no precipitation." When the subjective distribution is continuous, however, some approach to sketching its main features is a practical necessity if its probability information is to be conveyed succinctly in a publicly issued forecast. Discretizing a continuous subjective distribution is one approach to simplifying it in terms of one or a few easily expressible quantities. Alternatively, if the forecaster's subjective distribution on a given occasion can be reasonably well approximated by one of the parametric distributions described in Chapter 4, another approach to simplifying its communication could be to report the parameters of the approximating distribution. There is no guarantee, however, that subjective distributions will always (or even ever) correspond to a familiar parametric form.

One very attractive and workable alternative for introducing probability information into forecasts for continuous meteorological variables is the use of *credible interval forecasts*. This forecast format has been used operationally in Sweden (Ivarsson et al., 1986), but to date has been used only experimentally in the United States (Murphy and Winkler, 1974; Peterson et al., 1972; Winkler and Murphy, 1979). In unrestricted form, a credible interval forecast requires specification of three quantities: two points defining an interval for the continuous forecast variable, and a probability (according to the forecaster's subjective distribution) that the forecast quantity will fall in the designated interval. Usually the requirement is also made that the credible interval be located in the middle of the subjective distribution. In this case the specified probability is distributed equally on either side of the subjective median, and the forecast is called a *central credible interval* forecast.

There are two special cases of the central credible interval forecast format, each requiring that only two quantities be communicated. The first is the fixed-width central credible interval forecast. As the name implies, the width of the central credible interval is the same for all forecasting situations and is specified in advance for each predictand. Thus the forecast includes a location for the interval, generally specified as its midpoint, and a probability that the outcome will occur in the forecast interval. For example, the Swedish central credible interval forecasts for temperature are of the fixed-width type, with the interval size specified to be $\pm 3 \,°C$ around the midpoint temperature. These forecasts thus include a forecast temperature, together with a probability that the subsequently observed temperature will be within $3 \,°C$ of the forecast temperature. The two forecasts $15 \,°C$, 90% and $15 \,°C$, 60% would both indicate that the forecaster expects the temperature to be about $15 \,°C$, but the inclusion of

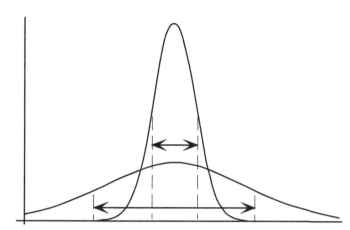

FIGURE 7.30 Two hypothetical subjective distributions shown as probability density functions. The two distributions have the same mean, but reflect different degrees of uncertainty. The tall, narrow distribution represents an easier (less uncertain) forecasting situation, and the broader distribution represents a more difficult forecast problem. Arrows delineate 75% central credible intervals in each case.

probabilities in the forecasts shows that much more confidence can be placed in the former as opposed to the latter of the two forecasts of 15 °C. Because the forecast interval is central, these two forecasts would also imply 5%, and 20% chances, respectively, for the temperature to be colder than 12° or warmer than 18°.

Some forecast users would find the unfamiliar juxtaposition of a temperature and a probability in a fixed-width central credible interval forecast to be somewhat jarring. The fixed-probability central credible interval forecast is an alternative forecast format that could be implemented more subtly. In this format, it is the probability contained in the forecast interval, rather than the width of the interval, that is specified in advance and is constant from forecast to forecast. This format makes the probability component of the credible interval forecast implicit, so the forecast consists of two numbers having the same physical dimensions as the quantity being forecast.

Figure 7.30 illustrates the relationship of 75% central credible intervals for two subjective distributions having the same mean. The shorter, broader distribution represents a relatively uncertain forecasting situation, where events fairly far away from the center of the distribution are regarded as having substantial probability. A relatively wide interval is therefore required to subsume 75% of this distribution's probability. On the other hand, the tall and narrow distribution describes considerably less uncertainty, and a much narrower forecast interval contains 75% of its density. If the variable being forecast is temperature, the 75% central credible interval forecasts for these two cases might be 10° to 20°, and 14° to 16°, respectively.

A strong case can be made for operational credible-interval forecasts (Grounds et al. 2017; Murphy and Winkler, 1974, 1979). Since nonprobabilistic temperature forecasts are already often specified as ranges, fixed-probability central credible interval forecasts could be introduced into forecasting operations quite unobtrusively. Forecast users not wishing to take advantage of the implicit probability information would notice little difference from the present forecast format, whereas those understanding the meaning of the forecast ranges would derive additional benefit. Even forecast users unaware that the forecast range is meant to define a particular interval of fixed probability might notice over time that the interval widths were related to the precision of the forecasts.

7.10.4. Assessing Discrete Probabilities

Experienced weather forecasters are able to formulate subjective probability forecasts that evidently quantify their uncertainty regarding future weather quite successfully. Examination of the error characteristics of such forecasts (see Chapter 9) reveals that they are largely free of the biases and inconsistencies sometimes exhibited in the subjective probability assessments made by less experienced individuals. Commonly, inexperienced forecasters produce probability forecasts exhibiting overconfidence (Murphy, 1985), or biases due to such factors as excessive reliance on recently acquired information (Spetzler and Staël von Holstein, 1975; Tversky, 1974).

Individuals who are experienced at assessing their subjective probabilities can do so in a seemingly subconscious or automatic manner. People who are new to the practice often find it helpful to use physical or conceptual devices that allow comparison of the uncertainty to be assessed with a situation that is more concrete and familiar (Garthwaite et al., 2005). For example, Spetzler and Staël von Holstein (1975) describe a physical device called a probability wheel, which consists of a spinner of the sort that might be found in a child's board game, on a background that has the form of a pie chart. This background has two colors, blue and orange, and the proportion of the background covered by each of the colors can be adjusted. The probability wheel is used to assess the probability of a dichotomous event (e.g., a PoP forecast) by adjusting the relative coverages of the two colors until the forecaster feels the probability of the event to be forecast is about equal to the probability of the spinner stopping in the orange sector. The subjective probability forecast is then read as the angle subtended by the orange sector, divided by 360°.

Conceptual devices can also be employed to assess subjective probabilities. For many people, comparison of the uncertainty surrounding the future weather is most easily assessed in the context of lottery games or betting games. Such conceptual devices translate the probability of an event to be forecast into more concrete terms by posing hypothetical questions such as "would you prefer to be given \$2 if precipitation occurs tomorrow, or \$1 for sure (regardless of whether or not precipitation occurs)?" Individuals preferring the sure \$1 in this lottery situation evidently feel that the relevant PoP is <0.5, whereas individuals who feel the PoP is >0.5 would generally prefer to receive \$2 on the chance of precipitation. A forecaster can use this lottery device by adjusting the variable payoff relative to the certainty equivalent (the sum to be received for sure) until the point of indifference, where either choice would be equally attractive. That is, the variable payoff is adjusted until the expected (i.e., probability-weighted average) payment is equal to the certainty equivalent. Denoting the subjective probability as p, the procedure can be written formally as

$$\text{Expected payoff} = p\,(\text{Variable payoff}) + (1-p)(\$0) = \text{Certainty equivalent} \qquad (7.65a)$$

which leads to

$$p = \frac{\text{Certainty equivalent}}{\text{Variable payoff}}. \qquad (7.65b)$$

The same kind of logic can be applied in an imagined betting situation. Here the forecasters ask themselves whether receiving a specified payment should the weather event to be forecast occurs, or suffering some other monetary loss if the event does not occur, is preferable. In this case the subjective probability is assessed by finding monetary amounts for the payment and loss such that the bet is a fair one, implying that the forecaster would be equally happy to be on either side of it. Since the expected payoff from a fair bet is zero, the betting game situation can be represented as

$$\text{Expected payoff} = p\,(\$\text{payoff}) + (1-p)(-\$\text{loss}) = 0, \qquad (7.66a)$$

leading to

$$p = \frac{\$\text{loss}}{\$\text{loss} + \$\text{payoff}}. \qquad (7.66b)$$

Many betting people think in terms of odds in this context. Equation 7.66a can be expressed alternatively as

$$\text{odds ratio} = \frac{p}{1-p} = \frac{\$\text{loss}}{\$\text{payoff}}. \qquad (7.67)$$

Thus a forecaster being indifferent to an even-money bet (1:1 odds) harbors an internal subjective probability of $p = 0.5$. Indifference to being on either side of a 2:1 bet implies a subjective probability of 2/3, and indifference at 1:2 odds is consistent with an internal probability of 1/3.

The same kind of thinking as these lottery and betting games for individual probability elicitation can also be applied to probability evaluation through "prediction markets" (Wolfers and Zitzewitz, 2004). However, in the case of prediction markets the tacit probability assessments are made through a *consensus forecasting* process, where the judgments of multiple individuals are aggregated. Experience in meteorological contexts (Baars and Mass, 2005; Hamill and Wilks, 1995; Sanders, 1963; Thompson, 1977; Vislocky and Fritsch, 1995a) has shown that such consensus forecasts typically improve on the performance of the individual forecasts that are combined into the consensus.

7.10.5. Assessing Continuous Distributions

The same kinds of lotteries or betting games just described can also be used to assess quantiles of a subjective continuous probability distribution using the *method of successive subdivision*. Here the approach is to identify quantiles of the subjective distribution by comparing event probabilities that they imply with the reference probabilities derived from conceptual money games. Use of this method in an operational setting is described in Krzysztofowicz et al. (1993).

The easiest quantile to identify is the median. Suppose the distribution to be identified pertains to tomorrow's maximum temperature. Since the median divides the subjective distribution into two equally probable halves, its location can be assessed by evaluating a preference between, say, $1 for sure and $2 if tomorrow's maximum temperature is warmer than 14 °C. The situation is the same as that described in Equation 7.65. Preferring the certainty of $1 implies a subjective probability for the event {maximum temperature warmer than 14 °C} that is smaller than 0.5. A forecaster preferring the chance at $2 evidently feels that the probability for this event is larger than 0.5. Since the cumulative probability, p, for the median is fixed at 0.5, we can locate the threshold defining the event {outcome above median} by adjusting it to the point of indifference between the certainty equivalent and a variable payoff equal to twice the certainty equivalent.

The quartiles can be assessed in the same way, except that the ratios of certainty equivalent to variable payoff must correspond to the cumulative probabilities of the quartiles, that is, 1/4 or 3/4. At what temperature T_{LQ} are we indifferent to the alternatives of receiving \$1 for sure, or \$4 if tomorrow's maximum temperature is below T_{LQ}? The temperature T_{LQ} then estimates the forecaster's subjective lower quartile. Similarly, the temperature T_{UQ}, at which we are indifferent to the alternatives of \$1 for sure or \$4 if the temperature is above T_{UQ}, estimates the upper quartile.

Especially when someone is inexperienced at probability assessments, it is a good idea to perform some consistency checks. In the method just described, the quartiles were assessed independently, but together define a range—the 50% central credible interval—in which half the probability should lie. Therefore a good check on their consistency would be to verify that we are indifferent to the choices between \$1 for sure, and \$2 if $T_{LQ} \leq T \leq T_{UQ}$. If we prefer the certainty equivalent in this comparison the quartile estimates T_{LQ} and T_{UQ} are apparently too close. If we prefer the chance at the \$2 they apparently subtend too much probability. Similarly, we could verify indifference between the certainty equivalent, and four times the certainty equivalent if the temperature falls between the median and one of the quartiles. Any inconsistencies discovered in checks of this type indicate that some or all of the previously estimated quantiles need to be reassessed.

7.11. EXERCISES

7.1 a. Derive a simple linear regression equation using the data in Table A.3, relating June temperature (as the predictand) to June pressure (as the predictor).
   b. Explain the physical meanings of the two parameters.
   c. Formally test whether the fitted slope is significantly different from zero.
   d. Compute the R^2 statistic.
   e. Estimate the probability that a predicted value corresponding to $x_0 = 1013$ mb will be within 1 °C of the regression line, using Equation 7.23.
   f. Repeat (e), assuming the prediction variance equals the MSE.

7.2 Consider the following partial ANOVA table, describing the results of a regression analysis:

Source	df	SS	MS
Total	26	318.2874	
Regression	_____	316.6065	_____
Residual	25	_____	_____

   a. Fill in the 4 blanks in the table.
   b. How many predictor variables are in the equation?
   c. What is the sample variance of the predictand?
   d. What is the R^2 value?
   e. Estimate the probability that a prediction made by this regression will be within ± 0.2 units of the actual value.
   f. Formally test whether the estimated slope $b = 0.69$ is significantly different from zero.

7.3 Derive an expression for the maximum likelihood estimate of the intercept b_0 in logistic regression (Equation 7.36), for the constant model in which $b_1 = b_2 = \ldots = b_K = 0$.

7.4 The 19 nonmissing precipitation values in Table A.3 can be used to fit the regression equation:

$$\ln\,[(\text{Precipitation}) + 1\,\text{mm}] = 499.4 - 0.512\,(\text{Pressure}) + 0.796\,(\text{Temperature})$$

The MSE for this regression is 0.701. (The constant 1 mm has been added to ensure that the logarithm is defined for all data values.)

a. Estimate the missing precipitation value for 1956 using this equation.

b. Construct a 95% prediction interval for the estimated 1956 precipitation.

7.5 Explain how to use cross-validation to estimate the prediction mean squared error, and the sampling distribution of the regression slope, for the problem in Exercise 7.1. If the appropriate computing resources are available, implement your algorithm.

7.6 Hurricane Zeke is an extremely late storm in a very busy hurricane season. It has recently formed in the Caribbean, the 500 mb height at gridpoint 37 (relative to the storm) is 5400 m, the 500 mb height at gridpoint 3 is 5500 m, and the 1000 mb height at gridpoint 51 is −200 m (i.e., the surface pressure near the storm is well below 1000 mb).

a. Use the NHC 67 model (see Table 7.7) to forecast the westward component of its movement over the next 12 h, if storm has moved 80 n.mi. due westward in the previous 12 hours.

b. What would the NHC 67 forecast of the westward displacement be if, in the previous 12 hours, the storm had moved 80 n.mi. westward *and* 30 n.mi. northward (i.e., $P_y = 30$ n. mi.)?

7.7 The fall (September, October, November) MOS equation for predicting maximum temperature (in °F) at Binghamton, New York, formerly used with a now-discontinued dynamical model, at the 60-h lead time was

$$\text{MAX } T = -363.2 + 1.541\,(850\,\text{mb T}) - .1332\,(\text{SFC} - 490\,\text{mb RH}) - 10.3\,(\text{COS DOY})$$

where:

(850 mb T) is the 48-h dynamical forecast of temperature (K) at 850 mb

(SFC-490 mb RH) is the 48-h forecast lower tropospheric RH in %

(COS DOY) is the cosine of the day of the year transformed to radians or degrees, that is,

$= \cos\,(2\pi t/365)$ or $= \cos\,(360°\,t/365)$

and t is the day number of the valid time (the day number for January 1 is 1, and for October 31 it is 304)

Calculate what the 60-h MOS maximum temperature forecast would be for the following:

	Valid time	48-h 850 mb *T* fcst	48-h mean RH fcst
a.	September 4	278 K	30%
b.	November 28	278 K	30%
c.	November 28	258 K	30%
d.	November 28	278 K	90%

7.8 A MOS equation for 12–24h PoP in the warm season might look something like
PoP $= 0.25 + 0.0063$(Mean RH) $- 0.163$(0–12 ppt [bin @ 0.1 in.]) $- 0.165$(Mean RH [bin @ 70%]).
where:
Mean RH (%) is the same variable as in Exercise 7.7 for the appropriate lead time
0–12 ppt is the model-forecast precipitation amount in the first 12h of the forecast
[bin @ xxx] indicates use as a binary variable: $= 1$ if the predictor is $\leq$ xxx
$$= 0 \text{ otherwise}$$
Evaluate the MOS PoP forecasts for the following conditions:

	12-h mean RH	0–12 ppt
a.	90%	0.00 in.
b.	65%	0.15 in.
c.	75%	0.15 in.
d.	75%	0.09 in.

7.9 Explain why the slopes of the solid lines decrease, from Figure 7.27 to Figure 7.28a, to Figure 7.28b. What would the corresponding MOS equation be for an arbitrarily long lead time into the future?

7.10 A forecaster is equally happy with the prospect of receiving $1 for sure, or $5 if freezing temperatures occur on the following night. What is the forecaster's subjective probability for frost?

7.11 A forecaster is indifferent between receiving $1 for sure and any of the following: $8 if tomorrow's rainfall is >55 mm, $4 if tomorrow's rainfall is >32 mm, $2 if tomorrow's rainfall is >12 mm, $1.33 if tomorrow's rainfall is >5 mm, and $1.14 if tomorrow's precipitation is >1 mm.
a. What is the median of this individual's subjective distribution?
b. What would be a consistent 50% central credible interval forecast? A 75% central credible interval forecast?
c. In this forecaster's view, what is the probability of receiving more than one but no more than 32 mm of precipitation?

Chapter 8

Ensemble Forecasting

8.1. BACKGROUND

8.1.1. Inherent Uncertainty of Dynamical Forecasts

It was noted in Section 1.3 that dynamical chaos ensures that the future behavior of the atmosphere cannot be known with certainty. Because the atmosphere can never be fully observed, either in terms of spatial coverage or accuracy of measurements, a fluid-dynamical model of its behavior will always begin calculating forecasts from a state at least slightly different from that of the real atmosphere. These models (and other nonlinear dynamical systems, including the real atmosphere) exhibit the property that solutions (forecasts) started from only slightly different initial conditions will yield quite different results for lead times sufficiently far into the future. For synoptic scale weather predictions, "sufficiently far" is a matter of (at most) weeks, and for mesoscale forecasts this window is even shorter, so that the problem of sensitivity to initial conditions is of practical importance.

Dynamical forecast models are the mainstay of weather and climate forecasting, and the inherent uncertainty of their results must be appreciated and quantified if their information is to be utilized most effectively. For example, a single deterministic forecast of the hemispheric 500 mb height field two days in the future is at best only one member of an essentially infinite collection of 500 mb height fields that could plausibly occur. Even if this deterministic forecast of the 500 mb height field is the best possible single forecast that can be constructed, its usefulness and value will be enhanced if aspects of the probability distribution of which it is a member can be estimated and communicated. It is the purpose of ensemble forecasting to characterize the inherent uncertainty of dynamical forecasts, in a quantitative yet understandable way. Although much more attention has been devoted to initial-condition sensitivity in weather forecasts, the issue is also important for longer range forecasts and climate change projections (e.g., Deser et al., 2012; Hawkins et al., 2016).

8.1.2. Stochastic Dynamical Systems in Phase Space

Understanding the conceptual and mathematical underpinnings of ensemble forecasting requires the concept of a *phase space*. A phase space is an abstract geometrical space, each of the coordinate axes of which corresponds to one of the forecast variables of the dynamical system. Within the phase space, the "state" of the dynamical system is defined by specification of particular values for each of these forecast variables, and therefore corresponds to a single point in this (generally high-dimensional) space.

To concretely introduce the phase space concept, consider the behavior of a swinging pendulum, which is a simple dynamical system that is commonly encountered in textbooks on physics or differential equations. The state of the dynamics of a pendulum can be completely described by two variables: its

Statistical Methods in the Atmospheric Sciences. https://doi.org/10.1016/B978-0-12-815823-4.00008-0

angular position and its velocity. At the extremes of the pendulum's arc, its angular position is maximum (positive or negative) and its velocity is zero. At the bottom of its arc the angular position of the swinging pendulum is zero and its speed (corresponding to either a positive or negative velocity) is maximum. When the pendulum finally stops, both its angular position and velocity are zero. Because the motions of a pendulum can be fully described by two variables, its phase space is two-dimensional. That is, its phase space is a phase plane. The changes through time of the state of the pendulum system can be described by a path, known as an *orbit*, or *trajectory*, on this phase plane.

Figure 8.1 shows the trajectory of a hypothetical pendulum in its phase space. That is, this figure is a graph in phase space that represents the motions of a pendulum and their changes through time. The trajectory begins at the single point corresponding to the initial state of the pendulum: it is dropped from the right with zero initial velocity (A). As it drops it accelerates and acquires leftward velocity, which increases until the pendulum passes through the vertical position (B). The pendulum then decelerates, slowing until it stops at its maximum left position (C). As the pendulum drops again it moves to the right, stopping short of its initial position because of friction (D). The pendulum continues to swing back and forth until it finally comes to rest in the vertical position (E).

The dynamics of a pendulum are simple both because the phase space has only two dimensions, but also because its behavior is not sensitive to the initial condition. Releasing the pendulum slightly further to the right or left relative to its initial point in Figure 8.1, or with a slight upward or downward push, would produce a very similar trajectory that would track the spiral in Figure 8.1 very closely, and arrive at the same place in the center of the diagram at nearly the same time.

The corresponding behavior of the atmosphere, or of a realistic mathematical model of it, would be quite different. The landmark paper of Lorenz (1963) demonstrated that solutions to systems of deterministic non-linear differential equations can exhibit sensitive dependence on initial conditions. That is, even though deterministic equations yield unique and repeatable solutions when integrated forward from a given initial condition, projecting systems exhibiting sensitive dependence forward in time from very slightly different

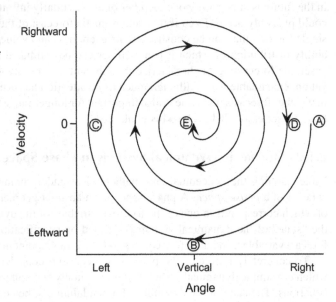

FIGURE 8.1 Trajectory of a swinging pendulum in its two-dimensional phase space or phase plane. The pendulum has been dropped from position (A) on the right, from which point it swings in arcs of decreasing angle. Finally, it slows to a stop, with zero velocity in the vertical position (E).

initial conditions eventually yields computed states that diverge strongly from each other. Sometime later Li and Yorke (1975) coined the name "chaotic" dynamics for this phenomenon, although this label is somewhat unfortunate in that it is not really descriptive of the sensitive-dependence phenomenon.

The system of three coupled ordinary differential equations used by Lorenz (1963) is deceptively simple:

$$\frac{dx}{dt} = -10x + 10y \tag{8.1a}$$

$$\frac{dy}{dt} = -xz + 28x - y \tag{8.1b}$$

$$\frac{dz}{dt} = xy - \frac{8}{3}z. \tag{8.1c}$$

This system is a highly abstracted representation of thermal convection in a fluid, where x represents the intensity of the convective motion, y represents the temperature difference between the ascending and descending branches of the convection, and z represents departure from linearity of the vertical temperature profile. Despite its low dimensionality and apparent simplicity, the system composed of Equation 8.1a–8.1c shares some key properties with the equations governing atmospheric flow. In addition to sensitive dependence, the simple Lorenz system and the equations governing atmospheric motion also both exhibit regime structure, multiple distinct timescales, and state-dependent variations in predictability (e.g., Palmer, 1993).

Because Equation 8.1a–8.1c involve three prognostic variables, the phase space of this system is a three-dimensional volume. However, not all points in its phase space can be visited by system trajectories once the system has settled into a steady state. Rather, the system will be confined to a subset of points in the phase space called the *attractor*. The attractor of a dynamical system is a geometrical object within the phase space, toward which trajectories are attracted in the course of time. Each point on the attractor represents a dynamically self-consistent state, jointly for all of the prognostic variables. The attractor for the simple pendulum system consists of the single point in the center of Figure 8.1, representing the motionless state toward which any undisturbed pendulum will converge. Figure 8.2, from

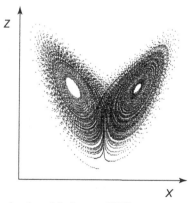

FIGURE 8.2   Projection of a finite approximation of the Lorenz (1963) attractor onto the x–z plane, yielding the Lorenz "butterfly." *From Palmer (1993). © American Meteorological Society. Used with permission.*

Palmer (1993), shows an approximate rendering of the Lorenz attractor, projected onto the x–z plane. The figure has been constructed by numerically integrating the Lorenz system forward for an extended time, with each dot representing the system state at a discrete time increment. The characteristic shape of this projection of the Lorenz attractor has come to be known as the Lorenz "butterfly." In a sense, the attractor can be thought of as representing the "climate" of its dynamical system, and each point on the attractor represents a possible instantaneous "weather" state. A sequence of these "weather" states then traces out a trajectory in the phase space, along the attractor.

Each wing of the attractor in Figure 8.2 represents a regime of the Lorenz system. Trajectories in the phase space consist of some number of clockwise circuits around the left-hand ($x < 0$) wing of the attractor, followed by a shift to the right-hand ($x > 0$) wing of the attractor where some number of counterclockwise cycles are executed, until the trajectory shifts again to the left wing, and so on. Circuits around one or the other of the wings occur on a faster timescale than residence times on each wing. The traces in Figure 8.3, which are example time series of the x variable, illustrate that the fast oscillations around one or the other wings are variable in number, and that transitions between the two wing regimes occur suddenly. The two traces in Figure 8.3 were initialized at very similar points, and the sudden difference between them that begins after the first regime transition illustrates the sensitive-dependence phenomenon.

The Lorenz system and the real atmosphere share the very interesting property of state-dependent variations in predictability. That is, forecasts initialized in some regions of the phase space (corresponding to particular subsets of the dynamically self-consistent states) may yield better predictions than others. Figure 8.4 illustrates this idea for the Lorenz system by tracing the trajectories of loops of initial conditions initialized at different parts of the attractor. The initial loop in Figure 8.4a, on the upper part of the left wing, illustrates extremely favorable forecast evolution. These initial points remain close together throughout the 10-stage forecast, although of course they would eventually diverge if the integration were to be carried further into the future. The result is that the forecast from any one of these

FIGURE 8.3 Example time series for the x variable in the Lorenz system. The two time series have been initialized at nearly identical values. *From Palmer (1993). © American Meteorological Society. Used with permission.*

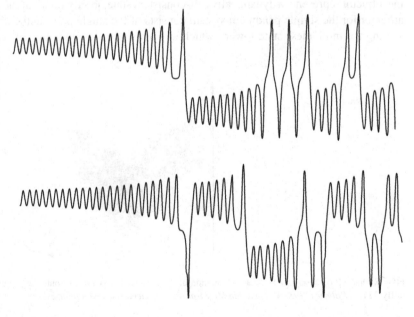

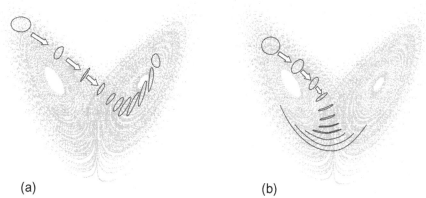

(a)   (b)

FIGURE 8.4 Collections of forecast trajectories for the Lorenz system, initialized at (a) a high-predictability portion of the attractor, and (b) a moderate-predictability portion of the attractor. Any of the forecasts in panel (a) would likely represent the unknown true future state well, whereas many of the results in panel (b) would correspond to poor forecasts. *From Palmer (1993). © American Meteorological Society. Used with permission.*

initial states would produce a good forecast of the trajectory from the (unknown) true initial condition, which might be located near the center of the initial loop. In contrast, Figure 8.4b shows forecasts for the same set of future times when the initial conditions are taken as the points on the loop that is a little lower on the left wing of the attractor. Here the dynamical predictions are reasonably good through the first half of the forecast period, but then diverge strongly toward the end of the period as some of the trajectories remain on the left-hand wing of the attractor while others undergo the regime transition to the right-hand wing. The result is that a broad range of the prognostic variables might be forecast from initial conditions near the unknown true initial condition, and there is no way to tell in advance which of the trajectories might represent good or poor forecasts.

The phase space of a realistic atmospheric model has many more dimensions than that of the Lorenz system. For example, the dimensions of the phase spaces of operational weather forecasting models are on the order of 10^9 (e.g., Lewis, 2014), each corresponding to one of the (horizontal gridpoints) $\times$ (vertical levels) $\times$ (prognostic variables) combinations represented. The trajectory of the atmosphere or a model of the atmosphere is more complicated than that of the Lorenz system, but the qualitative behavior is analogous in many ways, and the changes in the flow within a model atmosphere through time can still be imagined abstractly as a trajectory through its multidimensional phase space.

8.1.3. Uncertainty and Predictability in Dynamical Systems

The connection between uncertainty, probability, and dynamical forecasting can be approached using the phase space of the Lorenz attractor as a low-dimensional and comprehensible metaphor for the millions-dimensional phase spaces of realistic modern dynamical weather prediction models. Consider again the forecast trajectories portrayed in Figure 8.4. Rather than regarding the upper-left loops as collections of initial states, imagine that they represent boundaries containing most of the probability, perhaps the 99% probability ellipsoids, for probability density functions defined on the attractor. When initializing a dynamical forecast model we can never be certain of the true initial state, but we may be able to quantify that initial-condition uncertainty in terms of a probability distribution, and that distribution must be defined on the system's attractor if the possible initial states are to be dynamically consistent with the governing equations. In effect, those governing equations will operate on the probability distribution of initial-condition uncertainty, advecting it across the attractor and distorting its initial shape in the process. If the initial probability distribution is a correct representation of the initial-condition uncertainty, and if the governing equations are a correct representation of the dynamics of the true system, then the subsequent advected and distorted probability distributions will correctly quantify the forecast uncertainty at future times. This uncertainty may be larger (as represented by Figure 8.4b) or smaller (Figure 8.4a), depending on the intrinsic predictability of the states in the initial region of the attractor. To the extent that the forecast model equations are not complete and correct representations of the true dynamics, which is inevitable as a practical matter, then additional uncertainty will be introduced.

Using this concept of a probability distribution that quantifies the initial-condition uncertainty, Epstein (1969c) proposed the method of *stochastic-dynamic prediction*. The historical and biographical background leading to this important paper has been reviewed by Lewis (2014). Denoting the (multivariate) uncertainty distribution as φ and the vector $\dot{X}$ as containing the total derivatives with respect to time of the prognostic variables defining the coordinate axes of the phase space, Epstein (1969c) begins with the conservation equation for total probability, φ,

$$\frac{\partial \varphi}{\partial t} + \nabla \cdot \left(\dot{X}\varphi\right) = 0. \tag{8.2}$$

Equation 8.2, also known as the Liouville equation (Ehrendorfer, 1994a; Gleeson, 1970), is analogous to the more familiar continuity (i.e., conservation) equation for mass. As noted by Epstein (1969c),

> It is possible to visualize the probability density in phase space, as analogous to mass density (usually ρ) in three-dimensional physical space. Note that $\rho \geq 0$ for all space and time, and $\iiint(\rho/M)dxdydz = 1$ if M is the total mass of the system. The "total probability" of any system is, by definition, one.

Equation 8.2 states that any change in the probability contained within a small (hyper-) volume surrounding a point in phase space must be balanced by an equal net flux of probability through the boundaries of that volume. The governing physical dynamics of the system (e.g., Equations 8.1a–8.1c for the Lorenz system) are contained in the time derivatives $\dot{X}$ in Equation 8.2, which are also known as tendencies. Note that the integration of Equation 8.2 is deterministic, in the sense that there are no random terms introduced on the right-hand sides of the dynamical tendencies.

Epstein (1969c) considered that direct numerical integration of Equation 8.2 on a set of gridpoints within the phase space was computationally impractical, even for the idealized 3-dimensional dynamical system he used as an example. Instead he derived time-tendency equations for the elements of the mean vector and covariance matrix of φ, in effect, assuming multivariate normality (Chapter 12) for this distribution initially and at all forecast times. The result was a system of nine coupled differential equations (three each for the means, variances, and covariances), arrived at by assuming that the third and higher moments of the forecast distributions vanished. In addition to providing a (vector) mean forecast, the procedure characterizes state-dependent forecast uncertainty through the forecast variances and covariances that populate the forecast covariance matrix, the increasing determinant ("size") of which (Equation 12.6) at increasing lead times can be used to characterize the increasing forecast uncertainty. Concerning this procedure, Epstein (1969c) noted that "deterministic prediction implicitly assumes that all variances are zero. Thus the approximate stochastic equations are higher order approximations ... than have previously been used."

Stochastic-dynamic prediction in the phase space in terms of the first and second moments of the uncertainty distribution does not yield good probabilistic forecasts because of the neglect of the higher moment aspects of the forecast uncertainty. Even so, the method is computationally impractical when applied to realistic forecast models for the atmosphere. Full evaluation of Equation 8.2 (Ehrendorfer, 1994b; Thompson, 1985) is even further out of reach for practical problems.

8.2. ENSEMBLE FORECASTS

8.2.1. Discrete Approximation to Stochastic-Dynamic Forecasts

Even though the stochastic-dynamic approach to forecasting as proposed by Epstein (1969c) is out of reach computationally, it is theoretically sound and conceptually appealing. It provides the philosophical basis for addressing the problem of sensitivity to initial conditions in dynamical weather and climate models, which is currently best achieved through *ensemble forecasting*. Rather than computing the effects of the governing dynamics on the full continuous probability distribution of initial-condition uncertainty (Equation 8.2), ensemble forecasting proceeds by constructing a discrete approximation to this process. That is, a collection of individual initial conditions (each represented by a single point in the phase space) is chosen, and each is integrated forward in time according to the governing equations of the (modeled) dynamical system. Ideally, the distribution of these states in the phase space at future times, which can be mapped to physical space, will then represent a sample from the statistical distribution of forecast uncertainty. These Monte Carlo solutions bear the same relationship to stochastic-dynamic forecast equations as the Monte Carlo resampling tests introduced in Section 5.3.3 bear to the analytical tests they approximate. (Recall that resampling tests are appropriate and useful in situations where the underlying mathematics are difficult or impossible to evaluate analytically.) Lewis (2005) traces the history of this confluence of dynamical and statistical ideas in atmospheric prediction.

Ensemble forecasting is an instance of Monte Carlo integration (Metropolis and Ulam, 1949). Ensemble forecasting in meteorology appears to have been first proposed explicitly in a conference paper by Lorenz (1965):

The proposed procedure chooses a finite ensemble of initial states, rather than the single observed initial state. Each state within the ensemble resembles the observed state closely enough so that the differences might be ascribed to errors or inadequacies in observation. A system of dynamic equations previously deemed to be suitable for forecasting is then applied to each member of the ensemble, leading to an ensemble of states at any future time. From an ensemble of future states, the probability of occurrence of any event, or such statistics as the ensemble mean and ensemble standard deviation of any quantity, may be evaluated.

Ensemble forecasting was first implemented in a meteorological context by Epstein (1969c) as a means to provide representations of the true forecast distributions to which his truncated stochastic-dynamic calculations could be compared. He explicitly chose initial ensemble members as independent random draws from the initial-condition uncertainty distribution:

Discrete initial points in phase space are chosen by a random process such that the likelihood of selecting any given point is proportional to the given initial probability density. For each of these initial points (i.e. for each of the sample selected from the ensemble) deterministic trajectories in phase space are calculated by numerical integration... Means and variances are determined, corresponding to specific times, by averaging the appropriate quantities over the sample.

The ensemble forecast procedure begins in principle by drawing a finite sample from the probability distribution describing the uncertainty of the initial state of the atmosphere. Imagine that a few members of the point cloud surrounding the mean estimated atmospheric state in phase space are picked randomly. Collectively, these points are called the ensemble of initial conditions, and each represents a plausible initial state of the atmosphere consistent with the uncertainties in observation and analysis. Rather than explicitly predicting the movement of the entire initial-state probability distribution through the phase space of the dynamical model, that movement is approximated by the collective trajectories of the ensemble of sampled initial points. It is for this reason that the Monte Carlo approximation to stochastic-dynamic forecasting is known as ensemble forecasting. Each of the points in the initial ensemble provides the initial conditions for a separate dynamical integration. At this initial time, all the ensemble members are very similar to each other. The distribution in phase space of this ensemble of points after the forecasts have been advanced to a future time then approximates how the full true initial probability distribution would have been transformed by the governing physical laws that are expressed in the dynamics of the model.

Figure 8.5 illustrates the nature of ensemble forecasting in an idealized two-dimensional phase space. The circled X in the initial-time ellipse represents the single best initial value, from which a conventional deterministic dynamical integration would begin. Recall that, for a real model of the atmosphere, this initial point defines a full set of meteorological maps for all of the variables being forecast. The evolution of this single forecast in the phase space, through an intermediate forecast lead time and to a final forecast lead time, is represented by the heavy solid lines. However, the position of this point in phase space at the initial time represents only one of the many plausible initial states of the atmosphere consistent with errors in the analysis. Around it are other plausible states, which are meant to sample the probability distribution for states of the atmosphere at the initial time. This distribution is represented by the small

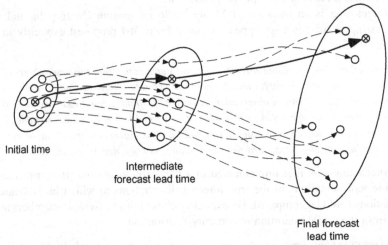

Initial time

Intermediate
forecast lead time

Final forecast
lead time

FIGURE 8.5 Schematic illustration of some concepts in ensemble forecasting, plotted in terms of an idealized two-dimensional phase space. The heavy line represents the evolution of the single best analysis of the initial state of the atmosphere, corresponding to the more traditional single deterministic forecast. The dashed lines represent the evolution of individual ensemble members. The ellipse in which they originate represents the probability distribution of initial atmospheric states, which are very close to each other. At the intermediate lead time, all the ensemble members are still reasonably similar. By the final lead time some of the ensemble members have undergone a regime change and represent qualitatively different flows. Any of the ensemble members, including the solid line, are plausible trajectories for the evolution of the real atmosphere, and there is no way of knowing in advance which will represent the real atmosphere most closely.

ellipse. The open circles in this ellipse represent eight other members of this distribution. This ensemble of nine initial states approximates the variations represented by the full distribution from which they were drawn.

The Monte Carlo approximation to a stochastic-dynamic forecast is constructed by repeatedly running the dynamical model, once for each of the members of the initial ensemble. The trajectories of each of the ensemble members through the phase space are only modestly different at first, indicating that all nine integrations represented in Figure 8.5 are producing fairly similar forecasts at the intermediate lead time. Accordingly, the probability distribution describing uncertainty about the state of the atmosphere at the intermediate lead time would not be a great deal larger than at the initial time. However, between the intermediate and final lead times the trajectories diverge markedly, with three (including the one started from the central value of the initial distribution) producing forecasts that are similar to each other, and the remaining six members of the ensemble predicting rather different atmospheric states at that time. The underlying distribution of uncertainty that was fairly small at the initial time has been stretched substantially, as represented by the large ellipse at the final lead time. The dispersion of the ensemble members at that time allows the nature of that distribution to be estimated, and is indicative of the uncertainty of the forecast, assuming that the dynamical model includes only negligible errors in the representations of the governing physical processes. If only the single forecast started from the best initial condition had been made, this information would not be available.

Reviews of recent operational use of the ensemble forecasting approach can be found in Buizza et al. (2005), Cheung (2001), and Kalnay (2003).

8.2.2. Choosing Initial Ensemble Members

Ideally, we would like to produce ensemble forecasts based on a large number of possible initial atmospheric states drawn randomly from the PDF of initial-condition uncertainty in phase space. However, each member of an ensemble of forecasts is produced by a complete rerunning of the dynamical model, each of which requires a substantial amount of computing. As a practical matter, computer time is a limiting factor at operational forecast centers, and each center must make a subjective judgment balancing the number of ensemble members to include in relation to the spatial resolution of the model used to integrate them forward in time. Consequently, the sizes of operational forecast ensembles are limited, and it is important that initial ensemble members be chosen well. Their selection is further complicated by the fact that the initial-condition PDF in phase space is unknown. Also, it presumably changes from day to day, so that the ideal of simple random samples from this distribution cannot be achieved in practice.

The simplest, and historically first, method of generating initial ensemble members was to begin with a best analysis, assuming it to be the mean of the probability distribution representing the uncertainty of the initial state of the atmosphere. Variations around this mean state can be easily generated, by adding random numbers characteristic of the errors or uncertainty in the instrumental observations underlying the analysis (Leith, 1974). For example, these random values might be Gaussian variates with zero mean, implying an unbiased combination of measurement and analysis errors. In practice, however, simply adding independent random numbers to a single initial field has been found to yield ensembles whose members are too similar to each other, probably because much of the joint variation introduced in this way is dynamically inconsistent, so that the corresponding energy is quickly dissipated in the model (Palmer et al., 1990). The consequence is that the dispersion of the resulting forecast ensemble underestimates the uncertainty in the forecast as the trajectories of the initial ensemble members quickly collapse onto the attractor.

As of the time of this writing (2018), there are three dominant methods of choosing initial ensemble members, although a definitively best method has yet to be demonstrated (e.g., Hamill and Swinbank, 2015). Until relatively recently, the United States National Centers for Environmental Prediction (NCEP) used the *breeding method* (Ehrendorfer, 1997; Kalnay, 2003; Toth and Kalnay, 1993, 1997; Wei et al., 2008). In this approach, differences in the three-dimensional patterns of the predicted variables, between the ensemble members and the single "best" (control) analysis, are chosen to look like differences between recent forecast ensemble members and the forecast from the corresponding previous control analysis. The underlying idea is that the most impactful initial-condition errors should resemble forecast errors in the most recent ("background") forecast. The patterns are then scaled to have magnitudes appropriate to analysis uncertainties. These bred patterns are different from day to day and emphasize features with respect to which the ensemble members are diverging most rapidly. The breeding method is relatively inexpensive computationally.

In contrast, the European Centre for Medium-Range Weather Forecasts (ECMWF) generates initial ensemble members using *singular vectors* (Bonavita et al., 2012; Buizza, 1997; Ehrendorfer, 1997; Kalnay, 2003; Molteni et al., 1996). Here the fastest growing characteristic patterns of differences from the control analysis in a linearized version of the full forecast model are calculated, again for the specific weather situation of a given day. Linear combinations (in effect, weighted averages) of these patterns, with magnitudes reflecting an appropriate level of analysis uncertainty, are then added to the control analysis to define the ensemble members. There is theoretical support for the use of singular vectors to choose initial ensemble members (Ehrendorfer and Tribbia, 1997), although its use requires substantially more computation than does the breeding method.

The Meteorological Service of Canada and the U.S. NCEP both currently generate their initial ensemble members using a method called the *ensemble Kalman filter* (EnKF) (Burgers et al., 1998; Houtekamer and Mitchell, 2005). This method is related to the multivariate extension of conjugate Bayesian updating of a Gaussian prior distribution (Section 6.3.4). Here the ensemble members from the previous forecast cycle define the Gaussian prior distribution, and the ensemble members are updated using a Gaussian likelihood function (i.e., data-generating process) for available observed data assuming known data variance (characteristic of the measurement errors), to yield new initial ensemble members from a Gaussian posterior distribution. The initial ensembles are relatively compact as a consequence of their (posterior) distribution being constrained by the observations, but the ensemble members diverge as each is integrated forward in time by the dynamical model, producing a more dispersed prior distribution for the next update cycle. The UK Met Office uses a related technique known as the ensemble transform Kalman filter, or ETKF (Bowler et al., 2008; Bowler and Mylne, 2009; Wang and Bishop, 2003). Expositions and literature reviews for the EnKF are provided by Evensen (2003) and Hamill (2006).

Choice of ensemble size is another important, but not yet definitively resolved, question regarding the structuring of an ensemble forecasting system (Leutbecher, 2018). Fundamentally, this issue involves a trade-off between the number of ensemble members versus the spatiotemporal resolution of the dynamical model used to integrate them (and therefore the required computational cost for each member), in the face of a finite computing resource. At present, addressing this question involves both balancing the priorities related to the needs of different users of the forecasts and also consideration of which lead times to favor in optimizing forecast accuracy. Richardson (2001) and Mullen and Buizza (2002) concluded that the appropriate ensemble size depends on user needs, and in particular that forecasts of rarer and more extreme events benefit from larger ensembles at the expense of dynamical model resolution. On the other hand, devoting more of the computational resource to improved model resolution is beneficial if forecast accuracy at shorter lead times (when overall uncertainty is lowest) has been prioritized, whereas the balance shifts toward larger ensembles of lower resolution integrations when medium- and long-range forecast accuracy is more important (Buizza, 2008, 2010; Ma et al., 2012; Machete and Smith, 2016; Raynaud and Bouttier, 2017). Ultimately, however, the balance should shift toward devoting new computational resources to increasing ensemble sizes. The reason is that error growth at the smallest scales (those newly resolved by an increased dynamical model resolution) both increases much faster than error growth at the larger scales of primary interest, and contaminates forecasts for the larger scales, so that a limit exists on the improvements to be achieved by increasing resolution (Lorenz, 1969; Palmer, 2014b).

8.2.3. Ensemble Mean and Ensemble Dispersion

One simple application of ensemble forecasting is averaging the members of the ensemble to obtain a single *ensemble mean* forecast. The motivation is to obtain a forecast that is more accurate than the single forecast initialized with the best estimate of the initial state of the atmosphere. Epstein (1969a) pointed out that the time-dependent behavior of the ensemble mean is different from the solution of forecast equations using the initial mean value, and concluded that in general the best forecast is not the single forecast initialized with the best estimate of initial conditions.

The first of these conclusions, at least, should not be surprising since a dynamical model is in effect a highly nonlinear function that transforms a set of initial atmospheric conditions to a set of forecast atmospheric conditions. In general, the average of a nonlinear function over some set of particular values of its argument is not the same as the function evaluated at the average of those values. That is, if the function $f(x)$ is nonlinear,

$$\frac{1}{n}\sum_{i=1}^{n} f(x_i) \neq f\left(\frac{1}{n}\sum_{i=1}^{n} x_i\right) \tag{8.3}$$

To illustrate simply, consider the three values $x_1 = 1$, $x_2 = 2$, and $x_3 = 3$. For the nonlinear function $f(x) = x^2 + 1$, the left side of Equation 8.3 is 5 2/3, and the right side of that equation is 5. We can easily verify that the inequality of Equation 8.3 holds for other nonlinear functions (e.g., $f(x) = \log(x)$ or $f(x) = 1/x$) as well. For the linear function $f(x) = 2x + 1$ the two sides of Equation 8.3 are both equal to 5.

Extending this idea to ensemble forecasting, we might like to know the atmospheric state corresponding to the center of the ensemble in phase space for some time in the future. Ideally, this central value of the ensemble will approximate the center of the stochastic-dynamic probability distribution at that future time, after the initial distribution has been transformed by the nonlinear forecast equations. The Monte Carlo approximation to this future value is the ensemble mean forecast. The ensemble mean forecast is obtained simply by averaging together the ensemble members for the lead time of interest, which corresponds to the left side of Equation 8.3. In contrast, the right side of Equation 8.3 represents the single forecast started from the average initial value of the ensemble members. Depending on the nature of the initial distribution and on the dynamics of the model, this single forecast may or may not be close to the ensemble average forecast. The benefits of ensemble averaging appear to derive primarily from averaging out elements of disagreement among the ensemble members, while emphasizing features that generally are shared by the members of the forecast ensemble. Empirically, ensemble means generally outperform single-integration forecasts, and theoretical support for this phenomenon is provided by Thompson (1977), Rougier (2016), and Christiansen, 2018).

Particularly for longer lead times, ensemble mean maps tend to be smoother than instantaneous states of the actual system, and so may seem unmeteorological, or more similar to smooth climatic averages. Palmer (1993) suggests that ensemble averaging will improve the forecast only until a regime change, or a change in the long-wave pattern, and he illustrates this concept nicely using the simple Lorenz (1963) model. This problem also is illustrated in Figure 8.5, where a regime change is represented by the bifurcation of the trajectories of the ensemble members between the intermediate and final lead times. At the intermediate lead time, before some of the ensemble members undergo this regime change, the center of the distribution of ensemble members is well represented by the ensemble average, which is a better central value than the single member of the ensemble started from the "best" initial condition. At the final forecast lead time the distribution of states has been distorted into two distinct groups. Here the ensemble average will be located somewhere in the middle, but near none of the ensemble members.

An especially important aspect of ensemble forecasting is its capacity to yield information about the magnitude and nature of the uncertainty in a forecast. In principle the forecast uncertainty is different on different forecast occasions, and this notion can be thought of as state-dependent predictability. The value to forecast users of communicating the different levels of forecast confidence that exist on different occasions was recognized early in the 20th century (Cooke, 1906b; Murphy, 1998). Qualitatively, we have more confidence that the ensemble mean is close to the eventual state of the atmosphere if the dispersion of the ensemble is small. Conversely, when the ensemble members are very different from each other the future state of the atmosphere may be more uncertain. One approach to "forecasting forecast skill" (Ehrendorfer, 1997; Kalnay and Dalcher, 1987; Palmer and Tibaldi, 1988) is to anticipate the accuracy of a forecast as being inversely related to the dispersion of the ensemble members. Operationally, forecasters do this informally when comparing the results from different dynamical models, or when comparing successive forecasts for a particular time in the future that were initialized on different days.

More formally, the *spread-skill relationship* for a collection of ensemble forecasts often is characterized by the correlation, over a collection of forecast occasions, between some measure of the ensemble spread such as the variance or standard deviation of the ensemble members around their ensemble mean on each occasion, and a measure of the predictive accuracy of the ensemble mean on that occasion. The idea is that forecasts entailing more or less uncertainty are then characterized by larger or smaller ensemble variances, respectively. The accuracy is often characterized using either the mean squared error (Equation 9.33) or its square root, although other measures have been used in some studies. These spread-skill correlations have sometimes been found to be fairly modest (e.g., Atger, 1999; Grimit and Mass, 2002; Hamill et al., 2004; Whittaker and Loughe, 1998), at least in part because of sampling variability of the chosen ensemble spread statistic, especially when the ensemble size is modest. Even so, some of the more recently reported spread-skill relationships (e.g., Sherrer et al. 2004; Stensrud and Yussouf, 2003) have been fairly strong. Figure 8.6 shows forecast accuracy, as measured by average root-mean squared error of ensemble members, as functions of ensemble spread measured by average root-mean squared differences among the ensemble members; for forecasts of 500 mb height over western Europe by the 51-member ECMWF ensemble prediction system for June 1997–December 2000. Clearly the more accurate forecasts (smaller RMSE) tend to be associated with smaller ensemble spreads, and vice versa, with this relationship being stronger for the shorter, 96-h lead time.

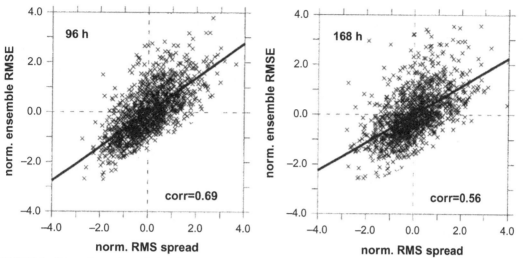

FIGURE 8.6   Scatterplots and correlations between forecast accuracy (vertical) and ensemble spread (horizontal) for ECMWF 500 mb height forecasts over western Europe, 1997–2000, at 96 h- and 168 h lead times. *Modified from Scherrer et al. (2004).* © *American Meteorological Society. Used with permission.*

8.2.4. Effects of Model Errors

Given a perfect dynamical model, integrating a random sample from the PDF of initial-condition uncertainty forward in time would yield a sample from the PDF characterizing forecast uncertainty. Of course dynamical models are not perfect, so that even if an initial-condition PDF could be known and correctly sampled from, the distribution of a forecast ensemble can at best be only an approximation to a sample from the true PDF for the forecast uncertainty (Hansen, 2002; Palmer, 2006; Smith, 2001).

Leith (1974) distinguished two kinds of model errors. The first derives from the models inevitably operating at a lower resolution than the real atmosphere or, equivalently, occupying a phase space of much lower dimension (Judd et al., 2008). Although still significant, this problem has been gradually addressed and partially ameliorated over the history of dynamical forecasting through progressive increases in model resolution. The second kind of model error derives from the fact that certain physical processes—prominently those operating at scales smaller than the model resolution—are represented incorrectly. In particular, such physical processes (known colloquially in this context as "physics") generally are represented using some relatively simple function of the explicitly resolved variables, known as a *parameterization.* Figure 8.7 shows a parameterization (solid curve) for the unresolved part of the tendency (dX/dt) of a resolved variable X, as a function of X itself, in the highly idealized Lorenz '96 (Lorenz, 2006) model. The individual points in Figure 8.7 are a sample of the actual unresolved tendencies, which are summarized by the regression function. In a realistic dynamical model there are a large number of such parameterizations for various unresolved physical processes, and the effects of these processes on the resolved variables are included in the model as functions of the resolved variables through these parameterizations. It is evident from Figure 8.7 that the parameterization (smooth curve) does not fully capture the range of behaviors for the parameterized process that are actually possible (scatter of points around the curve). Even if the large-scale dynamics have been modeled correctly, nature does not supply the value of the unresolved tendency given by "the" parameterized curve, but

FIGURE 8.7 Scatterplot of the unresolved time tendency, U, of a resolved variable, X, as a function of the resolved variable; together with a regression function representing the average dependence of the tendency on the resolved variable. *From Wilks (2005).*

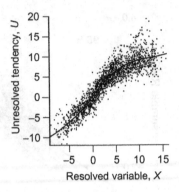

rather provides an effectively random realization from the point cloud around it. One way of looking at this kind of model error is that the effects of the parameterized physics are not fully determined by the resolved variables. That is, they are uncertain.

One way of representing the errors, or uncertainties, in the parameterized model physics is to extend the idea of the ensemble to include simultaneously a collection of different initial conditions *and* multiple dynamical models (each of which has a different collection of parameterizations). Harrison et al. (1999) found that ensemble forecasts using all four possible combinations of two sets of initial conditions and two dynamical model formulations differed significantly, with members of each of the four ensembles clustering relatively closely together, and distinctly from the other three, in the phase space. Other studies (e.g., Hansen, 2002; Houtekamer et al., 1996; Mullen et al., 1999; Mylne et al., 2002a; Stensrud et al., 2000) have found that using such *multimodel ensembles* (e.g., Kirtman et al., 2014) improves the resulting ensemble forecasts. The components of the Canadian Meteorological Center's operational multimodel ensemble share the same large-scale dynamical formulation, but differ with respect to the structure of various parameterizations (Houtekamer et al., 2009), in effect using different but similar parameterization curves of the kind represented in Figure 8.7, for different ensemble members.

A substantial part of the resulting improvement in ensemble performance derives from the multi-model ensembles exhibiting larger ensemble dispersion, so that the ensemble members are less like each other than if an identical dynamical model is used for all forecast integrations. Typically the dispersion of forecast ensembles is too small (e.g., Buizza, 1997; Stensrud et al., 1999; Toth and Kalnay, 1997), and so they express too little uncertainty about forecast outcomes (see Section 9.7).

Another approach to capturing uncertainties in the structure of dynamical models is suggested by the scatter around the regression curve in Figure 8.7. From the perspective of Section 7.2, the regression residuals that are differences between the actual (points) and parameterized (regression curve) behavior of the modeled system are random variables. Accordingly, the effects of parameterized processes can be more fully represented in a dynamical model if random numbers are added to the deterministic parame-terization function, making the dynamical model explicitly stochastic (e.g., Berner et al., 2017; Palmer, 2001, 2012; Palmer et al., 2005b). Even if the system being modeled truly does not contain random com-ponents, adopting the stochastic view of unresolved, parameterized processes in a dynamical model may improve the resulting forecasts in terms of both ensemble spread and climatological fidelity (e.g., Buizza et al., 1999; Leutbecher et al., 2017; Ollinaho et al., 2017; Palmer, 2012; Tenant et al., 2011; Wilks, 2005).

The idea of stochastic parameterizations in dynamical models is not new, having been proposed as early as the 1970s (Lorenz, 1975; Moritz and Sutera, 1981; Pitcher, 1974, 1977). However, its use in realistic atmospheric models has been relatively recent (Bowler et al., 2008; Buizza et al., 1999; Garratt et al., 1990; Leutbecher et al., 2017; Lin and Neelin, 2000, 2002; Sanchez et al., 2016; Williams et al., 2003). Particularly noteworthy is the first operational use of a stochastic representation of the effects of unresolved processes in the forecast model at ECMWF, which they called *stochastic physics*. Stochastic parameterization is still at a fairly early stage of development and continues to be the subject of ongoing research (e.g., Bengtsson et al., 2013; Berner et al., 2010, 2015; Christensen et al., 2017b; Frenkel et al., 2012; Neelin et al., 2010; Shutts, 2015).

Stochastic parameterizations also have been used in simplified climate models, to represent atmospheric variations on the timescale of weather, beginning the 1970s (e.g., Hasselmann, 1976; Lemke, 1977; Sutera, 1981), and in continuing work (e.g., Batté and Doblas-Reyes, 2015; Imkeller and von Storch, 2001; Imkeller and Monahan, 2002). Some papers applying this idea to prediction of the El Niño phenomenon are Christensen et al. (2017a), Penland and Sardeshmukh (1995), and Thompson and Battisti (2001).

8.3. UNIVARIATE ENSEMBLE POSTPROCESSING

8.3.1. Why Ensembles Need Postprocessing

In principle, initial ensemble members chosen at random from the PDF characterizing initial-condition uncertainty, and integrated forward in time with a perfect dynamical model, will produce an ensemble of future system states that is a random sample from the PDF characterizing forecast uncertainty. Ideally, then, the dispersion of a forecast ensemble characterizes the uncertainty in the forecast, so that small ensemble dispersion (all ensemble members similar to each other) indicates low uncertainty, and large ensemble dispersion (large differences among ensemble members) signals large forecast uncertainty.

In practice, the initial ensemble members are chosen in ways that do not randomly sample from the PDF of initial-condition uncertainty (Section 8.2.2), and errors in the dynamical models deriving mainly from processes operating on unresolved scales produce errors in ensemble forecasts just as they do in conventional single-integration forecasts. Accordingly, the dispersion of a forecast ensemble can at best only approximate the PDF of forecast uncertainty (Hansen, 2002; Smith, 2001). In particular, a forecast ensemble may reflect errors both in statistical location (most or all ensemble members being well away from the actual state of the atmosphere, but relatively nearer to each other) and dispersion (either under- or overrepresenting the forecast uncertainty). Often, operational ensemble forecasts are found to exhibit too little dispersion (e.g. Buizza, 1997; Buizza et al., 2005; Hamill, 2001; Toth et al., 2001; Wang and Bishop, 2005), which leads to overconfidence in probability assessment if ensemble relative frequencies are interpreted directly as estimating probabilities.

To the extent that ensemble forecast errors have consistent characteristics, they can be corrected through *ensemble-MOS* methods that summarize a historical database of these forecast errors, just as is done for single-integration dynamical forecasts. From the outset of ensemble forecasting (Leith, 1974) it was anticipated that use of finite ensembles would yield errors in the forecast ensemble mean that could be statistically corrected using a database of previous errors. MOS postprocessing (Section 7.9.2) is a more difficult problem for ensemble forecasts than for ordinary single-integration dynamical forecasts, or for the ensemble mean, because ensemble forecasts are susceptible to both the ordinary biases introduced by errors and inaccuracies in the dynamical model formulation, in addition to their usual underdispersion bias. Either or both of these kinds of problems in ensemble forecasts can be corrected using MOS methods.

Ultimately the goal of ensemble-MOS methods is to estimate a forecast PDF or CDF on the basis of the discrete approximation provided by a finite, m-member ensemble. If the effects of initial-condition errors and model errors were not important, this task could be accomplished by operating only on the ensemble members at hand, without regard to the statistical characteristics of past forecast errors. Probably the simplest such non-MOS approach is to regard the forecast ensemble as a random sample from the true forecast CDF and estimate cumulative probabilities from that CDF using a plotting-position estimator (Section 3.3.7). The so-called *democratic voting* method is a commonly used, although usually suboptimal, such estimator. Denoting the quantity being forecast, or verification, as V, and the distribution quantile whose cumulative probability is being estimated as q, this method computes

$$\Pr\{V \le q\} = \frac{1}{m}\sum_{k=1}^{m} I(q \le x_k) = \frac{\text{rank}(q) - 1}{m}, \tag{8.4}$$

where the indicator function $I(\bullet) = 1$ if its argument is true and is zero otherwise, and rank(q) indicates the rank of the quantile of interest in a hypothetical $m + 1$ member ensemble consisting of the ensemble members x_k and that quantile. Equation 8.4 is equivalent to the Gumbel plotting position estimator (Table 3.2) and has the unfortunate property of assigning zero probability to any quantile less than the smallest ensemble member, $x_{(1)}$, and full unit probability to any quantile greater than the largest ensemble member, $x_{(m)}$.

Other plotting position estimators do not have these deficiencies. For example, using the Tukey plotting position (Wilks, 2006b) yields

$$\Pr\{V \le q\} = \frac{\text{Rank}(q) - 1/3}{(m + 1) + 1/3}. \tag{8.5}$$

Katz and Ehrendorfer (2006) derive a cumulative probability estimator equivalent to the Weibull plotting position using a conjugate Bayesian analysis (Section 6.3.2) with a uniform prior distribution and a binomial likelihood for the ensemble members' binary forecasts of outcomes above or below q. However, the cumulative probability estimators in Equations 8.4 and 8.5 will still lead to inaccurate, overconfident results unless the ensemble size is large or the forecasts are reasonably skillful, even if the ensemble is free of bias errors and exhibits dispersion that is consistent with the actual forecast uncertainty (Richardson, 2001, see Section 9.7).

8.3.2. Nonhomogeneous Regression Methods

Direct transformation of a collection of ensemble forecasts using estimators such as Equation 8.5 will usually be inaccurate because of bias errors (e.g., observed temperatures warmer or cooler, on average, than the forecast temperatures) and/or dispersion errors (ensemble dispersion smaller or larger, on average, than required to accurately characterize the forecast uncertainty), which occur in general because of imperfect ensemble initialization and deficiencies in the structure of the dynamical model. Ordinary regression-based MOS postprocessing of single-integration dynamical forecasts through regression methods (Section 7.9.2) can be extended to compensate for ensemble dispersion errors also, by using an ensemble dispersion predictor in appropriately reformulated regression models. Adjusting the dispersion of the ensemble according to its historical error statistics can allow information on possible state- or flow-dependent predictability to be included also in an ensemble-MOS procedure.

In common with other regression methods, *nonhomogeneous regressions* represent the conditional mean of the predictive distribution as an optimized linear combination of the predictors, which in this setting are generally the ensemble members. However, unlike the ordinary regression models described in Sections 7.2 and 7.3, in which the "error" and predictive variances are assumed to be constant (homogeneous), in nonhomogeneous regressions these variances are formulated to depend on the ensemble variance. This property allows the predictive distributions produced by nonhomogeneous regression methods to exhibit more uncertainty when the ensemble dispersion is large, and less uncertainty when the ensemble dispersion is small.

Nonhomogeneous Gaussian Regression

Nonhomogeneous Gaussian regression (NGR) was independently proposed by Jewson et al. (2004), and Gneiting et al. (2005), where it was named EMOS (for ensemble MOS). However, it is only one of many ensemble-MOS methods, so that the more descriptive and specific name NGR was applied to it by Wilks (2006b). The distinguishing feature of NGR among other nonhomogeneous regression methods is that it yields predictive distributions that are Gaussian.

Specifically, for each forecasting occasion t the Gaussian predictive distribution has mean and variance that are particular to the ensemble for that occasion,

$$y_t \sim N[\mu_t, \sigma_t^2],$$ (8.6)

where y_t is the quantity being predicted. The Gaussian predictive distribution has mean

$$\mu_t = a + b_1 x_{t,1} + b_2 x_{t,2} + \cdots + b_m x_{t,m}$$ (8.7)

and variance

$$\sigma_t^2 = c + d\, s_t^2,$$ (8.8)

where $x_{t,k}$ is the kth ensemble member for the tth forecast, s_t^2 is the ensemble variance,

$$s_t^2 = \frac{1}{m} \sum_{k=1}^{m} (x_{t,k} - \bar{x}_t)^2,$$ (8.9)

m is the ensemble size, and the ensemble mean is

$$\bar{x}_t = \frac{1}{m} \sum_{k=1}^{m} x_{t,k}.$$ (8.10)

The $m+3$ regression constants a, b_k, c, and d are the same for each forecast occasion t and need to be estimated using past training data. (Division by $m-1$ in Equation 8.9 will yield identical forecasts because the larger sample variance estimates will be compensated by a smaller estimate for the parameter d.) Usually the $x_{t,k}$ are dynamical-model predictions of the same quantity as y_t, but these can be any useful predictors (e.g., Messner et al., 2017).

Equation 8.7 is a general specification for the predictive mean, which would be appropriate when the m ensemble members are nonexchangeable, meaning that they have distinct statistical characteristics.

This is the case, for example, when a multimodel ensemble (i.e., an ensemble where more than one dynamical model is being used) is composed of single integrations from each of its m constituent models. Very often a forecast ensemble comprises exchangeable members (necessarily from the same dynamical model) having the same statistical characteristics, so that the regression coefficients b_k must be the same apart from estimation errors. In this common situation of statistically exchangeable ensemble members, Equation 8.7 is replaced by

$$\mu_t = a + b\,\bar{x}_t,$$ (8.11)

in which case there are four regression parameters a, b, c, and d to be estimated. Mean functions of intermediate complexity are also sometimes appropriate. For example, a two-model ensemble in which the members within each model are exchangeable would require two b coefficients operating on the two within-model ensemble means. Sansom et al. (2016) extend NGR for seasonal forecasting, where time-dependent biases deriving from progressive climate warming are important (e.g., Wilks and Livezey, 2013), by specifying

$$\mu_t = a + b_1\,\bar{x}_t + b_2 t$$ (8.12)

rather than using Equation 8.11.

Probabilistic NGR forecasts for the predictand y_t are computed using

$$\Pr\{y_t \leq q\} = \Phi\left(\frac{q - \mu_t}{\sigma_t}\right),$$ (8.13)

where q is any quantile of interest in the predictive distribution, and $\Phi(\bullet)$ indicates the cumulative distribution function (CDF) of the standard Gaussian distribution. The predictive mean (Equation 8.7, or 8.11) corrects unconditional forecast bias (consistent over- or underforecasting) when $a \neq 0$, and corrects conditional forecast biases when the $b_k \neq 1/m$ in Equation 8.7 or $b \neq 1$ in Equation 8.11. The (square root of the) predictive variance (Equation 8.8) corrects dispersion errors, and further allows incorporation of any spread-skill relationship (i.e., positive correlation between ensemble spread and ensemble-mean error; e.g., Figure 8.6) into the forecast formulation. In Equation 8.8, ensembles exhibiting correct dispersion characteristics correspond to $c \approx 0$ and $d \approx 1$, larger values of d reflect stronger spread-skill relationships, and $d \approx 0$ indicates lack of a useful spread-skill relationship.

Unlike the situation for conventional linear regression, there is no analytic formula that can be used for NGR parameter fitting. An innovative idea proposed by Gneiting et al. (2005) is to estimate the NGR parameters a, b_k, c, and d by minimizing the average continuous ranked probability score (CRPS, Matheson and Winkler, 1976, see Section 9.5.1) over the training data. For Gaussian predictive distributions this is

$$\overline{\text{CRPS}}_G = \frac{1}{n} \sum_{t=1}^{n} \sigma_t \left\{ \frac{y_t - \mu_t}{\sigma_t} \left[2\Phi\left(\frac{y_t - \mu_t}{\sigma_t}\right) - 1 \right] + 2\phi\left(\frac{y_t - \mu_t}{\sigma_t}\right) - \frac{1}{\sqrt{\pi}} \right\},$$ (8.14)

where $\phi(\bullet)$ denotes the probability density function of the standard Gaussian distribution, n is the number of training samples, μ_t is defined by Equation 8.7 or Equation 8.11 as appropriate, and σ_t is the square root of the quantity in Equation 8.9. Both $c \geq 0$ and $d \geq 0$ are required mathematically, which can be achieved by setting $c = \gamma^2$ and $d = \delta^2$, and optimizing over γ and δ.

A more conventional parameter estimation approach, used by Jewson et al. (2004), is maximization of the Gaussian log-likelihood function

$$L_G = -\sum_{t=1}^{n} \left[\frac{(y_t - \mu_t)^2}{2\sigma_t^2} - \ln(\sigma_t) \right], \tag{8.15}$$

which formally assumes independence among the n training samples. Maximization of this objective function is equivalent to minimization of the average logarithmic score (Good, 1952), which is also known as the Ignorance score (Roulston and Smith, 2002, see Section 9.5.3). Gneiting et al. (2005) found that use of Equation 8.15 for estimation of NGR parameters yielded somewhat overdispersive predictive distributions, although Baran and Lerch (2016), Gebetsberger et al. (2017a), Prokosch (2013), Williams et al. (2014), and Williams (2016) have reported little difference in forecast performance between use of Equations 8.14 and 8.15 for estimating the NGR parameters.

Parameter estimates for many of the ensemble-MOS methods in addition to NGR can be computed using either CRPS or negative log-likelihood minimization. Both require iterative solutions. Maximum likelihood is typically less computationally demanding than CRPS minimization, although it is less robust to the influence of outlying extreme values (Gneiting and Raftery, 2007). Either approach can be modified by adding a penalty for lack of calibration (i.e., correspondence between forecast probabilities and subsequent predictand relative frequencies), Section 9.7.1, which may increase the economic value of the postprocessed forecasts to users (Wilks, 2018a).

Figure 8.8 shows an example NGR predictive distribution (solid), in relation to the raw $m = 35$-member ensemble (tick-marks on the horizontal axis, some of which represent multiple members), and a kernel density smoothing (Section 3.3.6) of them. The NGR distribution is shifted several degrees to the right relative to the raw ensemble, indicating that the training data used for fitting the parameters exhibited a cold bias. The dispersion of the NGR predictive distribution is wider than that of the raw ensemble, exemplifying correction of the typical underdispersion of raw ensemble forecasts. However, because the NGR predictive distribution is constrained to be Gaussian, it is incapable of representing the bimodality exhibited by the kernel density smoothing of the raw ensemble.

Wilks and Hamill (2007), Hagedorn et al. (2008), and Kann et al. (2009) have reported good results when postprocessing ensemble forecasts for surface temperatures (which are approximately Gaussian) using NGR, yielding substantial improvements in forecast skill over direct use of raw ensemble output (e.g., Equation 8.4 or 8.5).

More Flexible Nonhomogeneous Regression Distributions

By construction, the NGR formulation can yield only Gaussian predictive distributions. Some meteorological and climatological predictands are not Gaussian or approximately Gaussian, so that modifications to the nonhomogeneous regression framework are required to adequately represent such data. One possible approach is to subject the target predictand and its predictor counterparts in the ensemble to a Box-Cox ("power" or "normalizing") transformation (Equation 3.20), before fitting an NGR model (Hemri et al., 2015), which is intended to make the climatological distribution of the predictand y as nearly Gaussian as possible. The transformation exponent λ is then an additional parameter to be estimated.

In a special case of this approach, Baran and Lerch (2015, 2016) investigate use of nonhomogeneous lognormal regression, which is equivalent to NGR after the predictand y has been log-transformed. Predictive probabilities on the transformed scale are estimated using

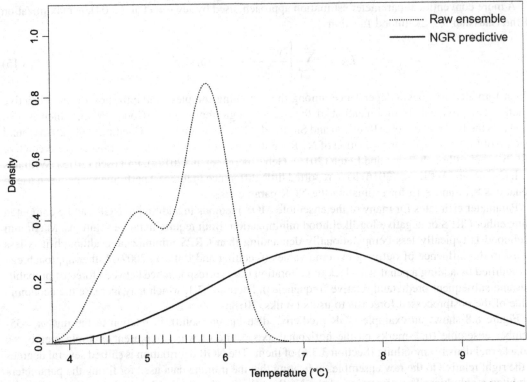

FIGURE 8.8 Example NGR (solid) predictive distribution, compared to the raw ensemble (tick-marks on horizontal axes) and smoothed predictive distribution based on it (dotted). *Modified from Taillardat et al. (2016).* © *American Meteorological Society. Used with permission.*

$$\Pr\{y_t \le q\} = \Phi\left[\left(\ln(q) - \ln\left(\frac{\mu_t^2}{\sqrt{\mu_t^2 + \sigma_t^2}}\right)\right) \bigg/ \sqrt{\ln\left(1 + \frac{\sigma_t^2}{\mu_t^2}\right)}\right], \tag{8.16}$$

where σ_t^2 and μ_t are parameterized according to Equations 8.8 and 8.11 (for an exchangeable ensemble), respectively. The counterpart of Equation 8.14 for average lognormal CRPS is given in Baran and Lerch (2015).

Another possible approach to nonhomogeneous regression is to specify non-Gaussian predictive distributions. Messner et al. (2014a) modeled square-root transformed wind speeds using nonhomogeneous regressions with logistic (Section 4.4.4) predictive distributions,

$$\Pr\{y_t \le q\} = \frac{\exp\left[(q - \mu_t)/\sigma_t\right]}{1 + \exp\left[(q - \mu_t)/\sigma_t\right]}. \tag{8.17}$$

Here the conditional means μ_t were modeled using Equation 8.11, and the corresponding (strictly positive) scale parameters were specified using

$$\sigma_t = \exp(c + d\,s_t). \tag{8.18}$$

Messner et al. (2014a) estimated the regression parameters in Equation 8.17 using maximum likelihood, and Taillardat et al. (2016) provide the expression necessary for minimum-CRPS estimation. Scheuerer and Möller (2015) investigate nonhomogeneous regressions with gamma-distributed predictive distributions and provide a closed-form CRPS equation.

In another approach to modeling positively skewed predictands in a nonhomogeneous ensemble regression setting, Lerch and Thorarinsdottir (2013) forecast probability distributions for maximum daily wind speed using generalized extreme-value (GEV) predictive distributions. In this case the prediction probabilities are given by

$$
\Pr(y_t \leq q) =
\begin{cases}
\exp\left\{-\left[1 + \xi\left(\dfrac{q - \mu_t}{\sigma_t}\right)\right]^{-1/\xi}\right\}, & \xi \neq 0 \\
\exp\left\{-\exp\left[-\left(\dfrac{q - \mu_t}{\sigma_t}\right)\right]\right\}, & \xi = 0
\end{cases}
\tag{8.19}
$$

Lerch and Thorarinsdottir (2013) used Equation 8.11 to specify the location parameter μ_t but in their application found better results when the scale parameter depended on the ensemble mean according to

$$
\sigma_t = \exp(c + d\,\overline{x}_t),
\tag{8.20}
$$

which ensures $\sigma_t > 0$ as required. Equation 8.20 does not allow explicit dependence on the ensemble spread, but does reflect the heteroscedastic nature of variables such as wind speeds and precipitation amounts which often exhibit greater variability for larger values. When $\xi > 0$ (found in the overwhelming majority of forecasts by Lerch and Thorarinsdottir, 2013), Equation 38.19 allows $y_t > \mu_t - \sigma_t/\xi$, so that GEV predictive distributions may yield nonzero probability for negative values of the predictand. These probabilities were quite low when maximum daily wind speeds in Lerch and Thorarinsdottir (2013) but could be more substantial if the predictand is daily precipitation.

Truncated Nonhomogeneous Regressions

The original NGR approach will often be appropriate when the predictand of interest exhibits a reasonably symmetric distribution, for example, temperature or sea-level pressure. When the distribution of the predictand is not symmetric but unimodal, and if any discontinuity in its distribution at zero is small, NGR for the transformed predictand or nonhomogeneous regressions with different predictive forms, as just described, may work well. However, strictly nonnegative predictands often exhibit large discontinuities in their probability densities at zero, which may prevent simple transformations from achieving adequate approximation to Gaussian shape, and which may be poorly represented by alternative conventional predictive distributions. This problem is especially acute in the case of relatively short-duration (such as daily) precipitation accumulations, which often feature a large probability "spike" at zero. Nonhomogeneous regression approaches based on truncation or censoring can be successful at addressing this issue.

Thorarinsdottir and Gneiting (2010) proposed using zero-truncated Gaussian predictive distributions (Equations 4.39–4.41) in a nonhomogeneous regression framework for forecasting positively skewed and nonnegative predictands. The average CRPS for the truncated Gaussian distribution (Thorarinsdottir and Gneiting, 2010; Taillardat et al., 2016), which is the counterpart to Equation 8.14, can again be used to fit the regression parameters. Having obtained these estimates, either Equation 8.7 or 8.11 is used to compute the location parameter μ_t for the current forecast, and Equation 8.8 is used to compute the scale parameter σ_t. Forecast probabilities can then be calculated using Equation 4.40.

Hemri et al. (2014) used zero-truncated Gaussian regressions following square-root transformation of the predictand. Scheuerer and Möller (2015) used nonhomogeneous regressions with truncated logistic predictive distributions (Equation 8.17). Junk et al. (2015) fit zero-truncated Gaussian regressions using training data restricted to near-analogs of the current forecast.

Observing that the heavier tails of the lognormal probability model better represent the distribution of stronger wind speeds whereas the truncated Gaussian model is better over the remainder of the distribution, Baran and Lerch (2016) investigated probability mixtures of nonhomogeneous lognormal and zero-truncated Gaussian predictive distributions, so that predictive probabilities were computed using

$$\Pr\{y_t \leq q\} = w\Pr_{TG}\{y_t \leq q\} + (1 - w)\Pr_{LN}\{y_t \leq q\}, \tag{8.21}$$

where the indicated truncated Gaussian (TG) and lognormal (LN) probabilities are specified by Equations 4.40 and 8.16, respectively. They found clearly better-calibrated predictive distributions derived from the mixtures as compared to the individual lognormal or truncated Gaussian distributions. A related model (Lerch and Thorarinsdottir, 2013) uses a regime-switching idea, where GEV predictive distributions (Equation 8.19) are used when the ensemble median is above a threshold θ, and truncated Gaussian predictive distributions (Equation 4.40) are used otherwise, with θ being an additional parameter requiring estimation. Baran and Lerch (2015) investigate a similar regime-switching model, where the predictive distribution is lognormal rather than GEV if the ensemble median is larger than θ.

Censored Nonhomogeneous Regressions

Censoring, in contrast to truncation, allows a probability distribution to represent values falling hypothetically below a censoring threshold, even though those values have not been observed (Section 4.4.3). In the context of ensemble postprocessing, the censoring threshold is generally zero, and any probability corresponding to negative predictand values is assigned to exactly zero, yielding a probability spike there. Specifying the parameters for these censored predictive distributions using regression models yields what is known in the economics literature as Tobit regression (Tobin, 1958). Censored formulations allow representation of a discontinuous probability spike at the lower limit of the predictive distribution, whereas truncated formulations are likely more appropriate when the variable in question is at least approximately continuous there.

Scheuerer (2014) describes a nonhomogeneous regression model having zero-censored GEV predictive distributions, so that that any nonzero probability for negative y_t is assigned to zero precipitation. Assuming $\xi \neq 0$ in Equation 8.19 (with an obvious modification for $\xi = 0$), predictive probabilities are computed using

$$\Pr(y_t \leq q) = \begin{cases} \exp\left\{-\left[1 + \xi\left(\dfrac{q - \mu_t}{\sigma_t}\right)\right]^{-1/\xi}\right\}, & q \geq 0 \\ 0, & q < 0 \end{cases}, \tag{8.22}$$

so that all probability for $y_t \leq 0$ is assigned to zero. Scheuerer (2014) linked the censored GEV parameters to the ensemble statistics using the relationships

$$\mu_t = a + b_1\bar{x}_t + b_2\sum_{k=1}^{m} I(x_{t,k} = 0) - \frac{\sigma_t}{\xi}[\Gamma(1 - \xi) - 1], \tag{8.23}$$

where $I(\bullet)$ is the indicator function whose value is 1 if its argument is true, and is zero otherwise, $\Gamma(\bullet)$ denotes the gamma function, and the predictive scale parameter is specified using

$$\sigma_t = c + \frac{d}{m^2} \sum_{k=1}^{m} \sum_{j=1}^{m} \left| x_{t,k} - x_{t,j} \right|. \tag{8.24}$$

Here division by $m(m-1)$ rather than m^2 might be preferred since the m terms in Equation 8.24 for which $k=j$ will be identically zero (Ferro et al., 2008; Wilks, 2018a). The distribution parameters were estimated by minimizing average CRPS, with the shape parameter ξ assumed independent of the ensemble statistics as was also done by Lerch and Thorarinsdottir (2013).

Using a similar approach, Scheuerer and Hamill (2015a) and Baran and Nemoda (2016) implement nonhomogeneous regressions using zero-censored shifted-gamma (i.e., Pearson Type III, Equation 4.55) predictive distributions for precipitation amounts, where both the shape α_t and scale β_t parameters are required to be positive, and the shift parameter $\zeta_t < 0$ in this application. As is also the case for the Scheuerer (2014) zero-censored GEV model, any nonzero probability for negative values of y_t is assigned to $y_t = 0$, so that

$$\Pr\{y_t \leq q\} = \begin{cases} F_{\gamma(\alpha_t)}\left(\dfrac{q-\zeta_t}{\beta_t}\right), & q \geq 0 \\ 0 & , q < 0 \end{cases}, \tag{8.25}$$

where $F_{\gamma(\alpha)}$ denotes the CDF for the standard ($\beta = 1$) 2-parameter gamma distribution (Section 4.4.5) with shape parameter α.

Messner et al. (2014b) proposed nonhomogeneous regression for square-root transformed wind speeds, using censored logistic predictive distributions. Probabilities on the transformed scale are then computed using

$$\Pr\{y_t \leq q\} = \begin{cases} \dfrac{\exp\left[(q-\mu_t)/\sigma_t\right]}{1+\exp\left[(q-\mu_t)/\sigma_t\right]}, & q \geq 0, \\ 0 & , \text{ otherwise,} \end{cases} \tag{8.26}$$

where the dependence of the mean and scale parameters on the ensemble statistics were formulated according to Equations 8.11 and 8.18, respectively. Stauffer et al. (2017) and Gebetsberger et al. (2017b) further extended this censored-logistic approach for (power-transformed) precipitation forecasts, where in some cases all ensemble members may forecast zero, by defining the logistic distribution parameters as

$$\mu_t = a + b_1 I\left(\sum_{k=1}^{m} x_k = 0\right) + b_2 \bar{x}_t \left[1 - I\left(\sum_{k=1}^{m} x_k = 0\right)\right], \tag{8.27a}$$

and

$$\ln(\sigma_t) = c + d \ln(s_t)\left[1 - I\left(\sum_{k=1}^{m} x_k = 0\right)\right]. \tag{8.27b}$$

Thus regression specifications for both mean and standard deviation parameters are invoked if at least one ensemble member forecasts nonzero precipitation, and fixed $\mu_t = a + b_1$ and $\sigma_t = \exp(c)$ are used if all ensemble members are dry. The logarithms in Equation 8.27b ensure that the predictive standard deviation is always positive. Equation 8.27b allows the logarithmic link for the standard deviation to be defined even when all ensemble members are zero, in which case $\sigma_t = 0$.

An alternative to censoring for representing the finite probability of zero precipitation is to construct a mixture model where one element of the mixture is a probability p_t for zero precipitation (Bentzien and Friederichs, 2012). In that paper this probability was specified using logistic regression (Section 7.6.2), and then combined with either a gamma or lognormal distribution F_t for the nonzero precipitation amounts, yielding the probability specification

$$\Pr\{y_t \leq q\} = p_t + (1 - p_t)F_t(q| q > 0).$$ (8.28)

Bentzien and Friederichs (2012) also proposed specifying probabilities for large extremes using generalized Pareto distributions, which arise in extreme-value theory (Section 4.4.8) for the distribution of values above a high threshold, appended to the right tails of the gamma or lognormal predictive distributions.

8.3.3. Logistic Regression Methods

Hamill et al. (2004) first proposed use of logistic regression (Section 7.6.2) for ensemble postprocessing, using the ensemble mean as the sole predictor, in an application where only the two climatological terciles were used as the prediction thresholds, q:

$$\Pr\{y_i \leq q\} = \frac{\exp\left[a_q + b_q \bar{x}_t\right]}{1 + \exp\left[a_q + b_q \bar{x}_t\right]}.$$ (8.29)

When fitting logistic regression parameters, the training-data predictands are binary (as indicated by the dots in Figure 7.20), being one if the condition in curly brackets on the left-hand side of Equation 8.29 is true and zero otherwise.

As indicated by the subscripts in Equation 8.29, separate regression coefficients must in general be estimated for each predictand threshold q in logistic regression. Especially when logistic regressions are estimated for large numbers of predictand thresholds, it becomes increasingly likely that probability specifications using Equation 8.29 may be inconsistent, implying negative probabilities for some outcomes. Wilks (2009) proposed extending ordinary logistic regression in a way that unifies the regression functions for all quantiles of the distribution of the predictand, by assuming equal regression coefficients b_q, and specifying the regression intercepts as a nondecreasing function of the target quantile,

$$\Pr\{y_t \leq q\} = \frac{\exp\left[a_q g(q) + a_0 + b\bar{x}_t\right]}{1 + \exp\left[a_q g(q) + a_0 + b\bar{x}_t\right]}.$$ (8.30)

The function $g(q) = \sqrt{q}$ was found to provide good results for the data used in that paper. Equation 8.30 ensures coherent probability specifications and requires estimation of fewer parameters than do conventional logistic regressions for multiple quantiles. The parameters are generally fit using maximum likelihood (Messner et al., 2014a) using a selected set of observed quantiles, but once fit Equation 8.30 can be applied for any value of q. This approach has come to be known as extended logistic regression (XLR).

The mechanism of XLR can most easily be appreciated by realizing that the regression function in Equation 8.30 is linear when expressed on the log-odds scale:

$$\ln\left(\frac{\Pr\{y_t \leq q\}}{1 - \Pr\{y_t \leq q\}}\right) = a_q g(q) + a_0 + b\bar{x}_t.$$ (8.31)

Figure 8.9a shows example XLR probability specifications for selected predictand quantiles q, compared with corresponding results in when logistic regressions (Equation 8.29) have been fit individually for the same predictand quantiles in Figure 8.9b. The common slope parameter b in Equations 8.30 and 8.31 force regressions for all quantiles to be parallel in log-odds in Figure 8.9a; whereas the individual logistic regressions in Figure 8.9b cross, leading in some cases to cumulative probability specifications for smaller precipitation amounts being larger than specifications for larger amounts.

Of course the logistic regression function inside the exponentials of Equations 8.29 or 8.30 can include multiple predictors, in the form $b_1x_{t,1} + b_2x_{t,2} + \dots + b_mx_{t,m}$ analogously to Equation 8.7, where the various x's may be nonexchangeable ensemble members or other covariates. Messner et al. (2014a) point out that even if one or more of these involves ensemble spread, these forms do not explicitly represent any spread-skill relationship exhibited by the forecast ensemble, but that the logistic regression framework can be further modified to do so (Equation 8.17).

Equations 3.30 and 3.31 can be seen as a continuous extension of the proportional-odds logistic regression approach (McCullagh, 1980) for specifying cumulative probabilities for an ordered set of discrete outcomes. Messner et al. (2014a) applied proportional-odds logistic regression to prediction of the climatological deciles of wind speed and precipitation, using both ordinary (homoscedastic) and nonconstant-variance (heteroscedastic) formulations. Hemri et al. (2016) applied the more conventional homoscedastic proportional-odds logistic regression to postprocessing ensemble forecasts of cloud cover. The cloud-cover predictand is measured in "octas," or discrete eighths of the sky hemisphere, so that the nine ordered predictand values are $y_t = 0/8$, $1/8$, $\dots$, $8/8$. The (homoscedastic) proportional-odds logistic regression forecasts are then formulated as

$$\ln\left(\frac{\Pr\{y_t \le q\}}{1 - \Pr\{y_t \le q\}}\right) = a_q + b_1x_{t,1} + b_2x_{t,2} + \dots + b_mx_{t,m}, \tag{8.32}$$

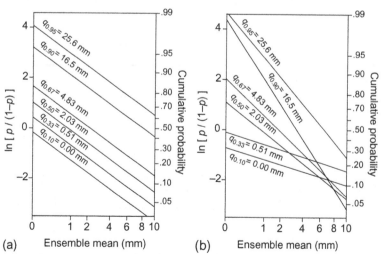

FIGURE 8.9 Logistic regressions plotted on the log-odds scale, for 28 November–2 December accumulated precipitation at Minneapolis, at 6–10 day lead time. Forecasts from Equation (8.30), evaluated at selected quantiles, are shown by the parallel lines in panel (a), which cannot yield logically inconsistent sets of forecasts. Regressions for the same quantiles, fitted separately using Equation (8.29), are shown in panel (b). Because these regressions are not constrained to be parallel, logically inconsistent forecasts are inevitable for sufficiently extreme values of the predictor. *From Wilks (2009).*

where $q = 0/8$, $1/8$, ...,$8/8$, as before the predictors $x_{t,k}$ might be nonexchangeable ensemble members and/or other statistics derived from the ensemble, and the intercepts are strictly ordered so that $a_{0/8} < a_{1/8} < \cdots < a_{8/8}$. The overall result is very much like that shown in Figure 8.9a, with regression functions that are parallel in the log-odd space and which have monotonically increasing intercepts a_q, but it differs in that intermediate functions between the plotted lines are not defined because of the discreteness of the predictand.

8.3.4. Bayesian Model Averaging and other "Ensemble Dressing" Methods

Bayesian model averaging (BMA), introduced as an ensemble postprocessing tool by Raftery et al. (2005), is the second of the two most commonly used ensemble-MOS methods, the other being nonhomogeneous regression (Section 8.3.2). Despite the name, it is not a fully Bayesian method in the sense of the approach presented in Chapter 6. In common with regression methods, BMA yields a continuous predictive distribution for the forecast variable y. However, rather than imposing a particular parametric form, BMA predictive distributions are mixture distributions (Section 4.4.9) or equivalently kernel density estimates (Section 3.3.6). That is, BMA predictive distributions are weighted sums of m component probability distributions, each centered at the corrected value of one of the m ensemble members being postprocessed. The BMA procedure is an example of the process sometimes referred to as "ensemble dressing," because it is the aggregate result of m probability distributions being metaphorically draped around each corrected ensemble member.

Construction of a BMA predictive distribution can be expressed in general as

$$f_{BMA}(y_t) = \sum_{k=1}^{m} w_k f_k(y_t), \tag{8.33}$$

where each w_k is the nonnegative weight associated with the component probability density $f_k(y_t)$ pertaining to the kth ensemble member $x_{t,k}$, and $\Sigma_k w_k = 1$. Figure 8.10 illustrates the process for a five-member ensemble of nonexchangeable members, where the component distributions are Gaussian, have been constrained to have equal variance, and the weights can be interpreted as probabilities that the respective members will provide the best forecast (Raftery et al., 2005). Figure 8.10 emphasizes that, even when the component densities are Gaussian, their weighted sum can take on a wide variety of shapes. In the case of Figure 8.10 the BMA density is bimodal, reflecting the bifurcation of this small ensemble into two groups.

In order for BMA and other ensemble dressing procedures to work correctly, the raw ensemble members must first be debiased, in order to correct systematic forecast errors exhibited in the available training data. Usually this initial debiasing step is accomplished using linear regressions, although more sophisticated methods could be used if appropriate. For nonexchangeable ensembles, separate regressions are fit for each ensemble member to reflect their different error statistics (Raftery et al., 2005), so that the bias correction is accomplished by centering each component distribution at mean

$$\mu_{t,k} = a_k + b_k x_{t,k}, k = 1,\ldots,m, \tag{8.34}$$

where as usual the regression coefficients minimize the average squared difference over the training period between the observed values y_t and the conditional regression specifications $\mu_{t,k}$. When the ensemble members are exchangeable, then these correction equations should be the same for each member, which can be accomplished by constraining the regression parameters in Equation 8.34 to be equal for all ensemble members (Wilson et al., 2007), or with a regression involving the ensemble mean as the sole predictor (Hamill, 2007; Williams et al., 2014),

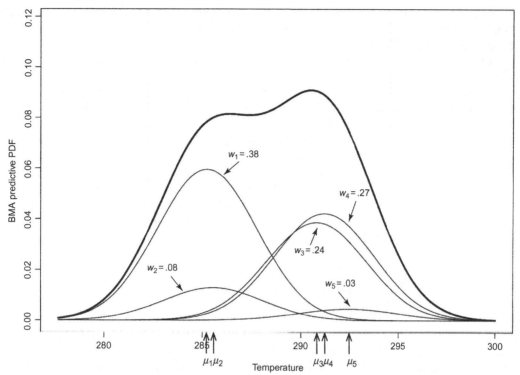

FIGURE 8.10 Illustration of a BMA predictive distribution (heavy curve) constructed from a bias-corrected ensemble of size $m = 5$. The magnitudes of the bias-corrected ensemble members are indicated by the arrows on the horizontal axis, and the weights for the five Gaussian components (light curves) correspond to their areas. The BMA predictive distribution is the weighted sum. *Modified from Raftery et al. (2005). © American Meteorological Society. Used with permission.*

$$\mu_{t,k} = (x_{t,k} - \bar{x}_t) + (a + b\bar{x}_t). \tag{8.35}$$

Very often Gaussian distributions are adopted for the component probability densities $f_k(y_t)$ in Equation 8.33, as illustrated in Figure 8.10. The corresponding component standard deviations σ_k and weights w_k are then usually estimated by maximizing the log-likelihood function

$$L_{\mathrm{BMA}} = \sum_{t=1}^{n} \sum_{k=1}^{m} \left[\ln(w_k) - \frac{(y_t - \mu_{t,k})^2}{2\sigma_k^2} - \ln(\sigma_k) \right] \tag{8.36}$$

over the n available training samples. In this context maximum likelihood estimation is preferably computed using the Expectation-Maximization (EM) algorithm (Section 4.6.3), which is particularly convenient for fitting parameters of mixture distributions such as Equation 8.33 (e.g., Example 4.14). The Gaussian BMA predictive probabilities are then computed as

$$\mathrm{Pr}\{y_t \leq q\} = \sum_{k=1}^{m} w_k \Phi\left(\frac{q - \mu_{t,k}}{\sigma_k} \right). \tag{8.37}$$

This equation reflects the fact that case-to-case differences in the spread of BMA predictive distributions derive from the dispersion of the corrected ensemble members, $\mu_{t,k}$, since the standard deviations σ_k of the component ("dressing") distributions are fixed.

When all ensemble members are exchangeable, then the standard deviations σ_k and weights w_k would be constrained to be equal (Wilks, 2006b; Wilson et al., 2007). Similarly if there are groups of exchangeable members within the ensemble (e.g., exchangeable members from each of several dynamical models), then these parameters (and also the debiasing means in Equation 8.34) would be equal within each group (Fraley et al., 2010).

As was also the case for the nonhomogeneous regression methods (Section 8.3.2), basing BMA post-processing on Gaussian component distributions may be inappropriate for predictands having skewed distributions, and/or those that can take on only nonnegative values. Duan et al. (2007) approach the first of these problems simply by forecasting Box-Cox (i.e., "power") transformed predictands (Equation 3.20).

The problem of nonnegative predictands, which is relevant especially for wind speed and precip-itation amount forecasting, is somewhat more difficult. Sloughter et al. (2010) describe BMA forecasts for wind speed using gamma distributions (Section 4.4.5) for the component probability densities in Equation 8.33. They link the ensemble statistics to the parameters of these component gamma den-sities using Equation 8.34 for the means, and

$$\sigma_{t,k} = c + d x_{t,k},$$
(8.39)

for the standard deviation, where $\mu_{t,k} = \alpha_{t,k}\beta_{t,k}$ and $\sigma_{t,k} = \beta_{t,k}\sqrt{\alpha_{t,k}}$ relate these regressions to the two gamma distribution parameters $\alpha_{t,k}$ and $\beta_{t,k}$ in Equation 4.45.

Baran (2014) proposed using truncated Gaussian component distributions in BMA for forecasting wind speeds, analogously to the nonhomogeneous regression approach of Thorarinsdottir and Gneiting (2010). The component probability density functions to be weighted in Equation 8.33 then have the same form as Equation 4.39, with the location parameters $\mu_{t,k}$ defined using Equation 8.34 and the scale parameter σ assumed to be the same for each ensemble member. Forecast probabilities are then computed using

$$\Pr\{y_t \leq q\} = \sum_{k=1}^{m} w_k \left[\Phi\left(\frac{q - \mu_{t,k}}{\sigma}\right) - \Phi\left(\frac{-\mu_{t,k}}{\sigma}\right) \Big/ \Phi\left(\frac{\mu_{t,k}}{\sigma}\right) \right], q > 0.$$
(8.40)

A forecast precipitation distribution is usually more difficult to model, as it often consists of a discrete probability for exactly zero, and a continuous probability density for the nonzero amounts. Sloughter et al. (2007) describe BMA for such precipitation distributions, specifying the probability of exactly zero precipitation with the logistic regression

$$p_{t,k} = \frac{\exp\left[a_{0,k} + a_{1,k}x_{t,k}^{1/3} + a_{2,k}I(x_{t,k} = 0)\right]}{1 + \exp\left[a_{0,k} + a_{1,k}x_{t,k}^{1/3} + a_{2,k}I(x_{t,k} = 0)\right]},$$
(8.41)

and a gamma distribution for transformed nonzero amounts, yielding the mixed discrete-continuous component distributions

$$f_k(y_t) = p_{t,k}I(y_t = 0) + (1 - p_{t,k})I(y_t > 0)\frac{(y_t/\beta_{t,k})^{\alpha_{t,k}-1}\exp(-y_t/\beta_{t,k})}{\beta_{t,k}\Gamma(\alpha_{t,k})}.$$
(8.42)

Schmeits and Kok (2010) modified this approach slightly, specifying equal probabilities of zero precip-itation for each ensemble member's component dressing distribution using the logistic regression

$$p_{t,k} = \frac{\exp\left[a_{0,k} + a_1 \sum_{k=1}^{m} x_{t,k}^{1/3}\right]}{1 + \exp\left[a_{0,k} + a_1 \sum_{k=1}^{m} x_{t,k}^{1/3}\right]}, \tag{8.43}$$

rather than Equation 8.41. Equations 8.41, and 3.43 indicate that Sloughter et al. (2007) and Schmeits and Kok (2010) found best results when working with cube-root transformed precipitation.

Figure 8.11 illustrates the construction of a BMA predictive distribution for precipitation using these mixed discrete and continuous component distributions. The heavy vertical bar at zero indicates that the weighted sum of the probabilities specified by $w_k\, p_{t,k}$ is approximately 0.37. The weighted component gamma distributions are shown as the light curves, each having area equal to $w_k\,(1 - p_{t,k})$, and their sum is the nonzero part of the predictive distribution indicated by the heavy curve. Note that this mixed discrete-continuous distribution form is different from both the truncated and censored distributions exemplified by Figs. 4.3 and 4.7, respectively, in that no part of the continuous distribution is defined for $y_{t,k} \leq 0$, but is similar in spirit to the regression mixture distribution (Equation 8.28) proposed by Bentzien and Friederichs (2012). Each $f_k(y_t)$ in Equation 8.42 will integrate to unit probability because the second term contains the scaling factor $(1 - p_{t,k})$. Probability calculations for this model are therefore

$$\Pr\{y_t \leq q\} = \sum_{k=1}^{m} w_k \left[p_{t,k} + (1 - p_{t,k})F_{\gamma\left(\alpha_{t,k}\right)}\left(q/\beta_{t,k}\right)\right], \; y_t \geq 0, \tag{8.44}$$

since $F_{\gamma(\alpha)}(y_t) = 0$ for $y_t = 0$, where $F_{\gamma(\alpha)}(y_t)$ is the CDF for the standard gamma distribution.

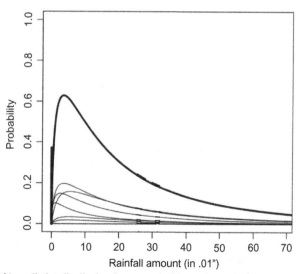

FIGURE 8.11   Example BMA predictive distribution (heavy vertical bar and curve) composed of component distributions that are mixtures of a discrete component at zero and gamma distributions for nonzero precipitation amounts (light curves). *Modified from Sloughter et al. (2007). © American Meteorological Society. Used with permission.*

Other Ensemble Dressing Methods

The BMA and other ensemble dressing approaches are closely allied to kernel density smoothing (Section 3.3.6), in which a probability distribution is estimated from a finite set of data by centering a characteristic shape (the kernel) at each data point (here, each corrected ensemble member), and summing or averaging these. Ensemble dressing as a statistical postprocessing idea was introduced by Roulston and Smith (2003), as a nonparametric method that will be described in Section 8.3.6. They proposed that the dressing kernels should represent the distribution of forecast errors around each ensemble member assuming that member provided the "best" of the individual members' forecasts on that occasion, in order that the dressing kernels should represent only uncertainty not already reflected by the ensemble dispersion.

Wang and Bishop (2005) extended the best-member dressing idea of Roulston and Smith (2003) as a parametric method by proposing use of continuous Gaussian kernels centered on each corrected ensemble member, all of which have the same variance

$$\sigma_D^2 = s_{\mu_t - y_t}^2 - \left(1 + \frac{1}{m}\right)\bar{s}_t^2 , \tag{8.45}$$

where the first term on the right-hand side indicates the error variance for the corrected ensemble-mean forecasts, and the second term approaches the average ensemble variance over the training period as the ensemble size m increases. Accordingly the second-moment constraint in Equation 8.45 can be viewed as reflecting a partition of the total error variance $s_{\mu_t - y_t}^2$ for the corrected ensemble-mean forecasts into uncertainty due to the average ensemble spread, $\bar{s}_t^2$, plus uncertainty σ_D^2 around each ensemble member. Probability forecasts are then computed using

$$\Pr\{y_t \leq q\} = \frac{1}{m}\sum_{k=1}^{m}\Phi\left(\frac{q - \mu_{t,k}}{\sigma_D}\right), \tag{8.46}$$

where $\mu_{t,k}$ denotes the kth corrected ensemble member (Equation 8.34). Equation 8.46 amounts to a simplification relative to the calculation for BMA predictive probabilities (Equation 8.37), with all weights $w_k = 1/m$, and the common dressing variance estimated using Equation 8.45 rather than Equation 8.36. This Gaussian ensemble dressing approach is effective for the usual case of underdispersed ensembles, but it cannot reduce the predictive variances of overdispersed ensembles, and will fail by specifying negative dressing variance if the ensembles are sufficiently overdispersed. If the difference of the two terms in Equation 8.45 is positive but small, the mixture distribution will be noisy and unrealistic (Bishop and Shanley, 2008).

Fortin et al. (2006) note that different predictive mixture-distribution weights for each ensemble member should be appropriate in the best-member setting, depending on each member's rank within the ensemble, even if the raw ensemble members are exchangeable. The basic idea is that more extreme ensemble members are more likely to be the best member when the dynamical forecast model is underdispersive, whereas the more central ensemble members are more likely to be best when the raw ensemble is overdispersive. Fortin et al. (2006) model the mixture probabilities using beta distributions and show that the resulting postprocessed forecasts can correct both over- and underdispersion of the raw ensemble. Furthermore, for overdispersed ensembles the dressing kernels may be centered between their corrected ensemble members and the ensemble mean, at least for the more extreme ensemble members. Their method also allows different component distributional forms to be associated with the corrected ensemble members depending on their ranks.

Bröcker and Smith (2008) derived an ensemble dressing extension that they call affine kernel dressing (AKD). They proposed centering component Gaussian dressing distributions at the corrected values

$$\mu_{t,k} = a + b_1 x_{t,k} + b_2 \bar{x}_t, \tag{8.47}$$

and setting the variances for these dressing distributions as

$$\sigma_t^2 = c + b_1^2 d s_t^2, \tag{8.48}$$

where the parameters a, b_1, b_2, c, and d must be estimated from the training data. Equation 8.48 reduces to Equation 8.8, for $b_1 = 1$. The linkage of Equations 8.47 and 8.48 through the parameter b_1 allows AKD to correct both over- and underdispersion in the raw ensembles because the variance of the resulting predictive mixture distribution is

$$\sigma_{y_t}^2 = c + (1 + d) b_1^2 s_t^2, \tag{8.49}$$

which can be smaller than the ensemble variance.

Bröcker and Smith (2008) also propose adding an additional "ensemble member," consisting of the climatological distribution for y, to the dressing procedure in order to make it more robust to the occasional particularly bad forecast ensemble. Including this climatological Gaussian distribution, with mean μ_C and standard deviation σ_C, AKD predictive probabilities are computed using

$$\Pr\{y_t \leq q\} = \frac{1 - w_c}{m} \sum_{k=1}^{m} \Phi\left(\frac{q - \mu_{t,k}}{\sigma_t}\right) + w_c \Phi\left(\frac{q - \mu_c}{\sigma_c}\right). \tag{8.50}$$

Thus each of the actual ensemble members is given equal weight and equal dressing variance, although this variance changes from forecast to forecast depending on the raw ensemble variance (Equation 8.48). Bröcker and Smith (2008) found values for the weighting parameter for the climatological distribution w_C ranging from approximately 0.02 to 0.06 in their example, with larger values chosen for longer lead times.

Unger et al. (2009) calculate regression-based ensemble-dressing probabilities for ensembles with exchangeable members using Equation 8.46, but compute corrections to the individual ensemble members using the same correction parameters for each member,

$$\mu_{t,k} = a + b_1 x_{t,k}, \tag{8.51}$$

where the regression parameters a and b are fit through regression between the ensemble mean and the observations. Using results from regression, they set the standard deviation for the component Gaussian distributions to be

$$\sigma_D = \left[\frac{n}{n-2} s_y^2 \left(1 - \frac{r_M^2}{r_x}\right)\right]^{1/2} \tag{8.52}$$

where s_y^2 is the (climatological) sample variance of the predictand, r_M is the correlation between the ensemble means and the predictand, and r_x is the correlation between the individual ensemble members and the predictand, over the training period. Glahn et al. (2009b) and Veenhuis (2013) present a similar approach, using empirically based formulations for σ_D.

8.3.5. Fully Bayesian Ensemble Postprocessing Approaches

Although BMA (Section 8.3.4) has a grounding in Bayesian ideas, it is not a fully Bayesian procedure (Di Narzo and Cocchi, 2010). Truly Bayesian analyses are based on the relationship in Equation 6.1. In that equation, θ denotes the target of inference (in the present context the quantity being forecast, y), and x represents the available relevant training data.

Most implementations of Bayesian forecast postprocessing have assumed Gaussian distributions for the prior and likelihood distributions. This assumption is convenient because, if the variance associated with the Gaussian likelihood can be specified externally to the Bayesian analysis as a single value, the posterior distribution is Gaussian also and its parameters can be specified analytically (Section 6.3.4). Krzysztofowicz (1983) was apparently the first to use this framework for postprocessing weather forecasts, using Gaussian distributions to forecast a temperature variable y by postprocessing a single (i.e., nonprobabilistic) dynamical forecast x, although x in Equation 6.1 can as easily be regarded as the ensemble mean in the context of ensemble forecasting. Krzysztofowicz and Evans (2008) extended this postprocessing framework to a much broader range of possible distribution forms, by transformation to Gaussian distributions.

When the variable y to be postprocessed is an observed weather quantity, a natural choice for the prior distribution $f(y)$ is its climatological distribution. Long climatological records for y are generally available, and one strength of the Bayesian approach in the context of forecast postprocessing is that these long records can be brought naturally into the analysis even if the training data relating x and y are much more limited. In this context the likelihood encodes probabilistic information about past forecast errors within the training sample, characterizing the degree to which a forecast x reduces uncertainty about the predictand y. Accordingly, Equation 6.1 expresses a modification or updating of the prior information $f(y)$ in light of the past observed performance of the forecasts. However, the prior information need not necessarily be provided by the climatological distribution. For example, Coelho et al. (2004) use the forecast distribution from a statistical model as the prior, together with a likelihood encoding performance of a dynamical model, to combine the two predictive information sources through Bayes' Theorem.

Because the likelihood denotes conditional distributions for the (often, ensemble mean) forecast, given particular values of the observed variable y, it characterizes the discrimination (Section 9.1.3) of the forecasts in the training data set. Coelho et al. (2004) and Luo et al. (2007) estimate Gaussian likelihoods using linear regressions where the predictand is the ensemble mean, and the predictor is the observation y, so that the mean function for the Gaussian likelihood is $\mu_L = a + by$, and the variance σ_L^2 is the regression prediction variance. Figure 8.12, from Coelho et al. (2004), illustrates the procedure for an example where the colder forecasts are nearly unbiased but the warmer forecasts exhibit a marked warm bias. Because the regression prediction variance σ_L^2 is specified as a point value external the Bayesian estimation, the resulting predictive distributions are also Gaussian with mean $\mu_{P,t}$ and standard deviation $\sigma_{P,t}$ for the tth forecast so that probabilities are calculated using

$$\Pr\{y_t \leq q\} = \Phi\left(\frac{q - \mu_{P,t}}{\sigma_{P,t}}\right). \tag{8.53}$$

Luo et al. (2008) formulate the Gaussian predictive parameters as

$$\mu_{P,t} = \sigma_P^2 \left[\frac{\mu_C}{\sigma_C^2} + \frac{b\,(\bar{x}_t - a)}{\sigma_L^2 + \bar{\sigma}_e^2}\right] \tag{8.54a}$$

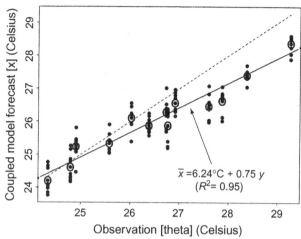

FIGURE 8.12   Individual ensemble members (small dots) and ensemble means (circles) as functions of observed December Niño 3.4 temperatures. Solid line is the weighted regression for the ensemble means defining the Bayesian likelihood. The dashed 1:1 line would correspond to perfect forecasts. *Modified from Coelho et al. (2004). © American Meteorological Society. Used with permission.*

and

$$\sigma_P = \left[\frac{\left(\upsilon_L^2 + \overline{\upsilon}_e^2 \right) \upsilon_C^2}{\sigma_L^2 + \overline{\sigma}_e^2 + b^2 \sigma_C^2} \right]^{1/2}, \tag{8.54b}$$

where μ_C and σ_C^2 are the (prior) climatological mean and variance, and $\overline{\sigma}_e^2$ is the average ensemble variance.

In a method they call Forecast Assimilation, Stephenson et al. (2005) extend Gaussian-based Bayesian calibration using a multivariate normal likelihood distribution for individual nonexchangeable ensemble members, rather than the ensemble mean. Reggiani et al. (2009) implement Gaussian-based Bayesian recalibration (after transformation of the positively skewed hydrological variables) individually for each ensemble member, and then construct the predictive distribution as the average of the m individual predictive density functions, analogously to Equation 8.33 with $w_k = 1/m$. Hodyss et al. (2016) accommodate nonexchangeability of individual ensemble members by representing the likelihood as the product of conditional distributions for each member, but at the cost of requiring the estimation of a large number of regression parameters to define the likelihood. Siegert et al. (2016) describe a more elaborate Bayesian framework for ensemble-mean recalibration which does not have an analytic result for the posterior distribution, and so requires resampling from the final predictive distribution.

Friederichs and Thorarinsdottir (2012) describe a Bayesian approach to postprocessing peak wind speed forecasts, using the GEV (Equation 4.63) as the form of the likelihood distribution, where the location (ζ_t) and scale (β_t) parameters are specified as linear functions of covariates, while the shape parameter κ is assumed to be the same for all forecasts. They use noninformative distributions (independent Gaussian distributions with very large variances) for the prior and a computationally intensive parameter estimation procedure.

A different approach to Bayesian ensemble calibration was proposed by Satterfield and Bishop (2014). Their target of inference is the forecast error variance as predicted by the current ensemble variance, so that the procedure specifically seeks to represent the spread-skill relationship of the ensemble forecasts.

The Bayesian predictions summarized so far in this section are analogous to the regression methods reviewed in Section 8.3.2, in that the output is a single predictive distribution that in most cases is of a known parametric form. In contrast, Bishop and Shanley (2008), DiNarzo and Cocchi (2010), and Marty et al. (2015) formulate full Bayesian analyses for the ensemble-dressing setting, which is an alternative approach to the methods described in Section 8.3.4. Bishop and Shanley (2008) propose using the BMA mixture-distribution formulation as the Bayesian likelihood rather than the predictive distribution. DiNarzo and Cocchi (2010) employ a hierarchical Bayesian model in which a latent variable represents the "best member." Marty et al. (2015) combine the Krzysztofowicz and Evans (2008) Bayesian approach with conventional BMA. These approaches incorporate ensemble-variance information into the predictive distribution by construction, and so allow features such as bimodality in the raw ensemble to carry through to the postprocessed predictive distribution.

8.3.6. Nonparametric Ensemble Postprocessing Methods

Nonparametric ensemble postprocessing methods are wholly or mostly data based, in contrast to the methods described in previous sections which rely on prespecified mathematical forms. Hamill and Colucci (1997) proposed the earliest of these nonparametric methods, which estimates postprocessed ensemble probabilities on the basis of the verification rank histogram (Section 9.7.1). The verification rank histogram tabulates relative frequencies p_j in the training data that the observed value y_t is larger than j of its forecast ensemble members $x_{t,j}$, plus 1. For example, p_1 is the proportion of training-sample forecasts for which the observation was smaller than all ensemble members, and p_{m+1} is the proportion of forecasts where the observation is larger than all m ensemble members.

The Hamill and Colucci (1997) recalibration procedure operates on the rank histogram for the unconditionally debiased ensemble members

$$\widetilde{x}_{t,k} = x_{t,k} - \sum_{t=1}^{n} (\overline{x}_t - y_t), \tag{8.55}$$

and aims to achieve flatness (which is the ideal result) of the recalibrated rank histogram. When a quantile of interest is not outside the range of the ensemble, probabilities are estimated by linear interpolation:

$$\Pr\{y_t \le q\} = \sum_{j=1}^{k} p_j + p_{j+1} \frac{q - \widetilde{x}_{t,(k)}}{\widetilde{x}_{t,(k+1)} \widetilde{x}_{t,(k)}}, \ \widetilde{x}_{t,(k)} \le q \le \widetilde{x}_{t,(k+1)}. \tag{8.56}$$

The parenthetical subscripts denote that the ensemble members have been sorted in ascending order. When a quantile of interest is outside the range of the ensemble, the probabilities represented in p_1 and p_{m+1} must be extrapolated in some way. Because rank histogram recalibration tends to underperform more recently developed methods (e.g., Wilks, 2006b; Ruiz and Saulo, 2012) it is rarely used.

Quantile Regression

Bremnes (2004) introduced the use of quantile regression (Section 7.7.3) for ensemble postprocessing of continuous predictands. The predictand considered by Bremnes (2004) was precipitation amount, which has a mixed discrete (finite probability mass at zero) and continuous (nonzero precipitation amounts) distribution. He separately forecast the probability of exactly zero precipitation using probit regression (which is very similar to logistic regression, Section 7.6.2), forecast selected nonzero precipitation quantiles with quantile regression, and combined the two according to the mathematical definition of conditional probability, yielding

$$\Pr\{y_t \le q_p\} = 1 - \frac{1 - p_{y>0}}{1 - \Pr\{y_t = 0\}}, \tag{8.57}$$

where $p_{y>0}$ is an appropriate cumulative probability derived from the quantile regression describing the nonzero amounts. For example, in order to calculate the median ($q_{.50}$) of the predictive distribution when the probability of zero precipitation is 0.2, the precipitation amount of interest would be the quantile of the predictive distribution for nonzero amounts corresponding to $p_{y>0} = 0.6$. When $p_{y>0} \le \Pr\{y_t = 0\}$ then $q_p = 0$ is implied. The method can of course be used, and indeed is more easily implemented, for strictly positive predictands where $\Pr\{y_t = 0\} = 0$.

For each preselected predictand quantile q_p an appropriate list of ensemble predictors x_t must be selected, as is also the case for ordinary least-squares regression, although Ben Bouallègue (2016) suggests use of lasso-penalization (Section 7.5.2) to choose predictor variables. Bremnes (2004) considered the five predictand quantiles $q_{.05}, q_{.25}, q_{.50}, q_{.75},$ and $q_{.95}$, and for each of the five quantile regressions he used the same $I = 2$ predictors, being the two quartiles of the ensemble. Figure 8.13 illustrates the resulting forecasts as functions of the 75th percentiles of the ensembles, for three levels of the ensemble 25th percentiles. These show greater forecast uncertainty (larger distances between the forecast quantiles) for increasing values of both but especially for $q_{.25}$.

Ensemble Dressing

Although ensemble dressing is usually implemented using parametric kernels (Section 8.3.4), it was originally proposed by Roulston and Smith (2003) as a nonparametric ensemble postprocessing approach. They "dressed" each corrected ensemble member using a random sample from the collection of errors that were defined relative to the ensemble member closest to the observation on each

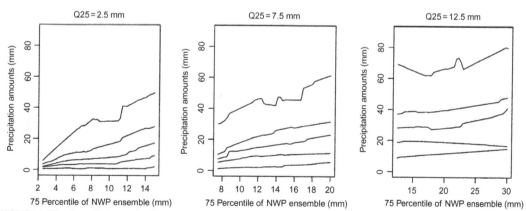

FIGURE 8.13 Quantile regression forecasts for (from top to bottom) the $q_{.95}, q_{.75}, q_{.50}, q_{.25},$ and $q_{.05}$ quantiles of the predictive distribution of nonzero precipitation, for three levels of the ensemble lower quartile, as functions of the ensemble upper quartile. *From Bremnes (2004). © American Meteorological Society. Used with permission.*

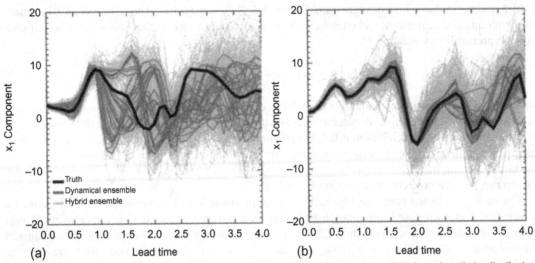

FIGURE 8.14 True evolution (heavy curves), 32 raw ensemble members (medium curves), and the dressed predictive distribution represented by 16 best-member trajectories around each raw ensemble member (light curves), for two forecast cases of the Lorenz '96 (Lorenz, 2006) system. Panel (a) shows a low-predictability case and panel (b) shows a high-predictability case. *From Roulston and Smith (2003).*

occasion (the "best member"), in an archive of past forecasts. Figure 8.14 shows example dressing distributions in the form of time trajectories for forecasts of the Lorenz '96 (Lorenz, 2006) system. Here there are 32 ensemble members (medium gray curves), each of which has been dressed using a sample of 16 best-member error trajectories (light curves), so that the predictive distribution is represented by a collection of 512 trajectories. Figure 8.14a shows a low-predictability case, Figure 8.14b shows a high-predictability case, and in both panels the heavy curves indicate the true trajectories. The same catalog of past best-member errors has been sampled in both examples, but the spread of the predictive distributions for the low-predictability case is larger because the underlying ensemble spread is larger. Messner and Mayr (2011) propose a similar resampling approach, where the discrete dressing kernels are derived from close analogs in the training data to the current ensemble members, following the method of Hamill and Whitaker (2006).

Individual Ensemble-Member Adjustments

Another nonparametric approach to ensemble postprocessing involves transforming each of the ensemble members individually, leading to a corrected ensemble of finite size m. This has been termed the member-by-member postprocessing (MBMP) approach (Van Schaeybroeck and Vannitsem, 2015). MBMP adjustments can be expressed as

$$y_{t,k} = (a + b\bar{x}_t) + \gamma_t(x_{t,k} - \bar{x}_t), \tag{8.58}$$

where each raw ensemble member $x_{t,k}$ maps to a distinct corrected counterpart, $y_{t,k}$. The parameters a and b define unconditional and conditional bias corrections, and the parameter γ_t controls a forecast-dependent expansion or contraction relative to the ensemble mean. Various definitions for these parameters have been proposed.

Eade et al. (2014) estimate the parameters in Equation 8.58 using

$$a = \bar{y} - b\bar{\bar{x}}, \tag{8.59a}$$

$$b = \frac{s_{y_t}}{s_{\bar{x}_t}} r_{y_t \bar{x}_t}, \tag{8.59b}$$

and

$$\gamma_t = \frac{s_{y_t}}{s_t} \sqrt{1 - r^2_{y_t, \bar{x}_t}}, \tag{8.59c}$$

where s_{y_t} is the climatological predictand standard deviation, the correlations in Equation 8.59 relate the predictand and the ensemble means, and

$$\bar{\bar{x}} = \frac{1}{n} \sum_{t=1}^{n} \bar{x}_t \tag{8.60}$$

is the average of the ensemble means over the training data. Equations 8.59a and 8.59b are the ordinary least-squares regression coefficients relating the ensemble mean to the predictand y, and Equation 8.59c varies from forecast to forecast according to the ensemble standard deviation s_t in the denominator. Doblas-Reyes et al. (2005) and Johnson and Bowler (2009) have proposed an equivalent model assuming that both the forecast ensembles and the observations have been centered, so that $a = 0$ in Equation 8.59a.

Van Schaeybroeck and Vannitsem (2015) allow the "stretch" coefficient γ_t to depend on the ensemble spread according to

$$\gamma_t = c + \frac{d}{\delta_t}, \tag{8.61}$$

where

$$\delta_t = \frac{1}{m(m-1)} \sum_{j=1}^{m} \sum_{k=1}^{m} |x_{t,j} - x_{t,k}| \tag{8.62}$$

is the average absolute difference among pairs of the uncorrected ensemble members, and both c and d are constrained to be nonnegative. Thus Equation 8.58 can be viewed as a nonparametric counterpart to the nonhomogeneous regressions described in Sections 8.3.2 (Schefzik, 2017). Van Schaeybroeck and Vannitsem (2015) describe estimation of the parameters a, b, c, and d in two ways. The first is using the method maximum likelihood, assuming the errors of the corrected ensemble-mean forecasts are exponentially distributed, with mean δ_t, and so maximizing

$$L_{\text{exp}} = - \sum_{t=1}^{n} \left[\left| \frac{\bar{y}_t - y_t}{\delta_t} \right| + \ln(\delta_t) \right] \tag{8.63}$$

with respect to a, b, c, and d. Here

$$\bar{y}_t = \frac{1}{m} \sum_{k=1}^{m} y_{t,k} \tag{8.64}$$

is the average of the individually corrected ensemble members for forecast case t. Alternatively, the parameters can be estimated by minimizing the ensemble CRPS (Equation 9.83).

Williams (2016) uses the nonparametric adjustment procedure in Equation 8.58, but defines the "stretch" coefficient as

$$\gamma_t = \frac{\sqrt{d + cs_t^2}}{s_t} = \sqrt{c + d/s_t^2}. \tag{8.65}$$

With this formulation, method-of-moments estimators can be derived for the required parameters, the computation of which will be relatively fast. The two bias-correction parameters a and b are again the least-squares regression parameters defined in Equations 8.59a and 8.59b. The method-of-moments estimates for the "stretch" parameters are

$$c = \frac{\text{cov}(s_t^2, y_t^2) - 2ab\,\text{cov}(s_t^2, \bar{x}_t) - b^2\,\text{cov}(s_t^2, \bar{x}_t)}{\text{Var}(s_t^2)} \tag{8.66a}$$

and

$$d = s_{y_t}^2 - c\bar{s}^2 - b^2 s_{\bar{x}_t}^2, \tag{8.66b}$$

where

$$\bar{s}^2 = \frac{1}{n} \sum_{t=1}^{n} s_t^2 \tag{8.67}$$

is the average ensemble variance over the training period, and

$$s_{\bar{x}_t}^2 = \frac{1}{n-1} \sum_{t=1}^{n} (\bar{x}_t - \bar{\bar{x}})^2 \tag{8.68}$$

is the variance of the ensemble means over the training period. Again both c and d are constrained to be nonnegative. Equations 8.61 and 8.65 were found to perform similarly in Wilks (2018a).

Figure 8.15 illustrates the MBMP adjustment method. Figure 8.15a (solid line) shows the debiasing function defined by the parameters a and b in Equations 8.59a and 8.59b, indicating that ensembles with moderate and large means were positively biased in the training data, and that ensembles having small ensemble means were negatively biased. Figure 8.15b shows a hypothetical $m = 5$ member ensemble (filled circles) to be corrected. The dotted arrows originating at the raw ensemble mean in Figure 8.15b locate the corrected ensemble mean in Figure 8.15c. The corrected ensemble members (open circles) in Figure 8.15c retain their distributional shape but have been expanded away from the corrected ensemble mean, by the factor $\gamma_t = 1.5$ relative to the raw ensemble in Figure 8.15b, so that the indicated corrected ensemble range in Figure 815c is larger than the ensemble range of the example underdispersive raw ensemble in Figure 8.15b by the factor 1.5.

Because the identities of the individual ensemble members are preserved in these adjustments, different forecast variables that have been subjected to independent postprocessing will continue to exhibit the rank correlation structures inherited from the raw ensemble. Thus correlations in the raw ensemble among different variables at a single location, and spatiotemporal correlations for a given variable, will all be carried forward to the postprocessed ensemble. Therefore MBMP can also serve as the basis for a multivariate postprocessing algorithm, as described in Section 8.4.3. On the other hand, postprocessing predictands such as wind speed or precipitation may be problematic without constraints requiring all

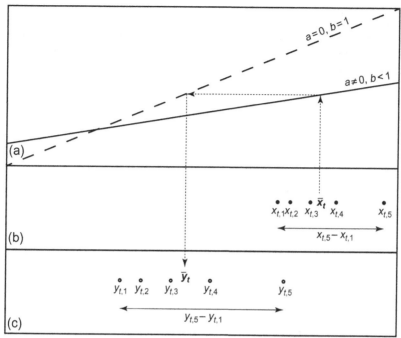

FIGURE 8.15 Illustration of the MBMP adjustment method, for (b) a hypothetical $m = 5$ member underdispersive and positively biased raw ensemble (filled circles), which has been transformed to (c) the corrected ensemble (open circles) using the debiasing function (solid line) in (a) and the "stretch" coefficient $\gamma_t = 1.5$. *From Wilks (2018a).*

postprocessed ensemble members to be nonnegative; because the smaller members of an underdispersed ensemble may be transformed to negative values, and Equations 8.61 or 8.65 will be undefined for precipitation ensembles in which all members are zero.

8.4. MULTIVARIATE ENSEMBLE POSTPROCESSING

The ensemble postprocessing methods outlined in Section 8.3 pertained to scalar predictands at single lead times. These univariate methods are adequate for many purposes, but in some settings forecasts of multiple predictands exhibiting realistic correlations structures are essential. Depending on the application, such multivariate forecasts might involve two or more different but correlated predictand variables, forecasts for a given variable at many spatially correlated locations simultaneously, forecasts for a given variable at multiple lead times into the future, or various combinations of these possibilities. For example, correct portrayal of the joint spatial and temporal correlation structure of forecast precipitation is essential for realistic streamflow and flood forecasting (e.g., Clark et al., 2004; Li et al., 2017), and realistic portrayal of the space-time correlation structure of wind speed forecasts is necessary in such applications as optimal wind power decision making (e.g. Pinson, 2013) and aircraft routing (Chaloulos and Lygeros, 2007). This section describes ensemble postprocessing approaches designed to capture these correlations in multivariate predictive distributions, which can be broadly classified as parametric methods, copula methods, and analog methods.

8.4.1. Parametric Methods

Schuhen et al. (2012) proposed extension of NGR (Section 8.3.2) to postprocessing of ensemble forecasts for northward (u) and eastward (v) wind components, jointly using bivariate normal (Equation 4.31) predictive distributions. Extending Equation 8.11, bivariate mean vectors are defined as linear functions of the respective ensemble means,

$$\mu_u = a_u + b_u \bar{u} \tag{8.69a}$$

and

$$\mu_v = a_v + b_v \bar{v}, \tag{8.69b}$$

and extending Equation 8.8 the respective postprocessed variances are defined as linear functions of the respective ensemble variances,

$$\sigma_u^2 = c_u + d_u s_u^2 \tag{8.70a}$$

and

$$\sigma_v^2 = c_v + d_v s_v^2. \tag{8.70b}$$

Schuhen et al. (2012) defined the correlation parameter as a function of the ensemble-mean angular wind direction θ, although not the correlation within the ensemble, using

$$\rho_{uv} = r \cos\left[\frac{2\pi}{360°}(k\theta + \varphi) \right] + s, \tag{8.71}$$

where r, s, φ, and k are parameters to be estimated. Thus even with the simplification of Equation 8.71, the approach requires estimation of 12 parameters in Equations 8.69–8.71. Figure 8.16a shows an example result from Equation 8.71 (dashed curve) for the Sea-Tac (Seattle) airport together with the training data. Figure 8.16b shows an example forecast, with the dark contours representing the postprocessed bivariate normal distribution, in relation to the original 8-member ensemble (dots) and the 90% probability ellipse of the bivariate normal distribution fit to the raw ensemble (gray curve). Bias, mainly in the u wind speed, and overall underdispersion of the raw ensemble have been corrected.

Baran and Möller (2017) extended the idea of bivariate NGR to jointly postprocess bivariate forecasts of (scalar) wind speed (x_w) and temperature (x_T), with the marginal distribution of wind speed represented as zero-truncated Gaussian (Equation 4.39). Accordingly, Equation 4.31 for the bivariate normal PDF is modified to read

$$f(x_w, x_T) = \frac{I(x_w \geq 0)}{\Phi\left(\frac{\mu_w}{\sigma_w}\right) 2\pi \sigma_w \sigma_T \sqrt{1 - \rho^2}}$$
$$\times \exp\left\{ -\frac{1}{2(1-\rho^2)}\left[\left(\frac{x_w - \mu_w}{\sigma_w}\right)^2 + \left(\frac{x_T - \mu_T}{\sigma_T}\right)^2 - 2\rho\left(\frac{x_w - \mu_w}{\sigma_w}\right)\left(\frac{x_T - \mu_T}{\sigma_T}\right) \right] \right\} \tag{8.72}$$

In this case, eight parameters were required to jointly correct the two means, extending Equation 8.11, and another eight parameters were required to specify the standard deviations, for a total parameter count of 16.

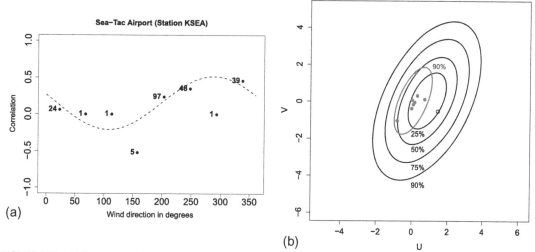

FIGURE 8.16 (a) Example correlation function (Equation 8.71) fit to bivariate normal wind component correlations as a function of ensemble-mean wind direction, with training data and sample sizes. (b) Example bivariate normal predictive distribution (dark contours), based on an 8-member ensemble (light dots). Gray ellipse is the 90% probability ellipse fit to the raw ensemble members, and open circle is the eventual verifying observation. *From Schuhen et al. (2012). © American Meteorological Society. Used with permission.*

Bivariate Gaussian distributions can also be used to extend the BMA and allied approaches (Section 8.3.4), in which case the PDFs f_k in Equation 8.33 are bivariate normal. As was also the case for univariate BMA, the means must first be debiased before being used to center the bivariate normal kernels that are used to build up the predictive distribution. Figure 8.17 shows an example of the result, where the circles locate the eight uncorrected ensemble members, the plusses locate their bias-corrected counterparts, and the triangle shows the corresponding observed wind vector. Because the extension of the debiasing Equation 8.34 to the bivariate setting required estimation of six parameters for each of the eight nonexchangeable ensemble members, the approach is highly parameterized. Schölzel and Hense (2011) employ this approach in a climate change setting. Baran and Möller (2015) extended the bivariate BMA idea to joint forecasts of temperature and scalar wind speed using the bivariate Gaussian distributions with zero-truncated wind speed component (Equation 8.72) as the dressing kernels. Berrocal et al. (2007) described a higher dimensional multivariate BMA approach, in which temperature forecasts at multiple locations are constructed using multivariate normal (Chapter 13) dressing kernels. The number of required parameters was kept to a manageable level by assuming particular mathematical forms for the spatial covariances (Gel et al., 2004). Hemri et al. (2013) represented (power-transformed) river runoff forecasts for a particular gage at multiple lead times using an equivalent approach.

8.4.2. Copula Methods

That large numbers of parameters need to be estimated for parametric multivariate postprocessing models, even in the low-dimensional settings described in Section 8.4.1, suggests that nonparametric methods may often be preferred for specifying the statistical dependence among the multiple predictands. *Empirical copulas* provide a flexible and attractive means to do so. For multivariate ensemble postprocessing, copulas allow each of multiple forecast variables to be postprocessed separately, using

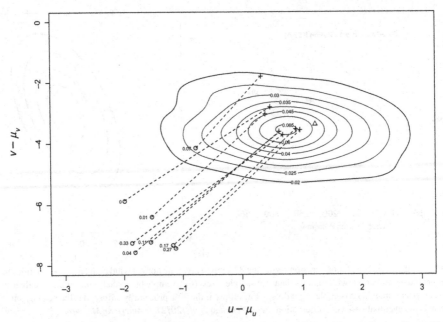

FIGURE 8.17 Example bivariate BMA forecast distribution for u and v wind components. The raw (nonexchangeable) ensemble members are shown as circles with associated weights, and the dashed lines connect to bias-corrected ensemble members (plusses). Triangle shows the eventual observed wind vector. *From Sloughter et al. (2013). © American Meteorological Society. Used with permission.*

a method such as NGR (Section 8.3.2) or BMA (Section 8.3.4), with the results assembled subsequently into a joint distribution with appropriate correlation structure.

Copulas are structures that connect multivariate distribution functions to their individual marginal distributions. In effect, a copula provides a "dependence template" onto which samples drawn independently from a collection of univariate marginal distributions can be arranged, in a way that will reflect a specified dependence structure. Empirical copulas (e.g., Bárdossy and Pegram, 2009; Rüschendorf, 2009) derive their structure from independent rank transformations of a sample of training data in each of the dimensions of the multivariate data space. Schefzik et al. (2013) proposed using uncorrected ensembles to provide the structure of the empirical copulas, calling the method *empirical copula coupling* (ECC), which allows the nature of the relationships among the prognostic variables to change from forecast to forecast. The underlying assumption is that, even though the raw ensemble may exhibit bias and dispersion errors, the dynamical model (at least approximately) correctly represents the statistical relationships among the forecast variables.

An empirical copula can be constructed for any data dimension K, but it is easiest to visualize and understand for bivariate ($K = 2$) data. Consider a hypothetical $m = 5$ member raw ensemble of bivariate data, x, shown in the matrix (or, equivalently the data table) in Figure 8.18a. The process of constructing the copula template begins by separately transforming each of the two variables to their ranks within the ensemble,

$$r_{i,k} = \sum_{j=1}^{m} I\left(x_{j,k} \leq x_{i,k}\right), \quad i = 1, \ldots, m; \; k = 1, \ldots, K, \tag{8.73}$$

where again $I(\bullet)$ denotes the indicator function. These ranks then populate the matrix of $[R]$ of ranks in Figure 8.18b. For example, the first ($i = 1$) ensemble member for the first ($k = 1$) variable, $x_{1,1}$ is the

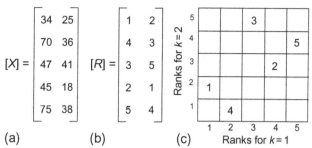

FIGURE 8.18   (a) Hypothetical bivariate ensemble [X] of size $m = 2$ that is transformed to ranks in (b) according to Equation 8.73. (c) Latin square representation of the hypothetical matrix of ranks in (b). Horizontal axis pertains to the first column and vertical axis pertains to the second column of [R]. The integers in the interior of the square are the row indices. *Modified from Wilks (2015).*

smallest of the five x's, and so transforms to $r_{1,1} = 1$. Similarly the ($i = 5$) fifth ensemble member for the second ($k = 2$) variable, $x_{5,2}$ is the fourth-smallest among the ensemble for the second variable, and so transforms to $r_{5,2} = 4$. Schefzik et al. (2013) motivate the dependence template of ranks [R] using a latin hypercube of dimension K, which for $K = 2$ is a Latin square, having one entry only in each row and column (Figure 8.18c). Another way to look at Figure 8.18c is as a scatterplot of the rank-transformed bivariate data. The arrangement of the five row index labels (the 'data points') within the Latin square suggests a positive correlation, and indeed the Spearman rank correlation (Equation 3.34) between the two columns of Figure 8.18a is 0.6.

Having constructed the empirical copula [R], it can be used to arrange independent samples from the K independently postprocessed univariate marginal distributions, each of size m, in a way that preserves the intervariable relationships encoded in the ranks of [R]. Denote these samples, which might be random draws from an NGR (Equation 8.6) or BMA (Equation 8.33) predictive distribution, or derived through some other postprocessing method (e.g., Flowerdew, 2014; Roulin and Vannitsem, 2012), as $y_{i,k}$, $i = 1, ..., m$; $k = 1, ..., K$. Adopting the conventional notation of parenthetical subscripts to indicate sorted data (so that, for example, $y_{(1),k}$ and $y_{(m),k}$, are the smallest and largest postprocessed samples, respectively, for the kth variable), the arrangement of these ranked values duplicating the multivariate relationships encoded in R is achieved by ordering the sorted data $y_{(i),k}$ according to the ranks $r_{i,k}$,

$$\tilde{y}_{i,k} = y_{(r_{i,k}),k}, \quad i = 1,...,m, \; k = 1,...,K, \tag{8.74}$$

which organizes the sorted samples $y_{(i),k}$ onto the empirical copula R in Figure 8.18b and c. The resulting matrix of postprocessed and correlated ensemble members for the example in Figure 8.18 is then

$$\left[\tilde{Y}\right] = \begin{bmatrix} y_{(r_{1,1}),1} & y_{(r_{1,2}),2} \\ y_{(r_{2,1}),1} & y_{(r_{2,2}),2} \\ y_{(r_{3,1}),1} & y_{(r_{3,2}),2} \\ y_{(r_{4,1}),1} & y_{(r_{4,2}),2} \\ y_{(r_{5,1}),1} & y_{(r_{5,2}),2} \end{bmatrix} = \begin{bmatrix} y_{(1),1} & y_{(2),2} \\ y_{(4),1} & y_{(3),2} \\ y_{(3),1} & y_{(5),2} \\ y_{(2),1} & y_{(1),2} \\ y_{(5),1} & y_{(4),2} \end{bmatrix}. \tag{8.75}$$

The rank correlation between the two columns in Equation 8.75 is also 0.6 because the reordered postprocessed data have exactly the same rank structure as those in Figure 8.18c. In higher dimensional settings the matrices in Figure 8.18 and Equation 8.75 are extended to include K columns, and all $K(K-1)/2$ pairwise correlations among the K columns of [R] are reproduced for the corresponding columns of $\left[\tilde{Y}\right]$.

Schefzik et al. (2013) proposed several ECC variants, which differ in the way the postprocessed variables are chosen before they are reordered by the copula. The first of these is based on equidistant quantiles (ECC-Q) from the individual univariate predictive distributions, F_k. That is, each of the univariate postprocessed predictive distributions is sampled systematically by evaluating the respective quantile functions at the points specified by

$$y_{i,k} = F_k^{-1}\left(\frac{i}{m+1}\right), \quad i = 1, \ldots, m, \quad k = 1, \ldots, K, \qquad (8.76)$$

the argument of which is the Weibull plotting position estimator (Table 3.2).

Because the sampling of each postprocessed distribution F_k in Equation 8.76 is systematic and not random, the ECC-Q method is limited to producing postprocessed ensembles of the same size m as the original ensemble. The second empirical copula coupler is based on independent random (ECC-R) samples from the K univariate postprocessed predictive distributions,

$$y_{i,k} = F_k^{-1}(u_{i,k}), \quad i = 1, \ldots, m, \quad k = 1, \ldots, K, \qquad (8.77)$$

where the $u_{i,k}$ are independent standard uniform random numbers. Because Equation 8.77 samples each univariate postprocessed distribution F_k randomly, repeated implementation of the ECC is not redundant, so that Equation 8.77 can be repeatedly reevaluated, reordered according to the copula template, and pooled. Accordingly the size of a multivariate ECC-R postprocessed ensemble can be any integer multiple of the original ensemble size m, and so can represent the forecast distribution more smoothly and is better able to sample its extremes.

Figure 8.19 illustrates the ECC approach, operating on an $m = 50$-member ensemble representing a $K = 4$-dimensional (temperatures and pressures at Berlin and Hamburg) multivariate forecast. Figure 8.19a shows the scatterplot matrix of the raw ensemble, with histograms on the diagonal representing the four marginal distributions. Figure 8.19b shows the corresponding plot after each of the four forecast variables has been independently postprocessed using BMA. The marginal distributions on the diagonal have been transformed, but the pairwise scatters indicate near zero correlation. Figure 8.19c shows the result after the BMA-postprocessed forecasts in Figure 8.19b are reordered using ECC. The marginal distributions in Figure 8.19b and c are identical, and the rank correlation structures in Figures 8.19c reproduce those in Figure 8.19a.

If the underlying dynamical model represents the statistical dependences among the forecast variables poorly, or if the ensemble size is too small to sample them adequately, then ECC methods may not perform well. Alternatively empirical copulas can be constructed using the *Schaake shuffle* (Clark et al., 2004). This method operates by drawing samples from the historical climatological record to populate the $x_{i,k}$ values used to compute the ranks in Equation 8.73, which in turn are used to reorder samples $y_{i,k}$ from the individually postprocessed forecast variables in Equation 8.74. Because this resampling from the climatological record is not constrained by the original ensemble size, the size of the final ensemble (corresponding to the number of rows in the matrices $[X]$ and $[R]$ in Figure 8.18) can be of any size, up to the size of the relevant climatology.

A disadvantage of the Schaake shuffle is that the climatology is sampled independently of the raw ensemble. To the extent that there may be case-to-case differences in the correlation structure among the K forecast variables (e.g., Clark et al., 2004; Demargne et al., 2014; Verkade et al., 2013) copulas derived from random sampling of the climatology will not reflect them. This problem can be addressed by sampling observations from the historical climatology that are most similar to the current forecast

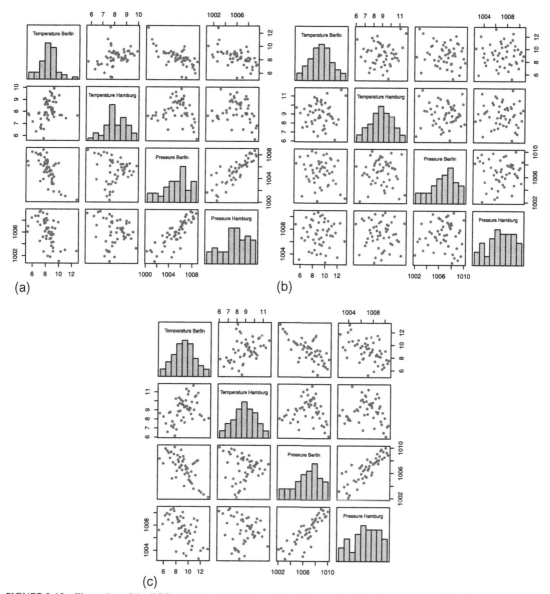

FIGURE 8.19 Illustration of the ECC approach, operating on a $K = 4$ dimensional (Berlin temperature, Hamburg temperature, Berlin pressure, Hamburg pressure) and $m = 50$-member ensemble forecast. (a) Scatterplot matrix of the raw, uncorrected ensemble, (b), scatterplot matrix after independent BMA postprocessing of the four forecast variables, and (c) the postprocessed forecasts after ECC reordering. *From Schefzik et al. (2013).*

(Clark et al., 2004), or sampling dates for which past forecasts are most similar to the current forecast (Schefzik, 2016; Scheuerer et al., 2017; Scheuerer and Hamill, 2018).

Application of empirical copula methods to precipitation fields, where many ensemble members may be exactly zero, may lead to inconsistent representation of the spatiotemporal variability that derives from random assignments to the lowest ranks (Wu et al., 2018).

This section has emphasized empirical copulas, which are computationally fast, and can accommodate very high dimension K. Copulas can also be constructed parametrically, most commonly on the basis of the multivariate Gaussian distribution (Chapter 12). These *Gaussian copulas* require estimation of a K-dimensional correlation matrix using training data, and then sampling from the resulting multivariate Gaussian distribution (Section 12.4.1) to define the values $x_{i,k}$ in Equation 8.73. The utility of this approach may be limited by the larger numbers of correlation parameters that must be estimated for large K, unless a theoretical model for the correlation structure (e.g., Equation 10.23 in a setting where the K forecast dimensions pertain to a time sequence) can be reasonably assumed. Hemri et al. (2015) and Möller et al. (2013) provide examples of the use of Gaussian copulas in ensemble postprocessing.

8.4.3. Member-by-Member Postprocessing

The nonparametric MBMP method presented in Equation 8.58 operates as a univariate postprocessing method that separately transforms individual members of an ensemble of scalar forecasts. However, because the procedure is a monotonic transformation, the resulting transformed ensemble members are ordered in the same way as their counterparts in the original raw ensemble. Accordingly, multivariate collections of MBMP-transformed ensembles inherit the rank correlation (Equation 3.34) structure of the original raw ensemble, so that the independently transformed ensembles can be assembled into a plausible multivariate forecast, in a way that is closely aligned with the ECC approach (Schefzik, 2017).

Figure 8.20 illustrates the result, for bivariate forecasts of 2 m temperatures, and the corresponding coldest temperatures during the previous 6 h. Figure 8.20a shows the result when these two temperature variables have been separately adjusted using Equation 8.58. The correlation between the two variables is realistically strong, and nearly all of the points (black) are above the dashed 1:1 line, with only a few physically implausible points (gray) below, even though the two MBMP adjustments were computed independently. Figure 8.20b shows the corresponding result when the points are drawn randomly from NGR (Section 8.3.2) distributions that were also fit independently to the two sets of temperature forecasts. In this case the correlation is nearly zero, and a large fraction of the points are below the 1:1 line.

Pinson (2012) describes a similar but less general approach, applied to bivariate (u, v) forecast wind vectors.

8.4.4. Analog Methods

A rather different approach to multivariate ensemble postprocessing searches an archive of past forecasts of the same kind that are currently available and chooses one or more of these previous forecasts as analogs for each ensemble member. The collection of observed weather states or events following these analogs is then assembled as a discrete ensemble representing a sample from the forecast PDF.

Roulston et al. (2003) used this approach to obtain probabilistic forecasts of time series of wind speeds having realistic temporal correlation characteristics, to support decision making in wind power management. Hamill et al. (2006, 2015) and Hamill and Whitaker (2006) implemented this idea for areal forecasts of precipitation amounts that appropriately capture the spatial correlations. However, the result is an estimate of a high-dimensional forecast distribution that is difficult to communicate concisely. Figure 8.21 shows one approach to conveying a portion of this information, namely, maps of exceedance probabilities ($\times 100$) for three precipitation amounts. The format is informative but incomplete, since for example it does not indicate or allow computation of such quantities as probability of at least 10 mm at

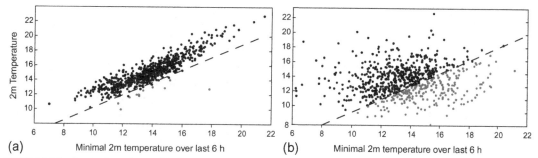

FIGURE 8.20   Scatterplots of ensemble-forecast surface temperatures versus corresponding coldest forecast temperatures in the previous 6 h (a) after MBM postprocessing, and (b) sampled from independent univariate NGR distributions. Values below the dashed 1:1 line, shown in gray, are physically implausible. *Modified from Van Schaeybroeck and Vannitsem (2015).*

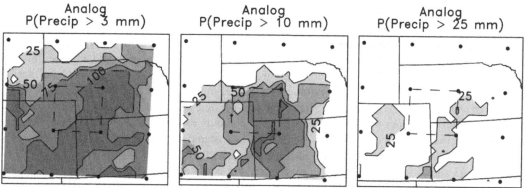

FIGURE 8.21   Exceedance probabilities (×100) for three precipitation amounts, derived from historical observations following forecasts analogous to a current ensemble forecast. *Modified from Hamill and Whitaker (2006). © American Meteorological Society. Used with permission.*

one location given at least 10 mm at another, or the probability that at least 3 mm will occur everywhere in the domain.

8.5. GRAPHICAL DISPLAY OF ENSEMBLE FORECAST INFORMATION

A prominent attribute of ensemble forecast systems is that they generate large amounts of multivariate information. As noted in Section 3.6, the difficulty of gaining even an initial understanding of a new multivariate data set can be reduced through the use of well-designed graphical displays. It was recognized early in the development of what is now ensemble forecasting that graphical display would be an important means of conveying the resulting complex information to forecasters (Epstein and Fleming, 1971; Gleeson, 1967), and operational experience is still accumulating regarding the most effective means of doing so. This section presents a selection of some graphical devices that are in current use for displaying ensemble forecast information.

Perhaps the most direct way to visualize an ensemble of forecasts is to plot them simultaneously. Of course, for even modestly sized ensembles each element (corresponding to one ensemble member) of such a plot must be small in order for all the ensemble members to be viewed simultaneously. Such collections are called *stamp maps*, because each of its individual component maps is sized approximately like a postage stamp, allowing only the broadest features to be discerned. For example, Figure 8.22 shows 51 stamp maps from the ECMWF ensemble prediction system, for surface pressure over western Europe ahead of a large and destructive winter storm that occurred in December 1999. The ensemble consists of 50 members, plus the control forecast begun at the "best" initial atmospheric state, labeled "deterministic predictions." The subsequently analyzed surface pressure field, labeled "verification," indicates a deep, intense surface low centered near Paris. The control forecast missed this important feature completely, as did many of the ensemble members. However, a substantial number of the ensemble members did portray a deep surface low, suggesting a potentially actionable probability for this destructive storm, 42 h in advance. Although fine details of the forecast are difficult if not impossible to discern from the small images in a stamp map, a forecaster with experience in the interpretation of this kind of display can get an overall sense of the outcomes that are plausible, according to this sample of ensemble members. A further step that sometimes is taken with a collection of stamp maps is to group them objectively into subsets of similar maps using a cluster analysis (see Chapter 16).

Part of the difficulty in interpreting a collection of stamp maps is that the many individual displays are difficult to comprehend simultaneously. Superposition of a set of stamp maps would alleviate this difficulty if not for the problem that the resulting plot would be too cluttered to be useful. However, seeing each contour of each map is not necessary to form a general impression of the flow, and indeed seeing only one or two well-chosen pressure or height contours is often sufficient to define the main features, since (especially away from the surface) typically the contours roughly parallel each other. Superposition of one or two well-selected contours from each of the stamp maps often does yield a sufficiently uncluttered composite to be interpretable, which is known as the *spaghetti plot*.

Figure 8.23 shows four spaghetti plots for the 5520-m contour of the 500 mb surface over North America, as forecast (a) 72, (b) 96, (c) 120 and (d) 144 h after the initial time of 0000 UTC, 15 March 2018. In Figure 8.23a the ensemble members generally agree quite well, and even with only the 5520-m contour shown the general nature of the flow is clear. At the 96-h lead time (Figure 8.23b) the ensemble members are still generally in close agreement about the forecast flow, except over the northeast Pacific. The 500 mb field over most of the domain would be regarded as fairly certain except in this area. At the 120- and 144-h lead times (Figure 8.23c and d) there is still substantial agreement about (and thus relatively high probability would be inferred for) the developing cutoff low over the eastern Pacific, but the forecasts throughout the domain have begun to diverge quite strongly, suggesting the pasta dish for which this kind of plot is named. Spaghetti plots have proven to be quite useful in visualizing the evolution of the forecast flow, simultaneously with the dispersion of the ensemble. The effect is even more striking when a series of spaghetti plots is animated, which can be appreciated at some operational forecast center websites.

It can be informative to condense the large amount of information from an ensemble forecast into a small number of summary statistics, and to plot maps of these. By far the most common such plot, suggested initially by Epstein and Fleming (1971), is simultaneous display of the ensemble mean and standard deviation fields. That is, at each of a number of gridpoints the average of the ensemble members is calculated, as well as the standard deviation of the ensemble members around this average.

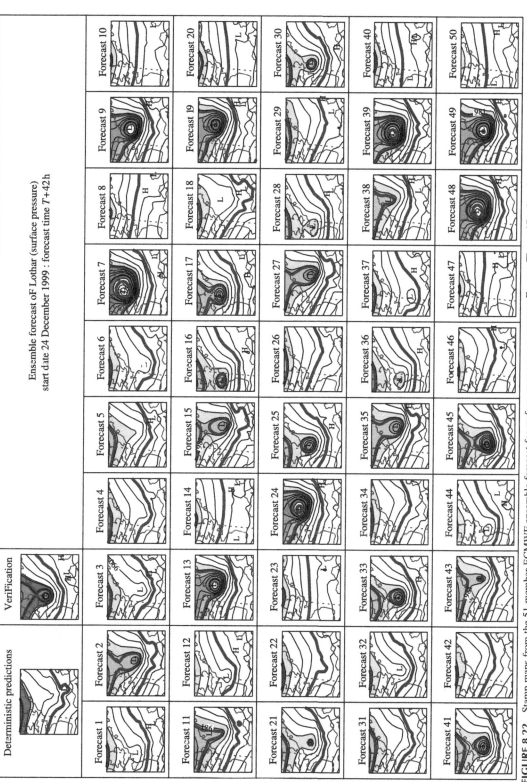

FIGURE 8.22 Stamp maps from the 51-member ECMWF ensemble forecast for surface pressure over western Europe. The verification shows the corresponding surface analysis 42 h later during winter storm Lothar. *From Palmer et al. (2005a).*

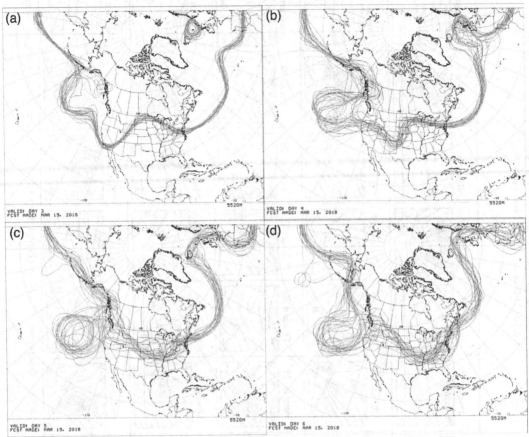

FIGURE 8.23　Spaghetti plots for the 5520-m contour of the 500 mb height field over North America, showing forecasts for (a) 72h, (b) 96h, (c) 120 and (d) 144h after the initial time of 0000 UTC, 15 March 2018. *From www.cpc.ncep.noaa.gov.*

Figure 8.24 is one such plot, for a 96-h forecast of 500 mb heights over much of the northern hemisphere, valid at 0000 UTC, 19 March 2018. Here the solid contours represent the ensemble-mean field, and the shading indicates the field of ensemble standard deviations. Large standard deviations indicate that the possible trough or cutoff low over the northeastern Pacific, corresponding to the dispersion in the spaghetti plot in Figure 8.23b, is fairly uncertain.

Gleeson (1967) suggested combining maps of forecast u and v wind components with maps of probabilities that the forecasts will be within 10 knots of the eventual observed values. Epstein and Fleming (1971) suggested that a probabilistic depiction of a horizontal wind field could take the form of Figure 8.25. Here the lengths and orientations of the arrows indicate the means of the forecast distributions of wind vectors, and the probability is 0.75 that the true wind vectors will terminate within the corresponding ellipse. It has been assumed in this figure that the uncertainty in the wind forecasts is described by the bivariate normal distribution, and the ellipses have been drawn as explained in Example 12.1.

Ensemble forecasts for surface weather elements at a single location can be concisely summarized by time series of boxplots for selected predictands, in a plot called an *ensemble meteogram*.

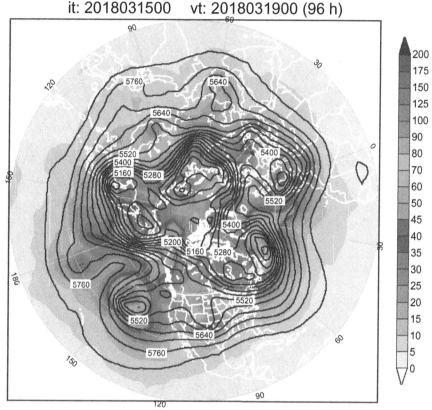

FIGURE 8.24 Ensemble mean (solid) and ensemble standard deviation (shading) for a 96-h forecast of hemispheric 500mb heights, valid 0000 UTC, 19 March 2018, corresponding approximately to Figure 8.23b. *From http://www.emc.ncep.noaa.gov/.*

Each of these boxplots displays the dispersion of the ensemble for one predictand at a particular forecast lead time, and jointly they show the time evolutions of the forecast central tendencies and uncertainties, through the forecast period. Figure 8.26 shows an example from the Japan Meteorological Agency, in which boxplots representing ensemble dispersion for four weather elements at Tsukuba are plotted at 6-hourly intervals. The plot indicates greater uncertainty in the cloud cover and precipitation forecasts, and the increasing uncertainty with increasing lead time is especially evident for the temperature forecasts.

Figure 8.27 shows an alternative to boxplots for portraying the time evolution of the ensemble distribution for a predictand. In this *plume graph* the shadings indicate the PDF of ensemble dispersion, as a function of time into the future for forecast surface temperatures at Hamburg, Germany. The ensemble can be seen to be quite compact early in the forecast, and expresses a large degree of uncertainty by the end of the period.

Finally, information from ensemble forecasts is very commonly displayed as maps of ensemble relative frequencies (Equation 8.4) for dichotomous events, which may be defined according to a threshold

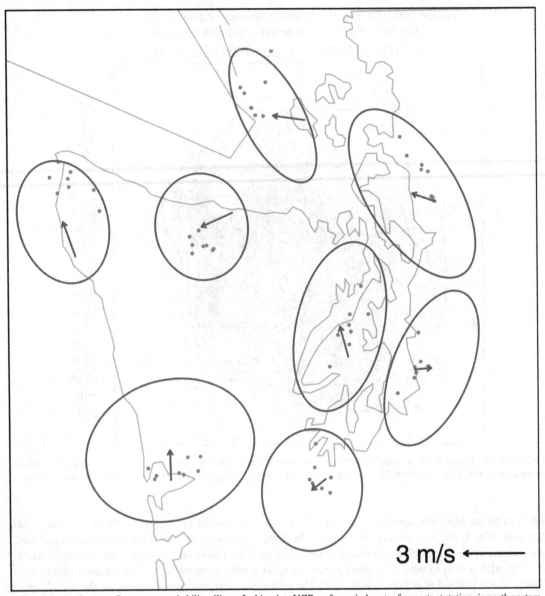

FIGURE 8.25 Seventy-five percent probability ellipses for bivariate NGR surface wind vector forecasts at stations in northwestern Washington State, valid at 0000 UTC 20 Oct 2008, at a lead time of 48 h. Arrows indicate the mean vectors, the bases of which locate the forecast locations. The eight raw ensemble members are shown as gray dots. *Modified from Schuhen et al. (2012). © American Meteorological Society. Used with permission.*

for a continuous variable. Such maps are often labeled as portraying "probability," but unless the underlying ensemble members have first been subjected to postprocessing using methods such as those described in Section 8.4.2 or 8.4.3, this description is generally not justified. Figure 8.28 shows an example of a very common plot of this kind, for ensemble relative frequency of >5 mm of precipitation at a lead time of 48 h.

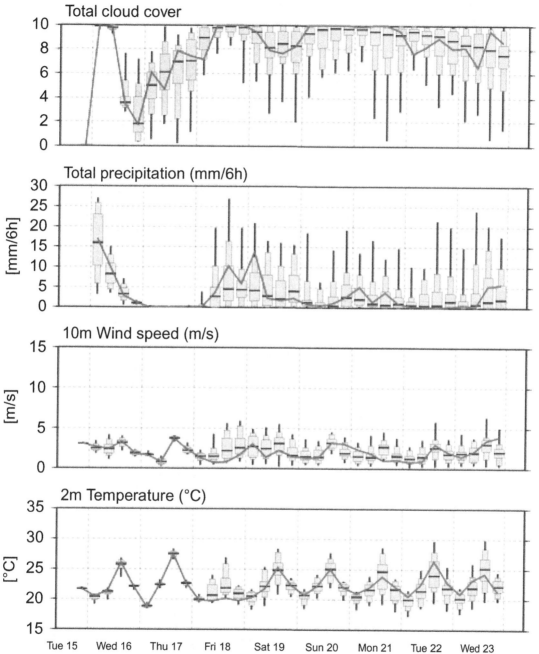

FIGURE 8.26   Ensemble meteogram for Tsukuba, Japan, from a Japan Meteorological Agency forecast ensemble begun on 1200 UTC, 15 June 2010. Wider portions of boxplots indicate the interquartile ranges, narrower box portions show middle 80% of the ensemble distributions, and whiskers extend to most extreme ensemble members. Solid line shows the control forecast. *From gpvjma.ccs.hpcc.jp.*

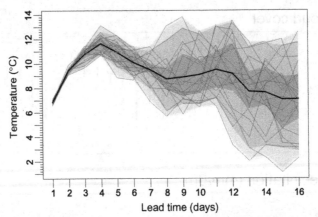

FIGURE 8.27　A plume graph, indicating probability density as a function of time, for a 16-day forecast of surface temperature at Hamburg, Germany, initiated 27 October 2010. The black line shows the ensemble mean, the light lines indicate the individual ensemble members, and the gray shadings indicate 50%, 90% and 99% probability intervals. *From Keune et al. (2014). © American Meteorological Society. Used with permission.*

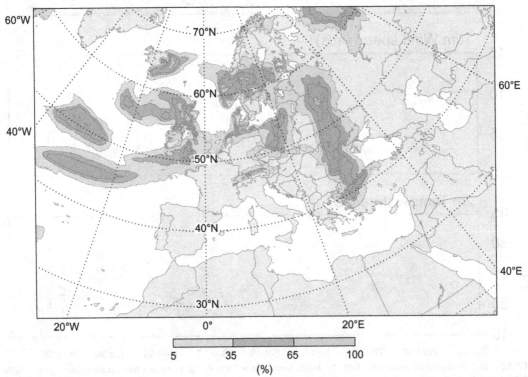

FIGURE 8.28　Ensemble relative frequency for accumulation of >5 mm precipitation over Europe, 48 h ahead of 27 July 2017. *From Buizza and Richardson (2017).*

8.6. EXERCISES

8.1. For the nonhomogeneous Gaussian regression model with $a=-0.5$, $b=1.4$, $c=1.5$, and $d=3.0$, compute the 10th and 90th percentiles of the predictive distribution if the ensemble consists of exchangeable members, and

(a) the ensemble mean is 5.25°C and ensemble variance is 0.25°C^2.

(b) the ensemble mean is 5.25°C and ensemble variance is 0.10°C^2.

8.2. The five bias corrected, nonexchangeable ensemble members shown in Figure 8.10 are 285.2, 285.6, 290.8, 291.2, and 292.5 K. Using the fact that each of the Gaussian BMA dressing kernels have standard deviation 2.6 K, compute $\Pr\{y \le 293\,K\}$, which was the verifying observation.

8.3. Figure 8.9a shows XLR functions in the form of Equation 8.31, where the right-hand side is $0.836\sqrt{q}-0.157-1.222\sqrt{\bar{x}_t}$. What is the probability of precipitation between 10 mm and 25 mm if the ensemble mean is

(a) 0 mm?

(b) 5 mm?

(c) 20 mm?

8.4. A five-member ensemble of temperature forecasts is NGR-transformed to yield a Gaussian predictive distribution with mean $\mu_1 = 27.0$°C and standard deviation $\sigma_1 = 1.5$°C. Wind speed forecasts for the same location and lead time are similarly postprocessed to yield a zero-truncated Gaussian predictive distribution with location parameter $\mu_2 = 1.5$ m/s and scale parameter $\sigma_2 = 1.5$ m/s. Use the ECC-Q approach together with the empirical copula structure in Figure 8.18 to compute a bivariate ensemble, jointly for the postprocessed temperatures and wind speeds.

Forecast Verification

9.1. BACKGROUND

9.1.1. Purposes of Forecast Verification

Forecast verification is the process of assessing the quality of forecasts. This process perhaps has been most fully developed in the atmospheric sciences, although parallel developments have taken place within other disciplines as well, such as finance and economics (e.g., Armstrong, 2001; Clark and McCracken, 2013), medical diagnosis (e.g., Pepe, 2003), and statistics (e.g., Gneiting, 2011a; Gneiting and Raftery, 2007; Schervish, 1989; Winkler, 1996). In the literature of those disciplines the activity is more typically called validation, or assessment, or evaluation. Verification of weather forecasts has been undertaken since at least 1884 (Muller, 1944; Murphy, 1996), and use of the term "verification" to mean evaluation of forecasts appears to have originated with the seminal paper of Finley (1884). In addition to this chapter, other reviews of forecast verification can be found in Jolliffe and Stephenson (2012a), Murphy (1997), and Stanski et al. (1989). Open-source software resources are also available (Pocernich, 2007).

Perhaps not surprisingly, there can be differing views of what constitutes a good forecast (Murphy, 1993). Many forecast verification procedures exist, but all involve statistics characterizing the relationship between a forecast or set of forecasts, and the corresponding observation(s) of the predictand. Any forecast verification method thus necessarily involves comparisons between matched pairs of forecasts and the observations to which they pertain.

On a fundamental level, forecast verification involves investigation of the properties of the *joint distribution of forecasts and observations* (Murphy and Winkler, 1987). That is, any given verification data set consists of a collection of forecast/observation pairs whose joint behavior can be characterized in terms of the relative frequencies of the possible combinations of forecast/observation pairs. A parametric joint distribution such as the bivariate normal (Section 4.4.2) can sometimes be useful in representing this joint distribution for a particular data set, but the empirical joint distribution of these quantities (more in the spirit of Chapter 3) usually forms the basis of forecast verification measures. Ideally, the association between forecasts and the observations to which they pertain will be reasonably strong, but in any case the nature and strength of this association will be reflected in their joint distribution.

Objective evaluations of forecast quality are undertaken for a variety of reasons. Brier and Allen (1951) categorized these as serving administrative, scientific, and economic purposes. In this view, administrative use of forecast verification pertains to ongoing monitoring of operational forecasts. For example, it is often of interest to examine trends of forecast performance through time (e.g., Stern and Davidson, 2015). Rates of forecast improvement, if any, for different predictands, locations, or lead times can be compared. Verification of forecasts from different sources for the same events can also be compared. Here forecast verification techniques allow comparison of the relative merits of competing forecasters or forecasting systems. For example, this is the purpose to which forecast verification is often put in scoring student forecast contests at colleges and universities.

Statistical Methods in the Atmospheric Sciences. https://doi.org/10.1016/B978-0-12-815823-4.00009-2

Analysis of verification statistics and their components can also help in the assessment of specific strengths and weaknesses of forecasters or forecasting systems. Although classified by Brier and Allen as scientific, this application of forecast verification is perhaps better regarded as *diagnostic verification* (Murphy et al., 1989; Murphy and Winkler, 1992). Here specific attributes of the relationship between forecasts and the subsequent events are investigated, highlighting strengths and deficiencies in a set of forecasts. Diagnostic verification allows human forecasters to be given feedback on the performance of their forecasts in different situations, which hopefully will lead to better forecasts in the future. Similarly, forecast verification measures can point to problems in forecasts produced by objective means, possibly leading to better forecasts through methodological improvements.

Ultimately, the justification for any forecasting enterprise is that it supports better decision making. The usefulness of forecasts to support decision making clearly depends on their error characteristics, which are elucidated through forecast verification methods. Thus the economic motivations for forecast verification are to provide the information necessary for users to derive full economic value from forecasts and to enable estimation of that value. However, since the economic value of forecast information in different decision situations must be evaluated on a case-by-case basis (e.g., Katz and Murphy, 1997a), forecast value cannot be computed from the verification statistics alone. Although it is sometimes possible to guarantee the economic superiority of one forecast source over another for all forecast users on the basis of a detailed verification analysis, which is a condition called *sufficiency* (Ehrendorfer and Murphy, 1988; Krzysztofowicz and Long, 1990, 1991; Murphy, 1997; Murphy and Ye, 1990), superiority with respect to a single verification measure does not necessarily imply superior forecast value for all users. Furthermore, actual as opposed to potential forecast value depends on psychosocial factors as well as purely economic ones (Millner, 2008; Stewart, 1997).

9.1.2. The Joint Distribution of Forecasts and Observations

The joint distribution of the forecasts and observations is of fundamental interest with respect to the verification of forecasts. In most practical settings, both the forecasts and observations are discrete variables. That is, even if the forecasts and observations are not already discrete quantities, they are typically rounded operationally to one of a finite set of values. Denote the forecast by y_i, which can take on any of the I values $y_1, y_2, \ldots, y_I$; and the corresponding observation as o_j, which can take on any of the J values $o_1, o_2, \ldots, o_J$. Then the joint distribution of the forecasts and observations is denoted

$$p\left(y_i, o_j\right) = \Pr\{y_i, o_j\} = \Pr\{y_i \cap o_j\}; \quad i = 1, \ldots, I; j = 1, \ldots, J. \tag{9.1}$$

This is a discrete bivariate probability distribution function, associating a probability with each of the $I \times J$ possible combinations of forecast and observation.

Even in the simplest cases, for which $I = J = 2$, this joint distribution can be difficult to use directly. From the definition of conditional probability (Equation 2.10) the joint distribution can be factored in two ways that are informative about different aspects of the verification problem. From a forecasting standpoint, the more familiar and intuitive of the two is

$$p\left(y_i, o_j\right) = p(o_j | y_i)\, p(y_i); \quad i = 1, \ldots, I; \ j = 1, \ldots J; \tag{9.2}$$

which is called the *calibration-refinement factorization* (Murphy and Winkler, 1987). One part of this factorization consists of a set of the I conditional distributions, $p(o_j | y_i)$, each of which consists of probabilities for all the J outcomes o_j, given one of the forecasts y_i. That is, each of these conditional distributions specifies how often each possible weather event occurred on those occasions when the single forecast y_i was issued, or how well each forecast y_i is calibrated. The other part of this factorization is the unconditional (marginal) distribution $p(y_i)$, which specifies the relative frequencies of use of each of the forecast values y_i, or how often each of the I possible forecast values were used. This marginal distribution is called the *refinement distribution* of the forecasts. The refinement of a set of forecasts refers to the dispersion of the distribution $p(y_i)$. A refinement distribution with a large spread implies refined forecasts, in that different forecasts are issued relatively frequently, and so have the potential to discern a broad range of conditions. Conversely, if most of the forecasts y_i are the same or very similar, $p(y_i)$ is narrow, which indicates a lack of refinement. This attribute of forecast refinement often is referred to as *sharpness*, in the sense that refined forecasts are called sharp.

The *likelihood-base rate factorization* (Murphy and Winkler, 1987) is the other factorization of the joint distribution of forecasts and observations,

$$p\left(y_i, o_j\right) = p\{y_i|\, o_j\} p\{o_j\}; \quad i = 1, \ldots, I; j = 1, \ldots, J. \tag{9.3}$$

Here the conditional distributions $p(y_i \mid o_j)$ express the likelihoods that each of the allowable forecast values y_i would have been issued in advance of each of the observed weather events o_j. Although this concept may seem logically reversed, it can reveal useful information about the nature of forecast performance. In particular, these conditional distributions relate to how well a set of forecasts is able to discriminate among the events o_j, in the same sense of the word used in Chapter 15. The unconditional distribution $p(o_j)$ consists simply of the relative frequencies of the J weather events o_j in the verification data set. This distribution usually is called the sample climatological distribution or simply the *sample climatology*.

Both the likelihood-base rate factorization (Equation 9.3) and the calibration-refinement factorization (Equation 9.2) can be calculated from the full joint distribution $p(y_i, o_j)$. Conversely, the full joint distribution can be reconstructed from either of the two factorizations. Accordingly, the full information content of the joint distribution $p(y_i, o_j)$ is included in either of the pairs of distributions, Equation 9.2 or Equation 9.3. Forecast verification approaches based on these distributions are sometimes known as *distributions-oriented* (Murphy, 1997) or *diagnostic verification* (Murphy et al., 1989; Murphy, 1991; Murphy and Winkler, 1992) approaches, in distinction to potentially incomplete summaries based on one or a few scalar verification measures, known as *measures-oriented* approaches. The name diagnostic verification is appropriate because portraying the full joint distribution of forecasts and observations allows detailed strengths and weaknesses of a set of forecasts to be diagnosed, potentially providing insights into the optimal use and value of the forecasts (e.g., Epstein, 1966), and into particular avenues for future forecast improvement (Murphy et al., 1989; Murphy and Winkler, 1992).

Although the two factorizations of the joint distribution of forecasts and observations can help organize the verification information conceptually, neither reduces the dimensionality (Murphy, 1991), or degrees of freedom, of the verification problem. That is, since all the probabilities in the joint distribution (Equation 9.1) must add to 1, it is completely specified by any $(I \times J) - 1$ of these probabilities. The factorizations of Equations 9.2 and 9.3 express this information differently and informatively, but $(I \times J) - 1$ distinct probabilities are still required to completely specify each factorization.

9.1.3. Scalar Attributes of Forecast Performance

Even in the simplest case of $I = J = 2$, complete specification of forecast performance requires a $(I \times J) - 1 = 3$-dimensional set of verification quantities. This minimum level of dimensionality is already sufficient to make understanding and comparison of forecast evaluation statistics less than straightforward. The difficulty is compounded in the many verification situations where $I > 2$ and/or $J > 2$, and such higher-dimensional verification situations may be further complicated if the sample size is not large enough to obtain good estimates for all of the required $(I \times J)-1$ probabilities. As a consequence, it is traditional to summarize forecast performance using one or several scalar (i.e., one-dimensional) verification measures. Many of the scalar summary statistics have been found through analysis and experience to provide very useful information about forecast performance, but some of the information in the full joint distribution of forecasts and observations is inevitably discarded when the dimensionality of the verification problem is reduced.

The following is a partial list of scalar aspects, or attributes, of forecast quality. These attributes are not uniquely defined, so that each of these concepts may be characterized using different functions of a verification data set.

(1) *Accuracy* refers to the average correspondence between individual forecasts and the events they predict. Scalar measures of accuracy are meant to summarize, in a single number, the overall quality of a set of forecasts. Several of the more common measures of accuracy will be presented in subsequent sections. The remaining forecast attributes in this list can often be interpreted as components, or aspects, of accuracy.

(2) *Bias*, or *unconditional bias*, or systematic bias, measures the correspondence between the average forecast and the average observed value of the predictand. This concept is different from accuracy, which measures the average correspondence between individual pairs of forecasts and observations. Temperature forecasts that are consistently too warm or precipitation forecasts that are consistently too wet both exhibit bias, whether or not the forecasts are otherwise reasonably accurate or quite inaccurate.

(3) *Reliability*, or *calibration*, or *conditional bias*, pertains to the relationship of the forecast to the distribution of observations, for specific values of (i.e., conditional on) the forecast. Reliability statistics sort the forecast/observation pairs into groups according to the value of the forecast variable, and characterize the conditional distributions of the observations given the forecasts. Thus measures of reliability summarize the I conditional distributions $p(o_j \mid y_i)$ of the calibration-refinement factorization (Equation 9.2). This attribute is also referred to as *validity* in some older literature (Bross, 1953; Sanders, 1963).

(4) *Resolution* refers to the degree to which the forecasts sort the observed events into groups that are different from each other. It is related to reliability, in that both are concerned with the properties of the conditional distributions of the observations given the forecasts, $p(o_j \mid y_i)$. Therefore resolution also relates to the calibration-refinement factorization of the joint distribution of forecasts and observations. However, resolution pertains to the differences between the conditional averages of the observations for different values of the forecast, whereas reliability compares the conditional distributions of the observations with the forecast values themselves. If average temperature outcomes following forecasts of, say, 10°C and 20°C are very different, the forecasts can resolve these different temperature outcomes, and are said to exhibit resolution. If the temperature outcomes following forecasts of 10°C and 20°C are nearly the same on average, the forecasts exhibit almost no resolution. Resolution can be regarded as a measure of the information content of forecasts, in the sense of characterizing the reduction in uncertainty about the predictand (e.g., Jolliffe and Stephenson, 2012b).

(5) *Discrimination* is the converse of resolution, in that it pertains to differences between the conditional distributions of the forecasts for different values of the observation. Measures of discrimination

summarize the J conditional distributions of the forecasts given the observations, $p(y_i \mid o_j)$, in the likelihood-base rate factorization (Equation 9.3). The discrimination attribute reflects the ability of the forecasting system to produce different forecasts for those occasions having different realized outcomes of the predictand. If a forecasting system forecasts $y =$ "snow" with equal frequency when $o =$ "snow" and $o =$ "sleet," the two conditional probabilities of a forecast of snow are equal, and the forecasts are not able to discriminate between snow and sleet events.

(6) *Sharpness*, or refinement, is an attribute of the forecasts alone, without regard to their corresponding observations. Measures of sharpness characterize the unconditional distribution (relative frequencies of use) of the forecasts, $p(y_i)$ in the calibration-refinement factorization (Equation 9.2). Forecasts that rarely deviate much from the climatological value of the predictand exhibit low sharpness. In the extreme, forecasts consisting only of the climatological value of the predictand exhibit no sharpness. By contrast, forecasts that are frequently much different from the climatological value of the predictand are sharp. Sharp forecasts exhibit the tendency to "stick their neck out." Sharp forecasts will be accurate only if they also exhibit good reliability, or calibration, and an important goal is to maximize sharpness without sacrificing calibration (Brier, 1950; Bross, 1953; Sanders, 1963; Gneiting et al., 2007; Murphy and Winkler, 1987; Winkler, 1996). Anyone can produce sharp forecasts, but the difficult task is to ensure that these forecasts correspond well to the subsequent observations.

9.1.4. Forecast Skill

Forecast *skill* refers to the relative accuracy of a set of forecasts, with respect to some set of standard *reference forecasts*. Common choices for the reference forecasts are climatological values of the predictand, persistence forecasts (values of the predictand in the previous time period), or random forecasts (with respect to the climatological relative frequencies of the observed events o_j). Yet other choices for the reference forecasts can be more appropriate in some cases. For example, when evaluating the performance of a new forecasting system, it could be appropriate to compute skill relative to the forecasts that this new system might replace.

Forecast skill is usually presented as a *skill score*, which is often interpreted as a percentage improvement over the reference forecasts. In generic form, the skill score for forecasts characterized by a particular measure of accuracy A, with respect to the accuracy A_{ref} of a set of reference forecasts, is given by

$$SS_{\text{ref}} = \frac{A - A_{\text{ref}}}{A_{\text{perf}} - A_{\text{ref}}} \times 100\%, \tag{9.4}$$

where A_{perf} is the value of the accuracy measure that would be achieved by perfect forecasts. Note that this generic skill score formulation gives consistent results whether the accuracy measure has a positive (larger values of A are better) or negative (smaller values of A are better) orientation. If $A = A_{\text{perf}}$ the skill score attains its maximum value of 100%. If $A = A_{\text{ref}}$ then $SS_{\text{ref}} = 0\%$, indicating no improvement over the reference forecasts. If the forecasts being evaluated are inferior to the reference forecasts with respect to the accuracy measure A, $SS_{\text{ref}} < 0\%$. Skill scores are known as U-statistics in the economics literature (Campbell and Diebold, 2005).

The use of skill scores often is motivated by a desire to equalize effects of intrinsically more or less difficult forecasting situations, when comparing forecasters or forecast systems. For example, forecasting precipitation in a very dry climate is generally relatively easy, since forecasts of zero, or the climatological average (which will be very near zero), will exhibit good accuracy on most days. If the accuracy of the reference forecasts (A_{ref} in Equation 9.4) is relatively high, a higher accuracy A is required to achieve a given skill level than would be the case in a more difficult forecast situation, in which A_{ref} would be smaller. Some of the effects of the intrinsic ease or difficulty of different forecast situations can be equalized through use of skill scores such as Equation 9.4, but unfortunately skill scores have not been found to be fully effective for this purpose (Glahn and Jorgensen, 1970; Winkler, 1994, 1996).

When skill scores are averaged over nonhomogeneous forecast-observation pairs (e.g., for a single location across a substantial fraction of the annual cycle, or for multiple locations with different climates), care must be taken to compute the skill scores consistently, so that credit is not given for correctly "forecasting" mere climatological differences (Hamill and Juras, 2006; Juras, 2000). For example, correctly forecasting summers being warmer than winters is not credited as a contribution to skill. When computing averaged skill scores, each of the three quantities on the right-hand side of Equation 9.4 should be computed separately for each homogeneous subset of the forecast-observation pairs, with the summary average skill calculated as the weighted average of the resulting component skills.

9.2. NONPROBABILISTIC FORECASTS FOR DISCRETE PREDICTANDS

Forecast verification is perhaps easiest to understand in the context of nonprobabilistic forecasts for discrete predictands. Nonprobabilistic indicates that the forecast consists of an unqualified statement that a single outcome will occur. Nonprobabilistic forecasts contain no expression of uncertainty, in distinction to probabilistic forecasts. A discrete predictand is an observable variable that takes on one and only one of a finite set of possible values. This is in distinction to a scalar continuous predictand, which (at least conceptually) may take on any value on the relevant portion of the real line.

Verification for nonprobabilistic forecasts of discrete predictands has been undertaken since the 19th century (Murphy, 1996), and during this considerable time a variety of sometimes conflicting terminology has been used. For example, nonprobabilistic forecasts have been called categorical, in the sense their being firm statements that do not admit the possibility of alternative outcomes. This usage of the term appears to date from early in the 20th century (Liljas and Murphy, 1994). However, more recently the term categorical has come to be understood as relating to a predictand belonging to one of a set of MECE categories, that is, a discrete variable. In an attempt to minimize confusion, the term categorical will be avoided here, in favor of the more explicit terms, nonprobabilistic and discrete. Other instances of the multifarious nature of forecast verification terminology will also be noted in this chapter.

9.2.1. The 2×2 Contingency Table

There is usually a one-to-one correspondence between allowable nonprobabilistic forecast values and values of the discrete observable predictand to which they pertain. In terms of the joint distribution of forecasts and observations (Equation 9.1), $I = J$. The simplest possible situation is for the dichotomous $I = J = 2$ case, or verification of nonprobabilistic yes/no forecasts. Here there are $I = 2$ possible forecasts, either that the event will ($i = 1$, or y_1) or will not ($i = 2$, or y_2) occur. Similarly, there are $J = 2$ possible outcomes: either the event subsequently occurs (o_1) or it does not (o_2). Despite the simplicity of this verification setting, a surprisingly large body of work on the 2×2 verification problem has developed.

Conventionally, nonprobabilistic verification data are displayed in an $I \times J$ contingency table of absolute frequencies, or counts, of the $I \times J$ possible combinations of forecast and event pairs. If these counts are transformed to relative frequencies, by dividing each tabulated entry by the sample size (total number of forecast and event pairs), the (sample estimate of the) joint distribution of forecasts and observations (Equation 9.1) is obtained. Figure 9.1 illustrates the essential equivalence of the contingency table and the joint distribution of forecasts and observations for the simple $I = J = 2$ case. The boldface portion in Figure 9.1a shows the arrangement of the four possible combinations of forecast and event pairs as a square contingency table, and the corresponding portion of Figure 9.1b shows these counts transformed to joint relative frequencies.

In terms of Figure 9.1, the event in question was successfully forecast to occur a times out of n total forecasts. These a forecast-observation pairs usually are called *hits*, and their relative frequency, a/n, is the sample estimate of the corresponding joint probability $p(y_1, o_1)$ in Equation 9.1. Similarly, on b occasions, called *false alarms*, the event was forecast to occur but did not, and the relative frequency b/n estimates the joint probability $p(y_1, o_2)$. There are also c instances of the event of interest occurring despite not being forecast, called *misses*, the relative frequency of which estimates the joint probability $p(y_2, o_1)$; and d instances of the event not occurring after a forecast that it would not occur, sometimes called a *correct rejection* or *correct negative*, the relative frequency of which corresponds to the joint probability $p(y_2, o_2)$.

It is also usual to include what are called *marginal totals* with a contingency table of counts. These are simply the row and column totals yielding, in this case, the numbers of times each yes or no forecast, or observation, respectively, occurred. These are shown in Figure 9.1a in normal typeface, as is the sample size,

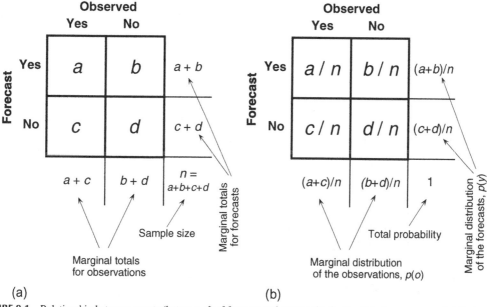

(a)   (b)

FIGURE 9.1   Relationship between counts (letters a–d) of forecast and event pairs for the dichotomous nonprobabilistic verification situation as displayed in a 2×2 contingency table (bold, panel a), and the corresponding joint distribution of forecasts and observations [$p(y, o)$] (bold, panel b). Also shown are the marginal totals, indicating how often each of the two events was forecast and observed in absolute terms; and the marginal distributions of the observations [$p(o)$] and forecasts [$p(y)$], which indicates the same information in relative frequency terms.

$n = a+b+c+d$. Expressing the marginal totals in relative frequency terms, again by dividing through by the sample size, yields the *marginal distribution* of the forecasts, $p(y_i)$, and the marginal distribution of the observations, $p(o_j)$. The marginal distribution $p(y_i)$ is the refinement distribution, in the calibration-refinement factorization (Equation 9.2) for the 2×2 joint distribution in Figure 9.1b. Since there are $I = 2$ possible forecasts, there are two calibration distributions $p(o_j | y_i)$, each of which consists of $J = 2$ probabilities. Therefore in addition to the refinement distribution $p(y_1) = (a+b)/n$ and $p(y_2) = (c+d)/n$, the calibration-refinement factorization in the 2×2 verification setting consists of the conditional probabilities

$$p(o_1 | y_1) = a/(a+b) \tag{9.5a}$$

$$p(o_2 | y_1) = b/(a+b) \tag{9.5b}$$

$$p(o_1 | y_2) = c/(c+d) \tag{9.5c}$$

and

$$p(o_2 | y_2) = d/(c+d). \tag{9.5d}$$

In terms of the definition of conditional probability (Equation 2.10), Equation 9.5a (for example) would be obtained as $[a/n]/[(a+b)/n] = a/(a+b)$.

Similarly, the marginal distribution $p(o_j)$, with elements $p(o_1) = (a+c)/n$ and $p(o_2) = (b+d)/n$, is the base-rate (i.e., sample climatological) distribution in the likelihood-base rate factorization (Equation 9.3). That factorization is completed by the four conditional probabilities

$$p(y_1 | o_1) = a/(a+c) \tag{9.6a}$$

$$p(y_2 | o_1) = c/(a+c) \tag{9.6b}$$

$$p(y_1 | o_2) = b/(b+d) \tag{9.6c}$$

and

$$p(y_2 | o_2) = d/(b+d). \tag{9.6d}$$

9.2.2. Scalar Attributes of the 2×2 Contingency Table

Even though the 2×2 contingency table summarizes verification data for the simplest possible forecast setting, its dimensionality is 3. That is, the forecast performance information contained in the contingency table cannot fully be expressed with fewer than three sample statistics. It is perhaps not surprising that a wide variety of these scalar attributes have been devised and used to characterize forecast performance, over the long history of the verification of forecasts of this type. Unfortunately, a similarly wide variety of nomenclature also has appeared in relation to these attributes, both within and outside the atmospheric sciences. This section lists scalar attributes of the 2×2 contingency table that have been most widely used, together with much of the synonymy associated with them. The organization follows the general classification of attributes in Section 9.1.3.

Accuracy

Accuracy statistics reflect correspondence between pairs of forecasts and the events they are meant to predict. Perfectly accurate forecasts in the 2×2 nonprobabilistic forecasting situation will clearly exhibit all "yes" forecasts for the event followed by the event and all "no" forecasts for the event followed by nonoccurrence, so that $b = c = 0$. For real, imperfect forecasts, accuracy measures characterize degrees of this correspondence. Several scalar accuracy measures are in common use, with each reflecting somewhat different aspects of the underlying joint distribution.

Perhaps the most direct and intuitive measure of the accuracy of nonprobabilistic forecasts for discrete events is the *proportion correct* proposed by Finley (1884). This is simply the fraction of the n forecast occasions for which the nonprobabilistic forecast correctly anticipated the subsequent event or nonevent. In terms of the counts Figure 9.1a, the proportion correct is given by

$$PC = \frac{a+d}{n}. \tag{9.7}$$

The proportion correct satisfies the principle of *equivalence of events*, since it credits correct "yes" and "no" forecasts equally. As Example 9.1 will show, however, this is not always a desirable attribute, particularly when the "yes" event is rare, so that correct "no" forecasts can be made fairly easily. The proportion correct also penalizes both kinds of errors (false alarms and misses) equally. The worst possible proportion correct is zero. The best possible proportion correct is one. Sometimes PC in Equation 9.7 is multiplied by 100%, and referred to as the *percent correct*, or percentage of forecasts correct. Because the proportion correct does not distinguish between correct forecasts of the event, a, and correct forecasts of the nonevent, d, this fraction of correct forecasts has also been called the hit rate. However, in current usage the term hit rate usually is reserved for the discrimination measure given in Equation 9.12.

The *threat score* (TS) or *critical success index* (CSI) is an alternative to the proportion correct that is particularly useful when the event to be forecast (as the "yes" event) occurs substantially less frequently than the nonoccurrence (the "no" event). In terms of Figure 9.1a, the threat score is computed as

$$TS = CSI = \frac{a}{a+b+c}, \tag{9.8}$$

which is the number of correct "yes" forecasts divided by the total number of occasions on which that event was forecast and/or observed. It can be viewed as a proportion correct for the quantity being forecast, after removing correct "no" forecasts from consideration. The worst possible threat score is zero, and the best possible threat score is one. When originally proposed (Gilbert, 1884) it was called the *ratio of verification*, and denoted as V, and so Equation 9.8 is sometimes called the Gilbert Score (as distinct from the Gilbert Skill Score, Equation 9.20). In the ecology literature TS is known as the Jaccard coefficient (Janson and Vegelius, 1981). Very often each of the counts in a 2×2 contingency table pertains to a different forecasting occasion (as illustrated in Example 9.1), but the threat score (and the skill score based on it, Equation 9.20) may also be used to assess simultaneously issued spatial forecasts, for example, heavy precipitation warnings (e.g., Ebert and McBride, 2000; Stensrud and Wandishin, 2000). In this setting, a represents the intersection of the areas over which the event was forecast and subsequently occurred, b represents the area over which the event was forecast but failed to occur, and c is the area over which the event occurred but was not forecast to occur. The threat score is convenient in this setting because often the relevant "no" forecast area is arbitrary or unknown.

A third approach to characterizing forecast accuracy in the 2×2 situation is in terms of odds, or the ratio of a probability to its complementary probability, $p/(1-p)$. In the context of forecast verification the ratio of the conditional odds of a hit, given that the event occurs, to the conditional odds of a false alarm, given that the event does not occur, is called the *odds ratio*,

$$\theta = \frac{p(y_1|o_1)/[1-p(y_1|o_1)]}{p(y_1|o_2)/[1-p(y_1|o_2)]} = \frac{p(y_1|o_1)/p(y_2|o_1)}{p(y_1|o_2)/p(y_2|o_2)} = \frac{a\,d}{b\,c}. \tag{9.9}$$

The conditional distributions making up the odds ratio are all likelihoods from Equation 9.6. In terms of the 2×2 contingency table, the odds ratio is the product of the numbers of correct forecasts divided by the product of the numbers of incorrect forecasts. Clearly, larger values of this ratio indicate more accurate forecasts. No-information forecasts, for which the forecasts and observations are statistically independent (i.e., $p(y_i, o_j) = p(y_i)\,p(o_j)$, Equation 2.12), yield $\theta = 1$. The odds ratio was introduced into meteorological forecast verification by Stephenson (2000), although it has a longer history of use in medical statistics.

Bias

The bias, or comparison of the average forecast with the average observation, usually is represented as a ratio for verification of contingency tables. In terms of the 2×2 table in Figure 9.1a the bias ratio is

$$B = \frac{a+b}{a+c}. \tag{9.10}$$

The bias is simply the ratio of the number of "yes" forecasts to the number of "yes" observations. Unbiased forecasts exhibit $B = 1$, indicating that the event was forecast the same number of times that it was observed. Note that bias provides no information about the correspondence between the individual forecasts and observations of the event on particular occasions, so that Equation 9.10 is not an accuracy measure. Bias greater than one indicates that the event was forecast more often than observed, which is called *overforecasting*. Conversely, bias less than one indicates that the event was forecast less often than observed, or was *underforecast*.

Reliability and Resolution

Equation 9.5 shows four calibration attributes for the 2×2 contingency table. That is, each quantity in Equation 9.5 is a conditional relative frequency for event occurrence or nonoccurrence, given either a "yes" or "no" forecast, in the sense of the calibration distributions $p(o_j \mid y_i)$ of Equation 9.2. Actually, Equation 9.5 indicates two calibration distributions, one conditional on the "yes" forecasts (Equations 9.5a and 9.5b), and the other conditional on the "no" forecasts (Equations 9.5c and 9.5d). Each of these four conditional probabilities is a scalar reliability statistic for the 2×2 contingency table, and all four have been given names (e.g., Doswell et al., 1990). By far the most commonly used of these conditional relative frequencies is Equation 9.5b, which is called the *false alarm ratio* (FAR). In terms of Figure 9.1a, the false alarm ratio is computed as

$$\text{FAR} = \frac{b}{a+b}. \tag{9.11}$$

That is, FAR is the fraction of "yes" forecasts that turn out to be wrong, or that proportion of the forecast events that fail to materialize. The FAR has a negative orientation, so that smaller values of FAR are to be preferred. The best possible FAR is zero, and the worst possible FAR is one. The FAR has also been called the false alarm *rate* (Barnes et al., 2009 sketch a history of the confusion), although this rather similar term is now generally reserved for the discrimination measure in Equation 9.13. 1–FAR is known as the *positive predictive value* in medical statistics.

Discrimination

Two of the conditional probabilities in Equation 9.6 are used frequently to characterize 2×2 contingency tables, although all four of them have been named (e.g., Doswell et al., 1990). Equation 9.6a is commonly known as the *hit rate*,

$$H = \frac{a}{a+c}. \tag{9.12}$$

Regarding only the event o_1 as "the" event of interest, the hit rate is the ratio of correct forecasts to the number of times this event occurred. Equivalently this statistic can be regarded as the fraction of those occasions when the forecast event occurred on which it was also forecast, and so is also called the *probability of detection* (POD). In medical statistics this quantity is known as the *true-positive fraction*, or the *sensitivity*.

Equation 9.6c is called the *false alarm rate*,

$$F = \frac{b}{b+d}, \tag{9.13}$$

which is the ratio of false alarms to the total number of nonoccurrences of the event o_1, or the conditional relative frequency of a wrong forecast given that the event does not occur. The false alarm rate is also known as the *probability of false detection* (POFD). Jointly the hit rate and false alarm rate provide both the conceptual and geometrical basis for the signal detection approach for verifying probabilistic forecasts (Section 9.4.6). In medical statistics this quantity is known as the *false-positive fraction*, or 1 minus the *specificity*.

Noting that many summary statistics degenerate to trivial values as the climatological probability, or base rate,

$$s = p(o_1) = (a+c)/n \tag{9.14}$$

becomes small, Ferro and Stephenson (2011) proposed the extremal dependence index,

$$EDI = \frac{\ln(F) - \ln(H)}{\ln(F) + \ln(H)} \tag{9.15}$$

to address 2×2 verification for extreme (and therefore rare) events. This measure takes on values between ± 1, with positive values indicating better performance than random forecasts.

9.2.3. Skill Scores for 2×2 Contingency Tables

Forecast verification data in contingency tables are often characterized using relative accuracy measures or skill scores in the general form of Equation 9.4. A large number of such skill scores have been developed for the 2×2 verification situation, and many of these have been presented by Muller (1944), Mason (2003), Murphy and Daan (1985), Stanski et al. (1989), and Woodcock (1976). Some of these skill measures date from the earliest literature on forecast verification (Murphy, 1996) and have been rediscovered and (unfortunately) renamed on multiple occasions. In general the different skill scores perform differently and sometimes inconsistently. This situation can be disconcerting if we hope to choose among alternative skill scores, but should not really be surprising given that all of these skill scores are scalar measures of forecast performance in what is intrinsically a higher-dimensional setting. Scalar skill scores are used because they are conceptually convenient, but they are necessarily incomplete representations of forecast performance.

One of the most frequently used skill scores for summarizing square contingency tables was originally proposed by Doolittle (1888), but because it is nearly universally known as the Heidke Skill Score (Heidke, 1926) this latter name will be used here. The *Heidke Skill Score* (HSS) follows the form of Equation 9.4, based on the proportion correct (Equation 9.7) as the basic accuracy measure. Thus perfect forecasts receive HSS = 1, forecasts equivalent to the reference forecasts receive zero scores, and forecasts worse than the reference forecasts receive negative scores.

The reference accuracy measure in the Heidke score is the proportion correct that would be achieved by random forecasts that are statistically independent of the observations. In the 2×2 situation, the marginal probability of a "yes" forecast is $p(y_1) = (a+b)/n$, and the marginal probability of a "yes" observation is $s = p(o_1) = (a+c)/n$. Therefore the probability of a correct "yes" forecast by chance is

$$p(y_1)p(o_1) = \frac{(a+b)}{n} \frac{(a+c)}{n} = \frac{(a+b)(a+c)}{n^2}, \tag{9.16a}$$

and similarly the probability of a correct "no" forecast by chance is

$$p(y_2)p(o_2) = \frac{(b+d)}{n} \frac{(c+d)}{n} = \frac{(b+d)(c+d)}{n^2}. \tag{9.16b}$$

Thus following Equation 9.4, for the 2×2 verification setting the Heidke Skill Score is

$$\text{HSS} = \frac{(a+d)/n - [(a+b)(a+c) + (b+d)(c+d)]/n^2}{1 - [(a+b)(a+c) + (b+d)(c+d)]/n^2}$$

$$= \frac{2(ad - bc)}{(a+c)(c+d) + (a+b)(b+d)}, \tag{9.17}$$

where the second equality is easier to compute. The HSS is referred to as Cohen's kappa (Cohen, 1960) in the social science literature. Hyvärinen (2014) provides an alternative, Bayesian derivation for HSS that does not invoke the notion of naive random forecasts.

Another popular skill score for contingency-table forecast verification has been rediscovered many times since being first proposed by Peirce (1884). The *Peirce Skill Score* is also commonly referred to as the *Hanssen-Kuipers discriminant* (Hanssen and Kuipers, 1965) or *Kuipers' performance index* (Murphy and Daan, 1985), and is sometimes also called the *true skill statistic* (TSS) (Flueck, 1987). It is known as *Youden's* (1950) index in the medical statistics literature. *Gringorten's* (1967) *skill score* contains equivalent information, as it is a linear transformation of the Peirce Skill Score. The Peirce Skill Score is formulated similarly to the Heidke score, except that the reference hit rate in the denominator is that for random forecasts that are constrained to be unbiased. That is, the imagined random reference forecasts in the denominator of Equation 9.4 have a marginal distribution that is equal to the (sample) climatology, so that $p(y_1) = p(o_1)$ and $p(y_2) = p(o_2)$. Again following Equation 9.4 for the 2×2 situation of Figure 9.1, the Peirce Skill Score is computed as

$$\text{PSS} = \frac{(a+d)/n - [(a+b)(a+c) + (b+d)(c+d)]/n^2}{1 - \left[(a+c)^2 + (b+d)^2\right]/n^2}$$

$$= \frac{ad - bc}{(a+c)(b+d)}, \tag{9.18}$$

where again the second equality is computationally more convenient. The PSS can also be understood as the difference between two conditional probabilities in the likelihood-base rate factorization of the joint distribution (Equation 9.6), namely, the hit rate (Equation 9.12) and the false alarm rate (Equation 9.13). That is, PSS $= H - F$. Perfect forecasts receive a score of one (because $b = c = 0$, or in an alternative view, $H = 1$ and $F = 0$), random forecasts receive a score of zero (because $H = F$), and forecasts inferior to the random forecasts receive negative scores. Constant forecasts (i.e., always forecasting one or the other of y_1 or y_2) are also accorded zero skill. Furthermore, unlike the Heidke score, the contribution made to the Peirce Skill Score by a correct "no" or "yes" forecast increases as the event is more or less likely, respectively. Thus forecasters are not discouraged from forecasting rare events on the basis of their low climatological probability alone.

The *Clayton* (1927, 1934) *Skill Score* can be formulated as the difference of the conditional probabilities in Equation 9.5a and 9.5c, relating to the calibration-refinement factorization of the joint distribution, that is,

$$\text{CSS} = \frac{a}{(a+b)} - \frac{c}{(c+d)} = \frac{ad - bc}{(a+b)(c+d)}. \tag{9.19}$$

The CSS indicates positive skill to the extent that the event occurs more frequently when forecast than when not forecast, so that the conditional relative frequency of the "yes" outcome given "yes" forecasts is larger than the conditional relative frequency given "no" forecasts. Clayton (1927) originally called this difference of conditional relative frequencies (multiplied by 100%) the percentage of skill, where he understood skill in the modern sense of accuracy relative to climatological expectancy. Perfect forecasts exhibit $b = c = 0$, yielding CSS $= 1$. Random forecasts (Equation 9.16) yield CSS $= 0$.

A skill score in the form of Equation 9.4 can also be constructed using the threat score (Equation 9.8) as the basic accuracy measure, using TS for random (Equation 9.16) forecasts as the reference. In particular, $\text{TS}_{ref} = a_{ref}/(a+b+c)$, where Equation 9.16a implies $a_{ref} = (a+b)(a+c)/n$. Since $\text{TS}_{perf} = 1$, the resulting skill score is

$$\text{GSS} = \frac{a/(a+b+c) - a_{ref}/(a+b+c)}{1 - a_{ref}/(a+b+c)} = \frac{a - a_{ref}}{a - a_{ref} + b + c}. \tag{9.20}$$

This skill score, called the *Gilbert Skill Score* (GSS) originated with Gilbert (1884), who referred to it as the *ratio of success*. It is also commonly (although erroneously, Hogan et al., 2010) called the *Equitable Threat Score* (ETS). Because the sample size n is required to compute a_{ref}, the GSS depends also on the number of correct "no" forecasts, unlike the TS.

The odds ratio (Equation 9.9) can also be used as the basis of a skill score,

$$Q = \frac{\theta - 1}{\theta + 1} = \frac{(ad/bc) - 1}{(ad/bc) + 1} = \frac{ad - bc}{ad + bc}. \tag{9.21}$$

This skill score originated with Yule (1900), and is called *Yule's Q* (Woodcock, 1976), or the *Odds Ratio Skill Score* (ORSS) (Stephenson, 2000). Random (Equation 9.15) forecasts exhibit $\theta = 1$, yielding $Q = 0$; and perfect forecasts exhibit $b = c = 0$, producing $Q = 1$. However, an apparently perfect skill of $Q = 1$ is also obtained for imperfect forecasts, if either one or the other of b or c is zero.

All the skill scores listed in this section depend only on the four counts $a, b, c,$ and d in Figure 9.1 and are therefore necessarily related. Notably, HSS, PSS, CSS, and Q are all proportional to the quantity $ad - bc$. Some specific mathematical relationships among the various skill scores are noted in Hogan and Mason (2012), Mason (2003), Murphy (1996), Stephenson (2000), and Wandishin and Brooks (2002).

Example 9.1. The Finley Tornado Forecasts

The Finley tornado forecasts (Finley, 1884) are historical 2×2 forecast verification data that are often used to illustrate evaluation of forecasts in this format. John Finley was a sergeant in the U.S. Army who, using telegraphed synoptic information, formulated yes/no tornado forecasts for 18 regions of the United States east of the Rocky Mountains. The data set and its analysis were instrumental in stimulating much of the early work on forecast verification (Murphy, 1996). The contingency table for Finley's $n = 2803$ forecasts is presented in Table 9.1a.

Finley chose to evaluate his forecasts using the proportion correct (Equation 9.7), which for his data is $PC = (28 + 2680)/2803 = 0.966$. On the basis of this result, Finley claimed 96.6% accuracy. However, the proportion correct for this data set is dominated by the correct "no" forecasts, since tornados are relatively rare. Very shortly after Finley's paper appeared, Gilbert (1884) pointed out that always forecasting "no" would produce an even higher proportion correct. The contingency table that would be obtained if tornados had never been forecast is shown in Table 9.1b. These hypothetical forecasts yield a proportion correct of $PC = (0 + 2752)/2803 = 0.982$, which is indeed higher than the proportion correct for the actual forecasts.

Employing the threat score (Equation 9.8) gives a more reasonable comparison, because the large number of easy, correct "no" forecasts are ignored. For Finley's original forecasts, the threat score is $TS = 28/(28 + 72 + 23) = 0.228$, whereas for the obviously useless "no" forecasts in Table 9.1b the threat score is $TS = 0/(0 + 0 + 51) = 0$. Clearly the threat score would be preferable to the proportion correct in this instance, but it is still not completely satisfactory. Equally useless would be a forecasting system that always forecast "yes" for tornados. For constant "yes" forecasts the threat score would be $TS = 51/(51 + 2752 + 0) = 0.018$, which is small, but not zero. The odds ratio for the Finley forecasts is $\theta = (28)(2680)/(72)(23) = 45.3 > 1$, suggesting better than random performance for the forecasts in Table 9.1a. The odds ratio is not computable for the forecasts in Table 9.1b.

The bias ratio for the Finley tornado forecasts is $B = 1.96$, indicating that approximately twice as many tornados were forecast as actually occurred, much of which might be attributable to the sparseness of the observation network at the time. The false alarm ratio is $FAR = 0.720$, which expresses the fact that a fairly large fraction of the forecast tornados did not eventually materialize. On the other hand, the hit rate is $H = 0.549$ and the false alarm rate is $F = 0.0262$; indicating that more than half of the actually observed tornados were forecast to occur, whereas a very small fraction of the nontornado cases falsely warned of a tornado.

TABLE 9.1 Contingency Tables for Verification of the Finley Tornado Forecasts, from 1884

(a)

		Tornados Observed	
		Yes	No
Tornados	Yes	28	72
Forecast	No	23	2680
			$n = 2803$

(b)

		Tornados Observed	
		Yes	No
Tornados	Yes	0	0
Forecast	No	51	2752
			$n = 2803$

The forecast event is occurrence of a tornado, with separate forecasts for 18 regions of the United States east of the Rocky Mountains. (a) The table for the forecasts as originally issued; and (b) data that would have been obtained if "no tornados" had always been forecast.
© American Meteorological Society. Used with permission.

The various skill scores yield a very wide range of results for the Finely tornado forecasts: HSS = 0.355, PSS = 0.523, CSS = 0.271, GSS = 0.216, EDI = 0.717, and $Q = 0.957$. Zero skill is attributed to the constant "no" forecasts in Table 9.1b by HSS, PSS, and GSS, but CSS, EDS, SEDS, and Q cannot be computed for $a = b = 0$. ◇

9.2.4. Which Score?

The wide range of skills attributed to the Finley tornado forecasts in Example 9.1 may be somewhat disconcerting, but should not be surprising. The root of the problem is that, even in this simplest of all possible forecast verification settings, the dimensionality (Murphy, 1991) of the problem is $I \times J - 1 = 3$, and the collapse of this three-dimensional information into a single number by any scalar verification measure necessarily involves a loss of information. Put another way, there are a variety of ways for forecasts to go right and for forecasts to go wrong, and mixtures of these are combined differently by different scalar attributes and skill scores. There is no single answer to the question posed in the heading for this section.

It is sometimes necessary to choose a single scalar summary of forecast performance, accepting that the summary will necessarily be incomplete. For example, competing forecasters in a contest must be evaluated in a way that produces an unambiguous ranking of their performances. Choosing a single score for such a purpose involves investigating and comparing relevant properties of competing candidate verification statistics, a process that is called *metaverification* (Murphy, 1996). Which property or properties might be most relevant may depend on the specific situation, but one reasonable criterion can be that a chosen verification statistic should be *equitable* (Gandin and Murphy, 1992). An equitable skill score rates random forecasts, and all constant forecasts (such as always forecasting "no tornados" in Example 9.1), equally. Usually this score for useless forecasts is set to zero, and equitable scores are scaled such that perfect forecasts have unit skill. Equitability also implies that correct forecasts of less frequent events (such as tornados in Example 9.1) are weighted more strongly than correct forecasts of more common events, which discourages distortion of forecasts toward the more common event in order to artificially inflate the resulting score.

The original Gandin and Murphy (1992) definition of equitability imposed the additional condition that any equitable verification measure must be expressible as a linear weighted sum of the elements of the contingency table, which leads to the use of the PSS (Equation 9.18) as the only equitable skill score for the 2×2 verification setting. However, Hogan et al. (2010) have argued persuasively that this second condition is not compelling, and if it is not required HSS (Equation 9.17) also equitable, in the sense of also yielding zero skill for random or constant forecasts. Interestingly Hogan et al. (2010) also show that GSS (also known as the "equitable" threat score, Equation 9.20) is not equitable because it does not yield zero skill for random forecasts. However, Hogan et al. (2010) find that GSS, CSS (Equation 9.19), and Q (Equation 9.21) are *asymptotically equitable*, meaning that they approach equitability as the sample size becomes very large.

Because the dimensionality of the 2×2 problem is 3, the full information in the 2×2 contingency table can be captured fully by three well-chosen scalar attributes. Using the likelihood-base rate factorization (Equation 9.6), the full joint distribution can be summarized by (and recovered from) the hit rate H (Equations 9.12 and 9.6a), the false alarm rate F (Equation 9.13 and 9.6c), and the base rate s (Equation 9.14). Similarly, using the calibration-refinement factorization (Equation 9.5), forecast performance depicted in a 2×2 contingency table can be fully captured using the false alarm ratio FAR (Equations 9.11 and 9.5b), its counterpart in Equation 9.5d, and the probability $p(y_1) = (a+b)/n$ defining the refinement distribution. Other triplets of verification measures can also be used jointly to illuminate the data in a 2 × 2 contingency table (although not any three scalar statistics calculated from a 2 × 2 table will fully represent its information content). For example, Stephenson (2000) suggests use of H and F together with the bias ratio B, calling this the BHF representation. He also notes that, jointly, the likelihood ratio θ and Peirce Skill Score PSS represent the same information as H and F, so that these two statistics together with either s or B will also fully represent the 2 × 2 table. The joint characterization using H, F, and s is also sometimes used in the medical literature (Pepe, 2003). Stephenson et al. (2008a) and Brill (2009) analyze properties of various 2 × 2 performance measures in terms of H, B, and s.

Consistent with the ideas presented in Chapter 3, the understanding of multivariable statistics can be enhanced through well-designed graphics. Roebber (2009) suggests summarizing 2 × 2 verification tables in a 2-dimensional diagram whose axes are $1-FAR$ and H, an example of which is shown in Figure 9.2. In this diagram the solid contours show isolines of constant TS (Equation 9.8) and dashed radial lines indicate forecast bias (Equation 9.10). The short line segments in this particular example of the diagram connect results for forecasts before (circles) and after (triangles) human intervention, highlighting the resulting improvements in both H and TS. The Roebber (2009) performance diagram is not a complete 3-dimensional summary of the joint distribution, since it does not reflect the number of correct "no" forecasts, but may nevertheless be informative when summarizing performance of forecasts of rare events, or spatial forecasts where the quantity d in Figure 9.1 may be ambiguous or unknown.

Wilks (2016a) proposed two diagnostic verification diagrams for the 2 × 2 forecast setting in Figure 9.1, based on one of the factorizations of the joint distribution of forecasts and observations. The first, called the $H–F$ diagram, plots verification results in the space defined by H (Equations 9.6a and 9.13) and F (Equations 9.6c and 9.13), in common with the ROC diagram (Section 9.4.6). Together with the base rate s (Equation 9.14) these quantities completely specify the likelihood-base rate factorization (Equation 9.3). Mason (2003) points out that isolines of the bias ratio B (Equation 9.10) exhibit slope

FIGURE 9.2 A Roebber (2009) performance diagram, showing paired results for convective storm initiation forecasts before (circles) and after (triangles) human intervention, as functions of $1-FAR$ and H. Curves show TS and dashed lines (with labels outside the figure frame) show B, as functions of the two coordinate axes. *Modified from Roberts et al. (2012). © American Meteorological Society. Used with permission.*

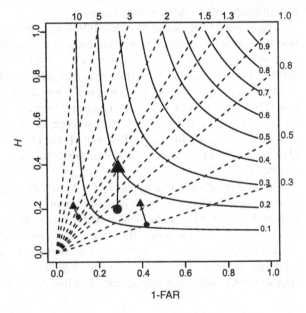

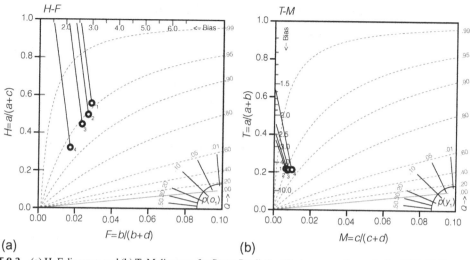

FIGURE 9.3   (a) H–F diagram, and (b) T–M diagram, for Storm Prediction Center convective outlooks, 2002–2012, at 6-h ("1"), 18.5-h ("2") 30-h ("3"), and 48-h ("4") lead times. Angles of vectors from the horizontal represent the event probability in panel (a), and relative frequencies of "yes" forecasts in panel (b), as indicated by the legends in the lower-right corners. Bias ratios are shown on the indicated diagram margins, and dashed curves show isopleths of the Odds Ratio Skill Score, Q (Equation 9.21). *From Wilks (2016a).*

$$\frac{dH}{dF} = \frac{-(1-s)}{s} \tag{9.22}$$

and intersect the H axis at $H = B$. Therefore s can be represented by the angle from the horizontal of a vector in the H–F space given by

$$\alpha = \tan^{-1}\left(\frac{1-s}{s}\right) \tag{9.23}$$

The full joint distribution for a set of forecasts can be represented as a point in the H–F plane locating the likelihood distribution probabilities, together with an arrow pointing upward and to the left at an angle defined by Equation 9.23. An example is shown in Figure 9.3a, where the legend in the lower-right corner indicates values of $s = p(o_1)$, and the upward arrows locate the values of B. The dashed contours show isolines of the Odds Ratio Skill Score Q (Equation 9.21), indicating that more skillful forecasts are located toward the upper-left portion of the diagram. Either *EDI* (Equation 9.15) or *PSS* (Equation 9.18) could be used instead since both may be expressed as functions of H and F only.

Figure 9.3b shows an example of the T–M diagram, which visually portrays the calibration-refinement factorization (Equation 9.2) for the 2×2 verification setting. In this diagram, a set of forecasts is located as a point defined by $T = a/(a+b)$ (Equation 9.5a) and $M = c/(c+d)$ (Equation 9.5c). Representation of the calibration-refinement factorization is completed by the relative frequency of "yes" forecasts, $p(y_1) = (a+b)/n$ which, analogously to Equations 9.22 and 9.23, can be represented by an angle in the T–M space. Arrows drawn at these angles from the plotted points locate the bias on the margin of the diagram, with the legend in the lower-right corner again indicating the correspondence between these directions and the frequency of "yes" forecasts. The dashed curves in Figure 9.3b show isolines of Q (Equation 9.21), indicating that more skillful forecasts are located toward the upper-left portion of the diagram.

9.2.5. Conversion of Probabilistic to Nonprobabilistic Forecasts

The MOS system from which the nonprobabilistic precipitation amount forecasts in Table 7.8 were taken actually produces probability forecasts for discrete precipitation amount classes. The publicly issued precipitation amount forecasts in the table were then derived by converting the underlying probabilities to the nonprobabilistic format by choosing one and only one of the possible categories. This unfortunate information degradation is distressing, but is sometimes advocated under the rationale that nonprobabilistic forecasts are easier to understand. However, the loss of information content is detrimental to the forecast users.

For a dichotomous predictand, the conversion from a probabilistic to a nonprobabilistic format requires selection of a threshold probability, above which the forecast will be "yes," and below which the forecast will be "no." This procedure seems simple enough; however, the proper threshold to choose depends on the user of the forecast and the particular decision problem(s) to which that user will apply the forecast. If a decision problem has undergone quantitative analysis, for example, as that described in Section 9.9 or by Manzato and Jolliffe (2017), then the choice of the appropriate threshold may be clear. However, different decision problems will require different threshold probabilities, and this is the crux of the information-loss issue. In a very real sense, the conversion from a probabilistic to a nonprobabilistic format amounts to the forecaster making decisions for the forecast users, but without knowing the particulars of their decision problems. Often the choice of threshold is made which optimizes the value of some verification score, but different scores will be optimized for different thresholds, and indeed a score can be constructed that corresponds to any chosen threshold (Mason, 1979). Necessarily, then, the conversion of a probabilistic forecast to a nonprobabilistic format is arbitrary.

Example 9.2. Effects of Different Thresholds on Conversion to Nonprobabilistic Forecasts

It is instructive to examine the procedures used to convert probabilistic to nonprobabilistic forecasts. Table 9.2 contains a verification data set of probability-of-precipitation forecasts, issued for the United States during the period October 1980 through March 1981. Here the joint distribution of the $I = 12$ possible forecasts and the $J = 2$ possible observations is presented in the form of the calibration-refinement factorization (Equation 9.2). For each allowable forecast probability, y_i, the conditional probability $p(o_1 \mid y_i)$ indicates the relative frequency of the event $j = 1$ (precipitation occurrence) for these $n = 12,402$ forecasts. The marginal probabilities $p(y_i)$ indicate the relative frequencies with which each of the $I = 12$ possible forecast values was used.

These precipitation occurrence forecasts were issued as probabilities. If it had been intended to convert them first to a nonprobabilistic rain/no rain format, a threshold probability would have been

TABLE 9.2 Verification data for subjective 12–24h lead time probability-of-precipitation forecasts for the United States during October 1980–March 1981, expressed in the form of the calibration-refinement factorization (Equation 9.2) of the joint distribution of these forecasts and observations.

y_i	0.00	0.05	0.10	0.20	0.30	0.40	0.50	0.60	0.70	0.80	0.90	1.00
$p(o_1 \mid y_i)$	.006	.019	.059	.150	.277	.377	.511	.587	.723	.799	.934	.933
$p(y_i)$	.4112	.0671	.1833	.0986	.0616	.0366	.0303	.0275	.0245	.0220	.0170	.0203

There are $I = 12$ allowable values for the forecast probabilities, y_i, and $J = 2$ events ($j = 1$ for precipitation and $j = 2$ for no precipitation). The sample climatological relative frequency is 0.162, and the sample size is $n = 12,402$.
From Murphy and Daan (1985).

chosen in advance. There are many possibilities for this choice, each of which gives different results. The two simplest approaches are used rarely, if ever, in operational practice. The first procedure is to forecast the more likely event, which corresponds to selecting a threshold probability of 0.50. The other simple approach is to use the climatological relative frequency of the event being forecast as the threshold probability. For the data set in Table 9.2 this relative frequency is $\Sigma_i \, p(o_j|y_i)p(y_i) = 0.162$ (Equation 2.14), although in practice this probability threshold would need to have been estimated in advance using historical climatological data, and likely would have been estimated separately for the different locations whose data are aggregated in the table. Forecasting the more likely event turns out to maximize the expected values of both the proportion correct (Equation 9.7) and the Heidke Skill Score (Equation 9.17), and using the climatological relative frequency for the probability threshold maximizes the expected Peirce Skill Score (Equation 9.18) (Mason, 1979).

The two methods for choosing the threshold probability that are most often used operationally are based on the threat score (Equation 9.8) and the bias ratio (Equation 9.10) for 2×2 contingency tables. For each possible choice of a threshold probability, a different 2×2 contingency table, in the form of Figure 9.1a, results, and therefore different values of TS and B are obtained. When using the threat score to choose the threshold, that threshold producing the maximum TS is selected. When using the bias ratio, that threshold producing, as nearly as possible, no bias ($B = 1$) is chosen.

Figure 9.4 illustrates the dependence of the bias ratio and threat score on the threshold probability for the data given in Table 9.2. Also shown are the hit rates H and false alarm ratios FAR that would be obtained. The threshold probabilities that would be chosen according to the climatological

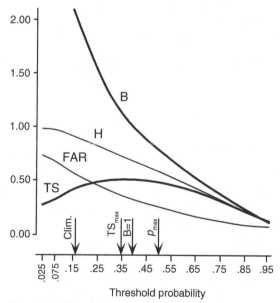

Threshold probability

FIGURE 9.4 Derivation of candidate threshold probabilities for converting the probability-of-precipitation forecasts in Table 9.2 to nonprobabilistic rain/no rain forecasts. The Clim threshold indicates a forecast of rain if the probability is higher than the climatological probability of precipitation, TS_{max} is the threshold that would maximize the threat score of the resulting nonprobabilistic forecasts, the $B=1$ threshold would produce unbiased forecasts, and the p_{max} threshold would produce nonprobabilistic forecasts of the more likely of the two events. Also shown (lighter lines) are the hit rates H and false alarm ratios FAR for the resulting 2×2 contingency tables.

relative frequency (Clim), the maximum threat score (TS_{max}), unbiased nonprobabilistic forecasts ($B = 1$), and maximum probability (p_{max}) are indicated by the arrows at the bottom of the figure. For example, choosing the overall relative frequency of precipitation occurrence, 0.162, as the threshold results in forecasts of PoP $= 0.00, 0.05$, and 0.10 being converted to "no rain," and the other forecasts being converted to "rain." This would have resulted in $n[p(y_1)+p(y_2)+p(y_3)] = 12,402[0.4112+ 0.0671+0.1833] = 8205$ "no" forecasts, and $12,402–8205 = 4197$ "yes" forecasts. Of the 8205 "no" forecasts, we can compute, using the multiplicative law of probability (Equation 2.11), that the proportion of occasions that "no" was forecast but precipitation occurred was $p(o_1|y_1)p(y_1)+p(o_1|y_2)p(y_2)+p(o_1|y_3)p(y_3) = (.006)(.4112)+(.019)(.0671)+(.059)(.1833) = 0.0146$. This relative frequency is c/n in Figure 9.1, so that $c = (0.0146)(12,402) = 181$, and $d = 8205–181 = 8024$. Similarly, we can compute that, for this cutoff, $a = 12,402[(0.150)(0.0986)+...+(0.933)(0.203)] = 1828$ and $b = 2369$. The resulting 2×2 table yields $B = 2.09$, and $TS = 0.417$. By contrast, the threshold maximizing TS is near 0.35, which also would have resulted in overforecasting of precipitation occurrence. ◇

9.2.6. Extensions for Multicategory Discrete Predictands

Nonprobabilistic forecasts for discrete predictands are not limited to the 2×2 format, although that simple situation is the most commonly encountered and the easiest to understand. In some settings it is natural or desirable to consider and forecast more than two discrete MECE events. The left side of Figure 9.5, in boldface type, shows a generic contingency table for the case of $I = J = 3$ possible forecasts and events. Here the counts for each of the nine possible forecast and event pair outcomes are denoted by the letters r through z, yielding a total sample size $n = r+s+t+u+v+w+x+y+z$. As before, dividing each of the nine counts in this 3×3 contingency table by the sample size yields a sample estimate of the joint distribution of forecasts and observations, $p(y_i, o_j)$.

Of the accuracy measures listed in Equations 9.7 through 9.9, only the proportion correct (Equation 9.7) generalizes directly to situations with more than two forecast and event categories. Regardless of the size of I and J, the proportion correct is still given by the number of correct forecasts

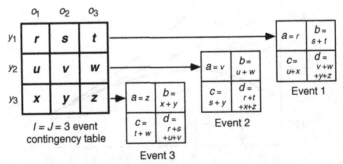

FIGURE 9.5 Contingency table for the $I = J = 3$ nonprobabilistic forecast verification situation (bold), and its reduction to three 2×2 contingency tables. Each 2×2 contingency table is constructed by regarding one of the three original events as "the" event being forecast, and the remaining two original events combined as complementary, that is, not the forecast event. For example, the 2×2 table for Event 1 lumps Event 2 and Event 3 as the single event "not Event 1." The letters a, b, c, and d are used in the same sense as in Figure 9.1a. Performance measures specific to the 2×2 contingency tables can then be computed separately for each of the resulting tables. This procedure generalizes easily to square forecast verification contingency tables with arbitrarily many forecast and event categories.

divided by the total number of forecasts, n. This number of correct forecasts is obtained by adding the counts along the diagonal from the upper left to the lower-right corners of the contingency table. In Figure 9.5, the numbers r, v, and z represent the numbers of occasions when the first, second, and third events were correctly forecast, respectively. Therefore in the 3×3 table represented in this figure, the proportion correct would be $PC = (r+v+z)/n$.

The other statistics listed in Section 9.2.2 pertain only to the dichotomous, yes/no forecast situation. In order to apply these to nonprobabilistic forecasts that are not dichotomous, it is necessary to collapse the $I = J > 2$ contingency table into a series of 2×2 contingency tables. Each of these 2×2 tables is constructed, as indicated in Figure 9.5, by considering "the" forecast event in distinction to the complementary, "not the forecast event." This complementary event simply is constructed as the union of the $J-1$ remaining events. For example, in Figure 9.5 the 2×2 contingency table for Event 1 lumps Events 2 and 3 as "not Event 1." Thus the number of times Event 1 is correctly forecast is still $a = r$, but the number of times it is incorrectly forecast is $b = s+t$. From the standpoint of this collapsed 2×2 contingency table, whether the incorrect forecast of Event 1 was followed by Event 2 or Event 3 is unimportant. Similarly, the number of times "not Event 1" is correctly forecast is $d = v+w+y+z$, and includes cases where Event 2 was forecast but Event 3 occurred, and Event 3 was forecast but Event 2 occurred.

Attributes for 2×2 contingency tables can be computed for any or all of the 2×2 tables constructed in this way from larger square tables. For the 3×3 contingency table in Figure 9.5, the bias (Equation 9.10) for forecasts of Event 1 would be $B_1 = (r+s+t)/(r+u+x)$, the bias for forecasts of Event 2 would be $B_2 = (u+v+w)/(s+v+y)$, and the bias for forecasts of Event 3 would be $B_3 = (x+y+z)/(t+w+z)$.

Example 9.3. A Set of Multicategory Forecasts

The left-hand side of Table 9.3 shows a 3×3 verification contingency table for forecasts of freezing rain (y_1), snow (y_2), and rain (y_3). These are nonprobabilistic MOS forecasts, conditional on the occurrence of some form of precipitation, for the Eastern region of the United States, for October through March of 1983/1984 through 1988/1989. For each of the three precipitation types, a 2×2 contingency table can be constructed, following Figure 9.5, that summarizes the performance of forecasts of that precipitation type in distinction to the other two precipitation types together. Table 9.3 also includes forecast attributes from Section 9.2.2 for each 2×2 decomposition of the 3×3 contingency table. These are reasonably consistent with each other for a given 2×2 table, and indicate that the rain forecasts were slightly superior to the snow forecasts, but that the freezing rain forecasts were substantially less successful, with respect to most of these measures. ◇

The Heidke and Peirce Skill Scores can be extended easily to verification problems where there are more than $I = J = 2$ possible forecasts and events. The formulae for these scores in the more general case can be written most easily in terms of the joint distribution of forecasts and observations, $p(y_i, o_j)$, and the marginal distributions of the forecasts, $p(y_i)$ and of the observations, $p(o_j)$. For the Heidke Skill Score this more general form is

$$HSS = \frac{\sum_{i=1}^{I} p(y_i, o_i) - \sum_{i=1}^{I} p(y_i)p(o_i)}{1 - \sum_{i=1}^{I} p(y_i)p(o_i)}, \tag{9.24}$$

and the higher-dimensional generalization of the Peirce Skill Score is

TABLE 9.3 Nonprobabilistic MOS Forecasts for Freezing Rain (y_1), Snow (y_2), and Rain (y_3), Conditional on Occurrence of Some Form of Precipitation, for the Eastern Region of the United States During Cool Seasons of 1983/1984 Through 1988/1989

Full 3 × 3 Contingency Table				Freezing Rain		Snow		Rain	
o_1	o_2	o_3		o_1	Not o_1	o_2	Not o_2	o_3	Not o_3
y_1 50	91	71	y_1	50	162	y_2 2364	217	y_3 3288	259
y_2 47	2364	170	Not y_1 101		6027	Not y_2 296	3463	Not y_3 241	2552
y_3 54	205	3288							
				TS = 0.160		TS = 0.822		TS = 0.868	
				$\theta = 18.4$		$\theta = 127.5$		$\theta = 134.4$	
				B = 1.40		B = 0.97		B = 1.01	
				FAR = 0.764		FAR = 0.084		FAR = 0.073	
				H = 0.331		H = 0.889		H = 0.932	
				F = 0.026		F = 0.059		F = 0.092	

The verification data are presented as a 3 × 3 contingency table on the left, and then as three 2 × 2 contingency tables for each of the three precipitation types. Also shown are scalar attributes from Section 9.2.2 for each of the 2 × 2 tables. The sample size is $n = 6340$. Data are from Goldsmith (1990).

$$PSS = \frac{\sum_{i=1}^{I} p(y_i, o_i) - \sum_{i=1}^{I} p(y_i)p(o_i)}{1 - \sum_{j=1}^{J} [p(o_j)]^2}. \tag{9.25}$$

Equation 9.24 reduces to Equation 9.17, and Equation 9.25 reduces to Equation 9.18, for $I = J = 2$.

Using Equation 9.24, the Heidke score for the 3 × 3 contingency table in Table 9.3 would be computed as follows. The proportion correct, PC $= \Sigma_i p(y_i, o_i) = (50/6340)+(2364/6340)+(3288/6340) = 0.8994$. The proportion correct for the random reference forecasts would be $\Sigma_i p(y_i)p(o_i) = (.0334)(.0238)+(.4071)(.4196)+(.5595)(.5566) = 0.4830$. Here, for example, the marginal probability $p(y_1) = (50+91+71)/6340 = 0.0344$. The proportion correct for perfect forecasts is of course one, yielding HSS $= (.8944 - .4830)/(1 - .4830) = 0.8054$. The computation for the Peirce Skill Score, Equation 9.25, is the same except that a different reference proportion correct is used in the denominator only. This is the unbiased random proportion $\Sigma_i[p(o_i)^2] = .0238^2 + .4196^2 + .5566^2 = 0.4864$. The Peirce Skill Score for this 3 × 3 contingency table is then PSS $= (.8944 - .4830)/(1 - .4864) = 0.8108$. The difference between the HSS and the PSS for these data is small, because the forecasts exhibit little bias.

There are many more degrees of freedom in the general $I \times J$ contingency table setting than in the simpler 2×2 problem. In particular $I \times J - 1$ elements are necessary to fully specify the contingency table, so that a scalar score must summarize much more even in the 3 x 3 setting as compared to the 2 x 2 problem. Accordingly, the number of possible scalar skill scores that are plausible candidates increases rapidly with the size of the verification table. The notion of *equitability* for skill scores describing performance of nonprobabilistic forecasts of discrete predictands was proposed by Gandin and Murphy (1992) to define a restricted set of these yielding equal (zero) scores for random or constant forecasts.

When three or more events having a natural ordering are being forecast, it is usually required in addition that multiple-category forecast misses are scored as worse forecasts than single-category misses. Equations 9.24 and 9.25 both fail this requirement, as they depend only on the proportion correct. Gerrity (1992) suggested a family of equitable (in the sense of Gandin and Murphy, 1992) skill scores that are also sensitive to distance in this way and appear to provide generally reasonable results for rewarding correct forecasts and penalizing incorrect ones (Livezey, 2003). The computation of Gandin-Murphy skill scores involves first defining a set of scoring weights $w_{i,j}$, $i = 1, \ldots, I$, $j = 1, \ldots, J$; each of which is applied to one of the joint probabilities $p(y_j, o_j)$, so that in general a *Gandin-Murphy Skill Score* is computed as linear weighted sum of the elements of the contingency table

$$\text{GMSS} = \sum_{i=1}^{I} \sum_{j=1}^{J} p(y_i, o_j) \, w_{i,j}. \tag{9.26}$$

As noted in Section 9.2.4 for the simple case of $I = J = 2$, when linear scoring weights are required as one of the equitability criteria in the 2 x 2 setting, the result is the Peirce Skill Score (Equation 9.18). More constraints are required for larger verification problems, and Gerrity (1992) suggested the following approach to defining the scoring weights based on the sample climatology $p(o_j)$. First, define the sequence of $J - 1$ odds ratios

$$D(j) = \frac{1 - \sum_{r=1}^{j} p(o_r)}{\sum_{r=1}^{j} p(o_r)}, \quad j = 1, \ldots, J - 1, \tag{9.27}$$

where r is a dummy summation index. The scoring weights for correct forecasts are then

$$w_{j,j} = \frac{1}{J-1} \left[\sum_{r=1}^{j-1} \frac{1}{D(r)} + \sum_{r=j}^{J-1} D(r) \right], \quad j = 1, \ldots, J; \tag{9.28a}$$

and the weights for the incorrect forecasts are

$$w_{i,j} = \frac{1}{J-1} \left[\sum_{r=1}^{i-1} \frac{1}{D(r)} + \sum_{r=j}^{J-1} D(r) - (j - i) \right], \quad 1 \le i < j \le J. \tag{9.28b}$$

The summations in Equation 9.28 are taken to be equal to zero if the lower index is larger than the upper index. These two equations fully define the $I \times J$ scoring weights when symmetric errors are penalized equally, that is, when $w_{i,j} = w_{j,i}$. Equation 9.28a gives more credit for correct forecasts of rarer events and less credit for correct forecasts of common events. Equation 9.28b also accounts for the intrinsic rarity of the J events, and increasingly penalizes errors for greater differences between the forecast category i and the observed category j, through the penalty term $(j-i)$. Each scoring weight in Equation 9.28 is used together with the corresponding member of the joint distribution $p(y_j, o_j)$ in Equation 9.26 to compute the skill score. When the weights for the Gandin-Murphy Skill Score are computed according to Equations 9.27 and 9.26, the result is sometimes called the *Gerrity skill score*.

Example 9.4. Gerrity Skill Score for a 3 x 3 Verification Table

Table 9.3 includes a 3 x 3 contingency table for nonprobabilistic forecasts of freezing rain, snow, and rain, conditional on the occurrence of precipitation of some kind. Since there is not an obvious ordering of these three categories, use of *HSS* (Equation 9.24) or *PSS* (Equation 9.25) might be preferred in this case, but for convenience these data will be used in this example to illustrate computation of the Gerrity skill score. Figure 9.6a shows the corresponding joint sample probability distribution $p(y_i, o_j)$, calculated by dividing the counts in the contingency table by the sample size, $n = 6340$. Figure 9.6a also shows the sample climatological distribution $p(o_j)$, computed by summing the columns of the joint distribution.

The Gerrity (1992) scoring weights for the Gandin-Murphy Skill Score (Equation 9.26) are computed from these sample climatological relative frequencies using Equations 9.27 and 9.28. First, Equation 9.27 yields the $J-1 = 2$ likelihood ratios $D(1) = (1-.0238)/.0238 = 41.02$, and $D(2) = [1-(.0238+.4196)]/(.0238+.4196) = 1.25$. The rather large value for $D(1)$ reflects the fact that freezing rain was observed rarely, on only approximately 2% of the precipitation days during the period considered. The scoring weights for the three possible correct forecasts, computed using Equation 9.28a, are

$$w_{1,1} = \frac{1}{2}(41.02 + 1.25) = 21.14, \tag{9.29a}$$

(a) Joint distribution

Forecast \ Observed	Frz Rain	Snow	Rain
Frz Rain	$p(y_1,o_1)$ = .0079	$p(y_1,o_2)$ = .0144	$p(y_1,o_3)$ = .0112
Snow	$p(y_2,o_1)$ = .0074	$p(y_2,o_2)$ = .3729	$p(y_2,o_3)$ = .0268
Rain	$p(y_3,o_1)$ = .0085	$p(y_3,o_2)$ = .0323	$p(y_3,o_3)$ = .5186
	$p(o_1)$ = .0238	$p(o_2)$ = .4196	$p(o_3)$ = .5566

(b) Scoring weights

$w_{1,1}$ = 21.14	$w_{1,2}$ = 0.13	$w_{1,3}$ = −1.00
$w_{2,1}$ = 0.13	$w_{2,2}$ = 0.64	$w_{2,3}$ = −0.49
$w_{3,1}$ = −1.00	$w_{3,2}$ = −0.49	$w_{3,4}$ = 0.41

FIGURE 9.6 (a) Joint distribution of forecasts and observations for the 3 × 3 contingency table in Table 9.3, with the marginal probabilities for the three observations (the sample climatological probabilities). (b) The Gerrity (1992) scoring weights computed from the sample climatological probabilities.

$$w_{2,2} = \frac{1}{2}\left(\frac{1}{41.02} + 1.25\right) = 0.64, \tag{9.29b}$$

and

$$w_{3,3} = \frac{1}{2}\left(\frac{1}{41.02} + \frac{1}{1.25}\right) = 0.41; \tag{9.29c}$$

and the weights for the incorrect forecasts are

$$w_{1,2} = w_{2,1} = \frac{1}{2}(1.25 - 1) = 0.13, \tag{9.30a}$$

$$w_{2,3} = w_{3,2} = \frac{1}{2}\left(\frac{1}{41.02} - 1\right) = -0.49 \tag{9.30b}$$

and

$$w_{3,1} = w_{1,3} = \frac{1}{2}(-2) = -1.00. \tag{9.30c}$$

These scoring weights are arranged in Figure 9.6b in positions corresponding to the joint probabilities in Figure 9.6a to which they pertain.

The scoring weight $w_{1,1} = 21.14$ is much larger than the others in order to reward correct forecasts of the rare freezing rain events. Correct forecasts of snow and rain are credited with much smaller positive values, with $w_{3,3} = 0.41$ for rain being smallest because rain is the most common event. The scoring weight $w_{2,3} = -1.00$ is the minimum value according to the Gerrity algorithm, produced because the $(j-i) = 2$-category error (cf. Equation 9.28b) is the most severe possible when there is a natural ordering among the three outcomes. The penalty for an incorrect forecast of snow when rain occurs, or for rain when snow occurs (Equation 9.30b), is moderately large because these two events are relatively common. Mistakenly forecasting freezing rain when snow occurs, or vice versa, actually receives a small positive score because the frequency $p(o_1)$ is so small.

Finally, the Gandin-Murphy Skill Score in Equation 9.26 is computed by summing the products of pairs of joint probabilities and scoring weights in corresponding positions in Figure 9.6. That is, GMSS = $(.0079)(21.14)+(.0144)(.13)+(.0112)(-1)+(.0074)(.13)+(.3729)(.64)+(.0268)(-.49)+(.0085)$ $(-1)+(.0323)(-.49)+(.5186)(.41) = 0.57.$ ◇

The Gerrity skill score is one of an essentially infinite number of skill scores that can be constructed using the framework of Equation 9.26. For example, Rodwell et al. (2010) constructed such a skill score, for three-category precipitation forecasts. Defining p_1, p_2, and p_3 as the climatological probabilities of "dry," "light precipitation," and "heavy precipitation" categories, respectively, they construct the matrix of scoring weights in Equation 9.26 as

$$[W] = \frac{1}{2}\begin{bmatrix} 0 & \dfrac{1}{1-p_1} & \dfrac{1}{p_3} + \dfrac{1}{1-p_1} \\[2ex] \dfrac{1}{p_1} & 0 & \dfrac{1}{p_3} \\[2ex] \dfrac{1}{p_1} + \dfrac{1}{1-p_3} & \dfrac{1}{1-p_3} & 0 \end{bmatrix}, \tag{9.31}$$

in which $w_{i,j}$ is the entry in the ith row and jth column of $[W]$. Rodwell et al. (2010) recommend choosing the boundary between the two wet categories to yield $p_2/p_3 = 2$.

9.3. NONPROBABILISTIC FORECASTS FOR CONTINUOUS PREDICTANDS

A different set of verification measures generally is applied to forecasts of continuous variables. Continuous variables in principle can take on any value in a specified segment of the real line, rather than being limited to a finite number of discrete events. Temperature is an example of a continuous variable. In practice, however, forecasts and observations for continuous atmospheric variables are made using a finite number of discrete values. For example, temperature forecasts usually are rounded to integer degrees. It would be possible to deal with this kind of forecast verification data in discrete form, but there are usually so many allowable values of forecast and observation pairs that the resulting contingency tables would become unwieldy and possibly quite sparse. Just as discretely reported observations of continuous variables were treated as continuous quantities in Chapter 4, it is convenient and useful to treat the verification of (operationally discrete) forecasts of continuous quantities in a continuous framework as well.

Conceptually, the joint distribution of forecasts and observations is again of fundamental interest. This distribution will be the continuous analog of the discrete joint distribution of Equation 9.1. Because of the finite nature of the verification data, however, explicitly using the concept of the joint distribution in a continuous setting generally requires that a parametric distribution such as the bivariate normal (Equation 4.31) be assumed and fit. Parametric distributions and other statistical models occasionally are assumed for the joint distribution of forecasts and observations or their factorizations (e.g., Bradley et al., 2003; Katz et al., 1982; Krzysztofowicz and Long, 1991; Murphy and Wilks, 1998), but it is far more common that scalar performance and skill measures, computed using individual forecast/observation pairs, are used in verification of continuous nonprobabilistic forecasts.

9.3.1. Scalar Accuracy Measures

Scalar accuracy measures for evaluation of nonprobabilistic forecasts of continuous predictands are generally functions of the differences between pairs of forecasts and observations. It is important that the function chosen for evaluation of forecasts of this type be *consistent* (Murphy and Daan, 1985) with the process used to develop the forecast. In the case of a human forecaster, whose underlying judgment about the future weather condition will be characterized by a subjective probability distribution (Section 7.10), the extraction of a single real-valued forecast quantity from this distribution will depend on a *directive* (Murphy and Daan, 1985). Typical directives in this context are to forecast a quantity such as the mean, or the median or some other quantile, of the subjective distribution. A scoring function that is optimized by the forecasts derived from a particular directive is consistent with that directive (Gneiting, 2011a; Murphy and Daan, 1985). Conversely, once a scoring function has been decided upon, the forecaster should be informed (or the nonhuman forecasting algorithm should be constructed in light) of either its mathematical structure or of the directive with which it is consistent (Gneiting, 2011a).

Only two scalar measures of forecast accuracy for continuous predictands are in common use in meteorology and climatology, although Gneiting (2011a) lists some others that can be found in the statistics, econometrics, and nonmeteorological forecasting literature. The first of the two is the *mean absolute error*,

$$\text{MAE} = \frac{1}{n} \sum_{k=1}^{n} |y_k - o_k|. \tag{9.32}$$

Here (y_k, o_k) is the kth of n pairs of forecasts and observations. The MAE is the arithmetic average of the absolute values of the differences between the members of each pair. Clearly the MAE is zero if the forecasts are perfect (each $y_k = o_k$), and increases as discrepancies between the forecasts and observations become larger. The MAE can be interpreted as a typical magnitude for the forecast error in a given verification data set.

The MAE is consistent with the directive to forecast the median. It is a symmetric piecewise linear function, in that over- and underforecasts of the same magnitude are penalized equally. Penalizing these two types of errors differently, with an asymmetric piecewise linear function, would be consistent with forecasting directives involving different distribution quantiles (Gneiting, 2011b).

The MAE often is used for verification of temperature forecasts in the United States. Figure 9.7 shows seasonally stratified MAE for MOS minimum temperature forecasts over a portion of the country, during December 2004-November 2005, as functions of lead time. Winter temperatures are least accurate, summer temperatures are most accurate, and accuracy for the transition seasons is intermediate.

The *mean squared error* is the other common accuracy measure for continuous nonprobabilistic forecasts,

$$\text{MSE} = \frac{1}{n} \sum_{k=1}^{n} (y_k - o_k)^2. \tag{9.33}$$

The MSE is the average of the squared differences between the forecast and observation pairs. It is consistent with the directive to forecast the mean. Since the MSE is computed by squaring the forecast errors, it will be more sensitive to larger errors than will the MAE, and so will also be more sensitive to outliers. Squaring the errors necessarily produces positive terms in Equation 9.33, so the MSE

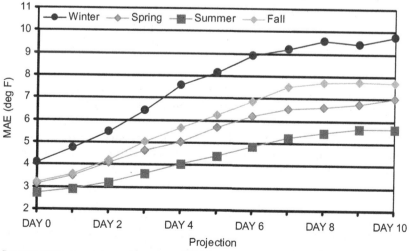

FIGURE 9.7 Seasonal MAEs for minimum temperature MOS forecasts over a portion of the United States, during December 2004 through November 2005, as functions of lead times. *From www.nws.noaa.gov/tdl/synop.*

increases from zero for perfect forecasts through larger positive values as the discrepancies between forecasts and observations become increasingly large. The similarity between Equations 9.33 and 3.6 indicates that forecasting the climatological mean on each of the n occasions being evaluated will yield MSE essentially equal to the climatological variance of the predictand o. On the other hand, forecasting a random draw from the climatological distribution yields MSE that is double the climatological variance (Hayashi, 1986). Sometimes the MSE is expressed as its square root, $\text{RMSE} = \sqrt{\text{MSE}}$, which has the same physical dimensions as the forecasts and observations, and like the MAE can also be thought of as a typical magnitude for forecast errors.

Initially, it might seem that the correlation coefficient (Equation 3.28) could be another useful accuracy measure for nonprobabilistic forecasts of continuous predictands. However, although the correlation does reflect linear association between two variables (in this case, forecasts and observations), it is sensitive to outliers, and is not sensitive to either conditional or unconditional biases that may be present in the forecasts. This latter problem can be appreciated by considering an algebraic manipulation of the MSE (Murphy, 1988):

$$\text{MSE} = (\bar{y} - \bar{o})^2 + s_y^2 + s_o^2 - 2 s_y s_o r_{yo}. \tag{9.34}$$

Here $r_{y,o}$ is the product-moment correlation between the forecasts and observations, s_y and s_o are the standard deviations of the marginal distributions of the forecasts and observations, respectively, and the first term in Equation 9.34 is the square of the *mean error*,

$$\text{ME} = \frac{1}{n} \sum_{k=1}^{n} (y_k - o_k) = \bar{y} - \bar{o}. \tag{9.35}$$

The Mean Error is simply the difference between the average forecast and average observation, and therefore expresses the unconditional bias of the forecasts. Equation 9.35 differs from Equation 9.33 in that the individual forecast errors are not squared before they are averaged. Forecasts that are, on average, too large will exhibit $\text{ME} > 0$ and forecasts that are, on average, too small will exhibit $\text{ME} < 0$. It is important to note that the bias gives no information about the typical magnitude of individual forecast errors, and indeed the second equality in Equation 9.35 shows that it need not relate individual forecast and observation pairs. It is therefore not in itself an accuracy measure.

Returning to Equation 9.34, it can be seen that forecasts that are more highly positively correlated with the observations will exhibit lower MSE, other factors being equal. However, since the MSE can be written with the correlation $r_{y,o}$ and the bias (ME) in separate terms, we can imagine forecasts that may be highly correlated with the observations, but with sufficiently severe bias that they would be useless at face value. A hypothetical set of temperature forecasts could, for example, be exactly half of the subsequently observed temperatures, and so exhibit conditional bias. For convenience, imagine that these temperatures are nonnegative. A scatterplot of the observed temperatures versus the corresponding forecasts would exhibit all points falling perfectly on a straight line ($r_{y,o} = 1$), but the slope of that line would be 2. The bias, or mean error, would be $\text{ME} = n^{-1} \Sigma_k (f_k - o_k) = n^{-1} \Sigma_k (0.5\, o_k - o_k)$, or the negative of half of the average observation. This bias would be squared in Equation 9.34, leading to a very large MSE. A similar situation would result if all the forecasts were exactly 10 degrees colder than the observed temperatures, and so would be unconditionally biased. The correlation $r_{y,o}$ would still be one, the points on the scatterplot would fall on a straight line (this time with unit slope), the ME would be $-10°$, and the MSE would be inflated by $(10°)^2$. The definition of correlation (Equation 3.29) shows clearly why these problems would occur: the means of the two variables being correlated are separately subtracted, and any differences in scale are removed by separately dividing by the two standard deviations, before calculating the correlation. Therefore any mismatches between either location or scale between the forecasts and observations are not reflected in the result. The Taylor diagram (e.g., - Figure 9.30) is an interesting graphical approach for separating the contributions of the correlation and the standard deviations in Equation 9.34 to the RMSE, when forecast biases are zero or are ignored.

9.3.2. Skill Scores

Skill scores, or relative accuracy measures of the form of Equation 9.4, can easily be constructed using the MAE, MSE, or RMSE as the underlying accuracy statistics. Usually the reference, or control, forecasts are provided either by the climatological values of the predictand or by persistence (i.e., the previous value in a sequence of observations). For the MSE, the accuracies of these two references are, respectively,

$$\text{MSE}_{\text{clim}} = \frac{1}{n} \sum_{k=1}^{n} (\bar{o} - o_k)^2 \tag{9.36a}$$

and

$$\text{MSE}_{\text{pers}} = \frac{1}{n-1} \sum_{k=2}^{n} (o_{k-1} - o_k)^2. \tag{9.36b}$$

Analogous equations can be written for the MAE, in which the squaring function would be replaced by the absolute value function.

In Equation 9.36a, it is implied that the climatological average value does not change from forecast occasion to forecast occasion (i.e., as a function of the index, k). If this implication is true, then MSE_{clim} in Equation 9.36a is an estimate of the sample variance of the predictand (compare Equation 3.6). In some applications the climatological value of the predictand will be different for different forecasts. For example, if daily temperature forecasts at a single location were being verified over the course of several months, the index k would represent time, and the climatological average temperature usually would change smoothly as a function of the date. In this case the quantity being summed in Equation 9.36a would be $(c_k - o_k)^2$, with c_k being the climatological value of the predictand on day k. Failing to account for a time-varying climatology would produce an unrealistically large MSE_{clim}, because the correct seasonality for the predictand would not be reflected (Hamill and Juras, 2006; Juras, 2000). The MSE for persistence in 9.36b implies that the index k represents time, so that the reference forecast for the observation o_k at time k is just the observation of the predictand during the previous time period, o_{k-1}.

Either of the reference measures for accuracy in Equations 9.36a or 9.36b, or their MAE counterparts, can be used in Equation 9.4 to calculate skill. Murphy (1992) advocates use of the more accurate reference forecasts to standardize the skill. For skill scores based on MSE, Equation 9.36a is more accurate (i.e., is smaller) if the lag-1 autocorrelation (Equation 3.36) of the time series of observations is smaller than 0.5, and Equation 9.36b is more accurate when the autocorrelation of the observations is larger than 0.5. For the MSE using climatology as the control forecasts, the skill score (in proportion rather than percentage terms) becomes

$$SS_{clim} = \frac{MSE - MSE_{clim}}{0 - MSE_{clim}} = 1 - \frac{MSE}{MSE_{clim}}. \qquad (9.37)$$

Notice that perfect forecasts have MSE or MAE $= 0$, which allows the rearrangement of the skill score in Equation 9.37. By virtue of this second equality in Equation 9.37, SS_{clim} based on MSE is sometimes called the *reduction of variance* (RV), because the quotient being subtracted is the average squared error (or residual, in the nomenclature of regression) divided by the climatological variance (cf. Equation 7.17). Equation 9.37 is known as the *Nash-Sutcliffe efficiency* in the hydrology literature.

The skill score for the MSE in Equation 9.37 can be manipulated algebraically in a way that yields some insight into the determinants of forecast skill as measured by the MSE, with respect to climatology as the reference (Equation 9.36a). Rearranging Equation 9.37, and substituting an expression for the Pearson product-moment correlation between the forecasts and observations, $r_{y,o}$ (Equation 3.29), yields (Murphy, 1988; Murphy and Epstein, 1989)

$$SS_{clim} = r_{y,o}^2 - \left[r_{y,o} - \frac{s_y}{s_o} \right]^2 - \left[\frac{\bar{y} - \bar{o}}{s_o} \right]^2. \qquad (9.38)$$

Equation 9.38 indicates that the skill in terms of the MSE can be regarded as consisting of a contribution due to the correlation between the forecasts and observations, and penalties relating to the calibration (reliability, or conditional bias) and unconditional bias of the forecasts. The first term in Equation 9.38 is the square of the Pearson correlation coefficient, and is a measure of the proportion of variability in the observations that is (linearly) accounted for by the forecasts. Here the squared correlation is similar to the R^2 in regression (Equation 7.17), although least-squares regressions are constrained to be unbiased by construction, whereas forecasts in general are not.

The second term in Equation 9.38 is a measure of reliability, or calibration, or conditional bias, of the forecasts. This is most easily appreciated by imagining a linear regression between the observations and the forecasts. The slope, b, of that linear regression equation can be expressed in terms of the correlation and the standard deviations of the predictor and predictand as $b = (s_o/s_y) r_{y,o}$. This relationship can be verified by substituting Equations 3.6 and 3.25 into Equation 7.7a. If this slope is smaller than $b = 1$, then the predictions made with this regression are too large (positively biased) for smaller forecasts, and too small (negatively biased) for larger forecasts. However, if $b = 1$, there will be no conditional bias, and substituting $b = (s_o/s_y) r_{y,o} = 1$ into the second term in Equation 9.38 yields a zero penalty for conditional bias.

The third term in Equation 9.38 is the square of the unconditional bias, as a fraction of the standard deviation of the observations, s_o. If the bias is small compared to the variability of the observations as measured by s_o the reduction in skill will be modest, whereas increasing bias of either sign progressively degrades the skill.

Thus if the forecasts are completely reliable and unbiased, the second and third terms in Equation 9.38 are both zero, and the skill score is exactly $r_{y,o}^2$. To the extent that the forecasts are biased or not correctly calibrated (exhibiting conditional biases), then the square of the correlation coefficient will overestimate skill. Squared correlation is accordingly best regarded as measuring potential skill.

9.3.3. Conditional Quantile Plots

It is possible and quite informative to graphically represent certain aspects of the joint distribution of nonprobabilistic forecasts and observations for continuous variables. The joint distribution contains a large amount of information that is most easily absorbed from a well-designed graphical presentation. For example, Figure 9.8 shows *conditional quantile plots* for a sample of daily maximum temperature forecasts issued during the winters of 1980/1981 through 1985/1986 for Minneapolis, Minnesota. Panel (a) illustrates the performance of objective (MOS) forecasts, and panel (b) illustrates the performance of the corresponding subjective forecasts. These diagrams contain two parts, representing the two factors in the calibration-refinement factorization of the joint distribution of forecasts and observations (Equation 9.2).

The conditional distributions of the observations given each of the forecasts are represented in terms of selected quantiles, in comparison to the 1:1 diagonal line representing perfect forecasts. These have been presented as slightly smoothed versions of the raw values, although they could instead be represented using quantile regressions (Section 7.7.3). Here it can be seen that the MOS forecasts (panel a) exhibit a small degree of overforecasting (the conditional medians of the observed temperatures are consistently colder than the forecasts), but that the subjective forecasts are essentially unbiased. The histograms in the lower parts of the panels represent the frequency of use of the forecasts, or $p(y_i)$. Here it can be seen that the subjective forecasts are somewhat sharper, or more refined, with more extreme temperatures being forecast more frequently, especially on the left tail.

Figure 9.8a shows the same data that is displayed in the glyph scatterplot in Figure 3.24, and the bivariate histogram in Figure 3.25. However, the two figures in Chapter 3 show the data in terms of their joint distribution, whereas the calibration-refinement factorization plotted in Figure 9.8a allows an easy visual separation between the frequencies of use of each of the possible forecasts, and the distributions of temperature outcomes conditional on each forecast. The conditional quantile plot is an example of a diagnostic verification technique, because it allows diagnosis of particular strengths and weakness of this set of forecasts through exposition of the full joint distribution of the forecasts and observations. In particular, Figure 9.8 shows that the subjective forecasts have improved over the MOS forecasts, both by correcting the overforecasting of the colder temperatures, and by exhibiting better sharpness for the coldest forecasts. Comparison of scalar scores such as MSE for the two forecast sources would indicate that the subjective forecasts were more accurate, but would yield no information about the specific nature of the improvements, or how even better subjective forecasts might be achieved in the future.

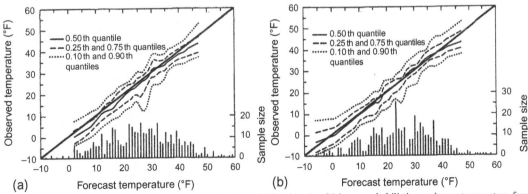

FIGURE 9.8 Conditional quantile plots for (a) objective and (b) subjective 24-h nonprobabilistic maximum temperature forecasts, for winter seasons of 1980 through 1986 at Minneapolis, Minnesota. Main body of the figures delineate smoothed quantiles from the conditional distributions $p(o_j \mid y_i)$ (i.e., the calibration distributions) in relation to the 1:1 line, and the lower parts of the figures show the unconditional distributions of the forecasts, $p(y_i)$ (the refinement distributions). *From Murphy et al. (1989). © American Meteorological Society. Used with permission.*

9.4. PROBABILITY FORECASTS FOR DISCRETE PREDICTANDS

9.4.1. The Joint Distribution for Dichotomous Events

Formulation and verification of probability forecasts for weather events have a long history, dating at least to Cooke (1906a) (Murphy and Winkler, 1984). Verification of probability forecasts is somewhat more subtle than verification of nonprobabilistic forecasts. Since nonprobabilistic forecasts contain no expression of uncertainty, it is clear whether an individual forecast is correct or not. However, unless a probability forecast is either 0.0 or 1.0, the situation is less clear-cut. For probability values between these two (certainty) extremes a single forecast is neither right nor wrong, so that meaningful assessments can only be made using collections of forecast and observation pairs. Again, it is the joint distribution of forecasts and observations that contains the relevant information for forecast verification.

The simplest setting for probability forecasts is in relation to dichotomous predictands, which are limited to $J = 2$ possible outcomes. The probability of precipitation (PoP) forecast is the most familiar example of probability forecasts for a dichotomous event. Here the event is either the occurrence (o_1) or nonoccurrence (o_2) of measurable precipitation. The joint distribution of forecasts and observations is more complicated than for the case of nonprobabilistic forecasts for binary predictands, however, because more than $I = 2$ probability values can allowably be forecast. In theory any real number between zero and one is an allowable probability forecast, but in practice the forecasts usually are rounded to one of a reasonably small number of values.

Table 9.4a contains a hypothetical joint distribution for probability forecasts of a dichotomous predictand, where the $I = 11$ possible forecasts might have been obtained by rounding continuous

TABLE 9.4 A Hypothetical Joint Distribution of Forecasts and Observations (a) for Probability Forecasts (Rounded to Tenths) for a Dichotomous Event, with (b) Its Calibration-Refinement factorization, and (c) Its Likelihood-Base Rate factorization

y_i	(a) Joint Distribution		(b) Calibration-Refinement		(c) Likelihood-Base Rate	
	$p(y_i, o_1)$	$p(y_i, o_2)$	$p(y_i)$	$p(o_1 \mid y_i)$	$p(y_i \mid o_1)$	$p(y_i \mid o_2)$
0.0	.045	.255	.300	.150	.152	.363
0.1	.032	.128	.160	.200	.108	.182
0.2	.025	.075	.100	.250	.084	.107
0.3	.024	.056	.080	.300	.081	.080
0.4	.024	.046	.070	.350	.081	.065
0.5	.024	.036	.060	.400	.081	.051
0.6	.027	.033	.060	.450	.091	.047
0.7	.025	.025	.050	.500	.084	.036
0.8	.028	.022	.050	.550	.094	.031
0.9	.030	.020	.050	.600	.101	.028
1.0	.013	.007	.020	.650	.044	.010
					$p(o_1) = .297$	$p(o_2) = .703$

probability assessments to the nearest tenth. Thus this joint distribution of forecasts and observations contains $I \times J = 22$ individual probabilities. For example, on 4.5% of the forecast occasions a zero forecast probability was nevertheless followed by occurrence of the event, and on 25.5% of the occasions zero probability forecasts were correct in that the event o_1 did not occur.

Table 9.4b presents the same joint distribution in terms of the calibration-refinement factorization (Equation 9.2). That is, for each possible forecast probability, y_i, Table 9.4b presents the relative frequency with which that forecast value was used, $p(y_i)$, and the conditional probability that the event o_1 occurred given the forecast y_i, $p(o_1|y_i)$, $i = 1, \ldots, I$. For example, $p(y_1) = p(y_1, o_1) + p(y_1, o_2) = .045 + .255 = .300$, and (using the definition of conditional probability, Equation 2.10) $p(o_1|y_1) = p(y_1, o_1)/p(y_1) = .045/.300 = .150$. Because the predictand is binary it is not necessary to specify the conditional probabilities for the complementary event, o_2, given each of the forecasts. That is, since the two predictand values represented by o_1 and o_2 constitute a MECE partition of the sample space, $p(o_2|y_i) = 1 - p(o_1|y_i)$. Not all the $J = 11$ probabilities in the refinement distribution $p(y_i)$ can be specified independently either, since $\Sigma_i p(y_j) = 1$. Thus the joint distribution can be completely specified with $I \times J - 1 = 21$ of the 22 probabilities given in either Table 9.4a or Table 9.4b, which is the dimensionality of this verification problem.

Similarly, Table 9.4c presents the likelihood-base Rate factorization (Equation 9.3) for the joint distribution in Table 9.4a. Since there are $J = 2$ MECE events, there are two conditional distributions $p(y_i|o_j)$, each of which includes $I = 11$ probabilities. Since these 11 probabilities must sum to 1, each conditional distribution is fully specified by any 10 of them. The refinement (i.e., sample climatological) distribution consists of the two complementary probabilities $p(o_1)$ and $p(o_2)$, and so can be completely defined by either of the two. Therefore the likelihood-base rate factorization is also fully specified by $10 + 10 + 1 = 21$ probabilities. The information in any of the three portions of Table 9.4 can be recovered fully from either of the others. For example, $p(o_1) = \Sigma_i p(y_i, o_1) = .297$, and $p(y_1|o_1) = p(y_1, o_1)/p(o_1) = .045/.297 = .152$.

9.4.2. The Brier Score

Given the generally high dimensionality of verification problems involving probability forecasts even for dichotomous predictands (e.g., $I \times J - 1 = 21$ for Table 9.4), it is not surprising that such forecasts are often evaluated using scalar summary measures. Although attractive from a practical standpoint, such simplifications necessarily will give incomplete pictures of forecast performance. Multiple scalar accuracy measures for verification of probabilistic forecasts of dichotomous events exist (e.g., Bröcker, 2012a; Murphy and Daan, 1985), but by far the most common is the *Brier score* (BS). The Brier score is essentially the mean squared error of the probability forecasts, considering that the observation is $o_1 = 1$ if the event occurs, and that the observation is $o_2 = 0$ if the event does not occur. The score averages the squared differences between pairs of forecast probabilities and the subsequent binary observations,

$$\text{BS} = \frac{1}{n}\sum_{k=1}^{n}(y_k - o_k)^2, \tag{9.39}$$

where the index k again denotes a numbering of the n forecast-event pairs. Comparing the Brier score with Equation 9.33 for the mean squared error, it can be seen that the two are completely analogous. As a mean-squared-error measure of accuracy, the Brier score is negatively oriented, with perfect forecasts exhibiting BS $= 0$. Less accurate forecasts receive higher Brier scores, but since individual

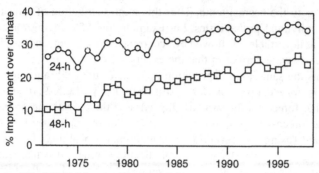

FIGURE 9.9 Trends in the skill of United States subjective PoP forecasts, measured in terms of the Brier score relative to climatological probabilities, April–September 1972–1998. *From www.nws.noaa.gov/tdl/synop.*

forecasts and observations are both bounded by zero and one, the score can take on values only in the range $0 \leq \text{BS} \leq 1$.

The Brier score as expressed in Equation 9.39 is nearly universally used, but it differs from the score as originally introduced by Brier (1950) in that it averages only the squared differences pertaining to one of the two binary events. The original Brier score also included squared differences for the complementary (or non-) event in the average, with the result that Brier's original score is exactly twice that given by Equation 9.39. The confusion is unfortunate, but the usual present-day understanding of the meaning of Brier score is that given in Equation 9.39. In order to distinguish this from the original formulation, the Brier Score in Equation 9.39 sometimes is referred to as the *half-Brier score*.

Skill scores of the form of Equation 9.4 often are computed for the Brier score, yielding the *Brier Skill Score*

$$\text{BSS} = \frac{\text{BS} - \text{BS}_{\text{ref}}}{0 - \text{BS}_{\text{ref}}} = 1 - \frac{\text{BS}}{\text{BS}_{\text{ref}}}, \tag{9.40}$$

since $\text{BS}_{\text{perf}} = 0$. This equation is analogous to Equation 9.37 for the MSE more generally. The BSS is the conventional skill-score form using the Brier score as the underlying accuracy measure. Usually the reference forecasts are the relevant climatological relative frequencies, which may vary with location and/or time of year (Hamill and Juras, 2006; Juras, 2000). Skill scores with respect to the climatological probabilities for subjective U.S. PoP forecasts during the warm seasons of 1972 through 1998 are shown in Figure 9.9. The labeling of the vertical axis as % improvement over climate indicates that it is the skill score in Equation 9.40, using climatological probabilities as the reference forecasts, that is plotted in the figure. According to this score, forecasts made for the 48-hour lead time in the 1990s exhibited skill comparable to 24-hour forecasts made in the 1970s.

9.4.3. Algebraic Decomposition of the Brier Score

An instructive algebraic decomposition of the Brier score (Equation 9.36) has been derived by Murphy (1973b). It relates to the calibration-refinement factorization of the joint distribution, Equation 9.2, in that it pertains to quantities that are conditional on particular values of the forecasts.

As before, consider that a verification data set contains forecasts taking on any of a discrete number, I, of forecast values y_i. For example, in the verification data set in Table 9.4, there are $I = 11$ allowable forecast values, ranging from $y_1 = 0.0$ to $y_{11} = 1.0$. Let N_i be the number of times each forecast y_i is used in the collection of forecasts being verified. The total number of forecast-event pairs is simply the sum of these subsample, or conditional sample, sizes,

$$n = \sum_{i=1}^{I} N_i. \tag{9.41}$$

The marginal distribution of the forecasts—the refinement in the calibration-refinement factorization—consists simply of the relative frequencies

$$p(y_i) = \frac{N_i}{n}. \tag{9.42}$$

The first column in Table 9.4b shows these relative frequencies for the data set represented there.

For each of the subsamples delineated by the I allowable forecast values there is a relative frequency of occurrence of the forecast event. Since the observed event is dichotomous, a single conditional relative frequency defines the conditional distribution of observations given each forecast y_i. It is convenient to think of this relative frequency as the subsample relative frequency, or conditional average observation,

$$\bar{o}_i = p(o_1 \mid y_i) = \frac{1}{N_i} \sum_{k \in N_i} o_k, \tag{9.43}$$

where $o_k = 1$ if the event occurs for the kth forecast-event pair, $o_k = 0$ if it does not, and the summation is over only those values of k corresponding to occasions when the forecast y_i was issued. The second column in Table 9.4b shows these conditional relative frequencies. Similarly, the overall (unconditional) relative frequency, or sample climatology, of the observations is given by

$$\bar{o} = \frac{1}{n} \sum_{k=1}^{n} o_k = \frac{1}{n} \sum_{i=1}^{I} N_i \bar{o}_i. \tag{9.44}$$

After some algebra, the Brier score in Equation 9.39 can be expressed in terms of the quantities just defined as the sum of the three terms

$$\mathrm{BS} = \underbrace{\frac{1}{n} \sum_{i=1}^{I} N_i (y_i - \bar{o}_i)^2}_{(\text{"Reliability"})} - \underbrace{\frac{1}{n} \sum_{i=1}^{I} N_i (\bar{o}_i - \bar{o})^2}_{(\text{"Resolution"})} + \underbrace{\bar{o}(1 - \bar{o})}_{(\text{"Uncertainty"})} . \tag{9.45}$$

As indicated in this equation, these three terms are known as Reliability, Resolution, and Uncertainty. Since more accurate forecasts are characterized by smaller values of BS, a forecaster would like the reliability term to be as small as possible, and the resolution term to be as large (in absolute value) as possible. Equation 9.44 shows that the uncertainty term depends only on the sample climatological relative frequency, and so is unaffected by the forecasts. The reliability and resolution terms in Equation 9.45 sometimes are used individually as scalar measures of these two aspects of forecast quality, and called REL and RES, respectively. If these two measures are normalized by dividing each by the uncertainty term, their difference equals the Brier skill score BSS, as in Equation 9.46.

The reliability term in Equation 9.45 summarizes the calibration, or conditional bias, of the forecasts. It consists of a weighted average of the squared differences between the forecast probabilities and the relative frequencies of the observed event in each subsample. For forecasts that are perfectly reliable, the subsample relative frequency is exactly equal to the forecast probability in each subsample. The relative frequency of the event should be small on occasions when $y_1 = 0.0$ is forecast, and should be large when $y_I = 1.0$ is forecast. On those occasions when the forecast probability is 0.5, the relative frequency of the event should be near 1/2. For reliable, or well-calibrated forecasts, all the squared differences in the reliability term will be near zero, and their weighted average will be small.

The resolution term in Equation 9.45 summarizes the ability of the forecasts to discern subsample forecast periods with relative frequencies of the event that are different from each other. The forecast probabilities y_i do not appear explicitly in this term, yet it still depends on the forecasts through the sorting of the events making up the subsample relative frequencies (Equation 9.43). Mathematically, the resolution term is a weighted average of the squared differences between these subsample relative frequencies, and the overall sample climatological relative frequency. If the forecasts sort the observations into subsamples having substantially different relative frequencies than the overall sample climatology, the resolution term will be large. This is a desirable situation, since the resolution term is subtracted in Equation 9.45. Conversely, if the forecasts sort the events into subsamples with very similar event relative frequencies, the squared differences in the summation of the resolution term will be small. In that case the forecasts resolve the event only weakly, and the resolution term will be small.

The uncertainty term in Equation 9.45 depends only on the variability of the observations and cannot be influenced by anything the forecaster may do. This term is identical to the variance of the Bernoulli (binomial, with $N = 1$) distribution (see Table 4.3), exhibiting minima of zero when the climatological probability is either zero or one, and a maximum when the climatological probability is 1/2. When the event being forecast almost never happens, or almost always happens, the uncertainty in the forecasting situation is small. In these cases, always forecasting the climatological probability will give generally good results. When the climatological probability is closer to 1/2 there is substantially more uncertainty inherent in the forecasting situation, and the third term in Equation 9.45 is commensurately larger.

Equation 9.45 is an exact decomposition of the Brier score when the allowable forecast values are only the I probabilities y_i. This is the case when, for example, human forecasters are required to round their judgments to fixed fractions such as tenths. When a richer set of probabilities, derived perhaps from relative frequencies within a large forecast ensemble, or a logistic regression, have been rounded into I bins, Equation 9.45 will not balance exactly if BS on the left-hand side has been computed using the unrounded values. However, the resulting discrepancy can be quantified using two additional terms (Stephenson et al., 2008b). The three terms on the right-hand side of Equation 9.45 exhibit biases that may be appreciable if the subsample sizes N_i are not large (Bröcker, 2012b, 2012c), although corrections are available (Ferro and Fricker, 2012; Siegert, 2014).

The algebraic decomposition of the Brier score in Equation 9.45 is interpretable in terms of the calibration-refinement factorization of the joint distribution of forecasts and observations (Equation 9.2), as will become clear in Section 9.4.4. Murphy and Winkler (1987) also proposed a different three-term algebraic decomposition of the mean-squared error (of which the Brier score is a special case), based on the likelihood-base rate factorization (Equation 9.3), which has been applied to the Brier score for the data in Table 9.2 by Bradley et al. (2003).

9.4.4. The Reliability Diagram

Single-number summaries of forecast performance such as the Brier score can provide a convenient quick impression, but a comprehensive appreciation of forecast quality can be achieved only through the full joint distribution of forecasts and observations. Because of the typically large dimensionality ($= I \times J - 1$) of these distributions their information content can be difficult to absorb from numerical tabulations such as those in Tables 9.2 or 9.4, but becomes conceptually accessible when presented in a well-designed graphical format. The *reliability diagram*, apparently introduced by Sanders (1963), is one such graphical device. It shows the full joint distribution of forecasts and observations for probability forecasts of a binary predictand, in terms of its calibration-refinement factorization (Equation 9.2). Accordingly, it is the counterpart of the conditional quantile plot (Section 9.3.3) for nonprobabilistic forecasts of continuous predictands. The fuller picture of forecast performance portrayed in the reliability diagram as compared to a scalar summary, such as BSS, allows diagnosis of particular strengths and weaknesses in a verification data set.

The two elements of the calibration-refinement factorization are the calibration distributions, or conditional distributions of the observation given each of the I allowable values of the forecast, $p(o_j|y_i)$; and the refinement distribution $p(y_i)$, expressing the frequency of use of each of the possible forecasts. Each of the calibration distributions is a Bernoulli (binomial, with $N = 1$) distribution, because there is a single binary outcome O on each forecast occasion, and for each forecast y_i the probability of the outcome o_1 is the conditional probability $p(o_1|y_i)$. This probability fully defines the corresponding Bernoulli distribution, because $p(o_2|y_i) = 1 - p(o_1|y_i)$. Taken together, these I calibration probabilities $p(o_1|y_i)$ define a *calibration function*, which expresses the conditional probability of the event o_1 as a function of the forecast y_i. In some settings, the forecasts to be evaluated have been rounded to a prespecified set of I probabilities before being issued. However, when the probability forecasts being evaluated are continuous, and so can take on any value on the unit interval, the number of bins I must be chosen in order to plot a reliability diagram. Bröcker (2008) suggests optimizing this choice by minimizing the cross-validated BS of the probability forecasts rounded to I discrete values. The forecasts y_i, $i = 1, ..., I$, corresponding to each bin are most consistently calculated as the average forecast probability within each bin (Bröcker, 2008; Bröcker and Smith, 2007b).

The first element of a reliability diagram is a plot of the calibration function, usually as I points connected by line segments for visual clarity. Figure 9.10a shows five characteristic forms for this portion of the reliability diagram, which allows immediate visual diagnosis of unconditional and conditional biases that may be exhibited by the forecasts in question. The center panel in Figure 9.10a shows the characteristic signature of well-calibrated forecasts, in which the conditional event relative frequency is essentially equal to the forecast probability. That is, $p(o_1|y_i) \approx y_i$, so that the I dots fall along the dashed 1:1 line except for deviations consistent with sampling variability. Well-calibrated probability forecasts "mean what they say," in the sense that subsequent event relative frequencies are essentially equal to the forecast probabilities. In terms of the algebraic decomposition of the Brier score (Equation 9.45), such forecasts exhibit excellent reliability, because the squared differences in the reliability term correspond to squared vertical distances between the dots and the 1:1 line in the reliability diagram. These distances are all small for well-calibrated forecasts, yielding a small reliability term, which is a weighted average of the I squared vertical distances.

The top and bottom panels in Figure 9.10a show characteristic forms of the calibration function for forecasts exhibiting unconditional biases. In the top panel, the calibration function is entirely to the right of the 1:1 line, indicating the forecasts are consistently too large relative to the conditional event relative frequencies, so that the average forecast is larger than the average observation (Equation 9.44). This pattern is the signature of overforecasting, or if the predictand is precipitation occurrence, a wet bias. Similarly, the bottom panel in Figure 9.10a shows the characteristic signature of underforecasting, or a dry bias, because the calibration function being entirely to the left of the 1:1 line indicates that the

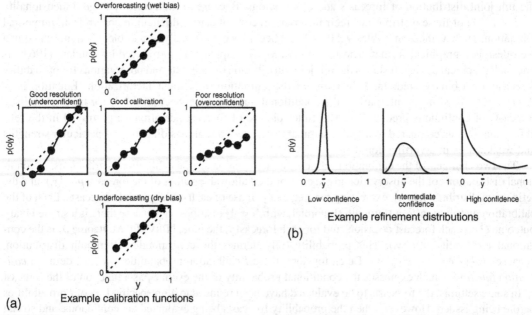

(a) Example calibration functions

(b) Example refinement distributions

FIGURE 9.10   Example characteristic forms for the two elements of the reliability diagram. (a) Calibration functions, showing calibration distributions $p(o|y)$ (i.e., conditional Bernoulli probabilities), as functions of the forecast y. (b) Refinement distributions, $p(y)$, reflecting aggregate forecaster confidence.

forecast probabilities are consistently too small relative to the corresponding conditional event relative frequencies given by $p(o_1|y_i)$, and so the average forecast is smaller than the average observation. Forecasts that are unconditionally biased in either of these two ways are miscalibrated, or not reliable, in the sense that the conditional event probabilities $p(o_1|y_i)$ do not correspond well to the stated probabilities y_i. The vertical distances between the points and the dashed 1:1 line are nonnegligible, leading to substantial squared differences in the first summation of Equation 9.45, and thus to a large reliability term in that equation.

The deficiencies in forecast performance indicated by the calibration functions in the left and right panels of Figure 9.10a are more subtle and indicate conditional biases. That is, the sense and/or magnitudes of the biases exhibited by forecasts having these types of calibration functions depend on the forecasts themselves. In the left ("good resolution") panel, there are overforecasting biases associated with smaller forecast probabilities and underforecasting biases associated with larger forecast probabilities, and the reverse is true of the calibration function in the right ("poor resolution") panel. The calibration function in the right panel of Figure 9.10a is characteristic of forecasts showing poor resolution in the sense that the conditional outcome relative frequencies $p(o_1|y_i)$ depend only weakly on the forecasts and are all near the climatological probability. (That the climatological relative frequency is somewhere near the center of the vertical locations of the points in this panel can be appreciated from the law of total probability (Equation 2.14), which expresses the unconditional climatology as a weighted average of these conditional relative frequencies.) Because the differences in this panel between the calibration probabilities $p(o_1|y_i)$ (Equation 9.43) and the overall sample climatology are small, the resolution term in Equation 9.45 is small, reflecting the fact that these forecasts resolve the event o_1

poorly. Because the sign of this term in Equation 9.45 is negative, poor resolution leads to larger (worse) Brier scores.

Conversely, the calibration function in the left panel of Figure 9.10a indicates good resolution, in the sense that the weighted average of the squared vertical distances between the points and the sample climatology in the resolution term of Equation 9.45 is large. Here the forecasts are able to identify subsets of forecast occasions for which the outcomes are quite different from each other. For example, small but nonzero forecast probabilities have identified a subset of forecast occasions when the event o_1 did not occur at all. However, the forecasts are conditionally biased, and so mislabeled, and therefore not well calibrated. Their Brier score would be penalized for this miscalibration through a substantial positive value for the reliability term in Equation 9.45.

The labels underconfident and overconfident in the left and right panels of Figure 9.10a can be understood in relation to the other element of the reliability diagram, namely, the refinement distribution $p(y_i)$. The dispersion of the refinement distribution reflects the overall confidence of the forecaster, as indicated in Figure 9.10b. Forecasts that deviate rarely and quantitatively little from their average value (left panel) exhibit little confidence. Forecasts that are frequently extreme—that is, specifying probabilities close to the certainty values $y_1 = 0$ and $y_I = 1$ (right panel)—exhibit high confidence. However, the degree to which a particular level of forecaster confidence may be justified will be evident only from inspection of the calibration function for the same forecasts. The forecast probabilities in the right-hand ("overconfident") panel of Figure 9.10a are mislabeled in the sense that the extreme probabilities are too extreme. Outcome relative frequencies following probability forecasts near 1 are substantially smaller than 1, and outcome relative frequencies following forecasts near 0 are substantially larger than 0. A calibration-function slope that is shallower than the 1:1 reference line is diagnostic for overconfident forecasts, because correcting the forecasts to bring the calibration function into the correct orientation would require adjusting extreme probabilities to be less extreme, thus shrinking the dispersion of the refinement distribution, which would connote less confidence. Conversely, the underconfident forecasts in the left panel of Figure 9.10a could achieve good reliability (calibration function aligned with the 1:1 line) by adjusting the forecast probabilities to be more extreme, thus increasing the dispersion of the refinement distribution and connoting greater confidence.

A reliability diagram consists of plots of both the calibration function and the refinement distribution, and so is a full graphical representation of the joint distribution of the forecasts and observations, through its calibration-refinement factorization. Figure 9.11 shows two reliability diagrams, for seasonal (three-month) forecasts for (a) average temperatures and (b) total precipitation above the climatological terciles (outcomes in the warm and wet 1/3 of the respective local climatological distributions), for global land areas equatorward of 30° (Mason et al., 1999). The most prominent feature of Figure 9.11 is the substantial cold (underforecasting) bias evident for the temperature forecasts. The period 1997 through 2000 was evidently substantially warmer than the preceding several decades that defined the reference climate. The relative frequency of the observed warm outcome was about 0.7 (rather than the long-term climatological value of 1/3), but Figure 9.11a shows clearly that that warmth was not anticipated by these forecasts, in aggregate. There is also an indication of conditional bias in the temperature forecasts, with the overall calibration slope being slightly shallower than 45°, and so reflecting some forecast overconfidence. The precipitation forecasts (Figure 9.11b) are better calibrated, showing only a slight overforecasting (wet) bias and a more nearly correct overall slope for the calibration function. The refinement distributions (insets, with logarithmic vertical scales) show much more confidence (more frequent use of more extreme probabilities) for the temperature forecasts.

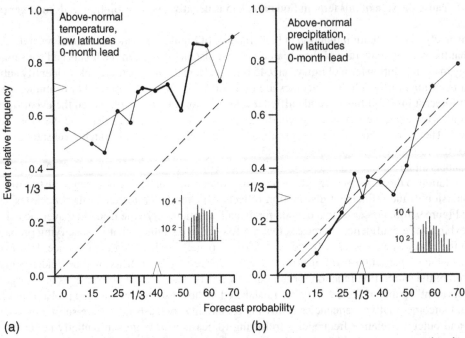

FIGURE 9.11   Reliability diagrams for seasonal (three-month) forecasts of (a) average temperature warmer than the climatological upper tercile, and (b) total precipitation wetter than the climatological upper tercile, for global land areas equatorward of 30°, during the period 1997–2000. *From Wilks and Godfrey (2002).*

The reliability diagrams in Figure 9.11 include some additional features that are not always plotted in reliability diagrams, which help interpret the results. The light lines through the calibration functions show weighted (to make points with larger subsample size N_i more influential) least-squares regressions (Murphy and Wilks, 1998), which help guide the eye through the irregularities that are due at least in part to sampling variations. In order to emphasize the better-estimated portions of the calibration function, the line segments connecting points based on larger sample sizes have been drawn more heavily. Finally, the average forecasts are indicated by the triangles on the horizontal axes, and the average observations are indicated by the triangles on the vertical axes, which emphasize the strong underforecasting of temperature in Figure 9.11a. Other modifications to the basic reliability diagram that have been suggested include plotting Brier Score contours on the diagram (Ehrendorfer, 1997) and plotting conditional boxplots in place of dots locating the conditional means (Bentzien and Friederichs, 2012). Use of nonuniformly spaced bins containing equal numbers of forecasts has also been suggested (Bröcker and Smith, 2007b), although the resulting Brier score may not be proper (Mitchell and Ferro, 2017) in the sense explained in Section 9.4.8.

An elaboration of the reliability diagram (Hsu and Murphy, 1986) includes reference lines related to the algebraic decomposition of the Brier score (Equation 9.45) and the Brier skill score (Equation 9.40), in addition to plots of the calibration function and the refinement distribution. This version of the reliability diagram is called the *attributes diagram*, an example of which (for the joint distribution in Table 9.2) is shown in Figure 9.12. The horizontal "no-resolution" line in the attributes diagram relates to the resolution term in Equation 9.45. Geometrically, the ability of a set of forecasts to identify event

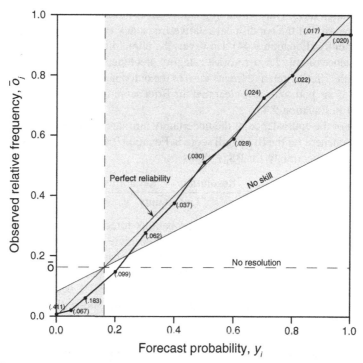

FIGURE 9.12 Attributes diagram for the $n = 12{,}402$ PoP forecasts summarized in Table 9.2. Solid dots show observed relative frequency of precipitation occurrence, conditional on each of the $I = 12$ possible probability forecasts. Forecasts not defining event subsets with different relative frequencies of the forecast event would exhibit all points on the dashed no-resolution line, which is plotted at the level of the sample climatological probability. Points in the shaded region bounded by the line labeled "no skill" contribute positively to forecast skill, according to Equation 9.40. Relative frequencies of use of each of the forecast values, $p(y_i)$, are shown parenthetically, although they could also have been indicated graphically.

subsets with different relative frequencies produces points in the attributes diagram that are well removed, vertically, from the level of the overall sample climatology, which is indicated by the no-resolution line. Points falling on the no-resolution line indicate forecasts y_i that are unable to resolve occasions where the event is more or less likely than the overall climatological probability. The weighted average making up the resolution term is of the squares of the vertical distances between the points (the subsample relative frequencies) and the no-resolution line. These distances will be large for forecasts exhibiting good resolution, in which case the resolution term will contribute to a small (i.e., good) Brier score. The forecasts summarized in Figure 9.12 exhibit a substantial degree of resolution, with forecasts that are most different from the sample climatological probability of 0.162 making the largest contributions to the resolution term.

Another interpretation of the uncertainty term in Equation 9.45 emerges from imagining the attributes diagram for climatological forecasts, that is, constant forecasts of the sample climatological relative frequency, Equation 9.44. Since only a single forecast value would ever be used in this case, there would be only $I = 1$ dot on such a diagram. The horizontal position of this dot would be at the constant forecast value, and the vertical position of the single dot would be at the same sample climatological relative frequency. This single point would be located at the intersection of the 1:1 (perfect reliability), no-skill and no-resolution lines. Thus climatological forecasts have perfect (zero, in Equation 9.45) reliability, since the forecast and the conditional relative frequency (Equation 9.43) are both equal to the climatological probability (Equation 9.44). However, the climatological forecasts also have zero resolution since the existence of only $I = 1$ forecast category precludes discerning different subsets of forecasting occasions with differing relative frequencies of the outcomes. Since the reliability and resolution terms in Equation 9.45 are both zero, it is clear that the Brier score for climatological forecasts is exactly the uncertainty term in Equation 9.45.

This observation of the equivalence of the uncertainty term and the BS for climatological forecasts has interesting consequences for the Brier skill score in Equation 9.40. Substituting Equation 9.45 for BS into Equation 9.40, and uncertainty for BS_{ref} yields

$$BSS = \frac{"Resolution" - "Reliability"}{"Uncertainty"}. \tag{9.46}$$

Since the uncertainty term is always positive, the probability forecasts will exhibit positive skill in the sense of Equation 9.40 if the resolution term is larger in absolute value than the reliability term. This means that subsamples of the forecasts identified by the forecasts y_i will contribute positively to the overall skill when their resolution term is larger than their reliability term. Geometrically, this corresponds to points on the attributes diagram being closer to the 1:1 perfect-reliability line than to the horizontal no-resolution line. This condition defines the no-skill line, which is midway between the perfect-reliability and no-resolution lines, and delimits the shaded region, in which subsamples contribute positively to forecast skill, according to BSS. In Figure 9.12 only the subsample for $y_4 = 0.2$, which is nearly equal to the climatological probability, fails to contribute positively to the overall BSS.

Note that forecasts whose calibration functions lie outside the shaded region in an attributes diagram are not necessarily useless. Zero or negative skill according to BSS or indeed any other scalar measure may still be consistent with positive economic value for some users, since it is possible for forecasts with lower BSS to be more valuable for some users (e.g., Murphy and Ehrendorfer, 1987). At minimum, forecasts exhibiting a calibration function with positive slope different from 1 have the potential for *recalibration*, which is most often achieved by relabeling each of the forecasts y_i with the corresponding conditional relative frequencies $\bar{o}_i$ (Equation 9.43) defining the calibration function in a previously analyzed training data set. For overconfident forecasts, the recalibration process comes at the expense of sharpness, although sharpness is increased when underconfident forecasts are recalibrated. Bröcker (2008) suggests recalibration using a kernel-smoothing (Section 3.3.6) estimate of the calibration function, which is an appealing approach for continuously varying probability forecasts. Van den Dool et al. (2017) suggest that such miscalibrations might be addressed using a linear regression postprocessing step.

9.4.5. The Discrimination Diagram

The joint distribution of forecasts and observations can also be displayed graphically through the likelihood-base rate factorization (Equation 9.3). For probability forecasts of dichotomous ($J = 2$) predictands, this factorization consists of two conditional likelihood distributions $p(y_i \mid o_j), j = 1, 2$; and a base rate (i.e., sample climatological) distribution $p(o_j)$ consisting of the relative frequencies for the two dichotomous events in the verification sample.

The *discrimination diagram* displays superimposed plots of the two likelihood distributions, as functions of the forecast probability y_i, together with a specification of the sample climatological probabilities $p(o_1)$ and $p(o_2)$. Together, these quantities completely represent the information in the full joint distribution. Therefore the discrimination diagram presents the same information as the reliability diagram, but in a different format.

Figure 9.13 shows an example discrimination diagram, for the probability-of-precipitation forecasts whose calibration-refinement factorization is presented in Table 9.2 and whose attributes diagram is shown in Figure 9.12. The probabilities in the two likelihood distributions calculated from their joint distribution are presented in Table 15.2. Clearly the conditional probabilities given the "no precipitation" event o_2 are greater for the smaller forecast probabilities, and the conditional probabilities given the "precipitation" event o_1 are greater for the intermediate and larger probability forecasts. Forecasts that discriminated perfectly between the two events would exhibit no overlap in their likelihoods. The two likelihood distributions in Figure 9.13 overlap somewhat, but exhibit substantial separation, indicating substantial discrimination by the forecasts of the dry and wet events.

The separation of the two likelihood distributions in a discrimination diagram can be summarized by the absolute difference between their means, called the *discrimination distance*,

$$d = \left| \mu_{y|o_1} - \mu_{y|o_2} \right|. \tag{9.47}$$

For the two conditional distributions in Figure 9.13 this difference is $d = |0.567 – 0.101| = 0.466$, which is also plotted in the figure. This distance is zero if the two likelihood distributions are the same (i.e., if the forecasts cannot discriminate the event at all), and increases as the two likelihood distributions become more distinct. In the limit $d = 1$ for perfect forecasts, which have all probability concentrated at $p(1|o_1) = 1$ and $p(0|o_2) = 1$.

There is a connection between the likelihood distributions in the discrimination diagram and statistical discrimination as discussed in Chapter 15. In particular, the two likelihood distributions in Figure 9.13 could be used together with the sample climatological probabilities, as in Section 15.3.3,

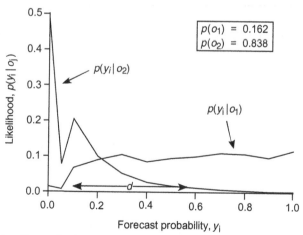

FIGURE 9.13   Discrimination diagram for the data in Table 9.2, which is shown in likelihood-base rate form in Table 15.2. The discrimination distance d (Equation 9.47) is also indicated.

to recalibrate these probability forecasts by calculating posterior probabilities for the two events given each of the possible forecast probabilities (cf. Exercise 15.5).

9.4.6. The ROC Diagram

The ROC (*relative operating characteristic*, or *receiver operating characteristic*) diagram is another discrimination-based graphical forecast verification display, although unlike the reliability diagram and discrimination diagram it does not include the full information contained in the joint distribution of forecasts and observations. The ROC diagram was first introduced into the meteorological literature by Mason (1982), although it has a longer history of use in such disciplines as psychology (Swets, 1973) and medicine (Pepe, 2003; Swets, 1979), after arising from signal detection theory in electrical engineering. The ROC diagram is a square plot, with the false alarm rate F (Equation 9.13) on the horizontal axis and the hit rate H (Equation 9.12) on the vertical.

One way to view the ROC diagram and the ideas behind it is in relation to the class of idealized decision problems outlined in Section 9.9.1. Here hypothetical decision makers must choose between two alternatives on the basis of a probability forecast for a dichotomous variable, with one of the decisions (say, action A) being preferred if the event o_1 occurs, and the other (action B) being preferable if the event does not occur. As explained in Section 9.9.1, the probability threshold determining which of the two decisions will be optimal depends on the decision problem, and in particular on the relative undesirability of having taken action B when the event occurs versus action A when the event does not occur. Therefore different probability thresholds for the choice between actions A and B will be appropriate for different decision problems.

If the forecast probabilities y_i have been issued as, or rounded to, I discrete values, there are $I - 1$ such thresholds, excluding the trivial cases of always taking action A or always taking action B. Operating on the joint distribution of forecasts and observations (e.g., Table 9.4a) consistent with each of these probability thresholds yields $I - 1$ contingency tables of dimension 2 x 2 of the kind treated in Section 9.2. A "yes" forecast is imputed if the probability y_i is above the threshold in question (sufficient probability to warrant a nonprobabilistic forecast of the event, for those decision problems appropriate to that probability threshold), and a "no" forecast is imputed if the forecast probability is below the threshold (insufficient probability for a nonprobabilistic forecast of the event). The mechanics of constructing these 2 x 2 contingency tables are exactly as illustrated in Example 9.2.

ROC diagrams are constructed by evaluating each of the $I - 1$ contingency tables using the hit rate H (Equation 9.12) and the false alarm rate F (Equation 9.13). As the hypothetical decision threshold is increased from lower to higher probabilities there are progressively more "no" forecasts and progressively fewer "yes" forecasts, yielding corresponding decreases in both H and F. In terms of the two likelihood distributions $p(o_1 \mid y_i)$ for the "yes" event and $p(o_2 \mid y_i)$ for the "no" event (e.g., Figure 9.13),

$$H = \int_{y^*}^{1} p(o_1 \mid y)dy \tag{9.48a}$$

and

$$F = \int_{y^*}^{1} p(o_2 \mid y)dy, \tag{9.48b}$$

where y^* is the decision threshold. The resulting $I-1$ point pairs (F_i, H_i) are then plotted and connected with line segments, and also connected to the point $(0, 0)$ corresponding to never forecasting the event (i.e., always choosing action A), and to the point $(1,1)$ corresponding to always forecasting the event (always choosing action B).

The ability of a set of probability forecasts to discriminate a dichotomous event can be easily appreciated from its ROC diagram. Consider first the ROC diagram for perfect forecasts, which use only $I=2$ probabilities, $y_1 = 0.00$ and $y_2 = 1.00$. For such forecasts there is only one probability threshold from which to calculate a 2 x 2 contingency table. That table for perfect forecasts exhibits $F = 0.0$ and $H = 1.0$, so its ROC curve consists of two line segments coincident with the left boundary and the upper boundary of the ROC diagram. At the other extreme of forecast performance, random forecasts consistent with the sample climatological probabilities $p(o_1)$ and $p(o_2)$ will exhibit $F_i = H_i$ regardless of how many or how few different probabilities y_i are used, and so their ROC curve will fall near (within sampling variability) the 45° diagonal connecting the points $(0, 0)$ and $(1,1)$. ROC curves for real forecasts usually fall between these two extremes, lying above and to the left of the 45° diagonal. Forecasts with better discrimination exhibit ROC curves approaching the upper-left corner of the ROC diagram more closely, whereas forecasts with very little ability to discriminate the event o_1 exhibit ROC curves very close to the $H = F$ diagonal.

It can be convenient to summarize a ROC diagram using a single scalar value, and the usual choice for this purpose is the area under the ROC curve, A. Since ROC curves for perfect forecasts pass through the upper-left corner, the area under a perfect ROC curve includes the entire unit square, so $A_{perf} = 1$. Similarly ROC curves for random forecasts lie along the 45° diagonal of the unit square, yielding the area $A_{rand} = 0.5$. The area A under a ROC curve of interest can therefore also be expressed in standard skill-score form (Equation 9.4), as

$$SS_{ROC} = \frac{A - A_{rand}}{A_{perf} - A_{rand}} = \frac{A - 1/2}{1 - 1/2} = 2A - 1. \tag{9.49}$$

Marzban (2004) describes some characteristics of forecasts that can be diagnosed from the shapes of their ROC curves, based on analysis of some simple idealized discrimination diagrams. Symmetrical ROC curves result when the two likelihood distributions $p(y_i | o_1)$ and $p(y_i | o_2)$ have similar dispersion, or widths, so the ranges of the forecasts y_i corresponding to each of the two outcomes are comparable. On the other hand, asymmetrical ROC curves, which might intersect either the vertical or horizontal axis at either $H \approx 0.5$ or $F \approx 0.5$, respectively, are indicative of one of the two likelihoods being substantially more concentrated than the other. Marzban (2004) also finds that A (or, equivalently, SS_{ROC}) is a reasonably good discriminator among relatively low-quality forecasts, but that relatively good forecasts tend to be characterized by quite similar (near-unit) areas under their ROC curves.

Example 9.5. Two Example ROC Curves

Example 9.2 illustrated the conversion of the probabilistic forecasts summarized by the joint distribution in Table 9.2 to nonprobabilistic yes/no forecasts, using a probability threshold y^* between $y_3 = 0.1$ and $y_4 = 0.2$. The resulting 2 x 2 contingency table consists of (cf. Figure 9.1a) $a = 1828, b = 2369, c = 181$, and $d = 8024$; yielding $F = 2369/(2369+8024) = 0.228$ and $H = 1828/(1828+181) = 0.910$. This point is indicated by the dot on the ROC curve for the Table 9.2 data in Figure 9.14. The entire ROC curve for the Table 9.2 data consists of this and all other partitions of these forecasts into yes/no forecasts using different probability thresholds. For example, the point just to the left of $(0.228, 0.910)$ on this ROC curve is obtained by moving the threshold between $y_4 = 0.2$ and $y_5 = 0.3$. This partition produces $a = 1644, b = 1330, c = 364$, and $d = 9064$, defining the point $(F, H) = (0.128, 0.819)$.

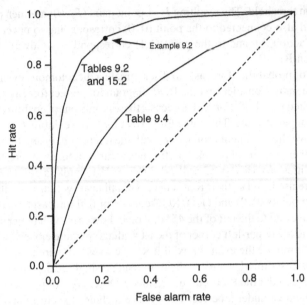

FIGURE 9.14 ROC diagrams for the PoP forecasts in Tables 9.2 and 15.2 (upper solid curve), and the hypothetical forecasts in Table 9.4 (lower solid curve). Solid dot locates the (F, H) pair corresponding to the probability threshold in Example 9.2.

Summarizing ROC curves according to the areas underneath them is usually accomplished through summation of the areas under each of the I trapezoids defined by the point pairs (F_i, H_i), $i = 1, \ldots, I-1$, together with the two endpoints (0, 0) and (1, 1). For example, the trapezoid defined by the dot in Figure 9.14 and the point just to its left has area $0.5(0.910+0.819)(0.228-0.128) = 0.08645$. This area, together with the areas of the other $I - 1 = 11$ trapezoids defined by the segments of the ROC curve for these data yield the total area $A = 0.922$.

The ROC curve, and the area under it, can also be computed directly from the joint probabilities in $p(y_i, o_j)$, that is, without knowing the sample size n. Table 9.5 summarizes the conversion of the hypothetical joint distribution in Table 9.4a to the $I - 1 = 10$ sets of 2 x 2 tables, by operating directly on the joint probabilities. Note that these data have one fewer forecast value y_i than those in Table 9.2, because in Table 9.2 the forecast $y_2 = 0.05$ has been allowed. For example, for the first probability threshold in Table 9.5, 0.05, only the forecasts $y_1 = 0.0$ are converted to "no" forecasts, so the entries of the resulting 2 x 2 joint distribution (cf. Figure 9.1b) are $a/n = 0.032+0.025+\cdots+0.013 = 0.252$, $b/n = 0.128+0.075+$ $+0.007 = 0.448$, $c/n = p(y_2, o_1) = 0.045$, and $d/n = p(y_2, o_2) = 0.255$. For the second probability threshold, 0.15, both the forecasts $y_1 = 0.0$ and $y_2 = 0.1$ are converted to "no" forecasts, so the resulting 2 x 2 joint distribution contains the four probabilities $a/n = 0.025+0.024+ . . +0.013 = 0.220$, $b/n =$ $0.075+0.056+\cdots+ 0.007 = 0.320$, $c/n = 0.045+0.032 = 0.077$, and $d/n = 0.255+0.128 = 0.383$.

Table 9.5 also shows the hit rate H and false alarm rate F for each of the 10 partitions of the joint distribution in Table 9.4a. These pairs define the lower ROC curve in Figure 9.14, with the points corresponding to the smaller probability thresholds occurring in the upper right portion of the ROC diagram, and points corresponding to the larger probability thresholds occurring in the lower left portion. Proceeding from left to right, the areas under the $I = 11$ trapezoids defined by these points

TABLE 9.5 The $I-1 = 10$ 2 × 2 Tables Derived From Successive Partitions of the Joint Distribution in Table 9.4, and the Corresponding Values for H and F

Threshold	a/n	b/n	c/n	d/n	H	F
0.05	.252	.448	.045	.255	.848	.637
0.15	.220	.320	.077	.383	.741	.455
0.25	.195	.245	.102	.458	.657	.348
0.35	.171	.189	.126	.514	.576	.269
0.45	.147	.143	.150	.560	.495	.203
0.55	.123	.107	.174	.596	.414	.152
0.65	.096	.074	.201	.629	.323	.105
0.75	.071	.049	.226	.654	.239	.070
0.85	.043	.027	.254	.676	.145	.038
0.95	.013	.007	.284	.696	.044	.010

together with the points at the corners of the ROC diagram are $0.5(0.044+0.000)(0.010 - 0.000) = 0.00022$, $0.5(0.145+0.044)(0.038-0.010) = 0.00265$, $0.5(0.239+0.145)(0.070-0.038) = 0.00614$, ... , $0.5(1.000+0.848)(1.000-0.637) = 0.33541$; yielding a total area of $A = 0.698$.

Figure 9.14 shows clearly that the forecasts in Table 9.2 exhibit greater event discrimination than those in Table 9.4, because the arc of the corresponding ROC curve for the former is everywhere above that for the latter, and approaches the upper left-hand corner of the ROC diagram more closely. This difference in discrimination is summarized by the differences in the areas under the two ROC curves, that is, $A = 0.922$ versus $A = 0.698$. ◇

Because ROC diagrams display the relationship between H (Equation 9.12) and F (Equation 9.13), which are two characteristics of the 2 × 2 contingency table, it is not surprising that others of the verification statistics in Sections 9.2.2 and 9.2.3 can be related to these plots. For example, since PSS (Equation 9.18) can be written as PSS = $H - F$, and the 45° diagonal line in the ROC diagram is exactly the $H = F$ line, the PSS for any threshold is exactly the vertical distance between the ROC curve and the $H = F$ diagonal. Sometimes the partition maximizing this vertical distance is chosen as "optimal" in medical statistics. Isolines of equal bias (Equation 9.10) having slope $-p(o_2)/p(o_1) = -(b+d)/(a+c)$ can also be drawn on the ROC diagram, which intersect the vertical axis at $H = B$ (Mason, 2003). Thus the partition yielding unbiased forecasts occurs at the intersection of the ROC curve and the equation $H = 1 - F(b+d)/(a+c)$.

ROC diagrams have been used increasingly in recent years to evaluate probability forecasts for binary predictands, so it is worthwhile to reiterate that (unlike the reliability diagram and the discrimination diagram) they do *not* provide a full depiction of the joint distribution of forecasts and observations. This deficiency of the ROC diagram can be appreciated by recalling the mechanics of its construction, as outlined in Example 9.5. In particular, the calculations behind the ROC diagrams are carried out without regard to the specific values for the probability labels, $p(y_i)$. That is, the forecast probabilities are used only to sort the elements of the joint distribution into a sequence of 2 x 2 tables, but otherwise their actual numerical values are immaterial. For example, Table 9.4b shows that the forecasts defining the lower

ROC curve in Figure 9.14 are poorly calibrated, and in particular they exhibit strong conditional (over-confidence) bias. However this and other biases are not reflected in the ROC diagram, because the specific numerical values for the forecast probabilities $p(y_i)$ do not enter into the ROC calculations, and so ROC diagrams are insensitive to such conditional and unconditional biases (e.g., Glahn, 2004; Jolliffe and Stephenson, 2005; Kharin and Zwiers, 2003; Wilks, 2001). In fact, if the forecast probabilities $p(y_i)$ had corresponded exactly to the corresponding conditional event probabilities $p(o_1|y_i)$, or even if the probability labels on the forecasts in Tables 9.2 or 9.4 had been assigned values that were allowed to range outside the [0, 1] interval (while maintaining the same ordering, and so the same groupings of event outcomes), the resulting ROC curves would be identical!

As illustrated in Example 9.5, numerical integration of the ROC curve using the trapezoidal rule is an easy and typical approach to estimating the area A under it. To the extent that the plotted points only sample an underlying smooth function, trapezoidal integration will underestimate the area because ROC curves generally exhibit negative second derivatives, although this underestimation will decrease as the number of points increases. An alternative is to represent the two likelihood distributions $p(o_1 | y_i)$ and $p(o_2 | y_i)$ as Gaussian, which is called the *bi-normal model*. This assumption is more robust that might initially be imagined, because as noted before the ROC analysis is insensitive to the numerical labels of the forecasts, so that the bi-normal model will yield good results to the extent that monotonic transformations of the two likelihood distributions yield approximately Gaussian distributions. The necessary Gaussian parameters are estimated from the data at hand, using either regression (Mason, 1982; Swets, 1979) or maximum likelihood (Dorfman and Alf, 1969). The result is a smooth continuous ROC curve that can be integrated without bias, but may yield inaccuracies to the extent that the bi-normal model is not fully adequate.

ROC-based statistics provide information that is similar to the "resolution" term in the Brier score decomposition (Equation 9.45), independently of forecast calibration or lack thereof. This insensitivity to calibration is typically not a problem for the widespread use of ROC diagrams in applications like medical statistics (Pepe, 2003), because there the "forecast" (perhaps the blood concentration of a particular protein) is incommensurate with the "observation" (the patient has disease or not): there is no expectation that the "forecasts" are calibrated to or even pertain to the same variable as the observations. In such cases what is required is evaluation of the mapping between increasing levels of the diagnostic measurement with the probability of disease, and for this purpose the ROC diagram is a natural tool.

The insensitivity of ROC diagrams and ROC areas to both conditional and unconditional forecast biases—that they are independent of calibration—is sometimes cited as an advantage. This property is an advantage only in the sense that ROC diagrams reflect potential skill (which would be actually achieved only if the forecasts were correctly calibrated), in much the same way that the correlation coefficient reflects potential skill (cf. Equation 9.38). However, this property is not an advantage for forecast users who do not have access to the historical forecast data necessary to correct miscalibrations, and who therefore have no choice but to take forecast probabilities at face value. On the other hand, when forecasts underlying ROC diagrams are correctly calibrated, dominance of one ROC curve over another (i.e., one curve lying entirely above and to the left of another) implies statistical sufficiency (DeGroot and Fienberg, 1982, Ehrendorfer and Murphy, 1988) for the dominating forecasts, so that these will be of greater economic value for all rational forecast users (Krzysztofowicz and Long, 1990).

9.4.7. The Logarithmic, or Ignorance Score

The *logarithmic score*, or *Ignorance score* (Good, 1952; Roulston and Smith, 2002; Winkler and Murphy, 1968) is an alternative to the Brier score (Equation 9.39) for probability forecasts for dichotomous events. On a given forecasting occasion, k, it is the negative of the logarithm of the forecast probability corresponding to the event that subsequently occurs:

$$I_k = \begin{cases} -\ln(y_k), & \text{if } o_k = 1 \\ -\ln(1-y_k), & \text{if } o_k = 0, \end{cases} \tag{9.50a}$$

with the average Ignorance over n forecasting occasions being

$$\bar{I} = \frac{1}{n}\sum_{k=1}^{n} I_k. \tag{9.50b}$$

The Ignorance score ranges from zero for perfect forecasts ($y = 1$ if the binary event occurs or $y = 0$ if it does not) to infinity for certainty forecasts that are wrong ($y = 0$ if the event occurs or $y = 1$ if it does not). Another way of looking at Equation 9.50 is as the negative of the log-likelihood function (Section 4.6.1) for a sequence of n independent Bernoulli (i.e., binomial, Section 4.2.1, with $N = 1$) distributions, with probabilities y_k.

Even a single wrong certainty forecast in Equation 9.50 implies that the average Ignorance score for the entire collections of forecasts in Equation 9.50 will also be infinite, regardless of the accuracy of the other $n-1$ forecasts considered. Accordingly the Ignorance score is not appropriate in settings where the forecasts must be rounded before being issued. When forecast probabilities evaluated using the Ignorance score are to be estimated on the basis of a finite sample (e.g., in the context of ensemble forecasting), it would be natural to estimate the probabilities using one of the plotting position formulae from Table 3.2, for example, the implementation of the Tukey plotting position in Equation 8.5, that cannot produce either $y = 0$ or $y = 1$.

A natural skill score in this context, or relative accuracy as measured by Ignorance relative to a set of reference forecasts y_{ref}, is provided by the relative (or, reduction in) Ignorance (Bröcker and Smith, 2008; Peirolo, 2011),

$$\begin{aligned} \text{RI} &= \bar{I}_{ref} - \bar{I}_k \\ &= \frac{1}{n}\sum_{k=1}^{n} \ln\left(\frac{y_k}{y_{ref}}\right). \end{aligned} \tag{9.51}$$

Terms in Equation 9.51 for which $y_k > y_{ref}$ (larger probability than the reference forecast for the "yes" event) contribute positively to the sum, and those for which $y_k < y_{ref}$ contribute negatively, so that a positive value for RI corresponds to positive skill. Murphy and Epstein (1967a) pointed out that a consistent measure of forecast sharpness in this context is provided by the average negative entropy, or Shannon (1948) information

$$\bar{E} = -\frac{1}{n}\sum_{k=1}^{n} y_k \ln(y_k). \tag{9.52}$$

The Ignorance score behaves generally similarly to the Brier score, except for extreme (near-certainty) probability forecasts, for which the behavior of the two scores diverge markedly (e.g., Benedetti, 2010). Accordingly Ignorance may be better able than the Brier score to resolve the accuracy of forecasts for extreme or otherwise rare, and therefore low-probability, events. The Ignorance score generalizes to probability forecasts for nonbinary discrete events (Section 9.4.9), and to full continuous probability distribution forecasts (Section 9.5.1).

The Ignorance score and the corresponding skill score RI have interesting connections to probability assessment in betting and insurance (Good, 1952; Hagedorn and Smith, 2009; Roulston and Smith, 2002). In these contexts, the base-2 logarithms are typically used in Equations 9.50–9.52. The Ignorance score can be interpreted as the expected returns to be obtained by placing bets in proportion to the forecast probabilities, and the RI describes the gambler's return relative to the reference forecasts of the "house."

9.4.8. Hedging and Strictly Proper Scoring Rules

It is natural for forecasters to want to achieve the best scores they can. Depending on the evaluation measure, it may be possible to improve scores by *hedging*, or "gaming," which in the context of forecasting implies reporting something other than our true beliefs about future weather events in order to produce a better score (Jolliffe, 2008; Murphy and Epstein, 1967b). For example, in the setting of a forecast contest in a college or university, if the evaluation of our performance can be improved by playing the score, then it is entirely rational to try to do so. Conversely, if we are responsible for assuring that forecasts are of the highest possible quality, evaluating those forecasts in a way that penalizes hedging is desirable.

A forecast evaluation procedure that awards a forecaster's best expected score only when his or her true beliefs are forecast is called a *strictly proper* scoring rule (Winkler and Murphy, 1968). A slightly weaker condition, that no better expected score than that earned by the forecaster's true beliefs, is called *proper*. Strictly proper scoring procedures cannot be hedged, and attempting to hedge a proper scoring rule can at best achieve only a comparable result. One very appealing attribute of both the Brier score (Equation 9.39) and the Ignorance score (Equation 9.50) is that they are strictly proper, and this is a strong motivation for using one or the other to evaluate the accuracy of probability forecasts for dichotomous predictands, although many other strictly proper scoring rules can also be formulated (Gneiting and Raftery, 2007; Merkle and Steyvers, 2013).

It is instructive to look at the strict propriety of the Brier score in more detail. Of course it is not possible to know in advance what score a given forecast will achieve, unless we can make perfect forecasts. However, it is possible on each forecasting occasion to calculate the expected, or probability-weighted, future score using our subjective probability for the forecast event. Suppose a forecaster's subjective probability for the event being forecast is y^*, and that the forecaster must publicly communicate a forecast probability, y. The expected Brier score is simply

$$E[BS] = y^*(y-1)^2 + (1-y^*)(y-0)^2, \tag{9.53}$$

where the first term is the score received if the event occurs multiplied by the subjective probability that it will occur, and the second term is the score received if the event does not occur multiplied by the subjective probability that it will not occur. Consider that the forecaster has decided on a subjective probability y^* and is weighing the problem of what forecast y to issue publicly. Regarding y^* as constant, it is easy to minimize the expected Brier score by differentiating Equation 9.53 by y, and setting the result equal to zero. Then,

$$\frac{\partial E[BS]}{\partial y} = 2y^*(y-1) + 2(1-y^*)y = 0, \tag{9.54a}$$

yielding

$$2y\,y^* - 2y^* + 2y - 2y\,y^* = 0$$
$$2y = 2y^* \tag{9.54b}$$
$$y = y^*.$$

That is, regardless of the forecaster's subjective probability, the minimum expected Brier score is achieved only when the publicly communicated forecast corresponds exactly to the subjective probability. A similar derivation demonstrating that the Ignorance score is strictly proper can be found in Winkler and Murphy (1968). By contrast, for example, the absolute error (linear) score, $LS = |y - o|$ is minimized by forecasting $y = 0$ when $y^* < 0.5$, and forecasting $y = 1$ when $y^* > 0.5$, and so is not proper.

The concept of a strictly proper scoring rule is easiest to understand and prove for a case such as the Brier score, since the probability distribution being forecast (Bernoulli) is so simple. Gneiting and Raftery (2007) show that the concept of strict propriety can be applied in more general settings, where the form of forecast distribution is not necessarily Bernoulli. It is also not necessary to invoke forecaster honesty in order to motivate the concept of strict propriety. Bröcker and Smith (2007a) and Gneiting and Raftery (2007) note that strictly proper scores are internally consistent, in the sense that a forecast probability distribution yields an optimal expected score when the verification is drawn from that same probability distribution.

Equation 9.54 proves that the Brier score is strictly proper. Often Brier scores are expressed in the skill-score format of Equation 9.40. Even though the Brier score itself is strictly proper, this standard skill score based upon it, and other skill scores as well (Winkler, 1996) are not. However, for moderately large sample sizes (perhaps $n > 100$) the BSS closely approximates a strictly proper scoring rule (Murphy, 1973a).

9.4.9. Probability Forecasts for Multiple-Category Events

Probability forecasts may be formulated for discrete events having more than two ("yes" vs. "no") possible outcomes. These events may be *nominal*, for which there is not a natural ordering; or *ordinal*, where it is clear which of the outcomes are larger or smaller than others. The approaches to verification of probability forecasts for nominal and ordinal predictands may differ, because the magnitude of the forecast error is not a meaningful quantity in the case of nominal events, but is potentially quite important for ordinal events. One approach to verifying forecasts for nominal predictands is to collapse them to a sequence of binary predictands. Having done this, Brier scores, reliability diagrams, and so on, can be used to evaluate each of the derived binary forecasting situations.

Verification of probability forecasts for multicategory ordinal predictands presents a more difficult problem. First, the dimensionality of the verification problem increases exponentially with the number of outcomes over which the forecast probability is distributed. For example, consider a $J = 3$-event situation for which the forecast probabilities are constrained to be one of the 11 values 0.0, 0.1, 0.2, ... , 1.0. The dimensionality of the problem is not simply $33 - 1 = 32$, as might be expected by extension of the formula for dimensionality for the dichotomous forecast problem, because the forecasts are now vector quantities. For example, the forecast vector (0.2, 0.3, 0.5) is a different and distinct forecast from the vector (0.3, 0.2, 0.5). Since the three forecast probabilities must sum to 1.0, only two of them can vary freely. In this situation there are $I = 66$ possible three-dimensional forecast vectors, yielding a dimensionality for the forecast problem of $(66 \times 3) - 1 = 197$ (Murphy, 1991). Similarly, the dimensionality for the four-category ordinal verification situation with the same restriction on the forecast probabilities would be $(286 \times 4)-1 = 1143$. As a practical matter, because of their high dimensionality, probability forecasts for ordinal predictands primarily have been evaluated using scalar performance measures, even though such approaches will necessarily be incomplete.

Ranked Probability Score

Verification measures that are *sensitive to distance* reflect at least some aspects of the magnitudes of forecast errors, and for this reason are often preferred for probability forecasts of ordinal predictands. That is, such verification statistics are capable of penalizing forecasts increasingly as more probability is assigned to event categories further removed from the actual outcome. In addition, we would like the verification measure to be strictly proper (see Section 9.4.8), so that forecasters are encouraged to report their true beliefs. Several strictly proper scalar scores that are sensitive to distance exist for this type of forecast (Murphy and Daan, 1985; Staël von Holstein and Murphy, 1978), but of these the *ranked probability score* (RPS) is almost universally chosen. The RPS was originally formulated by R.J. Thompson around 1967 (Murphy, 1971), although Epstein (1969b) independently proposed a more elaborate but equivalent score.

The ranked probability score is essentially an extension of the Brier score (Equation 9.39) to the many-event situation. That is, it is a squared-error score with respect to the observation $o_j = 1$ if the forecast event j occurs, and $o_j = 0$ if the event does not occur. However, in order for the score to be sensitive to distance, the squared errors are computed with respect to the cumulative probabilities in the forecast and observation vectors. This characteristic introduces some notational complications.

As before, let J be the number of event categories, and therefore also the number of probabilities included in each forecast. For example, a common format for seasonal forecasts (pertaining to average conditions over 3-month periods) is to allocate probability among three climatologically equiprobable classes (e.g., Mason et al., 1999; O'Lenic et al., 2008). If a precipitation forecast is 20% chance of "dry," 40% chance of "near-normal," and 40% chance of "wet," then $y_1 = 0.2$, $y_2 = 0.4$, and $y_3 = 0.4$. Each of these components y_j pertains to one of the J events being forecast. That is, y_1, y_2, and y_3 are the three components of a forecast vector $\mathbf{y}$, and if all probabilities were to be rounded to tenths this forecast vector would be one of $I = 66$ possible forecasts $\mathbf{y}_i$. Similarly, in this setting the observation vector has three components. One of these components, corresponding to the event that occurs, will equal 1, and the other $J - 1$ components will equal zero. For example, if the observed precipitation outcome is in the "wet" category, then $o_1 = 0$, $o_2 = 0$, and $o_3 = 1$.

The cumulative forecasts and observations, denoted Y_m and O_m, are defined as functions of the components of the forecast vector and observation vector, respectively, according to

$$Y_m = \sum_{j=1}^{m} y_j, \quad m = 1, \ldots, J; \tag{9.55a}$$

and

$$O_m = \sum_{j=1}^{m} o_j, \quad m = 1, \ldots, J. \tag{9.55b}$$

In terms of the foregoing hypothetical example, $Y_1 = y_1 = 0.2$, $Y_2 = y_1 + y_2 = 0.6$, and $Y_3 = y_1 + y_2 + y_3 = 1.0$; and $O_1 = o_1 = 0$, $O_2 = o_1 + o_2 = 0$, and $O_3 = o_1 + o_2 + o_3 = 1$. Notice that since Y_m and O_m are both cumulative functions of probability components that must add to one, the final sums Y_J and O_J are always both equal to one by definition.

The ranked probability score is the sum of squared differences between the components of the cumulative forecast and observation vectors in Equation 9.55a and 9.55b, given by

$$\text{RPS} = \sum_{m=1}^{J} (Y_m - O_m)^2,$$ (9.56a)

or, in terms of the forecast and observed vector components y_j and o_j,

$$\text{RPS} = \sum_{m=1}^{J} \left[\left(\sum_{j=1}^{m} y_j \right) - \left(\sum_{j=1}^{m} o_j \right) \right]^2.$$ (9.56b)

A perfect forecast would assign all the probability to the single y_j corresponding to the event that subsequently occurs, so that the forecast and observation vectors would be the same. In this case, RPS = 0. Forecasts that are less than perfect receive scores that are positive numbers, so the RPS has a negative orientation. Notice also that the final ($m = J$) term in Equation 9.56 is always zero, because the accumulations in Equations 9.55 ensure that $Y_J = O_J = 1$. Accordingly Equation 9.56 will have only $J–1$ nonzero terms so that the last of these need not be computed, and the worst possible score is $J - 1$. The net effect of the calculation in Equation 9.56 is to add Brier scores for the $J–1$ binary probability forecasts defined by the $J–1$ boundaries between the J categories. For $J = 2$, the ranked probability score reduces to the Brier score, Equation 9.39.

Equation 9.56 yields the ranked probability score for a single forecast-event pair. Jointly evaluating a collection of n forecasts using the ranked probability score requires nothing more than averaging the RPS values for each forecast-event pair,

$$\overline{\text{RPS}} = \frac{1}{n} \sum_{k=1}^{n} \text{RPS}_k.$$ (9.57)

Similarly, the skill score for a collection of RPS values relative to the RPS computed from the climatological probabilities can be computed as

$$\text{SS}_{\text{RPS}} = \frac{\overline{\text{RPS}} - \overline{\text{RPS}_{\text{clim}}}}{0 - \overline{\text{RPS}_{\text{clim}}}} = 1 - \frac{\overline{\text{RPS}}}{\overline{\text{RPS}_{\text{clim}}}}.$$ (9.58)

Bradley and Schwartz (2011) point out that the RPS skill score in Equation 9.58 can also be formulated as a weighted average of the Brier skill scores for the $J–1$ possible binary probability forecasts defined by the boundaries of the J categories.

Example 9.6. Illustration of the Mechanics of the Ranked Probability Score

Table 9.6 demonstrates the mechanics of computing the RPS, and illustrates the property of sensitivity to distance, for two hypothetical probability forecasts for precipitation amounts. Here the continuum of precipitation has been divided into $J = 3$ categories, < 0.01 in., 0.01 – 0.24 in., and ≥ 0.25 in. Forecaster 1 has assigned the probabilities (0.2, 0.5, 0.3) to the three events, and Forecaster 2 has assigned the probabilities (0.2, 0.3, 0.5). The two forecasts are similar, except that Forecaster 2 has allocated more probability to the ≥ 0.25 in. category at the expense of the middle category. If no precipitation falls on this occasion the observation vector will be that indicated in the table. Many forecasters and forecast users would intuitively feel that Forecaster 1 should receive a better score, because this forecaster has assigned more probability closer to the observed category than did Forecaster 2. The score for Forecaster 1 is

TABLE 9.6 Comparison of Two Hypothetical Probability Forecasts for Precipitation Amount, Divided into $J = 3$ Ordinal Categories

	Forecaster 1		Forecaster 2		Observed	
Event	y_j	Y_m	y_j	Y_m	o_j	O_m
< 0.01 in.	0.2	0.2	0.2	0.2	1	1
0.01–0.24 in.	0.5	0.7	0.3	0.5	0	1
≥ 0.25 in.	0.3	1.0	0.5	1.0	0	1

The three components of the observation vector indicate that the observed precipitation was in the smallest category.

$RPS = (0.2–1)^2 + (0.7–1)^2 = 0.73$, and for Forecaster 2 it is $RPS = (0.2–1)^2 + (0.5–1)^2 = 0.89$. The lower RPS for Forecaster 1 indicates a more accurate forecast according to RPS.

These RPS results are equivalent to adding the $J–1 = 2$ Brier scores for the two possible binary probability forecasts based on the forecast vectors that respect the ordinal nature of the categories, that is, Pr$\{< 0.01$ in.$\}$ (vs. $≥ 0.01$ in.) and Pr$\{< 0.25$ in.$\}$ (vs. $≥ 0.25$ in.). For Forecaster 1, these two probability forecasts are 0.2 and 0.7, respectively, and for Forecaster 2 they are 0.2 and 0.5, which are exactly the probabilities specified by Equation 9.55a. Since the observation indicated in Table 9.6 is < 0.01 in., the two Brier scores for Forecaster 1 are $(0.2–1)^2 = 0.64$ and $(0.7–1)^2 = 0.09$, which sum to 0.73, consistent with Equation 9.56. The corresponding results for Forecaster 2 are $(0.2–1)^2 = 0.64$ and $(0.5–1)^2 = 0.25$, which sum to 0.89.

Alternatively, if some amount of precipitation larger than 0.25 in. had fallen, Forecaster 2's probabilities would have been closer and would have received the better score. The score for Forecaster 1 would have been $RPS = (0.2–0)^2 + (0.7–0)^2 = 0.53$, and the score for Forecaster 2 would have been $RPS = (0.2–0)^2 + (0.5–0)^2 = 0.29$. Note that in both of these examples, only the first $J–1 = 2$ terms in Equation 9.56 were needed to compute the RPS. ◇

The Calibration Simplex

The reliability diagram (Section 9.4.4) for probability forecasts of binary events can be extended to the *Calibration Simplex* (Wilks, 2013), which is a graphical representation of the calibration-refinement factorization (Equation 9.2) for three-category probability forecasts, regardless of whether the categories are nominal or ordinal.

The reliability diagram plots the possible scalar probability forecasts along the unit interval on its horizontal axis. Each set of forecast probabilities for three categories can be plotted in two dimensions because they must sum to 1. The geometrically appropriate graph is the regular 2-simplex (e.g., Epstein and Murphy, 1965; Murphy, 1972), which takes the shape of an equilateral triangle. The area within the triangle can be partitioned to represent possible 3-dimensional forecast vectors in the same way that the reliability diagram partitions the unit interval. Figure 9.15 illustrates the plotting of discretized forecast vectors onto the simplex, which has been rendered as a tessellation of hexagons. Here the categories are labeled B (below normal), N (near-normal), and A (above normal), consistent with a typical forecast format for medium-range and seasonal forecasts. Each scalar forecast probability has been rounded to one of the ten values $0/9, 1/9, \ldots, 9/9$, yielding $I = 55$ distinct possible vector forecasts, each of which

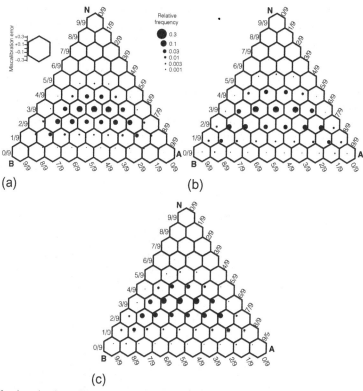

FIGURE 9.15 Calibration simplexes illustrating (a) overconfident forecasts, (b) underconfident forecasts, and (c) unconditionally biased forecasts. In each case the refinement distribution is represented by the sizes of the dots. *From Wilks (2013).*

is represented by one of the hexagons. The hexagons at the three vertices represent forecasts assigning all probability to the outcome labeled at that corner. Hexagons representing other forecast vectors are located at perpendicular distances from the respective opposite sides, which are indicated by the probability labels in the margins. The result, for example, is that the hexagon in the center of Figure 9.15 locates the climatological forecast [1/3, 1/3, 1/3].

The I conditional calibration distributions $p(o_j \mid y_i)$, for dichotomous probability forecasts are Bernoulli, each of which can be characterized by single probabilities that are indicated in the reliability diagram by the vertical positions of the I dots. In the case of three-element probability forecasts, the calibration distributions are multinomial (Equation 4.11) and are completely specified by two of the three probability parameters. The locations of dots within each hexagon of the calibration simplex indicate the magnitudes of these conditional multinomial probabilities. The calibration-refinement factorization is completed by specifying the frequencies of use of the I possible probability vectors (the refinement distribution), the elements of which are shown in proportion to the areas of the dots within each hexagon.

Figure 9.15a shows a calibration simplex reflecting overconfident forecasts, where the plotted dots have been displaced toward the center of the diagram, indicating that probabilities larger than 1/3 have been overforecast and probabilities smaller than 1/3 have been underforecast. Figure 9.15b shows a corresponding plot for underconfident forecasts, where the probabilities larger than 1/3 have been under-forecast and probabilities smaller than 1/3 have been overforecast, so that the dots have been displaced toward the margins of the diagram. Figure 9.15c shows an example plot for unconditionally biased fore-casts, where the "A" category has been overforecast at the expense of the "B" and "N" categories, so that the dots are displaced away from the "A" vertex.

A similar idea, proposed independently by Murphy and Daan (1985) and Jupp et al. (2012), is to draw line segments on the 2-simplex connecting points locating each distinct forecast vector with its corresponding average outcome vector, and indicate the corresponding subsample sizes with numerals printed at the mid-points of these line segments. Jupp et al. call the resulting plot the *ternary reliability diagram*. This graphical approach also expresses the full information content of the joint distribution, but overall the diagram may be somewhat difficult to read, unless very few of the possible forecast vectors have been used.

Logarithmic (Ignorance) Score for Multiple Categories

An alternative to RPS for evaluation of probability forecasts for multicategory events is provided by an extension of the Ignorance score (Equation 9.50a), which is also strictly proper. For a single forecast k, the Ignorance score is simply the negative logarithm of that element of the forecast vector corresponding to the event that actually occurred (i.e., for which $o_j = 1$), which can be expressed as

$$I_k = -\sum_{j=1}^{J} o_j \ln (y_j), \tag{9.59}$$

where it is understood that $0 \ln(0) = 0$. As was also the case for the 2-category Ignorance score (Equation 9.50a), incorrect certainty forecasts yield infinite Ignorance, so that the Ignorance score is usually not suitable for forecasts that have been rounded to a finite set of discrete allowable probabilities. The average Ignorance score over n forecasting occasions would again be given by Equation 9.50b.

Extending Equation 9.51, a natural skill measure is the Relative Ignorance

$$\mathrm{RI} = \frac{1}{n}\sum_{k=1}^{n}\sum_{j=1}^{J} o_{k,j} \ln \left(\frac{y_{k,j}}{y_{ref,j}}\right), \tag{9.60}$$

where $o_{k,j}$ is the binary observation variable for the jth category on the kth forecast occasion, $y_{k,j}$ is the corresponding forecast, and $y_{ref,j}$ is the appropriate (perhaps climatological) reference (Krakauer et al., 2013).

The Ignorance score is not sensitive to distance, and indeed exhibits a property known as *locality*, meaning that only the probability assigned to the event that occurs matters in its computation. For the forecasts in Table 9.6, $I = 1.61$ for both Forecaster 1 and Forecaster 2 because the distribution of forecast probabilities among outcome categories for which $o_j = 0$ is irrelevant. It should be clear that locality and sensitivity to distance are mutually incompatible, and some discussion of their relative merits is included in Bröcker and Smith (2007a) and Mason (2008), although as scalars both the ranked probability score and the Ignorance score are in any case incomplete measures of forecast quality. If locality is accepted as a desirable characteristic, then preference for the Ignorance score is indicated, since it is the only score that is both local and strictly proper. Use of the Ignorance score in preference to the RPS is also indicated when evaluating probability forecasts for nominal categories, in which case the concepts of ordering and distance not meaningful.

9.5. PROBABILITY DISTRIBUTION FORECASTS FOR CONTINUOUS PREDICTANDS

When the predictand is a continuous, real-valued quantity the most informative forecast format will involve communicating its predictive distribution. That is, each forecast will be a full probability distribution, which might be communicated as a PDF $f(y)$, or CDF $F(y)$. For some forecasts, such as the nonhomogeneous regressions described in Sections 8.3.2–8.3.4, these predictive distributions may be one the conventional parametric forms described in Section 4.4, in which case they might be specified in terms of a few distribution parameters. Forecasts formulated nonparametrically (e.g., Section 8.3.6), or through Bayesian methods solved using MCMC methods (Section 6.4), may be expressed as discrete approximations to continuous PDFs or CDFs. Although this latter class includes raw ensemble forecasts, and member-by-member postprocessed ensemble forecasts (Section 8.3.6), discussion of verification methods designed specifically for forecast ensembles will be deferred until Section 9.7.

9.5.1. Continuous Ranked Probability Score

Regardless of how a forecast probability distribution is expressed, providing a full forecast probability distribution is both a conceptual and a mathematical extension of multicategory probability forecasting (Section 9.4.9), to forecasts for an infinite number of predictand classes of infinitesimal width. A natural approach to evaluating this kind of forecast is to extend the ranked probability score to the continuous case, replacing the summations in Equation 9.56 with integrals. The result is the *Continuous Ranked Probability Score* (Hersbach, 2000; Matheson and Winkler, 1976; Unger, 1985),

$$\text{CRPS} = \int_{-\infty}^{\infty} [F(y) - F_o(y)]^2 dy, \tag{9.61a}$$

where

$$F_o(y) = \begin{cases} 0, & y < o \\ 1, & y \geq o \end{cases} \tag{9.61b}$$

is a cumulative-probability step function that jumps from 0 to 1 at the point where the forecast variable y equals the observation o. The squared difference between continuous CDFs in Equation 9.61a is analogous to the same operation applied to the cumulative discrete variables in Equation 9.56a for the Ranked Probability Score, and accordingly CRPS can be seen as an extension of RPS for infinitely many bins of infinitesimal width. Also, just as the RPS can be viewed as the sum of Brier scores for the binary variables defined by the bin boundaries, the CRPS can be viewed as the integral of the BS over all values of the predictand (Hersbach, 2000). In common with the discrete BS and RPS, the CRPS is also strictly proper (Matheson and Winkler, 1976).

The CRPS has a negative orientation (smaller values are better), and it rewards concentration of probability around the step function located at the observed value. Figure 9.16 illustrates the CRPS with a hypothetical example. Figure 9.16a shows three Gaussian forecast PDFs $f(y)$ in relation to a single observed value of the continuous predictand y. Forecast Distribution 1 is centered on the eventual observation and strongly concentrates its probability around the observation. Distribution 2 is equally sharp (i.e., expresses the same degree of confidence in distributing probability), but is centered well away from the observation. Distribution 3 is centered on the observation but exhibits low confidence (distributes probability more diffusely than the other two forecast distributions). Figure 9.16b shows the same three forecast distributions expressed as CDFs, $F(y)$, together with the step-function CDF $F_0(y)$ (thick line) that jumps from 0 to 1 at the observed value (Equation 9.61b). Since the CRPS is the integrated squared difference between the CDF and the step function, CDFs that approximate the step function (Distribution 1) produce relatively small integrated squared differences, and so good scores. Distribution 2 is equally sharp, but its displacement away from the observation produces large discrepancies with the step function, especially for values of the predictand slightly larger than the observation, and therefore a very large integrated squared difference. Distribution 3 is centered on the observation, but its diffuse assignment of forecast probability means that it is nevertheless a poor approximation to the step function, and so also yields a large integrated squared difference.

Alternatively, the CRPS can also be formulated as (Gneiting and Raftery, 2007)

$$\text{CRPS} = E_F|Y - o| - \frac{1}{2}E_F|Y - Y'|, \tag{9.62}$$

where E_F denotes statistical expectation with respect to the forecast distribution $F(y)$, and Y and Y' are independent realizations from $F(y)$. This alternative representation forms the basis of the Energy Score described in Section 9.5.2, and the discrete version of CRPS described in Section 9.7.3.

Equation 9.61 may be difficult to evaluate for arbitrary an forecast CDF, $F(y)$. However, if this forecast distribution is Gaussian with mean μ and variance σ^2 the CRPS when the observation o occurs is (Gneiting et al., 2005)

$$\text{CRPS}(\mu, \sigma^2, o) = \sigma \left\{ \frac{o-\mu}{\sigma}\left[2\Phi\left(\frac{o-\mu}{\sigma}\right) - 1\right] + 2\phi\left(\frac{o-\mu}{\sigma}\right) - \frac{1}{\sqrt{\pi}} \right\}, \tag{9.63}$$

where $\Phi(\cdot)$ and $\phi(\cdot)$ are the CDF and PDF (Equation 4.25) of the standard Gaussian distribution. In Figure 9.16, $f_1(y)$ has $\mu = 0$ and $\sigma^2 = 1$, $f_2(y)$ has $\mu = 2$ and $\sigma^2 = 1$, and $f_3(y)$ has $\mu = 0$ and $\sigma^2 = 9$. Using Equation 9.63 the observation $o = 0$ yields $\text{CRPS}_1 = .23$, $\text{CRPS}_2 = 1.45$, and $\text{CRPS}_3 = .70$.

References providing closed-form expressions for CRPS when the forecast distribution takes on other known parametric forms are listed in Table 9.7.

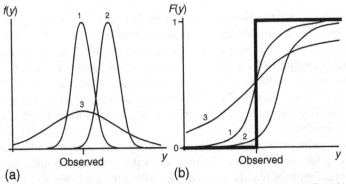

FIGURE 9.16 Schematic illustration of the continuous ranked probability score, as expressed in Equation 9.61. Three Gaussian forecast PDFs are shown in relation to the observed outcome in (a). The corresponding CDFs are shown in (b), together with the step-function CDF for the observation $F_0(y)$ (heavy line). Distribution 1 would produce a small (good) CRPS because its CDF is the closest approximation to the step function, yielding the smallest integrated squared difference. Distribution 2 concentrates probability away from the observation, and Distribution 3 is penalized for lack of sharpness even though it is centered on the observation.

Because CRPS can also be computed as the Brier score for dichotomous events integrated over all possible division points of the continuous variable y, the CRPS has an algebraic decomposition into reliability, resolution, and uncertainty components that is analogous to an integrated form of Equation 9.45 (Candille and Talagrand, 2005). Indeed, any strictly proper scoring rule can be expressed in that way (Bröcker, 2009). Hersbach (2000) also shows that for nonprobabilistic forecasts (all probability concentrated at y, with $F(y)$ also a step function in the form of Equation 9.61b), CRPS reduces to absolute error (because $0^2 = 0$ and $1^2 = 1$), in which case the average CRPS over n forecasts reduces to the MAE (Equation 9.32).

Analogously to the RPS in Equation 9.57, CRPS values for multiple occasions are typically averaged to characterize forecast quality overall in terms of a single number. Also analogously to the RPS in Equation 9.58, these average CRPS values are often expressed in the usual skill score format (Equation 9.4).

TABLE 9.7 Locations in the Literature for Analytic CRPS Formulas for Various Continuous Distributions

Distribution	References
Beta	Jordan et al. (2017), Taillardat et al. (2016)
Exponential	Jordan et al. (2017)
truncated	Jordan et al. (2017)
Gamma	Jordan et al. (2017), Scheuerer and Möller (2015), Taillardat et al. (2016)
Gaussian	
lognormal	Baran and Lerch (2015), Taillardat et al. (2016)
mixture	Gneiting et al. (2007), Grimit et al. (2006), Jordan et al. (2017)
square-root truncated	Hemri et al. (2014), Taillardat et al. (2016)
truncated	Gneiting et al. (2006), Jordan et al. (2017), Taillardat et al. (2016), Thorarinsdottir and Gneiting (2010)
Generalized Pareto	Friederichs and Thorarinsdottir (2012), Jordan et al. (2017)
GEV	Friederichs and Thorarinsdottir (2012), Jordan et al. (2017)
left-censored	Scheuerer (2014)
Laplace	Jordan et al. (2017)
log-Laplace	Jordan et al. (2017)
Logistic	Jordan et al. (2017), Taillardat et al. (2016)
censored	Taillardat et al. (2016)
log-Logistic	Jordan et al. (2017), Taillardat et al. (2016)
square-root censored	Taillardat et al. (2016)
truncated	Jordan et al. (2017), Scheuerer and Möller (2015), Taillardat et al. (2016)
t	Jordan et al. (2017)
Truncated	Jordan et al. (2017)
Uniform	Jordan et al. (2017)
Von Mises (circular)	Grimit et al. (2006)

9.5.2. Energy Score

The *Energy Score* (Gneiting and Raftery, 2007; Gneiting et al., 2008) is a multivariate extension of the CRPS, based on its representation in Equation 9.62. Defining Y and Y' as independent random vectors drawn from the multivariate forecast distribution $F(y)$, and o as the corresponding vector observation, the Energy Score is

$$\text{ES} = E_F\|Y - o\| - \frac{1}{2}E_F\|Y - Y'\|, \qquad (9.64)$$

where again E_F denotes statistical expectation with respect to F, and $\|\cdot\|$ indicates Euclidean distance (Equation 11.6). It is a strictly proper score (Gneiting and Raftery, 2007).

Often closed-form expressions for the expectations in Equation 9.63 will not be available, in which case Monte Carlo evaluation analogous to the scalar expression in Equation 9.83 using a large number of realizations m can be employed. If the elements of the forecast and observation vectors are measured in different physical units it may be advisable to standardize them before calculation of the score. Initial experience with the Energy Score has suggested that it may not be sufficiently sensitive to correlations among the elements of the forecast and observation vectors (Pinson and Girard, 2012; Scheuerer and Hamill, 2015b), and accordingly Scheuerer and Hamill (2015b) suggest that the Dawid-Sebastiani score (Section 9.5.3) may be preferable if the information required to compute it is available.

9.5.3. Ignorance, and Dawid-Sebastiani Scores

The strictly proper Ignorance score (Equations 9.50 and 9.59) also generalizes to probability density forecasts for a continuous predictand. When the forecast PDF is $f(y)$ and the observation is o, the Ignorance score for a single forecast is

$$I = -\ln[f(o)]. \qquad (9.65)$$

Equation 9.65 is written in terms of a univariate PDF and a scalar observation, although it is equally applicable to multivariate forecast PDFs and vector observations. The Ignorance score is local, since it is simply the negative logarithm of the forecast PDF evaluated at the observation, regardless of the behavior of $f(y)$ for other values of its argument. If $f(y)$ is a Gaussian forecast PDF with mean μ and variance σ^2, the Ignorance when the observation is o is therefore

$$I = \frac{\ln(2\pi\sigma^2)}{2} + \frac{(o - \mu)^2}{2\sigma^2}, \qquad (9.66)$$

where the first term penalizes lack of sharpness, independently of the observation, and the second term penalizes in proportion to the square of the standardized error in the location of the forecast distribution. Equation 9.66 is the negative of the Gaussian log-likelihood (Equation 4.82) for a single ($n = 1$) observation o.

Example 9.7. Comparison of CRPS and Ignorance for 2 Gaussian forecast PDFs

Figure 9.17b compares CRPS and the Ignorance scores for the two Gaussian forecast PDFs shown in Figure 9.17a. The forecast PDF $f_1(y)$ (solid curve) is standard Gaussian (zero mean, unit standard deviation), and $f_2(y)$ (dashed) has mean 1 and standard deviation 3. Since both forecast PDFs are Gaussian, their CRPS and Ignorance scores, as functions of the observation o, can be computed using Equations 9.63 and 9.66, respectively.

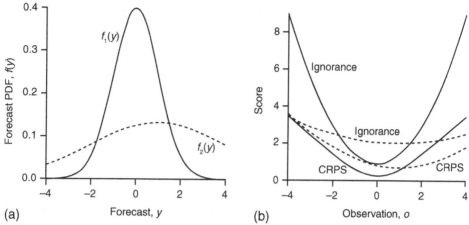

FIGURE 9.17 Comparison of the continuous ranked probability and Ignorance scores, for the two Gaussian PDFs shown in (a), as functions of the observed value (b). The solid PDF $f_1(y)$ is standard Gaussian, and the dashed $f_2(y)$ has mean 1 and standard deviation 3.

Since $f_1(y) = f_2(y)$ for $y \approx -1.7$ and $y \approx 1.5$, the Ignorance scores for the two forecasts are equal for these two values of the observation. On the other hand, CRPS yields a better score for $f_1(y)$ when $o = -1.7$, but a slightly worse score for $f_1(y)$ when $o = 1.5$. It is not immediately clear which of these two results should be preferable, and indeed the preference would generally depend on the circumstances of individual forecast users.

Even though the Ignorance score depends only on the local value $f(o)$ whereas CRPS is an integrated quantity, the two scores behave qualitatively similarly in Figure 9.17b. This similarity in behavior derives from the smoothness of the Gaussian forecast PDFs, which implicitly imparts a degree of sensitivity to distance for the Ignorance score in this example. More prominent differences for the two scores would be expected if the forecast PDFs were bi- or multimodal. The biggest differences between the two scores occurs for extreme values of the observation, 3 or 4 standard deviations away from the forecast mean, which the Ignorance score penalizes relatively more heavily than does the CRPS. ◇

The *Dawid-Sebastiani score* (Dawid and Sebastiani, 1999),

$$DSS = 2 \ln(\sigma) + \frac{(o - \mu)^2}{\sigma^2}, \tag{9.67}$$

is closely related to the Ignorance score for Gaussian predictive distributions (Equation 9.66), and indeed is simply a linear transformation of it. However, it can be applied to density forecasts in general when the mean and variance are both specified, in which case it is proper, but not strictly proper. For multivariate forecast PDFs the Dawid-Sebastiani score is

$$DSS = \ln(\det[\Sigma]) + (o - \mu)^T [\Sigma]^{-1} (o - \mu). \tag{9.68}$$

This is a linear transformation of the log-likelihood for the multivariate normal distribution, the PDF for which is Equation 12.1, but it can be used in connection with general multivariate forecasts PDFs and vector observations o when the covariance matrix $[\Sigma]$ is known or can be estimated well. In common with the Ignorance score for Gaussian predictive distributions (Equation 9.66) the DSS penalizes dispersion of the forecast distribution through the determinant of $[\Sigma]$ in the first term of Equation 9.68, and penalizes (Mahalanobis, Equation 11.91) distance between the forecast mean and observation vectors in the second term.

9.5.4. The PIT Histogram

It is sometimes of interest to characterize the calibration, as distinct from the accuracy, of a set of probability density forecasts. For this type of forecast, calibration implies that the forecasts and observations are consistent in the sense that, collectively, the observations can plausibly be regarded as a random draws from their respective forecast distributions. For each forecast occasion, k, the *probability integral transform* (PIT) is simply the value of the cumulative probability of the observation o_i within the forecast CDF F_k,

$$u_k = F_k(o_k). \tag{9.69}$$

The forecast distributions F_k may be different for different forecasts, but if they are calibrated then each cumulative probability in Equation 9.69 is a random variable that has been drawn from the standard uniform distribution, $f(u_k) = 1, 0 \le u_k \le 1$. Therefore if a collection of n density forecasts are calibrated, then a histogram of the corresponding collection of PIT values $u_k, k, = 1, \dots, n$ will be uniform on the unit interval, within the limits of sampling variability. This is the *PIT histogram* (Dawid, 1984; Diebold et al., 1998). Typically between ten and twenty bins are used to discretize the unit interval.

Figure 9.18 shows PIT histograms for a collection of temperature forecasts. The unit interval has been discretized into ten bins, and the histogram bars indicate the relative frequencies of PIT values in each bin. Figure 9.18a shows the PIT histogram for Gaussian distributions fit to individual raw forecast ensembles by computing the ensemble means and ensemble standard deviations, and Figure 9.18b shows corresponding results after NGR (Section 8.3.2) postprocessing. Interpretations of characteristic PIT histogram shapes are the same as for the closely allied Verification Rank Histogram (Section 9.7.1), and the

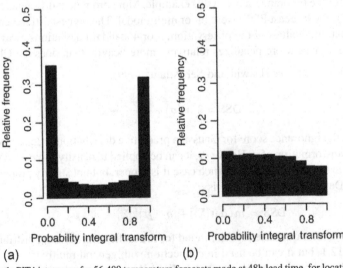

(a)   (b)

FIGURE 9.18   Example PIT histograms for 56,489 temperature forecasts made at 48h lead time, for locations in the northwestern U.S. and southwestern Canada, for (a) Gaussian distributions defined by raw ensemble means and standard deviations, and (b) NGR (Section 8.3.2) postprocessed forecasts. *From Gneiting et al. (2005). © American Meteorological Society. Used with permission.*

scalar metrics listed in Section 9.7.1 for characterizing rank histogram flatness are equally applicable to PIT histograms. The U-shaped PIT histogram in Figure 9.18a shows that the uncorrected forecast distributions are too narrow on average, so that the corresponding observations fall too frequently in the tails, corresponding to $u_k \approx 0$ or $u_k \approx 1$. The NGR-postprocessed results in Figure 9.18b are much more uniform, although they still reveal a tendency toward warm bias, so that the observations fall too frequently in the left tails of the forecast distributions.

9.6. QUANTILE FORECASTS

9.6.1. The Quantile Score

In some settings only selected quantiles of the predictive distribution are issued as forecasts. Examples include forecasts of the median, and quantile regression (Section 7.7.3) forecasts. In this setting a quantile q is forecast, with the implication that $\Pr\{o \leq q\} = \alpha$, where o is the observed value and α is the probability level for the forecast quantile. Such forecasts can be evaluated using the proper *Quantile Score* (Gneiting and Raftery, 2007; Taylor, 1999),

$$\text{QS} = \begin{cases} (q-o)(1-\alpha) \,, & o \leq q \\ (o-q)\alpha \,, & o > q \end{cases}. \tag{9.70}$$

The Quantile Score is based on a check function (Equation 7.55). The integral of QS over all forecast quantile probability levels yields the CRPS (Gneiting and Ranjan, 2011). Bentzien and Friederichs (2014) derive a reliability-resolution-uncertainty decomposition of the average QS that is analogous to Equation 9.45 for the Brier score, and propose new verification diagrams for QS. Ben Bouallègue et al. (2015) propose computing skill scores based on QS in the form of Equation 9.4, and draw connections with economic value of forecasts.

9.6.2. Central Credible Interval Forecasts

The burden of communicating a full probability distribution is reduced considerably if the forecast distribution is merely sketched, using the central credible interval (CCI) format (Section 7.10.3). In full form, a CCI forecast consists of a range of the predictand that is centered in a probability sense, together with the probability covered by that range within the forecast distribution. Usually CCI forecasts are abbreviated in one of two ways: either the interval width is constant on every forecast occasion but the location of the interval and the probability it subtends are allowed to vary (fixed-width CCI forecasts), or the probability within the interval is constant on every forecast occasion but the interval location and width may both change (fixed-probability CCI forecasts).

The ranked probability score (Equation 9.56) is an appropriate scalar accuracy measure for fixed-width CCI forecasts (Baker, 1981; Gordon, 1982). In this case there are three categories (below, within, and above the forecast interval) among which the forecast probability is distributed. The probability p pertaining to the forecast interval is specified as part of the forecast, and because the forecast interval is located in the probability center of the distribution, probabilities for the two extreme categories are each $(1-p)/2$. The result is that $\text{RPS} = (p-1)^2/2$ if the observation falls within the interval, or $\text{RPS} = (p^2+1)/2$ if the observation is outside the interval. The RPS thus reflects a balance between preferring a large p if the observation is within the interval, but preferring a smaller p if it is outside, and that balance is optimized when the forecasters report their true judgment.

The RPS is not an appropriate accuracy measure for fixed-probability CCI forecasts. For this forecast format, small (i.e., better) RPS can be achieved by always forecasting extremely wide intervals, because the RPS does not penalize vague forecasts that include wide central intervals. In particular, forecasting an interval that is sufficiently wide that the observation is nearly certain to fall within it will produce a smaller RPS than a verification outside the interval if $(p-1)^2/2 < (p^2+1)/2$. A little algebra shows that this inequality is satisfied for any positive probability p.

Fixed-probability CCI forecasts are appropriately evaluated using *Winkler's score* (Dunsmore, 1968; Winkler, 1972a; Winkler and Murphy, 1979),

$$W = \begin{cases} (b-a+1)+c(a-o), & o<a \\ (b-a+1), & a\leq o\leq b. \\ (b-a+1)+c(o-b), & b<o \end{cases} \tag{9.71}$$

Here the forecast interval ranges from a to b, inclusive, with $a \leq b$, and the value of the observed variable is o. Regardless of the actual observation, a forecast is assessed penalty points equal to the width of the forecast interval, which is $b-a+1$ to account for both endpoints when (as is usual) the interval is specified in terms of integer units of the predictand. An additional penalty is added if the observation falls outside the specified interval, and the magnitudes of these "miss" penalties are proportional to the distance from the interval. Winkler's score thus expresses a trade-off between short intervals to reduce the fixed penalty (and thus encouraging sharp forecasts), versus sufficiently wide intervals to avoid incurring the additional penalties too frequently. This trade-off is balanced by the constant c, which depends on the fixed probability to which the forecast CCI pertains, and increases as the implicit probability for the interval increases, because outcomes outside the interval should occur increasingly rarely for larger interval probabilities. In particular, $c = 4$ for 50% CCI forecasts, and $c = 8$ for 75% CCI forecasts. More generally, $c = 1/F(a)$, where $F(a) = 1 - F(b)$ is the cumulative probability associated with the lower interval boundary according to the forecast CDF. Equation 9.71 has also been called the *Interval Score* by Gneiting and Raftery (2007), who note that it is a proper score.

Winkler's score is equally applicable to fixed-width CCI forecasts, and to unabbreviated CCI forecasts for which the forecaster is free to choose both the interval width and the subtended probability. In these two cases, where the stated probability may change from forecast to forecast, the penalty function for observations falling outside the forecast interval will also change, according to $c = 1/F(a)$.

The calibration (reliability) of fixed-probability CCI forecasts can be evaluated simply, by tabulating the relative frequency over a sample of n forecasts with which the observation falls in the forecast interval, and checking numbers of observations falling above and below the interval for equality. Relative frequency of observations within the interval being less than the specified forecast probability suggests that improvements could be achieved by widening the forecast intervals, on average, and vice versa for the observed relative frequency being larger than the forecast probability. Of course good calibration does not guarantee skillful forecasts, as constant interval forecasts based on the central part of the predictand climatological distribution will also exhibit good calibration.

9.7. VERIFICATION OF ENSEMBLE FORECASTS

Chapter 8 presented the method of ensemble forecasting, in which the effects of initial-condition uncertainty on dynamical forecasts are represented by a finite collection, or ensemble, of very similar initial conditions. Ideally, an initial ensemble represents a random sample from the PDF quantifying initial-condition uncertainty, defined on the phase space of the dynamical model. Integrating the forecast model forward in time from each of these initial conditions individually thus is a Monte Carlo approach to estimating the effects of the initial-condition uncertainty on the forecast uncertainty for the quantities being predicted. That is, if the initial ensemble members have been chosen as a random sample from the initial-condition uncertainty PDF, and if the forecast model consists of correct and accurate representations of the physical dynamics, the ensemble after being integrated forward in time represents a random sample from the PDF of forecast uncertainty. If this ideal situation could be realized, the true state of the atmosphere would be just one more member of the ensemble, at the initial time and throughout the integration period, and should be statistically indistinguishable from the forecast ensemble. This condition, that the actual future atmospheric state behaves like a random draw from the same distribution that produced the ensemble, is called *ensemble consistency* (Anderson, 1997), and is closely related to the notion of exchangeability

In light of this background, it should be clear that ensemble forecasts are probability forecasts that are expressed as a discrete approximation to a full forecast PDF. According to this approximation, ensemble relative frequency should estimate actual probability. Depending on what the predictand(s) of interest may be, the formats for these probability forecasts can vary widely. Probability forecasts can be obtained for simple predictands, such as continuous scalars (e.g., temperature or precipitation at a single location), or discrete scalars (possibly constructed by thresholding a continuous variable, e.g., zero or trace precipitation vs. nonzero precipitation, at a given location), or quite complicated multivariate predictands such as entire fields (e.g., the joint distribution of 500 mb heights at the global set of horizontal gridpoints).

When ensemble forecasts are expressed as conventional probability forecasts, perhaps after transformation to a probability for a dichotomous outcome, or after kernel density smoothing of the ensemble to yield a continuous probability distribution (e.g., Roulston and Smith, 2003), or by fitting parametric probability distributions (e.g., Hannachi and O'Neill, 2001; Stephenson and Doblas-Reyes, 2000; Wilks, 2002b), then the probabilistic forecast verification methods presented in Sections 9.4–9.6 can be applied to evaluate them. The same is true for results of ensemble postprocessing methods that yield conventional probability forecasts, such as many of the methods outlined in Section 8.3.2–8.3.4. Similarly, ensemble-mean forecasts can be evaluated using methods described in Section 9.3. However, other verification tools have been developed for evaluation of raw ensemble forecasts, or postprocessed ensembles expressed as discrete approximations to underlying probability distributions (as described in Sections 8.4.2 and 8.4.3), and these methods are presented in the following subsections. Some of these ensemble verification methods have been designed to evaluate ensemble consistency, through calibration diagnostics, and others are accuracy measures that operate on the discrete samples provided by ensembles.

9.7.1. Assessing Univariate Ensemble Calibration

To the extent that the ensemble consistency condition has been met, the observation being predicted looks statistically like just another member of the forecast ensemble. The result is that the ensemble will be calibrated, so that ensemble relative frequency can be regarded as a good estimator of probability, regardless of whether those probability estimates are sharp or not. Several methods have been proposed to evaluate ensemble calibration.

Verification Rank Histogram

Construction of a *verification rank histogram*, sometimes called simply the *rank histogram*, is the most common approach to evaluating whether a collection of ensemble forecasts for a scalar predictand satisfy the consistency condition. That is, the rank histogram is used to evaluate whether the forecast ensembles apparently include the observations being predicted as equiprobable members. The rank histogram is a graphical approach that was devised independently by Anderson (1996), Hamill and Colucci (1998), and Talagrand et al. (1997), and is sometimes also called the *Talagrand diagram*.

Consider the evaluation of n ensemble forecasts, each of which consists of m ensemble members, in relation to the n corresponding observed values for the predictand. Within each of these n sets, if the m members and the single observation all have been drawn from the same distribution, then the rank of the observation within these $m+1$ values is equally likely to be any of the integers $i = 1$, 2, 3, ... , $m+1$. For example, if the observation is smaller than all m ensemble members, then its rank is $i = 1$. If it is larger than all the ensemble members then its rank is $i = m+1$. For each of the n forecasting occasions, the rank of the observation within this $m+1$-member distribution is tabulated. Collectively these n verification ranks are plotted in the form of a histogram to produce the verification rank histogram. (Equality of the observation with one or more of the ensemble members requires a slightly more elaborate procedure; see Hamill and Colucci, 1998). If the ensemble consistency condition has been met this histogram of verification ranks will be uniform, reflecting equiprobability of the observations within their ensemble distributions, except for departures that are small enough to be attributable to sampling variations. In the limit of infinite ensemble size, or if the ensemble distribution has been represented as a smooth, continuous PDF, the rank histogram is identical to the PIT histogram (Section 9.5.4).

Departures from the ideal of rank uniformity can be used to diagnose aggregate deficiencies of the ensembles (Hamill, 2001). Figure 9.19 shows four problems that can be discerned from the rank histogram, together with a rank histogram (center panel) that shows only small sampling departures from a uniform distribution of ranks, or rank uniformity. The horizontal dashed lines in Figure 9.19 indicate the relative frequency $[= (m+1)^{-1}]$ attained by a uniform distribution for the ranks, which is often plotted for reference as part of the rank histogram. The hypothetical rank histograms in Figure 9.19 each have $m+1 = 9$ bars, and so would pertain to ensembles of size $m = 8$.

Overdispersed ensembles produce rank histograms with relative frequencies concentrated in the middle ranks (left-hand panel in Figure 9.19). In this situation excessive dispersion produces ensembles that range beyond the observation more frequently than would occur by chance if the ensembles exhibited consistency. The verification is accordingly an extreme member of the $m+1$-member (ensemble + verification) collection too infrequently, so that the extreme ranks are underpopulated; and is near the center of the ensemble too frequently, producing overpopulation of the middle ranks. Conversely, a set of n underdispersed ensembles produce a U-shaped rank histogram (right-hand panel in Figure 9.19) because the ensemble members tend to be too similar to each other, and different from the verification. The result is that the verification is too frequently an outlier among the collection of $m+1$ values, so the extreme ranks are overpopulated; and occurs too rarely as a middle value, so the central ranks are underpopulated.

An appropriate degree of ensemble dispersion is a necessary condition for a set of ensemble forecasts to exhibit consistency, but it is not sufficient. It is also necessary for consistent ensembles not to exhibit unconditional biases. That is, consistent ensembles will not be centered either above or below their corresponding verifications, on average. Unconditional ensemble bias can be diagnosed from overpopulation of either the smallest ranks, or the largest ranks, in the verification rank histogram. Forecasts that are centered above the verification, on average, exhibit overpopulation of the smallest ranks (upper panel in Figure 9.19) because the tendency for overforecasting leaves the verification too frequently as the smallest or one of the smallest values of the $m+1$-member collection. Similarly, underforecasting bias (lower panel in Figure 9.19) produces overpopulation of the higher ranks, because a consistent

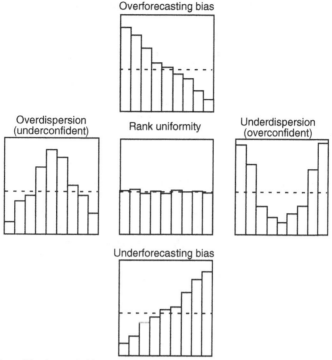

FIGURE 9.19 Example verification rank histograms for hypothetical ensembles of size $m = 8$, illustrating characteristic ensemble dispersion and bias errors. Perfect rank uniformity is indicated by the horizontal dashed lines. The arrangement of the panels corresponds to the calibration portions of the reliability diagrams in Figure 9.10a.

tendency for the ensemble to be below the verification leaves the verification too frequently as the largest or one of the largest members.

There are connections with the calibration function $p(o_j|y_i)$ that is plotted as part of the reliability diagram (Section 9.4.4), which can be appreciated by comparing Figures 9.19 and 9.10a. The five pairs of panels in these two figures bear a one-to-one correspondence for forecast ensembles yielding probabilities for a dichotomous variable defined by a fixed threshold applied to a continuous predictand. That is, the "yes" component of a dichotomous outcome occurs if the value of the continuous predictand y is at or above a threshold. For example, the event "precipitation occurs" corresponds to the value of a continuous precipitation variable being at or above a detection limit, such as 0.01 in. In this setting, forecast ensembles that would produce each of the five reliability diagrams in Figure 9.10a would exhibit rank histograms having the forms in the corresponding positions in Figure 9.19.

Correspondences between the unconditional bias signatures in these pairs of figures are easiest to understand. Ensemble overforecasting (upper panels) yields average probabilities that are larger than average outcome relative frequencies in Figure 9.10a, because ensembles that are too frequently centered above the verification will exhibit a majority of members above a given threshold more frequently than the verification is above that threshold (or, equivalently, more frequently than the corresponding probability of being above the threshold, according to the climatological distribution). Conversely, underforecasting (lower panels) simultaneously yields average probabilities for dichotomous events that are smaller than the corresponding average outcome relative frequencies in Figure 9.10a, and overpopulation of the higher ranks in Figure 9.19.

In underdispersed ensembles, most or all ensemble members will fall too frequently on one side or the other of the threshold defining a dichotomous event. The result is that probability forecasts from underdispersed ensembles will be excessively sharp and will use extreme probabilities more frequently than justified by the ability of the ensemble to resolve the event being forecast. The probability forecasts therefore will be overconfident. That is, too little uncertainty is communicated, so that the conditional event relative frequencies are less extreme than the forecast probabilities. Reliability diagrams reflecting such conditional biases, in the form of the right-hand panel of Figure 9.10a, are the result. Conversely, overdispersed ensembles will rarely have a large majority of members on one side or the other of the event threshold, so the probability forecasts derived from them will rarely be extreme. These probability forecasts will be underconfident and produce conditional biases of the kind illustrated in the left-hand panel of Figure 9.10a, namely, that the conditional event relative frequencies tend to be more extreme than the forecast probabilities.

Several single-number summaries with which to characterize the degree of rank histogram flatness are available. They are equally applicable to characterization of the flatness of the PIT histogram (Section 9.5.4). The χ^2 statistic is the first of these summaries, which essentially addresses the goodness of the fit (Section 5.2.5) of a rank histogram to a discretized uniform distribution,

$$\chi^2 = \frac{m+1}{n} \sum_{i=1}^{m+1} \left(n_i - \frac{n}{m+1} \right)^2, \tag{9.72}$$

where n_i is the number of counts in the ith rank histogram bin. An advantage to using of the χ^2 statistic in this context is that its sampling distribution is known, and so formal testing of the null hypothesis of a flat rank histogram can be undertaken (Section 9.11.5). The *Reliability Index* (Delle Monache et al., 2006) operates similarly to the χ^2 flatness statistic, but measures absolute rather than squared discrepancies from uniformity,

$$\mathrm{RI} = \frac{1}{n} \sum_{i=1}^{m+1} \left| n_i - \frac{n}{m+1} \right|. \tag{9.73}$$

This statistic is sometimes referred to as D, or Δ by some authors, and is sometimes expressed as a percentage after multiplication by 100%. Taillardat et al. (2016) propose characterizing rank histogram flatness with the entropy statistic

$$\varphi = \frac{-1}{\ln(m+1)} \sum_{i=1}^{m+1} \frac{n_i}{n} \ln \left(\frac{n_i}{n} \right), \tag{9.74}$$

which achieves its maximum of $\varphi = 1$ for calibrated forecasts.

Lack of uniformity in a rank histogram quickly reveals the presence of conditional and/or unconditional biases in a collection of ensemble forecasts, but unlike the reliability diagram it does not provide a complete picture of forecast performance in the sense of fully expressing the joint distribution of forecasts and observations. In particular, the rank histogram does not include a representation of the refinement, or sharpness, of the ensemble forecasts. Rather, it indicates only if the forecast refinement is appropriate, relative to the degree to which the ensemble can resolve the predictand. The nature of this incompleteness can be appreciated by imagining the rank histogram for ensemble forecasts constructed as random samples of size m from the historical climatological distribution of the predictand. Such ensembles would be consistent, by definition, because the value of the predictand on any future occasion will have been drawn from the same distribution that generated the finite sample in each ensemble. The resulting rank histogram would be accordingly flat, but would not reveal that these forecasts exhibited so little refinement as to be useless.

If these climatological ensembles were to be converted to probability forecasts for a discrete event according to a fixed threshold of the predictand, in the limit of $m \to \infty$ their reliability diagram would consist of a single point, located on the 1:1 diagonal, at the magnitude of the climatological relative frequency. This abbreviated reliability diagram immediately would communicate the fact that the forecasts underlying it exhibited no sharpness, because the same event probability would have been forecast on each of the n occasions. Of course real ensembles are of finite size, and climatological ensembles of finite size would exhibit sampling variations from forecast to forecast, yielding a refinement distribution $p(y_i)$ (Equation 9.2) with nonzero variance, but a reliability diagram exhibiting no resolution and therefore a horizontal calibration function (indicated by the "no resolution" line in Figure 9.12).

Even when a set of consistent ensemble forecasts can resolve the event better than does the climatological distribution, sampling variations in the resulting probability estimates will generally lead to reliability diagram calibration functions that suggest overconfidence. Richardson (2001) presents analytic expressions for this apparent overconfidence when the probability estimates are estimated using ensemble relative frequency (Equation 8.4), which indicate that this effect decreases with increasing ensemble size, and with forecast accuracy as measured by decreasing Brier scores. This same effect produces sampling-error-based degradations in the Brier and ranked probability scores for probability forecasts based on ensemble relative frequency (Ferro et al., 2008).

It is worth pointing out that a flat rank histogram is a necessary, but not sufficient condition for diagnosing ensemble calibration. In particular, a flat rank histogram can result from compensating effects of different types of miscalibration in subsets of the data set being examined (Hamill, 2001).

Relating Ensemble-Mean Errors and Ensemble Variances

Two consequences of the ensemble consistency condition are that forecasts from the individual ensemble members (and therefore also the ensemble-mean forecasts) are unbiased, and that the average (over multiple forecast occasions) MSE for the ensemble-mean forecasts should be equal to the average ensemble variance. If, on any given forecast occasion, the observation o is statistically indistinguishable from any of the ensemble members y_i, then clearly the bias is zero since $E[y_i] = E[o_i]$. Statistical equivalence of any ensemble member and the observation further implies that

$$E\left[(o - \bar{y})^2\right] = E\left[(y_i - \bar{y})^2\right], \tag{9.75}$$

where $\bar{y}$ denotes the ensemble mean. Realizing that $(o-\bar{y})^2 = (\bar{y}-o)^2$, it is easy to see that the left-hand side of Equation 9.75 is the MSE for the ensemble-mean forecasts, whereas the right-hand side expresses dispersion of the ensemble members y_i around the ensemble mean, as the average ensemble variance. It is important to realize that Equation 9.75 holds only for forecasts from a consistent ensemble, and in particular assumes unbiasedness in the forecasts. Forecast biases will inflate the ensemble-mean MSE without affecting the ensemble dispersion (because ensemble dispersion is computed relative to the sample ensemble mean), so that this diagnostic cannot distinguish forecast bias from ensemble underdispersion (Wilks, 2011). Therefore attempting to correct ensemble underdispersion by inflating ensemble variance to match ensemble-mean MSE will yield overdispersed ensembles if the underlying forecasts are biased.

Accounting for finite ensemble size m, the sample counterpart of Equation 9.75 is (e.g., Leutbecher and Palmer, 2008)

$$\frac{m}{m+1}\left[\frac{1}{n}\sum_{t=1}^{n}(\bar{x}_t - o_t)^2\right] = \frac{m}{m-1}\left[\frac{1}{n}\sum_{t=1}^{n}s_t^2\right], \tag{9.76}$$

where the ensemble mean $\bar{x}_t$ on the left-hand side is defined by Equation 8.10, and the ensemble variance s_t^2 on the right-hand side is defined by Equation 8.9. When the ensemble size is large the corrections outside the square brackets in Equation 9.76 are often neglected. If the ensemble variances have been computed using division by $m-1$ rather than by m as in Equation 8.9, then the correction $m/(m-1)$ on the right-hand side is not needed.

Equation 9.76 applies both to a full n-member verification data set and also to subsets. The *binned spread-error diagram* (Van den Dool, 1989) is a graphical device that displays both the extent to which a collection of ensemble forecasts to be verified is consistent with Equation 9.75 and the degree to which variations in ensemble spread may predict variations in forecast accuracy. It is constructed by collecting subsets of the verification data having similar values for the ensemble variance into discrete bins that define points on the horizontal axis and plotting the corresponding conditional average squared errors on the vertical. Although these MSE values reflect conditional accuracy, these diagrams are sometimes inaccurately called "spread-skill" diagrams. Often it is the square roots of the two sides of Equation 9.76 that are plotted in binned spread-error diagrams, in which case it is important that the square root of the MSE of the left-hand side is used, rather than the average RMSE (Fortin et al., 2014, 2015).

Figure 9.20 shows two examples, where twenty equally populated class intervals for the ensemble spread have been chosen. The binned spread-error plot for a collection of uncorrected ensemble forecasts in Figure 9.76a shows that Equation 9.76 has not been satisfied, because the points are well away from the desired 1:1 line. The conditional ensemble-mean errors are too large relative to the binned ensemble spreads, which could indicate underdispersion or bias or both. It also indicates that there is a predictive relationship between ensemble spread and average forecast accuracy. The rank histogram for these raw ensemble forecasts (Figure 9.21) shows both underdispersion and a strong negative bias, although it does not by itself give

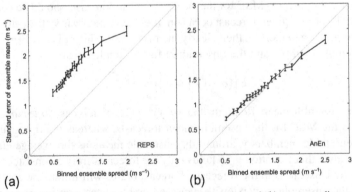

FIGURE 9.20 Binned spread-error diagrams for (a) raw ensemble forecasts and (b) corresponding postprocessed forecasts of surface wind speed at 42 h lead time. Error bars indicate 95% bootstrap confidence intervals. *From Delle Monache et al. (2013). © American Meteorological Society. Used with permission.*

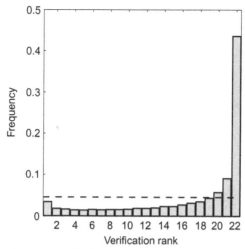

FIGURE 9.21 Rank histogram for the raw ensemble forecasts whose binned-spread error diagram is shown in Figure 9.20a. *Modified from Delle Monache et al. (2013). © American Meteorological Society. Used with permission.*

an indication of the accuracy of these forecasts. The spread-error diagram for the postprocessed ensembles in Figure 9.76b shows much better calibration, since the points falling close to the 1:1 line support Equation 9.76 having been satisfied approximately, suggesting that these postprocessed forecasts are both nearly unbiased and appropriately dispersive.

Another alternative for displaying the degree to which average ensemble spread and ensemble-mean accuracy correspond is to plot the two sides of Equation 9.76 (or their square roots) unconditionally, as functions of lead time. Figure 9.22 does this for the same forecasts underlying Figure 9.20, which shows again that the raw ensembles in Figure 9.22a are poorly calibrated whereas the postprocessed ensembles in Figure 9.22b satisfy Equation 9.76 to good approximation.

9.7.2. Assessing Multivariate Ensemble Calibration

Assessment of calibration for multivariate (i.e., vector) ensemble forecasts is more difficult, since multivariate calibration implies that both the marginal distributions of each of the elements of the forecast vectors, and their joint relationships, must be correctly represented. Several graphical approaches to assessing multivariate calibration have been proposed.

Minimum Spanning Tree (MST) Histogram

The concept behind the verification rank histogram (Section 9.7.1) can be extended to ensemble forecasts for multiple predictands, using the *minimum spanning tree* (MST) *histogram*, which allows investigation of simultaneous forecast calibration in multiple dimensions. This idea was proposed by Smith (2001) and explored more fully by Smith and Hansen (2004), Wilks (2004), and Gombos et al. (2007). The MST histogram is constructed from an ensemble of d-dimensional vector forecasts $y_i, i, = 1, \ldots, m$, and the corresponding vector observation o. Each of these vectors defines a point in a d-dimensional space, the coordinate axes of which corresponds to the d variables in the vectors y and o. In general, these vectors will not have a natural ordering in the same way that a set of $m+1$ scalars would, so

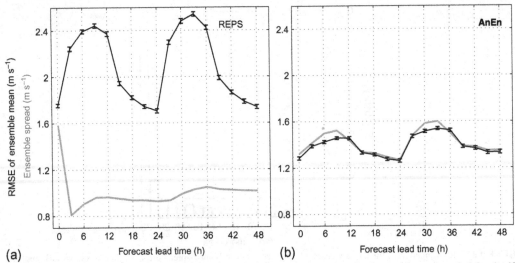

FIGURE 9.22 Square roots of the two sides of Equation 9.76 as a function of lead time, for the same forecasts underlying the 42-h lead forecasts in Figure 9.20. *From Delle Monache et al. (2013). © American Meteorological Society. Used with permission.*

the conventional verification rank histogram is not applicable to these multidimensional quantities. The minimum spanning tree for m the members y_i of a particular ensemble is the set of line segments (in the d-dimensional space of these vectors) that connect all the points y_i in an arrangement having no closed loops, and for which the sum of the lengths of these line segments is minimized. The solid lines in Figure 9.23 show a minimum spanning tree for a hypothetical $m = 10$-member forecast ensemble, labeled A–J.

If each (multidimensional) ensemble member is replaced in turn with the observation vector o, the lengths of the minimum spanning trees for each of these substitutions make up a set of m reference MST lengths. The dashed lines in Figure 9.23 show the MST obtained when ensemble member D is replaced by the observation, O. To the extent that the ensemble consistency condition has been satisfied,

FIGURE 9.23 Hypothetical example minimum spanning trees in $d = 2$ dimensions. The $m = 10$ ensemble members are labeled A–J, and the corresponding observation is O. Solid lines indicate MST for the ensemble as forecast, and dashed lines indicate the MST that results when the observation is substituted for ensemble member D. *From Wilks (2004).*

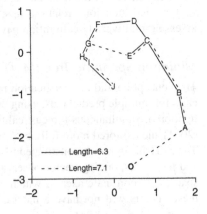

the observation vector is statistically indistinguishable from any of the forecast vectors y_i, implying that the length of the MST connecting only the m vectors y_i has been drawn from the same distribution of MST lengths as those obtained by substituting the observation for each of the ensemble members in turn. The MST histogram investigates the plausibility of this proposition, and thus the plausibility of ensemble consistency for the n d-dimensional ensemble forecasts, by tabulating the ranks of the MST lengths for the original ensembles within each group of $m+1$ MST lengths. A collection of consistent multivariate ensembles should yield a flat MST histogram, within reasonable sampling variations, and the degree of flatness can be characterized using Equations 9.72, 9.73, or 9.74.

Although the concept behind the MST histogram is similar to that of the univariate rank histogram, it is not a multidimensional generalization of the rank histogram, and the interpretations of MST histograms are different (Wilks, 2004). Some example MST histograms are shown in the bottom row of Figure 9.25. Underdispersed ensembles will yield negatively sloped (MSTs excluding the observation vector are typically shorter than those that include it) rather than U-shaped MST histograms, and overdispersed ensembles will exhibit the reverse. In raw form, the MST histogram is unable to distinguish between ensemble underdispersion and bias (the outlier observation O in Figure 9.23 could be the result of either of these problems), and deemphasizes variables in the forecast and observation vectors with small variance. However, useful diagnostics can be obtained from MST histograms of debiased and rescaled forecast and observation vectors. When the elements of the forecast vectors are measured in different physical units, or when some elements are intrinsically more variable than others, it will be advantageous to standardize each of the variables in order to avoid the result being dominated by the higher-variance elements (Wilks, 2004).

Multivariate Rank Histograms

Direct extension of the verification rank histogram for multivariate (i.e., vector) forecasts is not straightforward, because an unambiguous ranking of a collection of vectors along the real line must be defined before proceeding to assign a particular ensemble to one of the $m+1$ rank histogram bins. Gneiting et al. (2008) proposed the concept of the *prerank*, $\pi(y_k)$ to define one-dimensional orderings of a collection of vectors y_i that allow accumulation of counts into rank histogram bins. For notational convenience in this section, define the d-dimensional vector of ensemble members y_i, $i = 1, \dots, m$, and define the corresponding vector observation as y_0 rather than o. Having evaluated the preranks $\pi(y_i)$ for all $m+1$ of these vectors, a forecast ensemble contributes a count to rank histogram bin b according to

$$b = 1 + \sum_{i=1}^{m} I[\pi(y_i) < \pi(y_0)], \tag{9.77}$$

where as before the indicator function $I(\cdot)=1$ if its argument is true, and equals zero otherwise, so that $1 \leq b \leq m+1$. That is, Equation 9.77 defines the multivariate rank histogram bin to which an ensemble contributes by counting the number of ensemble members whose prerank is smaller than the prerank of the observation and adding 1. Equation 9.77 generalizes the scalar verification rank histogram (Section 9.7.1), to which it reduces when $d = 1$ and $\pi(y_i) = y_i$.

Several prerank functions have been proposed. The original (Gneiting et al., 2008) paper used the multivariate rank histogram prerank

FIGURE 9.24 Illustration of the prerank function in Equation 9.78 for a $d=2$ dimensional observation vector y_0 and a hypothetical $m=5$ member ensemble. Ensemble members within the gray shading contribute nonzero terms in Equation 9.78.

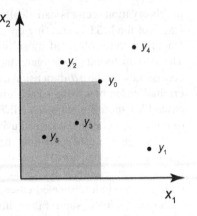

$$\pi_{\mathrm{MRH}}(y_i) = \sum_{j=0}^{m} I\left[y_j \preceq y_i\right], \tag{9.78}$$

where $y_j \preceq y_i$ if *each* of the d elements of y_j is less than or equal to its counterpart in y_i. Each prerank $\pi_{\mathrm{MRH}}(y_i)$, $i = 0, 1, m$, is an integer between 1 and $m+1$, and ties among the preranks are resolved at random. Figure 9.24 illustrates the operation of Equation 9.78 for a hypothetical $m = 5$ member ensemble and corresponding observation, of dimension $d = 2$. Here, the prerank $\pi_{\mathrm{MRH}}(y_0) = 3$ for the observation vector y_0 is equal to the number of ensemble members both below and to the left of y_0 (i.e., within the gray shading), and including y_0.

Equation 9.78 is not the only way to order a collection of vectors, and alternative prerank functions have been defined. Thorarinsdottir et al. (2016) proposed the *average-rank* histogram (ARH), computation of which begins with component-wise ranks for each of the d elements of the d-dimensional vectors y_k,

$$c_l(y_i) = \sum_{j=0}^{m} I\left[y_{j,l} \le y_{i,l}\right], \quad l = 1,\dots,d. \tag{9.79}$$

The $m+1$ preranks for the ARH are then simply the average of the d component-wise ranks in Equation 9.79,

$$\pi_{\mathrm{ARH}}(y_i) = \frac{1}{d}\sum_{l=1}^{d} c_l(y_i). \tag{9.80}$$

The ARH preranks need not be integer-valued. Ties among the ARH preranks are again resolved at random. Both Equation 9.78 and 9.80 are "ascending rank" methods in that in some sense they reflect distance from the smallest values.

Thorarinsdottir et al. (2016) also introduced the band-depth rank histogram (BDH), the computations for which also begin with the component-wise ranks (Equation 9.79). The BDH prerank function is

$$\pi_{\mathrm{BDH}}(y_i) = \frac{1}{d}\sum_{l=1}^{d} \left[m + 1 - c_l(y_i)\right]\left[c_l(y_i) - 1\right]. \tag{9.81}$$

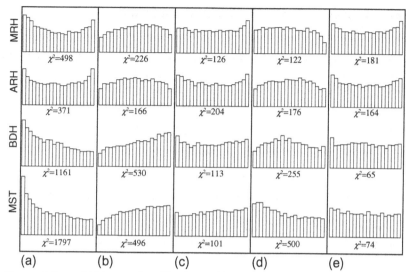

FIGURE 9.25 Representative rank histograms for unbiased ensembles according to MRH (Equation 9.78), ARH (Equation 9.80), BDH (Equation 9.81), and MST (Equation 9.82) prerank functions in an idealized situation with $m = 20$; when the variances are (a) underforecast, and (b) overforecast; when the correlations among the three variables are (c) underforecast, and (d) overforecast; and when (e) the forecast distribution is rotated $20°$ in the $d = 3$-dimensional forecast space. χ^2 values characterize corresponding histogram flatness according to Equation 9.72. *Modified from Wilks (2017).*

Again, prerank ties are resolved at random. Unlike the MRH and ARH prerank functions, which analogously to the familiar univariate rank histogram provide measures of "ascending rank" of the observation vector y_0 relative to the ensemble, the BDH prerank function in Equation 9.81 assesses "centrality," in the sense that $\pi_{BDH}(y_0)$ is large when y_0 is near the middle of the ensemble, and is small when y_0 is extreme relative to the ensemble.

Computation of the minimum spanning tree histogram can also be cast in the framework of preranks, where the MST prerank function is

$$\pi_{MST}(y_i) = \|MST[\{y_0, y_1, \dots, y_m\} \setminus \{y_i\}]\|, \tag{9.82}$$

where $\|\cdot\|$ denotes Euclidean length (Equation 11.14). That is, the MST prerank for the vector y_i is the minimum sum of lengths of $m-1$ line segments connecting the m points of the ensemble plus the observation in their d-dimensional space, when (as denoted by the backslash) the vector y_i is omitted. The MST histogram also assesses centrality of the observation within the ensemble.

Figure 9.25 shows that the various prerank functions that have been proposed react differently to different types of miscalibration, although the two ascending-rank methods MRH and ARH behave similarly to each other, as do the two center-outward methods BDH and MST. Thorarinsdottir et al. (2016) and Mirzargar and Anderson (2017) provide additional comparisons. The synthetic ensembles underlying the histograms in Figure 9.25 were constructed without bias, but biases are more sensitively detected using scalar verification rank histograms separately for each of the d forecast dimensions (Scheuerer and Hamill, 2015b; Wilks, 2017). None of the four methods dominate the others in terms of detecting all types and combinations of types of miscalibration, so that it is good practice to examine the results of multiple preranking methods (Thorarinsdottir et al., 2016; Wilks, 2017).

9.7.3. Assessing Univariate Ensemble Accuracy and Skill

Ensemble CRPS

Although the continuous ranked probability score is defined in Equation 9.61 as the integral over a continuous CDF, the alternative representation in Equation 9.62 provides the basis for a discrete, or "ensemble" CRPS. In particular, straightforward substitution of sample averages within an individual forecast ensemble into Equation 9.62 yields (Van Schaeybroeck and Vannitsem, 2015)

$$\text{eCRPS} = \frac{1}{m} \sum_{i=1}^{m} |y_i - o| - \frac{1}{m(m-1)} \sum_{i=1}^{m-1} \sum_{j=i+1}^{m} |y_i - y_j|. \tag{9.83}$$

The second term in Equation 9.83 is half of the λ_2, or L-scale dispersion statistic (Equation 5.33). Terms for which $x_i = x_j$ are not included in the double summation (Ferro et al., 2008, Wilks, 2018a) because these are not independent realizations from the underlying distribution with respect to which the expectations are calculated in Equation 9.62. Alternative formulations for the eCRPS have been derived by Bröcker (2012d), Candille and Talagrand (2005), and Hersbach (2000).

Dawid–Sebastiani Score

The Dawid–Sebastiani score, which is equivalent to the Ignorance score for a continuous Gaussian predictive distribution, was expressed in Equation 9.67 as a function of the mean and variance of a continuous predictive distribution. Substitution of a sample ensemble mean (Equation 8.10) and ensemble variance (Equation 8.11) yields the discrete counterpart,

$$\text{eDSS} = \ln\left(s^2\right) + \frac{(o - \bar{y})^2}{s^2}. \tag{9.84}$$

9.7.4. Assessing Multivariate Ensemble Accuracy and Skill

Ensemble Energy Score

Gneiting et al. (2008) proposed the ensemble Energy Score, as a multivariate extension of the eCRPS (Equation 9.83) to vector forecasts, and as a discretized version of the Energy Score for continuous multivariate predictive distributions (Equation 9.64),

$$\text{eES} = \frac{1}{m} \sum_{i=1}^{m} \|\boldsymbol{y}_i - \boldsymbol{o}\| - \frac{1}{m(m-1)} \sum_{i=1}^{m-1} \sum_{j=i+1}^{m} \|\boldsymbol{y}_i - \boldsymbol{y}_j\|. \tag{9.85}$$

Here the Euclidean distance in the d-dimensional space of the vector ensemble members x and observation o substitute for the absolute values in Equation 9.84, and both of these equations follow as discrete versions of the CRPS as expressed in Equation 9.62. As was also the case for calculating minimum spanning trees (Section 9.7.2), when the elements of the forecast and observed vectors are expressed in different physical units, or when their distributions differ substantially in range or spread, it may be advantageous to standardize each element before calculation of Equation 9.85, for example, by dividing by standard deviations. Although it is conceptually appealing as a direct multivariate generalization of the CRPS, experience with the ensemble Energy Score has shown it to exhibit weak sensitivity to misspecification of correlations among the elements of the forecast vectors (Pinson and Girard, 2012; Scheuerer and Hamill, 2015b). Accordingly its use as a sole metric for ensemble forecast accuracy may be problematic.

Ensemble Variogram Score

The ensemble Variogram Score (Scheuerer and Hamill, 2015b) is an alternative to the ensemble Energy Score that is more responsive to errors in the specification of correlations among the elements of the forecast and observation vectors. The ensemble Variogram Score of order λ is defined as

$$\text{eVS} = \sum_{k=1}^{d}\sum_{j=1}^{d} w_{k,j}\left(\left|o_k - o_j\right|^{\lambda} - \frac{1}{m}\sum_{i=1}^{m}\left|y_{k,i} - y_{j,i}\right|^{\lambda}\right)^2, \tag{9.86}$$

which is composed of the sum of weighted squared differences for all pairs of variogram (or *structure function*) approximations, among elements of the forecast and observed vectors. The ensemble Variogram score is proper, but not strictly proper. Both the nonnegative weights $w_{k,j}$ and the exponent order λ are adjustable parameters that must be selected. Scheuerer and Hamill (2015b) provide some guidance on criteria that could inform choices for the weights. They also found that selecting $\lambda = 0.5$ yielded good results in their example, but note that other choices are possible. Biases that are similar for all elements of the forecast vectors (e.g., in a spatial setting where gridpoint values all tend to be too wet or too warm) might not be penalized appreciably by eVS. Similarly to MST and eES calculation, standardization of the forecast and observation vector elements will be advantageous in settings where they have been expressed on incommensurate scales.

Ensemble Dawid–Sebastiani Score

In principle, the Dawid–Sebastiani score (Equation 9.68) could also be adapted for multivariate ensemble forecasts, by substituting the vector ensemble mean for $\boldsymbol{\mu}$ and an ensemble-based estimate of the covariance matrix for $[\Sigma]$. However, in practice implementation of this idea is problematic. First, unless the ensemble size m is larger than the forecast and observation vector dimension d, the estimated covariance matrix will be singular so that its inversion in Equation 9.68 will fail. Furthermore, unless $m \gg d$ the covariance matrices will be estimated poorly, so that the resulting score calculation may be extremely misleading (Feldmann et al., 2015; Scheuerer and Hamill, 2015b). Since ensemble sizes are strongly limited by computational constraints, application of the Dawid–Sebastiani score to ensemble forecasts is unlikely to be feasible unless the forecast and observation vector dimension d is perhaps 2 or 3.

9.8. NONPROBABILISTIC FORECASTS FOR FIELDS

9.8.1. General Considerations for Field Forecasts

Characterization of the quality of forecasts for atmospheric fields (spatial arrays of atmospheric variables) is an important problem in forecast verification. Forecasts for such fields as surface pressures, geopotential heights, temperatures, and so on, are produced routinely by weather forecasting centers worldwide. Often these forecasts are nonprobabilistic, without expressions of uncertainty as part of the forecast format. An example of this kind of forecast is shown in Figure 9.26a, which displays 24-h forecasts of sea-level pressures and 1000-500 mb thicknesses over a portion of North America, made 4 May 1993 by the U.S. National Meteorological Center. Figure 9.26b shows the same fields

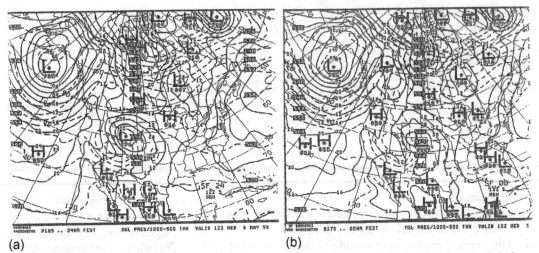

FIGURE 9.26   Forecast (a) and subsequently analyzed (b) sea-level pressures (solid) and 1000–500 mb thicknesses (dashed) over a portion of North America for 5 May 1993.

as analyzed 24 hours later. A subjective visual assessment of the two pairs of fields indicates that the main features correspond well, but that some discrepancies exist in their locations and magnitudes.

Objective, quantitative methods of verification for forecasts of atmospheric fields allow more rigorous assessments of forecast quality to be made. In practice, such methods operate on gridded fields, or collections of values of the field variable sampled at, interpolated to, or averaged over, a grid in the spatial domain. Usually this geographical grid consists of regularly spaced points either in distance, or in latitude and longitude. Figure 9.27 illustrates the gridding process for a hypothetical pair of forecast and observed fields in a small spatial domain. Each of the fields can be represented in map form as contours, or interpolated isolines, of the mapped quantity. The grid imposed on each map is a regular array of points at which the fields are represented. Here the grid consists of four rows in the north–south direction

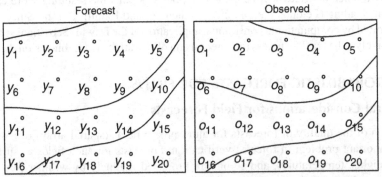

FIGURE 9.27   Hypothetical forecast (left) and observed (right) atmospheric fields represented as contour maps over a small rectangular domain. Objective assessments of the accuracy of the forecast operate on gridded versions of both the forecast and observed fields, that is, by representing them as discrete values on the same geographical grid (small circles). Here the grid has four rows in the north–south direction, and five columns in the east–west direction, so the forecast and observed fields are represented by the $M=20$ discrete values y_m and o_m, respectively.

and five columns in the east–west direction. Thus the gridded forecast field consists of the $M = 20$ discrete values y_m, which represent the smoothly varying continuous forecast field. The gridded observed field consists of the $M = 20$ discrete values o_m, which represent the smoothly varying observed field at these same locations.

The accuracy of a field forecast usually is assessed by computing measures of the correspondence between the values y_m and o_m. If a forecast is perfect, then $y_m = o_m$ for each of the M gridpoints. Of course there are many ways that gridded forecast and observed fields can be different, even when there are only a small number of gridpoints. Put another way, the verification of field forecasts is a problem of very high dimensionality, even for small grids. Although examination of the joint distribution of forecasts and observation is in theory the preferred approach to verification of field forecasts, its large dimensionality suggests that this ideal may not be practically realizable. Rather, the correspondence between forecast and observed fields generally has been characterized using scalar summary measures. These scalar accuracy measures are necessarily incomplete but can be useful in practice.

When comparing gridded forecasts and observations, it is assumed tacitly or otherwise that the two pertain to the same spatial scale. This assumption may not be valid in all cases, for example, when the grid-scale value of a dynamical model forecast represents an area average, but the observed field is an interpolation of point observations or even the irregularly spaced point observations themselves. In such cases discrepancies deriving solely from the scale mismatch are expected (e.g., Cavanaugh and Shen, 2015; Director and Bornn, 2015), and it may be better to upscale the point values (create area averages of the smaller-scale field consistent with the larger grid scale) before comparison (e.g., Gober et al., 2008; Osborn and Hulme, 1997; Tustison et al., 2001). Even different interpolation or gridding algorithms may affect spatial verification scores to a degree (Accadia et al., 2003).

9.8.2. The S1 Score

The *S1 score* is an accuracy measure that is primarily of historical interest. It was designed to reflect the accuracy of forecasts for gradients of pressure or geopotential height, in consideration of the relationship of these gradients to the wind field at the same level (Teweles and Wobus, 1954).

Rather than operating on individual gridded values, the S1 score operates on the differences between gridded values at adjacent gridpoints. Denote the differences between the gridded values at any particular pair of adjacent gridpoints as Δy for points in the forecast field, and Δo for points in the observed field. In terms of Figure 9.27, for example, one possible value of Δy is $y_3 - y_2$, which would be compared to the corresponding gradient in the observed field, $\Delta o = o_3 - o_2$. Similarly, the difference $\Delta y = y_9 - y_4$, would be compared to the observed difference $\Delta o = o_9 - o_4$. If the forecast field reproduces the signs and magnitudes of the gradients in the observed field exactly, each Δy will equal its corresponding Δo.

The S1 score summarizes the differences between the $(\Delta y, \Delta o)$ pairs according to

$$S1 = \frac{\sum_{\text{adjacent pairs}} |\Delta y - \Delta o|}{\sum_{\text{adjacent pairs}} \max\{|\Delta y|, |\Delta o|\}} \times 100. \tag{9.87}$$

Here the numerator consists of the sum of the absolute errors in forecast gradient over all adjacent pairs of gridpoints. The denominator consists of the sum, over the same pairs of points, of the larger of the absolute value of the forecast gradient, $|\Delta y|$, or the absolute value of the observed gradient, $|\Delta o|$. The resulting ratio is multiplied by 100 as a convenience. Equation 9.87 yields the S1 score for a single pair of

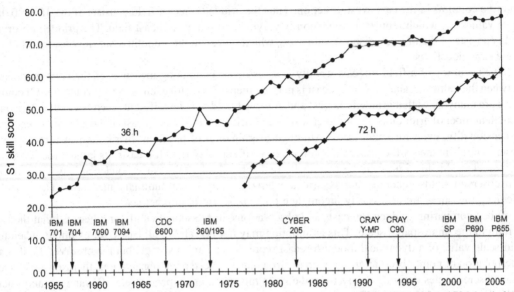

FIGURE 9.28 Average S1 score, converted to skill scores using S1 = 70 as the reference, for NCEP 36-h and 72-h hemispheric forecasts of 500 mb heights for 1955–2006. *From Harper et al. (2007). © American Meteorological Society. Used with permission.*

forecast-observed fields. When the aggregate performance of a series of field forecasts is to be assessed, the S1 scores for each forecast occasion are simply averaged. This averaging smooths sampling variations and allows trends through time of forecast performance to be assessed more easily.

Clearly, perfect forecasts will exhibit S1 = 0, with poorer gradient forecasts being characterized by increasingly larger scores. The S1 score exhibits some undesirable characteristics that have resulted in its going out of favor. The most obvious is that the actual magnitudes of the forecast pressures or heights are unimportant, since only pairwise gridpoint differences are scored. Thus the S1 score does not reflect bias. Summer scores tend to be larger (apparently worse) because of generally weaker gradients, producing a smaller denominator in Equation 9.87. The S1 score can also be improved by deliberately overforecasting the magnitudes of gradients (Thompson and Carter, 1972). Finally, the score depends on the size of the domain and the spacing of the grid, so that it is difficult to compare S1 scores not pertaining to the same domain and grid.

The S1 score has limited operational usefulness for current forecasts, but its continued tabulation has allowed forecast centers to examine very long-term trends in their field-forecast accuracy. Decades-old forecast maps may not have survived, but summaries of their accuracy in terms of the S1 score have often been retained. For example, Figure 9.28 shows S1 scores, converted to skill scores (Equation 9.4) using $S1_{ref} = 70$ (a traditional subjective rule of thumb for "useless" forecasts), for hemispheric 500 mb heights at 36- and 72-h lead times, over the period 1955-2006.

9.8.3. Mean Squared Error

The *mean squared error*, or MSE, is a much more common accuracy measure for field forecasts. The MSE operates on the gridded forecast and observed fields by spatially averaging the individual squared differences between the two at each of the M gridpoints. That is,

Chapter | 9   Forecast Verification

$$MSE = \frac{1}{M}\sum_{m=1}^{M}(y_m - o_m)^2. \qquad (9.88)$$

This formulation is mathematically the same as Equation 9.33, with the mechanics of both equations centered on averaging squared errors. The difference in application between the two equations is that the MSE in Equation 9.88 is computed over the gridpoints of a single pair of forecast/observation fields—that is, to $n = 1$ pair of maps—whereas Equation 9.33 pertains to the average over n pairs of scalar forecasts and observations. Clearly the MSE for a perfectly forecast field is zero, with larger MSE indicating decreasing accuracy of the forecast.

Often the MSE is expressed as its square root, the *root-mean squared error*, $RMSE = \sqrt{MSE}$. This transformation of the MSE has the advantage that it retains the units of the forecast variable and is thus more easily interpretable as a typical error magnitude. To illustrate, the solid line in Figure 9.29 shows RMSE in meters for 30-day forecasts of 500 mb heights initialized on 108 consecutive days during 1986–1987 (Tracton et al., 1989). There is considerable variation in forecast accuracy from day to day, with the most accurate forecast fields exhibiting RMSE near 45 m, and the least accurate forecast fields exhibiting RMSE around 90 m. Also shown in Figure 9.29 are RMSE values of 30-day forecasts of persistence, obtained by averaging observed 500 mb heights for the most recent 30 days prior to the forecast. Usually the persistence forecast exhibits slightly higher RMSE than the 30-day dynamical forecasts, but it is apparent from the figure that there are many days when the reverse is true, and that at this extended range the accuracy of these persistence forecasts was competitive with that of the dynamical forecasts.

The plot in Figure 9.29 shows accuracy of individual field forecasts, but it is also possible to express the aggregate accuracy of a collection of field forecasts by averaging the MSEs for each of a collection of paired comparisons. This average of MSE values across many forecast maps can then be converted to an average MSE as before, or expressed as a skill score in the same form as Equation 9.37. Since the MSE for perfect field forecasts is zero, the skill score following the form of Equation 9.4 is computed using

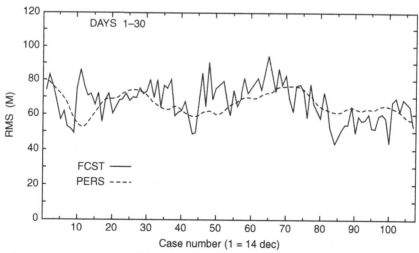

FIGURE 9.29   Root-mean squared error (RMSE) for dynamical 30-day forecasts of 500 mb heights for the northern hemisphere between 20° and 80° N (solid), and persistence of the previous 30-day average 500 mb field (dashed), for forecasts initialized 14 December 1986 through 31 March 1987. *From Tracton et al. (1989).* © *American Meteorological Society. Used with permission.*

$$SS = \frac{\sum\limits_{i=1}^{n} MSE(i) - \sum\limits_{i=1}^{n} MSE_{ref}(i)}{0 - \sum\limits_{i=1}^{n} MSE_{ref}(i)} = 1 - \frac{\sum\limits_{i=1}^{n} MSE(i)}{\sum\limits_{i=1}^{n} MSE_{ref}(i)}, \tag{9.89}$$

where the aggregate skill of n individual field forecasts is being summarized. When this skill score is computed, the reference field forecast is usually either the climatological average field (in which case it may be called the reduction of variance, in common with Equation 9.37) or individual persistence forecasts as shown in Figure 9.29.

The MSE skill score in Equation 9.89, when calculated with respect to climatological forecasts as the reference, allows an interesting interpretation for field forecasts when algebraically decomposed in the same way as in Equation 9.38. When applied to field forecasts, this decomposition is conventionally expressed in terms of the differences (anomalies) of forecasts and observations from the corresponding climatological values at each gridpoint (Murphy and Epstein, 1989),

$$y'_m = y_m - c_m \tag{9.90a}$$

and

$$o_m = o_m - c_m, \tag{9.90b}$$

where c_m is the climatological value at gridpoint m. The resulting MSE and skill scores are identical, because the climatological values c_m can be both added to and subtracted from the squared terms in Equation 9.88 without changing the result, that is,

$$MSE = \frac{1}{M}\sum_{m=1}^{M}(y_m - o_m)^2 = \frac{1}{M}\sum_{m=1}^{M}([y_m - c_m] - [o_m - c_m])^2 = \frac{1}{M}\sum_{m=1}^{M}(y'_m - o'_m)^2. \tag{9.91}$$

When expressed in this way, the algebraic decomposition of MSE skill score in Equation 9.38 becomes

$$SS_{clim} = \frac{r_{y'o'}^2 - [r_{y'o'} - (s_{y'}/s_{o'})]^2 - [(\overline{y'} - \overline{o'})/s_{o'}]^2 + (\overline{o'}/s_{o'})^2}{1 + (\overline{o'}/s_{o'})^2} \tag{9.92a}$$

$$\approx r_{y'o'}^2 - [r_{y'o'} - (s_{y'}/s_{o'})]^2 - [(\overline{y'} - \overline{o'})/s_{o'}]^2. \tag{9.92b}$$

The difference between this decomposition and Equation 9.38 is the normalization factor involving the average differences between the observed and climatological gridpoint values, in both the numerator and denominator of Equation 9.92a. This factor depends only on the observed field, and Murphy and Epstein (1989) note that it is likely to be small if the skill is being evaluated over a sufficiently large spatial domain, because positive and negative differences with the gridpoint climatological values will tend to balance. Neglecting this term leads to the approximate algebraic decomposition of the skill score in Equation 9.92b, which is identical to Equation 9.38 except that it involves the differences from the gridpoint climatological values, y' and o'. In particular, it shows that the (squared) correlation between the forecast and observed fields is at best a measure of potential rather than actual skill, which can be realized only in the absence of conditional (second term in Equation 9.92b) and unconditional (third term in Equation 9.92b) biases. It is worthwhile to work with these climatological anomalies when investigating skill of field forecasts in this way, in order to avoid ascribing spurious skill to forecasts for merely forecasting a correct climatology.

The Taylor Diagram

The joint contributions to the RMSE of the correlation between two fields, and the discrepancy between their standard deviations, can be visualized using a graphical device that is known as the *Taylor diagram* (Taylor, 2001), although a very similar diagram was proposed independently by Lambert and Boer (2001). The Taylor diagram is based on a geometrical representation of an algebraic decomposition of the debiased MSE, which is the MSE after subtraction of contributions due to overall bias errors:

$$MSE' = MSE - (\bar{y} - \bar{o})^2 \tag{9.93a}$$

$$= \frac{1}{M} \sum_{m=1}^{M} [(y_m - \bar{y}) - (o_m - \bar{o})]^2 = \sigma_y^2 + \sigma_o^2 - 2\sigma_y\sigma_o r_{yo}. \tag{9.93b}$$

Equation 9.93a defines the debiased MSE to be equal to MSE after subtraction of the squared bias. Clearly MSE = MSE' if the means over the M gridpoints, $\bar{y}$ and $\bar{o}$, are equal, and otherwise MSE' reflects only those contributions to MSE not deriving from the unconditional bias. The first equality in Equation 9.93b indicates that MSE' is equivalent to MSE calculated after subtracting the map means (i.e., averages over the M gridpoints) $\bar{y}$ and $\bar{o}$, so that both transformed fields have equal, zero, area averages. The second equality in Equation 9.93b suggests the geometrical representation of the relationship between the standard deviations over the M gridpoints, σ_y and σ_o, their correlation $r_{y,o}$, and RMSE', through the direct analogy to the *law of cosines*,

$$c^2 = a^2 + b^2 - 2ab\cos\theta. \tag{9.94}$$

The correspondence is between the three terms on the right-hand sides of Equation 9.93b and 9.94.

The Taylor diagram represents the two standard deviations σ_y and σ_o as the lengths of two legs of a triangle, which are separated by an angle θ whose cosine is equal to the correlation $r_{y,o}$ between the two fields. The length of the third side is then RMSE'. The diagram itself plots the vertices of these triangles in polar coordinates, where the angle from the horizontal is the cosine of the correlation, and the radial distances from the origin are defined by the standard deviations. The correlation of the observed field with itself is 1, so the corresponding vertex is at an angle $\cos^{-1}(1) = 0°$ from the horizontal, with radius σ_o. The vertices for each forecast field are represented as points, plotted at a radius σ_y and angle $\cos^{-1}(r_{y,o})$.

Taylor diagrams are most useful when multiple "y" fields are being compared simultaneously to a corresponding reference field "o." Figure 9.30 shows a superposition of three such Taylor diagrams, showing performance of 16 dynamical climate models in representing global fields of precipitation, surface temperatures, and sea-level pressures. Since the "o" fields are different in the three comparisons, all standard deviations have been divided by the appropriate σ_o, so that the "b" vertex (c.f. Equation 9.94) of each triangle is located at unit radius on the horizontal axis, at the point labeled "observed." The tight clustering near the "observed" of the points representing the "a" vertices for the simulated temperature fields indicate that these patterns have been best simulated of the three variables, with standard deviations nearly correctly simulated (all points are near unit radius from the origin), and correlations with the observed temperature field that all exceed 0.95. The distances from each of these points to the reference "observed" vertex are geometrically equal to RMSE'. In contrast, precipitation is the least well simulated variable of the three, with simulated standard deviations ranging from approximately 75% to 125% of the

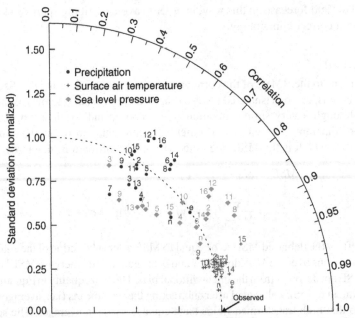

FIGURE 9.30 Taylor diagram comparing 16 climate-model generated fields of precipitation, surface air temperatures, and sea-level pressures to instrumental ("observed") values. Points labeled "n" and "e" are corresponding reanalysis values, which are model results that have been strongly constrained by actual data. All standard deviations have been divided by the appropriate σ_o in order to superimpose the Taylor diagrams for the three variables. Transformation of correlation to angle from the horizontal is indicated by the cosine scale along the curved edge. *From McAvaney et al. (2001).*

correct value, and correlations ranging from about 0.5 to 0.7, so that RMSE' (distances to "observed") is substantially larger in all cases than for the temperature simulations.

In order to emphasize that distances to the reference point indicates RMSE', Taylor diagrams are sometimes drawn with semicircles of equal RMSE', centered at radius σ_o on the horizontal axis. Negative correlations can be accommodated by extending the diagram counterclockwise to include an additional quadrant. While Taylor diagrams are most frequently used to illustrate the phase association (correlation) and amplitude (standard deviation) errors of spatial fields over M gridpoints, the mathematical decomposition and its geometrical representation are equally applicable to MSEs of nonprobabilistic forecasts for scalar predictands, over n forecasting occasions (Equation 9.33), again after removal of any contributions to the MSE from squared bias.

Recently, Koh et al. (2012) have proposed a suite if diagrams devised in a similar spirit, that portray biases, in addition to phase and amplitude errors.

9.8.4. Anomaly Correlation

The *anomaly correlation* (AC) is another commonly used measure of association that operates on pairs of gridpoint values in the forecast and observed fields. To compute the anomaly correlation, the forecast and observed values are first converted to anomalies in the sense of Equation 9.90: the climatological average value of the observed field at each of M gridpoints is subtracted from both forecasts y_m and observations o_m.

There are actually two forms of anomaly correlation in use, and it is unfortunately not always clear which has been employed in a particular instance. The first form, called the *centered* anomaly correlation, was apparently first suggested by Glenn Brier in an unpublished 1942 U.S. Weather Bureau mimeo (Namias, 1952). It is computed according to the usual Pearson correlation (Equation 3.28), operating on the M gridpoint pairs of forecasts and observations that have been referred to the climatological averages c_m at each gridpoint,

$$\text{AC}_C = \frac{\sum\limits_{m=1}^{M} \left(y'_m - \overline{y'} \right) \left(o'_m - \overline{o'} \right)}{\left[\sum\limits_{m=1}^{M} \left(y'_m - \overline{y'} \right)^2 \sum\limits_{m=1}^{M} \left(o'_m - \overline{o'} \right)^2 \right]^{1/2}}. \tag{9.95}$$

Here the primed quantities are the anomalies relative to the climatological averages (Equation 9.90), and the overbars refer to these anomalies averaged over a given map of M gridpoints. The square of Equation 9.95 is thus exactly $r^2_{y'o'}$ in Equation 9.92.

The other form for the anomaly correlation differs from Equation 9.95 in that the map-mean anomalies are not subtracted, yielding the *uncentered* anomaly correlation

$$\text{AC}_U = \frac{\sum\limits_{m=1}^{M} (y_m - c_m)(o_m - c_m)}{\left[\sum\limits_{m=1}^{M} (y_m - c_m)^2 \sum\limits_{m=1}^{M} (o_m - c_m)^2 \right]^{1/2}} = \frac{\sum\limits_{m=1}^{M} y'_m o'_m}{\left[\sum\limits_{m=1}^{M} (y'_m)^2 \sum\limits_{m=1}^{M} (o'_m)^2 \right]^{1/2}}. \tag{9.96}$$

This form was apparently first suggested by Miyakoda et al. (1972). Superficially, the AC_U in Equation 9.96 resembles the Pearson product-moment correlation coefficient (Equations 3.28 and 9.95), in that that both are bounded by ± 1, and that neither is sensitive to biases or scale errors in the forecasts. However, the centered and uncentered anomaly correlations are equivalent only if the averages over the M gridpoints of the two anomalies are zero, that is, only if $\Sigma_m (y_m - c_m) = 0$ and $\Sigma_m (o_m - c_m) = 0$. These conditions may be approximately true if the forecast and observed fields are being compared over a large (e.g., hemispheric) domain, but will almost certainly not hold if the fields are compared over a relatively small area. In this latter situation DelSole and Shukla (2006) recommend use of the uncentered anomaly correlation as conforming better to plausible subjective evaluation of forecast accuracy.

The anomaly correlation is designed to detect similarities in the patterns of departures (i.e., anomalies) from the climatological mean field and is sometimes therefore referred to as a *pattern correlation*. However, as Equation 9.92 makes clear, the anomaly correlation does not penalize either conditional or unconditional biases. Accordingly, it is reasonable to regard the anomaly correlation as reflecting potential skill (that might be achieved in the absence of conditional and unconditional biases), but it is incorrect to regard the anomaly correlation (or, indeed, any correlation) as measuring actual skill (e.g., Murphy, 1995).

The anomaly correlation often is used to evaluate extended-range (beyond a few days) forecasts. Figure 9.31 shows anomaly correlation values for the same 30-day dynamical and persistence forecasts of 500 mb height that are verified in terms of the RMSE in Figure 9.29. Since the anomaly correlation has

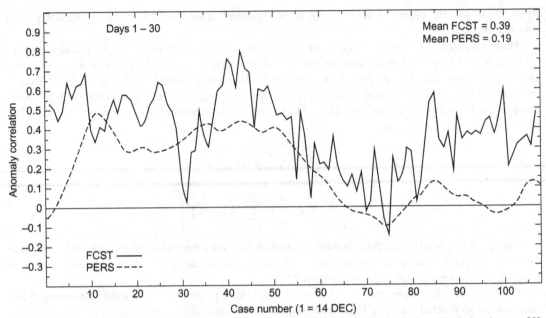

FIGURE 9.31   Anomaly correlations for dynamical 30-day forecasts of 500 mb heights for the Northern Hemisphere between 20° and 80° N (solid), and persistence of the previous 30-day average 500 mb field (dashed), for forecasts initialized 14 December 1986 through 31 March 1987. The performance of the same forecasts is characterized in Figure 9.29 using the RMSE. *From Tracton et al. (1989). © American Meteorological Society. Used with permission.*

a positive orientation (larger values indicate more accurate forecasts) and the RMSE has a negative orientation (smaller values indicate more accurate forecasts), we must mentally "flip" one of these two plots vertically in order to compare them. When this is done, it can be seen that the two measures usually rate a given forecast map similarly, although some differences are apparent. For example, in this data set the anomaly correlation values in Figure 9.31 show a more consistent separation between the performance of the dynamical and persistence forecasts than do the RMSE values in Figure 9.29, suggesting that the patterns of the fields may have been forecast better than the magnitudes.

As is also the case for the MSE, aggregate performance of a collection of field forecasts can be summarized by averaging anomaly correlations across many forecasts. However, skill scores of the form of Equation 9.4 usually are not calculated for the anomaly correlation. For the uncentered anomaly correlation, AC_U is undefined for climatological forecasts, because the denominator of Equation 9.96 is zero. Rather, AC skill generally is evaluated relative to the reference values $AC_{ref} = 0.6$ or $AC_{ref} = 0.5$. Individuals working operationally with the anomaly correlation have found, subjectively, that $AC_{ref} = 0.6$ seems to represent a reasonable lower limit for delimiting field forecasts that are synoptically useful (Hollingsworth et al., 1980). Murphy and Epstein (1989) have shown that if the average forecast and observed anomalies are zero, and if the forecast field exhibits a realistic level of variability (i.e., the two summations in the denominator of Equation 9.95 are of comparable magnitude), then $AC_C = 0.5$ corresponds to the skill score for the MSE in Equation 9.89 being zero. Under these same restrictions, $AC_C = 0.6$ corresponds to the MSE skill score being 0.20.

Figure 9.32 illustrates the use of the subjective $AC_{ref} = 0.6$ reference level. Figure 9.32a shows average AC values for 500 mb height forecasts made during the winters (December-February) of

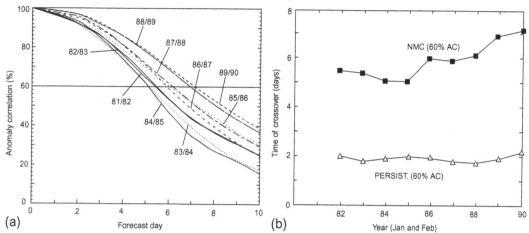

FIGURE 9.32 (a) Average anomaly correlations as a function of forecast lead time for 1981/1982 through 1989/1990 winter 500 mb heights between 20°N and 80°N. Accuracy decreases as forecast lead time increases, but there are substantial differences among winters. (b) Average lead time at which anomaly correlations for the dynamical and persistence forecasts cross the $AC_{ref} = 0.6$ level, for Januarys and Februarys of these nine winters. *From Kalnay et al. (1990). © American Meteorological Society. Used with permission.*

1981/1982 through 1989/1990. For lead time zero days into the future (i.e., initial time), AC = 1 since $y_m = o_m$ at all grid points. The average AC declines progressively for longer lead times, falling below $AC_{ref} = 0.6$ between days five and seven. The curves for the later years tend to lie above the curves for the earlier years, reflecting, at least in part, improvements made to the dynamical forecast model during the decade. One measure of this overall improvement is the increase in the average lead time at which an AC curve crosses the 0.6 line. These times are plotted in Figure 9.32b, and range from five days in the early- and mid-1980s, to seven days in the late 1980s. Also plotted in this panel are the average lead times at which anomaly correlations for persistence forecasts fall below 0.6. The crossover time at the $AC_{ref} = 0.6$ threshold for persistence forecasts is consistently about two days. Thus imagining the average correspondence between observed 500 mb maps separated by 48 hour intervals allows a qualitative appreciation of the level of forecast performance represented by the $AC_{ref} = 0.6$ threshold.

9.8.5. Field Verification Based on Spatial Structure

Because the numbers of gridpoints M typically used to represent meteorological fields is relatively large, and the numbers of allowable values for forecasts and observations of continuous predictands defined on these grids may also be large, the dimensionality of verification problems for field forecasts is typically huge. Using scalar scores such as GSS (Equation 9.20) or MSE (Equation 9.88) to summarize forecast performance in these settings may seem at times to be a welcome relief from the inherent complexity of the verification problem, but necessarily masks very much relevant detail. For example, forecast and observed precipitation fields are sometimes converted to binary (yes/no) fields according to whether the gridpoint values are above or below a threshold such as 0.25 in., with the resulting field forecast scored according to a 2 x 2 contingency table measure such as the GSS. But, as demonstrated by Ahijevych et al. (2009), all such forecasts exhibiting no spatial overlap with the observed "yes" area, but which have the same number of

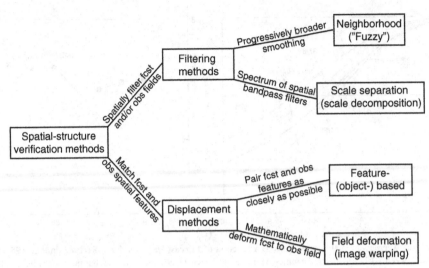

FIGURE 9.33   The taxonomy of spatial-structure verification methods proposed by Gilleland et al. (2009).

"yes" gridpoints, yield equal scores regardless of the distance between the forecast and observed features, or the similarity or dissimilarity of the shapes of the forecast and observed features. Similarly, a modest error in the advection speed of a relatively small-scale forecast feature may produce a large MSE as a consequence of the *"double penalty" problem*: the forecast will be penalized once for placing the feature where it did not occur, and penalized again for placing no feature where the actual event happened, even though the feature itself may have been well forecast with respect to its presence, shape, trajectory, and intensity.

Accordingly, forecasters and forecast evaluators often are dissatisfied with the correspondence between the traditional single-number performance summaries and their subjective perceptions about the goodness of a spatial field forecast. This dissatisfaction has stimulated work on field-verification methods that may be able to quantify aspects of forecast performance which better reflect human visual reactions to map features. Casati et al. (2008) and Gilleland et al. (2009) reviewed initial developments in this area, and Gilleland (2013) provides an extensive list of papers using these methods.

Gilleland et al. (2009) have proposed a taxonomy for these spatial verification methods. This taxonomy is shown in Figure 9.33, and Figure 9.34 illustrates the ideas behind the four types. Most of the spatial-structure verification methods that have been proposed to date can be classified either as filtering methods or displacement methods. For filtering methods, the forecast and/or observed fields (or, the difference field) are subjected to spatial filters before application of more conventional verification metrics at multiple spatial scales. In contrast, displacement methods operate on discrepancies between individual features in the forecast and observed fields, generally in terms of nontraditional metrics that describe the nature and degree of spatial manipulation necessary to achieve congruence between the manipulated forecast field and the corresponding observed field. In this context a "feature" is usually understood to be a collection of contiguous nonzero gridpoints or pixels in either the forecast or observed fields.

Software implementing many of the methods described in this section is described by Fowler et al. (2018).

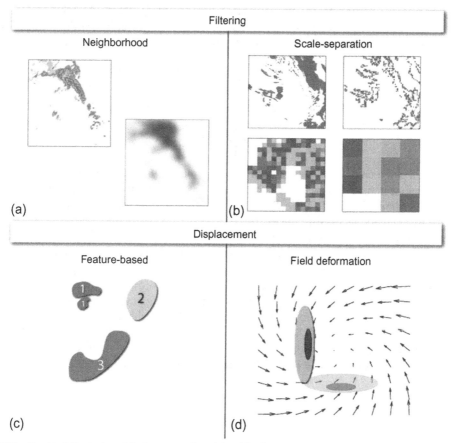

FIGURE 9.34 Graphical illustration of the four types of spatial verification methods named in Figure 9.33. *From Gilleland et al. (2009). © American Meteorological Society. Used with permission.*

Neighborhood Methods

Neighborhood (or "fuzzy") methods address the problem of excessive penalties for small displacement errors by expanding the comparisons from individual gridpoint pairs to larger spatial and/or temporal neighborhoods (Ebert, 2008). In effect these methods smooth the forecast and observation fields before computing a verification measure of interest, as suggested by Figure 9.34a. The underlying idea is particularly appealing for very high-resolution forecasts, for which exact matches between forecast and observed fields cannot be reasonably expected. In such cases forecasts that are in some sense close can be valuable, for example, in the construction of severe-weather warnings.

The *Fractions Skill Score* (Roberts and Lean, 2008) is perhaps the most widely used of the neighborhood methods. For it, and other neighborhood methods, a collection of gridpoints around each point in the domain to be verified is defined as its neighborhood. Whether these neighborhoods are defined in terms of radial distance from a base point, or as square domains centered on a base point, appears to have little effect on the results. Skok and Roberts (2016) investigated the effects of different treatments for neighborhoods that are near the edge of the domain. A binary outcome is defined for each of the M points in the neighborhood, separately for the forecast and observed fields, which might, for example, be occurrence or not of precipitation larger than some threshold. Defining $P_{y,m}$ and $P_{o,m}$ as the fractions of above-threshold points in the forecast and observed neighborhoods around the point m, respectively, the Fractions Skill Score is computed as

$$FSS = 1 - \frac{\sum_{m=1}^{M} (P_{y,m} - P_{o,m})^2}{\sum_{m=1}^{M} P_{y,m}^2 + \sum_{m=1}^{M} P_{o,m}^2}. \tag{9.97}$$

Equation 9.97 is based on the conventional skill score (Equation 9.37) for the MSE (Roberts and Lean, 2008), so to the extent that $P_{y,m}$ and $P_{o,m}$ can be regarded as probabilities for within-neighborhood grid-points exhibiting above-threshold values, the FSS is reminiscent of the Brier skill score (Equation 9.40), although the $P_{o,m}$ are not necessarily binary.

Because the results will be sensitive to the degree of smoothing (the sizes of the neighborhoods), the FSS and other neighborhood methods may be applied at a range of increasing scales, which may provide information on possible scale dependence of forecast performance. The score will generally be better for the larger scales (i.e., for increasing neighborhood size), as poorly- or unpredictable scales are progressively filtered. Figure 9.35 illustrates this point in a synthetic setting in which a 1-pixel-wide rain band is displaced by 1, 3, 11, and 21 pixels relative to the truth, and the neighborhood is expanded from 1 to 49 grid squares centered on each pixel. The FSS increases for increasing neighborhood size, and as expected is smaller for greater displacement between the forecast and observed features The dashed horizontal line labeled "uniform" locates the value FSS = 0.5 that is often interpreted as the threshold for a useful

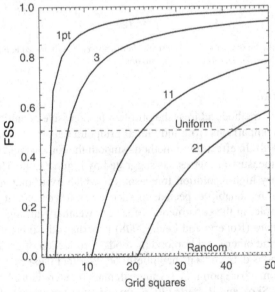

FIGURE 9.35 Illustration of the dependence of the FSS on the neighborhood size, in grid squares, for an idealized case in which a 1-pixel wide band of rain occupying 1% of the domain is displaced by 1, 3, 11, and 21 pixels. *Modified from Roberts and Lean (2008). © American Meteorological Society. Used with permission.*

forecast (e.g., Skok and Roberts, 2016), and accordingly the size of the spatial scale that might be considered as plausibly predictable with respect to a particular curve for FSS as a function of domain size, provided the domain size is large compared to the typical displacement error. Similarly, when the forecast and observation are converted to binary fields at a range of thresholds before application of neighborhood methods, information on variation of skill with event intensity (e.g., rainfall rate) can be extracted. Skok and Roberts (2018) study how a transformation of FSS can be used to diagnose overall average displacement errors, emphasizing errors for the largest objects in the field.

Scale Separation Methods

In contrast to the progressive blurring of the fields by neighborhood methods, scale separation methods apply mathematical spatial filters to the forecast and observed fields, allowing isolation and separation of the verification for features of different sizes. Unlike the results of the progressive smoothing produced by the neighborhood methods, the filtered fields generated by scale separation methods at particular spatial scales may not closely resemble the original fields, as exemplified by the maps at four spatial scales shown in Figure 9.34b.

Briggs and Levine (1997) proposed this general approach using *wavelets*, which are a particular kind of mathematical basis function (Section 10.6), applying it to forecasts of geopotential height fields. Weniger et al. (2017) review the subsequent use of wavelets to decompose forecast and observed fields into hierarchies of spatial scales for the purpose of forecast verification. Figure 9.36 illustrates the procedure, comparing a pair of example analyzed (upper left panel) and forecast (upper right panel) 500 mb fields, together with corresponding wavelet decompositions of those fields at five spatial scales. After the scale separation, conventional measures of agreement such as MSE are then applied to the pairs of forecast and observation fields at each decomposed scale. Because the wavelet basis functions are orthogonal, these scale-specific MSE values sum to the overall MSE for the unfiltered comparison, and the fractional contribution to error made by each of the scales can be easily computed

Casati (2010) extends this method to settings such as precipitation field forecasts, which include also an intensity dimension. Denis et al. (2002) and de Elia et al. (2002) consider a similar approach based on more conventional spectral basis functions. These methods allow investigation of scale dependence of forecast errors, possibly including a minimum spatial scale below which the forecasts do not exhibit useful skill.

Feature-Based Methods

The displacement methods operate by comparing the structure of specific features (contiguous gridpoints or pixels sharing a relevant property) in the forecast and observed fields, with respect to such attributes as position, shape, size, and intensity. Thus as suggested by Figure 9.34c, the methods operate by identifying discrete portions of interest of the spatial domain, which often correspond to forecast or observed precipitation amounts larger than some threshold. An early example of this class of methods was described by Ebert and McBride (2000). Differences among displacement methods relate to such issues as how features occurring in one but not the other of the forecast and observed fields are dealt with, whether two or more nearby but distinct features in one of the fields should be merged before comparison with the other field, and what summary diagnostics are used to characterize differences.

In the Method for Object-based Diagnostic Evaluation (MODE; Davis et al., 2006a, 2009), discrete objects of interest in the forecast and observation fields are identified by a combination of smoothing and thresholding operations. Figure 9.37 illustrates an example of the transformation by MODE of a

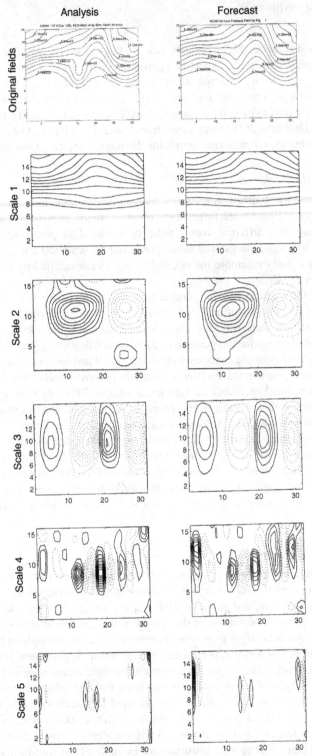

FIGURE 9.36 Example analyzed (upper left) and forecast (upper right) 500 mb height fields, and their wavelet decompositions at a sequence of five progressively finer spatial scales. *From Briggs and Levine (1997). © American Meteorological Society. Used with permission.*

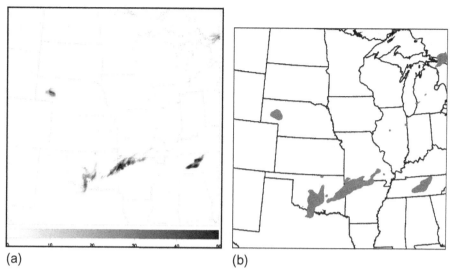

FIGURE 9.37 (a) Example field of 1-h rainfall accumulation (mm), and (b) its transformation to a collection of objects after MODE smoothing and thresholding operations. *From Davis et al. (2006b). © American Meteorological Society. Used with permission.*

radar-derived precipitation field to a collection of precipitation objects. These processed features are matched between the forecast and observed fields through a user-adjustable multiattribute agreement function that includes the separation of their centroids, minimum separation of their boundaries, orientation differences, relative sizes, and the areas of their intersections; and the degree of agreement is quantified, again with a user-adjustable multiattribute function. Mittermaier and Bullock (2013) describe extension of the 2-dimensional spatial MODE algorithm with a third dimension that represents the time evolution of the forecast and observed fields. Wolff et al. (2014) compare MODE and FSS (-Equation 9.97) for a sample of dynamical precipitation forecast fields.

The Structure-Amplitude-Location (SAL; Wernli et al., 2008, 2009) method characterizes agreement between forecast and observed fields with respect to the three characteristics that make up its name. It was designed with the verification of precipitation fields in mind, although Weniger and Friederichs (2016) apply the method to cloud fields. The SAL method begins by identifying objects that consist of contiguous gridpoints or pixels having precipitation amounts at least 1/15 of the 95th percentile of amounts larger than 0.1 mm, as evaluated over the domain. Each of these N objects in either the forecast or observed fields has center of mass s_n, and total (summed over points) precipitation amount R_n, $n = 1, \dots, N$.

The structure error is defined as

$$S = \frac{V(\mathbf{y}) - V(\mathbf{o})}{\frac{1}{2}[V(\mathbf{y}) + V(\mathbf{o})]}, \tag{9.98}$$

where

$$V(\bullet) = \frac{\sum\limits_{n=1}^{N} R_n^2 / R_n^{\max}}{\sum\limits_{n=1}^{N} R_n} \tag{9.99}$$

is the mass-weighted scaled volume, and R_n^{max} is the maximum precipitation in object n. The structure error S ranges from -2 to $+2$, with zero reflecting perfect structure, positive values indicating forecast fields that are too smooth, and negative values indicating forecast fields that are too noisy.

The amplitude error is

$$A = \frac{D(y) - D(o)}{\frac{1}{2}[D(y) + D(o)]}, \tag{9.100}$$

where $D(\cdot)$ is the domain-mean value. The amplitude error ranges from -2 to $+2$, with zero indicating unbiasedness.

The location error is

$$L = \frac{|s(y) - s(o)|}{d} + 2\frac{|r(y) - r(o)|}{d}, \tag{9.101}$$

where d is the maximum distance across the domain, and

$$r(\bullet) = \frac{\sum_{n=1}^{N} R_n |\bar{s} - s_n|}{\sum_{n=1}^{N} R_n} \tag{9.102}$$

is the mass-weighted mean distance of the center of mass of each object to the overall center of mass $\bar{s}$. The location error ranges from 0 to 2, with $L = 0$ indicating perfect locations for the objects.

The three dimensions of the SAL method can be displayed compactly on a diagram in which the horizontal axis is the structure error, the vertical axis is the amplitude error, and the shading of the plotted points indicates the location error. Figure 9.38a shows an example diagram for dynamical summer precipitation forecasts, and Figure 9.38b shows corresponding results for persistence forecasts. Ideally, one would like to see points with minimal shading that are concentrated near the origin.

Field Deformation Methods

Field deformation techniques mathematically manipulate the entire forecast field (rather than only individual objects within the field) in order to match the observed field to the extent possible. This approach was first proposed by Hoffman et al. (1995), who characterized forecast errors using a decomposition into displacement, amplitude, and residual components. Location error was determined by horizontal translation of the forecast field until the best match was obtained, where "best" may be interpreted through such criteria as minimum MSE, maximal area overlap, or alignment of the forecast and observed centroids. Alternatively, the required warping of the forecast field can be characterized with a field of deformation vectors, as illustrated in Figure 9.34d, which shows idealized forecast (lighter grays) and observed (darker grays) precipitation areas, with the heavier shading indicating a region of higher precipitation. The field deformation vectors warp the forecast into congruence with the observed area. This vector field indicates that the forecast error is with respect to rotation, and not displacement or aspect ratio.

Another approach to field deformation is to iteratively displace fields though a hierarchy of progressively increasing resolution by minimizing an error statistic such as RMSE, in a method known as "optical flow" (Han and Szunyogh, 2016; Keil and Craig, 2007). Keil and Craig (2009) propose

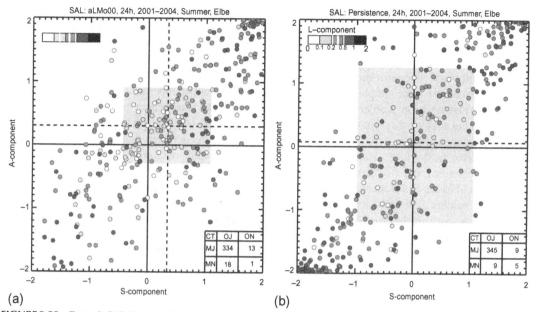

FIGURE 9.38 Example SAL diagrams for daily summer precipitation over a portion of Germany, for (a) dynamical forecasts, and (b) persistence forecasts. Median values for S and A are shown with dashed lines, and the shaded regions are defined by the quartiles of the respective distributions. *From Wernli et al. (2008). © American Meteorological Society. Used with permission.*

quantifying the agreement between the two original fields in terms of the magnitude of overall displacement required for the morphing and the amplitude errors of the morphed field.

9.9. VERIFICATION BASED ON ECONOMIC VALUE

9.9.1. Optimal Decision Making and the Cost/Loss Ratio Problem

The practical justification for effort expended in developing forecast systems and making forecasts is that these forecasts should result in better decision making in the face of uncertainty. Often such decisions have direct economic consequences, or their consequences can be mapped onto an economic (i.e., monetary) scale. There is a substantial literature in the fields of economics and statistics on the use and value of information for decision making under uncertainty (e.g., Clemen, 1996; Johnson and Holt, 1997), and the concepts and methods in this body of knowledge have been extended to the context of optimal use and economic value of weather forecasts (e.g., Katz and Murphy, 1997a; Winkler and Murphy, 1985). Forecast verification is an essential component of this extension, because it is the joint distribution of forecasts and observations (Equation 9.1) that will determine the economic value of forecasts (on average) for a particular decision problem. It is therefore natural to consider characterizing forecast goodness (i.e., computing forecast verification measures) in terms of the mathematical transformations of the joint distribution that define forecast value for particular decision problems. This notion may have first been proposed by Epstein (1962).

The reason that economic value of weather forecasts must be calculated for particular decision problems—that is, on a case-by-case basis—is that the economic value of a particular set of forecasts will be different for different decision problems (e.g., Roebber and Bosart, 1996; Wilks, 1997a). However, a useful and convenient prototype, or "toy," decision model is available, called the *cost/loss ratio* problem (e.g., Katz and Murphy, 1997b; Murphy, 1977; Thompson, 1962). This simplified decision model apparently originated with Anders Angstrom, in a 1922 paper (Liljas and Murphy, 1994), and has been frequently used since that time. Despite its simplicity, the cost/loss problem nevertheless can reasonably approximate some simple real-world decision problems (Roebber and Bosart, 1996).

The cost/loss decision problem relates to a hypothetical decision maker for whom some kind of adverse weather may or may not occur, and who has the option of either protecting or not protecting against the economic effects of the adverse weather. That is, this decision maker must choose one of two alternatives in the face of an uncertain dichotomous weather outcome. Because there are only two possible actions and two possible outcomes, this is the simplest possible decision problem: no decision would be needed if there was only one course of action, and no uncertainty would be involved if only one weather outcome was possible. The protective action available to the decision maker is assumed to be completely effective, but requires payment of a cost C, regardless of whether or not the adverse weather subsequently occurs. If the adverse weather occurs in the absence of the protective action being taken, the decision maker suffers a loss L. The economic effect is zero loss if protection is not taken and the event does not occur. Figure 9.39a shows the loss function for the four possible combinations of decisions and outcomes in this problem.

Probability forecasts for the dichotomous weather event are assumed to be available and, depending on their quality, better decisions (in the sense of improved economic outcomes, on average) may be possible. Taking these forecasts at face value (i.e., assuming that they are calibrated, so $p(o_1 \mid y_i) = y_i$ for all forecasts y_i), the optimal decision on any particular occasion will be the one yielding the smallest expected (i.e., probability-weighted average) expense. If the decision is made to protect, the expense will be C with probability 1, and if no protective action is taken the expected loss will be $y_i L$ (because no loss is incurred, with probability $1-y_i$). Therefore the smaller expected expense will be associated with the protection action whenever

$$C < y_i L, \tag{9.103a}$$

or

$$C/L < y_i. \tag{9.103b}$$

That is, protection is the optimal action when the probability of the adverse event is larger than the ratio of the cost C to the loss L, which is the origin of the name cost/loss ratio. Different decision makers face problems involving different costs and losses, and so their optimal thresholds for action will be different. Clearly this analysis can be relevant only if $C < L$, because otherwise the protective action offers no potential gains, so that meaningful cost/loss ratios are confined to the unit interval, $0 < C/L < 1$.

Mathematically explicit decision problems of this kind not only prescribe optimal actions, but also provide a way to calculate expected economic outcomes associated with forecasts having particular characteristics. For the simple cost/loss ratio problem these expected economic expenses are the probability-weighted average costs and losses, according the probabilities in the joint distribution of the forecasts and observations, $p(y_i, o_j)$. If only climatological forecasts are available (i.e., if the climatological relative frequency, $\bar{o}$, is forecast on each occasion), the optimal action will be to protect if this climatological probability is larger than C/L, and not to protect otherwise. Accordingly, the expected expense associated with the climatological forecast depends on its magnitude relative to the cost/loss ratio:

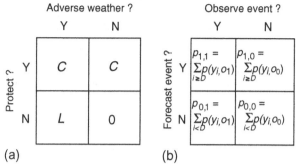

FIGURE 9.39 (a) Loss function for the 2×2 cost/loss ratio situation. (b) Corresponding 2×2 verification table resulting from probability forecasts characterized by the joint distribution $p(y_i, o_j)$ being transformed to nonprobabilistic forecasts according to a particular decision maker's cost/loss ratio. *Adapted from Wilks (2001).*

$$\text{EE}_{\text{clim}} = \begin{cases} C, & \text{if } C/L < \bar{o} \\ \bar{o}L, & \text{otherwise.} \end{cases} \tag{9.104}$$

Similarly, if perfect forecasts were available the hypothetical decision maker would incur the protection cost only on the occasions when the adverse weather was about to occur, so the corresponding expected expense would be

$$\text{EE}_{\text{perf}} = \bar{o}C. \tag{9.105}$$

The expressions for expected expenses in Equation 9.104 and 9.105 are simple because the joint distributions of forecasts and observations for climatological and perfect forecasts are also very simple. More generally, a set of probability forecasts for a dichotomous event would be characterized by a joint distribution of the kind shown in Table 9.4a. A cost/loss decision maker with access to probability forecasts that may range throughout the unit interval has an optimal decision threshold, D, corresponding to the cost/loss ratio, C/L. That is, the decision threshold D is that value of the index i corresponding to the smallest probability y_i that is larger than C/L. In effect, the hypothetical cost/loss decision maker transforms probability forecasts summarized by a joint distribution $p(y_i, o_j)$ into nonprobabilistic forecasts for the dichotomous event "adverse weather," in the same way that was described in Sections 9.2.5 and 9.4.6: probabilities y_i for which $i \geq D$ are transformed to "yes" forecasts and forecasts for which $i < D$ are transformed to "no" forecasts. Figure 9.39b illustrates the 2×2 joint distribution (corresponding to Figure 9.1b) for the resulting nonprobabilistic forecasts of the binary event, in terms of the joint distribution of forecasts and observations for the probability forecasts. Here $p_{1,1}$ is the joint frequency that the probability forecast y_i is above the decision threshold D and the event subsequently occurs, $p_{1,0}$ is the joint frequency that the forecast is above the probability threshold but the event does not occur, $p_{0,1}$ is the joint frequency of forecasts below the threshold and the event occurring, and $p_{0,0}$ is the joint frequency of the probability forecasts being below threshold and the event not occurring.

Because the hypothetical decision maker has constructed yes/no forecasts using the decision threshold D that is customized to a particular cost/loss ratio of interest, there is a one-to-one correspondence between the joint probabilities in Figure 9.39b and the loss function in Figure 9.39a. Combining these leads to the expected expense associated with the forecasts characterized by the joint distribution $p(y_i, o_j)$,

$$\text{EE}_{\text{f}} = (p_{1,1} + p_{1,0}) C + p_{0,1} L \tag{9.106a}$$

$$= C \sum_{j=0}^{1} \sum_{i \geq D} p(y_i, o_j) + L \sum_{i < D} p(y_i\, o_1). \qquad (9.106b)$$

This expected expense depends both on the particular nature of the decision maker's circumstances, through the cost C, the loss L, and their ratio that defines the decision threshold D; and on the quality of the probability forecasts available to the decision maker, as summarized in the joint distribution of forecasts and observations $p(y_i, o_j)$.

9.9.2. The Value Score

Economic value as calculated in the simple cost/loss ratio decision problem is, for a given cost/loss ratio, a rational and meaningful single-number summary of the quality of probabilistic forecasts for a dichotomous event. However, this measure of forecast quality is different for different decision makers (i.e., different values of C/L). Richardson (2000) proposed using economic value, plotted as a function of the cost/loss ratio, as a graphical verification device for probabilistic forecasts for dichotomous events, after a transformation that ensures retrospective calibration of the forecasts (i.e., $y_i \equiv p(o_1 | y_i)$). The ideas are similar to those behind the ROC diagram (see Section 9.4.6), in that forecasts are evaluated through a function that is based on reducing probability forecasts to yes/no forecasts at all possible probability thresholds y_D, and also because conditional and unconditional biases are not penalized. The result is a strictly nonnegative measure of potential (not necessarily actual) economic value in the simplified decision problem, as a function of C/L, for $0 < C/L < 1$.

This basic procedure can be extended to reflect potentially important forecast deficiencies by computing the economic expenses using the original, uncalibrated forecasts (Wilks, 2001). A forecast user without the information or sophistication necessary to recalibrate the forecasts would need to take them at face value and, to the extent that they might be miscalibrated (i.e., that the probability labels y_i might be inaccurate), make suboptimal decisions. Whether or not the forecasts are preprocessed to remove biases, the calculated expected expense (Equation 9.106) can be expressed in the form of a standard skill score (Equation 9.4), relative to the expected expenses associates with climatological (Equation 9.104) and perfect (Equation 9.105) forecasts, called the *value score*:

$$VS = \frac{EE_f - EE_{clim}}{EE_{perf} - EE_{clim}} \qquad (9.107a)$$

$$= \begin{cases} \dfrac{(C/L)(p_{1,1} + p_{1,0} - 1) + p_{0,1}}{(C/L)(\bar{o} - 1)}, & \text{if } C/L < \bar{o} \\[2mm] \dfrac{(C/L)(p_{1,1} + p_{1,0}) + p_{0,1} - \bar{o}}{\bar{o}\,[(C/L) - 1]}, & \text{if } C/L > \bar{o}. \end{cases} \qquad (9.107b)$$

The advantage of this rescaling of EE_f is that sensitivities to particular values of C and L are removed, so that (unlike Equations 9.104–9.106) Equation 9.107 depends only on their ratio, C/L. Perfect forecasts exhibit $VS = 1$, and climatological forecasts exhibit $VS = 0$, for all cost/loss ratios. If the forecasts are recalibrated before calculation of the value score, it will be nonnegative for all cost/loss ratios. Richardson (2001) called this score, for recalibrated forecasts, the potential value, V. However, in the more realistic case that the forecasts are scored at face value, $VS < 0$ is possible if some or all of the hypothetical decision makers would be better served on average by adopting the climatological decision rule, leading to EE_{clim} in Equation 9.104. Related ideas have been presented by Thompson and

Brier (1955), Murphy (1977), and Granger and Pesaran (2000). Mylne (2002) has extended this verification framework for 2×2 decision problems in which protection against the adverse event is only partially effective, and Diebold et al. (1998) present a similar approach for continuous density forecasts.

Figure 9.40 shows VS curves for binary frost forecasts derived from ensembles postprocessed using an MBMP (Equation 8.58) method, with (black solid and dashed) and without (gray) being constrained to maximize calibration on an independent training data set (Wilks, 2018a). The larger economic values occur for cost/loss ratios near the climatological probability of about 0.66, and the economic value is smallest for extreme C/L because in those cases the best decisions are usually obvious according to the climatological probabilities. Although the forecasts postprocessed using the MBMP method (Equation 8.58) are sufficiently well calibrated that no negative economic values occur in Figure 9.40, imposing the calibration constraint improves their economic value for nearly all C/L ratios (i.e., for nearly all forecast users), even though the improved calibration has come at the cost of a slightly reduced eCRPS (Equation 9.83). Comparing Figures 9.40a and 9.40b shows that the improvement in economic value deriving from the calibration constraint is similar to the difference in value between 24–48 h and 48–72 h lead times.

9.9.3. Connections Between VS and Other Verification Approaches

Just as ROC curves are sometimes characterized in terms of the area beneath them, value score curves also can be collapsed to scalar summary statistics. The simple unweighted integral of VS over the full unit interval of C/L is one such summary. This simple function of VS turns out to be equivalent to evaluation of the full set of forecasts using the Brier score, because the expected expense in the cost/lost ratio situation (Equation 9.106) is a linear function of BS (Equation 9.39) (Murphy, 1966). That is, ranking competing forecasts according to their Brier scores, or Brier skill scores, yields the same result as a ranking based on the unweighted integrals of their VS curves. To the extent that the forecast user community might have a nonuniform distribution of cost/loss ratios (e.g., a preponderance of forecast users for whom the protection option is relatively inexpensive), single-number weighted averages of VS also can be computed as statistical expectations of VS with respect to the probability density function for C/L across users of interest (Richardson, 2001; Wilks, 2001).

The VS curve is constructed through a series of 2×2 verification tables, and there are accordingly connections both with scores used to evaluate nonprobabilistic forecasts of binary predictands, and with the ROC curve. For correctly calibrated forecasts, maximum economic value in the cost/loss decision problem is achieved for decision makers for whom C/L is equal to the climatological event relative frequency, because for these individuals the optimal action is least clear from the climatological information alone. This explains the peaks in Figure 9.40 coinciding with C/L equal to the climatological frost probability. Gandin et al. (1992) called these *ideal users*, recognizing that such individuals will benefit most from forecasts. Interestingly, this maximum (potential, because calibrated forecasts are assumed) economic value is given by the Peirce skill score (Equation 9.18), evaluated for the 2×2 table appropriate to this "ideal" cost/loss ratio (Richardson, 2000; Wandishin and Brooks, 2002). Furthermore, the odds ratio (Equation 9.9) $\theta > 1$ for this table is a necessary condition for economic value to be imparted for at least one of the possible cost/loss ratio decision problems (Richardson, 2003; Wandishin and Brooks, 2002). The range of cost/loss ratios for which positive potential economic value can be realized for a given 2×2 verification table is given by its Clayton skill score (Equation 9.19) (Wandishin and Brooks, 2002). Semazzi and Mera (2006) show that the area between the ROC curve and a line that depends on C/L and the economic value of baseline forecasts is proportional to the

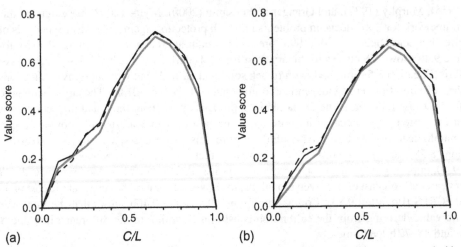

FIGURE 9.40 VS curves for binary January New York city frost forecasts derived from ensembles postprocessed without (gray) and with (light solid and dashed) calibration constraints, at (a) 24–48 h and (b) 48–72 h lead times. *From Wilks (2018a).*

potential economic value V of forecasts that are free of conditional and unconditional biases. Additional connections between VS and attributes of the ROC diagram are provided in Marzban (2012), Mylne (2002), and Richardson (2003).

9.10. VERIFICATION WHEN THE OBSERVATION IS UNCERTAIN

Forecast verification is traditionally undertaken under the tacit assumption that the verifying observation is a true and error-free representation of the predictand. This assumption may be reasonable when observation errors are small relative to forecast errors, but the true state of the predictand can never really be known with certainty because of measurement (or instrument) errors and representativeness errors, and as forecast systems improve the gap between forecast and observation error narrows (e.g., Bowler, 2008; Mittermaier and Stephenson, 2015). Measurement errors may be quite small when instruments such as ordinary thermometers and raingauges, which interact fairly directly with the process or quantity being measured, are used. *Representativeness errors* occur when the measured quantity differs from the predicted quantity. These occur in a spatial sense when there is a scale mismatch between that of the instrument (e.g., a raingage with area approximately 350 cm^2) and the predictand (which might be rainfall averaged over a 100 km^2 area), and in a temporal sense when there is a mismatch between observation time and prediction time.

Bowler (2006) considered the effects of observation error in the 2×2 contingency table setting. Given an (externally derived) characterization of the observational error characteristics, it is possible to reconstruct an expected 2×2 table for hypothetical error-free observations. The existence of observational error degrades apparent forecast skill relative to what might be achieved with error-free observations if that skill is positive, but errors in the observations tend to make negatively skillful forecasts appear less bad. Alternatively, one could consider randomly perturbing the binary observations according to misclassification probabilities for the "yes" and "no" events (Briggs et al., 2005).

Ciach and Krajewski (1999) partition the MSE for radar-derived area-averaged rainfall into two terms: MSE for instrument error (discrepancies between radar-estimated and gauge measurements at the gauge locations) and MSE for representativeness error (discrepancies between local gauge measurements and the true area average). The first of these contributions is straightforward to characterize, using prior data sets for radar-derived precipitation at raingauge locations. The second contribution is much more difficult to calculate without data from a very dense raingauge network, although a modeling-based estimate can be derived by assuming characteristics of the spatial correlation structure of the rainfall fields. Results of this study showed the representativeness error component to be most important at the shortest timescales, but that it remained a significant contribution to overall MSE even for 4-day rainfall accumulations.

The notion of uncertain observations fits well in the context of ensemble forecasting (Chapter 8). The conceptual basis for ensemble forecasting begins with the idea of uncertainty in the initial condition (i.e., the observation at initialization time) for a dynamical forecast model. Therefore a necessary condition for a forecast ensemble to be consistent (for the observation of the predictand to be statistically indistinguishable from the forecast ensemble members) is that the verification, which is the basis for initial conditions in the next forecast cycle, must also be subject to errors and uncertainty. When the magnitude of the observation error is a substantial fraction of the ensemble spread, ignoring observation errors produces overpopulation of the extreme bins of the rank histogram, which leads to an erroneous (or, at least, exaggerated) diagnosis of ensemble underdispersion. The most usual remedy in such cases is to simulate the effects of observational errors by adding random numbers with error characteristics mimicking the observational errors to each ensemble member (Anderson, 1996; Candille and Talagrand, 2008; Hamill, 2001; Saetra et al., 2004).

Replacing the observation with a probability distribution is a seemingly natural approach when uncertainty about the observation will have an important effect. Candille and Talagrand (2008) proposed treating verifying observations explicitly as probabilities (for discrete events) or probability distributions (for continuous predictands). This approach is mathematically tractable, while continuing to allow use of the reliability diagram, the Brier score (including its usual algebraic decomposition), the Brier skill score, the ranked probability score, the continuous ranked probability score, and the ROC diagram. Other studies proceeding along these lines include Pappenberger et al. (2009) and Santos and Ghelli (2012), who consider probability forecasts for binary observations evaluated with the Brier score; Friederichs and Thorarinsdottir (2012), who replace the step function in the CRPS (Equation 9.61) with a Gaussian distribution for the observational uncertainty; and Vannitsem and Hagedorn (2011) who account for observational uncertainty in the fitting of MOS regression equations. Ahrens and Jaun (2007) and Pinson and Hagedorn (2012) consider verification against Monte Carlo samples from observation distributions, Gorgas and Dorninger (2012) propose use of an "analysis ensemble," and Röpnack et al. (2013) and Weijs and van de Giesen (2011) outline Bayesian approaches. Of course, in order to implement the methods just outlined, appropriate distributions characterizing the observation errors must still be defined and estimated, either externally to the verification data being evaluated or subjectively (Briggs et al., 2005).

Ferro (2017) has introduced the notion of the unbiased and proper scoring rule for forecasts of uncertain observations. The expected value of an unbiased proper score S evaluated with respect to the uncertain observation o will be the same as the expected value of the score that would hypothetically be achieved by evaluating its counterpart proper score S_0 with respect to the true but unknown value x. He further defines two unbiased and proper scoring rules pertaining to the Dawid-Sebastiani score (Equation 9.67) for PDF forecasts of a continuous variable, which require knowledge of only the first two moments of the conditional distribution of the observation o given the truth x. The first of these assumes that the observational uncertainty can be represented as additive white noise, so that $E(o|x) = a + bx$ and $Var(o|x) = c^2$, where a, b, and c are constants that are assumed known. In this case the score that is unbiased and proper for the DSS is

$$DSS_A = 2 \ln(\sigma) + \frac{(o - a - b\mu)^2 - c^2}{b^2\sigma^2}, \tag{9.108}$$

where μ and σ^2 are the mean and variance of the forecast distribution. In the limit of no observational uncertainty ($a = c = 0$ and $b = 1$), Equation 9.108 reduces to the DSS in Equation 9.67. The second unbiased and proper score for the DSS pertains to a multiplicative error distribution for o, for which $E(o|x) = bx$ and $Var(o|x) = c^2x^2$, yielding the score

$$DSS_M = 2 \ln(\sigma) + \frac{(o - b\mu)^2 - o^2c^2/(b^2 + c^2)}{b^2\sigma^2}. \tag{9.109}$$

Naveau and Bessac (2018) extend these ideas to encompass also multiplicative gamma-distributed errors, applicable to such quantities as precipitation, and provide results for CRPS.

9.11. SAMPLING AND INFERENCE FOR VERIFICATION STATISTICS

Practical forecast verification is necessarily concerned with finite samples of forecast-observation pairs. The verification statistics that can be computed from a particular data set are no less subject to sampling variability than are any other sort of statistics. If a different sample of the same kind of forecasts and observations were hypothetically to become available, the value of verification statistic(s) computed from it likely would be at least somewhat different. To the extent that the sampling distribution for a verification statistic is known or can be estimated, confidence intervals around it can be obtained, and formal tests (e.g., against a null hypothesis of zero skill) can be constructed.

In many cases the procedures presented in Chapter 5 can be used to construct statistical inferences for forecast verification statistics, and therefore to assess statistical significance for sample assessments of accuracy, skill, and so on. Notably, when results for a large number of forecasts and observations have been averaged to form a sample mean score, the Central Limit Theorem would suggest use of a t test (Section 5.2.1–5.2.4) to make inferences about that mean. Of course, when such averages have been computed over a sequence of autocorrelated (e.g., daily) forecasts and observations, the effects of autocorrelation on their sampling distributions must be represented. Similarly, when different forecast sources are being compared with respect to the same sequence of observations, failing to represent the paired nature of these data when assessing the possible significance of their differences, whether the data are also autocorrelated (Section 5.2.4) or not (Section 5.2.3), will likely lead to erroneous inferences (DelSole and Tippett, 2014; Gilleland et al., 2018; Jarman and Smith, 2018; Siegert et al., 2017). When multiple inferences are being assessed simultaneously the inferential standards need to be adjusted appropriately (Geer, 2016), as outlined in Section 5.4.

9.11.1. Sampling Characteristics of Contingency Table Statistics

In principle, the sampling characteristics of many 2×2 contingency table statistics follow from a fairly straightforward application of independent binomial sampling (Agresti, 1996). For example, such measures as the false alarm ratio (Equation 9.11), the hit rate (Equation 9.12), and the false alarm rate

(Equation 9.13) are all proportions that estimate (conditional) probabilities. If the contingency table counts (see Figure 9.1a) have been produced independently from stationary (i.e., constant-p) forecast and observation systems, those counts are (conditional) binomial variables, and the corresponding proportions (such as FAR, H and F) are sample estimates of the corresponding binomial probabilities (Seaman et al., 1996).

A direct approach to finding confidence intervals for sample proportions x/N that estimate the binomial parameter p is to use the binomial probability distribution function (Equation 4.1). A $(1-\alpha)\cdot 100\%$ confidence interval for the underlying probability that is consistent with the observed proportion x/N can be defined by the extreme values of x on each tail that include probabilities of at least $1-\alpha$ between them, inclusive. Unfortunately the result, called the *Clopper-Pearson exact interval*, generally will be inaccurate to a degree (and, specifically, too wide) because of the discreteness of the binomial distribution (Agresti and Coull, 1998). Another simple approach to calculation of confidence intervals for sample proportions is to invert the Gaussian approximation to the binomial distribution (Equation 5.2). Since Equation 5.2b is the standard deviation σ_x for the number of binomial successes X, the corresponding variance for the estimated proportion $\hat{p} = x/N$ is $\sigma_p^2 = \sigma_x^2/N^2 = \hat{p}(1-\hat{p})/N$ (using Equation 4.17). The resulting $(1-\alpha)\cdot 100\%$ confidence interval is then

$$p = \hat{p} \pm z_{(1-\alpha/2)}[\hat{p}(1-\hat{p})/N]^{1/2}, \tag{9.110}$$

where $z_{(1-\alpha/2)}$ is the $(1-\alpha/2)$ quantile of the standard Gaussian distribution (e.g., $z_{(1-\alpha/2)} = 1.96$ for $\alpha = 0.05$).

Equation 9.110 can be quite inaccurate, in the sense that the actual probability of including the true p is substantially smaller than $1-\alpha$, unless N is very large. However, this bias can be corrected using the modification (Agresti and Coull, 1998) to Equation 9.110,

$$p = \frac{\hat{p} + \dfrac{z_{(1-\alpha/2)}^2}{2N} \pm z_{(1-\alpha/2)} \sqrt{\dfrac{\hat{p}(1-\hat{p})}{N} + \dfrac{z_{(1-\alpha/2)}^2}{4N^2}}}{1 + \dfrac{z_{(1-\alpha/2)}^2}{N}}. \tag{9.111}$$

The differences between Equations 9.111 and 9.110 are in the three terms involving $z_{(1-\alpha/2)}^2/N$, which approach zero in Equation 9.111 for large N. Standard errors according to Equation 9.110 are tabulated for ranges of $\hat{p}$ and N in Thornes and Stephenson (2001).

Marzban and Sandgathe (2008) derive the approximation to the standard deviation of the sampling distribution of the threat score (Equation 9.8),

$$\hat{\sigma}_{TS} \approx TS \sqrt{\frac{1}{a}\left(\frac{b}{a+b} + \frac{c}{a+c}\right)}. \tag{9.112}$$

Another relevant result from the statistics of contingency tables (Agresti, 1996), is that the sampling distribution of the logarithm of the odds ratio (Equation 8.9) is approximately Gaussian distributed for sufficiently large $n = a+b+c+d$, with estimated standard deviation

$$\hat{\sigma}_{\ln(\theta)} = \left[\frac{1}{a} + \frac{1}{b} + \frac{1}{c} + \frac{1}{d}\right]^{1/2}. \tag{9.113}$$

Thus a floor on the magnitude of the sampling uncertainty for the odds ratio is imposed by the smallest of the four counts in Table 8.1a. When the null hypothesis of independence between forecasts and observations (i.e., $\theta = 1$) is of interest, it could be rejected if the observed $\ln(\theta)$ is sufficiently far from $\ln(1) = 0$, with respect to Equation 9.113.

For large n, sampling distributions for the contingency-table statistics are expected to be approximately Gaussian, which suggests use of one-sample t tests as the basis of inferences about them, provided the underlying data are stationary and independent. Hogan and Mason (2012) tabulate expressions for the sampling variance of additional 2×2 contingency table verification measures that could be used in a similar way.

Radok (1988) shows that the sampling distribution for the multicategory Heidke skill score (Equation 9.24) is proportional to that of a chi-square variable.

Example 9.8. Inferences for Selected Contingency Table Verification Measures

The hit and false alarm rates for the Finley tornado forecasts in Table 1.1a are $H = 28/51 = 0.549$ and $F = 72/2752 = 0.026$, respectively. Because the underlying forecast situations were fairly widely separated in space and time, the underlying counts reasonably approximate independent sampling. These proportions are sample estimates of the conditional probabilities of tornados having been forecast, given either that tornados were or were not subsequently reported. Using Equation 9.111, $(1-\alpha)\cdot 100\% = 95\%$ confidence intervals for the true underlying conditional probabilities can be estimated as

$$H = \frac{.549 + \dfrac{1.96^2}{(2)(51)} \pm 1.96 \sqrt{\dfrac{.549(1-.549)}{51} + \dfrac{1.96^2}{(4)(51)^2}}}{1 + \dfrac{1.96^2}{51}} \tag{9.114a}$$

$$= .546 \pm .132 = \{.414, .678\}$$

and

$$F = \frac{.026 + \dfrac{1.96^2}{(2)(2752)} \pm 1.96 \sqrt{\dfrac{.026(1-.026)}{2752} + \dfrac{1.96^2}{(4)(2752)^2}}}{1 + \dfrac{1.96^2}{2752}} \tag{9.114b}$$

$$= .0267 \pm .00598 = \{.0207, .0326\}.$$

The precision of the estimated false alarm rate is much better (its standard error is much smaller) in part because the overwhelming majority of observations $(b+d)$ were "no tornado;" but also in part because $p(1-p)$ is small for extreme values, and larger for intermediate values of p. Assuming independence of the forecasts and observations (in the sense illustrated in Equation 9.16), plausible useless-forecast benchmarks for the hit and false alarm rates might be $H_0 = F_0 = (a+b)/n = 100/2803 = 0.0357$. Neither of the 95% confidence intervals in Equation 9.114 include this value, leading to the inference that H and F for the Finley forecasts are better than would have been achieved by chance.

Stephenson (2000) notes that, because the Peirce skill score (Equation 9.18) can be calculated as the difference between H and F, confidence intervals for it can be calculated using simple binomial sampling considerations because H and F are mutually independent. In particular, since on the strength of the Central Limit Theorem the sampling distributions of both H and F are approximately Gaussian for sufficiently large sample sizes, under these conditions the sampling distribution of the PSS will be Gaussian, with estimated standard deviation

$$\hat{\sigma}_{PSS} = \sqrt{\hat{\sigma}_H^2 + \hat{\sigma}_F^2}. \tag{9.115}$$

For the Finley tornado forecasts, PSS = 0.523, so that a 95% confidence interval around this value could be constructed as $0.523 \pm 1.96\, \hat{\sigma}_{PSS}$. Estimating $\hat{\sigma}_H = .132/1.96 = 0.673$ from Equation 9.114a and $\hat{\sigma}_F = .00698/1.96 = .00305$ from Equation 9.114b, or interpolating from the table in Thornes and Stephenson (2001), this interval would be $0.523 \pm 1.96\, (0.0673^2 + 0.00305^2)^{1/2} = 0.523 \pm 0.132 = \{\,0.391, 0.655\}$. Since this interval does not include zero, a reasonable inference would be that these forecasts exhibited significant skill according to the Peirce skill score. Hanssen and Kuipers (1965) and Woodcock (1976) derive the alternative expression for the sampling variance of the PSS,

$$\hat{\sigma}_{PSS}^2 = \frac{n^2 - 4(a+c)(b+d)\, PSS^2}{4n\,(a+c)(b+d)}. \tag{9.116}$$

Again assuming a Gaussian sampling distribution, Equation 9.116 estimates the 95% confidence interval for PSS for the Finley tornado forecasts as $0.523 \pm (1.96)(.070) = \{0.386, 0.660\}$.

Finally, the odds ratio for the Finley forecasts is $\theta = (28)(2680)/(23)(72) = 45.31$, and the standard deviation of the (approximately Gaussian) sampling distribution of its logarithm (Equation 9.113) is $(1/28 + 1/72 + 1/23 + 1/2680)^{1/2} = 0.306$. The null hypothesis that the forecasts and observations are independent (i.e., $\theta_0 = 1$) produces the t-statistic $[\ln(45.31) - \ln(1)]/0.306 - 12.5$, which would lead to emphatic rejection of that null hypothesis. $\diamond$

The calculations in this section have relied on the assumptions that the verification data are independent and, for the sampling distribution of proportions, that the underlying probability p is stationary (i.e., constant across forecasts). The independence assumption might be violated, for example, if the data set consists of a sequence of daily forecast-observation pairs. The stationarity assumption might be violated if the data set includes a range of locations with different climatologies for the forecast variable. When comparing the performance of two forecasting systems pertaining to the same set of observations, use of the equations presented in this section to formulate 2-sample t tests would be inappropriate because correlations between the paired forecasts would not be represented. In cases where these assumptions might be violated, inferences for contingency-table verification measures still can be made, by estimating their sampling distributions using appropriately constructed resampling approaches (see Section 9.11.6).

9.11.2. ROC Diagram Sampling Characteristics

Because confidence intervals around sample estimates for the hit rate H and the false alarm rate F can be calculated using Equation 9.111, confidence regions around individual (F, H) points in a ROC diagram can also be calculated and plotted. A complication is that, in order to define a joint, simultaneous $(1-\alpha) \cdot 100\%$ confidence region around a sample (F, H) point, each of the two individual confidence intervals must cover its corresponding true value with a probability that is somewhat larger than $(1-\alpha)$. Essentially, this adjustment is necessary in order to make valid simultaneous inference in a multiple testing situation (cf. Section 5.4.1). Since H and F are at least approximately independent, a reasonable approach to deciding the appropriate sizes of the two confidence intervals is to use the *Bonferroni inequality* (Equation 12.56). In the case of the ROC diagram, where the joint confidence region is $K = 2$-dimensional, Equation 12.56 says that the rectangular region defined by two $(1-\alpha/2) \cdot 100\%$ confidence intervals for F and H will jointly enclose the true (F, H) pair with probability at least as large as $1-\alpha$. For example, a joint 95% (at least) rectangular confidence region will be defined by two 97.5% confidence intervals, calculated using $z_{1-\alpha/4} = z_{.9875} = 2.24$, in Equation 9.111.

A test for the statistical significance of the area A under the ROC curve, against the null hypothesis that the forecasts and observations are independent (i.e., that $A_0 = 1/2$), is available. The sampling distribution of the ROC area computed using trapezoidal integration, given the null hypothesis of no relationship between forecasts and observations, is proportional to the distribution of the Mann-Whitney U-statistic (Equations 5.22 and 5.23) (Bamber, 1975; Mason and Graham, 2002), so that the test for the ROC area is equivalent to the Wilcoxon-Mann-Whitney test applied to the two likelihood distributions $p(y_i|o_1)$ and $p(y_i|o_2)$ (cf. Figure 9.13). To calculate this test, the ROC area A is transformed to a Mann-Whitney U variable according to

$$U = n_1 n_2 (1 - A). \tag{9.117}$$

Here $n_1 = a+c$ is the number of "yes" observations, and $n_2 = b+d$ is the number of "no" observations. Notice that, under the null hypothesis $A_0 = 1/2$, Equation 9.117 is exactly the mean of the Gaussian approximation to the sampling distribution of U in Equation 5.24a. This null hypothesis is rejected for sufficiently small U, or equivalently for sufficiently large ROC area A.

Example 9.9. Confidence and Significance Statements about a ROC Diagram

Figure 9.41 shows the ROC diagram for the Finley tornado forecasts (Table 9.1a), together with the 97.5% confidence intervals for F and H. These are $0.020 \leq F \leq 0.034$ and $0.396 \leq H \leq 0.649$, and were calculated from Equation 9.111 using $z_{1-\alpha/4} = z_{.9875} = 2.24$. The confidence interval for F is only about as wide as the dot locating the sample (F, H) pair, both because the number of "no tornado" observations is large, and because the proportion of false alarms is quite small. These two 97.5% confidence intervals define a 95% rectangular confidence region for the true (F, H) pair according to the Bonferroni inequality

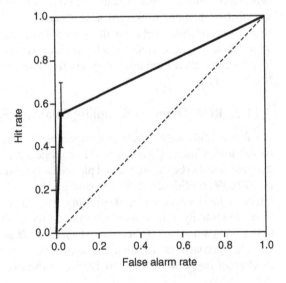

FIGURE 9.41 ROC diagram for the Finley tornado forecasts (Table 9.1a), with the 95% simultaneous Bonferroni (Equation 12.56) confidence intervals for the single (F, H) point, calculated using Equation 9.111.

(Equation 12.56). This region does not include the dashed 1:1 line, indicating that it is improbable for these forecasts to have been generated by a process that was independent of the observations.

The area under the ROC curve in Figure 9.41 is 0.761. If the true ROC curve for the process from which these forecast-observation pairs have been sampled is the dashed 1:1 diagonal line, what is the probability that a ROC area A this large or larger could have been achieved by chance, given $n_1 = 51$ "yes" observations and $n_2 = 2752$ "no" observations? Equation 9.117 yields $U = (51)(2752)(1-0.761) = 33544$, the unusualness of which can be evaluated in the context of the (null) Gaussian distribution with mean $\mu_U = (51)(2752)/2 = 70176$ (Equation 5.24a) and standard deviation $\sigma_U = [(51)(2752)(51+2752+1)/12]^{1/2} = 5727$ (Equation 5.24b). The resulting test statistic is $z = (33544-70176)/5727 = -6.4$, so that the null hypothesis of no association between the forecasts and observations would be strongly rejected. $\diamond$

As before, when two forecast systems have been used to predict the same observations it is necessary to account for the correlation between them in order to judge the possible statistical significance of the difference in the resulting ROC areas. Hanley and McNeil (1983) and DeLong et al. (1988) provide methods for doing so when the data are serially independent. Lahiri and Yang (2018) describe computation of confidence bands around a single ROC curve, even in the presence of serial correlation.

9.11.3. Brier Score and Brier Skill Score Inference

Bradley et al. (2008) have derived an expression for the variance of the sampling distribution of the Brier score (Equation 9.39), assuming that the n forecast observation pairs (y_i, o_i) are independent random samples from a homogeneous joint distribution of forecasts and observations. Their result can be expressed as

$$\hat{\sigma}_{BS}^2 = \frac{1}{n} \sum_{i=1}^{n} \left(y_i^4 - 4y_i^3 o_i + 6y_i^2 o_i - 4y_i o_i + o_i \right) - \frac{BS^2}{n}. \tag{9.118}$$

Similarly, Bradley et al. (2008) derive the approximate sampling variance for the Brier skill score (Equation 9.40),

$$\hat{\sigma}_{BSS}^2 \approx \left(\frac{n}{n-1}\right)^2 \frac{\hat{\sigma}_{BS}^2}{\bar{o}^2(1-\bar{o})^2} + \frac{n}{(n-1)^3} \frac{(1-BSS)^2}{\bar{o}(1-\bar{o})} [\bar{o}(1-\bar{o})(6-4n)+n-1]$$

$$+ \left(\frac{n}{n-1}\right)^2 \frac{(2-4\bar{o})(1-BSS)}{\bar{o}(1-\bar{o})} \left(1 + \frac{\sum_{i=1}^{n}(y_i^2 o_i - 2y_i o_i)}{n\bar{o}} + \frac{\sum_{i=1}^{n}(y_i^2 o_i - y_i^2)}{n(1-\bar{o})} \right). \tag{9.119}$$

Since Equation 9.118 and 9.119 require estimates of high (up to fourth) moments of the joint distribution of the forecasts and observations, the sample size n must be fairly large in order for these estimates to be usefully accurate. Figure 9.42 shows sample sizes necessary for Equations 9.118 and 9.119 to yield Gaussian 95% and 99% confidence intervals exhibiting approximately correct coverage. Quite large sample sizes ($n > 1000$) are required for higher-skill forecasts of relatively rare events, whereas much more modest sample sizes are adequate for low-skill forecasts of common events. Using too few samples yields estimated sampling variances, and therefore confidence intervals, that are too small. Bradley et al. (2008) also note that the sample estimate of BSS exhibits a negative bias that may be appreciable for small sample sizes and relatively rare events (small n and $\bar{o}$).

FIGURE 9.42 Sample sizes necessary for Equation 9.118 and 9.119 to yield approximately correct variance estimates, as a function of the sample climatological probability and sample Brier skill score. *From Wilks (2010).*

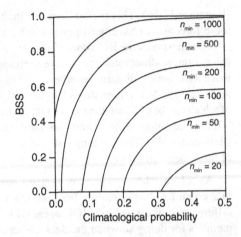

When the forecasts and binary events exhibit first-order autoregressive serial dependence (Section 10.3.1), Equations 9.118 and 9.119 can be used with "effective sample size" adjustments to that appropriately inflate these sampling variances (Wilks, 2010). For the sampling variance of the Brier score, the sample size n in Equation 9.118 is replaced by the effective sample size estimate

$$n' = n\frac{1-(1-\bar{o})[b(1-BS)r_1]^2}{1+(1-\bar{o})[b(1-BS)r_1]^2}, \tag{9.120}$$

and for the Brier skill score (Equation 9.119) the effective sample size can be estimated as

$$n' = n\frac{1-(1-\bar{o})[b(1-BS)r_1]^4}{1+(1-\bar{o})[b(1-BS)r_1]^4}. \tag{9.121}$$

In these two effective sample size equations r_1 is the lag-1 autocorrelation, and b is the slope of the calibration function in the reliability diagram, where $b = 1$ for calibrated forecasts, $b < 1$ for overconfident forecasts, and $b > 1$ for underconfident forecasts. Lahiri and Yang (2016) analyze inflation of the variance estimates in Equation 9.118 and 9.119 under more general time dependence conditions.

Siegert (2014) describes a method to estimate the sampling variances for each of the three components of the Brier score decomposition in Equation 9.45, assuming statistical independence among the n samples.

9.11.4. Reliability Diagram Sampling Characteristics

The calibration-function portion of the reliability diagram consists of I conditional outcome relative frequencies that estimate the conditional probabilities $p(o_1 | y_i), i = 1, ..., I$. If independence and stationarity are reasonably approximated, then confidence intervals around these points can be computed using either Equation 9.110 or Equation 9.111. To the extent that these intervals include the 1:1 perfect reliability diagonal, a null hypothesis that the forecaster(s) or forecast system has produced calibrated forecasts would not be rejected. To the extent that these intervals do not include the horizontal "no resolution" line, a null hypothesis that the forecasts are no better than climatological guessing would be rejected.

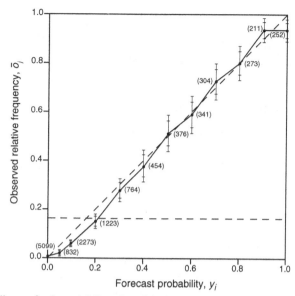

FIGURE 9.43   Reliability diagram for the probability-of-precipitation data in Table 9.2, with 95% confidence intervals on each conditional probability estimate, calculated using Equation 9.110. Inner confidence limits pertain to individual points, and outer bounds are joint Bonferroni (Equation 12.56) confidence limits. Raw subsample sizes N_i are shown parenthetically. The 1:1 perfect reliability and horizontal "no resolution" lines are dashed.

Figure 9.43 shows the reliability diagram for the forecasts summarized in Table 9.2, with 95% confidence intervals drawn around each of the $I = 12$ conditional relative frequencies. The stationarity assumption for these estimated probabilities is reasonable, because the forecasters have sorted the forecast-observation pairs according to their judgments about those probabilities. The independence assumption is less well justified, because these data are simultaneous forecast-observation pairs for about one hundred locations in the United States, so that positive spatial and temporal correlations among both the forecasts and observations would be expected. Accordingly the confidence intervals drawn in Figure 9.43 are possibly too narrow.

Because the sample sizes (shown parenthetically in Figure 9.43) are large, Equation 9.110 was used to compute the confidence intervals. For each point, two confidence intervals are shown. The inner, narrower intervals are ordinary individual confidence intervals, computed using $z_{1-\alpha/2} = 1.96$, for $\alpha = 0.05$ in Equation 9.110. An interval of this kind would be appropriate if confidence statements about a single one of these points are of interest. The outer, wider confidence intervals are joint $(1-\alpha) \cdot 100\% = 95\%$ Bonferroni (Equation 12.56) intervals, computed using $z_{1-[\alpha/12]/2} = 2.87$, again for $\alpha = 0.05$. The meaning of these outer, Bonferroni, intervals is that the probability is at least 0.95 that all $I = 12$ of the conditional probabilities being estimated are simultaneously within their respective individual confidence intervals. Thus a (joint) null hypothesis that all of the forecast probabilities are calibrated would be rejected if any one of them fails to include the diagonal 1:1 line (dashed), which in fact does occur for $y_1 = 0.0$, $y_2 = 0.05$, $y_3 = 0.1$, $y_4 = 0.2$, and $y_{12} = 1.0$. On the other hand, it is clear that these forecasts are overall much better than random climatological guessing, since the Bonferroni confidence intervals overlap the dashed horizontal no resolution line only for $y_4 = 0.2$, and are in general quite far from it.

An alternative approach to computation of confidence intervals around the points on the reliability diagram has been proposed by Bröcker and Smith (2007b). Using a bootstrap approach that accounts for the randomness of the number of forecasts in each of the I bins, they plot "consistency bars" around the vertical projections of the reliability diagram points onto the 1:1 diagonal line, in order to evaluate the likelihoods of the observed conditional relative frequencies under a null assumption of perfect reliability. Pinson et al. (2010) extend this approach to accommodate serial correlation in the verification data.

9.11.5. Assessing Ensemble Calibration

Ensemble calibration is typically diagnosed using verification rank histograms (for scalar, discrete ensembles), PIT histograms (for ensembles postprocessed to yield continuous predictive distributions) or MST histograms (for multivariate discrete ensembles). Distinguishing between true deviations from verification rank histogram (and also PIT and MST histogram) uniformity and mere sampling variations usually is approached through the chi-square goodness-of-fit test (Section 5.2.5). Here the null hypothesis is a uniform rank (or PIT or MST) histogram, so the expected number of counts in each bin is $n/(m+1)$, and the test is evaluated using the chi-square distribution with $v = m$ degrees of freedom (because there are $m+1$ bins). This approach assumes independence of the n ensembles being evaluated, and so may not be appropriate in unmodified form if, for example, the ensembles pertain to consecutive days. Additive corrections to tabulated critical chi-square values appropriate to serial correlation in the forecasts are given in Table 9.8, for rank- and PIT-histogram uniformity, and in Table 9.9 for MST histogram uniformity, assuming that the time dependence can be characterized by the lag-1 autocorrelation r_1. That is, given autocorrelation of the underlying data, larger values of the χ^2 test statistic are required in order to reject a null hypothesis of rank uniformity.

A potential drawback of the ordinary chi-square test in this context is that it assesses deviations from rank uniformity independently of the ordering of the bars in the histogram. That is, deviations from the ideal uniformity level are treated equally whether they occur randomly across the histogram bars, or if (for example) the positive deviations are concentrated toward the left, and the negative deviations are concentrated toward the right. Accordingly it is not focused toward detecting coherent patterns in the

TABLE 9.8 Additive Corrections to Tabulated χ^2 Critical Values to Test Uniformity of Conventional (Scalar) Rank Histograms and PIT Histograms, as Functions of lag-1 Autocorrelation r_1

	Test level, α			
r_1	0.10	0.05	0.01	0.001
0.1	0.3	0.3	0.6	1.1
0.2	0.8	0.9	1.4	2.4
0.3	1.5	1.8	2.8	4.6
0.4	2.6	3.1	4.9	8.3
0.5	4.1	5.1	8.4	14.6
0.6	6.6	8.6	14.3	25.3
0.7	11.2	14.8	25.2	44.3
0.8	20.9	28.1	48.6	85.1
0.9	50.5	69.0	121.7	214.2

From Wilks (2004).

TABLE 9.9 Additive Corrections to Tabulated χ^2 Critical Values to Test Uniformity of MST Histograms, as Functions of lag-1 Autocorrelation r_1

r_1	Test level, α			
	0.10	0.05	0.01	0.001
0.4	0.4	0.5	0.6	1.1
0.5	0.6	0.9	1.3	2.2
0.6	1.3	1.6	2.4	4.4
0.7	2.6	3.4	5.0	8.8
0.8	5.4	7.1	11.9	22.6
0.9	15.6	21.0	37.2	68.6

Corrections for $r_1 < 0.4$ are negligible.
From Wilks (2004).

rank histogram such as those illustrated in Figure 9.19. Alternative tests designed to be sensitive to these characteristic patterns, and which therefore exhibit greater power to detect them, are described in Elmore (2005) and Jolliffe and Primo (2008).

9.11.6. Resampling Verification Statistics

Often the uncertainty characteristics of verification statistics with unknown sampling distributions are of interest. Or, sampling characteristics of verification statistics discussed previously in this section are of interest, but assumptions of independent sampling or first-order autoregressive dependence cannot be supported. In such cases, statistical inference for forecast verification statistics can be addressed through resampling tests, as described in Sections 5.3.3 through 5.3.5. These procedures are very flexible, and the resampling algorithm used in any particular case will depend on the specific setting.

For problems where the sampling distribution of the verification statistic is unknown, but independence can reasonably be assumed, implementation of conventional permutation or bootstrap tests are straightforward. Illustrative examples of the bootstrap in forecast verification can be found, for example, in Bröcker and Smith (2007b) and Roulston and Smith (2003). Bradley et al. (2003) use the bootstrap to evaluate the sampling distributions of the reliability and resolution terms in Equation 9.4, using the probability-of-precipitation data in Table 9.2. Déqué (2003) illustrates permutation tests for a variety of verification statistics.

Special problems occur when the data to be resampled exhibit spatial and/or temporal correlation. A typical cause of spatial correlation is the occurrence of simultaneous data at multiple locations, that is, maps of forecasts and observations. Hamill (1999) describes a permutation test for a paired comparison of two forecasting systems, in which problems of nonindependence of forecast errors have been obviated by spatial pooling. Alternatively, the effects of spatial correlation on resampled verification statistics can be accounted for automatically if the resampled objects are entire maps, rather than individual locations resampled independently of each other. Similarly, the effects of time correlation in the forecast verification statistics can be accounted for using the moving-blocks bootstrap (see Section 5.3.5). The moving-blocks bootstrap is equally applicable to scalar data (e.g., individual forecast-observation pairs at single locations, which are autocorrelated), or to entire autocorrelated maps of forecasts and observations

(Wilks, 1997b). Pinson et al. (2010) apply a spectrally based (i.e., based mathematically on concepts from Chapter 10) resampling procedure to account for the effects of serial correlation on the reliability diagram.

9.12. EXERCISES

9.1. For the forecast verification data in Table 9.2,
   a. Reconstruct the joint distribution, $p(y_i, o_j)$, $i = 1, ..., 12, j = 1, 2$.
   b. Compute the unconditional (sample climatological) probability $p(o_1)$.

9.2. Construct the 2×2 contingency table that would result if the probability forecasts in Table 9.2 had been converted to nonprobabilistic rain/no rain forecasts, with a threshold probability of 0.25.

9.3. Using the 2×2 contingency table from Exercise 9.2, compute
   a. The proportion correct.
   b. The threat score.
   c. The Heidke skill score.
   d. The Peirce skill score.
   e. The Gilbert skill score.

9.4. For the event o_3 (3 to 4 in. of snow) in Table 9.10 find
   a. The threat score.
   b. The hit rate.
   c. The false alarm ratio.
   d. The bias ratio.

TABLE 9.10 A 4×4 Contingency Table for Snow Amount Forecasts in the Eastern Region of the United States During the Winters 1983/1984 Through 1988/1989

	o_1	o_2	o_3	o_4
y_1	35,915	477	80	28
y_2	280	162	51	17
y_3	50	48	34	10
y_4	28	23	185	34

The event o_1 is 0–1 in., o_2 is 2–3 in., o_3 is 3–4 in., and o_4 is ≥ 6 in.
From Goldsmith (1990).

9.5. Using the 4×4 contingency table in Table 9.10, compute
   a. The joint distribution of the forecasts and the observations.
   b. The proportion correct.
   c. The Heidke skill score.
   d. The Peirce skill score.

9.6. For the persistence forecasts for the January 1987 Ithaca maximum temperatures in Table A.1 (i.e., the forecast for 2 January is the observed temperature on 1 January, etc.), compute
   a. The MAE.
   b. The RMSE.

c. The ME (bias).
d. The skill, in terms of RMSE, with respect to the sample climatology.

9.7. Hypothetical forecasts and observations for three stations on three days are shown in Table 9.11.
   a. Compute the overall MSE skill score (Equation 9.37), correctly computing the skill with respect to the sample climatological means pertaining to the individual stations.
   b. Compute the overall MSE skill score, incorrectly based on the overall sample climatological mean.

TABLE 9.11 Hypothetical Nonprobabilistic Forecasts and Observations of Temperature on Three Days at Three Stations Having Different Climatological Values

	Station A		Station B		Station C	
	Forecast	Observed	Forecast	Observed	Forecast	Observed
Day 1	−2	2	3	7	8	11
Day 2	2	4	7	8	12	13
Day 3	9	6	13	11	18	16

9.8. Using the collection of hypothetical PoP forecasts summarized in Table 9.12,
   a. Calculate the Brier score.
   b. Calculate the Brier score for (the sample) climatological forecast.
   c. Calculate the skill of the forecasts with respect to the sample climatology.
   d. Draw the reliability diagram.
   e. Compute the average Ignorance score, assuming that $y_1 = 0.01$ and $y_{11} = 0.99$.

TABLE 9.12 Hypothetical Verification Data for 1000 Probability-of-Precipitation Forecasts

Forecast probability, y_i	0.00	0.10	0.20	0.30	0.40	0.50	0.60	0.70	0.80	0.90	1.00
Number of times forecast	293	237	162	98	64	36	39	26	21	14	10
Number of precipitation occurrences	9	21	34	31	25	18	23	18	17	12	9

9.9. For the hypothetical forecast data in Table 9.12,
   a. Compute the likelihood-base rate factorization of the joint distribution $p(y_i, o_j)$.
   b. Draw the discrimination diagram.
   c. Draw the ROC curve.
   d. Test whether the area under the ROC curve is significantly greater than 1/2.

9.10. Suppose you honestly forecast the probability of precipitation for tomorrow as 20%. What is your expected Brier Score for this forecast?

9.11. For the hypothetical probabilistic three-category precipitation amount forecasts in Table 9.13, where only five distinct forecast vectors have been used,
   a. Calculate the average RPS.

    b. Calculate the RPS skill of the forecasts with respect to the sample climatology.

    c. Calculate the average Ignorance score.

TABLE 9.13 Hypothetical Verification for 500 Probability Forecasts of Precipitation Amounts

Forecast Probabilities For			Number of Forecast Periods Verifying As		
<0.01 in.	0.01–0.24 in.	≥0.25 in.	<0.01 in.	0.01–0.24 in.	≥0.25 in.
.8	.1	.1	263	24	37
.5	.4	.1	42	37	12
.4	.4	.2	14	16	10
.2	.6	.2	4	13	6
.2	.3	.5	4	6	12

9.12. For the hypothetical five-member bivariate ensemble and corresponding forecast shown in Figure 9.24,

    a. Compute the preranks π_{MRH} for the observation and the five ensemble members.

    b. To which histogram bin will this observation contribute?

9.13. Suppose a particular forecaster's 75% central credible interval for the next day's maximum temperature is 27–32°F.

    a. If the verifying temperature is 34°F, evaluate the accuracy of this forecast using Winkler's score.

    b. Assuming that this forecaster's subjective distribution is Gaussian, what would that person's probability be for tomorrow's maximum temperature falling below 25°F?

9.14. Table 9.14 shows a set of 20 hypothetical ensemble forecasts, each with five members, and corresponding observations.

    a. Plot the verification rank histogram.

    b. Qualitatively diagnose the performance of this sample of forecast ensembles.

TABLE 9.14 A Set of 20 Hypothetical Ensemble Forecasts, of Ensemble Size 5, and Corresponding Observations

Case	Member 1	Member 2	Member 3	Member 4	Member 5	Observation
1	7.9	7.3	5.5	6.9	8.3	7.7
2	7.4	5.6	8.2	5.8	6.1	9.4
3	9.5	8.3	10.5	8.9	6.1	8.7
4	6.1	7.8	5.1	10.4	4.9	3.4
5	6.3	5.8	5.1	6.0	4.1	7.3
6	8.1	6.8	1.8	6.7	10.5	8.2
7	4.4	5.6	7.7	6.0	7.0	4.3
8	5.9	3.0	4.4	7.2	9.1	7.0
9	5.2	5.7	5.3	6.0	7.5	4.1

TABLE 9.14 A Set of 20 Hypothetical Ensemble Forecasts, of Ensemble Size 5, and Corresponding Observations—cont'd

Case	Member 1	Member 2	Member 3	Member 4	Member 5	Observation
10	2.7	6.6	5.8	7.5	5.1	8.3
11	6.6	5.2	5.3	5.5	3.2	4.7
12	6.7	6.0	8.6	7.7	4.8	8.7
13	8.9	1.3	5.9	7.3	6.3	8.5
14	8.5	5.0	4.6	7.6	1.4	4.8
15	9.2	4.4	8.9	5.3	6.5	9.5
16	2.7	8.7	3.4	7.6	5.1	4.3
17	4.1	7.0	7.5	7.2	7.0	5.4
18	7.7	4.7	5.7	5.7	6.8	2.1
19	6.7	7.4	6.2	5.3	5.8	3.3
20	4.4	3.3	1.9	5.4	6.6	7.4

9.15. For Case 1 in Table 9.14,
    a. Compute the eCRPS.
    b. Compute the Dawid-Sebastiani score.

9.16. For the hypothetical forecast and observed 500 mb fields in Figure 9.44,
    a. Calculate the S1 score, comparing the 24 pairs of gradients in the north-south and east-west directions.
    b. Calculate the MSE.
    c. Calculate the skill score for the MSE with respect to the climatological field.
    d. Calculate the centered AC.
    e. Calculate the uncentered AC.

9.17. Using the results from Exercise 9.1, construct the VS curve for the verification data in Table 9.2.

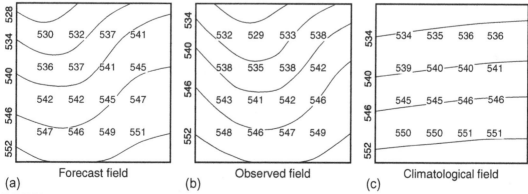

FIGURE 9.44 Hypothetical forecast (a), observed (b), and climatological average (c) fields of 500 mb heights (dam) over a small domain, and interpolations onto 16-point grids.

Chapter 10

Time Series

10.1. BACKGROUND

This chapter presents methods for characterizing and analyzing the time variations of data series. Often we encounter data sets consisting of consecutive realizations of atmospheric variables. When the ordering of the data in time is important to their information content, summarization and analysis using time series methods are appropriate.

 As has been illustrated earlier, atmospheric observations separated by relatively short times tend to be similar or correlated. Analyzing and characterizing the nature of these temporal correlations, or relationships through time, can be useful both for understanding the underlying physical processes and for forecasting future atmospheric events. Accounting for these correlations is also necessary if valid statistical inferences about time-series data are to be made (see Chapter 5).

10.1.1. Stationarity

Of course, we do not expect the future values of a data series to be identical to some past series of existing observations. However, in many instances it may be very reasonable to assume that their statistical properties will be the same. The idea that past and future values of a time series will be similar statistically is an informal expression of what is called *stationarity*. Usually, the term stationarity is understood to mean *weak stationarity* or *covariance stationarity*. In this sense, stationarity implies that the mean and auto-covariance function (Equation 3.39) of the data series do not change through time. Different time slices of a stationary data series (e.g., the data observed to date and the data to be observed in the future) can be regarded as having the same underlying mean, variance, and covariances. Furthermore, the correlations between variables in a stationary series are determined only by their separation in time (i.e., their lag, k, in Equation 3.37) and not their absolute positions in time. Qualitatively, different portions of a stationary time series look alike statistically, even though the individual data values may be very different. Covariance stationarity is a less restrictive assumption than *strict stationarity*, which implies that the full joint distribution of the variables in the series does not change through time. More technical expositions of the concept of stationarity can be found in, for example, Fuller (1996) or Kendall and Ord (1990).

 Most methods for analyzing time series assume stationarity of the data. However, many atmospheric processes are distinctly not stationary. Obvious examples of nonstationary atmospheric data series are those exhibiting annual or diurnal cycles. For example, temperatures typically exhibit very strong annual cycles in mid- and high-latitude climates, and we expect the average of the distribution of January temperature to be very different from that for July temperature. Similarly, time series of wind speeds often

Statistical Methods in the Atmospheric Sciences. https://doi.org/10.1016/B978-0-12-815823-4.00010-9

exhibit a diurnal cycle, which derives physically from the tendency for diurnal changes in static stability, imposing a diurnal cycle on the downward momentum transport.

There are two approaches to dealing with nonstationary series. Both aim to process the data in a way that will subsequently allow stationarity to be reasonably assumed. The first approach is to mathematically transform the nonstationary data to approximate stationarity. For example, subtracting a periodic mean function from data subject to an annual cycle would produce a transformed data series with constant (zero) mean. In order to produce a series with both constant mean and variance, it might be necessary to further transform these anomalies to standardized anomalies (Equation 3.27), that is, to divide the values in the anomaly series by standard deviations that also vary through an annual cycle. Not only do temperatures tend to be colder in winter, but the variability of temperature also tends to be higher. Data that become stationary after such cycles have been removed are said to exhibit *cyclostationarity*. A possible approach to transforming a monthly cyclostationary temperature series to (at least approximate) stationarity could be to compute the 12 monthly mean values and 12 monthly standard deviation values, and then to apply Equation 3.27 using the different means and standard deviations for the appropriate calendar month. This was the first step used to construct the time series of SOI values in Figure 3.16.

The alternative to data transformation is to stratify the data. That is, we can conduct separate analyses of subsets of the data record that are short enough to be regarded as nearly stationary. We might analyze daily observations for all available January records at a given location, assuming that each 31-day data record is a sample from the same physical process, but not necessarily assuming that process to be the same as for July, or even for February, data.

10.1.2. Time-Series Models

Characterization of the properties of a time series often is achieved by invoking a mathematical model for the data variations. Having obtained a time-series model for an observed data set, that model might then be viewed as a generating process, or algorithm, that could have produced the data. A mathematical model for the time variations of a data set can allow compact representation of the characteristics of those data in terms of a few parameters. This approach is entirely analogous to the fitting of parametric probability distributions, which constitute another kind of probability model, in Chapter 4. The distinction is that the distributions in Chapter 4 are used without regard to the ordering of the data, whereas the motivation for using time-series models is specifically to characterize the nature of the ordering. Time-series methods are thus appropriate when the ordering of the data values in time is important to a given application.

Regarding an observed time series as having been generated by a theoretical (i.e., model) process is convenient because it allows characteristics of future, and so still unobserved, values of a time series to be inferred from the inevitably limited data in hand. That is, characteristics of an observed time series are summarized by the parameters of a time-series model. Invoking the assumption of stationarity, future values of the time series should then also exhibit the statistical properties implied by the model, so that the properties of the model generating process can be used to infer characteristics of yet unobserved values of the series.

10.1.3. Time-Domain vs. Frequency-Domain Approaches

There are two fundamental approaches to time series analysis: *time-domain* analysis and *frequency-domain* analysis. Although these two approaches proceed very differently and may seem quite distinct, they are not mutually independent. Rather, they are complementary views that are linked mathematically.

Time-domain methods seek to characterize data series in the same terms in which they are observed and reported. The autocorrelation function (Section 3.5.5) is a primary tool for characterization of relationships among data values in the time-domain approach. Mathematically, time-domain analyses operate in the same space as the data values. Separate sections in this chapter describe different time-domain methods for use with discrete and continuous data. Here the terms discrete and continuous are used in the same sense as in Chapter 4: discrete random variables are allowed to take on only a finite (or possibly countably infinite) number of values, and continuous random variables may take on any of the infinitely many real values within their range.

Frequency-domain analyses represent data series in terms of contributions occurring at different timescales or characteristic frequencies. Most commonly, each timescale is represented by a pair of sine and cosine functions. In that case, the overall time series is regarded as having arisen from the combined effects of a collection of sine and cosine waves oscillating at different rates. The sum of these waves reproduces the original data, but it is often the relative strengths of the individual component waves that are of primary interest. Frequency-domain analyses take place in the mathematical space defined by this collection of sine and cosine waves. That is, such frequency-domain analyses involve transformation of the n original data values into coefficients that multiply an equal number of periodic (the sine and cosine) functions. At first exposure this process can seem very strange and possibly difficult to grasp. However, frequency-domain methods very commonly are applied to atmospheric time series, and important insights can be gained from frequency-domain analyses.

10.2. TIME DOMAIN—I. DISCRETE DATA

10.2.1. Markov Chains

Recall that a discrete random variable is one that can take on only values from among a defined, finite or countably infinite set. The most common class of model, or stochastic process, used to represent time series of discrete variables is known as the *Markov chain*. A Markov chain can be imagined as being based on collection of "states" of a model system. Each state corresponds to one of the elements of the MECE partition of the sample space describing the random variable in question.

For each time period, the length of which is equal to the time separation between observations in the time series, the Markov chain can either remain in the same state or change to one of the other states. Remaining in the same state corresponds to two successive observations of the same value of the discrete random variable in the time series, and a change of state implies two successive values of the time series that are different.

The behavior of a Markov chain is governed by a set of probabilities for these transitions, called the *transition probabilities*. The transition probabilities specify conditional probabilities for the system being in each of its possible states during the next time period. The simplest form is called a first-order Markov chain, for which the transition probabilities controlling the next state of the system depend only on the current state of the system. That is, knowing the current state of the system and the full sequence of states leading up to the current state, provides no more information about the probability distribution for the states at the next observation time than does knowledge of the current state alone. This characteristic of first-order Markov chains is known as the *Markovian property*, which can be expressed more formally as

$$Pr\{X_{t+1}|X_t, X_{t-1}, X_{t-2}, ..., X_1\} = Pr\{X_{t+1}|X_t\}. \tag{10.1}$$

The probabilities of future states depend on the present state, but they do not depend on the particular way that the model system arrived at the present state. In terms of a time series of observed data the Markovian property means, for example, that forecasts of tomorrow's data value can be made on the basis of today's observation, and that also knowing yesterday's data value provides no additional information.

The transition probabilities of a Markov chain are conditional probabilities. That is, there is a conditional probability distribution pertaining to each possible current state, and each of these distributions specifies probabilities for the states of the system in the next time period. To say that these probability distributions are conditional allows for the possibility that the transition probabilities can be different, depending on the current state. The fact that these distributions can be different is the essence of the capacity of a Markov chain to represent the serial correlation, or persistence, often exhibited by atmospheric variables. If probabilities for future states are the same, regardless of the current state, then the time series consists of independent values. In that case the probability of occurrence of any given state in the upcoming time period would not be affected by the occurrence or nonoccurrence of a particular state in the current time period. If the time series being modeled exhibits persistence, the probability of the system staying in a given state will be higher than the probabilities of arriving at that state from other states, and higher than the corresponding unconditional probability.

If the transition probabilities of a Markov chain do not change through time and none of them are zero, then the resulting time series will be stationary. Modeling nonstationary data series exhibiting, for example, an annual cycle can require allowing the transition probabilities to vary through an annual cycle as well. One way to capture this kind of nonstationarity is to specify that the probabilities vary according to some smooth periodic curve, such as a cosine function, as described in Section 10.4. Alternatively, separate transition probabilities can be used for nearly stationary portions of the cycle, for example, each of the 12 calendar months.

Certain classes of Markov chains are described more concretely, but relatively informally, in the following sections. More formal and comprehensive treatments can be found in, for example, Feller (1970), Karlin and Taylor (1975), or Katz (1985).

10.2.2. Two-State, First-Order Markov Chains

The simplest kind of discrete random variable pertains to dichotomous (yes/no) events. The behavior of a stationary sequence of independent (exhibiting no serial correlation) values of a dichotomous discrete random variable is described by the binomial distribution (Equation 4.1). That is, for serially independent events, the ordering in time is of no importance from the perspective of specifying probabilities for future events, so that a time-series model for their behavior does not provide more information than does the binomial distribution.

A two-state Markov chain is a statistical model for the persistence of binary events. The occurrence or nonoccurrence of rain on a given day is a simple meteorological example of a binary random event, and a sequence of daily observations of "rain" and "no rain" for a particular location would constitute a time series of that variable. Consider a series where the random variable takes on the values $x_t = 1$ if precipitation occurs on day t and $x_t = 0$ if it does not. For the January 1987 Ithaca precipitation data in Table A.1, this time series would consist of the values presented in Table 10.1. That is, $x_1 = 0$, $x_2 = 1$, $x_3 = 1$, $x_4 = 0$, ... , and $x_{31} = 1$. It is evident from looking at this series of numbers that the 1's and 0's

TABLE 10.1 Time Series of a Dichotomous Random Variable Derived from the January 1987 Ithaca Precipitation Data in Table A.1

Date, t	1 2 3 4 5 6 7 8 9 10 11 12 13 14 15 16 17 18 19 20 21 22 23 24 25 26 27 28 29 30 31
x_t	0 1 1 0 0 0 0 1 1 1 1 1 1 1 1 1 0 0 0 0 1 0 0 1 0 0 0 0 1 1 1

Days on which nonzero precipitation was reported yield $x_t = 1$, and days with zero precipitation yield $x_t = 0$.

tend to cluster in time. As was illustrated in Example 2.2, this clustering is an expression of the serial correlation present in the time series, which in turn reflects the fact that the daily measurement interval is shorter than the characteristic timescale of the underlying physical data-generating process. That is, the probability of a 1 following a 1 is apparently higher than the probability of a 1 following a 0, and the probability of a 0 following a 0 is apparently higher than the probability of a 0 following a 1.

A first-order, two-state Markov chain is a common and often quite good stochastic model for data of this kind. A two-state Markov chain is natural for dichotomous data since each of the two states will pertain to one of the two possible data values. A first-order Markov chain has the property that the transition probabilities governing each observation in the time series depend only on the value of the previous member of the time series.

Figure 10.1 illustrates schematically the nature of a two-state, first-order Markov chain. In order to help fix ideas, the two states are labeled in a manner consistent with the data in Table 10.1. For each value of the time series, the stochastic process is either in state 0 (no precipitation occurs and $x_t = 0$) or in state 1 (precipitation occurs and $x_t = 1$). At each time step the process can either stay in the same state or switch to the other state. Therefore four distinct transitions are possible, corresponding to a dry day following a dry day (p_{00}), a wet day following a dry day (p_{01}), a dry day following a wet day (p_{10}), and a wet day following a wet day (p_{11}). Each of these four transitions is represented in Figure 10.1 by arrows, labeled with the appropriate transition probabilities. Here the notation is such that the first subscript on the probability is the state at time t, and the second subscript is the state at time $t+1$.

The transition probabilities are conditional probabilities for the state at time $t+1$ (e.g., whether precipitation will occur tomorrow) given the state at time t (e.g., whether or not precipitation occurred today). That is,

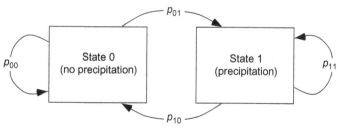

FIGURE 10.1   Schematic representation of a two-state, first-order Markov chain, illustrated in terms of daily precipitation occurrence or nonoccurrence. The two states are labeled "0" for no precipitation, and "1" for precipitation occurrence. For a first-order Markov chain, there are four transition probabilities controlling the state of the system in the next time period. Since these four probabilities are pairs of conditional probabilities, $p_{00} + p_{01} = 1$ and $p_{10} + p_{11} = 1$. For quantities like day-to-day precipitation occurrence that exhibit positive serial correlation, $p_{01} < p_{00}$, and $p_{01} < p_{11}$.

$$p_{00} = \text{Pr}\{X_{t+1} = 0 | X_t = 0\} \tag{10.2a}$$

$$p_{01} = \text{Pr}\{X_{t+1} = 1 | X_t = 0\} \tag{10.2b}$$

$$p_{10} = \text{Pr}\{X_{t+1} = 0 | X_t = 1\} \tag{10.2c}$$

and

$$p_{11} = \text{Pr}\{X_{t+1} = 1 | X_t = 1\}. \tag{10.2d}$$

Together, Equations 10.2a and 10.2b define the conditional probability distribution for the value of the time series at time $t+1$, given that $X_t = 0$ at time t. Similarly, Equations 10.2c and 10.2d express the conditional probability distribution for the next value of the time series given that the current value is $X_t = 1$.

Notice that the four probabilities in Equation 10.2 provide some redundant information. Given that the Markov chain is in one state or the other at time t, the sample space for X_{t+1} consists of only two MECE events. Therefore $p_{00} + p_{01} = 1$ and $p_{10} + p_{11} = 1$, so that it is really only necessary to focus on one of each of the pairs of transition probabilities, say p_{01} and p_{11}. In particular, it is sufficient to estimate only two parameters for a two-state first-order Markov chain, since each of the two pairs of conditional probabilities must sum to 1. The parameter estimation procedure usually consists simply of computing the conditional relative frequencies, which yield the maximum likelihood estimators (MLEs)

$$\hat{p}_{01} = \frac{\text{\# of 1's following 0's}}{\text{Total \# of 0's}} = \frac{n_{01}}{n_{0\bullet}} = \frac{n_{01}}{n_{00} + n_{01}} \tag{10.3a}$$

and

$$\hat{p}_{11} = \frac{\text{\# of 1's following 1's}}{\text{Total \# of 1's}} = \frac{n_{11}}{n_{1\bullet}} = \frac{n_{11}}{n_{10} + n_{11}}. \tag{10.3b}$$

Here n_{01} is the number of transitions from State 0 to State 1, n_{11} is the number of pairs of time steps in which there are two consecutive 1's in the series, $n_{0\bullet}$ is the number of 0's in the series followed by another data point, and $n_{1\bullet}$ is the number of 1's in the series followed by another data point. That is, the subscript "$\bullet$" indicates the total over all values of the index replaced by this symbol, so that $n_{1\bullet} = n_{10} + n_{11}$ and $n_{0\bullet} = n_{00} + n_{01}$. Equations 10.3 state that the parameter p_{01} is estimated by looking at the relative frequency of the event $X_{t+1} = 1$ considering only those points in the time series following data values for which $X_t = 0$. Similarly, p_{11} is estimated as the fraction of points for which $X_t = 1$ that are followed by points with $X_{t+1} = 1$. These somewhat labored definitions of $n_{0\bullet}$ and $n_{1\bullet}$ are necessary to account for the edge effects in a finite sample. The final point in the time series is not counted in the denominator of Equation 10.3a or 10.3b, whichever is appropriate, because there is no available data value following it to be incorporated into the counts in one of the numerators. These definitions also cover cases of missing values, and stratified samples such as 30 years of January data, for example.

Equation 10.3 suggests that parameter estimation for a two-state first-order Markov chain is equivalent to fitting two Bernoulli distributions (i.e., binomial distributions with $N = 1$). One of these binomial distributions pertains to points in the time series preceded by 0's, and the other describes the behavior of points in the time series preceded by 1's. Knowing that the process is currently in state 0 (e.g., no precipitation today), the probability distribution for the event $X_{t+1} = 1$ (precipitation tomorrow) is simply binomial (Equation 4.1) with $p = p_{01}$. The second binomial parameter is $N = 1$, because there is only one data point in the series for each time step. Similarly, if $X_t = 1$, then the distribution for the event $X_{t+1} = 1$ is binomial with $N = 1$ and $p = p_{11}$. The conditional dichotomous events of a stationary Markov chain satisfy the requirements listed in Section 4.2.1 for the binomial distribution. For a stationary process the probabilities do not change through time, and conditioning on the current value of the time series satisfies the independence assumption for the binomial distribution because of the Markovian property. It is the fitting of two Bernoulli distributions that allows the time dependence in the data series to be represented.

Certain properties are implied for a time series described by a Markov chain. These properties are controlled by the values of the transition probabilities and can be computed from them. First, the long-run relative frequencies of the events corresponding to the two states of the Markov chain are called the *stationary probabilities*. For a Markov chain describing the daily occurrence or nonoccurrence of precipitation, the stationary probability for precipitation, π_1, corresponds to the (unconditional) climatological probability of precipitation. In terms of the transition probabilities p_{01} and p_{11},

$$\pi_1 = \frac{p_{01}}{1 + p_{01} - p_{11}}, \tag{10.4}$$

with the stationary probability for state 0 being simply $\pi_0 = 1 - \pi_1$. The usual situation of positive serial correlation or persistence produces $p_{01} < \pi_1 < p_{11}$. Applied to daily precipitation occurrence, this relationship means that the conditional probability of a wet day following a dry day is less than the overall climatological relative frequency, which in turn is less than the conditional probability of a wet day following a wet day.

The transition probabilities also imply a specific degree of serial correlation, or persistence, for the binary time series. In terms of the transition probabilities, the lag-1 autocorrelation (Equation 3.36) of the binary time series is simply

$$r_1 = p_{11} - p_{01}. \tag{10.5}$$

In the context of Markov chains, r_1 is sometimes known as the *persistence parameter*. As the correlation r_1 increases, the difference between p_{11} and p_{01} widens, so that state 1 is more and more likely to follow state 1, and less and less likely to follow state 0. That is, there is an increasing tendency for 0's and 1's to cluster in time, or occur in runs. A time series exhibiting no autocorrelation would be characterized by $r_1 = p_{11} - p_{01} = 0$, and $p_{11} = p_{01} = \pi_1$. In that case the two conditional probability distributions specified by Equation 10.2 are the same, and the time series is simply a string of independent Bernoulli realizations. The Bernoulli distribution can be viewed as defining a two-state, zero-order Markov chain.

Once the state of a Markov chain has changed, the number of time periods it will remain in the new state is a random variable, with a probability distribution function. Because the conditional independence implies conditional Bernoulli distributions, this probability distribution function for numbers of consecutive time periods in the same state, or the "spell length," will be the geometric distribution (Equation 4.5), with $p = p_{01}$ for sequences of 0's (dry spells), and $p = p_{10} = 1 - p_{11}$ for sequences of 1's (wet spells).

The full autocorrelation function, Equation 3.37, for the first-order Markov chain follows easily from the lag-1 autocorrelation r_1. Because of the Markovian property, the autocorrelation between members of the time series separated by k time steps is simply the lag-1 autocorrelation multiplied by itself k times,

$$r_k = (r_1)^k. \tag{10.6}$$

A common misconception is that the Markovian property implies independence of values in a first-order Markov chain that are separated by more than one time period. Equation 10.6 shows that the correlation, and hence the statistical dependence, among elements of the time series tails off at increasing lags, but it is never exactly zero unless $r_1 = 0$. Rather, the Markovian property implies conditional independence of data values separated by more than one time period, as expressed by Equation 10.1. Given a particular value for x_t, the different possible values for x_{t-1}, x_{t-2}, x_{t-3}, and so on, do not affect the probabilities for

x_{t+1}. However, for example, $\Pr\{x_{t+1} = 1 \mid x_{t-1} = 1\} \neq \Pr\{x_{t+1} = 1 \mid x_{t-1} = 0\}$, indicating statistical dependence among members of a Markov chain separated by more than one time period. Put another way, it is not that the Markov chain has no memory of the past, but rather that it is only the recent past that matters.

10.2.3. Test for Independence vs. First-Order Serial Dependence

Even if a series of binary data is generated by a mechanism producing serially independent values, the sample lag-one autocorrelation (Equation 3.36) computed from a finite sample is unlikely to be exactly zero. A formal test, similar to the χ^2 goodness-of-fit test (Equation 5.14), can be computed to investigate the statistical significance of the sample autocorrelation for a binary data series. The null hypothesis for this test is that the data series is serially independent (i.e., the data are independent Bernoulli variables), with the alternative being that the series was generated by a first-order Markov chain.

The test is based on a contingency table of the observed transition counts n_{00}, n_{01}, n_{10}, and n_{11}, in relation to the numbers of transitions expected under the null hypothesis. The corresponding expected counts, e_{00}, e_{01}, e_{10}, and e_{11}, are computed from the observed transition counts under the constraint that the marginal totals of the expected counts are the same as for the observed transitions. The comparison is illustrated in Figure 10.2, which shows generic contingency tables for the observed transition counts (a) and those expected under the null hypothesis of independence (b). For example, the transition count n_{00} specifies the number of consecutive pairs of 0's in the time series. This is related to the joint probability $\Pr\{X_t = 0 \cap X_{t+1} = 0\}$. Under the null hypothesis of independence this joint probability is simply the product of the two event probabilities, or in relative frequency terms, $\Pr\{X_t = 0\} \Pr\{X_{t+1} = 0\} = (n_{0\bullet}/n)(n_{\bullet 0}/n)$. Thus the corresponding number of expected transition counts is simply this product multiplied by the sample size, or $e_{00} = (n_{0\bullet})(n_{\bullet 0})/n$. More generally,

$$e_{ij} = \frac{n_{i\bullet}\, n_{\bullet j}}{n}. \tag{10.7}$$

The test statistic is computed from the observed and expected transition counts using

$$\chi^2 = \sum_i \sum_j \frac{(n_{ij} - e_{ij})^2}{e_{ij}}, \tag{10.8}$$

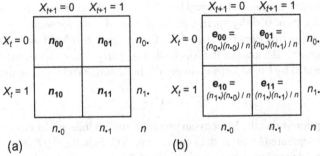

(a)   (b)

FIGURE 10.2   Contingency tables of observed transition counts n_{ij} (a) for a binary time series, and (b) transition counts e_{ij} expected if the time series actually consists of serially independent values with the same marginal totals. The transition counts are shown in boldface, and the marginal totals are in plain type.

where, for the 2×2 contingency table appropriate for dichotomous data, the summations are for $i = 0$ to 1 and $j = 0$ to 1. That is, there is a separate term in Equation 10.8 for each of the four pairs of contingency table cells in Figure 10.2. Note that Equation 10.8 is analogous to Equation 5.14, with the n_{ij} being the observed counts, and the e_{ij} being the expected counts. Under the null hypothesis, the test statistic follows the χ^2 distribution with $v = 1$ degree of freedom. This value of the degrees-of-freedom parameter is appropriate because, given that the marginal totals are fixed, arbitrarily specifying one of the transition counts completely determines the other three.

The fact that the numerator in Equation 10.8 is squared implies that values of the test statistic on the left tail of the null distribution are favorable to H_0, because small values of the test statistic are produced by pairs of observed and expected transition counts of similar magnitudes. Therefore the test is one-tailed. The p value associated with a particular test can be assessed using the χ^2 quantiles in Table B.3.

Example 10.1. Fitting a Two-State, First Order Markov Chain

Consider summarizing the time series in Table 10.1, derived from the January 1987 Ithaca precipitation series in Table A1, using a first-order Markov chain. The parameter estimates in Equation 10.3 are obtained easily from the transition counts. For example, the number of 1's following 0's in the time series of Table 10.1 is $n_{01} = 5$. Similarly, $n_{00} = 11$, $n_{10} = 4$, and $n_{11} = 10$. The resulting sample estimates for the transition probabilities (Equations 10.3) are $p_{01} = 5/16 = 0.312$, and $p_{11} = 10/14 = 0.714$. Note that these are identical to the conditional probabilities computed in Example 2.2.

Whether fitting a first-order Markov chain to the data in Table 10.1 is justified can be investigated using the χ^2 test in Equation 10.8. Here the null hypothesis is that these data resulted from an independent (i.e., Bernoulli) process, and the expected transition counts e_{ij} that must be computed are those consistent with this null hypothesis. These are obtained from the marginal totals $n_{0\bullet} = 11 + 5 = 16$, $n_{1\bullet} = 4 + 10 = 14$, $n_{\bullet 0} = 11 + 4 = 15$, and $n_{\bullet 1} = 5 + 10 = 15$. The expected transition counts follow easily as $e_{00} = (16)(15)/30 = 8$, $e_{01} = (16)(15)/30 = 8$, $e_{10} = (14)(15)/30 = 7$, and $e_{11} = (14)(15)/30 = 7$. Note that usually the expected transition counts will be different from each other and need not be integer values.

Computing the test statistic in Equation 10.8, we find $\chi^2 = (11-8)^2/8 + (5-8)^2/8 + (4-7)^2/7 + (10-7)^2/7 = 4.82$. The degree of unusualness of this result with reference to the null hypothesis can be assessed with the aid of Table B.3. Looking on the $v = 1$ row, the result lies between the 95th and 99th percentiles of the appropriate χ^2 distribution. Thus even for this rather small sample size, the null hypothesis of serial independence would be rejected at the 5% level, although not at the 1% level.

The degree of serial correlation exhibited by this data sample can be summarized using the persistence parameter, which is also the lag-one autocorrelation, $r_1 = p_{11} - p_{01} = 0.714 - 0.312 = 0.402$. This value could also be obtained by operating on the series of 0's and 1's in Table 10.1, using Equation 3.36. This lag-1 autocorrelation is fairly large, indicating substantial serial correlation in the time series. Assuming first-order Markov dependence, it also implies the full autocorrelation function, through Equation 10.6. Figure 10.3 shows that the implied theoretical correlation function for this Markov process, indicated by the dashed line, agrees very closely with the sample autocorrelation function shown by the solid line, for the first few lags. This agreement provides qualitative support for the first-order Markov chain as an appropriate model for the data series.

Finally, the stationary (i.e., climatological) probability for precipitation implied for these data by the Markov chain model is, using Equation 10.4, $\pi_1 = 0.312/(1 + 0.312 - 0.714) = 0.522$. This value agrees closely with the relative frequency $16/30 = 0.533$, obtained by counting the number of 1's in the last 30 values of the series in Table 10.1. ◇

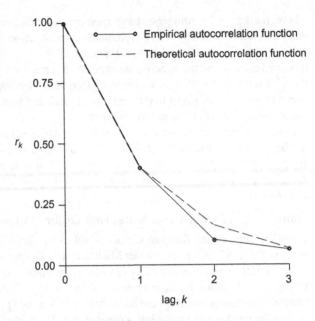

FIGURE 10.3 Sample autocorrelation function for the January 1987 Ithaca binary precipitation occurrence series, Table 10.1 (solid, with circles), and theoretical autocorrelation function (dashed) specified by the fitted first-order Markov chain model (Equation 10.6). The correlations are 1.00 for $k = 0$, since the unlagged data are perfectly correlated with themselves.

10.2.4. Some Applications of Two-State Markov Chains

One interesting application of the Markov chain model is in the computer generation of synthetic rainfall series. Time series of random binary numbers, statistically resembling real rainfall occurrence data, can be produced using a Markov chain as the generating algorithm. This procedure is an extension of the ideas presented in Section 4.7, to time-series data. To generate sequences of numbers statistically resembling those in Table 10.1, for example, the parameters $p_{01} = 0.312$ and $p_{11} = 0.714$, estimated in Example 10.1, would be used together with a uniform [0, 1] random number generator (see Section 4.7.1). The synthetic time series would begin using the stationary probability $\pi_1 = 0.522$. If the first uniform number generated were less than π_1, then $x_1 = 1$, meaning that the first simulated day would be wet. For subsequent values in the series, corresponding to day $t+1$, each new uniform random number would be compared to the appropriate transition probability, depending on whether the most recently generated number, corresponding to day t, was wet or dry. That is, the transition probability p_{01} would be used to generate x_{t+1} if $x_t = 0$, and p_{11} would be used if $x_t = 1$. A wet day ($x_{t+1} = 1$) is simulated if the next uniform random number is less than the transition probability, and a dry day ($x_{t+1} = 0$) is generated if it is not. Since typically $p_{11} > p_{01}$ for daily precipitation occurrence data, simulated wet days are more likely to follow wet days than dry days, as is the case in the real data series.

The Markov chain approach for simulating precipitation occurrence can be extended to include simulation of daily precipitation amounts. This is accomplished by adopting a statistical model for the nonzero rainfall amounts, yielding a sequence of random variables defined on the Markov chain, called a *chain-dependent process* (Katz, 1977; Todorovic and Woolhiser, 1975). Often a gamma distribution (Section 4.4.5) is fit to the precipitation amounts on wet days in the data record (e.g., Katz, 1977; Richardson, 1981; Stern and Coe, 1984), although the mixed exponential distribution (Equation 4.78) often provides a better fit to nonzero daily precipitation data (e.g., Foufoula-Georgiou and

Lettenmaier, 1987; Wilks, 1999a; Woolhiser and Roldan, 1982). Computer algorithms are available to generate random variables drawn from gamma distributions (e.g., Bratley et al., 1987; Johnson, 1987), or together Example 4.15 and Section 4.7.5 can be used to simulate from the mixed exponential distribution, to produce synthetic precipitation amounts on days when the Markov chain calls for a wet day. The tacit assumption that precipitation amounts on consecutive wet days are independent has turned out to be a reasonable approximation in most instances where it has been investigated (e.g., Katz, 1977; Stern and Coe, 1984), but may not adequately simulate extreme multiday precipitation events that could arise, for example, from a slow-moving landfalling hurricane (Wilks, 2002a). Generally both the Markov chain transition probabilities and the parameters of the distributions describing precipitation amounts change through the year. These seasonal cycles can be handled by fitting separate sets of parameters for each of the 12 calendar months (e.g., Wilks, 1989), or by representing them using smoothly varying sine and cosine functions (Stern and Coe, 1984).

Properties of longer term precipitation quantities resulting from simulated daily series (e.g., the monthly frequency distributions of numbers of wet days in a month, or of total monthly precipitation) can be calculated from the parameters of the chain-dependent process that governs the daily precipitation series. Since observed monthly precipitation statistics are computed from individual daily values, it should not be surprising that the statistical characteristics of monthly precipitation quantities will depend directly on the statistical characteristics of daily precipitation occurrences and amounts. Katz (1977, 1985) gives equations specifying some of these relationships, which can be used in a variety of ways (e.g., Katz and Parlange, 1993; Wilks, 1992, 1999b; Wilks and Wilby, 1999).

Finally, another interesting perspective on the Markov chain model for daily precipitation occurrence is in relation to forecasting precipitation probabilities. Recall that forecast skill is assessed relative to a set of benchmark, or reference forecasts (Equation 9.4). Usually one of two reference forecasts is used: either the climatological probability of the forecast event, in this case π_1; or persistence forecasts specifying unit probability if precipitation occurred in the previous period, or zero probability if the event did not occur. Neither of these reference forecasting systems is particularly sophisticated, and both are relatively easy to improve upon, at least for short-range forecasts. A more challenging, yet still fairly simple alternative is to use the transition probabilities of a two-state Markov chain as the reference forecasts. If precipitation did not occur in the preceding period, the reference forecast would be p_{01}, and the conditional reference forecast probability for precipitation following a day with precipitation would be p_{11}. Note that for quantities exhibiting persistence, $0 < p_{01} < \pi_1 < p_{11} < 1$, so that reference forecasts consisting of Markov chain transition probabilities constitute a compromise between the persistence (either 0 or 1) and climatological (π_1) probabilities. Furthermore, the balance of this compromise depends on the strength of the persistence exhibited by the climatological data on which the estimated transition probabilities are based. A weakly persistent quantity would be characterized by transition probabilities differing little from π_1, whereas strong serial correlation will produce transition probabilities much closer to 0 and 1.

10.2.5. Multiple-State Markov Chains

Markov chains are also useful for representing the time correlation of discrete variables that can take on more than two values. For example, a three-state, first-order Markov chain is illustrated schematically in Figure 10.4. Here the three states are arbitrarily labeled 1, 2, and 3. At each time t, the random variable in the series can take on one of the three values $x_t = 1$, $x_t = 2$, or $x_t = 3$, and each of these values corresponds

FIGURE 10.4 Schematic illustration of a three-state, first-order Markov chain. There are nine possible transitions among the three states, including the possibility that two consecutive points in the time series will be in the same state. First-order time dependence implies that the transition probabilities depend only on the current state of the system, or present value of the time series.

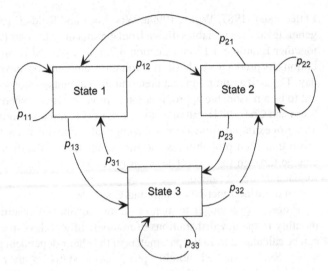

to a different state. First-order time dependence implies that the transition probabilities for x_{t+1} depend only on the state x_t, so that there are $3^2 = $ nine transition probabilities, p_{ij}. In general, for a first-order, s-state Markov chain, there are s^2 transition probabilities.

As is also the case for the two-state Markov chain, the transition probabilities for multiple-state Markov chains are conditional probabilities. For example, the transition probability p_{12} in Figure 10.4 is the conditional probability that state 2 will occur at time $t+1$, given that state 1 occurred at time t. Therefore in an s-state Markov chain it must be the case that the probabilities for the s transitions emanating from each state must sum to one, or $\Sigma_j p_{ij} = 1$ for each value of i.

Estimation of the transition probabilities for multiple-state Markov chains is a straightforward generalization of the formulas in Equations 10.3 for two-state chains. Each of these estimates is simply obtained from the conditional relative frequencies of the transition counts,

$$\hat{p}_{ij} = \frac{n_{ij}}{n_{i\bullet}}, \quad i,j=1,\ldots,s. \tag{10.9}$$

As before, the dot indicates summation over all values of the replaced subscript so that, for example, $n_{1\bullet} = \Sigma_j n_{1j}$. For the $s = 3$-state Markov chain represented in Figure 10.4, for example, $\hat{p}_{12} = n_{12}/(n_{11}+n_{12}+n_{13})$. In general, a contingency table of transition counts, corresponding to Figure 10.2a for the $s = 2$-state case, will contain s^2 entries.

Testing whether the observed degree of serial correlation is significantly different from zero in a multiple-state situation can be done using the χ^2 test in Equation 10.8. Here the summations are over all s possible states and will include s^2 terms. As before, the expected numbers of transition counts e_{ij} are computed using Equation 10.7. Under the null hypothesis of no serial correlation, the distribution of the test statistic in Equation 10.8 is χ^2 with $\nu = (s-1)^2$ degrees of freedom.

Three-state Markov chains have been used to characterize transitions between below-normal, near-normal, and above-normal months, as defined by the U.S. Climate Prediction Center (see Example 4.9), by Preisendorfer and Mobley (1984) and Wilks (1989). Mo and Ghil (1987) used a five-state Markov chain to characterize transitions between persistent hemispheric 500-mb flow types.

10.2.6. Higher-Order Markov Chains

First-order Markov chains often provide good representations of daily precipitation occurrences, but it is not obvious just from inspection of the series in Table 10.1, for example, that this simple model will be adequate to capture the observed correlation structure. More generally, an mth order Markov chain is one where the transition probabilities depend on the states in the previous m time periods. Formally, the extension of the Markovian property expressed in Equation 10.1 to the mth order Markov chain is

$$\Pr\{X_{t+1}|X_t, X_{t-1}, X_{t-2}, ..., X_1\} = \Pr\{X_{t+1}|X_t, X_{t-1}, ..., X_{t-m}\}. \tag{10.10}$$

Consider, for example, a second-order Markov chain. Second-order time dependence means that the transition probabilities depend on the states (values of the time series) at lags of both one and two time periods. Notationally, then, the transition probabilities for a second-order Markov chain require three subscripts: the first denotes the state at time $t-1$, the second denotes the state in time t, and the third specifies the state at (the future) time $t+1$. The notation for the transition probabilities of a second-order Markov chain can be defined as

$$p_{hij} = \{X_{t+1} = j | X_t = i, X_{t-1} = h\}. \tag{10.11}$$

In general, the notation for an mth order Markov chain requires $m+1$ subscripts on the transition counts and transition probabilities. If Equation 10.11 is being applied to a binary time series such as that in Table 10.1, the model would be a two-state, second-order Markov chain, and the indices h, i, and j could take on either of the $s = 2$ values of the time series, say 0 and 1. However, Equation 10.11 is equally applicable to discrete time series with larger numbers ($s > 2$) of states.

As is the case for first-order Markov chains, transition probability estimates are obtained from relative frequencies of observed transition counts. However, since data values further back in time now need to be considered, the number of possible transitions increases exponentially with the order, m, of the Markov chain. In particular, for an s-state, mth order Markov chain, there are $s^{(m+1)}$ distinct transition counts and transition probabilities. The arrangement of the resulting transition counts, in the form of Figure 10.2a, is presented in Table 10.2 for an $s = 2$ state, $m =$ second-order Markov chain. The transition counts are determined from the observed data series by examining consecutive groups of $m+1$ data points. For example, the first three data points in Table 10.1 are $x_{t-1} = 0$, $x_t = 1$, $x_{t+1} = 1$, and this triplet would contribute one to the transition count n_{011}. Overall the data series in Table 10.1 exhibits three transitions of this kind, so the final transition count $n_{011} = 3$ for this data set. The second triplet in

TABLE 10.2 Arrangement of the $2^{(2+1)} = 8$ Transition Counts for a Two-State, Second-Order Markov Chain in a Table of the Form of Figure 10.2a

X_{t-1}	X_t	$X_{t+1} = 0$	$X_{t+1} = 1$	Marginal Totals
0	0	n_{000}	n_{001}	$n_{00\cdot} = n_{000} + n_{001}$
0	1	n_{010}	n_{011}	$n_{01\cdot} = n_{010} + n_{011}$
1	0	n_{100}	n_{101}	$n_{10\cdot} = n_{100} + n_{101}$
1	1	n_{110}	n_{111}	$n_{11\cdot} = n_{110} + n_{111}$

Determining these counts from an observed time series requires examination of successive triplets of data values.

the data set in Table 10.1 would contribute one count to n_{110}. There is only one other triplet in this data for which $x_{t-1} = 1$, $x_t = 1$, $x_{t+1} = 0$, yielding the final count for $n_{110} = 2$.

The transition probabilities for a second-order Markov chain are obtained from the conditional relative frequencies of the transition counts

$$\hat{p}_{hij} = \frac{n_{hij}}{n_{hi\bullet}}. \tag{10.12}$$

That is, given the value of the time series at time $t-1$ was $x_{t-1} = h$ and the value of the time series at time t was $x_t = i$, the probability that the next value of the time series $x_{t+1} = j$ is p_{hij}, and the sample estimate of this probability is given in Equation 10.12. Just as the two-state first-order Markov chain consists essentially of two conditional Bernoulli distributions, a two-state second-order Markov chain amounts to four conditional Bernoulli distributions, with parameters $p = p_{hi1}$, for each of the four distinct combinations of the indices h and i.

Note that the small data set in Table 10.1 is really too short to fit a second-order Markov chain. Since there are no consecutive triplets in this series for which $x_{t-1} = 1$, $x_t = 0$, and $x_{t+1} = 1$ (i.e., a single dry day following and followed by a wet day) the transition count $n_{101} = 0$. This zero transition count would lead to the sample estimate for the transition probability $\hat{p}_{101} = 0$, even though there is no physical reason why that particular sequence of wet and dry days could not or should not occur.

10.2.7. Deciding Among Alternative Orders of Markov Chains

How are we to know what order m for a Markov chain is appropriate to represent a particular data series? One approach is to use a hypothesis test. For example, the χ^2 test in Equation 10.8 can be used to assess the plausibility of a first-order Markov chain model versus a null zero-order, or binomial model. The mathematical structure of this test can be modified to investigate the suitability of, say, a first-order versus a second-order, or a second-order versus a third-order Markov chain, but the overall statistical significance of a collection of such tests would be difficult to evaluate. This difficulty arises in part because of the issue of test multiplicity. As discussed in Section 5.4, the level of the strongest of a collection of simultaneous, correlated tests is difficult if not impossible to evaluate.

Two criteria are in common use for choosing among alternative orders of Markov chain models. These are the *Akaike Information Criterion* (AIC) (Akaike, 1974; Tong, 1975) and the *Bayesian Information Criterion* (BIC) (Schwarz, 1978; Katz, 1981). Both are based on the log-likelihood functions. In the present context these log-likelihoods depend on the transition counts and the estimated transition probabilities. The log-likelihoods for s-state Markov chains of order 0, 1, 2, and 3 are

$$L_0 = \sum_{j=0}^{s-1} n_j \ln \left(\hat{p}_j \right) \tag{10.13a}$$

$$L_1 = \sum_{i=0}^{s-1} \sum_{j=0}^{s-1} n_{ij} \ln \left(\hat{p}_{ij} \right) \tag{10.13b}$$

$$L_2 = \sum_{h=0}^{s-1} \sum_{i=0}^{s-1} \sum_{j=0}^{s-1} n_{hij} \ln \left(\hat{p}_{hij} \right) \tag{10.13c}$$

and

$$L_3 = \sum_{g=0}^{s-1}\sum_{h=0}^{s-1}\sum_{i=0}^{s-1}\sum_{j=0}^{s-1} n_{ghij} \ln\left(\hat{p}_{ghij}\right), \tag{10.13d}$$

with obvious extension for fourth and higher-order Markov chains. Here the summations are over all s states of the Markov chain, and so will include only two terms each for two-state (binary) time series. Equation 10.13a is simply the log-likelihood for the independent binomial model.

Example 10.2. Likelihood Ratio Test for the Order of a Markov Chain

To illustrate the application of Equations 10.13, consider a likelihood ratio test of first-order dependence of the binary time series in Table 10.1 versus the null hypothesis of zero serial correlation. The test involves computation of the log-likelihoods in Equations 10.13a and 10.13b. The resulting two log-likelihoods are compared using the test statistic given by Equation 5.20.

In the last 30 data points in Table 10.1, there are $n_0 = 15$ 0's and $n_1 = 15$ 1's, yielding the unconditional relative frequencies of rain and no rain $\hat{p}_0 = 15/30 = 0.500$ and $\hat{p}_1 = 15/30 = 0.500$, respectively. The last 30 points are used because the first-order Markov chain amounts to two conditional Bernoulli distributions, given the previous day's value, and the value for 31 December 1986 is not available in Table A.1. The log-likelihood in Equation 10.13a for these data is $L_0 = 15 \ln(0.500) + 15 \ln(0.500) = -20.79$. Values of n_{ij} and $\hat{p}_{ij}$ were computed previously and can be substituted into Equation 10.13b to yield $L_1 = 11 \ln(0.688) + 5 \ln(0.312) + 4 \ln(0.286) + 10 \ln(0.714) = -18.31$. Necessarily, $L_1 \geq L_0$ because the greater number of parameters in the more elaborate first-order Markov model provide more flexibility for a closer fit to the data at hand. The statistical significance of the difference in log-likelihoods can be assessed knowing that the null distribution of $\Lambda = 2 (L_1 - L_0) = 4.96$ is χ^2, with $v = (s^{m(H_A)} s^{m(H_0)}) (s-1)$ degrees of freedom. Since the time series being tested in binary, $s = 2$. The null hypothesis is that the time dependence is zero order, so $m(H_0) = 0$, and the alternative hypothesis is first-order serial dependence, or $m(H_A) = 1$. Thus $v = (2^1 - 2^0)(2-1) = 1$ degree of freedom. In general the appropriate degrees-of-freedom will be the difference in dimensionality (i.e., the parameter counts) between the competing models. This likelihood test result is consistent with the χ^2 goodness-of-fit test conducted in Example 10.1, where $\chi^2 = 4.82$, which is not surprising because the χ^2 test conducted there is an approximation to the likelihood ratio test. ◇

Both the AIC and BIC criteria attempt to find the most appropriate model order by striking a balance between goodness of fit, as reflected in the log-likelihoods, and a penalty that increases with the number of fitted parameters. The two approaches differ only in the form of the penalty function. The AIC and BIC statistics are computed for each trial order m, using

$$AIC(m) = -2L_m + 2s^m(s-1), \tag{10.14}$$

and

$$BIC(m) = -2L_m + s^m \ln(n), \tag{10.15}$$

respectively. The order m is chosen as appropriate that minimizes either Equation 10.14 or 10.15. The BIC criterion tends to be more conservative, generally picking lower orders than the AIC criterion when results of the two approaches differ. Use of the BIC statistic is generally preferable for sufficiently long time series, although "sufficiently long" may range from around $n = 100$ to over $n = 1000$, depending on the nature of the serial correlation (Katz, 1981).

10.3. TIME DOMAIN—II. CONTINUOUS DATA

10.3.1. First-Order Autoregression

The Markov chain models described in the previous section are not suitable for describing time series of data that are continuous, in the sense of the data being able to take on infinitely many values on all or part of the real line. As discussed in Chapter 4, atmospheric variables such as temperature, wind speed, geopotential height, and so on, are continuous variables in this sense. The correlation structure of such time series often can be represented successfully using a class of time series models known as *Box–Jenkins models*, after the classic text by Box and Jenkins (1976).

The *first-order autoregression*, or AR(1) model, is the simplest Box–Jenkins model. It is the continuous counterpart of the first-order Markov chain. As the name suggests, one way of viewing the AR(1) model is as a simple linear regression (Section 7.2.1), where the predictand is the value of the time series at time $t+1$, x_{t+1}, and the predictor is the current value of the time series, x_t. The AR(1) model can be written as

$$x_{t+1} - \mu = \phi(x_t - \mu) + \varepsilon_{t+1}, \tag{10.16}$$

where μ is the mean of the time series, ϕ is the autoregressive parameter, and ε_{t+1} is a random quantity corresponding to the residual in ordinary regression. The right-hand side of Equation 10.16 consists of a deterministic part in the first term and a random part in the second term. That is, the next value of the time series x_{t+1} is given by the function of x_t in the first term, plus the random shock or innovation ε_{t+1}.

The time series of x is assumed to be stationary, so that its mean μ is the same for each interval of time. The data series also exhibits a variance, σ_x^2, the sample counterpart of which is just the ordinary sample variance computed from the values of the time series by squaring Equation 3.6. The ε's are mutually independent random quantities having mean $\mu_\varepsilon = 0$ and variance σ_ε^2. Very often it is further assumed that the ε's follow a Gaussian distribution.

As illustrated in Figure 10.5, the autoregressive model in Equation 10.16 can represent the serial correlation of a time series. This is a scatterplot of minimum temperatures at Canandaigua, New York, during January 1987, from Table A.1. The first 30 data values, for 1–30 January, are plotted on the horizontal axis. The corresponding temperatures for the following days, 2–31 January, are plotted on the vertical axis. The serial correlation, or persistence, is evident from the appearance of the point cloud and from the positive slope of the regression line. Equation 10.16 can be viewed as a prediction equation for x_{t+1} using x_t as the predictor. Rearranging Equation 10.16 to more closely resemble the simple linear regression Equation 7.3 yields the intercept $a = \mu(1-\phi)$, and slope $b = \phi$.

Another way to look at Equation 10.16 is as an algorithm for generating synthetic time series of values of x, in the same sense as in Section 4.7. Beginning with an initial value, x_0, we would subtract the mean value (i.e., construct the corresponding anomaly), multiply by the autoregressive parameter ϕ, and then add a randomly generated variable ε_1 drawn from a Gaussian distribution (see Section 4.7.4) with mean zero and variance σ_ε^2. The first value of the time series, x_1, would then be produced by adding back the mean μ. The next time series value, x_2, would then be produced in a similar way, by operating on x_1 and adding a new random Gaussian quantity ε_2. For positive values of the parameter ϕ, synthetic time series constructed in this way will exhibit positive serial correlation because each newly generated data value x_{t+1} includes some information carried forward from the preceding value x_t. Since x_t was in turn generated in part from x_{t-1}, and so on, members of the time series separated by more than one time unit will be correlated, although this correlation becomes progressively weaker as the time separation increases.

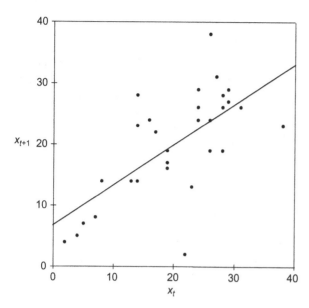

FIGURE 10.5 Scatterplot of January 1–30, 1987 minimum temperatures (°F) at Canandaigua, New York (x_t, horizontal) paired with minimum temperatures for the following days, January 2–31 (x_{t+1}, vertical). The data are from Table A.1. The regression line corresponding to the first term of the AR(1) time series model (Equation 10.16) is also shown.

The first-order autoregression is sometimes called the *Markov process* or Markov scheme. It shares with the first-order Markov chain the property that the full history of the time series prior to x_t provides no additional information regarding x_{t+1}, once x_t is known. This property can be expressed formally as

$$\Pr\{X_{t+1} \le x_{t+1} \mid X_t \le x_t, X_{t-1} \le x_{t-1}, \ldots, X_1 \le x_1\} = \Pr\{X_{t+1} \le x_{t+1} \mid X_t \le x_t\}. \tag{10.17}$$

Here the notation for continuous random variables has been used to express essentially the same idea as in Equation 10.1 for a sequence of discrete events. Again, Equation 10.17 does not imply that values of the time series separated by more than one time step are independent, but only that the influence of the prior history of the time series on its future values is fully contained in the current value x_t, regardless of the particular path by which the time series arrived at x_t.

Equation 10.16 is also sometimes known as a *red noise* process, because a positive value of the parameter ϕ averages or smooths out short-term fluctuations in the serially independent series of innovations, ε, while affecting the slower random variations much less strongly. The resulting time series is called red noise by analogy to visible light depleted in the shorter wavelengths, which appears reddish. This topic will be discussed more fully in Section 10.5, but the effect can be appreciated by looking at Figure 5.4. That figure compares a series of uncorrelated Gaussian values, ε_t (panel a), with an autocorrelated series generated from them using Equation 10.16 and the value $\phi = 0.6$ (panel b). It is evident that the most erratic point-to-point variations in the uncorrelated series have been smoothed out, but the slower random variations are essentially preserved. In the time domain this smoothing is expressed as positive serial correlation. From a frequency perspective, the resulting series is "reddened."

Parameter estimation for the first-order autoregressive model is straightforward. The estimated mean of the time series, μ, is simply the usual sample average (Equation 3.2) of the data set, provided that the series can be considered to be stationary. Nonstationary series must first be dealt with in one of the ways sketched in Section 10.1.1.

The estimated autoregressive parameter is simply equal to the sample lag-1 autocorrelation coefficient, Equation 3.36,

$$\hat{\phi} = r_1. \tag{10.18}$$

For the resulting probability model to be stationary, it is required that $-1 < \phi < 1$. As a practical matter this presents no problem for the first-order autoregression, because the correlation coefficient also is bounded by the same limits. For most atmospheric time series the parameter ϕ will be positive, reflecting persistence. Negative values of ϕ are possible, but correspond to very jagged (anticorrelated) time series with a tendency for alternating values above and below the mean. Because of the Markovian property, the full (theoretical, or population, or generating-process) autocorrelation function for a time series governed by a first-order autoregressive process can be written in terms of the autoregressive parameter as

$$\rho_k = \phi^k. \tag{10.19}$$

Equations 10.18 and 10.19 correspond directly to Equation 10.6 for the discrete first-order Markov chain. Thus the autocorrelation function for an AR(1) process with $\phi > 0$ decays exponentially from $\rho_0 = 1$, approaching zero as $k \to \infty$.

A series of truly independent data would have $\phi = 0$. However, a finite sample of independent data generally will exhibit a nonzero sample estimate for the autoregressive parameter. For a sufficiently long data series the sampling distribution of $\hat{\phi}$ is approximately Gaussian, with $\mu_\phi = \hat{\phi}$ and variance $\sigma_\phi^2 = (1-\phi^2)/n$. Therefore a test for the sample estimate of the autoregressive parameter, corresponding to Equation 5.3 with the null hypothesis that $\phi = 0$, can be carried out using the test statistic

$$z = \frac{\hat{\phi} - 0}{\left[\text{Var}\left(\hat{\phi}\right)\right]^{1/2}} = \frac{\hat{\phi}}{[1/n]^{1/2}}, \tag{10.20}$$

because $\phi = 0$ under the null hypothesis. Statistical significance would be assessed approximately using standard Gaussian probabilities. This test is virtually identical to the t test for the slope of a regression line (Section 7.2.5).

The final parameter of the statistical model in Equation 10.16 is the residual variance, or innovation variance, σ_ε^2. This quantity is sometimes also known as the *white-noise variance*, for reasons that are explained in Section 10.5. This parameter expresses the variability or uncertainty in the time series not accounted for by the serial correlation or, put another way, the uncertainty in x_{t+1} given that x_t is known. The brute-force approach to estimating σ_ε^2 is to estimate ϕ using Equation 10.18, compute the time series ε_{t+1} from the data using a rearrangement of Equation 10.16, and then to compute the ordinary sample variance of these ε values. Since the variance of the data is often computed as a matter of course, another way to estimate the white-noise variance is to use the relationship between the variances of the data series and the innovation series in the AR(1) model,

$$\sigma_\varepsilon^2 = \left(1 - \phi^2\right) \sigma_x^2. \tag{10.21}$$

Equation 10.21 implies $\sigma_\varepsilon^2 \leq \sigma_x^2$, with equality only for independent data, for which $\phi = 0$. Equation 10.21 also implies that knowing the current value of an autocorrelated time series decreases uncertainty about the next value of the time series. In practical settings we work with sample estimates of the autoregressive parameter and of the variance of the data series, so that the corresponding sample estimator of the white-noise variance is

$$\hat{s}_\varepsilon^2 = \frac{1 - \hat{\phi}^2}{n-2} \sum_{t=1}^{n} (x_t - \bar{x})^2 = \frac{n-1}{n-2}\left(1 - \hat{\phi}^2\right) s_x^2. \tag{10.22}$$

The difference between Equations 10.22 and 10.21 is appreciable only if the data series is relatively short.

Example 10.3. A First-Order Autoregression

Consider fitting an AR(1) process to the series of January 1987 minimum temperatures from Canandaigua, in Table A.1. As indicated in the table, the average of these 31 values is 20.23°F, and this would be adopted as the estimated mean of the time series, assuming stationarity. The sample lag-1 autocorrelation coefficient, from Equation 3.29, is $r_1 = 0.67$, and this value would be adopted as the estimated autoregressive parameter according to Equation 10.18.

The scatterplot of these data against themselves lagged by one time unit in Figure 10.5 suggests the positive serial correlation typical of daily temperature data. A formal test of the estimated autoregressive parameter versus the null hypothesis that it is really zero would use the test statistic in Equation 10.20, $z = 0.67/[1/31]^{1/2} = 3.73$. This test provides strong evidence that the observed nonzero sample autocorrelation did not arise by chance from a sequence of 31 independent values.

The sample standard deviation of the 31 Canandaigua minimum temperatures in Table A1 is 8.81°F. Using Equation 10.22, the estimated white-noise variance for the fitted autoregression would be $s_\varepsilon^2 = (30/29)(1-0.67^2)(8.81^2) = 44.24°F^2$, corresponding to a standard deviation of 6.65°F. By comparison, the brute-force sample standard deviation of the series of sample residuals, each computed from the rearrangement of Equation 10.16 as $e_{t+1} = (x_{t+1}-\mu)-\phi(x-\mu)$ is 6.55°F.

The computations in this example have been conducted under the assumption that the time series being analyzed is stationary, which implies that the mean value does not change through time. This assumption is not exactly satisfied by these data, as illustrated in Figure 10.6. Here the time series of the January 1987 Canandaigua minimum temperature data is shown together with the climatological average temperatures for the period 1961–1990 (dashed line), and the linear trend fit to the 31 data points for 1987 (solid line).

Of course, the dashed line in Figure 10.6 is a better representation of the long-term (population) mean minimum temperatures at this location, and it indicates that early January is slightly warmer than late January on average. Strictly speaking, the data series is not stationary, since the underlying mean value for the time series is not constant through time. However, the change through the month represented by the dashed line is sufficiently minor (in comparison to the variability around this mean function) that

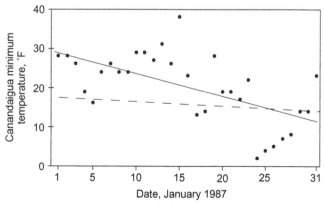

FIGURE 10.6 Time series of the January 1987 Canandaigua minimum temperature data. Solid line is the least-squares linear trend in the data, and the dashed line represents the climatological average minimum temperatures for the period 1961–1990.

generally we would be comfortable in pooling data from a collection of Januaries and assuming statio-narity. In fact, the preceding results for the 1987 data sample are not very much different if the January 1987 mean minimum temperature of 20.23°F, or the long-term climatological temperatures represented by the dashed line, are assumed. In this latter case, we find $\phi = 0.64$, and $s_\varepsilon^2 = 6.49°F$.

Because the long-term climatological minimum temperature declines so slowly, it is clear that the rather steep negative slope of the solid line in Figure 10.6 results mainly from sampling variations in this short example data record. Normally an analysis of this kind would be carried out using a much longer data series. However, if no other information about the January minimum temperature climate of this location were available, it would be sensible to produce a stationary series before proceeding further, by subtracting the mean values represented by the solid line from the data points, provided the estimated slope is significantly different from zero (after accounting for the serial correlation in the data). The regression equation for this line is $\mu(t) = 29.6-0.584\,t$, where t is the date, and the slope is indeed nominally significant. Hypothetically, the autoregressive process in Equation 10.16 would then be fit using the time series of the anomalies $x'_t = x_t - \mu(t)$. For example, $x'_1 = 28°F-(29.6-0.584) = -1.02°$ F. Since the average residual from a least-squares regression line is zero (see Section 7.2.2), the mean of this series of anomalies x'_t will be zero. Fitting Equation 10.16 to this anomaly series yields $\hat{\phi} = 0.47$, and $s_\varepsilon^2 = 39.95°F^2$.   ◇

10.3.2. Higher-Order Autoregressions

The first-order autoregression in Equation 10.16 generalizes readily to higher orders. That is, the regression equation predicting x_{t+1} can be expanded to include data values progressively further back in time as predictors. The general autoregressive model of order K, or AR(K) model is

$$x_{t+1} - \mu = \sum_{k=1}^{K} [\phi_k (x_{t-k+1} - \mu)] + \varepsilon_{t+1}. \tag{10.23}$$

Here the anomaly for the next time point, $x_{t+1} - \mu$, is a weighted sum of the previous K anomalies plus the random component ε_{t+1}, where the weights are the autoregressive coefficients ϕ_k. As before, the ε's are mutually independent, with zero mean and variance σ_ε^2. Stationarity of the process implies that μ and σ_ε^2 do not change through time. For $K = 1$, Equation 10.23 is identical to Equation 10.16.

Estimation of the K autoregressive parameters ϕ_k is perhaps most easily done using the set of equations relating them to the autocorrelation function, which are known as the *Yule–Walker equations*. These are

$$
\begin{aligned}
r_1 &= \hat{\phi}_1 &+ \hat{\phi}_2 r_1 &+ \hat{\phi}_3 r_2 &+ \cdots &+ \hat{\phi}_K r_{K-1} \\
r_2 &= \hat{\phi}_1 r_1 &+ \hat{\phi}_2 &+ \hat{\phi}_3 r_1 &+ \cdots &+ \hat{\phi}_K r_{K-2} \\
r_3 &= \hat{\phi}_1 r_2 &+ \hat{\phi}_2 r_1 &+ \hat{\phi}_3 &+ \cdots &+ \hat{\phi}_K r_{K-3} \cdot \\
\vdots &\quad \vdots &\vdots &\vdots &\cdots &\vdots \\
r_K &= \hat{\phi}_1 r_{K-1} &+ \hat{\phi}_2 r_{K-2} &+ \hat{\phi}_3 r_{K-3} &+ \cdots + &\hat{\phi}_K
\end{aligned}
\tag{10.24}
$$

Here $\phi_k = 0$ for $k > K$. The Yule–Walker equations arise from Equation 10.23, by multiplying by x_{t-k}, applying the expected value operator, and evaluating the result for different values of k (e.g., Box and

Jenkins, 1976). These equations can be solved simultaneously for the ϕ_k. Another approach to estimating the autoregressive parameters in Equation 10.23 is through maximum likelihood, assuming Gaussian-distributed random innovations ε_t (Miller, 1995). Alternatively, a method to use these equations recursively for parameter estimation—that is, to compute ϕ_1 and ϕ_2 to fit the AR(2) model knowing ϕ for the AR(1) model, and then to compute ϕ_1, ϕ_2, and ϕ_3 for the AR(3) model knowing ϕ_1 and ϕ_2 for the AR(2) model, and so on—is given in Box and Jenkins (1976) and Katz (1982). Constraints on the autoregressive parameters necessary for Equation 10.23 to describe a stationary process are given in Box and Jenkins (1976).

The theoretical autocorrelation function corresponding to a particular set of the ϕ_k's can be determined by solving Equation 10.24 for the first K autocorrelations and then recursively applying

$$\rho_m = \sum_{k=1}^{K} \phi_k \, \rho_{m-k}. \tag{10.25}$$

Equation 10.25 holds for lags $m \geq k$, with the understanding that $\rho_0 \equiv 1$. Finally, the generalization of Equation 10.21 for the relationship between the white-noise variance and the variance of the data values themselves is

$$\sigma_\varepsilon^2 = \left(1 - \sum_{k=1}^{K} \phi_k \rho_k\right) \sigma_x^2. \tag{10.26}$$

10.3.3. The AR(2) Model

The second-order autoregression, or AR(2) process is a common and important higher-order autoregressive model. It is reasonably simple, requiring the fitting of only two parameters in addition to the sample mean and variance of the series, yet it can describe a variety of qualitatively quite different behaviors of time series. The defining equation for AR(2) processes is

$$x_{t+1} - \mu = \phi_1 (x_t - \mu) + \phi_2 (x_{t-1} - \mu) + \varepsilon_{t+1}, \tag{10.27}$$

which is easily seen to be a special case of Equation 10.23. Using the first $K = 2$ of the Yule–Walker Equations (10.24),

$$r_1 = \hat{\phi}_1 + \hat{\phi}_2 r_1 \tag{10.28a}$$

and

$$r_2 = \hat{\phi}_1 r_1 + \hat{\phi}_2, \tag{10.28b}$$

the two autoregressive parameters can be estimated as

$$\hat{\phi}_1 = \frac{r_1 (1 - r_2)}{1 - r_1^2} \tag{10.29a}$$

and

$$\hat{\phi}_2 = \frac{r_2 - r_1^2}{1 - r_1^2}, \tag{10.29b}$$

by solving Equations 10.28 for $\hat{\phi}_1$ and $\hat{\phi}_2$.

The white-noise variance for a fitted AR(2) model can be estimated in several ways. For very large samples, Equation 10.26 with $K = 2$ can be used with the sample variance of the time series, s_x^2. Alternatively, once the autoregressive parameters have been fit using Equations 10.29 or some other means, the corresponding estimated time series of the random innovations ε can be computed from a rearrangement of Equation 10.27 and their sample variance computed, as was done in Example 10.3 for the fitted AR(1) process. Another possibility is to use the recursive equation given by Katz (1982),

$$s_\varepsilon^2(m) = \left[1 - \hat{\phi}_m^2(m)\right] s_\varepsilon^2(m-1). \tag{10.30}$$

Here the autoregressive models AR(1), AR(2), ... are fitted successively, $s_\varepsilon^2(m)$ is the estimated white-noise variance of the mth (i.e., current) autoregression, $s_\varepsilon^2(m-1)$ is the estimated white-noise variance for the previously fitted (one order smaller) model, and $\hat{\phi}_m(m)$ is the estimated autoregressive parameter for the highest lag in the current model. For the AR(2) model, Equation 10.30 can be used with the expression for $s_\varepsilon^2(1)$ in Equation 10.22 to yield

$$s_\varepsilon^2(2) = \left(1 - \hat{\phi}_2^2\right) \frac{n-1}{n-2} \left(1 - r_1^2\right) s_x^2, \tag{10.31}$$

since $\hat{\phi} = r_1$ for the AR(1) model.

For an AR(2) process to be stationary, its two parameters must satisfy the constraints

$$\left.\begin{array}{c} \phi_1 + \phi_2 < 1 \\ \phi_2 - \phi_1 < 1 \\ -1 < \phi_2 < 1 \end{array}\right\}, \tag{10.32}$$

which define the triangular region in the (ϕ_1, ϕ_2) plane shown in Figure 10.7. Note that substituting $\phi_2 = 0$ into Equation 10.32 yields the stationarity condition $-1 < \phi_1 < 1$ applicable to the AR(1) model. Figure 10.7 includes AR(1) models as special cases on the horizontal $\phi_2 = 0$ line, for which that stationarity condition applies.

The first two values of the theoretical autocorrelation function for a particular AR(2) process can be obtained by solving Equations 10.28 as

$$\rho_1 = \frac{\phi_1}{1 - \phi_2} \tag{10.33a}$$

and

$$\rho_2 = \phi_2 + \frac{\phi_1^2}{1 - \phi_2}, \tag{10.33b}$$

and subsequent values of the autocorrelation function can be calculated using Equation 10.25. Figure 10.7 indicates that a wide range of types of autocorrelation functions, and thus a wide range of time correlation behaviors, can be represented by AR(2) processes. First, AR(2) models include the simpler AR(1) models as special cases. Two AR(1) autocorrelation functions are shown in Figure 10.7. The autocorrelation function for the model with $\phi_1 = 0.5$ and $\phi_2 = 0.0$ decays exponentially toward zero, following Equation 10.19. Autocorrelation functions for many atmospheric time series exhibit this kind of behavior, at least approximately. The other AR(1) model for which an autocorrelation function is shown is for $\phi_1 = -0.6$ and $\phi_2 = 0.0$. Because of the negative lag-one autocorrelation, the autocorrelation function exhibits oscillations around zero that are progressively damped at longer lags (again, compare Equation 10.19).

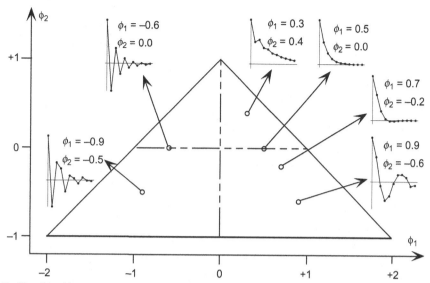

FIGURE 10.7 The allowable parameter space for stationary AR(2) processes, with insets showing autocorrelation functions for selected AR(2) models. The horizontal $\phi_2 = 0$ line locates the AR(1) models as special cases, and autocorrelation functions for two of these are shown. AR(2) models appropriate to atmospheric time series usually exhibit $\phi_1 > 0$.

That is, there is a tendency for the anomalies of consecutive data values to have opposite signs, so that data separated by even numbers of lags are positively correlated. This kind of behavior rarely is seen in atmospheric data series, and most AR(2) models for atmospheric data have $\phi_1 > 0$.

The second autoregressive parameter allows many other kinds of behaviors in the autocorrelation function. For example, the autocorrelation function for the AR(2) model with $\phi_1 = 0.3$ and $\phi_2 = 0.4$ exhibits a larger correlation at two lags than at one lag. For $\phi_1 = 0.7$ and $\phi_2 = -0.2$ the autocorrelation function declines very quickly and is almost zero for lags $k \geq 4$. The autocorrelation function for the AR (2) model with $\phi_1 = 0.9$ and $\phi_2 = -0.6$ is very interesting in that it exhibits a slow damped oscillation around zero. This characteristic reflects what are called *pseudoperiodicities* in the corresponding time series. That is, time series values separated by very few lags exhibit fairly strong positive correlation, those separated by a few more lags exhibit negative correlation, and values separated by a few more additional lags exhibit positive correlation again. The qualitative effect is for the corresponding time series to exhibit oscillations around the mean resembling an irregular cosine curve with an average period that is approximately equal to the number of lags at the first positive hump in the autocorrelation function. Thus AR(2) models can represent data that are approximately but not strictly periodic, such as barometric pressure variations resulting from the movement of midlatitude synoptic systems.

Some properties of autoregressive models are illustrated by the four example synthetic time series in Figure 10.8. Series (a) is simply a sequence of 50 independent Gaussian variates with $\mu = 0$. Series (b) is a realization of the AR(1) process generated using Equation 10.16 or, equivalently, Equation 10.27 with $\mu = 0$, $\phi_1 = 0.5$ and $\phi_2 = 0.0$. The apparent similarity between series (a) and (b) arises because series (a) has been used as the ε_{t+1} series forcing the autoregressive process in Equation 10.27. The effect of the parameter $\phi = \phi_1 > 0$ is to smooth out step-to-step variations in the white-noise series (a), and to give the resulting time series a bit of memory. The relationship of the series in these two panels is analogous to that in Figure 5.4, in which $\phi = 0.6$.

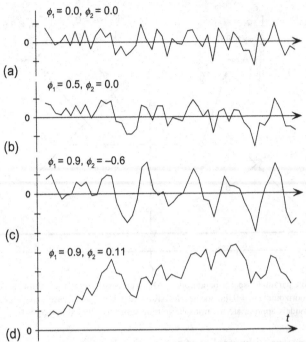

FIGURE 10.8 Four synthetic time series illustrating some properties of autoregressive models. Series (a) consists of independent Gaussian variables (white noise). Series (b) is a realization of the AR(1) process with $\phi_1 = 0.5$, and series (c) is a realization of the AR(2) process with $\phi_1 = 0.9$ and $\phi_2 = -0.6$, autocorrelation functions for both of which are shown in Figure 10.7. Series (d) is nonstationary because its parameters lie outside the triangle in Figure 10.7, and this nonstationarity can be seen as a drifting in the mean value. The series (b)–(d) were constructed using Equation 10.27 with $\mu = 0$ and the ε's from series (a).

Series (c) in Figure 10.8 is a realization of the AR(2) process with $\mu = 0$, $\phi_1 = 0.9$ and $\phi_2 = -0.6$. It resembles qualitatively some atmospheric series (e.g., midlatitude sea-level pressures), but has been generated using Equation 10.27 with series (a) as the forcing white noise. This series exhibits pseudoperiodicities. That is, peaks and troughs in this time series tend to recur with a period near six or seven time intervals, but these are not so regular that a cosine function or the sum of a few cosine functions would represent them very well. This feature is the expression in the data series of the positive hump in the autocorrelation function for this autoregressive model shown in the inset of Figure 10.7, which occurs at a lag interval of six or seven time periods. Similarly, the peak–trough pairs tend to be separated by perhaps three or four time intervals, corresponding to the minimum in the autocorrelation function at these lags shown in the inset in Figure 10.7.

The autoregressive parameters $\phi_1 = 0.9$ and $\phi_2 = 0.11$ for series (d) in Figure 10.8 fall outside the triangular region in Figure 10.7 that defines the limits of stationary AR(2) processes. The series is therefore not stationary, and the nonstationarity can be seen as a drifting of the mean value in the realization of this process shown in Figure 10.8.

Finally, series (a) through (c) in Figure 10.8 illustrate the nature of the relationship between the variance of the time series, σ_x^2, and the white-noise variance, σ_ε^2, of an autoregressive process. Series (a) consists simply of independent Gaussian variates, or white noise. Formally, it can be viewed as a special case of an autoregressive process, with all the ϕ_k's $= 0$. Using Equation 10.26 it is clear that

$\sigma_x^2 = \sigma_\varepsilon^2$ for this series. Since series (b) and (c) were generated using series (a) as the white-noise forcing ε_t +1, σ_ε^2 for all three of these series are equal. Time series (c) gives the visual impression of being more variable than series (b), which in turn appears to exhibit more variability than series (a). Using Equations 10.33 with Equation 10.26 it is easy to compute that σ_x^2 for series (b) is 1.33 times larger than the common σ_ε^2, and for series (c) it is 2.29 times larger. The equations on which these computations are based pertain only to stationary autoregressive series, and so cannot be applied meaningfully to the non-stationary series (d).

10.3.4. Order Selection Criteria

The Yule–Walker equations (10.24) can be used to fit autoregressive models to essentially arbitrarily high order. At some point, however, expanding the complexity of the model will not appreciably improve its representation of the data. Adding more terms in Equation 10.23 will eventually result in the model being overfit or excessively tuned to the data used for parameter estimation.

The BIC (Schwarz, 1978) and AIC (Akaike, 1974) statistics, applied to Markov chains in Section 10.2, are also often used to decide among potential orders of autoregressive models. Both statistics involve the log-likelihood plus a penalty for the number of parameters, with the two criteria differing only in the form of the penalty function. Here the likelihood function involves the estimated (assumed Gaussian) white-noise variance.

For each candidate order m, the order selection statistics

$$BIC(m) = n \ln \left[\frac{n}{n-m-1} s_\varepsilon^2(m) \right] + (m+1) \ln(n), \tag{10.34}$$

or

$$AIC(m) = n \ln \left[\frac{n}{n-m-1} s_\varepsilon^2(m) \right] + 2(m+1), \tag{10.35}$$

are computed, using $s_\varepsilon^2(m)$ from Equation 10.30. Better fitting models will exhibit smaller white-noise variance, implying less residual uncertainty. Arbitrarily adding more parameters (fitting higher- and higher-order autoregressive models) will not increase the white-noise variance estimated from the data sample, but neither will the estimated white-noise variance decrease much if the extra parameters are not effective in improving the description of the behavior of the data. Thus the penalty functions serve to guard against overfitting. That order m is chosen as appropriate, which minimizes either Equation 10.34 or 10.35.

Example 10.4. Order Selection among Autoregressive Models

Table 10.3 summarizes the results of fitting successively higher-order autoregressive models to the January 1987 Canandaigua minimum temperature data, assuming that they are stationary without removal of a trend. The second column shows the sample autocorrelation function up to seven lags. The estimated white-noise variance for autoregressions of orders one through seven, computed using the Yule–Walker equations and Equation 10.30, is shown in the third column. Notice that $s_\varepsilon^2(0)$ is simply the sample variance of the time series itself, or s_x^2. The estimated white-noise variances decrease progressively as more terms are added to Equation 10.23, but toward the bottom of the table adding yet more terms has little further effect.

The BIC and AIC statistics for each candidate autoregression are shown in the last two columns. Both indicate that the AR(1) model is most appropriate for this data, as $m = 1$ produces the minimum in both order selection statistics. Similar results also are obtained for the other three temperature series in

TABLE 10.3 Illustration of Order Selection for Autoregressive Models to Represent the January 1987 Canandaigua Minimum Temperature Series, Assuming Stationarity

Lag, m	r_m	$s_\varepsilon^2(m)$	BIC(m)	AIC(m)
0	1.000	77.58	138.32	136.89
1	0.672	42.55	125.20	122.34
2	0.507	42.11	129.41	125.11
3	0.397	42.04	133.91	128.18
4	0.432	39.72	136.76	129.59
5	0.198	34.39	136.94	128.34
6	0.183	33.03	140.39	130.35
7	0.161	33.02	145.14	133.66

Presented are the autocorrelation function for the first seven lags m, the estimated white-noise variance for each AR(m) model, and the BIC and AIC statistics for each trial order. For $m = 0$ the autocorrelation function is 1.00, and the white-noise variance is equal to the sample variance of the series. The AR(1) model is selected by both the BIC and AIC criteria.

Table A.1. Note, however, that with a larger sample size, higher-order autoregressions could be chosen by both criteria. For the estimated residual variances presented in Table 10.3, using the AIC statistic would lead to the choice of the AR(2) model for n greater than about 290, and the AR(2) model would minimize the BIC statistic for n larger than about 430. ◇

10.3.5. The Variance of a Time Average

Estimation of the sampling distribution of the average of a correlated time series is an important application of time series models in atmospheric data analysis. Recall that a sampling distribution characterizes the batch-to-batch variability of a statistic computed from a finite data sample. If the data values making up a sample average are independent, the variance of the sampling distribution of that average is given by the variance of the data, s_x^2, divided by the sample size (Equation 5.4).

When the underlying data are positively autocorrelated, using Equation 5.4 to calculate the variance of (the sampling distribution of) their time average leads to an underestimate. This discrepancy is a consequence of the tendency for nearby values of correlated time series to be similar, leading to less batch-to-batch consistency of the sample average. The phenomenon is illustrated in Figure 5.4. As discussed in Chapter 5, underestimating the variance of the sampling distribution of the mean can cause serious problems for statistical inference, leading for example, to unwarranted rejection of valid null hypotheses.

The effect of serial correlation on the variance of a time average over a sufficiently large sample can be accounted for through a variance inflation factor, V, modifying Equation 5.4:

$$\text{Var}[\bar{x}] = \frac{V\sigma_x^2}{n}. \tag{10.36}$$

If the data series is uncorrelated, $V = 1$ and Equation 10.36 corresponds to Equation 5.4. If the data exhibit positive serial correlation, $V > 1$ and the variance of the time average is inflated above what would be implied for independent data. Note, however, that even if the underlying data are correlated, the mean of the sampling distribution of the time average is the same as the underlying mean of the data being averaged,

$$E[\bar{x}] = \mu_{\bar{x}} = E[x_t] = \mu_x. \tag{10.37}$$

For arbitrarily large sample sizes, the variance inflation factor depends on the autocorrelation function according to

$$V = 1 + 2 \sum_{k=1}^{\infty} \rho_k. \tag{10.38}$$

However, the variance inflation factor can be estimated with greater ease and precision if a data series is well represented by an autoregressive model. In terms of the parameters of an AR(K) model, the large-sample variance inflation factor in Equation 10.38 is well approximated by

$$V = \frac{1 - \sum_{k=1}^{K} \phi_k \rho_k}{\left[1 - \sum_{k=1}^{K} \phi_k \right]^2}. \tag{10.39}$$

Note that the theoretical autocorrelations ρ_k in Equation 10.39 can be expressed in terms of the autoregressive parameters by solving the Yule–Walker Equations (10.24) for the correlations. In the special case of an AR(1) model being appropriate for a time series of interest, Equation 10.39 reduces to

$$V = \frac{1 + \phi}{1 - \phi}, \tag{10.40}$$

which was used to estimate the effective sample size in Equation 5.12 and the variance of the sampling distribution of a sample mean in Equation 5.13. Equations 10.39 and 10.40 are convenient large-sample approximations to the formula for the variance inflation factor based on sample autocorrelation estimates

$$V = 1 + 2 \sum_{k=1}^{n} \left(1 - \frac{k}{n} \right) r_k, \tag{10.41}$$

which more closely follows Equation 10.38. Equation 10.41 approaches Equations 10.39 and 10.40 for large sample size n, when the autocorrelations r_k are expressed in terms of the autoregressive parameters (Equation 10.24). Usually either Equation 10.39 or 10.40, as appropriate, would be used to compute the variance inflation factor, although the Diebold–Mariano test (Diebold and Mariano, 1995) uses Equation 10.41 (see Section 5.2.4).

Example 10.5. Variances of Time Averages of Different Lengths

The relationship between the variance of a time average and the variance of the individual elements of a time series in Equation 10.36 can be useful in surprising ways. Consider, for example, the average winter (December–February) geopotential heights, and the standard deviations of those averages, for the northern hemisphere shown in Figures 10.9a and 10.9b, respectively. Figure 10.9a shows the average

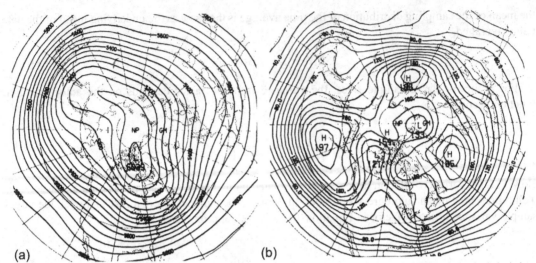

FIGURE 10.9 Average 500 mb height field for the northern hemisphere winter (a), and the field of standard deviations of that average, reflecting winter-to-winter variations (b). *From Blackmon (1976). © American Meteorological Society. Used with permission.*

field (Equation 10.37), and Figure 10.9b shows the standard deviation of 90-day averages of winter 500 mb heights, representing the interannual variability. That is, Figure 10.9b shows the square root of Equation 10.36, with s_x^2 being the variance of the daily 500 mb height measurements and $n = 90$. The maps in Figure 10.9 summarize statistics that have been computed for a large number of gridpoints in the hemispheric domain.

Suppose, however, that the sampling distribution of 500 mb heights averaged over a different length of time were needed. We might be interested in the variance of 10-day averages of 500 mb heights at selected locations, for use as a climatological reference when calculating the skill of forecasts of 10-day averages of this quantity, using Equation 9.37. (The variance of the climatological distribution is exactly the mean-squared error of the climatological reference forecast.) Assuming that time series of winter 500 mb heights are stationary, the variance of an average over some different time period can be approximated without explicitly knowing the variance inflation factor in either Equations 10.38 or 10.39, and therefore without necessarily having the daily data. The ratio of the variances of 10-day and 90-day averages can be constructed from Equation 10.36,

$$\frac{\mathrm{Var}[\bar{x}_{10}]}{\mathrm{Var}[\bar{x}_{90}]} = \frac{V\, s_x^2/10}{V\, s_x^2/90},$$

(10.42a)

leading to

$$\mathrm{Var}[\bar{x}_{10}] = \frac{90}{10}\mathrm{Var}[\bar{x}_{90}].$$

(10.42b)

Regardless of the averaging period, the variance inflation factor V and the variance of the daily observations s_x^2 are the same because they are characteristics of the underlying daily time series. Thus the variance of a 10-day average is approximately nine times larger than the variance of a 90-day average, and a map of hemispheric 10-day standard deviations of winter 500 mb heights would be qualitatively very similar to Figure 10.9b, but exhibiting magnitudes about $\sqrt{9} = 3$ times larger.   ◇

10.3.6. Autoregressive-Moving Average Models

Autoregressions actually constitute a subset of a broader class of the Box–Jenkins time-domain models, also known as *autoregressive-moving average*, or *ARMA*, *models*. The general ARMA(K, M) model has K autoregressive terms, as in the AR(K) process in Equation 10.23, and in addition contains M moving average terms that compose a weighted average of the M previous values of the ε's. The ARMA(K, M) model thus contains K autoregressive parameters ϕ_k and M moving average parameters θ_m that affect the time series according to

$$x_{t+1} - \mu = \sum_{k=1}^{K} \phi_k (x_{t-k+1} - \mu) + \varepsilon_{t+1} - \sum_{m=1}^{M} \theta_m \varepsilon_{t-m+1}. \tag{10.43}$$

The AR(K) process in Equation 10.23 is a special case of the ARMA(K, M) model in Equation 10.43, with all the $\theta_m = 0$. Similarly, a pure *moving average process* of order M, or MA(M) process, would be a special case of Equation 10.43, with all the $\phi_k = 0$. Parameter estimation and derivation of the autocorrelation function for the general ARMA(K, M) process is more difficult than for the simpler AR(K) models. Parameter estimation methods are given in Box and Jenkins (1976), and many time-series computer packages will fit ARMA models.

The ARMA(1,1) is an important and common ARMA model,

$$x_{t+1} - \mu = \phi_1 (x_t - \mu) + \varepsilon_{t+1} - \theta_1 \varepsilon_t. \tag{10.44}$$

Computing parameter estimates even for this simple ARMA model is somewhat complicated, although Box and Jenkins (1976) present an easy graphical technique that allows estimation of ϕ_1 and θ_1 using the first two sample lagged correlations r_1 and r_2. The autocorrelation function for the ARMA (1,1) process can be calculated from the parameters using

$$\rho_1 = \frac{(1 - \phi_1 \theta_1)(\phi_1 - \theta_1)}{1 + \theta_1^2 - 2\phi_1 \theta_1} \tag{10.45a}$$

and

$$\rho_k = \phi_1 \rho_{k-1}, \quad k > 1. \tag{10.45b}$$

The autocorrelation function of an ARMA(1, 1) process decays exponentially from its value at ρ_1, which depends on both ϕ_1 and θ_1. This differs from the autocorrelation function for an AR(1) process, which decays exponentially from $\rho_0 = 1$. The relationship between the time-series variance and the white-noise variance of an ARMA(1,1) process is

$$\sigma_\varepsilon^2 = \frac{1 - \phi_1^2}{1 + \theta_1^2 + 2\phi_1 \theta_1} \sigma_x^2. \tag{10.46}$$

Equations 10.45 and 10.46 also apply to the simpler AR(1) and MA(1) processes, for which $\theta_1 = 0$ or $\phi_1 = 0$, respectively.

10.3.7. Simulation and Forecasting with Continuous Time-Domain Models

An important application of time-domain models is in the simulation of synthetic (i.e., random number, as in Section 4.7) series having statistical characteristics that are similar to observed data series. Such

Monte Carlo simulations are useful for investigating impacts of atmospheric variability in situations where the record length of the observed data is known or suspected to be insufficient to include representative sequences of the relevant variable(s). Here it is necessary to choose the type and order of time-series model carefully, so that the simulated time series will well represent the variability of the real generating process.

Once an appropriate time series model has been identified and its parameters estimated, its defining equation can be used as an algorithm to generate synthetic time series. For example, if an AR(2) model is representative of the data, Equation 10.27 would be used, whereas Equation 10.44 would be used as the generation algorithm for ARMA(1,1) models. The simulation method is similar to that described earlier for sequences of binary variables generated using the Markov chain model. Here, however, the noise or innovation series, ε_{t+1}, usually is assumed to consist of independent Gaussian variates with $\mu_\varepsilon = 0$ and variance σ_ε^2, which is estimated from the data as described earlier.

At each time step, a new Gaussian ε_{t+1} is chosen (see Section 4.7.4) and substituted into the defining equation. The next value of the synthetic time series x_{t+1} is then computed using the previous K values of x (for AR models), the previous M values of ε (for MA models), or both (for ARMA models). The only real difficulty in implementing the process is at the beginning of each synthetic series, where there are no prior values of x and/or ε that can be used. A simple solution to this problem is to substitute the corresponding averages (expected values) for the unknown previous values. That is, $(x_t-\mu) = 0$ and $\varepsilon_t = 0$ for $t \le 0$ might be assumed.

A better procedure is to generate the first values in a way that is consistent with the structure of the time-series model. For example, with an AR(1) model we could choose x_0 from a Gaussian distribution with variance $\sigma_x^2 = \sigma_\varepsilon^2/(1-\phi^2)$ (cf. Equation 10.21). Another very workable solution is to begin with $(x_t-\mu) = 0$ and $\varepsilon_t = 0$, but generate a longer time series than needed. The first few members of the resulting time series, which are most influenced by the initial values, are then discarded.

Example 10.6. Statistical Simulation with an Autoregressive Model

The time series in Figures 10.8b–d were produced according to the procedure just described, using the independent Gaussian series in Figure 10.8a as the series of ε's. The first and last few values of this independent series, and of the two series plotted in Figures 10.8b and c, are presented in Table 10.4. For all three series, $\mu = 0$ and $\sigma_\varepsilon^2 = 1$. Equation 10.16 has been used to generate the values of the AR(1) series, with $\phi_1 = 0.5$, and Equation 10.27 was used to generate the AR(2) series, with $\phi_1 = 0.9$ and $\phi_2 = -0.6$.

Consider the more difficult case of generating the AR(2) series. Calculating x_1 and x_2 in order to begin the series presents an immediate problem, because x_0 and x_{-1} do not exist. This simulation was initialized by assuming the expected values $E[x_0] = E[x_{-1}] = \mu = 0$. Thus since $\mu = 0$, $x_1 = \phi_1 x_0 + \phi_2 x_{-1} + \varepsilon_1 = (0.9)(0) - (0.6)(0) + 1.526 = 1.562$. Having generated x_1 in this way, it is then used to obtain $x_2 = \phi_1 x_1 + \phi_2 x_0 + \varepsilon_2 = (0.9)(1.526) - (0.6)(0) + 0.623 = 1.996$. For values of the AR(2) series at times $t = 3$ and larger, the computation is a straightforward application of Equation 10.27. For example, $x_3 = \phi_1 x_2 + \phi_2 x_1 + \varepsilon_3 = (0.9)(1.996) - (0.6)(1.526) - 0.272 = 0.609$. Similarly, $x_4 = \phi_1 x_3 + \phi_2 x_2 + \varepsilon_4 = (0.9)(0.609) - (0.6)(1.996) + 0.092 = -0.558$. If this synthetic series were to be used as part of a larger simulation, the first portion would generally be discarded, so that the retained values would have negligible memory of the initial condition $x_{-1} = x_0 = 0$. ◇

Purely statistical forecasts of the future evolution of time series can be produced using time-domain models. These are accomplished by simply extrapolating the most recently observed value(s) into the future using the appropriate defining equation, on the basis of parameter estimates fitted from the

TABLE 10.4 Values of the Time Series Plotted in Figure 10.8a–c

t	Independent Gaussian Series, ε_t, (Figure 10.8a)	AR(1) Series, x_t, (Figure 10.8b)	AR(2) Series, x_t, (Figure 10.8c)
1	1.526	1.526	1.526
2	0.623	1.387	1.996
3	−0.272	0.421	0.609
4	0.092	0.302	−0.558
5	0.823	0.974	−0.045
⋮	⋮	⋮	⋮
49	−0.505	−1.073	−3.172
50	−0.927	−1.463	−2.648

The AR(1) and AR(2) series have been generated from the independent Gaussian series using Equations 10.16 and 10.27, respectively, as the algorithms.

previous history of the series. Since the future values of the ε's cannot be known, the extrapolations are usually made using their expected values, that is, $E[\varepsilon] = 0$. Probability bounds on these extrapolations can be calculated as well.

The nature of this kind of forecast is most easily appreciated for the AR(1) model, the defining equation for which is Equation 10.16. Assume that the mean μ and the autoregressive parameter ϕ have been estimated from past data, the most recent of which is x_t. A nonprobabilistic forecast for x_{t+1} could be made by setting the unknown future ε_{t+1} to zero, and rearranging Equation 10.16 to yield $x_{t+1} = \mu + \phi(x_t - \mu)$. Note that, in common with the forecasting of a binary time series using a Markov chain model, this forecast is a compromise between persistence ($x_{t+1} = x_t$, which would result if $\phi = 1$) and climatology ($x_{t+1} = \mu$, which would result if $\phi = 0$). Further projections into the future would be obtained by extrapolating the previously forecast values, for example, $x_{t+2} = \mu + \phi(x_{t+1} - \mu)$, and $x_{t+3} = \mu + \phi(x_{t+2} - \mu)$. For the AR(1) model and $\phi > 0$, this series of forecasts would exponentially approach $x_\infty = \mu$.

The same procedure is used for higher-order autoregressions, except that the most recent K values of the time series are needed to extrapolate an AR(K) process (Equation 10.23). Forecasts derived from an AR(2) model, for example, would be made using the previous two observations of the time series, or $x_{t+1} = \mu + \phi_1(x_t - \mu) + \phi_2(x_{t-1} - \mu)$. Forecasts using ARMA models are only slightly more difficult, requiring that the last M values of the ε series be backcalculated before the projections begin.

Forecasts made using time-series models are of course uncertain, and the forecast uncertainty increases for longer lead times into the future. This uncertainty also depends on the nature of the appropriate time-series model (e.g., the order of an autoregression and its parameter values) and on the intrinsic uncertainty in the random noise series that is quantified by the white-noise variance σ_ε^2. The variance of a forecast made only one time step into the future is simply equal to the white-noise variance. Assuming the ε's follow a Gaussian distribution, a 95% probability interval on a forecast of x_{t+1} is then approximately $x_{t+1} \pm 2\sigma_\varepsilon$. For very long extrapolations, the variance of the forecasts approaches the variance of the time series itself, σ_x^2, which for AR models can be computed from the white-noise variance using Equation 10.26.

For intermediate lead times, calculation of forecast uncertainty is more complicated. For a forecast j time units into the future, the variance of the forecast is given by

$$\sigma^2(x_{t+j}) = \sigma_\varepsilon^2 \left[1 + \sum_{i=1}^{j-1} \psi_i^2 \right]. \tag{10.47}$$

Here the weights ψ_i depend on the parameters of the time series model, so that Equation 10.47 indicates that the variance of the forecast increases with both the white-noise variance and the lead time, and that the increase in uncertainty at increasing lead times depends on the specific nature of the time-series model. For the $j = 1$ time step forecast, there are no terms in the summation in Equation 10.47, and the forecast variance is equal to the white-noise variance, as noted earlier.

For AR(1) models, the ψ weights are simply

$$\psi_i = \phi^i, \quad i > 0, \tag{10.48}$$

so that, for example, $\psi_1 = \phi$, $\psi_2 = \phi^2$, and so on. More generally, for AR(K) models, the ψ weights are computed recursively, using

$$\psi_i = \sum_{k=1}^{K} \phi_k \psi_{i-k}, \tag{10.49}$$

where it is understood that $\psi_0 = 1$ and $\psi_i = 0$ for $i < 0$. For AR(2) models, for example, $\psi_1 = \phi_1$, $\psi_2 = \phi_1^2 + \phi_2$, $\psi_3 = \phi_1(\phi_1^2 + \phi_2) + \phi_2\phi_1$, and so on. Equations that can be used to compute the ψ weights for MA and ARMA models are given in Box and Jenkins (1976).

Example 10.7. Forecasting with an Autoregressive Model

Figure 10.10 illustrates forecasts using the AR(2) model with $\phi_1 = 0.9$ and $\phi_2 = -0.6$. The first six points in the time series, shown by the circles connected by heavy lines, are the same as the final six points in the time series shown in Figure 10.8c. The extrapolation of this series into the future, using Equation 10.27 with all $\varepsilon_{t+1} = 0$, is shown by the continuation of the heavy line connecting the x's. Note that only the

FIGURE 10.10 The final six points of the AR(2) time series in Figure 10.8c (heavy line, with circles), and its forecast evolution (heavy line, with x's) extrapolated using Equation 10.27 with all the $\varepsilon = 0$. The $\pm 2\sigma$ limits describing the uncertainty of the forecasts are shown with dashed lines. These standard deviations depend on the forecast lead time. For the 1-step ahead forecast, the width of the confidence interval is a function simply of the white-noise variance, $\pm 2\sigma_\varepsilon$. For very long lead times, the forecast series converges to the mean, μ, of the process, and the width of the confidence interval increases to $\pm 2\sigma_x$. Three example realizations of the first five points of the future evolution of the time series, simulated using Equation 10.27 and particular random ε values, are also shown (thin lines connecting numbered points).

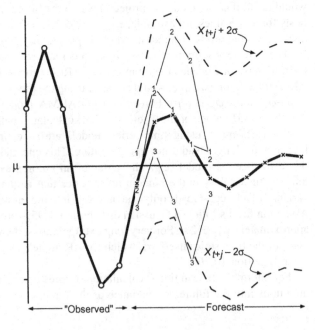

final two observed values are used to extrapolate the series. The forecast series continues to show the pseudoperiodic behavior characteristic of this AR(2) model, but its oscillations damp out at longer lead times as the forecast series approaches the mean μ.

The approximate 95% confidence intervals for the forecast time series values, given by $\pm 2\sigma(x_{t+j})$ as computed from Equation 10.47 are shown by the dashed lines in Figure 10.10. For the particular values of the autoregressive parameters $\phi_1 = 0.9$ and $\phi_2 = -0.6$, Equation 10.49 yields $\psi_1 = 0.90, \psi_2 = 0.21, \psi_3 = -0.35, \psi_4 = -0.44$, and so on. Note that the confidence band follows the oscillations of the forecast series, and broadens from $\pm 2\sigma_\varepsilon$ at a lead of one time unit to nearly $\pm 2\sigma_x$ at the longer lead times.

Finally, Figure 10.10 shows the relationship between the forecast time series mean, and the first five points of three realizations of this AR(2) process, shown by the thin lines connecting points labeled "1," "2," and "3." (In effect these represent a 3-member ensemble forecast for the future evolution of this time series.) Each of these three series was computed using Equation 10.27, starting from $x_t = -2.648$ and $x_{t-1} = -3.172$, but using different sequences of independent Gaussian ε's. For the first two or three projections these remain reasonably close to the (mean) forecasts. Subsequently the three series begin to diverge as the influence of the final two points from Figure 10.8c diminishes and the accumulated influence of the new (and different) random ε's increases. For clarity these series have not been plotted more than five time units into the future, although doing so would have shown each to oscillate irregularly, with progressively less relationship to the forecast mean series. $\diamond$

10.4. FREQUENCY DOMAIN—I. HARMONIC ANALYSIS

Analysis in the frequency domain involves representing data series in terms of contributions made at different timescales. For example, a time series of hourly temperature data from a midlatitude location usually will exhibit strong variations both at the daily timescale (corresponding to the diurnal cycle of solar heating) and at the annual timescale (reflecting the march of the seasons). In the time domain, these cycles would appear as large positive values in the autocorrelation function for lags at and near 24 hours for the diurnal cycle, and 24 x 365 = 8760 hours for the annual cycle. Thinking about the same time series in the frequency domain, we speak of large contributions to the total variability of the time series at periods of 24 and 8760 h, or at frequencies of 1/24 = 0.0417 h^{-1} and 1/8760 = 0.000114 h^{-1}.

Harmonic analysis consists of representing the fluctuations or variations in a time series as having arisen from the adding together of a series of sine and cosine functions. These trigonometric functions are "harmonic" in the sense that they are chosen to have frequencies exhibiting integer multiples of the fundamental frequency determined by the sample size (i.e., length) of the data series. A common physical analogy is the musical sound produced by a vibrating string, where the pitch is determined by the fundamental frequency, but the esthetic quality of the sound depends also on the relative contributions of the higher harmonics.

10.4.1. Cosine and Sine Functions

It is worthwhile to review briefly the nature of the cosine function $\cos(\alpha)$, and the sine function $\sin(\alpha)$. The argument in both is a quantity α, measured in angular units, which can be either degrees or radians. Figure 10.11 shows portions of the cosine (solid) and sine (dashed) functions, on the angular interval 0 to $5\pi/2$ radians ($0°$ to $450°$).

The cosine and sine functions extend through indefinitely large negative and positive angles. The same wave pattern repeats every 2π radians or $360°$, so that

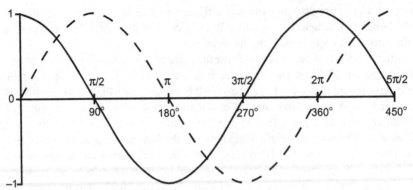

FIGURE 10.11   Portions of the cosine (solid) and sine (dashed) functions on the interval 0° to 450° or, equivalently, 0 to 5π/2 radians. Each executes a full cycle every 360°, or 2π radians, and extends left to −∞ and right to +∞.

$$\cos(2\pi k + \alpha) = \cos(\alpha), \qquad (10.50)$$

where k is any integer. An analogous equation holds for the sine function. That is, both the cosine and sine functions are periodic. Both functions oscillate around their average value of zero and attain maximum values of +1 and minimum values of −1. The cosine function is maximized at 0°, 360°, and so on, and the sine function is maximized at 90°, 450°, and so on.

These two functions have exactly the same shape but are offset from each other by 90°. Sliding the cosine function to the right by 90° produces the sine function, and sliding the sine function to the left by 90° produces the cosine function. That is,

$$\cos\left(\alpha - \frac{\pi}{2}\right) = \sin(\alpha) \qquad (10.51a)$$

and

$$\sin\left(\alpha + \frac{\pi}{2}\right) = \cos(\alpha). \qquad (10.51b)$$

10.4.2. Representing a Simple Time Series with a Harmonic Function

Even in the simple situation of time series having a sinusoidal character and executing a single cycle over the course of n observations, three small difficulties must be overcome in order to use a sine or cosine function to represent it. These are:

(1) The argument of a trigonometric function is an angle, whereas the data series is a function of time.
(2) Cosine and sine functions fluctuate between +1 and −1, but the data will generally fluctuate between different limits.
(3) The cosine function is at its maximum value for $\alpha = 0$ and $\alpha = 2\pi$, and the sine function is at its mean value for $\alpha = 0$ and $\alpha = 2\pi$. Both the sine and cosine may thus be positioned arbitrarily in the horizontal with respect to the data.

The solution to the first problem comes through regarding the length of the data record, n, as constituting a full cycle, or the fundamental period. Since the full cycle corresponds to 360° or 2π radians in angular measure, it is easy to proportionally rescale time to angular measure, using

$$\alpha = \left(\frac{360°}{\text{cycle}}\right)\left(\frac{t \text{ time units}}{n \text{ time units/cycle}}\right) = \frac{t}{n}360° \tag{10.52a}$$

or

$$\alpha = \left(\frac{2\pi}{\text{cycle}}\right)\left(\frac{t \text{ time units}}{n \text{ time units/cycle}}\right) = 2\pi\frac{t}{n}. \tag{10.52b}$$

These equations can be viewed as specifying the angle that subtends proportionally the same part of the angular distance between 0 and 2π, as the point t is located in time between 0 and n. The quantity

$$\omega_1 = \frac{2\pi}{n} \tag{10.53}$$

is called the *fundamental frequency*. This quantity is an angular frequency, having physical dimensions of radians per unit time. The fundamental frequency specifies the fraction of the full cycle, spanning n time units, that is executed during a single time unit. The subscript "1" indicates that ω_1 pertains to the wave that executes one full cycle over the whole data series.

The second problem is overcome by shifting a cosine or sine function up or down to the general level of the data, and then stretching or compressing it vertically until its range corresponds to that of the data. Since the mean of a pure cosine or sine wave is zero, simply adding the mean value of the data series to the cosine function assures that it will fluctuate around that mean value. The stretching or shrinking is accomplished by multiplying the cosine function by a constant, C_1, known as the *amplitude*. Again, the subscript indicates that this is the amplitude of the fundamental harmonic. Since the maximum and minimum values of a cosine function are ± 1, the maximum and minimum values of the function $C_1 \cos(\alpha)$ will be $\pm C_1$. Combining the solutions to these first two problems for a data series (call it y) yields

$$y_t = \bar{y} + C_1 \cos\left(\frac{2\pi t}{n}\right). \tag{10.54}$$

This function is plotted as the lighter curve in Figure 10.12. In this figure the horizontal axis indicates the equivalence of angular and time measure, through Equation 10.52, and the vertical shifting and stretching has produced a function fluctuating around the mean, with a range of $\pm C_1$.

Finally, it is usually necessary to shift a harmonic function laterally in order to have it match the peaks and troughs of a data series. This time shifting is most conveniently accomplished when the cosine function is used, because its maximum value is achieved when the angle on which it operates is zero. Shifting the cosine function to the right by the angle ϕ_1 results in a new function that is maximized at ϕ_1,

$$y_t = \bar{y} + C_1 \cos\left(\frac{2\pi t}{n} - \phi_1\right). \tag{10.55}$$

The angle ϕ_1 is called the *phase angle* or *phase shift*. Shifting the cosine function to the right by this amount requires subtracting ϕ_1, so that the argument of the cosine function is zero when $(2\pi t/n) = \phi_1$. Notice that by using Equation 10.51 it would be possible to rewrite Equation 10.55 using the sine function. However, the cosine usually is used as in Equation 10.55, because the phase angle can then be easily interpreted as corresponding to the time of the maximum of the harmonic function. That is, the function in Equation 10.55 is maximized at time $t = \phi_1 n/2\pi$.

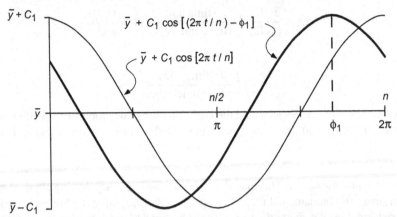

FIGURE 10.12   Transformation of a simple cosine function defined on 0 to 2π radians to a function representing a data series on the interval 0 to n time units. After changing from time to angular units, multiplying the cosine function by the amplitude C_1 stretches it so that it fluctuates through a range of $2C_1$. Adding the mean of the time series then shifts it to the proper vertical level, producing the lighter curve. The function can then be shifted laterally by subtracting the phase angle ϕ_1 that corresponds to the time of the maximum in the function (heavier curve).

Example 10.8. Transforming a Cosine Wave to Represent an Annual Cycle

Figure 10.13 illustrates the foregoing procedure using the 12 mean monthly temperatures (°F) for 1943–1989 at Ithaca, New York. Figure 10.13a is simply a plot of the 12 data points, with $t = 1$ indicating January, $t = 2$ indicating February, and so on. The overall annual average temperature of 46.1°F is located by the dashed horizontal line. These data appear to be at least approximately sinusoidal, executing a single full cycle over the course of the 12 months. The warmest mean temperature is 68.8°F in July and the coldest is 22.2°F in January.

The light curve at the bottom of Figure 10.13b is simply a cosine function with the argument transformed so that it executes one full cycle in 12 months. It is obviously a poor representation of the data. The dashed curve in Figure 10.13b shows this function lifted to the level of the average annual temperature, and stretched so that its range is similar to that of the data series (Equation 10.54). The stretching has been done only approximately, by choosing the amplitude C_1 to be half the difference between the July and January temperatures.

Finally, the cosine curve needs to be shifted to the right to line up well with the data. The maximum in the curve can be made to occur at $t = 7$ months (July) by introducing the phase shift, using Equation 10.52, of $\phi_1 = (7)(2\pi)/12 = 7\pi/6$. The result is the heavy curve in Figure 10.13b, which is of the form of Equation 10.55. This function lines up with the data points, albeit somewhat roughly. The correspondence between the curve and the data can be improved by using better estimators for the amplitude and phase of the cosine wave, as explained in the next section.   ◇

10.4.3. Estimation of the Amplitude and Phase of a Single Harmonic

The heavy curve in Figure 10.13 represents the associated temperature data reasonably well, but the correspondence will be improved with more refined choices for C_1 and ϕ_1. The easiest way to do this is to use the trigonometric identity

$$\cos(\alpha - \phi_1) = \cos(\phi_1)\cos(\alpha) + \sin(\phi_1)\sin(\alpha). \tag{10.56}$$

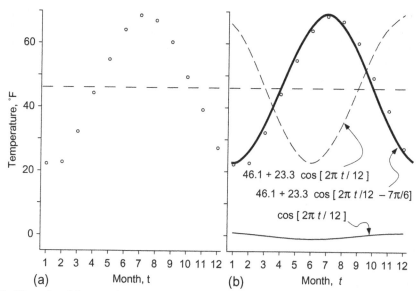

FIGURE 10.13   Illustration of the approximate matching of a cosine function to a data series. (a) Average monthly temperatures (° F) for Ithaca, New York for the years 1943–1989 (the data values are given in Table 10.5). The annual cycle of average temperature is evidently approximately sinusoidal. (b) Three cosine functions illustrating transformation from time to angular measure (light curve at bottom), vertical positioning and stretching (dashed curve), and lateral shifting (heavy curve) yielding finally the function matching the data approximately. The horizontal dashed lines indicate the average of the 12 data points, 46.1°F.

Substituting $\alpha = 2\pi\, t/n$ from Equation 10.52 and multiplying both sides by the amplitude C_1 yields

$$C_1 \cos\left(\frac{2\pi t}{n} - \phi_1\right) = C_1 \cos(\phi_1) \cos\left(\frac{2\pi t}{n}\right) + C_1 \sin(\phi_1) \sin\left(\frac{2\pi t}{n}\right)$$

$$= A_1 \cos\left(\frac{2\pi t}{n}\right) + B_1 \sin\left(\frac{2\pi t}{n}\right), \tag{10.57}$$

where

$$A_1 = C_1 \cos(\phi_1). \tag{10.58a}$$

and

$$B_1 = C_1 \sin(\phi_1). \tag{10.58b}$$

Equation 10.57 says that it is mathematically equivalent to represent a harmonic wave either as a cosine function with amplitude C_1 and phase ϕ_1, or as the sum of an unshifted cosine and unshifted sine wave with amplitudes A_1 and B_1.

For the purpose of estimating one or the other of these pairs of parameters from a set of data, the advantage of representing the wave using the second line of Equation 10.57 rather than Equation 10.55 derives from the fact that the former is a linear function of the parameters A_1 and B_1. Notice that making the variable transformations $x_1 = \cos(2\pi\, t/n)$ and $x_2 = \sin(2\pi\, t/n)$, and substituting these into the second line of Equation 10.57, produces what looks like a two-predictor regression equation with $A_1 = b_1$ and $B_1 = b_2$. In fact, given a data series y_t we can use this transformation together with ordinary regression software to find least-squares estimates of the parameters A_1 and B_1, with y_t as the predictand. Furthermore, the regression package will also produce the average of the predictand values as the intercept, b_0. Subsequently, the more convenient form of Equation 10.55 can be recovered by inverting Equations 10.58 to yield

$$C_1 = \left[A_1^2 + B_1^2\right]^{1/2} \tag{10.59a}$$

and

$$\phi_1 = \begin{cases} \tan^{-1}(B_1/A_1), & A_1 > 0 \\ \tan^{-1}(B_1/A_1) \pm \pi, & \text{or} \pm 180^\circ, \quad A_1 < 0. \\ \pi/2, \text{ or } 90^\circ, & A_1 = 0 \end{cases} \tag{10.59b}$$

Notice that since the trigonometric functions are periodic, effectively the same phase angle is produced by adding or subtracting a half-circle of angular measure if $A_1 < 0$. The alternative that yields $0 < \phi_1 < 2\pi$ is usually selected.

Finding the parameters A_1 and B_1 in Equation 10.57 using least-squares regression will work in the general case. For the special (although not too unusual) situation where the data are equally spaced in time with no missing values, the properties of the sine and cosine functions allow the same least-squares parameter values to be obtained more easily and efficiently using

$$A_1 = \frac{2}{n}\sum_{t=1}^{n} y_t \cos\left(\frac{2\pi t}{n}\right) \tag{10.60a}$$

and

$$B_1 = \frac{2}{n}\sum_{t=1}^{n} y_t \sin\left(\frac{2\pi t}{n}\right). \tag{10.60b}$$

Example 10.9. Harmonic Analysis of Average Monthly Temperatures

Table 10.5 presents the calculations necessary to obtain least-squares estimates for the parameters of the annual harmonic representing the Ithaca mean monthly temperatures plotted in Figure 10.13a, using Equations 10.60. The temperature data are shown in the column labeled y_t, and their average is easily computed as $552.9/12 = 46.1^\circ$F. The $n = 12$ terms of the sums in Equations 10.60a and 10.60b are shown in the last two columns. Applying Equations 10.60 to these yields $A_1 = (2/12)(-110.329) = -18.39$, and $B_1 = (2/12)(-86.417) = -14.40$.

Equation 10.59 transforms these two amplitudes to the parameters of the amplitude-phase form of Equation 10.55. This transformation allows easier comparison to the heavy curve plotted in Figure 10.13b. The amplitude is $C_1 = [-18.39^2 - 14.40^2]^{1/2} = 23.36^\circ$F, and the phase angle is $\phi_1 = \tan^{-1}(-14.40/-18.39) + 180^\circ = 218^\circ$. Here 180° has been added rather than subtracted, so that $0^\circ < \phi_1 < 360^\circ$. The least-squares amplitude of $C_1 = 23.36^\circ$F is quite close to the one used to draw Figure 10.13b, and the phase angle is 8° greater than the $(7)(360^\circ)/12 = 210^\circ$ angle that was eyeballed on the basis of the July mean being the warmest of the 12 months. The value of $\phi_1 = 218^\circ$ is a better estimate and implies a somewhat later (than mid-July) date for the time of the climatologically warmest temperature at this location. In fact, since there are very nearly as many degrees in a full cycle as there are days in one year, the results from Table 10.5 indicate that the heavy curve in Figure 10.13b should be

TABLE 10.5 Illustration of the Mechanics of Using Equations 10.60 to Estimate the Parameters of a Fundamental Harmonic

	y_t	$\cos(2\pi t/12)$	$\sin(2\pi t/12)$	$y_t \cos(2\pi t/12)$	$y_t \sin(2\pi t/12)$
1	22.2	0.866	0.500	19.225	11.100
2	22.7	0.500	0.866	11.350	19.658
3	32.2	0.000	1.000	0.000	32.200
4	44.4	−0.500	0.866	−22.200	38.450
5	54.8	−0.866	0.500	−47.457	27.400
6	64.3	−1.000	0.000	−64.300	0.000
7	68.8	−0.866	−0.500	−59.581	−34.400
8	67.1	−0.500	−0.866	−33.550	−58.109
9	60.2	0.000	−1.000	0.000	−60.200
10	49.5	0.500	−0.866	24.750	−42.867
11	39.3	0.866	−0.500	34.034	−19.650
12	27.4	1.000	0.000	27.400	0.000
Sums:	552.9	0.000	0.000	−110.329	−86.417

The data series y_t are the mean monthly temperatures at Ithaca for month t plotted in Figure 10.13a. Each of the 12 terms in Equations 10.60a and 10.60b, respectively, is shown in the last two columns.

shifted to the right by about one week. It is apparent that the result would be an improved correspondence with the data points. ◇

Example 10.10. Interpolation of the Annual Cycle to Average Daily Values

The calculations in Example 10.9 result in a smoothly varying representation of the annual cycle of mean temperature at Ithaca, based on the monthly values. Particularly if this were a location for which daily data were not available, it might be valuable to be able to use a function like this to represent the climatological average temperatures on a day-by-day basis. In order to employ the cosine curve in Equation 10.55 with time t in days, it would be necessary to use $n = 365$ days rather than $n = 12$ months. The amplitude can be left unchanged, although Epstein (1991) suggests a method to adjust this parameter that will produce a somewhat better representation of the annual cycle of daily values. In any case, however, it is necessary to make an adjustment to the phase angle.

Consider that the time $t = 1$ month represents all of January, and thus might be reasonably assigned to the middle of the month, perhaps the 15th. Thus the $t = 0$ months point of this function corresponds to the middle of December. Therefore when using $n = 365$ rather than $n = 12$, simply substituting the day number (1 January $= 1$, 2 January $= 2$, ..., 1 February $= 32$, etc.) for the time variable will result in a curve that is shifted too far left by about two weeks. What is required is a new phase angle, say ϕ_1', consistent with a time variable t' in days, that will position the cosine function correctly.

On 15 December, the two time variables are $t = 0$ months, and $t' = -15$ days. On 31 December, they are $t = 0.5$ month $= 15$ days, and $t' = 0$ days. Thus in consistent units, $t' = t - 15$ days, or $t = t' + 15$ days. Substituting $n = 365$ days and $t = t' + 15$ into Equation 10.55 yields

$$y_t = \bar{y} + C_1 \cos\left[\frac{2\pi t}{12} - \phi_1\right] = \bar{y} + C_1 \cos\left[\frac{2\pi (t' + 15)}{365} - \phi_1\right]$$

$$= \bar{y} + C_1 \cos\left[\frac{2\pi t'}{365} + 2\pi\frac{15}{365} - \phi_1\right]$$

$$= \bar{y} + C_1 \cos\left[\frac{2\pi t'}{365} - \left(\phi_1 - 2\pi\frac{15}{365}\right)\right] \quad (10.61)$$

$$= \bar{y} + C_1 \cos\left[\frac{2\pi t'}{365} - \phi_1'\right].$$

That is, the required new phase angle is $\phi_1' = \phi_1 - (2\pi)(15)/365$. ◇

10.4.4. Higher Harmonics

The computations in Example 10.9 produced a single cosine function passing quite close to the 12 monthly mean temperature values. This very good fit results because the shape of the annual cycle of temperature at this location is approximately sinusoidal, with a single full cycle being executed over the $n = 12$ points of the data series. We do not expect that a single harmonic wave will represent every time series this well. However, just as adding more predictors to a multiple regression will improve the fit to a set of training data, adding more cosine waves to a harmonic analysis will improve the fit to any time series.

Any data series consisting of n points can be represented exactly, meaning that a function can be found that passes through each of the points, by adding together a series of $n/2$ harmonic functions,

$$y_t = \bar{y} + \sum_{k=1}^{n/2}\left\{C_k \cos\left[\frac{2\pi kt}{n} - \phi_k\right]\right\} \quad (10.62a)$$

$$= \bar{y} + \sum_{k=1}^{n/2}\left\{A_k \cos\left[\frac{2\pi kt}{n}\right] + B_k \sin\left[\frac{2\pi kt}{n}\right]\right\}, \quad (10.62b)$$

which is sometimes called the *synthesis equation*. Equation 10.62b emphasizes that Equation 10.57 holds for any cosine wave, regardless of its frequency. The cosine wave that is the $k = 1$ term of Equation 10.62a is simply the fundamental, or first harmonic, that was the subject of the previous section. The other $n/2 - 1$ terms in the summation of Equation 10.62 are *higher harmonics*, or cosine waves with frequencies

$$\omega_k = \frac{2\pi k}{n} \quad (10.63)$$

that are integer multiples of the fundamental frequency ω_1.

For example, the second harmonic is that cosine function that completes exactly two full cycles over the n points of the data series. It has its own amplitude C_2 and phase angle ϕ_2. Notice that the factor k inside the cosine and sine functions in Equation 10.62a is of critical importance. When $k = 1$, the angle $\alpha = 2\pi kt/n$ varies through a single full cycle of 0 to 2π radians as the time index increased from $t = 0$ to $t = n$, as described earlier. In the case of the second harmonic where $k = 2$, $\alpha = 2\pi kt/n$ executes one full cycle as t increases from 0 to $n/2$, and then executes a second full cycle between $t = n/2$ and

$t = n$. Similarly, the third harmonic is defined by the amplitude C_3 and the phase angle ϕ_3, and varies through three cycles as t increases from 0 to n.

Equation 10.62b suggests that the coefficients A_k and B_k corresponding to particular data series y_t can be found using multiple regression methods, after the data transformations $x_1 = \cos(2\pi\ t/n)$, $x_2 = \sin(2\pi\ t/n)$, $x_3 = \cos(2\pi\ 2t/n)$, $x_4 = \sin(2\pi\ 2t/n)$, $x_5 = \cos(2\pi\ 3t/n)$, and so on. This is, in fact, the case in general, but if the data series is equally spaced in time and contains no missing values, Equation 10.60 generalizes to

$$A_k = \frac{2}{n} \sum_{t=1}^{n} y_t \cos\left(\frac{2\pi\ kt}{n}\right) \tag{10.64a}$$

and

$$B_k = \frac{2}{n} \sum_{t=1}^{n} y_t \sin\left(\frac{2\pi\ kt}{n}\right), \tag{10.64b}$$

which are sometimes called *analysis equations*. To compute a particular A_k, for example, these equations indicate than an n-term sum is formed, consisting of the products of the data series y_t with values of a cosine function executing k full cycles during the n time units. For relatively short data series these equations can be easily programmed or evaluated using spreadsheet software. For larger data series the A_k and B_k coefficients usually are computed using a more efficient method that will be mentioned in Section 10.5.3. Having computed these coefficients, the amplitude-phase form of the first line of Equation 10.62 can be arrived at by computing, separately for each harmonic,

$$C_k = \left[A_k^2 + B_k^2\right]^{1/2} \tag{10.65a}$$

and

$$\phi_k = \begin{cases} \tan^{-1}(B_k/A_k), & A_k > 0 \\ \tan^{-1}(B_k/A_k) \pm \pi,\ \text{or} \pm 180°, & A_k < 0. \\ \pi/2\ ,\text{or}\ 90°, & A_k = 0 \end{cases} \tag{10.65b}$$

Recall that a multiple regression function will pass through all the developmental data points, and exhibit $R^2 = 100\%$, if there are as many predictor values as data points. The series of cosine terms in Equation 10.62a is an instance of this overfitting principle, because there are two parameters (the amplitude and phase) for each harmonic term. Thus the $n/2$ harmonics in Equation 10.62 consist of n predictor variables, and any set of data, regardless of how untrigonometric it may look, can be represented exactly using Equation 10.62.

Since the sample mean in Equation 10.62 is effectively one of the estimated parameters, corresponding to the regression intercept b_0, an adjustment to Equation 10.62 is required if n is odd. In this case a summation over only $(n-1)/2$ harmonics is required to completely represent the function. That is, $(n-1)/2$ amplitudes plus $(n-1)/2$ phase angles plus the sample average of the data equals n. If n is even, there are $n/2$ terms in the summation, but the phase angle for the final and highest harmonic, $\phi_{n/2}$, is zero.

We may or may not want to use all $n/2$ harmonics indicated in Equation 10.62, depending on the context. Often for defining, say, an annual cycle of a climatological quantity, the first few harmonics may give a quite adequate representation from a practical standpoint, and will typically be more accurate than simpler sample averages (e.g., 12 discrete monthly mean values), in terms of representing future data values not used in the fitting (Narapusetty et al., 2009). If the goal is to find a function passing exactly through each of the data points, then all $n/2$ harmonics would be used. Recall that Section 7.4 warned against overfitting in the context of developing forecast equations, because the artificial skill exhibited on the developmental data does not carry forward when the equation is used to forecast future independent data. In this latter case the goal would not be to forecast but rather to represent the data, so that the over-fitting ensures that Equation 10.62 reproduces a particular data series exactly.

Example 10.11. A More Complicated Annual Cycle

Figure 10.14 illustrates the use of a small number of harmonics to smoothly represent the annual cycle of a climatological quantity. Here the quantity is the probability (expressed as a percentage) of five consecutive days without measurable precipitation, for El Paso, Texas. The irregular curve is a plot of the individual daily relative frequencies, computed using data for the years 1948–1983. These execute a regular but asymmetric annual cycle, with the wettest time of year being summer, and with dry springs and falls separated by a somewhat less dry winter. The figure also shows irregular, short-term fluctuations that have probably arisen mainly from sampling variations particular to the specific years analyzed. If a different sample of El Paso precipitation data had been used to compute the relative frequencies (say, 1984–2018), the same broad pattern would be evident, but the details of the individual "wiggles" would be different.

The annual cycle in Figure 10.14 is quite evident, yet it does not resemble a simple cosine wave. However, this cycle is reasonably well represented by the smooth curve, which is the sum of the first three harmonics. That is, the smooth curve is a plot of Equation 10.62 with three, rather than $n/2$, terms in the summation. The mean value for these data is 61.4%, and the parameters for the first two of these

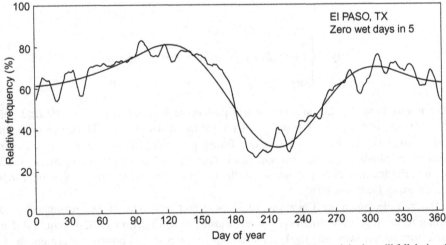

FIGURE 10.14   The annual cycle of the climatological probability that no measurable precipitation will fall during the five-day period centered on the date on the horizontal axis, for El Paso, Texas. Irregular line is the plot of the daily relative frequencies, and the smooth curve is a three-harmonic fit to the data. *From Epstein and Barnston (1988).*

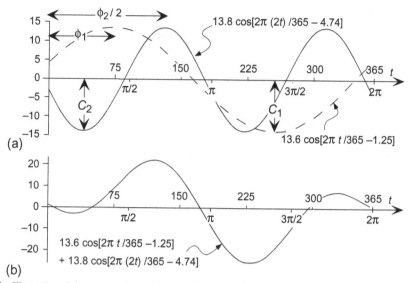

FIGURE 10.15 Illustration of the construction of the smooth curve in Figure 10.14. (a) The first (dashed) and second (solid) harmonics of the annual cycle plotted separately. These are defined by $C_1 = 13.6\%$, $\phi_1 = 72° = 0.4\,\pi$, $C_2 = 13.8\%$, and $\phi_2 = 272° = 1.51\,\pi$. The horizontal axis is labeled both in days and radians. (b) The smoothed representation of the annual cycle is produced by adding the values of the two functions in panel (a) for each time point. Subsequently adding the annual mean value of 61.4% produces a curve very similar to that in Figure 10.14. The small differences are accounted for by the third harmonic. Note that panels (a) and (b) have different vertical scales.

harmonics are $C_1 = 13.6\%$, $\phi_1 = 72° = 0.4\,\pi$, $C_2 = 13.8\%$, and $\phi_2 = 272° = 1.51\,\pi$. These values were computed from the underlying data using Equations 10.64 and 10.65. Computing and plotting the sum of all possible $(365{-}1)/2 = 182$ harmonics would result in a function identical to the irregular curve in Figure 10.14.

Figure 10.15 illustrates the construction of the smooth curve representing the annual cycle in Figure 10.14. Panel (a) shows the first (dashed) and second (solid) harmonics plotted separately, both as a function of time (t) in days and as a function of the corresponding angular measure in radians. Also indicated are the magnitudes of the amplitudes C_k in the vertical, and the correspondence of the phase angles ϕ_k to the maxima of the two functions. Note that since the second harmonic executes two cycles during the full 365 days of the year, there are two times of maximum, located at $\phi_2/2$ and $\pi + \phi_2/2$. (The maxima for the third harmonic would occur at $\phi_3/3$, $2\pi/3 + \phi_3/3$, and $4\pi/3 + \phi_3/3$, with a similar pattern holding for the higher harmonics.)

The curve in Figure 10.15b has been constructed by simply adding the values for the two functions in Figure 10.15a at each time point. (Note that the two panels in Figure 10.15 have been plotted using different vertical scales.) During times of the year where the two harmonics are of opposite sign but comparable magnitude, their sum is near zero. The maximum and minimum of the function in Figure 10.15b are achieved when its two components have relatively large magnitudes of the same sign. Adding the annual mean value of 61.4% to the lower curve results in a close approximation to the smooth curve in Figure 10.14, with the small differences between the two attributable to the third harmonic. ◇

10.5. FREQUENCY DOMAIN—II. SPECTRAL ANALYSIS

10.5.1. The Harmonic Functions as Uncorrelated Regression Predictors

Equation 10.62b suggests the use of multiple regression to find best-fitting harmonics for a given data series y_t. But for equally spaced data with no missing values Equations 10.64b will produce the same least-squares estimates for the coefficients A_k and B_k as will multiple regression software. Notice, however, that Equations 10.64 do not depend on any harmonic other than the one whose coefficients are being computed. That is, these equations depend on the current value of k, but not $k-1$, or $k-2$, or any other harmonic index. This fact implies that the coefficients A_k and B_k for any particular harmonic can be computed independently of those for any other harmonic.

Recall that usually regression parameters need to be recomputed each time a new predictor variable is entered into a multiple regression equation, and each time a predictor variable is removed from a regression equation. As noted in Section 7.3, this recomputation is necessary in the general case of sets of predictor variables that are mutually correlated, because correlated predictors carry redundant information to a greater or lesser extent. It is a remarkable property of the harmonic functions that (for equally spaced and complete data) they are uncorrelated so, for example, the parameters (amplitude and phase) for the first or second harmonic are the same whether or not they will be used in an equation with the third, fourth, or any other harmonics.

This attribute of the harmonic functions is a consequence of what is called the *orthogonality* property of the sine and cosine functions. That is, for integer harmonic indices k and j,

$$\sum_{t=1}^{n} \cos\left(\frac{2\pi kt}{n}\right) \sin\left(\frac{2\pi jt}{n}\right) = 0, \text{ for any integer values of } k \text{ and } j; \tag{10.66a}$$

and

$$\sum_{t=1}^{n} \cos\left(\frac{2\pi kt}{n}\right) \cos\left(\frac{2\pi jt}{n}\right) = \sum_{t=1}^{n} \sin\left(\frac{2\pi kt}{n}\right) \sin\left(\frac{2\pi jt}{n}\right) = 0, \text{ for } k \neq j. \tag{10.66b}$$

To illustrate, consider the two transformed predictor variables $x_1 = \cos[2\pi t/n]$ and $x_3 = \cos[2\pi(2t)/n]$. The Pearson correlation between these derived variables is given by

$$r_{x_1 x_3} = \frac{\sum_{t=1}^{n} (x_1 - \bar{x}_1)(x_3 - \bar{x}_3)}{\left[\sum_{t=1}^{n} (x_1 - \bar{x}_1)^2 \sum_{t=1}^{n} (x_3 - \bar{x}_3)^2\right]}, \tag{10.67a}$$

and since the averages $\bar{x}_1$ and $\bar{x}_3$ of cosine functions over integer numbers of cycles are zero,

$$r_{x_1 x_3} = \frac{\sum_{t=1}^{n} \cos\left(\frac{2\pi t}{n}\right) \cos\left(\frac{2\pi 2t}{n}\right)}{\left[\sum_{t=1}^{n} \cos^2\left(\frac{2\pi t}{n}\right) \cos^2\left(\frac{2\pi 2t}{n}\right)\right]^{1/2}} = 0, \tag{10.67b}$$

because the numerator is zero by Equation 10.66b.

Since the relationships between harmonic predictor variables and the data series y_t do not depend on what other harmonic functions are also being used to represent the series, the proportion of the variance of y_t accounted for by each harmonic is also fixed. Expressing this proportion as the R^2 statistic commonly computed in regression, the R^2 for the kth harmonic is simply

$$R_k^2 = \frac{(n/2)\,C_k^2}{(n-1)\,s_y^2}. \tag{10.68}$$

In terms of the regression ANOVA table, the numerator of Equation 10.68 is the regression sum of squares for the kth harmonic. The factor s_y^2 is simply the sample variance of the data series, so the denominator of Equation 10.68 is the total sum of squares, SST. Notice that the strength of the relationship between the kth harmonic and the data series can be expressed entirely in terms of the ratio C_k/s_y. The phase angle ϕ_k is necessary only to determine the positioning of the cosine curve in time. Furthermore, since each harmonic provides independent information about the data series, the joint R^2 exhibited by a regression equation with only harmonic predictors is simply the sum of the R_k^2 values for each of the harmonics,

$$R^2 = \sum_{k\text{ in the equation}} R_k^2. \tag{10.69}$$

If all the $n/2$ possible harmonics are used as predictors (Equation 10.62), then the total R^2 in Equation 10.69 will be exactly 1.

Equation 10.62 says that a data series y_t of length n can be specified completely in terms of the n parameters of $n/2$ harmonic functions. Equivalently, we can take the view that the data y_t are transformed into new set of quantities A_k and B_k according to Equations 10.64. For this reason, Equations 10.64 are called the *discrete Fourier transform*. The data series can be represented as the n quantities C_k and ϕ_k, obtained from the A_k's and B_k's using the transformations in Equations 10.65. According to Equations 10.68 and 10.69, this data transformation accounts for all of the variation in the series y_t. Another perspective on Equations 10.68 and 10.69 is that the variance of the time-series variable y_t can be apportioned among the $n/2$ harmonic functions, each of which represents a different timescale of variation.

10.5.2. The Periodogram or Fourier Line Spectrum

The foregoing suggests that a different way to look at a time series is as a collection of Fourier coefficients A_k and B_k that are a function of frequency ω_k (Equation 10.63), rather than as a collection of data points y_t measured as a function of time. The advantage of this new perspective is that it allows us to see separately the contributions to a time series that are made by processes varying at different speeds, that is, by processes operating at a spectrum of different frequencies. Panofsky and Brier (1958) illustrate this distinction with an analogy: "An optical spectrum shows the contributions of different wave lengths or frequencies to the energy of a given light source. The spectrum of a time series shows the contributions of oscillations with various frequencies to the variance of a time series." Even if the underlying physical basis for a data series y_t is not really well represented by a series of cosine waves, often much can still be learned about the data by viewing it from this perspective.

The characteristics of a time series that has been Fourier transformed into the frequency domain are most often examined graphically, using a plot known as the *periodogram*, or *Fourier line spectrum*. This plot sometimes is called the *power spectrum*, or simply the spectrum, of the data series. In simplest form, the plot of a spectrum consists of the squared amplitudes C_k^2 as a function of the frequencies ω_k. Note that information contained in the phase angles ϕ_k is not portrayed in the spectrum. Therefore the spectrum conveys the proportion of variation in the original data series accounted for by oscillations at the harmonic frequencies, but does not supply information about when in time these oscillations are expressed. Fisher (2006) suggests that this characteristic is similar to what would be achieved by representing the frequencies of use of the various pitches in a piece of music with a histogram. Such a histogram might well identify the musical key, but not the piece of music itself. A spectrum thus does not provide a full picture of the behavior of the time series from which it has been calculated and is not sufficient to reconstruct the time series.

The vertical axis of a plotted spectrum is sometimes numerically rescaled, in which case the plotted points are proportional to the squared amplitudes. One choice for this proportional rescaling is that in Equation 10.68. It is also common for the vertical axis of a spectrum to be plotted on a logarithmic scale. Plotting the vertical axis logarithmically is particularly useful if the variations in the time series are dominated by harmonics of only a few frequencies. In this case a linear plot might result in the remaining spectral components being invisibly small. A logarithmic vertical axis also regularizes the representation of confidence limits for the spectral estimates (Section 10.5.6).

The horizontal axis of the line spectrum consists of $n/2$ frequencies ω_k if n is even, and $(n-1)/2$ frequencies if n is odd. The smallest of these will be the lowest frequency $\omega_1 = 2\pi/n$ (the fundamental frequency), and this corresponds to the cosine wave that executes a single cycle over the n time points. The highest frequency, $\omega_{n/2} = \pi$, is called the *Nyquist frequency*. It is the frequency of the cosine wave that executes a full cycle over only two time intervals, and which executes $n/2$ cycles over the full data record. The Nyquist frequency depends on the time resolution of the original data series y_t, and imposes an important limitation on the information available from a spectral analysis.

The horizontal axis of a plotted spectrum is often simply the angular frequency, ω, with units of radians/time. A common alternative is to use the frequencies

$$f_k = \frac{k}{n} = \frac{\omega_k}{2\pi}, \tag{10.70}$$

which have dimensions of time^{-1}. Under this alternative convention, the allowable frequencies range from $f_1 = 1/n$ for the fundamental to $f_{n/2} = 1/2$ for the Nyquist frequency. The horizontal axis of a spectrum can also be scaled according to the reciprocal of the frequency, or the period of the kth harmonic,

$$\tau_k = \frac{n}{k} = \frac{2\pi}{\omega_k} = \frac{1}{f_k}. \tag{10.71}$$

The period τ_k specifies the length of time required for a cycle of frequency ω_k to be completed. Associating periods with the periodogram estimates can help visualize the timescales on which the important variations in the data are occurring.

Example 10.12. Discrete Fourier Transform of a Small Data Set

Table 10.6 presents a simple data set and its discrete Fourier transform. The leftmost columns contain the observed average monthly temperatures at Ithaca, New York, for the two years 1987 and 1988. This is such a familiar type of data that, even without doing a spectral analysis, we know in advance that the primary feature will be the annual cycle of cold winters and warm summers. This expectation is validated by the plot of the data in Figure 10.16a, which shows these temperatures as a function of time. The overall impression is of a data series that is approximately sinusoidal with a period of 12 months, but that a single cosine wave with this period would not pass exactly through all the points.

TABLE 10.6 Average Monthly Temperatures, °F, at Ithaca, New York, for 1987–1988, and Their Discrete Fourier Transform

Month	1987	1988	k	τ_k, months	A_k	B_k	C_k
1	21.4	20.6	1	24.00	−0.14	0.44	0.46
2	17.9	22.5	2	12.00	−23.76	−2.20	23.86
3	35.9	32.9	3	8.00	−0.99	0.39	1.06
4	47.7	43.6	4	6.00	−0.46	−1.25	1.33
5	56.4	56.5	5	4.80	−0.02	−0.43	0.43
6	66.3	61.9	6	4.00	−1.49	−2.15	2.62
7	70.9	71.6	7	3.43	−0.53	−0.07	0.53
8	65.8	69.9	8	3.00	−0.34	−0.21	0.40
9	60.1	57.9	9	2.67	1.56	0.07	1.56
10	45.4	45.2	10	2.40	0.13	0.22	0.26
11	39.5	40.5	11	2.18	0.52	0.11	0.53
12	31.3	26.7	12	2.00	0.79	—	0.79

Columns 4 to 8 of Table 10.6 show the same data after being subjected to the discrete Fourier transform. Since $n = 24$ is an even number, the data are completely represented by $n/2 = 12$ harmonics. These are indicated by the rows labeled by the harmonic index, k. Column 5 of Table 10.6 indicates the period (Equation 10.71) of each of the 12 harmonics used to represent the data. The period of the fundamental frequency, $\tau_1 = 24$ months, is equal to the length of the data record. Since there are two annual cycles in the $n = 24$ month record, it is the $k = 2^{nd}$ harmonic with period $\tau_2 = 24/2 = 12$ months that is expected to be most important. The Nyquist frequency is $\omega_{12} = \pi$ radians/month, or $f_{12} = 0.5$ month^{-1}, corresponding to the period $\tau_{12} = 2$ months.

The coefficients A_k and B_k that could be used in Equation 10.62 to reconstruct the original data are shown in the next columns of the table. These constitute the discrete Fourier transform of the data series of temperatures. Notice that there are only 23 Fourier coefficients, because 24 independent pieces of information are necessary to fully represent the $n = 24$ data points, including the sample mean of 46.1°F. To use the synthesis Equation 10.62 to reconstitute the data, we would substitute $B_{12} = 0$.

Column 8 in Table 10.6 shows the amplitudes C_k, computed according to Equation 10.65a. The phase angles could also be computed, using Equation 10.65b, but these are not needed to plot the spectrum. Figure 10.16b shows the spectrum for these temperature data, plotted in the form of a histogram. The vertical axis consists of the squared amplitudes C_k^2, normalized according to Equation 10.68 to show the R^2 attributable to each harmonic. The horizontal axis is linear in frequency, but the corresponding periods are also shown, to aid the interpretation. Clearly most of the variation in the data is described by the second harmonic, the R^2 for which is 97.5%. As expected, the variations of the annual cycle dominate these data, but the fact that the amplitudes of the other harmonics are not zero indicates that the data do not consist of a pure cosine wave with a frequency of $f_2 = 1$ year^{-1}. Notice, however, that the logarithmic vertical axis tends to deemphasize the smallness of these other harmonics. If the vertical axis had been scaled linearly, the plot would consist of a spike at $k = 2$ and a small bump at $k = 6$, with the rest of the points being essentially indistinguishable from the horizontal axis. ◇

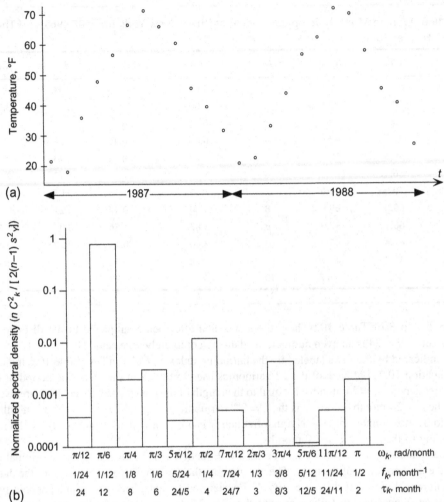

FIGURE 10.16   Illustration of the relationship between a simple time series and its spectrum. (a) Average monthly temperatures at Ithaca, New York for 1987–1988, from Table 10.6. The data are approximately sinusoidal, with a period of 12 months. (b) The spectrum of the data series in panel (a), plotted in the form of a histogram, and expressed in the normalized form of Equation 10.68. Clearly the most important variations in the data series are represented by the second harmonic, with period $\tau_2 = 12$ months, which is the annual cycle. Note that the vertical scale is logarithmic, so that the next most important harmonic accounts for barely more than 1% of the total variation. The horizontal scale is linear in frequency.

Example 10.13. Another Sample Spectrum

A less trivial example spectrum is shown in Figure 10.17. This is a portion of the (smoothed) spectrum of the monthly Tahiti minus Darwin sea-level pressure time series for 1951–1979. That time series resembles the (normalized) SOI index shown in Figure 3.16, including the tendency for quasiperiodic behavior. That the variations in the time series are not strictly periodic is evident from the irregular variations in Figure 3.16, and from the broad (i.e., spread over many frequencies) maximum in the spectrum.

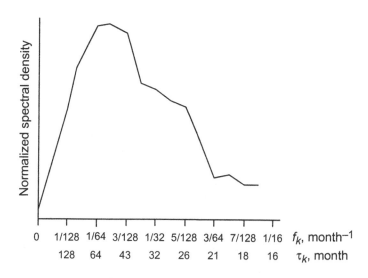

FIGURE 10.17 The low-frequency portion of the smoothed spectrum for the monthly time series of Tahiti minus Darwin sea-level pressures, 1951–1979. The underlying time series resembles that in Figure 3.16, and the tendency for oscillations to occur in roughly three- to seven-year cycles is reflected in the broad maximum of the spectrum in this range. *After Chen (1982). © American Meteorological Society. Used with permission.*

Figure 10.17 indicates that the typical length of one of these pseudocycles (corresponding to typical times between El Niño events) is something between $\tau = [(1/36) \text{ mo}^{-1}]^{-1} = 3$ years and $\tau = [(1/84) \text{ mo}^{-1}]^{-1} = 7$ years.

The vertical axis in Figure 10.17 has been plotted on a linear scale, but units have been omitted because they do not contribute to a qualitative interpretation of the plot. The horizontal axis is linear in frequency, and therefore nonlinear in period. Notice also that the labeling of the horizontal axis indicates that the full spectrum of the underlying data series has not been presented in the figure. Since the data series consists of monthly values, the Nyquist frequency must be 0.5 month^{-1}, corresponding to a period of two months. Only the left-most one-eighth of the spectrum has been shown because it is these lower frequencies that reflect the physical phenomenon of interest, namely, the ENSO cycle. The estimated spectral density function for the omitted higher frequencies would show only a long, irregular, and generally uninformative right tail exhibiting small spectral density estimates.   ◇

10.5.3. Computing Spectra

One way to compute the spectrum of a data series is simply to apply Equations 10.64 and then to find the amplitudes using Equation 10.65a. This is a reasonable approach for relatively short data series and can be programmed easily using, for example, spreadsheet software. These equations would be implemented only for $k = 1, 2, \ldots, (n/2-1)$. Because we want exactly n Fourier coefficients (A_k and B_k) to represent the n points in the data series, the computation for the highest harmonic, $k = n/2$, is done using

$$A_{n/2} = \begin{cases} \left(\dfrac{1}{2}\right)\left(\dfrac{2}{n}\right) \displaystyle\sum_{t=1}^{n} y_t \cos\left[\dfrac{2\pi(n/2)t}{n}\right] = \left(\dfrac{1}{n}\right) \displaystyle\sum_{t=1}^{n} y_t \cos[\pi t], & n \text{ even} \\[2em] 0, & n \text{ odd} \end{cases} \tag{10.72a}$$

and

$$B_{n/2} = 0, \quad n \text{ even or odd.} \tag{10.72b}$$

Although straightforward notationally, this method of computing the discrete Fourier transform is quite inefficient computationally. In particular, many of the calculations called for by Equation 10.64 are redundant. Consider, for example, the data for April 1987 in Table 10.6. The term for $t = 4$ in the summation in Equation 10.64b is $(47.7°F) \sin[(2\pi)(1)(4)/24] = (47.7°F)(0.866) = 41.31°F$. However, the term involving this same data point for $k = 2$ is exactly the same: $(47.7°F) \sin[(2\pi)(2)(4)/24] = (47.7°F)(0.866) = 41.31°F$. There are many other such redundancies in the computation of discrete Fourier transforms using Equations 10.64. These can be avoided through the use of clever algorithms known as *Fast Fourier Transforms* (FFTs) (Cooley and Tukey, 1965). Most scientific software packages include one or more FFT routines, which give very substantial speed improvements, especially as the length of the data series increases. In comparison to computation of the Fourier coefficients using a regression approach, an FFT is approximately $n/\log_2(n)$ times faster; or about 15 times faster for $n = 100$, and about 750 times faster for $n = 10000$.

It is worth noting that FFTs usually are documented and implemented in terms of the *Euler complex exponential notation*,

$$e^{i\omega t} = \cos(\omega t) + i \sin(\omega t), \tag{10.73}$$

where i is the unit imaginary number satisfying $i = \sqrt{-1}$, and $i^2 = -1$. Complex exponentials are used rather than sines and cosines purely as a notational convenience that makes some of the manipulations less cumbersome. The mathematics are still entirely the same. In terms of complex exponentials, Equation 10.62 becomes

$$y_t = \bar{y} + \sum_{k=1}^{n/2} H_k\, e^{i\,[2\pi\, k/n]t}, \tag{10.74}$$

where H_k is the complex Fourier coefficient

$$H_k = A_k + i B_k. \tag{10.75}$$

That is, the real part of H_k is the coefficient A_k, and the imaginary part of H_k is the coefficient B_k.

10.5.4. Aliasing

Aliasing is a hazard inherent in spectral analysis of discrete data. It arises because of the limits imposed by the sampling interval, or the time between consecutive pairs of data points. Since a minimum of two points are required to even sketch a cosine wave—one point for the peak and one point for the trough—the highest representable frequency is the Nyquist frequency, with $\omega_{n/2} = \pi$, or $f_{n/2} = 0.5$. A wave of this frequency executes one cycle every two data points, and thus a discrete data set can represent explicitly variations that occur no faster than this speed.

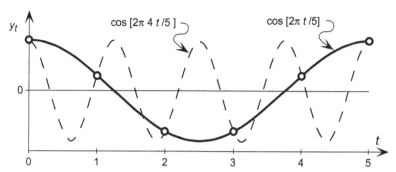

FIGURE 10.18   Illustration of the basis of aliasing. Heavy circles represent data points in a time series y_t. Fitting a harmonic function to them results in the heavy curve. However, if the data series actually had been produced by the process indicated by the light dashed curve, the fitted heavy curve would present the misleading impression that the source of the data was actually fluctuating at the lower frequency. The lighter curve has not been sampled densely enough because its frequency, $\omega = 8\pi/5$ (or $f = 4/5$), is higher than the Nyquist frequency of $\omega = \pi$ (or $f = 1/2$). Variations at the frequency of the dashed curve are said to be aliased to the frequency of the heavier curve.

It is worth wondering what happens to the spectrum of a data series if it includes an important physical process that varies faster than the Nyquist frequency. If so, the data series is said to be *under-sampled*, which means that the points in the time series are spaced too far apart to properly capture these fast variations. However, variations that occur at frequencies higher than the Nyquist frequency do not disappear. Rather, their contributions are spuriously attributed to some lower but representable frequency, between ω_1 and $\omega_{n/2}$. These high-frequency variations are said to be aliased.

Figure 10.18 illustrates the meaning of aliasing. Imagine that the physical data-generating process is represented by the dashed cosine curve. The data series y_t is produced by sampling this process at integer values of the time index t, resulting in the points indicated by the circles. However, the frequency of the dashed curve ($\omega = 8\pi/5$, or $f = 4/5$) is higher than the Nyquist frequency ($\omega = \pi$, or $f = 1/2$), meaning that it oscillates too quickly to be adequately sampled at this time resolution. If only the information in the discrete time points is available, these data look like the heavy cosine function, the frequency of which ($\omega = 2\pi/5$, or $f = 1/5$) is lower than the Nyquist frequency, and is therefore representable. Note that because the cosine functions are orthogonal, this same effect will occur whether or not variations of different frequencies are also present in the data.

The effect of aliasing on spectral analysis is that any energy (squared amplitudes) attributable to processes varying at frequencies higher than the Nyquist frequency will be erroneously added to that of one of the $n/2$ frequencies that are represented by the discrete Fourier spectrum. A frequency $f_A > 1/2$ will be aliased into one of the representable frequencies f (with $0 < f \le 1/2$) if it differs by an integer multiple of 1 time^{-1}, that is, if

$$f_A = j \pm f, j \text{ any positive integer.} \qquad (10.76a)$$

In terms of angular frequency, variations at the aliased frequency ω_A appear to occur at the representable frequency ω if

$$\omega_A = 2\pi j \pm \omega, j \text{ any positive integer.} \qquad (10.76b)$$

These equations imply that the squared amplitudes for frequencies higher than the Nyquist frequency will be added to the representable frequencies in an accordion-like pattern, with each "fold" of the accordion occurring at an integer multiple of the Nyquist frequency. For this reason the Nyquist frequency is sometimes called the *folding frequency*. An aliased frequency f_A that is just slightly higher than the Nyquist frequency of $f_{n/2} = 1/2$ is aliased to a frequency slightly lower than 1/2. Frequencies only slightly lower than twice the Nyquist frequency are aliased to frequencies only slightly higher than zero. The pattern then reverses for $2f_{n/2} < f_A < 3f_{n/2}$. That is, frequencies just higher than $2f_{n/2}$ are aliased to very low frequencies, and frequencies almost as high as $3f_{n/2}$ are aliased to frequencies near $f_{n/2}$.

Figure 10.19 illustrates the effects of aliasing on a hypothetical spectrum. The gray line represents the true spectrum, which exhibits a concentration of power at low frequencies, but also has a sharp peak at $f = 5/8$ and a broader peak at $f = 19/16$. These second two peaks occur at frequencies higher than the Nyquist frequency of $f = 1/2$, which means that the physical process that generated the data was not sampled often enough to resolve them explicitly. The variations actually occurring at the frequency $f_A = 5/8$ are aliased to (i.e., appear to occur at) the frequency $f = 3/8$. That is, according to Equation 10.76a, $f_A = 1 - f = 1 - 3/8 = 5/8$. In the spectrum, the squared amplitude for $f_A = 5/8$ is added to the (genuine) squared amplitude at $f = 3/8$ in the true spectrum. Similarly, the variations represented by the broader hump centered at $f_A = 19/16$ in the true spectrum are aliased to frequencies at and around $f = 3/16$ ($f_A = 1 + f = 1 + 3/16 = 19/16$). The dashed line in Figure 10.19 indicates the portions of the aliased spectral energy (the total area between the gray and black lines) contributed by frequencies between $f = 1/2$ and $f = 1$ (the area below the dashed line), and by frequencies between $f = 1$ and $f = 3/2$ (the area above the dashed line); emphasizing the fan-folded nature of the aliased spectral density.

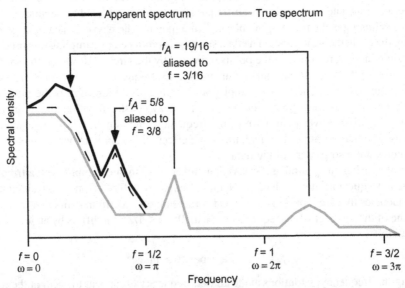

FIGURE 10.19   Illustration of aliasing in a hypothetical spectrum. The true spectrum (gray line) exhibits a sharp peak at $f = 5/8$, and a broader peak at $f = 19/16$. Since both of these frequencies are higher than the Nyquist frequency $f = 1/2$, they are aliased in the spectrum (erroneously attributed) to the frequencies indicated. The aliasing follows an accordion-like pattern, with the area between the gray line and the dashed line contributed by frequencies from $f = 1$ to $f = 1/2$, and the area between the dashed line and the heavy line contributed by frequencies between $f = 1$ and $f = 3/2$. The resulting apparent spectrum (heavy line) includes both the true spectral density values for frequencies between zero and 1/2, as well as the contributions from the aliased frequencies.

Aliasing can be particularly severe when isolated segments of a time series are averaged and then analyzed, for example, a time series of average January values in each of n years. This problem has been studied by Madden and Jones (2001), who conclude that badly aliased spectra are expected to result unless the averaging time is at least as large as the sampling interval. For example, a spectrum for January averages is expected to be heavily aliased since the one-month averaging period is much shorter than the annual sampling interval.

Unfortunately, once a data series has been collected, there is no way to "de-alias" its spectrum. That is, it is not possible to tell from the data values alone whether appreciable contributions to the spectrum have been made by frequencies higher than $f_{n/2}$, or how large these contributions might be. In practice, it is desirable to have an understanding of the physical basis of the processes generating the data series, so that it can be seen in advance that the sampling rate is adequate. Of course in an exploratory setting this advice is of no help, since the point of an exploratory analysis is exactly to gain a better understanding of an unknown or a poorly known generating process. In this latter situation, we would like to see the spectrum approach zero for frequencies near $f_{n/2}$, which would give some hope that the contributions from higher frequencies are minimal. A spectrum such as the heavy line in Figure 10.19 would lead us to expect that aliasing might be a problem, since its not being essentially zero at the Nyquist frequency could well mean that the true spectrum is nonzero at higher frequencies as well.

10.5.5. The Spectra of Autoregressive Models

Another perspective on the time-domain autoregressive models described in Section 10.3 is provided by their spectra. The types of time dependence produced by different autoregressive models produce characteristic spectral signatures that can be related to the autoregressive parameters.

The simplest case is the AR(1) process, Equation 10.16. Here positive values of the single autoregressive parameter ϕ induce a memory into the time series that tends to smooth over short-term (high-frequency) variations in the ε series, and emphasize the slower (low-frequency) variations. In terms of the spectrum, these effects lead to more density at lower frequencies and less density at higher frequencies. Furthermore, these effects are progressively stronger for ϕ closer to 1.

These ideas are quantified by the spectral density function for AR(1) processes,

$$S(f) = \frac{4\sigma_\varepsilon^2/n}{1 + \phi^2 - 2\phi \cos(2\pi f)}, 0 \leq f \leq 1/2. \tag{10.77}$$

This is a function that associates a spectral density with all frequencies in the range $0 \leq f \leq 1/2$. The shape of the function is determined by the denominator, and the numerator contains scaling constants that give the function numerical values that are comparable to the empirical spectrum of squared amplitudes, C_k^2. This equation also applies for negative values of the autoregressive parameter, which produce time series tending to oscillate quickly around the mean, and for which the spectral density is greatest at the high frequencies.

Note that, for zero frequency, Equation 10.77 is proportional to the variance of a time average. This can be appreciated by substituting $f = 0$, and Equations 10.21 and 10.39 into Equation 10.77, and comparing to Equation 10.36. Thus the extrapolation of the spectrum to zero frequency has been used to estimate variances of time averages (e.g., Madden and Shea, 1978).

Figure 10.20 shows spectra for the AR(1) processes having $\phi = 0.5, 0.3, 0.0$, and -0.6. The autocorrelation functions for the first and last of these are shown as insets in Figure 10.7. As might have been

FIGURE 10.20 Spectral density functions for four AR(1) processes, computed using Equation 10.77. Autoregressions with $\phi > 0$ are red-noise processes, since their spectra are enriched at the lower frequencies and depleted at the higher frequencies. The spectrum for the $\phi = 0$ process (i.e., serially independent data) is flat, exhibiting no tendency to emphasize either high- or low-frequency variations. This is a white-noise process. The autoregression with $\phi = -0.6$ tends to exhibit rapid variations around its mean, which results in a spectrum enriched in the high frequencies, or a blue-noise process. Autocorrelation functions for the $\phi = 0.5$ and $\phi = -0.6$ processes are shown as insets in Figure 10.7.

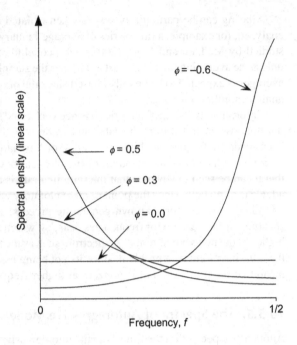

anticipated, the two processes with $\phi > 0$ show enrichment of the spectral densities in the lower frequencies and depletion in the higher frequencies, and these characteristics are stronger for the process with the larger autoregressive parameter. By analogy to the properties of visible light, AR(1) processes with $\phi > 0$ are sometimes referred to as *red-noise* processes.

AR(1) processes with $\phi = 0$ consist of series of temporally uncorrelated data values $x_{t+1} = \mu + \varepsilon_{t+1}$ (compare Equation 10.16). These exhibit no tendency to emphasize either low-frequency or high-frequency variations, so their spectrum is constant or flat. Again by analogy to visible light, this is called *white noise* because of the equal mixture of all frequencies. This analogy is the basis of the independent series of ε's being called the white-noise forcing, and of the parameter σ_ε^2 being known as the white-noise variance.

Finally, the AR(1) process with $\phi = -0.6$ tends to produce erratic short-term variations in the time series, resulting in negative correlations at odd lags and positive correlations at even lags. (This kind of correlation structure is rare in atmospheric time series.) The spectrum for this process is thus enriched at the high frequencies and depleted at the low frequencies, as indicated in Figure 10.20. Such series are accordingly known as *blue-noise* processes.

Expressions for the spectra of other autoregressive processes, and for ARMA processes as well, are given in Box and Jenkins (1976). Of particular importance is the spectrum for the AR(2) process,

$$S(f) = \frac{4\sigma_\varepsilon^2/n}{1 + \phi_1^2 + \phi_2^2 - 2\phi_1(1 - \phi_2)\cos(2\pi f) - 2\phi_2\cos(4\pi f)}, 0 \le f \le 1/2. \qquad (10.78)$$

This equation reduces to Equation 10.77 for $\phi_2 = 0$, since an AR(2) process (Equation 10.27) with $\phi_2 = 0$ is simply an AR(1) process.

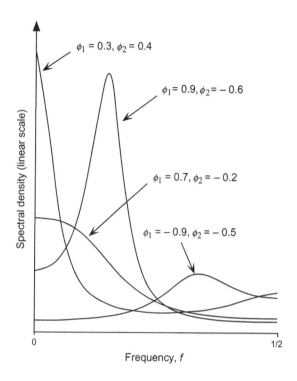

FIGURE 10.21 Spectral density functions for four AR(2) processes, computed using Equation 10.78. The diversity of the forms of the of spectra in this figure illustrates the flexibility of the AR(2) model. The autocorrelation functions for these autoregressions are shown as insets in Figure 10.7.

The AR(2) processes are particularly interesting because of their capacity to exhibit a wide variety of behaviors, including pseudoperiodicities. This diversity is reflected in the various forms of the spectra that are included in Equation 10.78. Figure 10.21 illustrates a few of these, corresponding to the AR(2) processes whose autocorrelation functions are shown as insets in Figure 10.7. The processes with $\phi_1 = 0.9$, $\phi_2 = -0.6$, and $\phi_1 = -0.9$, $\phi_2 = -0.5$, exhibit pseudoperiodicities, as indicated by the broad humps in their spectra at intermediate frequencies. The process with $\phi_1 = 0.3$, $\phi_2 = 0.4$ exhibits most of its variation at low frequencies, but also shows a smaller maximum at high frequencies. The spectrum for the process with $\phi_1 = 0.7$, $\phi_2 = -0.2$ resembles the red-noise spectra in Figure 10.20, although with a broader and flatter low-frequency maximum.

Example 10.14 Smoothing a Sample Spectrum Using an Autoregressive Model

The equations for spectra of autoregressive models can be useful in interpreting sample spectra from data series. The erratic sampling properties of the individual periodogram estimates, as described in Section 10.5.6, can make it difficult to discern features of the true spectrum that underlie a particular sample spectrum. However, if a well-fitting time-domain model can be estimated from the same data series, its spectrum can be used to guide the eye through the sample spectrum. Autoregressive models are sometimes fitted to time series for the sole purpose of obtaining smooth spectra. Chu and Katz (1989) show that the spectrum corresponding to a time-domain model fit using the SOI time series (see Figure 10.17) corresponds well to the spectrum computed directly from the data.

Consider the data series in Figure 10.8c, which was generated according to the AR(2) process with $\phi_1 = 0.9$ and $\phi_2 = -0.6$. The sample spectrum for this particular batch of 50 points is shown as the solid

FIGURE 10.22 Illustration of the use of the spectrum of a fitted autoregressive model to guide the eye in interpreting a sample spectrum. The solid curve is the sample spectrum for the $n = 50$ data points shown in Figure 10.8c, generated by the AR(2) process with $\phi_1 = 0.9$, $\phi_2 = -0.6$, and $\sigma_\varepsilon^2 = 1.0$. A fuller perspective on this spectrum is provided by the dashed line, which is the spectrum of the AR(2) process fitted to this same series of 50 data points.

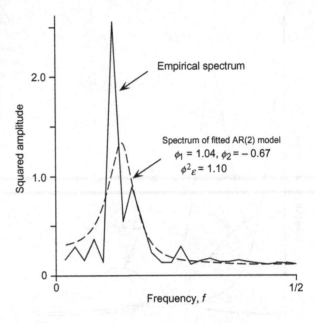

function in Figure 10.22. Apparently the series exhibits pseudoperiodicities in the frequency range around $f = 0.12$ through $f = 0.16$, but sampling variability makes the interpretation somewhat difficult. Although the empirical spectrum in Figure 10.22 somewhat resembles the theoretical spectrum for this AR(2) model shown in Figure 10.21, its nature might not be obvious from the empirical spectrum alone.

A fuller perspective on the spectrum in Figure 10.22 is gained when the dashed curve is provided to guide the eye. This is the spectrum for an AR(2) model fitted to the same data points from which the empirical spectrum was computed. The first two sample autocorrelations for these data are $r_1 = 0.624$ and $r_2 = -0.019$, which are near the true generating-process values that would be obtained from Equation 10.33. Using Equation 10.29, the corresponding estimated autoregressive parameters are $\phi_1 = 1.04$ and $\phi_2 = -0.67$. The sample variance of the $n = 50$ data points is 1.69, which leads through Equation 10.30 to the estimated white-noise variance $\sigma_\varepsilon^2 = 1.10$. The resulting spectrum, according to Equation 10.78, is plotted as the dashed curve. $\diamond$

10.5.6. Sampling Properties of Spectral Estimates

Since the data from which empirical spectra are computed are subject to sampling fluctuations, Fourier coefficients computed from these data will exhibit random batch-to-batch variations as well. That is, different data batches of size n from the same source will transform to somewhat different C_k^2 values, resulting in somewhat different sample spectra.

Each squared amplitude is an unbiased estimator of the true spectral density, which means that averaged over many batches the mean of the many C_k^2 values at any frequency would closely approximate their true generating-process counterpart. Another favorable property of raw sample spectra is that the periodogram estimates at different frequencies are uncorrelated with each other. Unfortunately, the sampling distribution for an individual C_k^2 is rather broad. In particular, the sampling distribution of suitably scaled squared amplitudes is the χ^2 distribution with $v = 2$ degrees of freedom, which is an exponential distribution, or a gamma distribution having $\alpha = 1$ (compare Figure 4.9).

The particular scaling of the raw spectral estimates that has this χ^2 sampling distribution is

$$\frac{v C_k^2}{S(f_k)} \sim \chi_v^2, \qquad (10.79)$$

where $S(f_k)$ is the (unknown, true) spectral density being estimated by C_k^2, and $v = 2$ degrees of freedom for a single spectral estimate C_k^2. Note that the various choices that can be made for multiplicative scaling of periodogram estimates will cancel in the ratio on the left-hand side of Equation 10.79. One way of appreciating the appropriateness of the χ^2 sampling distribution is to realize that the Fourier amplitudes in Equation 10.64 will be approximately Gaussian-distributed according to the Central Limit Theorem, because they are each derived from sums of n terms. Each squared amplitude C_k^2 is the sum of the squares of its respective pair of amplitudes A_k^2 and B_k^2, and the χ^2 is the distribution of the sum of v squared independent standard Gaussian variates (cf. Section 4.4.5). Because the sampling distributions of the squared Fourier amplitudes in Equation 10.65a are not standard Gaussian, the scaling constants in Equation 10.79 are necessary to produce a χ^2 distribution.

Because the sampling distribution of an individual periodogram estimate is exponential, these estimates are strongly positively skewed, and their standard errors (standard deviation of the sampling distribution) are equal to their means. An unhappy consequence of these properties is that the individual C_k^2 estimates represent the true spectrum rather poorly. This very erratic nature of raw spectral estimates is illustrated by the two sample spectra shown in Figure 10.23. The heavy and light lines are two sample spectra computed from different batches of $n = 30$ independent Gaussian random variables. Each of the two sample spectra varies rather wildly around the true spectrum, which is shown by the dashed horizontal line. In a real application, the true spectrum is, of course, not known in advance, and Figure 10.23 shows that the poor sampling properties of the individual spectral estimates can make it very difficult to discern much about the true spectrum if only a single sample spectrum is available.

Confidence limits for the underlying population quantities corresponding to raw spectral estimates are rather broad. Equation 10.79 implies that confidence interval widths are proportional to the raw periodogram estimates themselves, so that

$$\Pr\left[\frac{v C_k^2}{\chi_v^2 (1 - \alpha/2)} < S(f_k) \leq \frac{v C_k^2}{\chi_v^2 (\alpha/2)}\right] = 1 - \alpha, \qquad (10.80)$$

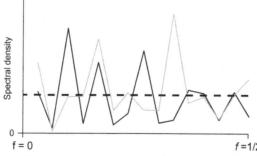

FIGURE 10.23 Illustration of the erratic sampling characteristics of estimated spectra. The solid and gray curves are two sample spectra, each computed using different batches of $n = 30$ independent Gaussian random variables. Both are quite erratic, with points of relative agreement being more fortuitous than meaningful. The true spectrum for the underlying serially independent data is shown by the horizontal dashed line. The vertical axis is linear.

where again $v = 2$ for a single raw periodogram estimate, and $\chi^2_v(\alpha)$ is the α quantile of the appropriate χ^2 distribution. For example, $\alpha = 0.05$ for a 95% confidence interval. The form of Equation 10.80 suggests one reason that it can be convenient to plot spectra on a logarithmic scale, since in that case the widths of the $(1-\alpha) \bullet 100\%$ confidence intervals are constant across frequencies, regardless of the magnitudes of the estimated C^2_k.

The usual remedy in statistics for an unacceptably broad sampling distribution is to increase the sample size. For spectra, however, simply increasing the sample size does not give more precise information about any of the individual frequencies, but rather results in equally imprecise information about more frequencies. For example, the spectra in Figure 10.23 were computed from $n = 30$ data points, and thus consist of $n/2 = 15$ squared amplitudes. Doubling the sample size to $n = 60$ data values would result in a spectrum at $n/2 = 30$ frequencies, each point of which would exhibit the same large sampling variations as the individual C^2_k values in Figure 10.23.

It is possible, however, to use larger data samples to obtain sample spectra that are more representative of the underlying population spectra. One approach is to compute replicate spectra from separate sections of the time series and then to average the resulting squared amplitudes. In the context of Figure 10.23, for example, a time series of $n = 60$ could be split into two series of length $n = 30$. The two spectra in Figure 10.23 might be viewed as having resulted from such a process. Here averaging each of the $n/2 = 15$ pairs of C^2_k values would result in a less erratic spectrum, that somewhat more faithfully represents the true spectrum. In fact the sampling distributions of each of these $n/2$ average spectral values would be proportional (Equation 10.79) to the χ^2 distribution with $v = 4$, or a gamma distribution with $\alpha = 2$, as each would be proportional to the sum of four squared Fourier amplitudes. This distribution is substantially less variable and less strongly skewed than the exponential distribution, having standard deviation of $1/\sqrt{2}$ of the averaged estimates, or about 70% of the previous individual ($v = 2$) estimates. If we had a data series with $n = 300$ points, 10 sample spectra could be computed whose average would smooth out a large fraction of the sampling variability evident in Figure 10.23. The sampling distribution for the averaged squared amplitudes in that case would have $v = 20$. The standard deviations for these averages would be smaller by the factor $1/\sqrt{10}$, or about one-third of the magnitudes of those for single squared amplitudes. Since the confidence interval widths are still proportional to the estimated squared amplitudes, a logarithmic vertical scale again results in plotted confidence interval widths not depending on frequency. A drawback to this procedure is that the lowest frequencies are no longer resolved.

An essentially equivalent approach to obtaining a smoother and more representative spectrum using more data begins with computation of the discrete Fourier transform for the longer data series. Although this results at first in more spectral estimates that are equally variable, their sampling variability can be smoothed out by adding (not averaging) the squared amplitudes for groups of adjacent frequencies. The spectrum shown in Figure 10.17 has been smoothed in this way. For example, if we wanted to estimate the spectrum at the 15 frequencies plotted in Figure 10.23, these could be obtained by summing consecutive pairs of the 30 squared amplitudes obtained from the spectrum of a data record that was $n = 60$ observations long. If $n = 300$ observations were available, the spectrum at these same 15 frequencies could be estimated by adding the squared amplitudes for groups of 10 of the $n/2 = 150$ original frequencies. Here again the sampling distribution is χ^2 with v equal to twice the number of pooled frequencies, or gamma with α equal to the number of pooled frequencies.

A variety of more sophisticated smoothing functions are commonly applied to sample spectra (e.g., Ghil et al., 2002; Jenkins and Watts, 1968; Von Storch and Zwiers, 1999). For example, a common approach to smoothing the sampling variability in a raw spectrum is to construct running weighted averages of the raw values,

$$\overline{C}_k^2 = \sum_{j=k-(m-1)/2}^{k+(m-1)/2} w_{j-k} C_j^2. \tag{10.81}$$

Here the averaging window covers an odd number m of adjacent frequencies; and the weights w, the indices for which run from $-(m-1)/2$ to $(m-1)/2$ are nonnegative, symmetric about the zero index, and sum to one. For example, running means of successive triplets would be achieved using Equation 10.81 with $m = 3$, and $w_{-1} = w_0 = w_1 = 1/3$. A $m = 3$-point triangular filter could be constructed using $w_{-1} = 1/4$, $w_0 = 1/2$, and $w_1 = 1/4$. Equation 10.80 for the confidence interval of one of these smoothed spectral estimates again applies, with degrees of freedom given by

$$v = \frac{2}{\sum\limits_{j=-(m-1)/2}^{(m-1)/2} w_j^2}, \tag{10.82}$$

so, for example, $v = 6$ for the three-point running mean with equal weights.

Unfortunately, spectra smoothed using overlapping averaging windows yields individual spectral estimates that are no longer statistically independent of each other. Note also that, regardless of the specific form of the smoothing procedure, the increased smoothness and representativeness of the resulting spectra come at the expense of decreased frequency resolution and introduction of bias. Essentially, stability of the sampling distributions of the spectral estimates is obtained by smearing spectral information from a range of frequencies across a frequency band. Smoothing across broader bands produces less erratic estimates, but hides sharp contributions that may be made at particular frequencies. In practice, there is always a compromise to be made between sampling stability and frequency resolution, which is resolved as a matter of subjective judgment.

It is sometimes of interest to investigate whether the largest C_k^2 among K such squared amplitudes is significantly different from a hypothesized population value. That is, has the largest periodogram estimate plausibly resulted from sampling variations in the Fourier transform of data arising from a purely random process, or does it reflect a real periodicity that may be partially hidden by random noise in the time series? Addressing this question is complicated by two issues: choosing a null spectrum that is appropriate to the data series, and accounting for test multiplicity if the frequency f_k corresponding to the largest C_k^2 is chosen according to the test data rather than on the basis of external, prior information.

Initially we might adopt the white-noise spectrum (Equation 10.77, with $\phi = 0$) to define the null hypothesis. This could be an appropriate choice if there is little or no prior information about the nature of the data series, or if we expect in advance that the possible periodic signal is embedded in uncorrelated noise. However, most atmospheric time series are positively autocorrelated, and usually a null spectrum reflecting this tendency is a preferable null reference function (Gilman et al., 1963). Commonly it is the AR(1) spectrum (Equation 10.77) that is chosen for the purpose, with ϕ and σ_ε^2 fit to the data whose spectrum is being investigated. Using Equation 10.79, the null hypothesis that the squared amplitude C_k^2 at frequency f_k is significantly larger than the null (possibly red-noise) spectrum at that frequency, $S_0(f_k)$, would be rejected at the α level if

$$C_k^2 \geq \frac{S_0(f_k)}{v} \chi_v^2 (1-\alpha), \tag{10.83}$$

where $\chi_v^2 (1-\alpha)$ denotes right-tail quantiles of the appropriate chi-square distribution, given in Table B.3. The parameter v may be greater than 2 if spectral smoothing has been employed.

The rejection rule given in Equation 10.83 is appropriate if the frequency f_k being tested has been chosen on the basis of prior, or external information, so that the choice is in no way dependent on the data used to calculate the C_k^2. When such prior information is lacking, testing the statistical significance of the largest squared amplitude is complicated by the problem of test multiplicity. When, in effect, K independent hypothesis tests are conducted in the search for the most significant squared amplitude, direct application of Equation 10.83 results in a test that is substantially less stringent than the nominal level, α. If the K spectral estimates being tested are uncorrelated, dealing with this multiplicity problem is reasonably straightforward, and involves choosing a nominal test level small enough that Equation 10.83 specifies the correct rejection rule when applied to the *largest* of the K squared amplitudes. Walker (1914) derived the appropriate level for computing the exact values,

$$\alpha = 1 - (1 - \alpha*)^{1/K}, \tag{10.84}$$

use of which is called the *Walker test* (Katz, 2002 provides more historical context). The derivation of Equation 10.84 is based on the sampling distribution of the smallest of K independent p values (Wilks, 2006a). The resulting individual test levels, α, to be used in Equation 10.83 to yield a true probability $\alpha*$ of falsely rejecting the null hypothesis that the largest of K periodogram estimates is significantly larger than the null spectral density at that frequency, are closely approximated by those calculated using the Bonferroni method (Section 12.5.3),

$$\alpha = \alpha*/K. \tag{10.85}$$

In order to account for the test multiplicity, Equation 10.85 chooses a nominal test level α that is smaller than the actual test level $\alpha*$, and that reduction is proportional to the number of frequencies (i.e., independent tests) being considered. Alternatively, the FDR approach (Section 5.4.2) could be used for this purpose. In either case the result is that a relatively large C_k^2 is required in order to reject the null hypothesis in the properly reformulated test.

Example 10.15. Statistical Significance of the Largest Spectral Peak Relative to a Red-Noise H_0

Imagine a hypothetical time series of length $n = 200$ for which the sample estimates of the lag-1 autocorrelation and white-noise variance are $r_1 = 0.6$ and $s_\varepsilon^2 = 1$, respectively. A reasonable candidate to describe the behavior of these data as a purely random series could be the AR(1) process with these two parameters. Substituting these values into Equation 10.77 yields the spectrum for this process, shown as the heavy curve in Figure 10.24. A sample spectrum, C_k^2, $k = 1, \ldots, 100$, could also be computed from this series. This spectrum would include squared amplitudes at $K = 100$ frequencies because $n = 200$ data points have been Fourier transformed. The sample spectrum will be rather erratic whether or not the series also contains one or more periodic components, and it may be of interest to calculate how large the largest C_k^2 must be in order to infer that it is significantly different from the null red spectrum at that frequency. Equation 10.83 provides the decision criterion.

Because $K = 100$ frequencies are being searched for the largest squared amplitude, the standard of proof must be much more stringent than if a particular single frequency had been chosen for testing in advance of seeing the data. In particular, Equation 10.83 and Equation 10.85 both show that a test at the $\alpha* = 0.10$ level requires that the largest of the 100 squared amplitudes trigger a test rejection at the nominal $\alpha = 0.10/100 = 0.001$ level, and a test at the $\alpha* = 0.01$ level requires the nominal test level $\alpha = 0.01/100 = 0.0001$. Each squared amplitude in the unsmoothed sample spectrum follows a χ^2 distribution with $\nu = 2$ degrees of freedom, so the relevant right-tail quantiles $\chi_2^2(1-\alpha)$ from the second line of Table B.3 are $\chi_2^2(0.999) = 13.816$ and $\chi_2^2(0.9999) = 18.421$, respectively. (Because $\nu = 2$ these limits

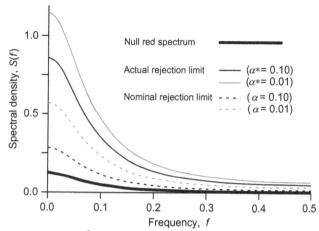

FIGURE 10.24   Red spectrum for $\phi_1 = 0.6$, $\sigma_\varepsilon^2 = 1.0$, and $n = 200$ (heavy curve) with minimum values necessary to conclude that the largest of $K = 100$ periodogram estimates is significantly larger (lighter solid curves) at the 0.10 (black) and 0.01 (gray) levels. Dashed curves show erroneous minimum values resulting when test multiplicity is not accounted for.

can also be calculated using the quantile function for the exponential distribution, Equation 4.94, with $\beta = 2$.) Substituting these values into Equation 10.83, and using Equation 10.77 with $\phi_1 = 0.6$ and $\sigma_\varepsilon^2 = 1$ to define $S_0(f_k)$, yields the two light solid lines in Figure 10.24. If the largest of the $K - 100$ C_k^2 values does not rise above these curves, the null hypothesis that the series arose from a purely random AR(1) process cannot be rejected at the specified α^* levels.

The dashed curves in Figure 10.24 are the rejection limits computed in the same way as the solid curves, except that the nominal test levels α have been taken to be equal to the overall test levels α^*, so that $\chi_2^2(0.90) = 4.605$ and $\chi_2^2(0.99) = 9.210$ have been used in Equation 10.83. These dashed curves would be appropriate thresholds for rejecting the null hypothesis that the estimated spectrum, at a single frequency that had been chosen in advance without reference to the data being tested, had resulted from sampling variations in the null red-noise process. If these thresholds were to be used to evaluate the largest among $K = 100$ squared amplitudes, the probabilities according to Equation 10.84 of falsely rejecting the null hypothesis if it were true would be $\alpha^* = 0.634$ and $\alpha^* = 0.99997$ (i.e., virtual certainty), at the nominal $\alpha = 0.01$ and $\alpha = 0.10$ levels, respectively.

Choice of the null spectrum can also have a large effect on the test results. If instead a white spectrum—Equation 10.77, with $\phi = 0$, implying $\sigma_x^2 = 1.5625$ (cf. Equation 10.21)—had been chosen as the baseline against which to judge potentially significant squared amplitudes, the null spectrum in Equation 10.83 would have been $S_0(f_k) = 0.031$ for all frequencies. In that case, the rejection limits would be parallel horizontal lines with magnitudes comparable to those at $f = 0.15$ in Figure 10.24.   ◇

10.6. TIME-FREQUENCY ANALYSES

Although spectral analysis can be very useful for characterizing the important timescales of variability of a data series, the fact that it does not use the phase information from the Fourier transform of the data implies that the locations of these variations in time cannot be represented. For example, consider the two artificial time series in Figure 10.25. Figure 10.25a shows the function $[\cos(2\pi\,11t/n) + \cos(2\pi\,19t/n)]/2$, which is superposition of the 11th and 19th harmonics for the full data record. Figure 10.25b plots $\cos(2\pi$

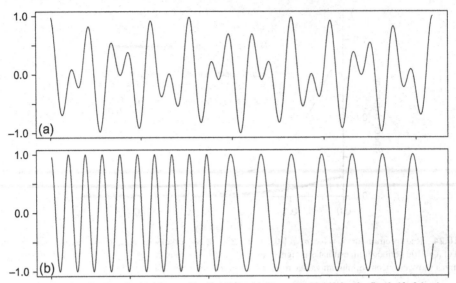

FIGURE 10.25   (a) The function [cos(2π 11t/n) + cos(2π 19t/n)]/2, and (b) cos(2π 19t/n)] for the first half of the time series and cos(2π 11t/n) for the last half. Fourier spectra computed from the two series would be identical.

19t/n) for the first $n/2$ points of the time series and cos(2π 11t/n) for the last $n/2$. The power spectra corresponding to these two series are identical.

10.6.1. Windowed Fourier Transforms

Windowed Fourier transforms offer one approach to time localization of spectral features. This method involves choosing a segment length $l < n$, and computing spectra for consecutive time-series subsets of length l as the "window" is moved along the series. That is, the first spectrum computed would be for the time points $t = 1$ to l, the second would be for $t = 2$ to $l + 1$, and so on. To construct the time-frequency plot, each of these $n - l + 1$ spectra is rotated 90° counterclockwise so that the low-frequency amplitudes are at the bottom and the high-frequency amplitudes are at the top, and the individual spectra are placed consecutively in time order. The result is a two-dimensional plot in which the vertical axis is frequency and the horizontal axis is time. Horizontal nonuniformity in the plot will indicate where different frequency components are more or less important during different portions of the time series, and the overlapping nature of the consecutive windowed time series will generally yield smooth variations in the horizontal.

Figure 10.26a illustrates the result for the SOI series for the years 1876–1995, a different segment of which is shown in Figure 3.16. Here the window length l is 21 years, the initial and final years of which have been tapered to zero using a cosine shape. Consistent with the usual understanding of ENSO, maximum power is exhibited in the range of about 0.14 to 0.33 cycles/per year (periods of about 3–7 years, as indicated also in Figures 10.7 and 10.26c). However, Figure 10.26a indicates that these strongest SOI variations do not occur uniformly through the record, but rather are stronger in the first and last quarters of the 20th century, with a more quiescent period in the middle of the century.

The time resolution provided by windowed Fourier transforms comes at the cost that the method is unable to see longer-period variations. In particular, it cannot resolve variations with periods longer than the window length l. Choosing a larger l improves the low-frequency resolution, but degrades the time resolution, so that a balance between these two desirable attributes must be struck.

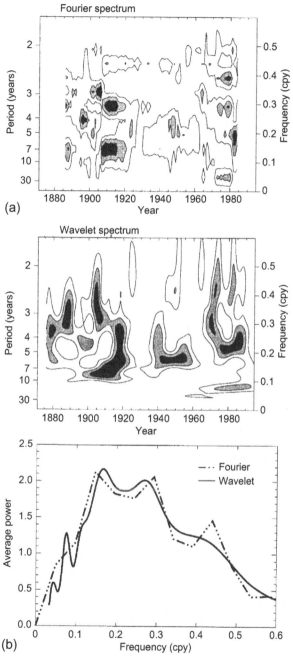

FIGURE 10.26 (a) Windowed Fourier transform, and (b) wavelet spectrum for the SOI series, 1876–1995. Contours indicate 20%, 40%, and 60% of maximum spectral power. (c) Time-integrated and standardized spectra from panel (a) (dash-dot) and (b) (solid). *From Kestin et al. (1998). © American Meteorological Society. Used with permission.*

10.6.2. Wavelets

Wavelet analysis is an alternative to windowed Fourier transforms that also yields a two-dimensional plot showing strengths of variations as a function of both period (or frequency) and time. Unlike Fourier analysis, which characterizes similarities between time series and trigonometric functions of infinite extent, wavelet analysis addresses similarities, over limited portions of the time series, to waves of limited time extent called wavelets.

A variety of choices for the basic wavelet shape are available (e.g., Torrance and Compo 1998), but all have zero mean and are nonzero over a limited portion of the real line. Often the Morlet wavelet is chosen in geophysical settings because of its similarity in shape to typical variations in those kinds of data. The Morlet wavelet consists of a sinusoidal wave that is damped for increasing absolute values of its arguments using a Gaussian shape,

$$\psi(t) = \pi^{-1/4} e^{i\omega_0 t} e^{-(t^2/2)}, \tag{10.86}$$

where often the constant $\omega_0 = 6$ is chosen. Figure 10.27 shows the shape of this wavelet function. It is nonzero for absolute values of its argument smaller than about 3.

The time localization in wavelet analysis is achieved through multiplication of the time series with a wavelet shape ψ that is centered at various points t' in the series, in effect sliding the wavelet shape along the time series and computing its similarity to that portion of the time series over which the wavelet is nonzero. Different frequencies are resolved by repeating this process for different dilations $s, s > 1$ of the basic wavelet shape, because larger values of s expand the portion of the time series for which the scaled wavelet function is nonzero, and its peaks and troughs are more fully separated. The result is a two-dimensional array of wavelet coefficients

$$W(s, t') = \sum_{t=1}^{n} y_t \frac{1}{\sqrt{s}} \psi^* \left(\frac{t - t'}{s} \right), \tag{10.87}$$

where the asterisk denotes complex conjugate. Although the necessary computations can be carried out using Equation 10.87, more efficient algorithms are available (e.g., Torrence and Compo, 1998). Figure 10.26b shows the resulting wavelet spectrum for the 1876-1995 SOI series, using the Morlet wavelet. Here the coefficients $W(s, t')$ have been plotted in two dimensions and smoothed. The basic features portrayed by the windowed Fourier transform of these data in Figure 10.26a are also present in their wavelet spectrum.

Wavelet analysis can be generalized to representing the variations in two-dimensional images in terms of a spectrum of spatial scales. One application of two-dimensional wavelets is in spatial field verification (Section 9.8.5). Another is in in the JPEG format for photographic image compression.

FIGURE 10.27 Basic shape of the Morlet wavelet, with real part solid and imaginary part dashed. *From Torrence and Compo (1998). © American Meteorological Society. Used with permission.*

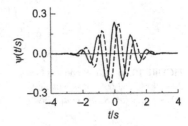

Both the windowed Fourier transforms and the wavelet spectra would be different for the two time series in Figure 10.25, and both analysis methods would allow the different characteristics of the two time series to be diagnosed. For the superposition of the two underlying waves in Figure 10.25a, both approaches would yield horizontal bands at the frequencies $\omega_{11} = 2\pi\,11/n$ and $\omega_{19} = 2\pi\,19/n$, that would extend across the full widths of the diagrams. For the series in Figure 10.25b, both would exhibit horizontal bands at $\omega_{19} = 2\pi\,19/n$ across the left halves of the diagrams, and separate horizontal bands at $\omega_{11} = 2\pi\,11/n$ extending across the right halves.

10.7. EXERCISES

10.1. Using the January 1987 precipitation data for Canandaigua in Table A.1,
    a. Fit a two-state, first-order Markov chain to represent daily precipitation occurrence.
    b. Test whether this Markov model provides a significantly better representation of the data than does the assumption of independence.
    c. Compare the theoretical stationary probability, π_1 with the empirical relative frequency.
    d. Graph the theoretical autocorrelation function, for the first three lags.
    e. Compute the probability according to the Markov model that a sequence of consecutive wet days will last at least three days.

10.2. Graph the autocorrelation functions up to five lags for
    a. The AR(1) process with $\phi = 0.4$.
    b. The AR(2) process with $\phi_1 = 0.7$ and $\phi_2 = -0.7$.

10.3. Computing sample lag correlations for a time series with $n = 100$ values, whose variance is 100, yields $r_1 = 0.80$, $r_2 = 0.60$, and $r_3 = 0.50$.
    a. Use the Yule–Walker equations to fit AR(1), AR(2), and AR(3) models to the data. Assume the sample size is large enough that Equation 10.26 provides a good estimate for the white-noise variance.
    b. Select the best autoregressive model for the series according to the BIC statistic.
    c. Select the best autoregressive model for the series according to the AIC statistic.

10.4. Given that the mean of the time series in Exercise 10.3 is 50, use the fitted AR(2) model to forecast the future values of the time series x_1, x_2, and x_3; assuming the current value is $x_0 = 76$ and the previous value is $x_{-1} = 65$.

10.5. The variance of a time series governed by the AR(1) model with $\phi = 0.8$, is 25. Compute the variances of the sampling distributions of averages of consecutive values of this time series, with lengths
    a. $n = 5$,
    b. $n = 10$,
    c. $n = 50$.

10.6. A square-root transformed time series of hourly wind speeds is well described using an AR(2) model with parameters $\phi_1 = 1.09$ and $\phi_2 = -0.21$, and having Gaussian residuals with variance $\sigma_\varepsilon^2 = 0.12\,\text{m/s}$. The mean of the transformed series is 2.3 $(\text{m/s})^{1/2}$.
    a. Calculate the variance of the transformed wind speeds.
    b. If the wind speed in the current hour is 15 m/s and the wind speed for the previous hour was 10 m/s, calculate a 50% central prediction interval for the wind speed (m/s) in the coming hour.

10.7. For the temperature data in Table 10.7,
a. Calculate the first two harmonics.
b. Plot each of the two harmonics separately.
c. Plot the function representing the annual cycle defined by the first two harmonics. Also include the original data points in this plot, and visually compare the goodness of fit.

TABLE 10.7 Average Monthly Temperature Data for New Delhi, India

Month	J	F	M	A	M	J	J	A	S	O	N	D
Average Temperature, °F	57	62	73	82	92	94	88	86	84	79	68	59

10.8. Use the two-harmonic equation for the annual cycle from Exercise 10.7 to estimate the mean daily temperatures for
a. April 10.
b. October 27.
10.9. The amplitudes of the third, fourth, fifth, and sixth harmonics, respectively, of the data in Table 10.7 are 1.4907, 0.5773, 0.6311, and 0.0001°F.
a. Plot a periodogram for these data. Explain what it shows.
b. What proportion of the variation in the monthly average temperature data is described by the first two harmonics?
c. Which of the harmonics have squared amplitudes that are significant, relative to the white-noise null hypothesis, not accounting for test multiplicity, with $\alpha = 0.05$?
d. Which of the harmonics have squared amplitudes that are significant, relative to the white-noise null hypothesis, using Bonferroni criterion, with $\alpha^* = 0.05$?
e. Which of the harmonics have squared amplitudes that are significant harmonics, relative to the white-noise hull hypothesis, using FDR criterion, with $\alpha_{global} = 0.05$?
10.10. How many tick-marks for frequency are missing from the horizontal axis of Figure 10.17?
10.11. Suppose the minor peak in Figure 10.17 at $f = 13/256 = 0.0508$ mo^{-1} resulted in part from aliasing.
a. Compute a frequency that could have produced this spurious signal in the spectrum.
b. How often would the underlying sea-level pressure data need to be recorded and processed in order to resolve this frequency explicitly?
10.12. Plot the spectra for the two autoregressive processes in Exercise 10.2, assuming unit white-noise variance, and $n = 100$.
10.13. The largest squared amplitude in Figure 10.23 is $C_{11}^2 = 0.413$ (in the gray spectrum). Compute a 95% confidence interval for the value of the underlying spectral density at this frequency.

Part III

Multivariate Statistics

Chapter 11

Matrix Algebra and Random Matrices

11.1. BACKGROUND TO MULTIVARIATE STATISTICS

11.1.1. Contrasts Between Multivariate and Univariate Statistics

Much of the material in the first 10 chapters of this book has pertained to analysis of univariate or one-dimensional data. That is, the analysis methods presented were oriented primarily toward scalar data values and their distributions. However, in many practical situations data sets are composed of vector observations. In such cases each data record consists of simultaneous values for multiple quantities. Such data sets are known as *multivariate*. Examples of multivariate atmospheric data include simultaneous observations of multiple variables at one location, or an atmospheric field as represented by a set of grid-point values at a particular time.

Univariate methods can be, and are, applied to individual scalar elements of multivariate data observations. The distinguishing attribute of multivariate methods is that both the joint behavior of the multiple simultaneous values, as well as the variations of the individual data elements, are considered. The remaining chapters of this book present introductions to some of the multivariate methods that are used most commonly with atmospheric data. These include approaches to data reduction and structural simplification, characterization and summarization of multiple dependencies, predictions of subsets of the variables from the remaining ones, and grouping and classification of the multivariate observations.

Multivariate methods are more difficult to understand and implement than univariate methods. Notationally, they require use of matrix algebra to make the presentation and mathematical manipulations tractable. The elements of matrix algebra that are necessary to understand the subsequent material are presented briefly in Section 11.3.

The complexities of multivariate data and the methods that have been devised to deal with them dictate that all but the very simplest multivariate analyses will be implemented using a computer. Enough detail is included here for readers comfortable with numerical methods to be able to implement the analyses themselves. However, many readers will use statistical software for this purpose, and the material in this portion of this book should help to understand what these computer programs are doing, and why.

11.1.2. Organization of Data and Basic Notation

In conventional univariate statistics, each datum or observation is a single number, or scalar. In multivariate statistics each datum is a collection of simultaneous observations of $K \geq 2$ scalar values. For both notational and computational convenience, these multivariate observations are arranged in an ordered

Statistical Methods in the Atmospheric Sciences. https://doi.org/10.1016/B978-0-12-815823-4.00011-0

list known as a *vector*, with a boldface single symbol being used to represent the entire collection, for example,

$$\mathbf{x}^{\mathrm{T}} = \left[x_1, x_2, x_3, \cdots, x_K\right]. \tag{11.1}$$

The superscript "T" on the left-hand side has a specific meaning that will be explained in Section 11.3, but for now we can safely ignore it. Because the K individual values are arranged horizontally, Equation 11.1 is called a *row vector*, and each of the positions within it corresponds to one of the K scalars whose simultaneous relationships will be considered. It can be convenient to visualize (for $K=2$ or 3) or imagine (for higher dimensions) a data vector geometrically, as a point in a K-dimensional space, or as an arrow whose tip position is defined by the listed scalars and whose base is at the origin. Depending on the nature of the data, this abstract geometric space may correspond to a phase- or state-space (see Section 8.1.2), or some subset of the dimensions (a *subspace*) of such a space.

A univariate data set consists of a collection of n scalar observations x_i, $i=1, \ldots, n$. Similarly, a multivariate data set consists of a collection of n data vectors $\mathbf{x}_i$, $i=1, \ldots, n$. Again for both notational and computational convenience this collection of data vectors can be arranged into a rectangular array of numbers having n rows, each corresponding to one multivariate observation, and with each of the K columns containing all n observations of one of the variables. This arrangement of the $n \times K$ numbers in the multivariate data set is called a *data matrix*,

$$[X] = \begin{bmatrix} \mathbf{x}_1^{\mathrm{T}} \\ \mathbf{x}_2^{\mathrm{T}} \\ \mathbf{x}_3^{\mathrm{T}} \\ \vdots \\ \mathbf{x}_n^{\mathrm{T}} \end{bmatrix} = \begin{bmatrix} x_{1,1} & x_{1,2} & \cdots & x_{1,K} \\ x_{2,1} & x_{2,2} & \cdots & x_{2,K} \\ x_{3,1} & x_{3,2} & \cdots & x_{3,K} \\ \vdots & \vdots & \cdots & \vdots \\ x_{n,1} & x_{n,2} & \cdots & x_{n,K} \end{bmatrix}. \tag{11.2}$$

Here n row-vector observations of the form shown in Equation 11.1 have been stacked vertically, or subjected to *row binding*, to yield a rectangular array, called a *matrix*, with n rows and K columns. An equally valid view is that the K univariate data sets in each column have been subjected to *column binding* to produce [X]. Conventionally, the first of the two subscripts of the scalar elements of a matrix denotes the row number, and the second indicates the column number so, for example, $x_{3,2}$ is the third of the n observations of the second of the K variables. In this book matrices, such as [X], will be denoted using square brackets, as a pictorial reminder that the symbol within represents a rectangular array.

The data matrix [X] in Equation 11.2 corresponds exactly to a conventional data table or spreadsheet display, in which each column pertains to one of the variables considered, and each row represents one of the n observations or cases. Its contents can also be visualized or imagined geometrically within an abstract K-dimensional space, with each of the n rows defining a single point. The simplest example is a data matrix for bivariate data, which has n rows and $K=2$ columns. The pair of numbers in each of the rows locates a point on the Cartesian plane. The collection of these n points on the plane defines a scatterplot of the bivariate data.

11.1.3. Multivariate Extensions of Common Univariate Statistics

Just as the data vector in Equation 11.1 is the multivariate extension of a scalar datum, multivariate sample statistics can be expressed using the notation of vectors and matrices. The most common of these is the multivariate sample mean, which is just a vector of the K individual scalar sample means (Equation 3.2), arranged in the same order as the elements of the underlying data vectors,

$$\bar{x}^T = \left[\frac{1}{n}\sum_{i=1}^{n} x_{i,1}, \; \frac{1}{n}\sum_{i=1}^{n} x_{i,2}, \; \cdots, \; \frac{1}{n}\sum_{i=1}^{n} x_{i,K} \right] = [\bar{x}_1, \; \bar{x}_2, \; \cdots, \; \bar{x}_K]. \tag{11.3}$$

As before, the boldface symbol on the left-hand side of Equation 11.3 indicates a vector quantity, and the double-subscripted variables in the first equality are indexed according to the same convention as in Equation 11.2.

The multivariate extensions of the sample standard deviation (Equation 3.6), or (much more commonly, its square) the sample variance, are a little more complicated because all pairwise relationships among the K variables need to be considered. In particular, the multivariate extension of the sample variance is the collection of covariances between all possible pairs of the K variables,

$$\text{Cov}(x_k, x_\ell) = s_{k,\ell} = \frac{1}{n-1}\sum_{i=1}^{n} (x_{i,k} - \bar{x}_k)(x_{i,\ell} - \bar{x}_\ell), \tag{11.4}$$

which is equivalent to the numerator of Equation 3.22. If the two variables are the same, that is, if $k=\ell$, then Equation 11.4 defines the sample variance, $s_k^2 = s_{k,k}$, or the square of Equation 3.6.

Although the notation $s_{k,k}$ for the sample variance of the kth variable may seem a little strange at first, it is conventional in multivariate statistics, and is also convenient from the standpoint of arranging the covariances calculated according to Equation 11.4 into a square array called the *sample covariance matrix*,

$$[S] = \begin{bmatrix} s_{1,1} & s_{1,2} & s_{1,3} & \cdots & s_{1,K} \\ s_{2,1} & s_{2,2} & s_{2,3} & \cdots & s_{2,K} \\ s_{3,1} & s_{3,2} & s_{3,3} & \cdots & s_{3,K} \\ \vdots & \vdots & \vdots & \ddots & \vdots \\ s_{K,1} & s_{K,2} & s_{K,3} & \cdots & s_{K,K} \end{bmatrix}. \tag{11.5}$$

That is, the covariance $s_{k,\ell}$ is displayed in the kth row and ℓth column of the covariance matrix. The sample covariance matrix, or sample *variance-covariance matrix*, is directly analogous to the sample (Pearson) correlation matrix (see Figure 3.29), with the relationship between corresponding elements of the two matrices being given by Equation 3.28, that is, $r_{k,l} = s_{k,l}/(s_{k,k}s_{l,l})^{1/2}$. The K covariances $s_{k,k}$ in the diagonal positions between the upper-left and lower-right corners of the sample covariance matrix are simply the K sample variances. The remaining, off-diagonal, elements are covariances among unlike variables, and the values below and to the left of the diagonal positions duplicate the values above and to the right.

The variance-covariance matrix is also known as the *dispersion matrix*, because it describes how the underlying data are dispersed around their (vector) mean in the K-dimensional space defined by the K variables. The diagonal elements are the individual variances, which index the degree to which the data are spread out in directions parallel to the K coordinate axes for this space, and the covariances in the off-diagonal positions describe the extent to which the cloud of data points is oriented at angles to these axes. The matrix $[S]$ is the sample estimate of the population dispersion matrix $[\Sigma]$, which appears in the probability density function for the multivariate normal distribution (Equation 12.1).

11.2. MULTIVARIATE DISTANCE

It was pointed out in the previous section that a data vector can be regarded as a point in the K-dimensional geometric space whose coordinate axes correspond to the K variables being simultaneously represented. Many multivariate statistical approaches are based on, and/or can be interpreted in terms of, distances within this K-dimensional space. Any number of distance measures can be defined (see Section 16.1.2), but two of these are of particular importance.

11.2.1. Euclidean Distance

The conventional *Euclidean distance* is perhaps the easiest and most intuitive distance measure, because it corresponds to our ordinary experience in the three-dimensional world. Euclidean distance is easiest to visualize in two dimensions, where it can easily be seen as a consequence of the Pythagorean theorem, as illustrated in Figure 11.1. Here two points, x and y, located by the dots, define the hypotenuse of a right triangle whose other two legs are parallel to the two data axes. The Euclidean distance $||y-x||=||x-y||$ is obtained by taking the square root of the sum of the squared lengths of the other two sides.

Euclidean distance generalizes directly to $K \geq 3$ dimensions even though the corresponding geometric space may be difficult or impossible to imagine. In particular,

$$||x-y|| = \sqrt{\sum_{k=1}^{K} (x_k - y_k)^2}. \tag{11.6}$$

Distance between a point x and the origin can also be calculated using Equation 11.6 by substituting a vector of K zeros (which locates the origin in the corresponding K-dimensional space) for the vector y.

It is often mathematically convenient to work in terms of squared distances. No information is lost in so doing, because distance ordinarily is regarded as necessarily nonnegative, so that squared distance is a monotonic and invertible transformation of ordinary dimensional distance (e.g., Equation 11.6). In

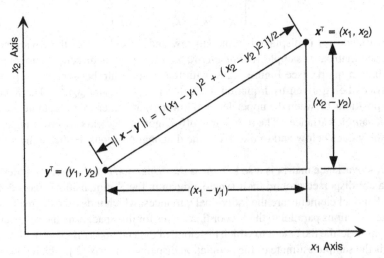

FIGURE 11.1   Illustration of the Euclidean distance between points x and y in $K=2$ dimensions using the Pythagorean theorem.

addition, the square-root operation is avoided. Points x at a constant squared distance $C^2 = ||x - y||^2$ from a fixed point y define a circle on the plane centered at y with radius C for $K = 2$ dimensions, a sphere in a volume with radius C for $K = 3$ dimensions, and a hypersphere with radius C within a K-dimensional hypervolume for $K > 3$ dimensions.

11.2.2. Mahalanobis (Statistical) Distance

Euclidean distance treats the separation of pairs of points in a K-dimensional space equally, regardless of their relative orientation. However, it will be very useful to interpret distances between points in terms of statistical dissimilarity or unusualness, and in this sense point separations in some directions are more unusual than others. The context for unusualness is established by a (K-dimensional, joint) probability distribution for the data points, which may be characterized using the scatter of a finite sample, or using a parametric probability density function.

Figure 11.2 illustrates the issues in $K = 2$ dimensions. Figure 11.2a shows a statistical context established by the scatter of points $x^T = (x_1, x_2)$. The distribution is centered at the origin, and the standard deviation of x_1 is approximately three times that of x_2, that is, $s_1 \approx 3s_2$. The orientation of the point cloud along one of the axes reflects the fact that the two variables x_1 and x_2 are essentially uncorrelated (the points in fact have been drawn from a bivariate Gaussian distribution with $\rho = 0$, see Section 4.4.2). Because of this difference in dispersion, a given Euclidean distance between a pair of points in the horizontal is less unusual than is the same distance in the vertical, relative to this data scatter. Although point A is closer to the center of the distribution according to Euclidean distance, it is more unusual than point B in the context established by the point cloud, and so is statistically further from the origin.

Because the points in Figure 11.2a are uncorrelated, a distance measure that reflects unusualness in the context of the data scatter can be defined simply as

$$D^2 = \frac{(x_1 - \bar{x}_1)^2}{s_{1,1}} + \frac{(x_2 - \bar{x}_2)^2}{s_{2,2}}, \tag{11.7}$$

which is a special case of the *Mahalanobis distance* between the point $x^T = (x_1, x_2)$ and the origin (because the two sample means are zero) when variations in the $K = 2$ dimensions are uncorrelated. For convenience Equation 11.7 is expressed as a squared distance, and it is equivalent to the ordinary squared Euclidean distance after the transformation that divides each element of the data vector by its respective standard deviation (recall that, e.g., $s_{1,1}$ is the sample variance of x_1). Another interpretation of

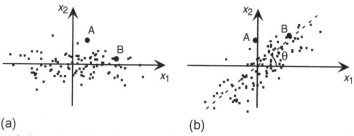

(a)   (b)

FIGURE 11.2   Distance in the context of data scatters centered at the origin. (a) The standard deviation of x_1 is approximately three times larger than the standard deviation of x_2. Point A is closer to the origin in terms of Euclidean distance, but point B is less unusual relative to the data scatter, and so is closer in statistical distance. (b) The same points rotated through an angle $\theta = 40$ degrees.

Equation 11.7 is as the sum of the two squared standardized anomalies, or z-scores (Equation 3.27). In either case, the importance ascribed to a distance along one of the axes is inversely proportional to the data scatter, or uncertainty, in that direction. Consequently point A is further from the origin than point B in Figure 11.2a when measured according to the Mahalanobis distance.

For a fixed Mahalanobis distance D^2, Equation 11.7 defines an ellipse of constant statistical distance on the plane, and that ellipse is also a circle if $s_{1,1}=s_{2,2}$. Generalizing Equation 11.7 to three dimensions by adding a third term for x_3, the set of points at a fixed distance D^2 constitute an ellipsoid that will be spherical if all three variances are equal, blimp-like if two variances are nearly equal but smaller than the third, and disk-like if two variances are nearly equal and larger than the third.

In general the variables within a multivariate data vector x will not be uncorrelated, and these correlations must also be accounted for when defining distances in terms of a data scatter or probability density. Figure 11.2b illustrates the situation in two dimensions, in which the points from Figure 11.2a have been rotated around the origin through an angle $\theta=40$ degrees, which results in the two variables being relatively strongly positively correlated. Again point B is closer to the origin in a statistical sense, although in order to calculate the actual Mahalanobis distances in terms of the variables x_1 and x_2 it would be necessary to use an equation of the form

$$D^2 = a_{1,1}(x_1 - \bar{x}_1)^2 + 2a_{1,2}(x_1 - \bar{x}_1)(x_2 - \bar{x}_2) + a_{2,2}(x_2 - \bar{x}_2)^2. \tag{11.8}$$

Analogous expressions of this kind for the Mahalanobis distance in K dimensions would involve $K(K+1)/2$ terms. Even in only two dimensions the coefficients $a_{1,1}$, $a_{1,2}$, and $a_{2,2}$ are fairly complicated functions of the rotation angle θ and the three covariances $s_{1,1}$, $s_{1,2}$, and $s_{2,2}$. For example,

$$a_{1,1} = \frac{\cos^2(\theta)}{s_{1,1}\cos^2(\theta) - 2s_{1,2}\sin(\theta)\cos(\theta) + s_{2,2}\sin^2(\theta)} \\ + \frac{\sin^2(\theta)}{s_{2,2}\cos^2(\theta) - 2s_{1,2}\sin(\theta)\cos(\theta) + s_{1,1}\sin^2(\theta)}. \tag{11.9}$$

Do not study this equation at all closely. It is here to help convince you, if that is even required, that conventional scalar notation is hopelessly impractical for expressing the mathematical ideas necessary to multivariate statistics. Matrix notation and matrix algebra, which will be reviewed in the next section, are practical necessities for taking the development further. Section 11.4 will resume the statistical development using matrix algebra notation, including revisiting the Mahalanobis distance in Section 11.4.4.

11.3. MATRIX ALGEBRA REVIEW

The mathematical mechanics of dealing simultaneously with multiple variables and their mutual correlations are greatly simplified by use of matrix notation, and a set of computational rules called *matrix algebra* or *linear algebra*. The notation for vectors and matrices was briefly introduced in Section 11.1.2. Matrix algebra is the toolkit used to mathematically manipulate these notational objects. A brief review of this subject, sufficient for the multivariate techniques described in the following chapters, is presented in this section. More complete introductions are readily available elsewhere (e.g., Golub and van Loan, 1996; Strang, 1988).

11.3.1. Vectors

The vector is a fundamental component of matrix algebra notation. It is essentially nothing more than an ordered list of scalar variables, or ordinary numbers, that are called the elements of the vector. The number of elements, also called the vector's dimension, will depend on the situation at hand. A familiar meteorological example is the two-dimensional horizontal wind vector, whose two elements are the eastward wind speed u, and the northward wind speed v.

Vectors already have been introduced in Equation 11.1, and as previously noted will be indicated using boldface type. A vector with only $K = 1$ element is just an ordinary number, or scalar. Unless otherwise indicated, vectors will be regarded as being *column vectors*, which means that their elements are arranged vertically. For example, the column vector x would consist of the elements $x_1, x_2, x_3, \ldots, x_K$; arranged as

$$x = \begin{bmatrix} x_1 \\ x_2 \\ x_3 \\ \vdots \\ x_K \end{bmatrix}. \tag{11.10}$$

These same elements can be arranged horizontally, as in Equation 11.1, which is a row vector. Column vectors are transformed to row vectors, and vice versa, through an operation called *transposing* the vector. The transpose operation is denoted by the superscript "T," so that we can write the vector x in Equation 11.10 as the row vector x^T in Equation 11.1, which is pronounced "x-transpose." The transpose of a column vector is useful for notational consistency within certain matrix operations. It is also useful for typographical purposes, as it allows a vector to be written on a horizontal line of text.

Addition of two or more vectors with the same dimension is straightforward. *Vector addition* is accomplished by adding the corresponding elements of the two vectors, for example

$$x + y = \begin{bmatrix} x_1 \\ x_2 \\ x_3 \\ \vdots \\ x_K \end{bmatrix} + \begin{bmatrix} y_1 \\ y_2 \\ y_3 \\ \vdots \\ y_K \end{bmatrix} = \begin{bmatrix} x_1 + y_1 \\ x_2 + y_2 \\ x_3 + y_3 \\ \vdots \\ x_K + y_K \end{bmatrix}. \tag{11.11}$$

Subtraction is accomplished analogously. This operation reduces to ordinary scalar addition or subtraction when the two vectors have dimension $K = 1$. Addition and subtraction of vectors with different dimensions are not defined.

Multiplying a vector by a scalar results in a new vector whose elements are simply the corresponding elements of the original vector multiplied by that scalar. For example, multiplying the vector x in Equation 11.10 by a scalar constant c yields

$$c x = \begin{bmatrix} c x_1 \\ c x_2 \\ c x_3 \\ \vdots \\ c x_K \end{bmatrix}. \tag{11.12}$$

Two vectors of the same dimension can be multiplied using an operation called the *dot product* or *inner product*. This operation consists of multiplying together each of the K like pairs of vector elements, and then summing these K products. That is,

$$x^\mathrm{T}y = [x_1, \; x_2, \; x_3, \; \cdots, \; x_K] \begin{bmatrix} y_1 \\ y_2 \\ y_3 \\ \vdots \\ y_K \end{bmatrix} = x_1 y_1 + x_2 y_2 + x_3 y_3 + \cdots + x_K y_K \tag{11.13}$$

$$= \sum_{k=1}^{K} x_k y_k.$$

This vector multiplication has been written as the product of a row vector on the left and a column vector on the right in order to be consistent with the operation of matrix multiplication, which will be presented in Section 11.3.2. As will be seen, the dot product is in fact a special case of matrix multiplication, and (unless $K=1$) the order of vector and matrix multiplication is important: in general the multiplications $x^\mathrm{T} y$ and $y\,x^\mathrm{T}$ and their matrix generalizations yield entirely different results. Equation 11.13 also shows that vector multiplication can be expressed in component form using summation notation. Expanding vector and matrix operations in component form can be useful if the calculation is to be programmed for a computer, depending on the programming language.

As noted previously, a vector can be visualized as a point in K-dimensional space. The Euclidean length of a vector in that space is the ordinary distance between the point and the origin. Length is a scalar quantity that can be computed using the dot product, as

$$\|x\| = \sqrt{x^\mathrm{T}x} = \left(\sum_{k=1}^{K} x_k^2 \right)^{1/2}. \tag{11.14}$$

Equation 11.14 is sometimes known as the *Euclidean norm* of the vector x. Figure 11.1, with $y=0$ as the origin, illustrates that this length is simply an application of the Pythagorean theorem. A common use of Euclidean length is in the computation of the total horizontal wind speed from the horizontal velocity vector $v^\mathrm{T} = [u, v]$, according to $v_\mathrm{H} = (u^2 + v^2)^{1/2}$. However, Equation 11.14 generalizes to arbitrarily high K as well.

The angle θ between two vectors is also computed using the dot product,

$$\theta = \cos^{-1}\left(\frac{x^\mathrm{T}y}{\|x\| \, \|y\|} \right). \tag{11.15}$$

This relationship implies that two vectors are perpendicular if their dot product is zero, since $\cos^{-1}(0) = 90$ degrees. Mutually perpendicular vectors are also called *orthogonal*.

The magnitude of the *projection* (or "length of the shadow") of a vector x onto a vector y is also a function of the dot product, given by

$$L_{x,y} = \frac{x^\mathrm{T}y}{\|y\|}. \tag{11.16}$$

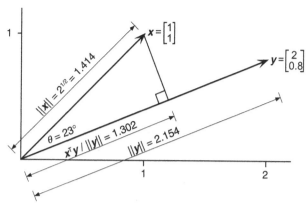

FIGURE 11.3 Illustration of the concepts of vector length (Equation 11.14), the angle between two vectors (Equation 11.15), and the projection of one vector onto another (Equation 11.16); for the two vectors $x^T = [1, 1]$ and $y^T = [2, 0.8]$.

The geometric interpretations of these three computations of length, angle, and projection are illustrated in Figure 11.3, for the vectors $x^T = [1, 1]$ and $y^T = [2, 0.8]$. The length of x is simply $||x|| = (1^2 + 1^2)^{1/2} = \sqrt{2}$, and the length of y is $||y|| = (2^2 + 0.8^2)^{1/2} = 2.154$. Since the dot product of the two vectors is $x^T y = (1)(2) + (1)(0.8) = 2.8$, the angle between them is $\theta = \cos^{-1}(2.8/(2.154\sqrt{2})) = 23$ degrees, and the length of the projection of x onto y is $2.8/2.154 = 1.302$.

11.3.2. Matrices

A matrix is a two-dimensional rectangular array of numbers having I rows and J columns. The *dimension* of a matrix is specified by these numbers of rows and columns. A matrix dimension is written $(I \times J)$, and pronounced "I by J." Matrices are denoted here by uppercase letters surrounded by square brackets. Sometimes, for notational clarity, a parenthetical expression for the dimension of a matrix will be written directly below it. The *elements* of a matrix are the individual variables or numerical values occupying the particular row-and-column positions. The matrix elements are identified notationally by two subscripts; the first of these identifies the row number and the second identifies the column number. Equation 11.2 shows a $(n \times K)$ data matrix, and Equation 11.5 shows a $(K \times K)$ covariance matrix, with the subscripting convention illustrated.

A vector is a special case of a matrix, having a single row or a single column, and matrix operations are applicable also to vectors. A K-dimensional row vector is a $(1 \times K)$ matrix, and a column vector is a $(K \times 1)$ matrix. Just as a $K = 1$ dimensional vector is also a scalar, so too is a (1×1) matrix.

A matrix with the same number of rows and columns, such as $[S]$ in Equation 11.5, is called a *square* matrix. The elements of a square matrix for which the subscript values $i = j$ are located on the diagonal between the upper left to the lower-right corners, and are called *diagonal* elements. Correlation matrices $[R]$ (see Figure 3.29) are square matrices having all 1's on the diagonal. A square matrix for which $a_{i,j} = a_{j,i}$ for all values of i and j is called *symmetric*. Correlation and covariance matrices are symmetric because the correlation between variable i and variable j is identical to the correlation between variable j and variable i. The *identity matrix* $[I]$, consisting of 1's on the diagonal and zeros everywhere else, is another important square, symmetric matrix,

$$[I] = \begin{bmatrix} 1 & 0 & 0 & \cdots & 0 \\ 0 & 1 & 0 & \cdots & 0 \\ 0 & 0 & 1 & \cdots & 0 \\ \vdots & \vdots & \vdots & \ddots & \vdots \\ 0 & 0 & 0 & \cdots & 1 \end{bmatrix}. \tag{11.17}$$

An identity matrix can be constructed for any (square) dimension. When the identity matrix appears in an equation it can be assumed to be of appropriate dimension for the relevant matrix operations to be defined. The identity matrix is a special case of a *diagonal matrix*, whose off-diagonal elements are all zeros.

The *transpose* operation is defined for any matrix, including the special case of vectors. The transpose of a matrix is obtained in general by exchanging row and column indices, not by a 90° rotation as might have been anticipated from a comparison of Equations 11.1 and 11.10. Geometrically, the transpose operation is like a reflection across the matrix diagonal, which extends downward and to the right from the upper, left-hand element. For example, the relationship between the (3×4) matrix $[B]$ and its transpose, the (4×3) matrix $[B]^T$, is illustrated by comparing

$$\underset{(3\times4)}{[B]} = \begin{bmatrix} \alpha & \beta & \gamma & \delta \\ \varepsilon & \varsigma & \eta & \theta \\ \iota & \kappa & \lambda & \mu \end{bmatrix} \tag{11.18a}$$

and

$$\underset{(4\times3)}{[B]}^T = \begin{bmatrix} \alpha & \varepsilon & \iota \\ \beta & \varsigma & \kappa \\ \gamma & \eta & \lambda \\ \delta & \theta & \mu \end{bmatrix}. \tag{11.18b}$$

Equation 11.18 also illustrates the convention of indicating the matrix dimension parenthetically, beneath the matrix symbol. If a square matrix $[A]$ is symmetric, then $[A]^T = [A]$.

Multiplication of a matrix by a scalar is the same as for vectors and is accomplished by multiplying each element of the matrix by the scalar,

$$c\,[D] = c \begin{bmatrix} d_{1,1} & d_{1,2} \\ d_{2,1} & d_{2,2} \end{bmatrix} = \begin{bmatrix} c\,d_{1,1} & c\,d_{1,2} \\ c\,d_{2,1} & c\,d_{2,2} \end{bmatrix}. \tag{11.19}$$

Similarly, matrix addition and subtraction are accomplished by performing these operations on the elements in corresponding row and column positions, and are defined only for matrices of identical dimension. For example, the sum of two (2×2) matrices would be computed as

$$[D] + [E] = \begin{bmatrix} d_{1,1} & d_{1,2} \\ d_{2,1} & d_{2,2} \end{bmatrix} + \begin{bmatrix} e_{1,1} & e_{1,2} \\ e_{2,1} & e_{2,2} \end{bmatrix} = \begin{bmatrix} d_{1,1}+e_{1,1} & d_{1,2}+e_{1,2} \\ d_{2,1}+e_{2,1} & d_{2,2}+e_{2,2} \end{bmatrix}. \tag{11.20}$$

Matrix multiplication is defined between two matrices if the number of columns in the left matrix is equal to the number of rows in the right matrix. Thus not only is matrix multiplication not commutative (i.e., $[A]\,[B] \neq [B]\,[A]$, in general), but multiplication of two matrices in reverse order is not even defined unless the two have complementary row and column dimensions. The product of a matrix multiplication is another matrix, the row dimension of which is the same as the row dimension of the left matrix, and the column dimension of which is the same as the column dimension of the right matrix. That is, multiplying a $(I \times J)$ matrix $[A]$ (on the left) and a $(J \times K)$ matrix $[B]$ (on the right) yields a $(I \times K)$ matrix $[C]$. In effect, the middle dimension J is "multiplied out."

Consider the case where $I = 2$, $J = 3$, and $K = 2$. In terms of the individual matrix elements, the matrix multiplication $[A][B] = [C]$ expands to

$$\underset{(2 \times 3)}{\begin{bmatrix} a_{1,1} & a_{1,2} & a_{1,3} \\ a_{2,1} & a_{2,2} & a_{2,3} \end{bmatrix}} \underset{(3 \times 2)}{\begin{bmatrix} b_{1,1} & b_{1,2} \\ b_{2,1} & b_{2,2} \\ b_{3,1} & b_{3,2} \end{bmatrix}} = \underset{(2 \times 2)}{\begin{bmatrix} c_{1,1} & c_{1,2} \\ c_{2,1} & c_{2,2} \end{bmatrix}}, \tag{11.21a}$$

where

$$[C] = \begin{bmatrix} c_{1,1} & c_{1,2} \\ c_{2,1} & c_{2,2} \end{bmatrix} = \begin{bmatrix} a_{1,1}b_{1,1} + a_{1,2}b_{2,1} + a_{1,3}b_{3,1} & a_{1,1}b_{1,2} + a_{1,2}b_{2,2} + a_{1,3}b_{1,3} \\ a_{2,1}b_{1,1} + a_{2,2}b_{2,1} + a_{2,3}b_{3,1} & a_{2,1}b_{1,2} + a_{2,2}b_{2,2} + a_{2,3}b_{3,2} \end{bmatrix}. \tag{11.21b}$$

The individual components of $[C]$ as written out in Equation 11.21b may look confusing at first exposure. In understanding matrix multiplication, it is helpful to realize that each element of the product matrix $[C]$ is simply the dot product, as defined in Equation 11.13, of one of the rows in the left matrix $[A]$ and one of the columns in the right matrix $[B]$. In particular, the number occupying the ith row and kth column of the matrix $[C]$ is exactly the dot product between the row vector comprising the ith row of $[A]$ and the column vector comprising the kth column of $[B]$. Equivalently, matrix multiplication can be written in terms of the individual matrix elements using summation notation,

$$c_{i,k} = \sum_{j=1}^{J} a_{i,j} b_{j,k}; i = 1, \ldots, I; k = 1, \ldots, K. \tag{11.22}$$

Figure 11.4 illustrates the procedure graphically, for one element of the matrix $[C]$ resulting from the multiplication $[A][B] = [C]$.

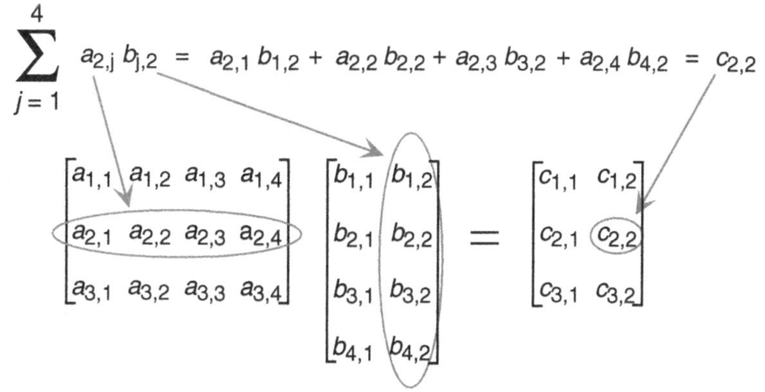

FIGURE 11.4   Graphical illustration of matrix multiplication as the dot product of the ith row of the left-hand matrix with the jth column of the right-hand matrix, yielding the element in the ith row and jth column of the matrix product.

The identity matrix (Equation 11.17) is so named because it functions as the multiplicative identity— that is, $[A] [I] = [A]$, and $[I] [A] = [A]$ regardless of the dimension of $[A]$—although in the former case $[I]$ is a square matrix with the same number of columns as $[A]$, and in the latter its dimension is the same as the number of rows in $[A]$.

The dot product, or inner product (Equation 11.13), is one application of matrix multiplication to vectors. But the rules of matrix multiplication also allow multiplication of two vectors in the opposite order, which is called the *outer product*. In contrast to the inner product, which is a $(1 \times K) \times (K \times 1)$ matrix multiplication yielding a (1×1) scalar, the outer product of two vectors of the same dimension K is a $(K \times 1) \times (1 \times K)$ matrix multiplication, yielding a $(K \times K)$ square matrix. For example, for $K = 3$,

$$\mathbf{xy}^{\mathrm{T}} = \begin{bmatrix} x_1 \\ x_2 \\ x_3 \end{bmatrix} [y_1, \ y_2, \ y_3] = \begin{bmatrix} x_1 y_1 & x_1 y_2 & x_1 y_3 \\ x_2 y_1 & x_2 y_2 & x_2 y_3 \\ x_3 y_1 & x_3 y_2 & x_3 y_3 \end{bmatrix}. \tag{11.23}$$

It is not necessary for two vectors forming an outer product to have the same dimension, because as vectors they have common ("inner") dimension 1. The outer product is sometimes known as the *dyadic product*, or *tensor product*, and the operation is sometimes indicated using a circled "x," that is, $\mathbf{xy}^{\mathrm{T}} = \mathbf{x} \otimes \mathbf{y}$.

The *trace* of a square matrix is simply the sum of its diagonal elements, that is,

$$\mathrm{tr}[A] = \sum_{k=1}^{K} a_{k,k}, \tag{11.24}$$

for the $(K \times K)$ matrix $[A]$. For the $(K \times K)$ identity matrix, $\mathrm{tr}[I] = K$.

The *determinant* of a square matrix is a scalar quantity defined as

$$\det[A] = |A| = \sum_{k=1}^{k} a_{1,k} |A_{1,k}| (-1)^{1+k}, \tag{11.25}$$

where $[A_{1,k}]$ is the $(K-1 \times K-1)$ matrix formed by deleting the first row and kth column of $[A]$, and $a_{1,k}$ is the element in the original matrix at the intersection of the deleted row and deleted column. The absolute value notation for the matrix determinant suggests that this operation produces a scalar that is in some sense a measure of the magnitude of the matrix. The definition in Equation 11.25 is recursive, so for example computing the determinant of a $(K \times K)$ matrix requires that K determinants of reduced $(K-1 \times K-1)$ matrices be calculated first, and so on, until reaching and $|A| = a_{1,1}$ for $K = 1$. Accordingly the process is quite tedious and is usually best left to a computer. However, in the (2×2) case,

$$\det_{(2 \times 2)} [A] = \det \begin{bmatrix} a_{1,1} & a_{1,2} \\ a_{2,1} & a_{2,2} \end{bmatrix} = a_{11} a_{2,2} - a_{1,2} a_{2,1}. \tag{11.26}$$

The matrix generalization of arithmetic division exists for square matrices that have a property known as *full rank*, or *nonsingularity*. This condition can be interpreted to mean that the matrix does not contain redundant information, in the sense that none of the rows can be constructed from linear combinations of the other rows. Considering each row of a nonsingular matrix as a vector, it is impossible to construct vector sums of rows multiplied by scalar constants that equal any one of the other rows. These same conditions applied to the columns also imply that the matrix is nonsingular. Nonsingular matrices have nonzero determinant.

Nonsingular square matrices are *invertible*. That a matrix $[A]$ is invertible means that another matrix $[B]$ exists such that

$$[A][B] = [B][A] = [I]. \tag{11.27}$$

It is then said that $[B]$ is the inverse of $[A]$, or $[B] = [A]^{-1}$; and that $[A]$ is the inverse of $[B]$, or $[A] = [B]^{-1}$. Loosely speaking, $[A][A]^{-1}$ indicates division of the matrix $[A]$ by itself, and so yields the (matrix) identity $[I]$. Inverses of (2×2) matrices are easy to compute by hand, using

$$[A]^{-1} = \frac{1}{\det[A]} \begin{bmatrix} a_{2,2} & -a_{1,2} \\ -a_{2,1} & a_{1,1} \end{bmatrix} = \frac{1}{a_{1,1}a_{2,2} - a_{2,1}a_{1,2}} \begin{bmatrix} a_{2,2} & -a_{1,2} \\ -a_{2,1} & a_{1,1} \end{bmatrix}. \tag{11.28}$$

The name of this matrix is pronounced "*A* inverse." Explicit formulas for inverting matrices of higher dimension also exist, but quickly become very cumbersome as the dimensions get larger. Computer algorithms for inverting matrices are widely available, and as a consequence matrices with dimension higher than two or three are rarely inverted by hand. An important exception is the inverse of a diagonal matrix, which is simply another diagonal matrix whose nonzero elements are the reciprocals of the diagonal elements of the original matrix. If $[A]$ is symmetric (frequently in statistics, symmetric matrices are inverted), then $[A]^{-1}$ is also symmetric.

Table 11.1 lists some additional properties of arithmetic operations with matrices that have not been specifically mentioned in the foregoing.

TABLE 11.1 Some Elementary Properties of Arithmetic Operations With Matrices

Distributive multiplication by a scalar	$c([A][B]) = (c[A])[B] = [A](c[B])$
Distributive matrix multiplication	$[A]([B]+[C]) = [A][B]+[A][C]$
	$([A]+[B])[C] = [A][C]+[B][C]$
Associative matrix multiplication	$[A]([B][C]) = ([A][B])[C]$
Inverse of a matrix product	$([A][B])^{-1} = [B]^{-1}[A]^{-1}, ([A][B][C])^{-1} = [C]^{-1}[B]^{-1}[A]^{-1}$, etc.
Transpose of a matrix product	$([A][B])^{\mathsf{T}} = [B]^{\mathsf{T}}[A]^{\mathsf{T}}, ([A][B][C])^{\mathsf{T}} = [C]^{\mathsf{T}}[B]^{\mathsf{T}}[A]^{\mathsf{T}}$, etc.
Combining matrix transpose and inverse	$([A]^{-1})^{\mathsf{T}} = ([A]^{\mathsf{T}})^{-1}$

Example 11.1. Computation of the Covariance and Correlation Matrices

The covariance matrix $[S]$ was introduced in Equation 11.5, and the correlation matrix $[R]$ was introduced in Figure 3.29 as a device for compactly representing the mutual correlations among K variables. The correlation matrix for the January 1987 data in Table A.1 (with the unit diagonal elements and the symmetry implicit) is presented in Table 3.5. The computation of the covariances in Equation 11.4 and of the correlations in Equation 3.29 can also be expressed in the notation of matrix algebra.

One way to begin the computation is with the $(n \times K)$ data matrix $[X]$ (Equation 11.2). Each row of this matrix is a vector, consisting of one observation for each of K variables. The number of these rows is the same as the sample size, n, so $[X]$ is just an ordinary data table such as Table A.1. In Table A.1 there are $K = 6$ variables (excluding the column containing the dates), each simultaneously observed on $n = 31$ occasions. An individual data element $x_{i,k}$ is the ith observation of the kth variable. For example, in Table A.1, $x_{4,6}$ would be the Canandaigua minimum temperature (19°F) observed on 4 January.

Define the $(n \times n)$ matrix $[1]$, whose elements are all equal to 1. The $(n \times K)$ matrix of anomalies (in the meteorological sense of variables with their means subtracted), or centered data $[X']$ is then

$$[X'] = [X] - \frac{1}{n}[1][X].$$

(11.29)

(Note that some authors use the prime notation to indicate matrix transpose, but the superscript "T" has been used for this purpose throughout this book, to avoid confusion.) The second term in Equation 11.29 is a $(n \times K)$ matrix containing the sample means. Each of its n rows is the same and consists of the K sample means in the same order as the corresponding variables appear in each row of $[X]$.

Multiplying $[X']$ by the transpose of itself, and dividing by $n-1$, yields the sample covariance matrix,

$$[S] = \frac{1}{n-1}[X']^{\mathrm{T}}[X'].$$

(11.30)

This is the same symmetric $(K \times K)$ matrix as in Equation 11.5, whose diagonal elements are the sample variances of the K variables, and whose other elements are the covariances among all possible pairs of the K variables. The operation in Equation 11.30 corresponds to the summation in the numerator of Equation 3.28.

Now define the $(K \times K)$ diagonal matrix $[D]$, whose diagonal elements are the sample standard deviations of the K variables. That is, $[D]$ consists of all zeros except for the diagonal elements, whose values are the square roots of the corresponding elements of $[S]$: $d_{k,k} = \sqrt{s_{k,k}}$, $k = 1, \ldots, K$. The correlation matrix can then be computed from the covariance matrix using

$$[R] = [D]^{-1}[S][D]^{-1}.$$

(11.31)

Since $[D]$ is diagonal, its inverse is the diagonal matrix whose elements are the reciprocals of the sample standard deviations on the diagonal of $[D]$. The matrix multiplication in Equation 11.31 corresponds to division by the standard deviations in Equation 3.29.

Note that the correlation matrix $[R]$ is equivalently the covariance matrix of the standardized variables (or standardized anomalies) z_k (Equation 3.27). That is, dividing the anomalies x_k' by their standard deviations $\sqrt{s_{k,k}}$ nondimensionalizes the variables and results in their having unit variance (1's on the diagonal of $[R]$) and covariances equal to their correlations. In matrix notation this can be seen by substituting Equation 11.30 into Equation 11.31 to yield

$$\begin{aligned} [R] &= \frac{1}{n-1}[D]^{-1}[X']^{\mathrm{T}}[X'][D]^{-1} \\ &= \frac{1}{n-1}[Z]^{\mathrm{T}}[Z] \end{aligned},$$

(11.32)

where $[Z]$ is the $(n \times K)$ matrix whose rows are the vectors of standardized variables z, analogously to the matrix $[X']$ of the anomalies. The first line of Equation 11.32 converts the matrix $[X']$ to the matrix $[Z]$ by dividing each element by its standard deviation, $d_{k,k}$. Comparing Equation 11.32 and 11.30 shows that $[R]$ is indeed the covariance matrix for the standardized variables z.

It is also possible to formulate the computation of the covariance and correlation matrices in terms of outer products of vectors. Define the ith of n (column) vectors of anomalies

$$x'_i = x_i - \overline{x}_i, \tag{11.33}$$

where the vector (sample) mean is the transpose of any of the rows of the matrix that is subtracted on the right-hand side of Equation 11.29 or, equivalently the transpose of Equation 11.3. Also let the corresponding standardized anomalies (the vector counterpart of Equation 3.27) be

$$z_i = [D]^{-1} x'_i, \tag{11.34}$$

where $[D]$ is again the diagonal matrix of standard deviations. Equation 11.34 is called the *scaling transformation*, and simply indicates division of all the values in a data vector by their respective standard deviations. The covariance matrix can then be computed in a way that is notationally analogous to the usual computation of the scalar variance (Equation 3.6, squared),

$$[S] = \frac{1}{n-1} \sum_{i=1}^{n} x'_i x'^{\mathrm{T}}_i, \tag{11.35}$$

and, similarly, the correlation matrix is

$$[R] = \frac{1}{n-1} \sum_{i=1}^{n} z_i z^{\mathrm{T}}_i. \tag{11.36}$$

$\diamond$

Example 11.2. Multiple Linear Regression Expressed in Matrix Notation

The discussion of multiple linear regression in Section 7.3.1 indicated that the relevant mathematics are most easily expressed and solved using matrix algebra. In this notation, the expression for the predictand y as a function of the predictor variables x_i (Equation 7.25) becomes

$$y = [X] b, \tag{11.37a}$$

or

$$\begin{bmatrix} y_1 \\ y_2 \\ y_3 \\ \vdots \\ y_n \end{bmatrix} = \begin{bmatrix} 1 & x_{1,1} & x_{1,2} & \cdots & x_{1,K} \\ 1 & x_{2,1} & x_{2,2} & \cdots & x_{2,K} \\ 1 & x_{3,1} & x_{3,2} & \cdots & x_{3,K} \\ \vdots & \vdots & \vdots & \cdots & \vdots \\ 1 & x_{n,1} & x_{n,2} & \cdots & x_{n,K} \end{bmatrix} \begin{bmatrix} b_0 \\ b_1 \\ b_2 \\ \vdots \\ b_K \end{bmatrix}. \tag{11.37b}$$

Here y is a $(n \times 1)$ matrix (i.e., a vector) of the n observations of the predictand, $[X]$ is a $(n \times K+1)$ data matrix containing the values of the predictors, and $b^{\mathrm{T}} = [b_0, b_1, b_2, ..., b_K]$ is a $(K+1 \times 1)$ vector of the regression parameters. The data matrix in the regression context is similar to that in Equation 11.2, except that it has $K+1$ rather than K columns. This extra column is the leftmost column of $[X]$ in Equation 11.37 and consists entirely of 1's. Thus Equation 11.37 is a vector equation, with dimension $(n \times 1)$ on each side. It is actually n repetitions of Equation 7.25, once each for the n data records.

The normal equations (presented in Equation 7.6 for the simple case of $K=1$) are obtained by left-multiplying each side of Equation 11.37 by $[X]^T$,

$$[X]^T y = [X^T][X] b, \tag{11.38a}$$

or

$$\begin{bmatrix} \Sigma y \\ \Sigma x_1 y \\ \Sigma x_2 y \\ \vdots \\ \Sigma x_K y \end{bmatrix} = \begin{bmatrix} n & \Sigma x_1 & \Sigma x_2 & \cdots & \Sigma x_K \\ \Sigma x_1 & \Sigma x_1^2 & \Sigma x_1 x_2 & \cdots & \Sigma x_1 x_K \\ \Sigma x_2 & \Sigma x_2 x_1 & \Sigma x_2^2 & \cdots & \Sigma x_2 x_K \\ \vdots & \vdots & \vdots & \cdots & \vdots \\ \Sigma x_K & \Sigma x_K x_1 & \Sigma x_K x_2 & \cdots & \Sigma x_K^2 \end{bmatrix} \begin{bmatrix} b_0 \\ b_1 \\ b_2 \\ \vdots \\ b_K \end{bmatrix}, \tag{11.38b}$$

where all the summations are over the n training data points. The symmetric $[X]^T[X]$ matrix has dimension $(K+1 \times K+1)$. Each side of Equation 11.38 has dimension $(K+1 \times 1)$, and this equation actually represents $K+1$ simultaneous equations involving the $K+1$ unknown regression coefficients. Matrix algebra very commonly is used to solve sets of simultaneous linear equations such as these. One (relatively computationally inefficient) way to obtain the solution is to left-multiply both sides of Equation 11.38 by the inverse of the $[X]^T[X]$ matrix. This operation is analogous to dividing both sides by this quantity, and yields

$$\begin{aligned} \left([X]^T[X]\right)^{-1}[X]^T y &= \left([X]^T[X]\right)^{-1}[X]^T[X] b \\ &= [X]^{-1}\left([X]^T\right)^{-1}[X]^T[X] b \\ &= [X]^{-1}[I][X] b \\ &= [I] b \\ &= b \end{aligned} \tag{11.39}$$

which is the solution for the vector of regression parameters. The result for the inverse of a matrix product from Table 11.1 has been used in the second line. If there are no linear dependencies among the predictor variables, then the matrix $[X]^T[X]$ is nonsingular, and its inverse will exist. Otherwise, regression software will be unable to compute Equation 11.39, and a suitable error message should be reported.

Variances and covariances for the joint sampling distribution of the $K+1$ regression parameters b^T, corresponding to Equations 7.18b and 7.19b, can also be calculated using matrix algebra. The $(K+1 \times K+1)$ covariance matrix, jointly for the intercept and the K regression coefficients, is

$$[S_b] = \begin{bmatrix} s_{b_0}^2 & s_{b_0,b_1} & \cdots & s_{b_0,b_K} \\ s_{b_1,b_0} & s_{b_1}^2 & \cdots & s_{b_1,b_K} \\ s_{b_2,b_0} & s_{b_2,b_1} & \cdots & s_{b_2,b_K} \\ \vdots & \vdots & \ddots & \vdots \\ s_{b_K,b_0} & s_{b_K,b_1} & \cdots & s_{b_K}^2 \end{bmatrix} = s_e^2 \left([X]^T[X]\right)^{-1}. \tag{11.40}$$

As before, s_e^2 is the estimated residual variance,

$$s_e^2 = \frac{1}{n-K-1}\left(\boldsymbol{y}^{\mathrm{T}}\boldsymbol{y} - \boldsymbol{b}^{\mathrm{T}}[X]^{\mathrm{T}}\boldsymbol{y}\right), \tag{11.41}$$

or MSE (as in Table 7.3). The diagonal elements of Equation 11.40 are the estimated variances of the sampling distributions of each of the elements of the parameter vector $\boldsymbol{b}$. The off-diagonal elements are the covariances among them, corresponding to (for covariances involving the intercept, b_0) the correlation in Equation 7.20. For sufficiently large sample sizes, the joint sampling distribution is multivariate normal (see Chapter 12) so Equation 11.40 fully defines its dispersion.

Similarly, the conditional variance of the sampling distribution of the multiple linear regression function, which is the multivariate extension of Equation 7.24, can be expressed in matrix form as

$$s_{\hat{y}|x_0}^2 = s_e^2\,\boldsymbol{x}_0^{\mathrm{T}}\left([X]^{\mathrm{T}}[X]\right)^{-1}\boldsymbol{x}_0, \tag{11.42}$$

and the prediction variance, corresponding to Equation 7.23 is

$$s_{\hat{y}|x_0}^2 = s_e^2\left\{1 + \boldsymbol{x}_0^{\mathrm{T}}\left([X]^{\mathrm{T}}[X]\right)^{-1}\boldsymbol{x}_0\right\}. \tag{11.43}$$

Equations 11.42 and 11.43 both depend on the values of the predictors for which the regression function is evaluated, $\boldsymbol{x}_0^{\mathrm{T}} = [1, x_1, x_2, \ldots, x_K]$.   ◇

A square matrix is called *orthogonal* if the vectors defined by its columns have unit lengths, and are mutually perpendicular (i.e., $\theta = 90$ degree according to Equation 11.15), and the same conditions hold for the vectors defined by its rows. In that case,

$$[A]^{\mathrm{T}} = [A]^{-1}, \tag{11.44a}$$

which implies that

$$[A][A]^T = [A]^T[A] = [I]. \tag{11.44b}$$

Orthogonal matrices are *unitary*, with this latter term encompassing also matrices that may have complex elements.

An *orthogonal transformation* is achieved by multiplying a vector by an orthogonal matrix. Considering a vector to define a point in K-dimensional space, an orthogonal transformation corresponds to a rigid rotation of the coordinate axes (and also a reflection, if the determinant is negative), resulting in a new basis (new set of coordinate axes) for the space. For example, consider $K = 2$ dimensions, and the orthogonal matrix

$$[T] = \begin{bmatrix} \cos(\theta) & -\sin(\theta) \\ \sin(\theta) & \cos(\theta) \end{bmatrix}, \tag{11.45}$$

The lengths of both rows and both columns of this matrix are $\sin^2(\theta) + \cos^2(\theta) = 1$ (Equation 11.14), and the angles between the two pairs of vectors are both 90 degrees (Equation 11.15), so $[T]$ is an orthogonal matrix.

Multiplication of a vector $\boldsymbol{x}$ by the transpose of this matrix corresponds to a rigid counterclockwise rotation of the coordinate axes through an angle θ. Consider the point $\boldsymbol{x}^{\mathrm{T}} = (1, 1)$ in Figure 11.5. Left-multiplying it by $[T]^{\mathrm{T}}$, with $\theta = 72$ degrees, yields the point in a new coordinate system (dashed axes)

FIGURE 11.5 The point $x^T=(1, 1)$, when subjected to an orthogonal rotation of the coordinate axes through an angle of $\theta=72$ degrees, is transformed to the point $\tilde{x}^T=(1.26, -0.64)$ in the new basis (dashed coordinate axes).

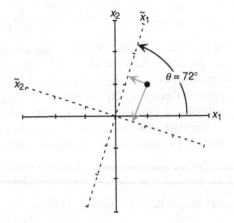

$$\tilde{x} = \begin{bmatrix} \cos(72^\circ) & \sin(72^\circ) \\ -\sin(72^\circ) & \cos(72^\circ) \end{bmatrix} x$$

$$= \begin{bmatrix} .309 & .951 \\ -.951 & .309 \end{bmatrix} \begin{bmatrix} 1 \\ 1 \end{bmatrix} = \begin{bmatrix} .309+.951 \\ -.951+.309 \end{bmatrix} = \begin{bmatrix} 1.26 \\ -0.64 \end{bmatrix}. \tag{11.46}$$

Because the rows and columns of an orthogonal matrix all have unit length, orthogonal transformations preserve length. That is, they do not compress or expand the (rotated) coordinate axes. In terms of (squared) Euclidean length (Equation 11.14),

$$\tilde{x}^T\tilde{x} = \left([T]^T x\right)^T \left([T]^T x\right)$$
$$= x^T[T][T]^T x$$
$$= x^T[I]x \tag{11.47}$$
$$= x^T x$$

The result for the transpose of a matrix product from Table 11.1 has been used in the second line, and Equation 11.44 has been used in the third.

11.3.3. Eigenvalues and Eigenvectors of a Square Matrix

An *eigenvalue* λ, and an *eigenvector*, e of a square matrix $[A]$ are a scalar and nonzero vector, respectively, satisfying the equation

$$[A]e = \lambda e, \tag{11.48a}$$

or equivalently

$$([A] - \lambda[I])e = 0, \tag{11.48b}$$

where 0 is a vector consisting entirely of zeros. For every eigenvalue and eigenvector pair that can be found to satisfy Equation 11.48, any scalar multiple of the eigenvector, ce, will also satisfy the equation together with that eigenvalue. Consequently, for definiteness it is usual to require that eigenvectors have unit length,

$$\|e\| = 1. \tag{11.49}$$

This restriction removes the ambiguity only up to a change in sign, since if a vector e satisfies Equation 11.48 then its negative, $-e$ will also.

If $[A]$ is nonsingular there will be K eigenvalue-eigenvector pairs λ_k and e_k with nonzero eigenvalues, where K is the number of rows and columns in $[A]$. Each eigenvector will be dimensioned $(K \times 1)$. If $[A]$ is singular at least one of its eigenvalues will be zero, with the corresponding eigenvector(s) being arbitrary. Synonymous terminology that is sometimes also used for eigenvalues and eigenvectors includes *characteristic values* and *characteristic vectors*, *latent values* and *latent vectors*, and *proper values* and *proper vectors*.

Because each eigenvector is defined to have unit length, the dot product of any eigenvector with itself is one. If, in addition, the matrix $[A]$ is symmetric, then its eigenvectors are mutually orthogonal, so that

$$e_i^T e_j = \begin{cases} 1, & i = j \\ 0, & i \neq j \end{cases}. \tag{11.50}$$

Orthogonal vectors of unit length are said to be *orthonormal*. (This terminology has nothing to do with the Gaussian or "normal" distribution.) The orthonormality property is analogous to Equation 10.66, expressing the orthogonality of the sine and cosine functions.

For many statistical applications, eigenvalues and eigenvectors are calculated for real (not containing complex or imaginary numbers) symmetric matrices, such as covariance or correlation matrices. Eigenvalues and eigenvectors of such matrices have a number of important and remarkable properties. The first of these properties is that their eigenvalues and eigenvectors are real valued. Also, as just noted, the eigenvectors of symmetric matrices are orthogonal. That is, their dot products with each other are zero, so that they are mutually perpendicular in K-dimensional space.

Often the $(K \times K)$ matrix $[E]$ is formed, the K columns of which are the eigenvectors e_k. That is,

$$[E] = [e_1, \ e_2, \ e_3, \ \cdots, \ e_K]. \tag{11.51}$$

Because of the orthogonality and unit length of the eigenvectors of symmetric matrices, the matrix $[E]$ is orthogonal, having the properties expressed in Equation 11.44. The orthogonal transformation $[E]^T x$ defines a rigid rotation of the K-dimensional coordinate axes of x, called an *eigenspace*. This space covers the same "territory" as the original coordinates, but using the different set of axes defined by the solutions to Equation 11.48. Metaphorically, and in two dimensions, the underlying landscape has not changed, but a compass would indicate that some different direction is north.

The K eigenvalue-eigenvector pairs contain the same information as the matrix $[A]$ from which they were computed, and so can be regarded as a transformation of $[A]$. This equivalence can be expressed, again for $[A]$ symmetric, as the *spectral decomposition*, or *Jordan decomposition*,

$$[A] = [E][\Lambda][E]^T \tag{11.52a}$$

$$= [E] \begin{bmatrix} \lambda_1 & 0 & 0 & \cdots & 0 \\ 0 & \lambda_2 & 0 & \cdots & 0 \\ 0 & 0 & \lambda_3 & \cdots & 0 \\ \vdots & \vdots & \vdots & \ddots & \vdots \\ 0 & 0 & 0 & \cdots & \lambda_K \end{bmatrix} [E]^T, \tag{11.52b}$$

so that $[\Lambda]$ denotes a diagonal matrix whose nonzero elements are the K eigenvalues of $[A]$. It is illuminating to consider also the equivalent of Equation 11.52 in summation notation,

$$[A] = \sum_{k=1}^{K} \lambda_k \, \boldsymbol{e}_k \boldsymbol{e}_k^{\mathrm{T}} \tag{11.53a}$$

$$= \sum_{k=1}^{K} \lambda_k [E_k]. \tag{11.53b}$$

The outer product of each eigenvector with itself in Equation 11.53a defines a matrix $[E_k]$. Equation 11.53b shows that the original matrix $[A]$ can be recovered as a weighted sum of these $[E_k]$ matrices, where the weights are the corresponding eigenvalues. Hence the spectral decomposition of a matrix is analogous to the Fourier decomposition of a function or data series (Equation 10.62a), with the eigenvalues playing the role of the Fourier amplitudes and the $[E_k]$ matrices corresponding to the cosine functions.

Other consequences of the equivalence of the information on the two sides of Equation 11.52 pertain to the eigenvalues. The first of these is

$$\mathrm{tr}[A] = \sum_{k=1}^{K} a_{k,k} = \sum_{k=1}^{K} \lambda_k = \mathrm{tr}[\Lambda]. \tag{11.54}$$

This relationship is especially important when $[A]$ is a covariance matrix, in which case its diagonal elements $a_{k,k}$ are the K variances. Equation 11.54 says the sum of these variances is equal to the sum of the eigenvalues of the covariance matrix.

The second consequence of Equation 11.52 for the eigenvalues is

$$\det[A] = \prod_{k=1}^{K} \lambda_k = \det[\Lambda], \tag{11.55}$$

which is consistent with the property that at least one of the eigenvalues of a singular matrix (having zero determinant) will be zero. A real symmetric matrix with all eigenvalues positive is called *positive definite*.

The matrix of eigenvectors $[E]$ has the property that it *diagonalizes* the original symmetric matrix $[A]$ from which the eigenvectors and eigenvalues were calculated. Left-multiplying Equation 11.52a by $[E]^{\mathrm{T}}$, right-multiplying by $[E]$, and using the orthogonality of $[E]$ yields

$$[E]^{\mathrm{T}}[A][E] = [\Lambda]. \tag{11.56}$$

That is, multiplication of $[A]$ on the left by $[E]^{\mathrm{T}}$ and on the right by $[E]$ produces the diagonal matrix of eigenvalues $[\Lambda]$.

There is also a strong connection between the eigenvalues λ_k and eigenvectors $\boldsymbol{e}_k$ of a nonsingular symmetric matrix, and the corresponding quantities λ_k^* and $\boldsymbol{e}_k^*$ of its inverse. The eigenvectors of matrix-inverse pairs are the same—that is, $\boldsymbol{e}_k^* = \boldsymbol{e}_k$ for each k—and the corresponding eigenvalues are reciprocals, $\lambda_k^* = \lambda_k^{-1}$. Therefore the eigenvector of $[A]$ associated with its largest eigenvalue is the same as the eigenvector of $[A]^{-1}$ associated with its smallest eigenvalue, and vice versa. One way to see this analytically is to subject both sides of Equation 11.52a to matrix inversion, and then use the property for the inverse of a matrix product in Table 11.1 and the orthonormality property (Equation 11.44) of an eigenvector matrix, to derive

$$([S])^{-1} = \left([E]^{\mathrm{T}}[\Lambda][E] \right)^{-1}, \tag{11.57}$$
$$= [E]^{\mathrm{T}}[\Lambda]^{-1}[E]$$

recalling that the inverse of a diagonal matrix is also diagonal with elements that are reciprocals of the original matrix.

The extraction of eigenvalue-eigenvector pairs from matrices is a computationally demanding task, particularly as the dimensionality of the problem increases. It is possible but very tedious to do the computations by hand if $K=2$, 3, or 4, using the equation

$$\det\left([A] - \lambda[I]\right) = 0. \tag{11.58}$$

This calculation requires first solving a Kth-order polynomial for the K eigenvalues, and then solving K sets of K simultaneous equations to obtain the eigenvectors. In general, however, widely available computer algorithms for calculating very close numerical approximations to eigenvalues and eigenvectors are used. These computations can also be done within the framework of the singular value decomposition (see Section 11.3.5).

Example 11.3. Eigenvalues and Eigenvectors of a (2×2) Symmetric Matrix and Its Inverse
The symmetric matrix

$$[A] = \begin{bmatrix} 185.47 & 110.84 \\ 110.84 & 77.58 \end{bmatrix} \tag{11.59}$$

has as its eigenvalues $\lambda_1=254.76$ and $\lambda_2=8.29$, with corresponding eigenvectors $e_1{}^T=[0.848, 0.530]$ and $e_2{}^T=[-0.530, 0.848]$. It is easily verified that both eigenvectors are of unit length. Their dot product is zero, which indicates that the two vectors are perpendicular, or orthogonal.

The matrix of eigenvectors is therefore

$$[E] = \begin{bmatrix} 0.848 & 0.530 \\ 0.530 & 0.848 \end{bmatrix}, \tag{11.60}$$

and the original matrix can be recovered using the eigenvalues and eigenvectors (Equations 11.52 and 11.53) as

$$[A] = \begin{bmatrix} 185.47 & 110.84 \\ 110.84 & 77.58 \end{bmatrix} = \begin{bmatrix} .848 & -.530 \\ .530 & .848 \end{bmatrix} \begin{bmatrix} 254.76 & 0 \\ 0 & 8.29 \end{bmatrix} \begin{bmatrix} .848 & .530 \\ -.530 & .848 \end{bmatrix} \tag{11.61a}$$

$$= 254.76 \begin{bmatrix} .848 \\ .530 \end{bmatrix} [.848 \quad .530] + 8.29 \begin{bmatrix} -.530 \\ .848 \end{bmatrix} [-.530 \quad .848] \tag{11.61b}$$

$$= 254.76 \begin{bmatrix} .719 & .449 \\ .449 & .281 \end{bmatrix} + 8.29 \begin{bmatrix} .281 & -.449 \\ -.449 & .719 \end{bmatrix} \tag{11.61c}$$

Equation 11.61a expresses the spectral decomposition of $[A]$ in the form of Equation 11.52, and Equations 11.61b and 11.61c show the same decomposition in the form of Equation 11.53.

The matrix of eigenvectors diagonalizes the original matrix $[A]$ according to

$$[E]^T[A][E] = \begin{bmatrix} .848 & .530 \\ -.530 & .848 \end{bmatrix} \begin{bmatrix} 185.47 & 110.84 \\ 110.84 & 77.58 \end{bmatrix} \begin{bmatrix} .848 & -.530 \\ .530 & .848 \end{bmatrix}$$

$$= \begin{bmatrix} 254.0 & 0 \\ 0 & 8.29 \end{bmatrix} = [\Lambda]. \tag{11.62}$$

The sum of the eigenvalues, $254.76+8.29=263.05$, equals the sum of the diagonal elements of the original $[A]$ matrix, $185.47+77.58=263.05$.

Applying Equation 11.29 to the matrix $[A]$ in Equation 11.59 yields its inverse

$$[A]^{-1} = \begin{bmatrix} .03688 & -.05270 \\ -.05270 & .08818 \end{bmatrix}. \tag{11.63}$$

The leading (i.e., largest) eigenvalue of $[A]^{-1}$ is then $\lambda_2{}^* = 1/8.29 = .1206$, and its last (i.e., smallest) eigenvalue is $\lambda_1{}^* = 1/254.76 = .003925$. The eigenvectors are the same as those shown in the columns of Equation 11.59, although conventionally the ordering of these columns would be reversed for $[A]^{-1}$ because the first column is paired with the largest eigenvalue. ◇

11.3.4. Square Roots of a Symmetric Matrix

Consider two square matrices of the same order, $[A]$ and $[B]$. If the condition

$$[A] = [B][B]^{\mathrm{T}} \tag{11.64}$$

holds, then $[B]$ multiplied by itself yields $[A]$, so $[B]$ is said to be a "square root" of $[A]$, or $[B] = [A]^{1/2}$. Unlike the square roots of scalars, the square root of a symmetric matrix is not uniquely defined. That is, there are any number of matrices $[B]$ that can satisfy Equation 11.64, although two algorithms are used most frequently to find solutions for it.

If $[A]$ is of full rank, a lower-triangular matrix $[B]$ satisfying Equation 11.64 can be found using the *Cholesky decomposition* of $[A]$. (A *lower-triangular* matrix has zeros above and to the right of the main diagonal, i.e., $b_{i,j} = 0$ for $i < j$.) Beginning with

$$b_{1,1} = \sqrt{a_{1,1}} \tag{11.65}$$

as the only nonzero element in the first row of $[B]$, the Cholesky decomposition proceeds iteratively, by calculating the nonzero elements of each of the subsequent rows, i, of $[B]$ in turn according to

$$b_{i,j} = \frac{a_{i,j} - \sum_{k=1}^{j-1} b_{i,k} b_{j,k}}{b_{j,j}}, j = 1, \ldots, i-1 \tag{11.66a}$$

and

$$b_{i,i} = \left[a_{i,i} - \sum_{k=1}^{i-1} b_{i,k}^2 \right]^{1/2}. \tag{11.66b}$$

It is a good idea to do these calculations in double precision in order to minimize the accumulation roundoff errors that can lead to a division by zero in Equation 10.64a for large matrix dimension K, even if $[A]$ is of full rank.

Some authors (e.g., Golub and van Loan, 1996) define the matrix square root more restrictively, requiring $[A] = [B][B] = [B]^2$ for $[B]$ to be a square root of $[A]$. A symmetric square-root matrix $[B]$ will satisfy both this condition and that in Equation 11.64. Such matrices can be found using the eigenvalues and eigenvectors of $[A]$ when $[A]$ is symmetric and is computable even if $[A]$ is not of full rank. Using the spectral decomposition (Equation 11.52) for $[B]$,

$$[B] = [A]^{1/2} = [E][\Lambda]^{1/2}[E]^{\mathrm{T}}, \tag{11.67}$$

where $[E]$ is the matrix of eigenvectors for both $[A]$ and $[B]$ (i.e., they are the same vectors). The matrix $[\Lambda]$ contains the eigenvalues of $[A]$, which are the squares of the eigenvalues of $[B]$ on the diagonal of

$[\Lambda]^{1/2}$. That is, $[\Lambda]^{1/2}$ is the diagonal matrix with elements $\lambda_k^{1/2}$, where the λ_k are the eigenvalues of $[A]$. Equation 11.67 is still defined even if some of these eigenvalues are zero, so this method can be used to find a square root for a matrix that is not of full rank. Note that $[\Lambda]^{1/2}$ also conforms to both definitions of a square-root matrix, since $[\Lambda]^{1/2}([\Lambda]^{1/2})^T = [\Lambda]^{1/2}[\Lambda]^{1/2} = ([\Lambda]^{1/2})^2 = [\Lambda]$. The square-root decomposition in Equation 11.67 is more tolerant of roundoff error than the Cholesky decomposition when the matrix dimension is large, because (computationally, as well as truly) zero eigenvalues do not produce undefined arithmetic operations.

Equation 11.67 can be extended to find the square root of a matrix inverse, $[A]^{-1/2}$, if $[A]$ is symmetric and of full rank. Because a matrix has the same eigenvectors as its inverse, so also will it have the same eigenvectors as the square root of its inverse. Accordingly,

$$[A]^{-1/2} = [E][\Lambda]^{-1/2}[E]^T, \tag{11.68}$$

where $[\Lambda]^{-1/2}$ is the diagonal matrix with elements $\lambda_k^{-1/2}$, the reciprocals of the square roots of the eigenvalues of $[A]$. The implications of Equation 11.68 are those that would be expected, that is, $[A]^{-1/2}([A]^{-1/2})^T = [A]^{-1}$, and $[A]^{-1/2}([A]^{1/2})^T = [I]$.

Example 11.4. Square Roots of a Matrix and Its Inverse

The symmetric matrix $[A]$ in Equation 11.59 is of full rank, since both of its eigenvalues are positive. Therefore a lower-triangular square-root matrix $[B] = [A]^{1/2}$ can be computed using the Cholesky decomposition. Equation 11.66 yields $b_{1,1} = (a_{1,1})^{1/2} = 185.47^{1/2} = 13.619$ as the only nonzero element of the first row ($i = 1$) of $[B]$. Because $[B]$ has only one additional row, Equations 11.66 need to be applied only once each. Equation 11.66a yields $b_{2,1} = (a_{1,1} - 0)/b_{1,1} = 110.84/13.619 = 8.139$. Zero is subtracted in the numerator of Equation 11.66a for $b_{2,1}$ because there are no terms in the summation. (If $[A]$ had been a (3×3) matrix, Equation 11.66a would be applied twice for the third ($i = 3$) row: the first of these applications, for $b_{3,1}$, would again have no terms in the summation; but when calculating $b_{3,2}$ there would be one term corresponding to $k = 1$.) Finally, the calculation indicated by Equation 11.66b is $b_{2,2} = (a_{2,2} - b_{2,1}^2)^{1/2} = (77.58 - 8.139^2)^{1/2} = 3.367$. The Cholesky lower-triangular square-root matrix for $[A]$ is thus

$$[B] = [A]^{1/2} = \begin{bmatrix} 13.619 & 0 \\ 8.139 & 3.367 \end{bmatrix}, \tag{11.69}$$

which can be verified as a valid square root of $[A]$ through the matrix multiplication $[B][B]^T$.

A symmetric square-root matrix for $[A]$ can be computed using its eigenvalues and eigenvectors from Example 11.3, and Equation 11.67:

$$\begin{aligned} [B] = [A]^{1/2} &= [E][\Lambda]^{1/2}[E]^T \\ &= \begin{bmatrix} .848 & -.530 \\ .530 & .848 \end{bmatrix} \begin{bmatrix} \sqrt{254.76} & 0 \\ 0 & \sqrt{8.29} \end{bmatrix} \begin{bmatrix} .848 & .530 \\ -.530 & .848 \end{bmatrix}. \\ &= \begin{bmatrix} 12.286 & 5.879 \\ 5.879 & 6.554 \end{bmatrix} \end{aligned} \tag{11.70}$$

This matrix also can be verified as a valid square root of $[A]$ by calculating $[B][B]^T = [B]^2$.

Equation 11.68 allows calculation of a square-root matrix for the inverse of $[A]$,

$$[A]^{-1/2} = [E][\Lambda]^{-1/2}[E]^{T}$$

$$= \begin{bmatrix} .848 & -.530 \\ .530 & .848 \end{bmatrix} \begin{bmatrix} 1/\sqrt{254.76} & 0 \\ 0 & 1/\sqrt{8.29} \end{bmatrix} \begin{bmatrix} .848 & .530 \\ -.530 & .848 \end{bmatrix}. \tag{11.71}$$

$$= \begin{bmatrix} .1426 & -.1279 \\ -.1279 & .2674 \end{bmatrix}$$

This is also a symmetric matrix. The matrix product $[A]^{-1/2} ([A]^{-1/2})^{T} = [A]^{-1/2} [A]^{-1/2} = [A]^{-1}$. The validity of Equation 11.71 can be checked by comparing the product $[A]^{-1/2} [A]^{-1/2}$ with $[A]^{-1}$ as calculated using Equation 11.28 or by verifying $[A][A]^{-1/2} [A]^{-1/2} = [A][A]^{-1} = [I]$.   ◇

11.3.5. Singular Value Decomposition (SVD)

Equation 11.52 expresses the spectral decomposition of a symmetric square matrix. This decomposition can be extended to any $(n \times m)$ rectangular matrix $[A]$ with at least as many rows as columns $(n \geq m)$ using the *singular value decomposition* (SVD),

$$\underset{(n\times m)}{[A]} = \underset{(n\times m)}{[L]} \ \underset{(m\times m)}{[\Omega]} \ \underset{(m\times m)}{[R]^{T}} , n \geq m. \tag{11.72}$$

The m columns of $[L]$ are called the left *singular vectors*, and the m columns of $[R]$ (not its transpose) are called the right singular vectors. (Note that, in the context of SVD, $[R]$ does not denote a correlation matrix.) Both sets of vectors are mutually orthonormal, so $[L]^{T}[L] = [R]^{T}[R] = [R][R]^{T} = [I]$, with dimension $(m \times m)$. The matrix $[\Omega]$ is diagonal, with nonnegative diagonal elements that are called the *singular values* of $[A]$. Equation 11.72 is sometimes called the "thin" SVD, in contrast to an equivalent expression in which the dimension of $[L]$ is $(n \times n)$, and the dimension of $[\Omega]$ is $(n \times m)$, but with the last $n - m$ rows of $[\Omega]$ containing all zeros so that the last $n - m$ columns of $[L]$ are arbitrary.

If $[A]$ is square and symmetric, then Equation 11.72 reduces to Equation 11.52, with $[L] = [R] = [E]$, and $[\Omega] = [\Lambda]$. It is therefore possible to compute eigenvalues and eigenvectors for symmetric matrices using an SVD algorithm from a package of matrix-algebra computer routines, which are widely available (e.g., Press et al., 1986). Analogously to Equation 11.53 for the spectral decomposition of a symmetric square matrix, Equation 11.72 can be expressed as a summation of weighted outer products of the left and right singular vectors,

$$[A] = \sum_{i=1}^{m} \omega_i \ell_i r_i^{T}. \tag{11.73}$$

Even if $[A]$ is not symmetric, there is a connection between the SVD and the eigenvalues and eigenvectors of both $[A]^{T}[A]$ and $[A][A]^{T}$, both of which matrix products are square (with dimensions $(m \times m)$ and $(n \times n)$, respectively) and symmetric. Specifically, the columns of $[R]$ are the $(m \times 1)$ eigenvectors of $[A]^{T}[A]$, the columns of $[L]$ are the $(n \times 1)$ eigenvectors of $[A][A]^{T}$. The respective singular values are the square roots of the corresponding eigenvalues, i.e., $\omega_i^2 = \lambda_i$.

Example 11.5. Eigenvalues and Eigenvectors of a Covariance Matrix Using SVD

Consider the (31×2) matrix $(30)^{-1/2}[X']$, where $[X']$ is the matrix of anomalies (Equation 11.29) for the minimum temperature data in Table A.1. The SVD of this matrix can be used to obtain the eigenvalues and eigenvectors of the sample covariance matrix for these data, without first explicitly computing $[S]$ (if $[S]$ is already known, SVD can also be used to compute the eigenvalues and eigenvectors, through the equivalence of Equations 11.72 and 11.52).

The SVD of $(30)^{-1/2}[X']$, in the form of Equation 11.72, is

$$\frac{1}{\sqrt{30}}[X'] = \begin{bmatrix} 1.09 & 1.42 \\ 2.19 & 1.42 \\ 1.64 & 1.05 \\ \vdots & \vdots \\ 1.83 & 0.51 \end{bmatrix}_{(31\times2)} = \begin{bmatrix} .105 & .216 \\ .164 & .014 \\ .122 & .008 \\ \vdots & \vdots \\ .114 & -.187 \end{bmatrix}_{(31\times2)} \begin{bmatrix} 15.961 & 0 \\ 0 & 2.879 \end{bmatrix}_{(2\times2)} \begin{bmatrix} .848 & .530 \\ -.530 & .848 \end{bmatrix}_{(2\times2)}. \qquad (11.74)$$

The reason for multiplying the anomaly matrix $[X']$ by $30^{-1/2}$ should be evident from Equation 11.30: the product $(30^{-1/2}\,[X']^T)\,(30^{-1/2}\,[X']) = (n-1)^{-1}\,[X']^T[X']$ yields the covariance matrix $[S]$ for these data, which is the same as the matrix $[A]$ in Equation 11.59. Because the matrix of right singular vectors $[R]$ contains the eigenvectors for the product of the matrix on the left-hand side of Equation 11.74, left-multiplied by its transpose, the matrix $[R]^T$ on the far right of Equation 11.74 is the same as the (transpose of) the matrix $[E]$ in Equation 11.60. Similarly the squares of the singular values in the diagonal matrix $[\Omega]$ in Equation 11.74 are the corresponding eigenvalues, for example, $\omega^2_1 = 15.961^2 = \lambda_1 = 254.7$.

The right-singular vectors of $(n-1)^{-1/2}\,[X']$ are the eigenvectors of the (2×2) covariance matrix $[S] = (n-1)^{-1}\,[X']^T[X']$. The left singular vectors in the matrix $[L]$ are eigenvectors of the (31×31) matrix $(n-1)^{-1}\,[X][X]^T$. This matrix actually has 31 eigenvectors, but only two of them (the two shown in Equation 11.74) are associated with nonzero eigenvalues. It is in this sense, of truncating the zero eigenvalues and their associated irrelevant eigenvectors, that Equation 11.74 is an example of a thin SVD. ◇

The SVD is a versatile tool with a variety of applications. One of these is maximum covariance analysis (MCA), to be described in Section 14.3. Sometimes MCA is confusingly called "SVD analysis," even though SVD is merely the computational tool used to calculate a MCA.

11.4. RANDOM VECTORS AND MATRICES

11.4.1. Expectations and Other Extensions of Univariate Concepts

Just as ordinary random variables are scalar quantities, a random vector (or random matrix) is a vector (or matrix) whose entries are random variables. The purpose of this section is to extend the rudiments of matrix algebra presented in Section 11.3 to include statistical ideas.

A vector x whose K elements are the random variables x_k is a random vector. The expected value of this random vector is also a vector, called the vector mean, whose K elements are the individual expected values (i.e., probability-weighted averages) of the corresponding random variables. If all the x_k are continuous variables,

$$\boldsymbol{\mu} = \begin{bmatrix} \int\limits_{-\infty}^{\infty} x_1 f_1(x_1)dx_1 \\ \int\limits_{-\infty}^{\infty} x_2 f_2(x_2)dx_2 \\ \vdots \\ \int\limits_{-\infty}^{\infty} x_K f_K(x_K)dx_K \end{bmatrix}. \tag{11.75}$$

If some or all of the K variables in x are discrete, the corresponding elements of $\boldsymbol{\mu}$ will be sums in the form of Equation 4.13.

The properties of expectations listed in Equation 4.15 extend also to vectors and matrices in ways that are consistent with the rules of matrix algebra. If c is a scalar constant, $[X]$ and $[Y]$ are random matrices with the same dimensions (and which may be random vectors if one of their dimensions is 1), and $[A]$ and $[B]$ are constant (nonrandom) matrices,

$$E(c[X]) = c\,E([X]), \tag{11.76a}$$

$$E([X]+[Y]) = E([X]) + E([Y]), \tag{11.76b}$$

$$E([A][X][B]) = [A]\,E([X])\,[B], \tag{11.76c}$$

$$E([A][X]+[B]) = [A]\,E([X]) + [B]. \tag{11.76d}$$

The (population, or generating-process) covariance matrix, corresponding to the sample estimate $[S]$ in Equation 11.5, is the matrix expected value

$$\underset{(K\times K)}{[\Sigma]} = E\left(\underset{(K\times 1)}{[x-\boldsymbol{\mu}]}\,\underset{(1\times K)}{[x-\boldsymbol{\mu}]^{\mathrm{T}}} \right) \tag{11.77a}$$

$$= E\left(\begin{bmatrix} (x_1-\mu_1)^2 & (x_1-\mu_1)(x_2-\mu_2) & \cdots & (x_1-\mu_1)(x_K-\mu_K) \\ (x_2-\mu_2)(x_1-\mu_1) & (x_2-\mu_2)^2 & \cdots & (x_2-\mu_2)(x_K-\mu_K) \\ \vdots & \vdots & \ddots & \vdots \\ (x_K-\mu_K)(x_1-\mu_1) & (x_K-\mu_K)(x_2-\mu_2) & \cdots & (x_K-\mu_K)^2 \end{bmatrix} \right) \tag{11.77b}$$

$$= \begin{bmatrix} \sigma_{1,1} & \sigma_{1,2} & \cdots & \sigma_{1,K} \\ \sigma_{2,1} & \sigma_{2,2} & \cdots & \sigma_{2,K} \\ \vdots & \vdots & \ddots & \vdots \\ \sigma_{K,1} & \sigma_{K,2} & \cdots & \sigma_{K,K} \end{bmatrix}. \tag{11.77c}$$

The diagonal elements of Equation 11.74 are the scalar (population or generating-process) variances, which would be computed (for continuous variables) using Equation 4.21 with $g(x_k)=(x_k-\mu_k)^2$ or, equivalently, Equation 4.22. The off-diagonal elements are the covariances, which would be computed using the double integrals

$$\sigma_{k,\ell} = \int_{-\infty}^{\infty} \int_{-\infty}^{\infty} (x_k - \mu_k)(x_\ell - \mu_\ell) f_{k,\ell}(x_k, x_\ell)\, dx_\ell\, dx_k, \tag{11.78}$$

which is analogous to the summation in Equation 11.4 for the sample covariances. Here $f_{k,\ell}(x_k, x_\ell)$ is the joint (bivariate) PDF for x_k and x_ℓ. Analogously to Equation 4.22b for the scalar variance, an equivalent expression for the (population) covariance matrix is

$$[\Sigma] = E(\boldsymbol{x}\boldsymbol{x}^T) - \boldsymbol{\mu}\boldsymbol{\mu}^T. \tag{11.79}$$

11.4.2. Partitioning Vectors and Matrices

In some settings it is natural to define collections of variables that segregate into two or more groups. Simple examples are one set of L predictands together with a different set of $K-L$ predictors, or sets of two or more variables, each observed simultaneously at some large number of locations or gridpoints. In such cases it is often convenient and useful to maintain these distinctions notationally, by partitioning the corresponding vectors and matrices.

Partitions are indicated by thin or dashed lines in the expanded representation of vectors and matrices. These indicators of partitions are imaginary lines, in the sense that they have no effect whatsoever on the matrix algebra as applied to the larger vectors or matrices. For example, consider a $(K \times 1)$ random vector $\boldsymbol{x}$ that consists of one group of L variables and another group of $K-L$ variables,

$$\boldsymbol{x}^T = \begin{bmatrix} x_1 & x_2 & \cdots & x_L & | & x_{L+1} & x_{L+2} & \cdots & x_K \end{bmatrix}, \tag{11.80a}$$

which would have expectation

$$E(\boldsymbol{x}^T) = \boldsymbol{\mu}^T = \begin{bmatrix} \mu_1 & \mu_2 & \cdots & \mu_L & | & \mu_{L+1} & \mu_{L+2} & \cdots & \mu_K \end{bmatrix}, \tag{11.80b}$$

exactly as Equation 11.75, except that both $\boldsymbol{x}$ and $\boldsymbol{\mu}$ are partitioned as (i.e., composed of a concatenation of) a $(L \times 1)$ vector and a $(K-L \times 1)$ vector.

The covariance matrix of $\boldsymbol{x}$ in Equation 11.80 would be computed in exactly the same way as indicated in Equation 11.77, with the partitions being carried forward:

$$[\Sigma] = E\left([\boldsymbol{x} - \boldsymbol{\mu}][\boldsymbol{x} - \boldsymbol{\mu}]^T\right) \tag{11.81a}$$

$$= \left[\begin{array}{cccc|cccc} \sigma_{1,1} & \sigma_{1,2} & \cdots & \sigma_{1,L} & \sigma_{1,L+1} & \sigma_{1,L+2} & \cdots & \sigma_{1,K} \\ \sigma_{2,1} & \sigma_{2,2} & \cdots & \sigma_{2,L} & \sigma_{2,L+1} & \sigma_{2,L+2} & \cdots & \sigma_{2,K} \\ \vdots & \vdots & \ddots & \vdots & \vdots & \vdots & \cdots & \vdots \\ \sigma_{L,1} & \sigma_{L,2} & \cdots & \sigma_{L,L} & \sigma_{L,L+1} & \sigma_{L,L+2} & \cdots & \sigma_{L,K} \\ \hline \sigma_{L+1,1} & \sigma_{L+1,2} & \cdots & \sigma_{L+1,L} & \sigma_{L+1,L+1} & \sigma_{L+1,L+2} & \cdots & \sigma_{L+1,K} \\ \sigma_{L+2,1} & \sigma_{L+2,2} & \cdots & \sigma_{L+2,L} & \sigma_{L+2,L+1} & \sigma_{L+2,L+2} & \cdots & \sigma_{L+2,K} \\ \vdots & \vdots & \cdots & \vdots & \vdots & \vdots & \ddots & \vdots \\ \sigma_{K,1} & \sigma_{K,2} & \cdots & \sigma_{K,L} & \sigma_{K,L+1} & \sigma_{K,L+2} & \cdots & \sigma_{K,K} \end{array}\right], \tag{11.81b}$$

$$= \left[\begin{array}{c|c} [\Sigma_{1,1}] & [\Sigma_{1,2}] \\ \hline [\Sigma_{2,1}] & [\Sigma_{2,2}] \end{array}\right], \tag{11.81c}$$

so that the covariance matrix $[\Sigma]$ for a data vector x partitioned into two segments as in Equation 11.80 is itself partitioned into four submatrices. The $(L \times L)$ matrix $[\Sigma_{1,1}]$ is the covariance matrix for the first L variables, $[x_1, x_2, ..., x_L]^T$, and the $(K-L \times K-L)$ matrix $[\Sigma_{2,2}]$ is the covariance matrix for the last $K-L$ variables, $[x_{L+1}, x_{L+2}, ..., x_K]^T$. Both of these matrices have variances on the main diagonal and covariances among the variables within its respective group in the other positions.

The $(K-L \times L)$ matrix $[\Sigma_{2,1}]$ contains the covariances among all possible pairs of variables consisting of one member in the second group and the other member in the first group. Because it is not a full covariance matrix it does not contain variances along the main diagonal even if it is square, and in general it is not symmetric. The $(L \times K-L)$ matrix $[\Sigma_{1,2}]$ contains the same covariances among all possible pairs of variables having one member in the first group and the other member in the second group. Because the full covariance matrix $[\Sigma]$ is symmetric, $[\Sigma_{1,2}]^T = [\Sigma_{2,1}]$.

11.4.3. Linear Combinations

A *linear combination* is essentially a weighted sum of two or more of the variables $x_1, x_2, ..., x_K$ in a data vector x. For example, the multiple linear regression in Equation 7.25 is a linear combination of the K regression predictors that yields a new variable, which in this case is the regression prediction. For simplicity, assume that the parameter $b_0 = 0$ in Equation 7.25 (this would be the case if the predictand and all predictors are expressed as anomalies). Then Equation 7.25 can be expressed in matrix notation as

$$y = b^T x, \tag{11.82}$$

where $b^T = [b_1, b_2, ..., b_K]$ is the vector of parameters that are the weights in the weighted sum.

Usually in regression the predictors x are considered to be fixed constants rather than random variables. But consider now the case where x is a random vector with mean μ_x and covariance $[\Sigma_x]$. The linear combination in Equation 11.82 will then also be a random variable. Extending Equation 4.14c for vector x, with $g_j(x) = b_j x_j$, the mean of y will be

$$\mu_y = \sum_{k=1}^{K} b_k \mu_k, \tag{11.83}$$

where $\mu_k = E(x_k)$. The variance of the linear combination is more complicated, both notationally and computationally, and involves the covariances among all pairs of the x's. For simplicity, suppose $K = 2$. Then,

$$\begin{aligned}
\sigma_y^2 = \text{Var}(b_1 x_1 + b_2 x_2) &= E\left\{ [(b_1 x_1 + b_2 x_2) - (b_1 \mu_1 + b_2 \mu_2)]^2 \right\} \\
&= E\left\{ [b_1(x_1 - \mu_1) + b_2(x_2 - \mu_2)]^2 \right\} \\
&= E\left\{ b_1^2(x_1 - \mu_1)^2 + b_2^2(x_2 - \mu_2)^2 + 2b_1 b_2(x_1 - \mu_1)(x_2 - \mu_2) \right\} \\
&= b_1^2 E\left\{ (x_1 - \mu_1)^2 \right\} + b_2^2 E\left\{ (x_2 - \mu_2)^2 \right\} + 2b_1 b_2 E\{ (x_1 - \mu_1)(x_2 - \mu_2) \} \\
&= b_1^2 \sigma_{1,1} + b_2^2 \sigma_{2,2} + 2b_1 b_2 \sigma_{1,2}
\end{aligned} \tag{11.84}$$

This scalar result is fairly cumbersome, even though the linear combination is composed of only two random variables, and the extension to linear combinations of K random variables involves $K(K+1)/2$ terms. More generally, and much more compactly, in matrix notation Equations 11.83 and 11.84 become

$$\mu_y = b^T \mu \tag{11.85a}$$

and

$$\sigma_y^2 = b^T [\Sigma_x] b. \tag{11.85b}$$

The quantities on the left-hand sides of Equations 11.85 are scalars, because the result of the single linear combination in Equation 11.82 is scalar. But consider simultaneously forming L linear combinations of the K random variables x,

$$
\begin{aligned}
y_1 &= b_{1,1} x_1 + b_{1,2} x_2 + \cdots + b_{1,K} x_K \\
y_2 &= b_{2,1} x_1 + b_{2,2} x_2 + \cdots + b_{2,K} x_K \\
\vdots \quad \vdots \quad \vdots & \qquad\qquad\qquad \vdots \\
y_L &= b_{L,1} x_1 + b_{L,2} x_2 + \cdots + b_{L,K} x_K
\end{aligned}
\tag{11.86a}
$$

or

$$
\underset{(L \times 1)}{y} = \underset{(L \times K)}{[B]^T} \ \underset{(K \times 1)}{x}. \tag{11.86b}
$$

Here each row of $[B]^T$ defines a single linear combination as in Equation 11.82, and collectively these L linear combinations define the random vector y. Extending Equations 11.85 to the mean vector and covariance matrix of this collection of L linear combinations of x,

$$
\underset{(L \times 1)}{\mu_y} = \underset{(L \times K)}{[B]^T} \ \underset{(K \times 1)}{\mu_x} \tag{11.87a}
$$

and

$$
\underset{(L \times L)}{\left[\Sigma_y \right]} = \underset{(L \times K)}{[B]^T} \ \underset{(K \times K)}{[\Sigma_x]} \ \underset{(K \times L)}{[B]}. \tag{11.87b}
$$

Note that by using Equations 11.87 it is not actually necessary to explicitly compute the transformed variables in Equation 11.86 in order to find their mean and covariance, if the mean vector and covariance matrix of the x's are known.

Example 11.6. Mean Vector and Covariance Matrix for a Pair of Linear Combinations

Example 11.5 showed that the matrix in Equation 11.59 is the covariance matrix for the Ithaca and Canandaigua minimum temperature data in Table A.1. The mean vector for these data is $\mu^T = (\mu_{Ith}, \mu_{Can}) = (13.0, 20.2)$. Consider now two linear combinations of these minimum temperature data in the form of Equation 11.45, with $\theta = 32$ degrees. That is, each of the two rows of $[T]^T$ defines a linear combination (Equation 11.82), which can be expressed jointly as in Equation 11.86b. Together, these two linear combinations are equivalent to a transformation that corresponds to a counterclockwise rotation of the coordinate axes through the angle θ. That is, each vector $y = [T]^T x$ would locate the same point, but in the framework of the rotated coordinate system.

One way to find the mean and covariance for the transformed points, μ_y and $[\Sigma_y]$, would be to carry out the transformation for all $n = 31$ point pairs, and then to compute the mean vector and covariance matrix for the transformed data set. However, knowing the mean and covariance of the underlying x's it is straightforward and much easier to use Equation 11.87 to obtain

$$\mu_y = \begin{bmatrix} \cos 32^\circ & \sin 32^\circ \\ -\sin 32^\circ & \cos 32^\circ \end{bmatrix} \mu_x = \begin{bmatrix} .848 & .530 \\ -.530 & .848 \end{bmatrix} \begin{bmatrix} 13.0 \\ 20.2 \end{bmatrix} = \begin{bmatrix} 21.7 \\ 10.2 \end{bmatrix} \qquad (11.88a)$$

and

$$\begin{bmatrix} \Sigma_y \end{bmatrix} = [T]^T [\Sigma_x][T] = \begin{bmatrix} .848 & .530 \\ -.530 & .848 \end{bmatrix} \begin{bmatrix} 185.47 & 110.84 \\ 110.84 & 77.58 \end{bmatrix} \begin{bmatrix} .848 & -.530 \\ .530 & .848 \end{bmatrix}$$

$$= \begin{bmatrix} 254.76 & 0 \\ 0 & 8.29 \end{bmatrix} \qquad (11.88b)$$

The rotation angle $\theta = 32$ degrees is evidently a special one for these data, as it produces a pair of transformed variables y that are uncorrelated. In fact this transformation is exactly the same as in Equation 11.62, which was expressed in terms of the eigenvectors of $[\Sigma_x]$. ◇

Just as the mean and variance of a linear combination can be expressed and computed without actually calculating the linear combinations, the covariance of two linear combinations can similarly be computed, using

$$\text{Cov}\left([A]^T x_1, [B]^T x_2 \right) = [A]^T [\Sigma_{1,2}] [B]. \qquad (11.89)$$

Here $[\Sigma_{1,2}]$ is the matrix of covariances between the vectors x_1 and x_2, which is the upper right-hand quadrant of Equation 11.81, when the vector x has been partitioned into the subvectors x_1 and x_2. If $[A]^T$ and $[B]^T$ are vectors (and so dimensioned $(1 \times L)$ and $(1 \times K - L)$, respectively), Equation 11.89 will yield the scalar covariance between the single pair of linear combinations. If $x_1 = x_2 = x$, then Equation 11.89 becomes

$$\text{Cov}\left([A]^T x, [B]^T x \right) = [A]^T [\Sigma] [B], \qquad (11.90)$$

where, as before $[\Sigma]$ is the covariance matrix for x.

11.4.4. Mahalanobis Distance, Revisited

Section 11.2.2 introduced the Mahalanobis, or statistical, distance as a way to measure differences or unusualness within the context established by an empirical data scatter or an underlying multivariate probability density. If the K variables in the data vector x are mutually uncorrelated, the (squared) Mahalanobis distance takes the simple form of the sum of the squared standardized anomalies z_k, as indicated in Equation 11.7 for $K = 2$ variables. When some or all of the variables are correlated the Mahalanobis distance accounts for the correlations as well, although as noted in Section 11.2.2 the notation is prohibitively complicated in scalar form. In matrix notation, the Mahalanobis distance between points x and y in their K-dimensional space is

$$D^2 = [x - y]^T [S]^{-1} [x - y], \qquad (11.91)$$

where $[S]$ is the covariance matrix in the context of which the distance is being calculated.

If the dispersion defined by $[S]$ involves zero correlation among the K variables, it is not difficult to see that Equation 11.91 reduces to Equation 11.7 (in two dimensions, with obvious extension to higher

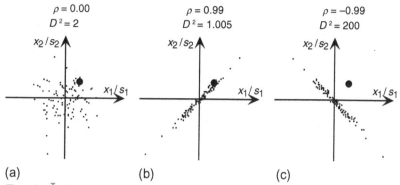

FIGURE 11.6   The point $z^T = (1, 1)$ (large dot) in the contexts of data scatters with (a) zero correlation, (b) correlation 0.99, and (c) correlation -0.99. Mahalanobis distances, D^2, to the origin are drastically different in these three cases.

dimensions). In that case, $[S]$ is diagonal, and its inverse is also diagonal with elements $(s_{k,k})^{-1}$, so Equation 11.91 would reduce to $D^2 = \Sigma_k \, (x_k - y_k)^2/s_{k,k}$. This observation underscores one important property of the Mahalanobis distance, namely, that different intrinsic scales of variability for the K variables in the data vector do not confound D^2, because each is divided by its standard deviation before squaring. If $[S]$ is diagonal, the Mahalanobis distance is the same as the Euclidean distance after dividing each variable by its standard deviation.

The second salient property of the Mahalanobis distance is that it accounts for the redundancy in information content among correlated variables. Again, this concept is easiest to see in two dimensions. Two strongly correlated variables provide very nearly the same information, and ignoring strong correlations when calculating statistical distance (i.e., using Equation 11.7 when the correlation is not zero), effectively double-counts the contribution of the (nearly) redundant second variable. The situation is illustrated in Figure 11.6, which shows the standardized point $z^T = (1, 1)$ in the contexts of three very different point clouds. In Figure 11.6a the correlation reflected by the circular point cloud is zero, so it is equivalent to use Equation 11.7 to calculate the Mahalanobis distance to the origin (which is also the vector mean of the point cloud), after having accounted for possibly different scales of variation for the two variables by dividing by the respective standard deviations. That distance is $D^2 = 2$ (corresponding to an ordinary Euclidean distance of $\sqrt{2} = 1.414$). The correlation between the two variables in Figure 11.6b is 0.99, so that one or the other of the two variables provides nearly the same information as both together: z_1 and z_2 are nearly the same variable. Using Equation 11.91 the Mahalanobis distance to the origin is $D^2 = 1.005$, which is only slightly more than if only one of the two nearly redundant variables had been considered alone, and substantially smaller than the distance appropriate to the context of the scatter in Figure 11.6a.

Finally, Figure 11.6c shows a very different situation, in which the correlation is -0.99. Here the point $(1, 1)$ is extremely unusual in the context of the data scatter, and using Equation 11.91 we find $D^2 = 200$. That is, it is extremely far from the origin relative to the dispersion of the point cloud, and this unusualness is reflected by the very large Mahalanobis distance. The point $(1, 1)$ in Figure 11.6c is a *multivariate outlier*. Visually it is well removed from the point scatter in two dimensions. But relative to either of the two univariate distributions it is a quite ordinary point that is relatively close to (one standard deviation from) each scalar mean, so that it would not stand out as unusual when applying standard scalar EDA methods to the two variables individually. It is an outlier in the sense that it does not behave like the scatter of the negatively correlated point cloud, in which large values of x_1/s_1 are associated with small values of x_2/s_2, and vice versa. The large Mahalanobis distance to the center (vector mean) of the point cloud identifies it as a multivariate outlier.

Equation 11.91 is an example of what is called a *quadratic form*. It is quadratic in the vector $x - y$, in the sense that this vector is multiplied by itself, together with scaling constants in the symmetric matrix $[S]^{-1}$. In $K = 2$ dimensions a quadratic form written in scalar notation takes the form of Equation 11.7 if the symmetric matrix of scaling constants is diagonal, and in the form of Equation 11.8 if it is not. Equation 11.91 emphasizes that quadratic forms can be interpreted as squared distances, and as such it is generally desirable for them to be nonnegative, and furthermore strictly positive if the vector being squared is not zero. This latter condition is met if the symmetric matrix of scaling constants is positive definite, so that all its eigenvalues are positive.

Finally, it was noted in Section 11.2.2 that Equation 11.7 describes ellipses of constant distance D^2. The ellipses described by Equation 11.7, corresponding to zero correlations in the matrix $[S]$ in Equation 11.91, have their axes aligned with the coordinate axes. Equation 11.91 also describes ellipses of constant Mahalanobis distance D^2, whose axes are rotated away from the directions of the coordinate axes to the extent that some or all of the correlations in $[S]$ are nonzero. In such cases the axes of the ellipses of constant D^2 are aligned in the directions of the eigenvectors of $[S]$, as will be seen in Section 12.1.

11.5. EXERCISES

11.1. Calculate the matrix product $[A][E]$, using the values in Equations 11.59 and 11.60.

11.2. Derive the regression equation produced in Example 7.1, using matrix notation.

11.3. Calculate the angle between the two eigenvectors of the matrix $[A]$ in Equation 11.59.

11.4. Verify through matrix multiplication that both $[T]$ in Equation 11.45, and its transpose, are orthogonal matrices.

11.5. Show that Equation 11.67 produces a valid square root.

11.6. Assuming all the relevant matrix and vector dimensions are compatible, simplify:

$$([C](\bar{x} - \mu))^{\mathrm{T}} \left(\frac{1}{n}[C][S][C]^{\mathrm{T}} \right)^{-1} ([C](\bar{x} - \mu))$$

11.7. The $(K \times K)$ square matrix $[E]$ contains eigenvectors of a covariance matrix as its columns. Solve for the $(K \times 1)$ vector x:

$$u = [E]^{\mathrm{T}} x$$

11.8. The eigenvalues and eigenvectors of the covariance matrix for the Ithaca and Canandaigua maximum temperatures in Table A.1 are $\lambda_1 = 118.8$ and $\lambda_2 = 2.60$, and $e_1^{\mathrm{T}} = [.700, .714]$ and $e_2^{\mathrm{T}} = [-.714, .700]$, where the first element of each vector corresponds to the Ithaca temperature.

    a. Find the covariance matrix $[S]$, using its spectral decomposition.

    b. Find $[S]^{-1}$ using its eigenvalues and eigenvectors.

    c. Find $[S]^{-1}$ using the result of part (a), and Equation 11.28.

    d. Find a symmetric $[S]^{1/2}$.

    e. Find the Mahalanobis distance between the observations for 1 January and 2 January.

11.9. a. Use the Pearson correlations in Table 3.5 and the standard deviations from Table A.1 to compute the covariance matrix $[S]$ for the four temperature variables in Table A.1.

b. Consider the average daily temperatures defined by the two linear combinations:

$y_1 = 0.5$ (Ithaca Max) $+ 0.5$ (Ithaca Min)

$y_2 = 0.5$ (Canandaigua Max) $+ 0.5$ (Canandaigua Min)

Find μ_y and $[S_y]$ without actually computing the individual y values.

Chapter 12

The Multivariate Normal Distribution

12.1. DEFINITION OF THE MVN

The *multivariate normal* (MVN) distribution is the natural generalization of the Gaussian, or normal distribution (Section 4.4.2) to multivariate, or vector data. The MVN is by no means the only known continuous parametric multivariate distribution (e.g., Johnson and Kotz, 1972; Johnson, 1987), but overwhelmingly it is the most commonly used. Some of the popularity of the MVN follows from its relationship to the multivariate Central Limit Theorem, although it is also used in other settings without strong theoretical justification because it enjoys a number of convenient properties that will be outlined in this section. This convenience is often sufficiently compelling to undertake transformation of non-Gaussian multivariate data to approximate multinormality before working with them (e.g., Kelly and Krzysztofowicz, 1997), which has been a strong motivation for development of the methods described in Section 3.4.

The univariate Gaussian PDF (Equation 4.24) describes the individual, or marginal, distribution of probability density for a scalar Gaussian variable. The MVN describes the joint distribution of probability density collectively for the K variables in a vector $\boldsymbol{x}$. The univariate Gaussian PDF is visualized as the bell curve defined on the real line (i.e., in a one-dimensional space). The MVN PDF is defined on the K-dimensional space whose coordinate axes correspond to the elements of $\boldsymbol{x}$, in which multivariate distances were calculated in Sections 11.2 and 11.4.4.

The probability density function for the MVN is

$$f(\boldsymbol{x}) = \frac{1}{(2\pi)^{K/2} \sqrt{\det[\Sigma]}} \exp\left[-\frac{1}{2}(\boldsymbol{x} - \boldsymbol{\mu})^T [\Sigma]^{-1} (\boldsymbol{x} - \boldsymbol{\mu})\right], \tag{12.1}$$

where $\boldsymbol{\mu}$ is the K-dimensional mean vector, and $[\Sigma]$ is the $(K \times K)$ covariance matrix for the K variables in the vector $\boldsymbol{x}$. In $K=1$ dimension, Equation 12.1 reduces to Equation 4.24, and for $K=2$ it reduces to the PDF for the bivariate normal distribution (Equation 4.31). The key part of the MVN PDF is the argument of the exponential function, and regardless of the dimension of $\boldsymbol{x}$ this argument is a squared, standardized distance (i.e., the difference between $\boldsymbol{x}$ and its mean, standardized by the (co-)variance). In the general multivariate form of Equation 12.1 this distance is the Mahalanobis distance, which is a positive-definite quadratic form when $[\Sigma]$ is of full rank, and is not defined otherwise because in that case $[\Sigma]^{-1}$ does not exist. The constants outside of the exponential in Equation 12.1 serve only to ensure that the integral over the entire K-dimensional space is 1,

$$\int_{-\infty}^{\infty} \int_{-\infty}^{\infty} \cdots \int_{-\infty}^{\infty} f(\boldsymbol{x}) \, dx_1 \, dx_2 \cdots dx_K = 1, \tag{12.2}$$

which is the multivariate extension of Equation 4.18.

If each of the K variables in $\boldsymbol{x}$ is separately standardized according to 4.26, the result is the standardized MVN density,

Statistical Methods in the Atmospheric Sciences. https://doi.org/10.1016/B978-0-12-815823-4.00012-2
© 2019 Elsevier Inc. All rights reserved.

$$\phi(z) = \frac{1}{(2\pi)^{K/2}\sqrt{\det[R]}} \exp\left[-\frac{z^T[R]^{-1}z}{2}\right], \tag{12.3}$$

where $[R]$ is the (Pearson) correlation matrix (e.g., Figure 3.29) for the K variables. Equation 12.3 is the multivariate generalization of Equation 4.25. The nearly universal notation for indicating that a random vector x follows a K-dimensional MVN with covariance matrix $[\Sigma]$ is

$$x \sim N_K(\mu, [\Sigma]) \tag{12.4a}$$

or, for standardized variables,

$$z \sim N_K(0, [R]), \tag{12.4b}$$

where 0 is the K-dimensional mean vector whose elements are all zero.

Because the only dependence of Equation 12.1 on the random vector x is through the Mahalanobis distance inside the exponential, contours of equal probability density are ellipsoids of constant Mahalanobis distance D^2 from μ. These ellipsoidal contours centered on the mean enclose the smallest regions in the K-dimensional space containing a given portion of the probability mass, and the link between the size of these ellipsoids and the enclosed probability is the χ^2 distribution:

$$\Pr\left\{D^2 = (x-\mu)^T[\Sigma]^{-1}(x-\mu) \le \chi_K^2(\alpha)\right\} = \alpha. \tag{12.5}$$

Here $\chi_K^2(\alpha)$ denotes the quantile of the χ^2 distribution with K degrees of freedom, associated with cumulative probability α (Table B.3). That is, the probability of an x being within a given Mahalanobis distance D^2 of the mean is equal to the area to the left of D^2 under the χ^2 distribution with degrees of freedom $\nu = K$. As noted at the end of Section 11.4.4 the orientations of these ellipsoids are defined by the eigenvectors of $[\Sigma]$, which are also the eigenvectors of $[\Sigma]^{-1}$. Furthermore, the elongation of the ellipsoids in the directions of each of these eigenvectors is given by the square root of the product of the respective eigenvalue of $[\Sigma]$ multiplied by the relevant χ^2 quantile. For a given D^2 the (hyper-) volume enclosed by one of these ellipsoids is proportional to the square root of the determinant of $[\Sigma]$,

$$V = \frac{2(\pi D^2)^{K/2}}{K\,\Gamma(K/2)}\sqrt{\det[\Sigma]}, \tag{12.6}$$

where $\Gamma(\cdot)$ denotes the gamma function (Equation 4.7). Here the determinant of $[\Sigma]$ functions as a scalar measure of the magnitude of the matrix, in terms of the volume occupied by the probability dispersion it describes. Accordingly, $\det[\Sigma]$ is sometimes called the *generalized variance*. The determinant, and thus also the volumes enclosed by constant-D^2 ellipsoids, increases as the K variances $\sigma_{k,k}$ increase; but also the determinant and these volumes decrease as the correlations among the K variables increase, because larger correlations result in the ellipsoids being less spherical and more elongated.

Example 12.1. Probability Ellipses for the Bivariate Normal Distribution

It is easiest to visualize multivariate ideas in two dimensions. Consider the MVN distribution fit to the Ithaca and Canandaigua minimum temperature data in Table A.1. Here $K = 2$, so this is a bivariate normal distribution with sample mean vector $[13.0, 20.2]^T$ and (2×2) covariance matrix as shown in Equation 11.59. Example 11.3 shows that this covariance matrix has eigenvalues $\lambda_1 = 254.76$ and $\lambda_2 = 8.29$, with corresponding eigenvectors $e_1^T = [0.848, 0.530]$ and $e_2^T = [-0.530, 0.848]$.

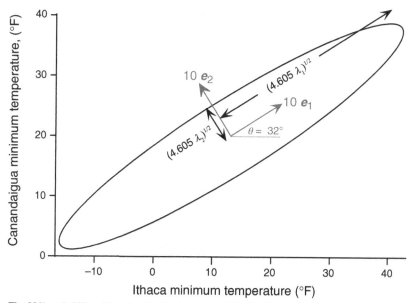

FIGURE 12.1   The 90% probability ellipse for the bivariate normal distribution representing the minimum temperature data in Table A.1, centered at the vector sample mean. Its major and minor axes are oriented in the directions of the eigenvectors *(gray)* of the covariance matrix in Equation 11.59, and stretched in these directions in proportion to the square roots of the respective eigenvalues. The constant of proportionality is the square root of the appropriate χ_2^2 quantile. The eigenvectors are drawn $10 \times$ larger than unit length for clarity.

Figure 12.1 shows the 90% probability ellipse for this distribution. All the probability ellipses for this distribution are oriented 32 degrees from the data axes, as shown in Example 11.6. (This angle between e_1 and the horizontal unit vector $[1, 0]^T$ can also be calculated using Equation 11.15.) The extent of this 90% probability ellipse in the directions of its two axes is determined by the 90% quantile of the χ^2 distribution with $v = K = 2$ degrees of freedom, which is $\chi_2^2(0.90) = 4.605$ from Table B.3. Therefore the ellipse extends to $(\chi_2^2(0.90)\lambda_k)^{1/2}$ in the directions of each of the two eigenvectors e_k; or the distances $(4.605 \cdot 254.67)^{1/2} = 34.2$ in the e_1 direction, and $(4.605 \cdot 8.29)^{1/2} = 6.2$ in the e_2 direction.

The volume enclosed by this ellipse is actually an area in two dimensions. From Equation 12.6 this area is $V = 2(\pi \, 4.605)^1 \sqrt{2103.26/(2 \cdot 1)} = 663.5$, since $\det[S] = 2103.26$. $\diamond$

Example 12.2. Approximate Confidence Region for Maximum Likelihood Estimates
Section 4.6.4 noted that when a training data sample is large, the sampling distribution of maximum likelihood parameter estimates is well approximated by a MVN distribution with the same dimensionality K as the number of parameters being estimated simultaneously. Accordingly K-dimensional confidence regions for the parameters can be represented by MVN distributions. Because the large-n bias for the maximum likelihood estimators is small, these confidence regions are centered at the maximum likelihood estimates, and the corresponding covariance matrix (Equation 4.91) is the inverse of the observed Fisher information matrix (Equation 4.92).

Equation 4.87 in Example 4.13 outlines use of Newton-Raphson iteration to estimate the two parameters α and β of a gamma distribution, which involves the observed Fisher information for this problem,

$$[I(\boldsymbol{\theta})] = - \begin{bmatrix} \partial^2 L/\partial\alpha^2 & \partial^2 L/\partial\alpha\partial\beta \\ \partial^2 L/\partial\beta\partial\alpha & \partial^2 L/\partial\beta^2 \end{bmatrix}$$

$$= \begin{bmatrix} n\Gamma''(\alpha) & n/\beta \\ n/\beta & -\dfrac{n\alpha}{\beta^2} + \dfrac{2\,\Sigma x}{\beta^3} \end{bmatrix},$$
(12.7)

where $\boldsymbol{\theta}$ represents the parameter vector $[\alpha, \beta]^T$. Applying the maximum likelihood algorithm in Example 4.13 to the 1933–1982 Ithaca January precipitation data in Table A.2 yields the parameter estimates $\hat{\alpha} = 3.76$ and $\hat{\beta} = 0.52$ in (Figure 4.16). Substituting these estimates into the second equality of Equation 12.7 yields

$$[I(\boldsymbol{\theta})] = \begin{bmatrix} 15.203 & 13.284 \\ 13.284 & 693.606 \end{bmatrix}$$
(12.8)

so that the covariance matrix appropriate to the sampling distribution of these two parameters is

$$Var\left(\hat{\boldsymbol{\theta}}\right) = \left[I\left(\hat{\boldsymbol{\theta}}\right)\right]^{-1} = \begin{bmatrix} 0.0669 & -0.00128 \\ -0.00128 & 0.00147 \end{bmatrix}.$$
(12.9)

Figure 12.2 shows the resulting 95% joint confidence region for the two parameters, which is bounded by an ellipse in the $K=2$-dimensional space. The two eigenvalues of the covariance matrix in Equation 12.9 are $\lambda_1 = 0.0669$ and $\lambda_2 = 0.00145$, and the corresponding eigenvectors (defining the directions of the dashed arrows locating the major and minor axes of the confidence ellipse) are $e_1 = [0.9998, -0.0196]^T$ and $e_2 = [0.0196, 0.9998]^T$. These eigenvectors are rotated away from the coordinate axes by only 1.1 degrees, indicating that sampling errors for the two parameters are only weakly correlated (Equation 12.9 implies a correlation of -0.129). Constructing the 95% confidence region requires finding the 95th percentile of the χ_2^2 distribution ($=5.991$, Table B.3), which together with the eigenvalues defines the extent of the confidence ellipse in the e_1 and e_2 directions. ◇

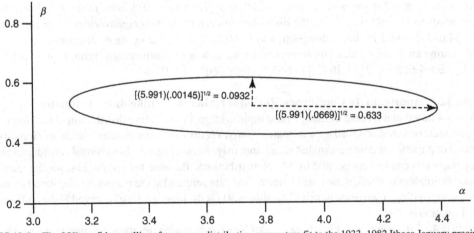

FIGURE 12.2 The 95% confidence ellipse for gamma distribution parameters fit to the 1933–1982 Ithaca January precipitation data in Table A.2.

12.2. FOUR HANDY PROPERTIES OF THE MVN

(1) *All subsets of variables from a MVN distribution are themselves distributed MVN.* Consider the partition of a $(K \times 1)$ MVN random vector x into the vectors $x_1 = (x_1, x_2, \ldots, x_L)$, and $x_2 = (x_{L+1}, x_{L+2}, \ldots, x_K)$, as in Equation 11.80a. Then each of these two subvectors themselves follows MVN distributions, with $x_1 \sim N_L (\mu_1, [\Sigma_{1,1}])$ and $x_2 \sim N_{K-L} (\mu_2, [\Sigma_{2,2}])$. Here the two mean vectors compose the corresponding partition of the original mean vector as in Equation 11.80b, and the covariance matrices are the indicated submatrices in Equations 11.81b and 11.81c. Note that the original ordering of the elements of x is immaterial, and that a MVN partition can be constructed from any subset. If a subset of the MVN x contains only one element (e.g., the scalar x_1) its distribution is univariate Gaussian: $x_1 \sim N_1 (\mu_1, \sigma_{1,1})$. That is, this first handy property implies that all the marginal distributions for the K elements of a MVN distribution x are univariate Gaussian. The converse may not be true: it is not necessarily the case that the joint distribution of an arbitrarily selected set of K Gaussian variables will follow a MVN distribution.

(2) *Linear combinations of a MVN x are Gaussian.* If x is a MVN random vector, then a single linear combination in the form of Equation 11.82 will be univariate Gaussian with mean and variance given by Equations 11.85a and 11.85b, respectively. This fact is a consequence of the property that sums of Gaussian variables are themselves Gaussian, as noted in connection with the sketch of the Central Limit Theorem in Section 4.4.2. Similarly the result of L simultaneous linear transformations, as in Equation 11.86, will have an L-dimensional MVN distribution, with mean vector and covariance matrix given by Equations 11.87a and 11.87b, respectively, provided the covariance matrix $[\Sigma_y]$ is invertible. This condition will hold if $L \leq K$, and if none of the transformed variables y_ℓ can be expressed as an exact linear combination of the others. In addition, the mean of a MVN distribution can be shifted without changing the covariance matrix. If c is a $(K \times 1)$ vector of constants then

$$x \sim N_K(\mu_x, [\Sigma_x]) \Rightarrow x + c \sim N_K(\mu_x + c, [\Sigma_x]). \tag{12.10}$$

(3) *Independence implies zero correlation, and vice versa, for Gaussian distributions.* Again consider the partition of a MVN x as in Equation 11.80a. If x_1 and x_2 are independent then the off-diagonal matrices of cross-covariances in Equation 11.81 contain only zeros: $[\Sigma_{1,2}] = [\Sigma_{2,1}]^T = [0]$. Conversely, if $[\Sigma_{1,2}] = [\Sigma_{2,1}]^T = [0]$ then the MVN PDF can be factored as $f(x) = f(x_1)f(x_2)$, implying independence (cf. Equation 2.12), because the argument inside the exponential in Equation 12.1 then breaks cleanly into two factors.

(4) *The conditional distribution of subset of a MVN x, given fixed values for other elements, is also MVN.* This is the multivariate generalization of Equations 4.35, which is illustrated in Example 4.7, expressing this idea for the bivariate normal distribution. Consider again the partition $x = [x_1, x_2]^T$ as defined in Equation 11.80b and used to illustrate properties (1) and (3) before. The conditional mean of one subset of the variables x_1 given particular values for the remaining variables $X_2 = x_2$ is

$$\mu_1 | x_2 = \mu_1 + [\Sigma_{12}][\Sigma_{22}]^{-1}(x_2 - \mu_2), \tag{12.11a}$$

and the conditional covariance matrix is

$$[\Sigma_{11} | x_2] = [\Sigma_{11}] - [\Sigma_{12}][\Sigma_{22}]^{-1}[\Sigma_{21}], \tag{12.11b}$$

where the submatrices of $[\Sigma]$ are again as defined in Equation 11.81. As was the case for the bivariate normal distribution, the conditional mean shift in Equation 12.11a depends on the particular value of the conditioning variable x_2, whereas the conditional covariance matrix in Equation 12.11b does not. If x_1 and x_2 are independent, then knowledge of one provides no additional information about the other. Mathematically, if $[\Sigma_{1,2}] = [\Sigma_{2,1}]^T = [0]$ then Equation 12.11a reduces to $\mu_1|x_2 = \mu_1$, and Equation 12.11b reduces to $[\Sigma_1|x_2] = [\Sigma_1]$.

Example 12.3. Three-Dimensional MVN Distributions as Cucumbers

Imagine a three-dimensional MVN PDF being represented by a cucumber, which is a solid, three-dimensional ovoid. Since the cucumber has a distinct edge, it would be more correct to imagine that it represents that part of a MVN PDF enclosed within a fixed-D^2 ellipsoidal surface. The cucumber would be an even better metaphor if its density increased toward the core and decreased toward the skin.

Figure 12.3a illustrates property (1), which is that all subsets of a MVN distribution are themselves MVN. Here are three hypothetical cucumbers floating above a kitchen cutting board in different orientations, and illuminated from above. Their shadows represent the joint distribution of the two variables whose axes are aligned with the edges of the board. Regardless of the orientation of the cucumber relative to the board (i.e., regardless of the covariance structure of the three-dimensional distribution) each of these two-dimensional joint shadow distributions for x_1 and x_2 is bivariate normal, with probability contours within fixed Mahalanobis distances of the means of the shadow ovals in the plane of the board.

Figure 12.3b illustrates property (4) that conditional distributions of subsets given particular values for the remaining variables in a MVN distribution are themselves MVN. Here portions of two cucumbers are lying on the cutting board, where the long axis of the left cucumber (indicated by the direction of the arrow or the corresponding eigenvector) is oriented parallel to the x_1 axis of the board, and the long axis of the right cucumber has been placed diagonally to the edges of the board. The three variables represented by the

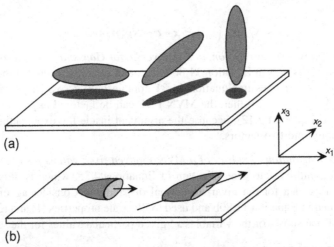

(a)

(b)

FIGURE 12.3 Three-dimensional MVN distributions as cucumbers on a kitchen cutting board. (a) Three cucumbers floating slightly above the cutting board and illuminated from above, illustrating that their shadows (the bivariate normal distributions representing the two-dimensional subsets of the original three variables in the plane of the cutting board) are ovals, regardless of the orientation (covariance structure) of the cucumber. (b) Two cucumbers resting on the cutting board, with faces exposed by cuts made perpendicularly to the x_1 coordinate axis; illustrating bivariate normality in the other two (x_2, x_3) dimensions, given the left-right location of the cut. Arrows indicate directions of the cucumber long-axis eigenvectors.

left cucumber are thus mutually independent, whereas the two horizontal (x_1 and x_2) variables for the right cucumber are positively correlated. Each cucumber has been sliced perpendicularly to the x_1 axis of the cutting board, and the exposed faces represent the joint conditional distributions of the remaining two (x_2 and x_3) variables. Both faces are ovals, illustrating that both of the resulting conditional distributions are bivariate normal. Because the cucumber on the left is oriented parallel to the cutting board edges (coordinate axes) it represents independent variables.

If parallel cuts had been made elsewhere on these cucumbers, the shapes of the exposed faces would have been the same, illustrating (as in Equation 12.11b) that the conditional covariance (shape of the exposed cucumber face) does not depend on the value of the conditioning variable (location left or right along the x_1 axis at which the cut is made). On the other hand, the conditional means (the centers of the exposed faces projected onto the $x_2 - x_3$ plane, Equation 12.11a) depend on the value of the conditioning variable (x_1), but only if the variables are correlated as in the right-hand cucumber. Making the cut further to the right shifts the location of the center of the exposed face of the right-hand cucumber toward the back of the board (the x_2 component of the conditional bivariate vector mean is greater). On the other hand, because the axes of the left cucumber ellipsoid are aligned with the coordinate axes, the location of the center of the exposed face in the $x_2 - x_3$ plane is the same regardless of where on the x_1 axis the cut has been made.   ◇

12.3. TRANSFORMING TO, AND ASSESSING MULTINORMALITY

It was noted in Section 3.4.1 that one strong motivation for transforming data to approximate normality is the ability to use the MVN distribution to describe the joint variations of a multivariate data set. Usually either the Box-Cox power transformations (Equation 3.20), or the Yeo and Johnson (2000) generalization to possibly nonpositive data (Equation 3.23), are used. The Hinkley statistic (Equation 3.21), which reflects the degree of symmetry in a transformed univariate distribution, is the simplest way to decide among power transformations. However, when the goal is specifically to approximate a Gaussian distribution, as is the case when we hope that each of the transformed distributions will form one of the marginal distributions of a MVN distribution, it is probably better to choose transformation exponents that maximize the Gaussian likelihood function (Equation 3.22). It is also possible to choose transformation exponents simultaneously for multiple elements of x, by choosing the corresponding vector of exponents $\boldsymbol{\lambda}$ that maximize the MVN likelihood function, although this approach requires substantially more computation than fitting the individual exponents independently, and in most cases is probably not worth the additional effort.

Choices other than the power transformations are also possible and may sometimes be more appropriate. For example, bimodal and/or strictly bounded data, such as might be well described by a beta distribution (see Section 4.4.6) with both parameters less than 1, will not power transform to approximate normality. However, if such data are adequately described by a parametric CDF $F(x)$, they can be transformed to approximate normality by matching cumulative probabilities, that is,

$$z_i = \Phi^{-1}[F(x_i)]. \tag{12.12}$$

Here $\Phi^{-1}[\cdot]$ is the quantile function for the standard Gaussian distribution, so Equation 12.12 transforms a data value x_i to the standard Gaussian z_i having the same cumulative probability as that associated with x_i within its CDF.

Methods for evaluating normality are necessary, both to assess the need for transformations, and to evaluate the effectiveness of candidate transformations. Mecklin and Mundfrom (2004) provide an extensive review and comparison of methods that have been proposed for assessing multivariate normality. There is no single best approach to this problem, and in practice we usually look at multiple indicators, which may include both quantitative formal tests and qualitative graphical tools.

Because all marginal distributions of a MVN distribution are univariate Gaussian, goodness-of-fit tests are often calculated for the univariate distributions corresponding to each of the elements of the x whose multinormality is being assessed. The Filliben test for the Gaussian Q-Q plot correlation (Table 5.5) is a good choice for the specific purpose of testing Gaussian distribution. Gaussian marginal distributions are a necessary consequence of joint multinormality, but are not sufficient to guarantee it. In particular, looking only at marginal distributions will not identify the presence of multivariate outliers (e.g., Figure 11.6c), which are points that are not extreme with respect to any of the individual variables, but are unusual in the context of the overall covariance structure.

Two tests for multinormality (i.e., jointly for all K dimensions of x) with respect to multivariate skewness and kurtosis are available (Mardia, 1970; Mardia et al., 1979). Both rely on the function of the point pair x_i and x_j given by

$$g_{i,j} = (x_i - \bar{x})^T [S]^{-1} (x_j - \bar{x}), \tag{12.13}$$

where $[S]$ is the sample covariance matrix. This function is used to calculate the multivariate skewness measure

$$b_{1,K} = \frac{1}{n^2} \sum_{i=1}^{n} \sum_{j=1}^{n} g_{i,j}^3, \tag{12.14}$$

which reflects high-dimensional symmetry and will be near zero for MVN data. This test statistic can be evaluated using

$$\frac{n\,b_{1,K}}{6} \sim \chi_\nu^2, \tag{12.15a}$$

where the degrees-of-freedom parameter is

$$\nu = \frac{K(K+1)(K+2)}{6}, \tag{12.15b}$$

and the null hypothesis of multinormality, with respect to its symmetry, is rejected for sufficiently large values of $b_{1,K}$.

Multivariate kurtosis (appropriately heavy tails for the MVN relative to probability density near the center of the distribution) can be tested using the statistic

$$b_{2,K} = \frac{1}{n} \sum_{i=1}^{n} g_{i,i}^2, \tag{12.16}$$

which is equivalent to the average of $(D^2)^2$ because for this statistic $i = j$ in Equation 12.13. Under the null hypothesis of multinormality,

$$\left[\frac{b_{2,K} - K(K+2)}{8K(K+2)/n} \right]^{1/2} \sim N[0, 1]. \tag{12.17}$$

Scatterplots of variable pairs are valuable qualitative indicators of multinormality, since all subsets of variables from a MVN distribution are jointly normal also, and two-dimensional graphs are easy to plot and grasp. Thus looking at a scatterplot matrix (see Section 3.6.5) is typically a valuable tool in assessing multinormality. Point clouds that are elliptical or circular are indicative of multinormality. Outliers away from the main scatter in one or more of the plots may be multivariate outliers, as in Figure 11.6c. Similarly, it can be valuable to look at rotating scatterplots (Section 3.6.3) of various three-dimensional subsets of x.

Absence of evidence for multivariate outliers in all possible pairwise scatterplots does not guarantee that none exist in higher-dimensional combinations. An approach to exposing the possible existence of high-dimensional multivariate outliers, as well as to detecting other possible problems, is to use Equation 12.5. This equation implies that if the data x are MVN, the (univariate) distribution for $D_i^2, i = 1, \ldots, n$, is χ_K^2. That is, the Mahalanobis distance D_i^2 from the sample mean for each x_i can be calculated, and the closeness of this distribution of D_i^2 values to the χ^2 distribution with K degrees of freedom can be evaluated. The easiest and most usual evaluation method is to visually inspect the Q-Q plot. It would also be possible to derive critical values to test the null hypothesis of multinormality according to the correlation coefficient for this kind of plot, using the method sketched in Section 5.2.5.

Because any linear combination of variables that are jointly multinormal will be univariate Gaussian, it can also be informative to look at and formally test linear combinations for Gaussian distribution. Often it is useful to look specifically at the linear combinations given by the eigenvectors of $[S]$,

$$y_i = e_k^T x_i. \tag{12.18}$$

It turns out that the linear combinations defined by the elements of the eigenvectors associated with the smallest eigenvalues can be particularly useful in identifying multivariate outliers, either by inspection of the Q-Q plots, or by formally testing the Q-Q correlations. (The reason behind linear combinations associated with the smallest eigenvalues being especially powerful in exposing outliers relates to principal component analysis, as explained in Section 13.1.5). Inspection of scatterplots of linear combinations in the rotated 2-dimensional spaces defined by pairs of eigenvectors of $[S]$ can also be revealing.

Example 12.4. Assessing Bivariate Normality for the Canandaigua Temperature Data

Are the January 1987 Canandaigua maximum and minimum temperature data in Table A.1 consistent with the proposition that they were drawn from a bivariate normal distribution? Figure 12.4 presents four plots indicating that this assumption is not unreasonable, considering the rather small sample size.

Figures 12.4a and b are Gaussian Q-Q plots for the maximum and minimum temperatures, respectively. The temperatures are plotted as functions of the standard Gaussian variables with the same cumulative probability, which has been estimated using a median plotting position (Table 3.2). Both plots are close to linear, supporting the notion that each of the two data batches was drawn from univariate Gaussian distributions. Somewhat more quantitatively, the correlations of the points in these two panels are 0.984 for the maximum temperatures and 0.978 for the minimum temperatures. If these data were serially independent, we could refer to Table 5.5 and find that both are larger than 0.970, which is the 10% critical value for $n = 30$. Since these data are serially correlated, the Q-Q correlations provide even weaker evidence against the null hypotheses that these two marginal distributions are Gaussian. Figure 12.4c shows the scatterplot for the two variables jointly. The distribution of points appears to be reasonably elliptical, with greater density near the sample mean, $[31.77, 20.23]^T$, and less density at the extremes. This assessment is supported by Figure 12.4d, which is the Q-Q plot for the Mahalanobis distances of each of the points from the sample mean. If the data are bivariate normal, the distribution of these D_i^2 values will be χ^2, with two degrees of freedom, which is an exponential distribution (Equations 4.52 and 4.53), with $\beta = 2$. Values of its quantile function on the horizontal axis of

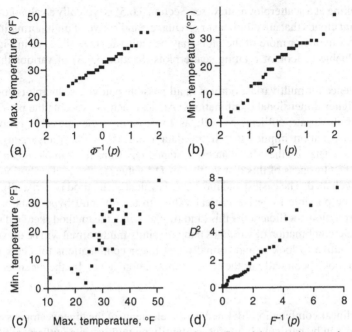

FIGURE 12.4 Graphical assessments of bivariate normality for the Canandaigua maximum and minimum temperature data. (a) Gaussian Q-Q plot for the maximum temperatures, (b) Gaussian Q-Q plot for the minimum temperatures, (c) scatterplot for the bivariate temperature data, and (D) Q-Q plot for Mahalanobis distances relative to the χ^2 distribution.

Figure 12.4d have been calculated using Equation 4.94. The points in this Q-Q plot are also reasonably straight, with the largest bivariate outlier ($D^2 = 7.23$) obtained for 25 January. This is the leftmost point in Figure 12.4c, corresponding to the coldest maximum temperature. The second-largest D^2 of 6.00 results from the data for 15 January, which is the warmest day in both the maximum and minimum temperature data.

The correlation of the points in Figure 12.4d is 0.989, but it would be inappropriate to use Table 5.5 to judge its unusualness relative to a null hypothesis that the data were drawn from a bivariate normal distribution, for two reasons. First, Table 5.5 was derived for Gaussian Q-Q plot correlations, and the null distribution (under the hypothesis of MVN data) for the Mahalanobis distance is χ^2. In addition, these data are not independent. However, it would be possible to derive critical values analogous to those in Table 5.5, by synthetically generating a large number of samples from a bivariate normal distribution with (bivariate) time correlations that simulate those in the Canandaigua temperatures, calculating the D^2 Q-Q plot for each of these samples, and tabulating the distribution of the resulting correlations. Methods appropriate to constructing such simulations are described in the next section. ◇

12.4. SIMULATION FROM THE MULTIVARIATE NORMAL DISTRIBUTION

12.4.1. Simulating Independent MVN Variates

Statistical simulation of MVN variates is accomplished through an extension of the univariate ideas presented in Section 4.7. Generation of synthetic MVN values takes advantage of the second property in Section 12.2 that linear combinations of MVN values are themselves MVN. In particular, realizations of K-dimensional MVN vectors $x \sim N_K(\mu, [\Sigma])$ are generated as linear combinations of K-dimensional standard MVN vectors $z \sim N_K(0, [I])$, each of the K elements of which are independent standard univariate Gaussian. These standard univariate Gaussian realizations are in turn generated on the basis of uniform variates (see Section 4.7.1) transformed according to an algorithm such as that described in Section 4.7.4.

Specifically, the linear combinations used to generate MVN variates with a given mean vector and covariance matrix are given by the rows of a square-root matrix (see Section 11.3.4) for $[\Sigma]$, with the appropriate element of the mean vector added:

$$x_i = [\Sigma]^{1/2} z_i + \mu. \tag{12.19}$$

As a linear combination of the K standard Gaussian values in the vector z, the generated vectors x will have a MVN distribution. It is straightforward to see that they will also have the correct mean vector and covariance matrix:

$$E(x) = E\left([\Sigma]^{1/2} z + \mu\right) = [\Sigma]^{1/2} E(z) + \mu = \mu \tag{12.20a}$$

because $E(z) = 0$, and

$$[\Sigma_x] = [\Sigma]^{1/2} [\Sigma_z] \left([\Sigma]^{1/2}\right)^T = [\Sigma]^{1/2} [I] \left([\Sigma]^{1/2}\right)^T$$
$$= [\Sigma]^{1/2} \left([\Sigma]^{1/2}\right)^T = [\Sigma] \tag{12.20b}$$

Different choices for the nonunique matrix $[\Sigma]^{1/2}$ will yield different simulated x vectors for a given input z, but Equation 12.20 shows that collectively, the resulting $x \sim N_K(\mu, [\Sigma])$ so long as $[\Sigma]^{1/2}$ $([\Sigma]^{1/2})^T = [\Sigma]$.

It is interesting to note that the transformation in Equation 12.19 can be inverted to produce standard MVN vectors $z \sim N_K(0, [I])$ corresponding to MVN vectors x of known distributions. Usually this manipulation is done to transform a sample of vectors x to the standard MVN according to their estimated mean vector and covariance matrix, analogously to the standardized anomaly (Equation 3.27),

$$z_i = [S]^{-1/2}(x_i - \bar{x}) = [S]^{-1/2} x_i'. \tag{12.21}$$

This relationship is called the *Mahalanobis transformation*. It is distinct from the scaling transformation (Equation 11.34), which produces a vector of standard Gaussian variates having unchanged correlation structure. It is straightforward to show that Equation 12.21 produces uncorrelated z_k values, each with unit variance:

$$[S_z] = [S_x]^{-1/2} [S_x] \left([S_x]^{-1/2}\right)^T$$
$$= [S_x]^{-1/2} [S_x]^{1/2} \left([S_x]^{1/2}\right)^T \left([S_x]^{-1/2}\right)^T = [I][I] = [I] \tag{12.22}$$

12.4.2. Simulating Multivariate Time Series

The autoregressive processes for scalar time series described in Sections 10.3.1 and 10.3.2 can be generalized to stationary multivariate, or vector, time series. In this case the variable x is a vector quantity observed at discrete and regularly spaced time intervals. The multivariate generalization of the AR(p) process in Equation 10.23 is

$$x_{t+1} - \mu = \sum_{i=1}^{p} [\Phi_i](x_{t-i+1} - \mu) + [B]\varepsilon_{t+1}. \tag{12.23}$$

Here the elements of the vector x consist of a set of K correlated time series, μ contains the corresponding mean vector, and the elements of the vector ε are mutually independent (and usually Gaussian) random variables with zero mean and unit variance. The matrices of autoregressive parameters $[\Phi_i]$ correspond to the scalar autoregressive parameters ϕ_k in Equation 10.23. The matrix $[B]$, operating on the vector ε_{t+1}, allows the random components in Equation 12.23 to have different variances, and to be mutually correlated at each time step (although they are uncorrelated in time). Note that the order, p, of the autoregression was denoted as K in Chapter 10 and does not indicate the dimension of a vector there. Multivariate autoregressive-moving average models, extending the scalar models in Section 10.3.6 to vector data, can also be defined.

The multivariate AR(1) process is the most common special case of Equation 12.20,

$$x_{t+1} - \mu = [\Phi](x_t - \mu) + [B]\varepsilon_{t+1}, \tag{12.24}$$

which is obtained from Equation 12.23 for the autoregressive order $p = 1$. It is the multivariate generalization of Equation 10.16 and will be stationary process if all the eigenvalues of $[\Phi]$ are between -1 and 1. Matalas (1967) and Bras and Rodríguez-Iturbe (1985) describe use of Equation 12.24 in hydrology, where the elements of x are typically simultaneously measured (possibly transformed) streamflows at different locations. This equation is also often used as part of a common synthetic *weather generator* formulation (Richardson, 1981). In this second application x usually has three elements, corresponding to daily maximum temperature, minimum temperature, and solar radiation at a given location.

The two parameter matrices in Equation 12.24 are most easily estimated using the simultaneous and lagged covariances among the elements of x. The simultaneous covariances are contained in the usual covariance matrix $[S]$, and the lagged covariances are contained in the matrix

$$[S_1] = \frac{1}{n-1} \sum_{t=1}^{n-1} x'_{t+1} x_t^{\prime T} \tag{12.25a}$$

$$= \begin{bmatrix} s_1(1 \to 1) & s_1(2 \to 1) & \cdots & s_1(K \to 1) \\ s_1(1 \to 2) & s_1(2 \to 2) & \cdots & s_1(K \to 2) \\ \vdots & \vdots & \ddots & \vdots \\ s_1(1 \to K) & s_1(2 \to K) & \cdots & s_1(K \to K) \end{bmatrix}. \tag{12.25b}$$

This equation is similar to Equation 11.35 for $[S]$, except that the pairs of vectors whose outer products are summed are data (anomalies) at pairs of successive time points. The diagonal elements of $[S_1]$ are the lag-1 autocovariances (the lagged autocorrelations in Equation 3.36 multiplied by the respective variances, as in Equation 3.39) for each of the K elements of x. The off-diagonal elements of $[S_1]$ are the lagged covariances among unlike elements of x. The arrow notation in this equation indicates the time sequence of the lagging of the variables. For example, $s_1(1 \to 2)$ denotes the correlation between x_1 at time t, and x_2 at time $t+1$, and $s_1(2 \to 1)$ denotes the correlation between x_2 at time t, and x_1 at time t+1. Notice that the matrix $[S]$ is symmetric, but that in general $[S_1]$ is not.

The matrix of autoregressive parameters $[\Phi]$ in Equation 12.24 is obtained from the lagged and unlagged covariance matrices using

$$[\Phi] = [S_1] [S]^{-1}.$$ (12.26)

Obtaining the matrix $[B]$ requires finding a matrix square root (Section 11.3.4) of

$$[B] [B]^T = [S] - [\Phi] [S_1]^T.$$ (12.27)

Chapman et al. (2015) provide a generalization of the Yule-Walker equations (Equation 10.24) for estimating the parameter matrices for the general vector autoregression (Equation 12.23) with order $p \geq 2$.

Having defined a multivariate autoregressive model, it is straightforward to simulate from it using the defining equation (e.g., Equation 11.24) together with an appropriate random number generator to provide time series of realizations for the random-forcing vector $\boldsymbol{\varepsilon}$. Usually these are taken to be standard Gaussian, in which case they can be generated using the algorithm described in Section 4.7.4. In any case the K elements of $\boldsymbol{\varepsilon}$ will have zero mean and unit variance, will be uncorrelated with each other at any one time t, and will be uncorrelated with other forcing vectors at different times $t+\tau$:

$$E[\boldsymbol{\varepsilon}_t] = \mathbf{0}$$ (12.28a)

$$E[\boldsymbol{\varepsilon}_t \boldsymbol{\varepsilon}_t^T] = [I]$$ (12.28b)

$$E[\boldsymbol{\varepsilon}_t \boldsymbol{\varepsilon}_{t+\tau}^T] = [0], \tau \neq 0.$$ (12.28c)

If the $\boldsymbol{\varepsilon}$ vectors contain realizations of independent Gaussian variates, then the resulting x vectors will have a MVN distribution, because they are linear combinations of (standard) MVN vectors $\boldsymbol{\varepsilon}$. If the original data that the simulated series are meant to emulate are clearly non-Gaussian, they may be transformed before fitting the time series model.

Example 12.5. Fitting and Simulating from a Bivariate Autoregression

Example 12.4 examined the Canandaigua maximum and minimum temperature data in Table A.1, and concluded that the MVN distribution is a reasonable model for their joint variations. The first-order autoregression (Equation 12.24) is a reasonable model for their time dependence, and fitting the parameter matrices $[\Phi]$ and $[B]$ will allow statistical simulation of synthetic bivariate series that statistically resemble these data. This process can be regarded as an extension of Example 10.3, which illustrated the univariate AR(1) model for the time series of Canandaigua minimum temperatures alone.

The sample statistics necessary to fit Equation 12.24 are easily computed from the Canandaigua temperature data in Table A.1 as

$$\bar{x} = [31.77 \; 20.23]^T$$ (12.29a)

$$[S] = \begin{bmatrix} 61.85 & 56.12 \\ 56.12 & 77.58 \end{bmatrix}$$ (12.29b)

and

$$[S_1] = \begin{bmatrix} s_{max \to max} & s_{min \to max} \\ s_{max \to min} & s_{min \to min} \end{bmatrix} = \begin{bmatrix} 37.32 & 44.51 \\ 42.11 & 51.33 \end{bmatrix}.$$ (12.29c)

The matrix of simultaneous covariances is the ordinary covariance matrix $[S]$, which is of course symmetric. The matrix of lagged covariances (Equation 12.29c) is not symmetric. Using Equation 12.26, the estimated matrix of autoregressive parameters is

$$[\Phi] = [S_1][S]^{-1} = \begin{bmatrix} 37.32 & 44.51 \\ 42.11 & 51.33 \end{bmatrix} \begin{bmatrix} .04705 & -.03404 \\ -.03404 & .03751 \end{bmatrix} = \begin{bmatrix} .241 & .399 \\ .234 & .492 \end{bmatrix}. \tag{12.30}$$

The matrix $[B]$ can be anything satisfying (c.f. Equation 12.27)

$$[B][B]^T = \begin{bmatrix} 61.85 & 56.12 \\ 56.12 & 77.58 \end{bmatrix} - \begin{bmatrix} .241 & .399 \\ .234 & .492 \end{bmatrix} \begin{bmatrix} 37.32 & 42.11 \\ 44.51 & 51.33 \end{bmatrix} = \begin{bmatrix} 35.10 & 25.49 \\ 25.49 & 42.47 \end{bmatrix}, \tag{12.31}$$

with one solution given by the Cholesky factorization (Equations 11.65 and 11.66),

$$[B] = \begin{bmatrix} 5.92 & 0 \\ 4.31 & 4.89 \end{bmatrix}. \tag{12.32}$$

Using the estimated values in Equations 12.30 and 12.32, and substituting the sample mean from Equation 12.29a for the mean vector, Equation 12.24 becomes an algorithm for simulating bivariate x_t series with the same (sample) first- and second-moment statistics as the Canandaigua temperatures in Table A.1. The Box-Muller algorithm (see Section 4.7.4) is especially convenient for generating the vectors ε_t in this case because it produces them in pairs. Figure 12.5a shows a 100-point realization of a bivariate time series generated in this way. Here the vertical lines connect the simulated maximum and minimum temperatures for a given day, and the light horizontal lines locate the two mean values (Equation 12.29a). These two time series statistically resemble the January 1987 Canandaigua temperature data to the extent that Equation 12.24 is capable of doing so. They are unrealistic in the sense that the generating-process statistics do not change through the 100 simulated days, since the underlying generating model is covariance stationary. That is, the means, variances, and covariances are constant throughout the 100 time points, whereas in nature these statistics change over the course of a winter. Also, the time series is potentially unrealistic in the sense that it is possible (although rare) to statistically simulate maximum temperatures that are colder than the simulated minimum temperature for the day. Recalculating the simulation, but starting from a different random number seed, would yield a different series, but with the same statistical characteristics.

Figure 12.5b shows a scatterplot for the 100 point pairs, corresponding to the scatterplot of the actual data in the lower-right panel of Figure 3.31. Since the points were generated by forcing Equation 12.24

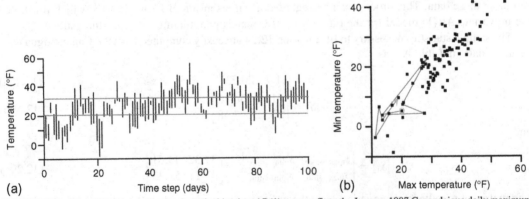

FIGURE 12.5　(a) A 100-point realization from the bivariate AR(1) process fit to the January 1987 Canandaigua daily maximum and minimum temperatures. Vertical lines connect the simulated maximum and minimum for each day, and light horizontal lines locate the two means. (b) Scatterplot of the 100 bivariate points. Light gray line segments connect the first 10 pairs of values.

with synthetic Gaussian variates for the elements of $\boldsymbol{\varepsilon}$, the resulting distribution for $\boldsymbol{x}$ is bivariate normal by construction. However, the points are not independent and exhibit time correlation mimicking that found in the original data series. The result is that successive points do not appear at random within the scatterplot, but rather tend to cluster. The light gray line illustrates this time dependence by tracing a path from the first point (circled) to the tenth point (indicated by the arrow tip).   ◇

Since the statistics underlying Figure 12.5a remained constant throughout the simulation, it is a realization of a stationary time series—in this case a perpetual January. Simulations of this kind can be made to be more realistic by allowing the parameters, based on the statistics in Equations 12.29, to vary periodically through an annual cycle. The result would be a *cyclostationary* autoregression whose statistics are different for different dates, but the same on the same date in different years. Cyclostationary autoregressions are described in Richardson (1981), Von Storch and Zwiers (1999), and Wilks and Wilby (1999), among others.

12.5. INFERENCES ABOUT A MULTINORMAL MEAN VECTOR

This section describes parametric multivariate hypothesis tests concerning mean vectors, based on the MVN distribution. There are many instances where multivariate nonparametric approaches are more appropriate. Some of these multivariate nonparametric tests have been described, as extensions to their univariate counterparts, in Sections 5.3 and 5.4. The parametric tests described in this section require the invertibility of the sample covariance matrix of $\boldsymbol{x}$, $[S_x]$, and so will be infeasible if $n \leq K$. In that case nonparametric tests would be indicated. Even if $[S_x]$ is invertible, the resulting parametric test may have disappointing power unless $n \gg K$, and this limitation can be another reason to choose a nonparametric alternative.

12.5.1. Multivariate Central Limit Theorem

The Central Limit Theorem for univariate data was described briefly in Section 4.4.2, and again more quantitatively in Section 5.2.1. It states that the sampling distribution of the average of a sufficiently large number, n, of random variables will be Gaussian, and that if the variables being averaged are mutually independent the variance of that sampling distribution will be smaller than the variance of the original variables by the factor $1/n$. The multivariate generalization of the Central Limit Theorem states that the sampling distribution of the mean of n independent random ($K \times 1$) vectors $\boldsymbol{x}$ with mean $\boldsymbol{\mu}_x$ and covariance matrix $[\Sigma_x]$ will be MVN with the same covariance matrix, again scaled by the factor $1/n$. That is,

$$\bar{\boldsymbol{x}} \sim N_K \left(\boldsymbol{\mu}_x, \frac{1}{n}[\Sigma_x] \right) \qquad (12.33a)$$

or, equivalently

$$\sqrt{n}(\bar{\boldsymbol{x}} - \boldsymbol{\mu}_x) \sim N_K(0, [\Sigma_x]). \qquad (12.33b)$$

If the random vectors $\boldsymbol{x}$ being averaged are themselves MVN, then the distributions indicated in Equations 12.33 are exact, because then the sample mean vector is a linear combination of the MVN vectors $\boldsymbol{x}$. Otherwise, the multinormality for the sample mean is approximate, and that approximation improves as the sample size n increases.

Multinormality for the sampling distribution of the sample mean vector implies that the sampling distribution for the Mahalanobis distance between the sample and population means will be χ^2. That is, assuming that $[\Sigma_x]$ is known, Equation 12.5 implies that

$$(\bar{x} - \mu)^T \left(\frac{1}{n}[\Sigma_x]\right)^{-1} (\bar{x} - \mu) \sim \chi_K^2, \tag{12.34a}$$

or

$$n\,(\bar{x} - \mu)^T [\Sigma_x]^{-1} (\bar{x} - \mu) \sim \chi_K^2. \tag{12.34b}$$

12.5.2. Hotelling's T^2

Usually inferences about means must be made without knowing the population variance, and this is true in both univariate and multivariate settings. Substituting the estimated covariance matrix into Equation 12.34 yields the one-sample *Hotelling T^2 statistic*,

$$T^2 = (\bar{x} - \mu_0)^T \left(\frac{1}{n}[S_x]\right)^{-1} (\bar{x} - \mu_0) = n\,(\bar{x} - \mu_0)^T [S_x]^{-1} (\bar{x} - \mu_0). \tag{12.35}$$

Here μ_0 indicates the unknown population mean (under the null hypothesis) about which inferences will be made. Equation 12.35 is the multivariate generalization of (the square of) the univariate one-sample t statistic that is obtained by combining Equations 5.3 and 5.4. The univariate t is recovered from the square root of Equation 12.35 for scalar (i.e., $K=1$) data. Both t and T^2 express differences between the sample mean being tested and its hypothesized true value under H_0, "divided by" an appropriate characterization of the dispersion of the null distribution. T^2 is a quadratic (and thus nonnegative) quantity, because the unambiguous ordering of univariate magnitudes on the real line that is expressed by the univariate t statistic does not generalize to higher dimensions. That is, the ordering of scalar magnitude is unambiguous (e.g., it is clear that $5 > 3$), whereas the ordering of vectors is not (e.g., is $(3, 5)^T$ larger or smaller than $(5, 3)^T$?).

The one-sample T^2 is simply the Mahalanobis distance between the vectors x and μ_0, within the context established by the estimated covariance matrix for the sampling distribution of the mean vector, $(1/n)[S_x]$. Since $\bar{x}$ is subject to sampling variations, a continuum of T^2 values is possible, and the probabilities for these outcomes are described by a PDF. Under the null hypothesis H_0: $E(x) = \mu_0$, an appropriately scaled version of T^2 follows what is known as the *F distribution*,

$$\frac{(n-K)}{(n-1)K} T^2 \sim F_{K,n-K}. \tag{12.36}$$

The F distribution is a two-parameter distribution whose quantiles are tabulated in most introductory statistics textbooks. Both parameters are referred to as degrees-of-freedom parameters, and in the context of Equation 12.36 they are $\nu_1 = K$ and $\nu_2 = n - K$, as indicated by the subscripts in Equation 12.36. Accordingly, a null hypothesis that $E(x) = \mu_0$ would be rejected at the α level if

$$T^2 > \frac{(n-1)K}{(n-K)} F_{K,n-K}(1 - \alpha), \tag{12.37}$$

where $F_{K,n-K}(1 - \alpha)$ is the $1 - \alpha$ quantile of the F distribution with K and $n - K$ degrees of freedom.

One way of looking at the F distribution is as the multivariate generalization of the t distribution, which is the null distribution for the t statistic in Equation 5.3. The sampling distribution of Equation 5.3 is t rather than standard univariate Gaussian, and the distribution of T^2 is F rather than χ^2 (as might have been expected from Equation 12.34), because the corresponding dispersion measures (s^2 and $[S]$, respectively) are sample estimates rather than known population values. Just as the univariate t distribution converges to the univariate standard Gaussian as its degrees-of-freedom parameter increases (and the variance s^2 is estimated increasingly more precisely), the F distribution approaches proportionality to the χ^2 with $v_1 = K$ degrees of freedom as the sample size (and thus also v_2) becomes large, because $[S]$ is estimated more precisely:

$$\chi_K^2(1-\alpha) = K F_{K,\infty}(1-\alpha). \tag{12.38}$$

That is, the $(1-\alpha)$ quantile of the χ^2 distribution with K degrees of freedom is exactly a factor of K larger than the $(1-\alpha)$ quantile of the F distribution with $v_1 = K$ and $v_2 = \infty$ degrees of freedom. Since $(n-1) \approx (n-K)$ for sufficiently large n, the large-sample counterparts of Equations 12.36 and 12.37 are, to good approximation,

$$T^2 \sim \chi_K^2 \tag{12.39a}$$

if the null hypothesis is true, leading to rejection at the α level if

$$T^2 > \chi_K^2(1-\alpha). \tag{12.39b}$$

Differences between χ^2 and (scaled) F quantiles are about 5% for $n-K=100$, so that this is a reasonable rule of thumb for appropriateness of Equations 12.39 as large-sample approximations to Equations 12.36 and 12.37.

The two-sample t-test statistic (Equation 5.5) is also extended in a straightforward way to inferences regarding the difference of two independent sample mean vectors:

$$T^2 = [(\bar{\mathbf{x}}_1 - \bar{\mathbf{x}}_2) - \boldsymbol{\delta}_0]^T [S_{\Delta\bar{\mathbf{x}}}]^{-1} [(\bar{\mathbf{x}}_1 - \bar{\mathbf{x}}_2) - \boldsymbol{\delta}_0], \tag{12.40}$$

where

$$\boldsymbol{\delta}_0 = E[\bar{\mathbf{x}}_1 - \bar{\mathbf{x}}_2] \tag{12.41}$$

is the difference between the two population mean vectors under H_0, corresponding to the second term in the numerator of Equation 5.5. If, is as often the case, the null hypothesis is that the two underlying means are equal, then $\boldsymbol{\delta}_0 = \mathbf{0}$ (corresponding to Equation 5.6). The two-sample Hotelling T^2 in Equation 12.40 is a Mahalanobis distance between the difference of the two sample mean vectors being tested and the corresponding difference of their expected values under the null hypothesis. If the null hypothesis is $\boldsymbol{\delta}_0 = \mathbf{0}$, Equation 12.40 reduces to a Mahalanobis distance between the two sample mean vectors.

The covariance matrix for the (MVN) sampling distribution of the difference of the two mean vectors is estimated differently, depending on whether the covariance matrices for the two samples, $[\Sigma_1]$ and $[\Sigma_2]$, can plausibly be assumed equal. If so, this matrix is estimated using a pooled estimate of that common covariance,

$$[S_{\Delta\bar{\mathbf{x}}}] = \left(\frac{1}{n_1} + \frac{1}{n_2}\right)[S_{pool}], \tag{12.42a}$$

where

$$[S_{pool}] = \frac{n_1 - 1}{n_1 + n_2 - 2}[S_1] + \frac{n_2 - 1}{n_1 + n_2 - 2}[S_2] \qquad (12.42b)$$

is a weighted average of the two sample covariance matrices for the underlying data. If these two matrices cannot plausibly be assumed equal, and if in addition the sample sizes are relatively large, then the dispersion matrix for the sampling distribution of the difference of the sample mean vectors may be estimated as

$$[S_{\Delta\bar{x}}] = \frac{1}{n_1}[S_1] + \frac{1}{n_2}[S_2], \qquad (12.43)$$

which is numerically equal to Equation 12.42 for $n_1 = n_2$.

If the sample sizes are not large, the two-sample null hypothesis is rejected at the α level if

$$T^2 > \frac{(n_1 + n_2 - 2)K}{(n_1 + n_2 - K - 1)} F_{K, n_1 + n_2 - K - 1}(1 - \alpha). \qquad (12.44)$$

That is, critical values are proportional to quantiles of the F distribution with $v_1 = K$ and $v_2 = n_1 + n_2 - K - 1$ degrees of freedom. For v_2 sufficiently large (> 100, perhaps), Equation 12.39b can be used, as before.

Finally, if $n_1 = n_2$ and corresponding observations of x_1 and x_2 are linked physically—and correlated as a consequence—it is appropriate to account for the correlations between the pairs of observations by computing a one-sample test involving their differences. Defining Δ_i as the difference between the ith observations of the vectors x_1 and x_2, analogously to Equation 5.10, the one-sample Hotelling T^2 test statistic, corresponding to Equation 5.11, and of exactly the same form as Equation 12.35, is

$$T^2 = (\bar{\Delta} - \mu_\Delta)^T \left(\frac{1}{n}[S_\Delta]\right)^{-1} (\bar{\Delta} - \mu_\Delta) = n (\bar{\Delta} - \mu_\Delta)^T [S_\Delta]^{-1} (\bar{\Delta} - \mu_\Delta). \qquad (12.45)$$

Here $n = n_1 = n_2$ is the common sample size, and $[S_\Delta]$ is the sample covariance matrix for the n vectors of differences Δ_i. The unusualness of Equation 12.45 in the context of the null hypothesis that the true difference of means is μ_Δ is evaluated using the F distribution (Equation 12.37) for relatively small samples, and the χ^2 distribution (Equation 12.39b) for large samples.

Example 12.6. Two-Sample, and One-Sample Paired T^2 Tests

Table 12.1 presents January averages of daily maximum and minimum temperatures at New York City and Boston, for the 30 years 1971 through 2000. Because these are annual values, their serial correlations are quite small. As averages of 31 daily values each, the univariate distributions of these monthly values are expected to closely approximate the Gaussian. Figure 12.6 shows scatterplots for the values at each location. The ellipsoidal dispersions of the two point clouds suggest bivariate normality for both pairs of maximum and minimum temperatures. The two scatterplots overlap somewhat, but the visual separation is sufficiently distinct to suspect strongly that their generating distributions are different.

The two vector means, and their difference vector, are

$$\bar{x}_N = \begin{bmatrix} 38.68 \\ 26.15 \end{bmatrix}, \qquad (12.46a)$$

$$\bar{x}_B = \begin{bmatrix} 36.50 \\ 22.13 \end{bmatrix}, \qquad (12.46b)$$

TABLE 12.1   Average January Maximum and Minimum Temperatures (°F) for New York City and Boston, 1971–2000, and the Corresponding Year-by-Year Differences

Year	New York		Boston		Differences	
	T_{max}	T_{min}	T_{max}	T_{min}	Δ_{max}	Δ_{min}
1971	33.1	20.8	30.9	16.6	2.2	4.2
1972	42.1	28.0	40.9	25.0	1.2	3.0
1973	42.1	28.8	39.1	23.7	3.0	5.1
1974	41.4	29.1	38.8	24.6	2.6	4.5
1975	43.3	31.3	41.4	28.4	1.9	2.9
1976	34.2	20.5	34.1	18.1	0.1	2.4
1977	27.7	16.4	29.8	16.7	−2.1	−0.3
1978	33.9	22.0	35.6	21.3	−1.7	0.7
1979	40.2	26.9	39.1	25.8	1.1	1.1
1980	39.4	28.0	35.6	23.2	3.8	4.8
1981	32.3	20.2	28.5	14.3	3.8	5.9
1982	32.5	19.6	30.5	15.2	2.0	4.4
1983	39.6	29.4	37.6	24.8	2.0	4.6
1984	35.1	24.6	32.4	20.9	2.7	3.7
1985	34.6	23.0	31.2	17.5	3.4	5.5
1986	40.8	27.4	39.6	23.1	1.2	4.3
1987	37.5	27.1	35.6	22.2	1.9	4.9
1988	35.8	23.2	35.1	20.5	0.7	2.7
1989	44.0	30.7	42.6	26.4	1.4	4.3
1990	47.5	35.2	43.3	29.5	4.2	5.7
1991	41.2	28.5	36.6	22.2	4.6	6.3
1992	42.5	28.9	38.2	23.8	4.3	5.1
1993	42.5	30.1	39.4	25.4	3.1	4.7
1994	33.2	17.9	31.0	13.4	2.2	4.5
1995	43.1	31.9	41.0	28.1	2.1	3.8
1996	37.0	24.0	37.5	22.7	−0.5	1.3
1997	39.2	25.1	36.7	21.7	2.5	3.4
1998	45.8	34.2	39.7	28.1	6.1	6.1
1999	40.8	27.0	37.5	21.5	3.3	5.5
2000	37.9	24.7	35.7	19.3	2.2	5.4

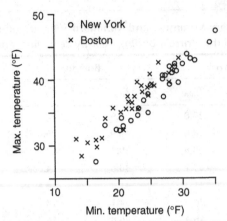

FIGURE 12.6   January average maximum and minimum temperatures, 1971–2000, for New York City (circles) and Boston (x's).

and

$$\overline{\Delta} = \overline{x}_N - \overline{x}_B = \begin{bmatrix} 2.18 \\ 4.02 \end{bmatrix}. \tag{12.46c}$$

As might have been expected from its lower latitude, the average temperatures at New York are warmer. The sample covariance matrix for all four variables jointly is

$$[S] = \begin{bmatrix} [S_N] & [S_{N-B}] \\ [S_{B-N}] & [S_B] \end{bmatrix} = \begin{bmatrix} 21.485 & 21.072 & 17.150 & 17.866 \\ 21.072 & 22.090 & 16.652 & 18.854 \\ 17.150 & 16.652 & 15.948 & 16.070 \\ 17.866 & 18.854 & 16.070 & 18.386 \end{bmatrix}. \tag{12.47}$$

Because the two locations are relatively close to each other and the data were taken in the same years, it is appropriate to treat them as paired values. This assertion is supported by the large cross-covariances in the submatrices $[S_{B-N}] = [S_{N-B}]^T$, corresponding to correlations ranging from 0.89 to 0.94: the data at the two locations are clearly not independent of each other. Nevertheless, it is instructive to first carry through T^2 calculations for differences of mean vectors as a two-sample test, ignoring these large cross-covariances for the moment.

Regarding the Boston and New York temperatures as mutually independent, the appropriate test statistic would be Equation 12.40. If the null hypothesis is that the underlying vector means of the two distributions from which these data were drawn are equal, $\delta_0 = 0$. Both the visual impressions of the two data scatters in Figure 12.6, and the similarity of the covariance matrices $[S_N]$ and $[S_B]$ in Equation 12.47, suggest that assuming equality of covariance matrices would be reasonable. The appropriate covariance for the sampling distribution of the mean difference would then be calculated using Equation 12.42, although because the sample sizes are equal the same numerical result is obtained with Equation 12.43:

$$[S_{\Delta \overline{x}}] = \left(\frac{1}{30} + \frac{1}{30} \right) \left(\frac{29}{58}[S_N] + \frac{29}{58}[S_B] \right) = \frac{1}{30}[S_N] + \frac{1}{30}[S_B] = \begin{bmatrix} 1.248 & 1.238 \\ 1.238 & 1.349 \end{bmatrix}. \tag{12.48}$$

The test statistic (Equation 12.40) can now be calculated as

$$T^2 = [2.18 \ \ 4.02] \begin{bmatrix} 1.248 & 1.238 \\ 1.238 & 1.349 \end{bmatrix}^{-1} \begin{bmatrix} 2.18 \\ 4.02 \end{bmatrix} = 32.34. \tag{12.49}$$

The $1-\alpha=0.9999$ quantile of the F distribution with $v_1=2$ and $v_2=57$ degrees of freedom is 10.9, so the null hypothesis is rejected at the $\alpha=0.0001$ level because $[(30+30-2)(2)/(30+30-2-1)]$ $10.9=22.2 \ll T^2=32.34$ (cf. Equation 12.44). The actual p-value is smaller than 0.0001, but more extreme F-distribution quantiles are not commonly tabulated. Using the χ^2 distribution will provide only a moderately close approximation (Equation 12.38) because $v_2=57$, but the cumulative probability corresponding to $\chi_2{}^2=32.34$ can be calculated using Equation 4.53 (because $\chi_2{}^2$ is the exponential distribution with $\beta=2$) to be 0.99999991, corresponding to a p value of 0.00000001 (Equation 12.39b).

Even though the two-sample T^2 test provides a definitive rejection of the null hypothesis, it underestimates the statistical significance, because it does not account for the positive covariances between the New York and Boston temperatures that are evident in the submatrices $[S_{N-B}]$ and $[S_{B-N}]$ in Equation 12.47. In effect, the estimate in Equation 12.48 has assumed $[S_{N-B}]=[S_{B-N}]=[0]$. One way to account for these correlations is to compute the differences between the maximum temperatures as the linear combination $\boldsymbol{b}_1{}^T = [1, 0, -1, 0]$; compute the differences between the minimum temperatures as the linear combination $\boldsymbol{b}_2{}^T = [0, 1, 0, -1]$; and then use these two vectors as the rows of the transformation matrix $[B]^T$ in Equation 11.87b to compute the covariance $[S_\Delta]$ of the $n=30$ vector differences, from the full covariance matrix $[S]$ in Equation 12.47. Equivalently, we could compute this covariance matrix from the 30 data pairs in the last two columns of Table 12.1. In either case the result is

$$[S_\Delta] = \begin{bmatrix} 3.133 & 2.623 \\ 2.623 & 2.768 \end{bmatrix}. \tag{12.50}$$

The null hypothesis of equal mean vectors for New York and Boston implies $\boldsymbol{\mu}_\Delta=\boldsymbol{0}$ in Equation 12.45, yielding the test statistic

$$T^2 = 30 \, [2.18 \ \ 4.02] \begin{bmatrix} 3.133 & 2.623 \\ 2.623 & 2.768 \end{bmatrix}^{-1} \begin{bmatrix} 2.18 \\ 4.02 \end{bmatrix} = 298. \tag{12.51}$$

Because these temperature data are spatially correlated, much of the variability that was ascribed to sampling uncertainty for the mean vectors separately in the two-sample test in Equation 12.49 is actually shared and does not contribute to sampling uncertainty about the temperature differences. The numerical consequence is that the variances in the matrix $(1/30)[S_\Delta]$ are much smaller than their counterparts in Equation 12.48 for the two-sample test. Accordingly, T^2 for the paired test in Equation 12.51 is much larger than for the two-sample test in Equation 12.49. In fact it is huge, leading to the rough (because the sample sizes are only moderate) estimate, through Equation 4.53, for the p value of 2×10^{-65}.

Both the (incorrect) two-sample test and the (appropriate) paired test yield strong rejections of the null hypothesis that the New York and Boston mean vectors are equal. But what can be concluded about the way(s) in which they are different? This question will be taken up in Example 12.8. ◇

The T^2 tests described so far are based on the assumption that the data vectors are mutually uncorrelated. That is, although the K elements of $\boldsymbol{x}$ may have nonzero correlations, the vector observations $\boldsymbol{x}_i$, $i=1, \ldots, n$, have been assumed to be mutually independent. As noted in Section 5.2.4, ignoring serial correlation may lead to large errors in statistical inference, typically because the sampling distributions of the test statistics have greater dispersion (the test statistics are more variable from batch to batch of data) than would be the case if the underlying data were independent.

A simple adjustment (Equation 5.13) is available for scalar t tests if the serial correlation in the data is consistent with a first-order autoregression (Equation 10.16). The situation is more complicated for the multivariate T^2 test because, even if the time dependence for each of K elements of x is reasonably represented by an AR(1) process, their autoregressive parameters ϕ may not be the same, and the lagged correlations among the different elements of x must also be accounted for. However, if the multivariate AR (1) process (Equation 12.24) can be assumed as reasonably representing the serial dependence of the data, and if the sample size is large enough to produce multinormality as a consequence of the Central Limit Theorem, the sampling distribution of the sample mean vector is

$$\bar{x} \sim N_K\left(\boldsymbol{\mu}_x, \frac{1}{n}[\Sigma_\phi]\right), \tag{12.52a}$$

where

$$[\Sigma_\phi] = ([I] - [\Phi])^{-1}[\Sigma_x] + [\Sigma_x]\left([I] - [\Phi]^T\right)^{-1} - [\Sigma_x]. \tag{12.52b}$$

Equation 12.52a corresponds to Equation 12.33a for independent data, and $[\Sigma_\Phi]$ reduces to $[\Sigma_x]$ if $[\Phi] = [0]$ (i.e., if the x's are serially independent). For large n, sample counterparts of the quantities in Equation 12.52 can be substituted, and the matrix $[S_\Phi]$ used in place of $[S_x]$ in the computation of T^2 test statistics.

12.5.3. Simultaneous Confidence Statements

As noted in Section 5.1.7, a confidence interval is a region around a sample statistic, containing values that would not be rejected by a test whose null hypothesis is that the observed sample value is the true value. In effect, confidence intervals are constructed by working hypothesis tests in reverse. The difference in multivariate settings is that a confidence interval defines a region in the K-dimensional space of the data vector x rather than an interval in the one-dimensional space (the real line) of the scalar x. That is, multivariate confidence intervals are K-dimensional hypervolumes, rather than one-dimensional line segments.

Consider the one-sample T^2 test, Equation 12.35. After the data x_i, $i = 1, \ldots, n$, have been observed and their sample covariance matrix $[S_x]$ has been computed, a $(1 - \alpha) \cdot 100\%$ confidence region for the true vector mean consists of the set of points satisfying

$$n(x - \bar{x})^T[S_x]^{-1}(x - \bar{x}) \leq \frac{K(n-1)}{n-K} F_{K,n-K}(1-\alpha), \tag{12.53}$$

because these are the x's that would not trigger a rejection of the null hypothesis that the true mean is the observed sample mean. For sufficiently large $n - K$, the right-hand side of Equation 12.53 would be well approximated by $\chi_K^2(1 - \alpha)$. Similarly, for the two-sample T^2 test (Equation 12.40) a $(1 - \alpha) \cdot 100\%$ confidence region for the difference of the two means consists of the points $\boldsymbol{\delta}$ satisfying

$$[\boldsymbol{\delta} - (\bar{x}_1 - \bar{x}_2)]^T[S_{\Delta\bar{x}}]^{-1}[\boldsymbol{\delta} - (\bar{x}_1 - \bar{x}_2)] \leq \frac{K(n_1 + n_2 - 2)}{n_1 + n_2 - K - 1} F_{K,n_1 + n_2 - K - 1}(1 - \alpha), \tag{12.54}$$

where again the right-hand side is approximately equal to $\chi_K^2(1 - \alpha)$ for large samples.

The points x satisfying Equation 12.53 are those whose Mahalanobis distance from $\bar{x}$ is no larger than the scaled $(1 - \alpha)$ quantile of the F (or χ^2, as appropriate) distribution on the right-hand side, and similarly for the points $\boldsymbol{\delta}$ satisfying Equation 12.54. Therefore the confidence regions defined by these equations are bounded by (hyper-) ellipsoids whose characteristics are defined by the covariance matrix for the sampling distribution of the respective test statistic; for example, by $(1/n)[S_x]$ for Equation 12.53. Because the sampling distribution of $\bar{x}$ approximates the MVN distribution on the strength of the Central Limit Theorem, the frontiers of the confidence regions defined by Equation 12.53 are probability ellipsoids for the MVN distribution with mean $\bar{x}$ and covariance $(1/n)[S_x]$ (cf. Equation 12.5). Similarly, the confidence regions defined by Equation 12.54 are bounded by hyper-ellipsoids centered on the vector mean difference between the two sample means.

As illustrated in Example 12.1, the properties of these confidence ellipses, other than their center, are defined by the eigenvalues and eigenvectors of the covariance matrix for the sampling distribution in question. In particular, each axis of one of these ellipsoids will be aligned in the direction of one of the eigenvectors, and each will be elongated in proportion to the square root of the corresponding eigenvalue. In the case of the one-sample confidence region, for example, the limits of x satisfying Equation 12.53 in the directions of each of the axes of the ellipse are

$$x = \bar{x} \pm e_k \sqrt{\lambda_k \frac{K(n-1)}{n-K} F_{K,n-K}(1-\alpha)}, \; k = 1,\ldots,K, \tag{12.55}$$

where λ_k and e_k are the kth eigenvalue-eigenvector pair of the matrix $(1/n)[S_x]$. Again, for sufficiently large n, the quantity under the radical would be well approximated by $\lambda_k \chi_K^2(1-\alpha)$. Equation 12.55 indicates that the confidence ellipses are centered at the observed sample mean $\bar{x}$, and extend further in the directions associated with the largest eigenvalues. They also extend further for smaller α because these produce larger cumulative probabilities for the distribution quantiles $F(1-\alpha)$ and $\chi_K^2(1-\alpha)$.

It would be possible and computationally simpler to conduct K univariate t tests, and to then compute K univariate confidence intervals separately for the means of each of the elements of x rather than the T^2 test examining the vector mean $\bar{x}$. What is the relationship between an ellipsoidal multivariate confidence region of the kind just described, and a collection of K univariate confidence intervals? Jointly, these univariate confidence intervals would define a hyper-rectangular region in the K-dimensional space of x; but the probability (or confidence) associated with outcomes enclosed by it will be substantially less than $1-\alpha$, if the lengths of each of its K sides are the corresponding $(1-\alpha)\cdot100\%$ scalar confidence intervals. The problem is one of test multiplicity: if the K tests on which the confidence intervals are based are independent, the joint probability of all the elements of the vector x being simultaneously within their scalar confidence bounds will be $(1-\alpha)^K$. To the extent that the scalar confidence interval calculations are not independent, the joint probability will be different, but difficult to calculate.

An expedient workaround for this multiplicity problem is to calculate the K one-dimensional *Bonferroni confidence intervals* and use these as the basis of a joint confidence statement:

$$\Pr\left\{ \bigcap_{k=1}^{K} \left[\bar{x}_k + z\left(\frac{\alpha/K}{2}\right)\sqrt{\frac{s_{k,k}}{n}} \leq \mu_k \leq \bar{x}_k + z\left(1 - \frac{\alpha/K}{2}\right)\sqrt{\frac{s_{k,k}}{n}} \right] \right\} \geq 1-\alpha. \tag{12.56}$$

The expression inside the square bracket defines a univariate, $(1-\alpha/K)\cdot100\%$ confidence interval for the kth variable in x. Each of these confidence intervals has been expanded relative to the nominal $(1-\alpha)\cdot100\%$ confidence interval, to compensate for the multiplicity in K dimensions simultaneously. For convenience, it has been assumed in Equation 12.56 that the sample size is adequate for standard Gaussian quantiles to be appropriate, although quantiles of the t distribution with $n-1$ degrees of freedom usually would be used for n smaller than about 30.

There are two problems with using Bonferroni confidence regions in this context. First, Equation 12.56 is an inequality rather than an exact specification. That is, the probability that all the

K elements of the hypothetical true mean vector $\boldsymbol{\mu}$ are contained simultaneously in their respective one-dimensional confidence intervals is at least $1-\alpha$, not exactly $1-\alpha$. That is, in general the K-dimensional Bonferroni confidence region is too large, but exactly how much more probability than $1-\alpha$ may be enclosed by it is not known.

The second problem is more serious. As a collection of univariate confidence intervals, the resulting K-dimensional hyper-rectangular confidence region ignores the covariance structure of the data. Bonferroni confidence statements can be reasonable if the correlation structure is weak, for example, in the setting described in Section 10.5.6. But Bonferroni confidence regions are inefficient when the correlations among elements of x are strong, in the sense that they will include large regions having very low plausibility. As a consequence they are too large in a multivariate sense and can lead to silly inferences.

Example 12.7. Comparison of Unadjusted Univariate, Bonferroni, and MVN Confidence Regions

Assume that the covariance matrix in Equation 11.59, for the Ithaca and Canandaigua minimum temperatures, had been calculated from $n=100$ independent temperature pairs. This many observations would justify large-sample approximations for the sampling distributions (standard Gaussian z and χ^2, rather than t and F quantiles), and assuming independence obviates the need for the nonindependence adjustments in Equation 12.52.

What is the best two-dimensional confidence region for the true climatological mean vector, given the sample mean $[13.00, 20.23]^T$, and assuming the sample covariance matrix for the data in Equation 11.59? Relying on the multivariate normality for the sampling distribution of the sample mean implied by the Central Limit Theorem, Equation 12.53 defines an elliptical 95% confidence region when the right-hand side is the χ^2 quantile $\chi_2^2(0.95) = 5.991$. The result is shown in Figure 12.7, centered on the sample mean (+). Compare this ellipse to Figure 12.1, which is centered on the same mean and based on the same covariance matrix (although drawn to enclose slightly less probability). Figure 12.7 has exactly the same shape and orientation, but it is much more compact, even though it encloses somewhat more probability. Both ellipses have the same eigenvectors, $e_1^T = [0.848, 0.530]$ and $e_2^T = [-0.530, 0.848]$, but the eigenvalues for Figure 12.7 are 100-fold smaller, that is, $\lambda_1 = 2.5476$ and $\lambda_2 = 0.0829$. The difference is that Figure 12.1 represents one contour of the MVN distribution for the data, with covariance $[S_x]$ given by Equation 11.59, but Figure 12.7 shows one contour of the MVN distribution with covariance $(1/n)[S_x]$, appropriate to Equation 12.53 and relevant to the sampling distribution of the mean rather than the distribution for the data. This ellipse is the smallest region enclosing 95% of the probability of this distribution for the sampling variations of the sample mean. Its elongation reflects the strong correlation between the minimum temperatures at the two locations, so that differences between the sample and true means due to sampling variations are much more likely to involve differences of the same sign for both the Ithaca and Canandaigua means.

The gray rectangle in Figure 12.7 outlines the 95% Bonferroni confidence region. It has been calculated using $\alpha=0.05$ in Equation 12.56, and so is based on the 0.0125 and 0.9875 quantiles of the standard Gaussian distribution, or $z=\pm2.24$. The resulting rectangular region encloses at least $(1-\alpha)\cdot100\%=95\%$ of the probability of the joint sampling distribution. It occupies much more area in the plane than does the confidence ellipse, because the rectangle includes large regions in the upper left and lower right that contain very little probability. However, from the standpoint of univariate inference—that is, confidence intervals for one location without regard to the other—the Bonferroni limits are narrower.

The dashed rectangular region results jointly from the two standard 95% confidence intervals. The length of each side has been computed using the 0.025 and 0.975 quantiles of the standard Gaussian

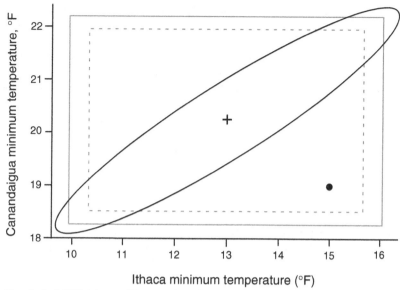

FIGURE 12.7 Hypothetical 95% joint confidence regions for the mean Ithaca and Canandaigua minimum temperatures, assuming that $n = 100$ independent bivariate observations had been used to calculate the covariance matrix in Equation 11.59. The ellipse encloses points within a Mahalanobis distance of $\chi^2 = 5.991$ of the sample mean (indicated by +) $[13.00, 20.23]^T$. Horizontal and vertical limits of the dashed rectangle are defined by two independent confidence intervals for the two variables, with $\pm z(0.025) = \pm 1.96$. Gray rectangle indicates corresponding Bonferroni confidence region, calculated with $\pm z(0.0125) = \pm 2.24$. The point $[15, 19]^T$ (large dot) is comfortably within both rectangular confidence regions, but is at Mahalanobis distance $\chi^2 = 1006$ from the mean relative to the joint covariance structure of the two variables, and is thus highly implausible.

distribution, which are $z = \pm 1.96$. They are, of course, narrower than the corresponding Bonferroni intervals, and according to Equation 12.56 the resulting rectangle includes at least 90% of the probability of this sampling distribution. Like the Bonferroni confidence region, it portrays large areas with very low probabilities as containing plausible values for the true mean.

The main difficulty with Bonferroni confidence regions is illustrated by the point $(15, 19)^T$, located by the large dot in Figure 12.7. It is comfortably within the gray rectangle delineating the Bonferroni confidence region, which carries the implication that this is a plausible value for the true mean vector. However, a Bonferroni confidence region is defined without regard to the multivariate covariance structure of the distribution that it purports to represent. In the case of Figure 12.7 the Bonferroni confidence region ignores the fact that sampling variations for these two positively correlated variables are much more likely to yield differences between the two sample and true means that are of the same sign. The Mahalanobis distance between the points $(15, 19)^T$ and $(13.00, 20.23)^T$, according to the covariance matrix $(1/n)[S_x]$, is 1006, implying an astronomically small probability for a separation this large and of this orientation for these two vectors (cf. Equation 12.34a). The vector $[15, 19]^T$ is an extremely implausible candidate for the true mean $\boldsymbol{\mu}_x$.   ◇

12.5.4. Interpretation of Multivariate Statistical Significance

What can be said about multivariate mean differences if the null hypothesis for a T^2 test is rejected, that is, if Equation 12.37 or 12.44 (or their large-sample counterpart, Equation 12.39b) is satisfied? This question is complicated by the fact that there are many ways for multivariate means to differ from one another, including but not limited to one or more of the pairwise differences between the elements that would be detected by the corresponding univariate tests.

If a T^2 test results in the rejection of its multivariate null hypothesis, the implication is that at least one scalar test for a linear combination $a^T x$ or $a^T(x_1 - x_2)$, for one- and two-sample tests, respectively, will be statistically significant. In any case, the scalar linear combination providing the most convincing evidence against the null hypothesis (regardless of whether or not it is sufficiently convincing to reject at a given test level) will satisfy

$$a \propto [S]^{-1}(\bar{x} - \boldsymbol{\mu}_0) \tag{12.57a}$$

for one-sample tests, or

$$a \propto [S]^{-1}[(\bar{x}_1 - \bar{x}_2) - \boldsymbol{\delta}_0] \tag{12.57b}$$

for two-sample tests. At minimum, then, if a multivariate T^2 calculation results in a null hypothesis rejection, then linear combinations corresponding to the K-dimensional direction defined by the vector a in Equation 12.57 will lead to significant results also. It can be very worthwhile to interpret the meaning, in the context of the data, of the direction a defined by Equation 12.57. Of course, depending on the strength of the overall multivariate result, other linear combinations may also lead to scalar test rejections, and it is possible that all linear combinations will be significant. The direction a also indicates the direction that best discriminates between the populations from which x_1 and x_2 were drawn (see Section 15.2.2).

The reason that any linear combination a satisfying Equation 12.57 yields the same test result can be seen most easily in terms of the corresponding confidence interval. Consider for simplicity the confidence interval for a one-sample T^2 test, Equation 12.53. Using the results in Equation 11.85, this scalar confidence interval is defined by

$$a^T \bar{x} - c \sqrt{\frac{a^T [S_x] a}{n}} \leq a^T \boldsymbol{\mu} \leq a^T \bar{x} + c \sqrt{\frac{a^T [S_x] a}{n}}, \tag{12.58}$$

where c^2 equals $[K(n-1)/(n-K)] F_{K, n-K}(1-\alpha)$, or χ_K^2, as appropriate. Even though the length of the vector a is arbitrary, so that the magnitude of the linear combination $a^T x$ is also arbitrary, the quantity $a^T \boldsymbol{\mu}$ is scaled identically.

Another remarkable property of the T^2 test is that valid inferences about any and all linear combinations can be made, even though they may not have been specified *a priori*. The price that is paid for this flexibility is that such inferences will be less precise than those made using conventional scalar tests for linear combinations that were specified in advance. This point can be appreciated in the context of the confidence regions shown in Figure 12.7. If a test regarding the Ithaca minimum temperature alone had been of interest, corresponding to the linear combination $a = [1, 0]^T$, the appropriate confidence interval would be defined by the horizontal extent of the dashed rectangle. The corresponding interval for this linear combination from the full T^2 test is substantially wider, being defined by the projection, or shadow, of the ellipse onto the horizontal axis. But what is gained from the multivariate test is the ability to make valid simultaneous probability statements regarding as many linear combinations as may be of interest.

Example 12.8. Interpreting the New York and Boston Mean January Temperature Differences
Return now to the comparisons made in Example 12.6, between the vectors of average January maximum and minimum temperatures for New York City and Boston. The difference between the

sample means was $[2.18, 4.02]^T$, and the null hypothesis was that the true means were equal, so the corresponding difference $\delta_0 = 0$. Even assuming, erroneously, that there is no spatial correlation between the two locations (or, equivalently for the purpose of the test, that the data for the two locations were taken in different years), T^2 in Equation 12.49 indicates that the null hypothesis should be strongly rejected.

Both means are warmer at New York, but Equation 12.49 does not necessarily imply significant differences between the average maxima or the average minima. Figure 12.6 shows substantial overlap between the data scatters for both maximum and minimum temperatures, with each scalar mean near the center of the corresponding univariate data distribution for the other city. Computing the separate univariate tests (Equation 5.8) yields $z = 2.18/\sqrt{1.248} = 1.95$ for the maxima and $z = 4.02/\sqrt{1.349} = 3.46$ for the minima. Even leaving aside the problem that two simultaneous comparisons are being made, the result for the difference of the average maximum temperatures is not quite significant at the 5% level, although the difference for the minima is stronger.

The significant result in Equation 12.49 ensures that there is at least one linear combination $a^T(x_1 - x_2)$ (and possibly others, although not necessarily the linear combinations resulting from $a^T = [1, 0]$ or $[0, 1]$) for which there is a significant difference. According to Equation 12.57b, the vectors producing the most significant linear combinations are proportional to

$$a \propto [S_{\Delta\bar{x}}]^{-1}\overline{\Delta} = \begin{bmatrix} 1.248 & 1.238 \\ 1.238 & 1.349 \end{bmatrix}^{-1} \begin{bmatrix} 2.18 \\ 4.02 \end{bmatrix} = \begin{bmatrix} -13.5 \\ 15.4 \end{bmatrix}. \tag{12.59}$$

This linear combination of the mean differences, and the estimated variance of its sampling distribution, are

$$a^T\overline{\Delta} = [-13.5 \quad 15.4]\begin{bmatrix} 2.18 \\ 4.02 \end{bmatrix} = 32.5, \tag{12.60a}$$

and

$$a^T[S_{\Delta\bar{x}}]a = [-13.5 \quad 15.4]\begin{bmatrix} 1.248 & 1.238 \\ 1.238 & 1.349 \end{bmatrix}\begin{bmatrix} -13.5 \\ 15.4 \end{bmatrix} = 32.6, \tag{12.60b}$$

yielding the univariate test statistic for this linear combination of the differences $z = 32.5/\sqrt{32.6} = 5.69$. This is, not coincidentally, the square root of Equation 12.49. The appropriate benchmark against which to compare the unusualness of this result in the context of the null hypothesis is not the standard Gaussian or t distributions (because this linear combination was derived from the test data, not *a priori*), but rather the square roots of either χ_2^2 quantiles or of appropriately scaled $F_{2,30}$ quantiles. The result is still very highly significant, with $p \approx 10^{-7}$.

Equation 12.59 indicates that the most significant aspect of the difference between the New York and Boston mean vectors is not the warmer temperatures at New York relative to Boston (which would correspond to $a \propto [1, 1]^T$). Rather, the elements of a are of opposite sign and of nearly equal magnitude, and so describe a *contrast*. Since $-a \propto a$, one way of interpreting this contrast is as the difference between the average maxima and minima, corresponding to the choice $a \approx [1, -1]^T$. That is, the most significant aspect of the difference between the two mean vectors is closely approximated by the difference in the average diurnal range, with the range for Boston being larger. The null hypothesis that that the two diurnal ranges are equal can be tested specifically, using the contrast vector $a = [1, -1]^T$ in Equation 12.60, rather than the linear combination defined by Equation 12.59. The result is $z = -1.84/\sqrt{0.121} = -5.29$. This test statistic is negative because the diurnal range at New York is smaller than the diurnal range at Boston. It is slightly smaller (in absolute value) than the result obtained when using $a = [-13.5, 15.4]$, because that is the most significant linear combination, although the result is almost the same because the two vectors are aligned in nearly the same direction. Comparing the result to the χ_2^2 distribution yields the very highly significant result $p \approx 10^{-6}$. Visually, the separation between the two point clouds in Figure 12.6 is consistent with this difference in diurnal range: The points for Boston tend to be closer to the upper left, and those for New York are closer to the lower right. On the other hand, the relative orientation of the two means is almost exactly opposite, with the New York mean closer to the upper right corner, and the Boston mean closer to the lower left. ◇

12.6. EXERCISES

12.1. Assume that the Ithaca and Canandaigua maximum temperatures in Table A.1 constitute a sample from a MVN distribution, and that their covariance matrix [S] has eigenvalues and eigenvectors as given in Exercise 11.8. Sketch the 50% and 95% probability ellipses of this distribution.

12.2. Assume that the four temperature variables in Table A.1 are MVN distributed, with the ordering of the variables in x being $[\text{Max}_{\text{Ith}}, \text{Min}_{\text{Ith}}, \text{Max}_{\text{Can}}, \text{Min}_{\text{Can}}]^T$. The respective means are also given in Table A.1, and the covariance matrix [S] is given in the answer to Exercise 11.9a. Assuming the true mean and covariance are the same as the sample values,

   a. Specify the conditional distribution of $[\text{Max}_{\text{Ith}}, \text{Min}_{\text{Ith}}]^T$, given that $[\text{Max}_{\text{Can}}, \text{Min}_{\text{Can}}]^T = [31.77, 20.23]^T$ (i.e., the average values for Canandaigua).

   b. Consider the linear combinations $b_1 = [1, 0, -1, 0]$, expressing the difference between the maximum temperatures, and $b_2 = [1, -1 -1, 1]$, expressing the difference between the diurnal ranges, as rows of a transformation matrix $[B]^T$. Specify the distribution of the transformed variables $[B]^T x$.

12.3. Consider the bivariate normal distribution for the random vector x, defined by:

$$\mu^T = [5.2 \ 16.1] \text{ and } [\Sigma] = \begin{bmatrix} 174.7 & 285.2 \\ 285.2 & 525.0 \end{bmatrix}.$$

   If x_1 and x_2 are Box-Cox ("power-") transformed versions of the variables y_1 and y_2, respectively, defined by $x_1 = (y_1^{1/2} - 1)/0.5$ and $x_2 = (y_2^{1/2} - 1)/0.5$, evaluate $\Pr\{y_1 \leq 10 | y_2 = 100\}$.

12.4. Use the mean vector for the New York January maximum and minimum temperatures in Equation 12.46a, and their covariance matrix in a portion of Equation 12.47, and assume that these define a bivariate normal distribution.

   a. Fully specify the distribution of the "mean" January temperature in a given year, which is computed as the average of the maximum and minimum temperatures.

   b. January 2012 was unusually warm in New York city, with $[x_{\text{max}}, x_{\text{min}}]^T = [44.2, 30.4]^T$. Evaluate the probability of seeing a January temperature vector at least as far removed from the long-term mean as that observed in 2012, assuming that the Chi-square distribution is an adequate approximation to the sampling distribution of the Mahalanobis distance.

12.5. The eigenvector associated with the smallest eigenvalue of the covariance matrix [S] for the January 1987 temperature data referred to in Exercise 11.2 is $e_4^T = [-0.665, 0.014, 0.738, -0.115]$. Assess the normality of the linear combination $e_4^T x$,

   a. Graphically, with a Q-Q plot. For computational convenience, evaluate $\Phi(z)$ using Equation 4.30.

   b. Formally, with the Filliben test (see Table 5.5), assuming no autocorrelation.

12.6. a. Compute the 1-sample T^2 testing the linear combinations $[B]^T \bar{x}$ with respect to H_0: $\mu_0 = 0$, where x and $[B]^T$ are defined as in Exercise 12.2. Ignoring the serial correlation, evaluate the plausibility of H_0, assuming that the χ^2 distribution is an adequate approximation to the sampling distribution of the test statistic.

b. Compute the most significant linear combination for this test.

12.7. Repeat Exercise 12.4, assuming spatial independence (i.e., setting all cross-covariances between Ithaca and Canandaigua variables to zero).

Chapter 13

Principal Component (EOF) Analysis

13.1. BASICS OF PRINCIPAL COMPONENT ANALYSIS

Principal component analysis, often denoted as PCA, is possibly the most widely used multivariate statistical technique in the atmospheric sciences. The technique was introduced into the atmospheric science literature by Obukhov (1947), and became popular for analysis of atmospheric data following the papers by Lorenz (1956), who called the technique *empirical orthogonal function* (EOF) analysis, and Davis (1976). Both the names PCA and EOF analysis are commonly used, and both refer to the same set of procedures. Sometimes the method is incorrectly referred to as *factor analysis*, which is a related but distinct multivariate statistical method. This chapter is intended to provide a basic introduction to what has become a very large subject. Book-length treatments of PCA are given in Preisendorfer (1988) and Navarra and Simoncini (2010), which are oriented specifically toward geophysical data; and in Jolliffe (2002), which describes PCA more generally. Hannachi et al. (2007) provide a comprehensive review. In addition, most textbooks on multivariate statistical analysis contain chapters on PCA.

13.1.1. Definition of PCA

PCA reduces a data set containing a large number of variables to a data set containing fewer (hopefully many fewer) new variables. These new variables are linear combinations of the original ones, and these linear combinations are chosen to represent the maximum possible fraction of the variability contained in the original data while being uncorrelated with each other. That is, given multiple observations of a $(K \times 1)$ data vector x, PCA finds $(M \times 1)$ vectors u whose elements are linear combinations of the elements of the x's, and which contain most of the information in the original collection of x's. PCA is most effective when this data compression can be achieved with $M \ll K$. This situation occurs when there are substantial correlations among the variables within x, in which case x contains redundant information. The elements of these new vectors u are called the *principal components* (PCs).

Data for atmospheric and other geophysical fields generally exhibit many large correlations among the variables x_k, and a PCA results in a much more compact representation of their variations. Beyond mere data compression, however, PCA can be a very useful tool for exploring large multivariate data sets, including those consisting of geophysical fields. Here PCA has the potential for yielding insights into both the spatial and temporal variations exhibited by the field or fields being analyzed, and new interpretations of the original data x can be suggested by the nature of the linear combinations that are most effective in compressing those data.

Usually it is convenient to calculate the PCs as linear combinations of the anomalies $x' = x - \bar{x}$. The first PC, u_1, is that linear combination of x' having the largest variance. The subsequent principal components u_m, $m = 2, 3, \ldots$, are the linear combinations having the largest possible variances, subject to the

Statistical Methods in the Atmospheric Sciences. https://doi.org/10.1016/B978-0-12-815823-4.00013-4

617

condition that they are uncorrelated with the principal components having lower indices. The result is that all the PCs are mutually uncorrelated.

The new variables or PCs—that is, the elements u_m of u that will account successively for the maximum amount of the joint variability of x' (and therefore also of x)—are uniquely defined (except for sign) by the eigenvectors of the covariance matrix of x, $[S]$. In particular, the mth principal component, u_m is obtained as the projection of the data vector x' onto the mth eigenvector, e_m,

$$u_m = e_m^T x' = \sum_{k=1}^{K} e_{k,m} x_k', \quad m = 1, \dots, M. \tag{13.1}$$

Notice that each of the M eigenvectors contains one element pertaining to each of the K variables, x_k. Similarly, each realization of the mth principal component in Equation 13.1 is computed from a particular set of observations of the K variables x_k. That is, each of the M principal components is a sort of weighted average of the x_k values that are the elements of a particular data vector x. Although the weights (the $e_{k,m}$'s) do not sum to 1, their squares do because of the scaling convention $\|e_m\| = 1$. (Note that a fixed scaling convention for the weights e_m of the linear combinations in Equation 13.1 allows the maximum variance constraint defining the PCs to be meaningful.) If the data sample consists of n observations (and therefore of n data vectors x, or n rows in the data matrix $[X]$), there will be n values for each of the principal components, or new variables, u_m. Each of these constitutes a single-number index of the resemblance between the eigenvector e_m and the corresponding individual data vector x.

Geometrically, the first eigenvector, e_1, points in the direction (in the K-dimensional space of x') in which the data vectors jointly exhibit the most variability. This first eigenvector is the one associated with the largest eigenvalue, λ_1. The second eigenvector e_2, associated with the second-largest eigenvalue λ_2, is constrained to be perpendicular to e_1 (Equation 11.50), but subject to this constraint it will align in the direction in which the x' vectors exhibit their next strongest variations. Subsequent eigenvectors e_m, $m = 3, 4, \dots, M$, are similarly numbered according to decreasing magnitudes of their associated eigenvalues, and in turn will be perpendicular to all the previous eigenvectors. Subject to this orthogonality constraint these eigenvectors will continue to locate directions in which the original data jointly exhibit maximum variability.

Put another way, the eigenvectors define a new coordinate system in which to view the data. In particular, the orthogonal matrix $[E]$ whose columns are the eigenvectors (Equation 11.51) defines the rigid rotation

$$u = [E]^T x', \tag{13.2}$$

which is the simultaneous matrix-notation representation of $M = K$ linear combinations of the form of Equation 13.1 (i.e., here the matrix $[E]$ is square, with K eigenvector columns). This new coordinate system is oriented such that each consecutively numbered axis is aligned along the direction of the maximum joint variability of the data, consistent with that axis being orthogonal to the preceding ones. These axes will turn out to be different for different data sets, because they are extracted from the sample covariance matrix $[S_x]$ particular to a given data set. That is, they are orthogonal functions, but are defined empirically according to the particular data set at hand. This observation is the basis for the eigenvectors being known in this context as empirical orthogonal functions (EOFs). The implied distinction is with theoretical orthogonal functions, such as Fourier harmonics or Tschebyschev polynomials, which also can be used to define alternative coordinate systems in which to view a data set.

It is a remarkable property of the principal components that they are uncorrelated. That is, the correlation matrix for the new variables u_m is simply $[I]$. This property implies that the covariances between pairs of the u_m's are all zero, so that the corresponding covariance matrix is diagonal. In fact, the covariance matrix for the principal components is obtained by the diagonalization of $[S_x]$ (Equation 11.56) and is thus simply the diagonal matrix $[\Lambda]$ of the eigenvalues of $[S]$:

$$[S_u] = \mathrm{Var}\left([E]^T x\right) = [E]^T [S_x] [E] = [E]^{-1} [S_x] [E] = [\Lambda]. \tag{13.3}$$

That is, the variance of the mth principal component u_m is the mth eigenvalue λ_m. Equation 11.54 then implies that each PC represents a share of the total variation in x that is proportional to its eigenvalue,

$$R_m^2 = \frac{\lambda_m}{\sum_{k=1}^{K} \lambda_k} \times 100\% = \frac{\lambda_m}{\sum_{k=1}^{K} s_{k,k}} \times 100\%. \tag{13.4}$$

Here R^2 is used in the same sense that is familiar from linear regression (Section 7.2.4). The total variation exhibited by the original data is completely represented in (or accounted for by) the full set of K u_m's, in the sense that the sum of the variances of the centered data x' (and therefore also of the uncentered variables x), $\Sigma_k s_{k,k}$, is equal to the sum of the variances $\Sigma_m \lambda_m$ of the principal component variables u.

Equation 13.2 expresses the transformation of a $(K \times 1)$ data vector x' to a vector u of PCs. If $[E]$ contains all K eigenvectors of $[S_x]$ (assuming $[S_x]$ is nonsingular) as its columns, the resulting vector u will also have dimension $(K \times 1)$. Equation 13.2 sometimes is called the *analysis formula* for x', expressing that the data can be analyzed, or summarized in terms of the principal components. Reversing the transformation in Equation 13.2, the data x' can be reconstructed from the principal components according to

$$\underset{(K \times 1)}{x'} = \underset{(K \times K)}{[E]} \underset{(K \times 1)}{u}, \tag{13.5}$$

which is obtained from Equation 13.2 by multiplying on the left by $[E]$ and using the orthogonality property of this matrix (Equation 11.44). The reconstruction of x' expressed by Equation 13.5 is sometimes called the *synthesis formula*. If the full set of $M = K$ PCs is used in the synthesis, the reconstruction is complete and exact, since $\Sigma_m R_m^2 = 1$ (cf. Equation 13.4). If $M < K$ PCs (usually those corresponding to the M largest eigenvalues) are used, the reconstruction is approximate,

$$\underset{(K \times 1)}{x'} \approx \underset{(K \times M)}{[E]} \underset{(M \times 1)}{u}, \tag{13.6a}$$

or

$$x_k' \approx \sum_{m=1}^{M} e_{k,m} u_m, \quad k = 1, \dots, K, \tag{13.6b}$$

but the approximation improves as the number M of PCs used (or, more precisely, as the sum of the corresponding eigenvalues, because of Equation 13.4) increases. Because $[E]$ in Equation 13.6a has only M columns, and operates on a truncated PC vector u of dimension $(M \times 1)$, Equation 13.6 is called the *truncated synthesis formula.* The original (in the case of Equation 13.5) or approximated (for Equation 13.6) uncentered data x can easily be obtained by adding back the vector of sample means, that is, by reversing Equation 11.33.

Because each principal component u_m is a linear combination of the original variables x_k (Equation 13.1), and vice versa (Equation 13.5), pairs of principal components and original variables will be correlated unless the eigenvector element $e_{k,m}$ relating them is zero. It can sometimes be informative to calculate these correlations, which are given by

$$r_{u,x} = \text{Corr}(u_m, x_k) = e_{k,m} \sqrt{\frac{\lambda_m}{s_{k,k}}}. \tag{13.7}$$

Example 13.1. PCA in Two Dimensions

The basics of PCA are most easily appreciated in a simple example where the geometry can be visualized. If $K = 2$ the space of the data is two-dimensional and can be graphed on a page. Figure 13.1a shows a scatterplot of centered (at zero) January 1987 Ithaca minimum temperatures (x_1') and Canandaigua minimum temperatures (x_2') from Table A.1. This is the same scatterplot that appears in the middle of the bottom row of Figure 3.31. It is apparent that the Ithaca temperatures are more variable than the Canandaigua temperatures, with the two standard deviations being $\sqrt{s_{1,1}} = 13.62°F$ and $\sqrt{s_{2,2}} = 8.81°F$, respectively. The two variables are clearly strongly correlated and have a Pearson correlation of $+0.924$ (see Table 3.5). The covariance matrix $[S]$ for these two variables is given as $[A]$ in Equation 11.59. The two eigenvectors of this matrix are $e_1^T = [0.848, 0.530]$ and $e_2^T = [-0.530, 0.848]$, so that the eigenvector matrix $[E]$ is that shown in Equation 11.60. The corresponding eigenvalues are $\lambda_1 = 254.76$ and $\lambda_2 = 8.29$. These are the same data used to fit the bivariate normal probability ellipses shown in Figures 12.1 and 12.7.

The orientations of the two eigenvectors are shown in Figure 13.1a, although their lengths have been exaggerated for clarity. It is evident that the first eigenvector is aligned in the direction in which the data jointly exhibit maximum variation. That is, the point cloud is inclined at the same angle as is e_1, which is $32°$ from the horizontal (i.e., from the vector $[1, 0]$, according to Equation 11.15). Since the data in this simple example exist in only $K = 2$ dimensions, the constraint that the second eigenvector must be perpendicular to the first determines its direction up to sign (i.e., it could as easily be $-e_2^T = [0.530, -0.848]$). This last eigenvector locates the direction in which data jointly exhibit their smallest variations.

Figure 13.1b illustrates another property of the first eigenvector, which is that it is the direction that minimizes the sum of squared distances (dashed lines) perpendicular to e_1, connecting it to the points. The leading eigenvector is thus different from the least-squares regression lines relating the two variables, which would minimize either the sum of squared vertical distances (if the Canandaigua temperatures were being predicted) or the squared horizontal distances (if the Ithaca temperatures were being predicted). In three dimensions, the second eigenvector would be the direction perpendicular to e_1, defining the e_1–e_2 plane minimizing the perpendicular squared distances from the plane to the points, and so on, in progressively higher dimensions.

The two eigenvectors determine an alternative coordinate system in which to view the data. This fact may become more clear if you rotate this book $32°$ clockwise while looking at Figure 13.1a. Within this rotated coordinate system, each point is defined by a principal component vector

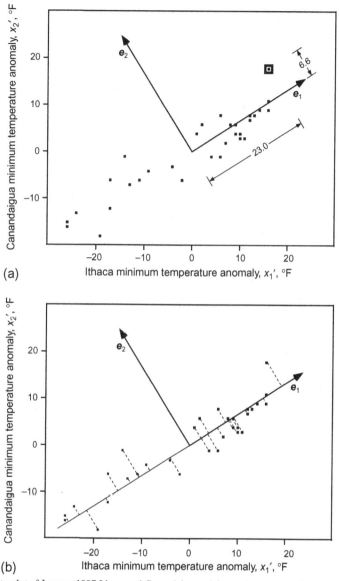

FIGURE 13.1 (a) Scatterplot of January 1987 Ithaca and Canandaigua minimum temperatures (converted to anomalies, or centered), illustrating the geometry of PCA in two dimensions. The eigenvectors e_1 and e_2 of the covariance matrix $[S]$ for these two variables, as computed in Example 11.3, have been plotted with lengths exaggerated for clarity. The data stretch out in the direction of e_1 to the extent that 96.8% of the joint variance of these two variables occurs along this axis. The coordinates u_1 and u_2, corresponding to the data point $x'^T [16.0, 17.8]$, recorded on January 15 and indicated by the large square symbol, are shown by lengths in the directions of the new coordinate system defined by the eigenvectors. That is, the vector $u^T = [23.0, 6.6]$ locates the same point as $x'^T = [16.0, 17.8]$. (b) The first eigenvector e_1 is also the direction that minimizes the sum of squared lengths of the dashed lines between the points and e_1, that are perpendicular to e_1.

$\boldsymbol{u}^{\mathrm{T}} = [u_1, u_2]$ of new transformed variables, whose elements consist of the projections of the original data onto the eigenvectors, according to the dot product in Equation 13.1. Figure 13.1a illustrates this projection for the 15 January data point $\boldsymbol{x}'^{\mathrm{T}} = [16.0, 17.8]$, which is indicated by the large square symbol. For this datum, $u_1 = (0.848)(16.0) + (0.530)(17.8) = 23.0$, and $u_2 = (-0.530)(16.0) + (0.848)(17.8) = 6.6$.

The sample variance of the new variable u_1 is an expression of the degree to which it spreads out along its axis (i.e., along the direction of $\boldsymbol{e}_1$). This dispersion is evidently greater than the dispersion of the data along either of the original axes, and indeed it is larger than the dispersion of the data in any other direction in this plane. This maximum sample variance of u_1 is equal to the eigenvalue $\lambda_1 = 254.76°\mathrm{F}^2$. The points in the data set tend to exhibit quite different values of u_1, whereas they have more similar values for u_2. That is, they are much less variable in the $\boldsymbol{e}_2$ direction, and the sample variance of u_2 is only $\lambda_2 = 8.29°\mathrm{F}^2$.

Since $\lambda_1 + \lambda_2 = s_{1,1} + s_{2,2} = 263.05°\mathrm{F}^2$, the new variables jointly retain all the variation exhibited by the original variables. However, the fact that the point cloud seems to exhibit no slope in the new coordinate frame defined by the eigenvectors indicates that u_1 and u_2 are uncorrelated. Their lack of correlation can be verified by transforming the 31 pairs of minimum temperatures in Table A.1 to principal components and computing the Pearson correlation, which is zero. The variance–covariance matrix for the principal components is therefore $[\Lambda]$, as shown in Equation 11.62.

The two original temperature variables are so strongly correlated that a very large fraction of their joint variance, $\lambda_1/(\lambda_1 + \lambda_2) = 0.968$, is represented by the first principal component alone. It would be said that the first principal component describes 96.8% of the total variance. The first principal component might be interpreted as reflecting the regional minimum temperature for the area including these two locations (they are about 50 miles, or about 80 km apart), with the second principal component describing local variations departing from the overall regional value.

Since so much of the joint variance of the two temperature series is captured by the first principal component, resynthesizing the series using only the first principal component will yield a good approximation to the original data. Using the synthesis Equation 13.6 with only the first ($M = 1$) principal component yields

$$\boldsymbol{x}'(t) = \begin{bmatrix} x_1'(t) \\ x_2'(t) \end{bmatrix} \approx \boldsymbol{e}_1 u_1(t) = \begin{bmatrix} .848 \\ .530 \end{bmatrix} u_1(t). \tag{13.8}$$

The temperature data x are time series, and therefore so are the principal components $\boldsymbol{u}$. The time dependence for both has been indicated explicitly in Equation 13.8. On the other hand, the eigenvectors are fixed by the covariance structure of the entire series and do not change through time. Figure 13.2 compares the original series (black) and the reconstructions using the first principal component $u_1(t)$ only (gray) for the (a) Ithaca and (b) Canandaigua anomalies. The discrepancies are small because $R^2_1 = 96.8\%$. The residual differences would be captured by u_2. The two gray series are exactly proportional to each other, since each is a scalar multiple of the same first principal component time series. Since $\mathrm{Var}(u_1) = \lambda_1 = 254.76$, the variances of the reconstructed series are $(0.848)^2\, 254.76 = 183.2$ and $(0.530)^2\, 254.76 = 71.6°\mathrm{F}^2$, respectively, which are close to but smaller than the corresponding diagonal elements of the original covariance matrix (Equation 11.59). The larger variance for the Ithaca temperatures is also visually evident in Figure 13.2. Using Equation 13.7, the correlations between the first principal component series $u_1(t)$ and the original temperature variables are $0.848(254.76/185.47)^{1/2} = 0.994$ for Ithaca, and $0.530 (254.76/77.58)^{1/2} = 0.960$ for Canandaigua. ◇

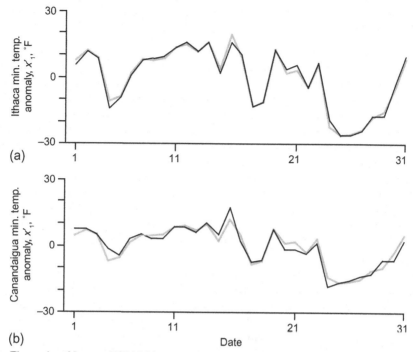

FIGURE 13.2   Time series of January 1987 (a) Ithaca and (b) Canandaigua minimum temperature anomalies (black), and their reconstruction using the first principal component only (gray), through the synthesis Equation 13.8.

13.1.2. PCA Based on the Covariance Matrix vs. the Correlation Matrix

A PCA can be computed as easily using the correlation matrix $[R]$ as it can on the covariance matrix $[S]$. The correlation matrix is the variance–covariance matrix of the vector of standardized variables z (Equation 11.32). The vector of standardized variables z is related to the vectors of original variables x and their centered counterparts x' according to the scaling transformation (Equation 11.34). Therefore PCA on the correlation matrix amounts to analysis of the joint variance structure of the standardized variables z_k, as computed using either Equation 11.34 or (in scalar form) Equation 3.27.

The difference between a PCA performed using the variance–covariance and correlation matrices will be one of emphasis. Since PCA seeks to find variables successively maximizing the proportion of the total variance ($\Sigma_k s_{k,k}$) represented, analyzing the covariance matrix $[S]$ results in principal components that emphasize the x_k's having the largest variances. Other things equal, the tendency will be for the first few eigenvectors to align near the directions of the variables having the biggest variances. In Example 13.1, the first eigenvector points more toward the Ithaca minimum temperature axis because the variance of the Ithaca minimum temperatures is larger than the variance of the Canandaigua minimum temperatures. Conversely, PCA applied to the correlation matrix $[R]$ weights all the standardized variables z_k equally, since all have equal (unit) variance.

If the PCA is computed using the correlation matrix, the analysis formula, Equations 13.1 and 13.2, will pertain to the standardized variables, z_k and z, respectively. Similarly the synthesis formulae, Equations 13.5 and 13.6 will pertain to z and z_k rather than to x' and x_k'. In this case the original data

x can be recovered from the result of the synthesis formula by reversing the standardization given by Equations 11.33 and 11.34, that is,

$$x = [D]z + \bar{x}. \tag{13.9}$$

Although z and x' can easily be obtained from each other using Equation 11.34, the eigenvalue–eigenvector pairs of $[R]$ and $[S]$ do not bear simple relationships to one another. In general, it is not possible to compute the eigenvectors and principal components of one knowing only the eigenvectors and principal components of the other. This fact implies that these two alternatives for PCA do not yield equivalent information and that an intelligent choice of one over the other must be made for a given application. If an important goal of the analysis is to identify or isolate the strongest variations in a data set, the better alternative usually will be PCA using the covariance matrix, although the choice will depend on the judgment of the analyst and the purpose of the study. For example, in analyzing gridded numbers of extratropical cyclones, Overland and Preisendorfer (1982) found that PCA on their covariance matrix better identified regions having the highest variability in cyclone numbers, and that correlation-based PCA was more effective at locating the primary storm tracks.

However, if the analysis is of unlike variables—variables not measured in the same units—it will almost always be preferable to compute the PCA using the correlation matrix. Measurement in unlike physical units yields arbitrary relative scalings of the variables, which results in arbitrary relative magnitudes of the variances of these variables. To take a simple example, the variance of a set of temperatures measured in °F will be $(1.8)^2 = 3.24$ times as large as the variance of the same temperatures expressed in °C. If the PCA has been done using the correlation matrix, the analysis formula, Equation 13.2, pertains to the vector z rather than x', and the synthesis in Equation 13.5 will yield the standardized variables z_k (or approximations to them if Equation 13.6 is used for the reconstruction). The summations in the denominators of Equation 13.4 will equal the number of standardized variables, since each has unit variance.

Example 13.2 Correlation-Versus Covariance-Based PCA for Arbitrarily Scaled Variables

The importance of basing a PCA on the correlation matrix when the variables being analyzed are not measured on comparable scales is presented in Table 13.1. This table summarizes PCAs of the January 1987 data in Table A.1 in (a) unstandardized (covariance matrix) and (b) standardized (correlation matrix) forms. Sample variances of the variables are shown, as are the six eigenvectors, the six eigenvalues, and the cumulative percentages of variance accounted for by the principal components. The (6x6) arrays in the upper-right portions of parts (a) and (b) of this table constitute the matrices $[E]$ whose columns are the eigenvectors.

The sample variances of each of the variables are shown, as are the six eigenvectors e_m arranged in decreasing order of their eigenvalues λ_m. The cumulative percentage of variance represented is calculated according to Equation 13.4. The much smaller variances of the precipitation variables in (a) are an artifact of the measurement units, but result in precipitation being unimportant in the first four principal components computed from the covariance matrix, which collectively account for 99.9% of the total variance of the data set. Computing the principal components from the correlation matrix ensures that variations of the temperature and precipitation variables are weighted equally.

Because of the different magnitudes of the variations of the data in relation to their measurement units, the variances of the unstandardized precipitation data are tiny in comparison to the variances of the temperature variables. This is purely an artifact of the measurement unit for precipitation (inches) being relatively large in comparison to the range of variation of the data (about 1 in.), and the

TABLE 13.1 Comparison of PCA Computed Using (a) the Covariance Matrix, and (b) the Correlation Matrix, of the Data in Table A.1

Variable	Sample Variance	e_1	e_2	e_3	e_4	e_5	e_6
(a) Covariance results:							
Ithaca ppt.	0.059 in.^{2}	.003	.017	.002	−.028	.818	−.575
Ithaca T_{max}	892.2 °F^2	.359	−.628	.182	−.665	−.014	−.003
Ithaca T_{min}	185.5 °F^2	.717	.527	.456	.015	−.014	.000
Canandaigua ppt.	0.028 in.^{2}	.002	.010	.005	−.023	.574	.818
Canandaigua T_{max}	61.8 °F^2	.381	−.557	.020	.737	.037	.000
Canandaigua T_{min}	77.6 °F^2	.459	.131	−.871	−.115	−.004	.003
	Eigenvalues, λ_k	337.7	36.9	7.49	2.38	0.065	0.001
	Cum. % variance	87.8	97.4	99.3	99.9	100.0	100.0
(b) Correlation results:							
Ithaca ppt.	1.000	.142	.677	.063	−.149	−.219	.668
Ithaca T_{max}	1.000	.475	−.203	.557	.093	.587	.265
Ithaca T_{min}	1.000	.495	.041	−.526	.688	−.020	.050
Canandaigua ppt.	1.000	.144	.670	.245	.096	.164	−.658
Canandaigua T_{max}	1.000	.486	−.220	.374	−.060	−.737	−.171
Canandaigua T_{min}	1.000	.502	−.021	−.458	−.695	−.192	−.135
	Eigenvalues, λ_k	3.532	1.985	0.344	0.074	0.038	0.027
	Cum. % variance	58.9	92.0	97.7	98.9	99.5	100.0

measurement unit for temperature (°F) being relatively small in comparison to the range of variation of the data (about 40°F). If the measurement units had been millimeters and °C, respectively, the differences in variances would have been much smaller. If the precipitation had been measured in micrometers, the variances of the precipitation variables would dominate the variances of the temperature variables.

Because the variances of the temperature variables are so much larger than the variances of the precipitation variables, the PCA calculated from the covariance matrix is dominated by the temperatures. The eigenvector elements corresponding to the two precipitation variables are negligibly small in the first four eigenvectors, so these variables make negligible contributions to the first four principal components. However, these first four principal components collectively describe 99.9% of the joint variance. An application of the truncated synthesis formula (Equation 13.6) with the leading $M = 4$ eigenvector therefore would result in reconstructed precipitation data very near their average values. That is, essentially none of the variation in precipitation would be represented.

Since the correlation matrix is the covariance matrix for comparably scaled variables z_k, each has equal variance. Unlike the analysis on the covariance matrix, this PCA does not ignore the precipitation variables when the correlation matrix is analyzed. Here the first (and most important) principal component represents primarily the closely intercorrelated temperature variables, as can be seen from the relatively larger elements of e_1 for the four temperature variables. However, the second principal component, which accounts for 33.1% of the total variance in the scaled data set, represents primarily the precipitation variations. The precipitation variations would not be lost in a truncated data representation including at least the first $M = 2$ eigenvectors, but rather would be very nearly completely reconstructed.   ◇

13.1.3. The Varied Terminology of PCA

The subject of PCA is sometimes regarded as a difficult and confusing one, but much of this confusion derives from a proliferation of the associated terminology, especially in writings by analysts of atmospheric data. Table 13.2 organizes the more common of these in a way that may be helpful in deciphering the PCA literature.

Lorenz (1956) introduced the term empirical orthogonal function (EOF) into the literature as another name for the eigenvectors of a PCA. The terms *modes of variation* and *pattern vectors* also are used primarily by analysts of geophysical data, especially in relation to analysis of fields, to be described in Section 13.2. The remaining terms for the eigenvectors derive from the geometric interpretation of the eigenvectors as basis vectors, or axes, in the K-dimensional space of the data. These terms are used in the literature of a broader range of disciplines.

The most common name for individual elements of the eigenvectors in the statistical literature is *loading*, connoting the weight of the kth variable x_k that is borne by the mth eigenvector e_m through the individual element $e_{k,m}$. The term "coefficient" is also a usual one in the statistical literature. The term *pattern coefficient* is used mainly in relation to PCA of field data, where the spatial patterns exhibited by the eigenvector elements can be illuminating. *Empirical orthogonal weights* is a term that is sometimes used to be consistent with the naming of the eigenvectors as EOFs.

The new variables u_m defined with respect to the eigenvectors are almost universally called "principal components." However, they are sometimes known as *empirical orthogonal variables* when the eigenvectors are called EOFs. There is more variation in the terminology for the individual values of the principal components $u_{i,m}$ corresponding to particular data vectors x_i'. In the statistical literature these are

TABLE 13.2 A Partial Guide to Synonymous Terminology Associated With PCA

Eigenvectors, e_m	Eigenvector elements, $e_{k,m}$	Principal Components, u_m	Principal Component Elements, $u_{i,m}$
EOFs	Loadings	Empirical Orthogonal Variables	Scores
Modes of Variation	Coefficients		Amplitudes
Pattern Vectors	Pattern Coefficients		Expansion Coefficients
Principal Axes	Empirical Orthogonal Weights		Coefficients
Principal Vectors			
Proper Functions			
Principal Directions			

most commonly called "scores," which has a historical basis in the early and widespread use of PCA in psychometrics. In atmospheric applications, the principal component elements are often called "amplitudes" by analogy to the amplitudes of a Fourier series, which multiply the (theoretical orthogonal) sine and cosine functions. Similarly, the term *expansion coefficient* is also used for this meaning. Sometimes expansion coefficient is shortened simply to "coefficient," although this can be the source of some confusion since it is more standard for the term coefficient to denote an eigenvector element.

13.1.4. Scaling Conventions in PCA

Another contribution to confusion in the literature of PCA is the existence of alternative scaling conventions for the eigenvectors. The presentation in this chapter assumes that the eigenvectors are scaled to unit length, that is, $||e_m|| \equiv 1$. Recall that vectors of any length will satisfy Equation 11.48 if they point in the appropriate direction, and as a consequence it is common for the output of eigenvector computations to be expressed with this scaling.

However, it is sometimes useful to express and manipulate PCA results using alternative scalings of the eigenvectors. When this is done, each element of an eigenvector is multiplied by the same constant, so their relative magnitudes and relationships remain unchanged. Therefore the qualitative results of an exploratory analysis based on PCA do not depend on the scaling selected, but if different, related analyses are to be compared it is important to be aware of the scaling convention used in each.

Rescaling the lengths of the eigenvectors changes the magnitudes of the principal components by the same factor. That is, multiplying the eigenvector e_m by a constant requires that the principal component scores u_m be multiplied by the same constant in order for the analysis formulas that define the principal components (Equations 13.1 and 13.2) to remain valid. The expected values of the principal component scores for centered data x' are zero, and multiplying the principal components by a constant will produce rescaled principal components whose means are also zero. However, their variances will change by a factor of the square of the scaling constant.

Table 13.3 summarizes the effects of three common scalings of the eigenvectors on the properties of the principal components. The first row indicates their properties under the scaling convention $||e_m|| \equiv 1$ adopted in this presentation. Under this scaling, the expected value (mean) of each of the principal components is zero (because it is the data anomalies x' that have been projected onto the eigenvectors), and the variance of each is equal to the respective eigenvalue, λ_m. This result is simply an expression of the diagonalization of the variance–covariance matrix (Equation 11.56) produced by adopting the rigidly rotated geometric coordinate system defined by the eigenvectors. When scaled in this way, the

TABLE 13.3 Three common eigenvector scalings used in PCA; their consequences for the properties of the principal components, u_m; and their relationship to the original variables, x_k; and the standardized original variables, z_k

Eigenvector Scaling	$E(u_m)$	$Var(u_m)$	$Corr(u_m, x_k)$	$Corr(u_m, z_k)$
$||e_m|| = 1$	0	λ_m	$e_{k,m}(\lambda_m)^{1/2}/s_k$	$e_{k,m}(\lambda_m)^{1/2}$
$||e_m|| = (\lambda_m)^{1/2}$	0	λ_m^2	$e_{k,m}/s_k$	$e_{k,m}$
$||e_m|| = (\lambda_m)^{-1/2}$	0	1	$e_{k,m}\lambda_m/s_k$	$e_{k,m}\lambda_m$

correlation between a principal component u_m and a variable x_k is given by Equation 13.7. The correlation between u_m and the standardized variable z_k is given by the product of the eigenvector element and the square root of the eigenvalue, since the standard deviation of a standardized variable is one.

The eigenvectors sometimes are rescaled by multiplying each element by the square root of the corresponding eigenvalue. This rescaling produces vectors of differing lengths, $||e_m|| \equiv (\lambda_m)^{1/2}$, but which point in exactly the same directions as the original eigenvectors having unit lengths. Consistency in the analysis formula implies that the principal components are also changed by the factor $(\lambda_m)^{1/2}$, with the result that the variance of each u_m increases to λ_m^2. A major advantage of this rescaling, however, is that the eigenvector elements are more directly interpretable in terms of the relationship between the principal components and the original data. Under this rescaling, each eigenvector element $e_{k,m}$ is numerically equal to the correlation $r_{u,z}$ between the mth principal component u_m and the kth standardized variable z_k.

The last scaling presented in Table 13.3, resulting in $||e_m|| \equiv (\lambda_m)^{-1/2}$, is less commonly used. This scaling is achieved by dividing each element of the original unit-length eigenvectors by the square root of the corresponding eigenvalue. The resulting expression for the correlations between the principal components and the original data is more awkward, but this scaling has the advantage that all the principal components have equal, unit variance. This property can be useful in the detection of outliers.

13.1.5. Connections to the Multivariate Normal Distribution

The distribution of the data x, whose sample covariance matrix $[S]$ is used to calculate a PCA, need not be multivariate normal in order for the PCA to be valid. Regardless of the joint distribution of x, the resulting principal components u_m will uniquely be those uncorrelated linear combinations that successively maximize the represented fractions of the variances on the diagonal of $[S]$. However, if in addition $x \sim N_K(\boldsymbol{\mu}_x, [\Sigma_x])$, then as linear combinations of the multinormal x's, the joint distribution of the principal components will also have a multivariate normal distribution,

$$u \sim N_M\left([E]^T \boldsymbol{\mu}_x, [\Lambda]\right). \tag{13.10}$$

Equation 13.10 is valid both when the matrix $[E]$ contains the full number $M = K$ of eigenvectors as its columns, or some fewer number $1 \le M < K$. If the principal components are calculated from the centered data x', then $\boldsymbol{\mu}_u = \boldsymbol{\mu}_{x'} = \mathbf{0}$.

If the joint distribution of x is multivariate normal, then the transformation of Equation 13.2 is a rigid rotation to the principal axes of the probability ellipses of the distribution of x, yielding the uncorrelated and mutually independent u_m. With this background it is not difficult to understand Equations 12.5 and 12.34, which say that the distribution of Mahalanobis distances to the mean of a multivariate normal distribution follows the χ_K^2 distribution. One way to view the χ_K^2 is as the distribution of K squared independent standard Gaussian variables z_k^2 (see Section 4.4.5). Calculation of the Mahalanobis distance (or, equivalently, the Mahalanobis transformation, Equation 12.21) produces uncorrelated values with zero mean and unit variance, and a (squared) distance involving them is then simply the sum of the squared values.

It was noted in Section 12.3 that an effective way to search for multivariate outliers when assessing multivariate normality is to examine the distribution of linear combinations formed using eigenvectors associated with the smallest eigenvalues of $[S]$ (Equation 12.18). These linear combinations are, of course, the last principal components. Figure 13.3 illustrates why this idea works, in the easily visualized

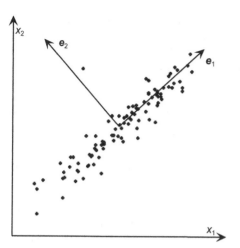

FIGURE 13.3   Identification of a multivariate outlier by examining the distribution of the last principal component. The projection of the single outlier onto the first eigenvector yields a quite ordinary value for its first principal component u_1, but its projection onto the second eigenvector yields a prominent outlier in the distribution of the u_2 values.

$K = 2$ situation. The point scatter shows a strongly correlated pair of Gaussian variables, with one multivariate outlier. The outlier is not especially unusual within either of the two univariate distributions, but it stands out in two dimensions because it is inconsistent with the strong positive correlation of the remaining points. The distribution of the first principal component u_1, obtained geometrically by projecting the points onto the first eigenvector e_1, is at least approximately Gaussian, and the projection of the outlier is a very ordinary member of this distribution. On the other hand, the distribution of the second principal component u_2, obtained by projecting the points onto the second eigenvector e_2, is concentrated near the origin except for the single large outlier. Other than the outlier, this distribution is also approximately Gaussian. This approach is effective in identifying the multivariate outlier because its existence has distorted the PCA only slightly, so that the leading eigenvector continues to be oriented in the direction of the main data scatter. Because a small number of outliers contribute only slightly to the full variability, it is the last (low-variance) principal components that represent them.

13.2. APPLICATION OF PCA TO GEOPHYSICAL FIELDS

13.2.1. PCA for a Single Field

The overwhelming majority of applications of PCA to atmospheric data have involved analyses of fields (i.e., spatial arrays of variables) such as geopotential heights, temperatures, precipitation, and so on. In these cases the full data set consists of multiple observations of a field or set of fields. Frequently these multiple observations take the form of time series, for example, a sequence of daily hemispheric 500 mb height maps. Another way to look at this kind of data is as a collection of K mutually correlated time series that have been sampled at each of K gridpoints or station locations. The goal of PCA as applied to this type of data is usually to explore, or to express succinctly, the joint space/time variations of the many variables in the data set.

Even though the locations at which the field is sampled are spread over a two-dimensional (or possibly three-dimensional) physical space, the data from these locations at a given observation time are arranged in the K-dimensional vector x. That is, regardless of their geographical arrangement, each location is assigned a number (as in Figure 9.27) from 1 to K, which refers to the appropriate element in the data vector $x = [x_1, x_2, x_3, \ldots , x_K]^T$. In this most common application of PCA to fields, the data matrices $[X]$ and $[X']$ are thus dimensioned $(n \times K)$, or (time $\times$ space), since data at K locations in space have been sampled at n successive times.

To emphasize that the original data consists of K time series, the analysis equation (13.1 or 13.2) is sometimes written with an explicit time index:

$$u(t) = [E]^T x'_t, \tag{13.11a}$$

or, in scalar form,

$$u_m(t) = \sum_{k=1}^{K} e_{k,m} x'_k(t), m = 1, \ldots, M. \tag{13.11b}$$

Here the time index t runs from 1 to n. The synthesis equations (13.5 or 13.6) can be written using the same notation, as was done in Equation 13.8. Equation 13.11 emphasizes that if the data x consist of a set of time series, then the principal components u are also time series. The time series of one of the principal components, $u_m(t)$, may very well exhibit serial correlation (correlation with itself through time), and the individual principal component time series are sometimes analyzed using the tools presented in Chapter 10. However, each of the time series of principal components will be uncorrelated with the time series of all the other principal components.

When the K elements of x are measurements at different locations in space, the eigenvectors can be displayed graphically in a quite informative way. Notice that each eigenvector contains exactly K elements, and that these elements have a one-to-one correspondence with each of the K locations in the dot product from which the corresponding principal component is calculated (Equation 13.11b). Each eigenvector element $e_{k,m}$ can be plotted on a map at the same location as its corresponding data value x'_k, and this field of eigenvector elements can itself be summarized using smooth contours in the same way as an ordinary meteorological field. Such maps depict clearly which locations are contributing most strongly to the respective principal components. Looked at another way, such maps indicate the geographic distribution of simultaneous data anomalies represented by the corresponding principal components. These geographic displays of eigenvectors sometimes also are interpreted as representing uncorrelated modes of variability of the fields from which the PCA was extracted. There are cases where this kind of interpretation can be reasonable (but see Section 13.2.4 for a cautionary counterexample), particularly for the leading eigenvector. However, because of the mutual orthogonality constraints on the eigenvectors, strong interpretations of this sort are often not justified for the subsequent EOFs (North, 1984).

Figure 13.4 shows the first four eigenvectors of a PCA of the correlation matrix for winter monthly mean 500 mb heights at gridpoints in the northern hemisphere. The percentages below and to the right of the panels show the fraction of the total hemispheric variance (Equation 13.4) represented by each of the corresponding principal components. Together, the first four principal components account for nearly half of the (normalized) hemispheric winter height variance. These patterns resemble the teleconnectivity patterns for the same data shown in Figure 3.33, and apparently reflect the same underlying physical processes in the atmosphere. For example, Figure 13.4b evidently reflects the PNA pattern of alternating height anomalies stretching from the Pacific Ocean through northwestern North America to southeastern North America. A positive value for the second principal component of this data set corresponds to negative 500 mb height anomalies (troughs) in the northeastern Pacific and in the southeastern United States, and to positive height anomalies (ridges) in the western part of the continent,

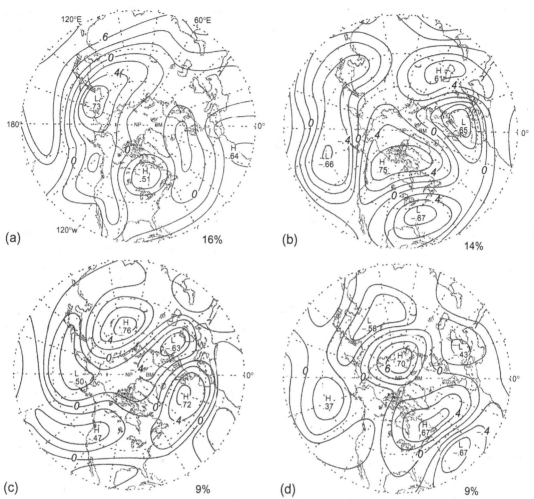

FIGURE 13.4 Spatial displays of the first four eigenvectors of gridded winter monthly mean 500 mb heights for the northern hemisphere, 1962–1977. This PCA was computed using the correlation matrix of the height data, and scaled so that $||e_m|| = \lambda_m^{1/2}$. Percentage values below and to the right of each map are proportion of total variance x100% (Equation 13.4). The patterns resemble the teleconnectivity patterns for the same data (Figure 3.33). *From Wallace and Gutzler (1981).* © *American Meteorological Society. Used with permission.*

and over the central tropical Pacific. A negative value of the second principal component yields the reverse pattern of anomalies, and a more zonal 500 mb flow over North America.

Principal component analyses are most frequently structured as just described, by computing the eigenvalues and eigenvectors from the $(K \times K)$ covariance or correlation matrix of the $(n \times K)$ data matrix $[X]$. However, this usual approach, known as *S-mode* PCA, is not the only possibility. An alternative, known as *T-mode* PCA is based on the eigenvalues and eigenvectors of the $(n \times n)$ covariance or correlation matrix of the data matrix $[X]^T$. Thus in a T-mode PCA the eigenvector elements correspond to the individual data samples (which often form a time series), and the principal components u relate to the K variables (which may be spatial points), so that the two approaches portray different aspects of a data set in complementary ways. Compagnucci and Richman (2008) compare these two approaches for representing atmospheric circulation fields. The eigenvalues and eigenvectors produced by the two approaches will be different, because the S-mode anomalies are computed by subtracting the K column means of $[X]$, whereas the T-mode anomalies are computed by subtracting the n row means, which will be the column means of $[X]^T$. Accordingly the number of nonzero eigenvalues in an S-mode analysis will be the smaller of $n-1$ and K, and the number of nonzero eigenvalues of a T-mode analysis will be the smaller of n and $K-1$.

13.2.2. Simultaneous PCA for Multiple Fields

It is also possible to apply PCA to vector-valued fields, which are fields with data for more than one variable at each location or gridpoint. This kind of analysis is equivalent to simultaneous PCA of two or more fields. If there are L such variables at each of the K gridpoints, then the dimensionality of the data vector x is given by the product KL. The first K elements of x are observations of the first variable, the second K elements are observations of the second variable, and the last K elements of x will be observations of the Lth variable. Since the L different variables generally will be measured in unlike units, it will almost always be appropriate to base the PCA of such data on the correlation matrix. The dimension of $[R]$, and of the matrix of eigenvectors $[E]$, will then be $(KL \times KL)$. Application of PCA to this kind of correlation matrix will produce principal components successively maximizing the joint variance of the L standardized variables in a way that considers the correlations both between and among these variables at the K locations. This joint PCA procedure is sometimes called *combined PCA*, (CPCA), or *extended EOF* (EEOF) analysis.

Figure 13.5 illustrates the structure of the correlation matrix (left) and the matrix of eigenvectors (right) for PCA of vector field data. The first K rows of $[R]$ contain the correlations between the first of the L variables at these locations and all of the KL variables. Rows $K+1$ to $2K$ similarly contain the correlations between the second of the L variables and all the KL variables, and so on. Another way to look at the correlation matrix is as a collection of L^2 submatrices, each dimensioned $(K \times K)$, which contain the correlations between sets of the L variables jointly at the K locations. The submatrices located on the diagonal of $[R]$ thus contain ordinary correlation matrices for each of the

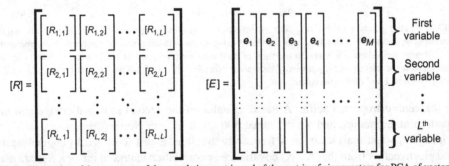

FIGURE 13.5 Illustration of the structures of the correlation matrix and of the matrix of eigenvectors for PCA of vector field data. The basic data consist of multiple observations of L variables at each of K locations, so the dimensions of both $[R]$ and $[E]$ are $(KL \times KL)$. The correlation matrix consists of $(K \times K)$ submatrices containing the correlations between sets of the L variables jointly at the K locations. The submatrices located on the diagonal of $[R]$ are the ordinary correlation matrices for each of the L variables. The off-diagonal submatrices contain correlation coefficients, but are not symmetric and will not contain 1's on the diagonals. Each eigenvector column of $[E]$ similarly consists of L segments, each of which contains K elements pertaining to the individual locations.

L variables. The off-diagonal submatrices contain correlation coefficients but are not symmetric and will not contain 1's on their diagonals. However, the overall symmetry of $[R]$ implies that $[R_{i,j}] = [R_{j,i}]^T$. Similarly, each column of $[E]$ consists of L segments, and each of these segments contains the K elements pertaining to each of the individual locations.

The eigenvector elements resulting from a PCA of a vector field can be displayed graphically in a manner that is similar to the maps drawn for ordinary scalar fields. Here, each of the L groups of K eigenvector elements is either overlaid on the same base map or plotted on separate maps. Figure 13.6, from the classic paper by Kutzbach (1967), illustrates this process for the case of $L = 2$ data values at each location. The two variables are average January surface pressure and average January temperature, measured at $K = 23$ locations in North America. The heavy lines are an analysis of the (first 23) elements of the first eigenvector that pertain to the pressure data, and the dashed lines with shading show an analysis of the temperature (second 23) elements of the same eigenvector. The corresponding principal component accounts for 28.6% of the joint variance of the $KL = 23 \times 2 = 46$ standardized variables.

In addition to effectively condensing very much information, the patterns shown in Figure 13.6 are consistent with the underlying atmospheric physical processes. In particular, the temperature anomalies are consistent with patterns of thermal advection implied by the pressure anomalies. If the first principal component u_1 is positive for a particular January, the solid contours imply positive pressure anomalies in the north and east, with lower than average pressures in the southwest. On the west coast, this pressure

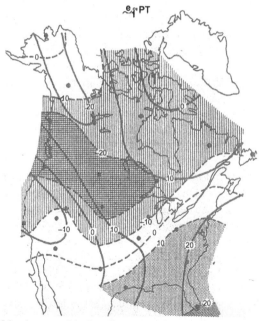

FIGURE 13.6   Spatial display of the elements of the first eigenvector of the (46×46) correlation matrix of average January sea-level pressures and temperatures at 23 locations in North America (dots). The first principal component of this correlation matrix accounts for 28.6% of the joint (standardized) variance of the pressures and temperatures. Heavy lines are a hand analysis of the sea-level pressure elements of the first eigenvector, and dashed lines with shading are a hand analysis of the temperature elements of the same eigenvector. The joint variations of pressure and temperature depicted are physically consistent with temperature advection in response to the pressure anomalies. *From Kutzbach (1967). © American Meteorological Society. Used with permission.*

pattern would result in weaker than average westerly surface winds and stronger than average northerly surface winds. The resulting advection of cold air from the north would produce colder temperatures, and this cold advection is reflected by the negative temperature anomalies in this region. Similarly, the pattern of pressure anomalies in the southeast would enhance southerly flow of warm air from the Gulf of Mexico, resulting in positive temperature anomalies as shown. Conversely, if u_1 is negative, reversing the signs of the pressure eigenvector elements implies enhanced westerlies in the west, and northerly wind anomalies in the southeast, which are consistent with positive and negative temperature anomalies, respectively. These temperature anomalies are indicated by the dashed contours and shading in Figure 13.6, when their signs are also reversed.

Figure 13.6 is a simple example involving familiar variables. Its interpretation is easy and obvious if we are conversant with the climatological relationships of pressure and temperature patterns over North America in winter. However, the physical consistency exhibited in this example (where the "right" answer is known ahead of time) is indicative of the power of this kind of PCA to uncover meaningful joint relationships among atmospheric (and other) fields in an exploratory setting, where clues about possibly unknown underlying physical mechanisms may be hidden in the complex relationships among several fields.

Example 13.3 Characterization of the Madden–Julian Oscillation

The Madden–Julian oscillation (MJO, Madden and Julian, 1972) is a travelling pattern of enhanced and suppressed tropical convection that propagates eastward from the western Indian to the eastern Pacific ocean basins on a one- to two-month timescale. It is a prominent element of subseasonal tropical atmospheric variability that exhibits characteristic signatures in satellite-observed outgoing longwave radiation (OLR, which is a proxy for cold, high convective cloud tops), coupled with upper- and lower-tropospheric zonal (east-west) wind convergence and divergence.

Monitoring of the MJO is typically done using a diagram derived from extended EOF analysis of OLR, 850 mb zonal winds and 200 mb zonal winds, averaged between 15°S and 15°N, as a function of longitude (Wheeler and Hendon, 2004). Figure 13.7 shows the two leading EOFs for the three combined variables. The traces for each of the three variables are shown as scalar functions of longitude rather than as maps because north–south variations have been collapsed by the meridional averaging.

It is notable that these two eigenvectors in Figure 13.7 have eigenvalues of comparable magnitude, which are well separated from the magnitudes of the third and subsequent eigenvalues (see Section 13.4), and that the traces for each of the three variables are in approximate quadrature (the loadings for EOF2 lag those in EOF1 by approximately a quarter cycle). These characteristics, which are spatial counterparts to the properties that can emerge from PCAs computed for time series (Section 13.7.1), allow the propagating nature of the MJO to be portrayed in a two-dimensional phase space defined by the corresponding principal components (Figure 13.8). In the context of the MJO diagram, these are conventionally denoted as RMM1 and RMM2, respectively. The diagram is divided into octants, with Phase 1 corresponding to convection over the western Indian ocean, and Phase 8 corresponding to convection over the eastern Pacific. An MJO cycle is portrayed as daily points in the diagram trace out a counterclockwise orbit in this phase space, an example of which for January through March of 2009 is shown in Figure 13.8. Figure 13.9 shows composites of December through February OLR and 850 mb zonal wind anomalies, averaged over the eight MJO phases jointly defined by the two principal components in the Wheeler–Hendon diagram (Figure 13.8), showing that the extended EOF analysis has been very effective at portraying this propagating phenomenon.   ◇

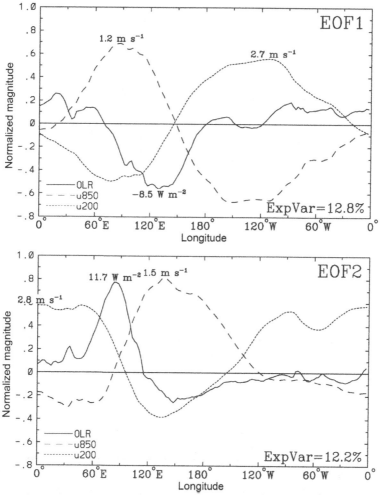

FIGURE 13.7 The leading two EOFs from an extended EOF analysis of OLR, 850 mb zonal wind, and 200 mb zonal wind, averaged within 15° of the equator, as functions of longitude. *From Wheeler and Hendon (2004). © American Meteorological Society. Used with permission.*

13.2.3. Scaling Considerations and Equalization of Variance

A complication arises in PCA of fields in which the geographical distribution of data locations is not uniform (Baldwin et al., 2009; Karl et al., 1982; North et al., 1982). The problem is that the PCA has no information about the spatial distributions of the locations, or even that the elements of the data vector x may pertain to different locations, but nevertheless finds linear combinations that maximize the joint variance. Regions that are overrepresented in x, in the sense that data locations are concentrated in that region, will tend to dominate the analysis, whereas data-sparse regions will be underweighted. In contrast, the goal of PCA on geophysical fields is usually to approximate the *intrinsic EOFs* (Baldwin et al., 2009; North et al., 1982; Stephenson, 1997), which are properties of the actual underlying continuous field(s), and are independent of any spatial sampling pattern.

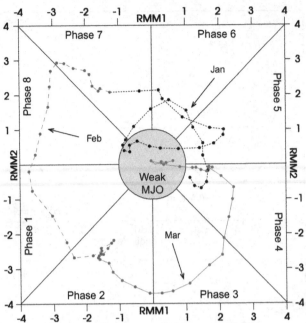

FIGURE 13.8 An example Wheeler–Hendon (2004) diagram, defined by the principal components RMM1 and RMM2, corresponding to the leading eigenvectors shown in Figure 13.7. The counterclockwise trace indicates progression of the MJO during January (dotted), February (dashed), and March (solid), 2009. *Modified from Peatman et al. (2015).*

Data available on a regular latitude–longitude grid is a common cause of this problem. In this case the number of gridpoints per unit area increases with increasing latitude because the meridians converge at the poles, so that a PCA for this kind of gridded data will emphasize high-latitude features and deemphasize low-latitude features. One approach to geographically equalizing the variances is to multiply the data by $\sqrt{\cos\phi}$, where ϕ is the latitude (North et al., 1982). The same effect can be achieved by multiplying each element of the covariance or correlation matrix being analyzed by $\sqrt{\cos\phi_k}\,\sqrt{\cos\phi_\ell}$, where k and ℓ are the indices for the two locations (or location/variable combinations) corresponding to that element of the matrix. The square roots are necessary even though the areas that are proportional to the cosines of the latitudes, because it is the variances and covariances of the analyzed quantities that need to be equalized for the PCA. Baldwin et al. (2009) formulate this process more generally by defining a weighting matrix that can concisely represent the effects of different spatial sampling arrays. Of course these rescalings must be reversed when recovering the original data from the principal components, as in Equations 13.5 and 13.6. An alternative procedure is to interpolate irregularly or nonuniformly distributed data onto an equal-area grid (Araneo and Compagnucci, 2004; Karl et al., 1982). This latter approach is also applicable when the data pertain to an irregularly spaced network, such as climatological observing stations.

Use of extended EOF analysis is not limited to settings involving multiple variables at a common set of locations. A slightly more complicated scaling problem arises when multiple fields with different spatial

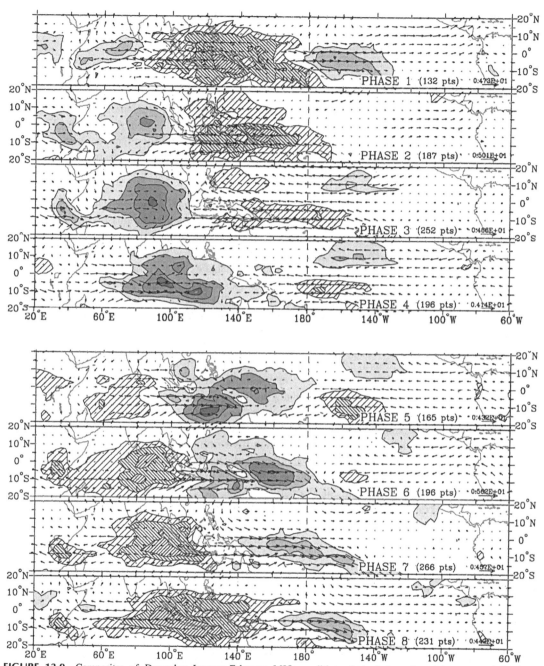

FIGURE 13.9 Composites of December–January–February MJO conditions derived from the eight sectors defined in Figure 13.8, showing regions of enhanced (shaded) and suppressed (hatched) convection, and corresponding 850 mb zonal wind anomalies, as the phenomenon propagates from west to east. *From Wheeler and Hendon (2004). © American Meteorological Society. Used with permission.*

resolutions or spatial domains are simultaneously analyzed with PCA. Here an additional rescaling is necessary to equalize the sums of the variances in each field. Otherwise fields with more gridpoints will dominate the PCA, even if all the fields pertain to the same geographic area.

13.2.4. Domain Size Effects: Buell Patterns

In addition to providing an efficient data compression, results of a PCA are sometimes interpreted in terms of underlying physical processes. For example, the spatial eigenvector patterns in Figure 13.4 have been interpreted as teleconnected modes of atmospheric variability, and the eigenvector displayed in Figure 13.6 reflects the connection between pressure and temperature fields that is expressed as thermal advection. The possibility that informative or at least suggestive interpretations may result can be a strong motivation for computing a PCA.

One problem that can occur when making such interpretations of a PCA for field data arises when the spatial scale (or one or more of the important spatial scales) of the data variations is comparable to or larger than the spatial domain being analyzed. In such cases the space/time variations in the data are still efficiently represented by the PCA, and PCA is still a valid approach to data compression. But the resulting spatial eigenvector patterns take on characteristic shapes that are nearly independent of the underlying variations in the data. These characteristic shapes are called *Buell patterns*, after the author of the paper that first pointed out their existence (Buell, 1979).

Consider, as an artificial but simple example, a 5×5 array of $K = 25$ points representing a square spatial domain. Define the correlations among data values observed at these points to be functions only of their spatial separation d, according to $r(d) = \exp(-d/2)$. The separations of adjacent points in the horizontal and vertical directions are $d = 1$, and so would exhibit correlation $r(1) = 0.61$; points adjacent diagonally would exhibit correlation $r(\sqrt{2}/2) = 0.49$, and so on. This correlation function is shown in Figure 13.10a. It is unchanging across the domain, and produces no spatially distinct features, or preferred patterns of variability. Its spatial scale is comparable to the domain size, which is 4×4 distance units vertically and horizontally, corresponding to $r(4) = 0.14$.

Even though there are no preferred regions of variability within the 5×5 domain, the eigenvectors of the resulting (25×25) correlation matrix $[R]$ appear to indicate that there are. The first of these eigenvectors, which accounts for 34.3% of the variance, is shown in Figure 13.10b. It appears to indicate generally in-phase variations throughout the domain, but with larger amplitude (greater magnitudes of variability) near the center. This first characteristic Buell pattern is an artifact of the mathematics behind the eigenvector calculation if all the correlations are positive and does not merit interpretation beyond its suggestion that the scale of variation of the data is comparable to or larger than the size of the spatial domain.

The dipole patterns in Figures 13.10c and 13.10d are also characteristic Buell patterns and result from the constraint of mutual orthogonality among the eigenvectors. They do not reflect "dipole oscillations" or "seesaws" in the underlying data, whose correlation structure (by virtue of the way this artificial example has been constructed) is homogeneous and isotropic. Here the patterns are oriented diagonally, because opposite corners of this square domain are further apart than opposite sides, but the characteristic dipole pairs in the second and third eigenvectors might instead have been oriented vertically and horizontally in a differently shaped domain. Notice that the second and third eigenvectors account for equal proportions of the variance, and so are actually oriented arbitrarily within the two-dimensional space that they span (see Section 13.4). Additional Buell patterns are sometimes seen in subsequent eigenvectors, the next of which typically suggest tripole patterns of the form $- + -$ or $+ - +$.

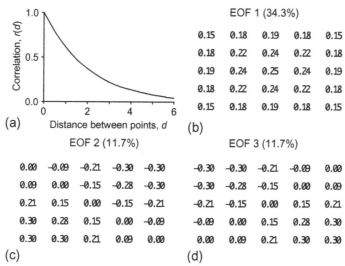

FIGURE 13.10   Artificial example of Buell patterns. Data on a 5×5 square grid with unit vertical and horizontal spatial separation exhibit correlations according to the function of their spatial separations shown in (a). Panels (b)–(d) show the first three eigenvectors of the resulting correlation matrix, displayed in the same 5×5 spatial arrangement. The resulting single central hump (b), and pair of orthogonal dipole patterns (c) and (d), are characteristic artifacts of the domain size being comparable to or smaller than the spatial scale of the underlying data variations.

13.3. TRUNCATION OF THE PRINCIPAL COMPONENTS

13.3.1. Why Truncate the Principal Components?

Mathematically, there are as many eigenvectors of [S] or [R] as there are elements of the data vector x, provided $K \leq n{-}1$. However, it is typical of atmospheric data that substantial covariances (or correlations) exist among the original K variables, and as a result there are few or no off-diagonal elements of [S] (or [R]) that are near zero. This situation implies that there is redundant information in x, and that the first few eigenvectors of its dispersion matrix will locate directions in which the joint variability of the data is greater than the variability of any single element of x. Similarly, the last few eigenvectors will point to directions in the K-dimensional space of x where the data jointly exhibit very little variation. This property was illustrated in Example 13.1 for daily temperature values measured at two nearby locations.

To the extent that there is redundancy in the original data x, it is possible to capture most of their variance by considering only the most important directions of their joint variations. That is, most of the information content of the data may be represented using some smaller number $M < K$ of the principal components u_m. In effect, the original data set containing the K variables x_k is approximated by the smaller set of new variables u_m. If $M << K$, retaining only the first M of the principal components results in a much smaller data set. This data compression capability of PCA is often a primary motivation for its use. If $M \approx K$ principal components are required to capture a usefully large proportion of the variance in the original data x there is probably little point to computing a PCA.

The truncated representation of the original data can be expressed mathematically by a truncated version of the analysis formula, Equation 13.2, in which the dimension of the truncated u is $(M \times 1)$, and [E] is the (nonsquare, $K \times M$) matrix whose columns consist only of the first M eigenvectors (i.e., those associated with the largest M eigenvalues) of $[S]$ or $[R]$. The corresponding synthesis formula, Equation 13.6, is then only approximately true because the original data cannot be exactly resynthesized without using all K eigenvectors.

Where is the appropriate balance between data compression (choosing M to be as small as possible) and avoiding excessive information loss (truncating only a small number, $K - M$, of the principal components)? There is no clear criterion that can be used to choose the number of principal components that are best retained in a given circumstance. The choice of the truncation level can be aided by one or more of the many available principal component selection rules, but it is ultimately a subjective choice that will depend in part on the data at hand and the purpose(s) of the analysis.

13.3.2. Subjective Truncation Criteria

Some approaches to truncating principal components are subjective, or nearly so. Perhaps the most basic criterion is to retain enough of the principal components to represent a "sufficient fraction" of the variances of the original x. That is, enough principal components are retained for the total amount of variability represented to be larger than some critical value,

$$\sum_{m=1}^{M} R_m^2 \geq R_{crit}^2, \tag{13.12}$$

where R_m^2 is defined as in Equation 13.4. Of course the difficulty comes in determining how large the fraction R_{crit}^2 must be in order to be considered "sufficient." Ultimately this will be a subjective choice, informed by the analyst's knowledge of the data at hand and the uses to which they will be put. Jolliffe (2002) suggests that $70\% \leq R_{crit}^2 \leq 90\%$ may often be a reasonable range.

Another essentially subjective approach to principal component truncation is based on the shape of the graph of the eigenvalues λ_m in decreasing order as a function of their index $m = 1, \ldots, K$, known as the *eigenvalue spectrum*. Since each eigenvalue measures the variance represented in its corresponding principal component, this graph is analogous to the power spectrum (see Section 10.5.2), further extending the parallels between EOF and Fourier analyses.

Plotting the eigenvalue spectrum with a linear vertical scale produces what is known as the *scree graph*. When using the scree graph qualitatively, the goal is to locate a point separating a steeply sloping portion to the left, and a more shallowly sloping portion to the right. The principal component number at which the separation occurs is then taken as the truncation cutoff, M. There is no guarantee that the eigenvalue spectrum for a given PCA will exhibit a single slope separation, or that it (or they) will be sufficiently abrupt to unambiguously locate a cutoff M. Sometimes this approach to principal component truncation is called the scree "test," although this name implies more objectivity and theoretical justification than is warranted: the scree-slope criterion does not involve quantitative statistical inference. Figure 13.11a shows the scree graph (circles) for the PCA summarized in Table 13.1b. This is a relatively well-behaved example, in which the last three eigenvalues are quite small, leading to a fairly distinct bend at $K = 3$, and so to a truncation after the first $M = 3$ principal components would be suggested.

An alternative but similar approach is based on the log-eigenvalue spectrum, or *log-eigenvalue (LEV) diagram*. Choosing a principal component truncation based on the LEV diagram is motivated by the idea that, if the last $K–M$ principal components represent uncorrelated noise, then the magnitudes of their eigenvalues should decay exponentially with increasing principal component number. This behavior should be identifiable in the LEV diagram as an approximately straight-line portion on its right-hand side. The M

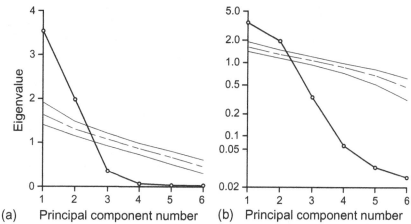

FIGURE 13.11   Graphical displays of eigenvalue spectra, that is, eigenvalue magnitudes as a function of the principal component number (heavier lines connecting circled points), for a $K = 6$-dimensional analysis (see Table 13.1b): (a) Linear scaling, or scree graph, (b) logarithmic scaling, or LEV diagram. Both the scree and LEV criteria would lead to retention of the first three principal components in this analysis. Lighter lines in both panels show results of the resampling tests necessary to apply Rule N of Priesendorfer et al. (1981). Dashed line is median of eigenvalues for 1000 (6×6) dispersion matrices of independent Gaussian variables, constructed using the same sample size as the data being analyzed. Solid lines indicate the 5th and 95th percentiles of these simulated eigenvalue distributions. Rule N would indicate retention of only the first two principal components, on the grounds that only these are significantly larger than what would be expected from data with no correlation structure.

retained principal components would then be the ones whose log-eigenvalues lie above the leftward extrapolation of this line. As before, depending on the data set there may no, or more than one, quasi-linear portions, and their limits may not be clearly defined. Figure 13.11b shows the LEV diagram for the PCA summarized in Table 13.1b. Here $M = 3$ would probably be chosen by most viewers of this LEV diagram, although the choice is not unambiguous.

13.3.3. Rules Based on the Size of the Last Retained Eigenvalue

Another class of principal-component selection rules involves focusing on how small an "important" eigenvalue can be. This set of selection rules can be summarized by the criterion

$$\text{Retain } \lambda_m \text{ if } \lambda_m > \frac{T}{K} \sum_{k=1}^{K} s_{k,k}, \tag{13.13}$$

where $s_{k,k}$ is the sample variance of the kth element of x, and T is a threshold parameter.

A simple application of this idea, known as *Kaiser's rule*, involves comparing each eigenvalue (and therefore the variance described by its principal component) to the amount of the joint variance reflected by the average eigenvalue. Principal components whose eigenvalues are above this threshold are retained. That is, Kaiser's rule uses Equation 13.13 with the threshold parameter $T = 1$. Jolliffe (1972, 2002) has argued that Kaiser's rule is too strict (i.e., typically seems to discard too many principal components). He suggests that the alternative $T = 0.7$ often will provide a roughly correct threshold, which allows for the effects of sampling variations.

A third alternative in this class of truncation rules is to use the *broken stick model*, so called because it is based on the expected length of the *m*th longest piece of a randomly broken unit line segment. According to this criterion, the threshold parameter in Equation 13.13 is taken to be

$$T(m) = \sum_{j=m}^{K} \frac{1}{j}. \tag{13.14}$$

This rule yields a different threshold for each candidate truncation level, that is, $T = T(m)$, so that the truncation is made at the smallest m for which Equation 13.13 is not satisfied, according to the threshold in Equation 13.14.

All of the three criteria described in this subsection would lead to choosing $M = 2$ for the eigenvalue spectrum in Figure 13.11.

13.3.4. Rules Based on Hypothesis Testing Ideas

Faced with a subjective choice among sometimes vague and possibly conflicting truncation criteria, it is natural to hope for a more objective approach based on the sampling properties of PCA statistics. Section 13.4 describes some large-sample results for the sampling distributions of eigenvalue and eigenvector estimates that have been calculated from multivariate normal samples. Based on these results, Mardia et al. (1979) and Jolliffe (2002) describe tests for the null hypothesis that the last $K-M$ eigenvalues are all equal, and so correspond to noise that should be discarded in the principal component truncation. One problem with this approach occurs when the data being analyzed do not have a multivariate normal distribution and/or are not independent, in which case inferences based on those assumptions may produce serious errors. But a more difficult problem with this approach is that it usually involves examining sequences of tests that are not independent: Are the last two eigenvalues plausibly equal, and if so, are the last three equal, and if so, are the last four equal…? The true test level for a random number of correlated tests will bear an unknown relationship to the nominal level at which each test in the sequence is conducted. This procedure can be used to choose a truncation level, but it will be as much a rule of thumb as the other possibilities already presented in this section, and not a quantitative choice based on a known small probability for falsely rejecting a null hypothesis.

Resampling counterparts to testing-based truncation rules have been used frequently with atmospheric data. The most common of these is known as *Rule N* (Overland and Preisendorfer, 1982; Preisendorfer et al., 1981). Rule N identifies the largest M principal components to be retained on the basis of a sequence of resampling tests involving the distribution of eigenvalues of randomly generated dispersion matrices. The procedure involves repeatedly generating sets of vectors of independent Gaussian random numbers with the same dimension (K) and sample size (n) as the data x being analyzed, and then computing the eigenvalues of their dispersion matrices. These randomly generated eigenvalues are then scaled in a way that makes them comparable to the eigenvalues λ_m to be tested, for example, by requiring that the sum of each set of randomly generated eigenvalues will equal the sum of the eigenvalues computed from the data. Each λ_m from the real data is then compared to the empirical distribution of its synthetic counterparts and is retained if it is larger than 95% of these.

The light lines in the panels of Figure 13.11 illustrate the use of Rule N to select a principal component truncation level. The dashed lines reflect the medians of 1000 sets of eigenvalues computed from 1000 (6×6) dispersion matrices of independent Gaussian variables, constructed using the same sample size as the data being analyzed. The solid lines show 95th and 5th percentiles of those distributions for each of the six eigenvalues. The first two eigenvalues λ_1 and λ_2 are larger than more than 95% of their synthetic counterparts, and accordingly, Rule N would choose $M = 2$ for this data.

A table of 95% critical values for Rule N, for selected sample sizes n and dimensions K, is presented in Overland and Preisendorfer (1982). Corresponding large-sample tables are given in Preisendorfer et al. (1981) and Preisendorfer (1988). Preisendorfer (1988) notes that if there is substantial temporal correlation present in the individual variables x_k, that it may be more appropriate to construct the resampling distributions for Rule N (or to use the tables just mentioned) using the smallest effective sample size (e.g., Bretherton et al., 1999; Preisendorfer et al., 1981)

$$n' \approx n \frac{1 - r_1^2}{1 + r_1^2} \tag{13.15}$$

appropriate to eigenvalues and other second moment quantities among the x_k, rather than using n independent vectors of Gaussian variables to construct each synthetic dispersion matrix. Equation 13.15 is analogous to Equation 5.12 pertaining to inferences about means, but the squaring of the lag-1 autocorrelation in Equation 13.15 renders the result much less sensitive to autocorrelation effects.

Another potential problem with Rule N, and other similar procedures, is that the data x may not be approximately Gaussian. For example, one or more of the x_k's could be precipitation variables. To the extent that the original data are not Gaussian, the random number generation procedure will not simulate accurately the underlying physical process, and the results of the test may be misleading. A possible remedy for the problem of non-Gaussian data might be to use a bootstrap version of Rule N, although this approach seems not to have been tried in the literature to date.

The primary weakness of the Rule N procedure derives from the fact that only its test for the leading eigenvalue is correct. The reason is that, having rejected the proposition that λ_1 is not different from the others, the Monte Carlo sampling distributions for the remaining eigenvalues are no longer meaningful because they are conditional on all K eigenvalues reflecting noise. That is, these synthetic sampling distributions will imply too much variance if it has been inferred that λ_1 has more than a random share, since the sum of the eigenvalues is constrained to equal the total variance. Accordingly, Preisendorfer (1988) notes that Rule N tends to retain too few principal components.

A better approach (Wilks, 2016c) is to test the sequence of scaled eigenvalues

$$\lambda_k^* = \frac{\lambda_k}{\dfrac{1}{N_{rank} - k + 1} \displaystyle\sum_{i=k}^{N_{rank}} \lambda_k}, \tag{13.16}$$

where $N_{rank} = \min(n-1, K)$ is the number of nonzero eigenvalues. Equation 13.16 is the fraction of variance represented by λ_k when the larger eigenvalues have been omitted from the denominator, for which analytic sampling distributions (Tracy and Widom, 1996) are available for sufficiently large n and K. Good approximations to the right tails of these sampling distributions, which are different for each λ^*_k, are provided by Pearson III distributions (Equation 4.55), with parameters

$$\alpha = 46.4 \tag{13.17a}$$

$$\beta_k = \frac{0.186 \, \sigma_k}{\max(n_k, K^*)} \tag{13.17b}$$

and

$$\zeta_k = \frac{\mu_k - 9.85 \, \sigma_k}{\max(n_k, K^*)}, \tag{13.17c}$$

where

$$\mu_k = \left(\sqrt{n_k - 1/2} + \sqrt{K^* - 1/2}\right)^2 \tag{13.18a}$$

and

$$\sigma_k = \sqrt{\mu_k}\left(\frac{1}{\sqrt{n_k - 1/2}} + \frac{1}{\sqrt{K^* - 1/2}}\right)^{1/3} \tag{13.18b}$$

are the Tracy–Widom location and scale parameters, $K^* = K-k+1$, and $n_k = n-k+1$. The sequence of tests, $k = 1, 2, \ldots, N_{rank}$ are computed, each pertaining to the null hypothesis $H_k:\{\lambda_k = \lambda_{k+1} = \lambda_{k+2} = \cdots = \lambda_{Nrank}\}$ that the eigenvalue λ_k and smaller eigenvalues represent only noise and so are equal. The truncation point M is chosen to be one smaller than the first k for which H_k is not rejected (i.e., for which λ^*_k is smaller than the $1-\alpha$ quantile of the relevant Pearson III distribution defined by Equations 13.17 and 13.18). That is,

$$M = \min(k \in \{1, 2, \ldots, N_{rank} - 1\} : p_k > \alpha) - 1, \tag{13.19}$$

where p_k is the p value pertaining to the kth test and the null hypothesis H_k. The first ($k=1$) of these tests will be equivalent to the Rule N test for the first eigenvalue, but the subsequent tests account for the larger-than-random fractions of variance in the lower indexed eigenvalues previously judged to be significant.

Example 13.4 Principal Component Truncation Using Sequential Testing

Table 13.1b presents the PCA computed from the standardized January 1987 data in Table A.1. The scree and LEV diagrams for the eigenvalues λ_k are plotted in Figure 13.11, together with the Monte Carlo distributions for Rule N. However, because the two leading eigenvalues clearly represent more of the variance than would their counterparts derived from purely random data, the Rule-N Monte Carlo distributions for the trailing eigenvalues are too large, which may lead to too few eigenvalues being retained according to this criterion.

Table 13.4 contains the information for this problem derived from Equations 13.16–13.18, leading to a sequence of hypothesis tests allowing estimation of the truncation point M. For the first ($k = 1$) of these tests, $\lambda_k = \lambda^*_k$ because the average over all six eigenvalues (denominator of Equation 13.16) is 1 for this PCA based on a correlation matrix. The quantity $(\lambda^*_1 - \zeta_1)/\beta_1$ should be distributed according to a gamma distribution with $\alpha = 46.4$ (Equation 13.17a) and $\beta = 1$ if the null hypothesis that all the (underlying generating-process) eigenvalues are equal, which is strongly rejected for the $k = 1$ test. Similarly, $(\lambda^*_2 - \zeta_2)/\beta_2$ and $(\lambda^*_3 - \zeta_3)/\beta_3$ are both on the far right tail of this gamma distribution, and so the null hypotheses H_2 and H_3 are strongly rejected as well. The p value p_4, for the $k = 4$ test, is not small, leading to H_4 being the first null hypothesis not rejected, so that the truncation point $M = 3$ is chosen by this procedure according to Equation 13.19. The more conservative Rule N procedure retains only the first two principal components. ◇

13.3.5. Rules Based on Structure in the Retained Principal Components

The truncation rules presented so far all relate to the magnitudes of the eigenvalues. The possibility that physically important principal components need not have the largest variances (i.e., eigenvalues) has

TABLE 13.4 Quantities from Equations 13.16–13.18 Applied to Truncation of the PCA Presented in Table 13.1b

k	n_k	K^*	μ_k	σ_k	β_k	ζ_k	λ_k	λ^*_k	$(\lambda^*_k - \zeta_k)/\beta_k$	p_k Value
1	31	6	61.90	6.663	.04000	−.1203	3.532	3.532	91.31	8×10^{-8}
2	30	5	57.04	6.561	.04068	−.2529	1.985	4.021	105.06	4×10^{-11}
3	29	4	51.97	6.467	.04148	−.4045	0.344	2.849	78.43	3.7×10^{-5}
4	28	3	46.58	6.396	.04249	−.5864	0.074	1.597	51.39	0.227
5	27	2	40.61	6.395	.04406	−.8289	0.038	1.169	45.34	0.543

motivated a class of truncation rules based on expected characteristics of physically important principal component series (Preisendorfer et al., 1981; Preisendorfer, 1988). Since most atmospheric data that are subjected to PCA are time series (e.g., time sequences of spatial fields recorded at K gridpoints), a plausible hypothesis may be that principal components corresponding to physically meaningful processes should exhibit time dependence, because the underlying physical processes are expected to exhibit time dependence. Preisendorfer et al. (1981) and Preisendorfer (1988) proposed several such truncation rules, which test null hypotheses that the individual principal component time series are uncorrelated, using either their power spectra or their autocorrelation functions. The truncated principal components are then those for which this null hypothesis is not rejected. This class of truncation rule seems to have been used very little in practice.

13.4. SAMPLING PROPERTIES OF THE EIGENVALUES AND EIGENVECTORS

13.4.1. Asymptotic Sampling Results for Multivariate Normal Data

Principal component analyses are calculated from finite data samples and are as subject to sampling variations as any other statistical estimation procedure. That is, we rarely if ever know the true covariance matrix $[\Sigma]$ for the population or underlying generating process, but rather estimate it using the sample counterpart $[S]$. Accordingly the eigenvalues and eigenvectors calculated from $[S]$ are also estimates based on the finite sample and are thus subject to sampling variations. Understanding the nature of these variations is quite important to correct interpretation of the results of a PCA.

The equations presented in this section must be regarded as approximate, as they are asymptotic (large-n) results, and are based also on the assumption that the underlying x's have a multivariate normal distribution. It is also assumed that no pair of the population eigenvalues are equal, implying (in the sense to be explained in Section 13.4.2) that all the population eigenvectors are well defined. The validity of these results is therefore approximate in most circumstances, but they are nevertheless quite useful for understanding the nature of sampling effects on the uncertainty about estimated eigenvalues and eigenvectors.

Eigenvalue Results

The basic result for the sampling properties of estimated eigenvalues is that, in the limit of very large sample size, their sampling distribution is unbiased, and multivariate normal,

$$\sqrt{n}(\hat{\boldsymbol{\lambda}}-\boldsymbol{\lambda}) \sim N_K\left(\mathbf{0}, 2[\varLambda]^2\right), \tag{13.20a}$$

or

$$\hat{\boldsymbol{\lambda}} \sim N_K\left(\boldsymbol{\lambda}, \frac{2}{n}[\varLambda]^2\right). \tag{13.20b}$$

Here $\hat{\boldsymbol{\lambda}}$ is the $(K \times 1)$ vector of estimated eigenvalues; $\boldsymbol{\lambda}$ is its true value; and the $(K \times \text{K})$ matrix $[\varLambda]^2$ is the square of the diagonal, population eigenvalue matrix, having elements $\lambda_k{}^2$. Because $[\varLambda]^2$ is diagonal the sampling distributions for each of the K estimated eigenvalues are (approximately) independent univariate Gaussian distributions,

$$\sqrt{n}\left(\hat{\lambda}_k - \lambda_k\right) \sim N\left(0, 2\lambda_k^2\right), \tag{13.21a}$$

or

$$\hat{\lambda}_k \sim N\left(\lambda_k, \frac{2}{n}\lambda_k^2\right). \tag{13.21b}$$

Equations 13.20 and 13.21 are large-sample approximations, and there is a bias in the sample eigenvalues for finite sample size. In particular, the largest eigenvalues will be overestimated (will tend to be larger than their population counterparts) and the smallest eigenvalues will tend to be underestimated, and these effects increase with decreasing sample size (Quadrelli et al., 2005; Von Storch and Hannoschock, 1985). These biases can be understood as a consequence of the sorting of the sample eigenvalues, so that the largest sample eigenvalue will be labeled as λ_1 regardless of the rank of its generating-process counterpart, and similarly the smallest sample eigenvalue will be labeled as λ_K. The Monte Carlo Rule N distributions in Figure 13.11 illustrate the results of this phenomenon for a small sample ($n = 31$, $K = 6$) situation where the true underlying eigenvalue spectrum is completely flat, with all generating-process $\lambda_k = 1$.

Using Equation 13.21a to construct a standard Gaussian variate provides an expression for the distribution of the relative error of the eigenvalue estimate,

$$z = \frac{\sqrt{n}\left(\hat{\lambda}_k - \lambda_k\right) - 0}{\sqrt{2}\lambda_k} = \sqrt{\frac{n}{2}}\left(\frac{\hat{\lambda}_k - \lambda_k}{\lambda_k}\right) \sim N(0, 1). \tag{13.22}$$

Equation 13.22 implies

$$\Pr\left\{\left|\sqrt{\frac{n}{2}}\left(\frac{\hat{\lambda}_k - \lambda_k}{\lambda_k}\right)\right| \leq z(1 - \alpha/2)\right\} = 1 - \alpha, \tag{13.23}$$

which leads to the $(1-\alpha) \cdot 100\%$ confidence interval for the kth eigenvalue,

$$\frac{\hat{\lambda}_k}{1 + z(1 - \alpha/2)\sqrt{2/n}} \leq \lambda_k \leq \frac{\hat{\lambda}_k}{1 - z(1 - \alpha/2)\sqrt{2/n}}. \tag{13.24}$$

Eigenvector Results

The elements of each sample eigenvector are approximately unbiased, and their sampling distributions are approximately multivariate normal. But the variances of the multivariate normal sampling distributions for each of the eigenvectors depend on all the other eigenvalues and eigenvectors in a somewhat complicated way. The sampling distribution for the kth eigenvector is

$$\hat{e}_k \sim N_K(e_k, [V_{e_k}]), \tag{13.25}$$

where the covariance matrix for this distribution is

$$[V_{e_k}] = \frac{\lambda_k}{n} \sum_{i \neq k}^{K} \frac{\lambda_i}{(\lambda_i - \lambda_k)^2} e_i e_i^T. \tag{13.26}$$

The summation in Equation 13.26 involves all K eigenvalue–eigenvector pairs, indexed here by i, *except* the kth pair, for which the covariance matrix is being calculated. It is a sum of weighted outer products of these eigenvectors, and so resembles the spectral decomposition of the true covariance matrix $[\Sigma]$ (cf. Equation 11.53). But rather than being weighted only by the corresponding eigenvalues, as in Equation 11.53, they are weighted also by the reciprocals of the squares of the differences between those eigenvalues, and the eigenvalue belonging to the eigenvector whose covariance matrix is being calculated. That is, the elements of the matrices in the summation of Equation 13.26 will be quite small, except for those that are paired with eigenvalues λ_i that are close in magnitude to the eigenvalue λ_k belonging to the eigenvector whose sampling distribution is being calculated.

13.4.2. Effective Multiplets

Equation 13.26, for the sampling uncertainty of the eigenvectors of a covariance matrix, has two important implications. First, the pattern of uncertainty in the estimated eigenvectors resembles a linear combination, or weighted sum, of all the *other* eigenvectors. Second, because the magnitudes of the weights in this weighted sum are inversely proportional to the squares of the differences between the corresponding eigenvalues, an eigenvector will be relatively precisely estimated (the sampling variances will be relatively small) if its eigenvalue is well separated from the other $K-1$ eigenvalues. Conversely, eigenvectors whose eigenvalues are similar in magnitude to one or more of the other eigenvalues will exhibit large sampling variations, and those variations will be larger for the eigenvector elements that are large in the eigenvectors with comparable eigenvalues.

The joint effect of these two considerations is that the sampling distributions of two (or more) eigenvectors having similar eigenvalues will be closely entangled. Their sampling variances will be large, and their patterns of sampling error will resemble the patterns of the eigenvector(s) with which they are entangled. The net effect will be that a realization of the corresponding sample eigenvectors will be a nearly arbitrary mixture of the true population counterparts. They will jointly represent the same amount of variance (within the sampling bounds approximated by Equation 13.21), but this joint variance will be arbitrarily mixed between (or among) them. Sets of such eigenvalue–eigenvector pairs are called effectively degenerate multiplets or *effective multiplets*. Attempts at physical interpretation of such sample eigenvectors will be frustrating if not hopeless.

The source of this problem can be appreciated in the context of a three-dimensional multivariate normal distribution, in which one of the eigenvectors is relatively large, and the two smaller ones are nearly equal. The resulting distribution has ellipsoidal probability contours resembling the cucumbers in Figure 12.3. The eigenvector associated with the single large eigenvalue will be aligned with the long axis of the ellipsoid. But this multivariate normal distribution has (essentially) no preferred direction in the plane perpendicular to the long axis (exposed face on the left-hand cucumber in Figure 12.3b). Any pair of perpendicular vectors that are also perpendicular to the long axis could as easily jointly represent variations in this plane. The leading eigenvector calculated from a sample covariance matrix from this distribution would be closely aligned with the true leading eigenvector (long axis of the cucumber) because its sampling variations will be small. In terms of Equation 13.26, both of the two terms in the summation would be small because $\lambda_1 >> \lambda_2 \approx \lambda_3$. On the other hand, each of the other two eigenvectors would be subject to large sampling variations: the term in Equation 13.26 corresponding to one or the other of them will be large, because $(\lambda_2 - \lambda_3)^{-2}$ will be large. The pattern of sampling error for e_2 will resemble the true generating-process e_3, and vice versa. That is, the orientation of the two sample eigenvectors in this plane will be arbitrary, beyond the constraints that they will be perpendicular to each other, and to e_1. The variations represented by each of these two sample eigenvectors will accordingly be an arbitrary mixture of the variations represented by their two population counterparts.

13.4.3. The North et al. Rule of Thumb

Equations 13.20 and 13.25, for the sampling distributions of the eigenvalues and eigenvectors, depend on the values of their true but unknown counterparts. Nevertheless, the sample estimates approximate the true values, so that large sampling errors are expected for those eigenvectors whose sample eigenvalues are close to other sample eigenvalues. The idea that it is possible to diagnose instances where sampling variations are expected to cause problems with eigenvector interpretation in PCA was expressed as a rule of thumb by North et al. (1982):

> "The rule is simply that if the sampling error of a particular eigenvalue λ [$\delta\lambda \sim \lambda(2/n)^{1/2}$] is comparable to or larger than the spacing between λ and a neighboring eigenvalue, then the sampling errors for the EOF associated with λ will be comparable to the size of the neighboring EOF. The interpretation is that if a group of true eigenvalues lie within one or two $\delta\lambda$ of each other, then they form an 'effectively degenerate multiplet,' and sample eigenvectors are a random mixture of the true eigenvectors."

However, caution is warranted in quantitatively interpreting the degree of overlap of the confidence intervals implied by the North et al. rule of thumb (see Section 5.2.2).

North et al. (1982) illustrated their rule of thumb with an instructive example. They constructed synthetic data from a set of known EOF patterns, the first four of which are shown in Figure 13.12a, together with their respective eigenvalues. Using a full set of such patterns, the covariance matrix $[\Sigma]$ from which they could be extracted was assembled using the spectral decomposition (Equation 11.53). Using $[\Sigma]^{1/2}$ (see Section 11.3.4), realizations of data vectors x from a distribution with covariance $[\Sigma]$ were generated as in Section 12.4. Figure 13.12b shows the first four eigenvalue–eigenvector pairs calculated from a sample of $n = 300$ such synthetic data vectors, and Figure 13.12c shows a realization of the leading eigenvalue–eigenvector pairs for $n = 1000$.

The leading four true eigenvector patterns in Figure 13.12a are visually distinct, but their eigenvalues are relatively close. Using Equation 13.21b and $n = 300$, 95% sampling intervals for the four eigenvalues are 14.02 ± 2.24, 12.61 ± 2.02, 10.67 ± 1.71, and 10.43 ± 1.67 (because $\Phi^{-1}(0.975) = 1.96$), all of which include the adjacent eigenvalues. Therefore it is expected according to the rule of thumb that the sample eigenvectors will be random mixtures of their population counterparts for this sample size, and Figure 13.12b bears out this expectation: the patterns in those four panels appear to be random mixtures of the four panels in Figure 13.12a. Even if the true eigenvectors were unknown, this conclusion would be expected from the North et al. rule of thumb, because adjacent sample eigenvectors in Figure 13.12b are within two estimated standard errors, or $2\,\delta\hat{\lambda} = 2\hat{\lambda}(2/n)^{1/2}$ of each other.

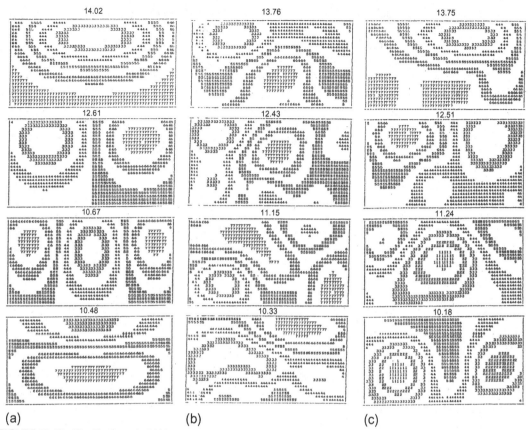

FIGURE 13.12 The North et al. (1982) example for effective degeneracy. (a) First four eigenvectors for the population from which synthetic data were drawn, with corresponding eigenvalues. (b) The first four eigenvectors calculated from a sample of $n = 300$, and the corresponding sample eigenvalues. (c) The first four eigenvectors calculated from a sample of $n = 1000$, and the corresponding sample eigenvalues. *From North et al. (1982). © American Meteorological Society. Used with permission.*

The situation is somewhat different for the larger sample size (Figure 13.12c). Again using Equation 13.21b but with $n = 1000$, the 95% sampling intervals for the four generating-process eigenvalues are 14.02 ± 1.22, 12.61 ± 1.10, 10.67 ± 0.93, and 10.43 ± 0.91. These intervals indicate that the first two sample EOFs should be reasonably distinct from each other and from the other EOFs, but that the third and fourth eigenvectors will probably still be entangled. Applying the rule of thumb to the sample eigenvalues in Figure 13.12c indicates that the separation between all adjacent pairs is close to $2\,\delta\hat{\lambda}$. The additional sampling precision provided by the larger sample size allows an approximation to the first two true EOF patterns to emerge, although an even larger sample still would be required before the sample eigenvectors would correspond well to their population counterparts.

The synthetic data realizations x in this artificial example were chosen independently of each other. If the data being analyzed are serially correlated, the unadjusted rule of thumb will imply better eigenvalue separation than is actually the case, because the variance of the sampling distribution of the sample eigenvalues will be larger than $2\,\lambda_k^2/n$ (as given in Equation 13.21). The cause of this discrepancy is that the sample eigenvalues are less consistent from batch to batch when calculated from autocorrelated data, so the qualitative effect is the same as was described for the sampling distribution of sample means, in Section 5.2.4. However, the effective sample size adjustment in Equation 13.15 would be appropriate in this case, which implies a much less extreme effect on the effective sample size than does Equation 5.12. Here r_1 would correspond to the lag-1 autocorrelation for the corresponding principal component time series when using Equation 13.21 or 13.24; and to the geometric mean of the autocorrelation coefficients for the two corresponding principal component series, when using Equation 13.26.

13.4.4. Bootstrap Approximations to the Sampling Distributions

The conditions specified in Section 13.4.1, of large sample size and/or underlying multivariate normal data, may be too unrealistic to be practical in some situations. In such cases it is possible to build good approximations to the sampling distributions of sample statistics using the bootstrap (see Section 5.3.5). Beran and Srivastava (1985) and Efron and Tibshirani (1993) specifically describe bootstrapping sample covariance matrices to produce sampling distributions for their eigenvalues and eigenvectors. The basic procedure is to repeatedly resample the underlying data vectors x with replacement; to produce some large number, n_B, of bootstrap samples, each of size n. Each of the n_B bootstrap samples yields a bootstrap realization of [S], whose eigenvalues and eigenvectors can be computed. Jointly these bootstrap realizations of eigenvalues and eigenvectors form reasonable approximations to the respective sampling distributions, which will reflect properties of the underlying data that may not conform to those assumed in Section 13.4.1.

Be careful in interpreting these bootstrap distributions. A (correctable) difficulty arises from the fact that the eigenvectors are determined up to sign only, so that in some bootstrap samples the counterpart of e_k may very well be $-e_k$. Failure to rectify such arbitrary sign switches will lead to large and unwarranted inflation of the computed sampling distributions for the eigenvector elements. Difficulties can also arise when resampling effective multiplets, because the random distribution of variance within a multiplet may be different from resample to resample, so the resampled eigenvectors may not bear one-to-one correspondences with their original sample counterparts. Finally, the bootstrap procedure destroys any serial correlation that may be present in the underlying data, which would lead to unrealistically narrow bootstrap sampling distributions. The moving-blocks bootstrap can be used for serially correlated data vectors (Wilks, 1997b) as well as scalars. Wang et al. (2014) provide an example using monthly surface pressure data.

13.5. ROTATION OF THE EIGENVECTORS

13.5.1. Why Rotate the Eigenvectors?

There is a strong tendency to try to ascribe physical interpretations to PCA eigenvectors and the corresponding principal components. The results shown in Figures 13.4 and 13.6 indicate that it can be both appropriate and informative to do so. However, the orthogonality constraint on the eigenvectors (Equation 11.48) can lead to problems with these interpretations, especially for the second and subsequent principal components. Although the orientation of the first eigenvector is determined solely by the direction of the maximum variation in the data, subsequent vectors must be orthogonal to each higher-variance eigenvector, regardless of the nature of the physical processes that may have given rise to the data. To the extent that those underlying physical processes are not independent, interpretation of the corresponding principal components as being independent modes of variability will not be justified (North, 1984). The first principal component may represent an important mode of variability or physical process, but it may well also include aspects of other correlated modes or processes. Thus the orthogonality constraint on the eigenvectors can result in the influences of several distinct physical processes being jumbled together in a single principal component.

When physical interpretation rather than data compression is a primary goal of PCA, it is often desirable to rotate a subset of the initial eigenvectors to a second set of new coordinate vectors. Usually it is some number M of the leading eigenvectors (i.e., eigenvectors with largest corresponding eigenvalues) of the original PCA that are rotated, with M chosen using a truncation criterion such as those discussed in Section 13.3. Rotated eigenvectors can be less prone to the artificial features resulting from the orthogonality constraint on the unrotated eigenvectors, such as Buell patterns (Richman, 1986). They also appear to exhibit better sampling properties (Richman, 1986; Cheng et al., 1995) than their unrotated counterparts. A large fraction of the review of PCA by Hannachi et al. (2007) is devoted to rotation.

Several procedures for rotating the original eigenvectors exist, but all seek to produce what is known as *simple structure* in the resulting analysis. Simple structure generally is understood to have been achieved if a large fraction of the elements of the resulting rotated vectors are near zero, and few of the remaining elements correspond to elements that are not near zero in the other rotated vectors. The desired result is that each rotated vector represents mainly the few original variables corresponding to the elements not near zero, and that the representation of the original variables is split between as few of the rotated principal components as possible. Simple structure aids interpretation of a rotated PCA to the extent that it allows association of each rotated eigenvector with a small number of the original K variables whose corresponding eigenvector elements are not near zero.

Following rotation of the eigenvectors, a second set of new variables is defined, called *rotated principal components*. The rotated principal components are obtained from the original data analogously to Equation 13.1 and 13.2, as the dot products of data vectors and the rotated eigenvectors. They can be interpreted as single-number summaries of the similarity between their corresponding rotated eigenvector and a data vector x. Depending on the method used to rotate the eigenvectors, the resulting rotated principal components may or may not be mutually uncorrelated.

A price is paid for the improved interpretability and better sampling stability of the rotated eigenvectors. One cost is that the dominant-variance property of PCA is lost. The first rotated principal component is no longer that linear combination of the original data with the largest variance. The variance represented by the original unrotated eigenvectors is spread more uniformly among the rotated eigenvectors, so that the corresponding eigenvalue spectrum is flatter. Also lost is either the orthogonality of the eigenvectors, or the uncorrelatedness of the resulting principal components, or both.

13.5.2. Rotation Mechanics

Rotated eigenvectors are produced as a linear transformation of a subset of M of the original K eigenvectors,

$$\left[\tilde{E}\right]_{(K \times M)} = [E]_{(K \times M)} \ [T]_{(M \times M)} \ , \tag{13.27}$$

where $[T]$ is the rotation matrix, and the matrix of rotated eigenvectors is denoted by the tilde. If $[T]$ is orthogonal, that is, if $[T][T]^T = [I]$, then the transformation Equation 13.27 is called an *orthogonal rotation*. Otherwise the rotation is called *oblique*.

Richman (1986) lists 19 approaches to defining the rotation matrix $[T]$ in order to achieve simple structure, although his list is not exhaustive. However, by far the most commonly used approach is the orthogonal rotation called the *varimax* (Kaiser, 1958). A varimax rotation is determined by choosing the elements of [T] to maximize

$$\sum_{m=1}^{M}\left[\sum_{k=1}^{K}e_{k,m}^{*}{}^{4}-\frac{1}{K}\left(\sum_{k=1}^{K}e_{k,m}^{*}{}^{2}\right)^{2}\right],\qquad(13.28a)$$

where

$$e_{k,m}^{*}=\frac{\widetilde{e}_{k,m}}{\left(\sum_{m=1}^{M}\widetilde{e}_{k,m}^{2}\right)^{1/2}},\qquad(13.28b)$$

are scaled versions of the rotated eigenvector elements. Together Equations 13.28a and 13.28b define the "normal varimax," whereas Equation 13.28a alone, using the unscaled eigenvector elements $\widetilde{e}_{k,m}$, is known as the "raw varimax." In either case the transformation is sought that maximizes the sum of the variances of the (either scaled or raw) squared rotated eigenvector elements, which tends to move them toward either their maximum or minimum (absolute) values (which are 0 and 1), and thus tends toward simple structure. The solution is iterative and is a standard feature of many statistical software packages.

The results of eigenvector rotation can depend on how many of the original eigenvectors are selected for rotation. That is, some or all of the leading rotated eigenvectors may be different if, say, $M+1$ rather than M eigenvectors are rotated (e.g., O'Lenic and Livezey, 1988). Unfortunately there is often not a clear answer to the question of what the best choice for M might be, and typically an essentially subjective choice is made. Some guidance is available from the various truncation criteria in Section 13.3, although these may not yield a unique answer. Sometimes a trial-and-error procedure is used, where M is increased slowly until the leading rotated eigenvectors are stable, that is, insensitive to further increases in M. In any case, however, it makes sense to include either all, or none, of the eigenvectors making up an effective multiplet, since jointly they carry information that has been arbitrarily mixed. Jolliffe (1987, 1989) suggests that it may be helpful to separately rotate groups of eigenvectors within effective multiplets in order to more easily interpret the information that they jointly represent.

Figure 13.13 compares unrotated and varimax-rotated PCAs for reconstructing spatial patterns that are independent (Figure 13.13a) and overlapping (Figure 13.13b). Both synthetic examples pertain to a 30×30-gridpoint square domain ($K = 900$), with $n = 256$. The leftmost columns in each panel show the three true generating-process eigenvectors, the nonzero features of which in Figure 13.13a are spatially disjoint. In this case both the unrotated (middle column) and rotated (rightmost column) recover the true patterns well. Because the underlying spatial patterns of variability in the leftmost column of Figure 13.13a already exhibit "simple structure," the unrotated and rotated solutions are equivalent because the rotation matrix [T] in Equation 13.27, implicitly resulting from Equation 13.28, very nearly equals the identity.

The features in the underlying "truth" eigenvectors in Figure 13.13b have overlapping spatial extents, and so cannot vary independently. The unrotated PCA in the middle column of Figure 13.13b is not able to separate the three. Here the first and third eigenvectors include portions of the underlying variability of the second mode, and the second eigenvector includes influences from the first and third underlying modes. In contrast, the varimax-rotated solution in the rightmost column of Figure 13.13b recovers the three true underlying modes well.

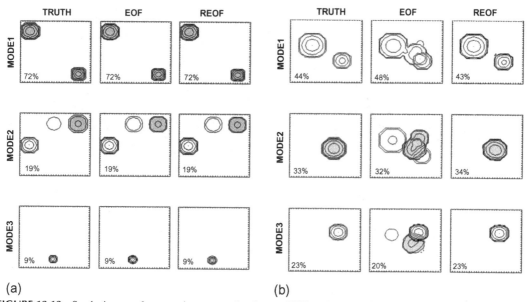

FIGURE 13.13   Synthetic example comparing unrotated and rotated PCAs when three underlying modes of variability are (a) spatially independent, and (b) spatially overlapping and thus nonindependent. The square spatial domain consists of 30 gridpoints in each direction, the contour interval is 0.25, and areas with negative eigenvector loadings are shaded. *Modified from Lian and Chen (2012). © American Meteorological Society. Used with permission.*

Figure 13.14 shows spatial displays of the first two rotated eigenvectors of monthly averaged hemispheric winter 500 mb heights. Using the truncation criterion of Equation 13.13 with $T = 1$, the first 19 eigenvectors of the correlation matrix for these data were rotated. The two patterns in Figure 13.14 are similar to the first two unrotated eigenvectors derived from the same data (see Figure 13.4a and 13.4b), although the signs have been (arbitrarily) reversed. However, the rotated vectors conform more to the idea of simple structure in that more of the hemispheric fields are fairly flat (near zero) in Figure 13.14, and each panel emphasizes more uniquely a particular feature of the variability of the 500 mb heights corresponding to the teleconnection patterns in Figure 3.33. The rotated vector in Figure 13.14a focuses primarily on height differences in the northwestern and western tropical Pacific, called the western Pacific teleconnection pattern. It thus represents variations in the 500 mb jet at these longitudes, with positive values of the corresponding rotated principal component indicating weaker than average westerlies, and negative values indicating the reverse. Similarly, the PNA pattern stands out exceptionally clearly in Figure 14.14b, where the rotation has separated it from the eastern hemisphere pattern evident in Figure 13.4b.

Figure 13.15 shows schematic representations of eigenvector rotation in two dimensions. The upper diagrams in each section represent the eigenvectors in the two-dimensional plane defined by the underlying variables x_1 and x_2, and the corresponding lower diagrams represent "maps" of the eigenvector elements plotted at the two "locations" x_1 and x_2 (these are meant to correspond to real-world maps such as those shown in Figures 13.4 and 13.14). Figure 13.15a illustrates the case of the original unrotated eigenvectors. The leading eigenvector e_1 is defined as the direction onto which a projection of the data

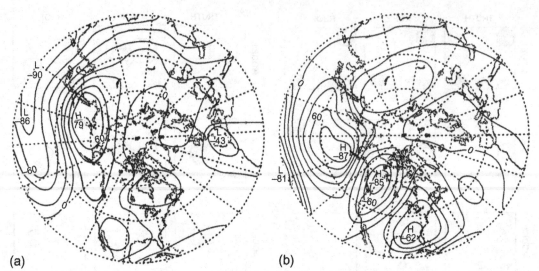

FIGURE 13.14   Spatial displays of the first two rotated eigenvectors of monthly averaged hemispheric winter 500 mb heights. The data are the same as those underlying Figure 13.4, but the rotation has better isolated the patterns of variability, allowing a clearer interpretation in terms of the teleconnection patterns in Figure 3.33. *From Horel (1981). © American Meteorological Society. Used with permission.*

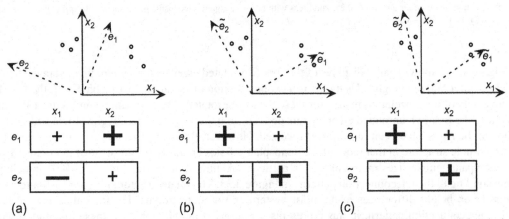

FIGURE 13.15   Schematic comparison of (a) unrotated, (b) orthogonally rotated, and (c) obliquely rotated unit-length eigenvectors in $K = 2$ dimensions. Left panels show eigenvectors in relation to scatterplots of the data, which exhibit two groups or modes. Right panels show schematic two-point maps of the two eigenvectors in each case. *After Karl and Koscielny (1982).*

points (i.e., the principal components) has the largest variance, which locates a compromise between the two clusters of points (modes). That is, it locates much of the variance of both groups, without really characterizing either. The leading eigenvector e_1 points in the positive direction for both x_1 and x_2, but is more strongly aligned toward x_2, so the corresponding e_1 map below shows a large positive "+" for x_2, and a smaller "+" for x_1. The second eigenvector is constrained to be orthogonal to the first, and so corresponds to large negative x_1, and mildly positive x_2, as indicated in the corresponding "map."

Figure 13.15b represents orthogonally rotated eigenvectors. Within the constraint of orthogonality they approximately locate the two point clusters, although the variance of the first rotated principal component is no longer maximum since the projections onto $\widetilde{e}_1$ of the three points with $x_1 < 0$ are quite small. However, the interpretation of the two features is enhanced in the maps of the two eigenvectors below, with $\widetilde{e}_1$ indicating large positive x_1 together with modest but positive x_2, whereas $\widetilde{e}_2$ shows large positive x_2 together with modestly negative x_1.

Finally, Figure 13.15c illustrates an oblique rotation, where the resulting rotated eigenvectors are no longer constrained to be orthogonal. Accordingly they have more flexibility in their orientations and can better accommodate features in the data that are not orthogonal.

13.5.3. Sensitivity of Orthogonal Rotation to Initial Eigenvector Scaling

An underappreciated aspect of orthogonal eigenvector rotation is that the orthogonality of the result depends strongly on the scaling of the original eigenvectors before rotation (Jolliffe, 1995, 2002; Mestas-Nuñez, 2000). This dependence is usually surprising because of the name "orthogonal rotation," which derives from the orthogonality of the transformation matrix $[T]$ in Equation 13.27, that is, $[T]^T[T] = [T][T]^T = [I]$. The confusion is compounded because of the incorrect assertion in a number of papers that an orthogonal rotation produces both orthogonal rotated eigenvectors and uncorrelated rotated principal components. At most one of these two results can be obtained by an orthogonal rotation, but neither will occur unless the eigenvectors are scaled correctly before the rotation matrix is applied. Because of the confusion about the issue, an explicit analysis of this counterintuitive phenomenon is worthwhile.

Denote as $[E]$ the possibly truncated $(K \times M)$ matrix of eigenvectors of $[S]$. Because these eigenvectors are orthogonal (Equation 11.50) and are originally scaled to unit length, the matrix $[E]$ is orthogonal, and so satisfies Equation 11.44b. The resulting principal components can be arranged in the matrix

$$\underset{(n \times M)}{[U]} = \underset{(n \times K)}{[X]} \underset{(K \times M)}{[E]} , \tag{13.29}$$

each of the n rows of which contain values for the M retained principal components, u_m^T. As before, $[X]$ is the original data matrix whose K columns correspond to the n observations on each of the original K variables. The uncorrelatedness of the unrotated principal components can be diagnosed by calculating their covariance matrix,

$$
\begin{aligned}
\underset{(M \times M)}{(n-1)^{-1}[U]^T[U]} &= (n-1)^{-1}([X][E])^T[X][E] \\
&= (n-1)^{-1}[E]^T[X]^T[X][E] \\
&= [E]^T \left([E][\Lambda][E]^T \right)[E] = [I][\Lambda][I] \\
&= [\Lambda].
\end{aligned} \tag{13.30}
$$

The u_m are uncorrelated because their covariance matrix $[\Lambda]$ is diagonal, and the variance for each u_m is λ_m. The steps on the third line of Equation 13.30 follow from the diagonalization of $[S] = (n-1)^{-1}[X]^T[X]$ (Equation 11.52a) and the orthogonality of the matrix $[E]$.

Consider now the effects of the three eigenvector scalings listed in Table 13.3 on the results of an orthogonal rotation. In the first case, the original eigenvectors are not rescaled from unit length, so the matrix of rotated eigenvectors is simply

$$\left[\widetilde{E}\right]_{(K \times M)} = \underset{(K \times M)}{[E]} \ \underset{(M \times M)}{[T]} .$$

(13.31)

That these rotated eigenvectors are still orthogonal, as expected, can be shown by

$$\left[\widetilde{E}\right]^{T}\left[\widetilde{E}\right] = ([E][T])^{T}[E][T] = [T]^{T}[E]^{T}[E][T]$$
$$= [T]^{T}[I][T] = [T]^{T}[T] = [I].$$

(13.32)

That is, the resulting rotated eigenvectors are still mutually perpendicular and of unit length. The corresponding rotated principal components are

$$\left[\widetilde{U}\right] = [X]\left[\widetilde{E}\right] = [X][E][T],$$

(13.33)

and their covariance matrix is

$$(n-1)^{-1}\left[\widetilde{U}\right]^{T}_{(M \times M)}\left[\widetilde{U}\right] = (n-1)^{-1}([X][E][T])^{T}[X][E][T]$$

$$= (n-1)^{-1}[T]^{T}[E]^{T}[X]^{T}[X][E][T]$$

$$= [T]^{T}[E]^{T}\left([E][\Lambda][E]^{T}\right)[E][T]$$

$$= [T]^{T}[\Lambda][T].$$

(13.34)

This matrix is not diagonal, reflecting the fact that the rotated principal components are no longer uncorrelated. This result is easy to appreciate geometrically, by looking at scatterplots such as Figure 13.1 or Figure 13.3. In each of these cases the point cloud is inclined relative to the original (x_1, x_2) axes, and the angle of inclination of the long axis of the cloud is located by the first eigenvector. The point cloud is not inclined in the (e_1, e_2) coordinate system defined by the two eigenvectors, reflecting the uncorrelatedness of the unrotated principal components (Equation 13.30). But relative to any other pair of mutually orthogonal axes in the plane, the points would exhibit some inclination, and therefore the projections of the data onto these axes would exhibit some nonzero correlation.

The second eigenvector scaling in Table 13.3, $\|\mathbf{e}_m\| = (\lambda_m)^{1/2}$, is commonly employed, and indeed is the default scaling in many statistical software packages for rotated principal components. In the notation of this section, employing this scaling is equivalent to rotating the scaled eigenvector matrix $[E][\Lambda]^{1/2}$, yielding the matrix of rotated eigenvectors

$$\left[\widetilde{E}\right] = \left([E][\Lambda]^{1/2}\right)[T].$$

(13.35)

The orthogonality of the rotated eigenvectors in this matrix can be checked with

$$\left[\widetilde{E}\right]^{T}\left[\widetilde{E}\right] = \left([E][\Lambda]^{1/2}[T]\right)^{T}[E][\Lambda]^{1/2}[T]$$

$$= [T]^{T}[\Lambda]^{1/2}[E]^{T}[E][\Lambda]^{1/2}[T]$$

$$= [T]^{T}[\Lambda]^{1/2}[I][\Lambda]^{1/2}[T] = [T]^{T}[\Lambda][T].$$

(13.36)

Here the equality on the second line is valid because the diagonal matrix $[\Lambda]^{1/2}$ is symmetric, so that $[\Lambda]^{1/2} = ([\Lambda]^{1/2})^{T}$. The rotated eigenvectors corresponding to the second, and frequently used, scaling in Table 13.3 are *not* orthogonal, because the result of Equation 13.36 is not a diagonal matrix. Neither are the corresponding rotated principal components independent. This can be seen by manipulating their covariance matrix, which is also not diagonal, that is,

$$
(n-1)^{-1} \underset{(M \times M)}{\left[\tilde{U}\right]^T \left[\tilde{U}\right]} = (n-1)^{-1} \left([X][E][\Lambda]^{1/2}[T]\right)^T [X][E][\Lambda]^{1/2}[T]
$$

$$
= (n-1)^{-1}[T]^T[\Lambda]^{1/2}[E]^T[X]^T[X][E][\Lambda]^{1/2}[T]
$$

$$
= [T]^T[\Lambda]^{1/2}[E]^T \left([E][\Lambda][E]^T\right)[E][\Lambda]^{1/2}[T] \tag{13.37}
$$

$$
= [T]^T[\Lambda]^{1/2}[I][\Lambda][I][\Lambda]^{1/2}[T]
$$

$$
= [T]^T[\Lambda]^{1/2}[\Lambda][\Lambda]^{1/2}[T]
$$

$$
= [T]^T[\Lambda]^2[T].
$$

The third eigenvector scaling in Table 13.3, $\|\mathbf{e}_m\| = (\lambda_m)^{-1/2}$, is used relatively rarely, although it can be convenient in that it yields unit variance for all the principal components u_m. The resulting rotated eigenvectors are not orthogonal, so that the matrix product

$$
\left[\tilde{E}\right]^T \left[\tilde{E}\right] = \left([E][\Lambda]^{-1/2}[T]\right)^T [E][\Lambda]^{-1/2}[T]
$$

$$
= [T]^T[\Lambda]^{-1/2}[E]^T[E][\Lambda]^{-1/2}[T] \tag{13.38}
$$

$$
= [T]^T[\Lambda]^{-1/2}[I][\Lambda]^{-1/2}[T] = [T]^T[\Lambda]^{-1}[T],
$$

is not diagonal. However, the resulting rotated principal components are uncorrelated since their covariance matrix,

$$
(n-1)^{-1} \underset{(M \times M)}{\left[\tilde{U}\right]^T \left[\tilde{U}\right]} = (n-1)^{-1} \left([X][E][\Lambda]^{-1/2}[T]\right)^T [X][E][\Lambda]^{-1/2}[T]
$$

$$
= (n-1)^{-1}[T]^T[\Lambda]^{-1/2}[E]^T[X]^T[X][E][\Lambda]^{-1/2}[T] \tag{13.39}
$$

$$
= [T]^T[\Lambda]^{-1/2}[E]^T \left([E][\Lambda][E]^T\right)[E][\Lambda]^{-1/2}[T]
$$

$$
= [T]^T[I][I][T] = [T]^T[T] = [I],
$$

is diagonal, and also reflects unit variances for all the rotated principal components.

Most frequently in meteorology and climatology, the eigenvectors in a PCA describe spatial patterns, and the principal components are time series reflecting the importance of the corresponding spatial patterns in the original data. When calculating orthogonally rotated principal components in this context, we can choose to have either orthogonal rotated spatial patterns but correlated rotated principal component time series (by using $\|\mathbf{e}_m\| = 1$), or nonorthogonal rotated spatial patterns whose time sequences are mutually uncorrelated (by using $\|\mathbf{e}_m\| = (\lambda_m)^{-1/2}$), but not both. It is not clear what the advantage of having neither property (using $\|\mathbf{e}_m\| = (\lambda_m)^{1/2}$, as is often done) might be. Differences in the results for the different scalings will be small if sets of effective multiplets are rotated separately, because their eigenvalues will necessarily be similar in magnitude, resulting in similar lengths for the scaled eigenvectors.

13.5.4. Simple Structure Through Regularization

A quite different approach to achieving simple structure, so that few of the eigenvector loadings in PCA are appreciably different from zero, is through Lasso regularization (Tibshirani, 1996). Use of the Lasso for regularization of least-squares regression, which imposes a budget on the sum of absolute values for the regression coefficients, was discussed in Section 7.5.2. The result is that, as the ceiling c on the sum of absolute values of the regression coefficients is decreased, more of them are progressively driven to zero.

A similar approach can be taken in PCA (Jolliffe et al., 2003), in which case the constraint can be expressed as

$$\sum_{k=1}^{K} |e_{k,m}| \leq c, c > 1, \tag{13.40}$$

for each eigenvector e_m. Jolliffe et al. (2003) named the approach Simplified Component Technique— LASSO, or SCoTLASS. For $c = 1$ the method yields exactly one nonzero loading in e_m, and so produces regularized eigenvectors that are exactly aligned with the original coordinate axes. Unconstrained ordinary PCA is produced when $c \geq \sqrt{K}$. For $1 < c \leq \sqrt{K}$ the resulting eigenvectors are orthogonal, and so still define a rigid rotation of the coordinate axes, but the data projections onto them (the regularized principal components) are not uncorrelated.

Of course the results will depend on the regularization parameter c, the choice of which is more ambiguous for PCA than in the regression setting because predictive cross-validation will in general not be available. Smaller values of the regularization parameter produce more loadings that are exactly zero, but simultaneously the fraction of variance represented decreases and the correlations among the regularized principal components increase. Jolliffe et al. (2003) suggest recomputation using multiple values of the regularization parameter, and then choosing c subjectively, perhaps on the basis of the problem-specific interpretability of the resulting regularized eigenvectors. Computation of regularized PCA is more difficult than its conventional counterpart and may involve multiple local minima in the numerical optimization. More details can be found in Hastie et al. (2015) and Jolliffe et al. (2003).

13.6. COMPUTATIONAL CONSIDERATIONS

13.6.1. Direct Extraction of Eigenvalues and Eigenvectors from [S]

The sample covariance matrix [S] is real and symmetric, and so will always have real-valued and nonnegative eigenvalues. Standard and stable algorithms are available to extract the eigenvalues and eigenvectors from real, symmetric matrices (e.g., Press et al., 1986), and this approach can be a very good one for computing a PCA. As noted earlier, it is sometimes preferable to calculate the PCA using the correlation matrix [R], which is also the covariance matrix for the standardized variables. The computational considerations presented in this section are equally appropriate to PCA based on the correlation matrix.

One practical difficulty that can arise is that the required computational time increases very quickly as the dimension of the covariance matrix increases. A typical application of PCA in meteorology or climatology involves a field observed at K grid- or other space-points, at a sequence of n times, where $K >> n$. The typical conceptualization is in terms of the $(K \times K)$ covariance matrix, which is very large— it is not unusual for K to include thousands of gridpoints. Hours of computation may be required to extract this many eigenvalue–eigenvector pairs. Yet since $K > n-1$ the sample covariance matrix is singular, implying that the last $K-n+1$ of its eigenvalues are exactly zero. It is pointless to calculate numerical approximations to these zero eigenvalues and their associated arbitrary eigenvectors.

In this situation fortunately it is possible to focus the computational effort on the n nonzero eigenvalues and their associated eigenvectors, using a computational trick (Von Storch and Hannoschöck, 1984). Recall that the $(K \times K)$ covariance matrix $[S]$ can be computed from the centered data matrix $[X']$ using Equation 11.30. Reversing the roles of the time and space points (although it is still the columns of the anomaly matrix $[X']$ that have zero mean), we also can compute the $(n \times n)$ covariance matrix

$$\underset{(n \times n)}{[S^*]} = \frac{1}{n-1} \underset{(n \times K)}{[X']} \underset{(K \times n)}{[X']}^{T} . \tag{13.41}$$

Both $[S]$ and $[S^*]$ have the same $\min(n-1, K)$ nonzero eigenvalues, $\lambda_k = \lambda^*_k$, so the required computational time may be much shorter if they are extracted from the smaller $(n \times n)$ matrix $[S^*]$, and this latter computation will be much faster in the usual situation where $K \gg n$.

The eigenvectors of $[S]$ and $[S^*]$ are different, but the leading $n-1$ (i.e., the meaningful) eigenvectors of $[S]$ can be computed from the eigenvectors e_k^* of $[S^*]$ using

$$e_k = \frac{[X']^T e_k^*}{\left\| [X']^T e_k^* \right\|} , k = 1, \dots, n-1 . \tag{13.42}$$

The dimensions of the multiplications in both numerator and denominator are $(K \times n)(n \times 1) = (K \times 1)$, and the role of the denominator is to ensure that the resulting e_k have unit length.

13.6.2. PCA via SVD

The eigenvalues and eigenvectors in a PCA can also be computed using the SVD (singular value decomposition) algorithm (Section 11.3.5), in two ways. First, as illustrated in Example 11.5, the eigenvalues and eigenvectors of a covariance matrix $[S]$ can be computed through SVD of the matrix $(n-1)^{-1/2}[X']$, where the centered $(n \times K)$ data matrix $[X']$ is related to the covariance matrix $[S]$ through Equation 11.30. In this case, the eigenvalues of $[S]$ are the squares of the singular values of $(n-1)^{-1/2}[X']$—that is, $\lambda_k = \omega_k^2$—and the eigenvectors of $[S]$ are the same as the right singular vectors of $(n-1)^{-1/2}[X']$—that is, $[E] - [R]$, or $e_k - r_k$.

An advantage of using SVD to compute a PCA in this way is that the left singular vectors (the columns of the $(n \times K)$ matrix $[L]$ in Equation 11.72) are proportional to the principal components (i.e., to the projections of the centered data vectors x'_i onto the eigenvectors e_k). In particular,

$$u_{i,k} = e_k^T x'_i = \sqrt{n-1} \, \ell_{i,k} \sqrt{\lambda_k}, \quad i = 1, \dots, n, \ k = 1, \dots, K; \tag{13.43a}$$

or

$$\underset{(n \times K)}{[U]} = \sqrt{n-1} \, \underset{(n \times K)}{[L]} \, \underset{(K \times K)}{[\Lambda]}^{1/2} . \tag{13.43b}$$

Here the matrix $[U]$ is used in the same sense as in Section 13.5.3, that is, each of its K columns contains the principal component series u_k corresponding to the sequence of n data values x_i, $i = 1, \dots, n$.

The SVD algorithm can also be used to compute a PCA by operating on the covariance matrix directly. Comparing the spectral decomposition of a square, symmetric matrix (Equation 11.52a) with its SVD (Equation 11.72), it is clear that these unique decompositions are the same. In particular, since a covariance matrix $[S]$ is square and symmetric, both the left and right matrices of its SVD are equal, and contain the eigenvectors, that is, $[E] = [L] = [R]$. In addition, the diagonal matrix of singular values is exactly the diagonal matrix of eigenvalues, $[\Lambda] = [\Omega]$.

Computation of PCA using the SVD algorithm is comparatively fast. However, be aware that particular software implementations of the SVD may not sort the eigenvalues in descending order, although each eigenvector will still be associated with the correct eigenvalue.

13.6.3. The Power Method

In some applications (e.g., Wilks, 2016c) only the leading eigenvalue and eigenvector are needed, so that computation of a full PCA is unnecessarily slow and wasteful of computing resources. In such cases it can be advantageous to find the leading eigenvalue–eigenvector pair using the *power method* (e.g., Golub and van Loan, 1996).

Beginning with an arbitrary initial guess for the leading eigenvector, e_1, with $\|e_1\| = 1$, the power method algorithm proceeds by iterating

$$v = [S]\, e \tag{13.44a}$$

$$\lambda_1 = \|v\| \tag{13.44b}$$

$$e_1 = v/\lambda_1 \tag{13.44c}$$

until convergence. Here v is an intermediate storage vector, and $\|v\|$ denotes its Euclidean length.

13.6.4. PCA and Missing Data

Computation of a sample covariance or correlation matrix using Equation 11.30 as input to a PCA, or alternatively computation of a PCA through SVD on the data matrix as in Example 11.5, both require that the input data anomaly matrix $[X']$ contains no missing values. Of course there are often missing value in real data sets, requiring that some accommodation be made before computation of a PCA. On the other hand, once the PCA has been computed, it can be used to estimate the missing values.

If there are very few missing values in a data set it may be reasonable to simply delete any incomplete data vectors before proceeding, but usually this approach will lead to an excessive portion of the data being lost. Alternatively, there are two approaches to dealing with the missing data before computation of a PCA. The first is to substitute the appropriate sample mean for any missing data, yielding a zero anomaly for the corresponding missing value $x'_{i,k}$, which is equally applicable when computing a covariance or correlation matrix, and when computing the SVD of a data matrix. The second approach is to estimate the elements of the covariance or correlation matrix using the numerator of Equation 3.28, including only terms for which both elements are nonmissing.

The two methods will differ only with respect to the divisor in the covariance calculation in the numerator of Equation 3.28, because the first approach will yield zero contribution to the sum when either of the two anomalies are zero, even though the divisor is $n-1$. Accordingly imputing zero anomaly for any missing values leads to negative bias in the absolute values of the resulting variance and covariance estimates. On the other hand, using the second approach yields different sample sizes for the estimated covariances or correlations, inconsistencies among which may result in a singular matrix. In that case some of the trailing eigenvalues may be negative which, as a consequence of Equation 11.54, will lead to an upward bias in the leading estimated eigenvalues.

Once a PCA involving missing data values has been computed, its eigenvectors can be used to estimate those missing data. First, the principal components in the matrix $[U]$ (Equation 13.29) are estimated using

$$u_{i,m} = \frac{\displaystyle\sum_{\substack{k=1 \\ x_{i,k} \text{ valid}}}^{K} x'_{i,k} e_{k,m}}{\displaystyle\sum_{\substack{k=1 \\ x_{i,k} \text{ valid}}}^{K} e^2_{k,m}}, \tag{13.45}$$

where the summations include only terms for which data values $x_{i,k}$ are nonmissing. When a data vector $\boldsymbol{x}_i$ contains no missing elements, the denominator of Equation 13.45 is 1 (Equation 11.49), in which case Equation 13.45 is equivalent to Equation 13.1. The missing anomaly values can then be estimated through application of a synthesis,

$$x'_{i,k} = \sum_{m=1}^{M} u_{i,m} e_{m,k}. \tag{13.46}$$

Equation 13.45 is the result of a least-squares approach that minimizes the error $([X]-[U][E^{\mathrm{T}}])^{\mathrm{T}}([X]-[U][E^{\mathrm{T}}])$, which is derived from Equation 13.29. It yields reasonable results when the proportion of missing data is not excessive. For larger (than perhaps 10% to 20%) fractions of missing data, more elaborate iterative approaches have been found to be more accurate (Taylor et al., 2013).

13.7. SOME ADDITIONAL USES OF PCA

13.7.1. Singular Spectrum Analysis (SSA): Time-Series PCA

Principal component analysis can also be applied to scalar or multivariate time series. This approach to time-series analysis is known both as *singular spectrum analysis* and *singular systems analysis* (SSA, in either case). Fuller developments of SSA than is presented here can be found in Elsner and Tsonis (1996), Ghil et al. (2002), and Vautard et al. (1992).

SSA is easiest to understand in terms of a scalar time series x_t, $t = 1, \ldots, n$; although the generalization to multivariate time series of a vector $\boldsymbol{x}_t$ is reasonably straightforward. As a variant of PCA, SSA involves extraction of eigenvalues and eigenvectors from a covariance matrix. This covariance matrix is calculated from a scalar time series by passing a *delay window*, or imposing an *embedding dimension*, of length K on the time series. The process is illustrated in Figure 13.16. For $K = 3$, the first K-dimensional data vector, $\boldsymbol{x}_{(1)}$ is composed of the first three members of the scalar time series, $\boldsymbol{x}_{(2)}$ is composed of the second three members of the scalar time series, and so on, yielding a total of $n - K + 1$ overlapping lagged data vectors.

If the time series x_t is covariance stationary, that is, if its mean, variance, and lagged correlations do not change through time, the $(K \times K)$ population covariance matrix of the lagged time-series vectors $\boldsymbol{x}_{()}$ takes on a special banded structure known as *Toeplitz*, in which the elements $\sigma_{i,j} = \gamma_{|i-j|} = \mathrm{E}[x'_t \, x'_{t+|i-j|}]$ are arranged in diagonal parallel bands. That is, the elements of the resulting covariance matrix are taken from the autocovariance function (Equation 3.39), with lags arranged in increasing order away from the main diagonal. All the elements of the main diagonal are $\sigma_{i,i} = \gamma_0$, that is, the variance. The elements on

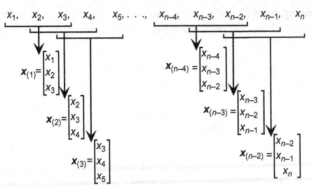

FIGURE 13.16 Illustration of the construction of the vector time series $x_{(t)}$, $t = 1, \ldots, n{-}M{+}1$, by passing a delay window of embedding dimension $M = 3$ over consecutive members of the scalar time series x_t.

the diagonal bands adjacent to the main diagonal are all equal to γ_1, reflecting the fact that, for example, the covariance between the first and second elements of the vectors $x_{(t)}$ in Figure 13.16 is the same as the covariance between the second and third elements. The elements separated from the main diagonal by one position are all equal to γ_2, and so on. Because of edge effects at the beginnings and ends of sample time series, the sample covariance matrix may be only approximately Toeplitz, although the diagonally banded Toeplitz structure is sometimes enforced before calculation of the SSA (Allen and Smith, 1996; Elsner and Tsonis, 1996; Groth and Ghil, 2015).

Since SSA is a PCA, the same mathematical considerations apply. In particular, the principal components are linear combinations of the data according to the eigenvectors (Equations 13.1 and 13.2). The analysis operation can be reversed to synthesize, or approximate, the data from all (Equation 13.20) or some (Equation 13.21) of the principal components. What makes SSA different follows from the different nature of the underlying data, and the implications of that different nature on interpretation of the eigenvectors and principal components. In particular, the data vectors are fragments of time series rather than the more usual spatial distribution of values at a single time, so that the eigenvectors in SSA represent characteristic time patterns exhibited by the data, rather than characteristic spatial patterns. Accordingly, the eigenvectors in SSA are sometimes called T-EOFs. Since the overlapping time series fragments x_t themselves occur in a time sequence, the principal components also have a time ordering, as in Equation 13.11. These temporal principal components u_k, or T-PCs, index the degree to which the corresponding time-series fragment x_t resembles the corresponding T-EOF, e_k. Because the data are consecutive fragments of the original time series, the principal components are linear combinations of these time-series segments, with the weights given by the T-EOF elements. The T-PCs are mutually uncorrelated, but in general an individual T-PC will exhibit temporal autocorrelations.

The analogy between SSA and Fourier analysis of time series is especially strong, with the T-EOFs corresponding to the sine and cosine functions, and the T-PCs corresponding to the amplitudes. However, there are two major differences. First, the orthogonal basis functions in a Fourier decomposition are the fixed harmonic functions, whereas the basis functions in SSA are the data-adaptive T-EOFs. Therefore an SSA may be more efficient than a Fourier analysis, in the sense of requiring fewer basis functions to represent a given fraction of the variance of a time series. Similarly, the Fourier amplitudes are time-independent constants, but their counterparts, the T-PCs, are themselves functions of time. Therefore similarly to wavelet analysis (Section 10.6), SSA can represent time variations that may be localized in time, and so not necessarily recurring throughout the time series.

Also in common with Fourier analysis, SSA can detect and represent oscillatory or quasi-oscillatory features in the underlying time series. A periodic or quasi-periodic feature in a time series is represented in SSA by pairs of T-PCs and their corresponding eigenvectors. These pairs have eigenvalues that are equal or nearly equal. The characteristic time patterns represented by these pairs of eigenvectors have the same (or very similar) shape, but are offset in time by a quarter cycle (as are a pair of sine and cosine functions). Unlike the sine and cosine functions these pairs of T-EOFs take on shapes that are determined by the time patterns in the underlying data. A common motivation for using SSA is to search, on an exploratory basis, for possible periodicities in time series, which periodicities may be intermittent and/or nonsinusoidal in form. Features of this kind are indeed identified by a SSA, but false periodicities arising only from sampling variations may also easily occur in the analysis (Allen and Robertson, 1996; Allen and Smith, 1996).

An important consideration in SSA is choice of the window length or embedding dimension, K. Obviously the analysis cannot represent variations longer than this length, although choosing too large a value results in a small sample size, $n-K+1$, from which to estimate the covariance matrix. Also, the computational effort increases quickly as K increases. Usual rules of thumb are that an adequate sample size may be achieved for $K < n/3$, and that the analysis will be successful in representing time variations with periods between $K/5$ and K.

Example 13.5 SSA for an AR(2) Series

Figure 13.17 shows an $n = 100$-point realization from the AR(2) process (Equation 10.27) with parameters $\phi_1 = 0.9$, $\phi_2 = -0.6$, $\mu = 0$, and $\sigma_\varepsilon = 1$. This is a purely random series, but the parameters ϕ_1 and ϕ_2 have been chosen in a way that allows the process to exhibit pseudoperiodicities. That is, there is a tendency for the series to oscillate, although the oscillations are irregular with respect to their frequency and phase. The spectral density function for this AR(2) process, included in Figure 10.21, shows a maximum centered near $f = 0.15$, corresponding to a typical period near $\tau = 1/f \approx 6.7$ time steps.

Analyzing the series using SSA requires choosing a delay window length, K, that should be long enough to capture the feature of interest yet short enough for reasonably stable covariance estimates to be calculated. Combining the rules of thumb for the window length, $K/5 < \tau < K < n/3$, a plausible choice is $K = 10$. This choice yields $n-K+1 = 91$ overlapping time series fragments $x_{(t)}$ of length $K = 10$.

Calculating the covariances for this sample of 91 data vectors $x_{(t)}$ in the conventional way yields the (10×10) matrix

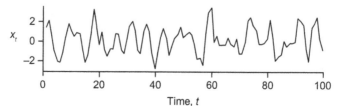

FIGURE 13.17   An $n = 100$-point realization from an AR(2) process with $\phi_1 = 0.9$ and $\phi_2 = -0.6$.

$$[S] = \begin{bmatrix} 1.792 \\ .955 & 1.813 \\ -.184 & .958 & 1.795 \\ -.819 & -.207 & .935 & 1.800 \\ -.716 & -.851 & -.222 & .959 & 1.843 \\ -.149 & -.657 & -.780 & -.222 & .903 & 1.805 \\ .079 & -.079 & -.575 & -.783 & -.291 & .867 & 1.773 \\ .008 & .146 & -.011 & -.588 & -.854 & -.293 & .873 & 1.809 \\ -.199 & .010 & .146 & -.013 & -.590 & -.850 & -.289 & .877 & 1.809 \\ -.149 & -.245 & -.044 & .148 & .033 & -.566 & -.828 & -.292 & .874 & 1.794 \end{bmatrix}. \tag{13.47}$$

For clarity, only the elements in the lower triangle of this symmetric matrix have been printed. Because of edge effects in the finite sample, this covariance matrix is approximately, but not exactly, Toeplitz. The 10 elements on the main diagonal are only approximately equal, and each is estimating the true lag-0 autocovariance $\gamma_0 = \sigma^2_x \approx 2.29$. Similarly, the nine elements on the second diagonal are approximately equal, with each estimating the lag-1 autocovariance $\gamma_1 \approx 1.29$, the eight elements on the third diagonal estimate the lag-2 autocovariance $\gamma_2 \approx -0.21$, and so on. The pseudoperiodicity in the data is reflected in the large negative autocovariance at three lags, and the subsequent damped oscillation in the autocovariance function, which can be seen easily by reading down the first column, or reading the bottom row from right to left.

Figure 13.18 shows the leading four eigenvectors of the covariance matrix in Equation 13.47 and their associated eigenvalues. The first two of these eigenvectors (Figure 13.18a), which are associated with nearly equal eigenvalues, are very similar in shape and are separated by approximately a quarter of the period τ corresponding to the middle of the spectral peak in Figure 10.21. Jointly they represent the dominant feature of the data series in Figure 13.17, namely the pseudoperiodic behavior, with successive peaks and crests tending to be separated by six or seven time units.

The third and fourth T-EOFs in Figure 13.18b represent other, nonperiodic aspects of the time series in Figure 12.17. Unlike the leading T-EOFs in Figure 13.18a, they are not offset images of each other and

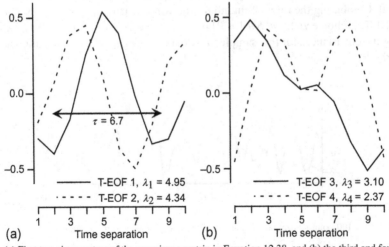

(a) Time separation

(b) Time separation

T-EOF 1, $\lambda_1 = 4.95$
T-EOF 2, $\lambda_2 = 4.34$

T-EOF 3, $\lambda_3 = 3.10$
T-EOF 4, $\lambda_4 = 2.37$

$\tau = 6.7$

FIGURE 13.18 (a) First two eigenvectors of the covariance matrix in Equation 12.38, and (b) the third and fourth eigenvectors.

do not have nearly equal eigenvalues. Jointly the four patterns in Figure 13.18 represent 83.5% of the variance within the 10-element time series fragments (but not including variance associated with longer timescales).

Ghil et al. (2002) present a similar extended example of SSA, using a time series of the Southern Oscillation Index (Figure 3.16).   ◇

It is conceptually straightforward to extend SSA to simultaneous analysis of multiple (i.e., vector) time series, which is called *multichannel SSA*, or MSSA (Ghil et al., 2002; Plaut and Vautard, 1994). The relationship between SSA and MSSA parallels that between an ordinary PCA for a single field and simultaneous PCA for multiple fields as described in Section 13.2.2. The multiple channels in a MSSA might be the L gridpoints representing a spatial field at time t, in which case the time series fragments corresponding to the delay window length K would be coded into a $(LK \times 1)$ vector $x_{(t)}$, yielding a $(LK \times LK)$ covariance matrix from which to extract space-time eigenvalues and eigenvectors (ST-EOFs). The dimension of such a matrix may become unmanageable, and one solution (Plaut and Vautard, 1994) can be to first calculate an ordinary PCA for the spatial fields, and then subject the first few principal components to the MSSA. In this case each channel corresponds to one of the spatial principal components calculated in the initial data compression step. Vautard et al. (1996, 1999) describe MSSA-based forecasts of fields constructed by forecasting the space-time principal components, and then reconstituting the forecast fields through a truncated synthesis.

13.7.2. Principal-Component Regression

A pathology that may occur in multiple linear regression (see Section 7.3.1) is that a set of predictor variables having strong mutual correlations can result in the calculation of an unstable regression relationship, in the sense that the sampling distributions of the estimated regression parameters may have very high variances. The problem can be appreciated in the context of Equation 11.40, for the covariance matrix of the joint sampling distribution of the estimated regression parameters. This equation depends on the inverse of the matrix $[X]^T [X]$, which is proportional to the covariance matrix $[S_x]$ of the predictors. Very strong intercorrelations among the predictors leads to their covariance matrix (and thus also $[X]^T [X]$) being nearly singular, or small in the sense that its determinant is near zero. The inverse, $([X]^T [X])^{-1}$ is then large, and inflates the covariance matrix $[S_b]$ in Equation 11.40. The result is that any specific set of estimated regression parameters may be very far from their correct values as a consequence of sampling variations, leading the fitted regression equation to perform poorly on independent data. The prediction intervals (based upon Equation 11.42) are also inflated.

An approach to remedying this problem is to first transform the predictors to their principal components, the correlations among which are zero. The resulting *principal-component regression* is convenient to work with, because the uncorrelated predictors can be added to or taken out of a tentative regression equation at will without affecting the contributions and parameter estimates of the other principal-component predictors. If all the principal components are retained in a principal-component regression, then nothing is gained over the conventional least-squares fit to the full predictor set. However, Jolliffe (2002) shows that multicollinearities, if present, are associated with the principal components having the smallest eigenvalues. As a consequence, the effects of the multicollinearities, and in particular the inflated covariance matrix for the estimated parameters, can in principle be removed by truncating the trailing principal components associated with the very small eigenvalues.

There are problems that may be associated with principal-component regression. Unless the principal components that are retained as predictors are interpretable in the context of the problem being analyzed, the insight to be gained from the regression may be limited. It is possible to reexpress the principal-component regression in terms of the original predictors using the synthesis equation (Equation 13.6), but the result will in general involve all the original predictor variables even if only one or a few principal component predictors have been used. This reconstituted regression will be biased, although often the variance is much smaller than for the least-squares alternative, resulting in a smaller MSE overall.

13.7.3. The Biplot

It was noted in Section 3.6 that graphical EDA for high-dimensional data is especially difficult. Since principal component analysis excels at data compression using the minimum number of dimensions, it is natural to think about applying PCA to EDA. The *biplot*, originated by Gabriel (1971), is such a tool. The "bi-" in biplot refers to the simultaneous representation of the n rows (the observations) and the K columns (the variables) of a data matrix $[X]$.

The biplot is a two-dimensional graph, whose axes are the first two eigenvectors of $[S_x]$. The biplot represents the n observations as their projections onto the plane defined by these two eigenvectors, that is, as the scatterplot of the first two principal components. To the extent that $(\lambda_1 + \lambda_2)/\Sigma_k \lambda_k \approx 1$, this scatterplot will be a close approximation to their higher-dimensional relationships, in a graphable two-dimensional space. Exploratory inspection of the data plotted in this way may reveal such aspects of the data as the points clustering into natural groups, or time sequences of points that are organized into coherent trajectories in the plane of the plot.

The other element of the biplot is the simultaneous representation of the K variables. Each of the coordinate axes of the K-dimensional data space defined by the variables can be thought of as a unit basis vector indicating the direction of the corresponding variable, that is, $b_1^T = [1, 0, 0, ..., 0], b_2^T = [0, 1, 0, ..., 0], ..., b_K^T = [0, 0, 0, ..., 1]$. These basis vectors can also be projected onto the two leading eigenvectors defining the plane of the biplot, that is,

$$e_1^T b_k = \sum_{k=1}^{K} e_{1,k} b_k \tag{13.48a}$$

and

$$e_2^T b_k = \sum_{k=1}^{K} e_{2,k} b_k. \tag{13.48b}$$

Since each of the elements of each of the basis vectors b_k is zero except for the kth, these dot products are simply the kth elements of the two eigenvectors. Therefore each of the K basis vectors b_k is located on the biplot by coordinates given by the corresponding eigenvector elements. Because the data values and their original coordinate axes are both projected in the same way, the biplot amounts to a projection of the full K-dimensional scatterplot of the data, including the coordinate axes, onto the plane defined by the two leading eigenvectors.

Figure 13.19 shows a biplot for the $K = 6$-dimensional January 1987 data in Table A.1, after standardization to zero mean and unit variance, so that the PCA pertains to their correlation matrix, $[R]$. The PCA for these data is given in Table 13.1b. The numbers indicate the calendar date for each plotted data point. The projections of the six original basis vectors (plotted longer than the actual projections in

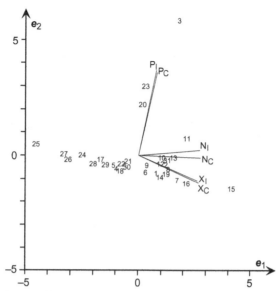

FIGURE 13.19   Biplot of the January 1987 data in Table A.1, after standardization. P = precipitation, X = maximum temperature, and N = minimum temperature. Numbered points refer to the corresponding calendar dates. The plot is a projection of the full six-dimensional scatterplot onto the plane defined by the first two principal components.

Equation 13.48 for clarity, but with the correct relative magnitudes) are indicated by the line segments diverging from the origin. "P," "N," and "X" indicate precipitation, minimum temperature, and maximum temperature, respectively, and the subscripts "I" and "C" indicate Ithaca and Canandaigua. It is immediately evident that the pairs of lines corresponding to like variables at the two locations are oriented nearly in the same directions, and that the temperature variables are oriented nearly perpendicularly to the precipitation variables. Approximately (because the variance described by the first two principal components is 92% rather than 100%), the correlations among these six variables are equal to the cosines of the angles between the corresponding lines in the biplot (compare Table 3.5), so the variables oriented in very similar directions form natural groupings.

The scatter of the n data points not only portrays their K-dimensional behavior in a potentially understandable way, but their interpretation is informed further by their relationship to the orientations of the variables. In Figure 13.19 most of the points are oriented nearly horizontally, with a slight inclination that is about midway between the angles of the minimum and maximum temperature variables, and perpendicular to the precipitation variables. These are the days corresponding to small or zero precipitation, whose primary variability characteristics relate to temperature differences. They are mainly located below the origin, because the mean precipitation is a bit above zero, and the precipitation variables are oriented nearly vertically (i.e., correspond closely to the second principal component). Points toward the right of the diagram, that are oriented similarly to the temperature variables, represent relatively warm days (with little or no precipitation), whereas points to the left are the cold days. Focusing on the dates for the coldest days, we can see that these occurred in a single run, toward the end of the month. Finally, the scatter of data points indicates that the few values in the upper portion of the biplot are different from the remaining observations, but it is the simultaneous display of the variables that allows us to see that these result from large positive values for precipitation.

13.8. EXERCISES

13.1. Using information from Exercise 11.8,

  a. Calculate the values of the first principal components for 1 January and for 2 January.

  b. Estimate the variance of all 31 values of the first principal component.

  c. What proportion of the total variability of the maximum temperature data is represented by the first principal component?

13.2. a. Compute the first two principal components for 1 January in the PCA in Table 13.1b.

  b. Reconstruct the six (standardized) weather variable values for 1 January, using the first 2 PCs only.

13.3. A principal component analysis of the data in Table A.3 yields the three eigenvectors $e_1^T = [.593, .552, -.587], e_2^T = [.332, -.831, -.446],$ and $e_3^T = [.734, -.069, .676]$, where the three elements in each vector pertain to the temperature, precipitation, and pressure data, respectively. The corresponding three eigenvalues are $\lambda_1 = 2.476$, $\lambda_2 = 0.356$, and $\lambda_3 = 0.169$.

  a. Was this analysis done using the covariance matrix or the correlation matrix? How can you tell?

  b. How many principal components should be retained according to Kaiser's rule, Jolliffe's modification, and the broken stick model?

  c. Estimate the missing precipitation value for 1956 using the first PC only.

13.4. Use the information in Exercise 13.3 to

  a. Compute 95% confidence intervals for the eigenvalues, assuming large samples and multinormal data.

  b. Examine the eigenvalue separation using the North et al. rule of thumb.

13.5. Using the information in Exercise 13.3, calculate the eigenvector matrix $[E]$ to be orthogonally rotated if

  a. The resulting rotated eigenvectors are to be orthogonal.

  b. The resulting principal components are to be uncorrelated.

13.6. Use the SVD in Equation 11.74 to find the first three values of the first principal component of the minimum temperature data in Table A.1.

13.7. Table 13.1b is the summary of a PCA on (the standardized) daily weather data at Ithaca and Canandaigua for January 1987. Using these results, the principal-component regression equation predicting the corresponding daily maximum temperatures at Central Park, NY city is:

$$\text{Max}_{NYC} = 37.5 + 7.15 u_1 - 2.81 u_3 \quad (R^2 = 75.9\%)$$

where u_1 and u_3 are the first and third principal components, respectively. Determine the corresponding regression equation in terms of the six original (standardized) predictor variables.

13.8. Construct a biplot for the data in Table A.3, using the information in Exercise 13.3.

Chapter 14

Multivariate Analysis of Vector Pairs

14.1. FINDING COUPLED PATTERNS: CCA, MCA, AND RA

This chapter describes three allied multivariate statistical methods: canonical correlation analysis (CCA), maximum covariance analysis (MCA), and redundancy analysis (RA). All three methods find sequences of pairwise relationships between two multivariate data sets. These relationships are constructed in terms of linear combinations, or patterns, in one data set that maximize interrelationships with linear combinations in the other data set. The three methods differ according to the criterion used to define the nature of the interrelationships that are maximized. These approaches thus bear some similarity to PCA (Chapter 13), which searches for patterns within a single multivariate data set that successively represent maximum amounts of its variations.

Denote data vectors from the two multivariate data sets whose relationships are to be characterized as x and y. The three methods CCA, MCA, and RA all extract relationships between pairs of data vectors x and y that are summarized in their joint covariance matrix. To compute this matrix, the two centered data vectors are concatenated into a single vector $c'^{\mathrm{T}} = [x'^{\mathrm{T}}, y'^{\mathrm{T}}]$. This partitioned vector contains $I + J$ elements, the first I of which are the elements of x', and the last J of which are the elements of y'. The $((I + J) \times (I + J))$ covariance matrix of c', $[S_C]$, is then partitioned into four blocks, in a manner similar to the partitioned covariance matrix in Equation 11.81 or the correlation matrix in Figure 13.5. That is,

$$[S_C] = \frac{1}{n-1}[C']^T[C'] = \begin{bmatrix} [S_{x,x}] & [S_{x,y}] \\ [S_{y,x}] & [S_{y,y}] \end{bmatrix}. \tag{14.1}$$

Each of the n rows of the $(n \times (I + J))$ matrix $[C']$ contains one observation of the vector x' and the corresponding observation of the vector y', with the primes indicating centering of the data by subtraction of each of the respective sample mean vectors. The $(I \times I)$ matrix $[S_{x,x}]$ is the variance–covariance matrix of the I variables in x. The $(J \times J)$ matrix $[S_{y,y}]$ is the variance–covariance matrix of the J variables in y. The matrices $[S_{x,y}]$ and $[S_{y,x}]$ contain the covariances between all combinations of the elements of x and the elements of y, and are related according to $[S_{x,y}] = [S_{y,x}]^{\mathrm{T}}$.

The three methods all find linear combinations of the original variables,

$$v_m = a_m^T x' = \sum_{i=1}^{I} a_{m,i} x'_i, \quad m = 1, \dots, \min(I, J); \tag{14.2a}$$

Statistical Methods in the Atmospheric Sciences. https://doi.org/10.1016/B978-0-12-815823-4.00014-6

and

$$w_m = \boldsymbol{b}_m^T \boldsymbol{y}' = \sum_{j=1}^{J} b_{m,j} y_j', \quad m = 1, \ldots, \min\ (I, J), \tag{14.2b}$$

by projecting them onto coefficient vectors $\boldsymbol{a}_m$ and $\boldsymbol{b}_m$. These coefficient vectors are defined differently for the three methods, on the basis of the relationships between the derived variables v_m and w_m, as characterized by the information in the joint covariance matrix in Equation 14.1.

CCA, MCA, and RA have been most widely applied to geophysical data in the form of spatial fields. In this setting the vector $\boldsymbol{x}$ often contains observations of one variable at a collection of gridpoints or locations, and the vector $\boldsymbol{y}$ contains observations of a different variable at a set of locations that may (in which case $I = J$) or may not be (in which case usually $I \ne J$) the same as those represented in $\boldsymbol{x}$. Typically the data consist of time series of observations of the two fields. When individual observations of the fields $\boldsymbol{x}$ and $\boldsymbol{y}$ are made simultaneously, these analyses can be useful in diagnosing aspects of the coupled variability of the two fields. When observations of $\boldsymbol{x}$ precede observations of $\boldsymbol{y}$ in time, the methods are natural vehicles for construction of linear statistical forecasts of the $\boldsymbol{y}$ field using the $\boldsymbol{x}$ field as a predictor.

14.2. CANONICAL CORRELATION ANALYSIS (CCA)

14.2.1. Properties of CCA

In CCA, the linear combination, or pattern, vectors $\boldsymbol{a}_m$ and $\boldsymbol{b}_m$ in Equation 14.2 are chosen such that each pair of the new variables v_m and w_m, called *canonical variates*, exhibit maximum correlation, while being uncorrelated with the projections of the data onto any of the other identified patterns. That is, CCA identifies new variables that maximize the interrelationships between two data sets in this sense. The vectors of linear combination weights, $\boldsymbol{a}_m$ and $\boldsymbol{b}_m$, are called the *canonical vectors*. The vectors $\boldsymbol{x}'$ and $\boldsymbol{a}_m$ each have I elements, and the vectors $\boldsymbol{y}'$ and $\boldsymbol{b}_m$ each have J elements. The number of pairs, M, of canonical variates that can be extracted from the two data sets is equal to the smaller of the dimensions of $\boldsymbol{x}$ and $\boldsymbol{y}$, that is, $M = \min\ (I, J)$.

The canonical vectors $\boldsymbol{a}_m$ and $\boldsymbol{b}_m$ are the unique choices that result in the canonical variates having the properties

$$\text{Corr}(v_1, w_1) \ge \text{Corr}(v_2, w_2) \ge \cdots \ge \text{Corr}(v_M, w_M) \ge 0, \tag{14.3a}$$

$$\text{Corr}(v_k, w_m) = \begin{cases} r_{C_m}, & k = m \\ 0, & k \ne m \end{cases}, \tag{14.3b}$$

$$\text{Corr}(v_k, v_m) = \text{Corr}(w_k, w_m) = 0, \quad k \ne m, \tag{14.3c}$$

and

$$\text{Var}(v_m) = \boldsymbol{a}_m^T [S_{x,x}]\,\boldsymbol{a}_m = \text{Var}(w_m) = \boldsymbol{b}_m^T [S_{y,y}]\,\boldsymbol{b}_m = 1, \quad m = 1, \ldots, M. \tag{14.3d}$$

Equation 14.3a states that each of the M successive pairs of canonical variates exhibits no greater correlation than the previous pair. These (Pearson product–moment) correlations between the pairs of canonical variates are called the *canonical correlations*, r_C. The canonical correlations can always be expressed as positive numbers, since either $\boldsymbol{a}_m$ or $\boldsymbol{b}_m$ can be multiplied by -1 if necessary. Equations 14.3b and 14.3c state that each canonical variate is uncorrelated with all the other canonical variates except its specific counterpart in the mth pair. Finally, Equation 14.3d states that each of the canonical variates has unit variance. Some restriction on the lengths of a_m and b_m is required for definiteness, and choosing these lengths to yield unit variances for the canonical variates turns out to be convenient for some applications. Accordingly, the joint $(2M \times 2M)$ covariance matrix for the resulting canonical variates then takes on the simple and interesting form

$$\text{Var}\left(\left[\frac{v}{w}\right]\right) = \begin{bmatrix} [S_{v,v}] & [S_{v,w}] \\ [S_{w,v}] & [S_{w,w}] \end{bmatrix} = \begin{bmatrix} [I] & [R_C] \\ [R_C] & [I] \end{bmatrix}, \tag{14.4a}$$

where $[R_C]$ is the diagonal matrix of the canonical correlations,

$$[R_C] = \begin{bmatrix} r_{C_1} & 0 & 0 & \cdots & 0 \\ 0 & r_{C_2} & 0 & \cdots & 0 \\ 0 & 0 & r_{C_3} & \cdots & 0 \\ \vdots & \vdots & \vdots & \ddots & \vdots \\ 0 & 0 & 0 & \cdots & r_{C_M} \end{bmatrix}. \tag{14.4b}$$

The definition of the canonical vectors is reminiscent of PCA, which finds a new orthonormal basis for a single multivariate data set (the eigenvectors of its covariance matrix), subject to a variance maximizing constraint. In CCA, two new bases are defined by the canonical vectors a_m and b_m. However, these basis vectors are neither orthogonal nor of unit length. The canonical variates are the projections of the centered data vectors x' and y' onto the canonical vectors and can be expressed in matrix form through the analysis formulae

$$\underset{(M\times 1)}{v} = \underset{(M\times I)}{[A]^{\mathrm{T}}} \underset{(I\times 1)}{x'} \tag{14.5a}$$

and

$$\underset{(M\times 1)}{w} = \underset{(M\times J)}{[B]^{T}} \underset{(J\times 1)}{y'} . \tag{14.5b}$$

Here the columns of the matrices $[A]$ and $[B]$ are the $M = \min(I, J)$ canonical vectors, a_m and b_m, respectively. Exposition of how the canonical vectors are calculated from the joint covariance matrix (Equation 14.1) will be deferred to Section 14.2.4.

Unlike the case of PCA, calculating a CCA on the basis of standardized (unit variance) variables yields results that are simple functions of the results from an unstandardized analysis. In particular, because in a standardized analysis the centered variables x_i' and y_j' in Equation 14.2 would be divided by their respective standard deviations, the corresponding elements of the canonical vectors would be larger by factors of those standard deviations. In particular, if a_m is the mth canonical $(I \times 1)$ vector for the x variables, its counterpart a_m^* in a CCA of the standardized variables would be

$$a_m^* = a_m [D_x], \tag{14.6}$$

where the $(I \times I)$ diagonal matrix $[D_x]$ (Equation 11.31) contains the standard deviations of the x variables, and a similar equation would hold for the canonical vectors b_m and the $(J \times J)$ diagonal matrix $[D_y]$ containing the standard deviations of the y variables. Regardless of whether a CCA is computed using standardized or unstandardized variables, the resulting canonical correlations are the same.

Correlations between the original and canonical variables can be calculated easily. The correlations between corresponding original and canonical variables, sometimes called *homogeneous correlations*, are given by

$$\text{Corr}(v_m, x) = \underset{(1 \times I)}{a_m^T} \underset{(1 \times I)}{[S_{x,x}]} \underset{(I \times I)}{[D_x]^{-1}} \qquad (14.7a)$$

and

$$\text{Corr}(w_m, y) = \underset{(1 \times J)}{b_m^T} \underset{(1 \times J)}{[S_{y,y}]} \underset{(J \times J)}{[D_y]^{-1}}. \qquad (14.7b)$$

These equations specify vectors of correlations, between the mth canonical variable v_m and each of the I original variables x_i; and between the canonical variable w_m and each of the J original variables y_k. Similarly, the vectors of *heterogeneous correlations* between the canonical variables and the "other" original variables are

$$\text{Corr}(v_m, y) = \underset{(1 \times J)}{a_m^T} \underset{(1 \times I)}{[S_{x,y}]} \underset{(J \times J)}{[D_y]^{-1}} \qquad (14.8a)$$

and

$$\text{Corr}(w_m, x) = \underset{(1 \times J)}{b_m^T} \underset{(1 \times J)}{[S_{y,x}]} \underset{(I \times I)}{[D_x]^{-1}}. \qquad (14.8b)$$

The homogeneous correlations indicate correspondence between each canonical variate and the underlying data field, whereas the heterogeneous correlations indicate how well the gridpoints in one field can be specified or predicted by the opposite canonical variate (Bretherton et al., 1992).

The canonical vectors a_m and b_m are chosen to maximize correlations between the resulting canonical variates v_m and w_m, but (unlike PCA) may or may not be particularly effective at summarizing the variances of the original variables x and y. If canonical pairs with high correlations turn out to represent small fractions of the underlying variability, their physical and practical significance may be limited. Therefore it is often worthwhile to calculate the variance proportions R_m^2 captured by each of the leading canonical variables for its underlying original variable.

How well the canonical variables represent the underlying variability is related to how accurately the underlying variables can be synthesized from the canonical variables. Solving the analysis equations (Equation 14.5) yields the CCA synthesis equations

$$\underset{(I \times 1)}{x'} = \underset{(I \times I)}{\left[\tilde{A}\right]^{-1}} \underset{(I \times 1)}{v} \qquad (14.9a)$$

and

$$\underset{(J \times 1)}{y'} = \underset{(J \times J)}{\left[\tilde{B}\right]^{-1}} \underset{(J \times 1)}{w}. \qquad (14.9b)$$

If $I = J$ (i.e., if the dimensions of the data vectors x and y are equal), then the matrices $[A]$ and $[B]$, whose columns are the corresponding M canonical vectors, are both square. In this case $\left[\tilde{A}\right] = [A]^T$ and $\left[\tilde{B}\right] = [B]^T$ in Equation 14.9, and the indicated matrix inversions can be calculated. If $I \neq J$ then one of the matrices $[A]$ or $[B]$ is not square, and so not invertible. In that case, the last $M - J$ columns of $[A]$ (if $I > J$), or the last $M - I$ columns of $[B]$ (if $I < J$), are filled out with the "phantom" canonical vectors corresponding to the zero eigenvalues, as described in Section 14.2.4.

Equation 14.9 describes the synthesis of individual observations of x and y on the basis of their corresponding canonical variables. In matrix form (i.e., for the full set of n observations), these become

$$[X']^T = [\widetilde{A}]^{-1} [V]^T \atop {(I \times n) \quad (I \times I) \quad (I \times n)}$$

(14.10a)

and

$$[Y']^T = [\widetilde{B}]^{-1} [W]^T . \atop {(J \times n) \quad (J \times J) \quad (J \times n)}$$

(14.10b)

Because the covariance matrices of the canonical variates are $(n-1)^{-1}[V]^T [V] = [I]$ and $(n-1)^{-1}$ $[W]^T[W] = [I]$ (Equation 14.4a), substituting Equation 14.10 into Equation 11.30 yields

$$[S_{x,x}] = \frac{1}{n-1}[X']^T[X'] = [\widetilde{A}]^{-1}\left([\widetilde{A}]^{-1}\right)^T = \sum_{m=1}^{I} \widetilde{a}_m \widetilde{a}_m^T$$

(14.11a)

and

$$[S_{y,y}] = \frac{1}{n-1}[Y']^T[Y'] = [\widetilde{B}]^{-1}\left([\widetilde{B}]^{-1}\right)^T = \sum_{m=1}^{I} \widetilde{b}_m \widetilde{b}_m^T,$$

(14.11b)

where the canonical vectors with tilde accents indicate columns of the *inverses* of the corresponding matrices. These decompositions are akin to the spectral decompositions (Equation 11.53a) of the two covariance matrices. Accordingly, the proportions of the variances of x and y represented by their mth canonical variables are

$$R_m^2(x) = \frac{\text{tr}\left(\widetilde{a}_m \widetilde{a}_m^T\right)}{\text{tr}([S_{x,x}])}$$

(14.12a)

and

$$R_m^2(y) = \frac{\text{tr}\left(\widetilde{b}_m \widetilde{b}_m^T\right)}{\text{tr}\left([S_{y,y}]\right)}.$$

(14.12b)

Example 14.1. CCA of the January 1987 Temperature Data

A simple illustration of the mechanics of a small CCA can be provided by again analyzing the January 1987 temperature data for Ithaca and Canandaigua, New York, given in Table A.1. Let the $I = 2$ Ithaca temperature variables be $x = [T_{max}, T_{min}]^T$, and similarly let the $J = 2$ Canandaigua temperature variables be y. The joint covariance matrix $[S_C]$ of these quantities is then the (4×4) matrix

$$[S_C] = \begin{bmatrix} 59.516 & 75.433 & 58.070 & 51.697 \\ 75.433 & 185.467 & 81.633 & 110.800 \\ 58.070 & 81.633 & 61.847 & 56.119 \\ 51.697 & 110.800 & 56.119 & 77.581 \end{bmatrix}.$$

(14.13)

This symmetric matrix contains the sample variances of the four variables on the diagonal and the covariances among the variables in the other positions. It is related to the corresponding elements of the

TABLE 14.1 The Canonical Vectors a_m (Corresponding to Ithaca Temperatures) and b_m (Corresponding to Canandaigua Temperatures) for the Partition of the Covariance Matrix in Equation 14.13 With $I = J = 2$

	a_1 (Ithaca)	b_1 (Canandaigua)	a_2 (Ithaca)	b_2 (Canandaigua)
T_{max}	0.0923	0.0946	−0.1618	−0.1952
T_{min}	0.0263	0.0338	0.1022	0.1907
λ_m		0.938		0.593
$r_{C_m} = \sqrt{\lambda_m}$		0.969		0.770

Also shown are the eigenvalues λ_m (cf. Example 14.3) and the canonical correlations, which are their square roots.

correlation matrix involving the same variables (see Table 3.5) through the square roots of the diagonal elements: dividing each element by the square root of the diagonal elements in its row and column produces the corresponding correlation matrix. This operation is shown in matrix notation in Equation 11.31.

Since $I = J = 2$, there are $M = 2$ canonical vectors for each of the two data sets being correlated. These are presented in Table 14.1, although the details of their computation will be left until Example 14.3. The first element of each pertains to the respective maximum temperature variable, and the second elements pertain to the minimum temperature variables. The correlation between the first pair of projections of the data onto these vectors, v_1 and w_1, is $r_{C_1} = 0.969$; and the second canonical correlation, between v_2 and w_2, is $r_{C_2} = 0.770$.

Each of the canonical vectors defines a direction in its two-dimensional (T_{max}, T_{min}) data space, but their absolute magnitudes are meaningful only in that they produce unit variances for their corresponding canonical variates. However, the relative magnitudes of the canonical vector elements can be interpreted in terms of which linear combinations of one underlying data vector are most correlated with which linear combination of the other. All the elements of a_1 and b_1 are positive, reflecting positive correlations among all four temperature variables; although the elements corresponding to the maximum temperatures are larger, reflecting the larger correlation between them than between the minima (Table 3.5). The pairs of elements in a_2 and b_2 are comparable in magnitude but opposite in sign, suggesting that the next most important pair of linear combinations with respect to correlation relate to the diurnal ranges at the two locations (recall that the signs of the canonical vectors are arbitrary, and chosen to produce positive canonical correlations so that reversing the signs on the second canonical vectors would put positive weights on the maxima and negative weights of comparable magnitudes on the minima).

The time series of the first pair of canonical variables is given by the dot products of a_1 and b_1 with the pairs of centered temperature values for Ithaca and Canandaigua, respectively, from Table A.1. The value of v_1 for 1 January would be constructed as $(33–29.87)(0.0923) + (19–13.00)(0.0263) = 0.447$. The time series of v_1 (pertaining to the Ithaca temperatures) would consist of the 31 values (one for each day): 0.447, 0.512, 0.249, −0.449, −0.686, ..., −0.041, 0.644. Similarly, the time series for w_1 (pertaining to the Canandaigua temperatures) is 0.474, 0.663, 0.028, −0.304, −0.310, ..., −0.283, 0.683. Each of this first pair of canonical variables is a scalar index of the general warmth at its respective location, with more emphasis on the maximum temperatures. Both series have unit sample variance.

The first canonical correlation coefficient, $r_{C_1} = 0.969$, is the correlation between this first pair of canonical variables, v_1 and w_1, and is the largest possible correlation between pairs of linear combinations of these two data sets.

Similarly, the time series of v_2 is 0.107, 0.882, 0.899, -1.290, -0.132, ..., -0.225, 0.354 and the time series of w_2 is 1.046, 0.656, 1.446, 0.306, -0.461, ..., -1.038, -0.688. Both of these series also have unit sample variance, and their correlation is $r_{C_2} = 0.770$. On each of the $n = 31$ days (the negatives of), these second canonical variates provide an approximate index of the diurnal temperature ranges at the corresponding locations.

The homogeneous correlations (Equation 14.7) for the leading canonical variates, v_1 and w_1, are

$$\text{Corr}(v_1, x^T) = [0.0923 \ 0.0263] \begin{bmatrix} 59.516 & 75.433 \\ 75.433 & 185.467 \end{bmatrix} \begin{bmatrix} 0.1296 & 0 \\ 0 & 0.0734 \end{bmatrix} = [0.969 \ 0.869] \quad (14.14a)$$

and

$$\text{Corr}(w_1, y^T) = [0.0946 \ 0.0338] \begin{bmatrix} 61.847 & 56.119 \\ 56.119 & 77.581 \end{bmatrix} \begin{bmatrix} 0.1272 & 0 \\ 0 & 0.1135 \end{bmatrix} = [0.985 \ 0.900]. \quad (14.14b)$$

All the four homogeneous correlations are strongly positive, reflecting the strong positive correlations among all four of the variables (see Table 3.5), and the fact that the two leading canonical variables have been constructed with positive weights on all four. The homogeneous correlations for the second canonical variates v_2 and w_2 are calculated in the same way, except that the second canonical vectors a_2^T and b_2^T are used in Equations 14.14a and 14.14b, respectively, yielding $\text{Corr}(v_2, x^T) = [-0.249, 0.495]$ and $\text{Corr}(w_2, y^T) = [-0.174, 0.436]$. The second canonical variables are less strongly correlated with the underlying temperature variables, because the magnitude of the diurnal temperature range is only weakly correlated with the overall temperatures: wide or narrow diurnal ranges can occur on both relatively warm and cool days. However, the diurnal ranges are evidently more strongly correlated with the minimum temperatures, with cooler minima tending to be associated with large diurnal ranges.

Similarly, the heterogeneous correlations (Equation 14.8) for the leading canonical variates are

$$\text{Corr}(v_1, y^T) = [.0923 \ .0263] \begin{bmatrix} 58.070 & 51.697 \\ 81.633 & 110.800 \end{bmatrix} \begin{bmatrix} .1272 & 0 \\ 0 & .1135 \end{bmatrix} = [.955 \ .872] \quad (14.15a)$$

and

$$\text{Corr}(w_1, x^T) = [.0946 \ .0338] \begin{bmatrix} 58.070 & 81.633 \\ 51.697 & 110.800 \end{bmatrix} \begin{bmatrix} .1296 & 0 \\ 0 & .0734 \end{bmatrix} = [.938 \ .842]. \quad (14.15b)$$

Because of the symmetry of these data (like variables at nearby locations), these correlations are very close to the homogeneous correlations in Equation 14.14. Similarly, the heterogeneous correlations for the second canonical vectors are also close to their homogeneous counterparts: $\text{Corr}(v_2, y^T) = [-0.132, 0.333]$ and $\text{Corr}(w_2, x^T) = [-0.191, 0.381]$.

Finally the variance fractions for the temperature data at each of the two locations that are described by the canonical variates depend, through the synthesis equations (Equation 14.9), on the matrices $[A]$ and $[B]$, whose columns are the canonical vectors. Because $I = J$,

$$\left[\tilde{A}\right]=[A]^T=\begin{bmatrix} 0.0923 & 0.0263 \\ -0.1618 & 0.1022 \end{bmatrix}, \text{ and } \left[\tilde{B}\right]=[B]^T=\begin{bmatrix} 0.0946 & 0.0338 \\ -0.1952 & 0.1907 \end{bmatrix}; \tag{14.16a}$$

so that

$$\left[\tilde{A}\right]^{-1}=\begin{bmatrix} 7.466 & -1.921 \\ 11.820 & 6.743 \end{bmatrix}, \text{ and } \left[\tilde{B}\right]^{-1}=\begin{bmatrix} 7.740 & -1.372 \\ 7.923 & 3.840 \end{bmatrix}. \tag{14.16b}$$

Contributions made by the canonical variates to the respective covariance matrices for the underlying data depend on the outer products of the columns of these inverse matrices (terms in the summations of Equations 14.11), that is,

$$\tilde{a}_1\tilde{a}_1^T=\begin{bmatrix} 7.466 \\ 11.820 \end{bmatrix}[7.466 \; 11.820]=\begin{bmatrix} 55.74 & 88.25 \\ 88.25 & 139.71 \end{bmatrix}, \tag{14.17a}$$

$$\tilde{a}_2\tilde{a}_2^T=\begin{bmatrix} -1.921 \\ 6.743 \end{bmatrix}[-1.921 \; 6.743]=\begin{bmatrix} 3.690 & -12.95 \\ -12.95 & 45.47 \end{bmatrix}, \tag{14.17b}$$

$$\tilde{b}_1\tilde{b}_1^T=\begin{bmatrix} 7.740 \\ 7.923 \end{bmatrix}[7.740 \; 7.923]=\begin{bmatrix} 59.91 & 61.36 \\ 61.36 & 62.77 \end{bmatrix}, \tag{14.17c}$$

$$\tilde{b}_2\tilde{b}_2^T=\begin{bmatrix} -1.372 \\ 3.840 \end{bmatrix}[-1.372 \; 3.840]=\begin{bmatrix} 1.882 & 5.279 \\ 5.279 & 14.75 \end{bmatrix}. \tag{14.17d}$$

Therefore the proportions of the Ithaca temperature variance described by its two canonical variates (Equation 14.12a) are

$$R_1^2(x)=\frac{55.74+139.71}{59.52+185.47}=0.798 \tag{14.18a}$$

and

$$R_2^2(x)=\frac{3.690+45.47}{59.52+185.47}=0.202, \tag{14.18b}$$

and the corresponding variance fractions for Canandaigua are

$$R_1^2(y)=\frac{59.91+62.77}{61.85+77.58}=0.880 \tag{14.19a}$$

and

$$R_2^2(y)=\frac{1.882+14.75}{61.85+77.58}=0.120, \tag{14.19b}$$

$\Diamond$

14.2.2. CCA Applied to Fields

Canonical correlation analysis is usually most interesting for atmospheric data when applied to fields. Here the spatially distributed observations (either at gridpoints or observing locations) are encoded into the vectors x and y in the same way as for PCA. That is, even though the data may pertain to a two- or three-dimensional field, each location is numbered sequentially and pertains to one element of the corresponding data vector. It is not necessary for the spatial domains encoded into x and y to be the same, and indeed in the applications of CCA that have appeared in the literature they are usually different.

As is the case when using PCA with spatial data, it is often informative to plot maps of the canonical vectors by associating the magnitudes of their elements with the geographic locations to which they pertain. In this context the canonical vectors are sometimes called *canonical patterns*, since the resulting maps show spatial patterns of the ways in which the original variables contribute to the canonical variables. Examining the pairs of maps formed by corresponding vectors a_m and b_m can be informative about the nature of the relationship between variations in the data over the domains encoded in x and y, respectively. Figures 14.2 and 14.3 show examples of maps of canonical vectors.

It can also be informative to plot pairs of maps of the homogeneous (Equation 14.7) or heterogeneous correlations (Equation 14.8). Each of these vectors contains correlations between an underlying data field and one of the canonical variables, and these correlations can also be plotted at the corresponding locations. Figure 14.1 shows one such pair of homogeneous correlation patterns. Figure 14.1a shows the spatial distribution of correlations between a canonical variable v, and the values of the corresponding data x that contains values of average December–February sea-surface temperatures (SSTs) in the north Pacific Ocean. This canonical variable accounts for 18% of the total variance of the SSTs in the data set analyzed (Equation 14.12). Figure 14.1b shows the spatial distribution of the correlations for the corresponding canonical variable w, pertaining to average hemispheric 500 mb heights y during the same winters included in the SST data in x. This canonical variable accounts for 23% of the total variance of the winter hemispheric height variations. The correlation pattern in Figure 14.1a corresponds to either cold water in the central north Pacific and warm water along the west coast of North America, or warm water in the central north Pacific and cold water along the west coast of North America. The pattern of 500 mb height correlations in Figure 14.1b is remarkably similar to the PNA pattern (cf. Figures 13.14b and 3.33).

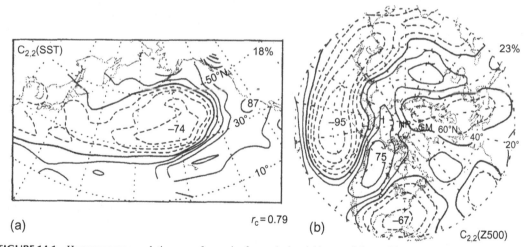

FIGURE 14.1   Homogeneous correlation maps for a pair of canonical variables pertaining to (a) average winter sea-surface temperatures (SSTs) in the northern Pacific Ocean, and (b) hemispheric winter 500 mb heights. The pattern of SST correlation in the left-hand panel (or its negative) is associated with the PNA pattern of 500 mb height correlations shown in the right-hand panel (or *its* negative). The canonical correlation for this pair of canonical variables is 0.79. *From Wallace et al. (1992). © American Meteorological Society. Used with permission.*

The correlation between the two time series v and w is the canonical correlation $r_C = 0.79$. Because v and w are well correlated, these figures indicate that cold SSTs in the central Pacific simultaneously with warm SSTs in the northeast Pacific (relatively large positive v) tend to coincide with a 500 mb ridge over northwestern North America and a 500 mb trough over southeastern North America (relatively large positive w). Similarly, warm water in the central north Pacific and cold water in the northwestern Pacific (relatively large negative v) are associated with the more zonal PNA flow (relatively large negative w).

The sampling properties of CCA may be poor when the available data are few relative to the dimensionality of the data vectors. The result can be that sample estimates for CCA parameters may be unstable (i.e., exhibit large variations from batch to batch) for small samples (e.g., Bretherton et al., 1992; Cherry, 1996). Friederichs and Hense (2003) describe, in the context of atmospheric data, both conventional parametric tests and resampling tests to help assess whether sample canonical correlations may be spurious sampling artifacts. These tests examine the null hypothesis that all the underlying population canonical correlations are zero.

Relatively small sample sizes are common when analyzing time series of atmospheric fields. In CCA, it is not uncommon for there to be fewer observations n than the dimensions I and J of the data vectors, in which case the necessary matrix inversions cannot be computed (see Section 14.2.4). However, even if the sample sizes are large enough to carry through the calculations, sample CCA statistics are erratic unless $n \gg M$. Barnett and Preisendorfer (1987) suggested that a remedy for this problem is to prefilter the two fields of raw data using separate PCAs before subjecting them to a CCA, and this has become a conventional procedure. Rather than directly correlating linear combinations of the fields x' and y', the CCA then operates on the vectors u_x and u_y, which consist of the leading principal components of x and y. The truncations for these two PCAs (i.e., the dimensions of the vectors u_x and u_y) need not be the same, but should be severe enough for the larger of the two to be substantially smaller than the sample size n. Livezey and Smith (1999) provide some guidance for the subjective choices that need to be made in this approach. This combined PCA/CCA approach is not always best and can be inferior if important information is discarded when truncating the PCA. In particular, there is no guarantee that the most strongly correlated linear combinations of x and y will be well related to the leading principal components of one field or the other.

An alternative approach to CCA when $n > M$ is to employ ridge regularization (Section 7.5.1) to the matrices $[S_{x,x}]$ and $[S_{y,y}]$, which allows the CCA computations to proceed because the regularized versions of these matrices are invertible (Cruz-Cano and Lee, 2014; Lim et al., 2012; Vinod, 1976). In addition (or instead, if $n < M$), L1 regularization (Section 7.5.2) can be applied in the computation of the canonical vectors a_m and b_m (Hastie et al., 2015; Witten et al., 2009).

14.2.3. Forecasting With CCA

When one of the fields, say x, is observed prior to y, and some of the canonical correlations between the two are large, it is natural to use CCA either as a purely statistical forecasting method (e.g., Barnston, 1994; Landman and Mason, 2001; Wilks, 2014a) or as a MOS implementation (e.g., Alfaro et al., 2018, Lim et al., 2012). In either case the entire $(I \times 1)$ field $x(t)$ is used to forecast the $(J \times 1)$ field $y(t + \tau)$, where τ is the time lag between the two fields in the training data, which becomes the forecast lead time. In applications with atmospheric data it is typical that there are too few observations n relative to the dimensions I and J of the fields for stable sample estimates (which are especially important for out-of-sample forecasting) to be calculated, even if the calculations can be performed because

$n > \max (I, J)$. It is therefore usual for both the x and y fields to be represented by separate series of truncated principal components, as described in the previous section. However, in order not to clutter the notation in this section, the mathematical development will be expressed in terms of the original variables x and y, rather than their principal components u_x and u_y.

The basic idea behind forecasting with CCA is straightforward: simple linear regressions are constructed that relate the predictand canonical variates w_m to the predictor canonical variates v_m,

$$w_m = \hat{\beta}_{0,m} + \hat{\beta}_{1,m} v_m, \quad m = 1,\ldots,M. \tag{14.20}$$

Here the estimated regression coefficients are indicated by the β's in order to distinguish clearly from the canonical vectors b, and the number of canonical pairs considered can be any number up to the smaller of the numbers of principal components retained for the x and y fields. These regressions are all simple linear regressions that can be computed individually, because canonical variables from different canonical pairs are uncorrelated (Equation 14.3b).

Parameter estimation for the regressions in Equation 14.20 is straightforward also. Using Equation 7.7a for the regression slopes,

$$\hat{\beta}_{1,m} = \frac{n \operatorname{Cov}(v_m, w_m)}{n \operatorname{Var}(v_m)} = \frac{n s_v s_w r_{v,w}}{n s_v^2} = r_{v,w} = r_{C_m}, \quad m = 1,\ldots,M. \tag{14.21}$$

That is, because the canonical variates are scaled to have unit variance (Equation 14.3c), the regression slopes are simply equal to the corresponding canonical correlations. Similarly, Equation 7.7b yields the regression intercepts

$$\hat{\beta}_{0,m} = \overline{w}_m - \hat{\beta}_{1,m} \overline{v}_m = b_m^T E(y') + \hat{\beta}_{1,m} a_m^T E(x') = 0, \quad m = 1,\ldots,M. \tag{14.22}$$

That is, because the CCA is calculated from the centered data x' and y' whose mean vectors are both 0, the averages of the canonical variables v_m and w_m are also both zero, so that all the intercepts in Equation 14.20 are zero. Equation 14.22 also holds when the CCA has been calculated from a principal component truncation of the original (centered) variables, because $E(u_x) = E(u_y) = 0$.

Once the CCA has been fit, the basic forecast procedure is as follows. First, centered values for the predictor field x' (or its first few principal components, u_x) are used in Equation 14.5a to calculate the M canonical variates v_m to be used as regression predictors. Combining Equations 14.20 through 14.22, the $(M \times 1)$ vector of predictand canonical variates is forecast to be

$$\hat{w} = [R_C] v = [R_C][A]^T x', \tag{14.23}$$

where $[R_C]$ is the diagonal $(M \times M)$ matrix of the canonical correlations (Equation 14.4b) and the $(I \times M)$ matrix $[A]$ contains the predictor canonical vectors a_m. In general, the forecast map $\hat{y}$ will need to be synthesized from its predicted canonical variates using Equation 14.9b, in order to see the forecast in a physically meaningful way. However, in order to be invertible, the matrix $[B]$, whose columns are the predictand canonical vectors b_m, must be square. This condition implies that the number of regressions M in Equation 14.20 needs to be equal to the dimensionality of y (or, more usually, to the number of predictand principal components that have been retained), although the dimension of x (or the number of predictor principal components retained) is not constrained in this way. If the CCA has been calculated using predictand principal components u_y, the centered predicted values $\hat{y}'$ are next recovered with the

PCA synthesis, Equation 13.6. Finally, the full predicted field is produced by adding back its mean vector. If the CCA has been computed using standardized variables, so that Equation 14.1 is a correlation matrix, the dimensional values of the predicted variables need to be reconstructed by multiplying by the appropriate standard deviation before adding the appropriate mean (i.e., by reversing Equation 3.27 or Equation 4.27 to yield Equation 4.29).

Example 14.2. An Operational CCA Forecast System

Canonical correlation is used as one of the elements contributing to the seasonal forecasts produced operationally at the U.S. Climate Prediction Center (Barnston et al., 1999). The predictands are seasonal (three-month) average temperature and total precipitation over the United States, made at lead times of 0.5 through 12.5 months.

The CCA forecasts contributing to this system are modified from the procedure described in Barnston (1994), whose temperature forecast procedure will be outlined in this example. The (59×1) predictand vector y represents temperature forecasts jointly at 59 locations in the conterminous United States. The predictors x consist of global sea-surface temperatures (SSTs) discretized to a 235-point grid, northern hemisphere 700 mb heights discretized to a 358-point grid, and previously observed temperatures at the 59 prediction locations. The predictors are three-month averages also, but in each of the four nonoverlapping three-month seasons for which data would be available preceding the season to be predicted. For example, the predictors for the January–February–March (JFM) forecast, to be made in mid-December, are seasonal averages of SSTs, 700 mb heights, and U.S. surface temperatures during the preceding September–October–November (SON), June–July–August (JJA), March–April–May (MAM), and December–January–February (DJF) seasons, so that the predictor vector x has $4 (235 + 358 + 59) = 2608$ elements. In principle, using sequences of four consecutive predictor seasons allows the forecast procedure to incorporate information about the time evolution of the predictor fields.

Since only $n = 37$ years of training data were available when this system was developed, drastic reductions in the dimensionality of both the predictors and predictands were necessary. Separate PCAs were calculated for the predictor and predictand vectors, which retained the leading six predictor principal components u_x and (depending on the forecast season) either five or six predictand principal components u_y. The CCAs for these pairs of principal component vectors yield either $M = 5$ or $M = 6$ canonical pairs. Figure 14.2 shows that portion of the first predictor canonical vector a_1 pertaining to the SSTs in the three seasons MAM, JJA, and SON, relating to the forecast for the following JFM. That is, each of these three maps expresses the SST contributions to the six elements of a_1 in terms of the original 235 spatial locations, through the corresponding elements of the eigenvector matrix $[E]$ for the predictor PCA. The most prominent feature in Figure 14.2 is the progressive evolution of increasingly negative values in the eastern tropical Pacific, which clearly represents an intensifying El Niño (warm) event when $v_1 < 0$, and development of a La Niña (cold) event when $v_1 > 0$, in the spring, summer, and fall before the JFM season to be forecast.

Figure 14.3 shows the first canonical predictand vector for the JFM forecast, b_1, again projected back to physical space at the 59 forecast locations. Because the CCA is constructed to have positive canonical correlations, a developing El Niño yielding $v_1 < 0$ results in a forecast $\hat{w}_1 < 0$ (Equation 14.23). The result is that the first canonical pair contributes a tendency toward relative warmth in the northern United States and relative coolness in the southern United States during El Niño winters. Conversely, this canonical pair forecasts cold in the north and warm in the south for La Niña winters. Evolving SSTs not resembling the patterns in Figure 14.2 would yield $v_1 \approx 0$, resulting in little contribution from the pattern in Figure 14.3 to the forecast.   ◇

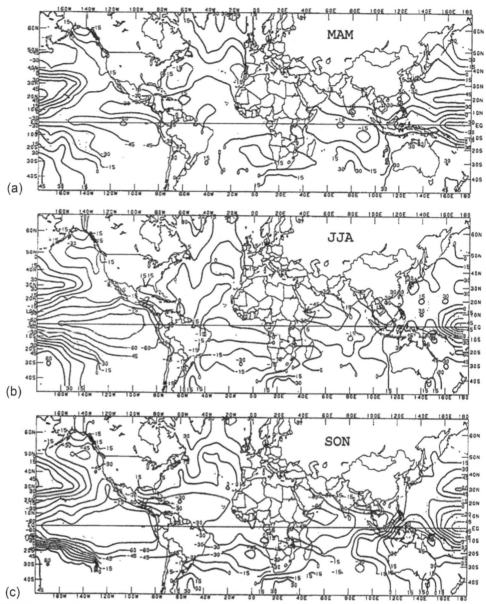

FIGURE 14.2 Spatial displays of portions of the first canonical vector for predictor sea-surface temperatures, in the three seasons preceding the JFM for which U.S. surface temperatures are forecast. The corresponding canonical vector for this predictand is shown in Figure 14.3. *From Barnston (1994). © American Meteorological Society. Used with permission.*

Extending CCA to probabilistic forecasts is relatively straightforward because the canonical variates are uncorrelated with all others except their counterparts in the mth pair. In particular, because each of the regressions in Equation 14.20 is independent of the others, the joint covariance matrix for the predictand canonical variates is

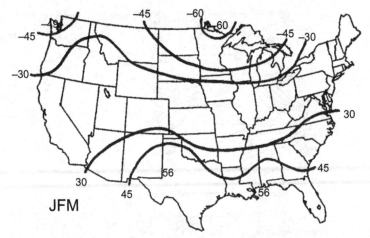

JFM

$$[S_{\hat{w}}] = \begin{bmatrix} s^2_{\hat{w}_1} & 0 & \cdots & 0 \\ 0 & s^2_{\hat{w}_2} & \cdots & 0 \\ \vdots & \vdots & \ddots & \vdots \\ 0 & 0 & \cdots & s^2_{\hat{w}_M} \end{bmatrix}, \tag{14.24}$$

where, using Equation 7.23 for the regression prediction variance,

$$s^2_{\hat{w}_m} = s^2_{e_m} \left[1 + \frac{1}{n} + \frac{(v_{0,m} - \bar{v}_m)^2}{\sum_{i=1}^{n}(v_{i,m} - \bar{v}_m)^2} \right] = \left(1 - r^2_{C_m} \right) \left[1 + \frac{1}{n} + \frac{v_{0,m}^2}{\sum_{i=1}^{n} v_{i,m}^2} \right]. \tag{14.25}$$

Here $v_{0,m}$ denotes the predictor linear combination in implementation, which is not one of the n training data values. The second equality follows because the fraction of variance represented by each regression residual is $1 - r^2_{C_m}$, and each predictand canonical variate has unit variance by construction. The centering of the predictor data implies that the average of the predictor canonical variates is zero.

If it is reasonable to invoke multivariate normality for the forecast errors, predictive distributions are fully defined by the vector mean in Equation 14.9b, and the covariance matrix

$$[S_{\hat{y}'}] = \left([B]^{-1} \right)^T [S_w] [B]^{-1}. \tag{14.26}$$

Figure 14.4 shows the 90% forecast probability ellipses derived from bivariate normal predictive distributions for the January–March 1995 SSTs, jointly in the Niño 3 and Niño 4 regions of the equatorial Pacific Ocean, which were computed on the basis of the $I = 4$ leading principal components of Indo-Pacific SSTs observed from 1 to 9 months prior. The forecast uncertainty (areas enclosed by the ellipses) decreases as the lead times become shorter, and the vector means of the distributions (numerals) move toward the observed January–March values (large dot).

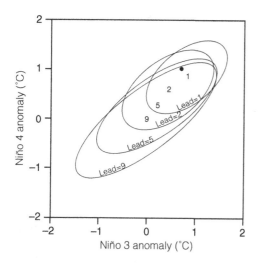

FIGURE 14.4   Ninety percent forecast probability ellipses for joint Niño 3 and Niño 4 forecasts for JFM 1995, at the 1-, 2-, 5-, and 9-month lead times. Numerals show respective forecast means, and the solid dot locates the observed 1995 value. *Modified from Wilks (2014a).*

In problems where a K-dimensional predictand vector has been represented by its leading J principal components, the corresponding predictive mean vector and covariance matrix are

$$\hat{y}' = \underset{(K \times 1)}{[E]} \underset{(K \times J)}{\left([B]^T\right)^{-1}} \underset{(J \times J)}{w} \underset{(J \times 1)}{} \tag{14.27}$$

and

$$\left[S_{\hat{y}'}\right] = \underset{(K \times K)}{[E]} \underset{(K \times J)}{\left([B]^{-1}\right)^T} \underset{(J \times J)}{[S_w]} \underset{(J \times J)}{[B]^{-1}} \underset{(J \times J)}{[E]^T}, \tag{14.28}$$

where it has been assumed that $I = J$, so that $\left[\tilde{B}\right] = [B]^T$.

Long-lead forecasts of monthly or seasonally averaged quantities are often made with linear multivariate statistical methods such as CCA. Although the dynamics of the climate system are nonlinear, in practice it has been found that nonlinear statistical methods perform no better than the traditional linear statistics for seasonal prediction (Tang et al., 2000; Van den Dool, 2007). Linear multivariate statistical methods also often perform comparably to fully nonlinear dynamical models in seasonal forecasting applications (e.g., Harrison, 2005, Quan et al., 2006, Toth et al., 2007, van den Dool 2007), at substantially reduced computational cost. Possibly this phenomenon occurs because the time averaging inherent in long-lead forecasting renders both the predictors and predictands Gaussian or quasi-Gaussian because of the Central Limit Theorem, which in turn induces linear or quasi-linear relationships among them (Hlinka et al., 2014; Hsieh, 2009; Yuval and Hsieh, 2002).

14.2.4. Computational Considerations

Finding canonical vectors and canonical correlations requires calculating pairs of eigenvectors e_m, corresponding to the x variables, and eigenvectors f_m, corresponding to the y variables; together with their corresponding eigenvalues λ_m, which are the same for each pair e_m and f_m. There are several computational approaches available to find these $e_m, f_m,$ and $\lambda_m, m = 1,\dots, M$.

Calculating CCA Through Direct Eigendecomposition

One approach is to find the eigenvectors e_m and f_m of the matrices

$$[S_{x,x}]^{-1} [S_{x,y}] [S_{y,y}]^{-1} [S_{y,x}] \tag{14.29a}$$

and

$$[S_{y,y}]^{-1} [S_{y,x}] [S_{x,x}]^{-1} [S_{x,y}], \tag{14.29b}$$

respectively. The factors in these equations correspond to the definitions in Equation 14.1. Equation 14.29a is dimensioned $(I \times I)$, and Equation 14.29b is dimensioned $(J \times J)$. The first $M = \min (I, J)$ eigenvalues of these two matrices are equal, and if $I \neq J$, the remaining eigenvalues of the larger matrix are zero. The corresponding "phantom" eigenvectors would fill the extra rows of one of the matrices in Equation 14.9. Equation 14.29 can be difficult computationally because in general these matrices are not symmetric, and algorithms to find eigenvalues and eigenvectors for general matrices are less stable numerically than routines designed specifically for real and symmetric matrices.

The eigenvalue–eigenvector computations are easier and more stable, and the same results are achieved, if the eigenvectors e_m and f_m are calculated from the symmetric matrices

$$[S_{x,x}]^{-1/2} [S_{x,y}] [S_{y,y}]^{-1} [S_{y,x}] [S_{x,x}]^{-1/2} \tag{14.30a}$$

and

$$[S_{y,y}]^{-1/2} [S_{y,x}] [S_{x,x}]^{-1} [S_{x,y}] [S_{y,y}]^{-1/2}, \tag{14.30b}$$

respectively. Equation 14.30a is dimensioned $(I \times I)$, and Equation 14.30b is dimensioned $(J \times J)$. Here the reciprocal square-root matrices must be symmetric (Equation 11.68), and not derived from Cholesky decompositions of the corresponding inverses or obtained by other means. The eigenvalue–eigenvector pairs for the symmetric matrices in Equation 14.30 can be computed using an algorithm specialized to the task, or through the singular value decomposition (Equation 11.72) operating on these matrices. In the latter case, the results are $[E] [\Lambda] [E]^T$ and $[F] [\Lambda] [F]^T$, respectively (compare Equations 11.72 and 11.52a), where the columns of $[E]$ are the e_m and the columns of $[F]$ are the f_m.

Regardless of how the eigenvectors e_m and f_m, and their common eigenvalues λ_m, are arrived at, the canonical correlations and canonical vectors are calculated from them. The canonical correlations are simply the positive square roots of the M nonzero eigenvalues,

$$r_{C_m} = \sqrt{\lambda_m}, \quad m = 1,\dots, M \tag{14.31}$$

The pairs of canonical vectors are calculated from the corresponding pairs of eigenvectors, using

$$\left. \begin{array}{l} a_m = [S_{x,x}]^{-1/2} e_m \\ b_m = [S_{y,y}]^{-1/2} f_m \end{array} \right\}, \quad m = 1,\dots, M. \tag{14.32}$$

Since $||e_m|| = ||f_m|| = 1$, this transformation ensures unit variances for the canonical variates, that is,

$$\text{Var}(v_m) = a_m^T [S_{x,x}] a_m = e_m^T [S_{x,x}]^{-1/2} [S_{x,x}] [S_{x,x}]^{-1/2} e_m = e_m^T e_m = 1, \tag{14.33}$$

because $[S_{x,x}]^{-1/2}$ is symmetric and the eigenvectors e_m are mutually orthogonal. An obvious analogous equation can be written for the variances $\text{Var}(w_m)$.

Extraction of eigenvalue–eigenvector pairs from large matrices can require large amounts of computing. However, the eigenvector pairs e_m and f_m are related in a way that makes it unnecessary to compute the eigendecompositions of both Equations 14.30a and 14.30b (or, both Equations 14.29a and 14.29b). For example, each f_m can be computed from the corresponding e_m using

$$f_m = \frac{[S_{y,y}]^{-1/2} [S_{y,x}] [S_{x,x}]^{-1/2} e_m}{\left\| [S_{y,y}]^{-1/2} [S_{y,x}] [S_{x,x}]^{-1/2} e_m \right\|}, \quad m = 1,...,M. \tag{14.34}$$

Here the Euclidean norm in the denominator ensures $||f_m|| = 1$. The eigenvectors e_m can be calculated from the corresponding f_m by reversing their roles, and reversing the matrix subscripts, in this equation.

CCA Via SVD

The special properties of the singular value decomposition (Equation 11.72) can be used to find both sets of the e_m and f_m pairs, together with the corresponding canonical correlations. This is achieved by computing the SVD

$$\underset{(I \times I)}{[S_{x,x}]^{-1/2}} \underset{(I \times J)}{[S_{x,y}]} \underset{(J \times J)}{[S_{y,y}]^{-1/2}} = \underset{(I \times J)}{[E]} \underset{(J \times J)}{[R_C]} \underset{(J \times J)}{[F]^T}. \tag{14.35}$$

The left-hand side of Equation 14.35 expresses the covariance between the vectors x and y after each has been subjected to the Mahalanobis transformation (Equation 12.21), which has also been called a *whitening transformation* (DelSole and Tippett, 2007; Swenson, 2015). This transformation both decorrelates the variables and scales them to unit variance. Equation 14.35 for the resulting covariance matrix is an instance of Equation 11.89 for the covariances among pairs of linear combinations because the square-root matrices are symmetric. As before the columns of $[E]$ are the e_m, the columns of $[F]$ are the f_m, and the diagonal matrix $[R_C]$ contains the canonical correlations. Here it has been assumed that $I \geq J$, but if $I < J$ the roles of x and y can be reversed in Equation 14.35. The canonical vectors are calculated as before, using Equation 14.32.

Example 14.3. The Computations behind Example 14.1

In Example 14.1 the canonical correlations and canonical vectors were given, with their computations deferred. Since $I = J$ in this example, the matrices required for these calculations are obtained by quartering $[S_C]$ (Equation 14.13) to yield

$$[S_{x,x}] = \begin{bmatrix} 59.516 & 75.433 \\ 75.433 & 185.467 \end{bmatrix}, \tag{14.36a}$$

$$[S_{y,y}] = \begin{bmatrix} 61.847 & 56.119 \\ 56.119 & 77.581 \end{bmatrix}, \tag{14.36b}$$

and

$$[S_{y,x}] = [S_{x,y}]^T = \begin{bmatrix} 58.070 & 81.633 \\ 51.697 & 110.800 \end{bmatrix}. \tag{14.36c}$$

The eigenvectors e_m and f_m, respectively, can be computed either from the pair of asymmetric matrices (Equation 14.29)

$$[S_{x,x}]^{-1} [S_{x,y}] [S_{y,y}]^{-1} [S_{y,x}] = \begin{bmatrix} 0.830 & 0.377 \\ 0.068 & 0.700 \end{bmatrix} \tag{14.37a}$$

and

$$[S_{y,y}]^{-1} [S_{y,x}] [S_{x,x}]^{-1} [S_{x,y}] = \begin{bmatrix} 0.845 & 0.259 \\ 0.091 & 0.686 \end{bmatrix}; \tag{14.37b}$$

or the symmetric matrices (Equation 14.30).

$$[S_{x,x}]^{-1/2} [S_{x,y}] [S_{y,y}]^{-1} [S_{y,x}] [S_{x,x}]^{-1/2} = \begin{bmatrix} 0.768 & 0.172 \\ 0.172 & 0.757 \end{bmatrix} \tag{14.38a}$$

and

$$[S_{y,y}]^{-1/2} [S_{y,x}] [S_{x,x}]^{-1} [S_{x,y}] [S_{y,y}]^{-1/2} = \begin{bmatrix} 0.800 & 0.168 \\ 0.168 & 0.726 \end{bmatrix}. \tag{14.38b}$$

The numerical stability of the computations is better if Equations 14.38a and 14.38b are used, but in either case the eigenvectors of Equations 14.37a and 14.38a are

$$e_1 = \begin{bmatrix} 0.719 \\ 0.695 \end{bmatrix} \text{ and } e_2 = \begin{bmatrix} -0.695 \\ 0.719 \end{bmatrix}, \tag{14.39}$$

with corresponding eigenvalues $\lambda_1 = 0.938$ and $\lambda_2 = 0.593$. The eigenvectors of Equations 14.37b and 14.38b are

$$f_1 = \begin{bmatrix} 0.780 \\ 0.626 \end{bmatrix} \text{ and } f_2 = \begin{bmatrix} -0.626 \\ 0.780 \end{bmatrix}, \tag{14.40}$$

again with eigenvalues $\lambda_1 = 0.938$ and $\lambda_2 = 0.593$. However, once the eigenvectors e_1 and e_2 have been computed it is not necessary to compute the eigendecomposition for either Equation 14.37b or Equation 14.38b, because their eigenvectors can also be obtained through Equation 14.34:

$$f_1 = \begin{bmatrix} 0.8781 & 0.1788 \\ 0.0185 & 0.8531 \end{bmatrix} \begin{bmatrix} 0.719 \\ 0.695 \end{bmatrix} \bigg/ \left\| \begin{bmatrix} 0.8781 & 0.1788 \\ 0.0185 & 0.8531 \end{bmatrix} \begin{bmatrix} 0.719 \\ 0.695 \end{bmatrix} \right\| = \begin{bmatrix} 0.780 \\ 0.626 \end{bmatrix} \tag{14.41a}$$

and

$$f_2 = \begin{bmatrix} 0.8781 & 0.1788 \\ 0.0185 & 0.8531 \end{bmatrix} \begin{bmatrix} -0.695 \\ 0.719 \end{bmatrix} \bigg/ \left\| \begin{bmatrix} 0.8781 & 0.1788 \\ 0.0185 & 0.8531 \end{bmatrix} \begin{bmatrix} -0.695 \\ 0.719 \end{bmatrix} \right\| = \begin{bmatrix} -0.626 \\ 0.780 \end{bmatrix}, \tag{14.41b}$$

since

$$[S_{y,y}]^{-1/2}[S_{y,x}][S_{x,x}]^{-1/2} = \begin{bmatrix} 0.1960 & -0.0930 \\ -0.0930 & 0.1699 \end{bmatrix} \begin{bmatrix} 58.070 & 81.633 \\ 51.697 & 110.800 \end{bmatrix} \begin{bmatrix} 0.1788 & -0.0522 \\ -0.0522 & 0.0917 \end{bmatrix}$$
$$= \begin{bmatrix} 0.8781 & 0.1788 \\ 0.0185 & 0.8531 \end{bmatrix}. \tag{14.41c}$$

The two canonical correlations are $r_{C_1} = \sqrt{\lambda_1} = 0.969$ and $r_{C_2} = \sqrt{\lambda_2} = 0.770$. The four canonical vectors are

$$a_1 = [S_{x,x}]^{-1/2}e_1 = \begin{bmatrix} .1788 & -.0522 \\ -.0522 & .0917 \end{bmatrix} \begin{bmatrix} .719 \\ .695 \end{bmatrix} = \begin{bmatrix} .0923 \\ .0263 \end{bmatrix}, \tag{14.42a}$$

$$a_2 = [S_{x,x}]^{-1/2}e_2 = \begin{bmatrix} .1788 & -.0522 \\ -.0522 & .0917 \end{bmatrix} \begin{bmatrix} -.695 \\ .719 \end{bmatrix} = \begin{bmatrix} -.1618 \\ .1022 \end{bmatrix}, \tag{14.42b}$$

$$b_1 = [S_{y,y}]^{-1/2}f_1 = \begin{bmatrix} .1960 & -.0930 \\ -.0930 & .1699 \end{bmatrix} \begin{bmatrix} .780 \\ .626 \end{bmatrix} = \begin{bmatrix} .0946 \\ .0338 \end{bmatrix}, \tag{14.42c}$$

and

$$b_2 = [S_{y,y}]^{-1/2}f_2 = \begin{bmatrix} .1960 & -.0930 \\ -.0930 & .1699 \end{bmatrix} \begin{bmatrix} -.626 \\ .780 \end{bmatrix} = \begin{bmatrix} -.1952 \\ .1907 \end{bmatrix}. \tag{14.42d}$$

Alternatively, the eigenvectors e_m and f_m can be obtained through the SVD (Equation 14.35) of the transpose of the matrix in Equation 14.41c (compare the left-hand sides of these two equations). The result is

$$\begin{bmatrix} .8781 & .0185 \\ .1788 & .8531 \end{bmatrix} = \begin{bmatrix} .719 & -.695 \\ .695 & .719 \end{bmatrix} \begin{bmatrix} .969 & 0 \\ 0 & .770 \end{bmatrix} \begin{bmatrix} .780 & .626 \\ -.626 & .780 \end{bmatrix}. \tag{14.43}$$

The canonical correlations are in the diagonal matrix $[R_C]$ in the middle of Equation 14.43. The eigenvectors are in the matrices $[E]$ and $[F]^T$ on either side of it, and can be used to compute the corresponding canonical vectors, as in Equation 14.42. ◇

14.3. MAXIMUM COVARIANCE ANALYSIS (MCA)

14.3.1. Definition of MCA

Maximum covariance analysis (MCA) is a similar technique to CCA, in that it finds pairs of linear combinations of two sets of vector data x and y (Equation 14.2) such that the squares of their covariances

$$\mathrm{Cov}(v_m, w_m) = a_m^T[S_{x,y}]b_m \tag{14.44}$$

(rather than their correlations, as in CCA) are maximized, subject to the constraint that the vectors a_m and b_m are orthonormal. Maximization of squared covariance allows for the possibility that a pair of vectors a_m and b_m may yield a negative covariance in Equation 14.44. As in CCA, the number of such pairs $M = \min$ (I, J) is equal to the smaller of the dimensions of the data vectors x and y, and each succeeding pair of projection vectors are chosen according to the maximization criterion, subject to the orthonormality constraint. In a typical application to atmospheric data, $x(t)$ and $y(t)$ are both time series of spatial fields, or the leading principal components of these fields, and so their projections in Equation 14.2 form time series also.

Computationally, the vectors a_m and b_m are found through a singular value decomposition (Equation 11.72) of the matrix $[S_{x,y}]$ in Equation 14.1, containing the cross-covariances between the elements of x and y,

$$\underset{(I \times J)}{[S_{x,y}]} = \underset{(I \times J)}{[A]} \; \underset{(J \times J)}{[\Omega]} \; \underset{(J \times J)}{[B]}^{T}.$$ (14.45)

The left singular vectors a_m are the columns of the matrix $[A]$ and the right singular vectors b_m are the columns of the matrix $[B]$ (i.e., the rows of $[B]^{T}$). The elements ω_m of the diagonal matrix $[\Omega]$ of singular values are the covariances (Equation 14.44) between the pairs of linear combinations in Equation 14.2. Comparison of Equation 14.45 with its CCA counterpart in Equation 14.35 shows that MCA is computed on the basis of the unwhitened (not subjected to Mahalanobis transformations) data vectors x' and y', and that the projection vectors a_m and b_m in MCA are not scaled subsequent to the SVD (cf. Equation 14.32 for CCA).

The proportions of the variances of the underlying variables represented by the projections v_m and w_m are

$$R_m^2(x) = \frac{a_m^T [S_{x,x}] a_m}{tr([S_{x,x}])}$$ (14.46a)

and

$$R_m^2(y) = \frac{b_m^T [S_{y,y}] b_m}{tr([S_{y,y}])},$$ (14.46b)

the numerators of which are $\text{Var}(v_m)$ and $\text{Var}(w_m)$, respectively. The homogeneous correlations are

$$\text{Corr}(v_m, x) = \frac{a_m^T [S_{x,x}] [D_x]^{-1}}{\left(a_m^T [S_{x,x}] a_m\right)^{1/2}}$$ (14.47a)

and

$$\text{Corr}(w_m, y) = \frac{b_m^T [S_{y,y}] [D_y]^{-1}}{\left(b_m^T [S_{y,y}] b_m\right)^{1/2}},$$ (14.47b)

which differ from their counterparts for CCA in Equation 14.7 because the square roots of the variances of the projection variables v_m and w_m in the denominators of Equations 14.47 are not equal to 1. Similarly, the heterogeneous correlations are

$$\text{Corr}(v_m, y) = \frac{a_m^T [S_{x,y}] [D_y]^{-1}}{\left(a_m^T [S_{x,x}] a_m\right)^{1/2}}$$ (14.48a)

and

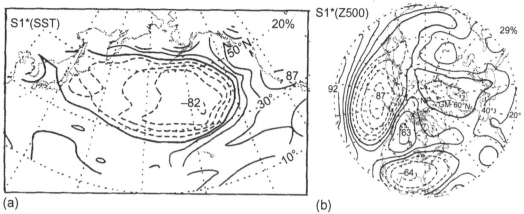

FIGURE 14.5   Homogeneous correlation maps of (a) average winter sea-surface temperatures in the northern Pacific Ocean, and (b) hemispheric winter 500 mb heights, derived from MCA. These are very similar to the corresponding CCA result in Figure 14.1. *From Wallace et al. (1992). © American Meteorological Society. Used with permission.*

$$\text{Corr}(w_m, x) = \frac{b_m^T [S_{y,x}][D_x]^{-1}}{\left(b_m^T [S_{y,y}] b_m\right)^{1/2}},$$

(14.48b)

which correspond to Equations 14.8.

Because the machinery of the singular value decomposition is used to find the vectors a_m and b_m, and the associated covariances ω_m, maximum covariance analysis sometimes unfortunately is known as SVD analysis. As illustrated earlier in this chapter and elsewhere in this book, the singular value decomposition has a rather broader range of uses (e.g., Golub and van Loan, 1996). In recognition of the parallels with CCA, the technique is also sometimes called *canonical covariance analysis* in which case the ω_m are called the canonical covariances. The method is also known as Co-inertia Analysis in the biology literature.

There are two main distinctions between CCA and MCA. The first is that CCA maximizes correlation, whereas MCA maximizes covariance. The leading CCA modes may capture relatively little of the corresponding variances (and thus yield small covariances even if the canonical correlations are high). On the other hand, MCA will find linear combinations with large covariances, which may result more from large variances than a large correlation. The second difference is that the vectors a_m and b_m in maximum covariance analysis are orthogonal, and the projections v_m and w_m of the data onto them are in general correlated, whereas the canonical variates in CCA are uncorrelated but the corresponding canonical vectors are not generally orthogonal. Bretherton et al. (1992), Cherry (1996), Tippett et al. (2008), and Van den Dool (2007) compare the two methods in greater detail.

It is not unusual to find similar results for CCA and MCA applied to the same data sets. For example, Figure 14.5 shows a pair of MCA-derived homogeneous correlation patterns for winter northern Pacific SSTs (a) and corresponding 500 mb heights (b), which are both very similar to their counterparts in Figure 14.1 that were based on CCA.

Example 14.4. Maximum Covariance Analysis of the January 1987 Temperature Data

Singular value decomposition of (the transpose of) the cross-covariance submatrix $[S_{x,y}]$ in Equation 14.36c yields

$$\begin{bmatrix} 58.07 & 51.70 \\ 81.63 & 110.8 \end{bmatrix} = \begin{bmatrix} .4876 & .8731 \\ .8731 & -.4876 \end{bmatrix} \begin{bmatrix} 157.4 & 0 \\ 0 & 14.06 \end{bmatrix} \begin{bmatrix} .6325 & .7745 \\ .7745 & -.6325 \end{bmatrix}. \tag{14.49}$$

The results are qualitatively similar to the CCA of the same data in Example 14.1. The first left and right vectors, $a_1 = [0.4876, 0.8731]^T$ and $b_1 = [0.6325, 0.7745]^T$, respectively, resemble the first pair of canonical vectors a_1 and b_1 in Example 14.1 in that both put positive weights on both variables in both data sets. Here the weights are closer in magnitude and emphasize the minimum temperatures rather than the maximum temperatures. The covariance between the linear combinations defined by these vectors is 157.4, which is larger than the covariance between any other pair of linear combinations for these data, subject to $\|a_1\| = \|b_1\| = 1$. The corresponding correlation is

$$\text{Corr}(v_1, w_1) = \frac{\omega_1}{(\text{Var}(v_1)\,\text{Var}(w_1))^{1/2}} = \frac{\omega_1}{\left(a_1^T [S_{x,x}]\, a_1\right)^{1/2} \left(b_1^T [S_{y,y}]\, b_1\right)^{1/2}}, \tag{14.50}$$

$$= \frac{157.44}{(219.8)^{1/2}(126.3)^{1/2}} = 0.945$$

which is large, but necessarily smaller than $r_{C_1} = 0.969$ for the CCA of the same data.

The second pair of vectors, $a_2 = [0.8731, -0.4876]^T$ and $b_2 = [0.7745, -0.6325]^T$, are also similar to the second pair of canonical vectors for the CCA in Example 14.1, in that they also describe a contrast between the maximum and minimum temperatures that can be interpreted as being related to the diurnal temperature ranges. The covariance of the second pair of linear combinations is ω_2, corresponding to a correlation of 0.772. This correlation is slightly larger than the second canonical correlation in Example 14.1, but has not been limited by the CCA constraint that the correlations between v_1 and v_2, and w_1 and w_2 must be zero.

The proportions of variability in the original variables captured by the MCA variables are

$$R_1^2(x) = \frac{[.4876\,.8731] \begin{bmatrix} 59.516 & 75.733 \\ 75.433 & 185.467 \end{bmatrix} \begin{bmatrix} .4876 \\ .8731 \end{bmatrix}}{59.516 + 185.467} = 0.897 \tag{14.51a}$$

$$R_2^2(x) = \frac{[.8731 - .4876] \begin{bmatrix} 59.516 & 75.733 \\ 75.433 & 185.467 \end{bmatrix} \begin{bmatrix} .8731 \\ -.4876 \end{bmatrix}}{59.516 + 185.467} = 0.103 \tag{14.52b}$$

$$R_1^2(y) = \frac{[.6325\,.7745] \begin{bmatrix} 61.847 & 56.119 \\ 56.119 & 77.581 \end{bmatrix} \begin{bmatrix} .6325 \\ .7745 \end{bmatrix}}{61.847 + 77.581} = 0.906 \tag{14.53c}$$

and

$$R_2^2(y) = \frac{[.7745 - .6325] \begin{bmatrix} 61.847 & 56.119 \\ 56.119 & 77.581 \end{bmatrix} \begin{bmatrix} .7745 \\ -.6325 \end{bmatrix}}{61.847 + 77.581} = 0.094 \tag{14.54d}$$

◇

The papers of Bretherton et al. (1992) and Wallace et al. (1992) have been influential advocates for the use of maximum covariance analysis. One advantage over CCA that sometimes is cited is that no matrix inversions are required, so that a maximum covariance analysis can be computed even if $n < \max (I, J)$. However, both techniques are subject to similar sampling problems in limited-data situations, so it is not clear that this advantage is of practical importance, and in any case dimension reduction through use of the leading principal components is often employed. Some cautions regarding maximum covariance analysis have been offered by Cherry (1997) and Hu (1997). Newman and Sardeshmukh (1995) emphasize that the a_m and b_m vectors may not represent physical modes of their respective fields, just as the eigenvectors in PCA do not necessarily represent physically meaningful modes.

14.3.2. Forecasting With MCA

The results of a MCA can be used to forecast one of the fields, say y, using the x field as the predictor, analogously to the CCA forecasts described in Section 14.2.3. However, since the projections in Equation 14.2 are not uncorrelated for different m in MCA, simultaneous application of M simple linear regressions, as in Equation 14.20 for CCA, will in general not yield optimal predictions. However, if the projection variables in Equation 14.2 have been computed from anomaly vectors x' and y' then the individual MCA prediction regressions will have zero intercept.

Because the v variables for the predictor fields are not independent of the w variables for the predictand fields, in principle any or all M of the v_i may be meaningful predictors for any of the w_j. The resulting M regressions can be represented jointly as

$$\underset{(J \times 1)}{\hat{w}} = \underset{(J \times J)}{[\beta]^T} \underset{(J \times 1)}{v} = \underset{(J \times J)}{[\beta]^T} \underset{(J \times I)}{[A]^T} \underset{(I \times 1)}{x'} , \tag{14.52}$$

where the jth column of the $(J \times J)$ matrix $[\beta]$ contains the regression coefficients predicting w_j. This equation is the counterpart of Equation 14.23 for CCA except that the parameter matrix $[\beta]$ is in general not diagonal. Wilks (2014b) estimated each column of $[\beta]$ using backward elimination (Section 7.4.2), stopping when nominal p values for all coefficients were no greater than 0.01, although other predictor selection strategies could be used instead.

Equation 14.52 is a multivariate multiple regression (e.g., Johnson and Wichern, 2007), for which the counterpart to Equation 11.37, but including the residuals, is

$$\underset{(J \times n)}{[\hat{W}]} = \underset{(J \times J)}{[\beta]^T} \underset{(J \times n)}{[V]} + \underset{(J \times n)}{[\varepsilon]} . \tag{14.53}$$

Here each of the columns of the matrices $[\hat{W}]$, $[V]$, and $[\varepsilon]$ contains one of the vectors $\hat{w}$, v and the residual vector ε for the n training samples. The prediction covariance matrix is then

$$[S_{\hat{w}}] = \frac{1 + v^T \left([V][V]^T \right)^{-1} v}{n - M - 1} [\varepsilon][\varepsilon]^T , \tag{14.54}$$

where the residual matrix $[\varepsilon]$ has been computed from the training data using Equation 14.53. Vector mean forecasts for the predictand anomalies y' and their covariance matrix can then be calculated using Equations 14.9 and 14.26 or, if the MCA has been computed using the leading principal components, Equations 14.27 and 14.28.

14.4. REDUNDANCY ANALYSIS (RA)

Both CCA and MCA treat the x and y variables symmetrically, in the sense that the resulting linear combinations and their correlations or covariances are the same if the roles of the two are reversed. In contrast, RA specifically computes the predictand projection vectors b_m in Equation 14.2b which yield maximal variances for the linear combinations w_m conditional on the predictor linear combinations v_m defined by the vectors a_m in Equation 14.2a (e.g., Tippett et al., 2008; Tyler, 1982; Von Storch and Zwiers, 1999). Redundancy analysis is thus specifically oriented toward predicting the y vectors on the basis of the x vectors.

The predictand linear combination vectors b_m are the eigenvectors of the real, symmetric $(J \times J)$ covariance matrix for the predictand vector given the predictors,

$$[S_{\hat{y}}] = [S_{y,x}][S_{x,x}]^{-1}[S_{x,y}], \tag{14.55}$$

after which the predictor linear combination vectors a_m can be computed using

$$a_j = \lambda_j^{1/2}[S_{x,x}]^{-1}[S_{x,y}]b_j, \tag{14.56}$$

where λ_j is the jth eigenvalue of the matrix in Equation 14.55 and b_j is its corresponding eigenvector.

As will be noted in Section 14.5, the paired pattern finding in RA operates on Mahalanobis-transformed (or whitened) anomaly predictor vectors x', but the predictand anomaly vectors y' are untransformed. Accordingly the predictor linear combinations v_j are uncorrelated with each other so that the regressions predicting the w_j can be considered independently of each other, as in CCA. The forecast vector mean is then computed using Equation 14.52, where the diagonal matrix of regression coefficients is

$$[\beta] = \left([B]^T[S_{y,x}][A]\right)^{1/2} = [\Lambda]^{1/2}, \tag{14.57}$$

where the columns of the matrices [A] and [B] are the vectors a_m and b_m, respectively, and the elements of the diagonal matrix $[\Lambda]$ are the eigenvalues of the matrix in Equation 14.55.

Because the predictor linear combinations v_m are uncorrelated, the prediction covariance matrix is of the same diagonal form as for CCA, Equation 14.24. The individual elements are

$$s_{\hat{w}_m}^2 = \left(-\lambda_m + \sum_{k=1}^{J} b_{k,m}^2\right)\left[1 + \frac{1}{n} + \frac{v_{0,m}^2}{\sum_{i=1}^{n}v_{i,m}^2}\right], \tag{14.58}$$

which differs from its CCA counterpart in Equation 14.25 only in the parenthetical expression for the residual variance. Again the λ_m are the eigenvalues of Equation 14.55. If the predictands are the principal components of the y vectors, the terms in the summation are replaced by $\gamma_m b_{k,m}^2$, where the γ_m are the eigenvalues associated with the eigenvectors onto which the predictands are projected.

Relatively little experience with RA for forecasting geophysical fields has yet accumulated, although the forecast skill for seasonal North American temperature forecasts was found to be comparable to those of CCA and MCA in Wilks (2014b).

14.5. UNIFICATION AND GENERALIZATION OF CCA, MCA, AND RA

Swenson (2015) notes that CCA, MCA, and RA are unified mathematically as singular value decompositions of the matrix form

$$[S_{x,x}]^{(\alpha-1)/2}[S_{x,y}][S_{y,y}]^{(\beta-1)/2}, \tag{14.59}$$

where α and β are parameters. For $\alpha = \beta = 0$, Equation 14.35 for CCA is obtained. For $\alpha = \beta = 1$, Equation 14.45 for MCA results, because $[S_{x,x}]^0 = [S_{y,y}]^0 = [I]$. Equation 14.59 also encompasses RA, when $\alpha = 0$ and $\beta = 1$, which indicates that the x vectors are subject to the Mahalanobis whitening transformation whereas the y vectors are not.

Intermediate, or *partial whitening*, forms of analysis are also allowed by Equation 14.59, when $0 < \alpha < 1$ and/or $0 < \beta < 1$. Partial whitening partially decorrelates the covariance matrices for x and y, so that the off-diagonal elements are reduced in absolute value but not shrunk fully to zero. Partial whitening also yields variances (diagonal elements) that are intermediate between the untransformed elements $s_{x,x}$ or $s_{y,y}$ when $\alpha = 0$ or $\beta = 0$, and unit variance when $\alpha = 1$ or $\beta = 1$. Accordingly the process can be seen as a regularization procedure. The partial whitening of $[S_{x,x}]$ or $[S_{y,y}]$ moves their eigenvalues toward 1 while leaving the eigenvectors unchanged.

Regarding the α and β parameters as free and tunable allows a 2-dimensional continuum of paired relationships to be considered. For example, Figure 14.6 shows contours of cross-validated specification skill in an artificial setting, as functions of α and β. The special cases of CCA, MCA, and RA are indicated by the large circles. For this synthetic example the best cross-validated result is obtained for $\alpha = \beta = 0.35$.

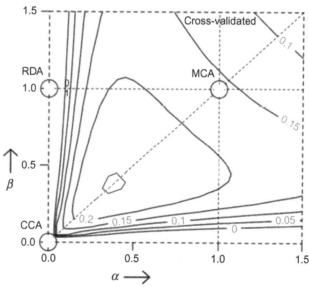

FIGURE 14.6 Dependence of specification skill on the partial-whitening parameters α and β in an artificial setting, showing best performance for $\alpha = \beta = 0.35$ in this case. The special cases of CCA, MCA, and RA are indicated by the circles. *Modified from Swenson (2015). © American Meteorological Society. Used with permission.*

14.6. EXERCISES

14.1. Using the information in Table 14.1 and the data in Table A.1, calculate the values of the canonical variables v_1 and w_1 for 6 January and 7 January.

14.2. The Ithaca maximum and minimum temperatures for 1 January 1988 were $x = (38°F, 16°F)^T$. Use the CCA in Example 14.1 to "forecast" the Canandaigua temperatures for that day.

14.3. Separate PCAs of the correlation matrices for the Ithaca and Canandaigua data in Table A.1 (after square-root transformation of the precipitation data) yields

$$[E_{\text{Ith}}] = \begin{bmatrix} 0.599 & 0.524 & 0.606 \\ 0.691 & 0.044 & -0.721 \\ 0.404 & -0.851 & 0.336 \end{bmatrix} \quad \text{and} \quad [E_{\text{Can}}] = \begin{bmatrix} 0.657 & 0.327 & 0.679 \\ 0.688 & 0.107 & -0.718 \\ 0.308 & -0.939 & 0.155 \end{bmatrix}, \quad (14.60)$$

with corresponding eigenvalues $\lambda_{\text{Ith}} = [1.883, 0.927, 0.190]^T$ and $\lambda_{\text{Can}} = (1.904, 0.925, 0.171)^T$. Given also the cross-correlations for these data.

$$[R_{I,C}] = \begin{bmatrix} 0.957 & 0.762 & 0.076 \\ 0.761 & 0.924 & 0.358 \\ 0.166 & 0.431 & 0.904 \end{bmatrix}, \quad (14.61)$$

compute the CCA after truncation to the two leading principal components for each of the locations (and notice that computational simplifications follow from using the principal components), by

a. Computing $[S_C]$, where c is the (4×1) vector $[u_{\text{Ith}}, u_{\text{Can}}]^T$, and then

b. Finding the canonical vectors and canonical correlations.

Chapter 15

Discrimination and Classification

15.1. DISCRIMINATION VS. CLASSIFICATION

This chapter deals with the problem of discerning membership among some number of groups, on the basis of a K-dimensional vector x of attributes that is observed for each member of each group. It is assumed that the number of groups G is known in advance; that this collection of groups constitutes a MECE partition of the sample space; that each data vector belongs to one and only one group; and that a set of training data is available, in which the group membership of each of the data vectors x_i, $i = 1, \ldots, n$, is known with certainty. The related problem, in which we know neither the group memberships of the data vectors nor the number of groups overall, is treated in Chapter 16.

The term *discrimination* refers to the process of estimating functions of the training data x_i that best describe the features separating the known group memberships of each x_i. In cases where this can be achieved well with three or fewer functions, it may be possible to express the discrimination graphically. The statistical basis of discrimination is the notion that each of the G groups corresponds to a different multivariate PDF for the data, $f_g(x)$, $g = 1, \ldots, G$. It is not necessary to assume multinormality for these distributions, but informative connections can be made with the material presented in Chapter 12 when that assumption is supported by the data.

Classification refers to use of the discrimination rule(s) to assign data that were not part of the original training sample to one of the G groups, or to the estimation of probabilities $p_g(x)$, $g = 1, \ldots, G$, that the vector x belongs to group g. If the groupings of x pertain to a time after x itself has been observed, then classification is a natural tool to use for forecasting discrete events. That is, a forecast can be made by classifying the current observation x as belonging to the group that is forecast to occur, or by computing the probabilities $p_g(x)$ for the occurrence of each of the G events.

Most of this chapter describes well-established, mainly linear, methods for discrimination and classification. More exotic and flexible, but also more computationally demanding, approaches are presented in the final sections.

15.2. SEPARATING TWO POPULATIONS

15.2.1. Equal Covariance Structure: Fisher's Linear Discriminant

The simplest form of discriminant analysis involves distinguishing between $G = 2$ groups on the basis of K-dimensional vectors of observations x. A training sample must exist, consisting of n_1 observations of x known to have come from Group 1, and n_2 observations of x known to have come from Group 2. That is, the basic data are the two matrices $[X_1]$, dimensioned $(n_1 \times K)$, and $[X_2]$, dimensioned $(n_2 \times K)$. The goal

Statistical Methods in the Atmospheric Sciences. https://doi.org/10.1016/B978-0-12-815823-4.00015-8

is to find a linear function of the K elements of the observation vector, that is, the linear combination $a^T x$, called the *discriminant function*, that will best allow a future K-dimensional vector of observations to be classified as belonging to either Group 1 or Group 2.

Assuming that the two populations corresponding to the groups have the same covariance structure, the approach to this problem taken by the statistician R.A. Fisher was to find the vector a as that direction in the K-dimensional space of the data that maximizes the separation in standard deviation units of the two vector means, when the data are projected onto a. This criterion is equivalent to choosing a to maximize

$$\frac{\left(a^T \bar{x}_1 - a^T \bar{x}_2\right)^2}{a^T \left[S_{\text{pool}}\right] a}. \tag{15.1}$$

Here the two mean vectors are calculated separately for each group, as would be expected, according to

$$\bar{x}_g = \frac{1}{n_g} \left[X_g\right]^T \mathbf{1} = \begin{bmatrix} \dfrac{1}{n_g} \displaystyle\sum_{i=1}^{n_g} x_{i,1} \\[2ex] \dfrac{1}{n_g} \displaystyle\sum_{i=1}^{n_g} x_{i,2} \\[1ex] \vdots \\[1ex] \dfrac{1}{n_g} \displaystyle\sum_{i=1}^{n_g} x_{i,K} \end{bmatrix}, \ g = 1,2; \tag{15.2}$$

where $\mathbf{1}$ is a $(n \times 1)$ vector containing only 1's, and n_g is the number of training-data vectors x in the gth group. The estimated common covariance matrix for the two groups, $[S_{\text{pool}}]$, is calculated using Equation 12.42b. If $n_1 = n_2$, the result is that each element of $[S_{\text{pool}}]$ is the simple average of the corresponding elements of $[S_1]$ and $[S_2]$. Note that multivariate normality has not been assumed for either of the groups. Rather, regardless of their distributions and whether or not those distributions are of the same form, all that has been assumed is that their underlying population covariance matrices $[\Sigma_1]$ and $[\Sigma_2]$ are equal.

Finding the direction a maximizing Equation 15.1 reduces the discrimination problem from one of sifting through and comparing relationships among the K elements of the data vectors, to looking at a single number. That is, the data vector x is transformed to a new scalar variable, $\delta_1 = a^T x$, known as *Fisher's linear discriminant function*. The groups of K-dimensional multivariate data are essentially reduced to groups of univariate data with different means (but equal variances), distributed along the a axis. The discriminant vector locating this direction of maximum separation is given by

$$a = \left[S_{\text{pool}}\right]^{-1} (\bar{x}_1 - \bar{x}_2), \tag{15.3}$$

so that Fisher's linear discriminant function is

$$\delta_1 = a^T x = (\bar{x}_1 - \bar{x}_2)^T \left[S_{\text{pool}}\right]^{-1} x. \tag{15.4}$$

As indicated in Equation 15.1, this transformation to Fisher's linear discriminant function maximizes the scaled distance between the two sample means in the training sample, which is

$$a^T (\bar{x}_1 - \bar{x}_2) = (\bar{x}_1 - \bar{x}_2)^T \left[S_{\text{pool}}\right]^{-1} (\bar{x}_1 - \bar{x}_2) = D^2. \tag{15.5}$$

That is, this maximum distance between the projections of the two sample means is exactly the Mahalanobis distance between them, according to $[S_{pool}]$.

A decision to classify a future observation x as belonging to either Group 1 or Group 2 can now be made according to the value of the scalar $\delta_1 = a^T x$. This dot product is a one-dimensional (i.e., scalar) projection of the vector x onto the direction of maximum separation, a. The discriminant function δ_1 is essentially a new variable, analogous to the new variable u in PCA and the new variables v and w in CCA, produced as a linear combination of the elements of a data vector x. The simplest way to classify an observation x is to assign it to Group 1 if the projection $a^T x$ is closer to the projection of the Group 1 mean onto the direction a, and assign it to Group 2 if $a^T x$ is closer to the projection of the mean of Group 2. Along the a axis, the midpoint between the means of the two groups is given by the projection of the average of these two mean vectors onto the vector a,

$$\hat{m} = \frac{1}{2}\left(a^T \bar{x}_1 + a^T \bar{x}_2\right) = \frac{1}{2}a^T\left(\bar{x}_1 + \bar{x}_2\right) = \frac{1}{2}\left(\bar{x}_1 - \bar{x}_2\right)^T \left[S_{pool}\right]^{-1}\left(\bar{x}_1 + \bar{x}_2\right). \tag{15.6}$$

Given an observation x_0 whose group membership is unknown, this simple midpoint criterion classifies it according to the rule

$$\text{Assign } x_0 \text{ to Group 1 if } a^T x_0 \geq \hat{m}, \tag{15.7a}$$

or

$$\text{Assign } x_0 \text{ to Group 2 if } a^T x_0 < \hat{m}. \tag{15.7b}$$

This classification rule divides the K-dimensional space of x into two regions, according to the (hyper-)plane perpendicular to a at the midpoint given by Equation 15.6. In two dimensions, the plane is divided into two regions according to the line perpendicular to a at this point. The volume in three dimensions is divided into two regions according to the plane perpendicular to a at this point, and so on for higher dimensions.

Example 15.1. Linear Discrimination in $K = 2$ Dimensions

Table 15.1 presents average July temperature and precipitation for cities in three regions of the United States. The data vectors include $K = 2$ elements each: one temperature element and one precipitation element. Consider the problem of distinguishing between membership in Group 1 vs. Group 2. This problem might arise if the locations in Table 15.1 represented the core portions of their respective climatic regions, and on the basis of these data we wanted to classify stations not listed in this table as belonging to one or the other of these two groups.

The mean vectors for the $n_1 = 10$ and $n_2 = 9$ data vectors in Groups 1 and 2 are

$$\bar{x}_1 = \begin{bmatrix} 80.6^\circ\text{F} \\ 5.67\,\text{in.} \end{bmatrix} \text{ and } \bar{x}_2 = \begin{bmatrix} 78.7^\circ\text{F} \\ 3.57\,\text{in.} \end{bmatrix}, \tag{15.8a}$$

and the two sample covariance matrices are

$$[S_1] = \begin{bmatrix} 1.47 & 0.65 \\ 0.65 & 1.45 \end{bmatrix} \text{ and } [S_2] = \begin{bmatrix} 2.01 & 0.06 \\ 0.06 & 0.17 \end{bmatrix}. \tag{15.8b}$$

Since $n_1 \neq n_2$ the pooled estimate for the common variance–covariance matrix is obtained by the weighted average specified by Equation 12.42b. The vector a pointing in the direction of maximum separation of the two sample mean vectors is then computed using Equation 15.3 as

TABLE 15.1 Average July Temperature (°F) and Precipitation (inches) for Locations in Three Regions of the United States, for the Period 1951–1980

Group 1: Southeast United States (o)			Group 2: Central United States (×)			Group 3: Northeast United States (+)		
Station	Temp.	Ppt.	Station	Temp.	Ppt.	Station	Temp.	Ppt.
Athens, GA	79.2	5.18	Concordia, KS	79.0	3.37	Albany, NY	71.4	3.00
Atlanta, GA	78.6	4.73	Des Moines, IA	76.3	3.22	Binghamton, NY	68.9	3.48
Augusta, GA	80.6	4.4	Dodge City, KS	80.0	3.08	Boston, MA	73.5	2.68
Gainesville, FL	80.8	6.99	Kansas City, MO	78.5	4.35	Bridgeport, CT	74.0	3.46
Huntsville, AL	79.3	5.05	Lincoln, NE	77.6	3.2	Burlington, VT	69.6	3.43
Jacksonville, FL	81.3	6.54	Springfield, MO	78.8	3.58	Hartford, CT	73.4	3.09
Macon, GA	81.4	4.46	St. Louis, MO	78.9	3.63	Portland, ME	68.1	2.83
Montgomery, AL	81.7	4.78	Topeka, KS	78.6	4.04	Providence, RI	72.5	3.01
Pensacola, FL	82.3	7.18	Wichita, KS	81.4	3.62	Worcester, MA	69.9	3.58
Savannah, GA	81.2	7.37						
Averages:	80.6	5.67		78.7	3.57		71.3	3.17

$$a = \begin{bmatrix} 1.73 & 0.37 \\ 0.37 & 0.84 \end{bmatrix}^{-1} \left(\begin{bmatrix} 80.6 \\ 5.67 \end{bmatrix} - \begin{bmatrix} 78.7 \\ 3.57 \end{bmatrix} \right)$$

$$= \begin{bmatrix} 0.640 & -.283 \\ -.283 & 1.309 \end{bmatrix} \begin{bmatrix} 1.9 \\ 2.10 \end{bmatrix} = \begin{bmatrix} 0.62 \\ 2.21 \end{bmatrix}.$$

(15.9)

Figure 15.1 illustrates the geometry of this problem. Here the data for the warmer and wetter southeastern stations of Group 1 are plotted as circles, and the central U.S. stations of Group 2 are plotted as ×'s. The vector means for the two groups are indicated by the heavy symbols. The direction a is not, and in general will not be, parallel to the line segment connecting the two group means. The projections of these two means onto a are indicated by the lighter dashed lines. The midpoint between these two projections locates the dividing point between the two groups in the one-dimensional discriminant space defined by a. The heavy dashed line perpendicular to the discriminant function δ_1 at this point divides the (temperature, precipitation) plane into two regions. Future points of unknown group membership falling above and to the right of this heavy dashed line would be classified as belonging to Group 1, and points falling below and to the left would be classified as belonging to Group 2.

Since the average of the mean vectors for Groups 1 and 2 is $[79.65, 4.62]^T$, the value of the dividing point is $\hat{m} = (0.62)(79.65) + (2.21)(4.62) = 59.59$. Of the 19 points in this training data, only that for Atlanta has been misclassified. For this station, $\delta_1 = a^T x = (0.62)(78.6) + (2.20)(4.73) = 59.18$. Since this value of δ_1 is slightly less than the midpoint value, Atlanta would be incorrectly classified as belonging to Group 2 (Equation 15.7). By contrast, the point for Augusta lies just to the Group 1 side of the heavy dashed line. For Augusta, $\delta_1 = a^T x = (0.62)(80.6) + (2.20)(4.40) = 59.70$, which is slightly greater than $\hat{m}$.

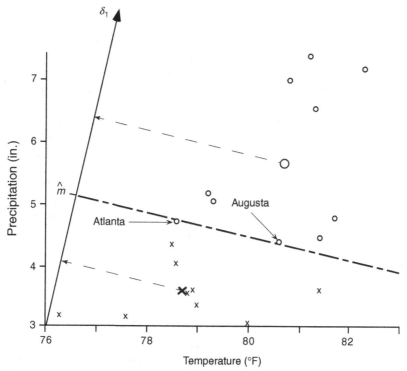

FIGURE 15.1 Illustration of the geometry of linear discriminant analysis applied to the southeastern (circles) and central (X's) U. S. data in Table 15.1. The (vector) means of the two groups of data are indicated by the heavy symbols, and their projections onto the discriminant function are indicated by the light dashed lines. The midpoint between these two projections, $\hat{m}$, defines the dividing line (heavier dashed line) used to assign future (temperature, precipitation) pairs to the groups. Of these training data, only the data point for Atlanta has been misclassified. Note that the discriminant function has been shifted to the right (i.e., does not pass through the origin, but is parallel to the vector a in Equation 15.9) in order to improve the clarity of the plot, but this does not affect the relative positions of the projections of the data points onto it.

Consider now the assignment to either Group 1 or Group 2 of two stations not listed in Table 15.1. For New Orleans, Louisiana, the average July temperature is 82.1°F, and the average July precipitation is 6.73 in. Applying Equation 15.7, we find $a^Tx = (0.62)(82.1)+(2.20)(6.73)=65.78 > 59.59$. Therefore New Orleans would be classified as belonging to Group 1. Similarly the average July temperature and precipitation for Columbus, Ohio, are 74.7°F and 3.37 in., respectively. For this station, $a^Tx = (0.62)(74.7)+(2.20)(3.37)=53.76 < 59.59$, which would result in Columbus being classified as belonging to Group 2. ◇

Example 15.1 was constructed with $K=2$ variables in each data vector in order to allow the geometry of the problem to be easily represented in two dimensions. However, it is not necessary to restrict the use of discriminant analysis to situations with only bivariate observations. In fact, discriminant analysis is potentially most powerful when allowed to operate on higher-dimensional data. For example, it would be possible to extend Example 15.1 to classifying stations according to average temperature and precipitation for all 12 months. If this were done, each data vector x would consist of $K=24$ values. The discriminant vector a would also consist of $K=24$ elements, but the dot product $\delta_1 = a^Tx$ would still be a single scalar that could be used to classify group memberships.

Usually high-dimensional vectors of atmospheric data exhibit substantial correlation among the K elements, and thus carry some redundant information. For example, the 12 monthly mean temperatures and 12 monthly mean precipitation values are not mutually independent. If only for computational economy, it can be a good idea to reduce the dimensionality of this kind of data before subjecting it to a discriminant analysis. This reduction in dimension is most commonly achieved through a principal component analysis (Chapter 13). When the groups in a discriminant analysis are assumed to have the same covariance structure, it seems natural to perform this PCA on the estimate of their common variance–covariance matrix, $[S_{pool}]$. However, if the shape of the dispersion of the group means (as characterized by Equation 15.18) is substantially different from $[S_{pool}]$, its leading principal components may not be good discriminators, and better results might be obtained from a discriminant analysis based on PCA of the overall covariance, $[S]$ (Jolliffe, 2002). If the data vectors are not of consistent units (some temperatures and some precipitation amounts, for example), it will make more sense to perform the PCA on the corresponding correlation matrix. The discriminant analysis can then be carried out using M-dimensional data vectors composed of elements that are the leading M principal components, rather than the original K-dimensional raw data vectors. The resulting discriminant function will then pertain to the principal components in the $(M \times 1)$ vector u, rather than to the original $(K \times 1)$ data, x. In addition, if the first two principal components account for a large fraction of the total variance, the data can effectively be visualized in a plot like Figure 15.1, in which the horizontal and vertical axes are the first two principal components.

15.2.2. Fisher's Linear Discriminant for Multivariate Normal Data

Use of Fisher's linear discriminant requires no assumptions about the specific nature of the distributions for the two groups, $f_1(x)$ and $f_2(x)$, except that they have equal covariance matrices. If in addition these are two multivariate normal distributions, or they are sufficiently close to multivariate normal for the sampling distributions of their means to be essentially multivariate normal according to the Central Limit Theorem, there are connections to the Hotelling T^2 test (Section 12.5) regarding differences between the two means.

In particular, Fisher's linear discriminant vector (Equation 15.3) identifies a direction that is identical to the linear combination of the data that is most strongly significant (Equation 12.57b), under the null hypothesis that the two population mean vectors are equal. That is, the vector a defines the direction maximizing the separation of the two means for both a discriminant analysis and the T^2 test. Furthermore, the distance between the two means in this direction (Equation 15.5) is their Mahalanobis distance with respect to the pooled estimate $[S_{pool}]$ of the common covariance $[\Sigma_1] = [\Sigma_2]$, which is proportional (through the factor $n_1^{-1} + n_2^{-1}$, in Equation 12.42a) to the 2-sample T^2 statistic itself (Equation 12.40).

In light of these relationships, one way to look at Fisher's linear discriminant, when applied to multivariate normal data, is as an implied test relating to the null hypothesis that $\mu_1 = \mu_2$. Even if this null hypothesis is true, the corresponding sample means in general will be different, and the result of the T^2 test is an informative necessary condition regarding the reasonableness of conducting the discriminant analysis. A multivariate normal distribution is fully defined by its mean vector and covariance matrix. Since $[\Sigma_1] = [\Sigma_2]$ already has been assumed, if in addition the two multivariate normal data groups are consistent with the condition $\mu_1 = \mu_2$, then there is no basis upon which to discriminate between them. Note, however, that rejecting the null hypothesis of equal means in the corresponding T^2 test is not a sufficient condition for good discrimination: arbitrarily small mean differences can be detected by this test as sample sizes increase, even though the scatter of the two data groups may overlap to such a limited degree that discrimination is completely pointless.

15.2.3. Minimizing Expected Cost of Misclassification

The point $\hat{m}$ on Fisher's discriminant function halfway between the projections of the two sample means is not always the best point at which to make a separation between groups. One might have prior information that the probability of membership in Group 1 is larger than that for Group 2, perhaps because Group 2 members are rather rare overall. If this is so, it would usually be desirable to move the classification boundary toward the Group 2 mean, with the result that more future observations x would be classified as belonging to Group 1. Similarly, if misclassifying a Group 1 data value as belonging to Group 2 were to be a more serious error than misclassifying a Group 2 data value as belonging to Group 1, again we would want to move the boundary toward the Group 2 mean.

One rational way to accommodate these considerations is to define the classification boundary based on the *expected cost of misclassification* (ECM) of a future data vector. Let p_1 be the prior probability (the unconditional probability according to previous information) that a future observation x_0 belongs to Group 1, and let p_2 be the prior probability that the observation x_0 belongs to Group 2. Define $P(2|1)$ to be the conditional probability that a Group 1 object is misclassified as belonging to Group 2, and $P(1|2)$ as the conditional probability that a Group 2 object is misclassified as belonging to Group 1. These conditional probabilities will depend on the two PDFs $f_1(x)$ and $f_2(x)$, respectively; and on the placement of the classification criterion, because these conditional probabilities will be given by the integrals of their respective PDFs over the regions in which classifications would be made to the other group. That is,

$$P(2|1) = \int_{R_2} f_1(x)dx \tag{15.10a}$$

and

$$P(1|2) = \int_{R_1} f_2(x)dx, \tag{15.10b}$$

where R_1 and R_2 denote the regions of the K-dimensional space of x in which classifications into Group 1 and Group 2, respectively, would be made. Unconditional probabilities of misclassification are given by the products of these conditional probabilities with the corresponding prior probabilities, that is, $P(2|1) p_1$ and $P(1|2) p_2$.

If $C(1|2)$ is the cost, or penalty, incurred when a Group 2 member is incorrectly classified as part of Group 1, and $C(2|1)$ is the cost incurred when a Group 1 member is incorrectly classified as part of Group 2, then the expected cost of misclassification will be

$$ECM = C(2|1) P(2|1) p_1 + C(1|2) P(1|2) p_2. \tag{15.11}$$

The classification boundary can be adjusted to minimize this expected cost of misclassification, through the effect of the boundary on the misclassification probabilities (Equations 15.10). The resulting classification rule is

$$\text{Assign } x_0 \text{ to Group 1 if } \frac{f_1(x_0)}{f_2(x_0)} \ge \frac{C(1|2) p_2}{C(2|1) p_1}, \tag{15.12a}$$

or

$$\text{Assign } x_0 \text{ to Group 2 if } \frac{f_1(x_0)}{f_2(x_0)} < \frac{C(1|2)\,p_2}{C(2|1)\,p_1}. \tag{15.12b}$$

That is, classification of x_0 depends on the ratio of its likelihoods according to the PDFs for the two groups (i.e., the ratio of the two density functions, evaluated at x_0), in relation to the ratios of the products of the misclassification costs and prior probabilities. Accordingly, it is not actually necessary to know the two misclassification costs specifically, but only their ratio, and likewise it is necessary only to know the ratio of the prior probabilities. If $C(1|2) \gg C(2|1)$—that is, if misclassifying a Group 2 member as belonging to Group 1 is especially undesirable—then the ratio of likelihoods on the left-hand side of Equation 15.12 must be quite large (x_0 must be substantially more plausible according to $f_1(x)$]) in order to assign x_0 to Group 1. Similarly, if Group 1 members are intrinsically rare, so that $p_1 \ll p_2$, a higher level of evidence must be met in order to classify x_0 as a member of Group 1. If both misclassification costs and prior probabilities are equal, or of the costs and priors compensate to make the right-hand sides of Equation 15.12 equal to 1, then classification is made according to the larger of $f_1(x_0)$ or $f_2(x_0)$.

Minimizing the ECM (Equation 15.11) does not require assuming that the distributions $f_1(x)$ or $f_2(x)$ have specific forms, or even that they are of the same parametric family. But it is necessary to know or assume a functional form for each of them in order to evaluate the left-hand side of Equation 15.12. Often it is assumed that both $f_1(x)$ and $f_2(x)$ are multivariate normal (possibly after data transformations for some or all of the elements of x), with equal covariance matrices that are estimated using $[S_{\text{pool}}]$. In this case Equation 15.12a, for the conditions under which x_0 would be assigned to Group 1, becomes

$$\frac{2\pi^{-K/2}\left|[S_{\text{pool}}]\right|^{-1/2} \exp\left(-\frac{1}{2}(x_0 - \bar{x}_1)^T [S_{\text{pool}}]^{-1}(x_0 - \bar{x}_1)\right)}{2\pi^{-K/2}\left|[S_{\text{pool}}]\right|^{-1/2} \exp\left(-\frac{1}{2}(x_0 - \bar{x}_2)^T [S_{\text{pool}}]^{-1}(x_0 - \bar{x}_2)\right)} \geq \frac{C(1|2)\,p_2}{C(2|1)\,p_1}, \tag{15.13a}$$

which, after some rearrangement, is equivalent to

$$(\bar{x}_1 - \bar{x}_2)^T [S_{\text{pool}}]^{-1} x_0 - \frac{1}{2}(\bar{x}_1 - \bar{x}_2)^T [S_{\text{pool}}]^{-1}(\bar{x}_1 + \bar{x}_2) \geq \ln\left(\frac{C(1|2)\,p_2}{C(2|1)\,p_1}\right). \tag{15.13b}$$

The left-hand side of Equation 15.13b looks elaborate, but its elements are familiar. In particular, its first term is exactly the linear combination $a^T x_0$ in Equation 15.7. The second term is the midpoint $\hat{m}$ between the two means when projected onto a, defined in Equation 15.6. Therefore if $C(1|2) = C(2|1)$ and $p_1 = p_2$ (or if other combinations of these quantities yield $\ln(1) = 0$ on the right-hand side of Equation 15.13b), the minimum ECM classification criterion for two multivariate normal populations with equal covariance is exactly the same as Fisher's linear discriminant. To the extent that the costs and/or prior probabilities are not equal, Equation 15.13 results in movement of the classification boundary away from the midpoint defined in Equation 15.6, and toward the projection of one of the two means onto a.

15.2.4. Unequal Covariances: Quadratic discrimination

Discrimination and classification are much more straightforward, both conceptually and mathematically, if equality of covariances for the G populations can be assumed. For example, it is the equality-of-covariance assumption that allows the simplification from Equation 15.13a to

Equation 15.13b for two multivariate normal populations. If it cannot be assumed that $[\Sigma_1]=[\Sigma_2]$, and instead these two covariance matrices are estimated separately by $[S_1]$ and $[S_2]$, respectively, minimum ECM classification for two multivariate populations leads to classification of x_0 as belonging to Group 1 if

$$\frac{1}{2}x_0^T\left([S_1]^{-1}-[S_2]^{-1}\right)x_0+\left(\bar{x}_1^T[S_1]^{-1}-\bar{x}_2^T[S_2]^{-1}\right)x_0-\text{const}\geq \ln\left(\frac{C(1|2)p_2}{C(2|1)p_1}\right), \qquad (15.14a)$$

where

$$\text{const}=\frac{1}{2}\left(\ln\frac{|[S_1]|}{|[S_2]|}+\bar{x}_1^T[S_1]^{-1}\bar{x}_1-\bar{x}_2^T[S_2]^{-1}\bar{x}_2\right) \qquad (15.14b)$$

contains scaling constants not involving x_0.

The mathematical differences between Equations 15.13b and 15.14 result because cancellations and recombinations that are possible in Equation 15.13 when the covariance matrices are equal, whereas additional terms in Equation 15.14 result when they are not. Classification and discrimination using Equation 15.14 are more difficult conceptually because the regions R_1 and R_2 are no longer necessarily contiguous. Equation 15.14, for classification with unequal covariances, is also less robust to non-Gaussian data than classification with Equation 15.13, when equality of covariance structure can reasonably be assumed.

Figure 15.2 illustrates quadratic discrimination and classification with a simple, one-dimensional example. Here it has been assumed for simplicity that the right-hand side of Equation 15.14a is $\ln(1)=0$, so the classification criterion reduces to assigning x_0 to whichever group yields the larger likelihood, $f_g(x_0)$. Because the variance for Group 1 is so much smaller, both very large and very small x_0 will be assigned to Group 2. Mathematically, this discontinuity for the region R_2 results from the first (i.e., the quadratic) term in Equation 15.14a, which in $K=1$ dimension is equal to $x_0^2(1/s_1^2-1/s_2^2)/2$. In higher dimensions the shapes of quadratic classification regions will usually be curved and more complicated.

Another approach to nonlinear discrimination within the straightforward framework of linear discriminant analysis is to extend the original data vector x to include non-linear-derived variables based on its elements, as is also done when computing support vector machine classifiers (Section 15.6.1). For example, if the original data vectors $x=[x_1,x_2]^T$ are $K=2$-dimensional, a quadratic discriminant analysis can be carried out in the $K^*=5$-dimensional space of the extended data vector $x^*=[x_1,x_2,x_1^2,x_2^2,x_1x_2]^T$. The resulting classification boundary can subsequently be mapped back to the original K-dimensional space of x, where in general it will be nonlinear.

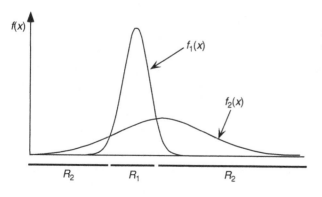

FIGURE 15.2 Discontinuous classification regions resulting from unequal variances for the populations described by two Gaussian PDFs $f_1(x)$ and $f_2(x)$.

15.3. MULTIPLE DISCRIMINANT ANALYSIS (MDA)

15.3.1. Fisher's Procedure for More Than Two Groups

Fisher's linear discriminant, described in Section 15.2.1, can be generalized for discrimination among $G = 3$ or more groups. This generalization is called *multiple discriminant analysis* (MDA). Here the basic problem is to allocate a K-dimensional data vector x to one of $G > 2$ groups on the basis $J = \min(G-1, K)$ discriminant vectors, a_j, $j = 1, \ldots, J$. The projection of the data onto these vectors yield the J discriminant functions

$$\delta_j = a_j^T x, \quad j = 1, \ldots, J. \tag{15.15}$$

The discriminant functions are computed on the basis of a training set of G data matrices $[X_1]$, $[X_2]$, $[X_3]$, ..., $[X_G]$, dimensioned, respectively, $(n_g \times K)$. A sample variance–covariance matrix can be computed from each of the G sets of data, $[S_1]$, $[S_2]$, $[S_3]$, ..., $[S_G]$, according to Equation 11.30. Assuming that the G groups represent populations having the same covariance matrix, the pooled estimate of this common covariance matrix is estimated by the weighted average

$$[S_{\text{pool}}] = \frac{1}{n-G} \sum_{g=1}^{G} (n_g - 1) [S_g], \tag{15.16}$$

where there are n_g observations in each group, and the total sample size is

$$n = \sum_{g=1}^{G} n_g. \tag{15.17}$$

Equation 12.42b is a special case of Equation 15.16, with $G = 2$.

Computation of multiple discriminant functions also requires calculation of the *between-groups covariance matrix*

$$[S_B] = \frac{1}{G-1} \sum_{g=1}^{G} (\bar{x}_g - \bar{x}_\bullet)(\bar{x}_g - \bar{x}_\bullet)^T, \tag{15.18}$$

where the individual group means are calculated as in Equation 15.2, and.

$$\bar{x}_\bullet = \frac{1}{n} \sum_{g=1}^{G} n_g \bar{x}_g \tag{15.19}$$

is the grand, or overall vector mean of all n observations. The between-groups covariance matrix $[S_B]$ is essentially a covariance matrix describing the dispersion of the G sample means around the overall mean (compare Equation 11.35).

The number J of discriminant functions that can be computed is the smaller of $G - 1$ and K. Thus for the two-group case discussed in Section 15.2, there is only $G - 1 = 1$ discriminant function, regardless of the dimensionality K of the data vectors. In the more general case, the discriminant functions are derived from the first J eigenvectors (corresponding to the nonzero eigenvalues) of the matrix

$$[S_{pool}]^{-1} [S_B]. \tag{15.20}$$

This $(K \times K)$ matrix in general is not symmetric. The discriminant vectors a_j are aligned with these eigenvectors, but are often scaled to yield unit variances for the data projected onto them, that is,

$$a_j^T [S_{pool}] a_j = 1, \ j = 1, ..., J. \tag{15.21}$$

Usually computer routines for calculating eigenvectors will scale eigenvectors to unit length, that is, $||e_j|| = 1$, but the condition in Equation 15.21 can be achieved by calculating

$$a_j = \frac{e_j}{\left(e_j^T [S_{pool}] e_j\right)^{1/2}}, \ j = 1, ..., J. \tag{15.22}$$

The first discriminant vector a_1, which is associated with the largest eigenvalue of the matrix in Equation 15.20, makes the largest contribution to separating the G group means, in aggregate; and a_J, which is associated with the smallest nonzero eigenvalue, makes the least contribution overall.

Many texts develop Fisher's method for MDA in terms of the eigenvectors of the product $[W]^{-1}[B]$, where $[W] = (n - G)[S_{pool}]$ is called the *within-groups covariance matrix*, and $[B] = (G - 1)[S_B]$ is called the sample between-groups matrix; rather than the eigenvectors of Equation 15.20. However, since the two matrix products are the same apart from a scalar multiple, their eigenvectors are also the same and so yield the same discriminant vectors a_j in Equation 15.22.

The J discriminant vectors a_j define a J-dimensional discriminant space, in which the G groups exhibit maximum separation. The projections δ_j (Equation 15.15) of the data onto these vectors are sometimes called the *discriminant coordinates* or canonical variates. This second name derives from the fact that equivalent results can be obtained through CCA using indicator variables for the groups (Hastie et al., 2009). As was also the case when distinguishing between $G = 2$ groups, observations x can be assigned to groups according to which of the G group means is closest in discriminant space. For the $G = 2$ case the discriminant space is one-dimensional, consisting only of a line. The group assignment rule (Equation 15.7) is then particularly simple. More generally, the Euclidean distances in discriminant space between the candidate vector x_0 and each of the G group means are evaluated in order to find which is closest. It is actually easier to evaluate these in terms of squared distances, yielding the classification rule:

$$\text{Assign } x_0 \text{ to group } g \text{ if } \sum_{j=1}^{J} [a_j(x_0 - \bar{x}_g)]^2 \leq \sum_{j=1}^{J} [a_j(x_0 - \bar{x}_h)]^2, \quad \text{for all } h \neq g. \tag{15.23}$$

That is, the sum of the squared distances between x_0 and each of the group means, along the directions defined by the vectors a_j, is compared in order to find the closest group mean.

Computing the discriminant vectors a_j allows one to define the discriminant space, which can lead to informative graphical displays of the data in this space or subspaces of it. However, if only a classification rule is needed, it can be computed without the eigendecomposition of the matrix product in Equation 15.20. Specifically, a candidate vector x_0 is assigned to the group g that maximizes

$$\bar{x}_g^T [S_{pool}]^{-1} x_0 - \frac{1}{2} \bar{x}_g^T [S_{pool}]^{-1} \bar{x}_g. \tag{15.24}$$

Example 15.2. Multiple Discriminant Analysis with G = 3 Groups

Consider discriminating among all three groups of data in Table 15.1. Using Equation 15.16 the pooled estimate of the assumed common covariance matrix is

$$[S_{pool}] = \frac{1}{28 - 3} \left(9 \begin{bmatrix} 1.47 & 0.65 \\ 0.65 & 1.45 \end{bmatrix} + 8 \begin{bmatrix} 2.08 & 0.06 \\ 0.06 & 0.17 \end{bmatrix} + 8 \begin{bmatrix} 4.85 & -0.17 \\ -0.17 & 0.10 \end{bmatrix} \right) = \begin{bmatrix} 2.75 & 0.20 \\ 0.20 & 0.61 \end{bmatrix}, \tag{15.25}$$

and using Equation 15.18 the between-groups covariance matrix is

$$[S_B] = \frac{1}{2} \left(\begin{bmatrix} 12.96 & 5.33 \\ 5.33 & 2.19 \end{bmatrix} + \begin{bmatrix} 2.89 & -1.05 \\ -1.05 & 0.38 \end{bmatrix} + \begin{bmatrix} 32.49 & 5.81 \\ 5.81 & 1.04 \end{bmatrix} \right) = \begin{bmatrix} 24.17 & 5.04 \\ 5.04 & 1.81 \end{bmatrix}. \qquad (15.26)$$

The directions of the two discriminant functions are specified by the eigenvectors of the matrix

$$[S_{pool}]^{-1}[S_B] = \begin{bmatrix} 0.373 & -.122 \\ -.122 & 1.685 \end{bmatrix} \begin{bmatrix} 24.17 & 5.04 \\ 5.04 & 1.81 \end{bmatrix} = \begin{bmatrix} 8.40 & 1.65 \\ 5.54 & 2.43 \end{bmatrix}, \qquad (15.27a)$$

which, when scaled according to Equation 15.22 are

$$a_1 = \begin{bmatrix} 0.542 \\ 0.415 \end{bmatrix} \quad \text{and} \quad a_2 = \begin{bmatrix} -0.282 \\ 1.230 \end{bmatrix}. \qquad (15.27b)$$

The discriminant vectors a_1 and a_2 define the directions of the first discriminant function $\delta_1 = a_1^T x$ and the second discriminant function $\delta_2 = a_2^T x$. Figure 15.3 shows the data for all three groups in Table 15.1 plotted in the discriminant space defined by these two functions. Points for Groups 1 and 2 are shown by circles and ×'s, as in Figure 15.1, and points for Group 3 are shown by +'s. The heavy symbols locate the respective vector means for the three groups. Note that the point clouds for Groups 1 and 2 appear to be stretched and distorted relative to their arrangement in Figure 15.1. This is because the matrix in Equation 15.27a is not symmetric so that its eigenvectors, and therefore the two discriminant vectors in Equation 15.27b, are not orthogonal.

The heavy dashed lines in Figure 15.3 divide the portions of the discriminant space that are assigned to each of the three groups by the classification criterion in Equation 15.23. These are the regions closest to each of the group means, and so define a Voronoi tessellation (or collection of Thiessen polygons) in the discriminant space. (Note, however, that the corresponding partition of the data space does not consist of

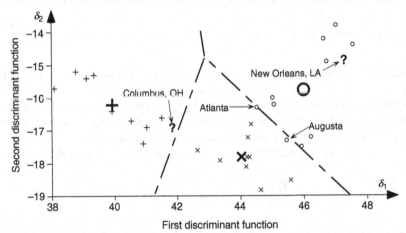

FIGURE 15.3   Illustration of the geometry of multiple discriminant analysis applied to the $G = 3$ groups of data in Table 15.1. Group 1 stations are plotted as circles, Group 2 stations are plotted as ×'s, and Group 3 stations are plotted as +'s. The three vector means are indicated by the corresponding heavy symbols. The two axes are the first and second discriminant functions, and the heavy dashed lines divide sections of this discriminant space allocated to each group. The data for Atlanta and Augusta are misclassified as belonging to Group 2. The two stations Columbus and New Orleans, which are not part of the training data in Table 15.1, are shown as question marks, and are allocated to Groups 3 and 1, respectively.

perpendicular bisectors of the line segments between group vector means, unless the pooled covariance matrix is proportional to [I].) Here the data for Atlanta and Augusta have both been misclassified as belonging to Group 2 rather than Group 1. For Atlanta, for example, the squared distance to the Group 1 mean is $[0.542(78.6–80.6)+0.415(4.73–5.67)]^2 +[-0.282(78.6–80.6)+1.230(4.73–5.67)]^2 =2.52$, and the squared distance to the Group 2 mean is $[0.542 (78.6–78.7)+0.415(4.73–3.57)]^2 +[-0.282 (78.6–78.7)+1.230(4.73–3.57)]^2 =2.31$. A line in this discriminant space could be drawn by eye that would include these two stations in the Group 1 region. That the discriminant analysis has not specified this line is probably a consequence of the assumption of equal covariance matrices not being well satisfied. In particular, the points in Group 1 appear to be more positively correlated in this discriminant space than the members of the other two groups.

The data points for the two stations Columbus and New Orleans, which are not part of the training data in Table 15.1, are shown by the question marks in Figure 15.3. The location in the discriminant space of the point for New Orleans, for which $x =[82.1\ 6.73]^T$, is $\delta_1 =(0.542)(82.1)+(0.415)(6.73)=47.3$ and $\delta_2 =(-0.282)(82.1)+(1.230)(6.73)=-14.9$, which is within the region assigned to Group 1. The coordinates in discriminant space for the Columbus data, $x =[74.7\ 3.37]^T$, are $\delta_1 =(0.542)(74.7)+(0.415)(3.37)=41.9$ and $\delta_2 =(-0.282)(74.7)+(1.230)(3.37)=-16.9$, which is within the region assigned to Group 3.

Drawing the discriminant space in Figure 15.3 required computation of the eigenvectors of the matrix product in Equation 15.20. If this diagram had not been of interest, the same group assignments could have been made instead using the computationally simpler Equation 15.24. Evaluating the data for New Orleans with respect to the three group means using Equation 15.24 yields

$$[80.6\ 5.67]\begin{bmatrix} .373 & -.122 \\ -.122 & 1.685 \end{bmatrix}\begin{bmatrix} 82.1 \\ 6.73 \end{bmatrix} -\frac{1}{2}[80.6\ 5.67]\begin{bmatrix} .373 & -.122 \\ -.122 & 1.685 \end{bmatrix}\begin{bmatrix} 80.6 \\ 5.67 \end{bmatrix} = 1226.7 \qquad (15.28a)$$

$$[78.7\ 3.57]\begin{bmatrix} .373 & -.122 \\ -.122 & 1.685 \end{bmatrix}\begin{bmatrix} 82.1 \\ 6.73 \end{bmatrix} -\frac{1}{2}[78.7\ 3.57]\begin{bmatrix} .373 & -.122 \\ -.122 & 1.685 \end{bmatrix}\begin{bmatrix} 78.7 \\ 3.57 \end{bmatrix} = 1218.6 \qquad (15.28b)$$

and

$$[71.3\ 3.17]\begin{bmatrix} .373 & -.122 \\ -.122 & 1.685 \end{bmatrix}\begin{bmatrix} 82.1 \\ 6.73 \end{bmatrix} -\frac{1}{2}[71.3\ 3.17]\begin{bmatrix} .373 & -.122 \\ -.122 & 1.685 \end{bmatrix}\begin{bmatrix} 71.3 \\ 3.17 \end{bmatrix} = 1200.1 \qquad (15.28c)$$

for Groups 1, 2, and 3, respectively, so that membership in Group 1 is chosen because the result in Equation 15.28a is largest. Similarly, the calculations for Columbus are

$$[80.6\ 5.67]\begin{bmatrix} .373 & -.122 \\ -.122 & 1.685 \end{bmatrix}\begin{bmatrix} 74.7 \\ 3.37 \end{bmatrix} -\frac{1}{2}[80.6\ 5.67]\begin{bmatrix} .373 & -.122 \\ -.122 & 1.685 \end{bmatrix}\begin{bmatrix} 80.6 \\ 5.67 \end{bmatrix} = 1010.2 \qquad (15.29a)$$

$$[78.7\ 3.57]\begin{bmatrix} .373 & -.122 \\ -.122 & 1.685 \end{bmatrix}\begin{bmatrix} 74.7 \\ 3.37 \end{bmatrix} -\frac{1}{2}[78.7\ 3.57]\begin{bmatrix} .373 & -.122 \\ -.122 & 1.685 \end{bmatrix}\begin{bmatrix} 78.7 \\ 3.57 \end{bmatrix} = 1016.6 \qquad (15.29b)$$

and

$$[71.3\ 3.17]\begin{bmatrix} .373 & -.122 \\ -.122 & 1.685 \end{bmatrix}\begin{bmatrix} 74.7 \\ 3.37 \end{bmatrix} -\frac{1}{2}[71.3\ 3.17]\begin{bmatrix} .373 & -.122 \\ -.122 & 1.685 \end{bmatrix}\begin{bmatrix} 71.3 \\ 3.17 \end{bmatrix} = 1219.2 \qquad (15.29c)$$

the largest of which is Equation 15.28c, so that the assignment is made to Group 3. ◇

Graphical displays of the discriminant space such as that in Figure 15.3 can be quite useful for visualizing the separation of data groups. If $J = \min(G-1, K) > 2$, we cannot plot the full discriminant space in only two dimensions, but it is still possible to calculate and plot its first two components, δ_1 and δ_2. The relationships among the data groups rendered in this reduced discriminant space will be a good approximation to those in the full J-dimensional discriminant space, if the corresponding eigenvalues of Equation 15.20 are large relative to the eigenvalues of the omitted dimensions. Similarly to the idea expressed in Equation 13.4 for PCA, the reduced discriminant space will be a good approximation to the full discriminant space, to the extent that $(\lambda_1 + \lambda_2)/\sum_j \lambda_j \approx 1$.

15.3.2. Minimizing Expected Cost of Misclassification

The procedure described in Section 15.2.3, accounting for misclassification costs and prior probabilities of group memberships, generalizes easily for MDA. Again, if equal covariances for each of the G populations can be assumed, there are no other restrictions on the PDFs $f_g(x)$ for each of the populations except that that these PDFs can be evaluated explicitly. The main additional complication is to specify cost functions for all possible $G(G-1)$ misclassifications of Group g members into Group h,

$$C(h|g); \quad g = 1,\ldots,G; \ h = 1,\ldots,G; \ g \neq h. \tag{15.30}$$

The resulting classification rule is to assign an observation x_0 to the group g for which

$$\sum_{\substack{h=1 \\ h \neq g}}^{G} C(g|h) p_h f_h(x_0) \tag{15.31}$$

is minimized. That is, the candidate Group g is selected for which the probability-weighted sum of misclassification costs, considering each of the other $G - 1$ groups h as the potential true home of x_0, is smallest. Equation 15.31 is the generalization of Equation 15.12 to $G \geq 3$ groups.

If all the misclassification costs are equal, minimizing Equation 15.31 simplifies to classifying x_0 as belonging to that group g for which

$$p_g f_g(x_0) \geq p_h f_h(x_0), \quad \text{for all } h \neq g. \tag{15.32}$$

If in addition the PDFs $f_g(x)$ are all multivariate normal distributions, with possibly different covariance matrices $[\Sigma_g]$, (the logs of) the terms in Equation 15.32 take on the form

$$\ln(p_g) - \frac{1}{2} \ln |[S_g]| - \frac{1}{2}(x_0 - \bar{x}_g)^T [S_g]^{-1} (x_0 - \bar{x}_g). \tag{15.33}$$

The observation x_0 would be allocated to the group whose multinormal PDF $f_g(x)$ maximizes Equation 15.33. The unequal covariances $[S_g]$ result in this classification rule being quadratic. If all the covariance matrices $[\Sigma_g]$ are assumed equal and are estimated by $[S_{pool}]$, the classification rule in Equation 15.33 simplifies to choosing that Group g maximizing the linear discriminant score

$$\ln(p_g) + \bar{x}_g^T [S_{pool}]^{-1} x_0 - \frac{1}{2}\bar{x}_g^T [S_{pool}]^{-1} \bar{x}_g. \tag{15.34}$$

This rule minimizes the total probability of misclassification. If the prior probabilities p_g are all equal, Equation 15.43 reduces to Equation 15.24.

15.3.3. Probabilistic Classification

The classification rules presented so far choose one and only one of the G groups in which to place a new observation x_0. Except for very easy cases, in which group means are well separated relative to the data scatter, these rules rarely will yield perfect results. Accordingly, probability information describing classification uncertainties is often useful.

Probabilistic classification, that is, specification of probabilities for x_0 belonging to each of the G groups, can be achieved through an application of Bayes' Theorem:

$$\Pr\{\text{Group } g \,|\, x_0\} = \frac{p_g f_g(x_0)}{\displaystyle\sum_{h=1}^{G} p_h f_h(x_0)}. \tag{15.35}$$

Here the p_g are the prior probabilities for group membership, which often will be the relative frequencies with which each of the groups is represented in the training data. The PDFs $f_g(x)$ for each of the groups can be of any form, so long as they can be evaluated explicitly for particular values of x_0.

Often it is assumed that each of the $f_g(x)$ is multivariate normal distribution. In this case, Equation 15.35 becomes

$$\Pr\{\text{Group } g \,|\, x_0\} = \frac{p_g\left(|[S_g]|^{-1/2} \exp\left(-\frac{1}{2}(x_0 - \bar{x}_g)^T [S_g]^{-1}(x_0 - \bar{x}_g)\right)\right)}{\displaystyle\sum_{h=1}^{G} p_h\left(|[S_h]|^{-1/2} \exp\left(-\frac{1}{2}(x_0 - \bar{x}_h)^T [S_h]^{-1}(x_0 - \bar{x}_h)\right)\right)}. \tag{15.36}$$

Equation 15.36 simplifies if all G of the covariance matrices are assumed to be equal, because in that case the factors involving determinants cancel. This equation also simplifies if all the prior probabilities are equal (i.e., $p_g = 1/G$, $g = 1, \ldots, G$), because these probabilities then cancel.

Example 15.3. Probabilistic Classification with $G = 3$ Groups

Consider probabilistic classification for Columbus, Ohio, into the three climate region groups of Example 15.2. The July mean vector for Columbus is $x_0 = [74.7\,°F, 3.37 \text{ in.}]^T$. Figure 15.3 shows that this point is near the boundary between the (nonprobabilistic) classification regions for Groups 2 (Central United States) and 3 (Northeastern United States) in the two-dimensional discriminant space, but the calculations in Example 15.2 do not quantify the certainty with which Columbus has been placed in Group 3.

Assume for simplicity that the three prior probabilities are equal, and that the three groups are all samples from multivariate normal distributions with a common covariance matrix. The pooled estimate for the common covariance is given in Equation 15.25, and its inverse is indicated in the middle equality of Equation 15.27a. The groups are then distinguished by their mean vectors, indicated in Table 15.1.

The differences between x_0 and the three group means are

$$x_0 - \bar{x}_1 = \begin{bmatrix} -5.90 \\ -2.30 \end{bmatrix}, \quad x_0 - \bar{x}_2 = \begin{bmatrix} -4.00 \\ -0.20 \end{bmatrix}, \text{ and } x_0 - \bar{x}_3 = \begin{bmatrix} 3.40 \\ 0.20 \end{bmatrix}; \tag{15.37a}$$

yielding the likelihoods

$$f_1(x_0) \propto \exp\left(-\frac{1}{2}[-5.90 \;\; -2.30]\begin{bmatrix} .373 & -.122 \\ -.122 & 1.679 \end{bmatrix}\begin{bmatrix} -5.90 \\ -2.30 \end{bmatrix}\right) = .000094, \tag{15.37b}$$

$$f_2(x_0) \propto \exp\left(-\frac{1}{2}[-4.00 \quad -0.20]\begin{bmatrix} .373 & -.122 \\ -.122 & 1.679 \end{bmatrix}\begin{bmatrix} -4.00 \\ -0.20 \end{bmatrix}\right) = .054, \tag{15.37c}$$

and

$$f_3(x_0) \propto \exp\left(-\frac{1}{2}[3.40 \quad 0.20]\begin{bmatrix} .373 & -.122 \\ -.122 & 1.679 \end{bmatrix}\begin{bmatrix} 3.40 \\ 0.20 \end{bmatrix}\right) = .122. \tag{15.37d}$$

Substituting these likelihoods into Equation 15.36 yields the three classification probabilities

$$\Pr(\text{Group } 1 | x_0) = .000094/(.000094 + .054 + .122) = .0005, \tag{15.38a}$$

$$\Pr(\text{Group } 2 | x_0) = .054/(.000094 + .054 + .122) = 0.31, \tag{15.38b}$$

and

$$\Pr(\text{Group } 3 | x_0) = .122/(.000094 + .054 + .122) = 0.69. \tag{15.38c}$$

Even though the group into which Columbus was classified in Example 15.2 is the most likely, there is still a substantial probability that it might belong to Group 2 instead. The possibility that Columbus is really a Group 1 station appears to be remote. ◇

15.4. FORECASTING WITH DISCRIMINANT ANALYSIS

Discriminant analysis is a natural tool to use in forecasting when the predictand consists of a finite set of discrete categories (groups), and vectors of predictors x are known sufficiently far in advance of the discrete observation that will be predicted. Apparently the first use of discriminant analysis for forecasting in meteorology was described by Miller (1962), who forecast airfield ceiling in five MECE categories at a lead time of 0–2h, and also made five-group forecasts of precipitation type (none, rain/freezing rain, snow/sleet) and amount (≤ 0.05 in., and > 0.05 in., if nonzero). Both of these applications today would be called *nowcasting*, because of the very short lead time. Some other examples of the use of discriminant analysis for forecasting can be found in Drosdowsky and Chambers (2001) and Ward and Folland (1991).

An informative case study in the use of discriminant analysis for forecasting is provided by Lehmiller et al. (1997). They consider the problem of forecasting hurricane occurrence (i.e., whether or not at least one hurricane will occur) during summer and autumn, within subbasins of the northwestern Atlantic Ocean, so that $G = 2$ for forecasts in each subbasin. They began with a quite large list of potential predictors and so needed to protect against overfitting in their $n = 43$-year training sample, 1950–1992. Their approach to predictor selection was computationally intensive, but statistically sound: different discriminant analyses were calculated for all possible subsets of predictors, and for each of these subsets the calculations were repeated 43 times, in order to produce leave-one-out cross-validations. The chosen predictor sets were those with the smallest number of predictors that minimized the number of cross-validated incorrect classifications.

Figure 15.4 shows one of the resulting discriminant analyses, for occurrence or nonoccurrence of hurricanes in the Caribbean Sea, using standardized African rainfall predictors that would be known as of 1 December in the preceding year. Because this is a binary forecast (two groups), there is only a single linear discriminant function, which would be perpendicular to the dividing line labeled "discriminant partition line" in Figure 15.4. This line compares to the long-short dashed dividing line in Figure 15.1. (The discriminant vector a would be perpendicular to this line and pass through the origin.)

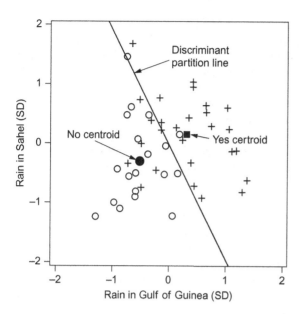

FIGURE 15.4 Binary (yes/no) forecasts for occurrence of at least one hurricane in the Caribbean Sea during summer and autumn, using two standardized predictors observed as of 1 December of the previous year to define a single linear discriminant function. Circles and plusses show the training data, and the two solid symbols locate the two group means (centroids). *From Lehmiller et al. (1997). © American Meteorological Society. Used with permission.*

The $n = 43$-year training sample is indicated by the open circles and plusses. Seven of the 18 hurricane years have been misclassified as "no" years, and only two of 25 nonhurricane years have been misclassified as "yes" years. Since there are more "yes" years, accounting for unequal prior probabilities would have moved the dividing line down and to the left, toward the "no" group mean (solid circle). Similarly, for some purposes it might be reasonable to assume that the cost of an incorrect "no" forecast would be greater than that of an incorrect "yes" forecast, and incorporating this asymmetry would also move the partition down and to the left, producing more "yes" forecasts.

15.5. CONVENTIONAL ALTERNATIVES TO CLASSICAL DISCRIMINANT ANALYSIS

Traditional discriminant analysis, as described in the first sections of this chapter, continues to be widely employed and extremely useful. Newer alternative approaches to discrimination and classification are also available. Two of these, based on conventional statistical methods, are described in this section. Two more modern "machine learning" alternatives are presented in Section 15.6.

15.5.1. Discrimination and Classification Using Logistic Regression

Section 7.6.2 described logistic regression, in which the nonlinear logistic function (Equation 7.36) is used to relate a linear combination of predictors, x, to the probability of one of the elements of a dichotomous outcome. Figure 7.20 shows a simple example of logistic regression, in which the probability of occurrence of precipitation at Ithaca has been specified as a logistic function of the minimum temperature on the same day.

Figure 7.20 could also be interpreted as portraying classification into $G = 2$ groups, with $g = 1$ indicating precipitation days, and $g = 2$ indicating dry days. The number densities (points per unit length) of the dots along the top and bottom of the figure suggest the locations and shapes of the two underlying PDFs, $f_1(x)$ and $f_2(x)$, respectively, as functions of the minimum temperature, x. The medians of these two conditional distributions for minimum temperature are near 23°F and 3°F, respectively. However, the classification function in this case is the logistic curve (solid), the equation for which is also given in the figure. Simply evaluating the function using the minimum temperature for a particular day provides an estimate of the probability that that day belonged to Group 1 (nonzero precipitation). A nonprobabilistic classifier could be constructed at the point of equal probability for the two groups, by setting the classification probability (= y in Figure 7.20) to 1/2. This probability is achieved when the argument of the exponential function is zero, implying a nonprobabilistic classification boundary of 15°F. In this case days could be classified as belonging to Group 1(wet) if the minimum temperature is warmer, and classified as belonging to Group 2 (dry) if the minimum temperature is colder. Seven days (the five warmest dry days, and the two coolest wet days) in the training data are misclassified by this rule. In this example the relative frequencies of the two groups are nearly equal, but logistic regression automatically accounts for relative frequencies of group memberships in the training sample (which estimate the prior probabilities) in the fitting process.

Figure 15.5 shows a forecasting example of two-group discrimination using logistic regression, with a $K = 2$-dimensional predictor vector x. The two groups are years with (solid dots) and without (open circles) landfalling hurricanes on the southeastern U.S. coast from August onward, and the two elements of x are July average values of sea-level pressure at Cape Hatteras, and 200–700 mb wind shear over southern Florida. The contour lines indicate the shape of the logistic function, which in this case is a surface deformed into an S shape, analogously to the logistic function in Figure 7.20 being a line deformed in the same way. High surface pressures and wind shears simultaneously result in large probabilities for hurricane landfalls, whereas low values for both predictors yield small probabilities. This surface could be calculated as indicated in Equation 7.42, except that the vectors would be dimensioned (3×1) and the matrix of second derivatives would be dimensioned (3×3).

Hastie et al. (2009, Section 4.4.5) compare logistic regression and linear discriminant analyses for the $G = 2$-group situation, concluding that logistic regression may be more robust, but that the two generally give very similar results.

FIGURE 15.5 Two-dimensional logistic regression surface, estimating the probability of at least one landfalling hurricane on the southeastern U.S. coastline from August onward, on the basis of July sea-level pressure at Cape Hatteras and 200–700 mb wind shear over south Florida. Solid dots indicate hurricane years, and open dots indicate nonhurricane years, in the training data. *Adapted from Lehmiller et al. (1997). © American Meteorological Society. Used with permission.*

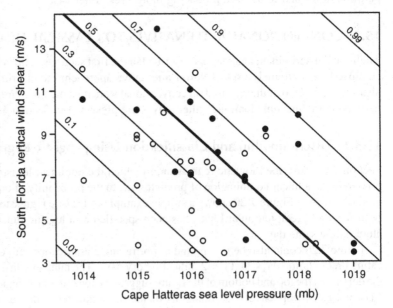

15.5.2. Discrimination and Classification Using Kernel Density Estimates

It was pointed out in Sections 15.2 and 15.3 that the G PDFs $f_g(x)$ need not be of particular parametric forms in order to implement Equations 15.12, 15.32, and 15.35, but rather it is necessary only that they can be evaluated explicitly. Gaussian or multivariate normal distributions often are assumed, but these and other parametric distributions may be poor approximations to data in some circumstances. Kernel density estimates (Section 3.3.6), which are nonparametric PDF estimates, provide viable alternatives. Indeed, nonparametric discrimination and classification motivated much of the early work on kernel density estimation (Silverman, 1986).

Nonparametric discrimination and classification are straightforward conceptually, but may be computationally demanding. The basic idea is to separately estimate the PDFs $f_g(x)$ for each of the G groups, using the methods described in Section 3.3.6. Somewhat subjective choices for appropriate kernel form and (especially) bandwidth are necessary. But having estimated these PDFs, they can be evaluated for any candidate x_0, and thus lead to specific classification results. Nearest-neighbor classification methods, where membership probabilities are derived from relative frequencies of group memberships for the data vectors nearest the datum x_0 to be classified, can be regarded as belonging to the class of kernel methods, where (hyper-) rectangular kernels are used to define the averaging.

Figure 15.6 illustrates the discrimination procedure for the same June Guayaquil temperature data (Table A.3) used in Figures 3.6 and 3.8. The distribution of these data is bimodal, as a consequence of four of the five El Niño years being warmer than 26°C whereas the warmest of the 15 non-El Niño years is 25.2°C. Discriminant analysis could be used to diagnose the presence or absence of El Niño, based on the June Guayaquil temperature, by specifying the two PDFs $f_1(x)$ for El Niño years and $f_2(x)$ for non-El Niño years. Parametric assumptions about the mathematical forms for these PDFs can be avoided through the use of kernel density estimates. The gray curves in Figure 15.6 show these two estimated PDFs. They exhibit fairly good separation, although $f_1(x)$, for El Niño years, is bimodal because the fifth El Niño year in the data set has a temperature of 24.8°C.

The posterior probability of an El Niño year as a function of the June temperature is calculated using Equation 15.35. The dashed curve is the result when equal prior probabilities $p_1 = p_2 = 1/2$ are assumed. Of course, El Niño occurs in fewer than half of all years, so it would be more reasonable to estimate the two prior probabilities as $p_1 = 1/4$ and $p_2 = 3/4$, which are the relative frequencies in the training sample. The resulting posterior probabilities are shown by the solid black curve in Figure 15.6.

Nonprobabilistic classification regions could be constructed using either Equation 15.12 or Equation 15.32, which would be equivalent if the two misclassification costs in Equation 15.12 were equal. If the two prior probabilities were also equal, the boundary between the two classification region

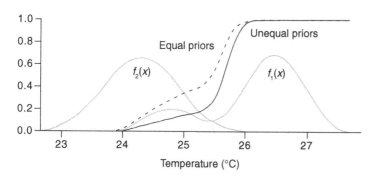

FIGURE 15.6 Separate kernel density estimates (quartic kernel, bandwidth$=0.92$) for Guayaquil June temperatures during El Nino $f_1(x)$ and non-El Niño years $f_2(x)$, 1951–1970 (gray PDFs); and posterior probabilities for an El Niño year according to Equation 15.32, assuming equal prior probabilities (dashed), and prior probabilities estimated by the training-sample relative frequencies (solid).

would occur at the point where $f_1(x) = f_2(x)$, or $x \approx 25.45°C$. This temperature corresponds to a posterior probability of 1/2, according to the dashed curve. For unequal prior probabilities the classification boundary would shift toward the less likely group (i.e., requiring a warmer temperature to classify as an El Niño year), occurring at the point where $f_1(x) = (p_2/p_1) f_2(x) = 3 f_2(x)$, or $x \approx 25.65$. Not coincidentally, this temperature corresponds to a posterior probability of 1/2 according to the solid black curve.

15.6. "MACHINE LEARNING" ALTERNATIVES TO CONVENTIONAL DISCRIMINANT ANALYSIS

15.6.1. Support Vector Machines (SVM)

Support Vector Machines (SVM) provide an approach to discrimination and classification that is similar to LDA, in that both define a separating hyperplane that partition a data space into regions assigned to one or the other of two groups. In SVM, this data space is generally expanded to include nonlinear transformations of the underlying data variables, which yields nonlinear classification boundaries in the original data space. (This strategy can also be used with MDA, as mentioned in Section 15.2.4.) The two methods differ primarily with respect to the criterion used to define the linear separator in the expanded data space. In addition, SVM is nonparametric in that it does not require specification or assumption of forms for the probability distribution(s) of the data, and so is an attractive alternative if the usual parametric assumptions in LDA seem doubtful. On the other hand, SVM does not provide a mechanism for probabilistic classification.

The idea behind SVM discrimination is most easily approached by considering the case of two linearly separable groups, as exemplified for $K = 2$ dimensions in Figure 15.7a. In such cases, any number of linear functions

$$f(\mathbf{x}) = b_0 + \mathbf{b}^T \mathbf{x} = 0 \qquad (15.39)$$

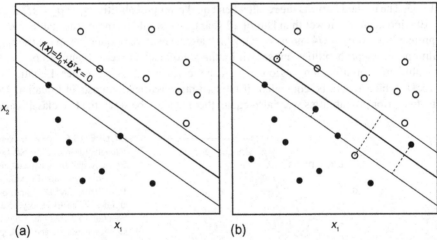

FIGURE 15.7   Illustration of (a) the maximum margin classifier, which is the support vector classifier when the two groups are linearly separable, and (b) the support vector classifier for two groups that are not linearly separable. The lengths of the light dashed lines in (b) indicate the magnitudes of the nonzero slack variables ε_i.

can be found to define lines (for $K=2$ dimensions, or surfaces for $K=3$, or hyperplanes for $K>3$) that fully separate the two groups, and thus can be used as classifiers. Classification of a data point x_i into one group or the other is coded as $y_i=\pm1$, depending on which side of the partition it occurs:

$$y_i = \text{sign}(b_0 + \boldsymbol{b}^T\boldsymbol{x}_i) = \begin{cases} 1 \text{ if } f(\boldsymbol{x}_i) > 0 \\ -1 \text{ if } f(\boldsymbol{x}_i) < 0 \end{cases}. \tag{15.40}$$

The *maximum margin classifier* differs from the LDA separator in the case of linearly separable groups, in that the parameters defining Equation 15.39 are chosen to maximize the width of the margin around it. In other words, the maximum margin classifier finds the function in Equation 15.39 that yields the broadest possible buffer zone between the two groups in the training data. The margins bound the symmetrical zone around the classification function in Equation 15.39 (lighter lines in Figure 15.7 that are parallel to the separating function) that contains data points only on its boundary. These defining data vectors are called the *support vectors*.

Although only the values of the support vectors define the parameters in Equation 15.39, all the data are used in determining which points are the support vectors. The criterion for finding these parameters can be expressed as

$$\min_{b_0, \boldsymbol{b}} \|\boldsymbol{b}\|, \text{ subject to } y_i(b_0 + \boldsymbol{b}^T\boldsymbol{x}_i) \geq 1, \ i=1,\ldots,n, \tag{15.41}$$

where $\|\bullet\|$ denotes Euclidean length and n is the training sample size. Because all points in a linearly separable case can be correctly classified, necessarily $y_i(b_0+\boldsymbol{b}^T\boldsymbol{x}_i)>0$. Support vectors exactly on a margin edge will have $y_i(b_0+\boldsymbol{b}^T\boldsymbol{x}_i)=1$, and $y_i(b_0+\boldsymbol{b}^T\boldsymbol{x}_i)>1$ for the other correctly classified points.

When two groups are not linearly separable, some points will necessarily lie on the wrong side of "their" margin, regardless of the parameters in Equation 15.39 and regardless of how thin the margin may be. In such cases the maximum margin classifier for linearly separable groups is relaxed to the *support vector classifier*, which allows some points in the training data to be misclassified, and also allows some correctly classified points to be inside their margin. These points are also support vectors, in addition to points that are exactly on their margin.

The degree of the intrusion into the margin of a correctly classified point, or degree of misclassification of a point on the wrong side of the separating hyperplane defined by Equation 15.39, is quantified by its corresponding *slack variable*, $\varepsilon_i \geq 0$. Any point on the correct side of its margin (i.e., those that are not support vectors), and support vectors that are exactly on their margin, have $\varepsilon_i=0$. Points within the margin but not misclassified have $0 < \varepsilon_i < 1$. Points exactly on the separating hyperplane have $\varepsilon_i=1$, and misclassified points have $\varepsilon_i > 1$. The slack variables thus quantify the proportional amount by which $f(\boldsymbol{x}_i)$ falls on the wrong side of its margin. Figure 15.7b illustrates the idea, where four of the seven support vectors have nonzero slack variables, the magnitudes of which are indicated by the light dashed lines.

Extending Equation 15.41, the support vector classifier optimizes the parameters in Equation 15.39 according to

$$\min_{b_0, \boldsymbol{b}} \|\boldsymbol{b}\|, \text{ subject to } \begin{cases} y_i(b_0 + \boldsymbol{b}^T\boldsymbol{x}_i) \geq 1 - \varepsilon_i, \ i=1,\ldots,n \\ \text{and} \qquad \sum_{i=1}^{n} \varepsilon_i \leq \text{constant} \end{cases}, \tag{15.42}$$

FIGURE 15.8 Illustration
of (a) nonlinear classification boundary and asso-
ciated margins, derived
from (b) a linear maximum
margin classifier in a
higher-dimensional space
that is rendered schematically in two dimensions.
*From https://commons.
wikimedia.org/wiki/File:
Kernel_Machine.svg.*

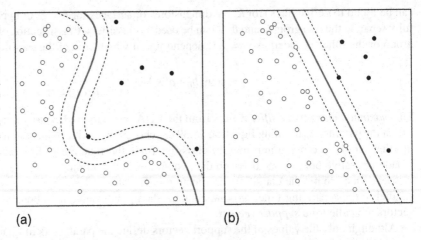

(a)   (b)

where the magnitude of the adjustable constant limiting the sum of the slack variables is typically optimized by minimizing the number of misclassified training points through cross-validation. The width of
the margin and the number of support vectors increase as this constant increases. Correctly classified
points that are far from the separating hyperplane have little influence in defining it, which is another
difference relative to LDA.

The support vector classifier is most often implemented by extending the dimension of the predictor
vectors x through the addition of (possibly very many) nonlinear transformations of the elements of x.
This extended implementation is the SVM. A simple and straightforward possibility is to expand the
predictor space using a polynomial basis. For example, in $K = 2$ elements the quadratic polynomial
expansion basis would consist of the $K^* = 5$-dimensional vector $x^* = [x_1, x_2, x_1^2, x_2^2, x_1 x_2]^T$. A linear
support vector classifier in this extended space then maps to a nonlinear classifier in the original space.
When $K^* + 1 \geq n$, a hyperplane fully separating two groups can always be found in the extended space, in
which case the maximum margin classifier would be implemented. Figs. 15.8 illustrates the result, in
which panel (a) shows a nonlinear classification boundary and associated margins in the original 2-
dimensional data space, which has been derived from a linear maximum margin classifier in a
higher-dimensional space that is shown schematically in two dimensions in panel (b).

Extension of SVM to multiclass ($G > 2$ groups) can be achieved by computing binary SVM classifiers for all $G (G–1)/2$ possible group pairs. Each datum x to be assigned to a group is classified by each
of these, and the datum is assigned to the group for which it received the largest number of "votes."

Computational aspects for SVM can be elaborate. Details can be found in such references as Efron
and Hastie (2016), Hastie et al. (2009), and Hsieh (2009).

15.6.2. Classification Trees and Neural Network Classifiers

Classification Trees

Classification trees can be viewed as an application of the regression trees described in Section 7.8.1, for
which the predictands are discrete categorical variables rather than continuous real numbers. That is, the
outcomes to be predicted are the integer-valued group indices $g = 1, 2, ..., G$. The two methods were
originally proposed together, under the name Classification And Regression Trees, or CART

(Breiman et al., 1984). As is also the case for a regression tree, a classification tree is built by recursive binary partitions based on the K elements of the predictor variables, where at each stage the split is made that best separates the groups to be discriminated.

For regression trees, each binary split is chosen as the one that minimizes the combined sum of squares in Equation 7.56. This metric is not appropriate in the case of classification trees because the group indices are arbitrarily chosen as consecutive integers, and in addition there may be no natural ordering among the groups. Instead, the splitting criteria are based on proportion of predictand values in a given group g at a current or potential terminal node ℓ,

$$p_{\ell,g} = \frac{1}{n_\ell} \sum_{x_i \in \ell} I(y_i \in g).\tag{15.43}$$

The next binary split is then chosen which minimizes the "node impurity," or lack of homogeneity of group membership within the resulting branches, according to either the overall cross-entropy

$$E[L] = -\sum_{\ell=1}^{L} \sum_{g=1}^{G} p_{\ell,g} \ln\left(p_{\ell,g}\right)\tag{15.44}$$

or the *Gini Index*

$$E[L] = \sum_{\ell-1}^{L} \sum_{g=1}^{G} p_{\ell,g}\left(1 - p_{\ell,g}\right),\tag{15.45}$$

where L is the number of branches or nodes after the currently contemplated binary split. The cross-entropy measure fails if one or more groups are absent from one or more of the nodes.

Nonprobabilistic classification is decided according to which group has the largest proportional membership in each terminal node or branch. Figure 15.9 shows a portion of an example classification

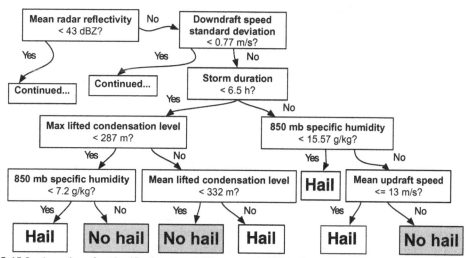

FIGURE 15.9 A portion of a classification tree yielding nonprobabilistic forecasts of hail occurrence or nonoccurrence. *From McGovern et al., 2017. © American Meteorological Society. Used with permission.*

tree, where the goal is to forecast one or the other of $G = 2$ future groups, pertaining to occurrence or nonoccurrence of hail. Alternatively, classification probabilities can be estimated using Equation 15.43 for each group $g = 1, 2, ..., G$, when the terminal node ℓ has been reached.

Because an individual classification tree is unstable, in the sense that small changes in the training data may yield substantial differences in the resulting tree structure, predictive performance can typically be improved by bootstrap aggregation, or bagging, of multiple trees, together with random predictor selection at each binary split. The result is a random forest, as described for regression trees in Section 7.8.1.

Neural Network Classifiers

Just as the framework of regression trees described in Section 7.8.1 can be used for classification by a discrete coding of the predictand variable, so also can neural networks (Section 7.8.2) be extended in an analogous way. As has already been noted in Section 7.8.2, use of a neural network for a classification problem usually involves choice of a suitable function for the output layer. In multicategory ($G > 2$) settings, often Equation 7.60 is used for this purpose. When only $G = 2$ groups are to be distinguished Equation 7.58 is another reasonable choice for the output function.

15.7. EXERCISES

15.1. Consider the two univariate PDFs $f_1(x) = 1 - |x|$, for $|x| \leq 1$; and $f_2(x) = 1 - |x - 0.5|$, for $-0.5 \leq x \leq 1.5$.
   a. Sketch the two PDFs.
   b. Identify the classification regions when $p_1 = p_2$ and $C(1|2) = C(2|1)$.
   c. Identify the classification regions when $p_1 = 0.2$ and $C(1|2) = C(2|1)$.

15.2. Use Fisher's linear discriminant to classify years in Table A.3 as either El Niño or non-El Niño, on the basis of the corresponding temperature and pressure data.
   a. What is the discriminant vector, scaled to have unit length?
   b. Which, if any, of the El Niño years have been misclassified?
   c. Assuming bivariate normal distributions, repeat part (b) accounting for unequal prior probabilities.

15.3. Figure 15.4 illustrates a discriminant analysis relating two standardized African precipitation variables, Gulf of Guinea rainfall (x_1) and Sahel rainfall (x_2), to the occurrence of at least one hurricane in the Caribbean sea in the following year (Y or N). The sample mean vectors and pooled variance estimates are

$$\bar{x}_Y = [0.29 \ 0.16]^T, \ \bar{x}_N = [-0.50 \ -0.32]^T, \ \text{and} \ [S_{pool}] = \begin{bmatrix} 1.000 & 0.154 \\ 0.154 & 1.000 \end{bmatrix}.$$

   a. Find the discriminant vector, using Fisher's procedure.
   b. Use Anderson's classification statistic to allocate the new observation $x_0 = [1/2 \ 1/2]^T$, assuming equal misclassification costs and equal priors.
   c. Assuming that both $f_Y(x)$ and $f_N(x)$ are bivariate normal with equal covariance matrices, evaluate the posterior probability of at least one Caribbean hurricane (i.e., find Pr{"yes"}) in the following year, given $x_0 = [1/2 \ 1/2]^T$.

TABLE 15.2 Likelihoods Calculated from the Forecast Verification Data for Subjective 12–24 h Projection Probability-of-Precipitation Forecasts for the United States During October 1980–March 1981, in Table 9.2

y_i	0.00	0.05	0.10	0.20	0.30	0.40	0.50	0.60	0.70	0.80	0.90	1.00
$p(y_i \mid o_1)$	0.0152	0.0079	0.0668	0.0913	0.1054	0.0852	0.0956	0.0997	0.1094	0.1086	0.0980	0.1169
$p(y_i \mid o_2)$	0.4877	0.0786	0.2058	0.1000	0.0531	0.0272	0.0177	0.0136	0.0081	0.0053	0.0013	0.0016

15.4. Average July temperature and precipitation at Ithaca, New York, are 68.6°F and 3.54 in.
   a. Classify Ithaca as belonging to one of the three groups in Example 14.2.
   b. Calculate probabilities that Ithaca is a member of each of the three groups, assuming bivariate normal distributions with common covariance matrix.

15.5. Using the forecast verification data in Table 9.2, we can calculate the likelihoods (i.e., conditional probabilities for each of the 12 possible forecasts, given either precipitation or no precipitation) in Table 15.2. The unconditional probability of precipitation is $p(o_1) = 0.162$. Considering the two precipitation outcomes as two groups to be discriminated between, calculate the posterior probabilities of precipitation if the forecast probability, y_i, is.
   a. 0.00.
   b. 0.10.
   c. 1.00.

Chapter 16

Cluster Analysis

16.1. BACKGROUND

16.1.1. Cluster Analysis vs. Discriminant Analysis

Cluster analysis deals with separating data into groups whose identities are not known in advance. This more limited state of knowledge is in contrast to the situation for discrimination methods, which require a training data set in which group memberships are known. In modern statistical parlance, cluster analysis is an example of unsupervised learning, whereas discriminant analysis is an instance of supervised learning. In general, in cluster analysis even the correct number of groups into which the data should be sorted is not known ahead of time. Rather, it is the degree of similarity and difference between individual observations x that is used to define the groups and to assign group membership. Examples of use of cluster analysis in the meteorological and climatological literature include grouping daily weather observations into synoptic types (Kalkstein et al., 1987), defining weather regimes from upper-air flow patterns (Mo and Ghil, 1988; Molteni et al., 1990), grouping members of forecast ensembles (Legg et al., 2002; Molteni et al., 1996; Tracton and Kalnay, 1993), and forecast evaluation (Kücken and Gerstengarbe, 2009; Marzban and Sandgathe, 2008). Gong and Richman (1995) have compared various clustering approaches in a climatological context and catalog the literature with applications of clustering to atmospheric data through 1993. Romesburg (1984) contains a general-purpose overview.

Given a sample of data vectors x defining the rows of a $(n \times K)$ data matrix $[X]$, a cluster analysis will define groups and assign group memberships at varying levels of aggregation. However, the method is primarily an exploratory data analysis tool, rather than an inferential tool. Indeed, the statistical underpinnings of most clustering approaches are vague. In most cases they are uncoupled from possible differences in underlying distributions, and useful associated inferential procedures are not generally available (e.g., Chacón, 2015). Unlike discriminant analysis, the procedure does not contain rules for assigning membership to future observations. However, a cluster analysis can bring out groupings in the data that might otherwise be overlooked, possibly leading to an empirically useful stratification of the data, or helping to suggest physical bases for observed structure in the data. For example, cluster analyses have been applied to geopotential height data in order to try to identify distinct atmospheric flow regimes (e.g., Cheng and Wallace, 1993; Mo and Ghil, 1988).

16.1.2. Distance Measures and the Distance Matrix

The idea of distance is central to the idea of the clustering of data points. Clusters should be composed of points separated by small distances, relative to the distances between clusters. However, there is a wide

Statistical Methods in the Atmospheric Sciences. https://doi.org/10.1016/B978-0-12-815823-4.00016-X

variety of plausible definitions for distance in this context, and the results of a cluster analysis may depend quite strongly on the distance measure chosen.

The most intuitive and commonly used distance measure in cluster analysis is Euclidean distance (Equation 11.6) in the K-dimensional space of the data vectors. Euclidean distance is by no means the only available choice for measuring distance between points or clusters, and in some instances may be a poor choice. In particular, if the elements of the data vectors are unlike variables with inconsistent measurement units, the variable with the largest values will tend to dominate the Euclidean distance. A more general alternative is the weighted Euclidean distance between two vectors x_i and x_j,

$$d_{i,j} = \left[\sum_{k=1}^{K} w_k \left(x_{i,k} - x_{j,k} \right)^2 \right]^{1/2}. \tag{16.1}$$

For $w_k = 1$ for each $k = 1, \ldots, K$, Equation 16.1 reduces to the ordinary Euclidean distance. If the weights are the reciprocals of the corresponding variances, that is, $w_k = 1/s_{k,k}$, the resulting function of the standardized variables is called the *Karl-Pearson distance*. Other choices for the weights are also possible. For example, if one or more of the K variables in x contains large outliers, it might be better to use weights that are reciprocals of the ranges of each of the variables.

Euclidean distance and Karl-Pearson distance are the most frequent choices in cluster analysis, but other alternatives are also possible. One alternative is to use the Mahalanobis distance (Equation 11.91), although deciding on an appropriate (pooled) dispersion matrix [S] may be difficult, since group memberships are not known in advance. A yet more general form of Equation 16.1 is the *Minkowski metric*,

$$d_{i,j} = \left[\sum_{k=1}^{K} w_k \left| x_{i,k} - x_{j,k} \right|^{\lambda} \right]^{1/\lambda}, \quad \lambda \geq 1. \tag{16.2}$$

Again, the weights w_k can equalize the influence of variables with incommensurate units. For $\lambda = 2$, Equation 16.2 reduces to the weighted Euclidean distance in Equation 16.1. For $\lambda = 1$, Equation 16.2 is known as the *city-block distance*.

The angles between pairs of vectors (Equation 11.15), or its cosine, are other possible choices for distance measures, as are the many alternatives presented in Mardia et al. (1979) or Romesburg (1984). Tracton and Kalnay (1993) have used the anomaly correlation (Equation 9.95) to group members of forecast ensembles, and the ordinary Pearson correlation sometimes is used as a clustering criterion as well. These latter two criteria are inverse distance measures, which should be maximized within groups and minimized between groups.

Having chosen a distance measure to quantify dissimilarity or similarity between pairs of vectors x_i and x_j, the next step in cluster analysis is to calculate the distances between all $n(n-1)/2$ possible pairs of the n observations. It can be convenient, either organizationally or conceptually, to arrange these into a $(n \times n)$ matrix of distances, [Δ], called the *distance matrix*. This symmetric matrix has zeros along the main diagonal, indicating zero distance between each x and itself.

16.2. HIERARCHICAL CLUSTERING

16.2.1. Agglomerative Methods Using the Distance Matrix

The most commonly implemented cluster analysis procedures are *hierarchical* and *agglomerative*. That is, they construct a hierarchy of sets of groups, each level of which is formed by merging one pair from the collection of previously defined groups. These procedures begin by considering that the n observations of x have no group structure or, equivalently, that the data set consists of n groups containing one observation each. The first step is to find the two groups (i.e., data vectors) that are closest in their K-dimensional space and to combine them into a new group. There are then $n-1$ groups, one of which has two members. On each subsequent step, the two groups that are closest are merged to form a larger group. Once a data vector x has been assigned to a group, it is not removed. Its group membership changes only when the group to which it has been assigned is merged with another group. This process continues until, at the final, $(n-1)^{st}$, step all n observations have been aggregated into a single group.

The n-group clustering at the beginning of this process and the one-group clustering at the end of this process are neither useful nor enlightening. Hopefully, however, a natural clustering of the data into a workable number of informative groups will emerge at some intermediate stage. That is, we hope that the n data vectors cluster or clump together in their K-dimensional space into some number G, $1 < G < n$, groups that reflect similar data-generating processes. The ideal result is a division of the data that both minimizes differences between members of a given cluster and maximizes differences between members of different clusters.

Distances between pairs of points can be unambiguously defined and stored in a distance matrix. However, even after choosing a distance measure and calculating a distance matrix there are alternative definitions for distances between groups of points if the groups contain more than a single member. The choice made for the distance measure, together with the criterion used to define cluster-to-cluster distances, essentially define the method of clustering. A few of the most common definitions and a less common but potentially interesting definition for intergroup distances based on the distance matrix are:

- *Single-linkage*, or *minimum-distance clustering*. Here the distance between clusters G_1 and G_2 is the smallest distance between one member of G_1 and one member of G_2. That is,

$$d_{G_1,G_2} = \min_{i \in G_1, j \in G_2} (d_{i,j}). \tag{16.3}$$

- *Complete-linkage*, or *maximum-distance clustering* groups data points on the basis of the largest distance between points in the two groups G_1 and G_2,

$$d_{G_1,G_2} = \max_{i \in G_1, j \in G_2} (d_{i,j}). \tag{16.4}$$

- *Average-linkage* clustering defines cluster-to-cluster distance as the average of distances between all possible pairs of points in the two groups being compared. If G_1 contains n_1 points and G_2 contains n_2 points, this measure for the distance between the two groups is

$$d_{G_1,G_2} = \frac{1}{n_1 n_2} \sum_{i=1}^{n_1} \sum_{j=1}^{n_2} d_{i,j}. \tag{16.5}$$

- *Centroid* clustering compares distances between the centroids, or vector averages, of pairs of clusters. According to this measure the distance between G_1 and G_2 is

$$d_{G_1,G_2} = \|\bar{x}_{G_1} - \bar{x}_{G_2}\|, \tag{16.6}$$

where the vector means are taken over all members of each of the groups separately.

- *Minimax linkage* (Bien and Tibshirani, 2011) is a relatively little known but potentially informative and useful alternative to the four more common distance measures,

$$dG_1,G_2 = \min_{x_i \in G_1 \cup G_2} \left[\max_{x_j \in G_1 \cup G_2} (d_{i,j}) \right]. \qquad (16.7)$$

For each of the $n_1 + n_2$ vectors x_i in a pair of groups that are candidates for merger, the expression inside the square brackets defines the maximum among distances to other members of the potentially merged group. The minimax distance in Equation 16.7 is then the smallest of these $n_1 + n_2$ maximized pairwise distances. Use of minimax linkage allows the potentially informative definition of prototype vectors, which provides a natural single-vector characterization of the group. These prototype vectors are exactly the x_i satisfying Equation 16.7 for each group.

Figure 16.1a illustrates single-linkage, complete-linkage, and centroid clustering for two hypothetical groups G_1 and G_2 in a $K = 2$-dimensional space. The open circles denote data points, of which there are $n_1 = 2$ in G_1 and $n_2 = 3$ in G_2. The centroids of the two groups are indicated by the solid circles. The single-linkage distance between G_1 and G_2 is the distance $d_{2,3}$ between the closest pair of points in the two groups. The complete-linkage distance is that between the most distant pair, $d_{1,5}$. The centroid distance is the distance between the two vector means $\|\bar{x}_{G_1} - \bar{x}_{G_2}\|$. The average-linkage distance can also be visualized in Figure 16.1 as the average of the six possible distances between individual members of G_1 and G_2, $(d_{1,5} + d_{1,4} + d_{1,3} + d_{2,5} + d_{2,4} + d_{2,3})/6$.

Figure 16.1b indicates that the minimax distance for the merger of the two groups is $d_{1,3}$, which is the largest distance between x_3 and any of the other points, but is the smallest of these maximum distances among the five points. Accordingly, the prototype vector for the merged group is x_3. The prototype vector within G_2 alone is x_4, because $d_{4,5}$ is smaller than the other two maximized distances within G_2, which are both $d_{3,5}$.

The results of a cluster analysis can depend strongly on which definition is chosen for the distances between clusters. Single-linkage clustering rarely is used, because it is susceptible to *chaining*, or the production of a few large clusters, which are formed by virtue of nearness to opposite edges of a cluster

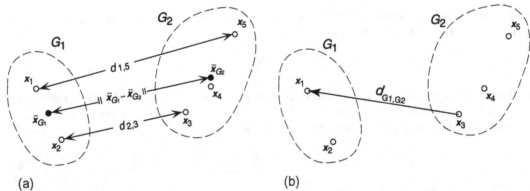

(a)   (b)

FIGURE 16.1   Illustration of four measures of the distance in $K = 2$-dimensional space, between a cluster G_1 containing the two elements x_1 and x_2, and a cluster G_2 containing the elements x_3, x_4, and x_5. The data points are indicated by open circles, and centroids of the two groups are indicated by the solid circles. (a) According to the maximum-distance, or complete-linkage criterion, the distance between the two groups is $d_{1,5}$, or the greatest distance between all of the six possible pairs of points in the two groups. The minimum-distance, or single-linkage criterion computes the distance between the groups as equal to the distance between the nearest pair of points, or $d_{2,3}$. According to the centroid method, the distance between the two clusters is the distance between the sample means of the points contained in each. (b) The minimax distance between the two groups is indicated by the arrow, which originates at the prototype vector for the merged groups, x_3.

of points to be merged at different steps. At the other extreme, complete-linkage clusters tend to be more numerous, as the criterion for merging clusters is more stringent. Average-distance clustering is usually intermediate between these two extremes and appears to be the most commonly used approach to hierarchical clustering based on the distance matrix. Hastie et al. (2009) argue that average-distance clustering is the only statistically consistent method of the three, meaning that as the sample size becomes arbitrarily large, the group-average dissimilarities approach true population values.

16.2.2. Ward's Minimum Variance Method

Ward's minimum variance method, or simply Ward's method, is a popular hierarchical clustering method that does not operate on a distance matrix. As a hierarchical method, it begins with n single-member groups, and merges two groups at each step, until all the data are in a single group after $n-1$ steps. However, the criterion for choosing which pair of groups to merge at each step is that, among all possible ways of merging two groups, the pair to be merged is chosen that minimizes the sum of squared distances between the points and the centroids of their respective groups, summed over the resulting groups. That is, among all possible ways of merging two of $G+1$ groups to make G groups, that merger is made that minimizes

$$W = \sum_{g=1}^{G} \sum_{i=1}^{n_g} \left\| \boldsymbol{x}_i - \bar{\boldsymbol{x}}_g \right\|^2 = \sum_{g=1}^{G} \sum_{i=1}^{n_g} \sum_{k=1}^{K} \left(x_{i,k} - \bar{x}_{g,k} \right)^2. \tag{16.8}$$

In order to implement Ward's method to choose the best pair from $G+1$ groups to merge, Equation 16.8 must be calculated for all of the $G(G+1)/2$ possible pairs of existing groups. For each trial pair, the centroid, or group mean, for the trial merged group is recomputed using the data for both of the previously separate groups, before the squared distances are calculated. In effect, Ward's method minimizes the sum, over the K dimensions of $\boldsymbol{x}$, of within-groups variances. At the first (n-group) stage this variance is zero, and at the last (1-group) stage this variance is tr$[S_x]$, so that $W = n$ tr$[S_x]$. For data vectors whose elements have incommensurate units, operating on nondimensionalized values (dividing by standard deviations) will prevent artificial domination of the procedure by one or a few of the K variables.

16.2.3. The Dendrogram or Tree Diagram

The progress and intermediate results of a hierarchical cluster analysis are conventionally illustrated using a *dendrogram* or tree diagram. Beginning with the "twigs" at the beginning of the analysis, when each of the n observations $\boldsymbol{x}$ constitutes its own cluster, one pair of "branches" is joined at each step as the closest two clusters are merged. The distances between these clusters before they are merged are also indicated in the diagram by the distances of the points of merger from the initial n-cluster stage of the twigs.

Figure 16.2 illustrates a simple dendrogram, reflecting the clustering of the five points plotted as open circles in Figure 16.1a. The analysis begins at the left of Figure 16.2, when all five points constitute separate clusters. At the first stage, the closest two points, $\boldsymbol{x}_3$ and $\boldsymbol{x}_4$, are merged into a new cluster. Their distance $d_{3,4}$ is proportional to the distance between the vertical bar joining these two points and the left edge of the figure. At the next stage, the points $\boldsymbol{x}_1$ and $\boldsymbol{x}_2$ are merged into a single cluster because the distance between them is smallest of the six distances among the four clusters that existed at the previous stage. The distance $d_{1,2}$ is necessarily larger than the distance $d_{3,4}$, since $\boldsymbol{x}_1$ and $\boldsymbol{x}_2$ were not chosen for merger on the first step, and the vertical line indicating the distance between them is plotted further to the

FIGURE 16.2   Illustration of a dendrogram or tree diagram, for a clustering of the five points plotted as open circles in Figure 16.1a. The results of the four clustering steps are indicated as the original five lines are progressively joined from left to right, with the distances between joined clusters indicated by the positions of the vertical lines.

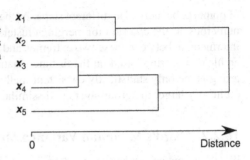

right in Figure 16.2 than the distance between x_3 and x_4. The third step merges x_5 and the pair (x_3, x_4), to yield the two-group stage indicated by the dashed ovals in Figure 16.1.

When minimax linkage is used to define distances, the interpretability of the dendrogram can be enhanced by indicating the prototype vector for each newly combined pair of groups at the point of the dendrogram indicating that merger. If the x vectors represent maps or other graphical objects, thumbnail plots of the prototypes can be shown. Otherwise, the name or case number of the prototype can be used. Showing prototypes only for the larger groups, or for the groups that are candidates for a final solution, might reduce clutter in the resulting plot.

16.2.4. How Many Clusters?

A hierarchical cluster analysis will produce a different grouping of n observations at each of the $n-1$ steps. At the first step each observation is in a separate group, and after the last step all the observations are in a single group. An important practical problem in cluster analysis is the choice of which intermediate stage will be chosen as the final solution. That is, we need to choose the level of aggregation in the tree diagram at which to stop further merging of clusters. The principle guiding this choice is to find that level of clustering that maximizes similarity within clusters and minimizes similarity between clusters, but in practice the best number of clusters for a given problem is usually not obvious. Generally the stopping point will require a subjective choice that will depend to some degree on the goals of the analysis.

One approach to the problem of choosing the best number of clusters is through summary statistics that relate to concepts in discrimination presented in Chapter 15. Several such criteria are based on the within-groups covariance matrix (Equation 15.16), either alone or in relation to the "between-groups" covariance matrix (Equation 15.18). Some of these objective stopping criteria are discussed in Jolliffe et al. (1986) and Fovell and Fovell (1993), who also provide references to the broader literature on such methods.

A traditional subjective approach to determination of the stopping level is to inspect a plot of the distances between merged clusters as a function of the stage of the analysis. When similar clusters are being merged early in the process, these distances are small and they increase relatively little from step to step. Late in the process there may be only a few clusters, separated by large distances. If a point can be discerned where the distances between merged clusters jump markedly, the process can be stopped just before these distances become large.

Wolter (1987) suggests a Monte Carlo approach, where sets of random numbers simulating the real data are subjected to cluster analysis. The distributions of clustering distances for the random numbers can be compared to the actual clustering distances for the data of interest. The idea here is that genuine clusters in the real data should be closer than clusters in the random data and that the clustering algorithm should be stopped at the point where clustering distances are greater than for the analysis of the random data. Similarly, Tibshirani et al. (2001) propose defining the stopping point as that exhibiting the maximum difference between the logs of the distances W in Equation 16.8 to those obtained by averaging the results of many cluster analyses of K-dimensional uniform random numbers.

Example 16.1. A Cluster Analysis in Two Dimensions

The mechanics of cluster analysis are easiest to see when the data vectors have only $K = 2$ dimensions. Consider the data in Table 15.1, where these two dimensions are average July temperature and average July precipitation. These data were collected into three groups for use in the discriminant analysis worked out in Example 15.2. However, the point of a cluster analysis is to try to discern group structure within a data set, without prior knowledge or information about the nature of that structure. Therefore for purposes of a cluster analysis, the data in Table 15.1 should be regarded as consisting of $n = 28$ observations of two-dimensional vectors x, whose natural groupings we would like to discern.

Because the temperature and precipitation values have different physical units, it is advisable to divide by the respective standard deviations before subjecting them to a clustering algorithm. That is, the temperature and precipitation values are divided by 4.42 °F and 1.36 in., respectively. The result is that the analysis is done using the Karl-Pearson distance, and the weights in Equation 16.1 are $w_1 = 4.42^{-2}$ and $w_2 = 1.36^{-2}$. The reason for this treatment of the data is to avoid the same kind of problem that can occur when conducting a principal component analysis using unlike data, where a variable with a much higher variance than the others will dominate the analysis even if that high variance is an artifact of the units of measurement. For example, if the precipitation had been expressed in millimeters there would be apparently more distance between points in the direction of the precipitation axis, and a clustering algorithm would focus on precipitation differences to define groups. If the precipitation were expressed in meters there would be essentially no distance between points in the direction of the precipitation axis, and a clustering algorithm would separate points almost entirely on the basis of the temperatures.

Figure 16.3 shows the results of clustering the data in Table 15.1, using the complete-linkage clustering criterion in Equation 16.4. On the left is a tree diagram for the process, with the individual stations listed at the bottom as the leaves. There are 27 horizontal lines in this tree diagram, each of which represents the merger of the two clusters it connects. At the first stage of the analysis the two closest points (Springfield and St. Louis) are merged into the same cluster, because their Karl-Pearson distance $d = [4.42^{-2}(78.8-78.9)^2 + 1.36^{-2}(3.58-3.63)^2]^{1/2} = 0.043$ is the smallest of the $(28)(28-1)/2 = 378$ distances between the possible pairs. This separation distance can be seen graphically in Figure 16.4: the distance $d = 0.043$ is the height of the leftmost dot in Figure 16.3b. At the second stage Huntsville and Athens are merged, because their Karl-Pearson distance $d = [4.42^{-2}(79.3-79.2)^2 + 1.36^{-2}(5.05-5.18)^2]^{1/2} = 0.098$ is the second-smallest separation of the points (cf. Figure 16.4), and this distance corresponds to the height of the second dot in Figure 16.3b. At the third stage, Worcester and Binghamton ($d = 0.130$) are merged, and at the fourth stage Macon and Augusta ($d = 0.186$) are merged. At the fifth stage, Concordia is merged with the cluster consisting of Springfield and St. Louis. Since the Karl-Pearson distance between Concordia and St. Louis is larger than the distance between Concordia and Springfield (but smaller than the distances between Concordia and the other 25 points), the complete-linkage criterion merges these three points at the larger distance $d = [4.42^{-2}(79.0-78.9)^2 + 1.36^{-2}(3.37-3.63)^2]^{1/2} = 0.193$ (height of the fifth dot in Figure 16.3b).

The heights of the horizontal lines in Figure 16.3a, indicating group mergers, also correspond to the distances between the merged clusters. Since the merger at each stage is between the two closest clusters, these distances become greater at later stages. Figure 16.3b shows these same distances between merged clusters as a function of the stage in the analysis. Subjectively, these distances climb gradually until perhaps stage 22 or stage 23, where the distances between combined clusters begin to become noticeably

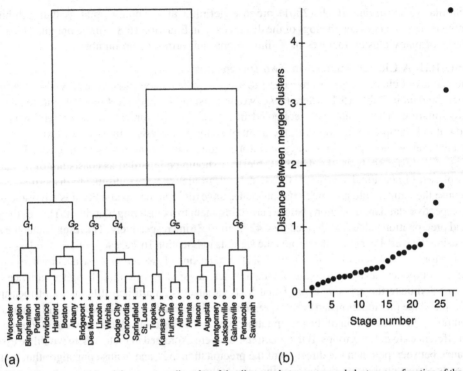

(a)   (b)

FIGURE 16.3 Dendrogram (a) and the corresponding plot of the distances between merged clusters as a function of the stage of the cluster analysis (b) for the data in Table 15.1. Standardized data (i.e., Karl-Pearson distances) have been clustered according to the complete-linkage criterion. The distances between merged groups appear to increase markedly at stage 22 or 23, indicating that the analysis should stop after 21 or 22 stages, which for these data would yield seven or six clusters, respectively. The six numbered clusters correspond to the grouping of the data shown in Figure 16.4. The seven-cluster solution would split Topeka and Kansas City from the Alabama and Georgia stations in G_5. The five-cluster solution would merge G_3 and G_4.

larger. A reasonable interpretation of this change in slope is that natural clusters have been defined at this point in the analysis, and that the larger distances at later stages indicate mergers of unlike clusters that should be distinct groups. Note, however, that a single change in slope does not occur in every cluster analysis, so that the choice of where to stop group mergers may not always be so clear-cut. It is possible, for example, for there to be two or more relatively flat regions in the plot of distance versus stage, separated by segments of larger slope. Different clustering criteria may also produce different breakpoints. In such cases the choice of where to stop the analysis is more ambiguous.

If Figure 16.3b is interpreted as exhibiting its first major slope increase between stages 22 and 23, a plausible point at which to stop the analysis would be after stage 22. This stopping point would result in the definition of the six clusters labeled $G_1 - G_6$ on the tree diagram in Figure 16.3a. This level of clustering assigns the nine northeastern stations (+ symbols) into two groups, assigns seven of the nine central stations (x symbols) into two groups, allocates the central stations Topeka and Kansas City to Group 5 with six of the southeastern stations (o symbols), and assigns the remaining four southeastern stations to a separate cluster.

Figure 16.4 indicates these six groups in the $K = 2$-dimensional space of the standardized data, by separating points in each cluster with dashed lines. If this solution seemed too highly aggregated on the basis of the prior knowledge and information available to the analyst, the seven-cluster solution produced after stage 21 could be chosen, which separates the central U.S. cities Topeka and Kansas City

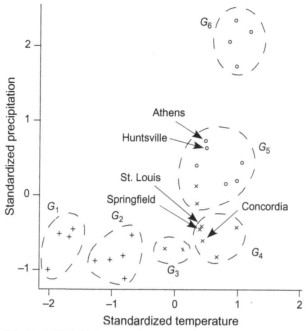

FIGURE 16.4 Scatterplot of the data in Table 15.1 expressed as standardized anomalies, with dashed lines showing the six groups defined in the cluster analysis tree diagram in Figure 16.3a. The five-group clustering would merge the central U.S. stations in Groups 3 and 4. The seven-group clustering would split the two central U.S. stations in Group 5 from six southeastern U.S. stations.

(x's) from the six southeastern cities in Group 5. If the six-cluster solution seemed too finely split, the five-cluster solution produced after stage 23 would merge the central U.S. stations in Groups 3 and 4. None of the groupings indicated in Figure 16.3a corresponds exactly to the group labels in Table 15.1, and we should not necessarily expect them to. It could be that limitations of the complete-linkage clustering algorithm operating on Karl-Pearson distances have produced some misclassifications, or that the group labels in Table 15.1 have been imperfectly defined, or both.

Finally, Figs. 16.5 and 16.6 illustrate the fact that different clustering algorithms will usually yield somewhat different results. Figure 16.5a shows distances at which groups are merged for the data in Table 15.1, according to single linkage operating on Karl-Pearson distances. There is a large jump after stage 21, suggesting a possible natural stopping point with seven groups. These seven groups are indicated in Figure 16.5b, which can be compared with the complete-linkage result in Figure 16.4. The clusters denoted G_2 and G_6 in Figure 16.4 occur also in Figure 16.5b. However, one long and thin group has developed in Figure 16.5b, composed of stations from G_3, G_4, and G_5. This result illustrates the chaining phenomenon to which single-linkage clusters are prone, as additional stations or groups are accumulated that are close to a point at one edge or another of a group, even though the added points may be quite far from other points in the same group.

Figure 16.6 shows the results when the same data are clustered using Ward's method. Here the results are very similar to those in Figure 16.3, which were derived using complete linkage. Ward's method groups the four southernmost stations on the far right of the dendrogram in Figure 16.6a somewhat differently than does complete linkage, and the first relatively large gap after stage 22 in Figure 16.6b suggests that a 5-group solution might be appropriate. ◇

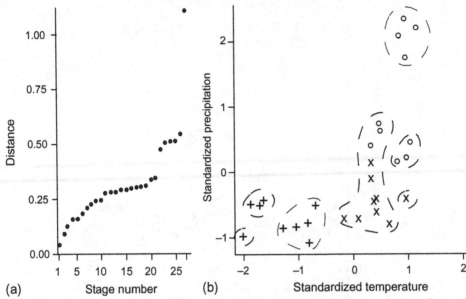

FIGURE 16.5   Clustering of the data in Table 15.1, using single linkage. (a) Merger distances as a function of stage, showing a large jump after 22 stages. (b) The seven clusters existing after stage 22, illustrating the chaining phenomenon.

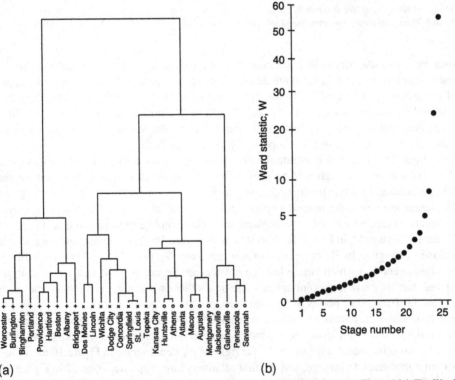

FIGURE 16.6   Clustering of the data in Table 15.1, using Ward's method, in the same format as Figure 16.4. The Ward sum-of-squares statistic W (Equation 16.8) is plotted in the vertical on a square-root scale.

Chapter 8 describes ensemble forecasting, in which the effects of uncertainty about the initial state of the atmosphere on the evolution of a forecast are addressed by calculating multiple dynamical forecasts beginning at an ensemble of similar initial conditions. The method has proved to be an extremely useful advance in forecasting technology, but requires extra effort to absorb the large amount of additional information produced. One way to summarize the information in a large collection of maps from a forecast ensemble is to group them according to a cluster analysis. If the smooth contours on each map have been interpolated from K gridpoint values, then each $(K \times 1)$ vector x included in the cluster analysis corresponds to one of the forecast maps.

Figure 16.7 shows the results of the use of Ward's method to group $n = 33$ ensemble members forecasting 500 mb heights over Europe, at a lead time of six days. An innovation in the approach is that it is

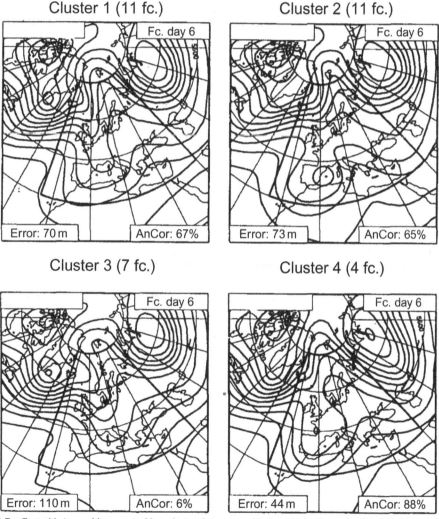

FIGURE 16.7   Centroids (ensemble means) of four clusters for an ensemble forecast for European 500 mb heights at a lead time of six days. Numbers of ensemble members in each cluster are indicated in the captions above each panel. *From Molteni et al. (1996).*

computed in a way that brings out the time trajectories of the forecasts, by simultaneously clustering maps for the five-, six-, and seven-day forecasts, although only the day 6 clusters are shown in Figure 16.7. That is, if each forecast map consists of K gridpoints, the x vectors being clustered would be dimensioned $(3 K \times 1)$, with the first K elements pertaining to day 5, the second K elements pertaining to day 6, and the last K elements pertaining to day 7. Each of the ensemble members remains in the same cluster throughout the forecast period, and clustering their joint behavior over the three-day period allows representation of the coherence in time of the flows that are represented. Because there are a large number of gridpoints underlying each map, the analysis actually was conducted using the first $K = 10$ principal components of the height fields, which was sufficient to capture 80% of the variance, so the clustered vectors had dimension (30×1). The currently operational counterpart of this method is described by Ferranti and Corti (2011).

Another interesting aspect of the example in Figure 16.7 is that the use of Ward's method provided an apparently natural stopping criterion for the clustering, which is related to forecast accuracy. Ward's method (Equation 16.8) is based on the sum of squared differences between the x's being clustered and their respective group means. Regarding the group means as forecasts, these squared differences would be contributions to the overall expected mean squared error if the ensemble members x were different realizations of plausible observed maps. The clustering in Figure 16.7 was stopped at the point where Equation 16.8 yields squared errors comparable to (the typically modest) 500 mb forecast errors obtained at the three-day lead time, so that their medium-range ensemble forecasts were grouped together if their differences were comparable to or smaller than typical short-range forecast errors.

16.2.5. Divisive Methods

In principle, hierarchical clustering can be achieved by reversing the agglomerative clustering process. That is, beginning with a single cluster containing all n observation vectors, this cluster can be split into the two most similar possible groups. At the third stage one of these groups could be split to yield the three most similar groups possible, and so on. The procedure would proceed, in principle, to the point of n clusters each populated by a single data vector, with an appropriate intermediate solution determined by a stopping criterion. This approach to clustering, which is opposite to agglomeration, is called *divisive clustering*.

Divisive clustering is almost never used, because it is computationally impractical for all except the smallest sample sizes. Agglomerative hierarchical clustering requires examination of all $G(G-1)/2$ possible pairs of G groups, in order to choose the most similar two for merger. In contrast, divisive clustering requires examination, for each group of size n_g members, all $2^{n_g - 1} - 1$ possible ways to make a split. This number of potential splits is 511 for $n_g = 10$, and rises to 524,287 for $n_g = 20$, and 5.4×10^8 for $n_g = 30$. Macnaughton-Smith et al. (1964) propose an alternative approach to divisive clustering that is much faster computationally but explores the possible splits less comprehensively.

16.3. NONHIERARCHICAL CLUSTERING

16.3.1. The K-Means Method

A potential drawback of hierarchical clustering methods is that once a data vector x has been assigned to a group it will remain in that group, and in groups with which it is merged. That is, hierarchical methods have no provision for reallocating points that may have been misclassified at an early stage. Clustering methods that allow reassignment of observations as the analysis proceeds are called *nonhierarchical*. Like hierarchical methods, nonhierarchical clustering algorithms also group observations according to some distance measure in the K-dimensional space of x.

The most widely used nonhierarchical clustering approach is called the *K-means* method. The "*K*" in K-means refers to the number of groups, called G in this text, and not to the dimension of the data vector. The K-means method is named for the number of clusters into which the data will be grouped, because this number must be specified in advance of the analysis, together with an initial guess for the group membership of each of the x_i, $i = 1, \dots, n$.

The K-means algorithm can begin either from a random partition of the n data vectors into the prespecified number G of groups, or from an initial selection of G seed points. The seed points might be defined by a random selection of G of the n data vectors or by some other approach that is unlikely to bias the results. Initial group memberships are then decided according to minimum distances to the seed points. Another possibility is to define the initial groups as the result of a hierarchical clustering that has been stopped at G groups, allowing reclassification of x's from their initial placement by the hierarchical clustering.

Having defined the initial membership of the G groups in some way, the K-means algorithm proceeds as follows:

(1) Compute the centroids (i.e., vector means) $\bar{x}_g$, $g = 1, \dots, G$; for each cluster.
(2) Calculate the distances between the current data vector x_i and each of the G $\bar{x}_g$'s. Usually Euclidean or Karl-Pearson distances are used, but distance can be defined by any measure that might be appropriate to the particular problem.
(3) If x_i is already a member of the group whose mean is closest, repeat step 2 for x_{i+1} (or for x_1, if $i = n$). Otherwise, reassign x_i to the group whose mean is closest, and return to step 1.

The algorithm is iterated until each x_i is closest to its group mean, that is, until a full cycle through all n data vectors produces no reassignments.

The need to prespecify the number of groups and their initial memberships can be a disadvantage of the K-means method, which may or may not compensate its ability to reassign potentially misclassified observations. Unless there is prior knowledge of the correct number of groups, and/or the clustering is a precursor to subsequent analyses requiring a particular number of groups, it is probably wise to repeat K-means clustering for a range of initial group numbers G. Because a particular set of initial guesses for group membership may yield a local rather than global minimum for the sum of distances to group centroids, K-means analyses should be repeated for different initial assignments of observations for each of the trial values of G. Hastie et al. (2009) suggest choosing that G minimizing an overall dissimilarity measure, such as the sum of squared distances between each x and its group mean in Equation 16.8.

16.3.2. Nucleated Agglomerative Clustering

Elements of agglomerative clustering and K-means clustering can be combined in an iterative procedure called *nucleated agglomerative clustering*. This method reduces somewhat the effects of arbitrary initial choices for group seeds in the K-means method and automatically produces a sequence of K-means clusters through a range of group sizes G.

The nucleated agglomerative method begins by specifying a number of groups G_{init} that is larger than the number of groups G_{final} that will exist at the end of the procedure. A K-means clustering into G_{init} groups is calculated, as described in Section 16.3.1. The following steps are then iterated:

(1) The two closest groups are merged according to Ward's method. That is, the two groups are merged that minimize the increase in Equation 16.8.
(2) K-means clustering is performed for the reduced number of groups, using the result of step 1 as the initial point. If the result is G_{final} groups, the algorithm stops. Otherwise, step 1 is repeated.

This algorithm produces a hierarchy of clustering solutions for the range of group sizes $G_{\text{init}} \geq G \geq G_{\text{final}}$, while allowing reassignment of observations to different groups at each stage in the hierarchy.

16.3.3. Clustering Using Mixture Distributions

The fitting of mixture distributions (see Section 4.4.9) (e.g., Everitt and Hand, 1981; McLachlan and Basford, 1988; Titterington et al., 1985) is another approach to nonhierarchical clustering. In the statistical literature, this approach to clustering is called "model-based," referring to the statistical model embodied in the mixture distribution (Banfield and Raftery, 1993; Fraley and Raftery, 2002). For multivariate data the most usual approach is to fit mixtures of multivariate normal distributions, for which maximum likelihood estimation using the EM algorithm (see Section 4.6.3) is straightforward (the algorithm is outlined in Hannachi and O'Neill, 2001, and Smyth et al., 1999). This approach to clustering has been applied to atmospheric data to identify large-scale flow regimes by Haines and Hannachi (1995), Hannachi (1997), and Smyth et al. (1999).

The basic idea in this approach to clustering is that each of the component PDFs $f_g(x)$, $g = 1, ..., G$, represents one of the G groups from which the data have been drawn. As illustrated in Example 4.14, using the EM algorithm to estimate a mixture distribution produces (in addition to the distribution parameters) posterior probabilities (Equation 4.88) for membership in each of the component PDFs, given each of the observed data values x_i. Using these posterior probabilities, a "hard" (i.e., nonprobabilistic) classification can be achieved by assigning each data vector x_i to that PDF $f_g(x)$ having the largest probability. However, in many applications retention of these probability estimates regarding the group memberships may be informative.

As is the case for other nonhierarchical clustering approaches, the number of groups G (in this case, the number of component PDFs $f_g(x)$) typically is specified in advance. However, the number of mixture densities to include can be cast as a model-selection exercise, using the BIC statistic (Equation 7.39) in connection with appropriate (e.g., multivariate normal) likelihood functions (Fraley and Raftery, 2002). Alternatively, Banfield and Raftery (1993) and Smyth et al. (1999) describe algorithms for choosing the number of groups, using a cross-validation approach.

16.4. SELF-ORGANIZING MAPS (SOM)

The method of Self-Organizing Maps (SOM) is a "machine learning" approach that is commonly used for clustering data sets in which the membership of the training data vectors in some prespecified number of groups G is not known. Accordingly, the method bears similarities to the K-means method described in Section 16.3.1.

The goal in a SOM analysis is to find a 1- or (more usually) 2-dimensional manifold within the K-dimensional data space that best conforms to the data distributions. Initially this manifold is represented by a regular planar grid containing G points, or nodes. As the analysis proceeds this grid is bent out of its initial planar arrangement and deformed away from its initial regular spacing. At the end of the

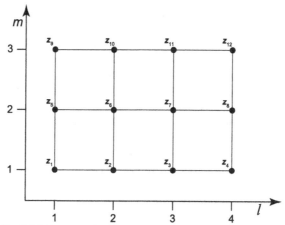

FIGURE 16.8   An example (4×3) initial rectangular grid containing $G = 12$ nodes, z_g.

procedure the data vectors x closest to each of the G nodes are allocated into a group associated with that node.

Figure 16.8 shows an example $(L \times M) = (4 \times 3)$ point grid, of nodes z_g, in which the grid index g might be defined as $g = (m-1)L + \ell$, for $m = 1, \ldots, M$, and $\ell = 1, \ldots, L$, so that $G = LM$. To initialize the calculations, this grid is often placed on the plane defined by the leading two principal components of the data set overall, in the K-dimensional space of the data vectors x. The initial grid collapses to a line, and the analysis will utilize a 1-dimensional manifold, for $M = 1$, in which case the initial placement might be along the axis of the leading principal component. In either case, the nodes z_g are $(K \times 1)$ vectors in the K-dimensional data space.

The SOM calculations proceed iteratively through the x_i, $i = 1, \ldots, n$, and generally more than one pass is made through the data. At each step the nodes are moved closer to the current data point x_i by first locating the node z_g that is closest (i.e., the node minimizing $\| x_i - z_g \|$), and then adjusting z_g and all other nodes $z_{g'}$ sufficiently close to x_i toward x_i according to

$$ z_{g'} \leftarrow z_{g'} + \alpha\, h\left(\left\| \begin{bmatrix} \ell \\ m \end{bmatrix} - \begin{bmatrix} \ell' \\ m' \end{bmatrix} \right\| \right) (x_i - z_g), \tag{16.9} $$

where α is a learning rate parameter, $h(\bullet)$ is a kernel function, the indices ℓ and m pertain to the closest node z_g, and the indices ℓ' and m' pertain to the node $z_{g'}$ being moved toward x_i. Notice that the proximity of a node $z_{g'}$ to the closest node z_g is defined in the argument of the kernel function with respect to distances between them in the original grid system, so that for example the distance between z_2 and z_7 in Figure 16.8 is always $\sqrt{2}$, regardless of how extremely the initial rectangular arrangement of nodes has been distorted by the iterative adjustments. Typical choices for the kernel function are circular, in which case only nodes $z_{g'}$ nearest to the closest node z_g are adjusted, and those adjustments are of equal magnitudes; or Gaussian, in which case all $z_{g'}$ are adjusted toward the current x_i but closer nodes are displaced more. In either case $z_{g'} = z_g$ in Equation 16.9 for the closest node, so that z_g is always adjusted toward the current data point x_i. As the iterations proceed, the learning rate parameter α is reduced from 1 to 0, and the width or standard deviation of the kernel function decreases, so that the adjustments toward each x_i tend to be smaller as the calculations proceed. These adjustments can be made either gradually and continuously as the iterations progress, or as step functions that change after each set of n iterations.

Figure 16.9 shows an example $G = 6$ result, obtained using an SOM based on a (2×3) grid. Here 104 locations in the northeastern United States have been clustered based on data vectors x_i, $i = 1, \ldots, 104$, containing the pairwise station covariances of daily precipitation data for the years 1900–1999. That is, each x_i is a row or column of the full (104×104) covariance matrix. The analysis has succeeded in regionalizing the data into six overlapping groups, which turn out to be spatially contiguous. The upper pair of diagrams in Figure 16.10 shows the arrangement of the 6-node grid, and the correspondence between nodes and the regions labeled A–F in Figure 16.9. The arrangement of the regions in the space of the grid mirrors their relative geographical positions, as a consequence of covariances between stations tending to decrease as the station separations increase. The lower pair of diagrams in Figure 16.10 shows the corresponding results from a $G = 12$-group clustering based on a (3×4) grid. The hatching patterns correspond to

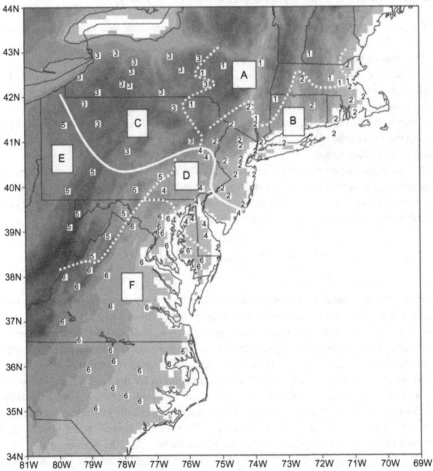

FIGURE 16.9   Clustering of 104 locations in the northeastern United States into $G = 6$ groups, according to daily precipitation data, 1900–1999, using a SOM based on a (2×3) grid. Shading indicates the topography. *From Crane and Hewitson (2003).*

the groups indicated in the upper diagram and in Figure 16.9, showing that the groups maintain their relative positions in the two grid systems, but their overlap in the lower-right diagram illustrates that the SOM clustering is nonhierarchical.

Although Figure 16.9 shows a geographic regionalization based on an SOM, the "map" in the name SOM usually refers to a display of representative data vectors which are those closest to each node z_g, on the original and undistorted grid. Figure 16.11 shows one such map, depicting fields of scatterometer-

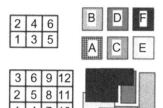

FIGURE 16.10   The relationship between the $G = 6$-group regionalization in Figure 16.9 and the (2×3) SOM grid is indicated in the upper pair of panels. The lower pair of panels shows corresponding results for the $G = 12$-group SOM, where the overlapping shading illustrates that the SOM clustering is nonhierarchical. *From Crane and Hewitson (2003).*

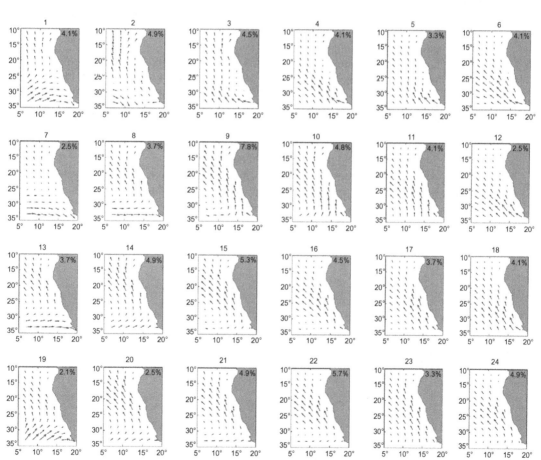

FIGURE 16.11   A SOM depicting a $G = 24$-group clustering of satellite-derived sea-surface winds in a portion of the southeastern Atlantic Ocean. The edge of the gray shading locates the west coast of southern Africa. *From Richardson et al. (2003).*

derived surface wind vectors for a region of the southeast Atlantic Ocean. Here the SOM has been computed using a (6×4) grid, and the proportion of the underlying (u, v) wind pair fields classified into each node are indicated in the upper-right corners. A close study of the individual panels reveals that nearby nodes are very similar to each other, reflecting the fact that the number density of SOM nodes in the K-dimensional data space will tend to follow the underlying density of data vectors. On the other hand, the strong similarities among nearby panels in Figure 16.11 suggest that a smaller initial grid might yield a more parsimonious clustering of the underlying data.

16.5. EXERCISES

16.1. Compute the distance matrix $[\Delta]$ for the Guayaquil temperature and pressure data in Table A.3 for the six years 1965–1970, using Karl-Pearson distance.

16.2. From the distance matrix computed in Exercise 16.1, cluster the six years using
   a. Single linkage.
   b. Complete linkage.
   c. Average linkage.

16.3. Cluster the Guayaquil pressure data (Table A.3) for the six years 1965–1970, using
   a. The centroid method and Euclidean distance.
   b. Ward's method operating on the raw data.

16.4. Cluster the Guayaquil temperature data (Table A.3) for the six years 1965–1970 into two groups using the K-means method, beginning with $G_1 = \{1965, 1966, 1967\}$ and $G_2 = \{1968, 1969, 1970\}$.

16.5. Consider the mixture distribution $f(x) = w f_1(x) + (1-w) f_2(x)$, where x is a nonnegative scalar, and each of the two constituent PDFs is a univariate exponential distribution. Applying the E-M algorithm to a collection of rainfall data, you have estimated the parameters as $\mu_1 = 2$ mm, $\mu_2 = 20$ mm, and $w = 0.7$, for the purpose of "model-based" clustering.
   a. Make a "hard" classification of a new datum, $x_0 = 5$ mm, into one of the two clusters.
   b. Calculate probabilities of membership for $x_0 = 5$ mm in each of the two clusters.

Example Data Sets

In real applications of climatological data analysis we would hope to use much more data (e.g., all available January daily data, rather than data for just a single year) and would have a computer perform the calculations. This small data set is used in a number of examples in this book so that the calculations can be performed by hand, and a clearer understanding of procedures can be achieved.

TABLE A.1 Daily Precipitation (in.) and Temperature (°F) Observations at Ithaca and Canandaigua, New York, for January 1987

Date	Ithaca			Canandaigua		
	Precipitation	Max Temp.	Min Temp.	Precipitation	Max Temp.	Min Temp.
1	0.00	33	19	0.00	34	28
2	0.07	32	25	0.04	36	28
3	1.11	30	22	0.84	30	26
4	0.00	29	−1	0.00	29	19
5	0.00	25	4	0.00	30	16
6	0.00	30	14	0.00	35	24
7	0.00	37	21	0.02	44	26
8	0.04	37	22	0.05	38	24
9	0.02	29	23	0.01	31	24
10	0.05	30	27	0.09	33	29
11	0.34	36	29	0.18	39	29
12	0.06	32	25	0.04	33	27
13	0.18	33	29	0.04	34	31
14	0.02	34	15	0.00	39	26
15	0.02	53	29	0.06	51	38
16	0.00	45	24	0.03	44	23
17	0.00	25	0	0.04	25	13
18	0.00	28	2	0.00	34	14
19	0.00	32	26	0.00	36	28
20	0.45	27	17	0.35	29	19
21	0.00	26	19	0.02	27	19
22	0.00	28	9	0.01	29	17
23	0.70	24	20	0.35	27	22
24	0.00	26	−6	0.08	24	2
25	0.00	9	−13	0.00	11	4
26	0.00	22	−13	0.00	21	5
27	0.00	17	−11	0.00	19	7

Continued

TABLE A.1 Daily Precipitation (in.) and Temperature (°F) Observations at Ithaca and Canandaigua, New York, for January 1987—Cont'd

	Ithaca			Canandaigua		
Date	Precipitation	Max Temp.	Min Temp.	Precipitation	Max Temp.	Min Temp.
28	0.00	26	−4	0.00	26	8
29	0.01	27	−4	0.01	28	14
30	0.03	30	11	0.01	31	14
31	0.05	34	23	0.13	38	23
Sum/avg.	3.15	29.87	13.00	2.40	31.77	20.23
Std. dev.	0.243	7.71	13.62	0.168	7.86	8.81

TABLE A.2 January Precipitation at Ithaca, New York, 1933–1982, Inches

1933	0.44	1945	2.74	1958	4.90	1970	1.03
1934	1.18	1946	1.13	1959	2.94	1971	1.11
1935	2.69	1947	2.50	1960	1.75	1972	1.35
1936	2.08	1948	1.72	1961	1.69	1973	1.44
1937	3.66	1949	2.27	1962	1.88	1974	1.84
1938	1.72	1950	2.82	1963	1.31	1975	1.69
1939	2.82	1951	1.98	1964	1.76	1976	3.00
1940	0.72	1952	2.44	1965	2.17	1977	1.36
1941	1.46	1953	2.53	1966	2.38	1978	6.37
1942	1.30	1954	2.00	1967	1.16	1979	4.55
1943	1.35	1955	1.12	1968	1.39	1980	0.52
1944	0.54	1956	2.13	1969	1.36	1981	0.87
		1957	1.36			1982	1.51

TABLE A.3 June Climate Data for Guayaquil, Ecuador, 1951–1970 (Asterisks Indicate El Niño Years)

Year	Temperature (°C)	Precipitation (mm)	Pressure (mb)
1951*	26.1	43	1009.5
1952	24.5	10	1010.9
1953*	24.8	4	1010.7
1954	24.5	0	1011.2
1955	24.1	2	1011.9
1956	24.3	Missing	1011.2
1957*	26.4	31	1009.3
1958	24.9	0	1011.1
1959	23.7	0	1012.0
1960	23.5	0	1011.4

TABLE A.3 June Climate Data for Guayaquil, Ecuador, 1951–1970 (Asterisks Indicate El Niño Years)—Cont'd

Year	Temperature (°C)	Precipitation (mm)	Pressure (mb)
1961	24.0	2	1010.9
1962	24.1	3	1011.5
1963	23.7	0	1011.0
1964	24.3	4	1011.2
1965*	26.6	15	1009.9
1966	24.6	2	1012.5
1967	24.8	0	1011.1
1968	24.4	1	1011.8
1969*	26.8	127	1009.3
1970	25.2	2	1010.6

Probability Tables

This Appendix contains tables for selected common probability distributions, for which closed-form expressions for the cumulative distribution functions do not exist.

TABLE B.1 Left-Tail Cumulative Probabilities for the Standard Gaussian Distribution, $\Phi(z) = \Pr\{Z \le z\}$

z	.09	.08	.07	.06	.05	.04	.03	.02	.01	.00	z
−4.0	.00002	.00002	.00002	.00002	.00003	.00003	.00003	.00003	.00003	.00003	−4.0
−3.9	.00003	.00003	.00004	.00004	.00004	.00004	.00004	.00004	.00005	.00005	−3.9
−3.8	.00005	.00005	.00005	.00006	.00006	.00006	.00006	.00007	.00007	.00007	−3.8
−3.7	.00008	.00008	.00008	.00008	.00009	.00009	.00010	.00010	.00010	.00011	−3.7
−3.6	.00011	.00012	.00012	.00013	.00013	.00014	.00014	.00015	.00015	.00016	−3.6
−3.5	.00017	.00017	.00018	.00019	.00019	.00020	.00021	.00022	.00022	.00023	−3.5
−3.4	.00024	.00025	.00026	.00027	.00028	.00029	.00030	.00031	.00032	.00034	−3.4
−3.3	.00035	.00036	.00038	.00039	.00040	.00042	.00043	.00045	.00047	.00048	−3.3
−3.2	.00050	.00052	.00054	.00056	.00058	.00060	.00062	.00064	.00066	.00069	−3.2
−3.1	.00071	.00074	.00076	.00079	.00082	.00084	.00087	.00090	.00094	.00097	−3.1
−3.0	.00100	.00104	.00107	.00111	.00114	.00118	.00122	.00126	.00131	.00135	−3.0
−2.9	.00139	.00144	.00149	.00154	.00159	.00164	.00169	.00175	.00181	.00187	−2.9
−2.8	.00193	.00199	.00205	.00212	.00219	.00226	.00233	.00240	.00248	.00256	−2.8
−2.7	.00264	.00272	.00280	.00289	.00298	.00307	.00317	.00326	.00336	.00347	−2.7
−2.6	.00357	.00368	.00379	.00391	.00402	.00415	.00427	.00440	.00453	.00466	−2.6
−2.5	.00480	.00494	.00508	.00523	.00539	.00554	.00570	.00587	.00604	.00621	−2.5
−2.4	.00639	.00657	.00676	.00695	.00714	.00734	.00755	.00776	.00798	.00820	−2.4
−2.3	.00842	.00866	.00889	.00914	.00939	.00964	.00990	.01017	.01044	.01072	−2.3
−2.2	.01101	.01130	.01160	.01191	.01222	.01255	.01287	.01321	.01355	.01390	−2.2
−2.1	.01426	.01463	.01500	.01539	.01578	.01618	.01659	.01700	.01743	.01786	−2.1
−2.0	.01831	.01876	.01923	.01970	.02018	.02068	.02118	.02169	.02222	.02275	−2.0
−1.9	.02330	.02385	.02442	.02500	.02559	.02619	.02680	.02743	.02807	.02872	−1.9
−1.8	.02938	.03005	.03074	.03144	.03216	.03288	.03362	.03438	.03515	.03593	−1.8
−1.7	.03673	.03754	.03836	.03920	.04006	.04093	.04182	.04272	.04363	.04457	−1.7
−1.6	.04551	.04648	.04746	.04846	.04947	.05050	.05155	.05262	.05370	.05480	−1.6
−1.5	.05592	.05705	.05821	.05938	.06057	.06178	.06301	.06426	.06552	.06681	−1.5
−1.4	.06811	.06944	.07078	.07215	.07353	.07493	.07636	.07780	.07927	.08076	−1.4
−1.3	.08226	.08379	.08534	.08692	.08851	.09012	.09176	.09342	.09510	.09680	−1.3
−1.2	.09853	.10027	.10204	.10383	.10565	.10749	.10935	.11123	.11314	.11507	−1.2
−1.1	.11702	.11900	.12100	.12302	.12507	.12714	.12924	.13136	.13350	.13567	−1.1
−1.0	.13786	.14007	.14231	.14457	.14686	.14917	.15151	.15386	.15625	.15866	−1.0
−0.9	.16109	.16354	.16602	.16853	.17106	.17361	.17619	.17879	.18141	.18406	−0.9
−0.8	.18673	.18943	.19215	.19489	.19766	.20045	.20327	.20611	.20897	.21186	−0.8
−0.7	.21476	.21770	.22065	.22363	.22663	.22965	.23270	.23576	.23885	.24196	−0.7
−0.6	.24510	.24825	.25143	.25463	.25785	.26109	.26435	.26763	.27093	.27425	−0.6
−0.5	.27760	.28096	.28434	.28774	.29116	.29460	.29806	.30153	.30503	.30854	−0.5
−0.4	.31207	.31561	.31918	.32276	.32636	.32997	.33360	.33724	.34090	.34458	−0.4
−0.3	.34827	.35197	.35569	.35942	.36317	.36693	.37070	.37448	.37828	.38209	−0.3
−0.2	.38591	.38974	.39358	.39743	.40129	.40517	.40905	.41294	.41683	.42074	−0.2
−0.1	.42465	.42858	.43251	.43644	.44038	.44433	.44828	.45224	.45620	.46017	−0.1
−0.0	.46414	.46812	.47210	.47608	.48006	.48405	.48803	.49202	.49601	.50000	0.0

Values of the standardized Gaussian variable, z, are listed to tenths in the rightmost and leftmost columns. Remaining column headings index the hundredth place of z. Right-tail probabilities are obtained using $\Pr\{Z > z\} = 1 - \Pr\{Z \le z\}$. Probabilities for $Z > 0$ are obtained using the symmetry of the Gaussian distribution, $\Pr\{Z \le z\} = 1 - \Pr\{Z \le -z\}$.

TABLE B.2 Quantiles of the Standard ($\beta = 1$) Gamma Distribution

α	Cumulative Probability														
	.001	.01	.05	.10	.20	.30	.40	.50	.60	.70	.80	.90	.95	.99	.999
0.05	0.0000	0.0000	0.0000	0.000	0.000	0.000	0.000	0.000	0.000	0.000	0.007	0.077	0.262	1.057	2.423
0.10	0.0000	0.0000	0.0000	0.000	0.000	0.000	0.000	0.001	0.004	0.018	0.070	0.264	0.575	1.554	3.035
0.15	0.0000	0.0000	0.0000	0.000	0.000	0.000	0.001	0.006	0.021	0.062	0.164	0.442	0.820	1.894	3.439
0.20	0.0000	0.0000	0.0000	0.000	0.000	0.002	0.007	0.021	0.053	0.122	0.265	0.602	1.024	2.164	3.756
0.25	0.0000	0.0000	0.0000	0.000	0.001	0.006	0.018	0.044	0.095	0.188	0.364	0.747	1.203	2.395	4.024
0.30	0.0000	0.0000	0.0000	0.000	0.003	0.013	0.034	0.073	0.142	0.257	0.461	0.882	1.365	2.599	4.262
0.35	0.0000	0.0000	0.0001	0.001	0.007	0.024	0.055	0.108	0.192	0.328	0.556	1.007	1.515	2.785	4.477
0.40	0.0000	0.0000	0.0004	0.002	0.013	0.038	0.080	0.145	0.245	0.398	0.644	1.126	1.654	2.958	4.677
0.45	0.0000	0.0000	0.0010	0.005	0.022	0.055	0.107	0.186	0.300	0.468	0.733	1.240	1.786	3.121	4.863
0.50	0.0000	0.0001	0.0020	0.008	0.032	0.074	0.138	0.228	0.355	0.538	0.819	1.349	1.913	3.274	5.040
0.55	0.0000	0.0002	0.0035	0.012	0.045	0.096	0.170	0.272	0.411	0.607	0.904	1.454	2.034	3.421	5.208
0.60	0.0000	0.0004	0.0057	0.018	0.059	0.120	0.204	0.316	0.467	0.676	0.987	1.556	2.150	3.562	5.370
0.65	0.0000	0.0008	0.0086	0.025	0.075	0.146	0.240	0.362	0.523	0.744	1.068	1.656	2.264	3.698	5.526
0.70	0.0001	0.0013	0.0123	0.033	0.093	0.173	0.276	0.408	0.579	0.811	1.149	1.753	2.374	3.830	5.676
0.75	0.0001	0.0020	0.0168	0.043	0.112	0.201	0.314	0.455	0.636	0.878	1.227	1.848	2.481	3.958	5.822
0.80	0.0003	0.0030	0.0221	0.053	0.132	0.231	0.352	0.502	0.692	0.945	1.305	1.941	2.586	4.083	5.964
0.85	0.0004	0.0044	0.0283	0.065	0.153	0.261	0.391	0.550	0.749	1.010	1.382	2.032	2.689	4.205	6.103
0.90	0.0007	0.0060	0.0353	0.078	0.176	0.292	0.431	0.598	0.805	1.076	1.458	2.122	2.790	4.325	6.239
0.95	0.0010	0.0080	0.0432	0.091	0.199	0.324	0.471	0.646	0.861	1.141	1.533	2.211	2.888	4.441	6.373
1.00	0.0014	0.0105	0.0517	0.106	0.224	0.357	0.512	0.694	0.918	1.206	1.607	2.298	2.986	4.556	6.503
1.05	0.0019	0.0133	0.0612	0.121	0.249	0.391	0.553	0.742	0.974	1.270	1.681	2.384	3.082	4.669	6.631
1.10	0.0022	0.0166	0.0713	0.138	0.275	0.425	0.594	0.791	1.030	1.334	1.759	2.469	3.177	4.781	6.757
1.15	0.0023	0.0202	0.0823	0.155	0.301	0.459	0.636	0.840	1.086	1.397	1.831	2.553	3.270	4.890	6.881
1.20	0.0024	0.0240	0.0938	0.173	0.329	0.494	0.678	0.889	1.141	1.460	1.903	2.636	3.362	4.998	7.003
1.25	0.0031	0.0271	0.1062	0.191	0.357	0.530	0.720	0.938	1.197	1.523	1.974	2.719	3.453	5.105	7.124
1.30	0.0037	0.0321	0.1192	0.210	0.385	0.566	0.763	0.987	1.253	1.586	2.045	2.800	3.544	5.211	7.242
1.35	0.0044	0.0371	0.1328	0.230	0.414	0.602	0.806	1.036	1.308	1.649	2.115	2.881	3.633	5.314	7.360
1.40	0.0054	0.0432	0.1451	0.250	0.443	0.639	0.849	1.085	1.364	1.711	2.185	2.961	3.722	5.418	7.476
1.45	0.0066	0.0493	0.1598	0.272	0.473	0.676	0.892	1.135	1.419	1.773	2.255	3.041	3.809	5.519	7.590
1.50	0.0083	0.0560	0.1747	0.293	0.504	0.713	0.935	1.184	1.474	1.834	2.324	3.120	3.897	5.620	7.704
1.55	0.0106	0.0632	0.1908	0.313	0.534	0.750	0.979	1.234	1.530	1.896	2.392	3.199	3.983	5.720	7.816
1.60	0.0136	0.0708	0.2070	0.336	0.565	0.788	1.023	1.283	1.585	1.957	2.461	3.276	4.068	5.818	7.928
1.65	0.0177	0.0780	0.2238	0.359	0.597	0.826	1.067	1.333	1.640	2.018	2.529	3.354	4.153	5.917	8.038
1.70	0.0232	0.0867	0.2411	0.382	0.628	0.865	1.111	1.382	1.695	2.079	2.597	3.431	4.237	6.014	8.147
1.75	0.0306	0.0958	0.2588	0.406	0.661	0.903	1.155	1.432	1.750	2.140	2.664	3.507	4.321	6.110	8.255

1.80	0.0360	0.1041	0.2771	0.430	0.693	0.942	1.199	1.431	1.805	2.200	2.731	3.584	4.405	6.207	8.362
1.85	0.0406	0.1145	0.2958	0.454	0.726	0.980	1.244	1.531	1.860	2.261	2.798	3.659	4.487	6.301	8.469
1.90	0.0447	0.1243	0.3142	0.479	0.759	1.020	1.288	1.580	1.915	2.321	2.865	3.735	4.569	6.396	8.575
1.95	0.0486	0.1361	0.3338	0.505	0.790	1.059	1.333	1.630	1.969	2.381	2.931	3.809	4.651	6.490	8.679
2.00	0.0525	0.1514	0.3537	0.530	0.823	1.099	1.378	1.680	2.024	2.442	2.997	3.883	4.732	6.582	8.783
2.05	0.0565	0.1637	0.3741	0.556	0.857	1.138	1.422	1.729	2.079	2.501	3.063	3.958	4.813	6.675	8.887
2.10	0.0657	0.1751	0.3949	0.583	0.891	1.178	1.467	1.779	2.133	2.561	3.129	4.032	4.894	6.767	8.989
2.15	0.0697	0.1864	0.4149	0.610	0.925	1.218	1.512	1.829	2.188	2.620	3.195	4.105	4.973	6.858	9.091
2.20	0.0740	0.2002	0.4365	0.637	0.959	1.258	1.557	1.879	2.242	2.680	3.260	4.179	5.053	6.949	9.193
2.25	0.0854	0.2116	0.4584	0.664	0.994	1.298	1.603	1.928	2.297	2.739	3.325	4.252	5.132	7.039	9.294
2.30	0.0898	0.2259	0.4807	0.691	1.029	1.338	1.648	1.978	2.351	2.799	3.390	4.324	5.211	7.129	9.394
2.35	0.0945	0.2378	0.5023	0.718	1.064	1.379	1.693	2.028	2.405	2.858	3.455	4.396	5.289	7.219	9.493
2.40	0.0996	0.2526	0.5244	0.747	1.099	1.420	1.738	2.078	2.459	2.917	3.519	4.468	5.367	7.308	9.592
2.45	0.1134	0.2680	0.5481	0.775	1.134	1.460	1.784	2.127	2.514	2.976	3.584	4.540	5.445	7.397	9.691
2.50	0.1184	0.2803	0.5754	0.804	1.170	1.500	1.829	2.178	2.568	3.035	3.648	4.612	5.522	7.484	9.789
2.55	0.1239	0.2962	0.5978	0.833	1.205	1.539	1.875	2.227	2.622	3.093	3.712	4.683	5.600	7.572	9.886
2.60	0.1297	0.3129	0.6211	0.862	1.241	1.581	1.920	2.277	2.676	3.152	3.776	4.754	5.677	7.660	9.983
2.65	0.1468	0.3255	0.6456	0.890	1.277	1.622	1.966	2.327	2.730	3.210	3.840	4.825	5.753	7.746	10.079
2.70	0.1523	0.3426	0.6705	0.920	1.314	1.663	2.011	2.376	2.784	3.269	3.903	4.896	5.830	7.833	10.176
2.75	0.1583	0.3561	0.6938	0.950	1.350	1.704	2.058	2.427	2.838	3.328	3.967	4.966	5.906	7.919	10.272
2.80	0.1647	0.3735	0.7188	0.980	1.386	1.746	2.103	2.476	2.892	3.386	4.030	5.040	5.982	8.004	10.367
2.85	0.1861	0.3919	0.7441	1.009	1.423	1.787	2.149	2.526	2.946	3.444	4.093	5.120	6.058	8.090	10.461
2.90	0.1919	0.4056	0.7697	1.040	1.460	1.829	2.195	2.576	2.999	3.502	4.156	5.190	6.133	8.175	10.556
2.95	0.1982	0.4242	0.7936	1.070	1.497	1.871	2.241	2.626	3.054	3.560	4.220	5.260	6.208	8.260	10.649
3.00	0.2050	0.4388	0.8193	1.101	1.534	1.913	2.287	2.676	3.108	3.618	4.283	5.329	6.283	8.345	10.743
3.05	0.2123	0.4577	0.8454	1.134	1.571	1.954	2.333	2.726	3.161	3.676	4.346	5.398	6.357	8.429	10.837
3.10	0.2385	0.4778	0.8717	1.165	1.607	1.996	2.378	2.776	3.215	3.734	4.408	5.468	6.432	8.513	10.930
3.15	0.2447	0.4922	0.8982	1.197	1.645	2.038	2.425	2.825	3.268	3.792	4.471	5.537	6.506	8.596	11.023
3.20	0.2514	0.5125	0.9251	1.227	1.682	2.080	2.471	2.875	3.322	3.850	4.533	5.605	6.580	8.680	11.113
3.25	0.2588	0.5278	0.9498	1.259	1.720	2.123	2.517	2.925	3.376	3.907	4.595	5.675	6.654	8.763	11.205
3.30	0.2667	0.5483	0.9767	1.291	1.758	2.165	2.563	2.975	3.430	3.965	4.658	5.743	6.727	8.845	11.298
3.35	0.2995	0.5704	1.0039	1.323	1.796	2.207	2.610	3.025	3.483	4.022	4.720	5.811	6.801	8.928	11.389
3.40	0.3057	0.5850	1.0313	1.354	1.834	2.250	2.656	3.075	3.537	4.079	4.782	5.879	6.874	9.010	11.480
3.45	0.3126	0.6072	1.0590	1.386	1.872	2.292	2.702	3.125	3.590	4.137	4.843	5.948	6.947	9.093	11.570
3.50	0.3201	0.6228	1.0870	1.418	1.910	2.334	2.748	3.175	3.644	4.194	4.905	6.015	7.020	9.174	11.660
3.55	0.3282	0.6450	1.1152	1.451	1.948	2.377	2.795	3.225	3.697	4.252	4.967	6.084	7.092	9.255	11.749
3.60	0.3370	0.6614	1.1405	1.483	1.985	2.420	2.841	3.274	3.750	4.309	5.028	6.152	7.165	9.337	11.840

Tabulated elements are values of the standardized random variable ξ corresponding to the cumulative probabilities $F(\xi)$ given in the column headings, for values of the shape parameter (α) given in the first column. To find quantiles for distributions with other scale parameters, enter the table at the appropriate row, read the standardized value in the appropriate column, and multiply the tabulated value by the scale parameter. To extract cumulative probabilities corresponding to a given value of the random variable, divide the value by the scale parameter, enter the table at the row appropriate to the shape parameter, and interpolate the result from the column headings.

TABLE B.2 Quantiles of the Standard ($\beta = 1$) Gamma Distribution—Cont'd

						Cumulative Probability									
α	.001	.01	.05	.10	.20	.30	.40	.50	.60	.70	.80	.90	.95	.99	.999
3.65	0.3767	0.6837	1.1687	1.516	2.024	2.462	2.887	3.324	3.804	4.366	5.091	6.219	7.237	9.418	11.929
3.70	0.3830	0.7084	1.1972	1.549	2.062	2.505	2.934	3.374	3.858	4.423	5.152	6.286	7.310	9.499	12.017
3.75	0.3900	0.7233	1.2259	1.582	2.101	2.547	2.980	3.425	3.911	4.480	5.214	6.354	7.381	9.579	12.107
3.80	0.3978	0.7480	1.2549	1.613	2.140	2.590	3.027	3.474	3.964	4.537	5.275	6.420	7.454	9.659	12.195
3.85	0.4064	0.7637	1.2843	1.646	2.179	2.633	3.073	3.524	4.018	4.594	5.336	6.488	7.525	9.740	12.284
3.90	0.4157	0.7883	1.3101	1.680	2.218	2.676	3.120	3.574	4.071	4.651	5.397	6.555	7.596	9.820	12.371
3.95	0.4259	0.8049	1.3393	1.713	2.257	2.719	3.163	3.624	4.124	4.708	5.458	6.622	7.668	9.900	12.459
4.00	0.4712	0.8294	1.3687	1.746	2.295	2.762	3.209	3.674	4.177	4.765	5.519	6.689	7.739	9.980	12.546
4.05	0.4779	0.8469	1.3984	1.780	2.334	2.805	3.256	3.724	4.231	4.822	5.580	6.755	7.811	10.059	12.634
4.10	0.4853	0.8714	1.4285	1.814	2.373	2.848	3.302	3.774	4.284	4.879	5.641	6.821	7.882	10.137	12.721
4.15	0.4937	0.8999	1.4551	1.848	2.413	2.891	3.350	3.823	4.337	4.936	5.701	6.888	7.952	10.216	12.807
4.20	0.5030	0.9141	1.4850	1.882	2.451	2.935	3.396	3.874	4.390	4.992	5.762	6.954	8.023	10.295	12.894
4.25	0.5133	0.9424	1.5150	1.916	2.491	2.978	3.443	3.924	4.444	5.049	5.823	7.020	8.093	10.374	12.981
4.30	0.5244	0.9575	1.5454	1.950	2.531	3.021	3.489	3.974	4.497	5.105	5.883	7.086	8.170	10.453	13.066
4.35	0.5779	0.9856	1.5762	1.985	2.572	3.065	3.537	4.024	4.550	5.162	5.944	7.153	8.264	10.531	13.152
4.40	0.5842	1.0016	1.6034	2.017	2.612	3.108	3.584	4.074	4.603	5.218	6.005	7.219	8.334	10.609	13.238
4.45	0.5916	1.0294	1.6339	2.051	2.653	3.152	3.630	4.123	4.656	5.274	6.065	7.284	8.405	10.687	13.324
4.50	0.6001	1.0463	1.6646	2.085	2.691	3.195	3.677	4.173	4.709	5.331	6.126	7.350	8.475	10.765	13.410
4.55	0.6096	1.0739	1.6956	2.120	2.731	3.239	3.724	4.223	4.762	5.387	6.186	7.415	8.544	10.843	13.495
4.60	0.6202	1.0917	1.7271	2.155	2.771	3.283	3.771	4.273	4.815	5.443	6.246	7.480	8.615	10.920	13.578
4.65	0.6319	1.1191	1.7547	2.190	2.812	3.326	3.817	4.323	4.868	5.501	6.306	7.546	8.684	10.998	13.663
4.70	0.6978	1.1378	1.7857	2.225	2.852	3.369	3.864	4.373	4.921	5.557	6.366	7.611	8.754	11.075	13.748
4.75	0.7031	1.1649	1.8170	2.260	2.890	3.412	3.911	4.423	4.974	5.613	6.426	7.676	8.823	11.152	13.832
4.80	0.7095	1.1844	1.8487	2.295	2.930	3.456	3.958	4.474	5.027	5.669	6.486	7.742	8.892	11.229	13.916
4.85	0.7172	1.2113	1.8809	2.330	2.970	3.500	4.005	4.524	5.081	5.725	6.546	7.807	8.962	11.306	14.000
4.90	0.7262	1.2465	1.9088	2.366	3.011	3.544	4.052	4.573	5.134	5.781	6.606	7.872	9.031	11.382	14.084
4.95	0.7365	1.2582	1.9403	2.398	3.051	3.588	4.099	4.623	5.186	5.837	6.665	7.937	9.100	11.457	14.168
5.00	0.7482	1.2931	1.9722	2.434	3.091	3.632	4.146	4.673	5.239	5.893	6.725	8.002	9.169	11.534	14.251
5.5	0.917	1.527	2.287	2.789	3.494	4.074	4.619	5.170	5.765	6.449	7.316	8.638	9.838	12.362	15.632
6.0	1.107	1.785	2.613	3.152	3.904	4.517	5.091	5.670	6.292	7.006	7.906	9.275	10.513	13.108	16.455
6.5	1.309	2.053	2.946	3.521	4.317	4.963	5.565	6.170	6.818	7.559	8.492	9.906	11.181	13.844	17.264
7.0	1.520	2.330	3.285	3.895	4.734	5.411	6.039	6.670	7.343	8.111	9.075	10.532	11.842	14.571	18.062
7.5	1.741	2.615	3.630	4.273	5.153	5.861	6.515	7.169	7.867	8.661	9.655	11.154	12.498	15.289	18.849
8.0	1.971	2.906	3.981	4.656	5.576	6.312	6.991	7.669	8.390	9.209	10.233	11.771	13.148	16.000	19.626
8.5	2.208	3.204	4.336	5.043	6.001	6.765	7.469	8.169	8.912	9.756	10.807	12.385	13.794	16.704	20.395
9.0	2.452	3.507	4.695	5.432	6.428	7.220	7.947	8.669	9.434	10.301	11.380	12.995	14.435	17.403	21.156

α															
9.5	2.703	3.816	5.059	5.825	6.858	7.676	8.425	9.169	9.955	10.845	11.950	13.602	15.072	18.095	21.910
10.0	2.961	4.130	5.425	6.221	7.289	8.133	8.904	9.669	10.476	11.387	12.519	14.206	15.705	18.783	22.657
10.5	3.223	4.449	5.796	6.620	7.722	8.591	9.384	10.169	10.996	11.929	13.086	14.808	16.335	19.466	23.399
11.0	3.491	4.771	6.169	7.021	8.157	9.050	9.864	10.669	11.515	12.470	13.651	15.407	16.962	20.145	24.134
11.5	3.765	5.098	6.545	7.424	8.593	9.511	10.345	11.168	12.034	13.009	14.214	16.003	17.586	20.819	24.864
12.0	4.042	5.428	6.924	7.829	9.031	9.972	10.826	11.668	12.553	13.548	14.777	16.598	18.208	21.490	25.589
12.5	4.325	5.762	7.306	8.237	9.470	10.434	11.308	12.168	13.072	14.086	15.338	17.191	18.826	22.157	26.310
13.0	4.611	6.099	7.690	8.646	9.910	10.896	11.790	12.668	13.589	14.623	15.897	17.782	19.443	22.821	27.026
13.5	4.901	6.439	8.076	9.057	10.351	11.360	12.272	13.168	14.107	15.160	16.456	18.371	20.057	23.481	27.738
14.0	5.195	6.782	8.464	9.470	10.794	11.824	12.755	13.668	14.624	15.695	17.013	18.958	20.669	24.139	28.446
14.5	5.493	7.128	8.854	9.884	11.238	12.288	13.238	14.168	15.141	16.231	17.570	19.544	21.278	24.794	29.151
15.0	5.794	7.477	9.246	10.300	11.682	12.754	13.721	14.668	15.658	16.765	18.125	20.128	21.886	25.446	29.852
15.5	6.098	7.828	9.640	10.717	12.128	13.220	14.204	15.168	16.174	17.299	18.680	20.711	22.493	26.096	30.549
16.0	6.405	8.181	10.036	11.135	12.574	13.686	14.688	15.668	16.690	17.832	19.233	21.292	23.097	26.743	31.244
16.5	6.715	8.537	10.433	11.555	13.021	14.153	15.172	16.168	17.206	18.365	19.786	21.873	23.700	27.388	31.935
17.0	7.028	8.895	10.832	11.976	13.469	14.621	15.657	16.668	17.722	18.898	20.338	22.452	24.301	28.030	32.624
17.5	7.344	9.254	11.233	12.398	13.918	15.089	16.141	17.168	18.237	19.430	20.889	23.029	24.901	28.671	33.309
18.0	7.662	9.616	11.634	12.822	14.367	15.558	16.626	17.668	18.752	19.961	21.439	23.606	25.499	29.310	33.993
18.5	7.983	9.980	12.037	13.246	14.818	16.027	17.111	18.168	19.267	20.492	21.989	24.182	26.096	29.946	34.673
19.0	8.306	10.346	12.442	13.671	15.269	16.496	17.596	18.658	19.782	21.023	22.538	24.756	26.692	30.581	35.351
19.5	8.631	10.713	12.848	14.098	15.720	16.966	18.081	19.158	20.297	21.553	23.087	25.330	27.286	31.214	36.027
20.0	8.958	11.082	13.255	14.525	16.172	17.436	18.567	19.658	20.811	22.082	23.634	25.903	27.879	31.845	36.701
25.0	12.337	14.853	17.382	18.844	20.725	22.157	23.432	24.667	25.946	27.361	29.082	31.584	33.752	38.077	43.330
30.0	15.869	18.742	21.594	23.229	25.320	26.905	28.310	29.667	31.067	32.613	34.486	37.198	39.541	44.190	49.804
35.0	19.518	22.721	25.870	27.664	29.949	31.673	33.198	34.667	36.179	37.845	39.857	42.764	45.266	50.213	56.159
40.0	23.260	26.770	30.196	32.139	34.603	36.458	38.094	39.667	41.283	43.060	45.203	48.289	50.940	56.164	62.420
45.0	27.078	30.877	34.563	36.646	39.279	41.256	42.996	44.667	46.381	48.262	50.527	53.782	56.573	62.058	68.604
50.0	30.959	35.032	38.965	41.179	43.973	46.064	47.904	49.667	51.473	53.453	55.833	59.249	62.171	67.903	74.725
55.0	34.895	39.229	43.396	45.736	48.681	50.883	52.816	54.667	56.561	58.634	61.125	64.693	67.740	73.707	80.790
60.0	38.878	43.462	47.852	50.312	53.403	55.709	57.732	59.667	61.645	63.808	66.403	70.116	73.284	79.475	86.809
65.0	42.902	47.725	52.331	54.905	58.136	60.543	62.652	64.667	66.725	68.974	71.670	75.523	78.805	85.212	92.786
70.0	46.963	52.017	56.830	59.515	62.879	65.383	67.575	69.667	71.802	74.134	76.927	80.913	84.306	90.920	98.725
80.0	55.180	60.673	65.878	68.773	72.392	75.079	77.428	79.667	81.949	84.438	87.414	91.655	95.258	102.26	110.51
90.0	63.506	69.410	74.984	78.076	81.934	84.794	87.290	89.667	92.087	94.723	97.872	102.35	106.15	113.53	122.19
100.0	71.921	78.216	84.139	87.418	91.501	94.524	97.160	99.669	102.22	104.99	108.30	113.01	117.00	124.72	133.77

Tabulated elements are values of the standardized random variable ξ corresponding to the cumulative probabilities $F(\xi)$ given in the column headings, for values of the shape parameter (α) given in the first column. To find quantiles for distributions with other scale parameters, enter the table at the appropriate row, read the standardized value in the appropriate column, and multiply the tabulated value by the scale parameter. To extract cumulative probabilities corresponding to a given value of the random variable, divide the value by the scale parameter, enter the table at the row appropriate to the shape parameter, and interpolate the result from the column headings.

TABLE B.3 Right-Tail Quantiles of the Chi-Square Distribution

ν	Cumulative Probability					
	0.50	0.90	0.95	0.99	0.999	0.9999
1	0.455	2.706	3.841	6.635	10.828	15.137
2	1.386	4.605	5.991	9.210	13.816	18.421
3	2.366	6.251	7.815	11.345	16.266	21.108
4	3.357	7.779	9.488	13.277	18.467	23.512
5	4.351	9.236	11.070	15.086	20.515	25.745
6	5.348	10.645	12.592	16.812	22.458	27.855
7	6.346	12.017	14.067	18.475	24.322	29.878
8	7.344	13.362	15.507	20.090	26.124	31.827
9	8.343	14.684	16.919	21.666	27.877	33.719
10	9.342	15.987	18.307	23.209	29.588	35.563
11	10.341	17.275	19.675	24.725	31.264	37.366
12	11.340	18.549	21.026	26.217	32.910	39.134
13	12.340	19.812	22.362	27.688	34.528	40.871
14	13.339	21.064	23.685	29.141	36.123	42.578
15	14.339	22.307	24.996	30.578	37.697	44.262
16	15.338	23.542	26.296	32.000	39.252	45.925
17	16.338	24.769	27.587	33.409	40.790	47.566
18	17.338	25.989	28.869	34.805	42.312	49.190
19	18.338	27.204	30.144	36.191	43.820	50.794
20	19.337	28.412	31.410	37.566	45.315	52.385
21	20.337	29.615	32.671	38.932	46.797	53.961
22	21.337	30.813	33.924	40.289	48.268	55.523
23	22.337	32.007	35.172	41.638	49.728	57.074
24	23.337	33.196	36.415	42.980	51.179	58.613
25	24.337	34.382	37.652	44.314	52.620	60.140
26	25.336	35.563	38.885	45.642	54.052	61.656
27	26.336	36.741	40.113	46.963	55.476	63.164
28	27.336	37.916	41.337	48.278	56.892	64.661
29	28.336	39.087	42.557	49.588	58.301	66.152
30	29.336	40.256	43.773	50.892	59.703	67.632
31	30.336	41.422	44.985	52.191	61.098	69.104
32	31.336	42.585	46.194	53.486	62.487	70.570
33	32.336	43.745	47.400	54.776	63.870	72.030
34	33.336	44.903	48.602	56.061	65.247	73.481
35	34.336	46.059	49.802	57.342	66.619	74.926
36	35.336	47.212	50.998	58.619	67.985	76.365
37	36.336	48.363	52.192	59.892	69.347	77.798
38	37.335	49.513	53.384	61.162	70.703	79.224
39	38.335	50.660	54.572	62.428	72.055	80.645
40	39.335	51.805	55.758	63.691	73.402	82.061
41	40.335	52.949	56.942	64.950	74.745	83.474
42	41.335	54.090	58.124	66.206	76.084	84.880
43	42.335	55.230	59.304	67.459	77.419	86.280
44	43.335	56.369	60.481	68.710	78.750	87.678
45	44.335	57.505	61.656	69.957	80.077	89.070
46	45.335	58.641	62.830	71.201	81.400	90.456
47	46.335	59.774	64.001	72.443	82.721	91.842
48	47.335	60.907	65.171	73.683	84.037	93.221
49	48.335	62.038	66.339	74.919	85.351	94.597
50	49.335	63.167	67.505	76.154	86.661	95.968

TABLE B.3 Right-Tail Quantiles of the Chi-Square Distribution—Cont'd

ν	Cumulative Probability					
	0.50	0.90	0.95	0.99	0.999	0.9999
55	54.335	68.796	73.311	82.292	93.168	102.776
60	59.335	74.397	79.082	88.379	99.607	109.501
65	64.335	79.973	84.821	94.422	105.988	116.160
70	69.334	85.527	90.531	100.425	112.317	122.754
75	74.334	91.061	96.217	106.393	118.599	129.294
80	79.334	96.578	101.879	112.329	124.839	135.783
85	84.334	102.079	107.522	118.236	131.041	142.226
90	89.334	107.565	113.145	124.116	137.208	148.626
95	94.334	113.038	118.752	129.973	143.344	154.989
100	99.334	118.498	124.342	135.807	149.449	161.318

For large ν, the Chi-square distribution is approximately Gaussian, with mean ν and variance 2ν.

Symbols and Acronyms

Roman Symbols	Meaning	Section First Introduced
A	area under a histogram	4.5.1
A	generic accuracy statistic	9.1.4
A	area under the ROC curve	9.4.6
A	amplitude error	9.8.5
A	amplitude of unshifted cosine	10.4.3
a	number of occurrences	2.3.1
a	plotting position constant	3.3.7
a	linear congruential generator multiplier parameter	4.7.1
a	regression intercept	7.2.1
a	hit (correct nonprobabilistic forecast)	9.2.1
$\boldsymbol{a}$	most significant linear combination, T^2 test	12.5.4
$\boldsymbol{a}$	canonical vector	14.2.1
$\boldsymbol{a}$	discriminant vector	15.2.1
$\hat{a}$	BC_a acceleration parameter	5.3.5
B	bias ratio	9.2.2
B	amplitude of unshifted sine	10.4.3
b	moving-blocks bootstrap block length	5.3.5
b	regression parameter	7.2.1
b	false alarm (nonprobabilistic forecast)	9.2.1
b	rank histogram bin	9.7.2
b^*	regularized regression parameter	7.5
$\boldsymbol{b}$	canonical vector	14.2.1
b_1, b_2	multivariate skewness and kurtosis statistics	12.3
C	(as superscript) complement	2.4.1
C	cost associated with adverse weather protection	9.9.1
C	amplitude of shifted cosine	10.4.2
C	misclassification cost	15.2.3
C_α	approximate critical value for K-S test	5.2.5
Corr	correlation	13.1.1
Cov	covariance	3.5.2
c	histogram binwidth scaling constant	3.3.5
c	generic constant	4.3.2
c	linear congruential generator increment parameter	4.7.1
c	threshold, or cutoff, level	7.3.2
c	missed nonprobabilistic forecast	9.2.1
c	climatological gridpoint value	9.8.3

Continued

Roman Symbols	Meaning	Section First Introduced
D	Thom's statistic for gamma distribution	4.4.5
D	decision threshold	9.9.1
$[D]$	diagonal matrix of standard deviations	11.3.2
D_i	difference in ranks	3.5.3
D_n	Kolmogorov-Smirnov test statistic	5.2.5
D_S	Smirnov (2-sample K-S test) test statistic	5.2.5
D	Gerrity odds ratio	9.2.6
D^2	Mahalanobis distance	11.2.2
d	Durbin-Watson statistic	7.2.6
d	correct negative (nonprobabilistic forecast)	9.2.1
d	discrimination distance	9.4.5
d	domain size	9.8.5
d	vector dimension	9.7.2
d	distance	13.2.4
det	determinant (of a matrix)	11.3.2
d_λ	Hinkley statistic	3.4.1
E	Event	2.3.1
E	statistical expectation	4.3.2
$[E]$	matrix of eigenvector columns	11.3.3
e	base of natural logarithms	4.2.5
e	regression residual	7.2.1
e	expected transition count	10.2.3
$\mathbf{e}$	eigenvector	11.3.3
F	generic cumulative distribution function	4.4.1
F	a specific distribution and hypothesis test name	5.3.4
F	false alarm rate	9.2.2
F_0	a step function	9.5.1
F_n	empirical CDF for Kolmogorov-Smirnov test	5.2.5
F^{-1}	inverse cumulative distribution (quantile) function	4.4.1
f	probability density function	4.4.1
f	regression function	7.9.1
$\hat{f}$	kernel density function	3.3.6
f_A	aliased frequency	10.5.4
f_k	frequency of kth harmonic	10.5.2
G	number of groups, discriminant analysis	15.1
g	multinormality comparison statistic	12.3
H	hit rate	9.2.2
H_0	null hypothesis	5.1.3
H_A	alternative hypothesis	5.1.3
H_k	complex Fourier coefficient	10.5.3
h	kernel function	3.3.6
h	ANN activation function	7.8.2
I	information matrix	4.6.4
I	indicator function	8.3.1
I	Ignorance (logarithmic) score	9.4.7
I	number of matrix rows	11.3.2
$[I]$	identity matrix	11.3.2
I_x	incomplete beta function	4.4.6
i	rank of order statistic	3.3.7
i	unit imaginary number	10.5.3

Roman Symbols	Meaning	Section First Introduced
J	number of matrix columns	11.3.2
K	dimensionality of multivariate observation	3.6.1
K	number of regression predictors	7.2.6
K	order of autoregressive model	10.3.2
K	number of periodogram amplitudes	10.5.6
K_α	constant for approximate K-S test	5.2.5
k	time lag	3.5.4
k	negative binomial distribution waiting parameter	4.2.3
k	harmonic index	10.4.4
k_0	number of restricted parameters, likelihood ratio test	5.2.6
k_A	number of alternative parameters, likelihood ratio test	5.2.6
L	log-likelihood function	3.4.1
L	bootstrap confidence interval lower bound	5.3.5
L	time-series block length	5.3.5
L	location error	9.8.5
L	loss associated with adverse weather	9.9.1
L	length of projection vector	11.3.1
L	number of multiple variables in combined principal components	13.2.2
$[L]$	matrix of "left" singular vectors	11.3.5
M	linear congruential generator modulus parameter	4.7.1
M	number of candidate regression predictors	7.4.2
M	number of gridpoints	9.8.1
M	order of moving-average model	10.3.6
M	number of retained principal components	13.1.1
M	embedding dimension (delay window)	13.7.1
M	number of canonical pairs	14.2.1
m	extreme-value block size	4.4.7
m	number of withheld points in cross-validation	7.4.4
m	ensemble size	8.3
m	order of a Markov chain	10.2.6
$\hat{m}$	discriminant function midpoint	15.2.1
N	number of binomial trials	4.2.1
N	Marshall-Palmer distribution function	4.4.5
N	subsample size	9.4.3
N_C, N_D	numbers of concordant or discordant pairs	3.5.3
n	sample size	2.3.1
n'	effective sample size	5.2.4
n^*	number of nonequal pairs, sign-rank test	5.3.1
n_B	number of bootstrap samples	5.3.5
O	cumulative observation vector	9.4.9
o	observation value	9.1.2
P	incomplete gamma function	4.4.5
P	probability function for group membership	4.6.3
p	a specific probability	3.1.2
p	binomial distribution parameter, (Bernoulli trial probability)	4.2.1
p	probability of evidence in a hypothesis test	5.1.4
p	order of autoregressive process	12.4.2
Pr	probability	2.3.1
Q	odds ratio skill score (Yule's Q)	9.2.3
q	generic distribution quantile	8.3.1
q_p	sample quantile associated with probability p	3.1.2

Continued

Roman Symbols	Meaning	Section First Introduced
$[R]$	sample correlation matrix	3.6.4
$[R]$	matrix of "right" singular vectors	11.3.5
R	average Return period function	4.4.7
R	Metropolis-Hastings ratio	6.4.2
R^2	coefficient of determination	7.2.4
R_1, R_2	sums of ranks, Wilcoxon test	5.3.1
R_1, R_2	discriminant classification regions	15.2.3
R_n	integrated precipitation amount	9.8.5
RI	relative (or reduction in) ignorance	9.4.6
r	Rayleigh distribution variate	4.7.4
r	mass-weighted average distance	9.8.5
r_1	sample lag-1 autocorrelation	3.5.4
r_C	canonical correlation	14.1.2
r_k	sample lag-k autocorrelation	3.5.4
r_{rank}	Spearman rank correlation	3.5.3
r_{xy}	Pearson correlation coefficient	3.5.2
S	Sample space	2.2.2
S	pseudo-random number seed	4.7.1
S	Mann-Kendall trend test statistic	5.3.2
S	generic sample statistic	5.3.5
S	structure error	9.8.5
S	spectral density function	10.5.5
$[S]$	Gandin-Murphy scoring matrix	9.2.6
$[S]$	sample covariance matrix	11.1.3
SS	generic skill score	8.3.3
S1	field gradient score	9.8.2
s	sample standard deviation	3.2.2
s	base rate (sample climatology)	8.2.3
s	number of states in a Markov chain	10.2.5
s^2	sample variance	3.2.2
s_{ens}	ensemble standard deviation	8.3.1
$s_{k,\ell}$	(co)variance between two variables	11.1.3
s_n	precipitation object center of mass	9.8.5
T	threshold parameter	13.3.3
$[T]$	transformation, or rotation matrix	11.3.2
T^2	Hotelling statistic	12.5.2
T_1, T_2, T_3	power transformation functions	3.4.1
T_i	teleconnectivity	3.6.7
T_i	rank of data differences, signed-rank test	5.3.1
T_{max}	maximum temperature	2.5
T_{min}	minimum temperature	2.4.4
T_0	time between effectively independent samples	5.2.4
T^+, T^-	sums of ranks of differences, signed-rank test	5.3.1
t	t-test statistic	5.2.1
t	time	7.3.2
t_j	number of repeated values, Wilcoxon-Mann-Whitney test	5.3.1
tr	trace (of a matrix)	11.3.2
U	bootstrap confidence interval upper bound	5.3.5
U	unresolved tendency	8.2.4
U, U_1, U_2	Mann-Whitney statistic	5.3.1
u	horizontal velocity component	3.5.2

Roman Symbols	Meaning	Section First Introduced
u	POT sampling threshold	4.4.7
u	standard uniform variate	4.7
$\boldsymbol{u}$	principal component vector	13.1.1
V	verification (observed value)	8.3.1
V	ratio of verification	9.2.2
V	mass-weighted scaled volume	9.8.5
V	potential economic value	9.9.2
V	variance inflation factor	10.3.5
V	volume	12.1
Var	variance function	4.3.2
v	horizontal velocity component	3.5.2
v	canonical variate	14.2.1
W	Winkler's score	9.6.2
W	Wavelet coefficient	10.6
W	Ward clustering statistic	16.2.2
w	histogram binwidth	3.3.5
w	mixture distribution weighting parameter	4.4.9
w	Gerrity scoring weight	9.2.6
w	weighting function for spectral smoothing	10.5.6
w	canonical variate	14.2.1
$x_{(\ )}$	order statistic (indicated by parenthetical subscript)	3.1.2
$\bar{x}$	sample mean of x	3.2.1
$\bar{x}_{ens}$	ensemble mean	8.3.1
x'	anomaly of x	3.4.3
Y	cumulative forecast vector	9.4.9
y	generic predictand, or forecast	7.2.1
Z	Fisher correlation transformation statistic	5.4.3
z	standardized anomaly	3.4.3
z	standard Gaussian variate	4.4.2
z	ANN hidden-layer variable	7.8.2
$\hat{z}_0$	BC_a bias correction parameter	5.3.5

Greek Symbols	Meaning	Section First Introduced
α	proportion of tail observations omitted in trimmed mean or variance	3.2.1
α	beta distribution parameter	3.3.7
α	gamma distribution shape parameter	4.4.5
α	Weibull distribution shape parameter	4.4.7
α	type I error rate (hypothesis level)	5.1.5
α_j	treatment effect coefficient	5.6.1
α	generic angle	10.4.1
α	learning rate parameter	16.4
α_{global}	global test level	5.4.1
β	beta distribution parameter	3.3.7
β	gamma distribution scale parameter	4.4.5

Continued

Greek Symbols	Meaning	Section First Introduced
β	exponential distribution mean	4.4.5
β	GEV distribution scale parameter	4.4.7
β	Weibull distribution scale parameter	4.4.7
β	type II error	5.1.5
β_1, β_2	mixed exponential scale parameters	4.4.9
Γ	gamma (factorial) function	4.2.3
γ	sample skewness coefficient	3.2.3
γ	Euler's constant (0.57721 …)	4.4.7
γ_k	autocovariance function	3.5.5
γ_t	member-by-member "stretch" parameter	8.3.6
γ_{YK}	Yule-Kendall index	3.2.3
Δ	difference	3.4.3
δ	difference	12.5.2
δ_1	Fisher's linear discriminant function	15.2.1
ε	arbitrarily small number	2.3.1
ε	generic random number	10.3.1
ζ	Pearson III distribution shift parameter	4.4.5
ζ	GEV distribution location parameter	4.4.7
η	local regression bandwidth	7.7.1
θ	generic parameter value or vector	4.6.2
θ	odds ratio	9.2.2
θ	moving-average parameter	10.3.6
θ	generic angle	11.2.2
κ	(excess) Kurtosis	3.2.4
κ	GEV distribution shape parameter	4.4.7
κ	generalized Pareto distribution shape parameter	4.4.7
Λ	likelihood function	4.6.1
Λ^*	likelihood ratio test statistic (deviance)	5.2.6
λ	Box-Cox ("power") transformation parameter	3.4.1
λ	regularization parameter	7.5.1
λ	nearest-neighbor fraction	7.7.1
λ	eigenvalue	11.3.3
λ_2	L-scale	5.3.4
μ	generic distribution mean	4.2.3
μ	Gaussian mean parameter	4.1.2
μ	Poisson distribution mean parameter	4.2.5
μ_0	mean under null hypothesis	5.1.6
ν	degrees-of-freedom parameter for Chi-square distribution	4.4.5
ν	degrees-of-freedom parameter for t distribution	5.2.1
ξ	standard gamma distribution variate	4.4.5
π	3.14159 …	4.4.2
$\pi(\mathbf{y})$	prerank of the vector $\mathbf{y}$	9.7.2
π	Markov chain stationary probability	10.2.2
ρ	bivariate Gaussian correlation parameter	
ρ_1	population (generating process) lag-1 autocorrelation	5.2.4
ρ_p	check function	7.7.3
ρ_k	lag-k autocorrelation	10.3.1
$[\Sigma]$	covariance matrix	4.6.4
σ	Gaussian standard deviation parameter	4.1.2
σ^2	generic distribution variance	4.2.3
σ^*	generalized Pareto distribution scale parameter	4.4.7
τ	Kendall correlation coefficient	3.5.3

Greek Symbols	Meaning	Section First Introduced
τ_k	period of kth harmonic	10.5.2
Φ	standard Gaussian cumulative distribution function	4.4.2
$[\Phi]$	matrix of autoregressive parameters	12.4.2
Φ^{-1}	standard Gaussian quantile function	4.4.2
ϕ	standard Gaussian probability density function	4.4.2
ϕ	autoregressive parameter	10.3.1
ϕ	phase shift, phase angle	10.4.2
ϕ	latitude	13.2.3
χ^2	Chi-square distribution variate	4.4.5
ψ	autoregressive forecast variance weight	10.3.7
ψ	wavelet function	10.6
$[\Omega]$	matrix of singular values	11.3.5
ω	average sampling frequency	4.4.7
ω	angular frequency	10.4.2
ω	singular value (in SVD)	11.3.5

Mathematical Operators	Meaning	Section First Introduced
$\cup$	union	2.4.1
$\cap$	intersection	2.4.1
$\mid$	conditional probability	2.4.3
Σ	series summation	2.4.5
Π	series product	3.3.6
$\binom{N}{x}$	combinatorial operator	4.2.1
$!$	factorial	4.2.1
$[\ ]^T$	vector or matrix transpose	11.1.2
$\|\ \|$	Euclidean distance	11.2.1
$\otimes$	dyadic (tensor) product	11.3.2
$\|\ \|$	matrix determinant	11.3.2
$[\ ]^{-1}$	matrix inverse	11.3.2
$[\ ]^{1/2}$	matrix "square root"	11.3.4

Acronyms	Section First Introduced
A	
AC: Anomaly Correlation	9.8.4
AIC: Akaike Information Criterion	7.6.2
AKD: Affine Kernel Dressing	8.3.4
ANN: artificial neural network	7.8.2
ANOVA: ANalysis Of VAriance	5.6
AR: AutoRegressive	10.3.1
ARH: Average-Rank Histogram	9.7.2

Continued

Acronyms	Section First Introduced
ARMA: AutoRegressive-Moving Average	10.3.6
ASA: American Statistical Association	5.1.4
B	
BC_a Bias-Corrected and Accelerated	5.3.5
BHF: Bias/Hit rate/False alarm rate	9.2.4
BIC: Bayesian Information Criterion	7.6.2
BMA: Bayesian Model Averaging	8.3.4
BS: Brier Score	9.4.2
BSS: Brier Skill Score	9.4.2
BUGS Bayesian inference Using Gibbs Sampling	6.4.3
C	
CART: Classification And Regression Trees	15.6.2
CCA: Canonical Correlation Analysis	14.1
CCI: Central Credible Interval	6.2.2
CDF: Cumulative Distribution Function	4.4.1
CPCA: Combined Principal Component Analysis	13.2.2
CRPS: Continuous Ranked Probability Score	8.3.2
CSI: Critical Success Index	9.2.2
CSS: Clayton Skill Score	9.2.3
CTI: Cold Tongue Index	6.4.3
D	
df: Degrees of Freedom	7.2.3
DSS: Dawid-Sebastiani Score	9.5.3
E	
ECC: Empirical Copula Coupling	8.4.2
ECM: Expected Cost of Misclassification	15.2.3
ECMWF: European Centre For Medium-range Weather Forecasts	8.2.2
eCRPS: ensemble Continuous Ranked Probability Score	9.7.3
eES: ensemble Energy Score	9.7.4
EDA: Exploratory Data Analysis	3.1
EDI: Extremal Dependence Index	9.2.2
EE: Expected Expense	9.9.1
EEOF: Extended Empirical Orthogonal Function	13.2.2
EM: Expectation-Maximization (algorithm)	4.4.9
EMOS: Ensemble Model Output Statistics	8.3.2
EnKF: Ensemble Kalman Filter	8.2.2
ENSO: El Niño-Southern Oscillation	3.4.2
EOF: Empirical Orthogonal Function	13.1
erf: error function	4.4.2
ES: Energy Score	9.5.2
ETS: Equitable Threat Score	9.2.3
eVS: ensemble Variogram Score	9.7.4
F	
FAR: False Alarm Ratio	9.2.2
FDR: False Discovery Rate	5.4.2
FFT: Fast Fourier Transform	10.5.3
FSS: Fractions Skill Score	9.8.5
G	
GED: Gaussian Ensemble Dressing	8.3.4
GEV: Generalized Extreme Value (distribution)	4.4.7
GLM: Generalized Linear Model	7.6.1
GMSS: Gandin-Murphy Skill Score	9.2.6
GSS: Gilbert Skill Score	9.2.3

Acronyms	Section First Introduced
H	
HPD: Highest Posterior Density	6.2.2
HSS: Heidke Skill Score	9.2.3
I	
IQR: Interquartile Range	3.2.2
J	
JAGS: Just Another Gibbs Sampler	6.4.3
K	
K-S: Kolmogorov-Smirnov	5.2.5
L	
LAD: Least Absolute Deviation	7.2.1
LEV: Log-EigenValue	13.3.2
LQ: Lower Quartile	3.1.2
LS: Linear Score	9.4.8
M	
MA: Moving Average	10.3.6
MAD: Median Absolute Deviation	3.2.2
MAE: Mean Absolute Error	9.3.1
MBMP: Member-by-Member Postprocessing	8.3.6
MCA: Maximum Covariance Analysis	11.3.5
MCMC: Markov Chain Monte Carlo	6.4.1
MDA: Multiple Discriminant Analysis	15.3
ME: Mean Error	9.3.1
MECE: Mutually Exclusive and Collectively Exhaustive	2.2.2
MJO: Madden-Julian Oscillation	12.2.2
MLE: Maximum Likelihood Estimator	4.6.1
MODE: Method for Object-based Diagnostic Evaluation	9.8.5
MOS: Model Output Statistics	7.9.2
MRH: Multivariate Rank Histogram	9.7.2
MS: Mean Squared	7.2.3
MSA: treatment mean squares	5.6.1
MSSA: Multichannel Singular Spectrum Analysis	13.7.1
MSE: Mean Squared Error	5.6.1
MSR: Regression Mean Square	7.2.3
MST: Minimum Spanning Tree	9.7.3
MVN: MultiVariate Normal distribution	12.1
N	
NAO: North Atlantic Oscillation	6.4.3
NCEP: National Centers for Environmental Prediction	8.1.5
NGR: Nonhomogeneous Gaussian Regression	8.3.2
NHC: National Hurricane Center	7.9.1
O	
OLR: Outgoing Longwave Radiation	12.2.2
OLS: Ordinary Least Squares	7.2.1
ORSS: Odds Ratio Skill Score	9.2.3
P	
PC: Proportion Correct	9.2.2
PC: Principal Component	13.1.1
PCA: Principal Component Analysis	13.1
PDF: Probability Density Function	4.4.1
PIT: Probability Integral Transform	4.7.2
POD: Probability Of Detection	9.2.2
POFD: Probability Of False Detection	9.2.2

Continued

Acronyms	Section First Introduced
PoP: Probability of Precipitation	7.10.2
POT: Peaks Over Threshold	4.4.7
PNA: Pacific-North America	3.6.7
P-P: Probability-Probability	4.5.2
PSS: Peirce Skill Score	9.2.3
Q	
Q-Q: Quantile-Quantile	4.5.2
QS: Quantile Score	9.6.1
R	
RA: Redundancy Analysis	14.4
REEP: Regression Estimation of Event Probabilities	7.6.2
REL: RELiability	9.4.3
RES: RESolution	9.4.3
RI: Reliability Index	9.7.1
RMSE: Root Mean Squared Error	8.2.3
ROC: Relative (or, Receiver) Operating Characteristic	9.4.6
RPS: Ranked Probability Score	9.4.9
RV: Reduction of Variance	9.3.2
S	
SAL: Structure-Amplitude-Location method	9.8.5
SAT: Scholastic Aptitude Test	7.4.1
s.e.: standard error	7.2.5
SOI: Southern Oscillation Index	5.4.3
SOM: Self-Organizing Map	16.4
SPI: Standardized Precipitation Index	4.4.5
SS: Sum of Squares	7.2.3
SSA Treatment sum of squares	5.6.1
SSB Block-effect sum of squares	5.6.2
SSA: Singular Spectrum Analysis	13.7.1
SSE: Error Sum of Squares	5.6.1
SSR: Regression Sum of Squares	7.2.2
SST: Total Sum of Squares	5.6.1
SST: Sea-Surface Temperature	14.2.3
ST-EOF: Space-Time Empirical Orthogonal Function	13.7.1
SVD: Singular Value Decomposition	11.3.5
SVM: Support Vector Machine	15.6.1
T	
T-EOF: Time-Empirical Orthogonal Function	13.7.1
T-PC: Time-Principal Component	13.7.1
TS: Threat Score	9.2.2
TSS: True Skill Statistic	9.2.3
U	
UNC: UNCertainty	9.4.3
UTC: Universal Time Coordinated	7.9.3
UQ: Upper Quartile	3.1.2
V	
VS: Value Score	9.9.2
W	
W Winkler score	9.6.2
X	
XLR: Extended Logistic Regression	8.3.3

Answers to Exercises

CHAPTER 2

2.1. b. Pr $\{A \cup B\} = 0.7$
    c. Pr $\{A \cap B^C\} = 0.1$
    d. Pr$\{A^C \cap B^C\} = 0.3$
2.2. b. Pr$\{A\} = 9/31$, Pr$\{B\} = 15/31$, Pr$\{A,B\} = 9/31$
    c. Pr$\{A|B\} = 9/15$
    d. No: Pr$\{A\} \neq$ Pr$\{A|B\}$
2.3. a. 18/22
    b. 22/31
2.4. b. Pr$\{E_1, E_2, E_3\} = .000125$
    c. Pr$\{E_1^C, E_2^C, E_3^C\} = .857$
2.5. a. 0.82
    b. Smaller: since Pr$\{B|A\}$>Pr$\{B\}$, Pr$\{A \cap B\}$=Pr$\{B|A\}$Pr$\{A\}$ > Pr$\{B\}$Pr$\{A\}$
2.6. 0.20

CHAPTER 3

3.1. Median $= 2$ mm, trimean $= 2.75$ mm, mean $= 12.95$ mm
3.2. MAD $= 0.4$ mb, IQR $= 0.8$ mb, $s = 0.88$ mb
3.4. $\gamma_{YK} = 0.158$, $\gamma = 0.877$
3.7. $\lambda = 0$
3.9. $z = 1.36$
3.10. $r_0 = 1.000$, $r_1 = 0.652$, $r_2 = 0.388$, $r_3 = 0.281$

3.12. Pearson: $\begin{bmatrix} 1.000 & 0.703 & -0.830 \\ 0.703 & 1.000 & -0.678 \\ -0.830 & -0.678 & 1.000 \end{bmatrix}$, Spearman: $\begin{bmatrix} 1.000 & 0.606 & -0.688 \\ 0.606 & 1.000 & -0.632 \\ -0.688 & -0.632 & 1.000 \end{bmatrix}$

CHAPTER 4

4.1. 0.163
4.2. a. 0.0364
    b. 0.344
4.3. a. $\mu_{\text{drought}} = 0.056$, $\mu_{\text{wet}} = 0.565$
    b. 0.054
    c. 0.432
4.4. $280 million, $2.825 billion

4.5. $Np(1-p) + (Np)^2$

4.6. a. 1/2

b. 0.00694

4.7. a. $\mu = 24.8°C$, $\sigma = 0.98°C$

b. $\mu = 76.6°F$, $\sigma = 1.76°F$

4.8. a. 0.278

b. 22.9°C

4.9. 0.043

4.10. a. $\alpha = 3.785$, $\beta = 0.934''$

b. $\alpha = 3.785$, $\beta = 23.7$ mm

4.11. a. $q_{30} = 2.41'' = 61.2$ mm; $q_{70} = 4.22'' = 107.2$ mm

b. 0.30'', or 7.7 mm

c. $\cong 0.05$

4.12. a. $q_{30} = 2.30'' = 58.3$ mm; $q_{70} = 4.13'' = 104.9$ mm

b. 0.46'', or 11.6 mm

c. $\cong 0.07$

4.13. 4.32''

4.14. a. $\beta = 35.1$ cm, $\zeta = 59.7$ cm

b. $x = \zeta - \beta \ln[-\ln(F)]$; $Pr\{X \le 221 \text{ cm}\} = 0.99$

4.15. a. 5.33''

b. 0.264

4.16. a. $\mu_{max} = 31.8°F$, $\sigma_{max} = 7.86°F$, $\mu_{min} = 20.2°F$, $\sigma_{min} = 8.81°F$, $\rho = 0.810$

b. 0.728

4.18. a. $\beta = \Sigma x/n$

b. $-I^{-1}(\hat{\beta}) = \hat{\beta}^2/n$

4.19. 0.201

4.20. $x(u) = \beta[-\ln(1-u)]^{1/\alpha}$

CHAPTER 5

5.1. a. $z = 4.88$, reject H_0

b. [1.10°C, 2.56°C]

5.2. 6.53 days (Ithaca), 6.08 days (Canandaigua)

5.3. $p = 0.40$

5.4. $z = -4.00$

$p = 0.000063$

$p = 0.000032$

5.5. $n \ge 86$

5.6. a. 0.5

b. 10.95°F

5.7. $|r| \ge 0.377$

5.8. a. $D_n = 0.152$ (reject at 10%, not at 5% level)

b. For classes: [<2, 2–3, 3–4, 4–5, ≥5], $\chi^2 = 0.33$ (do not reject)

$r = 0.971$ (do not reject)

5.9. $\Lambda = 21.86$, reject ($p < .001$)

5.10. a. $U_1 = 1$, reject ($p < .005$)

b. $z = -3.18$, reject ($p = .0007$)

5.11. $\approx [1.02, 3.59]$

5.12. a. Observed ($s^2_{E-N}/s^2_{non-E-N}$) = 329.5; permutation distribution critical value (1%, 2-tailed) $\approx$ 141, reject H_0 ($p < 0.01$)

b. 15/10000 members of bootstrap sampling distribution for $s^2_{E-N}/s^2_{non-E-N} \le 1$; 2-tailed p = 0.003

5.13. a. Counting method, no (need ≥ 3 locally significant); FDR, yes

b. $p = .007$ and $p = .009$ significant according to FDR

5.14. a

	df	SS	MS	F
Total	43	6777.73		
Treatment	3	2302.82	767.61	6.86
Error	40	4447.91	111.87	

b.

	df	SS	MS	F
Total	43	6777.73		
Blocks	10	3721.73	372.17	15.9
Treatment	3	2302.82	767.61	32.8
Error	30	701.18	23.37	

CHAPTER 6

6.1. a. $\alpha = 14.8$, $\beta = 7.41$

b. Beta distribution, with $\alpha' = 29.8$, $\beta' = 17.4$

c. $\Pr\{X^+ = 0\} = .0094$, $\Pr\{X^+ = 1\} = .0656$, $\Pr\{X^+ = 2\} = .1982$, $\Pr\{X^+ = 3\} = .3248$, $\Pr\{X^+ = 4\} = .2895$, $\Pr\{X^+ = 5\} = .1125$

6.2. a. $\beta = 190.8$, $\zeta = 162.3$

b. $\beta = 155.9$, $\zeta = 180.0$

c. 1040.0, 897.2

6.3. a. $\alpha = 1.5$, $\beta = 0.1$

b. .157

6.4. a. $\mu'_h = 455.6$, $\sigma'_h = 33.3$

b. $\mu_+ = 455.6$, $\sigma_+ = 60.1$

6.5. a. $\mu'_h = 427.4$, $\sigma'_h = 28.6$

b. $\mu_+ = 427.4$, $\sigma_+ = 57.6$

6.6. a. 462.3

b. 400

c. 450

CHAPTER 7

7.1. a. $a = 959.8°C$, $b = -0.925°C/mb$

c. $z = -6.33$

d. 0.690

e. 0.876

f. 0.925

7.2. a. Total   26   318.2874
     Regression   1   316.6065   316.6065
     Residual   25   1.6809   0.06724

    b. 1

    c. 12.73

    d. 0.9947

    e. 0.56

    f. $t = 68.6$

7.3. $\ln [\bar{y}/(1-\bar{y})]$

7.4. a. 1.74 mm

    b. [0 mm, 13.1 mm]

7.5. MSE = 0.369

    slopes: -.926,-.926,-.928,-.924,-.940,-.921,-.909,-.928,-.917,-.897,-.928,-.919,-.921, -.921,-.854,-1.095,-.927,-.952,.850,-.922

7.6. a. −59 n.m.

    b. −66 n.m.

7.7. a. 65.8°F

    b. 52.5°F

    c. 21.7°F

    d. 44.5°F

7.8. a. 0.65

    b. 0.49

    c. 0.72

    d. 0.56

7.9. $f_{MOS} = 30.8°F + (0) (Th)$

7.10. 0.20

7.11. a. 12 mm

    b. [5 mm, 32 mm], [1 mm, 55 mm]

    c. 0.625

CHAPTER 8

8.1. a. [4.9°C, 8.8°C]

    b. [5.1°C, 8.5°C]

8.2. 0.873

8.3. a. 0.059

    b. 0.345

    c. 0.143

8.4. (25.5°C, 1.3 m/s), (27.5°C, 1.8 m/s), (27.0°C, 3.1 m/s), (26.4°C, 0.7 m/s), (28.4°C, 2.4 m/s)

CHAPTER 9

9.1. a. .0025 .0013 .0108 .0148 .0171 .0138 .0155 .0161 .0177 .0176 .0159 .0189
     .4087 .0658 .1725 .0838 .0445 .0228 .0148 .0114 .0068 .0044 .0011 .0014

    b. 0.162

9.2. 1644   1330
      364   9064

9.3. a. 0.863
      b. 0.493
      c. 0.578
      d. 0.691
      e. 0.407

9.4. a. 0.074
      b. 0.097
      c. 0.761
      d. 0.406

9.5. a. .9597   .0127   .0021   .0007
         .0075   .0043   .0014   .0005
         .0013   .0013   .0009   .0003
         .0007   .0006   .0049   .0009
      b. 0.966
      c. 0.371
      d. 0.336

9.6. a. 5.37°F
      b. 7.54°F
      c. −0.03°F
      d. 1.95%

9.7. a. −118%
      b. 60.0%

9.8. a. 0.1215
      b. 0.1699
      c. 28.5%
      e. 0.392

9.9. a. .0415   .0968   .1567   .1428   .1152   .0829   .1060   .0829   .0783   .0553   .0415
          .3627   .2759   .1635   .0856   .0498   .0230   .0204   .0102   .0051   .0026   .0013
      c. H = .958, .862, .705, .562, .447, .364, .258, .175, .097, .042
          F = .637, .361, .198, .112, .062, .039, .019, .009, .004, .001
      d. $A = 0.831$, $z = -14.9$

9.10. 0.16

9.11. a. 0.298
       b. 16.4%
       c. 0.755

9.12. a. $\pi_{MRH}(y_0)=3$, $\pi_{MRH}(y_1)=1$, $\pi_{MRH}(y_2)=2$, $\pi_{MRH}(y_3)2$, $\pi_{MRH}(y_4)=5$, $\pi_{MRH}(y_5)=1$
       b. $b = 5$

9.13. a. 22
       b. 0.0192

9.14. a. 5 rank 1, 2 rank 2, 3 rank 3, 2 rank 4, 2 rank 5, 6 rank 6
       b. underdispersed

9.15. a. 0.140
       b. 0.224

9.16. a. 30.3

      b. 5.31 dam^2

      c. 46.9%

      d. 0.726

      e. 0.714

9.17. .352, .509, .673, .598, .504, .426, .343, .275, .195, .128, −.048

CHAPTER 10

10.1. a. $p_{01} = 0.45$, $p_{11} = 0.79$

      b. $\chi^2 = 3.51$, $p \approx 0.064$

      c. $\pi_1 = 0.682$, $n_{\bullet 1}/n = 0.667$

      d. $r_0 = 1.00$, $r_1 = 0.34$, $r_2 = 0.12$, $r_3 = 0.04$

      e. 0.624

10.2. a. $r_0 = 1.00$, $r_1 = 0.40$, $r_2 = 0.16$, $r_3 = 0.06$, $r_4 = 0.03$, $r_5 = 0.01$

      a. $r_0 = 1.00$, $r_1 = 0.41$, $r_2 = -0.41$, $r_3 = -0.58$, $r_4 = -0.12$, $r_5 = 0.32$

10.3. a. AR(1): $\phi = 0.80$; $s_\varepsilon^2 = 36.0$

       AR(2): $\phi_1 = 0.89$, $\phi_2 = -0.11$; $s_\varepsilon^2 = 35.5$

       AR(3): $\phi_1 = 0.91$, $\phi_2 = -0.25$, $\phi_3 = 0.16$; $s_\varepsilon^2 = 34.7$

      b. AR(1): BIC = 369.6

      c. AR(1): AIC = 364.4

10.4. $x_1 = 71.5$, $x_2 = 66.3$, $x_3 = 62.1$

10.5. a. 28.6

      b. 19.8

      c. 4.5

10.6. a. 0.67 m/s

      b. [12.9 m/s, 16.5 m/s]

10.7. a. $C_1 = 16.92°F$, $\phi_1 = 199°$; $C_2 = 4.16°F$, $\phi_2 = 256°$

10.8. a. 82.0°F

      b. 74.8°F

10.9. b. 0.990

      c. C_1^2, $p = .00593$

      d. C_1^2, $p = .00593 < 0.05/6$

      e. C_1^2

10.10. 56

10.11. a. e.g., $f_A = 1 - .0508\ \text{mo}^{-1} = .9492\ \text{mo}^{-1}$

       b. $\approx$ twice monthly

10.13. [0.11, 16.3]

CHAPTER 11

11.1. $\begin{bmatrix} 216.0 & -4.32 \\ 135.1 & 7.04 \end{bmatrix}$

11.2. $([X]^T y)^T = [627, 11475]$, $[X^T X]^{-1} = \begin{bmatrix} .06263 & -.002336 \\ -.00236 & .0001797 \end{bmatrix}$, $b^T = [12.46, 0.60]$

11.3. 90°

11.6. $\frac{1}{n}(\bar{x} - \mu)^T [S]^{-1} (\bar{x} - \mu)$

11.7. $[E]u$

11.8. a. $\begin{bmatrix} 59.5 & 58.1 \\ 58.1 & 61.8 \end{bmatrix}$

b. $\begin{bmatrix} .205 & -.193 \\ -.193 & .197 \end{bmatrix}$

c. $\begin{bmatrix} .205 & -.193 \\ -.193 & .197 \end{bmatrix}$

d. $\begin{bmatrix} 6.16 & 4.64 \\ 4.64 & 6.35 \end{bmatrix}$

e. 1.765

11.9. a. $\begin{bmatrix} 59.52 & 75.43 & 58.07 & 51.70 \\ 75.43 & 185.47 & 81.63 & 110.80 \\ 58.07 & 81.63 & 61.85 & 56.12 \\ 51.70 & 110.80 & 56.12 & 77.58 \end{bmatrix}$

b. $\mu_y^T = [21.4, 26.0]$

$[S_y] = \begin{bmatrix} 98.96 & 75.77 \\ 75.55 & 62.92 \end{bmatrix}$

CHAPTER 12

12.2. a. $\mu = [29.87, 13.00]^T$, $[S] = \begin{bmatrix} 4.96 & 0.15 \\ 0.15 & 27.12 \end{bmatrix}$

b. $N_2(\mu \ [\Sigma])$; $\mu = [-1.90, 5.33]^T$ $[\Sigma] = \begin{bmatrix} 5.23 & 7.01 \\ 7.01 & 50.24 \end{bmatrix}$

12.3. 0.334

12.4. a. $N_1(31.4, 21.4)$

b. 0.306

12.5. $r = 0.974 > r_{\text{crit}} (10\%) = 0.970$; do not reject

12.6. a. $T^2 = 68.5 >> 18.421 = \chi_2^2(.9999)$; reject

b. $a \propto [-.6217, .1929]^T$

12.7. a. $T^2 = 7.80$, reject @ 5%

b. $a \propto [-.0120, .0429]^T$

CHAPTER 13

13.1. a. 3.78, 4.51
     b. 118.8
     c. 0.979

13.2. a. 0.430, −0.738
     b. −.439, .354, .183, −.433, .371, .231

13.3. a. Correlation matrix: $\Sigma\, \lambda_k = 3$
     b. 1, 1, 1
     c. 2.3 mm

13.4. a. [1.51, 6.80], [0.22, 0.98], [0.10, 0.46]
     b. λ_2 and λ_3 may be entangled

13.5. a. $\begin{bmatrix} .593 & .332 & .734 \\ .552 & -.831 & -.069 \\ -.587 & -.446 & .676 \end{bmatrix}$

     b. $\begin{bmatrix} .377 & .556 & 1.785 \\ .351 & -1.39 & -.168 \\ -.373 & -.747 & 1.644 \end{bmatrix}$

13.6. 9.18, 14.34, 10.67

13.7. 37.5 + .838 IPpt + 1.831 IMax + 5.017 IMin + .341 CPpt + 2.42 CMax + 4.876 CMin

CHAPTER 14

14.1. 6 Jan: $v_1 = .038$, $w_1 = .433$; 7 Jan: $v_1 = .868$, $w_1 = 1.35$

14.2. 39.0°F, 23.6°F

14.3. a. $\begin{bmatrix} 1.883 & 0 & 1.838 & -.212 \\ 0 & .927 & .197 & .791 \\ 1.838 & .197 & 1.904 & 0 \\ -.212 & .791 & 0 & .925 \end{bmatrix}$

     b. $a_1 = [.728, .032]^T$, $b_1 = [.718, -.142]^T$, $r_{C_1} = 0.984$
        $a_2 = [-.023, 1.038]^T$, $b_2 = [.099, 1.030]^T$, $r_{C_2} = 0.867$

CHAPTER 15

15.1. b. $R_1: -1 \leq x \leq 0.25$
       $R_2: 0.25 < x \leq 1.5$
     c. $R_1: -1 \leq x \leq -0.33$
       $R_2: -0.33 < x \leq 1.5$

15.2. a. $a_1^T = [0.83, -0.56]$
     b. 1953
     c. 1953

15.3. a. $[.734\ .367]^T$
     b. w = 0.66 > 0 => "yes" group
     c. 0.66

15.4. a. $\delta_1 = 38.65$, $\delta_2 = -14.99$; Group 3
       b. 5.2 x 10^{-12}, 2.8 x 10^{-9}, 0.99999997
15.5. a. 0.006
       b. 0.059
       c. 0.934

CHAPTER 16

16.1.
$$
\begin{bmatrix}
0 & & & & & \\
3.59 & 0 & & & & \\
2.29 & 1.59 & 0 & & & \\
3.12 & 0.82 & 0.89 & 0 & & \\
0.71 & 4.27 & 2.89 & 3.75 & 0 & \\
1.64 & 2.24 & 0.71 & 1.59 & 2.20 & 0
\end{bmatrix}
$$

16.2. a. 1967+1970, $d = 0.71$; 1965+1969, $d = 0.71$; 1966+1968, $d = 0.82$;
          (1967+1970) + (1966+1968), $d = 1.59$; all, $d = 1.64$.
       b. 1967+1970, $d = 0.71$; 1965+1969, $d = 0.71$; 1966+1968, $d = 0.82$;
          (1967+1970) + (1966+1968), $d = 2.24$; all, $d = 4.27$.
       c. 1967+1970, $d = 0.71$; 1965+1969, $d = 0.71$; 1966+1968, $d = 0.82$;
          (1967+1970) + (1966+1968), $d - 1.58$; all, $d = 2.97$.
16.3. a. 1967+1970, $d = 0.50$; 1965+1969, $d = 0.60$; 1966+1968, $d = 0.70$;
          (1967+1970) + (1965+1969), $d = 1.25$; all, $d = 1.925$.
       b. 1967+1970, $d = 0.125$; 1965+1969, $d = 0.180$; 1966+1968, $d = .245$;
          (1967+1970) + (1965+1969), $d = 1.868$; all, $d = 7.053$.
16.4. {1966, 1967}, {1965, 1968, 1969, 1970}; {1966, 1967, 1968}, {1965, 1969, 1970};
       {1966, 1967, 1968, 1970}, {1965, 1969}.
16.5. a. f_1
       b. $\Pr\{f_1\}$=0.71, $\Pr\{f_2\}$=0.29

References

Abramowitz, M., Stegun, I.A. (Eds.), 1984. Pocketbook of Mathematical Functions. Frankfurt, Verlag Harri Deutsch. 468 pp.

Accadia, C., Mariani, S., Casaioli, M., Lavagnini, A., 2003. Sensitivity of precipitation forecast skill scores to bilinear interpolation and a simple nearest-neighbor average method on high-resolution verification grids. Weather Forecast. 18, 918–932.

Agresti, A., 1996. An Introduction to Categorical Data Analysis. Wiley. 290 pp.

Agresti, A., Coull, B.A., 1998. Approximate is better than "exact" for interval estimation of binomial proportions. Am. Stat. 52, 119–126.

Ahijevych, D., Gilleland, E., Brown, B.G., Ebert, E.E., 2009. Application of spatial verification methods to idealized and NWP-gridded precipitation forecasts. Weather Forecast. 24, 1485–1497.

Ahrens, B., Jaun, S., 2007. On evaluation of ensemble precipitation forecasts with observation-based ensembles. Adv. Geosci. 10, 139–144.

Akaike, H., 1974. A new look at the statistical model identification. IEEE Trans. Autom. Control 19, 716–723.

Alfaro, E.J., Chourio, X., Muñoz, A.G., Mason, S.J., 2018. Improved seasonal prediction skill of rainfall for the Primera season in Central America. Int. J. Climatol. 38 (Suppl. 1), e255–e268.

Allen, M.R., Robertson, A.W., 1996. Distinguishing modulated oscillations from coloured noise in multivariate datasets. Clim. Dyn. 12, 775–784.

Allen, M.R., Smith, L.A., 1996. Monte Carlo SSA: detecting irregular oscillations in the presence of colored noise. J. Clim. 9, 3373–3404.

Ambaum, M.H.P., 2010. Significance tests in climate science. J. Clim. 23, 5927–5932.

Anderson, J.L., 1996. A method for producing and evaluating probabilistic forecasts from ensemble model integrations. J. Clim. 9, 1518–1530.

Anderson, J.L., 1997. The impact of dynamical constraints on the selection of initial conditions for ensemble predictions: low-order perfect model results. Mon. Weather Rev. 125, 2969–2983.

Andrews, D.F., Gnanadesikan, R., Warner, J.L., 1971. Transformations of multivariate data. Biometrics 27, 825–840.

Andrews, D.F., Bickel, P.J., Hampel, F.R., Huber, P.J., Rogers, W.N., Tukey, J.W., 1972. Robust Estimates of Location—Survey and Advances. Princeton University Press.

Anscombe, F.J., 1973. Graphs in statistical analysis. Am. Stat. 27, 17–21.

Applequist, S., Gahrs, G.E., Pfeffer, R.L., 2002. Comparison of methodologies for probabilistic quantitative precipitation forecasting. Weather Forecast. 17, 783–799.

Araneo, D.C., Compagnucci, R.H., 2004. Removal of systematic biases in S-mode principal components arising from unequal grid spacing. J. Clim. 17, 394–400.

Armstrong, J.S., 2001. Evaluating forecasting methods. In: Armstrong, J.S. (Ed.), Principles of Forecasting: A Handbook for Researchers and Practitioners. Kluwer, pp. 443–472.

Atger, F., 1999. The skill of ensemble prediction systems. Mon. Weather Rev. 127, 1941–1953.

Azcarraga, R., and A.J. Ballester G., 1991. Statistical system for forecasting in Spain. In: H.R. Glahn, A.H. Murphy, L.J. Wilson, and J.S. Jensenius, Jr., eds., Programme on Short- and Medium-Range Weather Prediction Research. World Meteorological Organization WM/TD No. 421, XX, 23–25.

Baars, J.A., Mass, C.F., 2005. Performance of National Weather Service forecasts compared to operational, consensus, and weighted model output statistics. Weather Forecast. 20, 1034–1047.

Baker, D.G., 1981. Verification of fixed-width, credible interval temperature forecasts. Bull. Am. Meteorol. Soc. 62, 616–619.

Baldwin, M.P., Stephenson, D.B., Jolliffe, I.T., 2009. Spatial weighting and iterative projection methods for EOFs. J. Clim. 22, 234–243.

Bamber, D., 1975. The area above the ordinal dominance graph and the area below the receiver operating characteristic graph. J. Math. Psychol. 12, 387–415.

Banfield, J.D., Raftery, A.E., 1993. Model-based Gaussian and non-Gaussian clustering. Biometrics 49, 803–821.

Baran, S., 2014. Probabilistic wind speed forecasting using Bayesian model averaging with truncated normal components. Comput. Stat. Data Anal. 75, 227–238.

Baran, S., Lerch, S., 2015. Log-normal distribution based ensemble model output statistics models for probabilistic wind-speed forecasting. Q. J. R. Meteorol. Soc. 141, 2289–2299.

Baran, S., Lerch, S., 2016. Mixture EMOS model for calibrating ensemble forecasts of wind speed. Environmetrics 27, 116–130.

Baran, S., Möller, A., 2015. Joint probabilistic forecasting of wind speed and temperature using Bayesian model averaging. Environmetrics 26, 120–132.

Baran, S., Möller, A., 2017. Bivariate ensemble model output statistics approach for joint forecasting of wind speed and temperature. Meteorol. Atmos. Phys. 129, 99–112.

Baran, S., Nemoda, D., 2016. Censored and shifted gamma distribution based EMOS model for probabilistic quantitative precipitation forecasting. Environmetrics 27, 280–292.

Bárdossy, A., Pegram, G.G.S., 2009. Copula based multisite model for daily precipitation simulation. Hydrol. Earth Syst. Sci. 13, 2299–2314.

Bárdossy, A., Plate, E.J., 1992. Space-time model for daily rainfall using atmospheric circulation patterns. Water Resour. Res. 28, 1247–1259.

Barnes, L.R., Schultz, D.M., Gruntfest, E.C., Hayden, M.H., Benight, C.C., 2009. False alarm rate or false alarm ratio? Weather Forecast. 24, 1452–1454.

Barnett, T.P., Preisendorfer, R.W., 1987. Origins and levels of monthly and seasonal forecast skill for United States surface air temperatures determined by canonical correlation analysis. Mon. Weather Rev. 115, 1825–1850.

Barnston, A.G., 1994. Linear statistical short-term climate predictive skill in the northern hemisphere. J. Clim. 7, 1513–1564.

Barnston, A.G., van den Dool, H.M., 1993. A degeneracy in cross-validated skill in regression-based forecasts. J. Clim. 6, 963–977.

Barnston, A.G., Glantz, M.H., He, Y., 1999. Predictive skill of statistical and dynamical climate models in SST forecasts during the 1997–1998 El Niño episode and the 1998 La Niña onset. Bull. Am. Meteorol. Soc. 80, 217–243.

Barnston, A.G., Mason, S.J., Goddard, L., DeWitt, D.G., Zebiak, S.E., 2003. Multimodel ensembling in seasonal climate forecasting at IRI. Bull. Am. Meteorol. Soc. 84, 1783–1796.

Batté, L., Doblas-Reyes, F.J., 2015. Stochastic atmospheric perturbations in the EC-Earth3 global coupled model: impact of SPPT on seasonal forecast quality. Clim. Dyn. 45, 3419–3439.

Baughman, R.G., Fuquay, D.M., Mielke Jr., P.W., 1976. Statistical analysis of a randomized lightning modification experiment. J. Appl. Meteorol. 15, 790–794.

Ben Bouallègue, Z., 2016. Statistical postprocessing of ensemble global radiation forecasts with penalized quantile regression. Meteorol. Z. https://doi.org/10.1127/metz/2016/0748.

Ben Bouallégue, Z., Pinson, P., Friederichs, P., 2015. Quantile forecast discrimination and value. Q. J. R. Meteorol. Soc. 141, 3415–3424.

Benedetti, R., 2010. Scoring rules for forecast verification. Mon. Weather Rev. 138, 203–211.

Bengtsson, L., Steinheimer, M., Bechtold, P., Geleyn, J.-F., 2013. A stochastic parametrization for deep convection using cellular automata. Q. J. R. Meteorol. Soc. 139, 1533–1543.

Benjamini, Y., Hochberg, Y., 1995. Controlling the false discovery rate: a practical and powerful approach to multiple testing. J. R. Stat. Soc. Ser. B Methodol. 57, 289–300.

Bentzien, S., Friederichs, P., 2012. Generating and calibrating probabilistic quantitative precipitation forecasts from the high-resolution NWP model COSMO-DE. Weather Forecast. 27, 988–1002.

Bentzien, S., Friederichs, P., 2014. Decomposition and graphical portrayal of the quantile score. Q. J. R. Meteorol. Soc. 140, 1924–1934.

Beran, R., Srivastava, M.S., 1985. Bootstrap tests and confidence regions for functions of a covariance matrix. Ann. Stat. 13, 95–115.

Berner, J., Doblas-Reyes, F.J., Palmer, T.N., Shutts, G.J., Weisheimer, A., 2010. Impact of a quasi-stochastic cellular automaton backscatter scheme on the systematic error and seasonal prediction skill of a global climate model. In: Palmer, T., Williams, P. (Eds.), Stochastic Physics and Climate Modeling. Cambridge University Press, Cambridge, pp. 375–395.

Berner, J., Fossell, K.R., Ha, S.-Y., Hacker, J.P., Snyder, C., 2015. Increasing the skill of probabilistic forecasts: understanding performance improvements from model-error representations. Mon. Weather Rev. 143, 1295–1320.

Berner, J., et al., 2017. Stochastic parameterization: toward a new view of weather and climate models. Bull. Am. Meteorol. Soc. 98, 565–587.

Berrocal, V.J., Raftery, A.E., Gneiting, T., 2007. Combining spatial statistical and ensemble information in probabilistic weather forecasts. Mon. Weather Rev. 135, 1386–1402.

Beyth-Marom, R., 1982. How probable is probable? A numerical translation of verbal probability expressions. J. Forecast. 1, 257–269.

Bickel, P.J., Bühlmann, P., 1999. A new mixing notion and functional central limit theorems for a sieve bootstrap in time series. Bernoulli 5, 413–446.

Bien, J., Tibshirani, R., 2011. Hierarchical clustering with prototypes via minimax linkage. J. Am. Stat. Assoc. 106, 1075–1084.

Bishop, C.H., Shanley, K.T., 2008. Bayesian model averaging's problematic treatment of extreme weather and a paradigm shift that fixes it. Mon. Weather Rev. 136, 4641–4652.

Bjerknes, J., 1969. Atmospheric teleconnections from the equatorial Pacific. Mon. Weather Rev. 97, 163–172.

Blackmon, M.L., 1976. A climatological spectral study of the 500 mb geopotential height of the northern hemisphere. J. Atmos. Sci. 33, 1607–1623.

Bloomfield, P., Steiger, W., 1980. Least absolute deviations curve-fitting. SIAM J. Sci. Stat. Comput. 1, 290–301.

Blunden, J., Arndt, D.S. (Eds.), 2015. State of the climate in 2014. Bull. Am. Meteorol. Soc. 96, S1–S267.

Bonavita, M., Isaksen, L., Hólm, E., 2012. On the use of EDA background error variances in the ECMWF 4D-Var. Q. J. R. Meteorol. Soc. 138, 1540–1559.

Boswell, M.T., Gore, S.D., Patil, G.P., Taillie, C., 1993. The art of computer generation of random variables. In: Rao, C.R. (Ed.), Handbook of Statistics. In: vol. 9. Elsevier, pp. 661–721.

Bowler, N.E., 2006. Explicitly accounting for observation error in categorical verification of forecasts. Mon. Weather Rev. 134, 1600–1606.

Bowler, N.E., 2008. Accounting for the effect of observation errors on verification of MOGREPS. Meteorol. Appl. 15, 199–205.

Bowler, N.E., Mylne, K.R., 2009. Ensemble transform Kalman filter perturbations for a regional ensemble prediction system. Q. J. R. Meteorol. Soc. 135, 757–766.

Bowler, N.E., Arribas, A., Mylne, K.R., Robertson, K.B., Beare, S.E., 2008. The MOGREPS short-range ensemble prediction system. Q. J. R. Meteorol. Soc. 134, 703–722.

Box, G.E.P., Cox, D.R., 1964. An analysis of transformations. J. R. Stat. Soc. Ser. B Methodol. 26, 211–243.

Box, G.E.P., Jenkins, G.M., 1976. Time Series Analysis: Forecasting and Control. Holden-Day, San Francisco, CA. 575 pp.

Bradley, A.A., Schwartz, S.S., 2011. Summary verification measures and their interpretation for ensemble forecasts. Mon. Weather Rev. 139, 3075–3089.

Bradley, A.A., Hashino, T., Schwartz, S.S., 2003. Distributions-oriented verification of probability forecasts for small data samples. Weather Forecast. 18, 903–917.

Bradley, A.A., Schwartz, S.S., Hashino, T., 2008. Sampling uncertainty and confidence intervals for the Brier score and Brier Skill score. Weather Forecast. 23, 992–1006.

Bras, R.L., Rodríguez-Iturbe, I., 1985. Random Functions and Hydrology. Addison-Wesley. 559 pp.

Bratley, P., Fox, B.L., Schrage, L.E., 1987. A Guide to Simulation. Springer. 397 pp.

Braverman, A., Cressie, N., Teixeira, J., 2011. A likelihood-based comparison of temporal models for physical processes. Stat. Anal. Data Min. 4, 247–258.

Breiman, L., 1996. Bagging predictors. Mach. Learn. 24, 123–140.

Breiman, L., 2001. Random forests. Mach. Learn. 45, 5–32.

Breiman, L., Friedman, J., Stone, C.J., Olshen, R.A., 1984. Classification and Regression Trees. CRC Press. 368 pp.

Brelsford, W.M., Jones, R.H., 1967. Estimating probabilities. Mon. Weather Rev. 95, 570–576.

Bremnes, J.B., 2004. Probabilistic forecasts of precipitation in terms of quantiles using NWP model output. Mon. Weather Rev. 132, 338–347.

Bretherton, C.S., Smith, C., Wallace, J.M., 1992. An intercomparison of methods for finding coupled patterns in climate data. J. Clim. 5, 541–560.

Bretherton, C.S., Widmann, M., Dymnikov, V.P., Wallace, J.M., Bladé, I., 1999. The effective number of spatial degrees of freedom of a time-varying field. J. Clim. 12, 1990–2009.

Brier, G.W., 1950. Verification of forecasts expressed in terms of probabilities. Mon. Weather Rev. 78, 1–3.

Brier, G.W., Allen, R.A., 1951. Verification of weather forecasts. In: Malone, T.F. (Ed.), Compendium of Meteorology. American Meteorological Society, pp. 841–848.

Briggs, W.M., Levine, R.A., 1997. Wavelets and field forecast verification. Mon. Weather Rev. 125, 1329–1341.

Briggs, W., Pocernich, M., Ruppert, D., 2005. Incorporating misclassification error in skill assessment. Mon. Weather Rev. 133, 3382–3392.

Brill, K.F., 2009. A general analytic method for assessing sensitivity to bias of performance measures for dichotomous forecasts. Weather Forecast. 24, 307–318.

Bröcker, J., 2008. Some remarks on the reliability of categorical probability forecasts. Mon. Weather Rev. 136, 4488–4502.

Bröcker, J., 2009. Reliability, sufficiency, and the decomposition of proper scores. Q. J. R. Meteorol. Soc. 135, 1512–1519.

Bröcker, J., 2010. Regularized logistic models for probabilistic forecasting and diagnostics. Mon. Weather Rev. 138, 592–604.

Bröcker, J., 2012a. Probability forecasts. In: Jolliffe, I.T., Stephenson, D.B. (Eds.), Forecast Verification, A Practitioner's Guide in Atmospheric Science, second ed. Wiley-Blackwell, pp. 119–139.

Bröcker, J., 2012b. Estimating reliability and resolution of probability forecasts through decomposition of the empirical score. Clim. Dyn. 39, 655–667.

Bröcker, J., 2012c. Erratum to: Estimating reliability and resolution of probability forecasts through decomposition of the empirical score. Clim. Dyn. 39, 3123.

Bröcker, J., 2012d. Evaluating raw ensembles with the continuous ranked probability score. Q. J. R. Meteorol. Soc. 138, 1611–1617.

Bröcker, J., Smith, L.A., 2007a. Scoring probabilistic forecasts: the importance of being proper. Weather Forecast. 22, 382–388.

Bröcker, J., Smith, L.A., 2007b. Increasing the reliability of reliability diagrams. Weather Forecast. 22, 651–661.

Bröcker, J., Smith, L.A., 2008. From ensemble forecasts to predictive distribution functions. Tellus A 60A, 663–678.

Brooks, C.E.P., Carruthers, N., 1953. Handbook of Statistical Methods in Meteorology. Her Majesty's Stationery Office, London. 412 pp.

Brooks, H.E., Doswell III, C.A., Kay, M.P., 2003. Climatological estimates of local daily tornado probability for the United States. Weather Forecast. 18, 626–640.

Bross, I.D.J., 1953. Design for Decision. Macmillan, New York. 276 pp.

Brown, B.G., Katz, R.W., 1991. Use of statistical methods in the search for teleconnections: past, present, and future. In: Glantz, M., Katz, R.W., Nicholls, N. (Eds.), Teleconnections Linking Worldwide Climate Anomalies. Cambridge University Press, 371–400.

Brunet, N., Verret, R., Yacowar, N., 1988. An objective comparison of model output statistics and "perfect prog" systems in producing numerical weather element forecasts. Weather Forecast. 3, 273–283.

Buell, C.E., 1979. On the physical interpretation of empirical orthogonal functions. In: Preprints, 6th Conference on Probability and Statistics in the Atmospheric Sciences. American Meteorological Society, 112–117.

Bühlmann, P., 1997. Sieve bootstrap for time series. Bernoulli 3, 123–148.

Bühlmann, P., 2002. Bootstraps for time series. Stat. Sci. 17, 52–72.

Buizza, R., 1997. Potential forecast skill of ensemble prediction and ensemble spread and skill distributions of the ECMWF Ensemble Prediction System. Mon. Weather Rev. 125, 99–119.

Buizza, R., 2008. Comparison of a 51-member low-resolution (T_L399L62) ensemble with a 6- member high-resolution (T_L799L91) lagged-forecast ensemble. Mon. Weather Rev. 136, 3343–3362.

Buizza, R., 2010. Horizontal resolution impact on short- and long-range forecast error. Q. J. R. Meteorol. Soc. 136, 1020–1035.

Buizza, R., Leutbecher, M., 2015. The forecast skill horizon. Q. J. R. Meteorol. Soc. 141, 3366–3382.

Buizza, R., Richardson, D., 2017. 25 years of ensemble forecasting at ECMWF. ECMWF Newsl. 153, 18–31.

Buizza, R., Miller, M., Palmer, T.N., 1999. Stochastic representation of model uncertainties in the ECMWF Ensemble Prediction System. Q. J. R. Meteorol. Soc. 125, 2887–2908.

Buizza, R., Houtekamer, P.L., Toth, Z., Pellerin, G., Wei, M., Zhu, Y., 2005. A comparison of the ECMWF, MSC, and NCEP global ensemble prediction systems. Mon. Weather Rev. 133, 1076–1097.

Burgers, G., van Leeuwen, P.J., Evensen, G., 1998. Analysis scheme in the ensemble Kalman filter. Mon. Weather Rev. 12, 420–436.

Burman, P., Chow, E., Nolan, D., 1994. A cross-validatory method for dependent data. Biometrika 81, 351–358.

Cabilio, P., Zhang, Y., Chen, X., 2013. Bootstrap rank tests for trend in time series. Environmetrics 24, 537–549.

Campbell, S.D., Diebold, F.X., 2005. Weather forecasting and weather derivatives. J. Am. Stat. Assoc. 100, 6–16.

Candille, G., Talagrand, O., 2005. Evaluation of probabilistic prediction systems for a scalar variable. Q. J. R. Meteorol. Soc. 131, 2131–2150.

Candille, G., Talagrand, O., 2008. Impact of observational error on the validation of ensemble prediction systems. Q. J. R. Meteorol. Soc. 134, 959–971.

Carreau, J., Bengio, Y., 2009. A hybrid Pareto model for asymmetric fat-tailed data: the univariate case. Extremes 12, 53–76.

Carter, G.M., Dallavalle, J.P., Glahn, H.R., 1989. Statistical forecasts based on the National Meteorological Center's numerical weather prediction system. Weather Forecast. 4, 401–412.

Casati, B., 2010. New developments of the intensity-scale technique within the spatial verification methods intercomparison project. Weather Forecast. 25, 113–143.

Casati, B., Wilson, L.J., Stephenson, D.B., Nurmi, P., Ghelli, A., Pocernich, M., Damrath, U., Ebert, E.E., Brown, B.G., Mason, S., 2008. Forecast verification: current status and future directions. Meteorol. Appl. 15, 3–18.

Casella, G., 2008. Statistical Design. Springer. 307 pp.

Casella, G., George, E.I., 1992. Explaining the Gibbs sampler. Am. Stat. 46, 167–174.

Cavanaugh, N.R., Gershunov, A., 2015. Probabilistic tail dependence of intense precipitation on spatiotemporal scale in observations, reanalyses, and GCMs. Clim. Dyn. 45, 2965–2975.

Cavanaugh, N.R., Shen, S.S.P., 2015. The effects of gridding algorithms on the statistical moments and their trends of daily surface air temperature. J. Clim. 28, 9188–9205.

Chacón, J.E., 2015. A population background for nonparametric density-based clustering. Stat. Sci. 30, 518–523.

Chaloulos, G., Lygeros, J., 2007. Effect of wind correlation on aircraft conflict probability. J. Guid. Control. Dyn. 30, 1742–1752.

Chapman, D., Cane, M.A., Henderson, N., Lee, D.E., Chen, C., 2015. A vector autoregressive ENSO prediction model. J. Clim. 28, 8511–8520.

Charney, J.G., Eliassen, A., 1949. A numerical method for predicting the perturbations of the middle latitude westerlies. Tellus 1, 38–54.

Chen, W.Y., 1982. Assessment of southern oscillation sea-level pressure indices. Mon. Weather Rev. 110, 800–807.

Chen, Y.R., Chu, P.-S., 2014. Trends in precipitation extremes and return levels in the Hawaiin Islands under a changing climate. Int. J. Climatol. 34, 3913–3925.

Cheng, W.Y.Y., Steenburgh, W.J., 2007. Strengths and weaknesses of MOS, running-mean bias removal, and Kalman filter techniques for improving model forecasts over the western United States. Weather Forecast. 22, 1304–1318.

Cheng, X., Wallace, J.M., 1993. Cluster analysis of the northern hemisphere wintertime 500- hPa height field: spatial patterns. J. Atmos. Sci. 50, 2674–2696.

Cheng, X., Nitsche, G., Wallace, J.M., 1995. Robustness of low-frequency circulation patterns derived from EOF and rotated EOF analyses. J. Clim. 8, 1709–1713.

Cherry, S., 1996. Singular value decomposition and canonical correlation analysis. J. Clim. 9, 2003–2009.

Cherry, S., 1997. Some comments on singular value decomposition. J. Clim. 10, 1759–1761.

Cheung, K.K.W., 2001. A review of ensemble forecasting techniques with a focus on tropical cyclone forecasting. Meteorol. Appl. 8, 315–332.

Choi, E., Hall, P., 2000. Bootstrap confidence regions computed from autoregressions of arbitrary order. J. R. Stat. Soc. Ser. B Methodol. 62, 461–477.

Chowdhury, J.U., Stedinger, J.R., Lu, L.-H., 1991. Goodness-of-fit tests for regional GEV flood distributions. Water Resour. Res. 27, 1765–1776.

Christensen, H.M., Lock, S.-J., Moroz, I.M., Palmer, T.N., 2017a. Introducing independent patterns into the stochastically perturbed parameterisation tendencies (SPPT) scheme. Q. J. R. Meteorol. Soc. 143, 2168–2181.

Christensen, H.M., Berner, J., Coleman, D., Palmer, T.N., 2017b. Stochastic parameterization and the El Niño–Southern Oscillation. J. Clim. 30, 17–38.

Christiansen, B., 2018. Ensemble averaging and the curse of dimensionality. J. Clim. 31, 1587–1596.

Chu, P.-S., Katz, R.W., 1989. Spectral estimation from time series models with relevance to the southern oscillation. J. Clim. 2, 86–90.

Ciach, G.J., Krajewski, W.F., 1999. On the estimation of radar rainfall error variance. Adv. Water Resour. 22, 585–595.

Clark, T., McCracken, M., 2013. Advances in forecast evaluation. In: Elliot, G., Timmermann, A. (Eds.), Economic Forecasting. North-Holland, 1107–1201.

Clark, M., Gangopadhyay, S., Hay, L., Rajagopalan, B., Wilby, R., 2004. The Schaake shuffle: a method for reconstructing space–time variability in forecasted precipitation and temperature fields. J. Hydrometeorol. 5, 243–262.

Clayton, H.H., 1927. A method of verifying weather forecasts. Bull. Am. Meteorol. Soc. 8, 144–146.

Clayton, H.H., 1934. Rating weather forecasts. Bull. Am. Meteorol. Soc. 15, 279–283.

Clemen, R.T., 1996. Making Hard Decisions: an Introduction to Decision Analysis. Duxbury. 664 pp.

Cleveland, W.S., 1994. The Elements of Graphing Data. Hobart Press, 297 pp.

Coelho, C.A.S., Pezzulli, S., Balmaseda, M., Doblas-Reyes, F.J., Stephenson, D.B., 2004. Forecast calibration and combination: a simple Bayesian approach for ENSO. J. Clim. 17, 1504–1516.

Cohen, J., 1960. A coefficient of agreement for nominal scales. Educ. Psychol. Meas. 20, 213–220.

Coles, S., 2001. An Introduction to Statistical Modeling of Extreme Values. Springer. 208 pp.

Compagnucci, R.H., Richman, M.B., 2008. Can principal component analysis provide atmospheric circulation or teleconnection patterns? Int. J. Climatol. 28, 703–726.

Conover, W.J., 1999. Practical Nonparametric Statistics. Wiley. 584 pp.

Conover, W.J., Iman, R.L., 1981. Rank transformations as a bridge between parametric and nonparametric statistics. Am. Stat. 35, 124–129.

Conradsen, K., Nielsen, L.B., Prahm, L.P., 1984. Review of Weibull statistics for estimation of wind speed distributions. J. Clim. Appl. Meteorol. 23, 1173–1183.

Conte, M., DeSimone, C., Finizio, C., 1980. Post-processing of numerical models: forecasting the maximum temperature at Milano Linate. Rev. Meteor. Aeronautica 40, 247–265.

Cooke, W.E., 1906a. Forecasts and verifications in western Australia. Mon. Weather Rev. 34, 23–24.

Cooke, W.E., 1906b. Weighting forecasts. Mon. Weather Rev. 34, 274–275.

Cooley, D., 2009. Extreme value analysis and the study of climate change. Clim. Chang. 97, 77–83.

Cooley, J.W., Tukey, J.W., 1965. An algorithm for the machine calculation of complex Fourier series. Math. Comput. 19, 297–301.

Crane, R.G., Hewitson, B.C., 2003. Clustering and upscaling of station precipitation records to regional patterns using self-organizing maps (SOMs). Clim. Res. 25, 95–107.

Crochet, P., 2004. Adaptive Kalman filtering of 2-metre temperature and 10-metre wind-speed forecasts in Iceland. Meteorol. Appl. 11, 173–187.

Crutcher, H.L., 1975. A note on the possible misuse of the Kolmogorov-Smirnov test. J. Appl. Meteorol. 14, 1600–1603.

Cruz-Cano, R., Lee, M.-L.T., 2014. Fast regularized canonical correlation analysis. Comput. Stat. Data Anal. 70, 88–100.

Cui, B., Toth, Z., Zhu, Y., Hou, D., 2012. Bias correction for global ensemble forecast. Weather Forecast. 27, 396–410.

Cunnane, C., 1978. Unbiased plotting positions—a review. J. Hydrol. 37, 205–222.

D'Agostino, R.B., 1986. Tests for the normal distribution. In: D'Agostino, R.B., Stephens, M.A. (Eds.), Goodness-of-Fit Techniques. Marcel Dekker, 367–419.

D'Agostino, R.B., Stephens, M.A., 1986. Goodness-of-Fit Techniques. Marcel Dekker. 560 pp.

Dabernig, M., Mayr, G.J., Messner, J.W., Zeileis, A., 2017. Spatial ensemble post-processing with standardized anomalies. Q. J. R. Meteorol. Soc. 143, 909–916.

Dagpunar, J., 1988. Principles of Random Variate Generation. Clarendon Press, Oxford. 228 pp.

Daniel, W.W., 1990. Applied Nonparametric Statistics. Kent. 635 pp.

Davis, R.E., 1976. Predictability of sea level pressure anomalies over the north Pacific Ocean. J. Phys. Oceanogr. 6, 249–266.

Davis, C., Brown, B.G., Bullock, R., 2006a. Object-based verification of precipitation forecasts. Part I: Methodology and application to mesoscale rain areas. Mon. Weather Rev. 134, 1772–1784.

Davis, C., Brown, B.G., Bullock, R., 2006b. Object-based verification of precipitation forecasts. Part II: Application to convective rain systems. Mon. Weather Rev. 134, 1785–1795.

Davis, C.A., Brown, B.G., Bullock, R., Halley-Gotway, J., 2009. The method for object-based diagnostic evaluation (MODE) applied to numerical forecasts from the 2005 NSSL/SPC spring program. Weather Forecast. 24, 1252–1267.

Dawid, A.P., 1984. Present position and potential developments: some personal views: statistical theory: the prequential approach. J. R. Stat. Soc. Ser. A 147, 278–292.

Dawid, A.P., Sebastiani, P., 1999. Coherent dispersion criteria for optimal experimental design. Ann. Stat. 27, 65–81.

De Elia, R., Laprise, R., 2005. Diversity in interpretations of probability: implications for weather forecasting. Mon. Weather Rev. 133, 1129–1143.

De Elia, R., Laprise, R., Denis, B., 2002. Forecasting skill limits of nested, limited-area models: a perfect-model approach. Mon. Weather Rev. 130, 2006–2023.

DeGroot, M.W., Fienberg, S.E., 1982. Assessing probability assessors: calibration and refinement. Stat. Decis. Theory Relat. Top. 1, 291–314.

Delle Monache, L., Eckel, F.A., Rife, D.L., Nagarajan, B., Searight, K., 2013. Probabilistic weather prediction with an analog ensemble. Mon. Weather Rev. 141, 3498–3516.

Delle Monache, L., Hacker, J.P., Zhou, Y., Deng, X., Stull, R.B., 2006. Probabilistic aspects of meteorological and ozone regional ensemble forecasts. J. Geophys. Res. 111,. 15 pp. https://doi.org/10.1029/2005JD006917.

DeLong, E.R., DeLong, D.M., Clarke-Pearson, D.L., 1988. Comparing the areas under two or more correlated receiver operating characteristic curves: a nonparametric approach. Biometrics 44, 837–845.

DelSole, T., Shukla, J., 2006. Specification of wintertime North American surface temperature. J. Clim. 19, 2691–2716.

DelSole, T., Shukla, J., 2009. Artificial skill due to predictor selection. J. Clim. 22, 331–345.

DelSole, T., Tippett, M.K., 2007. Predictability: recent insights from information theory. Rev. Geophys. 45. https://doi.org/10.1029/2006RG000202.

DelSole, T., Tippett, M.K., 2014. Comparing forecast skill. Mon. Weather Rev. 142, 4658–4678.

DelSole, T., Tippett, M.K., 2018. Predictability in a changing climate. Clim. Dyn. 51, 531–545.

Demargne, J., Wu, L., Regonda, S.K., Brown, J.D., Lee, H., He, M., Seo, D.J., Hartman, R., Herr, H.D., Fresch, M., Schaake, J., Zhu, Y., 2014. The science of NOAA's operational hydrologic ensemble forecast service. Bull. Am. Meteorol. Soc. 95, 79–98.

Dempster, A.P., Laird, N.M., Rubin, D.B., 1977. Maximum likelihood from incomplete data via the EM algorithm. J. R. Stat. Soc. Ser. B Methodol. 39, 1–38.

Denis, B., Côté, J., Laprise, R., 2002. Spectral decomposition of two-dimensional atmospheric fields on limited-area domains using the discrete cosine transform (DCT). Mon. Weather Rev. 130, 1812–1829.

Déqué, M., 2003. Continuous variables. In: Jolliffe, I.T., Stephenson, D.B. (Eds.), Forecast Verification, first ed. Wiley, 97–119.

Deser, C., Phillips, A.S., Bourdette, V., Teng, H., 2012. Uncertainty in climate change projections: the role of internal variability. Clim. Dyn. 38, 527–546.

Devine, G.W., Norton, H.J., Barón, A.E., Juarez-Colunga, E., 2018. The Wilcoxon-Mann-Whitney procedure fails as a test of medians. Am. Stat. 72, 278–286.

Devroye, L., 1986. Non-Uniform Random Variate Generation. Springer. 843 pp.

Di Narzo, A.F., Cocchi, D., 2010. A Bayesian hierarchical approach to ensemble weather forecasting. J. R. Stat. Soc.: Ser. C: Appl. Stat. 59, 405–422.

Diebold, F.X., Mariano, R.S., 1995. Comparing predictive accuracy. J. Bus. Econ. Stat. 13, 253–263.

Diebold, F.X., Gunther, T.A., Tay, A.S., 1998. Evaluating density forecasts with applications to financial risk management. Int. Econ. Rev. 39, 863–883.

Director, H., Bornn, L., 2015. Connecting point-level and gridded moments in the analysis of climate data. J. Clim. 28, 3496–3510.

Doblas-Reyes, F.J., Hagedorn, R., Palmer, T.N., 2005. The rationale behind the success of multi-model ensembles in seasonal forecasting—II. Calibration and combination. Tellus A 57, 234–252.

Doolittle, M.H., 1888. Association ratios. Bull. Philos. Soc. Wash. 7, 122–127.

Dorfman, D.D., Alf, E., 1969. Maximum-likelihood estimation of parameters of signal-detection theory and determination of confidence intervals—rating-method data. J. Math. Psychol. 6, 487–496.

Doswell, C.A., 2004. Weather forecasting by humans—heuristics and decision making. Weather Forecast. 19, 1115–1126.

Doswell, C.A., Davies-Jones, R., Keller, D.L., 1990. On summary measures of skill in rare event forecasting based on contingency tables. Weather Forecast. 5, 576–585.

Downton, M.W., Katz, R.W., 1993. A test for inhomogeneous variance in time-averaged temperature data. J. Clim. 6, 2448–2464.

Draper, N.R., Smith, H., 1998. Applied Regression Analysis. Wiley. 706 pp.

Drosdowsky, W., Chambers, L.E., 2001. Near-global sea surface temperature anomalies as predictors of Australian seasonal rainfall. J. Clim. 14, 1677–1687.

Duan, Q., Ajami, N.K., Gao, X., Sorooshian, S., 2007. Multi-model ensemble hydrologic prediction using Bayesian model averaging. Adv. Water Resour. 30, 1371–1386.

Dunn, G.E., 1951. Short-range weather forecasting. In: Malone, T.F. (Ed.), Compendium of Meteorology. American Meteorological Society, 747–765.

Dunsmore, I.R., 1968. A Bayesian approach to calibration. J. R. Stat. Soc. Ser. B Methodol. 30, 396–405.

Durban, J., Watson, G.S., 1971. Testing for serial correlation in least squares regression. III. Biometrika 58 (1), 19.

Eade, R., Smith, D., Scaife, A., Wallace, E., Dunstone, N., Hermanson, L., Robinson, N., 2014. Do seasonal-to-decadal climate predictions underestimate the predictability of the real world? Geophys. Res. Lett. 41, 5620–5628.

Eady, E., 1951. The quantitative theory of cyclone development. In: Malone, T. (Ed.), Compendium of Meteorology. American Meteorological Society, 464–469.

Ebert, E.E., 2008. Fuzzy verification of high-resolution gridded forecasts: a review and proposed framework. Meteorol. Appl. 15, 51–64.

Ebert, E.E., McBride, J.L., 2000. Verification of precipitation in weather systems: determination of systematic errors. J. Hydrol. 239, 179–202.

Ebisuzaki, W., 1997. A method to estimate the statistical significance of a correlation when the data are serially correlated. J. Clim. 10, 2147–2153.

Efron, B., 1979. Bootstrap methods: another look at the jackknife. Ann. Stat. 7, 1–26.

Efron, B., 1982. The Jackknife, the Bootstrap and Other Resampling Plans. Society for Industrial and Applied Mathematics. 92 pp.

Efron, B., 1987. Better bootstrap confidence intervals. J. Am. Stat. Assoc. 82, 171–185.

Efron, B., Gong, G., 1983. A leisurely look at the bootstrap, the jackknife, and cross-validation. Am. Stat. 37, 36–48.

Efron, B., Hastie, T., 2016. Computer Age Statistical Inference. Cambridge University Press, Cambridge. 475 pp.

Efron, B., Tibshirani, R.J., 1993. An Introduction to the Bootstrap. Chapman and Hall. 436 pp.

Efron, B., Gous, A., Kass, R.E., Datta, G.S., Lahiri, P., 2001. Scales of evidence for model selection: Fisher versus Jeffreys. In: Lahiri, P. (Ed.), Model Selection. IMS, Hayward, CA, 208–256.

Ehrendorfer, M., 1994a. The Liouville equation and its potential usefulness for the prediction of forecast skill. Part I: Theory. Mon. Weather Rev. 122, 703–713.

Ehrendorfer, M., 1994b. The Liouville equation and its potential usefulness for the prediction of forecast skill. Part II: Applications. Mon. Weather Rev. 122, 714–728.

Ehrendorfer, M., 1997. Predicting the uncertainty of numerical weather forecasts: a review. Meteorol. Zeitschrift 6, 147–183.

Ehrendorfer, M., 2006. The Liouville equation and atmospheric predictability. In: Palmer, T., Hagedorn, R. (Eds.), Predictability of Weather and Climate. Cambridge University Press, Cambridge, 59–98.

Ehrendorfer, M., Murphy, A.H., 1988. Comparative evaluation of weather forecasting systems: sufficiency, quality, and accuracy. Mon. Weather Rev. 116, 1757–1770.

Ehrendorfer, M., Tribbia, J.J., 1997. Optimal prediction of forecast error covariances through singular vectors. J. Atmos. Sci. 54, 286–313.

Elmore, K.L., 2005. Alternatives to the chi-square test for evaluating rank histograms from ensemble forecasts. Weather Forecast. 20, 789–795.

Elsner, J.B., Bossak, B.H., 2001. Bayesian analysis of U.S. hurricane climate. J. Clim. 14, 4341–4350.

Elsner, J.B., Jagger, T.H., 2004. A hierarchical Bayesian approach to seasonal hurricane modeling. J. Clim. 17, 2813–2827.

Elsner, J.B., Schmertmann, C.P., 1993. Improving extended-range seasonal predictions of intense Atlantic hurricane activity. Weather Forecast. 8, 345–351.

Elsner, J.B., Schmertmann, C.P., 1994. Assessing forecast skill through cross validation. J. Clim. 9, 619–624.

Elsner, J.B., Tsonis, A.A., 1996. Singular Spectrum Analysis, A New Tool in Time Series Analysis. Plenum. 164 pp.

Epstein, E.S., 1962. A Bayesian approach to decision making in applied meteorology. J. Appl. Meteorol. 1, 169–177.

Epstein, E.S., 1966. Quality control for probability forecasts. Mon. Weather Rev. 94, 487–494.

Epstein, E.S., 1969a. The role of initial uncertainties in prediction. J. Appl. Meteorol. 8, 190–198.

Epstein, E.S., 1969b. A scoring system for probability forecasts of ranked categories. J. Appl. Meteorol. 8, 985–987.

Epstein, E.S., 1969c. Stochastic dynamic prediction. Tellus 21, 739–759.

Epstein, E.S., 1985. Statistical Inference and Prediction in Climatology: A Bayesian Approach. Meteorological Monograph 20(42), American Meteorological Society. 199 pp.

Epstein, E.S., 1991. On obtaining daily climatological values from monthly means. J. Clim. 4, 365–368.

Epstein, E.S., Barnston, A.G., 1988. A Precipitation Climatology of Five-Day Periods. NOAA Tech. Report NWS 41, Climate Analysis Center, National Weather Service, Camp Springs, MD. 162 pp.

Epstein, E.S., Fleming, R.J., 1971. Depicting stochastic dynamic forecasts. J. Atmos. Sci. 28, 500–511.

Epstein, E.S., Murphy, A.H., 1965. A note on the attributes of probabilistic predictions and the probability score. J. Appl. Meteorol. 4, 297–299.

Erickson, M.C., Bower, J.B., Dagostaro, V.J., Dallavalle, J.P., Jacks, E., Jensenius Jr., J.S., Su, J.C., 1991. Evaluating the impact of RAFS changes on the NGM-based MOS guidance. Weather Forecast. 6, 142–147.

Evensen, G., 2003. The ensemble Kalman filter: theoretical formulation and practical implementation. Ocean Dyn. 53, 343–367.

Everitt, B.S., Hand, D.J., 1981. Finite Mixture Distributions. Chapman and Hall. 143 pp.

Faes, C., Molenberghs, G., Aerts, M., Verbeke, G., Kenward, M.G., 2009. The effective sample size and an alternative small-sample degrees-of-freedom method. Am. Stat. 63, 389–399.

Farrugia, P.S., Micallef, A., 2006. Comparative analysis of estimators for wind direction standard deviation. Meteorol. Appl. 13, 29–41.

Feldmann, K., Scheuerer, M., Thorarinsdottir, T.L., 2015. Spatial postprocessing of ensemble forecasts for temperature using non-homogeneous Gaussian regression. Mon. Weather Rev. 143, 955–971.

Feller, W., 1970. An Introduction to Probability Theory and its Applications. Wiley. 509 pp.

Ferranti, L., Corti, S., 2011. New clustering products. ECMWF Newsl. 127, 6–11. www.ecmwf.int/sites/default/files/elibrary/2011/14596-newsletter-no127-spring-2011.pdf.

Ferro, C.A.T., 2017. Measuring forecast performance in the presence of observation error. Q. J. R. Meteorol. Soc. 143, 2665–2676.

Ferro, C.A.T., Fricker, T.E., 2012. A bias-corrected decomposition of the Brier score. Q. J. R. Meteorol. Soc. 138, 1954–1960.

Ferro, C.A.T., Stephenson, D.B., 2011. Extremal dependence indices: improved verification measures for deterministic forecasts of rare binary events. Weather Forecast. 26, 699–713.

Ferro, C.A.T., Richardson, D.S., Weigel, A.P., 2008. On the effect of ensemble size on the discrete and continuous ranked probability scores. Meteorol. Appl. 15, 19–24.

Filliben, J.J., 1975. The probability plot correlation coefficient test for normality. Technometrics 17, 111–117.

Finley, J.P., 1884. Tornado prediction. Am. Meteorol. J. 1, 85–88.

Fisher, R.A., 1925. Statistical Methods for Research Workers. Oliver & Boyd. 239 pp.

Fisher, R.A., 1935. The Design of Experiments. Oliver & Boyd. 252 pp.

Fisher, M., 2006. "Wavelet" J_b–A new way to model the statistics of background errors. ECMWF Newsl. 106, 23–28.

Flowerdew, J., 2014. Calibrating ensemble reliability whilst preserving spatial structure. Tellus A. 66, https://doi.org/10.3402/tellusa.v66.22662. 20 pp.

Flueck, J.A., 1987. A study of some measures of forecast verification. In: Preprints, Tenth Conference on Probability and Statistics in Atmospheric Sciences. American Meteorological Society, 69–73.

Folland, C., Anderson, C., 2002. Estimating changing extremes using empirical ranking methods. J. Clim. 15, 2954–2960.

Fortin, V., Favre, A.-C., Said, M., 2006. Probabilistic forecasting from ensemble prediction systems: improving upon the best-member method by using a different weight and dressing kernel for each member. Q. J. R. Meteorol. Soc. 132, 1349–1369.

Fortin, V., Abaza, M., Anctil, F., Turcotte, R., 2014. Why should ensemble spread match the RMSE of the ensemble mean? J. Hydrometeorol. 15, 1708–1713.

Fortin, V., Abaza, M., Anctil, F., Turcotte, R., 2015. Corrigendum. J. Hydrometeorol. 16, 484.

Foufoula-Georgiou, E., Lettenmaier, D.P., 1987. A Markov renewal model for rainfall occurrences. Water Resour. Res. 23, 875–884.

Fovell, R.G., Fovell, M.-Y., 1993. Climate zones of the conterminous United States defined using cluster analysis. J. Clim. 6, 2103–2135.

Fowler, T., Gotway, J.H., Newman, K., Jensen, T., Brown, B., Bullock, R., 2018. Model Evaluation Tools Version 7.0 (METv7.0) User's Guide. Developmental Testbed Center, Boulder CO. dtcenter.org/met/users/docs/users_guide/MET_Users_Guide_v7.0.pdf. 408 pp.

Fraley, C., Raftery, A.E., 2002. Model-based clustering, discriminant analysis, and density estimation. J. Am. Stat. Assoc. 97, 611–631.

Fraley, C., Raftery, A.E., Gneiting, T., 2010. Calibrating multimodel forecast ensembles with exchangeable and missing members using Bayesian model averaging. Mon. Weather Rev. 138, 190–202.

Francis, P.E., Day, A.P., Davis, G.P., 1982. Automated temperature forecasting, an application of Model Output Statistics to the Meteorological Office numerical weather prediction model. *Meteorological Magazine*. 111, 73–87.

Frenkel, Y., Majda, A.J., Khouider, B., 2012. Using the stochastic multicloud model to improve tropical convective parameterization: a paradigm example. J. Atmos. Sci. 69, 1080–1105.

Friederichs, P., Hense, A., 2003. Statistical inference in canonical correlation analyses exemplified by the influence of North Atlantic SST on European climate. J. Clim. 16, 522–534.

Friederichs, P., Thorarinsdottir, T.L., 2012. Forecast verification for extreme value distributions with an application to probabilistic peak wind prediction. Environmetrics 23, 579–594.

Friedman, R.M., 1989. Appropriating the Weather: Vilhelm Bjerknes and the Construction of a Modern Meteorology. Cornell University Press. 251 pp.

Fuller, W.A., 1996. Introduction to Statistical Time Series. Wiley. 698 pp.

Furrer, E.M., Katz, R.W., 2008. Improving the simulation of extreme precipitation events by stochastic weather generators. Water Resour. Res. 44. https://doi.org/10.1029/2008WR007316.

Gabriel, K.R., 1971. The biplot—graphic display of matrices with application to principal component analysis. Biometrika 58, 453–467.

Galanis, G., Anadranistakis, M., 2002. A one-dimensional Kalman filter for the correction of near surface temperature forecasts. Meteorol. Appl. 9, 437–441.

Gandin, L.S., Murphy, A.H., 1992. Equitable skill scores for categorical forecasts. Mon. Weather Rev. 120, 361–370.

Gandin, L.S., Murphy, A.H., Zhukovsky, E.E., 1992. Economically optimal decisions and the value of meteorological information. In: Preprints, 5th International Meeting on Statistical Climatology, 22–26 June, Toronto, Canada. 1992, J64–J71.

Garratt, J.R., Pielke Sr., R.A., Miller, W.F., Lee, T.J., 1990. Mesoscale model response to random, surface-based perturbations—a sea-breeze experiment. Bound.-Layer Meteorol. 52, 313–334.

Garthwaite, P.H., Kadane, J.B., O'Hagan, A., 2005. Statistical methods for eliciting probability distributions. J. Am. Stat. Assoc. 100, 680–700.

Gebetsberger, M., Messner, J.W., Mayr, G.J., Zeileis, A., 2017a. Estimation methods for non- homogeneous regression—minimum CRPS vs. maximum likelihood. Geophys. Res. Abstr. 19. EGU 2017-5573.

Gebetsberger, M., Messner, J.W., Mayr, G.J., Zeileis, A., 2017b. Fine-tuning nonhomogeneous regression for probabilistic precipitation forecasts: unanimous predictions, heavy tails, and link functions. Mon. Weather Rev. 145, 4693–4708.

Geer, A.J., 2016. Significance of changes in medium-range forecast scores. Tellus 68. https://doi.org/10.3402/tellusa.v68.30229. 21 pp.

Gel, Y., Raftery, A.E., Gneiting, T., 2004. Calibrated probabilistic mesoscale weather field forecasting: The geostatistical output perturbation method. J. Am. Stat. Assoc. 99, 575–583.

Gerrity Jr., J.P., 1992. A note on Gandin and Murphy's equitable skill score. Mon. Weather Rev. 120, 2709–2712.

Ghil, M., Allen, M.R., Dettinger, M.D., Ide, K., Kondrashov, D., Mann, M.E., Robertson, A.W., Saunders, A., Tian, Y., Varadi, F., Yiou, P., 2002. Advanced spectral methods for climatic time series. Rev. Geophys. 40, 1003–1044. https://doi.org/10.1029/2000RG000092.

Gilbert, G.K., 1884. Finley's tornado predictions. Am. Meteorol. J. 1, 166–172.

Gill, J., Rubiera, J., Martin, C., Cacic, I., Mylne, K., Chen, D., Gu, J., Tang, X., Yamaguchi, M., Foamouhoue, A.K., Poolman, E., Guiney, J., 2008. Guidelines on Communicating Forecast Uncertainty. World Meteorological Organization. WMO/TD No.1422, 22 pp.

Gilleland, E., 2013. Testing competing precipitation forecasts accurately and efficiently: the spatial prediction comparison test. Mon. Weather Rev. 141, 340–355.

Gilleland, E., Ahijevych, D., Brown, B.G., Casati, B., Ebert, E.E., 2009. Intercomparison of spatial forecast verification methods. Weather Forecast. 24, 1416–1430.

Gilleland, E., Hering, A.S., Fowler, T.L., Brown, B.G., 2018. Testing the tests: what are the impacts of incorrect assumptions when applying confidence intervals or hypothesis tests to compare competing forecasts? Mon. Weather Rev. 146, 1685–1703.

Gillies, D., 2000. Philosophical Theories of Probability. Routledge. 223 pp.

Gilman, D.L., Fuglister, F.J., Mitchell Jr., J.M., 1963. On the power spectrum of "red noise". J. Atmos. Sci. 20, 182–184.

Glahn, H.R., 1985. Statistical weather forecasting. In: Murphy, A.H., Katz, R.W. (Eds.), Probability, Statistics, and Decision Making in the Atmospheric Sciences. Westview Press, Boulder, CO, 289–335.

Glahn, H.R., 2004. Discussion of "verification concepts in forecast verification: a practitioner's guide in atmospheric science". Weather Forecast. 19, 769–775.

Glahn, B., 2014. A nonsymmetric logit model and grouped predictand category development. Mon. Weather Rev. 142, 2991–3002.

Glahn, H.R., Jorgensen, D.L., 1970. Climatological aspects of the Brier p-score. Mon. Weather Rev. 98, 136–141.

Glahn, H.R., Lowry, D.A., 1972. The use of Model Output Statistics (MOS) in objective weather forecasting. J. Appl. Meteorol. 11, 1203–1211.

Glahn, B., Gilbert, K., Cosgrove, R., Ruth, D.P., Sheets, K., 2009a. The gridding of MOS. Weather Forecast. 24, 520–529.

Glahn, B., Peroutka, M., Wiedenfeld, J., Wagner, J., Zylstra, G., Schuknecht, B., 2009b. MOS uncertainty estimates in an ensemble framework. Mon. Weather Rev. 137, 246–268.

Gleeson, T.A., 1967. Probability predictions of geostrophic winds. J. Appl. Meteorol. 6, 355–359.

Gleeson, T.A., 1970. Statistical-dynamical predictions. J. Appl. Meteorol. 9, 333–344.

Gneiting, T., 2011a. Making and evaluating point forecasts. J. Am. Stat. Assoc. 106, 746–762.

Gneiting, T., 2011b. Quantiles as optimal point forecasts. Int. J. Forecast. 27, 197–207.

Gneiting, T., Raftery, A.E., 2007. Strictly proper scoring rules, prediction, and estimation. J. Am. Stat. Assoc. 102, 359–378.

Gneiting, T., Ranjan, R., 2011. Comparing density forecasts using threshold- and quantile-weighted scoring rules. J. Bus. Econ. Stat. 29, 411–422.

Gneiting, T., Raftery, A.E., Westveld III, A.H., Goldman, T., 2005. Calibrated probabilistic forecasting using ensemble model output statistics and minimum CRPS estimation. Mon. Weather Rev. 133, 1098–1118.

Gneiting, T., Larson, K., Westrick, K., Genton, M.G., Aldrich, E., 2006. Calibrated probabilistic forecasting at the Stateline wind energy center: the regime-switching space-time method. J. Am. Stat. Assoc. 101, 968–979.

Gneiting, T., Balabdaoui, F., Raftery, A.E., 2007. Probabilistic forecasts, calibration and sharpness. J. R. Stat. Soc. Ser. B Methodol. 69, 243–268.

Gneiting, T., Stanberry, L.I., Grimit, E.P., Held, L., Johnson, N.A., 2008. Assessing probabilistic forecasts of multivariate quantities, with an application to ensemble predictions of surface winds. Test 17, 211–235.

Gober, M., Zsoter, E., Richardson, D.S., 2008. Could a perfect model ever satisfy a naive forecaster? On grid box mean versus point verification. Meteorol. Appl. 15, 359–365.

Godfrey, C.M., Wilks, D.S., Schultz, D.M., 2002. Is the January Thaw a statistical phantom? Bull. Am. Meteorol. Soc. 83, 53–62.

Goldsmith, B.S., 1990. NWS verification of precipitation type and snow amount forecasts during the AFOS era. NOAA Technical Memorandum NWS FCST 33. National Weather Service. 28 pp.

Golub, G.H., van Loan, C.F., 1996. Matrix Computations. Johns Hopkins Press. 694 pp.

Gombos, D., Hansen, J.A., Du, J., McQueen, J., 2007. Theory and applications of the minimum spanning tree rank histogram. Mon. Weather Rev. 135, 1490–1505.

Gong, X., Richman, M.B., 1995. On the application of cluster analysis to growing season precipitation data in North America east of the Rockies. J. Clim. 8, 897–931.

Good, I.J., 1952. Rational decisions. J. R. Stat. Soc. Ser. A 14, 107–114.

Good, P., 2000. Permutation Tests. Springer. 270 pp.

Goodall, C., 1983. M-Estimators of location: an outline of the theory. In: Hoaglin, D.C., Mosteller, F., Tukey, J.W. (Eds.), Understanding Robust and Exploratory Data Analysis. Wiley, New York, 339–403.

Gordon, N.D., 1982. Comments on "verification of fixed-width credible interval temperature forecasts" Bull. Am. Meteorol. Soc. 63, 325.

Gorgas, T., Dorninger, M., 2012. Concepts for a pattern-oriented analysis ensemble based on observational uncertainties. Q. J. R. Meteorol. Soc. 138, 769–784.

Graedel, T.E., Kleiner, B., 1985. Exploratory analysis of atmospheric data. In: Murphy, A.H., Katz, R.W. (Eds.), Probability, Statistics, and Decision Making in the Atmospheric Sciences. Westview Press, Boulder, CO, 1–43.

Granger, C.W.J., Pesaran, M.H., 2000. Economic and statistical measures of forecast accuracy. J. Forecast. 19, 537–560.

Gray, W.M., 1990. Strong association between West African rainfall and U.S. landfall of intense hurricanes. Science 249, 1251–1256.

Greenwood, J.A., Durand, D., 1960. Aids for fitting the gamma distribution by maximum likelihood. Technometrics 2, 55–65.

Griffis, V.W., Stedinger, J.R., 2007. Log-Pearson Type 3 distribution and its application in flood frequency analysis. I: Distribution characteristics. J. Hydraul. Eng. 12, 482–491.

Grimit, E.P., Mass, C.F., 2002. Initial results of a mesoscale short-range ensemble forecasting system over the Pacific Northwest. Weather Forecast. 17, 192–205.

Grimit, E.P., Gneiting, T., Berrocal, V.J., Johnson, N.A., 2006. The continuous ranked probability score for circular variables and its application to mesoscale forecast ensemble verification. Q. J. R. Meteorol. Soc. 132, 1–17.

Gringorten, I.I., 1949. A study in objective forecasting. Bull. Am. Meteorol. Soc. 30, 10–15.

Gringorten, I.I., 1967. Verification to determine and measure forecasting skill. J. Appl. Meteorol. 6, 742–747.

Groth, A., Ghil, M., 2015. Monte Carlo singular spectrum analysis (SSA) revisited: detecting oscillator clusters in multivariate datasets. J. Clim. 28, 7873–7893.

Grounds, M.A., LeClerc, J.E., Joslyn, S., 2018. Expressing flood likelihood: return period versus probability. Wea. Clim. Soc. 10, 5–17.

Gumbel, E.J., 1958. Statistics of Extremes. Columbia University Press. 375 pp.

Guttman, N.B., 1999. Accepting the standardized precipitation index: a calculation algorithm. J. Am. Water Resour. Assoc. 35, 311–322.

Hagedorn, R., 2008. Using the ECMWF reforecast data set to calibrate EPS reforecasts. ECMWF Newsl. 117, 8–13.

Hagedorn, R., Smith, L.A., 2009. Communicating the value of probabilistic forecasts with weather roulette. Meteorol. Appl. 16, 143–155.

Hagedorn, R., Hamill, T.M., Whitaker, J.S., 2008. Probabilistic forecast calibration using ECMWF and GFS ensemble reforecasts. Part I: Two-meter temperatures. Mon. Weather Rev. 136, 2608–2619.

Haines, K., Hannachi, A., 1995. Weather regimes in the Pacific from a GCM. J. Atmos. Sci. 52, 2444–2462.

Hall, T.M., Jewson, S., 2008. Comparison of local and basinwide methods for risk assessment of tropical cyclone landfall. J. Appl. Meteorol. Climatol. 47, 361–367.

Hall, P., Wilson, S.R., 1991. Two guidelines for bootstrap hypothesis testing. Biometrics 47, 757–762.

Hamed, K.H., 2009. Exact distribution of the Mann-Kendall trend test statistic for persistent data. J. Hydrol. 365, 86–94.

Hamill, T.M., 1999. Hypothesis tests for evaluating numerical precipitation forecasts. Weather Forecast. 14, 155–167.

Hamill, T.M., 2001. Interpretation of rank histograms for verifying ensemble forecasts. Mon. Weather Rev. 129, 550–560.

Hamill, T.M., 2006. Ensemble-based atmospheric data assimilation: a tutorial. In: Palmer, T.N., Hagedorn, R. (Eds.), Predictability of Weather and Climate. Cambridge University Press, Cambridge, 124–156.

Hamill, T.M., 2007. Comments on "Calibrated surface temperature forecasts from the Canadian ensemble prediction system using Bayesian Model Averaging". Mon. Weather Rev. 135, 4226–4230.

Hamill, T.M., Bates, G.T., Whitaker, J.S., Murray, D.R., Fiorino, M., Galarneau, T.J., Zhu, Y., Lapenta, W., 2013. NOAA's second-generation global medium-range ensemble reforecast dataset. Bull. Am. Meteorol. Soc. 94, 1553–1565.

Hamill, T.M., Colucci, S.J., 1997. Verification of Eta-RSM short-range ensemble forecasts. Mon. Weather Rev. 125, 1312–1327.

Hamill, T.M., Colucci, S.J., 1998. Evaluation of Eta–RSM ensemble probabilistic precipitation forecasts. Mon. Weather Rev. 126, 711–724.

Hamill, T.M., Juras, J., 2006. Measuring forecast skill: is it real skill or is it the varying climatology? Q. J. R. Meteorol. Soc. 132, 2905–2923.

Hamill, T.M., Scheuerer, M., Bates, G.T., 2015. Analog probabilistic precipitation forecasts using GEFS reforecasts and climatology-calibrated precipitation analyses. Mon. Weather Rev. 143, 3300–3309.

Hamill, T.M., Swinbank, R., 2015. Stochastic forcing, ensemble prediction systems, and TIGGE. In: Seamless Prediction of the Earth System: From Minutes to Months. World Meteorological Organization, 187–212.

Hamill, T.M., Whitaker, J.S., 2006. Probabilistic quantitative precipitation forecasts based on reforecast analogs: theory and application. Mon. Weather Rev. 134, 3209–3229.

Hamill, T.M., Whitaker, J.S., Mullen, S.L., 2006. Reforecasts: an important new dataset for improving weather predictions. Bull. Am. Meteorol. Soc. 87, 33–46.

Hamill, T.M., Whitaker, J.S., Wei, X., 2004. Ensemble re-forecasting: improving medium- range forecast skill using retrospective forecasts. Mon. Weather Rev. 132, 1434–1447.

Hamill, T.M., Wilks, D.S., 1995. A probabilistic forecast contest and the difficulty of assessing short-range uncertainty. Weather Forecast. 10, 620–631.

Han, F., Szunyogh, I., 2016. A morphing-based technique for the verification of precipitation forecasts. Mon. Weather Rev. 144, 295–313.

Hanley, J.A., McNeil, B.J., 1983. A method of comparing the areas under receiver operating characteristic curves derived from the same cases. Radiology 148, 839–843.

Hannachi, A., 1997. Low-frequency variability in a GCM: three dimensional flow regimes and their dynamics. J. Clim. 10, 1357–1379.

Hannachi, A., Jolliffe, I.T., Stephenson, D.B., 2007. Empirical orthogonal functions and related techniques in atmospheric science: a review. Int. J. Climatol. 27, 1119–1152.

Hannachi, A., O'Neill, A., 2001. Atmospheric multiple equilibria and non-Gaussian behavior in model simulations. Q. J. R. Meteorol. Soc. 127, 939–958.

Hansen, J.A., 2002. Accounting for model error in ensemble-based state estimation and forecasting. Mon. Weather Rev. 130, 2373–2391.

Hanssen, A.W., Kuipers, W.J.A., 1965. On the relationship between the frequency of rain and various meteorological parameters. Meded. Verh. 81, 2–15.

Harper, K., Uccellini, L.W., Kalnay, E., Carey, K., Morone, L., 2007. 50th anniversary of operational numerical weather prediction. Bull. Am. Meteorol. Soc. 88, 639–650.

Harrison, M., 2005. The development of seasonal and inter-annual climate forecasting. Clim. Chang. 70, 201–220.

Harrison, M.S.J., Palmer, T.N., Richardson, D.S., Buizza, R., 1999. Analysis and model dependencies in medium-range ensembles: two transplant case-studies. Q. J. R. Meteorol. Soc. 125, 2487–2515.

Harter, H.L., 1984. Another look at plotting positions. Commun. Stat. Theory Methods 13, 1613–1633.

Hasselmann, K., 1976. Stochastic climate models. Part I: Theory. Tellus 28, 474–485.

Hastenrath, S., Sun, L., Moura, A.D., 2009. Climate prediction for Brazil's Nordeste by empirical and numerical modeling methods. Int. J. Climatol. 29, 921–926.

Hastie, T., Tibshirani, R., Wainright, M., 2015. Statistical Learning with Sparsity: The Lasso and Generalizations. CRC Press. 351 pp.

Hawkins, E., Smith, R.S., Gregory, J.M., Stainforth, D.A., 2016. Irreducible uncertainty in near-term climate projections. Clim. Dyn. 46, 3807–3819.

Hayashi, Y., 1986. Statistical interpretations of ensemble-time mean predictability. J. Meteorol. Soc. Jpn. 64, 167–181.

He, Y., Monahan, A.H., Jones, C.G., Dai, A., Biner, S., Caya, D., Winger, K., 2010. Probability distributions of land surface wind speeds over North America. J. Geophys. Res. 115. https://doi.org/10.1029/2008JD010708.

Healy, M.J.R., 1988. Glim: An Introduction. Clarendon Press, Oxford. 130 pp.

Heidke, P., 1926. Berechnung des Erfolges und der Güte der Windstärkevorhersagen im Sturmwarnungsdienst. Geogr. Ann 8, 301–349.

Hemri, S., Haiden, T., Pappenberger, F., 2016. Discrete postprocessing of total cloud cover ensemble forecasts. Mon. Weather Rev. 144, 2565–2577.

Hemri, S., Lisniak, D., Klein, B., 2015. Multivariate postprocessing techniques for probabilistic hydrological forecasting. Water Resour. Res. 51, 7436–7451.

Hemri, S., Fundel, F., Zappa, M., 2013. Simultaneous calibration of ensemble river flow predictions over an entire range of lead times. Water Resour. Res. 49, 6744–6755.

Hemri, S., Scheuerer, M., Pappenberger, F., Bogner, K., Haiden, T., 2014. Trends in the predictive performance of raw ensemble weather forecasts. Geophys. Res. Lett. 41, 9197–9205.

Heo, J.-H., Kho, Y.W., Shin, H., Kim, S., Kim, T., 2008. Regression equations of probability plot correlation coefficient test statistics from several probability distributions. J. Hydrol. 355, 1–15.

Hering, A.S., Genton, M.C., 2011. Comparing spatial predictions. Technometrics 53, 414–425.

Herman, G.L., Schumacher, R.S., 2018. Money doesn't grow on trees, but forecasts to: forecasting extreme precipitation with random forests. Mon. Weather Rev. 146, 1571–1600.

Hersbach, H., 2000. Decomposition of the continuous ranked probability score for ensemble prediction systems. Weather Forecast. 15, 559–570.

Hilliker, J.L., Fritsch, J.M., 1999. An observations-based statistical system for warm-season hourly probabilistic precipitation forecasts of low ceiling at the San Francisco international airport. J. Appl. Meteorol. 38, 1692–1705.

Hingray, B., Mezghana, A., Buishand, T.A., 2007. Development of probability distributions for regional climate change from uncertain global mean warming and an uncertain scaling relationship. Hydrol. Earth Syst. Sci. 11, 1097–1114.

Hinkley, D., 1977. On quick choice of power transformation. Appl. Stat. 26, 67–69.

Hintze, J.L., Nelson, R.D., 1998. Violin plots: a box plot-density trace synergism. Am. Stat. 52, 181–184.

Hirsch, R.M., Slack, J.R., Smith, R.A., 1982. Techniques of trend analysis for monthly water quality data. Water Resour. Res. 18, 107–121.

Hlinka, J., Hartman, D., Vejmelka, M., Novotná, D., Palus, M., 2014. Non-linear dependence and teleconnections in climate data: sources, relevance, nonstationarity. Clim. Dyn. 42, 1873–1886.

Hodyss, D., Satterfield, E., McLay, J., Hamill, T.M., Scheuerer, M., 2016. Inaccuracies with multi-model post-processing methods involving weighted, regression-corrected forecasts. Mon. Weather Rev. 144, 1649–1668.

Hoerl, A.E., Kennard, R.W., 1970. Ridge regression: biased estimation for nonorthogonal problems. Technometrics 12, 55–67.

Hoffman, R.N., Liu, Z., Louis, J.-F., Grassotti, C., 1995. Distortion representation of forecast errors. Mon. Weather Rev. 123, 2758–2770.

Hogan, R.J., Ferro, C.A.T., Jolliffe, I.T., Stephenson, D.B., 2010. Equitability revisited: why the "equitable threat score" is not equitable. Weather Forecast. 25, 710–726.

Hogan, R.J., Mason, I.B., 2012. Deterministic forecasts of binary events. In: Jolliffe, I.T., Stephenson, D.B. (Eds.), Forecast Verification: A Practitioner's Guide in Atmospheric Science, second, ed. Wiley-Blackwell, 31–59.

Hollingsworth, A., Arpe, K., Tiedtke, M., Capaldo, M., Savijärvi, H., 1980. The performance of a medium range forecast model in winter—impact of physical parameterizations. Mon. Weather Rev. 108, 1736–1773.

Homleid, M., 1995. Diurnal corrections of short-term surface temperature forecasts using the Kalman filter. Weather Forecast. 10, 689–707.

Horel, J.D., 1981. A rotated principal component analysis of the interannual variability of the Northern Hemisphere 500 mb height field. Mon. Weather Rev. 109, 2080–2902.

Hosking, J.R.M., 1990. L-moments: analysis and estimation of distributions using linear combinations of order statistics. J. R. Stat. Soc. Ser. A 52, 105–124.

Hosking, J.R.M., Wallis, J.R., 1987. Parameter and quantile estimation for the generalized Pareto distribution. Technometrics 29, 339–349.

Hothorn, T., Leisch, F., Zeileis, A., Hornik, K., 2005. The design and analysis of benchmark experiments. J. Comput. Graph. Stat. 14, 675–699.

Houtekamer, P.L., Lefaivre, L., Derome, J., Ritchie, H., Mitchell, H.L., 1996. A system simulation approach to ensemble prediction. Mon. Weather Rev. 124, 1225–1242.

Houtekamer, P.L., Mitchell, H.L., 2005. Ensemble Kalman filtering. Q. J. R. Meteorol. Soc. 131, 3269–3289.

Houtekamer, P.L., Mitchell, H.L., Deng, X., 2009. Model error representation in an operational ensemble Kalman filter. Mon. Weather Rev. 137, 2126–2143.

Hsieh, W.W., 2009. Machine Learning Methods in the Environmental Sciences. Cambridge University Press, Cambridge. 349 pp.

Hsu, W.-R., Murphy, A.H., 1986. The attributes diagram: a geometrical framework for assessing the quality of probability forecasts. Int. J. Forecast. 2, 285–293.

Hu, Q., 1997. On the uniqueness of the singular value decomposition in meteorological applications. J. Clim. 10, 1762–1766.

Hurvich, C.M., Tsai, C.-L., 1989. Regression and time series model selection in small samples. Biometrika 76, 297–307.

Huth, R., Pokorná, L., 2004. Parametric versus non-parametric estimates of climatic trends. Theor. Appl. Climatol. 77, 107–112.

Hyndman, R.J., Fan, Y., 1996. Sample quantiles in statistical packages. Am. Stat. 50, 361–365.

Hyvärinen, O., 2014. A probabilistic derivation of the Heidke skill score. Weather Forecast. 29, 177–181.

Iglewicz, B., 1983. Robust scale estimators and confidence intervals for location. In: Hoaglin, D.C., Mosteller, F., Tukey, J.W. (Eds.), Understanding Robust and Exploratory Data Analysis. Wiley, 404–431.

Imkeller, P., Monahan, A., 2002. Conceptual stochastic climate models. Stochastic Dyn. 2, 311–326.

Imkeller, P., von Storch, J.-S. (Eds.), 2001. Stochastic Climate Models. Birkhäuser. 398 pp.

Ivarsson, K.-I., Joelsson, R., Liljas, E., Murphy, A.H., 1986. Probability forecasting in Sweden: some results of experimental and operational programs at the Swedish Meteorological and Hydrological Institute. Weather Forecast. 1, 136–154.

Jacks, E., Bower, J.B., Dagostaro, V.J., Dallavalle, J.P., Erickson, M.C., Su, J., 1990. New NGM-based MOS guidance for maximum/minimum temperature, probability of precipitation, cloud amount, and surface wind. Weather Forecast. 5, 128–138.

Jagger, T.H., Elsner, J.B., 2009. Modeling tropical cyclone intensity with quantile regression. Int. J. Climatol. 29, 1351–1361.

Janson, S., Vegelius, J., 1981. Measures of ecological association. Oecologia 49, 371–376.

Jarman, A.S., Smith, L.A., 2018. Quantifying the predictability of a predictant: demostrating the diverse roles of serial dependence in the estimation of forecast skill. Q. J. R. Meteorol. Soc. https://doiorg/10.1002/qj.3384.

Jenkins, G.M., Watts, D.G., 1968. Spectral Analysis and its Applications. Holden-Day, San Francisco, CA. 523 pp.

Jewson, S., Brix, A., Ziehmann, C., 2004. A new parametric model for the assessment and calibration of medium-range ensemble temperature forecasts. Atmos. Sci. Lett. 5, 96–102.

Johnson, C., Bowler, N., 2009. On the reliability and calibration of ensemble forecasts. Mon. Weather Rev. 137, 1717–1720.

Johnson, M.E., 1987. Multivariate Statistical Simulation. Wiley. 230 pp.

Johnson, N.L., Kotz, S., 1972. Distributions in Statistics. Continuous Multivariate Distributions, vol. 4. Wiley, New York. 333 pp.

Johnson, N.L., Kotz, S., Balakrishnan, N., 1994. Continuous Univariate Distributions. vol. 1. Wiley. 756 pp.

Johnson, N.L., Kotz, S., Balakrishnan, N., 1995. Continuous Univariate Distributions. vol. 2. Wiley. 719 pp.

Johnson, N.L., Kotz, S., Kemp, A.W., 1992. Univariate Discrete Distributions. Wiley. 565 pp.

Johnson, R.A., Wichern, D.W., 2007. Applied Multivariate Statistical Analysis, sixth ed. Prentice Hall. 773 pp.

Johnson, S.R., Holt, M.T., 1997. The value of weather information. In: Katz, R.W., Murphy, A.H. (Eds.), Economic Value of Weather and Climate Forecasts. Cambridge University Press, Cambridge, 75–107.

Jolliffe, I.T., 1972. Discarding variables in a principal component analysis, I: Artificial data. Appl. Stat. 21, 160–173.

Jolliffe, I.T., 1987. Rotation of principal components: some comments. Int. J. Climatol. 7, 507–510.

Jolliffe, I.T., 1989. Rotation of ill-defined principal components. Appl. Stat. 38, 139–147.

Jolliffe, I.T., 1995. Rotation of principal components: choice of normalization constraints. J. Appl. Stat. 22, 29–35.

Jolliffe, I.T., 2002. Principal Component Analysis, second ed. Springer. 487 pp.

Jolliffe, I.T., 2007. Uncertainty and inference for verification measures. Weather Forecast. 22, 637–650.

Jolliffe, I.T., 2008. The impenetrable hedge: a note on propriety, equitability, and consistency. Meteorol. Appl. 15, 25–29.

Jolliffe, I.T., Jones, B., Morgan, B.J.T., 1986. Comparison of cluster analyses of the English personal social services authorities. J. R. Stat. Soc. Ser. A 149, 254–270.

Jolliffe, I.T., Primo, C., 2008. Evaluating rank histograms using decompositions of the chi-square test statistic. Mon. Weather Rev. 136, 2133–2139.

Jolliffe, I.T., Stephenson, D.B., 2005. Comments on "discussion of verification concepts in forecast verification: a practitioner's guide in atmospheric science". Weather Forecast. 20, 796–800.

Jolliffe, I.T., Stephenson, D.B., 2012a. Forecast Verification: A Practitioner's Guide in Atmospheric Science, second ed. Wiley-Blackwell. 274 pp.

Jolliffe, I.T., Stephenson, D.B., 2012b. Epilogue: new directions in forecast verification. In: Jolliffe, I.T., Stephenson, D.B. (Eds.), Forecast Verification: A Practitioner's Guide in Atmospheric Science, second ed. Wiley-Blackwell, 221–230.

Jolliffe, I.T., Trendafilov, N.T., Uddin, M., 2003. A modified principal component technique based on the LASSO. J. Comput. Graph. Stat. 12, 531–547.

Jones, R.H., 1975. Estimating the variance of time averages. J. Appl. Meteorol. 14, 159–163.

Jordan, A., Krüger, F., Lerch, S., 2017. Evaluating probabilistic forecasts with the R package scoringRules. arXiv:1709.04743.

Joslyn, S., Savelli, S., 2010. Communicating forecast uncertainty: public perception of weather forecast uncertainty. Meteorol. Appl. 17, 180–195.

Judd, K., Reynolds, C.A., Rosmond, T.E., Smith, L.A., 2008. The geometry of model error. J. Atmos. Sci. 65, 1749–1772.

Junk, C., Delle Monache, L., Alessandrini, S., 2015. Analog-based ensemble model output statistics. Mon. Weather Rev. 143, 2909–2917.

Jupp, T.E., Lowe, R., Coelho, C.A.S., Stephenson, D.B., 2012. On the visualization, verification and recalibration of ternary probabilistic forecasts. Philos. Trans. R. Soc. Lond. A 370, 1100–1120.

Juras, J., 2000. Comments on "Probabilistic predictions of precipitation using the ECMWF ensemble prediction system". Weather Forecast. 15, 365–366.

Justus, C.G., Hargraves, W.R., Mikhail, A., Graber, D., 1978. Methods for estimating wind speed frequency distributions. J. Appl. Meteorol. 17, 350–353.

Kaiser, H.F., 1958. The varimax criterion for analytic rotation in factor analysis. Psychometrika 23, 187–200.

Kalkstein, L.S., Tan, G., Skindlov, J.A., 1987. An evaluation of three clustering procedures for use in synoptic climatological classification. J. Clim. Appl. Meteorol. 26, 717–730.

Kalnay, E., 2003. Atmospheric Modeling, Data Assimilation and Predictability. Cambridge University Press, Cambridge. 341 pp.

Kalnay, E., Dalcher, A., 1987. Forecasting the forecast skill. Mon. Weather Rev. 115, 349–356.

Kalnay, E., Kanamitsu, M., Baker, W.E., 1990. Global numerical weather prediction at the National Meteorological Center. Bull. Am. Meteorol. Soc. 71, 1410–1428.

Kann, A., Wittmann, C., Wang, Y., Ma, X., 2009. Calibrating 2-m temperature of limited-area ensemble forecasts using high-resolution analysis. Mon. Weather Rev. 137, 3373–3387.

Karl, T.R., Koscielny, A.J., 1982. Drought in the United States, 1895–1981. Int. J. Climatol. 2, 313–329.

Karl, T.R., Koscielny, A.J., Diaz, H.F., 1982. Potential errors in the application of principal component (eigenvector) analysis to geophysical data. J. Appl. Meteorol. 21, 1183–1186.

Karlin, S., Taylor, H.M., 1975. A First Course in Stochastic Processes. Academic Press. 557 pp.

Katz, R.W., 1977. Precipitation as a chain-dependent process. J. Appl. Meteorol. 16, 671–676.

Katz, R.W., 1981. On some criteria for estimating the order of a Markov chain. Technometrics 23, 243–249.

Katz, R.W., 1982. Statistical evaluation of climate experiments with general circulation models: a parametric time series modeling approach. J. Atmos. Sci. 39, 1446–1455.

Katz, R.W., 1985. Probabilistic models. In: Murphy, A.H., Katz, R.W. (Eds.), Probability, Statistics, and Decision Making in the Atmospheric Sciences. Westview Press, Boulder, CO, 261–288.

Katz, R.W., 2002. Sir Gilbert Walker and a connection between El Niño and statistics. Stat. Sci. 17, 97–112.

Katz, R.W., 2013. Statistical methods for nonstationary extremes. In: AghaKouchak, A., et al. (Eds.), Extremes in a Changing Climate. Springer, 15–36.

Katz, R.W., Ehrendorfer, M., 2006. Bayesian approach to decision making using ensemble weather forecasts. Weather Forecast. 21, 220–231.

Katz, R.W., Murphy, A.H., 1997a. Economic Value of Weather and Climate Forecasts. Cambridge University Press, Cambridge. 222 pp.

Katz, R.W., Murphy, A.H., 1997b. Forecast value: prototype decision-making models. In: Katz, R.W., Murphy, A.H. (Eds.), Economic Value of Weather and Climate Forecasts. Cambridge University Press, Cambridge, 183–217.

Katz, R.W., Murphy, A.H., Winkler, R.L., 1982. Assessing the value of frost forecasts to orchardists: a dynamic decision-making approach. J. Appl. Meteorol. 21, 518–531.

Katz, R.W., Parlange, M.B., 1993. Effects of an index of atmospheric circulation on stochastic properties of precipitation. Water Resour. Res. 29, 2335–2344.

Katz, R.W., Parlange, M.B., Naveau, P., 2002. Statistics of extremes in hydrology. Adv. Water Resour. 25, 1287–1304.

Katz, R.W., Zheng, X., 1999. Mixture model for overdispersion of precipitation. J. Clim. 12, 2528–2537.

Keil, C., Craig, G.C., 2007. A displacement-based error measure applied in a regional ensemble forecasting system. Mon. Weather Rev. 135, 3248–3259.

Keil, C., Craig, G.C., 2009. A displacement and amplitude score employing an optical flow technique. Weather Forecast. 24, 1298–1308.

Kelly, K.S., Krzysztofowicz, R., 1997. A bivariate meta-Gaussian density for use in hydrology. Stoch. Hydrol. Hydraul. 11, 17–31.

Kendall, M., Ord, J.K., 1990. Time Series. Edward Arnold, p. 296.

Kestin, T.S., Karoly, D.J., Yano, J.-I., Raynor, N.A., 1998. Time-frequency variability of ENSO and stochastic simulations. J. Clim. 11, 2258–2272.

Keune, J., Ohlwein, C., Hense, A., 2014. Multivariate probabilistic analysis and predictability of medium-range ensemble weather forecasts. Mon. Weather Rev. 142, 4074–4090.

Kharin, V.V., Zwiers, F.W., 2003. On the ROC score of probability forecasts. J. Clim. 16, 4145–4150.

Kharin, V.V., Zwiers, F.W., 2005. Estimating extremes in transient climate change simulations. J. Clim. 18, 1156–1173.

Kirtman, B.P., et al., 2014. The North American multimodel ensemble. Phase-1 seasonal-to-interannual prediction; Phase-2 toward developing intraseasonal prediction. Bull. Am. Meteorol. Soc. 95, 585–601.

Klein, W.H., Lewis, B.M., Enger, I., 1959. Objective prediction of five-day mean temperature during winter. J. Meteorol. 16, 672–682.

Knaff, J.A., Landsea, C.W., 1997. An El Niño-southern oscillation climatology and persistence (CLIPER) forecasting scheme. Weather Forecast. 12, 633–647.

Koenker, R.W., Bassett, B., 1978. Regression quantiles. Econometrica 46, 33–49.

Koh, T.-Y., Wang, S., Bhatt, B.C., 2012. A diagnostic suite to assess NWP performance. J. Geophys. Res. 117. https://doi.org/10.1029/2011JD017103. 20 pp.

Krakauer, N.Y., Grossberg, M.D., Gladkova, I., Aizenman, H., 2013. Information content of seasonal forecasts in a changing climate. Adv. Meteorol. https://doi.org/10.1155/2013/480210. 12 pp.

Krzysztofowicz, R., 1983. Why should a forecaster and a decision maker use Bayes' theorem. Water Resour. Res. 19, 327–336.

Krzysztofowicz, R., Drzal, W.J., Drake, T.R., Weyman, J.C., Giordano, L.A., 1993. Probabilistic quantitative precipitation forecasts for river basins. Weather Forecast. 8, 424–439.

Krzysztofowicz, R., Evans, W.B., 2008. Probabilistic forecasts from the National Digital Forecast database. Weather Forecast. 23, 270–289.

Krzysztofowicz, R., Long, D., 1990. Fusion of detection probabilities and comparison of multisensor systems. IEEE Trans. Syst. Man Cybern. 20, 665–677.

Krzysztofowicz, R., Long, D., 1991. Beta probability models of probabilistic forecasts. Int. J. Forecast. 7, 47–55.

Kücken, M., Gerstengarbe, F.-W., 2009. A combination of cluster analysis and kappa statistic for the evaluation of climate model results. J. Appl. Meteorol. Climatol. 48, 1757–1765.

Künsch, H.R., 1989. The jackknife and the bootstrap for general stationary observations. Ann. Stat. 17, 1217–1241.

Kutzbach, J.E., 1967. Empirical eigenvectors of sea-level pressure, surface temperature and precipitation complexes over North America. J. Appl. Meteorol. 6, 791–802.

Kysely, J., 2008. A cautionary note on the use of nonparametric bootstrap for estimating uncertainties in extreme-value models. J. Appl. Meteorol. Climatol. 47, 3226–3251.

Lahiri, S.N., 2003. Resampling Methods for Dependent Data. Springer. 374 pp.

Lahiri, K., Yang, L., 2016. Asymptotic variance of Brier (skill) score in the presence of serial correlation. Econ. Lett. 141, 125–129.

Lahiri, K., Yang, L., 2018. Confidence bands for ROC curves with serially dependent data. J. Bus. Econ. Stat. 36, 115–130.

Lall, U., Sharma, A., 1996. A nearest neighbor bootstrap for resampling hydrologic time series. Water Resour. Res. 32, 679–693.

Lambert, S.J., Boer, G.J., 2001. CMIP1 evaluation and intercomparison of coupled climate models. Clim. Dyn. 17, 83–106.

Landman, W.A., Mason, S.J., 2001. Forecasts of near-global sea surface temperatures using canonical correlation analysis. J. Clim. 14, 3819–3833.

Lanzante, J.R., 2005. A cautionary note on the use of error bars. J. Clim. 18, 3699–3703.

Leadbetter, M.R., Lindgren, G., Rootzen, H., 1983. Extremes and Related Properties of Random Sequences and Processes. Springer. 336 pp.

Lee, J., Li, S., Lund, R., 2014. Trends in extreme U.S. temperatures. J. Clim. 27, 4209–4225.

Lee, P.M., 1997. Bayesian Statistics: An Introduction, second ed. Wiley. 344 pp.

Leger, C., Politis, D.N., Romano, J.P., 1992. Bootstrap technology and applications. Technometrics 34, 378–398.

Legg, T.P., Mylne, K.R., Woodcock, C., 2002. Use of medium-range ensembles at the Met Office I: PREVIN—a system for the production of probabilistic forecast information from the ECMWF EPS. Meteorol. Appl. 9, 255–271.

Lehmiller, G.S., Kimberlain, T.B., Elsner, J.B., 1997. Seasonal prediction models for North Atlantic basin hurricane location. Mon. Weather Rev. 125, 1780–1791.

Leith, C.E., 1973. The standard error of time-average estimates of climatic means. J. Appl. Meteorol. 12, 1066–1069.

Leith, C.E., 1974. Theoretical skill of Monte-Carlo forecasts. Mon. Weather Rev. 102, 409–418.

Lemcke, C., Kruizinga, S., 1988. Model output statistics forecasts: three years of operational experience in the Netherlands. Mon. Weather Rev. 116, 1077–1090.

Lemke, P., 1977. Stochastic climate models. Part 3. Application to zonally averaged energy models. Tellus 29, 385–392.

Lerch, S., Thorarinsdottir, T.L., 2013. Comparison of non-homogeneous regression models for probabilistic wind speed forecasting. Tellus A. 65, https://doi.org/10.3402/tellusa.v65i0.21206. 13 pp.

Lepore, C., Tippett, M.K., Allen, J.T., 2017. ENSO-based probabilistic forecasts of March–May U.S. tornado and hail activity. Geophys. Res. Lett. 44, 9093–9101.

Lettenmaier, D.P., 1976. Detection of trends in water quality data from records with dependent observations. Water Resour. Res. 12, 1037–1046.

Lettenmaier, D.P., Wood, E.F., Wallis, J.R., 1994. Hydro-climatological trends in the continental United States, 1948-88. J. Clim. 7, 586–607.

Leutbecher, M., 2018. Ensemble size: how suboptimal is less than infinity? Q. J. R. Meteorol. Soc. https://doi.org/10.1002/qj.3387.

Leutbecher, M., Palmer, T.N., 2008. Ensemble forecasting. J. Comput. Phys. 227, 3515–3539.

Leutbecher, M., et al., 2017. Stochastic representations of model uncertainties at ECMWF: state of the art and future vision. Q. J. R. Meteorol. Soc. 143, 2315–2339.

Lewis, J.M., 2005. Roots of ensemble forecasting. Mon. Weather Rev. 133, 1865–1885.

Lewis, J.M., 2014. Edward Epstein's stochastic-dynamic approach to ensemble weather prediction. Bull. Am. Meteorol. Soc. 95, 99–116.

Li, C., Singh, V.P., Mishra, A.K., 2012. Simulation of the entire range of daily precipitation using a hybrid probability distribution. Water Resour. Res. 48. https://doi.org/10.1029/2011WR011446.

Li, T.Y., Yorke, J.A., 1975. Period three implies chaos. Am. Math. Mon. 82, 985–992.

Li, W., Duan, Q., Miao, C., Ye, A., Gong, W., Di, Z., 2017. A review on statistical postprocessing methods for hydrometeorological ensemble forecasting. WIREs Water 4. https://doi.org/10.1002/wat2.1246. 24 pp.

Lian, T., Chen, D., 2012. An evaluation of rotated EOF analysis and its application to tropical Pacific SST variability. J. Clim. 25, 5361–5373.

Light, A., Bartlein, P.J., 2004. The end of the rainbow? Color schemes for improved data graphics. Eos 85, 385, 391.

Liljas, E., Murphy, A.H., 1994. Anders Angstrom and his early papers on probability forecasting and the use/value of weather forecasts. Bull. Am. Meteorol. Soc. 75, 1227–1236.

Lilliefors, H.W., 1967. On the Kolmogorov-Smirnov test for normality with mean and variance unknown. J. Am. Stat. Assoc. 62, 399–402.

Lim, Y., Jo, S., Lee, J., Oh, H.-S., Kang, H.-S., 2012. An improvement of seasonal climate prediction by regularized canonical correlation analysis. Int. J. Climatol. 32, 1503–1512.

Lin, J.W.-B., Neelin, J.D., 2000. Influence of a stochastic moist convective parameterization on tropical climate variability. Geophys. Res. Lett. 27, 3691–3694.

Lin, J.W.-B., Neelin, J.D., 2002. Considerations for stochastic convective parameterization. J. Atmos. Sci. 59, 959–975.

Lindgren, B.W., 1976. Statistical Theory. MacMillan. 614 pp.

Lindsay, B.G., Kettenring, J., Siegmund, D.O., 2004. A report on the future of Statistics. Stat. Sci. 19, 387–413.

Little, R.J., 2006. Calibrated Bayes: a Bayes/frequentist roadmap. Am. Stat. 60, 213–223.

Livezey, R.E., 2003. Categorical events. In: Jolliffe, I.T., Stephenson, D.B. (Eds.), Forecast Verification, first ed. Wiley, 77–96.

Livezey, R.E., Chen, W.Y., 1983. Statistical field significance and its determination by Monte Carlo techniques. Mon. Weather Rev. 111, 46–59.

Livezey, R.E., Smith, T.M., 1999. Considerations for use of the Barnett and Preisendorfer (1987) algorithm for canonical correlation analysis of climate variations. J. Clim. 12, 303–305.

Loader, C., 1999. Local Regression and Likelihood. Springer. 290 pp.

Lorenz, E.N., 1956. Empirical Orthogonal Functions and Statistical Weather Prediction. Science Report 1, Statistical Forecasting Project, Department of Meteorology, MIT. NTIS AD 110268, 49 pp.

Lorenz, E.N., 1963. Deterministic nonperiodic flow. J. Atmos. Sci. 20, 130–141.

Lorenz, E.N., 1965. On the possible reasons for long-period fluctuations of the general circulation. In: *Proceedings of the WMO-IUGG Symposium on Research and Development Aspects of Long-Range Forecasting*. World Meteorological Organization, Boulder, CO, 203–211. WMO Tech. Note 66.

Lorenz, E.N., 1969. The predictability of a flow which possesses many scales of motion. Tellus 3, 290–307.

Lorenz, E.N., 1975. Climate predictability. In: The Physical Basis of Climate and Climate Modelling. GARP Publication Series, vol. 16. WMO, 132–136.

Lorenz, E.N., 1982. Atmospheric predictability experiments with a large numerical model. Tellus 34, 505–513.

Lorenz, E.N., 1993. The Essence of Chaos. University of Washington Press. 227 pp.

Lorenz, E.N., 2006. Predictability—a problem partly solved. In: Palmer, T., Hagedorn, R. (Eds.), Predictability of Weather and Climate. Cambridge University Press, Cambridge, 40–58.

Loucks, D.P., Stedinger, J.R., Haith, D.A., 1981. Water Resource Systems Planning and Analysis. Prentice-Hall. 559 pp.

Lu, R., 1991. The application of NWP products and progress of interpretation techniques in China. In: Glahn, H.R., Murphy, A.H., Wilson, L.J., Jensenius Jr., J.S. (Eds.), Programme on Short- and Medium-Range Weather Prediction Research, XX19–22. World Meteorological Organization WM/TD No. 421.

Lund, R., Liu, G., Shao, Q., 2016. A new approach to ANOVA methods for autocorrelated data. Am. Stat. 70, 55–62.

Luo, L., Wood, E.F., Pan, M., 2007. Bayesian merging of multiple climate model forecasts for seasonal hydrological predictions. J. Geophys. Res. D: Atmos. 112. https://doi.org/10.1029/2006JD007655.

Ma, J., Zhu, Y., Wobus, R., Wang, P., 2012. An effective configuration of ensemble size and horizontal resolution for the NCEP GEFS. Adv. Atmos. Sci. 29, 782–794.

Machete, R.L., Smith, L.A., 2016. Demonstrating the value of larger ensembles in forecasting physical systems. Tellus A 68, 28393. 21 pp.

Macnaughton-Smith, P., Williams, W.T., Dale, M.B., Mockett, L.G., 1964. Dissimilarity analysis: a new technique of hierarchical sub-division. Nature 202, 1034–1035.

Madden, R.A., 1979. A simple approximation for the variance of meteorological time averages. J. Appl. Meteorol. 18, 703–706.

Madden, R.A., Jones, R.H., 2001. A quantitative estimate of the effect of aliasing in climatological time series. J. Clim. 14, 3987–3993.

Madden, R.A., Julian, P.R., 1972. Description of global-scale circulation cells in the tropics with a 40-50 day period. J. Atmos. Sci. 29, 1109–1123.

Madden, R.A., Shea, D.J., 1978. Estimates of the natural variability of time-averaged temperatures over the United States. Mon. Weather Rev. 106, 1695–1703.

Madsen, H., Rasmussen, P.F., Rosbjerg, D., 1997. Comparison of annual maximum series and partial duration series methods for modeling extreme hydrologic events. 1. At-site modeling. Water Resour. Res. 33, 747–757.

Mann, H.B., Whitney, D.R., 1947. On a test of whether one of two random variables is stochastically larger than the other. Ann. Math. Stat. 18, 50–60.

Manzato, A., Jolliffe, I., 2017. Behaviour of verification measures for deterministic binary forecasts with respect to random changes and thresholding. Q. J. R. Meteorol. Soc. 143, 1903–1915.

Mao, Q., McNider, R.T., Mueller, S.F., Juang, H.-M.H., 1999. An optimal model output calibration algorithm suitable for objective temperature forecasting. Weather Forecast. 14, 190–202.

Mao, Y., Monahan, A., 2018. Linear and nonlinear regression prediction of surface wind components. Clim. Dyn. 51, 3291–3309.

Mardia, K.V., 1970. Measures of multivariate skewness and kurtosis with applications. Biometrika 57, 519–530.

Mardia, K.V., Kent, J.T., Bibby, J.M., 1979. Multivariate Analysis. Academic. 518 pp.

Marquardt, D.W., 1970. Generalized inverses, ridge regression, biased linear estimation, and nonlinear estimation. Technometrics 12, 591–612.

Marquardt, D.W., Snee, R.D., 1975. Ridge regression in practice. Am. Stat. 29, 3–20.

Marty, R., Fortin, V., Kuswanto, H., Favre, A.-C., Parent, E., 2015. Combining the Bayesian processor of output with Bayesian model averaging for reliable ensemble forecasting. Appl. Stat. 64, 75–92.

Marzban, C., 2004. The ROC curve and the area under it as performance measures. Weather Forecast. 19, 1106–1114.

Marzban, C., 2012. Displaying economic value. Weather Forecast. 27, 1604–1612.

Marzban, C., Leyton, S., Colman, B., 2007. Ceiling and visibility forecasts via neural networks. Weather Forecast. 22, 466–479.

Marzban, C., Sandgathe, S., 2008. Cluster analysis for object-oriented verification fields: a variation. Mon. Weather Rev. 136, 1013–1025.

Mason, I.B., 1979. On reducing probability forecasts to yes/no forecasts. Mon. Weather Rev. 107, 207–211.

Mason, I.B., 1982. A model for assessment of weather forecasts. Aust. Meteorol. Mag. 30, 291–303.

Mason, I.B., 2003. Binary events. In: Jolliffe, I.T., Stephenson, D.B. (Eds.), Forecast Verification, first ed. Wiley, 37–76.

Mason, S.J., 2008. Understanding forecast verification statistics. Meteorol. Appl. 15, 31–40.

Mason, S.J., Goddard, L., Graham, N.E., Yulaleva, E., Sun, L., Arkin, P.A., 1999. The IRI seasonal climate prediction system and the 1997/98 El Niño event. Bull. Am. Meteorol. Soc. 80, 1853–1873.

Mason, S.J., Graham, N.E., 2002. Areas beneath the relative operating characteristics (ROC) and relative operating levels (ROL) curves: statistical significance and interpretation. Q. J. R. Meteorol. Soc. 128, 2145–2166.

Mason, S.J., Mimmack, G.M., 1992. The use of bootstrap confidence intervals for the correlation coefficient in climatology. Theor. Appl. Climatol. 45, 229–233.

Mason, S.J., Mimmack, G.M., 2002. Comparison of some statistical methods of probabilistic forecasting of ENSO. J. Clim. 15, 8–29.

Matalas, N.C., 1967. Mathematical assessment of synthetic hydrology. Water Resour. Res. 3, 937–945.

Matalas, N.C., Sankarasubramanian, A., 2003. Effect of persistence on trend detection via regression. Water Resour. Res. 39, 1342–1348.

Matheson, J.E., Winkler, R.L., 1976. Scoring rules for continuous probability distributions. Manag. Sci. 22, 1087–1096.

Matsumoto, M., Nishimura, T., 1998. Mersenne twister: a 623-dimensionally equidistributed uniform pseudorandom number generator. ACM Trans. Model. Comput. Simul. 8, 3–30.

McAvaney, B.J., et al., 2001. Model evaluation. In: Houghton, J.T., et al. (Eds.), Climate Change 2001: The Scientific Basis. Cambridge University Press, 471–523.

McCullagh, P., 1980. Regression models for ordinal data. J. R. Stat. Soc. Ser. B Methodol. 42, 109–142.

McCullagh, P., Nelder, J.A., 1989. Generalized Linear Models. Chapman and Hall. 511 pp.

McCulloch, W.S., Pitts, W.H., 1943. A logical calculus of the ideas immanent in nervous activity. Bull. Math. Phys. 5, 115–137.

McGill, R., Tukey, J.W., Larsen, W.A., 1978. Variations of boxplots. Am. Stat. 32, 12–16.

McGovern, A., Elmore, K.L., Gagne, D.J., Haupt, S.E., Karstens, C.D., Lagerquist, R., Smith, T., Williams, J.K., 2017. Using artificial intelligence to improve real-time decision- making for high-impact weather. Bull. Am. Meteorol. Soc. 98, 2073–2090.

McKee, T.B., Doeskin, N.J., Kleist, J., 1993. The relationship of drought frequency and duration to time scales. In: Proceedings, 8th Conference on Applied Climatology. American Meteorological Society, 179–184.

McLachlan, G.J., Basford, K.E., 1988. Mixture Models: Inference and Application to Clustering. Dekker. 253 pp.

McLachlan, G.J., Krishnan, T., 1997. The EM Algorithm and Extensions. Wiley. 274 pp.

McLachlan, G.J., Peel, D., 2000. Finite Mixture Models. Wiley. 419 pp.

Mecklin, C.J., Mundfrom, D.J., 2004. An appraisal and bibliography of tests for multivariate normality. Int. Stat. Rev. 72, 123–138.

Meinshausen, N., 2006. Quantile regression forests. Journal of Machine Learning Research 7, 983–999.

Merkle, E.C., Steyvers, M., 2013. Choosing a strictly proper scoring rule. Decision Analysis 10, 292–304.

Messner, J.W., Mayr, G.J., 2011. Probabilistic forecasts using analogs in the idealized Lorenz'96 setting. Mon. Weather Rev. 139, 1960–1971.

Messner, J.W., Mayr, G.J., Zeileis, A., 2017. Nonhomogeneous boosting for predictor selection in ensemble postprocessing. Mon. Weather Rev. 145, 137–147.

Messner, J.W., Mayr, G.J., Wilks, D.S., Zeileis, A., 2014a. Extending extended logistic regression: extended versus separate versus ordered versus censored. Mon. Weather Rev. 142, 3003–3014.

Messner, J.W., Mayr, G.J., Zeileis, A., Wilks, D.S., 2014b. Heteroscedastic extended logistic regression for postprocessing of ensemble guidance. Mon. Weather Rev. 142, 448–456.

Mestas-Nuñez, A.M., 2000. Orthogonality properties of rotated empirical modes. Int. J. Climatol. 20, 1509–1516.

Metropolis, N., Ulam, S., 1949. The Monte-Carlo method. J. Am. Stat. Assoc. 44, 335–341.

Michaelson, J., 1987. Cross-validation in statistical climate forecast models. J. Clim. Appl. Meteorol. 26, 1589–1600.

Mielke, P.W., 1975. Convenient beta distribution likelihood techniques for describing and comparing meteorological data. J. Appl. Meteorol. 14, 985–990.

Mielke, P.W., 1991. The application of multivariate permutation methods based on distance functions in the earth sciences. Earth-Science Reviews 31, 55–71.

Mielke Jr., P.W., Berry, K.J., Brier, G.W., 1981. Application of multi-response permutation procedures for examining seasonal changes in monthly mean sea-level pressure patterns. Mon. Weather Rev. 109, 120–126.

Mielke Jr., P.W., Berry, K.J., Landsea, C.W., Gray, W.M., 1996. Artificial skill and validation in meteorological forecasting. Weather Forecast. 11, 153–169.

Miller, B.I., Hill, E.C., Chase, P.P., 1968. A revised technique for forecasting hurricane movement by statistical methods. Mon. Weather Rev. 96, 540–548.

Miller, J.M., 1995. Exact maximum likelihood estimation in autoregressive processes. Journal of Time Series Analysis 16, 607–615.

Miller, R.G., 1962. Statistical prediction by discriminant analysis. MeteorologicalMonographs, vol. 4, No. 25. American Meteorological Society. 53 pp.

Miller, R.G., 1964. Regression Estimation of Event Probabilities. Travelers Research Center, Hartford, CN. Tech Rept No. 1, Contract CWB-107704. 153 pp.

Millner, A., 2008. Getting the most out of ensemble forecasts: a valuation model based on user-forecast interactions. J. Appl. Meteorol. Climatol. 47, 2561–2571.

Mirzargar, M., Anderson, J.L., 2017. On evaluation of ensemble forecast calibration using the concept of data depth. Mon. Weather Rev. 145, 1679–1690.

Mitchell, K., Ferro, C.A.T., 2017. Proper scoring rules for interval probabilistic forecasts. Q. J. R. Meteorol. Soc. 143, 1597–1607.

Mittermaier, M.P., Bullock, R., 2013. Using MODE to explore the spatial and temporal characteristics of cloud cover forecasts from high-resolution NWP models. Meteorol. Appl. 20, 187–196.

Mittermaier, M.P., Stephenson, D.B., 2015. Inherent bounds on forecast accuracy due to observation uncertainty caused by temporal sampling. Mon. Weather Rev. 143, 4236–4243.

Miyakoda, K., Hembree, G.D., Strikler, R.F., Shulman, I., 1972. Cumulative results of extended forecast experiments. I: Model performance for winter cases. Mon. Weather Rev. 100, 836–855.

Mo, K.C., Ghil, M., 1987. Statistics and dynamics of persistent anomalies. J. Atmos. Sci. 44, 877–901.

Mo, K.C., Ghil, M., 1988. Cluster analysis of multiple planetary flow regimes. J. Geophys. Res. D: Atmos. 93, 10927–10952.

Möller, A., Lenkoski, A., Thorarinsdottir, T.L., 2013. Multivariate probabilistic forecasting using ensemble Bayesian model averaging and copulas. Q. J. R. Meteorol. Soc. 139, 982–991.

Molteni, F., Buizza, R., Palmer, T.N., Petroliagis, T., 1996. The new ECMWF ensemble prediction system: methodology and validation. Q. J. R. Meteorol. Soc. 122, 73–119.

Molteni, F., Tibaldi, S., Palmer, T.N., 1990. Regimes in wintertime circulation over northern extratropics. I: Observational evidence. Q. J. R. Meteorol. Soc. 116, 31–67.

Montgomery, D.C., 2013. Design and Analysis of Experiments. Wiley. 724 pp.

Moritz, R.E., Sutera, A., 1981. The predictability problem: effects of stochastic perturbations in multiequilibrium systems. Rev. Geophys. 23, 345–383.

Morrison, J.E., Smith, J.A., 2002. Stochastic modeling of flood peaks using the generalized extreme value distribution. Water Resour. Res. 38, 1305. https://doi.org/10.1029/2001WR000502.

Moura, A.D., Hastenrath, S., 2004. Climate prediction for Brazil's Nordeste: performance of empirical and numerical modeling methods. J. Clim. 17, 2667–2672.

Muhlbauer, A., Spichtinger, P., Lohmann, U., 2009. Application and comparison of robust linear regression methods for trend estimation. J. Appl. Meteorol. Climatol. 48, 1961–1970.

Mullen, S.L., Buizza, R., 2002. The impact of horizontal resolution and ensemble size on probabilistic forecasts of precipitation by the ECMWF ensemble prediction system. Weather Forecast. 17, 173–191.

Muller, R.H., 1944. Verification of short-range weather forecasts (a survey of the literature). Bull. Am. Meteorol. Soc. 25, 18–27. 47–53, 88–95.

Murphy, A.H., 1966. A note on the utility of probabilistic predictions and the probability score in the cost-loss ratio situation. J. Appl. Meteorol. 5, 534–537.

Murphy, A.H., 1971. A note on the ranked probability score. J. Appl. Meteorol. 10, 155–156.

Murphy, A.H., 1972. Scalar and vector partitions of the probability score: Part II. N-state situation. J. Appl. Meteorol. 11, 1183–1192.

Murphy, A.H., 1973a. Hedging and skill scores for probability forecasts. J. Appl. Meteorol. 12, 215–223.

Murphy, A.H., 1973b. A new vector partition of the probability score. J. Appl. Meteorol. 12, 595–600.

Murphy, A.H., 1977. The value of climatological, categorical, and probabilistic forecasts in the cost-loss ratio situation. Mon. Weather Rev. 105, 803–816.

Murphy, A.H., 1985. Probabilistic weather forecasting. In: Murphy, A.H., Katz, R.W. (Eds.), Probability, Statistics, and Decision Making in the Atmospheric Sciences. Westview Press, Boulder, CO, 337–377.

Murphy, A.H., 1988. Skill scores based on the mean square error and their relationships to the correlation coefficient. Mon. Weather Rev. 116, 2417–2424.

Murphy, A.H., 1991. Forecast verification: its complexity and dimensionality. Mon. Weather Rev. 119, 1590–1601.

Murphy, A.H., 1992. Climatology, persistence, and their linear combination as standards of reference in skill scores. Weather Forecast. 7, 692–698.

Murphy, A.H., 1993. What is a good forecast? An essay on the nature of goodness in weather forecasting. Weather Forecast. 8, 281–293.

Murphy, A.H., 1995. The coefficients of correlation and determination as measures of performance in forecast verification. Weather Forecast. 10, 681–688.

Murphy, A.H., 1996. The Finley affair: a signal event in the history of forecast verification. Weather Forecast. 11, 3–20.

Murphy, A.H., 1997. Forecast verification. In: Katz, R.W., Murphy, A.H. (Eds.), Economic Value of Weather and Climate Forecasts. Cambridge University Press, Cambridge, 19–74.

Murphy, A.H., 1998. The early history of probability forecasts: some extensions and clarifications. Weather Forecast. 13, 5–15.

Murphy, A.H., Brown, B.G., 1983. Forecast terminology: composition and interpretation of public weather forecasts. Bull. Am. Meteorol. Soc. 64, 13–22.

Murphy, A.H., Brown, B.G., Chen, Y.-S., 1989. Diagnostic verification of temperature forecasts. Weather Forecast. 4, 485–501.

Murphy, A.H., Daan, H., 1985. Forecast evaluation. In: Murphy, A.H., Katz, R.W. (Eds.), Probability, Statistics, and Decision Making in the Atmospheric Sciences. Westview Press, Boulder, CO, 379–437.

Murphy, A.H., Ehrendorfer, M., 1987. On the relationship between the accuracy and value of forecasts in the cost-loss ratio situation. Weather Forecast. 2, 243–251.

Murphy, A.H., Epstein, E.S., 1967a. Verification of probabilistic predictions: a brief review. J. Appl. Meteorol. 6, 748–755.

Murphy, A.H., Epstein, E.S., 1967b. A note on probability forecasts and "hedging". J. Appl. Meteorol. 6, 1002–1004.

Murphy, A.H., Epstein, E.S., 1989. Skill scores and correlation coefficients in model verification. Mon. Weather Rev. 117, 572–581.

Murphy, A.H., Wilks, D.S., 1998. A case study in the use of statistical models in forecast verification: precipitation probability forecasts. Weather Forecast. 13, 795–810.

Murphy, A.H., Winkler, R.L., 1974. Credible interval temperature forecasting: some experimental results. Mon. Weather Rev. 102, 784–794.

Murphy, A.H., Winkler, R.L., 1979. Probabilistic temperature forecasts: the case for an operational program. Bull. Am. Meteorol. Soc. 60, 12–19.

Murphy, A.H., Winkler, R.L., 1984. Probability forecasting in meteorology. J. Am. Stat. Assoc. 79, 489–500.

Murphy, A.H., Winkler, R.L., 1987. A general framework for forecast verification. Mon. Weather Rev. 115, 1330–1338.

Murphy, A.H., Winkler, R.L., 1992. Diagnostic verification of probability forecasts. Int. J. Forecast. 7, 435–455.

Murphy, A.H., Ye, Q., 1990. Comparison of objective and subjective precipitation probability forecasts: the sufficiency relation. Mon. Weather Rev. 118, 1783–1792.

Mylne, K.R., 2002. Decision-making from probability forecasts based on forecast value. Meteorol. Appl. 9, 307–315.

Mylne, K.R., Evans, R.E., Clark, R.T., 2002a. Multi-model multi-analysis ensembles in quasi-operational medium-range forecasting. Q. J. R. Meteorol. Soc. 128, 361–384.

Mylne, K.R., Woolcock, C., Denholm-Price, J.C.W., Darvell, R.J., 2002b. Operational calibrated probability forecasts from the ECMWF ensemble prediction system: implementation and verification. In: Preprints, Symposium on Observations, Data Analysis, and Probabilistic Prediction (Orlando, Florida). American Meteorological Society, 113–118.

Namias, J., 1952. The annual course of month-to-month persistence in climatic anomalies. Bull. Am. Meteorol. Soc. 33, 279–285.

Narapusetty, B., DelSole, T., Tippett, M.K., 2009. Optimal estimation of the climatological mean. J. Clim. 22, 4845–4859.

Narula, S.C., Wellington, J.F., 1982. The minimum sum of absolute errors regression: a state of the art survey. Int. Stat. Rev. 50, 317–326.

National Bureau of Standards, 1959. Tables of the Bivariate Normal Distribution Function and Related Functions. Applied Mathematics Series, 50U.S. Government Printing Office. 258 pp.

National Research Council, 2006. Completing the Forecast: Characterizing and Communicating Uncertainty for Better Decisions Using Weather and Climate Forecasts. National Academy Press, Washington DC. ISBN 0-309066327-X, www.nap.edu/catalog/11699.html.

Navarra, A., Simoncini, V., 2010. A Guide to Empirical Orthogonal Functions for Climate Data Analysis. Springer. 151 pp.

Naveau, P., Bessac, J., 2018. Forecast evaluation with imperfect observations and imperfect models. arXiv:1806.03745v1, 21 pp.

Neelin, J.D., Peters, O., Lin, J.W.-B., Hales, K., Holloway, C.E., 2010. Rethinking convective quasi-equilibrium: observational constraints for stochastic convective schemes in climate models. In: Palmer, T., Williams, P. (Eds.), Stochastic Physics and Climate Modeling. Cambridge University Press, Cambridge, 396–423.

Neilley, P.P., Myers, W., Young, G., 2002. Ensemble dynamic MOS. In: Preprints, 16th Conference on Probability and Statistics in the Atmospheric Sciences (Orlando, Florida). American Meteorological Society, 102–106.

Neter, J., Wasserman, W., Kutner, M.H., 1996. Applied Linear Statistical Models. McGraw-Hill. 1408 pp.

Neumann, C.J., Jarvinen, B.R., McAdie, C.J., Hammer, G.R., 1999. Tropical Cyclones of the North Atlantic Ocean, 1871-1998, 5th Revision. National Climatic Data Center, Asheville NC. 206 pp.

Neumann, C.J., Lawrence, M.B., Caso, E.L., 1977. Monte Carlo significance testing as applied to statistical tropical cyclone prediction models. J. Appl. Meteorol. 16, 1165–1174.

Newman, M., Sardeshmukh, P., 1995. A caveat concerning singular value decomposition. J. Clim. 8, 352–360.

Nicholls, N., 2001. The insignificance of significance testing. Bull. Am. Meteorol. Soc. 82, 981–986.

Nielsen, H.A., Madsen, H., Nielsen, T.S., 2006. Using quantile regression to extend an existing wind power forecasting system with probabilistic forecasts. Wind Energy 9, 95–108.

North, G.R., 1984. Empirical orthogonal functions and normal modes. J. Atmos. Sci. 41, 879–887.

North, G.R., Bell, T.L., Cahalan, R.F., Moeng, F.J., 1982. Sampling errors in the estimation of empirical orthogonal functions. Mon. Weather Rev. 110, 699–706.

Northrop, P.J., Chandler, R.E., 2014. Quantifying sources of uncertainty in projections of future climate. J. Clim. 27, 8793–8808.

Obukhov, A.M., 1947. Statistically homogeneous fields on a sphere. Usp. Mathematic. Nauk 2, 196–198.

O'Lenic, E.A., Livezey, R.E., 1988. Practical considerations in the use of rotated principal component analysis (RPCA) in diagnostic studies of upper-air height fields. Mon. Weather Rev. 116, 1682–1689.

O'Lenic, E.A., Unger, D.A., Halpert, M.S., Pelman, K.S., 2008. Developments in operational long-range climate prediction at CPC. Weather Forecast. 23, 496–515.

Ollinaho, P., Lock, S.-J., Leutbecher, M., Bechtold, P., Beljaars, A., Bozzo, A., Forbes, R.M., Haiden, T., Hogan, R.J., Sandu, I., 2017. Towards process-level representation of model uncertainties: stochastically perturbed parameterizations in the ECMWF ensemble. Q. J. R. Meteorol. Soc. 143, 408–422.

Osborn, T.J., Hulme, M., 1997. Development of a relationship between station and grid-box rainday frequencies for climate model evaluation. J. Clim. 10, 1885–1908.

Overland, J.E., Preisendorfer, R.W., 1982. A significance test for principal components applied to a cyclone climatology. Mon. Weather Rev. 110, 1–4.

Paciorek, C.J., Risbey, J.S., Ventura, V., Rosen, R.D., 2002. Multiple indices of Northern Hemisphere cyclone activity, winters 1949–99. J. Clim. 15, 1573–1590.

Palmer, T.N., 1993. Extended-range atmospheric prediction and the Lorenz model. Bull. Am. Meteorol. Soc. 74, 49–65.

Palmer, T.N., 2001. A nonlinear dynamical perspective on model error: A proposal for non-local stochastic-dynamic parameterization in weather and climate prediction models. Q. J. R. Meteorol. Soc. 127, 279–304.

Palmer, T.N., 2006. Predictability of weather and climate: from theory to practice. In: Palmer, T., Hagedorn, R. (Eds.), Predictability of Weather and Climate. Cambridge University Press, Cambridge, 1–29.

Palmer, T.N., 2012. Towards the probabilistic Earth-system simulator: a vision for the future of climate and weather prediction. Q. J. R. Meteorol. Soc. 138, 841–861.

Palmer, T.N., 2014a. More reliable forecasts with less precise computations: a fast-track route to cloud-resolved weather and climate simulators? Phil. Trans. R. Soc. A 372. https://doi.org/10.1098/rsta.2013.0391. 14 pp.

Palmer, T.N., 2014b. The real butterfly effect. Nonlinearity 27, R123–R141.

Palmer, T.N., Doblas-Reyes, F.J., Hagedorn, R., Weisheimer, A., 2005a. Probabilistic prediction of climate using multi-model ensembles: from basics to applications. Phil. Trans. R. Soc. B 360, 1991–1998.

Palmer, T.N., Mureau, R., Molteni, F., 1990. The Monte Carlo forecast. Weather 45, 198–207.

Palmer, T.N., Shutts, G.J., Hagedorn, R., Doblas-Reyes, F.J., Jung, T., Leutbecher, M., 2005b. Representing model uncertainty in weather and climate prediction. Annu. Rev. Earth Planet. Sci. 33, 163–193.

Palmer, T.N., Tibaldi, S., 1988. On the prediction of forecast skill. Mon. Weather Rev. 116, 2453–2480.

Panofsky, H.A., Brier, G.W., 1958. Some Applications of Statistics to Meteorology. Pennsylvania State University. 224 pp.

Papalexiou, S.M., Koutsoyiannis, D., 2013. Battle of extreme value distributions: A global survey on extreme daily rainfall. Water Resour. Res. 49, 187–201.

Pappenberger, F., Ghelli, A., Buizza, R., Bódis, K., 2009. The skill of probabilistic precipitation forecasts under observational uncertainties within the generalized likelihood uncertainty estimation framework for hydrological applications. J. Hydrometeorol. 10, 807–819.

Parisi, F., Lund, R., 2008. Return periods of continental U.S. hurricanes. J. Clim. 21, 403–410.

Peatman, S.C., Matthews, A.J., Stephens, D.P., 2015. Propagation of the Madden-Julian oscillation and scale interaction with the diurnal cycle in a high-resolution GCM. Clim. Dyn. 45, 2901–2918.

Peirce, C.S., 1884. The numerical measure of the success of predictions. Science 4, 453–454.

Peirolo, R., 2011. Information gain as a score for probabilistic forecasts. Meteorol. Appl. 18, 9–17.

Penland, C., Sardeshmukh, P.D., 1995. The optimal growth of tropical sea surface temperatures anomalies. J. Clim. 8, 1999–2024.

Pepe, M.S., 2003. The Statistical Evaluation of Medical Tests for Classification and Prediction. Oxford University Press. 302 pp.

Peterson, C.R., Snapper, K.J., Murphy, A.H., 1972. Credible interval temperature forecasts. Bull. Am. Meteorol. Soc. 53, 966–970.

Pinson, P., 2012. Adaptive calibration of (u,v)-wind ensemble forecasts. Q. J. R. Meteorol. Soc. 138, 1273–1284.

Pinson, P., 2013. Wind energy: forecasting challenges for its operational management. Stat. Sci. 28, 564–585.

Pinson, P., Girard, R., 2012. Evaluating the quality of scenarios of short-term wind power generation. Appl. Energy 96, 12–20.

Pinson, P., Hagedorn, R., 2012. Verification of the ECMWF ensemble forecasts of wind speed against analyses and observations. Meteorol. Appl. 19, 484–500.

Pinson, P., McSharry, P., Madsen, H., 2010. Reliability diagrams for non-parametric density forecasts of continuous variables: accounting for serial correlation. Q. J. R. Meteorol. Soc. 136, 77–90.

Pitcher, E.J., 1974. Stochastic Dynamic Prediction Using Atmospheric Data. Ph.D. dissertation, University of Michigan.

Pitcher, E.J., 1977. Application of stochastic dynamic prediction to real data. J. Atmos. Sci. 34, 3–21.

Pitman, E.J.G., 1937. Significance tests which may be applied to samples from any populations. J. R. Stat. Soc. Ser. B Methodol. 4, 119–130.

Plaut, G., Vautard, R., 1994. Spells of low-frequency oscillations and weather regimes in the Northern Hemisphere. J. Atmos. Sci. 51, 210–236.

Pocernich, M., 2007. Verification: forecast verification utilities. R package version 1.20. http://www.r-project.org.

Politis, D., Romano, J.P., 1992. A circular block resampling procedure for stationary data. In: Lepage, R., Billard, L. (Eds.), Exploring the Limits of Bootstrap. Wiley, 263–270.

Politis, D.N., Romano, J.P., Wolf, M., 1999. Subsampling. Springer, p. 347.

Preisendorfer, R.W., 1988. Mobley, C.D. (Ed.), Principal Component Analysis in Meteorology and Oceanography. Elsevier. 425 pp.

Preisendorfer, R.W., Barnett, T.P., 1983. Numerical-reality intercomparison tests using small-sample statistics. J. Atmos. Sci. 40, 1884–1896.

Preisendorfer, R.W., Mobley, C.D., 1984. Climate forecast verifications, United States Mainland, 1974–83. Mon. Weather Rev. 112, 809–825.

Preisendorfer, R.W., Zwiers, F.W., Barnett, T.P., 1981. Foundations of Principal Component Selection Rules. SIO Reference Series 81-4, Scripps Institution of Oceanography. 192 pp.

Press, W.H., Flannery, B.P., Teukolsky, S.A., Vetterling, W.T., 1986. Numerical Recipes: The Art of Scientific Computing. Cambridge University Press. 818 pp.

Prokosch, J., 2013. Bivariate Bayesian Model Averaging and Ensemble Model Output Statistics. M.S. Thesis, Norwegian University of Science and Technology. 85 pp. http://www.diva-portal.org/smash/get/diva2:656466/FULLTEXT01.pdf.

Quadrelli, R., Bretherton, C.S., Wallace, J.M., 2005. On sampling errors in empirical orthogonal functions. J. Clim. 18, 3704–3710.

Quan, X., Hoerling, M., Whitaker, J., Bates, G., Xu, T., 2006. Diagnosing sources of U.S. seasonal forecast skill. J. Clim. 19, 3279–3293.

R Development Core Team, 2017. An Introduction to R. R Foundation for Statistical Computing, http://www.R-project.org/.

Radok, U., 1988. Chance behavior of skill scores. Mon. Weather Rev. 116, 489–494.

Raftery, A.E., Gneiting, T., Balabdaoui, F., Polakowski, M., 2005. Using Bayesian model averaging to calibrate forecast ensembles. Mon. Weather Rev. 133, 1155–1174.

Räisänen, J., 2001. CO_2-induced climate change in CMIP2 experiments: quantification of agreement and role of internal variability. J. Clim. 14, 2088–2104.

Rajagopalan, B., Lall, U., Tarboton, D.G., 1997. Evaluation of kernel density estimation methods for daily precipitation resampling. Stoch. Hydrol. Hydraul. 11, 523–547.

Raynaud, L., Bouttier, F., 2017. The impact of horizontal resolution and ensemble size for convective-scale probabilistic forecasts. Q. J. R. Meteorol. Soc. 143, 3037–3047.

Razali, N.M., Wah, Y.B., 2011. Power comparisons of Shapiro-Wilk, Kolmogorov-Smirnov, Lilliefors and Anderson-Darling tests. J. Stat. Model. Anal. 2, 21–33.

Reggiani, P., Renner, M., Weerts, A.H., van Gelder, P.A.H.J.M., 2009. Uncertainty assessment via Bayesian revision of ensemble streamflow predictions in the operational river Rhine forecasting system. Water Resour. Res. 45. https://doi.org/10.1029/2007WR006758.

Retchless, D.P., Brewer, C.A., 2016. Guidance for representing uncertainty on global temperature change maps. Int. J. Climatol. 36, 1143–1159.

Richardson, A.J., Risien, C., Shillington, F.A., 2003. Using self-organizing maps to identify patterns in satellite imagery. Prog. Oceanogr. 59, 223–239.

Richardson, C.W., 1981. Stochastic simulation of daily precipitation, temperature, and solar radiation. Water Resour. Res. 17, 182–190.

Richardson, D.S., 2000. Skill and economic value of the ECMWF ensemble prediction system. Q. J. R. Meteorol. Soc. 126, 649–667.

Richardson, D.S., 2001. Measures of skill and value of ensemble predictions systems, their interrelationship and the effect of ensemble size. Q. J. R. Meteorol. Soc. 127, 2473–2489.

Richardson, D.S., 2003. Economic value and skill. In: Jolliffe, I.T., Stephenson, D.B. (Eds.), Forecast Verification, first ed. Wiley, 165–187.

Richman, M.B., 1986. Rotation of principal components. Int. J. Climatol. 6, 293–335.

Roberts, N.M., Lean, H.W., 2008. Scale-selective verification of rainfall accumulations from high-resolution forecasts of convective events. Mon. Weather Rev. 136, 78–97.

Roberts, R.D., Anderson, A.R.S., Nelson, E., Brown, B.G., Wilson, J.W., Pocernich, M., Saxen, T., 2012. Impacts of forecaster involvement on convective storm initiation and evolution nowcasting. Weather Forecast. 27, 1061–1089.

Rodwell, M.J., Richardson, D.S., Hewitson, T.D., Haiden, T., 2010. A new equitable score suitable for verifying precipitation in numerical weather prediction. Q. J. R. Meteorol. Soc. 136, 1344–1363.

Roebber, P.J., 2009. Visualizing multiple measures of forecast quality. Weather Forecast. 24, 601–608.

Roebber, P.J., Bosart, L.F., 1996. The complex relationship between forecast skill and forecast value: a real-world analysis. Weather Forecast. 11, 544–559.

Romanic, D., Curic, M., Jovicic, I., Lompar, M., 2015. Long-term trends of the 'Koshava' wind during the period 1949-2010. Int. J. Climatol. 35, 288–302.

Romesburg, H.C., 1984. Cluster Analysis for Researchers. Wadsworth/Lifetime Learning Publications. 334 pp.

Ropelewski, C.F., Jones, P.D., 1987. An extension of the Tahiti-Darwin Southern Oscillation index. Mon. Weather Rev. 115, 2161–2165.

Röpnack, A., Hense, A., Gebhardt, C., Majewski, D., 2013. Bayesian model verification of NWP ensemble forecasts. Mon. Weather Rev. 141, 375–387.

Rosenberger, J.L., Gasko, M., 1983. Comparing location estimators: trimmed means, medians, and trimean. In: Hoaglin, D.C., Mosteller, F., Tukey, J.W. (Eds.), Understanding Robust and Exploratory Data Analysis. Wiley, New York, 297–338.

Rougier, J., 2016. Ensemble averaging and mean squared error. J. Clim. 29, 8865–8870.

Roulin, E., Vannitsem, S., 2012. Postprocessing of ensemble precipitation predictions with extended logistic regression based on hindcasts. Mon. Weather Rev. 140, 874–888.

Roulston, M.S., Bolton, G.E., Kleit, A.N., Sears-Collins, A.L., 2006. A laboratory study of the benefits of including uncertainty information in weather forecasts. Weather Forecast. 21, 116–122.

Roulston, M.S., Kaplan, D.T., Hardenberg, J., Smith, L.A., 2003. Using medium-range weather forecasts to improve the value of wind energy production. Renew. Energy 28, 585–602.

Roulston, M.S., Smith, L.A., 2002. Evaluating probabilistic forecasts using information theory. Mon. Weather Rev. 130, 1653–1660.

Roulston, M.S., Smith, L.A., 2003. Combining dynamical and statistical ensembles. Tellus A 55, 16–30.

Ruiz, J.J., Saulo, C., 2012. How sensitive are probabilistic precipitation forecasts to the choice of calibration algorithms and the ensemble generation method? Part I: sensitivity to calibration methods. Meteorol. Appl. 19, 302–313.

Rüschendorf, L., 2009. On the distributional transform, Sklar's theorem, and the empirical copula process. J. Stat. Plan. Inference 139, 3921–3927.

Saetra, O., Hersbach, H., Bidlot, J.-R., Richardson, D.S., 2004. Effects of observation errors on the statistics for ensemble spread and reliability. Mon. Weather Rev. 132, 1487–1501.

Sain, S.R., Nychka, D., Mearns, L., 2011. Functional ANOVA and regional climate experiments: a statistical analysis of dynamic downscaling. Environmetrics 22, 700–711.

Sanchez, C., Williams, K.D., Collins, M., 2016. Improved stochastic physics schemes for global weather and climate models. Q. J. R. Meteorol. Soc. 142, 147–159.

Sanders, F., 1963. On subjective probability forecasting. J. Appl. Meteorol. 2, 191–201.

Sansom, J., Thomson, P.J., 1992. Rainfall classification using breakpoint pluviograph data. J. Clim. 5, 755–764.

Sansom, P.G., Ferro, C.A.T., Stephenson, D.B., Goddard, L., Mason, S.J., 2016. Best practices for postprocessing ensemble climate forecasts. Part I: Selecting appropriate calibration methods. J. Clim. 29, 7247–7264.

Sansom, P.G., Stephenson, D.B., Ferro, C.A.T., Zappa, G., Shaffrey, L., 2013. Simple uncertainty frameworks for selecting weighting schemes and interpreting multimodel ensemble climate change experiments. J. Clim. 26, 4017–4037.

Santer, B.D., Wigley, T.M.L., Boyle, J.S., Gaffen, D.J., Hnilo, J.J., Nychka, D., Parker, D.E., Taylor, K.E., 2000. Statistical significance of trends and trend differences in layer-average atmospheric temperature series. J. Geophys. Res. 105, 7337–7356.

Santos, C., Ghelli, A., 2012. Observational probability methods to assess ensemble precipitation forecasts. Q. J. R. Meteorol. Soc. 138, 209–221.

Satterfield, E.A., Bishop, C.H., 2014. Heteroscedastic ensemble postprocessing. Mon. Weather Rev. 142, 3484–3502.

Sauvageot, H., 1994. Rainfall measurement by radar: a review. Atmos. Res. 35, 27–54.

Schefzik, R., 2016. A similarity-based implementation of the Schaake shuffle. Mon. Weather Rev. 144, 1909–1921.

Schefzik, R., 2017. Ensemble calibration with preserved correlations: unifying and comparing ensemble copula coupling and member-by-member postprocessing. Q. J. R. Meteorol. Soc. 143, 999–1008. https://doi.org/10.1002/qj.2984.

Schefzik, R., Thorarinsdottir, T.L., Gneiting, T., 2013. Uncertainty quantification in complex simulation models using ensemble copula coupling. Stat. Sci. 28, 616–640.

Schenker, N., Gentleman, J.F., 2001. On judging the significance of differences by examining the overlap between confidence intervals. Am. Stat. 55, 182–186.

Scherrer, S.C., Appenzeller, C., Eckert, P., Cattani, D., 2004. Analysis of the spread-skill relations using the ECMWF ensemble prediction system over Europe. Weather Forecast. 19, 552–565.

Schervish, M.J., 1989. A general method for comparing probability assessors. Ann. Stat. 17, 1856–1879.

Scheuerer, M., 2014. Probabilistic quantitative precipitation forecasting using ensemble model output statistics. Q. J. R. Meteorol. Soc. 140, 1086–1096.

Scheuerer, M., Hamill, T.M., 2015a. Statistical postprocessing of ensemble precipitation forecasts by fitting censored, shifted gamma distributions. Mon. Weather Rev. 143, 4578–4596.

Scheuerer, M., Hamill, T.M., 2015b. Variogram-based proper scoring rules for probabilistic forecasts of multivariate quantities. Mon. Weather Rev. 143, 1321–1334.

Scheuerer, M., Hamill, T.M., 2018. Generating calibrated ensembles of physically realistic high-resolution precipitation forecast fields based on GEFS model output. J. Hydromet. 19, 1651–1670.

Scheuerer, M., Hamill, T.M., Whitin, B., He, M., Henkel, A., 2017. A method for preferential selection of dates in the Schaake shuffle approach to constructing spatiotemporal forecast fields of temperature and precipitation. Water Resour. Res. 53, 3029–3046.

Scheuerer, M., Möller, D., 2015. Probabilistic wind speed forecasting on a grid based on ensemble model output statistics. Ann. Appl. Stat. 9, 1328–1349.

Schmeits, M.J., Kok, K.J., 2010. A comparison between raw ensemble output, (modified) Bayesian Model Averaging, and extended logistic regression using ECMWF ensemble precipitation forecasts. Mon. Weather Rev. 138, 4199–4211.

Schölzel, C., Hense, A., 2011. Probabilistic assessment of regional climate change in southwest Germany by ensemble dressing. Clim. Dyn. https://doi.org/10.1007/s00382-010-0815-1. 12 pp.

Schuhen, N., Thorarinsdottir, T.L., Gneiting, T., 2012. Ensemble model output statistics for wind vectors. Mon. Weather Rev. 140, 3204–3219.

Schwarz, G., 1978. Estimating the dimension of a model. Ann. Stat. 6, 461–464.

Scott, D.W., 1992. Multivariate Density Estimation. Wiley. 317 pp.

Seaman, R., Mason, I., Woodcock, F., 1996. Confidence intervals for some performance measures of yes-no forecasts. Aust. Meteorol. Mag. 45, 49–53.

Semazzi, F.H.M., Mera, R.J., 2006. An extended procedure for implementing the relative operating characteristic graphical method. J. Appl. Meteorol. Climatol. 45, 1215–1223.

Sen, P.K., 1968. Estimates of the regression coefficient based on Kendall's tau. J. Am. Stat. Assoc. 63, 1379–1389.

Serinaldi, F., Kilsby, C.G., 2014. Rainfall extremes: toward reconciliation after the battle of distributions. Water Resour. Res. 50, 336–352.

Shannon, C.E., 1948. A mathematical theory of communication. Bell Syst. Tech. J. 27, 379–423, 623–656.

Shapiro, S.S., Wilk, M.B., 1965. An analysis of variance test for normality (complete samples). Biometrika 52, 591–610.

Sharma, A., Lall, U., Tarboton, D.G., 1998. Kernel bandwidth selection for a first order nonparametric streamflow simulation model. Stoch. Hydrol. Hydraul. 12, 33–52.

Sheets, R.C., 1990. The National Hurricane Center—past, present and future. Weather Forecast. 5, 185–232.

Shongwe, M.E., Landman, W.A., Mason, S.J., 2006. Performance of recalibration systems for GCM forecasts for southern Africa. Int. J. Climatol. 26, 1567–1585.

Shutts, G., 2015. A stochastic convective backscatter scheme for use in ensemble prediction systems. Q. J. R. Meteorol. Soc. 141, 2602–2616.

Siegert, S., 2014. Variance estimation for Brier score decomposition. Q. J. R. Meteorol. Soc. 140, 1771–1777.

Siegert, S., Bellprat, O., Ménégoz, M., Stephenson, D.B., Doblas-Reyes, F.J., 2017. Detecting improvements in forecast correlation skill: statistical testing and power analysis. Mon. Weather Rev. 145, 437–450.

Siegert, S., Stephenson, D.B., Sansom, P.G., Scaife, A.A., Eade, R., Arribas, A., 2016. A Bayesian framework for verification and recalibration of ensemble forecasts: how uncertain is NAO predictability? J. Clim. 29, 995–1012.

Silver, N., 2012. The Signal and the Noise. Penguin Books. 534 pp.

Silverman, B.W., 1986. Density Estimation for Statistics and Data Analysis. Chapman and Hall. 175 pp.

Simmons, A.J., Hollingsworth, A., 2002. Some aspects of the improvement in skill of numerical weather prediction. Q. J. R. Meteorol. Soc. 128, 647–677.

Skok, G., Roberts, N., 2016. Analysis of fractions skill score properties for random precipitation fields and ECMWF forecasts. Q. J. R. Meteorol. Soc. 142, 2599–2610.

Skok, G., Roberts, N., 2018. Estimating the displacement in precipitation forecasts using the fractions skill score. Q.J.R. Meteorol. Soc. 144, 414–425.

Sloughter, J.M., Gneiting, T., Raftery, A.E., 2010. Probabilistic wind speed forecasting using ensembles and Bayesian model averaging. J. Am. Stat. Assoc. 105, 25–35.

Sloughter, J.M., Gneiting, T., Raftery, A.E., 2013. Probabilistic wind vector forecasting using ensembles and Bayesian model averaging. Mon. Weather Rev. 141, 2107–2119.

Sloughter, J.M., Raftery, A.E., Gneiting, T., Fraley, C., 2007. Probabilistic quantitative precipitation forecasting using Bayesian model averaging. Mon. Weather Rev. 135, 3209–3220.

Smith, L.A., 2001. Disentangling uncertainty and error: on the predictability of nonlinear systems. In: Mees, A.I. (Ed.), Nonlinear Dynamics and Statistics. Birkhauser, 31–64.

Smith, L.A., 2007. Chaos, A Very Short Introduction. Oxford University Press. 180 pp.

Smith, L.A., Hansen, J.A., 2004. Extending the limits of ensemble forecast verification with the minimum spanning tree. Mon. Weather Rev. 132, 1522–1528.

Smith, R.E., Schreiber, H.A., 1974. Point process of seasonal thunder-storm rainfall: 2. Rainfall depth probabilities. Water Resour. Res. 10, 418–423.

Smyth, P., Ide, K., Ghil, M., 1999. Multiple regimes in Northern Hemisphere height fields via mixture model clustering. J. Atmos. Sci. 56, 3704–3723.

Solow, A.R., Moore, L., 2000. Testing for a trend in a partially incomplete hurricane record. J. Clim. 13, 3696–3710.

Spetzler, C.S., Staël von Holstein, C.-A.S., 1975. Probability encoding in decision analysis. Manag. Sci. 22, 340–358.

Sprent, P., Smeeton, N.C., 2001. Applied Nonparametric Statistical Methods. Chapman and Hall. 461 pp.

Stacy, E.W., 1962. A generalization of the Gamma distribution. Ann. Math. Stat. 33, 1187–1192.

Staël von Holstein, C.-A.S., Murphy, A.H., 1978. The family of quadratic scoring rules. Mon. Weather Rev. 106, 917–924.

Stanski, H.R., Wilson, L.J., Burrows, W.R., 1989. Survey of Common Verification Methods in Meteorology. World Weather Watch Technical Report No. 8, World Meteorological Organization. TD No. 358, 114 pp.

Stauffer, R., Mayr, G.J., Dabernig, M., Zeileis, A., 2015. Somewhere over the rainbow: how to make effective use of colors in meteorological visualizations. Bull. Am. Meteorol. Soc. 96, 203–215.

Stauffer, R., Umlauf, N., Messner, J.W., Mayr, G.J., Zeileis, A., 2017. Ensemble postprocessing of daily precipitation sums over complex terrain using censored high-resolution standardized anomalies. Mon. Weather Rev. 145, 955–969.

Stedinger, J.R., Vogel, R.M., Foufoula-Georgiou, E., 1993. Frequency analysis of extreme events. In: Maidment, D.R. (Ed.), Handbook of Hydrology. McGraw-Hill. 66 pp.

Steel, R.G.D., Torrie, J.H., 1960. Principles and Procedures of Statistics. McGraw-Hill. 481 pp.

Steinskog, D.J., Tjostheim, D.B., Kvamsto, N.G., 2007. A cautionary note on the use of the Kolmogorov-Smirnov test for normality. Mon. Weather Rev. 135, 1151–1157.

Stensrud, D.J., Bao, J.-W., Warner, T.T., 2000. Using initial conditions and model physics perturbations in short-range ensemble simulations of mesoscale convective systems. Mon. Weather Rev. 128, 2077–2107.

Stensrud, D.J., Brooks, H.E., Du, J., Tracton, M.S., Rogers, E., 1999. Using ensembles for short-range forecasting. Mon. Weather Rev. 127, 433–446.

Stensrud, D.J., Wandishin, M.S., 2000. The correspondence ratio in forecast evaluation. Weather Forecast. 15, 593–602.

Stensrud, D.J., Yussouf, N., 2003. Short-range ensemble predictions of 2-m temperature and dewpoint temperature over New England. Mon. Weather Rev. 131, 2510–2524.

Stephens, M., 1974. E.D.F. statistics for goodness of fit. J. Am. Stat. Assoc. 69, 730–737.

Stephenson, D.B., 1997. Correlation of spatial climate/weather maps and the advantages of using the Mahalanobis metric in predictions. Tellus A 49, 513–527.

Stephenson, D.B., 2000. Use of the "odds ratio" for diagnosing forecast skill. Weather Forecast. 15, 221–232.

Stephenson, D.B., Casati, B., Ferro, C.A.T., Wilson, C.A., 2008a. The extreme dependency score: a non-vanishing measure for forecasts of rare events. Meteorol. Appl. 15, 41–50.

Stephenson, D.B., Coelho, C.A.S., Doblas-Reyes, F.J., Balmaseda, M., 2005. Forecast assimilation: a unified framework for the combination of multi-model weather and climate predictions. Tellus A 57, 253–264.

Stephenson, D.B., Coelho, C.A.S., Jolliffe, I.T., 2008b. Two extra components in the Brier score decomposition. Weather Forecast. 23, 752–757.

Stephenson, D.B., Collins, M., Rougier, J.C., Chandler, R.E., 2012. Statistical problems in the probabilistic prediction of climate change. Environmetrics 23, 364–372.

Stephenson, D.B., Doblas-Reyes, F.J., 2000. Statistical methods for interpreting Monte-Carlo ensemble forecasts. Tellus A 52, 300–322.

Stern, H., Davidson, N.E., 2015. Trends in the skill of weather prediction at lead times of 1–14 days. Q. J. R. Meteorol. Soc. 141, 2726–2736.

Stern, R.D., Coe, R., 1984. A model fitting analysis of daily rainfall data. J. R. Stat. Soc. Ser. A 147, 1–34.

Strang, G., 1988. Linear Algebra and its Applications. Harcourt, 505 pp.

Stockdale, T.N., Anderson, D.L.T., Balmaseda, M.A., Doblas-Reyes, F., Ferranti, L., Mogensen, K., Palmer, T.N., Molteni, F., Vitart, F., 2011. ECMWF seasonal forecast system 3 and its prediction of sea surface temperature. Clim. Dyn. 37, 455–471.

Stewart, T.R., 1997. Forecast value: descriptive decision studies. In: Katz, R.W., Murphy, A.H. (Eds.), Economic Value of Weather and Climate Forecasts. Cambridge University Press, 147–181.

Stuart, N.A., Schultz, D.M., Klein, G., 2007. Maintaining the role of humans in the forecast process. Bull. Am. Meteorol. Soc. 88, 1893–1898.

Stull, R.B., 1988. An Introduction to Boundary Layer Meteorology. Kluwer. 666 pp.

Sutera, A., 1981. On stochastic perturbation and long-term climate behaviour. Q. J. R. Meteorol. Soc. 107, 137–151.

Swenson, E., 2015. Continuum power CCA: a unified approach for isolating coupled modes. J. Clim. 28, 1016–1030.

Swets, J.A., 1973. The relative operating characteristic in psychology. Science 182, 990–1000.

Swets, J.A., 1979. ROC analysis applied to the evaluation of medical imaging techniques. Investig. Radiol. 14, 109–121.

Taillardat, M., Mestre, O., Zamo, M., Naveau, P., 2016. Calibrated ensemble forecasts using quantile regression forests and ensemble model output statistics. Mon. Weather Rev. 144, 2375–2393.

Taillardat, M., Fougères, A.-L., Naveau, P., Mestre, O., 2017. Forest-based methods and ensemble model output statistics for rainfall ensemble forecasting. arXiv:1711.10937v1, 20 pp.

Talagrand, O., Vautard, R., Strauss, B., 1997. Evaluation of probabilistic prediction systems. In: Proceedings, ECMWF Workshop on Predictability. ECMWF, 1–25.

Taleb, N.N., 2001. Fooled by Randomness. Texere, New York. 203 pp.

Tang, B., Hsieh, W.W., Monahan, A.H., Tangang, F.T., 2000. Skill comparisons between neural networks and canonical correlation analysis in predicting the equatorial Pacific sea surface temperatures. J. Clim. 13, 287–293.

Tareghian, R., Rasmussen, P., 2013. Analysis of Arctic and Antarctic sea ice extent using quantile regression. Int. J. Climatol. 33, 1079–1086.

Taylor, J.W., 1999. Evaluating volatility and interval forecasts. J. Forecast. 18, 111–128.

Taylor, K.E., 2001. Summarizing multiple aspects of model performance in a single diagram. J. Geophys. Res. D: Atmos. 106, 7183–7192.

Taylor, K.E., Stouffer, R.J., Meehl, G.A., 2012. An overview of CMIP5 and the experiment design. Bull. Am. Meteorol. Soc. 93, 485–498.

Taylor, M.H., Losch, M., Wenzel, M., Schröter, J., 2013. On the sensitivity of field reconstruction and prediction using empirical orthogonal functions derived from gappy data. J. Clim. 26, 9194–9205.

Tenant, W.J., Shutts, G.J., Arribas, A., Thompson, S.A., 2011. Using a stochastic kinetic energy backscatter scheme to improve MOGREPS probabilistic forecast skill. Mon. Weather Rev. 139, 1190–1206.

Tezuka, S., 1995. Uniform Random Numbers: Theory and Practice. Kluwer, 209 pp.

Teweles, S., Wobus, H.B., 1954. Verification of prognostic charts. Bull. Am. Meteorol. Soc. 35, 455–463.

Theil, H., 1950. A rank-invariant method of linear and polynomial regression analysis. K. Nederlansdse Akad. Wet. 53, 386–392, 521–525, 1397–1412.

Theus, M., Urbanek, S., 2009. Interactive Graphics for Data Analysis. CRC Press. 280 pp.

Thiébaux, H.J., Pedder, M.A., 1987. Spatial Objective Analysis: with Applications in Atmospheric Science. Academic Press, London. 299 pp.

Thiébaux, H.J., Zwiers, F.W., 1984. The interpretation and estimation of effective sample size. J. Clim. Appl. Meteorol. 23, 800–811.

Thom, H.C.S., 1958. A note on the gamma distribution. Mon. Weather Rev. 86, 117–122.

Thompson, C.J., Battisti, D.S., 2001. A linear stochastic dynamical model of ENSO. Part II: Analysis. J. Clim. 14, 445–466.

Thompson, J.C., 1962. Economic gains from scientific advances and operational improvements in meteorological prediction. J. Appl. Meteorol. 1, 13–17.

Thompson, J.C., Brier, G.W., 1955. The economic utility of weather forecasts. Mon. Weather Rev. 83, 249–254.

Thompson, J.C., Carter, G.M., 1972. On some characteristics of the S1 score. J. Appl. Meteorol. 11, 1384–1385.

Thompson, P.D., 1977. How to improve accuracy by combining independent forecasts. Mon. Weather Rev. 105, 228–229.

Thompson, P.D., 1985. Prediction of the probable errors of prediction. Mon. Weather Rev. 113, 248–259.

Thorarinsdottir, T.L., Gneiting, T., 2010. Probabilistic forecasts of wind speed: ensemble model output statistics by using heteroscedastic censored regression. J. R. Stat. Soc. Ser. A 173, 371–388.

Thorarinsdottir, T.L., Scheuerer, M., Heinz, C., 2016. Assessing the calibration of high-dimensional ensemble forecasts using rank histograms. J. Comput. Graph. Stat. 25, 105–122.

Thornes, J.E., Stephenson, D.B., 2001. How to judge the quality and value of weather forecast products. Meteorol. Appl. 8, 307–314.

Tibshirani, R., 1996. Regression shrinkage and selection via the Lasso. J. R. Stat. Soc. Ser. B Methodol. 58, 267–288.

Tibshirani, R., Walther, G., Hastie, T., 2001. Estimating the number of clusters in a dataset via the gap statistic. J. R. Stat. Soc. Ser. B Methodol. 32, 411–423.

Tippet, M.K., Camargo, S.C., Sobel, A.H., 2011. A Poisson regression index for tropical cyclone genesis and the role of large-scale vorticity in genesis. J. Clim. 24, 2335–2357.

Tippett, M.K., DelSole, T., Mason, S.J., Barnston, A.G., 2008. Regression-based methods for finding coupled patterns. J. Clim. 21, 4384–4398.

Titterington, D.M., Smith, A.F.M., Makov, U.E., 1985. Statistical Analysis of Finite Mixture Distributions. Wiley. 243 pp.

Tobin, J., 1958. Estimation of relationships for limited dependent data. Econometrica 26, 24–36.

Todorovic, P., Woolhiser, D.A., 1975. A stochastic model of n-day precipitation. J. Appl. Meteorol. 14, 17–24.

Tong, H., 1975. Determination of the order of a Markov chain by Akaike's Information Criterion. J. Appl. Probab. 12, 488–497.

Torrence, C., Compo, G.P., 1998. A practical guide to wavelet analysis. Bull. Am. Meteorol. Soc. 79, 61–78.

Toth, Z., Kalnay, E., 1993. Ensemble forecasting at NMC: the generation of perturbations. Bull. Am. Meteorol. Soc. 74, 2317–2330.

Toth, Z., Kalnay, E., 1997. Ensemble forecasting at NCEP the breeding method. Mon. Weather Rev. 125, 3297–3318.

Toth, Z., Zhu, Y., Marchok, T., 2001. The use of ensembles to identify forecasts with small and large uncertainty. Weather Forecast. 16, 463–477.

Toth, Z., Peña, M., Vintzileos, A., 2007. Bridging the gap between weather and climate forecasting: research priorities for intra-seasonal prediction. Bull. Am. Meteorol. Soc. 88, 1427–1429.

Tracton, M.S., Kalnay, E., 1993. Operational ensemble prediction at the National Meteorological Center: practical aspects. Weather Forecast. 8, 379–398.

Tracton, M.S., Mo, K., Chen, W., Kalnay, E., Kistler, R., White, G., 1989. Dynamical extended range forecasting (DERF) at the National Meteorological Center. Mon. Weather Rev. 117, 1604–1635.

Tracy, C.A., Widom, H., 1996. On orthogonal and symplectic matrix ensembles. Commun. Math. Phys. 177, 727–754.

Tufte, E.R., 1983. The Visual Display of Quantitative Information. Graphics Press. 197 pp.

Tufte, E.R., 1990. Envisioning Information. Graphics Press. 126 pp.

Tukey, J.W., 1977. Exploratory Data Analysis. Addison-Wesley, Reading, MA. 688 pp.

Tustison, B., Harris, D., Foufoula-Georgiou, E., 2001. Scale issues in verification of precipitation forecasts. J. Geophys. Res. D: Atmos. 106, 11775–11784.

Tversky, A., 1974. Judgement under uncertainty: heuristics and biases. Science 185, 1124–1131.

Tyler, D.E., 1982. On the optimality of the simultaneous redundancy transformations. Psychometrika 47, 77–86.

Unger, D.A., 1985. A method to estimate the continuous ranked probability score. In: Preprints, 9th Conference on Probability and Statistics in the Atmospheric Sciences. American Meteorological Society, 206–213.

Unger, D.A., van den Dool, H., O'Lenic, E., Collins, D., 2009. Ensemble regression. Mon. Weather Rev. 137, 2365–2379.

Valée, M., Wilson, L.J., Bourgouin, P., 1996. New statistical methods for the interpretation of NWP output and the Canadian Meteorological Centre. In: Preprints, 13th Conference on Probability and Statistics in the Atmospheric Sciences (San Francisco, California). American Meteorological Society, 37–44.

Van den Dool, H.M., 1989. A new look at weather forecasting through analogues. Mon. Weather Rev. 117, 2230–2247.

Van den Dool, H., 2007. Empirical Methods in Short-Term Climate Prediction. Oxford University Press, Oxford. 215 pp.

Van den Dool, H.M., Becker, E., Chen, L.-C., Zhang, Q., 2017. The probability anomaly correlation and calibration of probabilistic forecasts. Weather Forecast. 32, 199–206.

Vannitsem, S., Hagedorn, R., 2011. Ensemble forecast post-processing over Belgium: comparison of deterministic-like and ensemble regression methods. Meteorol. Appl. 18, 94–104.

Van Schaeybroeck, B., Vannitsem, S., 2015. Ensemble post-processing using member-by-member approaches: theoretical aspects. Q. J. R. Meteorol. Soc. 141, 807–818.

Vautard, R., Pires, C., Plaut, G., 1996. Long-range atmospheric predictability using space-time principal components. Mon. Weather Rev. 124, 288–307.

Vautard, R., Plaut, G., Wang, R., Brunet, G., 1999. Seasonal prediction of North American surface air temperatures using space-time principal components. J. Clim. 12, 380–394.

Vautard, R., Yiou, P., Ghil, M., 1992. Singular spectrum analysis: a toolkit for short, noisy and chaotic series. Physica D 58, 95–126.

Veenhuis, B.A., 2013. Spread calibration of ensemble MOS forecasts. Mon. Weather Rev. 141, 2467–2482.

Velleman, P.F., Hoaglin, D.C., 1981. Applications, Basics, and Computing of Exploratory Data Analysis. Duxbury Press, Boston, MA. 354 pp.

Ventura, V., Paciorek, C.J., Risbey, J.S., 2004. Controlling the proportion of falsely rejected hypotheses when conducting multiple tests with climatological data. J. Clim. 17, 4343–4356.

Verkade, J.S., Brown, J.D., Reggiani, P., Weerts, A.H., 2013. Post-processing ECMWF precipitation and temperature ensemble reforecasts for operational hydrologic forecasting at various spatial scales. J. Hydrol. 501, 73–91.

Vigaud, N., Robertson, A.W., Tippett, M.K., 2017. Multimodel ensembling of subseasonal precipitation forecasts over North America. Mon. Weather Rev. 145, 3913–3928.

Vinod, H.D., 1976. Canonical ridge and econometrics of joint production. J. Econ. 4, 147–166.

Vislocky, R.L., Fritsch, J.M., 1995a. Improved model output statistics forecasts through model consensus. Bull. Am. Meteorol. Soc. 76, 1157–1164.

Vislocky, R.L., Fritsch, J.M., 1995b. Generalized additive models versus linear regression in generating probabilistic MOS forecasts of aviation weather parameters. Weather Forecast. 10, 669–680.

Vislocky, R.L., Fritsch, J.M., 1997. An automated, observations-based system for short-term prediction of ceiling and visibility. Weather Forecast. 12, 31–43.

Vogel, R.M., 1986. The probability plot correlation coefficient test for normal, lognormal, and Gumbel distributional hypotheses. Water Resour. Res. 22, 587–590.

Vogel, R.M., Kroll, C.N., 1989. Low-flow frequency analysis using probability-plot correlation coefficients. J. Water Resour. Plan. Manag. 115, 338–357.

Vogel, R.M., McMartin, D.E., 1991. Probability-plot goodness-of-fit and skewness estimation procedures for the Pearson type III distribution. Water Resour. Res. 27, 3149–3158.

Von Storch, H., 1982. A remark on Chervin-Schneider's algorithm to test significance of climate experiments with GCMs. J. Atmos. Sci. 39, 187–189.

Von Storch, H., 1995. Misuses of statistical analysis in climate research. In: von Storch, H., Navarra, A. (Eds.), Analysis of Climate Variability. Springer, 11–26.

Von Storch, H., Hannoschöck, G., 1984. Comments on "empirical orthogonal function analysis of wind vectors over the tropical Pacific region". Bull. Am. Meteorol. Soc. 65, 162.

Von Storch, H., Hannoschock, G., 1985. Statistical aspects of estimated principal vectors (EOFs) based on small samples sizes. J. Clim. Appl. Meteorol. 24, 716–724.

Von Storch, H., Zwiers, F.W., 1999. Statistical Analysis in Climate Research. Cambridge University Press, Cambridge. 484 pp.

Vrac, M., Naveau, P., 2007. Stochastic downscaling of precipitation: from dry events to heavy rainfalls. Water Resour. Res. 43. https://doi.org/10.1029/2006WR005308.

Walker, G.T., 1914. Correlation in seasonal variations of weather. III. On the criterion for the reality of relationships or periodicities. Mem. Indian Meteorol. Dep. 21 (9), 13–15.

Wallace, J.M., Blackmon, M.L., 1983. Observations of low-frequency atmospheric variability. In: Hoskins, B.J., Pearce, R.P. (Eds.), Large-Scale Dynamical Processes in the Atmosphere. Academic Press, 55–94.

Wallace, J.M., Gutzler, D.S., 1981. Teleconnections in the geopotential height field during the northern hemisphere winter. Mon. Weather Rev. 109, 784–812.

Wallace, J.M., Smith, C., Bretherton, C.S., 1992. Singular value decomposition of wintertime sea surface temperature and 500-mb height anomalies. J. Clim. 5, 561–576.

Wallsten, T.S., Budescu, D.V., Rapoport, A., Zwick, R., Forsyth, B., 1986. Measuring the vague meanings of probability terms. J. Exp. Psychol. 115, 348–365.

Walshaw, D., 2000. Modeling extreme wind speeds in regions prone to hurricanes. Appl. Stat. 49, 51–62.

Wandishin, M.S., Brooks, H.E., 2002. On the relationship between Clayton's skill score and expected value for forecasts of binary events. Meteorol. Appl. 9, 455–459.

Wang, Q.J., Shrestha, D.L., Robertson, D.E., Pokhrel, P., 2012. A log-sinh transformation for data normalization and variance stabilization. Water Resour. Res. 48. https://doi.org/10.1029/2011WR010973.

Wang, X., Bishop, C.H., 2003. A comparison of breeding and ensemble transform Kalman filter ensemble forecast schemes. J. Atmos. Sci. 60, 1140–1158.

Wang, X., Bishop, C.H., 2005. Improvement of ensemble reliability with a new dressing kernel. Q. J. R. Meteorol. Soc. 131, 965–986.

Wang, X.L., Zwiers, F.W., 1999. Interannual variability of precipitation in an ensemble of AMIP climate simulations conducted with the CCC GCM2. J. Clim. 12, 1322–1335.

Wang, Y.-H., Magnusdottir, G., Stern, H., Tian, X., Yu, Y., 2014. Uncertainty estimates of the EOF-derived North Atlantic oscillation. J. Clim. 27, 1290–1301.

Ward, M.N., Folland, C.K., 1991. Prediction of seasonal rainfall in the north Nordeste of Brazil using eigenvectors of sea-surface temperature. Int. J. Climatol. 11, 711–743.

Wasko, C., Sharma, A., 2014. Quantile regression for investigating scaling of extreme precipitation with temperature. Water Resour. Res. 50, 3608–3614.

Wasserstein, R.L., Lazar, N.A., 2016. The ASA's statement on p-values: context, process, and purpose. Am. Stat. 70, 129–133.

Waymire, E., Gupta, V.K., 1981. The mathematical structure of rainfall representations. 1. A review of stochastic rainfall models. Water Resour. Res. 17, 1261–1272.

Wei, M., Toth, Z., Wobus, R., Zhu, Y., 2008. Initial perturbations based on the ensemble transform (ET) technique in the NCEP global operational forecast system. Tellus A 60, 62–79.

Weijs, S.V., van de Giesen, N., 2011. Accounting for observational uncertainty in forecast verification: An information-theoretical view on forecasts, observations, and truth. Mon. Weather Rev. 139, 2156–2162.

Weniger, M., Friederichs, P., 2016. Using the SAL technique for spatial verification of cloud processes: a sensitivity analysis. J. Appl. Meteorol. Climatol. 55, 2091–2108.

Weniger, M., Kapp, F., Friederichs, P., 2017. Spatial verification using wavelet transforms: a review. Q. J. R. Meteorol. Soc. 143, 120–136.

Wernli, H., Hofmann, C., Zimmer, M., 2009. Spatial forecast verification methods intercomparison project: application of the SAL technique. Weather Forecast. 24, 1472–1484.

Wernli, H., Paulat, M., Hagen, M., Frei, C., 2008. SAL—A novel quality measure for the verification of quantitative precipitation forecasts. Mon. Weather Rev. 136, 4470–4487.

Westfall, P.H., 2014. Kurtosis as peakedness, 1905-2014. R.I.P. Am. Stat. 68, 191–195.

Whan, K., Schmeits, M., 2018. Comparing area-probability forecasts of (extreme) local precipitation using parametric and machine learning statistical post-processing methods. Mon. Weather Rev. https://doi.org/10.1175/MWR-D-17-0290.1. in press.

Wheeler, M.C., Hendon, H.H., 2004. An all-season real-time multivariate MJO index: development of an index for monitoring and prediction. Mon. Weather Rev. 132, 1917–1932.

Whitaker, J.S., Loughe, A.F., 1998. The relationship between ensemble spread and ensemble mean skill. Mon. Weather Rev. 126, 3292–3302.

Wickham, H., Hofmann, H., Wickham, C., Cook, D., 2012. Glyph-maps for visually exploring temporal patterns in climate data and models. Environmetrics 23, 382–393.

Wigley, T.M.L., 2009. The effect of changing climate on the frequency of absolute extreme events. Clim. Chang. 97, 67–76.

Wilkinson, L., 2005. The Grammar of Graphics, second ed. Springer. 690 pp.

Wilks, D.S., 1989. Conditioning stochastic daily precipitation models on total monthly precipitation. Water Resour. Res. 25, 1429–1439.

Wilks, D.S., 1990. Maximum likelihood estimation for the gamma distribution using data containing zeros. J. Clim. 3, 1495–1501.

Wilks, D.S., 1992. Adapting stochastic weather generation algorithms for climate change studies. Clim. Chang. 22, 67–84.

Wilks, D.S., 1993. Comparison of three-parameter probability distributions for representing annual extreme and partial duration precipitation series. Water Resour. Res. 29, 3543–3549.

Wilks, D.S., 1997a. Forecast value: prescriptive decision studies. In: R.W. Katz and A.H. Murphy (eds.), Economic Value of Weather and Climate Forecasts. Cambridge University Press, 109–145.

Wilks, D.S., 1997b. Resampling hypothesis tests for autocorrelated fields. J. Clim. 10, 65–82.

Wilks, D.S., 1998. Multisite generalization of a daily stochastic precipitation generation model. J. Hydrol. 210, 178–191.

Wilks, D.S., 1999a. Interannual variability and extreme-value characteristics of several stochastic daily precipitation models. Agric. For. Meteorol. 93, 153–169.

Wilks, D.S., 1999b. Multisite downscaling of daily precipitation with a stochastic weather generator. Clim. Res. 11, 125–136.

Wilks, D.S., 2001. A skill score based on economic value for probability forecasts. Meteorol. Appl. 8, 209–219.

Wilks, D.S., 2002a. Realizations of daily weather in forecast seasonal climate. J. Hydrometeorol. 3, 195–207.

Wilks, D.S., 2002b. Smoothing forecast ensembles with fitted probability distributions. Q. J. R. Meteorol. Soc. 128, 2821–2836.

Wilks, D.S., 2004. The minimum spanning tree histogram as a verification tool for multidimensional ensemble forecasts. Mon. Weather Rev. 132, 1329–1340.

Wilks, D.S., 2005. Effects of stochastic parametrizations in the Lorenz '96 system. Q. J. R. Meteorol. Soc. 131, 389–407.

Wilks, D.S., 2006a. On "field significance" and the false discovery rate. J. Appl. Meteorol. Climatol. 45, 1181–1189.

Wilks, D.S., 2006b. Comparison of ensemble-MOS methods in the Lorenz '96 setting. Meteorol. Appl. 13, 243–256.

Wilks, D.S., 2008. Improved statistical seasonal forecasts using extended training data. Int. J. Climatol. 28, 1589–1598.

Wilks, D.S., 2009. Extending logistic regression to provide full-probability-distribution MOS forecasts. Meteorol. Appl. 16, 361–368.

Wilks, D.S., 2010. Sampling distributions of the Brier score and Brier skill score under serial dependence. Q. J. R. Meteorol. Soc. 136, 2109–2118.

Wilks, D.S., 2011. On the reliability of the rank histogram. Mon. Weather Rev. 139, 311–316.

Wilks, D.S., 2013. The calibration simplex: a generalization of the reliability diagram for three-category probability forecasts. Weather Forecast. 28, 1210–1218.

Wilks, D.S., 2014a. Probabilistic canonical correlation analysis forecasts, with application to tropical Pacific sea-surface temperatures. Int. J. Climatol. 34, 1405–1413.

Wilks, D.S., 2014b. Comparison of probabilistic statistical forecast and trend adjustment methods for North American seasonal temperatures. J. Appl. Meteorol. Climatol. 53, 935–949.

Wilks, D.S., 2015. Multivariate ensemble model output statistics using empirical copulas. Q. J. R. Meteorol. Soc. 141, 945–952.

Wilks, D.S., 2016a. Three new diagnostic verification diagrams. Meteorol. Appl. 23, 371–378.

Wilks, D.S., 2016b. "The stippling shows statistically significant gridpoints": how research results are routinely overstated and overinterpreted, and what to do about it. Bull. Am. Meteorol. Soc. 97, 2263–2273.

Wilks, D.S., 2016c. Modified "Rule N" procedure for principal component (EOF) truncation. J. Clim. 29, 3049–3056.

Wilks, D.S., 2017. On assessing calibration of multivariate ensemble forecasts. Q. J. R. Meteorol. Soc. 143, 164–172.

Wilks, D.S., 2018a. Enforcing calibration in ensemble postprocessing. Q. J. R. Meteorol. Soc. 144, 76–84.

Wilks, D.S., 2018b. Univariate ensemble postprocessing. In: Vannitsem, S., Wilks, D.S., Messner, J.W. (Eds.), Statistical Postprocessing of Ensemble Forecasts. Elsevier, 49–89.

Wilks, D.S., Godfrey, C.M., 2002. Diagnostic verification of the IRI new assessment forecasts, 1997–2000. J. Clim. 15, 1369–1377.

Wilks, D.S., Hamill, T.M., 2007. Comparison of ensemble-MOS methods using GFS reforecasts. Mon. Weather Rev. 135, 2379–2390.

Wilks, D.S., Livezey, R.E., 2013. Performance of alternative "normals" for tracking climate changes, using homogenized and non-homogenized seasonal U.S. surface temperatures. J. Clim. Appl. Meteorol. 52, 1677–1687.

Wilks, D.S., Wilby, R.L., 1999. The weather generation game: a review of stochastic weather models. Prog. Phys. Geogr. 23, 329–357.

Williams, P.D., Read, P.L., Haine, T.W.N., 2003. Spontaneous generation and impact of inertia-gravity waves in a stratified, two-layer shear flow. Geophys. Res. Lett. 30, 2255–2258.

Williams, R.M., 2016. Statistical Methods for Post-Processing Ensemble Weather Forecasts. PhD. Dissertation, University of Exeter. 197 pp. https://ore.exeter.ac.uk/repository/bitstream/handle/10871/21693/WilliamsR.pdf.

Williams, R.M., Ferro, C.A.T., Kwasniok, F., 2014. A comparison of ensemble post-processing methods for extreme events. Q. J. R. Meteorol. Soc. 140, 1112–1120.

Wilson, L.J., Beauregard, S., Raftery, A.E., Verret, R., 2007. Reply. Mon. Weather Rev. 135, 4231–4236.

Wilson, L.J., Vallée, M., 2002. The Canadian updateable model output statistics (UMOS) system: design and development tests. Weather Forecast. 17, 206–222.

Wilson, L.J., Vallée, M., 2003. The Canadian updateable model output statistics (UMOS) system: validation against perfect prog. Weather Forecast. 18, 288–302.

Winkler, R.L., 1972a. A decision-theoretic approach to interval estimation. J. Am. Stat. Assoc. 67, 187–191.

Winkler, R.L., 1972b. Introduction to Bayesian Inference and Decision. Holt, Rinehart and Winston, New York. 563 pp.

Winkler, R.L., 1994. Evaluating probabilities: asymmetric scoring rules. Manag. Sci. 40, 1395–1405.

Winkler, R.L., 1996. Scoring rules and the evaluation of probabilities. Test 5, 1–60.

Winkler, R.L., Murphy, A.H., 1968. "Good" probability assessors. J. Appl. Meteorol. 7, 751–758.

Winkler, R.L., Murphy, A.H., 1979. The use of probabilities in forecasts of maximum and minimum temperatures. Meteorol. Mag. 108, 317–329.

Winkler, R.L., Murphy, A.H., 1985. Decision analysis. In: Murphy, A.H., Katz, R.W. (Eds.), Probability, Statistics and Decision Making in the Atmospheric Sciences. Westview Press, Boulder, CO, 493–524.

Witten, D.A., Tibshirani, R., Hastie, T., 2009. A penalized matrix decomposition, with applications to sparse principal components and canonical correlation analysis. Biostatistics 10, 515–534.

Wolfers, J., Zitzewitz, E., 2004. Prediction markets. J. Econ. Perspect. 18, 107–126.

Wolff, J.K., Harrold, M., Fowler, T., Gotway, J.H., Nance, L., Brown, B.G., 2014. Beyond the basics: Evaluating model-based precipitation forecasts using traditional, spatial, and object-based methods. Weather Forecast. 29, 1451–1472.

Wolter, K., 1987. The southern oscillation in surface circulation and climate over the tropical Atlantic, eastern Pacific, and Indian Oceans as captured by cluster analysis. J. Clim. Appl. Meteorol. 26, 540–558.

Woodcock, F., 1976. The evaluation of yes/no forecasts for scientific and administrative purposes. Mon. Weather Rev. 104, 1209–1214.

Woolhiser, D.A., Roldan, J., 1982. Stochastic daily precipitation models, 2. A comparison of distributions of amounts. Water Resour. Res. 18, 1461–1468.

Wu, C.F.J., Hamada, M.S., 2009. Experiments: Planning, Analysis and Optimization. Wiley. 760 pp.

Wu, L., Zhang, Y., Adams, T., Lee, H., Liu, Y., Schaake, J., 2018. Comparative evaluation of three Schaake shuffle schemes in postprocessing GEFS precipitation ensemble forecasts. J. Hydromet. 19, 575–598.

Ye, Z.S., Chen, N., 2017. Closed-form estimators for the gamma distribution derived from likelihood equations. Am. Stat. 71, 177–181.

Yeo, I.-K., Johnson, R.A., 2000. A new family of power transformations to improve normality or symmetry. Biometrika 87, 954–959.

Yip, S., Ferro, C.A.T., Stephenson, D.B., Hawkins, E., 2011. A simple, coherent framework for partitioning uncertainty in climate predictions. J. Clim. 24, 4634–4643.

Youden, W.J., 1950. Index for rating diagnostic tests. Cancer 3, 32–35.

Yue, S., Wang, C.-Y., 2002. The influence of serial correlation in the Mann-Whitney test for detecting a shift in median. Adv. Water Resour. 25, 325–333.

Yue, S., Wang, C.-Y., 2004. The Mann-Kendall test modified by effective sample size to detect trend in serially correlated hydrological series. Water Resour. Manag. 18, 201–218.

Yule, G.U., 1900. On the association of attributes in statistics. Philos. Trans. R. Soc. Lond. A 194, 257–319.

Yuval, Hsieh, W.W., 2002. The impact of time-averaging on the detectability of nonlinear empirical relations. Q. J. R. Meteorol. Soc. 128, 1609–1622.

Yuval, Hsieh, W.W., 2003. An adaptive nonlinear MOS scheme for precipitation forecasts using neural networks. Weather Forecast. 18, 303–310.

Zhang, X., Zwiers, F.W., Li, G., 2004. Monte Carlo experiments on the detection of trends in extreme values. J. Clim. 17, 1945–1952.

Zheng, X., Basher, R.E., Thomson, C.S., 1997. Trend detection in regional-mean temperature series: maximum, minimum, mean, diurnal range, and SST. J. Clim. 10, 317–326.

Zheng, X., Straus, D.M., Frederiksen, C.S., 2008. Variance decomposition approach to the prediction of the seasonal mean circulation: comparison with dynamical ensemble prediction using NCEP's CFS. Q. J. R. Meteorol. Soc. 134, 1997–2009.

Zwiers, F.W., 1987a. Statistical considerations for climate experiments. Part II: Multivariate tests. J. Clim. Appl. Meteorol. 26, 477–487.

Zwiers, F.W., 1987b. A potential predictability study conducted with an atmospheric general circulation model. Mon. Weather Rev. 115, 2957–2974.

Zwiers, F.W., 1990. The effect of serial correlation on statistical inferences made with resampling procedures. J. Clim. 3, 1452–1461.

Zwiers, F.W., Kharin, V.V., 1998. Intercomparison of interannual variability and potential predictability: an AMIP diagnostic subproject. Clim. Dyn. 14, 517–528.

Zwiers, F.W., Thiébaux, H.J., 1987. Statistical considerations for climate experiments. Part I: scalar tests. J. Clim. Appl. Meteorol. 26, 465–476.

Zwiers, F.W., von Storch, H., 1995. Taking serial correlation into account in tests of the mean. J. Clim. 8, 336–351.

Index

Note: Page numbers followed by *f* indicate figures, and *t* indicate tables.